MOTHS OF WESTERN NORTH AMERICA

Moths of Western North America

JERRY A. POWELL

PAUL A. OPLER

UNIVERSITY OF CALIFORNIA PRESS

University of California Press, one of the most distinguished university presses in the United States, enriches lives around the world by advancing scholarship in the humanities, social sciences, and natural sciences. Its activities are supported by the UC Press Foundation and by philanthropic contributions from individuals and institutions. For more information, visit www.ucpress.edu.

University of California Press
Oakland, California

Photo/image credits: T. Arcand: 61.25–61.27, 61.31, 61.34, 62.10, 62.20, 62.23, 62.24, 63.12, 63.31, 63.36, 64.1, 64.24–64.26, 64.29. E. Buckner-Opler: 63.3, 63.5, 63.21, 63.28–63.30. California Department of Food & Agriculture: 61.6. Canadian National Collection (CNC): 61.28, 61.32, 62.13, 62.25, 62.32, 63.19, 63.20, 63.26, 63.27, 63.33, 63.35, 64.3–64.6, 64.9, 64.13, 64.20–64.23, 64.27, 64.28, 64.32–64.34. R. Coville: Last page color section, 59.35, 60.4, 60.5, 60.7, 60.17, 60.21–60.24, 60.27, 61.2. H. V. Daly: 60.1. L. Dyer: 61.7. D. C. Ferguson: 61.11, 61.29, 61.30, 61.33, 61.36, 62.2, 62.3, 62.5–62.8, 62.11, 62.14, 62.16, 62.26, 62.27, 62.32, 62.34, 64.7, 64.8, 64.12, 64.14–64.16. C. Hansen: 61.13. P. G. Johnson II: First page color section, 59.2, 60.6, 60.9, 60.19, 60.30, 62.34, 63.22–63.25. N. McFarland: 61.9. L. Minor-Penland: 61.13. P. A. Opler: 61.23, 63.8, 63.9, 64.10, 64.11. S. Passoa: 64.19. D. J. Powell: 59.25, 59.33, 60.10, 60.15, 61.4, 61.5, 61.14, 61.15–61.22. J. A. Powell: 59.3, 59.4–59.22, 59.26–59.32, 59.34, 59.36, 60.2, 60.3, 60.8, 60.11–60.16, 60.18, 60.20, 60.25, 60.26, 60.28, 60.29, 60.31–60.36, 61.1, 61.3, 61.24, 62.33, 63.2, 64.17, 64.18, 64.35. E. S. Ross 59.23. B. Scaccia 61.35, 62.1, 62.4, 62.9, 62.12, 62.15, 62.17–62.19, 62.21, 62.22. D. L. Wagner 59.1, 61.8, 61.10. B. Walsh 62.28–62.31, 62.35, 62.36, 63.1, 63.4, 63.6, 63.7, 63.10, 63.11, 63.13–63.18, 64.2, 64.36.

Library of Congress Cataloging-in-Publication Data

Powell, Jerry A.
  Moths of western North America / Jerry A. Powell, Paul A. Opler.
    p. cm.
  Includes bibliographical references and index.
  ISBN 978-0-520-25197-7 (case : alk. paper) 1. Moths—West (U.S.)
2. Moths—Northwest, Canadian. I. Opler, Paul A. II. Title.
  QL548.P69 2009
  595.780970—dc22
                                                    2008048605

26 25 24 23 22 21
10 9 8 7 6 5 4 3

The paper used in this publication meets the minimum requirements of ANSI/NISO Z 39.48-1992 (R 1997)(*Permanence of Paper*).

Cover: *Arctia caja* (Linnaeus), California, Sierra Co., San Francisco State U., Sierra Nevada Field Campus, July 2007. Species is also shown in plate 47.8, which is from Elko Co., Nevada. The Sierra Nevada population moths differ in color and pattern from the Rocky Mountain and Basin Ranges form. Photograph by Rollin Coville.

CONTENTS

LIST OF FIGURES

PREFACE

Few people appreciate the diversity of insect life, and most naturalists, including professionally trained biologists who work with vertebrates or flowering plants, are surprised, even skeptical, when they hear estimated numbers of insect species. There are believed to be about 1.5 million named and described species of insects worldwide, which is more than five times the number of all other animals combined. Estimates of the number in collections remaining to be distinguished and named vary widely—3 to 5 million or more—and of course most of those, along with countless species never collected (estimates range from 10 to 30 million), will have been exterminated as a result of human population growth within the next few decades. Of the total described, an estimated 11 to 12%, or 160,000 to 180,000 species, are Lepidoptera, most of which occur in tropical regions. About 13,000 live in North America north of Mexico, of which 94% are moths, the remainder butterflies. We believe there are 7,500 to 8,000 named species of moths in the western states and provinces, along with at least another 3,000 that have been collected but not yet named.

It has been more than a century since a volume treating the moths of North America was published, *The Moth Book*, by W. J. Holland. In its era, when there were about half as many named species as now, Holland did an admirable job of summarizing knowledge and illustrating the larger moths with fine color photos, but his representation of small species, the so-called microlepidoptera, was sparse, only 6% of the described species. That volume, long unavailable, was reprinted in paperback form by Dover Press in 1968 with no additions and scant updating of the names. In 1926, E. O. Essig, in *Insects of Western North America*, gave accounts of more than 250 moth species and their larval biologies, especially those of concern in agriculture. This was a small selection of the western fauna, but Essig summarized life histories of species in many families, with greater emphasis on some of the microlepidoptera (Gracillariidae, Sesiidae, Tortricidae) and pyraloids than on many of the macromoths, so it served as a good general introduction. In 1984, Covell's *Field Guide to Moths of Eastern North America* provided a much-needed illustrated manual to the macromoths, which inspired many lepidopterists to expand their horizons beyond butterflies and begin developing local groups of moth enthusiasts. Again, however, the "micros," which make up about 45% of the described Lepidoptera in North America, were slighted, with only token representatives included, especially in the leaf-mining families, which had only 2% of their species illustrated.

In this text we summarize the diversity, taxonomy, and larval biology of western North American moths. The introductory sections give general features of the morphology of adults and immature stages and summarize adult behavior, larval biology, and life cycles. In this book, inspired by I. F. B. Common's *Moths of Australia*, we have taken an innovative approach as compared to previous treatments of Nearctic Lepidoptera. We attempt comparable treatment of all moth families, irrespective of the size of the moths or their traditional popularity with collectors. We present a summary of morphology, larval biology, geographical distribution, and diversity for superfamilies and families and discuss about 25% of the species in each family and larger genus known to occur in western North America. Obviously, therefore, this is not an identification manual that illustrates all moth species. On the other hand, we think entomologists, ecologists, and recreational collectors will find it useful in determining taxa to which species of interest belong, and we provide an entry into the literature—general references are given at the end of the introduction, and major taxonomic and larval biology references are cited at the end of each superfamily. The text summarizes host-plant relationships and life history patterns, which we hope will be valuable to ecologists and other biologists. We believe our treatment can open doors to the world of microlepidoptera for amateur collectors and others interested in biological studies, a field that has lagged in attracting specialists in North America. Moreover, we provide estimates of total species numbers in North America and the West and thereby give insights into the challenge remaining in descriptive taxonomy in each family, subfamily, and larger genus.

Between us we have more than 100 years' experience collecting moths and investigating their life history stages in the western United States, primarily in California, where many of the pioneers of moth study in the western Nearctic worked (e.g., Henry Edwards, Lord Walsingham, Albert Koebele, W. S. Wright, F. X. Williams, J. A. Comstock, H. H. Keifer, C. M. Dammers, and J. W. Tilden). Our experience has dictated an obvious Californicentric bias in species discussed, but we have also lived and studied for varying periods in Colorado, Virginia, Arizona, Mexico, Costa Rica, and Australia, all experiences that have contributed to the background of this book.

A great deal of progress has been made during the past half century toward understanding North American Lepidoptera, enabling development of this book. This era coincided with the growth of the Lepidopterists' Society, founded in 1947 by C. L. Remington and H. K. Clench, and progress was accelerated by modern advances in travel and easy access to areas that were remote, and by the use of black light and mercury vapor light traps. It must be difficult for young students today to imagine the state of knowledge of western North American moths when we started. One had to depend upon Forbes' *Lepidoptera of New York;* Holland's *Moth Book,* if you could find a copy; and Essig's *Insects of Western North America* to even get started, and there were no comprehensive museum collections of microlepidoptera and pyraloids in the western states.

Acknowledgments

Several persons served as our mentors, helping to show the way when the world of systematics and moths seemed an imponderable labyrinth, by providing insights and encouragement, including Richard Bohart, Charles Harbison, Paul Hurd, Jr., Hartford Keifer, Don MacNeill, Douglas Ferguson, and Edward Ross.

We owe much of our knowledge to lepidopterists who paved the way to writing this book by their collections, publications, and providing identifications during the past 50 years: Ron Hodges orchestrated the publication of *The Moths of America North of Mexico* series, including his treatments of the Sphingidae and parts of the Gelechioidea; and he spent several summers collecting in Arizona, Colorado, Utah, and western Texas. Don Davis wrote monographs of primitive moths, Psychidae, and Carposinidae. A publication in 1987 of *Immature Insects, Volume 1,* orchestrated by Fred Stehr, was an important advance, as have been monographs by Annette Braun (Bucculatricidae, Tischeriidae), Frank Hasbrouck (Acrolophidae), David Adamski (Blastobasidae), Don Duckworth (Stenomidae) and with Tom Eichlin (Sesiidae), Jean-Francois Landry (Coleophoridae and Scythrididae); Lauri Kaila (Elachistidae); Dalibor Povolný, many contributions in gnorimoschemine Gelechiidae; R. Gaedike (Acrolepiidae, Epermeniidae, Douglasiidae); John Heppner (Glyphipterigidae); Yu-Feng Hsu (Heliodinidae); and C. Gielis and Deborah Matthews (Pterophoridae). Margaret MacKay, T. N. Freeman, N. S. Obraztsov, John Brown, Richard L. Brown, and Bill Miller contributed greatly to our knowledge of Tortricidae; Marc Epstein to Zygaenoidea (especially Limacodidae); and Scott Miller (Dalceridae). Eugene Munroe, Alma Solis, and H. H. Neunzig have contributed extensively to Pyraloidea; W. C. McGuffin, Douglas Ferguson, and Klaus Bolte to Geometridae; and David Hardwick, Don Lafontaine, and Bob Poole to Noctuoidea. Jack Franclemont, who trained many of our contemporary lepidopterists during his long career at Cornell University, published the MONA treatment of bombycoid taxa and spent many summers collecting in Arizona and Texas. Fred Rindge lived in California, where he completed his graduate studies, and later was curator of Lepidoptera at the American Museum of Natural History, New York, he spent several summers collecting in the Rocky Mountains and published numerous revisions of geometrid genera. Doug Ferguson lived in the Northeast but made many collecting trips to the West and published revisions of geometrids, saturniids, and arctiid genera. The singularly comprehensive volume on Lepidoptera in *Handbook of Zoology,* orchestrated and edited by Niels Kristensen, published in 1999, pooled knowledge of specialists worldwide and provided the systematics framework upon which our text is based. With the increased literature access have come a legion of collectors, some of whom have posted lists and images on the internet.

We, along with Marc Epstein and James Adams, who wrote the Zygaenoidea, owe special thanks to the following people for providing helpful reviews of parts of the manuscript: David Adamski, Val Albu, Richard L. Brown, Charles Covell, Cliff Ferris, Chuck Harp, John Heppner, Ron Hodges, Boris Kondratieff, Jean-Francois Landry, Deborah Matthews, Noel McFarland, Don Lafontaine, Tomas Mustelin (who also provided his manuscript on southern California noctuids), Ted Poling, Amanda Roe, Dan Rubinoff, Alma Solis, Bruce Walsh, and Shen-Horn Yen. In addition to systematists listed in the preceding paragraph, many other taxonomists have provided specimens, identifications, and collection records for our respective museums, aiding the development of our understanding of western moths, including Bill Bauer, Norris Bloomfield, Don Bowman, Terry Dickel, Cliff Ferris, Todd Gilligan, Paul Johnson, Peter Jump, Ed Knudson, Ron Leuschner, John Nordin, Jack Powers, Ron Robertson, Dan Rubinoff, J. B. Sullivan, Catherine Toschi (Tauber), Bruce Walsh, Ron Wielgus, and Don Wright.

We are grateful for the efforts by generations of students and others at the University of California, Berkeley, who assisted by collecting larvae and helping to maintain the Lepidoptera rearing program and its database. This effort has been one of the most comprehensive of its kind in the United States, including collections and maintenance of nearly 15,000 rearing lots which successfully produced adults and field host data for about 1,350 species during 1957–2006. Participants included John Brown, Don Burdick, John Chemsak, John De Benedictis, Robert Dietz IV, Deborah Green, Yu-Feng Hsu, John Lawrence, Paul Rude, Brian Scaccia, Don Veirs, David Wagner, Jim Whitfield, and Jo Wolf. Much of this work was supported by research grants from governmental agencies: The National Science Foundation, "Comparative biology of Microlepidoptera " in 1965–1970; "Systematics of Tortricine Moths" in 1985–1990; the U.S.D.A., Forest Service, "Effects of Dylox on non-target organisms during a Modoc budworm outbreak" in 1974 and the CANUSA/West Research Agreement, "Taxonomic relationships and pheromone isolation among western spruce budworms populations" in 1978–1982; Bureau of Land Management, "Survey of Lepidoptera inhabiting three dune systems in the California desert" in 1977-1978. The Agricultural Experiment Station, University of California, Berkeley, funded the research project "Taxonomy and biology of Microlepidoptera," 1962–1994.

Color images were provided by Evi Buckner-Opler, Rollin Coville, Lee Dyer, Scott Kinnee, Jean-Francois Landry, David Powell, Brian Scaccia, David Wagner, and Bruce Walsh.

David Furth provided access and use of images made by the late Douglas Ferguson, in the Smithsonian Institution, Department of Entomology illustration archives. We are indebted to Steve Lew, Tina Mendez, Ainsley Seago, and Kip Will for assistance with a Microptics Digital Imaging System, enabling production of the specimen images of microlepidoptera and pyraloids by Powell, to Jocelyn Gill for production of macro moth images in Ottawa, and to Don Lafontaine for providing access and use of specimens in the CNC, Agriculture Canada. Jocelyn edited and composed the images on Plates 1–58. The line drawings were made by Powell (Figs. 1–27) and by Ainsley Seago (Figs. 28–252), redrawn from Powell drafts and literature sources acknowledged and funded by the P. C. Powell Trust Fund and the Essig Museum of Entomology. Figures 63, 64, and 86 are original drawings by Celeste Green and Figure 117 is by Tina Jordan; we regret that neither lived to see their use here.

Lastly, and most importantly, we sincerely acknowledge the support provided by our wives, Liz Randal and Evi Buckner-Opler, enabling completion of this project: To Liz for 30 years of incomparable companionship with an admirable balance of encouragement, tolerance, forbearance, and irreverence; and Evi exhibited almost infinite patience over a three year-period when most of the text was written, including taking an extra job to provide more quiet time for writing. There may be no way to adequately repay their kindness.

ABOUT THIS BOOK

Definition of Western North America

We define "the West" to include the states and Canadian provinces and parts thereof west of a line drawn along the eastern base of the western cordillera, from the British and Richardson mountains, Yukon Territory; and Mackenzie Mountains, Northwest Territory; along the Rocky Mountains from northeastern British Columbia, western Alberta, Montana, Wyoming, and Colorado; to New Mexico along the Sacramento Mountains; and western Texas along the Guadalupe, Davis, and Chisos mountains (Fig. 1). We mention transcontinental and other extralimital ranges, but we do not treat species that are restricted to regions east of the defined part of western North America.

Selection of Species

For each family, subfamily, and larger genus we aimed to discuss and illustrate about 25% of the named moth species known to occur in the West. Criteria that we used to select the species included, not necessarily in order of priority, are that they be (1) widespread, well-known species, especially those of economic importance in agriculture, forestry, garden and turf management, and stored products; (2) species for which we have information on life cycle and biology, particularly the larval foods; (3) distinctive or showy species and those otherwise liable to be easily recognized, especially in populated areas; (4) species with special adaptations, behavioral or morphological, even if they are obscure or local in occurrence; (5) a generic representation: ideally all genera might have been treated, but including at least one species from every genus would have exceeded 25% of the total in some taxa, and many of those would be poorly known species. Therefore, we excluded many monotypic genera or those with only one or a few species in the West for which scant information is available. Some families with only one or a few species are proportionately better represented (e.g., essentially all Micropterigidae, Acanthopteroctetidae, Incurvariidae, Adelidae, Schreckensteiniidae, Zygaenoidea, and Doidae). In summary, we selected the better-known species for which we have biological or other interesting information.

Scientific and Common Names

Animal names are governed by the *International Code of Zoological Nomenclature* (ICZN), which is among the very few laws agreed upon by all nations. The primary value of our scientific name system, and the reason it has been used in the same format for 250 years, is its universality. The genus name must be unique, different from every other animal's generic name, and the names are used in the same Latinized form worldwide. As our knowledge and the number of described species increase, species names change, especially because they are moved from one genus to another, or a species is found to be the same as an earlier described one, whose name takes priority.

Animals are classified by a hierarchy of categories called taxa (singular: taxon), with each higher taxon containing one to many of the next subordinate category. Within the animal kingdom, there are six obligate ranks that define the position of every species: phylum, class, order, family, genus, and species. The scientific name of every animal is based on the two lowest-rank categories, genus (plural: genera) and species (plural: species). Hence the name has two parts, the generic name, the first letter of which is always capitalized, followed by the specific name, the first letter of which is not capitalized. Animal names usually are given in italics. Thus an animal's name is more than a name because it is informative on relationships; the species is more closely related to other members of its genus than it is to those of other genera. Similarly, knowing the name of the family to which a species belongs tells us much about its relationships; genera in the same family are closer relatives than any of them is to genera in other families.

As an example, the largest moth in western North America, *Ascalapha odorata* (Linnaeus), is classified as

Phylum Arthropoda, which includes all animals with a hard exoskeleton and jointed appendages

Class Insecta, arthropods with three body sections and six pairs of legs

Order Lepidoptera, the moths and butterflies; having a covering of scales and sucking proboscis formed from the maxillary laciniae

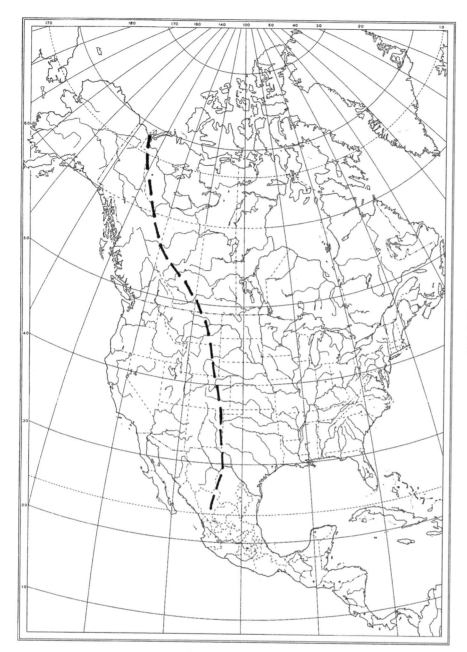

FIGURE 1. Map of North America with the western portion defined for faunal representation in ths book delineated by the bold, broken line.

Family Noctuidae, moths with thoracic tympana

Genus *Ascalapha*

Species *odorata*, a name proposed by Linnaeus in 1758, in the tenth edition of his *Systema Naturae*, which was later selected as the starting base of our binomial nomenclature system.

The author who originally proposed a species name usually is listed with the genus and species, to complete the name; the author's name is placed in parentheses if the species has been assigned to a genus different from its original combination.

In addition, several intermediate ranks are commonly employed for insects because there are vast numbers of species. In Lepidoptera, more than 120 families are defined, which are grouped into about 44 superfamilies. Many families are divided into subfamilies, and larger subfamilies into tribes, each of which contains one to many genera. Typically these intermediate ranks are used primarily by specialists, enabling communication about additional levels of relationships. For example, there are 9,000 described species of the family Tortricidae worldwide, and these are assigned to three subfamilies, which have 11, three, and eight tribes in which their genera are classified. Without the intermediate categories, comprehension of relationships among the more than a thousand genera would be exceedingly difficult, even for the most dedicated specialist. Moreover, endings for the names of ranks between genus and order are spelled out in the ICZN. Thus, -oidea for superfamily, -idae for family, -inae for subfamily, and -ini for tribe are consistently used and provide immediate recognition of the relative position of the group within its lineage.

Common or vernacular names for species vary from place to place for the same animal, particularly but not limited to species that occur in countries using different languages, and there are no rules governing their proposal or priority. Thus

Ascalapha odorata, mentioned above, is called the "black witch" in the United States but has a different common name, and probably more than one, in Mexico and other Spanish-speaking countries. A species often has more than one common name, depending upon where it is applied or in what circumstance. For example the noctuid moth *Heliocoverpa zea* is called the "corn earworm", "tomato fruitworm," and "cotton bollworm" in different places, depending on which crop it is infesting. In this book, we used the common names approved by the Entomological Society of America and those established in various books for some other species, which, however, sometimes contradict one another. We did not propose new common names for any of the vast majority of moth species that lack them.

Literature

We refer to authors of major works, such as the chapters in Kristensen's 1999 volume in *Handbook of Zoology*, or a systematic revision of a particular group, as in the *Moths of America North of Mexico*, but do not cite references in the text. General references that treat many families (e.g., Prentice and others in *Forest Lepidoptera of Canada*) appear in a bibliography at the end of this introduction. Taxonomically restricted publications on individual families or larger genera are cited at the end of each chapter. These contain the sources of much of our text and will provide an entry into the literature for most taxa. Of course, we have referred to hundreds of other publications in scientific journals that treat individual species or specific topics, citation of which is beyond the practical scope of this book.

Figures and Color Plates

Most of the line drawing figures depict moth genitalia, detailed comparison of which often is essential for identifying genera and species. We provide exemplar illustrations representing both sexes of one to several genera for most families. In some cases, these depict species for which there are no accounts in our text because better illustrations were available, from which our drawings were made, than was true for discussed species.

See color section for an explanation of how to use the plates.

Measurements

Size of adults is estimated by fore-wing length, measured from base to apex including the fringe. Using a micrometer disk, we measured five or more of the largest and smallest specimens available to obtain a range of fore-wing lengths, to 0.1 mm accuracy for tiny moths, and to 0.5 mm accuracy for intermedi-

ate-sized moths (ca. 6 to 20 mm). Measurements in centimeters accurate to 0.2 cm were made for larger moths, using vernier calipers.

Abbreviations

Generic names of both plants and insects often are abbreviated by their first letter if already spelled out in the same paragraph. Body parts with abbreviations are illustrated in Figs. 2, 3, and 18 to 24. Abdominal segments are numbered 1 to 10 (or A1 to A10), basal to caudal. In addition to morphological terms cited in the figures, the following abbreviations are used:

FW	fore wing(s)
HW	hind wing(s)
Sc, R, M, Cu, A	wing vein systems: subcostal, radial, medial, cubital, anal (Figs. 4–17)
cm	centimeter (1 inch = 2.54 cm)
mm	millimeter (10 mm = 1 cm)
m	meter (100 cm = 1 m = 3.28 feet)
km	kilometer (1,000 m = 1 km = 0.61 miles)
mi	mile
Co.	County
MONA	*Moths of America North of Mexico*
Mt., Mtn., Mtns.	mount, mountain, mountains
MyBP	million years before the present
'	feet (elevation)

Institutional collection acronyms:

AMNH	American Museum of Natural History, New York
CNC	Canadian National Collection, Ottawa
CSAC	California State Arthropod Collection, Department of Food and Agriculture, Sacramento
CSUC	C. P. Gillette Museum of Arthropod Diversity, Colorado State University, Ft. Collins
EME	Essig Museum of Entomology, University of California, Berkeley
LACM	Los Angeles County Museum of Natural History, Los Angeles
PMY	Peabody Museum, Yale University, New Haven, Conn.
SBNHM	Santa Barbara Natural History Museum, Santa Barbara, Calif.
SDMNH	San Diego Museum of Natural History, San Diego
UCD	Bohart Museum of Entomology, University of California, Davis
USNM	U.S. National Museum of Natural History, Washington, D.C.

PART ONE

INTRODUCTION TO LEPIDOPTERA AND MOTHS

Moths and butterflies make up the order Lepidoptera, and they are among the most familiar and easily recognized insects. The Lepidoptera is defined as a single evolutionary lineage (monophyletic) by a suite of more than 20 derived features, the most obvious of which are scales and proboscis. The scales are modified, flattened hairs that cover the body and wings—shinglelike—and are the source of the extraordinary variety of color patterns typical of these insects. In all but the most primitive forms, feeding by adults is accomplished by pumping in liquid via a tubular proboscis (haustellum), which usually is elongate and coiled under the head. The sister group of Lepidoptera, the Trichoptera, known as caddisflies, lack this development of mouthparts, and its members are covered with unmodified hairs rather than scales. Larval caddisflies are aquatic, whereas only a few species in the Lepidoptera have secondarily adapted to life underwater, primarily the Acentropiinae (=Nymphulinae)(Pyraloidea).

The life cycle of primitive kinds of insects, such as roaches and grasshoppers, includes egg, nymphal, and adult stages. The newly hatched nymph possesses body segments and appendages like those of the adult, and growth occurs though a series of stages (instars). The wings and reproductive organs appear in the final instar. By contrast, the more-derived lineages of insects, including lepidopterans, develop through four stages: egg, larva, pupa, and adult (a sequence called "complete metamorphosis" or "holometabolous"). Holometabolous insects make up more than 85% of extant insect species. Mating and egg deposition (oviposition) are carried out by the adult moths and butterflies. Within the egg, the embryo develops to a fully formed larva, which chews through the eggshell to hatch. The larva, commonly called a caterpillar, feeds and grows, usually through five or six instars. When fully grown, it transforms into a pupa, often within a silken cocoon spun by the larva, although many species pupate without a cocoon. Metamorphosis to the adult occurs during the pupal stage, and the fully developed adult breaks the pupal shell to emerge. Adults of most species feed, but they do not grow. Diapause, an arrested state of development, may occur in any of these stages, prolonging life and enabling the insect to bypass seasons that are unsuitable for growth and reproduction.

The Lepidoptera is one of the two or three largest orders of insects, with an estimated 160,000 to 180,000 named species. Based on specimens in collections and extrapolating from recent studies of Central American moths, we believe that fewer than half the species for which specimens are available in collections have been described and named by taxonomists; even in North America, an estimated one-third of the fauna is undescribed. Thus a realistic projection of the total world Lepidoptera is not possible, but probably the species number exceeds 350,000 and may be much larger. Much of this diversity can be attributed to the radiation of species in association with flowering plants. Lepidoptera constitutes the most species-rich lineage of organisms to have evolved in primary dependence upon angiosperm plants, rivaled only by the coleopteran (beetle) clade Phytophaga (Chrysomeloidea and Curculionoidea).

Morphology

Adult

The adult body framework (Fig. 2) consists of a hardened (sclerotized) exoskeleton made up of a head capsule with appendages; three fused thoracic segments, each with legs and two pairs of wings, on the middle (mesothoracic) and third (metathoracic) segments; and the abdomen, which has 10 segments and is less sclerotized than the thorax and movable by intersegmental membranes. Complex genital structures of external origin arise from abdominal segments 8 to 10, and often there are accessory structures (pouches, glands, hair brushes, etc.) associated with sound reception, courtship, or other functions.

Head

The head (Figs. 2, 3) is more or less globose, with relatively large compound eyes on the sides, antennae between them, and the mouthparts below. The crown or vertex is covered with scales, some or all of which may be hairlike (filiform), sometimes forming tufts. Above the eye and behind the base of the antenna there is a small ocellus and a small patch of sensory setae radiating from a scaleless, raised spot (chaetosema), although one or both are lost in many taxa. The antenna consists of a large basal segment (scape), often elongate second segment (pedicel), and a many-segmented filament (flagellum).

There is enormous variation in form of the antennae, filiform or with the segments variously enlarged or branched, with sensory setae of differing lengths. Often there are obvious differences between the sexes of a species. Antennae of butterflies are enlarged distally, forming apical clubs, while those of moths are not, although some moths (Sphingidae, Sesiidae) have distally enlarged antennae that are tapered or hooked to the tip.

The front of the head capsule (frons) often appears smooth, clothed in very short scales, or the lower part may be bare. The mouthparts of the most primitive moth families retain functional mandibles as in mecopteroid ancestors, or nonfunctional mandibular lobes. In the vast majority of moths the mandibles are lost and mouthparts consist of labrum with a pair of lateral pilifers, labial palpi, and maxillary palpi. The prominent labial palpi usually have three segments, the middle and terminal of which vary in curvature and length, affecting the orientation, decumbent, porrect, or turned or curved upward, but they are not folded. The maxillary palpi consist of one to five segments and in primitive moths are conspicuous, often folded. In most Lepidoptera the maxillary galeae are elongate and joined to form a tubular proboscis (haustellum) with musculature that enables it to be coiled under the head when not used, and other segments of the maxillae are reduced. Nectar from flowers or other fluids are sucked into the digestive tract by a pumping action.

Thorax

The three segments (Fig. 2), pro-, meso-, and metathorax, are fused, each consisting of a series of sclerotized plates (sclerites) that are connected and not movable. The prothorax is small in all Lepidoptera. Each segment gives rise to a pair of legs, and the fore and hind wings arise from the meso- and metathorax. In primitive groups the latter two and their wings are similar in size, but in derived lineages the mesothorax is larger and has more powerful musculature, and the fore wing has more rigid vein structure on the leading edge. The dorsal sclerite of each segment is known as the notum. At the anterior edge of the pronotum there is a pair of articulated plates (patagia), and on the lateral margins of the mesonotum there is another pair (tegulae) that covers the base of the fore wing. The mesonotum consists of a small prescutum, large mesoscutum, and much smaller mesoscutellum, while the metanotum has a bilobed metascutum and a smaller metascutellum. Generally all the surfaces are covered with scales, sometimes hairlike or forming tufts, except the metanotum, which is unscaled, usually with lateral hair brushes. In the largest superfamily, Noctuoidea, the metathorax is modified posteriorly into a pair of tympanal organs.

The legs (Fig. 2) have five segments: coxa, trochanter, femur, tibia, and tarsus. The tibia of the foreleg has an articulated lobe (epiphysis) on the inner surface, usually with a comb of stout setae, a uniquely derived feature in Lepidoptera, that is used to clean the antennae and proboscis by drawing them through the gap between the comb and tibia. The tibiae of males in

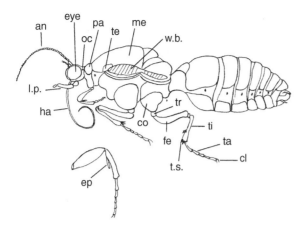

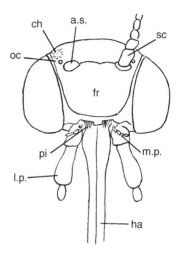

FIGURE 2. Schematic representation of the exoskeletal anatomy of a ditrysian moth, with prothoracic leg enlarged below. Head: an, antenna; eye, compound eye; ha, haustellum (proboscis); l.p., labial palpus; oc, ocellus. Thorax: cl, tarsal claws; co, coxa; ep, epiphysis; fe, femur; me, mesoscutum; pa, patagium; ta, tarsomeres; te, tegula; ti, tibia; tr, trochanter; t.s., tibial spurs; w.b., wing base. Abdomen: tergites and sternites 1 to 7 and spiracles shown. [Powell 2003]

FIGURE 3. Descaled lepidopteran head, frontal aspect. a.s., antennal socket; ch, chaetosema; fr, frons; ha, haustellum, consisting of fused galeae; l.p., labial palpus; m.p., maxillary palpus; oc, ocellus; pi, pilifer; sc, scape. [Powell 2003]

some families have various expandable brushes or tufts of specialized scales. The tarsi are five-segmented, usually with numerous setae and a pair of terminal claws, a central arolium, and sometimes a pair of ventral pads (pulvilli).

The wings are membranous, envelope-like, with a system of tubular veins that provides structure. The wings are tiny and soft upon the moth's eclosion from the pupa, then rapidly expand by circulation of blood (haemolymph) pumped into the flaccid veins, causing them to extend, stretching the wing membranes to full size, after which they rapidly harden, with the membranes pressed closely together, and blood ceases to flow through the veins. There are six series of veins (Figs. 4–15): costal (C), subcostal (Sc), radial (R), medial (M), cubital (Cu), and anal (A). Their homologies are discernible across all families of Lepidoptera, and the configuration of veins has been used extensively in classification. In the most primitive moths the fore and hind wings (FW, HW) are similar in shape and venation (homoneurous)(Figs. 4–5), whereas the more-derived groups have lost parts of the vein systems and have fewer remaining in the HW than the FW (heteroneurous)(Figs. 6–17). In particular, the basal sector of the M vein has been lost, leaving the central area defined by the R and Cu veins open (discal cell), from which the distal branches of the R, M, and Cu veins radiate. There are various wing-coupling mechanisms by which the FW and HW are linked to facilitate flight. Primitive homoneurous moths have an enlarged lobe at the base of the FW (jugum) that folds under the HW when the insect is at rest but extends over the HW in flight, which does not couple the wings efficiently. Most moths have one or more strong bristles (frenulum) at the base of the HW that hooks under a flap (retinaculum) on the underside of the FW, and the development of this varies among taxa and between the sexes of most species.

In a few groups (e.g., Psychidae, Lymantriinae) females of most species are flightless, having very reduced wings (brachypterous), or are apterous and may not even shed the pupal skin. Brachyptery has evolved many times independently, such as in high montane species of various families in Europe, North America, and Australia and in winter-active species of Ethmiidae and several unrelated genera of Geometridae. Both sexes are flightless in species of several families on remote southern oceanic islands and in one species of Scythrididae that lives on windswept coastal sand dunes in California.

Abdomen

There are 10 segments to the abdomen (A1 through A10), with segments 8 through 10 or 9 and 10 usually greatly modified to form external parts of the genitalia. Each segment 2 through 8 consists of a dorsal tergite and a ventral sternite that are joined at the sides by pleural membranes. The sternum of A1 in homoneurous families is small and is lost in other Lepidoptera. Intersegmental membranes join the successive segments, which are movable except where fused by genital structures. Functional spiracles leading to internal tracheae are located in the pleural membranes of A1 to A7. Articulation of the thorax and abdomen in derived families is accomplished by musculature attached to sclerotized struts (apodemes) that usually project from abdominal sternite 2, which is under tergite 1. There are paired tympanal organs at the base of the abdomen in Pyraloidea and Geometroidea. Various male glandular organs, associated with courtship, occur on the abdomen of Tortricidae, Arctiinae, other Noctuidae, and other families. Usually these are developed as expandable hair brushes or tufts, or as thin-walled, eversible sacs (coremata), and they occur at the base of the abdomen, from the intersegmental membrane at the base of the genitalia, or on other segments.

The genitalia of Lepidoptera are highly complex and provide the basis for taxonomic species discrimination in most families and often generic or family-defining characteristics. In the male (Fig. 18), segment A9 forms a sclerotized, dorsal hood-like cover (tegumen); it articulates with a ventral U-shaped vinculum, which sometimes has a short or long apodeme

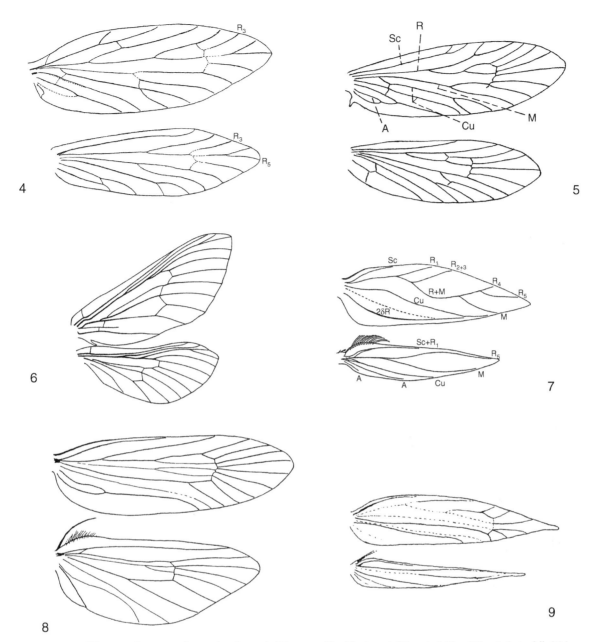

FIGURES 4–9. Wing venation: exemplar species of ancestral lineages of Lepidoptera. **4**, Micropterigidae: *Epimartyria pardella* Walsingham [Issiki 1931]. **5**, Eriocraniidae: *Eriocraniella* [Davis 1978]. **6**, Hepialidae: *Paraphymatopus californicus* (Boisduval) [Wagner 1985, unpublished Ph.D. thesis, U.C. Berkeley]. **7**, Nepticulidae: *Ectoedemia nyssaefoliella* (Chambers) [Braun 1917]. **8**, Prodoxidae: *Tegeticula yuccasella* (Riley) [Davis 1967]. **9**, Tischeriidae: *Tischeria omissa* Braun [Braun 1972].

extending into the body cavity (saccus). A sclerotized structure (uncus) extends caudally from the tegumen, usually rod- or hoodlike, sometimes forked, T-shaped, reduced, or absent. On the ventral side of the tegumen there are lateral "arms" (gnathos), sometimes joined, and often there are setate, membranous lobes or pads (socii) arising caudally to the gnathos. Laterally, there are broad lobes (valvae) articulating with the vinculum, which are thought to provide clasping stability during mating. The valvae usually are large, along with the tegumen more or less covering the other structures in repose, often densely setate on the inner surface, scaled exteriorly and the most visible part of the genitalia externally. The posterior margin (costa) and anterior margin (sacculus) of the valva often

are sclerotized, the sacculus sometimes heavily ornate. The distal portion of the valva is differentiated and heavily setate (cucullus). Between the bases of the valvae there is a transverse membrane (diaphragma), which often bears sclerotized bands posteriorly (fultura superior or transtilla) or plates anteriorly (fultura inferior or juxta). The phallus is separately articulated and passes through the diaphragma; it is sclerotized and consists of a basal lobe (phallobase) and a tubular aedeagus, which contains the membranous vesica, the intromittent organ. The vesica often is armed with various sclerotized spurs or other structures (cornuti), which sometimes are deciduous and deposited in the female. Sperm is produced in paired testes and passes through a duct leading to the vesica, entering the phallus above the phallobase. The precise functions of most of the external, sclerotized parts of the genitalia are unknown, and they vary independently in form, each being uniform in some taxa, highly variable in others, and thus differing in taxonomic value from one family, genus, or species to another.

In the female (Fig. 19) there are three fundamental types of genitalia. Primitive moths possess a single genital aperture near the posterior end of the abdomen, through which both copulation and oviposition occur (monotrysian). All other Lepidoptera have separate genital apertures for copulation and ovposition; in Hepialidae and related families the spermatozoa are conveyed from the gonopore (ostium bursae) to the ovipore via an external groove (exoporian)(Fig. 32). The vast majority of Lepidoptera have internal ducts that carry the sperm from the copulatory tract to oviduct (ditrysian). This feature defines the infraorder Ditrysia, containing most of the superfamilies and 98% of the species.

Females have paired ovaries, and the eggs pass through a common oviduct to a broader chamber where fertilization takes place just preceding the ovipositor. The ovipore is flanked by a pair of lobes (papillae anales) that typically are soft and covered with sensory setae but in many taxa are modified for various kinds of oviposition, such as piercing. The ostium bursae opens to the bursa copulatrix, consisting of a tube (ductus bursae) leading to a sac (corpus bursae) where the spermatophore secreted by the male is deposited. Both the ductus and corpus bursae are variously modified in different taxa, the corpus often with one or more thornlike, sclerotized structures (signa) that may aid in anchoring the spermatophore. Sperm are transported from the corpus bursae through a slender duct (ductus seminalis) usually to a diverticulum (bulla seminalis) and ultimately to the oviduct. The musculature that controls the ovipositor and papillae anales often involves extension and telescoping of the abdomen. The copulatory orifice is anchored to paired rods that extend internally from A10 (posterior apophyses) and A9 (anterior apophyses). The ostium bursae is located on the intersegmental membrane between A7 and A8, on sternite A7 or A8, and externally often is preceded by a sclerotized plate (lamella postvaginalis) or surrounded by various sclerotized folds or wrinkles (sterigma).

Internal Anatomy

Lepidoptera possess the same fundamental internal systems for breathing, blood circulation, digestion, excretion, central nerves, and endocrine functions as do most other insects. Oxygen enters through points of invaginated cuticle (spiracles), located laterally on the meso- and metathorax and first eight abdominal segments. These invaginations form cuticle-lined, air-conducting tubes (tracheae) that connect by longitudinal trunks between the body segments and branch to form fine tracheoles, which are the principle sites for gas exchange, throughout the body (a caterpillar has more than a million tracheoles). Blood (haemolymph) circulates through a musculated dorsal vessel, the functional equivalent to the mammalian heart. The abdominal part (heart) has segmental valves (ostia) that aspirate haemolymph from the body cavity (haemocoel) and pump it forward through a valveless portion in the thorax (aorta) that conducts it to the head, where the blood is released near the brain. From there it percolates back through the haemocoel, providing oxygen to the musculature and exchanging carbon dioxide for oxygen at the tracheoles, until it is again taken in by the heart.

The nervous system consists of the dorsal brain in the head, which is connected by a pair of nerves around the gut to the ventral nerve cord, where there are pairs of segmental ganglia connected by pairs of connective nerves. The three pairs of ganglia from the mouthpart segments are fused, and there are fused ganglia in the abdomen, so the maximum number of discrete ganglia pairs is 12, and usually there are fewer. Separate series of nerve fibers connect the brain and ganglia to the eyes, antennae, and other functional body parts. A small set of ganglia on the surface of the foregut is connected to the brain and to endocrine organs (corpora allata). These and the closely associated prothoracic gland produce hormones important in governing developmental stages.

The gut is a tube that runs through the body from mouth to anus. Food passes through the gut, where nutrients are digested and absorbed into the haemocoel. Food is covered with saliva from labial glands as it enters, is sucked in by the muscular pharynx, and passes through the foregut to the proventriculus, which is musculated. In the abdomen it enters the midgut, the principal center of digestion and absorption, aided by enzymes. Undigested material passes to the hindgut, which contains thick cells called rectal pads, the chief centers of water and ion absorption. Once through the hindgut, undigested wastes form feces and leave through the anus, as fluid in adult Lepidoptera and pellets (frass) in most caterpillars. Excretion is the removal of waste products from cellular metabolism. In insects this is accomplished by the Malpighian tubules, which absorb wastes from the haemolymph and deposit them in the hindgut, from which they are passed out with undigested wastes.

Egg

With few exceptions, female Lepidoptera produce eggs that are deposited externally after fertilization in the oviduct. Individual females produce 200 to 600 eggs or more, usually within a few days. Moth eggs vary enormously in size, shape, surface sculpture, and arrangement during oviposition. In general, within lineages such as families, larger species produce larger eggs, but depending upon the family, the size and numbers differ greatly. For example, females of Hepialidae, including some of the largest moths in the world, produce vast numbers of tiny eggs (20,000 to 30,000 or more by a single female) that are broadcast in the habitat. Conversely some small moths produce fewer (60 to 80), relatively large eggs that may be successively matured within the female over several days or weeks (e.g., Ethmiidae, Scythrididae).

The shell (chorion) is soft during development and quickly hardens after oviposition, assuming a regular form consistent for the species and often characteristic for genera or other groups. The chorion may be smooth or strengthened by raised longitudinal ribs or transverse ridges or both. At one end there is a tiny pore (micropyle), through which the sperm has entered, surrounded by a rosette of radiating lines or ridges. Two types of egg form are defined: those laid horizontally, with the micropyle at one end, which are usually more or less flat and may take the from of the underlying substrate; and those that are upright, with the micropyle at the top. In general, flat eggs are prevalent in the more ancestral lineages, "microlepidoptera," while most derived groups, larger moths and butterflies, have upright eggs with a more rigid and ornamented chorion. Eggs of either type are laid singly or in groups, varying with the species or family; flat eggs are sometimes deposited shinglelike, with the micropylar

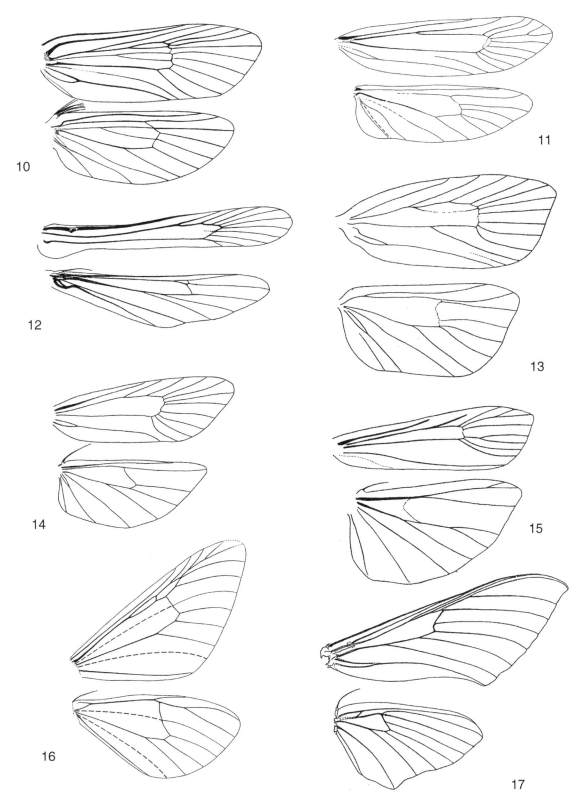

FIGURES 10–17. Wing venation: exemplars of ditrysian moths. **10,** Acrolophidae: *Acrolophus popeanellus* (Clemens) [Davis 1999]. **11,** Ethmiidae: *Ethmia charybdis* Powell [Powell 1973]. **12,** Sesiidae: *Zenodoxus palmii* (Neumoegen) [Eichlin and Duckworth 1988]. **13,** Tortricidae: *Anopina triangulana* (Kearfott) [Powell 1964]. **14,** Carposinidae: *Bondia comonana* (Kearfott) [Davis 1968]. **15,** Pyralidae: *Yosemitia graciella* (Hulst) [Heinrich 1939]. **16,** Geometridae: *Tescalsia giulianiata* Ferguson [Powell and Ferguson 1994]. **17,** Sphingidae: *Eumorpha achemon* (Drury) [Hodges 1970].

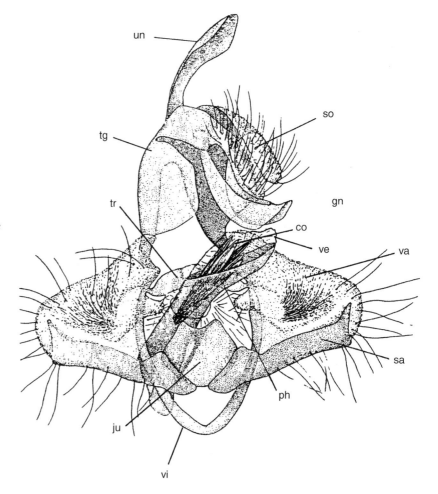

FIGURE 18. Male genitalia of a ditrysian moth (Tortricidae), ventral aspect with valvae reflexed. co, cornuti; gn, gnathos; ju, juxta; ph, phallus (aedeagus); sa, sacculus; so, socii; tg, tegumen; tr, transtilla; un, uncus; va, valva; ve, vesica; vi, vinculum. [Powell 2003]

ends protruding part way over the preceding row, while upright eggs may be laid singly or in groups arranged side by side, like rows of miniature barrels. Usually the eggs are glued to the substrate by a secretion of the female accessory (colleterial) glands, applied within the oviduct, and sometimes the colleterial fluid forms a thick, paintlike covering to egg masses. They may be covered with debris collected by the female or hairs or scales from her abdomen or wings, or surrounded by fences of upright scales. However, lepidopteran eggs are not tended or guarded by the adults.

Embryonic development is related to temperature, proceeding more rapidly in warmer conditions, but the rate is physiologically and hormonally controlled in many instances. It requires seven to 14 days in most Lepidoptera but may be delayed for many weeks or months in species that undergo diapause in the egg stage. In some flat eggs the chorion is translucent and development from embryo to first-instar larva is visible externally.

Larva

The **HEAD** (Figs. 20, 21) is a sclerotized, usually rounded (flattened in leaf-mining species) structure characterized by large lateral lobes, each bearing an ellipse of six simple eyes (stemmata) ventrolaterally, and systematically arranged primary setae; the lobes are joined by a median suture between two narrow adfrontal sclerites. The mouthparts include labrum, lateral, dentate mandibles, and small maxilla bearing a sensory palpus. The mouthparts may be directed downward (hypognathous) or forward (prognathous). The labium is weak but carries a spinneret behind the mouthparts ventrally, which distributes the silk produced by modified salivary glands. The antennae are short, usually three-segmented, located laterally to the lower ends of the adfrontal sutures.

The **THORAX** (Fig. 20) is three-segmented with well developed, segmented true legs in most Lepidoptera. The legs usually are five-segmented as in adults, with a terminal claw, but they may be variously modified or reduced. The prothorax usually has a dorsal, sclerotized, saddlelike area (thoracic shield). Spiracles are located on the meso- and metathoracic segments, except in some aquatic pyraloids, which have external gills.

The **ABDOMEN** (Fig. 20) has 10 segments, usually with spiracles on segments A1 through A8, restricted to segments A1 through A3, or absent in some aquatic pyraloids. There are paired, ventral, leglike organs on all segments in the most primitive moths, while on others they are restricted to segments A3 through A6 (ventral prolegs) and A10 (anal prolegs), which are equipped with circles or bands of tiny hooks (crochets)(Fig. 23) that aid in grasping and walking. The crochets may be arranged in a row so their bases are in line (uniserial), or they may be in two rows (biserial). They may be all similar

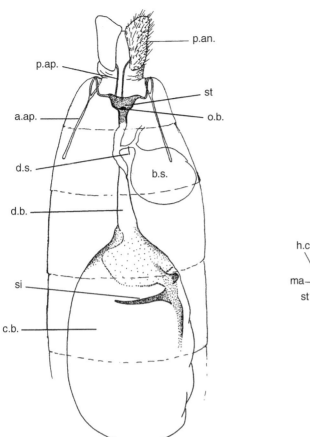

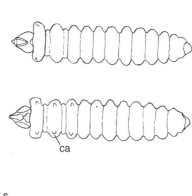

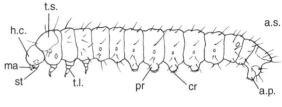

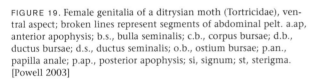

FIGURE 19. Female genitalia of a ditrysian moth (Tortricidae), ventral aspect; broken lines represent segments of abdominal pelt. a.ap, anterior apophysis; b.s., bulla seminalis; c.b., corpus bursae; d.b., ductus bursae; d.s., ductus seminalis; o.b., ostium bursae; p.an., papilla anale; p.ap., posterior apophysis; si, signum; st, sterigma. [Powell 2003]

FIGURE 20. Body forms of moth larvae: top two, Flattened, leaf-mining larva (Tischeriidae), dorsal aspect above, ventral below; middle, typical ditrysian caterpillar (Cossidae), lateral aspect; bottom, Geometridae ("inchworm"), lateral aspect, lacking prolegs on abdominal segments 3 to 5. a.s., anal shield; a.pr., anal proleg; ca, ambulatory calli that represent vestigial remnants of the thoracic legs; cr, crochets; h.c., head capsule; ma, mandible; pr, abdominal proleg; sp, spiracle; st, spinneret; t.l., thoracic leg; t.s., thoracic shield. [Powell 2003]

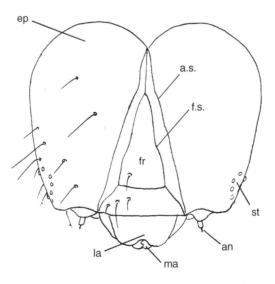

FIGURE 21. Schematic representation of the head capsule of a larval ditrysian moth, frontal aspect. an, antenna; a.s., adfrontal suture; ep, epicranial lobe; fr, frons; f.s., frontal suture; la, labrum; ma, mandible; st, stemmata. [Powell 2003]

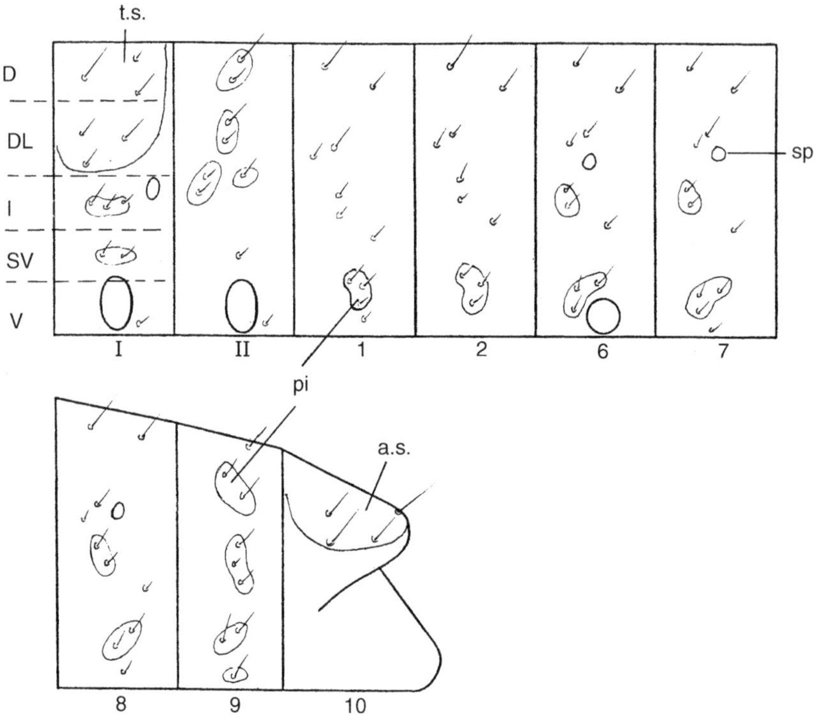

FIGURE 22. Chaetotaxy (setal map) of a larval ditrysian moth (Tortricidae); each rectangle represents one body segment from middorsum (upper border) to midventer (lower border). I, II, pro- and mesothoracic segments; 1, 2, and so on, abdominal segments. Setal groups: D, dorsal; DL, dorsolateral; L, lateral; SV, subventral; V, ventral. a.s., anal shield; pi, pinacula, which are raised and often pigmented; sp, spiracle; t.s., thoracic shield. [Powell 2003]

in size (uniordinal) or of two or three alternating lengths (bi- or triordinal). A specialized form occurs in some Noctuoidea, which have a series of crochets along the inner edge of the proleg with small or rudimentary crochets at both ends (heteroideous). The prolegs are reduced in number in Geometridae and some other groups and are lost in some borers (e.g., Prodoxinae) and sand-dwelling larvae (a few Noctuidae). In Tortricidae and some unrelated groups, A10 has a musculated comblike structure (anal fork) used to flip frass away from the larval shelter.

There are sensory setae on the head and body integument, and the homology of their primary arrangements (chaetotaxy)(Fig. 22) can be compared in all but the most primitive families, but only in the first or early instars of many macrolepidoptera. Their patterns have been valuable to understanding evolutionary trends and to identification of larvae, even though the primary arrangement is lost or replaced by numerous secondary setae in later instars of many taxa.

The adfrontal sclerites, arrangement of stemmata, and prolegs with crotchets distinguish Lepidoptera from other insect larvae.

Pupa

The head, thorax, and abdomen of the pupa (Fig. 24) resemble those of the adult and can be recognized externally. The mandibles of the most primitive families are functional and used to cut open the cocoon preceding eclosion of the adult. In other moths the head is sometimes provided with

a beak or other armature that assists in the eclosion process. The appendages of the head and thorax are each encased in cuticle and in most Lepidoptera are fused to the venter of the body, with the wing cases pulled around to the venter and lying adjacent to the antennae and mouthparts. Abdominal segments 7 to 10 are fused. In the more ancestral families some of the other segments are movable, usually provided with posteriorly directed spines or spurs, and the pupa wriggles forward to protrude from the cocoon or burrow just before moth eclosion. Gelechioidea and macrolepidoptera (Obtectomera, Fig. 24), possess fused abdominal segments (obtect), and their pupae are immobile. Adult eclosion from obtect pupae occurs along a silken track prepared by the larva or by other means of exit from the cocoon, or directly from the pupa in groups that do not spin cocoons. Many species have a group of hooked setae at the tip of the abdomen (cremaster) that anchors the pupa inside the cocoon or at the terminus of a silk emergence track, enabling pressure from the emerging adult to break the pupal shell. Others lack the cremaster but are held within a tight cocoon, within an earthen cell, or by a silk girdle.

The pupal integument is smooth and green or whitish when first formed but soon turns brown in most Lepidoptera. Some have numerous secondary setae. Those that pupate exposed, including Heliodinidae, Pterophoridae, some Gelechioidea, and a few other moths, are mottled green or brownish and often have prominent spines, ridges, or other projections that aid in camouflage. Pupae of Dioptidae are exposed and colorful, presumably aposematic, as are the larvae.

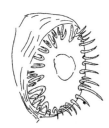 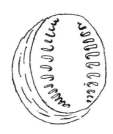 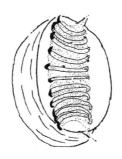

FIGURE 23. Abdominal proleg crochet patterns: left, *Ethmia* (partially biordinal circle) [Powell 1973]; middle, Sesiidae: *Synanthedon* (transverse bands) [modified from Heppner 1987]; right, Noctuidae: Agrotis [Crumb 1956].

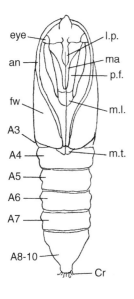 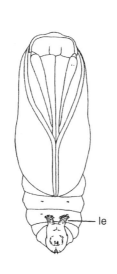 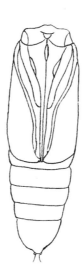

eye
an
fw
A3
A4
A5
A6
A7
A8-10
Cr
l.p.
ma
p.f.
m.l.
m.t.

le

FIGURE 24. Pupae of ditrysian moths, ventral aspect: left, Tortricidae, with abdominal segments 4 to 7 movable, enabling pupal movement forward at emergence; middle, Ethmiidae, with pupal movement restricted to flexible segments 5 and 6, and the pupa remains in place at emergence, a characteristic of Gelechioidea; right, Noctuidae (Obtectomera) with all segments immobile. A3–10, abdominal segments 3 to 10; an, antenna; cr, cremaster; fw, fore wing; le, leglike extensions of the ninth abdominal segment bearing hooked setae that anchor the pupa in lieu of a cremaster in some ethmiids and other Gelechioidea; l.p., labial palpus; ma, maxilla including galeae (haustellum); m.l., mesothoracic leg; m.t., metathoracic tarsus; p.f., prothoracic femur. [Powell 2003; a, c from Mosher 1916]

Biology

Success of Lepidoptera populations is dependent upon several factors in the climatic and biotic environment, interrelated with the insects' behavior. First, larval foods, and for most species adult nourishment, must be available. Climatic conditions suitable for mating and oviposition, larval feeding, and pupation are necessary. Females must find appropriate places for deposition of eggs. Larvae must sense proper foods, eat, molt, grow, and pupate. Pupae need to avoid desiccation and other factors that might prevent successful adult eclosion. Finally, egg, larval, and pupal parasites and predators have to combine on average to take all but two of the offspring of each female (whose eggs may number 200 to 600 or more) that survive physical dangers, but they cannot exceed that, in order to maintain stable population levels.

Adult Behavior

In any given population, males typically begin emergence and peak in numbers a few days ahead of females. Both are sexually mature upon eclosion, and males of nearly all moths are attracted by chemical signals (pheromones) emitted by "calling" females. Hence, in most Lepidoptera, mating takes place soon after female eclosion, and she has mature eggs ready to be fertilized and deposited within the first 24 hours. Mate seeking may involve primarily visual cues in most diurnal moths, although there may be short-range pheromones produced by one or both sexes that mitigate courtship. Males, and females too in most species, mate more than once. Older spermatophore remains are pushed to the distal end of the corpus bursae by the most recent one, and it is assumed that sperm precedence prevails, wherein the most recent male's sperm is effective.

Adults of both sexes of most Lepidoptera feed and in confinement die quickly if water is not available. There is experimental evidence that feeding on honey-enriched fluids extends the life of some moths and increases fecundity. Most macromoths and butterflies feed at flowers, imbibing amino acid–rich nectar, whereas most micromoths do not and gain nourishment from extrafloral nectaries, sap flows, and honeydew secreted by aphids or other Homoptera. Curious exceptions occur in diurnal microlepidoptera, which visit flowers, often other than the larval hosts (e.g., Adelidae, Sesiidae, Heliodinidae,

Scythrididae, Plutellidae, and Tortricidae), but nocturnal species of the same families, or even genera, rarely do so. The mouthparts are nonfunctional in a few families (e.g., Lasiocampidae, Saturniidae, Lymantriidae) and in some specialized species such as winter-active Geometridae and Ethmiidae. In such groups females possess a full complement of mature eggs upon eclosion.

Host-plant selection is made primarily by the female, which uses chemical and tactile cues to seek the proper substrate or habitat for oviposition. This choice is made by instinct, inherited genetically, and the newly hatched larvae also require specific stimuli, detected by chemoreceptors on the antennae and mouthparts; in host-specific species, they starve if the proper plant is not available, ignoring plants or synthetic diets that are quite acceptable and sufficient for nourishment of generalist species.

Larval Development

The newly formed larva, or caterpillar, first bites its way out of the eggshell with its strong mandibles, leaving a crescentic slit or ragged hole at the micropylar end. Some species then eat the remainder of the eggshell as a first meal. All growth takes place during the larval stages, so caterpillars consume enough nutrients to carry through cocoon formation, pupation, and metamorphosis to the adult. It must be sufficient for the moth to move on to its first feeding or, in species with nonfeeding adults, enough to provide for complete egg development of the next generation. To accommodate growth, the larva molts its skin (cuticle) several times, through successively larger stages (instars). Most Lepidoptera undergo five or six instars, but there may be many molts in larvae that feed on detritus or dry plant material.

Silk is produced by paired labial glands. It is composed of two proteins secreted in a viscous fluid in two strands, which consolidate as they leave the spinneret and contact the air. Its functions are many: first-instar larvae of many species are dispersed by air currents on silk strands; many or most species lay down a silk line as they move, enabling them to cling to substrates; silk is used by most external feeding micromoths to form shelters among foliage or other food sources, and some construct portable cases from which they feed (e.g., Adelidae,

Incurvariidae, Tineidae, Psychidae, Coleophoridae, aquatic Crambidae); others line tunnels with silk in fruits, stem, roots, or soil, from which they forage to feed; finally, perhaps the most ubiquitous use of silk is in cocoon formation preceding pupation, which sometimes is within the larval shelter or gallery, but often is separate and an elaborate structure, characteristic in shape.

Larval habits vary widely and often are quite specific for a family, genus, or species. These include leaf-mining, in which a larva spends it entire life within a leaf, and the depth and form of the mines are consistent, such that the moth family or genus often is recognizable from the mine (e.g., blotches just under the upper or lower epidermis, serpentine linear mines just under the surface or deeper, etc.). Other types of endophagous feeding include stem-mining, boring in seed, stems, and roots, or in galls developed by plants, stimulated by the larvae. Many external-feeding caterpillars avoid adverse conditions by seeking shelter in leaf litter at the base of the plant or in tunnels during the day and emerge at night to feed, when temperatures are cooler, humidity is higher, and diurnal predators are not active. Many macromoth larvae remain exposed, motionless, protected by cryptic coloration, body form, and behavior, or even camouflaged by a coat of flower bits or debris that collect on hooked body setae.

The duration of larval development varies greatly with the feeding and life-cycle types, even within families and genera. The time required to reach maturity also is dependent upon temperature and varies within species, such as between seasonal generations. Most Lepidoptera grow slowly in early instars, increasing in size much more rapidly in later instars, particularly the last. Growth after eclosion from the egg to maturity usually takes 30 to 50 days but sometimes is more rapid—as few as 18 or 19 days. Larval life can extend much longer, particularly in species that enter quiescent phases at lower temperatures, intermittently feeding when warmer, or in detritus feeders, which can simply wait for long periods when food is not suitable. Such species may live 100 to 140 days before pupation. Those that enter obligate diapause, usually as first or last instars, typically spend nine or 10 months as inactive larvae in addition to their feeding and growth period.

Larval Foods

The nutritional requirements of many caterpillars are generally similar. Synthetic diets that contain the same basic elements (casein, sucrose, salt, cellulose, wheat germ, amino acids, and vitamins, incorporated in an agar base) are successfully used for rearing many kinds of Lepidoptera. However, other species that are specific to certain plants fail to feed on a synthetic diet. Hence, nutritional value alone sometimes is not sufficient to elicit feeding, and natural plant chemicals act either as cues for feeding or as deterrents, often the same chemical in both roles with different larval species.

The vast majority of Lepidoptera caterpillars consume living plants (phytophagous), almost exclusively flowering plants, primarily angiosperms. All parts of plants are consumed—leaves, flowers, fruit, stems, roots—each kind of caterpillar specializing on its particular niche, feeding internally (endophagous) as leaf miners, root borers, and so forth, or externally (exophagous), either concealed in shelters constructed with silk, or exposed. Larvae of the most primitive family, Micropterigidae, consume liverworts and mosses or are general feeders on green plants, fern sporangia, fungal spores, and so on in damp habitats. Some other groups of moths do not feed on flowering plants (e.g., Tineidae) but specialize on wood rot fungi (Polyporaceae); or are detritivores on the ground, under bark of dead tree limbs, and in abandoned insect and spider nests; or feed on animal products in mammal burrows, bird nests, scats. A few can digest wool. Many species feed on fallen leaves, notably Oecophoridae and Tortricidae on *Eucalyptus* (Myrtaceae) in Australia, and several groups of Noctuidae in wet forest habitats. Rarely, Lepidoptera specialize on lichens (some Psychidae, Xylorictidae, Arctiinae), mosses (some Crambidae), or ferns (unrelated species, mainly oceanic islands). A few Lepidoptera are predaceous on scale insects or other Homoptera or in ant nests. A Hawaiian geometrid moth (*Eupithecia*) is predaceous on adult flies, which it catches by seizing the fly with elongate prothoracic legs. Other members of the worldwide genus *Eupithecia* are plant feeders.

Virtually every kind of flowering plant is eaten by one or more species of caterpillar. Food preferences vary enormously among families; they are summarized in the accounts of the families that follow. Internal feeders, such as leaf miners, stem and root borers, and gall inducers, and most other microlepidoptera are specialists to one or a few related plants. By contrast, the vast majority of macromoths are external feeders, and half or more of them are generalists within habitats. Examples include ground-dwelling cutworms feeding on low-growing herbaceous plants or shrub- and tree-feeding species either on broadleaf angiosperms or conifers.

Pupal Development

The duration of pupation, during which metamorphosis to the adult occurs, varies with temperature, usually requiring about 10 to 14 days, but many species require several weeks or undergo diapause as pupae for several months.

Pupal movement is an important adaptation in primitive moths and the ancestral Ditrysia. The pupa moves forward just preceding adult eclosion and either anchors by the cremaster to silk or wedges in the emergence aperture, which is prepared by the larva to be slightly narrower than the pupal abdomen.

Gelechioidea and the Obtectomera (Fig. 24) have independently derived fusion of abdominal segments that restricts movement to at most the posterior part of the abdomen, enabling turning within the cocoon but not forward movement, and the adult emerges directly from the pupation site. Pupae respond to tactile stimuli, including potential predators and probing by a parasitoid wasp ovipositor, by turning or wriggling. Many moth pupae have special structures on the abdomen that produce clicking or rattling sounds when the wriggling abdomen strikes the walls of the pupal cells or parchmentlike cocoon, or sounds are produced by rubbing fine pegs or rasplike surfaces on adjacent segments. Such sounds may aid in pupal defense.

Life Cycle

Most Lepidoptera in temperate climates undergo a single annual generation (univoltine), although many have two discrete seasonal broods (bivoltine), and some produce continuous generations so long as favorable temperature conditions prevail (multivoltine). Diapause, an arrested state of development regulated by hormones, controls the life-cycle pattern and enables populations to survive during unfavorable times (winter, dry season, etc.) when necessary resources are not available.

Diapause may be the single most important adaptation leading to species radiation of Lepidoptera in northern climates and high mountains, in the world's deserts and tropical dry season habitats, and other places where insects could not grow and reproduce continuously. Diapause occurs either in eggs, in first- or last-instar larvae, pupae, or as a reproductive delay in adults. The stage in which it occurs is constant for each species but varies between species, sometimes even within a genus. In Mediterranean climates, larval feeding typically occurs in spring when foliation peaks, and diapause lasts through the dry season in summer and hibernation in winter. Some species aestivate in diapause as prepupal larvae or pupae, fly in autumn, then hibernate as adults or eggs. Multivoltine species either enter diapause at the end of the growing season, often triggered by decreasing day length (photoperiod), or the larvae simply wait in a quiescent state, feeding slowly on warm days through winter, and adults of the spring generation metamorphose and eclose with warmer temperature in spring.

Many Lepidoptera are capable of maintaining the diapause to a second or later season if appropriate climatic conditions do not occur. This happens as a regular phenomenon in species adapted to seed-feeding on plants with biennial crops such as conifers or sporadically in species that depend upon resources limited to a specific season but erratic in abundance, such as flowering and fruiting by desert plants. Prepupal larvae of yucca moths (Prodoxidae) have successfully metamorphosed synchronously in large numbers after four to 30 years in diapause in experimental conditions.

Significance in Natural and Human Communities

Larvae of plant-feeding insects play a critical role in the biosphere by converting complex chemical energy in plants to digestible food for other members of the food chain, and their biomass far exceeds that of vertebrate herbivores. Hence the major role of Lepidoptera in natural communities is primary consumer of plants. Moths and butterflies make up the largest single evolutionary lineage adapted to depend upon living plants, in terms of species numbers and in many communities in biomass as well. Females of most species produce 200 to 600 eggs within a few days, vastly more in some species (1,000 to 30,000), releasing a potentially enormous load of caterpillars onto particular plant species or plant guilds such as herbs, woody shrubs, or trees. The caterpillars become an important food resource for specialized parasitoid wasps and flies; for general invertebrate predators such as spiders, mites, ants, social wasps; and for vertebrate predators, especially birds. There have been estimates of 80,000 caterpillars of several species feeding on a single oak tree and many times that number during outbreaks of single species that defoliate forest trees. Thus caterpillars comprise a major component of biological communities, affecting foraging by birds, buildup of yellow jacket colonies, and insect disease epidemics. A secondary role as decomposers also is filled by Lepidoptera. Tineidae, several groups of Gelechioidea, and some Noctuidae and other moths are detritivores and assist in reducing fallen leaves and fruit, fungi, and animal products (hair, feathers, predator scats) to humus. Finally, a few species are secondary consumers, predaceous on scale insects or other Homoptera in natural communities.

Economic Importance to Humans

Lepidoptera larvae damage plants grown for human use (food, lumber, cotton, garden ornamentals) and our stored products (grain, flour, nuts, woolen clothes and carpets). Most agricultural damage occurs because monoculture crops are grown in places distant from the natural enemies of the pest species, which themselves usually have been introduced by human activities to a new region. Wide-scale insecticide suppression of pest species has further increased problems because local parasites and invertebrate predators are eliminated, and the pest species become resistant to the insecticides by selection for survivors of repeated treatments. Similarly, pests of stored food and wool products have been transported worldwide by human activities. Lepidoptera probably are the most important insect group as plant defoliators (e.g., spruce budworm, the economically most important insect in Canada; larch budworm in Europe), and they cause huge losses by damage to fruits (e.g., codling moth, the "worm" in apples), corn (corn earworm, European corn borer), potatoes (potato tuberworm), cotton (pink bollworm), and many other crops and garden plants. They are a major problem in stored meal, grain, and nuts (Angoumois grain moth, Indian meal moth, Mediterranean flour moth, etc.) and in woolen products (casemaking clothes moth, webbing clothes moth, tapestry moth, and others). Still others infest bee nests, eating the combs (greater and lesser wax moths).

Conversely, some moths (e.g., Sphingidae and Noctuidae) are believed to play significant roles in pollination in natural communities, and they may aid in crop pollination in some instances. Several Lepidoptera have been purposefully introduced to act as biological control agents against noxious plants. Notable examples include a pyralid, the cactus moth, from Argentina, used to successfully suppress millions of acres of introduced prickly-pear cactus in Australia; an arctiid, the cinnabar moth from Europe, on tansy ragwort in the Pacific states of North America; and several Mexican species, to suppress lantana in Hawaii.

Fossil Record and Evolution

A widely accepted phylogenetic hypothesis of relationships among lepidopteran evolutionary lineages, based on morphological characteristics in living forms, primarily of the adults, is shown in Figure 25. The problem in such analysis is that we do not know what kinds of species might have preceded and interceded with the primitive extant lineages, each of which is now represented by one or a few relicts that have divergent larval features not shared with other Lepidoptera. Moreover, the fossil record is of little use in revealing clues to "missing links," and the paleontological preservation usually fails to provide information on critical characteristics, particularly of the immature stages.

Fossil Record

There are fossils of Triassic age (Fig. 26) assigned to Trichoptera (caddisflies), the presumed sister group of Lepidoptera, so branching of the two lineages may have occurred in the early Mesozoic. The earliest fossil recognized as lepidopteran is that of a small, scaled wing from the Lower Jurassic, from Dorset, England, nearly 200 MyBP. It was placed in a separate family, Archaeolepidae, suggested as a sister group to the Micropterigidae, but without characters available that might establish its relationships. Four genera were described from Russian Upper Jurassic tuffites (nearly 150 MyBP). Among these, two were assigned to Micropterigidae and two to Glossata and Ditrysia, but only one of them, *Protolepis*, possesses visible mouthpart structures. They were interpreted as a siphon formed of maxillary galeae, which would imply existence of Glossata 20 to 30 million years prior to the radiation of angiosperm plants during the early Cretaceous. That interpretation has been questioned; the structures may be maxillary palpi, and therefore the fossil may represent an extinct lineage of Aglossata. By the early Cretaceous there are well preserved Micropterigidae and an incurvariid (Heteroneura) in amber, and by the late Cretaceous several kinds of leaf mines representing modern families and host-plant associations, both Heteroneuran (Nepticulidae) and Ditrysian (Phyllocnistidae, Gracillariidae), as well as a Ditrysian larval head capsule of a free-living form such as Tineidae. That is, the fundamental clades of Lepidoptera are all represented before the beginning of the Tertiary. Hence, although Lepidoptera is the most recently evolved major insect order, its radiation has been relatively rapid, paralleling that of the angiosperms, the major lineages having evolved between about 140 and 90 MyBP.

Morphological Evolution

Major changes in morphological adaptation in adult feeding, oviposition mode, wing structure, and larval locomotion are indicated by Figure 25. The relict moths of ancient lineages (Micropterigidae, Agathiphagidae, Heterobathmiidae) share features of ancestral mecopteroids, functional mandibles in adults and pupae and similar fore and hind wings with complete venation, and a single female genital aperture. However, larvae of their extant species differ greatly from one another, each adapted for a particular lifestyle. Micropterigid larvae are free-living ground dwellers in moist situations, with well-developed thoracic legs, no crochet-bearing abdominal prolegs, and fluid-filled chambers in the cuticle. Agathiphagids are legless borers in primitive gymnosperm seeds, with reduced head sclerotization and sutures and few stemmata. Heterobathmiids are flattened leaf miners of southern beech, having a prognathous head with prominent adfrontal ridges, as well as seven stemmata laterally and thoracic legs with large, subdivided trochanters (unique in Lepidoptera), but no abdominal prolegs.

Adult Glossata (Eriocraniidae and all subsequent lineages) lack functional mandibles and feed by a proboscis formed of the maxillary galeae. Basal glossatan lineages have a piercing ovipositor and retain functional mandibles in the pupa, used to cut the cocoon at eclosion. The larvae have a spinneret. Several derived features occur beginning with the Exoporia (Hepialidae): The ovipore and gonopore are separate, connected by an external groove for sperm transfer; the larvae have differentiated prolegs on abdominal segments 3 to 6 and 10, with circles of crochets, and silk is used for various activities, not just cocoon formation, the ancestral condition in Lepidoptera. Functional pupal mandibles are lost, and there is no piercing ovipositor. Differentiated size, shape, and venation between fore and hind wings appear in the Heteroneura. The thoracic legs, crochet-bearing larval prolegs, and silk webbing are lost by larvae of Nepticuloidea, which are severely modified for leaf-mining. An independently derived piercing ovipositor occurs in Incurvarioidea, some of which have secondarily legless larvae.

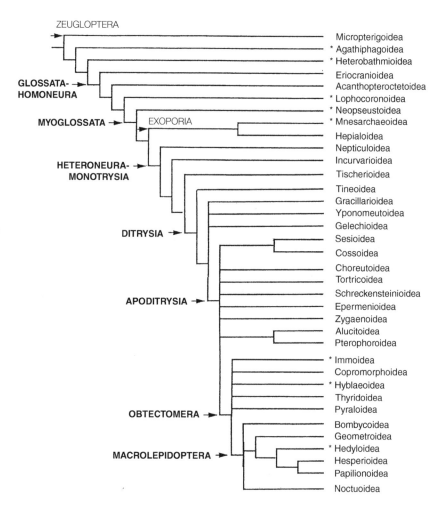

FIGURE 25. Two-dimensional portrayal of phylogenetic relationships of extant lepidopteran superfamilies, based on parsimony analysis. Successively more derived clades representing major morphological changes indicated in **boldface** to left. Asterisks (*) indicate superfamilies that do not occur in North America. [Based on Kristensen and Skalski 1999]

The last fundamental change, leading to the Ditrysia, is the internal system for storage and transfer of sperm from the gonopore to oviduct. Presumably this had evolved by the mid-Cretaceous, when larval mines of Gracillarioidea appear in the fossil record. The most successful lineages, in terms of extant diversity, Pyraloidea, Geometroidea, and Noctuoidea, which are defined by independently derived tympanal organs, presumably originated coincident with radiation of the bats during the late Paleocene and early Eocene.

Ecological Scenario

Questions remain concerning origins of angiosperm-feeding in basal lepidopteran lineages that led to major radiations of Lepidoptera. The ground-dwelling larvae of Micropterigidae are generalists, either as detritivores or fungivores in leaf litter or feeding on low-growing green plants in moist habitats, including bryophytes and soft angiosperm leaves. Similar habits occur in Exoporia (Mnesarchaeidae in New Zealand and Hepialidae, except many hepialids feed on roots or burrow into stems of woody angiosperms) and in basal Ditrysia (Tineidae, except none feeds on green plants). By contrast, extant larvae of the other lower Lepidoptera are internal (endophagous) feeders that specialize on particular flowering plants. We assume ground-dwelling, generalist habits are similar to those of mecopteroid ancestors of the Trichoptera-Lepidoptera clade, but we do not know if that mode of life persisted in basal members of all lineages through to the

Ditrysia. If so, adaptation to endophagy and to specialist angiosperm-feeding might have occurred at least four times: in heterobathmiids; in a Eriocraniid+Acanthopteroctetid+Lophocoronid+Neopseustid lineage; in nepticuloids, and probably independently in incurvarioids, when a piercing ovipositor reappears; and finally in a Palaephatid+Tischeriid lineage (Fig. 27). If an unknown angiosperm-feeding lineage was the common ancestor, at least two reversals to ground-dwelling, external-feeding, generalist caterpillars characterized by multiple morphological reversals must be postulated for Exoporia and again for Tineidae. In either scenario, there were independent origins of a piercing ovipositor (at least twice) and endophagous larval feeding accompanied by numerous derived morphological specializations in larvae (several times). Repeated shifts to angiosperm-feeding seem likely to have been facultative, as in extant micropterigids, rather than a single event. Multiple adaptations to endophagy implies parallel evolutionary trends, a more parsimonious scenario than multiple reversals to an ancestral morphological and behavioral ground plan.

Evolutionary Origin of Western Nearctic Floras

Painting with a broad brush, the geological scenario leading to the present-day flora of western North America, according to hypotheses of Chaney and Axelrod, included influence from three geofloras: The Neotropical-Tertiary, which developed from Cretaceous elements; the Madro-Tertiary, which arose in

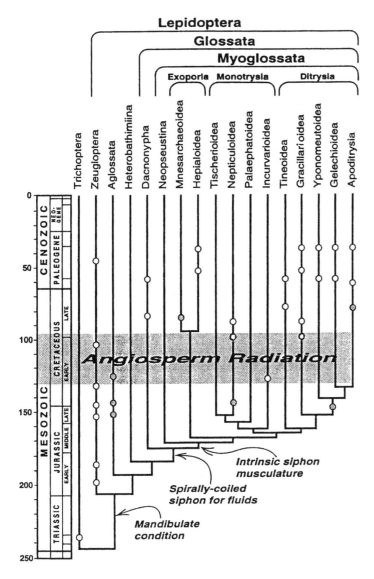

FIGURE 26. Phylogenetic hypothesis of major lepidopteran lineages superimposed on the geologic time scale, with fossil occurrences indicated. Open dots, reliable identifications; shaded dots, questionable assignments. Angiosperm radiation spans 130 to 95 Myra, from the earliest recognized occurrence of pollen to the time when angiosperms became the dominant vegetation. [Powell 2003, modified from Labandeira et al. 1994]

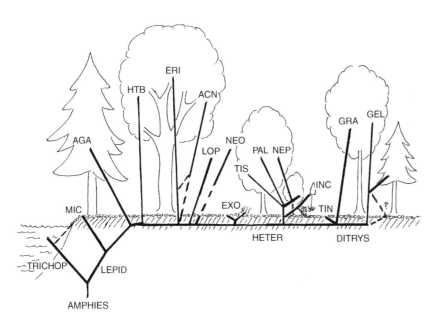

FIGURE 27. Pictorial representation of a theoretical scenario for the origins of angiosperm-feeding that led to the radiation of lepidopterans during the Cretaceous. A ground-dwelling mecopteroid-like ancestor gave rise to the Trichoptera-Lepidoptera split, then successively to the ancestor of extant Micropterigidae (MIC) and several specialized, radically differing, angiosperm-feeding lineages. The ancestral ground-dwelling caterpillar form is presumed to have been retained in Exoporia (EXO, Mnesarchaeoidea, Hepialoidea) and basal Ditrysia (Tineoidea). ACN, Acanthopteroctetoidea; AGA, Agathiphagoidea; DITRYS, Ditrysia; ERI, Eriocranioidea; GEL, Gelechioidea; GRA, Gracillarioidea; HETER, Heteroneura; HTB, Heterobathmioidea; INC, Incurvarioidea; LOP, Lophocoronoidea; NEO, Neopseustoidea; NEP, Nepticuloidea; PAL, Palaephatoidea; TIN, Tineoidea; TIS, Tischerioidea. [Powell 2003]

the Sierra Madre Occidental of Mexico and adjacent south-western United States during the Eocene and Oligocene from the Neotropical-Terriary; and the Arcto-Tertiary, which was widespread in northern latitudes during the Eocene. Each maintained its identity through time and contributed elements to the western flora at varying times through the Cenozoic.

There was a shift from humid, warm climate over a West Coast lowland dominated by subtropical forests in early Tertiary times to cooler, drier floras as the Pacific cordillera arose during the late Tertiary. As inferred by Axelrod, in the early Tertiary a broad coastal plain of low to moderate topography supported rich, mesic forests that included numerous hardwoods, many of which are related to descendants that live in deciduous forests of the eastern United States, and subtropical areas of abundant summer rainfall. The interior cordilleran axis (present-day Rocky Mountains area) had conifer-hardwood forest. The mesic lowland was gradually replaced by generally drier climatic regimes superimposed on increasingly complex geographic diversity, particularly the rise of the Cascades and Sierra Nevada. Fossil records of equivalents of modern conifer species extend back to the mid-Eocene (40 MyBP). A drier climate had spread over the region by the mid-Oligocene (32 MyBP), and forests were dominated by modern conifer genera, with associated deciduous shrubs. In early to mid-Miocene (25–15 MyBP) temperatures were warmer in the Northern Hemisphere than now. The Cascades and Sierra Nevada were still relatively low; their major uplifts occurred in the Pliocene and Pleistocene. Hence there was broad marine influence and more homogeneous vegetation in northwestern North America. Ocean temperatures shifted in the mid-Miocene (16–15 MyBP), causing greater temperature ranges in the interior, and this trend was accompanied by uplift of the mountains. Middle and late Miocene floras from lowland areas of the present-day Sierra Nevada were dominated by subtropical hardwoods. By early Pliocene (10–8 MyBP), rainfall, especially summer rain, is inferred to have decreased in the lowlands east of the cordillera, and redwood forests disappeared to be replaced by oak-juniper woodland. Douglas-fir forests spread to the Pacific Coast and into California during the late Pliocene. The progressive confinement of boreal forest conifers northward and coastalward and rise of the Sierra Nevada continued into the Pleistocene, and residual floras of the Great Basin took on modern aspects. Madrone, tanbark oak, and Catalina ironwood disappeared, and white fir was restricted to higher elevations of the Basin Ranges.

Quaternary forests in the Southwest formed as a result of increased precipitation in the early Pleistocene (3–2 MyBP), presaging the buildup of major ice sheets, which alternately shifted the floras southward and northward. Reconstruction of the extent of mountain glaciation in the western United States and associated floristic events is less precise than is true for the fluctuations of the major ice sheets in the Midwest and East. Owing to the topographic diversity, the situation in the West is more complex, with the mountain ranges oriented on north-south axes, and the fossil record is more fragmentary. The Cordilleran Ice Sheet extended from Alaska nearly to the Columbia River, where it attained maximum extent 14,000 years BP. Southward in the Cascades–Sierra Nevada, Basin Ranges, and northern Rocky Mountains, the montane glaciers extended down to low elevations, 610 m in the northern Cascades and 2,240 m in the Sierra Nevada. During each glacial cycle periglacial zones of permafrost formed, characterized by tundralike vegetation above the spruce-fir forests. Forests of the western United States survived in refugia at lower elevations and in some areas well south of their present ranges. Evidence from macrofossils and wood rat middens in the Southwest place spruce-fir forests about 450 km south and 415 m downslope from their modern counterparts. Fossil pollen analyses indicate that the last northern retreat of Douglas-fir forests reached southern Idaho by 11,000 years BP, and contact by its Pacific and Rocky Mountain races in northern Idaho occurred by 7,000 years BP, with each interglacial retreat, forest dominants, including many Madro-Tertiary elements, again moved northward, ultimately reaching their present-day positions.

A History of Moth Collectors in Western North America

As would be expected, collecting insects in North America lagged behind such activities in Europe, and in the West behind the eastern United States. Early in the nineteenth century the conspicuous species were collected and exported to be named by European workers. In North America, resident entomologists such as Fitch and Harris began publishing descriptions of insects in connection with agricultural concerns by the 1840s, and even the microlepidoptera specialists Clemens and Chambers were describing species in the 1860s.

Early Pioneers

THOMAS SAY visited the Rocky Mountains with the C. H. Long Expedition in 1819–1820 and later described many insects from his collections, but there are no known extant voucher specimens of moths from the Long expedition. The first insect collections on the Pacific Coast were made by Russians associated with the fur trade in the 1840s, but their emphasis was on beetles and butterflies. The earliest moth collections in the West were made by Europeans who migrated to California following the discovery of gold in California with the mass immigration of fortune seekers. P. J. M. LORQUIN (1797–1873) came to California in 1850–1856 and collected Lepidoptera in the Sierra Nevada and elsewhere. He is best known for having discovered many of the most common butterflies, which were described by Boisduval, his family physician in France. He also collected the first specimens of several endemic western moths, primarily Hepialidae, Sphingidae, and Saturniidae. JAMES RIDINGS (1803–1880) was an Englishman who came to Philadelphia in 1830. He collected insects extensively in the East and was instrumental in founding the Entomological Society of Philadelphia, forerunner of the American Entomological Society. Ridings was one of the earliest collectors to visit the Rocky Mountains, when he made a trip to Kansas and the Colorado Territory in 1864.

Additional collectors in the West arrived in the 1860s, including R. H. STRETCH (1837–1926), who came to North America from England in 1863 and migrated to Virginia City, Nevada, where he became state mineralogist. Later, as a mining engineer, he lived in San Francisco and two years at Havilah, Kern County. He sent specimens to Behr, Grote, and Scudder, but much of his material was given to Henry Edwards,

and it appears that many of the species credited by taxonomists to Edwards, whose specimens bear the label "Hy. Edwards Collection" without specified collectors, were collected by Stretch. He specialized on diurnal moths, and in the 1870s published on zygaenids and bombycids, which included the arctiids. He moved to Seattle and collected there after 1888 but is said to have lost interest in insects after Edwards died in 1891. According to Essig, Stretch's private collection was given to the University of California, Berkeley, but it was transferred to the California Academy of Sciences as a loan in 1919.

HENRY EDWARDS (1830–1891) was a British theater actor who traveled to Australia, Mexico, and South America before settling in San Francisco from 1865 to 1877, when he moved to New York. While in California he collected extensively and received specimens from Stretch and others. Edwards described many moths, especially sesiids and sphingids. His collection, said to have contained a quarter million specimens, was purchased by friends and given to the American Museum of Natural History.

Early collections were made in the Rocky Mountains by T. L. MEAD, who traveled with the G. M. Wheeler Geographical Survey expeditions in 1871–1874 and visited Nevada and territories later to become Colorado, Utah, New Mexico, and Arizona, before there were resident collectors in the Southwest. Moths he collected were described by Packard and Henry Edwards.

Far and away the most important expedition to the Pacific Coast for Lepidoptera was made in 1871–1872 by the English nobleman THOMAS DE GREY, SIXTH LORD BARON OF WALSINGHAM (1843–1919). Lord Walsingham, at the age of 27, accompanied by Thomas Eedle and Amos Carrier from his estate, sailed to New York. After examining collections of the American Museum, Philadelphia Academy of Sciences, and Smithsonian Institution, they traveled to San Francisco, just two years after completion of the transcontinental railroad. They spent about a week in San Francisco, during which Walsingham hired another assistant, obtained horses and mules, had a wagon outfitted as a field laboratory, and collected the type series of *Argyrotaenia franciscana* "amongst plants of wild lupin on the sand hills."

The party departed May 17, 1871, by boat to Petaluma, north of San Francisco Bay, via the Petaluma River. From there, they traveled overland for the ensuing 13 months, moving, often daily, from campsite to campsite. Walsingham hunted birds and mammals, for which he was renown in England. Bird skins were dried and shipped back to England, presumably along with moth specimens. He kept a diary of the travels, including insights into his collecting, but unfortunately the family destroyed the original diaries of his travels, except for the first five weeks. Records of this marvelous expedition consist of handwritten copies, evidently made after 1900 by Durrant, Walsingham's curator, which are deposited in the Natural History Museum, London (photocopy in Essig Museum, University of California, Berkeley). Included are the account by Walsingham of the first weeks, a daily itinerary of the full expedition made by Carrier, and numerical lists of species and of species bred from larvae. There are comments such as this one, made on May 25, at the head of Dry Creek: "On again; a short march after a long morning's setting" [spreading]; and at Camp 15, head of the Noyo River, south of Little Lake, June 8 to 11: "good collecting ground, several new species, . . . collected and set at this camp 400 specimens, many new." (Actually, nearly all of them represented new species.)

In all, 98 camps were established during 1871, as they traveled up the Coast Range, crossed over the northern Sacramento Valley to the Cow Creek drainage, Burney, Pit River, and Mt. Shasta. The most important bivouac was Camp 45 near the present McCloud at the foot of Mt. Shasta, where they stayed all of August, and Walsingham forayed widely on the mountain hunting bears and collecting moths. The party was in Oregon from September 13, 1871, until June 26, 1872. They traveled in southern and eastern Oregon in severe winter conditions before finally settling at Camp Watson on the John Day River in December and stayed over the winter. In 1872 the party left Camp Watson on April 14, traveled to The Dalles, shipped by boat to Portland, then resumed their trip southward, establishing 36 camps. They spent three weeks at Dutcher's Creek near Jacksonville in May, and a week at a camp in the Siskiyou Mountains before reentering California near Crescent City. They reached Eureka on June 30, where they spent 10 days, hunting and presumably packing all the specimens for the return trip, and left by ship to San Francisco July 11. "Homeward bound to England," Carrier wrote.

For more than 40 years Lord Walsingham published results of his expedition, describing hundreds of species, more than any subsequent student of western Nearctic microlepidoptera, many of them represented by long series or bred from larvae. Publication began in 1879 with *North American Tortricidae* and in 1880, *Pterophoridae of California and Oregon,* in which he described many of the now well-known species of the Pacific Coast. Later works included adelids, prodoxids, tineids, gelechioids, plutellids, heliodinids, and tortricids. Some of his species were not rediscovered for 90 to 100 years. Walsingham purchased the Zeller and other important collections, and his collection, numbering 260,000 microlepidoptera, was presented to the British Museum in 1910.

L. E. RICKSECKER (1841–1913) worked as railroad engineer in Utah in 1868 and later at Spokane, Washington, before settling in California in 1873. He lived in San Francisco and owned a ranch near the present-day Camp Meeker, Sonoma County, which was used as a field headquarters by biologists from the California Academy of Sciences. He reared and sold specimens to eastern and foreign taxonomists, and several Lepidoptera were named in his honor by Behr, Pearsall, Klots, and others. His private collection was destroyed by a fire at Santa Rosa following the 1906 earthquake, after which he and his wife moved to San Diego. She provided Essig with an informative account of Ricksecker's industrious collecting activities, through which they aimed to make expenses and usually did. *Hemileuca* and butterflies were reared by the 100s to be sold as pupae. Moths were trapped for Barnes, who paid half a cent apiece, and they often got 1,500 in a night. She also reported that Ricksecker tethered female beetles and moths to attract the males, probably the first use of pheromone-based collecting in western North America.

JAMES BEHRENS (1824–1898), a German, was one of the early pioneer collectors in California, beginning about 1874 in the San Francisco area. He collected all orders of insects and described a few species. He sent Lepidoptera to Grote, Hulst, Packard, Chambers, and others, and several species are named in his honor. Behrens's private collection went to the Museum of Lübeck, Germany, which was destroyed in 1943.

C. F. MCGLASHAN (1847–1931) came to California from Wisconsin by ox wagon as a small boy in 1854. He was the school principal in Placerville when he took up an interest in Lepidoptera and became superintendent of schools in Truckee in 1872, where he collected extensively for many years, the first to do so in the high Sierra Nevada. From 1875 to 1887 he sent most of the moths to Henry Edwards, and much Edwards collection material is labeled "Truckee," without a collector's name. Later he and his seventh daughter, Ximena, operated a butterfly farm for a few years beginning in 1912. They issued catalogs and sold specimens to many lepidopterists, especially Barnes, enabling Ximena to earn enough money to pay for her college education at the University of California and Stanford. There are extensive series of noctuids and other moths from the Barnes collection now housed in the Smithsonian, with the characteristic Barnes labels, lacking year and the collector's name. The remainder of McGlashan's collection is maintained in a small museum in Truckee.

H. K. MORRISON (1854–1885) did taxonomic work on Noctuidae as a young man and traveled to Georgia to explore the area where Abbot had made his famous eighteenth-century collections and paintings of insects. Morrison later undertook expeditions to the West as a professional collector, visiting Colorado in 1874–1877, Utah and Nevada in 1878, the Washington Territory in 1879, Arizona and southern California in 1880, Oregon in 1880–1881, New Mexico in 1883, and Nevada and Florida in 1884. But in Florida in 1885 he contracted a dysentery, which proved fatal, and he died at his home in North Carolina at the age of 31. He sold specimens to Henry Edwards, Grote, Walsingham, and others, from which many species were described. Unfortunately, he provided only the state name as data, although some records published by other taxonomists provide localities. Some of his specimens labeled North Carolina almost certainly originated elsewhere.

J. J. RIVERS (1824–1913) came to the United States from England in 1867 and settled in Kansas, where he was associated with F. H. Snow, the pioneer entomologist at the University of Kansas. Rivers moved to California about 1880 and was curator of natural history at the University of California, Berkeley, until 1895, when he moved to Santa Monica. He collected all kinds of natural history specimens and published several papers describing new species. In Lepidoptera he was chiefly interested in butterflies and *Malacosoma* and provided moth specimens to Behr, Dyar, and others. His collection, largely beetles, was sold in Europe to museums in Germany.

Two important contributors to western moth collection history in the late nineteenth century were **D. W. COQUILLETT** and **ALBERT KOEBELE,** who coincidentally lived in California in the 1880s. Coquillett (1856–1911) moved from Illinois to southern California in 1882 because of failing health due to tuberculosis. After recovery he resumed study of insects and published on diverse groups, especially Diptera. His collections of microlepidoptera, including the yucca moths described by Riley, mostly labeled "Los Angeles Co.," were the first in southern California and included many rearing records. His work led to an appointment as a field agent for the U.S. Department of Agriculture, Division of Entomology in 1885 by C. V. Riley. Coquillett was appointed curator of Diptera at the U.S. National Museum in Washington, D.C., in 1893. Koebele (1852–1924) came to the United States from Germany in 1880, met Riley and other entomologists, and was appointed to the Department of Agriculture, working in Georgia, Florida, and Brazil in 1882–1883. He requested a transfer from Washington and moved to Alameda, California, in 1885. He replaced Coquillett in southern California for a year in 1886, then returned to Alameda, where he collected extensively, reared many moth species, and did experimental spraying for pest species. When importation of ladybird beetles was proposed to replace fumigation control methods for scale insects on citrus, Riley sent Koebele to Australia, with Coquillett taking care of the beetle shipments and distributing them to orchards. The arrangement proved successful, and the three men were made famous for initiation of classical biological control. Koebele lived in Alameda intermittently but worked in Australia and New Zealand from 1888 until 1893, sending back large numbers of coccinellid beetles. Then he accepted a position with the Hawaiian Board of Agriculture and carried out similar work for many years. Koebele returned to Germany in 1910 because of ill health, survived World War I, but was too weak to travel again, and died there. His early California collection, which remained in his home in Alameda, was given to the California Academy of Sciences by Mrs. Koebele and contains the most important sample of nineteenth-century microlepidoptera of the San Francisco Bay area.

Early-Twentieth-Century Residents and Visitors

C. P. GILLETTE (1859–1941) was a pioneer entomologist in Colorado. After earning B.S. and M.S. degrees from Michigan State Agricultural College, he was employed for three years in the Experiment Station of Iowa State College. In 1891 he was placed in charge of the Department of Zoology and Entomology at the Colorado Agriculture College (now Colorado State University), and for more than 40 years he was associated with that institution and its growth. He was the Colorado state entomologist from 1907 to 1931, in addition to his university duties. He was said to have been a tireless collector, and although primarily interested in cynipid wasps and Homoptera, he motivated growth of the collection at Colorado State University of all insects. In 1898 he compiled a list of Lepidoptera captured in Colorado, including more than 600 species, annotated with locality, date, and ecological circumstances. The paper was illustrated by a photograph of his light trap. Most of the specimens were collected by Gillette, but he was assisted by others, especially C. F. Baker, who was his assistant for several years after he too graduated from Michigan State University. Though most records were from Ft. Collins and vicinity, the list was the first attempt to compile occurrence of all Lepidoptera in any of the western states.

H. G. DYAR (1866–1929) was a truly remarkable man in many ways. Described by his biographers, Epstein and Henson, as a person who wanted real life to conform to his idealistic expectations of efficiency and logic, he had little patience with the inaccurate work of colleagues and with bureaucratic procedures. Thus dedicated, his accomplishments stagger the imagination. Harrison Dyar was born to a wealthy family in New York, began an interest in insects as a youth, and became well known for studies on lepidopterous larvae. His seminal paper on the number of instars in lepidopterous larvae, in which he demonstrated that head capsule widths show a geometric progression in growth—Dyar's rule—was published before he began graduate school. While a M.S. student he proposed a classification of Lepidoptera based on the primary setal patterns of larvae, which was much advanced over the wing vein–based classification existing at the time. After obtaining his M.S. degree, then a Ph.D. in bacteriology at Columbia, he taught bacteriology there for three years. Beginning in 1897 he served as Custodian of Lepidoptera at the Smithsonian Institution for 31 years, without salary. Dyar is best known for his work with Lepidoptera larvae, his *Catalogue of North American Lepidoptera* in 1902, later monographic studies of mosquitoes, and founding the journal *Insecutor Inscitiae Menstruus*, which he edited and published for 14 years. Yet incredibly, he also managed two households simultaneously, wrote short stories, described large numbers of new Lepidoptera species, as well as larvae of sawflies and moths, and continued a lifelong practice of spending summers collecting and rearing insects. He made an amazing number of trips across the country for that era. Dyar worked in Yosemite Valley in May to June 1891, and around Portland, Oregon, in 1892. Later, accompanied by A. N. Caudell, who was appointed his assistant by the U.S. Department of Agriculture, he spent several summers in the West, including Colorado (1901) and British Columbia (1903), visited the Grand Canyon in 1906, and California in 1915 and 1916. In 1917 he was back in British Columbia, Seattle, and Montana, and in 1918 collected in Alberta. Dyar described larvae of about 90 species obtained in Colorado and published a list of Lepidoptera recorded in the Kootenai Valley area, British Columbia, compiled from his field work and that of **J. W. COCKLE,** a resident of Kaslo, and others. The work listed more than 650 species, 167 of which included larval descriptions. His specimen labels often were incomplete, locality names abbreviated, lacking larval host or year, but they included rearing lot numbers, referenced to notebooks at the Smithsonian. His collection was given to the Smithsonian in 1917.

Probably the most important contribution to our early knowledge of Great Basin moths was generated by **TOM SPALDING** (1866–1929), who lived and collected in Utah during the early 1900s. He left England in 1885 and migrated to Salt Lake City. He began to collect insects about 1900 when he learned that students in the East would buy specimens. He was a night watchman at mines southwest of Salt Lake City, in Stockton, Eureka, and later Dividend, where he harvested specimens from lights at the mine entrances. In 1905 Spalding established residency and built a "collecting cabin" at Vineyard, about five miles northwest of Provo, and he traveled to various other parts of Utah to collect moths and beetles. He sold specimens to Barnes, Dyar, Grossbeck, Kearfott, and others, from which a great many species were described, including more than a dozen named in his honor. Many taxonomists have used the Barnes material at the Smithsonian and benefited from Spalding's efforts.

G. H. FIELD and **W. S. WRIGHT** were collectors in the San Diego area around the turn of the century. Field (1850–1937) farmed in Kansas as a young man, then moved to San Diego in 1889, where he became head janitor of the city schools. He began collecting insects in 1895 through an association with a local coleopterist and later turned to Lepidoptera. He and Wright collected extensively in San Diego County, especially at the Stephens Ranch in La Puerta (now Mason) Valley at the edge of the desert. Field sent specimens to Barnes, Dyar, Grossbeck, and others, who described many new species including about a dozen patronyms for Field. After retirement he was curator of insects at the San Diego Museum of Natural History from 1920 to 1922, and he donated his collection and library to the museum in 1934. Wright (1866–1933) came to California from the Midwest, where he began a career as a school teacher that continued in San Diego until 1928. He too began collecting Lepidoptera about 1895, mainly around San Diego, but he made several trips to the Sierra Nevada and Arizona. Wright became a specialist on Geometridae and phycitine Pyralidae and published descriptions of many new species. He donated his collection of insects, including 40,000 Lepidoptera, to the San Diego Museum of Natural History where he was the curator of entomology from 1923 to 1933.

OTHO C. POLING (1871–1929) collected extensively in the Southwest, but much of his life remains enigmatic, and there appears to have been no published account of his field expeditions. He was born and raised in Quincy, Illinois, in a well-to-do family. As a youth he began collecting moths and butterflies and later became acquainted with William Barnes, of Decatur, Illinois, who described many species received from Poling. After college, Poling worked for some years with his father's firm, dealing in farm loans. Later, however, he developed his hobby into his profession, and during his last 12 years he devoted full time to collecting. Poling, who is said to have been better known as an ornithologist than entomologist, was married as a young man. The family included a daughter and son; a third child died young. According to Barnes, by 1904 Poling had made a number of trips to little-known parts of Arizona and Utah, including at least once with Barnes, and had collected in the Huachuca, Chiricahua, and Santa Catalina mountains and in Pinal and Pima counties. Poling's wife, Mary, died in 1911, and he moved to the Southwest several years later. He married again, and according to a 1920s article in a Laguna Beach, California, newspaper, he and his wife spent several months each year in the field, usually starting in early spring in the desert and working their way into the mountains as the season progressed. Poling sold specimens to Barnes and most of the larger museums—Smithsonian, Academy of Natural Sciences in Philadelphia, American Museum of Natural History—and institutions in Europe, especially England. During part or most of the 1920s the Polings spent winters in Laguna Beach. He collected in the California deserts, Arizona, New Mexico, and western Texas, and he was quoted as stating they always tried to find sites that had been as little disturbed as possible, spending weeks in then-remote places such as Big Bend, Texas, and the Baboquivari Mountains, Arizona, where he spent several months in 1923 and 1924.

Fourteen species and one genus of Lepidoptera were named in Poling's honor, mostly in 1900–1906 and 1922–1929, reflecting the two primary eras of his travels. In addition to Lepidoptera, at least 10 insects of other orders were named for Poling and/or his wife, including a camel cricket from Quincy, Illinois, in 1897, a plant bug, a blister beetle, a long-horned

beetle, two crane flies, two wasps, and a bee. Poling died in Tucson, Arizona, from complications following influenza. His collection, which was said to have contained 30,000 specimens, was sold to Barnes.

JAMES MCDUNNOUGH (1877–1962) is well known for his many publications on Lepidoptera and Ephemeroptera, author of about 1,500 species names and the widely used checklists of North American Lepidoptera in 1917 and 1938–1939. He was born in Ontario, educated there in private schools and in Germany, preparing for a career in music, but he abandoned it and completed a Ph.D. in zoology in Berlin and a M.A. at Queen's University in Ontario. He worked with Wm. Barnes as curator and wrote their joint publications from 1910 to 1919, after which he was appointed chief of the newly created Division of Systematic Entomology at Ottawa, where he remained for 28 years as curator of Lepidoptera in the Canadian National Collection. During both phases of his career McDunnough collected in the West, although none of his Barnes collection specimens bears his name. Among other western ventures financed by Barnes, McDunnough visited Mt. Rainier in 1911. He spent the summer of 1913 in Colorado, and in 1915 he collected in the vicinity of Shasta Retreat (a resort at Dunsmuir) and Soda Springs, south of Mt. Shasta, California, and visited Crater Lake, Oregon. Evidently he worked at Lake Tahoe (Deer Park Springs, now in South Lake Tahoe) and Truckee in 1911 and 1917. From Ottawa he made several trips to British Columbia, collecting there in 1924–1933, often with A. N. Gartrell, W. C. Carter, and Blackmore, such that Lepidoptera of British Columbia were better documented by the 1920s than those of any of the western United States. After retirement in 1946, McDunnough lived in New York for three years, then in Halifax, Nova Scotia, and continued to collect, rear, and publish on microlepidoptera. His specimens from his post-Barnes years are in the Canadian National Collection in Ottawa and the Halifax Museum, Nova Scotia.

E. H. BLACKMORE (1878–1929) was an early collector in British Columbia, from about 1895 until his untimely death. He published a check list of macrolepidoptera of British Columbia in 1927 and devoted time to the study of micros, while acting as honorary curator of the Lepidoptera at the Provincial Museum in Victoria.

JOHN A. COMSTOCK (1881–1970) is best known for his work on butterflies, particularly his influential book *Butterflies of California*, published in 1927, but he also collected and reared moths extensively in southern California and Arizona and published more than 150 articles on life histories of Lepidoptera, many in collaboration with **C. M. DAMMERS**. Comstock was born in Illinois and developed an early interest in natural history, especially Lepidoptera. He moved to Santa Rosa, California, in 1907, where he worked for Luther Burbank as a designer of publications. In 1910 he moved to southern California, earned an M.D., and ultimately became director of science of the Los Angeles County Museum of Natural History and was the editor of its *Bulletin* for 40 years. Among his achievements, he encouraged the development of numerous lepidopterists, including Dammers, **JOHN** and **GRACE SPERRY**, **CHRIS HENNE**, and **LLOYD MARTIN**, who collected extensively in southern California and the Southwest. After retirement in 1948, Comstock moved to Del Mar and continued Lepidoptera life-history studies. His collections are deposited in the Los Angeles County Museum.

F. X. WILLIAMS (1882–1967) was a native Californian who collected in the San Francisco Bay area and at Lake Tahoe around the turn of the century. He was born in Martinez, and the family moved to San Francisco when Francis was five years old. He developed an early interest in insects and collected in San Francisco when there were open areas with sand dunes and native vegetation. In 1905–1906, while Williams was a student at Stanford, he was a member of the California Academy of Sciences expedition to the Galapagos Islands, and his collection had been housed at the family home in Martinez, so it escaped the fire that destroyed much of San Francisco and the California Academy of Sciences collections. Williams's material, later given to the academy, comprises the best sample of the now-extinct San Francisco sand dunes Lepidoptera fauna. Williams published several papers on moths and butterflies from 1905 to 1910 but then entered graduate school at the University of Kansas, where he took up a lifelong study of Hymenoptera, particularly aculeate wasps. He obtained a D.Sc. at Harvard in 1915 and later spent most of his amazingly productive career with the Hawaiian Sugar Planters Association, working in the Philippines and other countries, investigating parasites and predators of pest species. After retirement, he returned to California in 1948, continued to publish on wasps, and served as a volunteer curator at the San Diego Museum of Natural History until 1966, but he did not revive his early interest in Lepidoptera.

ANNETTE BRAUN (1884–1978) was a pioneer microlepidopterist, especially fond of the leaf-mining groups, and published prolifically, including monographs of lithocolletine Gracillariidae, Nepticulidae, Elachistidae, Bucculatricidae, and Tischeriidae, over a period of nearly 70 years. She lived in Cincinnati and in 1911 was the first woman to receive a Ph.D. degree from the University of Cincinnati. During her early career Annette collected primarily in Ohio and Kentucky, traveling by hired horse and buggy with her sister, E. Lucy Braun, a renowned botanist. According to her biographer, Lisa Stern, the Braun sisters bought their first car in 1934. Nonetheless they had already made trips to Glacier National Park, Montana, and Grand Lake, Colorado, in 1920, and to northern Utah–southern Idaho and Zion Canyon in 1924, from which Annette published lists of species of microlepidoptera and described numerous species. Later they visited the West several times, including California in 1925 and 1930, the Grand Tetons National Park, Wyoming, in 1929, and Arizona in 1939. Moreover, Annette enlisted **G. R. PILATE** to collect for her in California and to send active leaf mines, from which she reared and described many species. Pilate's collection localities included Mills College in Oakland in 1908; Rivera, Los Angeles County, in 1909; Fredalba and Loma Linda, near San Bernardino, June to October, 1912; Palm Springs in March and Monache Meadows (Tulare County, now within a U.S. Forest Service wilderness area) in the high Sierra Nevada, where he also collected for Barnes in 1913 and 1917. After publication of her Tischeriidae monograph at age 88, with her eyesight failing, Annette Braun had to give up study of the tiny leaf miners, and she donated her collection to the Philadelphia Academy of Sciences.

DON MEADOWS (1897–1994) was born in Indiana and came to southern California when the family moved to Orange County when he was five years old. At an early age he became interested in natural history, especially birds and butterflies, and he went on to obtain a degree in biology at Pomona College. He was a high school teacher in the Long Beach school system beginning in 1925. Soon Don conceived the idea of doing a biological survey of Santa Catalina Island and in 1927 requested a transfer to Avalon High School, which was part of the Long Beach school district. He collected Lepidoptera, primarily butterflies and macromoths, extensively there for three

seasons. In 1931 he took a leave and completed requirements for an M.S. degree in entomology at the University of California, Berkeley, and although his thesis research was on the biology of a salt marsh horse fly (Tabanidae), he began collecting moths using a light trap for the first time. Evidently influenced by H. H. Keifer, who was in Sacramento but made frequent visits to Berkeley and identified Meadows's specimens, he began collecting microlepidoptera. Upon return to Catalina, his light trapping produced a dramatic increase in the numbers of collection records, emphasizing micros, in 1931–1933, before he returned to Long Beach. He recorded more than 250 species of moths on the island, the largest local inventory of Lepidoptera in California up to that time. In the 1930s, Meadows published lists of butterflies and macromoths of Catalina Island, described a few species, and proposed to the authorities of the Los Angeles County Museum a biological survey of the eight California Channel Islands. This was approved, and in February 1939 the first of 14 expeditions was undertaken, with Meadows as field leader. Each of the islands was visited at least twice, and the survey continued until December 1941, when the last trip was terminated by the bombing of Pearl Harbor. Exceedingly few Lepidoptera had been known from the islands other than Catalina, and each of the Los Angeles County Museum expeditions included one or more experienced lepidopterists: Meadows, Comstock, Lloyd Martin, and/or Chris Henne. Meadows estimated that 40,000 insect specimens were collected, of which about 5,000 were Lepidoptera. Thus his 1927–1934 Catalina survey and the 1939–1941 expeditions laid the foundation for all subsequent inventory of the island insects, and Meadows's visionary proposal was pivotal in the development of our knowledge of the Channel Islands fauna. Don Meadows published a few articles on Lepidoptera in the 1940s, but at about age 50 he terminated his interest in insects in favor of his other passion, the history of southern California and Baja California. Lepidoptera's loss was that field's gain, because between 1951 and 1980 Meadows published more than 40 articles in periodicals and magazines, as well as four books, on Orange County and the Spanish Missions era. The Meadows library, probably the best collection of southern Californiana ever assembled, was purchased by the University of California, Irvine, in 1972. His macrolepidoptera and other insects were bought by the Smithsonian Institution in 1950, and the micros were given or sold the Los Angeles County Museum.

J. F. G. CLARKE and **H. H. KEIFER** were the first resident lepidopterists to study microlepidoptera extensively on the Pacific Coast. Keifer (1902–1986) became interested in natural history and insects at an early age in Oroville, California, and attended the University of California, Berkeley, in 1920–1924, receiving a B.S. degree in entomology. He took a position as assistant to the curator at the California Academy of Sciences, 1925–1928, where he began to study microlepidoptera, possibly simply because nobody else in California had done so. He reared micros in San Francisco and described the early stages in detail, an approach he followed thereafter. He was appointed as systematic entomologist in the California Department of Agriculture in 1928 and spent the remainder of his career in Sacramento. During 1925–1937, Keifer published a series of excellent papers on California microlepidoptera, in which he described 46 new taxa, primarily Gelechioidea, 87% of them from specimens that he reared. He described the early stages of these and about 40 previously described species in painstaking detail, profusely illustrated. However, after an outbreak of citrus bud mite occurred in 1937, Keifer was obligated to identify eriophyid mites, and he soon perceived a wide-open field with no "experts" in the East upon whom he would need to depend, and at age 35 he abruptly terminated his work on microlepidoptera. Because his descriptive work on mites spanned more than 30 years and produced more than 630 new taxa, we can only speculate the impact he might have had on our knowledge of western Lepidoptera had his decision been otherwise. Clarke (1905–1990) collected around Bellingham, Washington, in his youth, then in southeastern Washington and Idaho while a student and instructor at Washington State University in 1927–1935. Quite likely inspired by Keifer's fine work, Clarke began publishing new species in 1932, many of which were reared from larval collections. He entered graduate school at Cornell but in 1935 accepted a position with the U.S. Department of Agriculture as systematist at the U.S. National Museum. In later years Clarke made several productive field trips to his native Northwest, published extensively, and became the hub of microlepidoptera studies in North America for many years. His collections are at the Smithsonian.

Most of the species in western North America were named during the descriptive era, 1870s to 1930s, which laid the foundation for our present-day studies. We owe a great deal to the persons mentioned above, who worked in adventurous surroundings, with travel difficult, without black lights (or any lights) to establish the pioneer collections.

The Modern Era

Probably 80% or more of the specimens in institutional collections that we use in systematics and biogeographical studies have been collected during the modern era, which might be defined as post–World War II. As the American population grew and shifted westward, travel by freeways and air became readily accessible to more people, who visited more formerly remote places. Aided by efficient blacklight, mercury vapor, and pheromone trapping, moth collectors have enlarged the available resources exponentially. Of course a great many persons have contributed to our knowledge of western moths through recreational collecting, by-products of research on particular plant or insect groups, teaching field courses, and concerted survey efforts. We cannot mention them all, but there has been a shift in recent decades toward comprehensive, year-round surveys at selected sites, and several of these inventories are worthy of special mention. They have provided much of the geographical, seasonal, and biological information that we attempt to summarize.

The CANADIAN FOREST INSECT SURVEY has been of singular importance in developing our knowledge of the distribution and larval hosts of northern and widespread moths. This project began in 1936 and by 1940 was expanded to cover the more accessible forested areas of Canada, operating from regional centers across Canada. By the 1950s nearly 25,000 collections were processed annually. *Forest Lepidoptera of Canada*, compiled by R. M. PRENTICE and others from 1958 to 1966, records larval hosts of nearly 1,000 species, a great many of them in Alberta and British Columbia.

JACK FRANCLEMONT trained many of our contemporary lepidopterists during his long career at Cornell University. Although he did not conduct year-round inventory at one site in the West, he spent a total of 31 months in Arizona and western Texas in nine extended summer periods between 1959 and 1974, often accompanied by a graduate student. Most notable among them was RON HODGES, who complemented the effort by collecting all the tiny species, which had been ignored by most lepidopterists in Arizona. In later years, Hodges continued the practice of intensive summer collecting at selected localities, visiting Colorado, Utah, and elsewhere.

NOEL MCFARLAND carried out surveys of macromoths at the family home in the Santa Monica Mountains, Los Angeles County, and at the McDonald State Forest near Corvallis, Oregon, in the 1960s, then for many years at Ash Canyon in the Huachuca Mountains, Arizona (more than 900 species of macromoths), rearing many species at each locality.

ANDRE BLANCHARD began a general survey of moths in Texas in the 1960s, including the mountains of western Texas. He and ED KNUDSON described many new species, and in recent years Knudson and C. BORDELON have conducted several in-depth surveys, including published lists from Big Bend, Davis Mountains, and Guadalupe national parks.

OPLER and EVI BUCKNER-OPLER, assisted by volunteers, particularly SARA SIMONSON, have conducted an inventory of moths at the Rocky Mountain National Park at many localities above 8,800(elevation (1988–1998, 200 trap nights, more than 1,200 species)(CSU). TERHUNE DICKEL has collected macromoths for more than 25 years on the west slope of Colorado, making his most-extensive collections in the Upper Colorado River drainage and at Dinosaur National Monument.

Several extensive local inventories have been carried out in California: CHARLES REMINGTON, his son, ERIC, SCOTT MILLER, CHARLES DROST, POWELL, and others have surveyed the California Islands for moths intermittently but comprehensively during 1966–2005 (EME, LACM, PMY, SMNHM). NILS WESTERLAND sampled moths for the LACM at black light near Ward Creek at Lake Tahoe continuously from May to September in the 1970s and 1980s, providing the finest survey of Sierra Nevada moths since that of McGlashan at Truckee.

There has been a long-term survey at Inverness Ridge, Marin County, California, 1940–2008, by BILL BAUER, STEVE BUCKETT, CATHERINE TOSCHI (TAUBER), BILL PATTERSON, C. W. O'BRIEN, OPLER, and POWELL (more than 600 sampling dates, more than 730 species). Powell carried out local inventories of all moths at suburban Walnut Creek, Contra Costa County in the 1960s (more than 500 species); at the University of California Big Creek Reserve in coastal Monterey County, 1980s and 1990s, assisted by Y.-F. HSU, BRIAN SCACCIA, and others (more than 200 dates, more than 950 species recorded, 380 of them reared); and in urban Berkeley (20 years, more than 5,000 black light dates, more than 250 microlep/pyraloid species). JOHN DE BENEDICTIS conducted a several-year inventory in the 1980s of microlepidoptera at San Bruno Mountain on the San Francisco Peninsula, primarily by rearing (EME), and an ongoing 20-year survey of all moths at the Cold Canyon Reserve, Solano County (UCD). JOHN BROWN and NORRIS BLOOMFIELD carried out the most intensive single locality survey by blacklight traps in the West, at the Miramar Marine Corps Air Station, San Diego County, a 23,000-acre island of coastal sage scrub and chaparral in a sea of urban sprawl, involving 364 nights of sampling (more than 1,000 trap/nights) in all months, 1995–1998 (more than 650 species). The specimens are divided among the EME, SDMNH, and USNM collections.

In recent years, RON WIELGUS has devoted a considerable effort to year-round collection of moths at Kneeland at 2,100′ in the coastal hills of Humboldt County, providing the first comprehensive sample of the Vancouveran Province fauna of northern California, including hundreds of new county and several new state records (EME).

References for Part 1

Axelrod, D. J. 1976. History of the coniferous forests, California and Nevada. University of California Publications in Botany 70; 62 pages.

Bosik, J. J. 1997. Common Names of Insects and Related Organisms. Entomological Society of America, College Park, MD; 232 pages.

Brown, J. B., and K. Bash, 2000. The Lepidoptera of Marine Corps Air Station Miramar: Calculating faunal similarity among sampling sites and estimating total species richness. Journal of Research on Lepidoptera 36: 15–78.

Chaney, R. W. 1947. Tertiary centers and migration routes. Ecological Monographs 17: 139–148.

Common, I. F. B. 1990. Moths of Australia. Melbourne University Press, Carlton, Victoria; distributed by E. J. Brill, Leiden, New York, Copenhagen; 535 pages.

Clarke, J. F. G. 1942. Notes and new species of microlepidoptera from Washington State. Proceedings of the U.S. National Museum 92: 267–276, plates 27–32.

Covell, C. V., Jr. 1984. A Field Guide to the Moths of Eastern North America. Houghton Mifflin Co., Boston; xv + 496 pages.

Covell, C. V., Jr. 1990. The status of our knowledge of North American Lepidoptera, pages 211–230, in: Kosztarab, M., and C. W. Schaefer (eds.) Systematics of the North American Insects and Arachnids: Status and Needs. Virginia Agricultural Experiment Station, Virginia Polytechnic Institute and State University, Blacksburg.

De Benedictis, J. A., D. L. Wagner, and J. B. Whitfield 1990. Larval hosts of Microlepidoptera of the San Bruno Mountains, California. Atala 16: 14–35.

Duncan, R. W. 2006. Conifer Defoliators of British Columbia. Canada Forest Service, Victoria, British Columbia; 359 pages.

Essig, E. O. 1926. Chapter 27: Lepidoptera, pages 629–755, in: Insects of Western North America. MacMillan Co., New York.

Essig, E. O. 1931. A History of Entomology. MacMillan Co., New York; vii + 1029 pages.

Ferguson, D. F. 1955. The Lepidoptera of Nova Scotia. Part 1: Macrolepidoptera. Nova Scotia Museum of Science, Halifax, Bulletin 2: 161–375.

Ferguson, D. F. 1977. Host records for Lepidoptera reared in eastern North America. Technical Bulletin 1521. U.S. Department of Agriculture, Washington DC; 49 pages.

Forbes, W. T. M. 1923–1960. The Lepidoptera of New York and Neighboring States. Vol. 1: Primitive Forms, Microlepidoptera, Pyraloids, Bombyces; 729 pages. Part 2: Geometridae, Sphingidae, Notodontidae, Lymantriidae; 263 pages. Part 4: Agaristidae through Nymphalidae; 188 pages. Cornell University Agricultural Experiment Station, Ithaca, NY.

Furniss, R. L., and V. M. Carolin 1977. Western Forest Insects. Miscellaneous Publication 1339. U.S. Department of Agriculture, Forest Service, Washington DC; vii + 654 pages.

Handfield, L. 1999. Les guide des papillons du Québec. Vol. 1. Broquet, Boucherville, Quebec, Canada; 982 pages.

Heppner, J. B. 1998. Classification of Lepidoptera. Part 1: Introduction. Holarctic Lepidoptera. Vol. 5: Supplement. Gainesville, FL; iv + 148 pages.

Hering, E. M. 1951. Biology of Leaf Miners. W. Junk, Gravenhage, Netherlands; iv + 420 pages.

Hodges, R. W, T. Dominick, D. R. Davis, D. C. Ferguson, J. G. Franclemont, E. G. Munroe, J. A. Powell. (eds.) 1971–2005. The Moths of America North of Mexico. E. W. Classey Ltd. and Wedge Entomology Research Foundation, Eugene, OR; printed by Allen Press, Lawrence, KS. (Check list and more than 25 volumes have been published, treating Sesioidea, Sphingoidea, and parts of the Gelechioidea, Pyraloidea, Geometroidea, Mimallonoidea, Bombycoidea, and Noctuoidea.)

Holland, W. J. 1908. The Moth Book. A Popular Guide to a Knowledge of the Moths of North America. Doubleday, Page, and Co., New York; xxiv + 479 pages, 48 color plates.

Holloway, J. D., J. D. Bradley, and D. J. Carter 1987. Lepidoptera. CIE Guides to Insects of Importance to Man. Commonwealth Agricultural Bureau, International Institute of Entomology, Wallingford, UK; 262 pages.

Johnson, W. T., and H. H. Lyon 1988. Insects That Feed on Trees and Shrubs. 2nd Edition. Cornell University Press, Ithaca; 556 pages. (Discusses a large and diverse array of moth species, including fine color photos of early stages and adults of many.)

Knudson, E., and C. Bordelon 1999. Checklist of Lepidoptera of Big Bend National Park, Texas. Texas Lepidoptera Survey 3; 63 pages.

Knudson, E., and C. Bordelon 1999. Checklist of Lepidoptera of Texas. Texas Lepidoptera Survey 6; 74 pages.

Knudson, E., and C. Bordelon 1999. Checklist of Lepidoptera of the Davis Mountains, Texas. Texas Lepidoptera Survey 7; pages unnumbered.

Knudson, E., and C. Bordelon 1999. Checklist of Lepidoptera of the Guadalupe Mountains National Park, Texas. Texas Lepidoptera Survey 4; 81 pages.

Kristensen, N. (ed.) 1999. Lepidoptera, Moths, and Butterflies. Vol. 1: Evolution, Systematics, and Biogeography. Handbook of Zoology. W. de Gruyter, Berlin, New York; x + 491 pages.

Kristensen, N., and A. W. Skalski 1999. Phylogeny and palaeontology, pages 7–25, in: Kristensen, N. (ed.) Lepidoptera, Moths, and Butterflies. Vol. 1: Evolution, Systematics, and Biogeography. Handbook of Zoology. W. de Gruyter, Berlin, New York; x + 491 pages.

Labandeira, C. C., D. L. Dilcher, D. R. Davis, and D. L. Wagner 1994. Ninety-seven million years of angiosperm-insect association: Paleobiological insights into the meaning of coevolution. Proceedings of the National Academy of Sciences 91: 12278–12282.

Lafontaine, J. D., and D. M. Wood. 1997. Butterflies and moths (Lepidoptera) of the Yukon, pages 723–785, in: Danks, H. V., and J. A. Downes (eds.) Insects of the Yukon. Biological Survey of Canada Monograph Series. Entomological Society of Canada, Ottawa.

Maier, C. T., C. R. Lemmon, and J. M. Fengler 2004. Caterpillars on the foliage of conifers in the northeastern United States. U.S. Department of Agriculture, Forest Service, Forest Health Technology Enterprise Team, Morgantown, WV; 151 pages.

Mallis, A. 1971. American Entomologists. Rutgers University Press, New Brunswick, NJ; xvii + 549 pages.

Mann, J. 1969. Cactus-feeding insects and mites. U.S. National Museum Bulletin 256; x + 158 pages.

McFarland, N. 1963. The Macroheterocera (Lepidoptera) of a mixed forest in western Oregon. M.S. Thesis, Oregon State University, Corvallis; 152 pages.

McFarland, N. 1965. The moths (Macroheterocera) of a chaparral plant association in the Santa Monica Mountains of southern California. Journal of Research on Lepidoptera 4: 43–74.

McFarland, N. 1975. Larval food plant records for 106 species of North American moths. Journal of the Lepidopterists' Society 29: 112–125.

Miller, J. C. 1995. Caterpillars of Pacific Northwest Forests and Woodlands. U.S. Department of Agriculture, Forest Service, Corvallis, OR; 80 pages.

Miller, J. C., and P. C. Hammond 2000. Macromoths of Northwest Forests and Woodlands. U.S. Department of Agriculture, Forest Service, Corvallis, OR; 133 pages.

Miller, J. C., and P. C. Hammond 2003. Lepidoptera of the Pacific Northwest: Caterpillars and Adults. FHTET-2003-03. U.S. Department of Agriculture, Forest Service, Forest Health Management Enterprise Team, Corvallis, OR; 324 pages.

Opler, P. A., and J. S. Buckett 1971. Seasonal distribution of "Macrolepidoptera" in Santa Clara County, California. Journal of Research on Lepidoptera 9: 75–88 ["1970"].

Opler, P. A., H. Pavulaan, R. E. Stanford, and M. Pogue (coordinators) 2002. Butterflies and Moths of North America. Northern Prairie Wildlife Research Center, Division of Biological Resources, U.S. Geological Survey, Bozeman, MT. Additional information available at www.butterfliesandmoths.org.

Powell, J. A. 1980. Evolution of larval food preferences in Microlepidoptera. Annual Review of Entomology 25: 133–157.

Powell, J. A. 1987. Records of prolonged diapause in Lepidoptera. Journal of Research on Lepidoptera 25: 83–109 [1986].

Powell, J. A. 2002. Landels-Hill Big Creek Reserve, Monterey County, California: Lepidoptera Checklist. Available at http://ucreserve.ucsc.edu/bigcreek/fauna/lepidoptera/index.html. (A check list of approximately 940 species, of which 38% have been reared from larval collections at Big Creek.)

Powell, J. A. 2003. Lepidoptera (Moths and Butterflies), pages 631–664, in: Resh, V. H., and R. T. Cardé (eds.) Encyclopedia of Insects. Academic Press, Amsterdam, San Diego; 116 figs., incl. 99 color photos.

Powell, J. A. 2005. Lepidoptera (moths and butterflies) at Inverness Ridge in central coastal California and their recovery following a wildfire; 29 pages. Available at http://essig.berkeley.edu/ Regional lists. (Includes history of Lepidoptera collections at Inverness Ridge and list of 650 species and their larval hosts.)

Powell, J. A., and J. A. De Benedictis 1995. Foliage-feeding Lepidoptera of *Abies* and *Pseudotsuga* associated with *Choristoneura* in California, Chap. 6, pages 167–215, *in:* Powell, J. A. (ed.) Biosystematic Studies of Conifer-feeding *Choristoneura* in the Western United States (Lepidoptera: Tortricidae). University of California Publications in Entomology 115, Berkeley and Los Angeles; vii + 275 pages.

Powell, J. A., and C. L. Hogue 1979. California Insects. California Natural History Guides No. 44. University of California Press, Berkeley; 388 pages, 16 photo plates, 459 figs. (Moths and butterflies, order Lepidoptera, pages 186–258.)

Powell, J. A., and Y.-F. Hsu 1999 (revised 2002). Annotated List of California Microlepidoptera. Available at www.berkeley.edu/essig/ Lepidoptera surveys.

Powell, J. A., C. Mitter, and B. Farrell 1998. Evolution of larval feeding preferences in Lepidoptera. Chapter 20, pages 403–422, *in:* Kristensen, N. (ed.) Lepidoptera. Vol. 1: Traite de Zoologie (new series), Teil. 75, Band 1.

Prentice, R. M. (compiler) 1958–1966. Forest Lepidoptera of Canada Recorded by the Forest Insect Survey, Volumes 1–4. Publication 1034, Bulletin 128, Publications 1013, 1142. Canada Department of Forestry, Ottawa; 840 pages.

Richers, K. 2006. County List of Moths of California. Available at www.berkeley.edu/essig/ Regional lists, California moths. (Updated annually.)

Robinson, G. S., P. R. Ackery, I. J. Kitching, G. W. Beccaloni, and L. M. Hernández 2002. Hostplants of the moth and butterfly caterpillars of America north of Mexico. Memoirs of the American Entomological Institute 69; 824 pages.

Sbordoni, V., and S. Forestiero 1984. Butterflies of the World. Times Books. A. Mondadori Editore, Milano, Italy; 312 pages. (Despite its title, this includes a magnificently illustrated survey of moth and butterfly families of the world.)

Scoble, M. J. 1992. The Lepidoptera. Form, Function, and Diversity. Oxford University Press, Oxford, New York, Toronto; xi + 404 pages.

Stehr, F. W. 1987. Lepidoptera, pages 288–596 *in:* Stehr, F. W. (ed.) Immature Insects. Kendall/Hunt Co., Dubuque, IA.

Tietz, H. M. 1972. An Index to the Described Life Histories, Early Stages, and Hosts of the Macrolepidoptera of the Continental United States and Canada. 2 volumes. A. C. Allyn, Sarasota, FL; 1041 pages.

Wagner, D. L. 2005. Caterpillars of Eastern North America: A Guide to Identification and Natural History. Princeton Field Guides. Princeton University Press, Woodstock, Oxfordshire, UK; 512 pages.

Wagner, D. L., V. Giles, R. C. Reardon, and M. L. McManus 1997. Caterpillars of Eastern Forests. Forest Health Technology Enterprise Team, U.S. Department of Agriculture, Forest Service 34; 113 pages.

Winter, W. D., Jr. 2000. Basic Techniques for Observing and Studying Moths and Butterflies. Memoir 5. The Lepidopterists' Society, Los Angeles; xvii + 444 pages.

Young, M. 1997. The Natural History of Moths. T. and A. D. Poyser, Ltd., London; xiv + 271 pages.

Zimmerman, E. C. 1978. Insects of Hawaii. Vol. 9: Microlepidoptera, Parts 1 and 2. University Press of Hawaii, Honolulu; xxx + 1903 pages.

PART TWO

CLASSIFICATION AND NATURAL HISTORY OF THE MOTHS OF WESTERN NORTH AMERICA

Historically the Lepidoptera have been classified in four or five suborders, all but one of which are primitive moths that retain ancestral characteristics as relict, morphologically dissimilar groups. All the more-derived moths and butterflies, more than 98% of the described species, comprise one evolutionary lineage, or clade, the Ditrysia. In recent decades much progress has been made in detailed analyses of the relationships of the primitive groups, research led by N. P. Kristensen in Denmark, aided by discoveries of new taxa and the previously unknown larvae and pupae of some of the families.

The resultant phylogenetic analyses have shown each of the primitive lineages to be paraphyletic with respect to the rest of the Lepidoptera (Fig. 25), and use of the Linnaean higher taxa (suborder and other ranks between order and superfamily) has generally been abandoned by lepidopterists. On the other hand, we continue to recognize the obligate categories (family, genus, species) for purposes of names and communication. Moreover, as discussed in Part 1, there is an inherent value in understanding the relative levels within lineages provided by the consistent endings to Linnaean rank taxa. Thus the name of the rank tells us its relative position within that lineage, and these ranks continue to be used. That is, we understand that families and genera in one superfamily (e.g., Tineoidea) are not comparable to their counterparts in another superfamily (e.g., Noctuoidea) in terms of evolutionary age, but they are ranks that enable efficient communication about their relationships within each of these lineages. Historically, the family level has been the common denominator for communication among entomologists, but in recent decades there has been inflation and a proliferation of family and superfamily divisions such that the superfamily has become a commonly used and understood rank for lepidopterists, and it is the primary division in the text that follows.

Recent authors have treated more than 120 families of Lepidoptera, and these are assigned to 44 to 48 superfamilies. There is considerable discrepancy between analyses within some superfamilies. In general we follow the hypothesis of phylogenetic relationships proposed by Kristensen and Skalski, as summarized in Figure 25. Six superfamilies and 11 families of moths in the primitive lineages are represented in the western United States.

Primitive Lineages

Zeugloptera Clade

Superfamily Micropterigoidea

FAMILY MICROPTERIGIDAE

These are the most primitive extant lepidopterans—they are living fossils. There are micropterigids recognizable as modern genera preserved in amber dating back to dinosaur times in the early Cretaceous period, 125 MyBP.

Adult Adults are small (FW length 3–6 mm), often colorful, with metallic sheens of bronze or purple and yellow FW markings, usually active in the daytime (Plate 59.1). They are characterized by numerous ancestral traits not shared by other moths, most notably retention of functional mandibles, which are used to feed on pollen of various trees in Europe, and more

primitive plants, sedges, Winteraceae, and fern spores in New Caledonia and Madagascar. A more complete, Mecoptera-like wing venation, similar in FW and HW, led to proposal of this group as a separate order, the Zeugloptera, but overall the evidence points to the combined Zeugloptera and other Lepidoptera as a sister group to the caddisflies (Trichoptera).

Larva Wholly unlike caterpillars of other Lepidoptera, they are plump, somewhat hexagonal in cross-section, with long antennae, short thoracic legs, and lacking the abdominal prolegs with crochets typical of most Lepidoptera. The larvae live in leaf litter or among mosses, in rotting wood, with high moisture conditions, and the cuticle has specializations unique among arthropods, with exo- and endocuticle separated by a fluid-filled space leading via pores to chambers in the exocuticle, overlaid by sticky pellicle to which particles of debris adhere. The pattern of primary setae on the body is unlike that of other moth larvae, and homologies of the two types have not been determined.

Larval Foods Some species feed on liverworts, which led to generalizations in the literature, perhaps based on a temptation to ascribe a primitive moth–primitive plant association to the ground plan for Lepidoptera. Many micropterigid larvae, however, are generalists, feeding on detritus, fungal hyphae, or angiosperm plant leaves. Fungus-feeding may have been the ancestral trait.

Diversity There are about 120 described species, occurring worldwide but in a disjunct, relictual distribution pattern. More than half the named species are in the genus *Micropteryx* in the Palaearctic Region, while only two are known in North America *(Epimartyria);* there is a greater diversity of genera in the Orient and southwestern Pacific, particularly New Zealand, eastern Australia, and New Caledonia, which has about 50 species, most of them undescribed.

Just one species, ***Epimartyria pardella*** Walsingham (Plate 1.1, Plate 59.1, Fig. 28), lives in moist habitats in and near the redwood forests along the coast of southern Oregon and northern California. The adults are small (FW 4.5–5.5 mm), diurnal, and active in bright sunshine, with iridescent purplish bronze FW marked by three variable, pale yellow patches, on the middorsal margin, from the costa distally, and at the tornus. They perch and fly short distances in fern-dominated canyons and marshes in late May to July. Larvae were reared by Tuskes and Smith from eggs, on a diet of liverworts, but they did not report whether other plant material was provided. Larvae preferred *Pellia* to *Conocephalum* and developed over a two-year period.

Glossata, Homoneura Clade

All the nonzeuglopteran Lepidoptera comprise the Glossata, the monophyly of which is well supported by a suite of derived characters. The most obvious traits that distinguish glossatans are in the adult mouthparts: mandibles nonfunctional (persisting as functional in the pupa in basal groups) and maxillary galeae elongated, forming a proboscis that is coiled in repose, accompanied by reduction of the head capsule and its cuticular thickening associated with mandibular musculature. The basal lineages retain ancestral features of the wings: similarly shaped FW and HW with relatively complete venation (homoneurous) and the jugal lobe at the base of the FW but no functional wing-coupling mechanism. Females in some ancestral families have a flattened, sclerotized abdominal apex with serrate edges, forming a "saw" that is everted to cut a pocket into host-plant leaves into which she deposits the egg.

Superfamily Eriocranioidea

FAMILY ERIOCRANIIDAE

This family is a Holarctic counterpart to the South American Heterobathmiidae, resembling them superficially as adults, and the larvae of both groups mine leaves of birch and oak (Fagales) in early spring.

Adult These are small moths (FW length 3.5–9 mm) with relatively narrow wings covered by iridescent, simple ("primitive type") scales and hairs, often golden with purplish markings. They fly in early spring just as the host trees are beginning to leaf out, and most species are primarily diurnal. Genitalia, Figs. 29, 30.

Larva Legless, obligate leaf miners throughout life, they mine in newly expanded, soft leaves, forming "baggy" full-depth mines, which dry and deteriorate after the leaf hardens. The larvae mature quickly and enter the soil for pupation. The pupae are mandibulate, and the following spring the adult within the pupal shell uses the mandibles to cut through the cocoon and reach the soil surface.

Larval Foods The larvae feed on birch (Betulaceae), oak (Fagaceae), and other Fagales, or Rosaceae (one species).

Diversity There are 25 described species assigned to five genera, with about half the species in Europe and Asia and half in North America.

Davis's monograph of North American Eriocraniidae in 1978 treated 15 species in five genera, and there have been no new species described since then, but *Acanthopteroctetes* has been treated as a separate family and superfamily. Nine species of Eriocraniidae occur in the West, one in Arizona, eight in California, one of which ranges to coastal Alaska. Three species that feed on oaks *(Quercus)* in California are the most commonly collected eriocraniids: *Dyseriocrania auricyanea* (Walsingham) (FW 4.5–7 mm, males larger than females; Plate 1.2, Plate 59.2) has mottled golden FW distinctly marked by shining purplish, most conspicuously by three regular, transverse bands slanted outward from the dorsal margin. Adults are diurnal or crepuscular and fly around newly foliating coast live oak *(Q. agrifolia)*,

February to April, depending upon tree exposure and phenology, to May in the Sierra Nevada, where they are associated with black oak, *(Q. kelloggii)*. We have found presumed mines of *D. auricyanea* on other inland oaks *(Q. wislizenii, Q. douglasii, Q. berberidifolia,* and *Q. lobata)*. Larvae mine the distal portion of new leaves in March to May. This species occurs from the northern coast of California and northern Sierra Nevada to southern California and Santa Cruz Island.

Eriocraniella aurosparsella (Walsingham) and *E. xanthocara* Davis are smaller (FW 3.5–5.5 mm; Plate 1.3, Plate 1.4, Figs. 29, 30) with dark, uniform, shining bronze to purplish FW in *E. aurosparsella*, deep blue in *E. xanthocara*, which has bright yellow to orange head vestiture in contrast to grayish in *E. aurosparsella*. Adults are diurnal and those of *E. aurosparsella* sometimes swarm about newly foliating *Quercus kelloggii*; a sample taken along the Stanislaus River, Tuolumne County, in early April consisted entirely of males, but another in Plumas County in June had 90% females. They fly from late February near the coast to mid-June in the Sierra Nevada. *Eriocraniella xanthocara* has been recorded primarily in association with *Q. agrifolia* but sometimes occurs on *Q. kelloggii*.

Eriocrania semipurpurella (Stephens)(Plate 1.5) is widespread in the northeastern United States, southern Canada, and along the Pacific Coast from Alaska near Anchorage to the Puget Sound, and it occurs at San Bruno Mountain on the San Francisco Peninsula, California. It is a sparsely scaled species with dark FW (4.5–6.5 mm) reflecting brassy to purplish, and obscure silvery spots distally on the costa and at the tornus, extended through the fringe. The moths are diurnal, active in May in Alaska, April in Washington, March in California, closely associated with oceanspray, *Holodiscus discolor* (Rosaceae), when it is leafing out. The larva forms a blotch mine along the leaf margin in April and May. *Neocrania bifasciata* Davis is a beautiful Californian species having shining purple FW (4.5–6.2 mm; Plate 1.6) with two broad, distinct golden bands. The moths are found around the new foliage of canyon oak *(Quercus chrysolepis)* April to early June in the Coast Ranges, Monterey south to San Diego County.

FAMILY ACANTHOPTEROCTETIDAE

This family, with three species in the western United States, has been included in the Eriocranioidea, or has been accorded superfamily status based on its more derived type of scales and first thoracic spiracle.

Acanthopteroctetes was proposed in 1921 for **tripunctata** from Glacier National Park by Annette Braun, who, we presume, never imagined the name would become a family and superfamily name bearer. The type species is known only from one male specimen. It is small (FW 5 mm), dark, FW with a coppery luster and three poorly defined pale spots, at basal one-third and one above the other at end of cell. A smaller species, *A. unifascia* Davis (FW 3.5–4.5 mm; Plate 1.7), in coastal California has dark bronzy FW with a white, transverse band at end of cell, usually reduced to streaks at the costa and tornus in northern populations. This is the only *Acanthopteroctetes* for which larvae are known, but the life cycle remains a mystery. Larvae mine leaves of *Ceanothus* (Rhamnaceae) in spring and typically undergo diapause until late summer, a peculiar adaptation to the California dry season, and puzzling because the type specimen and others were collected in April and May in southern California. We found the larval mines

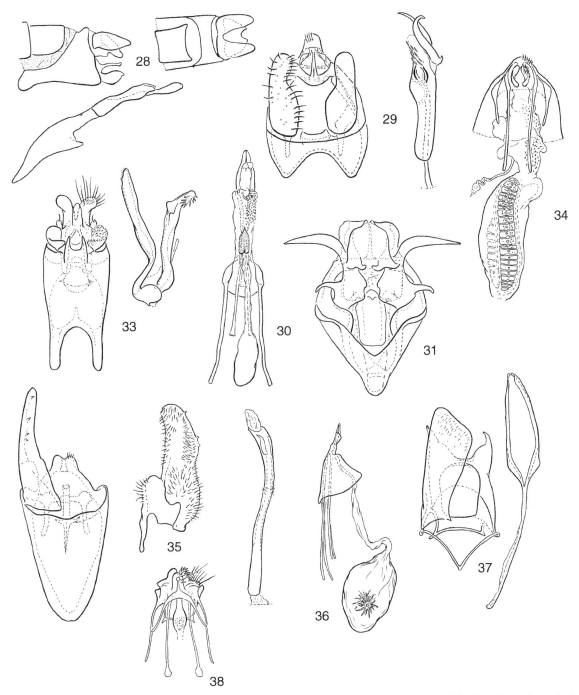

FIGURES 28–31, 33–38. Genitalia structures of ancestral lineages. **28,** Micropterigidae: *Epimartyria pardella* Walsingham ♂, terminal abdominal segments. upper left, lateral aspect; upper right, dorsal aspect; aedeagus below, lateral aspect [Issiki 1931]. **29,** Eriocraniidae: *Eriocraniella xanthocara* Davis ♂, aedeagus, lateral aspect to right. **30,** Eriocraniidae: *Eriocraniella aurosparsella* (Walsingham) ♀ [Davis 1978]. **31,** Hepialidae: *Paraphymatopus hectoides* (Boisduval) ♂, caudal aspect [Wagner 1985]. **33,** Nepticulidae: *Microcalyptris punctulata* (Braun) or species near ♂, aedeagus to right. **34,** Nepticulidae: *Microcalyptris punctulata* ♀ [Wilkinson 1979]. **35,** Prodoxidae: *Prodoxus phylloryctus* Wagner and Powell ♂, right valva removed, inner aspect in middle, aedeagus to right [Wagner and Powell 1988]. **36,** Prodoxidae: *Prodoxus coloradensis* Riley ♀, lateral aspect [Davis 1967]. **37,** Tischeriidae: *Tischeria distinca* Braun ♂, right valva omitted, aedeagus to right. **38,** Tischeriidae: *Tischeria omissa* Walsingham ♀, ductus bursae omitted [Braun 1972].

on Santa Cruz Island in March 1966, but the unusual diapause pattern fooled us, and the adults emerged after we stored the mines for future reference. Subsequently, Don Frack and David Wagner reared *A. unifascia* from mines collected in March, with adults emerging late June to August. The mode of overwintering and occurrence of fall and spring flights need explanation.

Exoporia Clade

Within the homoneurous Glossata, two superfamilies comprise the Exoporia, namely the Hepialoidea and the Mnesarchaeoidea, their monophyly established by the unique configuration of the female genital system, which is interpreted as homologous in these otherwise quite dissimilar moths. The copulatory orifice in these superfamilies is separate from the

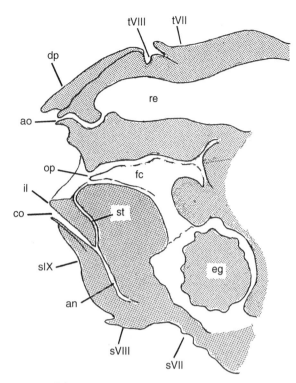

tVIII tVII
dp
re
ao
op
fc
il
co
st
sIX
eg
an
sVIII
sVII

FIGURE 32. Schematic representation of hepialid female terminal segments, saggital section. ao, anal opening; co, copulatory orifice; eg, eggs; fc, fertilization chamber; il, intergenital lobe; op, ovipore; re, rectal chamber; st, sperm tract, which is formed by invagination of ectoderm. Stippled areas represent sclerites and musculature. Roman numerals denote abdominal segments, tergal and sternal. [Modified from Nielsen and Kristensen 1989]

ovipore, but there is no internal connection between the two (Fig. 32). Sperm is transferred via an external groove in the body wall below the ovipore. Mnesarchaeids are eriocraniid-like moths in New Zealand, whose larval habits are similar to those of micropterigids.

Superfamily Hepialoidea

This is the most successful of the homoneuran and more primitive lineages in terms of extant diversity. The superfamily is characterized by having reduced mouthparts, with the proboscis absent or short and evidently nonfunctional. Hepialidae are large moths, even enormous in some genera, well represented on all nonpolar continents. Four other hepialoid families, Anomosetidae, Neotheoridae, Palaeosetidae and Prototheoridae, are Southern Hemisphere relicts represented by one to a few species and are smaller moths.

FAMILY HEPIALIDAE

Adult Adults are large to very large moths (wingspan 2–25 cm) and include some of the largest Lepidoptera in the world, such as *Trichophassus* in Central and South America and *Zelotypia* in Australia, with a 25 cm wingspan and often beautifully colored in greens and pinks. Genitalia, Figs. 31, 32. The females carry enormous numbers of eggs—one female of *Trictena* in Australia laid 29,000 eggs and had another 15,000 in her ovaries when dissected—and therefore are bulky, heavy-bodied creatures, surpassing in weight the largest sphingid and saturniid moths.

Hepialids are the so-called ghost moths, based on the habit of the males to form groups that fly together at dusk, as a ritual of courtship behavior; especially representative of the common name is one European species that has white FW and forms ghostlike clouds.

Larva The caterpillars are elongate, cylindrical (eruciform), with fully developed thoracic legs and abdominal prolegs that bear bi- or multiserial rings of crochets in later instars. They have primary setae distributed in patterns that are homologous with those of ditrysian larvae, and they lack secondary setae. Hepialid larvae are concealed feeders, living in tunnels in the soil or boring into stems of living trees and shrubs.

Larval Foods Many species tunnel in soil or live in silken galleries in leaf litter and feed on roots, leaves adjacent to the shelter, indiscriminately on pteridophyte, gymnosperm, or angiosperm plants. Early instars of some species feed on decaying wood and fungi associated with it and bore into tree trunks in later instars. Other hepialids feed on roots from galleries in the soil in grasslands.

Diversity More than 600 named species in 67 genera are found worldwide, best developed in Australia and Africa. There are about 20 species in America north of Mexico, fewer than half of which occur in the West. Although some hepialids are conspicuous and sometimes abundant, most biologists, naturalists, and even Lepidoptera enthusiasts never see these moths owing to their elusive, crepuscular flight behavior and because they rarely are attracted to lights.

Sthenopis quadriguttatus (Grote)(FW 3.5–6 cm; Plate 13.1), which ranges from northeastern North America to Alberta and the northern Rocky Mountain states, is the largest and most spectacular western species. The FW is pale ochreous gray with variable darker rust tan bands; the inner one has one or two silvery dots in the cell, bordered by black. Adults fly in July and occasionally come to black light early in the evening.

Paraphymatopus californicus (Boisduval)(Plate 13.4, Plate 13.5) is a common species along the Pacific Coast from Vancouver Island to California and in the Sierra Nevada to 5,000' elevation. Adults are active January to March along the coast, to June and July in Oregon and northward. They are moderately large (FW 1.5–2.5 cm), females usually appreciably larger than males, and males have remarkably variable FW patterns, gray with outwardly oblique brown or reddish bands, to dark brownish with white markings. David Wagner studied the behavior and likened their flight to leks formed by vertebrates, males gathering in groups as a function of courtship. Males select open spots adjacent to shrubs and perform arclike "dances" at dusk, just as the last daylight is fading. Mating lasts several hours, and pairs of this and the following species can be found after dark by diligently searching bushes in areas where the males were observed flying. The pair hangs, female clinging to a branch, male hanging head downward, connected only by the genitalia. The eggs are broadcast randomly on the ground, and newly hatched larvae burrow into the soil and find roots in which to feed. Late-instar larvae are large grubs (length 3.5–4.5 cm) that tunnel upward in the woody stems of bush lupines and other shrubs and are believed to require two years to reach maturity. *Paraphymatopus hectoides* (Boisduval) is smaller (FW 1.3–1.6 cm; Plate 13.6, Fig. 31), tan with reddish costa and usually oblique fasciae, rose reddish to whitish edged by dark gray. It is more widespread, occurring in Colorado, Utah, and Oregon. It flies with *P. californicus* along the California coast and in foothill and montane areas of the Coast Ranges and Sierra Nevada to 7,700' elevation, in a wide variety of habitats, often oak woodland. The flight habits are similar, and sometimes *P. hectoides* is very numerous—a group of lepidopterists observed more than 90 specimens during the brief flight at the University of California Forest Camp, in Plumas County, one evening in early June. A third species, *P. behrensi* (Stretch) is similar to *P. californicus*, but larger, FW (2–2.5 cm; Plate 13.2, Plate 13.3) ochreous to rose pink, often with two short rows of silvery white spots, obliquely outward in and beyond cell; HW pale rose pink. This species occurs in redwoods and other mesic, forested habitats from southern British Columbia in the Coast Ranges south to the Santa Cruz Mountains, California, and in the Sierra Nevada to 7,800′. *Paraphymatopus behrensi* flies in the summer, June to September, primarily July to August.

Gazorycta mcglashani (Henry Edwards) is a beautiful species having yellowish tan FW (14–15 mm) with a silvery white line from base to end of the cell, thence obliquely to the inner margin and apex. *Gazorycta lemberti* (Dyar)(Plate 13.7) is similar but has a series of white spot rows between the veins. Both are confined to high elevations of the Sierra Nevada and are particularly elusive owing to the improbable chance of short-term visitors encountering ideal conditions for the dusk flight. These species were discovered in the late 1800s but have been rarely collected since. McGlashan and Lembert reported that the flight occurs in grassy meadows and lasts only 20 minutes at dusk. Many recent visits to alpine habitats of the Sierra Nevada by hopeful lepidopterists have proved futile.

Heteroneura, Monotrysia Clade

All the remaining primitive and the derived (ditrysian) Lepidoptera have different venation in the FW and HW, with a reduced Rs system in the HW, differing in shape, and the HW usually smaller. Most possess a wing-coupling mechanism consisting of the frenulum (stiff setae at the base of the HW costal margin) that extends under the retinaculum (modified scales on the underside of the FW); and they have lost the first abdominal sternite. The five most basal superfamilies of the Heteroneura retain the ancestral monotrysian female reproductive system, but they share no derived characteristics that would unite them as monophyletic. The three most diverse of these are Nepticuloidea, Incurvarioidea, and Tischerioidea.

Superfamily Nepticuloidea

These are tiny moths whose larvae are leaf and stem miners. Specialization on diverse flowering plants has led this group to become the most species rich of the primitive Lepidoptera.

FAMILY NEPTICULIDAE

Adult Nepticulids include the smallest Lepidoptera (FW length 1.5–4.5 mm), characterized by having the basal antennal segment (scape) greatly enlarged, forming a cap over the upper half of the relatively large eye. The head is rough scaled, and the mouthparts are primitive, with long, folded maxillary palpi, rudimentary proboscis with galeae not joined, used to lap up moisture and honeydew secreted by aphids. The FW is relatively broad with long scale fringes. Genitalia, Figs. 33, 34.

Larva These legless, obligate leaf miners typically form a serpentine track beginning just below the egg, which is cemented to the leaf surface, gradually enlarging to a full-depth terminal tube (Plate 59.3) or irregular blotch. At maturity the larva cuts a crescentic slit in the upper cuticle and drops to the ground to form a tough silken cocoon.

Larval Foods They usually feed on mature leaves of woody plants, although a few mine stems or cause petiole galls. Individual nepticulid species are specialists, using more than 40 families of angiosperms, with a preponderance of records for Fagaceae and Rosaceae in the Holarctic. Some smaller lineages are specialists on one plant family, Anacardiaceae, Polygonaceae, or Fabaceae.

Diversity There are nearly 800 described species, placed in 11 genera, occurring in all nonpolar regions. No accurate estimate is available, but the total described probably includes fewer than 10% of the species in tropical regions and less than half the North American species. There are more than 80 described species of Nepticulidae in North America, the majority from the eastern deciduous forests, where Chambers and Braun worked. There are about 25 species identified in the West, as well as numerous unnamed, based on observed mines. Each species is restricted to a particular plant species or group of plants within a genus or closely related genera, almost exclusively woody plants, where the larvae make serpentine mines in mature leaves and require a lengthy time to reach maturity, often over winter. Similar looking mines found in herbaceous plants are usually those of agromyzid flies.

Ectoedemia platanella (Clemens)(Plate 1.8) is widespread in North America and occurs throughout much of the low-elevation West, forming mines at the base of the large leaves of sycamore, *Platanus racemosa* (Platanaceae). This is a tiny species (FW 1.8–2.3 mm), having ochreous orange head vestiture, FW dark bronze gray, usually with a transverse silver band at end of cell, rarely reduced to a trace at the dorsal margin. We have found larvae in May, August, and September, which developed without diapause.

Stigmella diffasciae (Braun), a tiny species (FW 2.2–2.4 mm; Plate 1.10), was originally reared from an unidentified shrub in the Sierra Nevada, but a species superficially matching the description of *S. diffasciae* often is abundant on coffeeberry, *Rhamnus californicum* (Rhamnaceae), in the Coast Ranges of California. Specimens are rare in collections because the larvae mature early, usually by February, so only abandoned mines are commonly seen, and even when collected in early spring, often most are parasitized. Newton and Wilkinson recorded the host as *Ceanothus* (Rhamnaceae), citing Braun's notes, but that identification has not been confirmed. *Stigmella heteromelis* Newton & Wilkinson (Plate 1.11, Plate 59.4) was described from specimens we reared from toyon, *Heteromeles arbutifolia* (Rosaceae), on the Berkeley campus in 1961. The FW (2.3–2.7 mm) is dark bronzy colored with a silvery band beyond the cell and rarely a second band basally. The mines are present, sometimes abundantly, throughout the range of toyon in California and the Channel Islands. Often they are completed and abandoned by February, but we have reared *S. heteromelis* from mines collected February to April. *Stigmella variella* (Braun)(Plate 1.12, Plate 1.13, Plate 59.3) is a variable species having whitish or pale tan head vesture and slate gray FW (2.5–3.5 mm) that varies from silvery white with the distal area dark gray to uniform slate gray or dark gray, sometimes with a faintly to well-developed silvery white subterminal band. The HW in the male is distinctive, gray with the basal half purplish black. The larvae feed in mature leaves of coast live oak, *Quercus agrifolia*, and *Q. wislizenii* (Fagaceae), and *S. variella* is multivoltine, mature mines appearing from

February to August. The mines are the most consistently present of the17 species of leaf miners occurring on *Q. agrifolia*. A quite similar appearing species creates huge mines on tanbark-oak *(Lithocarpus densiflorus)*.

Stigmella braunella (Jones) was described from specimens reared from cultivated Catalina cherry *(Prunus lyonii*, Rosaceae) in Berkeley, California. The head scaling is white, and FW (2.5–3.1 mm) is dark bronzy gray with a thin, transverse white line at end of cell. The species occurs on the larger Channel Islands of southern California and may have been an insular endemic, as it has not been recorded from native *Prunus* on the mainland. However, a similar *Stigmella* mining holly-leaf cherry *(P. ilicifolia)* may be the same species. On *P. lyonii* the larvae form conspicuous mines, much larger than most *Stigmella*, maturing in February to March, and adults emerge in March to June, evidently without diapause

Microcalyptris lotella Wagner (Plate 1.14) is a peculiar nepticulid in that the larvae form subcuticular mines in the newly developed stems of deerweed *(Lotus scoparius*, Fabaceae) during spring. The moth is tiny (FW 2.0–3.5 mm), gray, head scaling whitish to tan, FW pale gray, speckled by darker scale tips. There are dark androconial scales on basal one-third of FW venter and HW dorsum. Mines collected in March produced adults in April.

FAMILY OPOSTEGIDAE

Adult These moths are mostly larger than nepticulids (FW 1.8–8.3 mm), with enormous eye caps, completely obscuring the eyes from a frontal view. The wings are relatively broad with markedly reduced venation, the FW with apices often strongly bent upward, usually white with sparse dark markings.

Larva The larvae are extremely slender, flattened, legless, with callosities on second and third thoracic venter, adapted for subcutaneous mining. They are primarily cambium miners in woody stems and fruit.

Larval Foods They feed on a broad spectrum of angiosperms, including Betulaceae, Ranunculaceae, Polygonaceae, Saxifragaceae in the Holarctic, Rutaceae in Hawaii, and Fagaceae in Chile.

Diversity About 200 described species exist worldwide, placed in seven genera, certainly a small proportion of the actual fauna. The ALAS inventory at the La Selva Biological Reserve in lowland Costa Rica collected 17 species of opostegids (20% of the known Neotropical fauna), of which only one had been previously named. The described New World fauna has been greatly increased and elegantly treated in a 2007 monograph by Davis and Stonis. They found 87% of the described species occur in tropical regions, and they described 75% of the 91 species in the Western Hemisphere as new. Only 11 species are recorded for America north of Mexico, several of them only in the Florida Keys or southern Texas, and there are only three named species in the West.

Pseudopostega albogaleriella (Clemens)(FW 3–6 mm; Plate 1.15, Plate 1.16, Plate 1.17) is a shining, snow white moth; its FW varies from having a tiny black dot and delicate brown lines at the base of the fringe to having two transverse, brown bands and the fringe broadly brownish basally. We collected one specimen on Santa Rosa Island that has a large, brown smudge from the middorsal margin and distally two parallel, transverse lines angled at the middle. This species occurs in the northeastern states to Florida and eastern Texas, and there is a disjunct western component in California and Arizona, originally described as **P. bistrigulella** Braun, from southern California, which is now treated as a subjective synonym of *P. albogaleriella*. In California, it is recorded along the immediate coast from Monterey County southward and inland at low elevations. In the Miramar survey and earlier collections in San Diego, this species was recorded many times from February through May and twice in August, suggesting a diminished second generation during the dry season.

R. G. Rosenstiel found larvae of an unidentified opostegid mining fruit or cambium of *Lonicera* (Saxifragaceae) in Oregon, and it has been identified recently by Davis as **Opostegoides scioterma** (Meyrick).

Superfamily Incurvarioidea

These are tiny to small, mostly diurnal moths. The antennal scape is not enlarged into an eye cap. Adult and larval habits are diverse, but females all have a piercing ovipositor specialized for inserting the eggs into plant tissue, often ovules or young seed. Worldwide, there are six included families, of which the four most diverse, Heliozelidae, Incurvariidae, Adelidae, and Prodoxidae, occur in western North America. The other incurvarioid families are Cecidosidae in southern South America and South Africa, with seven species whose larvae cause stem galls on Anacardiaceae; and Crinopterigidae, with a single species in southern Europe that has a coleophorid-like larval biology, feeding on Cistaceae.

FAMILY HELIOZELIDAE

Species of this family occur worldwide but because of their minute size and diurnal habits they are rarely seen, and many more species are known from the characteristic abandoned larval mines than there are named.

Adult These moths are tiny to small, with FW 1.5 to 5 mm, but our species are less than 3 mm. Typically heliozelids have iridescent, metallic-appearing scaling. The eyes are small, characteristic of diurnal microlepidoptera. The mouthparts are well developed, maxillary palpi reduced, not folded, proboscis functional.

Larva They are flattened, usually legless leaf miners, but thoracic legs are well developed in some *Heliozela*; thorax in other genera with paired ventral and dorsal movable calli; abdominal prolegs absent. Early-instar larvae of *Heliozela* species in Europe and two species in the eastern United States mine at first in twigs or petioles (sometimes causing a gall in which they feed), or down a leaf midrib, then form a leaf mine. Females of *Antispila* and *Coptodisca* oviposit directly into the leaf, and the larva forms a short, serpentine mine, then enlarges it to a full-depth blotch. The last instar constructs a lenticular, portable case by cutting disks from the upper and lower epidermis and joining them with silk, giving rise to the common name "shield-bearers." The abandoned mines, with their distinctive "shot holes" are highly characteristic of heliozelids (Plate 59.6). The case-bearing larva crawls away and descends by a silken thread to attach to a lower leaf or bark, where pupation occurs.

Larval Foods Species are believed to be host specific, using at least 17 families of usually woody angiosperms, with preponderance in Myrtaceae in Australia and Cornaceae and Vitaceae in the Holarctic, primarily species of *Antispila*, among the few lepidopteran genera to specialize on Vitaceae.

Diversity There are more than 100 described species in about 12 genera, distributed in all major faunal realms except New Zealand but poorly known in tropical regions. Two genera with about 12 species occur in western North America.

Larvae of *Antispila voraginella* Braun form large blotch mines in leaves of wild grape (Vitaceae) in Utah and Arizona; mines sometimes coalesce and are occupied by two or more larvae, which cut out their relatively large, oval cases at the mine periphery (Plate 59.7). Braun found a few larvae in late July in Zion Canyon, but most of the mines were already abandoned, and adults emerged the following April. Although mature larvae form cases that are much larger than those of *Coptodisca*, the moths are not (FW 2.3–3.0 mm). Adults of *A. aurirubra* Braun in southern California and Utah are minute jewels, shining dark coppery or reddish with snow white FW markings. The larvae mine dogwood (*Cornus*, Cornaceae) in July and August, and adults emerge the following May

Species of *Coptodisca* are more prevalent in the West than are *Antispila* and have diversified on a wide array of host trees and shrubs, including Betulaceae *(Alnus, Ostrya)*, Rosaceae *(Cercocarpus, Malus, Prunus)*, Ericaceae *(Arbutus, Arctostaphylos, Kalmia, Vaccinium)*, Myricaceae *(Myrica)*, Juglandaceae *(Juglans)*, Salicaceae *(Populus, Salix)*, and Fagaceae *(Quercus)*. Adults of western Nearctic species are tiny (FW 1.5–2.35 mm), about the size of the smallest nepticulids. The species are superficially similar, shining metallic lead or silver colored, with the FW distally pale to dark orange with opposing costal and dorsal white triangles outlined in black and black blotches in the tornal area. Adults of *Coptodisca arbutiella* Busck (Plate 1.24, Plate 59.5, Plate 59.6) sometimes are numerous on madrone *(Arbutus menziesii)* along the coast from Vancouver Island to central California, running about during the daytime on the new leaves in spring. The larvae grow slowly after the leaves have hardened and do not reach maturity until early the following spring, in late February or early March in California. The mines sometimes are abundant, 30 or 40 per leaf on madrone and up to eight or 10 on a leaf of manzanita *(Arctostaphylos)*(Plate 59.6). *Coptodisca powellella* Opler (Plate 1.23) is similar and has the same habits and life cycle, feeding on coast live oak *(Q. agrifolia)* in California. *Coptodisca saliciella* (Clemens) is transcontinental in distribution. The mines often are numerous on coastal willow *(S. lasiolepis)*, as are those presumably of the same species on poplar *(P. fremontii)* in inland areas, the Central Valley of California south to Durango, Mexico.

FAMILY INCURVARIIDAE

Adult These are small moths (FW 3.5–9 mm), with rough head scaling, relatively small eyes, antennae shorter than FW, mandibles vestigial and proboscis short, half the palpi length, maxillary palpi elongate and folded. They are somber moths with dark, unicolorous wings, sometimes iridescent brown, bronze, or bluish.

Larva Larvae are moderately flattened, thoracic legs well developed but abdominal prolegs reduced, crochets uniserial, transverse bands. The early instars form blotch mines, after which they cut through the upper and lower epidermis to remove oval sections, which they sew together to form a portable case. Later instars graze leaf surfaces, forming skeletonized patches in the form of the case; they cut out larger circular sections to form larger cases as they grow. Larvae of a few genera remain in the mines throughout feeding, until ready to pupate,

then cut out a case in which they pupate, in the manner of heliozelids.

Larval Foods Oviposition is host specific. The ancestral, southern continent genera feed on Myrtaceae or Proteaceae, while Holarctic incurvariids use about 10 unrelated angiosperm families, none predominately.

Diversity There are about 100 described species in 11 genera, mostly Australian and Palaearctic, poorly represented in Africa and the Western Hemisphere. *Lampronia*, formerly considered to belong to this group, has been relocated to Prodoxidae.

Vespina quercivora (Davis) is our only western Nearctic incurvariid, occurring in southern and central California. It is a small (FW 3.4–5.3 mm; Plate 1.25), nondescript gray brown moth with pale tan head vestiture and a faint pale spot at the base of the tornal fringe in most specimens. Mines and shelters of the larvae sometimes are abundant, but there are few collection records of the adults in the field; they apparently are not attracted to lights and may be crepuscular. The larval biology is unlike any other western Nearctic moth, with a mining phase followed by two distinct case-bearing stages. The young larvae begin mining leaves of live oaks *(Quercus agrifolia* and *Q. wislizenii)* by September, ultimately forming a rectangular, full-depth mine with the frass plastered externally on its undersurface by silk; mines are occupied until February or March. The larva then cuts out an oval piece from the upper epidermis and a shorter one from the lower surface, and the two are sewn together forming a portable case (Plate 59.8). Its occupant feeds by grazing on the lower side of the leaf without exposing itself beyond the perimeter of the larger shield. In late March or April, the last-instar larva cuts out a larger, broadly oval or nearly circular disc, which is turned upside down and under which it grazes on the upper or lower surface of the leaves, including the new foliage. The resultant round cutouts and grazed areas are characteristic indicators of this insect. Pupation occurs in the last case in May or early June, and in the lab adults emerge between May and August, mostly June and July.

FAMILY ADELIDAE

Adult Adelids are best known for their enormously long antennae, often 2.5 or three times the FW length (Plate 59.10, Plate 59.11, Plate 59.12, Plate 59.13). Usually they are much longer in the male, which in many species possess greatly enlarged eyes, apparently correlated with swarming behavior. The eyes are small in both sexes of some species, irrespective of antennal length, and both sexes of those species are found on the flowers of the host plant. The antennae are fully scaled; *Nemophora* and some *Adela* species possess stout, hooklike spines on flagellar segments 7 to 10 or 10 to 13, but these are lacking in *Adela* in the western United States. The wing-locking mechanism is peculiar; the male retinaculum is a long costal fold extending under the base of the Sc, and the furculum is a stout seta accompanied by several setae along the HW costa. Females have three or four bristles in a row along the costa. Holarctic and Neotropical species (Adelinae) are small (FW 3.5–8.5 mm) diurnal moths, often brightly colored, iridescent green, blue, or purplish, with white antennae, while the primarily African Nematopogoninae are crepuscular or nocturnal and dull colored. Both sexes have a well-developed proboscis and take nectar, often from flowers other than the larval food plant. During mate-seeking, males of the large-eyed species form small, dancing groups, to which females evidently react. Females insert the eggs into the base of the ovaries of unopened flowers or into leaves.

Larva First-instar larvae of the few species studied in detail feed in the developing ovules; after molting they drop to the ground and construct flat, portable cases of silk covered by debris. They feed on fallen leaves or the lower leaves of the host plants, which often are short-lived annuals. Mature larvae are somewhat flattened, white or greenish with a darkly pigmented head and dorsal plates. The prolegs are reduced, with crochets arranged in several parallel rows. Pupation occurs in the flat, rectangulate case, with the long antennae free and coiled several times around the abdomen (Plate 59.15). The larval biology of these relatively conspicuous moths is surprisingly poorly known. *Nematopogon metaxella* (Hübner) in Europe long ago was reported to oviposit into leaves of any convenient plant, from which the larvae exit without feeding to construct portable cases on the ground, then feed indiscriminately on living or dead leaves. In Japan *Nemophora raddei* Rebel flies in spring and selects *Salix* flowers, and the first-instar larvae feed on the ovules before making their cases in the second instar; they complete development by autumn, and the pupae hibernate in diapause. *Adela* in California insert eggs into buds of Rosaceae (Plate 59.13), Polemoniaceae (Plate 59.9), Scrophulariaceae, or Papaveraceae (Plate 59.11).

Larval Foods Oviposition by each species for which hosts are recorded is restricted to one or a few closely related plants, which include members of at least 18 angiosperm families, with no strong emphasis on any one.

Diversity More than 300 species are described in five genera; adelids occur in all faunal regions except New Zealand. There are 13 species of adelids in Canada and the United States, of which nine occur in the West.

Nemophora bellela (Walker)(Plate 2.7) is the largest (FW 7.3–8.5 mm) and is a boreal species ranging from Quebec across Canada north of the plains to the mountains of British Columbia and Alaska. The antennae are thickly scaled, FW ochreous with the veins black, reflecting metallic blue on fresh specimens; there is a broad, transverse yellow band bordered by metallic blue at the end of the cell. The male antenna is about 1.9 times the FW length, that of the female 1.4 times. The host plant is unknown.

The other western species are members of the genus *Adela*, and all occur in California. They are beautiful diurnal moths, often resplendent in metallic colors, but the integument is black, and if handled carelessly, specimens lose their colorful scaling. They fly in spring when their annual host plants are in bloom. Sometimes four or five species are found in a serpentine grassland habitat, each closely associated with colonies of its respective host plants. *Adela trigrapha* Zeller (male FW 5.5–6.2 mm; Plate 2.9, Plate 59.9) is perhaps the most commonly seen species, ranging from Vancouver Island to southern California, including Santa Rosa and Santa Cruz islands. The males are seen in small dancing groups in open flowery meadows. They have large eyes, long white antennae (3 X FW length); FW black crossed by three white lines, which are reduced to two or only traces in some populations. Some inland populations consist of smaller individuals (FW 4.5–5.5 mm). Females are smaller than males and have small eyes and shorter antennae (1.5 X FW length), and head has bright orange vesture, FW bright, metallic blue, HW purple. They oviposit into flower heads of *Linanthus androsaceus* and *L. bicolor* (Polemoniaceae)(Plate 2.9). Both sexes visit other plants for nectar,

especially Asteraceae, late March to mid-May in the foothills, mid-April to mid-June north of San Francisco. *Adela eldorada* Powell (Plate 2.8) is similar, larger (male FW 6.5–7.5 mm, female 5.5–6.2 mm), and more brightly colored; males have some orange scaling on the head, and both sexes have broader, cream-colored bands on the FW than do *A. trigrapha*. This species flies in sheltered, shrubby sites rather than open meadows, in the central and southern Sierra Nevada foothills from about 1,500 to 6,000' elevation. Oviposition has not been observed. Two additional large-eyed species, *A. flammeusella* Chambers (Plate 2.11, Plate 59.10) and *A. thorpella* Powell (Plate 2.10), have bronzy FW and fly in open meadows. *Adela flammeusella* is associated with owl's clover, *Orthocarpus* (Scrophulariaceae), some species of which are now assigned to *Castilleja,* but the *Adela* do not care. Both sexes have dark coppery bronze FW (5–6 mm) and metallic purple HW; males usually have three to five cream-colored FW spots, but both sexes, especially females, have reduced or lack markings in some populations. By contrast, *A. thorpella* has shorter antennae (2.5 X FW length in males) and dull brassy bronze wings with a pale fringe and at most a broken pale band represented by a tornal spot (FW 6–7 mm). Both sexes tend to stay near the larval host plant, cream cups, *Platystemon californicus* (Papaveraceae). Both of these species are widespread in foothill situations in California, with *A. flammeusella* ranging to northeastern California and southern Washington at the Columbia River. *Adela oplerella* Powell, which also is closely tied to *P. californicus*, is smaller (FW 4.5–5.5 mm; Plate 2.5, Plate 59.11) and darker, sometimes with two faint whitish spots, the body with dense black vestiture. This uniquely derived species has small eyes and short antennae in both sexes, about equal to FW length in males, shorter in females. It is known only from about a dozen colonies in the San Francisco Bay area, almost exclusively serpentine grasslands. "Opler's longhorn moth" is under review as a threatened species.

Adela septentrionella Walsingham (Plate 2.12, Plate 2.13, Plate 59.13, Plate 59.14, Plate 59.15) ranges from southern British Columbia to central California north of the Transverse Ranges, and there is one record on Santa Cruz Island in 1966. It occurs from the coast to more boreal situations than other *Adela,* 6,000 to 8,000' in the Sierra Nevada. Adults are black, FW usually with two thin, transverse lines and a row of white dots around the terminal margin, but in coastal populations the white is reduced, and some individuals are immaculate black (FW 4.5–5.7 mm). The eyes are small in both sexes, and the antennae are longer than in some large-eyed species (2.5–3 X FW length in males; 2 X in females). Females have orange head scaling. This species flies later in the season, May and June, into July at higher elevations, in association with the bud stage of oceanspray, *Holodiscus discolor* (Rosaceae). Mature or nearly mature larvae overwinter in cases on the ground (Plate 59.14) and pupate in late March or April. *Adela singulella* Walsingham (Plate 2.14, Plate 2.15, Plate 59.12) also is unique, in having small eyes and very long, white antennae in both sexes (>4 X FW in males, >3 X in females). The FW (3.7–5 mm) is relatively narrow, dark bronzy purplish, with the costal margin white and usually crossed by a single white line. This typical form occurs north of San Francisco Bay, whereas populations with similar morphology range along the foothills of the Sierra Nevada on the same host plant but average smaller, with dull bronze FW and lacking the white transverse line. Oviposition occurs into flower heads of *Gilia capitata* and *G. achillaefolia* (Polemoniaceae), which also provide the nectar source. Lastly, *A. punctiferella* Walsingham (Plate 2.6) is an austral species, distributed in the Mojave Desert and northward along inland coastal foothills to Lake County, California. Adults resemble Sierran populations of *A. singulella* but are smaller (FW 3.8–4.6 mm), with much shorter antennae (3 X FW length in males, less in females), and the FW lacks the white costa. A white dot at the end of the cell, which gave the species its name, is present in only about 30% of the specimens; rarely is expressed as a thin, transverse line. Both sexes occur in close association with *G. tricolor* and *G. scopulorum*.

Some *Adela* flourish in lightly grazed areas and open habitats on poor soil, such as serpentine grasslands, where alien annual grasses are not dominant, and their annual host plants persist. *Adela trigrapha*, *A. flammeusella*, and *A. thorpella* were numerous on Santa Cruz Island in 1966, having survived many decades of feral sheep. However, their larval host plants became scarce following removal of the sheep in the 1980s and the resultant massive growth of alien grasses, and *A. flammeusella* occurred only sparsely, *A. trigrapha* was very rare, *A. thorpella* was not found from 2000 to 2006.

Species of **Cauchas** are smaller (FW 4–5.5 mm) than most *Adela*, slender moths with the wings rounded distally, and lack the dense hairiness of *Adela*. They are diurnal, mostly bronze colored, with short antennae, found in close association with native mustards (Brassicaceae), their larval hosts. *Cauchas simpliciella* (Walsingham)(Plate 2.4) has dark bronze to bright purplish FW with orange head vestiture and is widely distributed from Montana and the Pacific Northwest southward along the coast; often associated with *Erysimum*, more rarely with *Arabis* or *Thelypodium*, where females oviposit into young seed pods. Adults are active from late March near the coast to June at 5,000′ in montane California and July in Glacier National Park. There are several undescribed species found on *Arabis*, *Streptanthus*, *Thysanocarpus*, and other mustards and one on *Camissonia* (Onagraceae) in southern California.

The phylogenetic placement of **Tridentaforma** is problematic, with conflicting evidence in both morphological and molecular data, indicating a basal placement in Incurvarioidea, related to either Adelidae or Prodoxidae. *Tridentaforma fuscoleuca* (Braun), which was described from the Sierra Nevada and ranges north to Mt. Shasta, is a relatively large species (FW 8.5–10.5 mm; Plate 1.26) with shining, faintly purplish wings, FW overlaid with slate gray, usually with an indistinct black discal dot. Similar but appreciably smaller moths (FW 4.5–6.8; Plate 1.27, Plate 1.28), which likely comprise a complex of species, occur widely in the western United States. They fly around manzanita (*Arctostaphylos*, Ericaceae) in early spring, March and April in the foothills to June at Mt. Shasta, and presumably oviposit into developing seed.

FAMILY PRODOXIDAE

Prodoxids are renown for the close symbiotic relationship between species of *Tegeticula* and yucca plants (Agavaceae). Females possess enormous "tentacles," appendages of the maxillary palpi, unique among all insects, that are used to gather pollen (Plate 59.17), which is then transferred to the stigma of other flowers visited for oviposition, thus ensuring cross-pollination. Other kinds of insects are not attracted to yucca flowers to collect pollen. Females are believed to leave a pheromone signal at the oviposition sites that deters later females so that only a few larvae feed in any given seed pod, and many unaffected seeds are produced (Plate 59.18).

Adult The adults are small moths (FW 4–16 mm), generally dull colored, white or gray, although a few *Prodoxus* and

Greya species have the FW iridescent bronze colored and white or yellow patterned. The maxillary and labial palpi are relatively prominent but usually shorter than the proboscis, which is functional in the more ancestral genera of Prodoxidae. The derived, Agavaceae-feeding prodoxines apparently do not seek nectar, yet individuals of *Tegeticula* we marked were recaptured up to nine days later in the field. Genitalia, Figs. 35, 36.

Larva Early instars of the more ancestral genera (*Lampronia, Greya*) feed in young ovules of the host plant, then leave to spin overwintering shelters on the ground. In early spring they climb back onto the newly foliating host plants to feed in flower buds or foliage shoots. These caterpillars and those of the pollen-carrying genera are stout, highly mobile, with well-developed thoracic legs, lacking abdominal prolegs, while those of the "bogus yucca moths" (*Prodoxus*) are completely legless and apparently blind, living their entire life within the gallery and pupating there. Prepupal larvae of yucca moths are capable of maintaining the diapause for several to many years if optimal winter conditions are not experienced, followed by seasonally synchronized development.

Larval Foods Holarctic Lamproniinae and species of the basal prodoxine genus *Greya* specialize on Rosaceae, Ericaceae, Apiaceae, or Saxifragaceae, while the more-derived prodoxines are Agavaceae specialists that feed internally in seeds or sterile tissue of the inflorescence scapes and fruit.

Diversity Prodoxidae are predominately Holarctic, with Lamproniinae mainly Palaearctic and Prodoxinae largely Nearctic, with a few species ranging into southern Mexico. About 100 species in 12 genera are known worldwide, of which five genera and more than 50 species occur in western North America.

Although named for the English nobleman Lord Thomas de Grey of Walsingham, members of **Greya** live up to the image cast by the name, being primarily gray or gray brown. They are slender moths with relatively broad wings (FW 3–12.5 mm). There are 16 species distributed widely over western North America, particularly the Pacific Northwest. The genus was monographed by Davis and Pellmyr in 1992. *Greya politella* (Walsingham)(Plate 1.29) is the most frequently observed; it ranges from Montana and southern British Columbia to central California and Santa Cruz Island. It is a relatively large species (FW 5–9 mm) with dull bronze-colored FW, slightly shining; the adults are invariably found in close association with their host plant, species of *Lithophragma* (Saxifragaceae), especially *L. affine*, often in remarkably small, isolated colonies of the plant. Adults frequently are found with their heads protruding well down into the flowers and are believed to be effective in mechanical transfer of pollen, an association first reported by a plant systematist, R. L. Taylor, in the 1960s. Females possess a very elongate abdomen used to probe deeply into the base of the flower during oviposition. *Greya variabilis* Davis and Pellmyr (Plate 1.30, Plate 1.31) is the most northern species of Incurvarioidea and most widespread *Greya*, ranging from northern Oregon to Alaska beyond the Arctic Circle at 69° N and the Queen Charlotte, Pribilof, and St. Lawrence islands. Befitting its name, the FW (5–8 mm) pattern ranges from pale ochreous to dark bronzy brown, weakly or strongly marked with variable, distinct ochreous spots. Despite the wide distribution and numerous collections, the host plant is unknown. Adults are found from late June to August. Unlike nearly all other prodoxids, *G. reticulata* (Riley) exhibits marked sexual dimorphism, with the males gray and females' FW reticulate in metallic golden and white (Plate 59.16). Both sexes are represented in the original series collected at Los Angeles in the

1880s by Koebele. But evidently he sent only females to Riley, who described *G. reticulata* as a *Prodoxus*, believing it to be a yucca moth, with the FW pattern similar to some *Prodoxus*. It remained a "lost" species for more than 70 years, until we discovered a population and associated the sexes, in Santa Clara County in 1966, and observed its host plant, *Osmorhiza chilensis* (Apiaceae). Recently we found a colony on Albany Hill adjacent to San Francisco Bay, more than a century after the only previous Alameda County collection, also by Koebele. At Albany Hill, *G. reticulata* uses *Sanicula bipinnatifida* (Apiaceae) for oviposition (Plate 59.16) and larval feeding. Adults are active March through May to early June in the Northwest, and there is one record in western Colorado in August.

Members of the genus *Tegeticula* represent the most widely acclaimed classic example of plant–insect symbiosis. Female moths gather pollen and transfer it to the stigma of flowers on other yucca plants following their oviposition into the floral ovaries, which produces larvae that feed on some of the seeds. This was first observed by the botanist Engelmann in 1872. C. V. Riley, then the state entomologist for Missouri, carried out extensive studies on the relationship and used the wonderful story to help promote the importance of insects in his role of first entomologist in the U.S. Department of Agriculture and founder of the American Association of Economic Entomologists. Among other accomplishments, Riley commissioned Coquillett and Koebele to carry out entomological investigations in California. They collected the type series of most western yucca moths for Riley, along with the first specimens and host-plant records of many other microlepidoptera.

This biological coevolution has been repeated in textbooks almost unchanged from Riley's accounts, often becoming increasingly simplified, but in reality the relationships have been discovered to be much more complex. Populations of the white moths Riley named *Pronuba* (now *Tegeticula*) *yuccasella* were shown by Davis in the 1960s to possess appreciable morphological differences in FW shape and genitalia from one yucca species to another. Moreover, some females lack development of the maxillary tentacles, and Busck named them as a separate species, *Prodoxus intermedia*. Davis found 13 to 70% of females with this condition in samples from Texas and New Mexico. Later, Addicott and others proposed that these individuals constitute separate species that act as "cheaters" in the system, natural selection having favored their activity later in the season when developing fruit are available for oviposition. Thus they bypass the pollination process and depend upon other *Tegeticula* to have accomplished it. All of this knowledge was used by Olle Pellmyr in a series of elegant studies based on molecular and morphological evidence to demonstrate that the cheater species are independently derived at least three times, and in 1999 he described a complex of 13 species in the western United States among populations formerly treated as *T. yuccasella*.

Tegeticula yuccasella (Riley), the "yucca moth," is a relatively large species (FW 9–11.5 mm) with white FW, dark to pale brown HW. It ranges widely over the eastern United States, into southern Ontario, and westward into Montana, Wyoming, eastern Colorado, and New Mexico, acting as pollinator of several *Yucca* species. Populations show substantial variation in darkening of the integument and in wing breadth, and sympatric species are distinguished by genitalic features. Western species of the *T. yuccasella* complex that Pellmyr described include *T. altiplanella,* which occurs in the mountains and high plains of Colorado, New Mexico, southern Utah,

and northern Arizona, in association with the capsular yuccas (fruits that dehisce and open on the plant) of that region. Both this and *T. baccatella*, which lives on baccate yuccas (nondehiscent fruits)*(Y. baccata, Y. confinis, Y. thornberi),* have white FW over much of their ranges but tan FW at higher elevations and northern latitudes. *Tegeticula maderae* is a smaller (FW 6.5–10.5 mm; Plate 1.36), narrow-winged species confined to the mountains of southern Arizona on the summer-flowering *Y. schottii* in the oak forests above 5,000'. It thus is isolated spatially and seasonally from the adjacent broad-winged species, *T. baccatella* (on *Y. baccata*) and *T. elatella* (on *Y. elata*), which are larger species (FW 8–12 mm; Plate 1.36) that fly in May and June at lower elevations. *Tegeticula mojavella* is a narrow-winged species (Plate 1.38) that as the name suggests, occurs in the Mojave Desert to coastal southern California on Mojave yucca *(Y. schidigera).* The most widespread cheater species is *T. corruptrix* Pellmyr, which is associated with the pollinator species of numerous yuccas from western Nebraska and western Texas to the San Jacinto Mountains of southern California. It is the largest species in the complex (FW 10–14.8 mm).

Joshua tree *(Yucca brevifolia)* in the Mojave Desert hosts two *Tegeticula* species, which are very different from typical *Tegeticula*, being flattened insects, sawflylike in appearance. *Tegeticula synthetica* (Riley)(FW 7.5–9.5 mm; Plate 1.32) is dark brown with scattered pale gray scaling when freshly emerged, but most of the scaling is lost during the moths' activities among the nearly closed blossoms of *Y. brevifolia*. A sister species, *T. antithetica* Pellmyr, is smaller (FW 5.5–7.5 mm) with a paler gray, arrow-shaped mark in the discal cell, and there are distinct differences in genitalia of both sexes. The latter species occurs in the northeastern parts of the Joshua tree range, from Clark Mountain, San Bernardino County, California, to southwestern Utah, in association with *Y. brevifolia* var. *jaegeriana*, whereas *T. synthetica* is restricted to southwestern parts of the Joshua tree range, and the two are not known to overlap. Both species are active in March to early May at higher elevations.

Hesperoyucca whipplei in California and Baja California also harbors a distinctive species, *Tegeticula* **maculata** (Riley)(FW 6–10 mm). The moth is white, FW with black spotting distally in northern populations (Plate 1.33, Plate 1.34, Plate 59.17), but south of the Transverse Ranges it is black (*T. extranea* Henry Edwards)(Plate 1.35). Intermediates in the color shift occur in the Cajon Pass area, where some individuals have a brownish cast over the wings. Diurnal activity and black scaling may be a thermoregulatory adaptation in response to the early spring (February to March) flowering of *H. whipplei*.

A second kind of pollen-carrying prodoxid, **Parategeticula pollenifera** Davis (Plate 1.39), occurs along with *Tegeticula* on *Yucca schottii* in southern Arizona, and there are three other *Parategeticula* in Mexico. *Parategeticula pollenifera* (FW 11–15.5 mm) is larger and more bulky than *Tegeticula*, with broader, somewhat translucent, pale tan wings. It was first collected at Madera Canyon, Santa Rita Mountains, in 1947, but its distinctiveness was not recognized until Davis studied the yucca moths many years later. When newly emerged, the FW has a speckled appearance, with scattered black scales, which are shed in the flowers. The females possess maxillary tentacles and gather pollen in the same manner as *Tegeticula*, but the oviposition behavior and larval biology differ markedly. The female employs a sclerotized, hooklike extension of the abdomen to dig pits in the surface of the scape or petals, into which the round eggs are nested (Plate 59.19).

Larvae emerge outward, crawl to the young fruit, and burrow into the immature seed, causing a cyst of abnormal tissue to develop in place of a few seeds. The larva feeds within this cyst (Plate 59.20) at an earlier stage of seed development than do larvae of *Tegeticula*, which wait until the seeds are full sized. We conducted an exclusion experiment with bagged inflorescences demonstrating that *P. pollenifera* females are effective pollinators in absence of *Tegeticula*. But, if *T. maderae* females have been successful, survival of *P. pollenifera* is not dependent upon their own pollination behavior. Moreover, the number of *Parategeticula* larvae per fruit is not limited by pheromone-controlled oviposition. Therefore, *Parategeticula* ought to have a selective advantage. That appears to be the case in the Chiricahua Mountains, where *T. maderae* is rare, but numbers of the two are more or less equal in the Santa Rita Mountains.

Members of the genus *Prodoxus*, the "bogus yucca moths," are smaller (FW mostly 4.2–6.4 mm) and are also dependent upon the pollination system—the larvae feed in sterile tissue of the inflorescence, either the rapidly elongating scape (Plate 59.21) or in the fleshy fruit outside the seed tiers. All western yuccas have one or two scape-inhabiting species, and most harbor a different *Prodoxus* that feeds in the fruit. Adults are found in the flowers, along with *Tegeticula*, where mating occurs. The larvae evidently secrete a substance that causes the plant tissue surrounding the gallery to discolor; later it hardens, forming a woody cast around the cell containing the dormant larva over summer, fall, and winter. This capsule and a tough cocoon with a tightly closed lid prevent desiccation and enable larvae to maintain the diapause in years when climatic conditions, particularly temperature and rainfall, are not optimal. Transformation to the adult may be delayed many years, often four to eight years in our experimental trials with several species of *Prodoxus*, up to 30 years in *P. y-inversus* Riley, followed by synchronous mass emergence seasonally timed with the yucca flowering. In our experiment with *P. y-inversus* in dry fruits of *Yucca baccata* from southern Nevada, the vast majority of adults reared (*n*=520) emerged after more than 15 years in diapause—35% of them in years 16 and 17, 29% in year 20, 25% in year 25, and 3% (14 individuals) after 30 years. Recently, Pellmyr and others revised the genus *Prodoxus*, based on morphological and molecular analysis. They included the former *Agavenema* species, which feed on century plant *(Agave)*, and described nine new species, bringing the total to 22 species, of which 18 occur in the western United States.

Prodoxus quinquepunctellus (Chambers)(Plate 1.40) is the largest and most widespread species, occurring on the capsular (dehiscent) yuccas of the eastern United States and from the Rocky Mountains to Arizona. The moths are white, FW immaculate or with up to 18 small black dots, varying within populations. The adults (FW 6–10.5 mm) approach the size of male *Tegeticula* in the same flowers. Several *Prodoxus* have black-banded FW similar to *P. coloradensis* Riley (4.5–6 mm; Plate 1.43, Fig. 36), which ranges from western Texas, New Mexico, and southern Colorado to southern Nevada and the Mojave Desert in California, associated with baccate (nondehiscent) yuccas, including *Yucca baccata*, *Y. schidigera*, and *Y. arizonica*. The FW is white with four variable, dark gray or black, transverse bands, the middle two usually joined to form a Y or V shape connected at the tornus to a terminal band. The maculation is reduced, especially on the basal half, in some individuals and populations, varying to nearly immaculate. Populations in coastal southern California on *Y. schidigera* were identified as a sister species, *P. californicus*

Pellmyr & Balcázar-Lara. The superficially similar *P. phylloryctus* Wagner & Powell (Plate 1.44, Fig. 35) is unique as a leaf miner, the only prodoxine species not dependent on inflorescence growth. The larvae mine communally, and the hardened tissue surrounding the galleries coalesces to form large brownish blotches from which the pupae protrude at emergence. The adults resemble *P. coloradensis*, which occupies stalks on the same *Y. baccata* plants in Colorado and New Mexico, but is smaller (FW 4–5.5 mm), with the FW pattern consistently broader and darker.

Prodoxus y-inversus Riley, which bores in fruits of *Yucca baccata* and *Y. arizonensis*, has white FW (4.5–6.5 mm; Plate 1.45) with variable dark markings, forming an inverted Y in the distal half. The superficially quite similar *P. sonorensis* Pellmyr & Balcázar-Lara, which feeds in fruits of *Y. schottii*, has white FW with variable dark markings, typically with the inverted Y on the distal half broken. *Yucca schottii* in southern Arizona also has a scape borer, *P. ochrocarus* Davis (Plate 1.46), which has dark gray FW with a shining bronzy tint and pale yellow markings. Joshua tree *(Y. brevifolia)* has both a scape borer—*P. sordidus* Riley, which has cream-colored FW and white HW with the costal area dark gray (Plate 1.47)—and a fruit borer, *P. weethumpi* Pellmyr & Balcázar-Lara, with pale to dark tan FW and pale gray HW (Plate 1.48).

Hesperoyucca whipplei in California has the most complex guild, three species of *Prodoxus*, one feeding in the fruit and two in the scape, all three of which vary geographically in phenotype. *Prodoxus cinereus* Riley (Plate 2.1, Plate 59. 21), which is absent from the Coast Ranges, has ash gray FW and dark gray HW in the southern Sierra Nevada, dark gray FW south of the Transverse Ranges. It emerges first in spring, when the scapes are lengthening, and females tend to oviposit below the inflorescence, preceding its development. *P. aenescens* Riley (Plate 2.2), with dark bronze-colored FW in the Coast Ranges and Sierra Nevada, shining steel gray in southern populations, oviposits mostly in the inflorescence area, so the larvae occupy a higher portion of the scape and make shallower galleries. *P. marginatus* Riley (Plate 1.49, Plate 1.50) is smaller and has white FW with a terminal black band in northern populations (FW 4.2 mm) but is variable within southern populations, some individuals becoming diffusely gray speckled (Plate 1.50 shows *P. pulverulentus* Riley, treated as a separate species by Davis). The larvae feed at the base of the small dehiscent capsules. All three plus *Tegeticula maculata* are found together in the enormous panicles of *H. whipplei*, often in great numbers, from March to May. Although the moths are quite dissimilar, the analysis by Pellmyr et al. showed this guild forms a monophyletic clade with the three former *Agavenema* species. The latter are elongate, gray moths, larger than most *Prodoxus* (FW 5–15 mm); they share the same larval biology as the scape-feeding *Prodoxus* but feed on century plant *(Agave)*. *Prodoxus barberellus* Busck (FW 9–12.5 mm; Plate 1.41) occurs in southern Arizona on *A. palmeri* and other agaves, whereas *P. pallidus* (Davis)(FW 12–15 mm; Plate 1.42) is a larger and paler moth on *A. deserti* in the Colorado Desert of California and Baja California. Adults emerge from cocoons in the scapes after one to four years in confinement.

Mesepiola specca Davis is a tiny prodoxid (FW 3.8–5.6 mm; Plate 2.3) having whitish FW speckled with dark gray. The female of *M. specca* was collected in 1939 in New Mexico by Annette Braun, but its systematic position was not revealed until males were studied much later. The larvae feed in seeds of beargrass *(Nolina)*, discovered by Don Frack in the 1970s, and on *Dasylirion wheeleri*, according to Pellmyr. The species is recorded in Texas to Arizona, and in California based on one female collected on *Nolina* at El Toro, Orange County, in 1938.

Superfamily Tischerioidea

FAMILY TISCHERIIDAE

Adult The adults are very small moths (FW 2.7–5 mm), with lanceolate wings, usually dull colored, white, gray, or yellow, without distinct maculation. The proboscis is well developed. Genitalia, Figs. 37, 38. Adults are nocturnal with large eyes and when at rest they perch with the head appressed to the substrate and tail end lifted at a 45° angle.

Larva Larvae are slightly flattened leaf miners with thoracic legs reduced to two vestigial segments or ambulatory knobs (calli), abdominal prolegs rudimentary with crochets in multiserial bands or incomplete ellipses. They spin excessive silk, and the linear or blotch mines are characterized by a heavily silk-lined nest within which the larva retreats when not feeding.

Larval Foods Larvae are recorded from nine angiosperm families, most commonly Fagaceae, Rosaceae, and Asteraceae, more diverse on the last than is true of other lepidopterous miners, presumably a later evolutionary radiation.

Diversity This is a small, primarily Holarctic group with about 80 described species in one genus.

Tischeria is represented in the western United States and Canada by about 25 described species, and no doubt there are many yet to be discovered, as there has been no additional taxonomic study of them since Braun's monograph more than 30 years ago. The characteristic linear larval mines are known from a wide variety of trees and shrubs. Three oak-mining, similar, predominately yellow species are among the most often observed western species. *Tischeria mediostriata* Braun (Plate 1.19) is widespread in the southwestern states to eastern Washington, mining leaves of deciduous oaks, especially *Quercus gambellii* and *Q. kelloggii*. The FW (3.7–4.2 mm) is yellow orange with a broad lemon yellow stripe from base to two-thirds, its upper edge diverging from the costa. *Tischeria consanguinea* Braun (Plate 1.18) is similar, having bright yellow orange FW with a poorly defined darker subterminal line and tornal area, discal area paler yellow. The mines occur on several species of live oaks of the white oak group in arid parts of central and southern California. *Tischeria discreta* Braun is restricted to the live oaks, *Q. agrifolia* in coastal areas and *Q. wislizenii* in inland areas of California and on Santa Cruz Island. It is smaller (FW 3.2–3.5 mm) with pale ochreous FW basally becoming lustrous brownish orange distally, the apical area and termen darker brown.

Tischeria splendida Braun (FW 2.7–3 mm; Plate 1.20) creates conspicuous mines on native blackberry, *Rubus ursinus* (Plate 59.22), in coastal California. The moth, which originally had been collected at San Leandro in 1909 but confused with the eastern species, *T. aenea* Frey & Boll, was rediscovered in March 1961, when we found the mines at Mt. Diablo. It has a brilliant metallic bronze FW basally, shading to lustrous reddish apically. Walsingham collected *T. ceanothi* (Plate 1.22) in Mendocino County in 1871, and it is widespread in parts of California and western Nevada. The larvae mine several species of *Ceanothus*. This is a tiny species (FW 2.5–3 mm) with pale ashy to dark gray FW. Mines are often abundant on *C. thyrsiflorus* along the coast, and there are several generations per year. *Tischeria* **omissa** Braun (FW 3.5–4.7 mm; Plate 1.21, Fig. 38) was first described from garden hollyhock *(Althea rosea)* at Berkeley in 1927. Since then the species has been reared from various Malvaceae, including hollyhock in southern Arizona and *Sphaeralcea* in Texas, and we found the mines on alkali mallow, *Malvella leprosa* (=*Sida hederacea*), in the Sacramento Valley, on cheeseweed *(Malva parviflora)* in the San Francisco Bay area, and recently a population of this species or a closely similar one on *Malacothamnus* on San Clemente Island off southern California. The FW varies from pale straw color to ochreous gray with dark scaling nearly obscuring the pale ground color, and/or variably gray spotted; apical area and fringe pink or pinkish orange. The mines are obscured by leaf crimping on the soft-leaved mallows.

References for Primitive Lineages

Addicott, J. F., J. Bronstein, and F. Kjellberg 1990. Evolution of mutualisticlife-cycles: Yucca moths and fig wasps, pages 143–161, *in*: Gilbert, F. (ed.) Genetics, Evolution, and Coordination of Insect Life Cycles. Springer-Verlag, Berlin and New York.

Aker, C. L., and D. Udovik 1981. Oviposition and pollination behavior of the yucca moth, *Tegeticula maculata* (Lepidoptera: Prodoxidae), and its relation to the reproductive biology of *Yucca whipplei* (Agavaceae). Oecologia 49: 96–101.

Braun, A. 1917. Nepticulidae of North America. Transactions of the American Entomological Society, Philadelphia 48: 155–209.

Braun, A. 1972. Tischeriidae of America north of Mexico. Memoirs of the American Entomological Society, Philadelphia 28; 148 pages.

Davis, D. R. 1969. A revision of the moths of the subfamily Prodoxinae (Lepidoptera: Incurvariidae). U.S. National Museum Bulletin 255; 170 pages.

Davis, D. R. 1978. A revision of the North American moths of the superfamily Eriocranioidea with proposal of a new family, Acanthopteroctetidae (Lepidoptera). Smithsonian Contributions to Zoology 251; 131 pages.

Davis, D. R. 1987. Micropterigidae (Micropterigoidea), Eriocraniidae and Acanthopteroctetidae (Eriocranioidea), pages 341–347, *in*: Stehr, F. (ed.) Immature Insects. Vol. 1. Kendall/Hunt, Dubuque, IA.

Davis, D. R. 1987. Nepticulidae and Opostegidae (Nepticuloidea), Tischeriidae (Tischerioidea), Heliozelidae, Incurvariidae, Adelidae, Prodoxidae (Incurvarioidea), pages 350–362, *in*: Stehr, F. (ed.) Immature Insects. Vol. 1. Kendall/Hunt, Dubuque, IA.

Davis, D. R. 1999. The monotrysian Heteroneura, pages 65–90, *in*: Kristensen, N. (ed.) Lepidoptera, Moths and Butterflies, Vol. 1: Evolution, Systematics, and Biogeography. Handbook of Zoology. W. de Gruyter, Berlin, New York; x + 491 pages.

Davis, D. R., and D. C. Frack 1987. Micropterigidae (Micropterigoidea), Eriocraniidae, and Acanthopteroctetidae (Eriocranioidea), pages 341–347, *in*: Stehr, F. (ed.) Immature Insects. Vol. 1. Kendall/Hunt, Dubuque, IA.

Davis, D. R., O. Pellmyr, and J. N. Thompson 1992. Biology and systematics of *Greya* Busck and *Tetragma* new genus (Lepidoptera: Prodoxidae). Smithsonian Contributions to Zoology 524; 88 pages.

Davis, D. R., and J. R. Stonis 2007. A revision of the New World plant-mining moths of the family Opostegidae (Lepidoptera: Nepticuloidea). Smithsonian Contributions to Zoology 625; v + 212 pages.

Issiki, S. T., 1931. On the morphology and systematics of Micropterygidae (Lepidoptera, Homoneura) of Japan and Formosa, with some consideration on the Australian, European and North American forms. Proceedings of the Zooological Society of London 1931: 999–1039.

Kristensen, N. P. 1999. The homoneurous Glossata, pages 51–63, *in*: Kristensen, N. (ed.) Lepidoptera, Moths and Butterflies. Vol. 1: Evolution, Systematics, and Biogeography. Handbook of Zoology. W. de Gruyter, Berlin, New York; x + 491 pages.

Kristensen, N. P. 1999. The non-glossatan moths, pages 41–49, *in*: Kristensen, N. (ed.) Lepidoptera, Moths and Butterflies. Vol. 1: Evolution, Systematics, and Biogeography. Handbook of Zoology. W. de Gruyter, Berlin, New York; x + 491 pages.

Kristensen, N. P., and A. W. Skalski 1999. Phylogeny and paleontology, pages 8–25, *in*: Kristensen, N. (ed.) Lepidoptera, Moths and Butterflies.

Vol. 1: Evolution, Systematics, and Biogeography. Handbook of Zoology. W. de Gruyter, Berlin, New York; x + 491 pages.

Nielsen, E. S. 1980. A cladistic analysis of the Holarctic genera of adelid moths (Lepidoptera: Incurvarioidea). Entomologica Scandinavica 11: 161–178.

Nielsen, E. S., G. S. Robinson, and D. L. Wagner 2000. Ghost-moths of the world: A global inventory and bibliography of the Exoporia (Mnesarchaeoidea and Hepialoidea)(Lepidoptera). Journal of Natural History 34: 823–878.

Nieukerken, E. J. van 1986. Systematics and phylogeny of Holarctic genera of Nepticulidae (Lepidoptera, Heteroneura, Monotrysia). Zoologisches Verhandelingen, Leiden 236; 93 pages.

Pellmyr, O. 1996. Non-mutualistic yucca moths and their evolutionary consequences. Nature 380: 155–156.

Pellmyr, O. 1999. Systematic revision of yucca moths in the *Tegeticula yuccasella* complex (Lepidoptera: Prodoxidae) north of Mexico. Systematic Entomology 24: 243–271.

Pellmyr, O. 2003. Yuccas, yucca moths, and coevolution: A review. Annals of the Missouri Botanical Garden 90: 35–55.

Pellmyr, O., and M. Balcázar-Lara 2000. Systematics of the yucca moth genus *Parategeticula* (Lepidoptera: Prodoxidae), with description of three Mexican species. Annals of the Entomological Society of American 93: 432–439.

Pellmyr, O., M. Balcázar-Lara, D. M. Althoff, and K. A. Segraves 2000. Phylogeny and life history evolution of *Prodoxus* yucca moths (Lepidoptera: Prodoxidae). Systematic Entomology 31: 1–20.

Pellmyr, O., M. Balcázar-Lara, D. M. Althoff, K. A. Segraves, and J. Leebens-Mack 2005. Phylogeny and life history evolution of Prodoxus yucca moths (Lepidoptera: Prodoxidae). Systematic Zoology 31: 1–20.

Pellmyr, O., and K. A. Segraves 2003. Pollinator divergence within an obligate mutualism: Two yucca moth species (Lepidoptera: Prodoxidae: *Tegeticula*) on the Joshua tree (*Yucca brevifolia*: Agavaceae). Annals of the Entomological Society of American 93: 432–439.

Powell, J. A. 1969. A synopsis of Nearctic adelid moths, with descriptions of new species (Incurvariidae). Journal of the Lepidopterists' Society 23(4): 211–240.

Powell, J. A. 1984. Biological interrelationships of moths and *Yucca schottii* (Lepidoptera: Incurvariidae, Cochylidae, Blastobasidae). University of California Publications in Entomology 100; 93 pages.

Powell, J. A. 1992. Interrelationships of yuccas and yucca moths. Trends in Ecology and Evolution 7(1): 10–15.

Powell, J. A. 2001. Longest insect dormancy: Yucca moth larvae (Lepidoptera: Prodoxidae) metamorphose after 20, 25, and 30 years in diapause. Annals of the Entomological Society of America 94: 677–680.

Powell, J. A., and R. A. Mackie 1966. Biological interrelationships of moths and *Yucca whipplei* (Lepidoptera: Gelechiidae, Blastobasidae, Prodoxidae). University of California Publications in Entomology 42: 146.

Tuskes, P. M., and N. J. Smith 1984. The life history and behavior of *Epimartyria pardella* (Micropterigidae). Journal of the Lepidopterists' Society 38: 40–46.

Wagner, D. L. 1985. The biosystematics of *Hepialus* F. s. *lato*, with special emphasis on the *californicus-hectoides* species group. Ph.D. Dissertation, University of California, Berkeley; 391 pages.

Wagner, D. L. 1987. Hepialidae (Hepialoidea), pages 347–349, *in:* Stehr, F. (ed.) Immature Insects, Vol. 1. Kendall/Hunt, Dubuque, IA.

Ditrysia, Nonapoditrysian Superfamilies

The vast majority of living Lepidoptera make up the Ditrysia, a derived lineage that includes 98% or more of the described species, most of the superfamilies and families, almost all of the external plant-feeding caterpillars, and most of the special adaptations for prey avoidance. All possess reproductive systems based on separate female copulatory and oviposition orifices and internal ducts for transfer of the sperm.

Superfamily Tineoidea

The tineoids are generally recognized as the most ancestral living group of the Ditrysia. Most tineoids have erect, roughened head scaling and elongate, five-segmented maxillary palpi that are folded, usually longer than the labial palpi, while the haustellum has short, unfused galeae, used to lap up surface moisture from detritus or fungi. Females of most species possess elongate

apophyses of abdominal segments 9 and 10 that anchor musculature, enabling the ovipositor to be telescoped outward to insert the eggs in crevices or other niches in the habitat.

Six families are regarded as comprising the superfamily, two of which, Tineidae and Psychidae, are worldwide and more species rich. Among the others, Acrolophidae are found exclusively in the Western Hemisphere, are primarily Neotropical, and are well represented in the western United States. Arrhenophanidae are Indonesian and Neotropical, with several species in southern Mexico. Eriocottidae are mostly Old World, with one genus in South America.

FAMILY TINEIDAE

Adult   Tineids are slender, small to moderately large moths (FW length 2.5 to 25 mm), usually brown, tan, or whitish, with a shiny appearance and FW patterns of black on tan or ochreous on dark brown. The FW are held tentlike over the dorsum in repose. Tineids are most easily recognized by the lack of ocelli, rough head vestiture, maxillary palpi usually long and folded, labial palpi with stiff, lateral bristles, and the proboscis short or absent. They lack bipectination of the male antennae, characteristic of other tineoid families. Genitalia, Figs. 39–41.

Larva Slender, usually lacking integumental color pattern, the larvae often live within silken tubes or portable cases. The number of stemmata is sometimes reduced to one or none. All instars have well-developed thoracic legs and abdominal prolegs with a single circle of crochets.

Larval Foods Tineids do not feed on flowering plants, although one genus in the Palaearctic mines in ferns; mostly they are generalist detritivores or fungivores, with a wide range of biologies—from feeding on vegetable refuse, guano in bat caves, and detritus in spider webs to specialists on animal products such as scats, owl pellets, or feathers (Plate 59.24)(Tineinae)—some are capable of digesting wool, including several cosmopolitan species that feed on woolen clothes (Plate 59.24) and other manmade products. Others are primarily fungus feeders (e.g., Nemapogoninae, Scardiinae), especially in sporophores of wood rot fungi (Polyporaceae) or decaying wood permeated by the hyphae, sometimes quite specialized in host preference. Fungivory rather than detritivory may have been the evolutionary ground plan for the family and therefore for the Ditrysia. Larvipary, wherein eggs mature and first-instar larvae emerge within an enlarged oviduct in the female, is known in numerous Andean and Indo-Australian Tineinae. They produce many fewer eggs than do tineids with conventional reproductive systems. The pupa is mobile and wriggles to the surface of the feeding substrate to protrude at emergence of the adult.

Diversity There are more than 3,000 described species worldwide, probably less than half the number known in collections, especially from tropical regions. These are assigned to more than 300 genera in 15 subfamilies. There are more than 110 described species in America north of Mexico, of which an estimated 50 or more occur in the West, in addition to many not yet named.

SUBFAMILY TINEINAE

Several Tineinae are among the most notorious of all moths in the eyes of the general public because they eat holes in clothes and carpets. The larvae are capable of digesting keratin, and they consume material of vertebrate origin, especially fur and feathers, including wool products. The clothes moths that infest stored woolen goods in western North America are cosmopolitan species that have been transported by human activities since prehistoric times. Most of them are found only indoors and probably cannot survive our winters in native habitats except in warmer and immediate coastal areas. *Tineola bisselliella* (Hummel), the "webbing clothes moth," probably is the most ubiquitous, but the small (FW 4.5–6.2 mm; Plate 2.22), golden tan moths are rarely seen because they hide in dark places and are not attracted to lights. The larvae are elongate, pale without integumental color pattern, with a dark head that has no stemmata. They live in silken galleries among the fibers. *Tinea pellionella* Linnaeus (Plate 2.23, Plate 59.23), the "case making clothes moth," is a common species, often infesting carpets, especially in damp places. The larvae construct flat, oval, portable cases of silk and fibers, sometimes blends of many colors, depending upon the richness of the

scarves and sweaters they have encountered. The adults are small (FW 4.3–7 mm), brown, sometimes with a dark discal dot, and with ochreous or pale orange head vestiture. *Tinea pallescentella* Stainton (Plate 2.17, Plate 2.24) is closely related to South American species and is thought to have been introduced to Europe with sheep from Argentina prior to the midnineteenth century when it was first described. Koebele collected it at Alameda, California, in the 1880s. The adults are larger than *T. pellionella* (FW 6–11 mm) and generally shining pale tan with diffuse dark brownish markings. The larvae feed on keratinous materials if they are sufficiently moist, including mammal carcasses and hair and detritus in bird nests and poultry houses. The species is recorded from granaries in Europe and Australia, but it is not a primary stored-food pest.

Native species of *Tinea* occur in association with bird nests, mammal scats and carcasses, owl pellets, and ant nests. *Tinea occidentella* Chambers (Plate 2.18, Plate 59.24) and *T. niveocapitella* Chambers (Plate 2.19) are common species along the Pacific Coast, often attracted to lights. The moths are relatively large (FW 7.5–11.5 mm, rarely from 6.2 mm when reared on dry material), FW dark gray, almost black in the latter, with white head vestiture, as the name indicates. The FW is gray with whitish median area in *T. occidentella*. The larvae live on natural keratinous materials such as feathers of bird carcasses and coyote scats (Plate 59.24). Their size varies greatly with quality of diet, fresh scats producing large individuals, whereas older, dry scats yield only small moths. We found them in large numbers infesting owl pellets on Santa Barbara Island, California, 25 miles from the nearest other island.

Monopis crocicapitella (Clemens) has dark brown FW (4–6, rarely 8 mm; Plate 2.28, Fig. 39) with a golden tan dorsal margin. It was described originally in the 1860s from the eastern United States but later has been recorded from widespread parts of the Palaearctic, British Isles, and Mediterranean region to Japan and Australia, so its native land is unknown. Like the preceding species, *M. crocicapitella* frequently is found out of doors in coastal California, at least in urban areas, and occasionally comes to lights. The larvae live in portable, oval cases primarily on refuse in bird nests, owl pellets, mammal burrows, and bat guano. They are recorded feeding on dried bulbs, potatoes, and cotton bolls, in addition to animal materials, but *M. crocicapitella* is not a woolen products pest.

Elatobia carbonella Dietz is a small moth with charcoal black FW (6–8 mm; Plate 2.25) that lives in close association with *Polyporus volvatus*, the puff-ball fungus that blooms on dead conifers. The larvae, at least in later instars, are free-living inside the spore-bearing cavity that opens to the underside of the sporophore, and apparently they eat the spores. This species is numerous in early years following forest fires and occasionally comes to lights, April to August.

The "plaster moth," *Phereoeca praecox* Gozmany & Vari (Plate 2.26), is an interesting species in that the larvae make flat, oval, or slightly figure-eight-shaped cases covered with debris, which are commonly seen clinging to plaster walls in garages and other buildings in southern California. The adults are secretive, dark gray moths, FW (5.5–6.5 mm) with black scale tufts and were not identified until specimens were reared in the 1990s, 20 years after the larval cases began appearing in California. The larvae presumably are detritus feeders.

SUBFAMILY NEMAPOGONINAE

Fungus-feeding Tineidae include most Nemapogoninae and Scardiinae. Several native species of **Nemapogon** depend

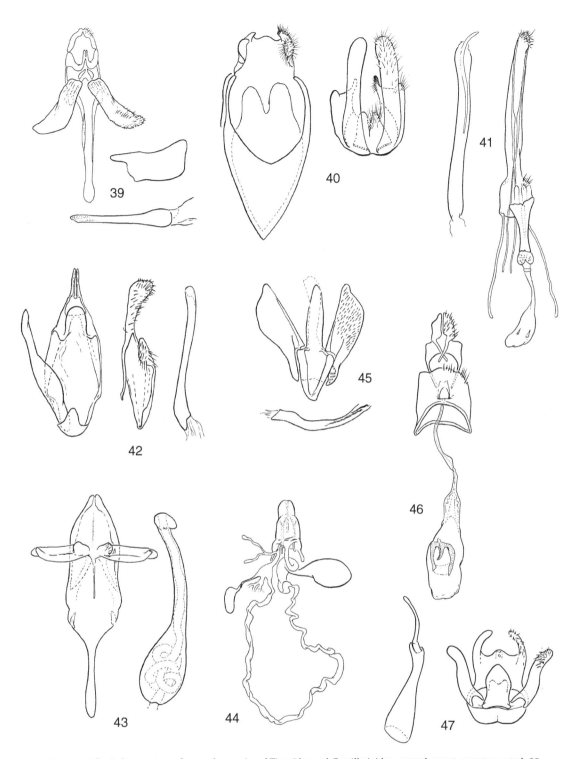

FIGURES 39–47. Genitalia structure of exemplar species of Tineoidea and Gracillarioidea; ventral aspect except as noted. **39,** Tineidae: *Monopis mycetophilella* Powell ♂, right valva, flattened, inner aspect, to right, aedeagus below. **40,** Tineidae: *Morophogoides berkeleyella* Powell ♂, ventral portion (diaphragm, valvae) removed and shown to at middle, aedeagus to right. **41,** Tineidae: *Morophogoides berkeleyella* Powell ♀ [Powell 1967]. **42,** Acrolophidae: *Acrolophus spilotus* Davis ♂, right valva removed, shown at middle, inner aspect, aedeagus to right [Davis 1990]. **43,** Psychidae: *Thyridopteryx alcora* Barnes ♂, aedeagus to right. **44,** Psychidae: *Thyridopteryx ephemeraeformis* (Haworth) ♀ [Davis 1964]. **45,** Gracillariidae: *Acrocercops affinis* Braun ♂, aedeagus below. **46,** Gracillariidae: *Neurobathra bohartiella* Opler ♀ [Opler 1971]. **47,** Bucculatricidae: *Bucculatrix variabilis* Braun ♂, aedeagus to left [Braun 1963].

primarily on Polyporaceae and use various species but tend to be habitat restricted. They are small moths (FW 5–8.5 mm, females larger than males) with gray or white FW patterned with black or brown. The larvae feed in the cortex of fresh sporophores before they dry or harden. *Nemapogon defectellus* (Zeller)(Plate 2.36) has white FW with black markings and is widespread in semiarid regions of the southwestern states, especially in riparian situations in association with fungi on willow and poplar. Hosts include *Polyporus munzii*, *P. rheadea*, *P. sulphureus*, and *P. vulpinus*, and *Trametes hispidus*, and we had one collection on *Armillaria mellea* (Agaricaceae) growing in association with *Trametes*. Colonies in the lab are capable of producing a second generation on dry fungus over a several-month period, whereas other *Nemapogon* require more moist conditions. *Nemapogon granellus* (Linnaeus)(Plate 2.33), known in the United States as the "European grain moth," is an introduced species that feeds on diverse vegetable matter such as dried fruit, mushrooms, and cork, and it is frequent around homes in coastal California. Larvae have been collected from various polypore fungi in numerous situations on hardwoods. A similar, native species, **N. molybdanellus** (Dietz) (Plate 2.34), is distinguished by its tan, rather than white, head vestiture. It is widespread in California, occurring up to 6,000' in the Sierra Nevada, feeding on several polypore fungi on oaks and other hardwoods, particularly *P. gilvus*, but also *Poria*, *Ganoderma*, and infrequently *Fomes* on conifers. The larvae also feed in fresh sporophores of the ascomycete *Hypoxylon occidentale* (Xylariaceae), where it grows adjacent to *Polyporus*. *Nemapogon oregonellus* (Busck)(Plate 2.35) has silvery white FW with longitudinal rust streaks and appears to be a specialist, feeding on *P. amarus* on incense cedar *(Calocedrus)* in Oregon and California.

SUBFAMILY SCARDIINAE

Among Scardiinae, the association with Polyporaceae also varies from generalist to specialist. Species feed on various fungi, and larvae often are found in the decaying wood subtending the sporophores. These are larger moths, FW 13–15.5 mm (but smaller, from 7.5 mm, in moths reared on dry fungus). *Daviscardia coloradella* (Dietz)(Plate 2.42) is the largest western species (FW 22–32 mm); its distinctly bicolored FW, dark on the costal half, pale yellowish dorsally, is diagnostic. This species is widespread in high-elevation coniferous forests of Colorado, Utah, and Arizona, south to Durango, associated with *Fomes*, *Polyporus*, and *Ganoderma* species on conifers and hardwoods. *Morophagoides burkerella* (Busck) has golden tan FW (7.5–13 mm; Plate 2.20, Plate 2.21) heavily mottled by dark brown. It is widely distributed along the Pacific Coast at elevations up to 6,200'; larvae feed in polypore fungi in a wide variety of situations, especially *P. gilvus* and *P. versicolor* on oaks and bay (*Umbellularia*) and associated rotten wood, even arborescent lupine, but also *Hypoxylon* on oak and *Fomes* on fir *(Abies)*. A similar but smaller species, **M. berkeleyella** (Powell)(FW 6–9.5 mm; Plate 2.27; Figs. 40, 41) is dark brown with obscure maculation. It has similar larval habits, living in redwood canyons and other mesic situations in the San Francisco Bay area. *Diataga leptosceles* Walsingham was described from Jalapa, Veracruz, Mexico, and ranges north to southern Arizona, where the larvae have been found in *P. vulpinus* on poplar.

Amorophaga cryptophori (Clarke)(Plate 2.40) is remarkable for its narrow host specialization, based on numerous collections. The larvae feed in the cortex of sporophores of *Polyporus*

volvatus (which was known as *Cryptophorus volvatus* when Clarke named the moth) on dead conifers in British Columbia and Idaho, south to the San Jacinto Mountains in southern California. The adults are large (FW 9.5–14.5 mm) and have elongate, narrow, gray FW with variable black longitudinal markings. Almost all specimens in museums were collected as larvae, but we have taken a few in blacklight traps in the Sierra Nevada. Larvae are capable of maintaining diapause for a year after sibs have emerged, presumably in response to dry conditions. They are found May to July, sometimes abundantly.

Dyotopasta yumaella (Kearfott) is a large desert species, FW (7.5–15.5 mm; Plate 2.41) pale gray with variable dark gray and rust, transverse maculation with raised, black scale tufts. The taxonomic relationships are not well understood, and *D. yumaella* may not belong with the Myrmecozelinae where now placed. The larvae feed in dead and dying cactus stems (*Opuntia* and other genera) and in Baja California in dead trunks of boojum tree (*Idria columnaris*, Fouqueriaceae). *Dyotopasta yumaella* occurs from Texas to southern California, south to Puebla, Mexico.

SUBFAMILY MEESSIINAE

Members of Meessiinae are small, often narrow winged with reduced venation. The larvae feed in silken galleries on lichens. *Homosetia marginimaculella* (Chambers)(Plate 2.30) is a widespread species in North America. The adults, which often come to lights in coastal areas in California, are small (FW 4.8–7.5 mm), characterized by slender wings, indistinctly spotted FW, and pale tan or rust head vestiture. They have been reared from fruiting bodies of *Poria incrassata*, a fungus that infests structural wood in southern California.

SUBFAMILY HIEROXESTINAE

The subfamily Hieroxestinae is pantropical, primarily Old World, but one species was described from Arizona, and two introduced species are among our most abundant urban moths in coastal California. They are characterized by having a flattened head, smooth scaled, unlike other tineids, by prominent, strongly divergent labial palpi, and by the elongate antennal scape. *Oinophila v-flava* (Haworth) is a small, narrow-winged moth (FW 3.5–4.5 mm; Plate 2.31) with shiny brown FW marked by variable ochreous, usually as outwardly oblique dashes forming a V-shaped median band and an outer line. It was described originally in England in the 1820s, where presumably it had been introduced. One specimen was collected at Los Angeles by Koebele in the 1880s, but there are no other western records until 1933, when *O. v-flava* appeared at Sacramento. By the 1940s this species had become widely distributed in coastal cities, and it has been a common urban species for the past half century. The larvae live in moldy situations and have been found under bark of dead limbs, in decaying acorns, leaf litter, polypore fungi, and debris associated with caterpillar-infested inflorescences. The adults often are numerous in houses, likely associated with indoor plants. *Opogona omoscopa* (Meyrick)(Plate 2.43) was introduced much later. It was described originally from Australia but is widely distributed in pantropical regions, and the native origin is unknown. In North America *O. omoscopa* was detected first near Santa Barbara in 1969, and by 1971 it was widespread in southern California. Larvae were found in Berkeley in ginger roots from Hawaii in 1972. The adults began appearing at urban lights in the San Francisco Bay area in 1978, and the species has been abundant since that time. Like *Oinophila*, this species flies throughout the year, appearing prevalently during the fall months. The larvae feed in decaying, usually moist plant material including wood, bark, leaves, and roots, whence they are easily transported. The adults have dark brown, almost black FW (6.5–10 mm), sometimes with faint ochreous subterminal and tornal spots, and the wing tips are upcurved. *Opogona arizonensis* Davis (Plate 2.32) was described from the mountains of southern Arizona and western Texas and evidently is a native species. It has straw yellow head scaling and FW (5–7.5 mm), with a prominent dark, triangular spot at the middorsal margin. A similar, undescribed species, FW with a dark bar along the costa basally, has appeared recently in chaparral habitat in southern California, at Miramar in the late 1990s, and at Rancho Santa Fe and on Santa Catalina and San Nicolas islands in 2004–2005.

SUBFAMILY DRYADAULINAE

The subfamily Dryadaulinae is represented in western North America by a curious tineid, **Dryadaula terpsichorella** (Busck), the so-called "Hawaiian dancing moth" (Plate 2.29). The common name is derived from characteristic circular gyrations, with a crablike sideways gait, which the moth performs each time it alights. Adults of this tiny species (FW 3.5–4 mm) have drooping labial palpi with recurved, spatulate apices. The FW is rounded distally and slightly bowed downward and is streaked with pale rust and dark gray brown, with a row of delicate metallic spots at the end of the cell. Adults were coincidentally collected within a few days at two sites in San Diego in 1998 and subsequently have been found at Manhattan Beach, California. This species was originally described from Hawaii, where it was believed to have been introduced about 1900, and other records suggest Polynesia as the source. The larvae were found in dead leaves of various plants in Hawaii and may be fungus feeders.

Phryganeopsis brunnea Walsingham is an unusual, large species (FW 8.3–10.5 mm; Plate 2.16) with broad, dark brown wings. The maxillary palpi are small but plicate, the labial palpi large and porrect, unlike those of most tineids. This species occurs in northwestern California, the Coast Ranges, and Sierra Nevada and is most often collected during the day, although the moths are also attracted to lights. Adults are recorded June to July, and one in April at the coast. Nothing is known about the larval biology, and its subfamily relationships are uncertain.

FAMILY ACROLOPHIDAE

Adult These small to moderately large (FW 6–16 mm), mostly brown moths resemble tineids but do not have a glossy appearance. They were long included in Tineidae but differ in several ways: antennae often are bipectinate in male, filiform in female (filiform in both sexes in Tineidae); the adults evidently are nonfeeding, with the proboscis reduced to a tiny, setose lobe or absent, maxillary palpi three- or four-segmented, rather than five-segmented; and the labial palpi are large, usually with second segment bushy scaled, and greatly elongated, curved over the head in males of some species of *Acrolophus*. FW vein R4 terminates on the apex or termen, rather than on the costa as in Tineidae. The anterior apophyses are lost in the female.

Larva Larvae are structurally similar to those of the tineids but usually more extensively pigmented, brownish dorsally and head, dorsal plates, pinacula brown to black; the

prespiracular plate is partially separated from pronotal plate. Genitalia, Fig. 42.

Larval Foods Larvae of acrolophids typically construct silken tubes in the ground or in detritus such as in mammal burrows and feed primarily on plant debris, rarely coprophagous or fungivorous.

Diversity The family is exclusively Western Hemisphere, primarily Neotropical, with about 275 described species, 80 in America north of Mexico, of which about 50 occur in the West.

Acrolophus includes the largest moths in the family and has about 250 described species, 37 described from the western United States. The adults are stout with rough head vesture and often very elongate labial palpi in males, curling over the head. Most are dark brown with obscure to distinct blackish markings; a few are primarily gray or tan. Genitalia, Fig. 42. The larvae typically create silken tubes in the soil and feed at the surface on grass thatch or roots, but two species in Florida and the West Indies are coprophagous in tortoise burrows. *Acrolophus variabilis* (Walsingham)(FW 10–16 mm; Plate 2.44, Plate 59.25) is our most widespread and largest species. Adults are attracted to lights in desert and semiarid regions, from western Kansas and Texas to southern Colorado, Utah, and central coastal and southern California. As the name implies, the FW color pattern varies considerably, from whitish gray mottled with dark gray to dark gray with darker patches or with the darker markings obsolete. The HW is brown with a lavender tinge. The male antennae are serrate and the labial palpi relatively short, erect, not curved over the head. *Acrolophus pyramellus* (Barnes & McDunnough)(FW 8.5–11 mm; Plate 2.47) is sometimes very abundant at lights in desert areas and ranges from western Nevada, Arizona, northern Baja California to central coastal California. It has dark brown to tan FW with distinct dark dots around the costa and termen, a spot at the end of the cell, and is mottled with dark to rust brown in some populations. Flight records span late March to early September. By contrast, *A. laticapitanus* (Walsingham)(Plate 2.37, Plate 2.45, Plate 2.46) is small and slender, superficially resembling some tortricids. The FW (6.5–9.5 mm) is pale yellow to tan or whitish gray, usually contrasting with dark HW. The species is widespread, from northern California to southern Arizona, and some of the variation is geographically limited, which resulted in several varietal names that have been called subspecies, even though not geographically distinct. There are genitalia differences as well, and the complex needs detailed study.

Species of *Amydria* and *Psilopsaltis* are smaller (FW 5.5–10 mm), superficially indistinguishable from one another, pale to dark brown, FW with delicate markings and smudges of black. The genitalia are fundamentally different, male *Psilopsaltis* having the aedeagus reduced and fused with the diaphragma, correlated in the female with a sclerotized cone projecting posteriorly at the ostium. This unusual structure is absent in *Amydria* and is analogous to that of some scythridids (Gelechioidea), for example, *Landryia*, and was termed the sterigmal cone by J.-F. Landry. *Amydria arizonella* Dietz is a mostly dark brown species with FW (6–9 mm; Plate 2.38) pattern obscure, more distinctly mottled in southern California populations. We have seen adults swarming around a compost heap at sundown in May, and Paul Rude reared them from composted sod collected in December at Berkeley and Oakland, California. Davis reported *A. arizonella* to be the most common moth in North American caves, particularly abundant in Carlsbad

Caverns, New Mexico. ***Amydria effrentella*** Clemens was described originally from Pennsylvania but is widespread over the western United States, from Colorado, Utah, and Washington to southern Arizona, the eastern Sierra Nevada, and southern California. The FW (7.5–9.3 mm) is generally pale tan, distinctly spotted to having dull, indistinct markings. The larvae were reported abundant in outer portions of older, underground nests of the mountain beaver (*Aplodontia*, a primitive rodent), in Washington. The nests contain large amounts of leaves, the outer of which are moist. ***Amydria obliquella*** Dietz has pale FW (5.5–8.5 mm; Plate 2.39) with distinct dark brown markings forming two outwardly oblique bands and spotting in the terminal area. It occurs in coastal southern California, north to Reno and south to Baja California and Nayarit, Mexico. We reared adults from root crowns of a desert shrub, *Isocoma menziesii* (Asteraceae), infested with larvae of *Eucosma* (Tortricidae). ***Ptilopsaltis confusella*** (Dietz) resembles *A. arizonella* but usually is paler brown or tan, the larger females with more strongly speckled FW (6.2–10 mm). This species occurs from Pennsylvania to the Pacific Coast, but specimens often are not distinguished in collections from species of *Amydria*. Adults are attracted to lights throughout the year in southern California and on the Channel Islands, and from May to October in central California, ranging to 5,000' on Mt. Pinos.

FAMILY PSYCHIDAE

The common name "bagworms" applies to psychids because the larvae live in portable cases, which they construct from silk plastered with debris, often symmetrically arranged pieces of plants on which the larvae feed (Plate 59.26).

Adult   Small and slender bodied to rather large and heavy bodied (FW 4–28 mm), males are fully winged, whereas females of different species may be fully winged, short winged (brachypterous), or wingless, even larviform and never leave the larval case. Some species are female only (parthenogenetic) or bisexual in some populations. The head vestiture is roughened, with long, slender scales directed forward, and the antennae often are strongly bipectinate in males, particularly in species with flightless females, but are filiform in both sexes of species having winged females, or not developed in larviform females. The wings of nearly all psychids are gray or brown, typically without color patterns. Genitalia, Figs. 43, 44.

Larva   Larvae are stout compared to tineid larvae, with the head and thorax larger and more heavily sclerotized, variously pigmented, and the abdomen tapered posteriorly. The thoracic legs are well developed and are used to pull their cases along on the host plant; the abdominal prolegs are present but reduced.

Larval Foods   Bagworms feed on lichens, grasses, conifer foliage, or leaves of angiosperm trees and shrubs, sometimes as specialists but more often as generalists on flowering plants. At maturity the larva attaches the case to a substrate, then inverts itself, and pupates in the case with its head toward the distal (older) end, whence the moth emerges.

Diversity   There are nearly 1,000 described species from all faunal regions, 90% of them in the Old World. Psychids are generally better studied than most microlepidoptera, owing to their fascinating behavior, biologies, and genetic complexity associated with the larviform females and parthenogenesis in five unrelated genera. About 80 species are named

from the Americas, including several described from just the larval cases, for which adult males are unknown. Only 10 occur in the West north of Mexico.

The male of ***Hyaloscotes fumosa*** Butler (Plate 2.48) is a slender moth with broad, rounded, translucent wings thinly clothed with fine, hairlike gray scales (FW 12–14 mm). It was described in 1881 from Mt. Shasta in northern California, based on males collected by Lord Walsingham, then was not seen again for more than a century. Subsequently males have been reared from larval cases found on manzanita (*Arctostaphylos*, Ericaceae) at Hat Creek, Lassen County. We found presumably the same species' case on blackberry (*Rubus*, Rosaceae) in the Trinity Mountains and collected males and larval cases on *Purshia* (Rosaceae) in the Sierra Nevada near Sierraville. Henry Edwards described three additional species of *Hyaloscotes* in 1877, based on larval cases, two on pine from Siskiyou County and the central Sierra Nevada, and one supposedly from Mt. Diablo in the central Coast Range on chinquapin (*Chrysolepis*, Fagaceae). If any of these can be shown to be the same species as *H. fumosa*, its name has priority. A mystery surrounds the Mt. Diablo species, **H. davidsonii**, because *Chrysolepis* does not occur on Mt. Diablo, so either the host was misidentified or the locality was in error. Moreover, according to Davis the case resembles that of some Old World *Clania* species, and because no more have been discovered on the Pacific Coast, he speculated that the 1870s collection represented an introduced species that did not persist; a mislabeled specimen seems more likely.

Thyridopteryx meadii Henry Edwards (Plate 2.49, Plate 59.26) is a desert species, ranging from western Texas to the Mojave Desert in California. Males have more robust bodies than *Hyaloscotes*, with black vestiture and transparent wings, very sparsely scaled when fresh (FW 8.5–11 mm). This species is primarily a specialist on creosote bush (*Larrea*, Zygophyllaceae), and it has been recorded from mesquite (*Prosopis*, Fabaceae) in Texas. Larvae are reported to seal the cases and remain dormant for long periods, then resume feeding when *Larrea* responds to rainfall with new vegetative growth. Thus the life cycle may be completed in a few weeks or last many months. The cases (Plate 59.26) are 30 to 40 mm in length and covered with small, flat leaf fragments and/or sections of twigs laid obliquely and spirally on the bag. The leaf material turns yellow with age, rendering the cases extremely cryptic in the foliage during dry periods.

Apterona helix (Siebold) is a widespread Palaearctic species that has both bisexual and asexual (parthenogenetic) populations, and the latter was introduced into California. The curious, spiral larval cases, which resemble tiny snails covered with mud, were first discovered in the Sierra Nevada (El Dorado County) in 1940. The larvae feed on all kinds of low-growing herbaceous plants, particularly favoring mule ears (*Wyethia*) and balsam root (*Balsamorhiza*)(Asteraceae). Dispersal is carried out by first-instar larvae ballooning in the wind on silk strands, and the species quickly spread northward through the eastern Sierra Nevada and intermountain region, becoming a pest in hedgerows in northeastern California and reaching southern Idaho and northern Utah by the early 1960s. Subsequently, *A. helix* has spread to the Coast Ranges west of the Sacramento Valley. Females oviposit in the cases, and first-instar larvae emerge in August. The early-instar larvae overwinter and begin feeding, mining out circular spots in the manner of *Coleophora*, and reach maturity early the following summer.

References for Superfamily Tineoidea

Davis, D. R. 1964. Bagworm moths of the Western Hemisphere (Lepidoptera: Psychidae). U.S. National Museum Bulletin 255; 233 pages.

Davis, D. R. 1987. Tineidae, Psychidae (Tineoidea), pages 362–370, *in*: Stehr, F. (ed.) Immature Insects. Vol. 1. Kendall/Hunt, Dubuque, IA.

Davis, D. R., and G. S. Robinson 1999. The Tineoidea and Gracillarioidea, pages 91–117, *in*: Kristensen, N. (ed.) Lepidoptera, Moths, and Butterflies. Vol. 1: Evolution, Systematics, and Biogeography. Handbook of Zoology. W. de Gruyter, Berlin, New York.

Hasbrouck, F. F. 1964. Moths of the family Acrolophidae in America north of Mexico (Microlepidoptera). Proceedings of the U.S. National Museum 114: 487–706.

Hinton, H. E. 1956. The larvae of the species of Tineidae of economic importance. Bulletin of Entomological Research 47: 251–346.

Kuznetsov, V. I., and A. A. Stekol'nikov 1997. Phylogenetic relations of some subfamilies of true moths (Lepidoptera: Tineidae) based on a new study of functional morphology of male genitalia. Entomological Review 76: 362–376.

Lawrence, J. F., and J. A. Powell 1969. Host relationships in North American fungus feeding moths (Lepidoptera: Oecophoridae, Oinophilidae, Tineidae). Museum of Comparative Zoology Bulletin, Harvard 138(2): 29–51.

Robinson, G. S., and E. S. Nielsen 1993. Tineid Genera of Australia. Monographs on Australian Lepidoptera 2. Commonwealth Scientific and Industrial Research Organization, Melbourne; 344 pages.

Superfamily Gracillarioidea

This is the major clade of Lepidoptera adapted for larval mining in leaves. Gracillarioids primarily feed on woody trees, shrubs, and vines, including both angiosperms and conifers. The larvae are obligate leaf, stem, or fruit miners in early instars; but later-instar larvae of many genera leave the mines to feed exposed or in webs externally on foliage and often assume a more caterpillar-like form (eruciform). Such change with more than one form in successive instars is termed hypermetamorphic. The adults lack the lateral bristles of the labial palpi possessed by tineoids and have a smoothly scaled frons and usually a well-developed, elongate, coiled proboscis. Like Tineidae, the antennae of many gracillariids are long, nearly equal to or longer than the FW.

FAMILY GRACILLARIIDAE

Adult These are tiny or small, slender moths (FW 2–10 mm), head with smooth scaling directed forward over the front, tufts of erect scales on the crown in Lithocolletinae; antennae 0.8 to longer than FW, filiform; labial palpus usually upturned, sometimes straight and drooping; maxillary palpi short to minute. The HW is lanceolate with scale fringe much broader than wing. Although nocturnal or crepuscular, the moths often are brightly colored, with the FW patterned in metallic orange, silver, bronze, purple, or yellow.

Larva Typically hypermetamorphic, early instars are modified for mining, flattened with prognathous head adapted for sap-feeding, legs reduced or lacking, transforming in the third or later instar to a caterpillar with chewing mouthparts and fully developed legs. The last instar uses silk to buckle the mine into a tentlike shelter or leaves the mine to feed externally, often folding a leaf into a tightly closed shelter. In some genera a variously modified, nonfeeding instar spins the cocoon. Pupation occurs outside the mine in most genera.

Larval Foods Nearly all gracillariids are specialists on one or a few closely related plants, typically woody angiosperms, encompassing an enormous variety of taxa—more than 80 plant families.

Diversity More than 2,000 species have been described from all major faunal regions, assigned to about 75 genera, and a much greater number of species remain to be identified, especially in tropical forests, judging from the diversity of observed mines. In North America there are about 275 named species of Gracillariidae, most of which were described from the eastern United States long ago. In the West there are numerous undescribed species. For example, about 70 unnamed species are thought to occur in California, based on mines and assumed host specificity. Most of these are in need of study of reared adults to confirm their status. Probably 300 or more species of gracillariids occur in the western Nearctic, more than half of them awaiting discovery through painstaking search for their larval mines, rearing, and correlated study of the reared adults.

SUBFAMILY GRACILLARIINAE

Species of the subfamily Gracillariinae are mostly elongate, slender moths that perch in an upright posture, with the prolegs extended and the abdomen and wings draped onto the substrate (Plate 59.27). The head is smooth scaled, and the antennae are longer than the FW. Many are quite colorful. *Caloptilia* is the largest genus in our fauna, with more than 30 described species and numerous others unnamed, known from reared adults or abandoned larval shelters. The larvae mine leaves in early instars, then molt to a cylindrical form not adapted for mining, vacate the mine, and create a closed shelter of a leaf roll or fold, in which they graze in the final instar. The mature larva leaves the shelter for pupation. *Caloptilia acerifoliella* (Chambers) is a small species (FW 5.0–7.5 mm; Plate 3.1) that occurs in Colorado and ranges widely, feeding on maple (*Acer*, Aceraceae). Populations in Oregon and California may comprise a complex of species. In the Willamette River Valley, the FW is uniform dull rust red, whereas for adults in the Cascade Range in Lane County on *A. circinatum* the FW is dull ochreous patterned with rust brown. In coastal central California, where *A. macrophyllum* is the larval host, the FW is straw yellow with a bright to dark rust basal band and along the dorsal margin, expanding to the terminal one-third of the wing (Plate 3.1), whereas in the Sierra Nevada the rust markings are reduced to a subterminal spot and sometimes faintly along the dorsal margin. The late-instar larval shelter is an inconspicuous fold along the distal edge of a maple leaf lobe. *Caloptilia alnivorella* (Chambers)(Plate 3.2, Plate 3.3) also was described from Colorado and ranges to the Pacific Coast, feeding on alders (*Alnus*, Betulaceae). The moths are relatively large (FW 6.5–9 mm) and strikingly handsome, having bright rust red FW with the costa beyond a basal patch pale yellow (north coastal California) and a well-defined costal patch into the cell (central coastal California). The shelters are leaf-edge rolls, not easily distinguished from those of other leaf rollers on *Alnus*.

Caloptilia azaleella (Brants), the "azalea leafminer," is a small species (FW 4.5–6 mm; Plate 3.4) with brown FW reflecting purplish and a pale yellow pattern. It was described originally from Belgium on greenhouse azaleas (*Rhododendron*, Ericaceae) imported in 1912 from Japan, and it was found in New York in 1911. This species was considered to be primarily a greenhouse pest but later became established on native azaleas in the eastern United States and Pacific Northwest. *Caloptilia azaleella* became established out of doors in the San Francisco Bay area by the 1960s. Late-instar larvae form shelters by curling a leaf edge or tying new leaves into a bundle.

The larval shelters become abundant on garden azaleas, turn the leaves brown, often a serious nuisance to gardeners, but we have no record of this species inhabiting native azalea. *Caloptilia azaleella* has three annual generations in Oregon and is multivoltine in California. Adults fly at Berkeley in every month except December and January, most often in April and November. *Caloptilia ferruginella* (Braun) was described from Yosemite Valley, mining *Rhododendron occidentale*, and from Mt. Tamalpais and the San Bernardino Mountains in southern California. It is larger than *C. azaleella* and has purplish and ochreous FW (6.5–7 mm) with rust red terminal fringe crossed by dark brown bands. The larvae mine the underside of the leaf, later rolling the leaf tip down into a cone, in July, and adults are recorded in July and August.

Two species of *Caloptilia* are described from coast live oak (*Quercus agrifolia*, Fagaceae) in California: *C. reticulata* (Braun)(Plate 3.5, Plate 59.27) has brown and ochreous reticulated FW (6–8 mm, rarely 4.5 mm), and the larva forms a shelter of a single fold at the tip of oak leaves, whereas *C. agrifoliella* Opler (Plate 3.6, Plate 3.7), which tends to occur more toward the immediate coast, is variable in color from rust to tan, FW (5.5–7.5 mm) indistinctly mottled with pale or blackish, the black sometimes pronounced on tan specimens along the basal 0.4 of the dorsal margin. The larval shelter of this species is a variable roll on new leaves. Like *C. azaleella*, both oak-feeding *Caloptilia* have been recorded at Berkeley in every month except December and January, *C. reticulata* most frequently in March, *C. agrifoliella* in May and June. A similar species with paler, washed-out FW pattern, occurs on *Q. pacifica* and *Q. parvula* on the California Channel Islands. Several other *Caloptilia* are known from other oaks, both evergreen (*Q. durata*) and deciduous (*Q. douglasii, Q. garryana, Q. kelloggii*).

Two species of *Caloptilia* occur on Anacardiaceae in California: *C. ovatiella* Opler (Plate 3.8) on sugar bush (*Rhus ovata*) in southern California, and *C. diversilobiella* Opler (Plate 3.9) on poison oak (*Toxicodendron diversilobum*) in the Coast Ranges of central California to southern Oregon. *Caloptilia ovatiella* (FW 6–8 mm) has dull brown FW reflecting purplish with the costal area diffuse white distally, while the FW pattern of *C. diversilobiella* (FW 6.5–7.5 mm) is resplendent in shining brown reflecting purplish, with costal pale yellow beyond the basal patch, dotted with black. The larval shelters are sometimes numerous on new foliage of poison oak but unfortunately not enough to suppress the abundant host.

Caloptilia nondeterminata (Braun)(Plate 3.10), which may be the same species as *C. ribesella* (Chambers), a species described from Colorado from the larva only, has bright orange FW (6–7.5 mm). This species occurs in southern Oregon and along the coast in central California, where the larvae feed on Grossulariaceae, both currants (*Ribes divaricatum, R. sanguineum*) and gooseberry (*R. menziesii*); larvae collected April to July, and August in Oregon. *Caloptilia palustriella* (Braun) has rust to gray brown FW (7–9 mm; Plate 3.11, Plate 3.12) with a sharply defined white triangle from the costa to the Cu crease; it is common on willow, especially *Salix lasiolepis* (Salicaceae) in coastal California. There is some question as to whether this is the same as the European *C. stigmatella* (Fabricius). An apparently undescribed species of *Caloptilia* feeds on California bay (*Umbellularia californica*, Lauraceae). It is large (FW 9.5–10 mm) with dark burnt sienna brown FW with the costal edge and fringe gray. Larvae mature on new foliage in April and secondary growth in summer. Adults overwinter and are seen sporadically October to March.

Micrurapteryx salicifoliella (Chambers) is a very slender moth with black FW (4.5–5.5 mm; Plate 3.14) delicately patterned with white along the costal and dorsal margins. It was described from Kentucky and is believed to occur across the country. The larvae form shallow, upperside blotch mines in leaves of willows (*Salix*). This species probably is multivoltine, with adults having been collected in June and July, larvae in July, September, and October in California and northern Baja California, occurs from the high Sierra to desert canyons. Species of *Parectopa* adults are superficially similar to *Micrurapteryx*, but the larvae are miners in soft-leafed legumes including *Lupinus* in Oregon and *Lathyrus* and *Melilotus* in the Sierra Nevada. Oddly, a *Parectopa* occurs on San Nicolas, the most distant California island, mining *Melilotus*, but has not been found on the larger islands. The FW is black with whitish markings, a series of outwardly oblique lines, and a subapical eyespot preceding a tail-like extension beyond the normal fringe scaling. The early instars are sap feeders in a subcutaneous, serpentine mine on the underside of the leaf. Later the larva migrates to the upper side and forms a spindle-shaped mine above the midrib, which gradually enlarges to oval, with short, digitate lobes radiating laterally. Mature mines are found in August and September. *Parectopa albicostella* Braun (Plate 3.15) was described from northern Utah, mining vetch (*Vicia*), distinguished from related species by its dark brownish gray head and frontal scaling and FW (4.5–5 mm) costal edge white. Apparently the same species mines wild pea (*Lathyrus nevadensis*) in the Sierra Nevada. Mines we collected in August and September produced adults in 3 to 4 weeks.

Neurobathra bohartiella Opler (Plate 3.13) is sometimes abundant, forming large blotch mines in not yet hardened leaves of *Quercus agrifolia* in coastal California and on Santa Cruz Island. The larva at first forms a thin, meandering, serpentine mine in the lower epidermis, then enters the midrib, migrates to the leaf tip, and makes an inflated, full-depth mine in the distal portion of the leaf. The adults are very slender, FW (3.5–5 mm) brown with parallel, brownish tan, black, and whitish chevrons. The adults fly April to early November at Berkeley, most numerous in July.

Parornix alta Braun is black, FW (3–3.8 mm) with silvery white costal and dorsal margin streaks and a transverse fascia from two-thirds to the tornus. It was described from northern Utah mining service berry (*Amelanchier*) and occurs in California, where six or more undescribed species of *Parornix* mine leaves of Rosaceae, David Wagner believes. The larvae form lower-surface, tentiform mines. Normal head capsule and mouthparts are achieved after the second molt, and the larva continues to mine until fully grown, when it emerges to form a cocoon within a turned-down border of a leaf. Western species may have a life cycle similar to the "tentiform leaf miner" of apple, *P. geminatella* (Packard), which is multivoltine in the eastern United States, with several summer generations.

Acrocercops too is well represented by undescribed species, known from either reared adults or abandoned mines, mostly on Fagaceae, but others on Asteraceae. The larvae feed entirely as miners, in the first two instars as a flat, sap feeder, forming a serpentine track, then in an upperside blotch in the middle of the leaf. Pupation occurs either within or outside the mine, varying with species. *Acrocercops* species are thought to be bivoltine, with adults active in California in May and June and again in August and September. *Acrocercops affinis* Braun is an oak miner in the foothills and mountains of California, and a similar species, *A. insulariella* Opler (Plate 3.16), a small, slender moth having rust-tinged brown FW with indistinct white along the costal and dorsal margins, occurs on Santa Cruz Island on *Quercus agrifolia*. In Arizona, **A. arbutusella** Braun is larger than other western *Acrocercops* (FW 4.5–5 mm; Plate 3.17), pale rust brown, beautifully patterned with distinct white chevrons edged by black. In the Huachuca Mountains adults appear in late April along with active mines that produce adults in May. The larvae mine leaves of *Arbutus arizonica* (Ericaceae), forming large, tentiform, upperside blotches similar to those of *Cameraria*, in which pupation occurs.

Species of the genus **Marmara** are tiny black moths, FW (2.0–3.5 mm) with fine transverse, white lines. Adults are superficially similar among species, as are the mines, and we do not know if some species are polyphagous or if mines on unrelated plants represent different species. The mines are subcutaneous, long, winding tracks (Plate 59.30). Often the larvae migrate from leaf to stem to leaf. **Marmara arbutiella** Busck (Plate 3.19) occurs from British Columbia to California on madrone (*Arbutus menziesii*, Ericaceae). Based on habitat occurrence, we suspect that the same species feeds on toyon (*Heteromeles*, Rosaceae) and perhaps other plants. **Marmara opuntiella** Busck mines the broad, modified stems of prickly pear (*Opuntia*, Cactaceae)(Plate 59.31) in Texas and the California Channel Islands. **Marmara gulosa** Guillén & Davis, a miner in citrus (Rutaceae) fruit skins and on many other hosts, was recently named from the Coachella Valley, California, after having been treated for decades as **M. pomonella** Busck, the "apple fruit miner," which was described from Oregon and is known only from the original female specimen.

SUBFAMILY LITHOCOLLETINAE

Members of the subfamily Lithocolletinae are smaller moths than most Gracillariinae and are miners throughout larval life. The genera *Cameraria* and *Phyllonorycter* are particularly diverse, with 140 described species in America north of Mexico. More than 50 occur in the western states, in addition to a like number known but not yet named. Adults of the two genera and *Cremastobombycia* are nearly all superficially similar, having orange- or rust-colored FW with white markings, often accentuated by black scaling, yet the mining habits differ markedly.

Protolithocolletis lathryi Braun is a tiny species with black FW (3.4–4.2 mm; Plate 3.20) having three opposed pairs of white, triangular spots along the costa and dorsal margin, the inner pair sometimes joined to form a chevronlike fascia. This species was described from Manitoba and occurs widely in the western states. Larvae of the first two instars are sap feeders, and later instars form upperside blotch mines that become inflated, bubblelike, on wild pea, *Lathyrus* (Fabaceae). Mines collected by Y.-F. Hsu in the central and northern Sierra Nevada, California, on *L. nevadensis* matured in late August and September.

Cremastobombycia are tiny moths (FW 2.5–4.5 mm) whose larvae mine in leaves of herbaceous and woody composites (Asteraceae), and the patterns of host specificity are not well understood. They feed in the first three instars on sap, then spin silk to buckle the leaf epidermis, separating it, and thereafter are tissue feeders, ultimately forming hollow, bubblelike mines, usually on lower, older leaves of the plant. Adults of **C. grindeliella** (Walsingham) have dark rust FW with thin, distinct white markings (FW 2.8–4.4 mm; Plate 3.21). The larvae mine leaves of gumplant (*Grindelia robusta*) in central coastal California, in November to December and February to March. Taxonomic relationships to populations mining *Artemisia*

douglasiana, Gnaphalium, and *Baccharis pilularis* in the same region are unclear. The FW patterns differ (Plate 3.22, Plate 3.23). An apparently undescribed species occurs on Lamiaceae, *Lespedezia* and *Salvia*, on Santa Catalina Island and mainland southern California.

Cameraria create blotch mines primarily in the upper leaf tissue (palisades layer) that are visible externally from the upper side of the leaf and usually not from below. During early instars the larva is flattened with a prognathous head capsule. When the full area of the mine is formed, the larva molts to a cylindrical, caterpillar form, with small legs. Using silk to buckle the mine roof, the final instar creates a space, the so-called tentiform mine, in which it feeds and pupates after forming a silken track for the adult emergence. The adults of western species are generally larger (FW 3.0–5.5 mm) than those of *Phyllonorycter*. **Cameraria gaultheriella** (Walsingham) is one of the largest western species (FW 3.5–5.5 mm; Plate 3.24), and its larvae create the largest mines—enormous brown blotches that sometimes encompass half or more the area of the large leaf of salal (*Gaultheria shallon*, Ericaceae)(Plate 59.29). Salal occurs in mixed evergreen and redwood forest habitats along the immediate Pacific Coast from Alaska to the Santa Ynez Mountains of southern California, but the moth is known only from Washington to the San Francisco Bay area. Adults have dark rust brown FW with usually short, broad, white markings along the costal and dorsal margins. They have been reared from April to August, and mines may be occupied throughout the year. In the same coastal habitats, **C. nemoris** (Walsingham)(Plate 3.25) is very common mining California huckleberry (*Vaccinium ovatum*, Ericaceae). Its mine encompasses the entire leaf, abundantly at times (Plate 59.28).

An impressive array of *Cameraria* species has evolved with western oaks. Opler and Davis recorded 17 species associated with California Fagaceae, a polyphyletic assemblage representing three species groups. Mines of **C. agrifoliella** (Braun)(Plate 3.26) are perhaps the most frequently observed lepidopterous leaf mines on any western oak, occurring conspicuously and almost ubiquitously on coast live oak (*Quercus agrifolia*) throughout its range in coastal California and northern Baja California. The FW (3.8–4.2 mm) is patterned white on dark rust, accentuated by black scaling more strongly so than in most lithocolletines. Adults have been collected in every month from February to November, and active mines can be found any time, often on sucker growth under the canopy and at the base of weakened or damaged trees. The large blotch mines are pale green when larvae are young, turning to tan with buckled creases toward maturity. A sister species, **C. wisizeniella** Opler (Plate 3.27), also feeds on *Q. agrifolia* but occurs at more inland sites and is more widespread than *C. agrifoliella*, in the inner Coast Ranges and foothills of the Sierra Nevada, following the distribution of *Q. wislizenii*. Similar mines occur on other oaks, created by species of *Cameraria* that are specific to one or a few oak species, in a complex pattern of allopatric and sympatric, shared host use among 14 species of *Quercus* and *Lithocarpus*. **Cameraria lobatiella** Opler & Davis (Plate 3.28) ranges widely at lower elevations west of the Sierra Nevada in association with deciduous hosts, valley oak (*Q. lobata*) in the Coast Ranges and Central Valley, black oak (*Q. kelloggii*) in southern California, and rarely on blue oak (*Q. douglasii*). This *Cameraria* has three or four annual generations followed by diapause as pupae in mines in fallen leaves until the new foliage has expanded the following spring. Other oak miners include **C. pentekes** Opler & Davis (Plate 3.29) on deciduous oaks, primarily *Q. douglasii*, rarely on

Q. lobata, and possibly on *Q. garryana* in the North Coast Ranges. Its life cycle is similar to that of *C. lobatiella*, with adult emergences in April, June to July, and September to October. **Cameraria jacintoensis** Opler & Davis (Plate 3.30) occurs in association with scrub oaks, *Q. berberidifoliella*, along the Coast Ranges and southern California coast, and on *Q. pacifica* and *Q. parvula* on Santa Cruz Island. This species is bivoltine, with larval feeding cohorts in summer and winter to early spring, and adult flights in late spring and fall.

Cameraria superimposita (Braun) was described from Logan Canyon, Utah, mining leaves of *Acer grandidentata*, distinguished from **C. macrocarpella** (Frey & Boll) from Texas by its golden ochreous FW with more oblique dorsal extensions of the fasciae and union of the apices of the three white streaks before the apex. Mines of this or a related species are common on *A. grandidentata* in canyons of the mountains in southern Arizona.

By contrast to *Cameraria*, larvae of **Phyllonorycter** usually mine in the lower mesophyll, at first shallowly as a sap feeder, then late instars eat the parenchyma and excavate deeper, forming mines seen from the underside of the leaf. By the last instar the mine of most species becomes visible on the upper side by an oval, speckled, paler area. Pupation occurs in the mine. Several species specialize on Salicaceae, forming a polyphyletic assemblage of at least three species groups in North America. They are all small species (FW 2.5–4.5 mm), mostly having orange FW with parallel, white chevron markings that are variable within populations, and genitalia dissections should be employed to confirm identifications. **Phyllonorycter nipigon** (Freeman)(Plate 3.33) has distinctly to poorly defined FW markings, sometimes obscured by ground color. It is widespread, from the northern Great Lakes states and southern Canada through British Columbia to central Alaska and the Rocky Mountains, eastern Washington, and along the eastern Sierra Nevada in California, feeding on aspen, *Populus tremuloides*, and several poplars, apparently preferring balsam poplar, *P. balsamifera*, in northern areas, and aspen in the Sierra Nevada. A closely related species, **Phyllonorycter deserticola** Davis & Deschka, is tiny (FW 2.8–3.5 mm; Plate 3.31), with highly variable extent of white FW maculation. As the name suggests, this species occurs widely in more arid parts of the Southwest, from northern Mexico through New Mexico, southern Utah to eastern California and the California Central Valley, feeding primarily on *Populus fremontii*. Mines of this species become extremely abundant in some years in localized areas, such as the northern Owens Valley in 2001–2004. With multiple mines per leaf, trees are severely affected with varying degrees of defoliation. *Phyllonorycter deserticola* is bivoltine, with flights in June and September. There are two century-old records for **P. erugatus** Davis & Deschka (Plate 3.32) in the San Francisco Bay area, which otherwise is not known south of the Willamette Valley near Portland, and a 1947 record for *P. nipigon* at Berkeley, but neither has been discovered west of the Sierra Nevada in the past half century. **Phyllonorycter salicifoliella** (Chambers)(Plate 3.34) mines primarily willows but occasionally poplar and aspen. Numerous *Salix* species are used over the broad distribution of this insect, the eastern United States and southern Canada into northern British Columbia, the Rocky Mountains from Wyoming to New Mexico, and the Pacific states. The FW color is less variable than in most *Phyllonorycter* and characteristically has the pale markings peppered with dark scales. A distinctive lithocolletine, **P. apicinigrella** (Braun)(Plate 3.35) has tan or ochreous FW with a variable black smudge at the base of the costa and black apical spot and subterminal band. The larvae mine *S. sitchensis* and *S. lasiandra*

in western Washington and *S. lasiolepis* in coastal California, in late summer and fall.

Phyllonorycter felinelle Heinrich is one of our most recognizable species, FW (3.0–4.3 mm; Plate 3.36) pale orange, having distinct, slender, silvery markings. It mines western sycamore (*Platanus racemosa*, Platanaceae) in coastal California, often abundantly, with several active and abandoned mines on most of the large leaves. This species is an adept colonizer and can maintain a colony on an isolated tree, such as on Santa Cruz and Santa Catalina islands, where sycamores are introduced. Several *Phyllonorycter* inhabit western Ericaceae, including two on madrone (*Arbutus menziesii*), one an upperside, one a lowerside miner, and one or more species on manzanita (*Arctostaphylos*): **P. arbutusella** (Braun)(Plate 3.37, Plate 3.38, Plate 3.39) appears to be the lowerside miner, but whether it or the upperside species is the same as **P. manzanita** (Braun)(Plate 3.40), which was described from Mt. Wilson in southern California, is unknown. **Phyllonorycter ribefoliae** (Braun)(Plate 3.41) mines leaves of red-flowering currant (*Ribes sanguineum*) and gooseberries (*R. divaricatum* and *R. menziesii*, Grossulariaceae), often becoming numerous late in the season on older leaves. This species colonizes readily and can maintain a colony on an isolated garden currant.

A complex of *Phyllonorycter* species occurs on Pacific states oaks, including three species on coast live oak (*Quercus agrifolia*) in California. Two are univoltine, flying in early spring, **P. antiochella** Opler, and late spring, **P. inusitatella** (Braun), and one bivoltine with spring and fall emergences, **P. sandraella** Opler. *Phyllonorycter inusitatella* is relatively large (FW 3.5–5 mm; Plate 3.44, Plate 3.45, Plate 3.46), shining golden orange with the white FW maculation reduced to weakly discernible costal chevrons, accompanied by an apical black dot; *P. sandraella* (Plate 3.47, Plate 3.48, Plate 3.49) is similar but with a bold, white, outwardly oblique chevron from the costa to end of the cell and sometimes a white line through the basal half; and *P. antiochella* is smaller (FW 3–4 mm; Plate 3.42, Plate 3.43), dull rust brown with distinct white chevrons, emphasized by dark brown margins.

Leucanthiza dircella (Braun), which was described from Ohio, or a closely related new species, occurs in the Berkeley Hills, California. The adults are tiny (FW 2.5–4.0 mm; Plate 3.18), beautiful, narrow-winged creatures having shining ochreous brown FW with metallic lead silver–colored markings and a white preapical patch. They fly around western leatherwood, *Dirca occidentialis* (Thymelaeaceae), in early spring, and the larvae make large, circular, upperside blotch mines in March and April.

FAMILY PHYLLOCNISTIDAE

Adult These are tiny, slender moths (FW 2–3 mm) with long antennae, often with shining white or silvery FW delicately banded with gray and rust distally; HW lanceolate with the fringe much broader.

Larva Greatly modified as a sap feeder, the larva creates an extremely long, meandering mine, often in the new, still-soft leaves, causing them to curl conspicuously, or regularly zigzag mines in mature leaves (Plate 59.32). The last instar is nonfeeding, lacking mandibles.

Larval Foods More than 20 angiosperm families have been recorded, but many additional trees of diverse families are used in Central America, judging from the mines. Phyllocnistid mines have been described from mid-Cretaceous (97 MyBP) Magnoliidae, providing the earliest known existence of any ditrysian.

Diversity Fewer than 100 species are described, a small fragment of the fauna; mines are found on more kinds of plants at one lowland forest locality in Costa Rica than there are described New World species. There are about a dozen named phyllocnistids in America north of Mexico, and only a few are documented in the western states and provinces.

Phyllocnistis populiella Chambers (Plate 3.50) was described from Colorado and ranges widely to central Alaska and California. The larva creates an upperside, subcutaneous mine on aspen (*Populus tremuloides*, Salicaceae)(Plate 59.32) and other poplars. The same or an undescribed species inhabits willows, including *Salix lasiolepis* in coastal California and Santa Rosa and Santa Catalina islands in April and May. Its mines may extend from leaf to leaf along the stems. Similar mines occur on *Prunus ilicifolia* and *P. lyonii* at Berkeley and on the California Channel Islands, the work of a presumably undescribed species (Plate 3.51). *Phyllocnistis vitifolia* Chambers (Plate 3.52, Plate 3.53) mines leaves of grape (*Vitis*, Vitaceae). It was originally described from Kentucky but has been found in desert wild grape *(V. girdiana)* in Riverside County, California, and recently in a greenhouse in Davis.

FAMILY BUCCULATRICIDAE

Adult Tiny to small moths (FW 2.5 to 7 mm), they are most easily recognized by their elongate frons and large, erect tuft of scales on the vertex. The appendages are short, antennae 0.6 to 0.9 the FW length, proboscis short but functional, and labial palpi vestigial. The wings are lanceolate, FW often with tufts of upraised scales.

Larva Larvae are hypermetamorphic legless leaf miners in the first two instars, then with well-developed legs, usually leave the mine to graze externally on leaf surfaces. Some species are stem miners or cause stem galls in which they feed, and a few are leaf miners throughout larval feeding. In such examples, at the last molt a caterpillar with functional spinneret and legs emerges and leaves the gall or mine to form a cocoon externally. *Bucculatrix* are known as the "ribbed cocoon makers" because larvae of nearly all species construct elaborate, elongate-oval cocoons having parallel, longitudinal ridges (Plate 3.54), and sometimes the cocoons are surrounded by a stockade of erect, silken posts. The shape of the cocoon varies from slender to rather stout and the number of ridges from about five to nine, with shape, color, and the rib number characteristic for individual species. The ridges are weakly developed or lacking in a few species, or in occasional specimens, such as parasitized individuals. Upon emergence, the pupa thrusts itself through a weakened fissure at the anterior end of the cocoon and protrudes. Some species are multivoltine and become more abundant as the season progresses, as do their parasites.

Larval Foods Each species is a host-plant specialist. There are more than 20 angiosperm plant families recorded, with Asteraceae and Fagaceae dominant in the Holarctic.

Diversity This is a small family of about 250 described species, distributed on all continents except New Zealand, most numerous in the Holarctic. There are about 100 named species of *Bucculatrix* in America north of Mexico, half of which were described by Annette Braun in her 1963 monograph. About 55 species are recorded in the western states, and no doubt many others await discovery.

Bucculatrix longula Braun, a large, shining white species, FW (5–5.5 mm; Plate 4.1) with pale ochreous markings, is an example of the gall-causing species. It has been reared from *Helianthus annuus* (Asteraceae) in northeastern Washington, Utah (Salt Lake), and eastern California (Bishop). *Bucculatrix angustata* Frey & Boll (FW 3.5–3.8 mm; Plate 4.2) is widespread, from the eastern United States to Texas, Utah, and Washington. The moths are pale to dark brown with a narrow, white, longitudinal streak from the base to half or more the FW. The larvae mine throughout their growth in leaves of *Aster* or *Solidago* (Asteraceae). The early instars form a long, linear, gradually widening mine, later instars form a second, linear, blotchlike mine.

Coyote brush (*Bacccharis pilularis*, Asteraceae) is inhabited by three species of *Bucculatrix* in California: *Bucculatrix variabilis* Braun, perhaps the most often seen and abundant western *Bucculatrix*, has mottled brownish FW (2.8–3.4 mm; Plate 4.3) with a pair of curved, white streaks that meet just before the middle of the cell. It was described in 1910, and the original series included individuals of the species later recognized as *B. separabilis* Braun (Plate 4.4), which has more extensive whitish areas of whitish on the FW (hence the name *variabilis*). Braun discovered the oversight after study of genitalia had become important in taxonomy of microlepidoptera, and she named *B. separabilis* in 1963. The latter species tends to live in more inland areas than *variabilis*. The third *Baccharis* species, *B. dominatrix* Rubinoff & Osborne, is larger (FW 4–5 mm; Plate 4.5) and features a prominent white, longitudinal FW mark. It occurs at several localities in the immediate San Francisco Bay area yet was not discovered until 1983. This species cohabits *Baccharis* with one or the other of its congeners, but all three are not known at one site.

A *Bucculatrix* in the Sierra Nevada has been identified as *B. sexnotata* Braun, which is a widespread, *Aster*-feeding species in the East, Ontario to Kentucky and North Carolina. The California species (Plate 4.6) feeds on *Wyethia*, so probably the similarity is superficial. *Bucculatrix ericameriae* Braun has white FW with pale brownish ochreous costa giving rise to oblique streaks into the cell. It occurs in the foothills of the Sierra Nevada and near the coast in Monterey County, California. The larvae feed on *Ericameria arborescens* (Asteraceae), and although *Ericameria* had been included in the genus *Haplopappus* in 1963 when Braun described the moth, she used the old plant name from the 1916 collection label as the basis for her species. Fittingly, 30 years later botanists reversed their decision and resurrected *Ericameria* from synonymy, validating Braun's decision. In the Great Basin, *Artemisia tridentata* (Asteraceae) harbors at least three species of *Bucculatrix*: *Bucculatrix salutatoria* Braun (FW 4.2–4.5 mm; Plate 4.7) is a pale whitish moth, FW finely dusted with ochreous and with indistinct darker, oblique streaks on the distal half and two distinct black dots in and below the cell. It occurs in Colorado, Wyoming, and Utah, generally in more northern, less xeric sites than *B. tridenticola* Braun, which has the FW (3.5 mm; Plate 4.8) densely black speckled with a gradually widening black streak from the base through the cell to the tornus. It is widespread from central Colorado and Utah to eastern Washington, Oregon, and northeastern California, whereas *B. seorsa* Braun has the FW white with two black dots as in *B. salutatoria,* and black-tipped scales grouped into poorly defined markings. It occurs from southeastern British Columbia to northeastern California. Other Asteraceae feeders include *B. koebelella* Busck, a small (FW 2.8–3.5 mm; Plate 4.9), ochreous gray moth that lives on *A. californica* along the coastal mainland, and *A. nesiotica* on the southern Channel Islands of California. The pale gray green larvae exactly match the foliage color of the host plants and sometimes are very abundant. Adults and larvae are found from February to July. ***Bucculatrix enceliae*** Braun lives on the widespread desert bush sunflower, *Encelia farinosa*, from southern Arizona through the low Colorado Desert of California. The moth is white with brown-tipped scales forming obscure markings and a general dusting on the distal half of the FW that defines two white streaks from the costa and a white patch before the tornus.

Among non–Asteraceae feeders, several *Bucculatrix* feed on oaks (Fagaceae). One eastern species feeds on various oaks and chestnut, but western species appear to be restricted to a single or closely related *Quercus* species. A complex of species is associated with both evergreen and deciduous oaks along the Pacific Coast: ***Bucculatrix albertiella*** Busck (Plate 4.9, Plate 4.10) occurs on *Q. agrifolia* and *Q. wislizenii* along the length of coastal California and northern Baja California. The adults are primarily ochreous, with FW (3.5–4 mm) pattern obscure, whitish, and an upraised tuft of black scales on the dorsal margin near the tornus. A presumably undescribed species occurs on deciduous oaks *(Q. douglasii, Q. lobata)* in central coastal hills of California; it has a similar phenotype, but the FW pattern is more extensive and pale brown, and the upraised black scaling of the dorsal margin is reduced to the inner edge of a broad, brown patch. ***Bucculatrix zophopasta*** Braun (Plate 4.12), described from Oregon on *Q. garryana*, has a dark ochreous brown to blackish brown FW pattern rather than ochreous. Braun assumed specimens from the southern Sierra Nevada to be the same species, but these may represent additional species, and another population at Salem, Oregon, on *Q. palustris* appears to be distinctive.

A complex of species also may be involved among Rhamnaceae feeders. ***Bucculatrix ceanothiella*** Braun (Plate 4.14) was described in 1918 from one female collected at Colton, San Bernardino County, California, and was still known only from the type in 1963. Populations in central California, originating from several species of *Ceanothus*, vary in phenotype and may not be the same as Braun's species. Those along the coast on an ornamental *Ceanothus* hybrid and native *C. thyrsiflorus*, although variable, tend to have the whitish FW with pale, delicate brown distal streaks, whereas those in Alameda County on ornamental *C. arboreus* and in the Sierra Nevada on *C. integerrimus* have more conspicuous, dark brown to blackish markings.

Bucculatrix quadrigemina Braun (Plate 4.13) was described from southern California in 1918; and a year later Busck, misled by its larger and darker phenotype, described *B. altheae* from the San Francisco Bay area, but it proved to be a synonym. The FW (3–5.5 mm) is cream white to pale ochreous, peppered with black-tipped scales that coalesce to form four evenly spaced, large costal patches alternating with four white marks, and there is a patch of upraised, black scales at the mid-dorsal margin. The cocoon of this species is unusual in lacking the typical longitudinal ribs, having instead a series of diagonal ridges from the cocoon to the substrate on each side, or sometimes no defined ridges. On garden hollyhock *(Alcea)* and *Lavatera* (Malvaceae), larvae are multivoltine and active through winter. In native communities *B. quadrigemina* has been reared from *Malacothamnus* in the central Coast Ranges and on San Clemente Island.

FAMILY DOUGLASIIDAE

Adult These are small moths (FW 3–6.5 mm) with smooth head vestiture consisting of broad scales; eyes relatively small, correlated with diurnal behavior; ocelli well developed and antennal scape not broadened into an eye cap, in contrast to other families of Gracillarioidea. Wings are narrow, lanceolate with reduced venation; Rs4 and M1 of FW forked from a common stem. Male genitalia lack uncus, gnathos, and transtilla, and the tegumen forms a triangular hood. Female genitalia with apophyses moderately long, ovipositor telescoping, but the oviposition behavior is not recorded. Genitalia, Figs. 111, 112.

Larva Larvae are stout, body whitish or yellowish with pigmented head capsule, pronotal, and anal shields, and long primary setae.

Larval Foods European species are leaf miners *(Klimeschia)* or petiole and stem miners *(Tinagma),* in Boraginaceae, Lamiaceae, or Rosaceae.

Diversity This is a small family, with 17 species recorded in the Palaearctic, seven in the Nearctic, and one in Australia. There are six described species in western North America, four of them named in 1990, and most likely others remain to be discovered.

Tinagma californicum Gaedike is a tiny (FW 3–3.5 mm; Plate 4.15, Plate 4.16), slate gray species with faint whitish powdering on the distal half of the FW. Adults are numerous on *Phacelia heterophylla* and *P. egena* (Hydrophyllaceae) in northern California (late May to July) and on *P. californica* in the San Francisco Bay area (late March to May), flying about the inflorescences during the daytime. ***Tinagma powelli*** Gaedike is a similar, slightly larger species (FW 3.5–4.3 mm; Plate 4.17) with more distinctly defined distal whitish. It flies in association with *P. distans* and *Cryptantha* (Boraginaceae) in the San Francisco Bay area and mountains of southern California. Presumably the larvae are miners in the flower petioles and/or inflorescence scapes. ***Tinagma giganteum*** Braun and ***T. pulverilineum*** Braun were described from Glacier National Park, Montana; both feature a pale distal area well defined by a strongly curved line, and they occur in Arctic-alpine situations of Colorado. The latter species ranges into Alberta, Nebraska, and northern California. Braun noted that adults of *T. giganteum* fly at dusk. Flight records are June to August, reaching 12,000' in Colorado.

References for Superfamily Gracillarioidea

Braun, A. F. 1908. A revision of the North American species of the genus *Lithocolletis* Hübner. Transactions of the American Entomological Society 34: 269–357, 5 plates.

Braun, A. F. 1963. The genus *Bucculatrix* in American North of Mexico (Microlepidoptera). Memoir 18. American Entomological Society, Philadelphia; iii + 208 pages, 45 plates.

Davis, D. R. 1987. Lyonetiidae, Gracillariidae (Tineoidea), pages 370–378, *in:* Stehr, F. (ed.) Immature Insects. Vol. 1. Kendall/Hunt, Dubuque, IA.

Davis, D. R., and G. Deschka 2001. Biology and systematics of the North American *Phyllonorycter* leafminers on Salicaceae, with a synoptic catalog of the Palearctic species (Lepidoptera: Gracillariidae). Smithsonian Contributions to Zoology 614; 89 pages.

Davis, D. R., and G. S. Robinson 1999. The Tineoidea and Gracillarioidea, pages 91–117, *in:* Kristensen, N. (ed.) Lepidoptera, Moths, and Butterflies. Vol. 1: Evolution, Systematics, and Bbiogeography. Handbook of Zoology. W. de Gruyter, Berlin, New York; x + 491 pages.

Gaedike, R. 1990. Revision der nearktischen Douglasiidae (Lepidoptera). Beitrag Entomologische (Berlin) 2: 287–300.

Opler, P. A. 1971. Seven new lepidopterous leaf-miners associated with *Quercus agrifolia* (Heliozelidae, Gracillariidae). Journal of the Lepidopterists' Society 25: 194–211.

Opler, P. A., and D. R. Davis 1981. The leafmining moths of the genus *Cameraria* associated with Fagaceae in California (Lepidoptera: Gracillariidae). Smithsonian Contributions to Zoology 333; 58 pages.

OECOPHORID LINEAGE

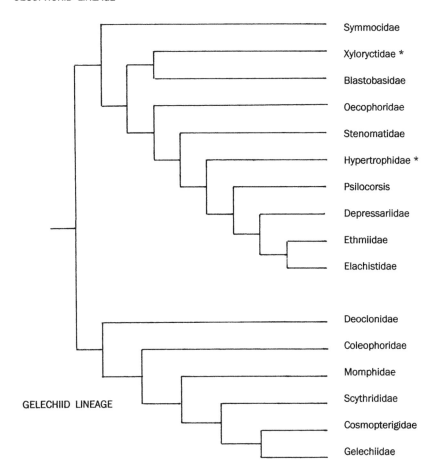

Symmocidae

Xyloryctidae *

Blastobasidae

Oecophoridae

Stenomatidae

Hypertrophidae *

Psilocorsis

Depressariidae

Ethmiidae

Elachistidae

Deoclonidae

Coleophoridae

Momphidae

GELECHIID LINEAGE

Scythrididae

Cosmopterigidae

Gelechiidae

FIGURE 48. Two-dimensional portrayal of phylogenetic relationships in Gelechioidea. Successively more derived clades within the Oecophorid and Gelechiid lineages indicated to right [based on Kaila 2001]. Asterisks (*) indicate families that do not occur in North America north of Mexico.

Superfamily Gelechioidea

This is the largest superfamily of micromoths by far, and because vast numbers of species remain undescribed, Gelechioidea may even surpass Noctuoidea as the most species rich and biologically diverse group of Lepidoptera. For example, in North America we estimate that only about half of the gelechioid species known in collections are named, and a total of nearly 4,000 species is projected, appreciably more than the much more thoroughly catalogued Noctuoidea. In tropical regions the inventory is imponderably incomplete—we have counted more than 1,100 species of Gelechioidea at the Estación Biologia La Selva, a 16 km² (about 4,000 acres) rain forest reserve in Costa Rica—and gelechioids comprise an estimated 20% of all Lepidoptera occurring there, the vast majority unnamed. In Australia, there are 4,000 described gelechioid species, and I. F. B. Common estimated a total fauna of twice that number, making up 35 to 40% of the Lepidoptera species. Besides the enormous species richness, gelechioids vary greatly in size, form, and biology, from tiny moths, such as *Chrysoesthia* (Gelechiidae) and *Ithome* (Cosmopterigidae)(FW 3–4 mm) whose larvae are leaf or seed miners, to large, bulky species in the tropics, such as *Thioscelis directrix* Meyrick (Stenomatidae) in Costa Rica and Peru (FW 2.5 cm) with hind legs longer than the wings, and tiger-moth-like *Maroga* (Xyloryctidae)(FW 3.0 cm) in Australia, whose larvae bore in tree limbs. In western North America we have several broad-winged depressariids and stenomatids with FW lengths of 12 to 13 mm, that is, more than an inch in wingspan on a spread specimen.

Metapleura potosi Busck, a Mexican species recently discovered in southern California, probably is our largest gelechioid (FW 15–16 mm).

The larval biology also varies enormously among gelechioids. Whereas feeding on living plants (phytophagy) is the mode of life for the majority of Lepidoptera, particularly ditrysian lineages other than the Tineidae, detrivory/saprophagy (feeding on decaying organic matter) and fungivory occur extensively in Gelechioidea and are prevalent in some taxa. Within the Oecophorid Lineage, detritivores and fungus feeders are numerous in the three major clades proposed as ancestral by Kaila (Fig. 48), including a massive radiation of Oecophoridae feeding on fallen leaves of *Eucalyptus* in Australia. Therefore detrivory/fungivory may represent the evolutionary groundplan of Gelechioidea. In the presumed more-derived Gelechiid Lineage some cosmopterigids are saprophagous, feeding in leaf litter or other dead plant matter, presumably an independent derivation.

Living plant feeders, especially xyloryctids, depressariids, coleophorids, scythridids, and gelechiids, use an enormous array of plants, more than 80 families of angiosperms by Gelechiidae alone, and probably every family-group of gymnosperms and angiosperms includes hosts for gelechioid larvae. There are specialists whose larvae burrow under bark or in roots, stems, buds, or seeds, as well as leaf miners and a diverse array of external-feeding foliage feeders.

Classification Gelechioids all have overlapping scales on the dorsal surface of the haustellum basally, up to one-third its length, and most have a smooth-scaled head, four-segmented maxillary palpi, and upwardly arched labial palpi with the third segment long and acute (Fig. 85). The most compelling synapomorphy is the nonmotile pupa. It remains in the cocoon at emergence of the moth, in contrast to other nonobtectomeran microlepidoptera. There is general agreement on the phylogenetic unity of this superfamily, but there have been wide differences of opinion on the number and relationships of the included families and subfamilies. About 25 groups have been treated as families, with considerable disagreement in their hypothesized relationships. Several authors have attempted phylogenetic treatments in recent decades, but using different approaches and character sets, they have presented fundamentally differing classifications. In the 1970s, Kuznetsov and Stekol'nikov based a revolutionary arrangement on functional musculature of the male genitalia, and Minet employed an array of morphological characters in the 1980s and early 1990s, as did Nielsen and Common in 1990, all without detailed assessment of relationships. Hodges (in Kristensen's, *Handbook of Zoology*) attempted the first comprehensive cladistic analysis, with weighted emphasis on male genitalia, and proposed two major lineages, the Elachistidae, including Ethmiidae and related families, which was defined by modified abdominal articulation in the pupa on abdominal segments 5 and 6 and on 6 and 7 that prevents lateral movement (Fig. 24 middle), although those of Elachistidae sensu stricto are polymorphic and questionably homologous. The sister clade included Oecophoridae in the strict sense, Coleophoridae and related families, and the Gelechiidae. In 2004, Kaila produced a phylogenetic hypothesis based on features of adults, larvae, and pupae. He too generated two major lineages, but they differ from the Hodges scheme, with Oecophoridae sensu stricto and others in one clade, including the Elachistid Lineage of Hodges, whereas Coleophoridae, Gelechiidae, and related families comprise the sister clade.

Recently, Bucheli and Wenzel produced a phylogenetic analysis of gelechioids emphasizing molecular data. Their study was based on 453 parsimony-informative characters for 42 species (38 North American species and four of Palaearctic origin), representing a broad array of gelechioid taxa. They abstracted 72 morphological characters from prior studies by Hodges and Passoa, with the remainder derived from mitochondrial DNA (cytochrome oxidase I and II), in effect swamping out the morphological component. A consensus phylogeny of the two most parsimonious trees defined two primary sister groups: an odd congregation of taxa including Pterolonchinae, a teleiodine gelechiid, *Scythris*, and *Mompha* formed a sister clade to the rest of the superfamily, which in their analysis included several gelechiids and another scythridid, *Landrya* (=*Asymmetrura*). Within the larger clade, this study confirmed the monophyly of Cosmopterigidae, Gelechiidae other than Teleiodini, Ethmiinae and Depressariinae, Blastobasidae, and Oecophoridae sensu stricto, but linked *Landrya* and the only stenomatine *(Antaeotricha)* examined with Batrachedrinae and Coleophorinae. Bucheli and Wenzel believed the results suggest traditional morphological characters that have defined taxa may not be homologous structures. No doubt the phylogeny will be reconfigured when additional genes, including nuclear, are examined.

We adopt the phylogeny proposed by Kaila, which has strength in use of a broad array of characters from the world fauna but has weaknesses in lack of data for larvae and/or pupae of nearly half the taxa and in its dependence on an exemplar approach, with few species representing huge assemblages of taxa, which are almost entirely Holarctic and

temperate Australian. Hence, the vast diversity of South Asian, African, and Neotropical regions is not included. A parsimony approach with 192 characters resulted in a phylogenetic hypothesis in which neither of the two fundamental lineages nor any of the major assemblages within the Oecophorid Lineage is defined by a uniquely derived synapomorphy.

Linnaean ranks (subfamilies, families) have been applied subjectively and variously to paraphyletic and monophyletic lineages by different authors, and we treat some of Kaila's subfamilies as families. Relationships of the North American families thus derived are illustrated in Figure 48.

Superfamily Gelechioidea, Oecophorid Lineage

The presumed ancestral clade of Gelechioidea is defined by three homoplastic features of the larva: the ecdysal line of the adfrontal area does not reach the cranial suture; seta L3 of prothoraxis is situated venterolateral of L2; and seta SD1 on A9 is more hairlike than the other setae. The pupae lack modified lateral articulation on the abdominal segments but are not defined as a lineage by a uniquely derived feature shared by all taxa.

FAMILY SYMMOCIDAE

This is a diverse, primarily Old World family with 42 genera that has been treated as a subfamily of Blastobasidae or Autostichidae, but blastobasines are placed in the Xyloryctid Assemblage by Kaila's phylogeny (see below). There are two introduced, widely distributed species of Symmocidae in North America, along with the New World genus, *Glyphidocera*.

Symmoca signatella (Herrich-Schäffer) is a gray moth, FW (6–8 mm; Plate 4.19) pale gray with poorly defined black median and distal, incomplete fasciae; HW dark gray. The species is believed to be native to southern Europe, where the larvae are reported to feed on lichens. It became established in urban situations around Paris and London in the late 1800s. *Symmoca signatella* was first detected in North America in Pasadena and Orange County, California, in 1942–1943, and by 1960 it had spread to in the San Francisco Bay area. Larvae are general scavengers in diverse situations, reported from lemon mummies and arbor vitae trash and under bark of dead tree limbs. Adults come to lights. In England this species and *Oegoconia quadripuncta* are univoltine, with adults active in July and August, but in California they fly from June through September, with rare records April to May and October to November, so probably there are two or more generations. *Oegoconia quadripuncta* (Haworth) is distinctive in the western Nearctic fauna, having dark brown FW (5.5–7 mm; Plate 4.20) with yellow spots. It was introduced to the Atlantic states nearly a century ago and described by Busck as *O. novimundi* in 1915. *Oegoconia quadripuncta* was collected in southern California in 1938–1940 and by the 1960s had become widespread in urban areas, including the Central Valley and San Francisco Bay area. Larvae of *O. quadripuncta* feed on decaying vegetable matter and have been reared from oak leaf litter in California.

The genus *Glyphidocera* has been included in the Symmocinae, or treated as a separate subfamily, and as an unrelated family by Hodges. *Glyphidocera* are drab moths, tan to brown, having somewhat truncated FW, superficially resembling dichomeriidine Gelechiidae, with faint darker markings. The labial palpi are thickly scaled, and males possess irregular rows of specialized, toothlike scales on abdominal terga 1 to 4, which are hidden in the folded intersegmental membranes on intact specimens. *Glyphidocera* is a New World genus with more than 100 described species, mostly Neotropical, in addition to many unnamed. There are 12 species in North America, mostly from Texas eastward. One species in the Gulf states feeds on bark of nursery junipers, causing twig dieback. *Glyphidocera septentrionella* Busck, which has pale rust brown or tan FW (7–8.5 mm; Plate 4.21), is widespread on the Pacific Coast from British Columbia to California, flying in July through September.

FAMILY XYLORYCTIDAE

The Xyloryctid Assemblage consists of the family Xyloryctidae sensu lato including in Kaila's phylogeny the Blastobasinae, which we retain as a separate family. There are more than 500 described species of Xyloryctidae assigned to 60 or more genera, in Africa, the Indo-Australian region, and Polynesia, with greatest numbers in Australia. Included are some of the largest, heavy-bodied Gelechioidea (FW 10–33 mm). They are nocturnal moths, often brightly colored, shining white or yellow, patterned with black or brown. The labial palpi usually are long, strongly curved, and slender. The larvae form silken tubes or shelters in lichens, on bark, or among foliage, and some species feed in bark or tunnel in trunks or stems and drag leaves back to the gallery at night. About half the known species feed on Myrtaceae and Proteaceae, or dead eucalypt leaves.

FAMILY BLASTOBASIDAE

Adult These are small (FW 4–15 mm), nocturnal moths with narrow wings and a short abdomen having conspicuous, transverse rows of stout, rust-colored or black, spiniform setae dorsally on segments 1 to 8 in males, 1 to 7 or 1 to 6 in females (Fig. 50). The labial palpi are short to moderately long, strongly curved upward, and appressed to the head. The first flagellar segment of the antenna is enlarged and deeply notched dorsomedially in males of some genera (Fig. 49). Blastobasids are uniformly dull colored, usually gray, sometimes tan or yellowish; FW with whitish or black, smudged maculation. Genitalia, Figs. 51, 52.

Larva Larvae are slender, cylindrical, often with darkly pigmented integument, at least the pinacula; labium with submental pit; abdominal SD1 setae often with sclerotized rings.

Larval Foods Some blastobasids feed on living plant material, but many species reared from plants are detritivores, living in a wide variety of situations: abandoned nests of insect larvae, galleries of stem and root borers or galls induced by other insects, and detritus associated with aphid and scale insect colonies, sometimes feeding on the associated insects.

Diversity Owing to their consistently drab appearance and relatively homogeneous genitalia structures, this family has been neglected in systematics studies. Blastobasines occur worldwide, but are most diverse in the Nearctic and Neotropical regions, with more than 275 described species and possibly five to 10 times that many awaiting study. There are about 70 species recognized in America north of Mexico, along with a large roster of synonyms, mostly described from the Atlantic states a century or more ago. Probably there are 120 to 140 Nearctic species, with only a small number of a great variety in collections from the western states validly identified.

Holcocera males possess a strong notch in the first flagellar segment of the antenna (Fig. 49), and the male genitalia have

setose anellus and a pointed apex on the proximal flange of the valva. *Holcocera gigantella* (Chambers) is our largest species (FW 11–15 mm; Plate 4.22, Fig. 51), FW whitish with well defined longitudinal black lines or gray with the lines obscure. The larvae are specialist feeders on living Agavaceae. The eggs are laid at the base of green fruit of *Yucca*, and the larvae bore into the fleshy cortex outside the seed tiers. Pupation occurs outside the fruit in a moderately dense silken cocoon. *Holcocera gigantella* was described originally from Colorado and has been collected widely in the Southwest, north to Monterey County, California, associated with several *Yucca* species. At a seaside locality in San Diego County there appear to be continuous generations, with mature larvae and cocoons on *Y. whipplei* from February to October. A similar, smaller species, *H. paradoxa* Powell (FW 9.5–10.5 mm; Fig. 52), is known only in the Santa Rita Mountains of southern Arizona, with the same larval habits, on *Y. schottii*.

Holcocera iceryaeella (Riley)(Plate 4.23, Plate 4.24, Plate 60.4) was so named because it was originally detected feeding on "cottony cushion scale" *(Icerya purchasi)*, a serious pest of citrus in southern California in the 1880s. Subsequently, *H. iceryaeella* has been detected in a wide variety of situations as a general detritivore. We reared this species at Dune Lakes in coastal California from bacterial cankers on willow, wood rot fungi (Polyporaceae) infested with ciid beetles, abandoned nests of Lepidoptera caterpillars on coyote brush *(Baccharis)*, and occupied ones on buckwheat *(Eriogonum)*. The adult is small (FW 5.5–9 mm) with pale gray FW marked by an oblique, whitish fascia at 0.4, edged by dark gray outwardly, and an incomplete fascia beyond the cell. There are two or more annual generations, with adults numerous at Berkeley from April to July or early August and September to October, rarely November.

Calosima species are similar to *Holcocera*, with narrow wings but lack the antennal notch and often have a satiny shine. In the male genitalia the valva has a secondary articulated process, and the aedeagus is spheroidal basally. There are six species described in North America, and the genus probably is better represented in the West than recognized. *Calosima munroei* Adamski has relatively long, narrow FW (7.5–9 mm; Plate 4.25) with fairly distinct transverse, black markings at midwing and end of the cell. It occurs in central coastal California (Marin County), where larvae feed in cones of Sargent cypress *(Cupressus sargenti)*, a serpentine soil endemic tree, after debris accumulates during feeding by larvae of cerambycid beetles and tortricids *(Cydia cupressana)*.

Blastobasis is the largest blastobasid genus worldwide, estimated by Adamski to include more than 100 species in Costa Rica alone, but there are only 10 named species in America north of Mexico. These are mostly larger moths than other blastobasids, and males are characterized by having a subconical process basally on the first flagellomere, which forms a notch with palmate scales on its inner surface. Females have a posterior lobe from the corpus bursae and a short, membranous antrum. *Blastobasis glandulella* Riley is relatively large (FW 8–11 mm; Plate 4.26), gray with paler basal area of FW. This species occurs in the Midwest and widely in the western states. Females seek acorns for oviposition, and larvae of this species, along with those of acorn weevils, constitute some of the food value to woodpeckers and other vertebrates that eat acorns. In coastal California adults fly July to September, rarely late June, probably univoltine.

Species of *Hypatopa* (Plate 4.34) lack modification of the antennal first flagellomere. The aedeagus is bulbous and sclerotized basally, and the valva has a dense patch of mixed setae and spines above the venterolateral margin. This may be the most species rich blastobasid genus in the Western Hemisphere and along with *Blastobasis* includes half the family's described species. There are 23 species recorded in North America, but many others have not been studied in detail or even named.

Members of the genus *Pigritia* are smaller (FW 4.5–6 mm; Plate 4.18), generally brown to tan rather than gray, nondescript moths, with reduced HW venation and with variable but often very short labial palpi. *Pigritia* has five valid species in America north of Mexico, with about 25 synonyms, nearly all of them from the eastern United States, named by Dietz a century ago. *Pigritia arizonella* Dietz occurs in the arid Southwest, and we reared specimens of this species in eastern California (Mono County) from root crowns of rabbit brush *(Chrysothamnus)* that had been tunneled by larvae of cerambycid beetles and tortricids. Larval habits of other *Pigritia* seem not to have been reported, perhaps because they are secluded in debris at or below ground level.

FAMILY OECOPHORIDAE SENSU STRICTO

This is a worldwide assemblage of dissimilar moths, with a tremendous radiation of forms in Australia.

Adult Small (FW 4–23 mm) with narrow to broad wings, they are mostly dull colored in Holarctic genera but wildly variable and colorful in Australia; FW patterned in yellow, rose, rust, and browns. The abdominal terga usually lack spiniform setal bands. Genitalia, Figs. 53, 54.

Larva Larvae are cylindrical, head often darkly sclerotized, sometimes with reduced numbers of stemmata, lack a submental pit, and the integument usually is not pigmented; thoracic legs and abdominal prolegs are well developed.

Larval Foods Most feed on dead plant material, leaf litter, and other vegetative refuse, and the rich fauna in Australia depends mainly on *Eucalyptus* (Myrtaceae), with about 60% feeding on fallen leaves, 25% on living foliage.

Diversity There are more than 3,000 described species in more than 500 genera worldwide; this is the dominant group in Australia, with more than 2,200 named species in 250 genera, which are estimated to comprise only half of the oecophorid fauna there. Several species are cosmopolitan household moths whose larvae feed in stored meal, potted plant humus, or other vegetative matter. In North America oecophorids are poorly represented but well known compared to most microlepidoptera, having been monographed by Clarke in 1941 and by Hodges in 1974. Several species are introduced, as scavengers or stored products pests. There are 42 Nearctic species, at least 20 of which occur in the West.

Species of *Inga* are broad winged, stout, rather bulky moths relative to other oecophorids, mostly drab, unicolorous brown, tan, or yellowish, although some tropical species are colorful. Five of the six known species in our fauna occur in Arizona and adjacent states. *Inga concolorella* (Beutenmüller)(FW 7–9 mm, rarely to 11 mm; Plate 4.27) varies from uniformly dark brown to pale yellowish gray. It is widespread in arid regions, from New Mexico to Reno, Nevada, and California's Central Valley north to Antioch in the San Joaquin delta. Its larva is peculiar, very slender, shaped like a wireworm (beetle family Elateridae), with quite short legs. We found them in roots of composite shrubs, burro-weed *(Isocoma tenuisecta)* in southern Arizona, and snakeweed *(Gutierrezia)* in southern California, which were infested with larvae of cerambycid beetle and tortricid moths. The *Inga* larvae live in narrow galleries

abandoned by other insects and form silken cocoons aboveground at the base of the stems. Presumably they feed on detritus and fungal mycelium.

Decantha are narrow-winged moths represented by two similar-looking species in western montane areas. *Decantha stonda* Hodges is widespread, from Wyoming to Arizona, the Pacific states, and British Columbia. It is small (FW 4–6.5 mm; Plate 4.37) with ochreous yellow FW delicately checkered with bluish black. Adults have been reared from the charcoal briquette–like sporophores of *Hypoxylon* (Xylariaceae) on dead oak at Santa Cruz Island, from pine bark in Yuba County, California, and from *Polyporus volvatus* (Polyporaceae) on lodgepole pine *(Pinus contorta)* in British Columbia. This species occurs primarily in montane conifer forest situations but ranges to low elevations in coastal California. *Decantha tistra* Hodges in northern Arizona averages a little larger, with darker markings, especially on the costal half of the FW.

Batia lunaris (Haworth)(Plate 4.38) is a European species that was first collected in North America along the Columbia River in 1931 by Clarke, but it was not recognized by the time of his oecophorid revision in 1941. It occurs in the Willamette Valley and Puget Sound areas of Oregon and Washington and appeared in California by 1956 in Marin County. *Batia lunaris* is small (FW 4–5.5 mm) but colorful, having bright ochreous FW suffused with reddish brown along the costa and termen, with a distinct dark blue mark at the tornus. The larvae feed in refuse and/or fungal hyphae under bark of dead limbs, mostly angiosperm shrubs, and there is a single annual generation; in California adults fly late April to July, primarily May and June.

Esperia sulphurella (Fabricius)(Plate 4.32, Plate 59.35) is another European immigrant, adding color to our early spring suite of dayflying microlepidoptera. Adults have shining black FW (6–8 mm), usually with a gold streak along the Cu crease, and a white triangle at the tornus; the HW is bright ochreous orange, bordered distally in black. Males have enlarged antennae. The moths are diurnal and can be seen fluttering in dappled sunlight in the morning around fallen trees or stacks of oak firewood. This species is univoltine and flies from the last days of February to mid- or late April. It was discovered in California by Don MacNeill, who found an adult in his urban yard in El Cerrito in April 1966. The following year, about five kilometers away in the Berkeley Hills, we reared *E. sulphurella* from black larvae found under bark of fallen live oak *(Quercus agrifolia)* and subsequently from cut fire wood. This colonist has been regularly seen in Berkeley and detected north of San Francisco Bay, in Marin, Sonoma, and Napa counties.

Polix coloradella (Walsingham)(Plate 4.29) is widespread in boreal North America and mountains of the West, to northern Arizona and central coastal California. The larvae live in decaying wood of various shrubs and trees, probably feeding on fungal hyphae. *Polix coloradella* has been reared from sporophores of *Fomes, Poria, Polyporus, Stereum* (Polyporaceae), *Hypoxylon* (Xylariaceae), and bark beetle-infested stems of Douglas-fir *(Pseudotsuga)*. The adults are dark gray to black, FW (7.5–9.3 mm, rarely 6.7 mm) powdered with pale yellow scales when fresh and with the dorsal margin yellow on its basal half, extended upward before the tornus. They are nocturnal, and most specimens have been collected at lights.

Borkhausenia nefrax Hodges (Plate 4.36) has a strange and not fully explained history. This species was first reported in North America in 1952 by the California Department of Agriculture as *Borkhausenia* species near *minutella* (Linnaeus), the latter a widespread Palaearctic species. In the 1960s we identified it as another Palaearctic species, *B. fuscescens* (Haworth),

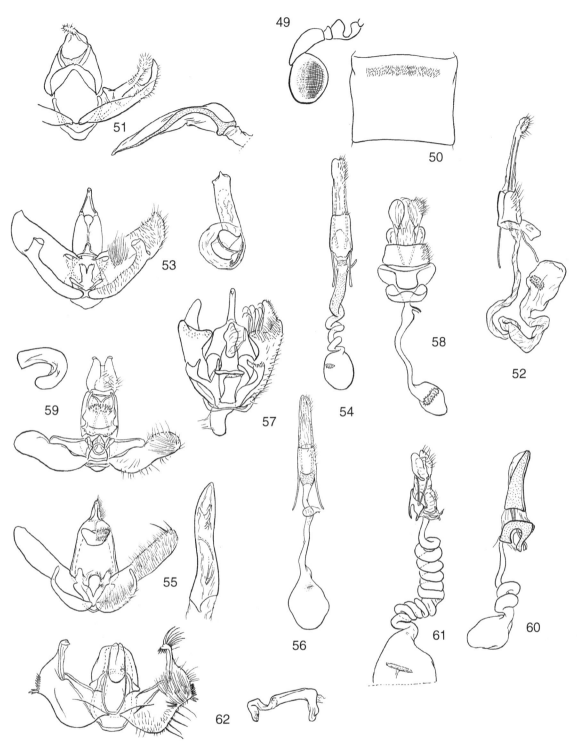

FIGURES 49–62. Antennal and abdominal structures, and genitalia of exemplar species of Gelechioidea; genitalia ventral aspect, except Fig. 60 lateral aspect. **49**, Blastobasidae: outline of eye and basal segments of antenna of *Holcocera gigantella* (Chambers). **50**, Blastobasidae: abdominal segment dorsum with band of spiculae characteristic of blastbasids [Powell 1976]. **51**, Blastobasidae: *Holcocera gigantella* ♂, aedeagus to right. **52**, Blastobasidae: *Holcocera paradoxa* Powell ♀ [Powell 1976]. **53**, Oecophoridae: *Hofmannophila pseudospretella* (Stainton) ♂, aedeagus to right. **54**, *Hofmannophila pseudospretella* ♀. **55**, Depressariidae: *Himmacia huachucella* (Busck) ♂, aedeagus to right. **56**, Depressariidae: *Himmacia huachucella* ♀ [Clarke 1941]. **57**, Stenomatidae: *Antaeotricha schlaegeri* (Zeller) ♂, aedeagus in situ. **58**, Stenomatidae: *Antaeotricha schlaegeri* ♀ [Duckworth 1964]. **59**, Ethmiidae: *Ethmia albistrigella* (Walsingham) ♂, aedeagus to left. **60**, Ethmiidae: *Ethmia minuta* Powell ♀. **61**, Ethmiidae: *Ethmia arctostaphylella* (Walsingham) ♀. **62**, Ethmiidae: *Ethmia marmorea* (Walsingham) ♂, aedeagus to right, lateral aspect [Powell 1973].

and documented its extensive occurrence in urban areas, dating from 1936. Both the recent history and geographic range indicated an introduced species. However, Hodges found specimens that seem to be the same species from native habitats in northern Arizona, New Mexico, and Colorado and described *B. nefrax* as a new, evidently native species. Its distribution in California as an urban moth, rarely collected in native habitats after 70 years, still supports the notion that *B. nefrax* had been introduced to the Pacific Coast. Moreover, soon after its description, *B. nefrax* was discovered in France, where it had long been present but unrecognized, so now we do not know if it was native in Europe and adventive in the western United States, or if a reverse colonization may have taken place. The adults are small (FW 4.5–8 mm), shiny tan with a variable pair of diffuse dark spots near the base and one near the end of the cell. In California they are often found in houses and other structures. *Borkhausenia nefrax* has multiple generations and flies in every month of the year, with a preponderance of records in September to November. The larvae are detritivores or fungivores in vegetative debris, small mammal burrows, wood rot fungi, and so on.

The "brown house moth," **Hoffmanophila pseudospretella** (Stainton)(Plate 4.30, Plate 4.31, Figs. 53, 54), is cosmopolitan, having been transported by activities of humans so long ago that neither the native origin nor timing of introduction into North America is known. It was originally described from England, where it is thought to have been introduced about 1840, and *H. pseudospretella* was known in British Columbia and California by 1871. Populations require high humidity for normal development, so *H. pseudospretella* is absent from inland, hot, dry areas. The adults have elongate, slender labial palpi, broad wings, are extraordinarily variable in size (FW 6.5–13 mm), and vary in color from dark brown to pale gray tan, although most are brown, with irregular, longitudinal black streaks and blotches. The larvae feed on a wide variety of plant and animal materials, in bird nests, dried fruit, insect pupae, vegetative detritus, and so forth, and the species is reported as a stored products pest in some circumstances. There are two or more generations, and colonies may be active throughout the year, especially in buildings. However, the brown house moth apparently is becoming less frequent with increasing prevalence of central heating and reduced humidity in buildings. At Berkeley adults are recorded outdoors from late April through October, rarely in November.

The "white-shouldered house moth," **Endrosis sarcitrella** (Linnaeus), is smaller than *Hoffmanophila*, also quite variable in size (FW 4.5–8.5 mm; Plate 4.28, Plate 4.35), and has narrower wings (HW as narrow as its fringe) with pale FW variably patterned with gray brown. The prothorax, patagia, and tegulae are contrastingly white, distinctive for this species and the source of the common name. The origin of *E. sarcitrella* is unknown. It was widely established in European urban situations by the 1700s, when formal zoological nomenclature began, and it was named several times, including *E. lacteella* by Denis and Schiffermüller, by which it was known in the literature for nearly 200 years. *Endrosis sarcitrella* was found on most continents before 1900 and probably came to North America long before it was recorded in California in 1902. Now it is established on the Pacific Coast from Alaska to California and in Nevada cities, and separately in Illinois and the Northeast. The larvae feed in bird nests and other vegetative refuse, wood rot fungi, compost, dry seeds, and occasionally stored meal. The adults are very common in and around houses and are rarely seen in native habitats. They fly in coastal California year-round and are seen most commonly March through September.

Pleurota albastrigulella (Kearfott)(Plate 4.33) is distinctive among all our gelechioids in having very long, porrect labial palpi. This deceived Kearfott into describing it as a species of *Dorata*, a tineid genus with similar palpi. This species is further interesting as the sole Nearctic representative of an Old World, primarily Mediterranean group, as though it had been introduced to the Pacific Coast long ago. *Pleurota albastrigulella* is widely distributed in the coastal counties of California and foothills of the Sierra Nevada, and there are a few records from Arizona and British Columbia. The moths are gray, FW (7–10 mm, males larger than females) with white costa and a whitish area through the middle of the FW enclosing two basal black dots and one at the end of the cell. This oecophorid often is observed diurnally, especially in grasslands, even grazed areas. The females have shorter wings than males but fly to lights. Hodges reported *P. albastrigulella* reared from chamise, *Adenostoma* (Rosaceae), but that was based on adults we collected in association with chamise, not larvae. **Pleurota rostrella** (Hübner) is recorded feeding in grass roots in Europe, and that seems a likely association for *P. albastrigulella*. Other European species are recorded on *Calluna* (Ericaceae) and *Salvia* (Lamiaceae).

Superfamily Gelechioidea, Elachistid Lineage

This group of taxa is defined by having modified abdominal articulation in the pupa, lateral condyles on abdominal segments 5 and 6 and on 6 and 7 that prevent lateral movement (Fig. 24), although those of Elachistidae sensu stricto are polymorphic and questionably homologous. The first three groups usually have been treated as families or subfamilies of Oecophoridae or Elachistidae. They are broad-winged moths with strongly curved labial palpi, often exceeding the top of the head.

FAMILY STENOMATIDAE

Adult Adults are small to moderately large (FW 5–25 mm), with rectangular to nearly oval FW. The valvae of male genitalia have setae with prominent, multilobed apices. The male antennae usually have long cilia. Genitalia, Figs. 57, 58.

Larva Only a small proportion of described species are known; they are relatively stout caterpillars, often with heavily pigmented integument.

Larval Foods The larvae are external feeders in concealed shelters, on diverse angiosperms (16 or more families), predominately Myrtaceae in the Southern Hemisphere and Fagales in the Nearctic.

Diversity More than 30 genera and 1,200 species are named, 90% from the Neotropical region, where there are countless other species not yet described. This group and the Ethmiidae tend to be mutually exclusive on a broad geographical scale. Stenomatids are species rich in the southeastern United States and lowland wet forests of Central America and northern South America, whereas ethmiids are speciose in arid parts of western North America and thorn forest regions of Mexico, Central America, and the Antilles. There are 25 described stenomatid species in America north of Mexico, most of which occur in the West, including several that range northward only to Texas or Arizona. A few others are known but not yet named. Larval host plants are recorded for more than half of the Nearctic species and in the western United States include Fagaceae, Fabaceae, Malvaceae, Ericaceae, and Poaceae.

Ten species occur in southern Arizona, including six *Antaeotricha*, several of them similar in appearance, having rectangular FW (7–15 mm), whitish clouded with variable gray and much broader, dark gray, fanlike HW. *Antaeotricha lindseyi* (Barnes & Busck). (Plate 4.40), which ranges from western Texas through New Mexico and Arizona, has the gray clouding restricted to the dorsal half of the FW (11.5–13.5 mm). *A. schlaegeri* Zeller (Plate 4.41, Plate 59.33, Figs. 57, 58) is widespread in the Northeast and Midwest, ranges into southern Arizona, and has more extensive gray markings, darker, almost black dorsal blotch on the basal half of the FW. Both species are double-brooded; adults fly in April and late July to September, just ahead of and during the rainy season. Larvae of *A. schlaegeri* and *A. lindseyi* probably feed on oaks *(Quercus)*. On the Pacific Coast, *A. manzanitae* Keifer is similar, with FW (12.5–14.5 mm; Plate 4.42) markings restricted to the dorsal area. It ranges from British Columbia to southern California in foothill areas, where the larvae feed on manzanita *(Arctostaphylos*, Ericaceae). The life cycle appears to be univoltine; larvae collected in April produce adults in May, and adults are recorded in May and June in the Coast Ranges and central Sierra Nevada, July at Mt. Shasta, to early August in southern Oregon.

Gonioterma crambitella Walsingham has FW apex slightly acute (ca. 10 mm; Plate 4.43), whitish with a few dark scales along the costa basally and a dot at end of the cell. This species ranges from central Texas to southeastern Arizona. The larval host plant is unknown, but a sister species, *G. mistrella* Busck, is recorded feeding in roots of a grass, *Phleum* (Poaceae). *Rectiostoma fernaldella* Riley, is the most distinctive western Nearctic stenomatid. This moth is small (FW 5–6.5 mm; Plate 4.39), black with basal 0.4 of FW bright greenish yellow, enclosing a black costal crescent, and there is a weakly raised, median transverse band of iridescent bronze to purplish scaling. The adults are diurnal and evidently have a short flight period. They are rarely seen in nature, although related species in the eastern United States and Mexico come to lights. *Rectiostoma fernaldella* occurs in southern California and along the Coast Ranges foothills and in the Sierra Nevada to mid-elevations and in southern Arizona. Most records have originated from collections of larvae, which are long-lived. The young larvae are found in the fall and winter on live oaks, including *Quercus agrifolia*, *Q. wislizenii*, probably *Q. arizonica* and chinquapin, *Chrysolepis*, in abandoned shelters of the parental generation. The larvae are distinctive, with thoracic area swollen, dark purplish, and abdomen abruptly tapered posteriorly, segmentally banded purplish. After foliation in spring, the partially grown larvae move to the new foliage and create shelters by plastering a leaf to the surface of another with tough silk; then they skeletonize the leaf surfaces from within. Pupation occurs in the shelter in April or May, and adults are active May to July in California and September in southern Arizona.

FAMILY DEPRESSARIIDAE

Adult Adults are small, FW (7–16 mm) elongate-oval or rectangular, labial palpi slender and strongly curved upward; male genitalia lack the multilobed setae of the valvae characteristic of stenomatids, and the strongly recurved phallobase and special pupal anchoring mechanisms of ethmiids. Genitalia, Figs. 55, 56.

Larva Most are leaf tiers, but some bore into stems or seeds, using at least 17 families of dicots, with strong specialization on Apiaceae and Asteraceae in the Holarctic.

Diversity Worldwide, more than 600 species are described, assigned to more than 80 genera, occurring in all major faunal realms, best represented in north temperate and tropical regions. There are more than 100 described species in America north of Mexico, assigned to 12 genera. About two-thirds of the named species occur in the West, and there are several other recently discovered species. Depressariids have been the subject of detailed studies on host-plant coevolution because many specialize on Apiaceae, which produce coumarins and other defensive chemicals.

Agonopterix is the largest genus, with about 20 species in our region, especially rich in the Northwest. Most are generally similar, having the second segment of the labial palpus with spreading scaling anteriorly and rectangular FW, brown or tan to rust, often with a poorly defined dark smudge at the end of the cell. The larvae are slender, green caterpillars that tie the terminals of the host plants, mostly perennial, herbaceous Apiaceae (carrot family) or Asteraceae (sunflower family). All are specialists. The adults, although nocturnal, are attracted to lights only sporadically, usually one or two at a time, so the best approach for obtaining specimens is through larval collections and rearing. Many species overwinter as adults in reproductive diapause and are secretive, hiding under loose bark or debris, in sheds, and so on. *Agonopterix alstroemeriana* (Clerck)(Plate 4.44) is one of our most distinctive, widespread, and often-seen species, distinguished from our native species by having white FW (7.5–9.5 mm) with a distinct, blood red discal spot. It is a Palaearctic species that was first detected in North America in New York in 1973, then suddenly appeared in 1983–1985 scattered over Washington, Oregon, Idaho, Utah, and coastal California. Its spread may have been aided by accidental transport of dormant moths, and the colonization was facilitated by the ubiquitous occurrence of the larval host, the weedy umbell, poison hemlock *(Conium maculatum)*. Adults of *A. alstroemeriana* are long-lived, from June to March, and sporadically come to lights, most often in June and during the winter. They are found inside outlying buildings and wood piles in rural and suburban areas. The larvae are bright green and form tight shelters in the leaflets of *Conium* in April and May, sometimes in great numbers, as local parasites seem not to have adopted this species extensively, and predators may be repelled by plant chemicals sequestered by the larvae. Among native species, *A. oregonensis* Clarke is small (FW 6–8.5 mm; Plate 4.45) and has dark brown to rust brown FW with a gray basal patch. The species ranges from British Columbia to central California, and adults are taken infrequently at lights in May and June and October to January. The larvae feed on various Apiaceae, including several species of *Lomatium* and *Sanicula*.

Agonopterix fusciterminella Clarke has pale tan to rust tan FW (8.5–10.5 mm; Plate 4.46) with moderately well defined discal spot and subterminal dots. It ranges from British Columbia to central coastal California and feeds on *Senecio* and *Cacaliopsis* (Asteraceae) in the Northwest. H. H. Keifer recorded *A. fusciterminella* feeding on *Cynoglossum grande* (Boraginaceae) in California, but we have failed to confirm that association, whereas we have found this species on *Senecio aronicoides*. The two plants grow in the same shaded habitats and have similar-appearing basal rosette foliage in early spring, so we think *Cynoglossum* was misidentified. *Agonopterix canadensis* (Busck)(Plate 4.47) is very widespread, from the northeastern United States, across southern Canada, south in the Rocky Mountains to central Colorado and the Basin Ranges, and along the Cascades and Sierra Nevada to central California. The adult ranges from pale gray to rust FW (8.5–10.5 mm) with well-developed black markings, similar to *A. fusciterminella*. The larval host plant is *S. serra* (Asteraceae) in Washington and likely other *Senecio* species over the broad distribution. *Agonopterix cajonensis* Clarke is similar, with tan to brownish FW, and occurs in the southern California Transverse Ranges at moderate elevations. An island endemic, *A. toega* Hodges is its sister species, occurring on the southernmost California islands. *Agonopterix toega* is variable from pale to dark brown FW (9.5–11 mm). We found the larvae on *Sanicula arguta* (Apiaceae) at three sites on San Clemente Island in 2004, so the species survived the long period of feral goats prior to their removal in the 1980s. We also found *A. toega* on tiny Santa Barbara Island, 36 miles to the north, where there is no *Sanicula*, and the larval host may be wild celery *(Apiastrum)*.

Two western species depart in larval host specialization, feeding on legumes formerly called *Psoralea* (Fabaceae), which have been fragmented into several genera by the industrious botanists, whom the *Agonopterix* ignore. *Agonopterix psoraliella* (Walsingham) is moderately large (FW 9–11 mm; Plate 4.48), with dark brown, almost black to rust brown FW with a basal gray patch that is well defined only in paler individuals. In contrast, *A. posticella* (Walsingham) is smaller (FW 5.5–8.5 mm; Plate 4.49), with yellow tan, rust, or orange brown FW, having the ground color covered by dark speckling, more concentrated distally, and a darker shade in the tornal area. Both species are widespread, from southern California to British Columbia, east to South Dakota, Arizona *(A. psoraliella)*, and Colorado *(A. posticella)*. Adults of both have been collected from May or June to September, and there is no circumstantial evidence of overwintering as adults.

Agonopterix nervosa (Haworth)(Plate 5.1), an introduced European species, is perhaps our most often seen western *Agonopteryx* because adults seem to come to lights more readily than native species and because the larvae commonly feed on invasive brooms (*Genista*, Fabaceae), which occur in human-disturbed rural and urban areas. The moth is moderately large (FW 7–10 mm) and is distinctive in having the FW apex acute, subtended by a slight indentation, and by having a tuft of pink scaling below the eye. The FW color varies considerably within populations, from pale tan to dark rust brown, sometimes showing a conspicuous dark discal smudge. There appear to be spring and fall generations; adults fly in May and July and more abundantly late August to mid-October, but not during winter. Larval hosts in western North America include gorse *(Ulex)*, French and Scotch brooms *(Genista)*, and *Liburnum* (Fabaceae). This species was found on Vancouver Island between 1915 and 1920 and later spread through the Puget Sound area and along the Pacific Coast to California or was separately introduced to the San Francisco Bay area, before 1957; it occurred in western Nevada by 1960.

Species of **Exaeretia** are similar to those of *Agonopterix* but are generally smaller and have relatively narrower FW and slender labial palpi, without expanded scaling. Each species is a specialist on plants of one family, Malvaceae, Salicaceae, or Asteraceae. *Exaeretia nechlys* Hodges, along with *E. thoracenigraeella* (Chambers) and *E. thoracefasciella* (Chambers), feed on Malvaceae, including bush mallow *(Malacothamnus)*, checkerbloom *(Sidalcea)*, and globemallow *(Sphaeralcea)*, in the Southwest and Pacific states. All have brown

FW, variably mottled darker, and a gray basal band. The FW of *E. nechlys* (6.5–8 mm; Plate 5.2) is uniformly dark brown with distinct, contrasting basal band and two white dots in the cell. The larvae are facultative leaf miners and can move from leaf to leaf, then pupate on the ground. *Exaeretia nechlys* ranges from southern Arizona to desert areas of southern California, the Sierra San Pedro Martir in northern Baja California, and northward to central coastal California. *Exaeretia gracilis* (Walsingham) is distinctive, having pale yellow tan FW (6.7–9 mm; Plate 5.3) with the base and prothorax dark brown. *Exaeretia gracilis* is widespread, from North Dakota to Texas and California. Its larvae are dark, almost black, and mine leaves of Asteraceae May to July and September in California. Hosts include western ragweed *(Ambrosia psilostachya)*, gumplant *(Grindelia stricta)*, and on San Clemente Island, telegraph weed *(Heterotheca grandiflora)*. This species is bi- or multivoltine; adults have been collected in all months except December and January in southern California. Another Asteraceae-feeding species, *E. umbraticostella* (Walsingham)(Plate 5.4), ranges from British Columbia and South Dakota to western Texas and southern California. It is small (FW 6–8.5 mm), with tan or pale brown FW having a dark brown basal band and a large, midcostal smudge extended distally. Larvae are recorded feeding on arrowleaf balsamroot *(Balsamorhiza sagittata)* and a sunflower, *Helianthus pumilus*.

Semioscopis species fly in early spring, too early for most collectors, and as a result our knowledge of them is fragmentary, although all are widespread across boreal North America. They are large (FW mostly 10–15 mm) with a strongly arched costa, giving the FW an oval appearance. **Semioscopis merriccella** Dyar has mottled pale gray FW (Plate 5.5) marked by a bold, black, serpentine bar in the cell. There are isolated records from British Columbia, Colorado, Oregon, and California. Specimens have been taken as early as March in Pennsylvania but mostly in May, including one in the central Sierra Nevada at 4,500' elevation. The larval habits are unknown. Similarly, *S. inornata* Walsingham (FW 10.5–15 mm, females consistently smaller than males; Plate 5.6, Plate 5.7) occurs in the northern United States and southern Canada, with peripheral records in Colorado, the White Mountains of Arizona, and northern California. The moths are gray, FW faintly mottled with whitish and black scaling. Adults fly from March to June, depending upon latitude and elevation, early April even in Alberta. Larvae have been collected by the Canadian Forest Insect Survey from several poplars and aspen (*Populus* species, Salicaceae), less frequently willow *(Salix)*.

Depressaria has more than 100 described Holarctic species, 24 of them in North America, all but two of which occur in the West. Larval host plants are primarily Apiaceae and are known for most of the western species, in large part due to the efforts of J. F. G. Clarke. The moths are similar to *Agonopterix* but are mostly larger and more heavy bodied in appearance and lack the anteriorly directed scale tuft of the second labial palpus segment. They are consistently drab with FW rust or shades of brown. The "parsnip webworm," **Depressaria pastinacella** (Duponchel), is our largest species (FW 9.5–13 mm; Plate 5.8) and perhaps the most often seen, particularly by its larval damage, which consists of webbed inflorescences of garden parsnip *(Pastinaca sativa)* and other umbels. The larvae eat the flowers and developing seeds, primarily in June. Adults emerge from June to August and overwinter. This species was first recorded in North America in southern Ontario in 1869, and with the St. Lawrence seaway as a likely port of entry, was believed to have been introduced from Europe. It was not mentioned in early-twentieth-century literature for the western United States, but by 1925 *D. pastinacella* was recorded in Washington, Oregon, Utah, and Arizona and subsequently has been regarded as natively Holarctic rather than introduced. **Depressaria artemisiae** Nickerl, a Holarctic species, is very small for a *Depressaria* (FW 5–8.5 mm; Plate 5.7). Its FW is dark brown or variably grayish brown, with a short whitish patch at base. In North America, this species had been known from southwestern Manitoba to British Columbia and eastern Washington, but we found larvae of *D. artemisiae* feeding on *Artemisia dracunculus* in the Huachuca Mountains, Arizona, in April 1986 and 1988. More than 50 years earlier, Clarke reared it from *A. dracunculoides* in Washington.

During a 1960s nomenclatural housecleaning of misapplied eighteenth- and nineteenth-century names, *D. pastinacella* replaced *heraclina* L. as the valid name because the latter is an *Agonopterix*, and *A. costosa* (Haworth) was replaced by *A. nervosa* (Haworth), which in turn had been wrongly applied to a *Depressaria*, and that species is now called *D. daucella* (Denis & Schiffermüller). So, three of the depressariid names most widely used in the literature for more than a century were shuffled, reminding us of how fortunate we are to work with a fauna that was mostly described much later than in Europe and does not suffer as much from incompetence preserved by the law of priority in the International Code of Zoological Nomenclature.

Depressaria daucella (Denis & Schiffermüller)(Plate 5.9) is a European species believed to have been introduced to the Puget Sound area of Washington after 1940, and it spread to the Portland, Oregon, area by 1950; along the Pacific coast to Mendocino County, California, by 1957; and the San Francisco Bay area by 1997. The moth is distinctive, having numerous fine, black, longitudinal striae on the rust FW (8–10.5 mm). The larvae feed on native umbells, Pacific waterdrop *(Oenanthe sarmentosa)*, and Douglas water hemlock *(Cicuta douglasii)* in Washington. The only other western Nearctic species with a similar striate FW pattern is **D. artemisiella** McDunnough; the lines are bolder on *D. artemisiella*, and the basal area is dark brown. It occurs inland, in southern British Columbia to the Wasatch Mountains, Utah, and the larvae feed on sagebrush (*Artemisia*, Asteraceae). **Depressaria alieniella** Busck (Plate 5.10), another *Artemisia* feeder, is very widespread, from Nova Scotia to British Columbia, Northwest Territories, and south to Mt. Shasta, California. The FW (8–10 mm) is dark chocolate to rust brown or pale rust, usually with a tiny but distinct white dot at the end of the cell. The life cycle appears to be similar to that of *D. pastinacella*.

Apachea barbarella (Busck), a moderately large species (FW 9–13 mm; Plate 5.11), has a generic name that calls up images of southern Arizona and rugged desert mountain retreats. But this species is widespread in the Southwest, from Colorado east of the Rocky Mountains, Utah, to Arizona and central California. The distribution follows that of hoptree, *Ptelea* (Rutaceae), a larval host documented in Arizona and California. *Prunus* was reported as a food plant by Clarke, but that was based on a specimen labeled "Prunus?," probably a misidentification. *Apachea barbarella* is broad winged, with a pronounced curvature in the HW posterior margin and has a strong scale tuft of the second segment of the labial palpus. The FW is dark to bright rust brown with multiple longitudinal black striae. We collected larvae in May; adults emerged in June and probably live until the following spring, as specimens have been taken through the winter.

Himmacia has three western species, one in western Texas known only from the type specimen collected before 1928, and two in southern Arizona. Of the latter, **H. huachucella** (Busck) has bright to pale rust rose–colored FW (8.5–11 mm; Plate 5.12, Figs. 55, 56), whereas **H. stratia** Hodges has pale rust tan to ochreous FW (Plate 5.13). We reared both species from *Quercus hypoleucoides* in the Santa Rita and Huachuca mountains and *H. huachucella* from *Q. arizonica* in Oak Creek Canyon. Both fly in summer, ahead of and during the rainy season, but the life cycle and overwintering stage are not documented.

Psilocorsis has been placed in the Amphisbatidae as a sister group to the Gelechiid Lineage by Hodges or in the Oecophorid Lineage as sister to the Elachistid Assemblage by Kaila. Its species are easily differentiated from those of Depressariidae sensu stricto and related taxa by having very slender, strongly arched labial palpi, narrow FW with acute rather than rounded apex, and by differences in the genitalia, especially the signum, which is many branched, like the pinnae of a fern leaf. There are about a dozen species known in North America, including several not yet named, and all are similar in habitus and FW pattern. Five occur from western Texas to Arizona, three of them restricted to southern Arizona: **Psilocorsis arguta** Hodges (FW 8.5–10.5 mm; Plate 5.14) and **P. amydra** Hodges, both described from Madera Canyon, Arizona, are similar in size and dark rust FW. *Psilocorsis arguta* has numerous distinct, fine, transverse, black lines, typical of many *Psilocorsis*, whereas *P. amydra* lacks them. We reared *P. arguta* from larvae collected in June on *Quercus hypoleucoides*. By contrast, **P. cirrhoptera** Hodges, described from the Chiricahua Mountains, has pale yellow FW (8.5–9.5 mm; Plate 5.15). All three fly in July and August, ahead of and during the rainy season.

FAMILY ETHMIIDAE

Adult These are mostly small moths (FW 4–16 mm, rarely 24 mm), having strongly upcurved labial palpi (Fig. 63) and elongate, narrow FW, often dark with a sinuate, pale band along the dorsal edge that renders a bird-dropping appearance when the moth is perched (Plate 59.36), or white with black spotting, superficially resembling *Yponomeuta*, and some tropical species are colorful with blotches of purple or blue and a terminal band of gold. Most are nocturnal, but some high-montane species are diurnal, as are a group of species in the southwestern Nearctic that fly in early spring. The family is characterized by a strongly recurved phallobase of the aedeagus, secondary subventral setae of the larva, and by two separately derived pupal anchoring mechanisms: either by development of "anal legs," ventral, setiferous, forward-directed extensions of the ninth segment (Fig. 24), or by grasping the exuvial head capsule or cocoon silk between abdominal segments 6 and 7. Genitalia, Figs. 59–62.

Larva Typical ethmiid larvae are slender, tapered toward both ends, and often colorful, living in slight webs exteriorly on new foliage of the host plant. Others have dull integumental colors and live within the scorpioid inflorescences of their borage and hydrophyll hosts.

Larval Foods About 80% of the species for which larvae are known worldwide feed on Boraginales (Boraginaceae, Ehretiaceae, Hydrophyllaceae), including species from the Palaearctic, Nearctic, Indo-Australian, and Neotropical regions and South Africa.

Diversity More than 300 species have been named worldwide, with the greatest richness in areas of seasonal drought,

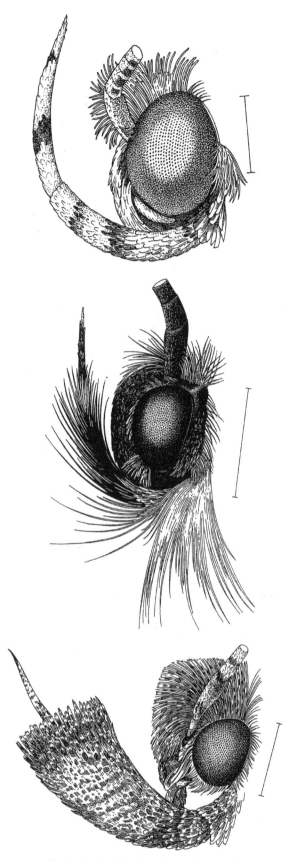

FIGURE 63. Heads of ethmiid moths, showing the strongly upcurved labial palpus that is characteristic of most Gelechioidea. top, *Ethmia confusella* (Walker); middle, *Ethmia coquillettella* Busck; bottom, *Pyramidobela quinquecristata* (Braun) [Celeste Green].

especially microphyllous thorn forests of the northern Neotropical region. This group is better studied than most micromoths, and most New World species in collections are described, having been monographed by Powell in 1973. About 50 species occur in America north of Mexico, most of which occur in the Southwest, 27 of them in California.

Ethmia are interesting on account of the diverse strategies they have evolved to cope with seasonal drought. Most western Nearctic species are univoltine, adapted to feed on annual plants in early spring. Members of the Albitogata Group fly early, January in coastal California and southern Arizona, other species in February, March, to early May. They are diurnal and darkly pigmented, evidently a thermoregulatory adaptation. Females oviposit into immature inflorescences of specific borage or hydrophyll annuals or herbaceous perennials, which appear with winter rains. The larvae feed within the gradually elongating inflorescences, ensuring survival that might not be afforded if they ate the foliage, which is ephemeral in drier seasons. A second mode is exhibited by species of the Geranella Group, which fly in late autumn in desert regions. The nonfeeding adults can live up to two weeks, fly in cold conditions (41–44°F), and produce eggs in November that hibernate until early spring. A third strategy involves nocturnal winter activity. The female of ***Ethmia charybdis*** Powell (Fig. 64) is flightless and active during winter in the Mojave Desert. Eggs are deposited at the onset of winter germination so that larvae can begin feeding as soon as plant growth begins. Finally, estivation and/or hibernation takes place by pupae in all western Nearctic species of *Ethmia* for which life cycles are known. During periods of drought, when annual plants may be very scarce or absent, pupae can delay diapause development and wait until a more favorable season. We obtained emergence of adults after two to four years by several species under lab conditions.

At least three species of *Ethmia* are of potential conservation concern, as they are known only from the original type specimen or from quite limited geographic areas, where they are threatened by human activities: ***Ethmia minuta*** Powell is known from several sites in Riverside and San Diego counties that are within the area of intense urban sprawl of southern coastal California. A similar species, ***E. tricula*** Powell, is known only from a unique type collected in 1963 near Moreno in Riverside County, California, an area that is under pressure of urban development. ***Ethmia monachella*** Busck is known only from the unique type specimen collected at Boulder, Colorado, in March 1908. No doubt it has been neglected owing to its early flight period, but sporadic searches in recent years near Boulder and nearby foothill and prairie habitats have been unsuccessful, and the original habitat may have been destroyed.

ALBITOGATA GROUP

The Albitogata Group of 10 species is restricted to the southwestern United States and adjacent Mexico. All are early-spring, diurnal moths and therefore have been overlooked by most collectors. ***Ethmia umbrimarginella*** Busck (Plate 5.19) was described from Mesilla, New Mexico, and has been collected in southern Arizona, but only a few times because it flies in January and February. The adult is small (FW 9.5 mm), dark gray, FW with a bright red orange basal spot and narrow line to nearly the end of the cell. Nothing is known of the larval biology, but a sister species, ***E. lassenella*** Busck

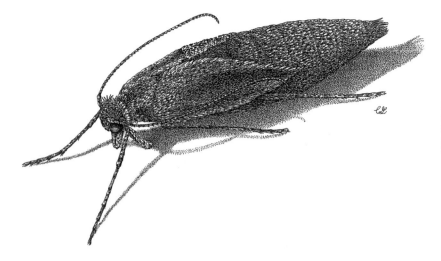

FIGURE 64. Habitus drawing of flightless female *Ethmia charybdis* Powell (Ethmiidae), a winter moth, which is active at freezing temperatures in the Mojave Desert [Celeste Green].

(Plate 5.20), feeds on *Phacelia procera* (Hydrophyllaceae) in northeastern California. *Ethmia coquillettella* Busck is similar but smaller (FW 6.5–8 mm; Plate 5.21; Fig. 63 middle), with pale yellow to yellow orange rather than orange or red FW spots. This species flies in early spring, February at the coast, to April inland in California, and April to May in British Columbia. We have associated adults with and reared larvae from eggs on *P. cicutaria* and *P. distans* (Hydrophyllaceae). *Ethmia brevistriga* Clarke (Plate 5.18) is the most often observed species of this complex because it flies later, mid-March to early May, in association with *P. distans*, which is an annual but is more shrub-like and persists into late spring. *Ethmia brevistriga* is small, dark brown, with variable white pattern on the posterior half of the FW (4.7–6.0 mm). It is endemic to California, ranging along the coast from Sonoma County to San Diego County and inland in the Coast Ranges, foothills of the Sierra Nevada, and the Owens Valley. *Ethmia scylla* Powell (Plate 5.22) diverges in plant selection, feeding on Chinese houses, *Collinsia heterophylla* (Scrophulariaceae). The moths are dull brownish gray with whitish scaling over the posterior half of the FW (5.8–7.2 mm) that defines weak darker markings distally. They fly in late February to early April in the California Coast Ranges, from Colusa County south to Los Angeles County. *Ethmia minuta* Powell is our smallest species (FW 4.3–5 mm; Plate 5.16, Plate 5.17, Fig. 60) and the only North American ethmiid exhibiting sexual dimorphism in wing color. The FW is black with a median longitudinal white line, HW white in males, dark gray in females. The papillae anales are elongated and sclerotized, enabling females to thrust the eggs into the densely bristled, unopened buds of *Cryptantha intermedia* (Boraginaceae). The eggs are smooth, rather than having a sculptured chorion as in most *Ethmia*, and larvae hatch directly into the interior of the bud.

Although genitalia morphology links *E. charybdis* Powell (Plate 5.23, Fig. 64) with the Albitogata Group, this species has diverged remarkably in adapting to winter behavior. Both sexes lack functional mouthparts, and males have long, narrow wings (FW 10.5–12.5 mm) and fly in very cold temperatures (31–35°F). The legs of both sexes are long and thin. *Ethmia charybdis* is active from mid-November through December (documented by pitfall-trap sampling at the Nevada Test Site) in the Mojave Desert and foothills west of California's San Joaquin Valley. We found the larvae on *Amsinckia tessellata* (Boraginaceae) in western Fresno County and at Joshua Tree National Monument in April.

Ethmia discostrigella (Chambers)(Plate 5.24, Plate 5.25) is by far the most abundantly observed ethmiid in North America—the adults swarm to lights throughout the West and in most collections make up most of the *Ethmia* holdings. This may be the typical image of an ethmiid to most lepidopterists, but *E. discostrigella* and its sister species, *E. semitenebrella* Dyar, are atypical, feeding on mountain mahogany (*Cercocarpus*, Rosaceae) rather than a borage. The moths have ash gray FW (10–14 mm) with short, black longitudinal lines, and *E. semitenebrella* (Plate 5.26, Plate 5.27) consistently has two oblique black dashes into the whitish dorsal area of the FW, whereas there are none or just one in *E. discostrigella*. The abdomen is ochreous yellow in both. For decades *E. semitenebrella* was treated as a synonym, but males of *E. semitenebrella* are easily distinguished by a strong, dark scale tuft from the base of the HW costa (Plate 5.27), which is lacking in *E. discostrigella*. It is concealed under the FW on classical spread specimens so was overlooked by tidy lepidopterists dedicated to the appearance of spread specimens in drawers. *Ethmia discostrigella* is the most widespread western species, ranging from western Texas and Chihuahua through the southwestern states to northern Utah, Wyoming, and northern Oregon and the foothills and mountains of California on both sides of the Sierra Nevada. *Ethmia semitenebrella* is largely sympatric, ranging from Nuevo León to eastern California, but not as far north and west as *E. discostrigella*. Both undergo at least two generations per season, and *E. discostrigella* has been regarded as an important defoliator of mountain mahogany in Idaho by the U.S. Forest Service. Although less widely distributed than the preceding species, *E. arctostaphylella* (Walsingham)(Plate 5.28, Plate 5.29, Fig. 61) occurs almost everywhere that mountain balm or yerba santa, *Eriodictyon* (Hydrophyllaceae), flourishes, most often in disturbed sites such as along rural roadsides. The FW (8.5–14 mm) is gray, paler on the dorsal half, and as the moths typically rest by day along the midrib of the *Eriodictyon* leaf, the image is that of bird droppings (Plate 59.36). This species ranges from southern Oregon to northern Baja California and northern Arizona. There are seasonal forms, differing in morphology and wing color. In central California the phenotype in early spring in association with *E. californicum* has narrower FW that are dark with an obscured whitish pattern, and the adults are primarily crepuscular, whereas later generations in the same populations have broader FW with an extensive whitish pattern and are nocturnal with larger eyes. The variation is also partly geographically related, with populations in arid areas very pale, especially in summer generations, and these are associated with *E. crassifolium*, which has wooly-white foliage.

Ethmia albistrigella (Walsingham)(Plate 5.30, Fig. 60) is a diurnal moth that flies in montane areas of Utah, Nevada, the northwestern states, and British Columbia, to the southern Sierra Nevada. The moth is black with a distinct, white longitudinal FW streak (FW 7–9 mm), and the abdomen is bright yellow ventrally and laterally. A race, *E. icariella* Powell, occurs above timberline (12,000′) in the southern Sierra Nevada and may be a separate species, and there is a comparable species, *E. orestella* Powell in high-elevation Colorado, known from two specimens. Various *Phacelia* species have been implicated as possible hosts in Oregon and California, and Braun suggested an association with *Hackelia floribunda* (Boraginaceae) in Utah. *Ethmia semilugens* (Zeller) is morphologically similar to *E. albistrigella* but is larger, FW (8.5–12 mm; Plate 5.31, Fig. 63 top) brown with a sinuate-margined white patch along the dorsal margin. In contrast to the phenotypically similar Trifurcella Group species, its abdomen is bright ochreous ventrally. *Ethmia semilugens* is widespread, from Texas and Chihuahua to northern Colorado, central Utah, and eastern California. We have reared it from *P. calthifolia* and *P. crenulata* in the Mojave Desert. Populations are facultatively bivoltine, with larval feeding in spring and adult emergences either the same year, July to September, or not until the following spring or a later year, presumably depending on rainfall. In the lab, emergences occurred from pupae in diapause after 33 and 52 months.

Ethmia geranella Barnes & Busck (FW 9–10.5 mm) is a member of a complex of similar, gray species in desert areas of Arizona and southern California. All feed on *Phacelia* in spring and fly in late fall, and the eggs enter diapause until winter germination begins. The proboscis is short and weakly coiled distally and may be nonfunctional, as captive moths lived two weeks without feeding. They are pale to dark gray, FW with a variably developed longitudinal, black streak that emphasizes a white spot at the end of the cell. *Ethmia macneilli* Powell (Plate 5.32, Plate 5.33) is quite similar superficially. We found its larvae on *P. ramosissima* (Hydrophyllaceae) at Keene in the Tehachapi Mountains in April, and those of *E. timberlakei* Powell on the same plant at Riverside, California. Larvae of two undescribed species of this group were numerous on *P. tanacetifolia* in the western foothills of the Central Valley in March.

Ethmia monticola (Walsingham) is a moderately large (FW 11–14.5 mm; Plate 5.34, Plate 5.35), gray species that ranges through boreal parts of western North America, from northern Arizona and New Mexico through Colorado, northern Utah, and the northwestern states to northern California. The FW is dark slate to pale ash gray, with longitudinal black lines and dots in the cell, and in Colorado, New Mexico, Arizona, and Nevada the costal half is mostly black (Plate 5.35). The abdomen is sharply contrasting, bright ochreous yellow. The adults are nocturnal but come to lights only sporadically; in southern Oregon large numbers of males were attracted to a pheromone lure for *Choristoneura* (Tortricidae), which was a trial recipe, ineffective for its target species, with principal components 96:4% Z:E tetradecenyl acetate. The larval hosts are unknown, but *Cynoglossum, Lithospermum, Symphytum*, and/or *Mertensia* (Boraginaceae) are possible. *Ethmia monticola* appears to be univoltine, emerging May to July, varying with elevation and latitude. A closely related species, *E. caliginosella* Busck, is known only from Arctic-alpine situations above 11,500′ in

Colorado. It is dark gray with bushy palpi and a partially dull ochreous abdomen.

Members of a complex of species related to ***Ethmia hagenella*** (Chambers) are concentrated in the Southwest from western Texas to eastern Arizona. They have white to pale yellowish FW heavily patterned with black spots and varying gray to brownish on the costal half (FW 9–12 mm). *Ethmia hagenella* occurs from southern and central Texas to southern New Mexico and in Chihuahua. Western populations have more extensive black smudging and were named *E. josephinella* Dyar (Plate 5.36). The species is bi- or multivoltine, with flight records January to April and September to October. Host plants for none of the southwestern species are known, but another member of the group, *E. zelleriella* (Chambers), which is widespread in the eastern United States to eastern Texas, feeds on *Phacelia dubia* and *P. bipinnatifida*.

TRIFURCELLA GROUP

The six species of the Trifurcella Group occurring in the western United States represent the northern tip of a large assemblage in Mexico and Central America. These moths are characterized by absence of the male uncus and "anal legs" of the pupa, and by having brown FW with the dorsal area paler, white to gray. ***Ethmia marmorea*** (Walsingham)(8.5–12 mm; Plate 5.37, Fig. 62) is our most widespread and frequently collected species. It resembles *E. semilugens* but has a white costal patch near the end of the FW cell that encloses a black dot (area all black in *E. semilugens*), and the abdomen in *E. marmorea* is brownish gray, not ochreous. *Ethmia marmorea* occurs from southern British Columbia through the Northwest and Basin Ranges, northeastern California, and Rocky Mountains to New Mexico, Chihuahua, and southern California. Nevertheless, nothing is known of the larval biology. A related eastern species, ***E. trifurcella*** (Chambers), which occurs west to Arizona, was reared from *Cynoglossum virginianum* (Boraginaceae) in Ohio and Kentucky by Annette Braun. The FW of ***E. sphenisca*** Powell (FW 10–13 mm; Plate 5.39), which ranges from Michoacan, Mexico, to northern Arizona, is even more similar to *E. semilugens*, and specimens masqueraded in collections as *E. semilugens* for 40 years, its bright ochreous anal scaling having been overlooked, before dissections revealed genitalic distinction. ***Ethmia albicostella*** (Beutenmüller)(Plate 5.38) is unusual in morphology and has a peculiar geographic distribution, from Manitoba to Durango along the Rocky Mountain–Sierra Madre Occidental axis, occurring in southern Alberta, the Rocky Mountain states, Kansas, and Arizona. The adult is moderately large (FW 10.5–13 mm) with the relatively narrow FW pale brownish gray on the costal, white on the dorsal half, divided by a straight line. Oddly, the labial palpus length, especially the third segment, varies geographically, from short in Canada to 1.5 times longer in Mexico. The larva was discovered on *Lithospermum* (Boraginaceae) in Wyoming by Braun.

Pseudethmia protuberans Clarke (Plate 5.40) was proposed as a monotypic genus because both sexes have the front produced into a strongly protuberant, rounded cone. Comparable and frontal modifications are developed in desert moths of unrelated taxa, often with elaborate differences between species (e.g., Gelechiidae, evergestiine Crambidae, stiriine Noctuidae). The labial palpi of *P. protuberans* are short and blunt, but genital morphology is not much derived from that of *E. trifurcella* and related species. Recognizing *Pseudethmia* as valid renders *Ethmia* paraphyletic. The adults are small (FW 8–9 mm) and drab, having pale gray FW with a weakly to well-defined longitudinal black streak through the cell. They are active in early spring, mid-February to early April. This species is limited to the deserts of southeastern California north to the Panamint Valley, Inyo County, and no doubt eastern Baja California. The larval host is unknown, but based on association of adults at sites we have seen (abundant at Rice and Kelso Dunes, San Bernardino County), *P. protuberans* is suspected to feed on *Tiquilia plicata* (Boraginaceae) and related plants formerly known as *Coldenia*. Males are numerous at lights, but females are rare, having been taken only by sweeping at night.

The assignment of **Pyramidobela** to Ethmiidae is based primarily on its split valva and recurved phallobase in the male genitalia and is equivocal. The labial palpi have a broad scale brush on segment 2 and very slender segment 3 (Fig. 63 bottom), and the FW has upraised scale tufts, unlike *Ethmia*. The larva lacks subventral secondary setae; and the pupa is densely covered with secondary setae, a condition partially developed by some Indo-Australian *Ethmia*. The moths are slender with narrow FW compared to *Ethmia*, FW (7.5–9.5 mm) tan, brown, or gray with two or three tufts of upraised, darker scales. **Pyramidobela quinquecristata** (Braun)(Plate 5.41, Fig. 63 bottom) occurs from southern British Columbia and Montana to Utah, northern California east of the Sierra Nevada, and Mt. Palomar in southern California. The FW is dark brown with tan shades distally, scale tufts black. The larvae feed on *Penstemon confertus* in Montana and *P. deustus* (Scrophulariaceae) in California. A similar species, **P. angelarum** Keifer (Plate 5.42), was discovered in Los Angeles in 1934 feeding on ornamental butterfly bush, *Buddleia* (Loganiaceae), a tropical plant possessing poisonous alkaloids. Later, *P. angelarum* spread to the San Francisco Bay area and the Russian River, as a specific feeder on Buddleia, and is multivoltine. Its origin is suspected to be Mexico, where there are several apparently related species.

FAMILY ELACHISTIDAE

Adult These are tiny or small moths (FW 2.5–6.5 mm) with narrow wings, HW lanceolate, fringe much wider; FW usually white, gray, or black with white markings. Maxillary palpi very reduced, one- or two-segmented, labial palpi slender, strongly curved upward. Genitalia, Figs. 65–67.

Larva Lavae are dorsoventrally compressed, with short legs; head prognathous, flattened, and recessed into the first body segment, adapted for mining.

Larval Foods Elachista mine monocots (Poaceae, Cyperaceae), but a few other elachistids are miners in dicots.

Diversity There are more than 400 described species, worldwide, mainly Holarctic. Following extensive taxonomic work in recent years by C. Kaila, there are about 140 described species in America north of Mexico, of which 86 were described as new in the 1990s (75% of them western), but as Kaila points out, the real number is no doubt much higher, as the elachistid fauna is virtually unexplored in large areas of the continent.

There are several reasons for the relative scarcity of specimens in collections: the small size and drab appearance of many species, brief seasonal occurrence of the adults, restriction of their activities to crepuscular and matinal flights, and elachistids rarely come to lights. Finally, the difficulty of finding mines in grasses and in rearing the larvae successfully deters one of the primary approaches for discovering microlepidoptera. We have found the best means of collecting elachistids is to lightly sweep with a fine-meshed net in grassy places, just before and after sundown and sunrise, or in shady wooded areas during the daytime. Kaila believes elachistids that are attracted to lights often keep their distance and are best collected by lightly sweeping foliage at a distance of a few to 10 m from the light source.

We had known **Perittia passula** Kaila (Plate 5.46) as a curiosity for 25 years prior to its recent naming, owing to its odd larval behavior and life cycle. The larvae form large, full-depth blotch mines in leaves of the native honeysuckle, *Lonicera hispidula* (Caprifoliaceae)(Plate 59.34), February to April. At maturity the larva exits the mine and burrows in the outer bark of a twig to aestivate. Adults emerge in late August to September, during the dry season, an atypical pattern for California microlepidoptera. The adult is tiny (FW 3–4 mm), dark gray brown, FW with a cream-colored pretornal mark. This species is known only from central coastal California, Marin County to Monterey County, and on Santa Catalina Island. Kaila illustrated a cocoon we provided from Marin Island but erroneously cited it from "California, Strawberry Co." There is no such county and no record of *P. passula* from Strawberry, Tuolumne County, or elsewhere in the Sierra Nevada. The type locality is in the Santa Lucia Range west of Jolon (misspelled as "Solon"), Monterey County. **Perittia cygnodiella** (Busck)(Plate 5.47, Figs. 65, 66) feeds on deciduous snowberry, *Symphoricarpos* (Caprifoliaceae), which the larvae mine in early spring. At Albany Hill in the urban San Francisco East Bay area, larvae leave the mines in April and pupae remain in diapause until the following February or early March. The moths are small (FW 3.5–5.5 mm), gray, FW with a patch of pale blue gray scaling on the distal half defining medial and pretornal, variable dark smudges, and the termen is black. *Perittia cygnodiella* ranges from Alberta and British Columbia to central California.

The genus **Elachista** includes most of the North American elachistids, and there are about 85 named species in the western United States and Canada. They are more species rich in northern and montane regions but also occur in lowland, arid habitats. Larvae of the reared species are miners in monocotyledons, mostly grasses (Poaceae), a few in sedges (Cyperaceae). The FW phenotype is one of three forms, black with well-defined white spots, whitish to gray with dark lines or smudges, or all white with tiny black markings. However, the FW pattern does not correlate with species groups defined by Kaila, so superficial appearance does not necessarily indicate relationships, and genitalia dissections should be made for reliable identification.

Elachista cucullata Braun (Plate 6.7) has black FW with white median and distal transverse bands, partial or complete. California specimens are larger (FW 4–4.6 mm) than those found elsewhere (3.3–3.7 mm). This species exemplifies the fragmentary state of Nearctic elachistid knowledge, having been recorded in the Northeast, from Newfoundland to Pennsylvania, and disjunct along the coast in California in Humboldt and Monterey counties. Adults have been taken in June. **Elachista leucosticta** Braun was described originally from Ontario but is widespread in the West, in Utah (Wasatch Range), northern Arizona, and northern California. The FW is black with three evenly spaced white spots along the dorsal margin, the outer one opposed by a triangular mark on the costa near the end of the cell, and there is a terminal spot preceding the fringe. As in *E. cucullata*, specimens from Utah and California are appreciably larger than those from other areas (FW 4.2 mm vs. 3 mm). Adults have been collected primarily in June, but records in June and August at one locality in Arizona the same year suggest a possible second generation. **Elachista coniophora** Braun (Plate 6.1, Plate 6.2, Fig. 67) is perhaps the most frequently collected western species, although

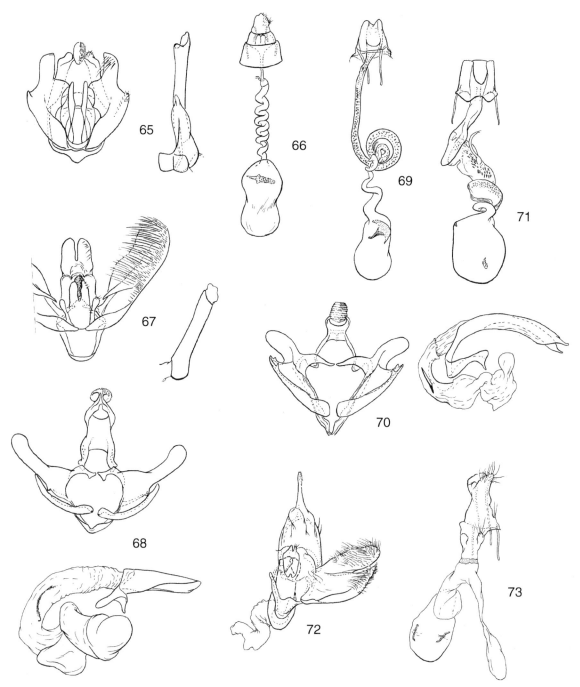

FIGURES 65–73. Genitalia structure of exemplar species of Gelechioidea, ventral aspect, except Fig. 73 lateral aspect. **65,** Elachistidae: *Perittia cygnodiella* (Busck) ♂, aedeagus to right. **66,** Elachistidae: *C. cygnodiella* ♀ [Braun 1948]. **67,** Elachistidae: *Elachista coniophora* Braun ♂, aedeagus to right [Braun 1948]. **68,** Coleophoridae: *Coleophora accordella* Walsingham ♂, aedeagus below. **69,** Coleophoridae: *Coleophora accordella* ♀. **70,** Coleophoridae: *Coleophora viscidiflorella* Walsingham ♂, aedeagus to right. **71,** Coleophoridae: *Coleophora viscidiflorella* ♀ [J.-F. Landry]. **72,** Momphidae: *Mompha franclemonti* Hodges ♂, aedeagus in situ. **73,** Momphidae: *Mompha franclemonti* ♀ [Hodges 1992].

restricted to low-elevation sites in California from the San Francisco Bay area and the Central Valley to San Diego and on the Channel Islands. The adults are often netted diurnally in serpentine grasslands in April and May, and we have seen them flying in large numbers at sundown in sand-based grasslands at Antioch on the San Joaquin delta and in western Fresno County. We have not observed *E. coniophora* attracted to lights. The FW is pale gray with a poorly defined whitish streak through the cell, expanded to cover the distal area, sprinkled with black dots. A similar species, **E. thelma** Kaila, occurs on the same serpentine grassland sites, as well as at several non-serpentine habitats in central California, and on San Clemente Island. Its FW is mostly dull white, sometimes with gray shading along the costa basally, and with short, black longitudinal marks distally on the Cu crease and at the end of the cell. The holotype was taken in late March, most of the others in April, and a specimen in Glenn County on July 1 at about 5,000' elevation.

The Argentella Group of Kaila includes many species that have white FW, usually with one or a few black dots. Several species are widespread in western North America, including the northwestern states and southern Canada. Larval hosts are unknown for any of the western species, but two members of this group in the eastern United States and others in the Palaearctic all mine in grasses (Poaceae). *Elachista controversa* Braun (Plate 6.3) is white, FW sometimes lightly dusted with gray, and has three short, weak, gray lines in the dorsal area near base and beyond in the cell. This species ranges through the mountains of California, from Mt. Shasta to the southern Sierra Nevada and in the White Mountains at 10,000′. *Elachista hololeuca* Braun usually is all white or cream tinged and is widespread in the Northwest, from British Columbia to Colorado, northern California, and northern Arizona. Both species fly primarily in June, but we collected *E. controversa* in late August in the Sierra Nevada, so there may be a second generation in May in some populations. *Elachista ossuaria* Kaila occurs from Dawson, Yukon, through Alberta, the Rocky Mountains to northern Arizona, flying in July. The scape, notum, and FW are pale yellowish white, base of the costa dark gray with a small dark gray spot basally in the cell (FW 5–6 mm). The FW of *E. ischnella* Kaila is speckled with gray and has two inconsistently developed lines on the Cu crease and end of the cell. It is widespread in the Southwest, from western Texas and southern Colorado, to the mountains of southern Arizona. The moths are active in April and May and both April and July in Arizona, so at least those populations may be double-brooded. *Elachista argillacea* Kaila was collected in both the Mojave Desert above 4,000′ (Ord Mountain) and the low Colorado Desert (near Palm Desert) in April. The FW (3.7–4.8 mm) is dull white with the basal one-third of the costa gray. The labial palpi are exceptionally short, a distinguishing feature for this species.

The Bifasciella Group of Kaila consists of about 40 described species, most of them similar in phenotype, FW black or gray with pale markings, usually a median transverse band and paired, preapical, and pretornal spots. A few species are pale gray or whitish, and many show considerable variation in hue of ground color, extent of pale spots, and sometimes a pale form. More than 30 species of these tiny moths (FW 3–5 mm) occur in our western area. We have reared two species in California, and larvae of at least eight others were recorded by Braun in the eastern United States, all miners of grasses (Poaceae); two Holarctic species have been reared from sedges (Cyperaceae) in Europe, so sedges need investigation in our region. *Elachista dagnirella* Kaila (Plate 6.4) is typical of the group, FW gray with a bronzy sheen, mottled by paler scale bases, forming a median band, the pale markings relatively narrow. This is a widespread, often-collected species, ranging from the Yukon, Alaska (Anchorage), and South Dakota to Colorado, Washington, and northern California. Despite the great latitudinal range, all collections have been in late June and July. Flight records for *E. stramineola* Braun are July and early August. This species is paler than most in the group, with mottled gray to dark gray FW having variable whitish maculation, often poorly defined, along the dorsal margin basally, fairly distinct transverse band and distal spots. It is widespread in the western United States, from Montana to Utah and northern Arizona, Nevada, and northeastern California. There are several species in this group from coastal California. *Elachista marachella* Kaila (Plate 6.5) is remarkable because it has adapted to an introduced African grass, which contradicts our usual assumption that leaf miners tend to be host specific and do not feed on exotic plants. During a several-year span, University of California entomologists reared *E. marachella*, including the type specimens, from mines in a veldtgrass, *Ehrharta erecta*, which has been naturalized in Strawberry Canyon on the University of California, Berkeley, campus since the 1950s. We also have found adults in relatively undisturbed habitats in Marin and Mendocino counties, where presumably *Elachista marachella* depends on native grasses. Field-collected adults are recorded in March, April, and May. The moths are black, FW with a complete submedial, transverse band and opposing preapical and pretornal white spots. The maculation is distinct but variable in extent. Similarly, *E. telcharella* Kaila (Plate 6.10) is highly variable in FW pattern, from pale to dark gray, scales pale basally, with median band and preapical pale maculation weakly developed, to black with basal half creamy white, except the costal area basally. This species is known from a few sites on Inverness Ridge, Marin County, and on California's northern coast at Inglenook Fen, Mendocino County, active March to May.

Elachista finarfinella Kaila (Plate 6.6) also occurs on the northern coast of California, having been collected by Lord Walsingham in June 1872, at Crescent City, and not again for 97 years, when we found numerous males 60 miles to the south at Arcata, in June. The FW median band is an outwardly directed chevron, in contrast to related species. Walsingham also collected the first *E. calegormella* Kaila "in redwoods nr. Navarro River, S. Mendocino Co." in May 1871, and we found it 90 years later, April 1961, and subsequently in Marin County. The FW is black with complete, narrow, transverse, ante- and postmedial lines. *Elachista indisella* Kaila (Plate 6.8) is known only from the Big Creek Reserve, Monterey County, where we have collected adults in May and June and reared them from mines in sweetgrass, *Hierochloe occidentalis*, which grows in the redwoods and ranges north along the coast to Washington. Larval collections made in March and April produced adults from late March to mid-May. The FW has a narrow, complete to broken and jagged median band, and the preapical spot is more distal, offset from the pretornal. By contrast, *E. gildorella* Kaila (Plate 6.9) has mottled gray FW with the maculation only weakly developed. It occurs in the San Francisco Bay area and was described from Albany Hill and San Bruno Mountain, city and county reserves, each isolated in a sea of urban housing for more than 50 years, and at Pt. Reyes National Seashore. *Elachista caranthirella* Kaila (Plate 6.11) is endemic to the Channel Islands so far as known; we collected the type series on Santa Catalina Island in 1968 and one specimen on Santa Rosa Island in 1997. The FW is nondescript mottled, dark gray with weakly defined pale median band and preapical spot. *Elachista excelsicola* Braun (Plate 6.12) is a nondescript, gray moth having poorly defined darker costal and tornal spots that form a weak transverse fascia. This is a Holarctic species, recorded in North America only in Alaska (Cantwell and Eagle Summit) and New Hampshire (Mt. Washington). It occurs widely in the Palaearctic in moist, northern conifer forests, bogs, and tundra, but the larval hosts are unknown.

Two sedge-feeding, holarctic species are among several *Elachista* that range across boreal North America: *E. alpinella* Stainton and *E. kilmunella* Stainton. The former is larger than most American *Elachista* (FW 4.5–5.2 mm). It is recorded from northern Manitoba (Churchill), Northwest Territories, and Yukon, and has been reared from several *Carex* species (Cyperaceae) in Europe. The FW variation exceeds that of other species, pale, mottled gray with poorly defined maculation to black with well-defined transverse band, preapical, and sometimes pretornal spots; *E. kilmunella* has a similar distribution. Cotton grass,

Eriophorum (Cyperaceae), is recorded as a host plant in Denmark. The FW is black with poorly defined maculation.

Superfamily Gelechioidea, Gelechiid Lineage

FAMILY DEOCLONIDAE

The genus *Deoclona* was described by Busck in 1903, and *D. yuccasella* Busck remains its only species. It was placed in the Gelechiidae, where it remained for nearly a century, even though in 1936 H. H. Keifer illustrated the adult, larva, and pupa in detail and called attention to several differences from any known gelechiid.

Adult The HW margin lacks the indentation of gelechiids and is more like that of depressariids; *D. yuccasella* has abdominal setal bands like a blastobasid, which the habitus of the moth generally resembles; the male genitalia are yponomeutid-like; and the pupa has peculiar features, pronotum with a pair of central, toothlike protuberances, abdominal segments 5 to 7 movable. The characteristics combined in parsimony-driven analysis place *Deoclona*, *Proclesis* in Argentina, and the Palaearctic *Syringopais* as a sister clade to the rest of the Gelechiid Lineage by Hodges, who defined the family.

Larva Larvae are stout, head brown, thoracic shield pale yellow, widely divided, integument unpigmented, without secondary setae; crochets tineid-like, stout, slightly biordinal in an elliptical oval.

Larval Foods Larvae eat dry seed of *Hesperoyucca* (Liliaceae sensu lato), create fruit mummies of Rosaceae, or are leaf miners of Poaceae.

Diversity There are four species in three genera, of which one occurs in western North America.

Deoclona yuccasella Busck is a small moth (FW 5.5–9.5 mm, females appreciably larger than males; Plate 5.43) having uniformly tan head, thorax, and FW; HW white. We have collected *D. yuccasella* adults from dry yucca inflorescences in October near San Diego, but they are rarely seen in the field, apparently are not attracted to lights, and may be diurnal. They fly after the yucca seed pods begin to dehisce late in the season. This species occurs on *Hesperoyucca whipplei* throughout most of its range in California from Monterey and Tulare counties southward. Fruit of *H. whipplei* is dehiscent, meaning the capsule breaks open, and the three carpels flare outward, enabling discharge of seed not eaten by the larvae of the pollinator, *Tegeticula* (Prodoxidae). Seeds infested and hollowed out by *T. maculata* are held in place by silk, trapping some viable seed at the base of the carpels. The open capsules remain on the dry inflorescence for several years, and *D. yuccasella* is a specialist on this remaining food resource. Collections of dry capsules made in summer, fall, or the following spring yield adults from May through summer and fall. Mating and oviposition take place in lab collections, so adults may emerge for two or three years. We do not have evidence of prolonged diapause.

FAMILY COLEOPHORIDAE

The cladistic hypotheses of Hodges and Kaila differ in interpretation of the Coleophoridae, particularly in placement of *Coelopoeta*, *Batrachedra*, and *Stathmopoda*. The *Batrachedra* group is included in an informal Coleophoridae by Kaila, as it had been traditionally (e.g., in Hodges's MONA Check List), whereas Hodges's 1999 analysis treated Batrachedridae as a family-level clade, sister to all the rest of the Gelechiid Lineage. *Coelopoeta*, which had traditionally been listed in Elachistidae,

is placed as a sister clade to the rest of the Coleophoridae by Kaila but was not mentioned by Hodges. *Stathmopoda* is a sister to *Coleophora* in Kaila's system but is retained as a subfamily of Oecophoridae sensu stricto by Hodges.

Adult Coleophoridae sensu Kaila are typically tiny to small (FW 3.5–13 mm), slender moths with lanceolate wings; the HW fringe much wider than the wing. *Coleophora* have rather long, nearly straight labial palpi, which project forward, often slightly drooping. *Batrachedra* have somewhat porrect labial palpi but with the third segment upturned, not curved as in most gelechioids. *Coleophora* have paired, elongate patches of spiniform setae situated longitudinally on abdominal terga 2 to 7; in *Batrachedra* the setae are reduced to minute spicules. These are mostly dull colored, yellowish, white, gray, or brownish moths, FW often with linear, pale or dark streaks. Genitalia, Figs. 68–71.

Larva Slender, with very reduced abdominal prolegs in *Coleophora*, the larvae are leaf miners in the first instar, then nearly always live in a portable case they construct, often of silk covered with frass, sand grains, or bits of plant material. They feed by mining outward as far as possible without leaving the affixed case, resulting in characteristic, round mines with a central hole, as they move from spot to spot. Larvae of *Batrachedra* are slender, well pigmented, free-living, with fully developed prolegs.

Larval Foods Most *Coleophora* species are specialists, as a group feeding on more than 30 plant families, including monocots, especially Juncaceae, and diverse dicots. *Batrachedra* are mostly phytophagous, feeding primarily in flowers and seed of monocots (Arecaceae, Cyperaceae, Juncaceae), but a few feed on Pinaceae or Salicaceae or are predators in leaf galls induced by sawflies (Tenthredinidae) or in scale insect colonies, even social spider webs.

Diversity There are more than 1,100 described species found worldwide, but they are mainly Holarctic *Coleophora*, which are lacking in tropical rain forests An estimated 500 to 1,000 unnamed species are known in the Nearctic. *Batrachedra* and stathmopodines are diverse in the Southern Hemisphere.

Coelopoeta glutinosi Walsingham is a small (FW 4.2–6.5 mm; Plate 5.44), somewhat broader winged moth than typical *Coleophora*, though the wings are still lanceolate, FW pale brownish gray to cream whitish, faintly mottled with pale rust; HW dark gray. The larvae feed as specialists on mountain balm (*Eriodictyon*, Hydrophyllaceae), mining the leaves and causing a gall-like swelling. *Coelopoeta glutinosi* ranges from northern to southern California, and in southern, especially desert areas, where the densely white-wooly *E. crassifolium* is the host plant, the moth is whitish. Larvae are active in May, adults primarily in June and July. A related species, *C. phaceliae* Kaila (Plate 5.45), mines foliage of *Phacelia californica* (Hydrophyllaceae) in the San Francisco Bay area (San Bruno Mountain) and *P. procera*, *P. hastata*, and *P. mutabilis* northeastern California. The moths are smaller than *C. glutinosi* (FW 4.5–6.0 mm) and vary from pale tan to brown, FW with a powdering of white scales, white dorsal margin, and triangular tornal patch. The larvae cause the leaves to curl, concealing the mines, which are found April to May at the coast, late May to early July in the mountains. Adults emerge May to July but have not been observed in the field.

Batrachedra and related genera are species rich in pantropical and temperate regions. The group is in need of rigorous phylogenetic study to ascertain relationships, but the 18 species described in America north of Mexico appear to be monophyletic, sharing possession of coremata and both dorsal and ventral condyles in the pupal abdomen. About a dozen species occur in the West, mostly in Arizona. ***Batrachedra praeangusta*** (Haworth)(FW 4.5–7 mm; Plate 6.15) is thought to be natively Holarctic in distribution and is widespread in the Nearctic, from Ontario and Illinois to British Columbia, Utah, California, and Arizona. Specimens were collected in Colorado before 1877, but not before the 1930s in other western states. The adults are sporadically attracted to lights and have been collected from trunks of aspen and poplars (*Populus*, Salicaceae) and reared from larvae feeding on the catkins and seed. The FW is dark gray to black with variable whitish clouding, and in Rocky Mountain populations there are two pale yellow patches along the Cu crease. Flight records in western North America are July to September. The similar but smaller ***B. striolata*** (Zeller)(FW 4–6 mm; Plate 6.13) has gray or mostly black FW, evenly speckled with darker and whitish scaling, and two brown spots along the Cu crease. This species is even more widespread and poorly documented than *B. praeangusta*, from Connecticut, Texas, Washington, and California. Most adults in collections have been reared from willow, feeding in galls caused by sawflies (Tenthredinidae), on the leaves *(Pontania)* or stems galls caused by *Euura*, and the moths are rarely are attracted to lights. *Batrachedra striolata* has an amazing ability to locate isolated stands of willow and colonize even sparse colonies of the host galls. The larvae feed on gall tissue, exuding frass from an entry hole, and probably destroy eggs or young larvae of the sawfly indiscriminately as they move from one gall to another. In coastal California the larvae of *B. striolata* are found almost anyplace one searches *Salix lasiolepis*, from the San Joaquin River delta to Los Angeles and on Catalina Island. The life cycle is multivoltine, adults having emerged from galls collected in March, May to October, and December. We reared ***B. illusor*** Hodges (FW 4–4.5 mm) from recently cut willow branches collected in March at the San Joaquin River delta. This is another widespread but poorly recorded species, recorded from Ohio, Illinois, New Mexico, and California. The FW is brown uniformly mottled with whitish, the scale bases pale. ***Batrachedra enormis*** Meyrick is a large species (FW 12–14 mm; Plate 6.14) having pale tan FW with dark brown dots at the middorsal margin and beyond the cell and a dark brown HW fringe. This species occurs in Mexico, western Texas, and the mountains of southern Arizona, where adults are recorded in April and August. Its larval habits are unknown.

The genus *Coleophora* is represented by about 150 described species in America north of Mexico, most of them from the Northeast, in large part due to the efforts of McDunnough in the Maritime Provinces. There are only about 30 named in the western states and provinces, but the *Coleophora* fauna as represented in collections is thought to be not more than 10% described. These moths are easily recognized by having porrect, somewhat decumbent labial palpi, in contrast to the arched palpi of most gelechioids, and by the paired patches of black or rust spiniform setae on abdominal terga, usually visible without removal of the scaling. *Coleophora* are so named because the larvae construct portable cases (from the Greek, *koleos* meaning "sheath" and *phor* meaning "carry"). The architecture of the larval cases varies considerably and has been used to support taxonomic species grouping as well as in recognition of species. The shelter is made primarily of silk, and nine general types are defined, depending upon whether it is covered with bits of leaf, debris, or sand; is parchmentlike; or is covered by cutouts from host-plant leaves or flowers. A larva may cut out the upper and lower epidermis of the mined portion to incorporate into the case, or it may use the cutout to

form a tubular case and abandon the smaller, earlier one, so the cutouts become increasingly larger. There is a valvelike aperture for frass ejection at the posterior end of the case, and it may be bivalve or trivalve. The architecture behavior is consistent within species, but similar architectures are not necessarily correlated with the proposed relationships based on genital characters. Larvae of a few Palaearctic species are borers and do not construct a portable case, but this habit has not been discovered in North American species.

Seed cases have been proposed as the ancestral type, although we do not have a rigorous phylogenetic study to confirm the hypothesis. Seed-feeding coleophorids feed within a single seed without a separate case until they have exhausted its contents, and therefore the size of the *Coleophora* larva and the size of the host-plant seed determine timing of constructing a case. Most species make a simple silken case from which they feed externally on seeds, whereas others create the case within a seed capsule. If the husk is large enough, there is no external modification other than feeding- and frass-ejection apertures, and such seed cases are extremely cryptic. If the capsule is too small, the case may project from it at the anal end. ***Coleophora mayrella*** (Hübner), the "clover case bearer" (FW 4.5–6.5 mm; Plate 6.16), is perhaps the most widespread and frequently collected coleophorid in North America. It is thought to have been introduced from Europe by early settlers and was recorded in the northeastern United States by 1860 (as *C. coruscipennella* Clemens, a synonym). The larvae feed in flower heads of clovers (*Trifolium*, Fabaceae), where they are difficult to see. This species now occurs throughout the northeastern United States, the Mississippi Valley, the Rocky Mountain states, across southern Canada, and abundantly in the Northwest. In California it ranges from the coast to the Central Valley, where it has been a pest to white and red clovers (as *C. spissicornis* Haworth, a synonym). In contrast to its univoltine life cycle in Europe, its North American records indicate *C. mayrella* is double-brooded in some populations. The adults are dark metallic, brassy green, with the basal half of the antenna markedly thickened by dense scaling, and the flagellum beyond has brown and white alternating bands. No doubt the colorful appearance compared to other coleophorids is in part responsible for its abundant collection records. ***Coleophora viridicuprella*** Walsingham is similar, pale brassy green with banded antennae that lack the thick scale brush basally. It is known only from scattered sites in eastern Washington to central coastal California. Walsingham found the cases affixed to a "grass or rush" and *C. viridicuprella* is inferred to feed on *Juncus*, based on its close relationship to ***C. maritella*** McDunnough (Plate 6.17) of the Rocky Mountains, larvae of which feed in the seeds of *Juncus*. *Coleophora maritella* is smaller (FW 3.2–4.3 mm), has unbanded antennae, olive brown FW, with the apical one-fourth of the flagellum white.

In the Pistol Case Group, silk is used to form a tubular section, turned downward, and enlarged into a broad basal section, like the silhouette of a pistol when viewed laterally. ***Coleophora malivorella*** Riley is a small species (FW 4 mm) with all-white FW, dark gray HW. This is the most widespread member of this group in the West. It was described originally from Pennsylvania, from apple *(Malus)*, and Kentucky (given three different names in 1878!) and was reared from *Salix* by Lord Walsingham in 1871 in California, where it occurs from Mt. Shasta to the San Francisco Bay area. The larger but quite similar *C. sacramenta* Heinrich has a large, expandable scale tuft on the antennal scape; FW (6–8 mm; Plate 6.18) white with variable brownish dusting distally, HW dark gray. The larvae construct similar cases, pistol shaped in outline. This species was described in 1914 as a pest of plum trees (*Prunus domestica*, Rosaceae) in California, where it may have been introduced. Subsequently, it has been reported from commercial stone fruits (prune, cherry, apricot) and apple in the San Francisco Bay area and on aspen and poplar (*Populus*, Salicaceae) and Tiliaceae in the eastern United States and Canada.

The Annulate Cases are elongate, very slender, attenuate tubes that are formed of a series of rings, like a mollusc shell. *Coleophora annulicola* Braun (Plate 6.19) was described from Logan, Utah, cases collected in June on *Aster* and *Solidago* (Asteraceae), adults emerging in July and August. The adults are dimorphic in color, very unusual for a coleophorid, males FW (6–8 mm) dark brownish, females pale grayish to whitish ochreous. Braun also observed the cases in Glacier National Park, Montana. A smaller, closely related, undescribed species, has whitish FW with fine longitudinal, ochreous lines and occurs on *Aster chilense* in the San Francisco Bay area, California. *Coleophora wyethiae* Walsingham (FW 7–9 mm; Plate 6.20) has white FW with faint linear streaks of pale ochreous, broader and more conspicuous on the Cu crease. It was discovered in July 1871 near the Pit River, California, on *Wyethia angustifolia* (Asteraceae), and Walsingham found it again in early April 1872 at the Rogue River in southern Oregon, so the species may be double-brooded.

Composite Leaf, or Lobe Cases, are created by the larva cutting out discs from the mined portions of leaves or flower parts and stacking them around itself, reinforced by silk. *Coleophora accordella* Walsingham is one of our most beautiful and easily recognizable species, having the FW (5.5–7.5 mm; Plate 6.21, Plate 6.22, Figs. 68, 69) yellow at the base, shading to ochreous or rust brown distally, with the costa broadly snow white, and there is a white streak along the Cu crease. The larvae feed on legumes (Fabaceae), *Lotus* in coastal California, and *Hedysarum* in Utah. The cases are constructed from pieces of flowers, so they are extremely cryptic while perched on the inflorescences. *Coleophora pruniella* Clemens (Plate 6.23, Plate 60.5), which was described originally from the larval case only on wild cherry (*Prunus*, Rosaceae) in Pennsylvania, is extremely widespread and is polyphagous. It is reported on poplar (*Populus*, Salicaceae), on crab apple (*Crataegus*, Rosaceae) in southeastern Canada, as a pest on apple trees *(Malus)* in the east and in California, and from birch (*Betula*, Betulaceae) and *Salix* in Utah. We have found *C. pruniella* on willow at several localities in California, where adults fly in June and July. The case is at first tubular, made of silk, in which the young larva overwinters; in spring it is attached to a larger, irregularly oval section formed by cutting out a portion of the mine, and the early section is discarded. The FW (5.5–7 mm) is dark gray.

The Sheath or Spatulate Leaf Case is constructed from a single leaf cutout strengthened by silk. It differs from other *Coleophora* mines in that the larva makes a new, larger case when it outgrows an earlier one, which it abandons. *Coleophora glaucella* Walsingham is drab, uniform pale gray to brownish gray, FW (5–6.5 mm; Plate 6.24) sometimes darker distally. The larva mines Ericaceae and makes a linear, flat case (Plate 60.6), which can be seen perched erectly at the end of a mine and later abandoned at the end of a linear hole where the upper and lower epidermis of the mined section have been cut out to cover the new case. The larval cases were discovered by Walsingham near Santa Rosa, California, on *Arctostaphylos glauca* (Ericaceae) in May, and adults emerged in July. We have found the mines on *A. manzanita* and *A. glandulosa*, on madrone *(Arbutus menziesii)* from northern and central California, and on *Arctostaphylos insularis* and *Comarostaphylis* on Santa Rosa and Santa Cruz islands. Larvae are active from March to May, adults July to August.

Several kinds of tubular cases are recognized, with or without external ornamentation. Tubular Leaf shelters are formed when the larva plasters leaflets or pieces of leaves onto a silken tube that gradually increases in diameter. *Coleophora irroratella* Walsingham (Plate 6.25) was described from Mt. Shasta, California, from adults only, and larval cases have been discovered subsequently on *Crataegus* (Rosaceae). The antennae are distinctly annulated with brown and white, and the FW is white with scattered dark brown scaling, more dense distally and concentrated into two elongate spots above the Cu crease. Adults fly in late July to August. Tubular Silk or "Cigar" cases often incorporate bits of frass or other debris or are plastered with sand grains, sometimes ribbed, having either a two- or three-flap closure of the opening for frass ejection. They are constructed by numerous species. Several western Asteraceae feeders make trivalve cases. One of the largest western species, *Coleophora lynosyridella* Walsingham (FW 11–11.5 mm; Plate 6.26), has pale ochreous FW with brown streaks, one through the cell that gives off oblique lines between the veins toward the costa and a broken line along the Cu crease. The cases occur on rabbit brush *(Chrysothamnus viscidiflorus)* and are pale tan, elongate (20 mm or more) cylindrical, smooth walled without incorporated debris, active in August. We have reared a closely related species from *Ericameria nauseosus* at Truckee in the Sierra Nevada from cases collected in late July. *Coleophora viscidiflorella* Walsingham is a similar but much smaller species (FW 6.5–7 mm; Figs. 70, 71) also on *Chrysothamnus viscidiflorus* at Mt. Shasta. Its case is elongate (13 mm), slender, and tapered, with the anterior end turned down. These two *Coleophora* names were applied to two as yet undescribed, tubular silk case species on coyote brush, *Baccharis pilularis* (Plate 60.7), in the San Francisco Bay area, by J. W. Tilden, who described their life histories in detail.

Two Eurasian *Coleophora* were introduced into the southwestern United States for biological control of Russian thistle (*Salsola tragus*, Asteraceae) and have become widely established. *Coleophora parthenica* Meyrick (Plate 6.28) differs from all known native species in its biology: Eggs are laid on foliage of young *Salsola*, and newly hatched larvae bore into the leaves and then the stems where they complete development and pupate. No larval case is constructed. There are at least two annual generations. The adults are white with the basal portion of the antennae thickly scaled. Field releases of moths reared from overwintering larvae were made at many sites in the San Joaquin Valley and southern California and in Nevada, Utah, and Idaho in 1973–1975, and by the late 1970s populations were established. In contrast, **C. klimeschiella** Toll (Plate 6.27) is a case bearer that feeds on the foliage of Russian thistle and was approved by the U.S. Department of Agriculture for release in 1978. The case is a straight, slender tube unornamented with leaf fragments. The moth is whitish, speckled with gray.

FAMILY MOMPHIDAE

Adult In North America, the family Momphidae is restricted primarily to the genus *Mompha*, which are tiny, stout moths with thick, diverging, upward-turned palpi, and usually upraised scale rows on the FW. Momphids lack the gnathos in the male genitalia, unusual for gelechioids. Genitalia, Figs. 72, 73.

Larva Larvae are stout, grublike, with fully functional but short thoracic legs, able to migrate from one mined site to another, feeding as leaf miners, within meristem of growing tips or stems, or in galls they induce. The L group on T1 has only two setae, in contrast to most microlepidoptera.

Larval Foods *Mompha* specialize on Onagraceae (70% of species for which larvae are known), the only Lepidopteran group to do so.

Diversity Momphidae are mainly Holarctic, with six genera and 60 described species, very few in tropical areas. There are about 40 named species in America north of Mexico, and at least that many more are unnamed in collections, especially in the West.

Mompha eloisella (Clemens) is larger than most momphids (FW 6.5–7.8 mm; Plate 6.29) and is a distinctive species, having white FW with bold, oblique, rust markings distally that converge into a V-shaped apex in a tail-like extension of the fringe, and there are two conspicuous black dots on the dorsal margin. The larvae tunnel in the flowering stalks of evening primroses (*Oenothera*, Onagraceae), without causing obvious dieback or other damage, and overwinter in the dry stems. This species is widespread, recorded in the northeastern United States and Midwest to Texas, Colorado, Oregon, and California, where it occurs in the Owens Valley and Central Valley, in association with *O. deltoides*. The life cycle appears to be multivoltine, as we have taken adults from early April to late August at Antioch on the San Joaquin River delta. **Mompha murtfeldtella** (Chambers)(Plate 6.30) also is widespread and may represent a complex of species. The adult is smaller than *M. eloisella* (FW 5–5.7 mm) with a white crown, thorax, and FW basally and two triangular patches on the dorsal margin that contrast with the gray or rust brown FW ground color. Each dorsal triangle has a weakly upraised scale tuft bordered inwardly by black. In California this species sometimes occurs with *M. eloisella*, such as at Antioch, where *M. murtfeldtella* is a tip borer in *O. deltoides* in sand dunes and in *O. elata hookeri* along the river edge. *Mompha murtfeldtella* appears to be multivoltine, as adults have emerged in March, April, July, and September from larvae taken in March, June, and August.

Mompha franclemonti Hodges (FW 3–4 mm; Plate 6.31, Plate 6.32, Figs. 72, 73) is a member of a complex of mostly undescribed species related to **M. metallifera** (Walsingham) of the eastern United States, black moths with silvery white markings. *Mompha franclemonti* is an elegant, diurnal, heliodinid-like species, metallic dark gray to black uniformly speckled with pale scales; there are incomplete golden silvery fasciae before the midwing and at the end of the cell, separated and followed by strong upraised scale tufts. This species is known only from the Big Creek Reserve in coastal Monterey County, collected in February, March, and May. We suspect the larval host to be *Epilobium canum* (Onagraceae), by its association in the field and because we reared a sister species, **M. powelli** Hodges, in the Oakland Hills, Alameda County, from *E. canum*. The latter species is smaller (FW 2.5–3.8 mm) with more distinct, chrome silver markings. There are several additional species of *Mompha* feeding on *E. canum* in coastal California and its offshore islands. Two of them cause stem and tip galls.

Annette Braun described **Mompha sexstrigella** from one specimen she reared from a leaf miner on the fireweed *Epilobium* (=*Chamaenerion*) *angustifolium* (Onagraceae) at Glacier National Park, Montana, in 1921. The species has brown FW (4 mm) with metallic leaden scales, basal one-fifth purplish, orange spots at one-fifth below the Cu and another at the tornus, and white streaks radiating before the apex. *Mompha sexstrigella* was later discovered in Russia, Finland, and Sweden, having been reared from the same plant in Finland. Recently, two closely related species were given cacophonous names by Koster and Harrison, **M. cleidarotrypa** from northern Arizona, and **M. achlynognoma**, which had been collected for several decades in California, Oregon, Washington, and Colorado and reared by Y.-F. Hsu from *E. brachycarpum* in the San Francisco Bay area. These species have slender, black and dark lead gray FW with ochreous markings along the dorsal margin and distally in the cell and beyond; *M. cleidarotrypa* is paler and has a row of four white spots along the costa.

Mompha cephalanthiella (Chambers) is peculiar both by its phenotype and host plant. The moths are tiny, FW (2.5–4 mm; Plate 6.33) dark brown fused with black basally, rust brown delicately patterned with black distally, and upraised scale tufts at middle and distally along the posterior margin. This species is widely distributed across the United States to California, where it occurs in the Central Valley and San Joaquin River delta in association with button willow, *Cephalanthus occidentalis* (Rubiaceae). The larvae make round, full-depth mines at the leaf margins and leave to pupate. Evidently *M. cephalanthiella* is double-brooded, as we have found abandoned mines on the deciduous host trees in June and have reared specimens from mines collected late July to October.

FAMILY SCYTHRIDIDAE

Adult These are tiny to small (FW 3–12 mm), slender moths, many of which are diurnal and visit flowers for nectar. They have narrow wings that wrap around the body in repose, rendering a tapered appearance from thorax to wing tips, comet shaped. Most are somber colored, gray or brown, with darker or white FW markings, the nocturnal species tending to be white or pale gray with larger eyes, while diurnal scythrids are mostly dark brown with small eyes. The male genitalia display an astonishing array of forms, from a relatively unmodified gelechioid plan in most nocturnal species to extremely reduced and modified by fusion, often asymmetrical and defying interpretation of homologies of the structures, mostly in diurnal forms. Moreover, there are many secondary sexual characteristics, often ochreous, brown, or black scaling of specialized form on the HW or abdomen of the females. Genitalia, Figs. 74–78.

Larva Slender, with a small head, tapered toward both ends, these larvae are usually gray or greenish without integumental markings but with sclerotized rings around setae SD1 on the abdomen. Head capsule with submental pit in most genera, at least in early instars, and both the setal rings and submental pit are shared with Blastobasidae. Most species live in frail webs and feed on growing tips of herbaceous plants; some are leaf miners in early instars.

Larval Foods At least 20 families of angiosperms are used, mostly dicots, with Cistaceae and Asteraceae prevalent in the Palaearctic; a few eat lichens, mosses, grasses, or cacti. There are too few rearing records for New World species to evaluate host specificity, with host records for any given species often based on only one collection or population.

Diversity There are more than 700 described species, 26 genera, with vast numbers awaiting study and naming. They

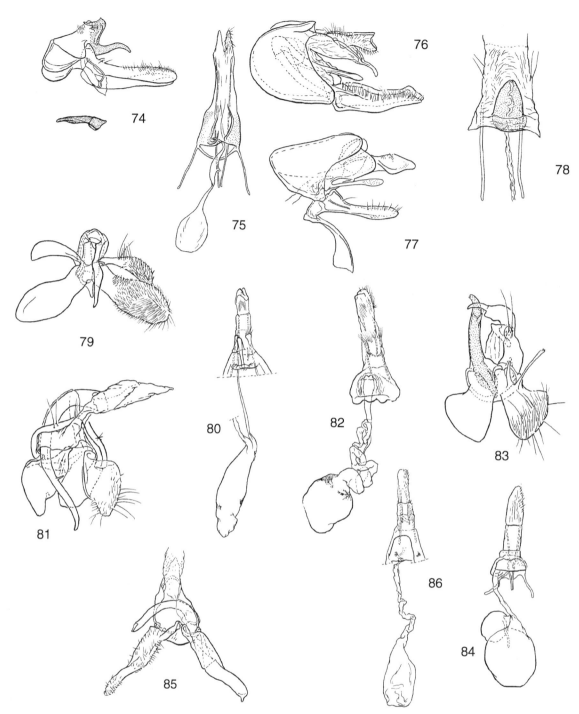

FIGURES 74–86. Genitalia structure of exemplar species of Gelechioidea, aedeagus in situ, except Fig. 74. **74,** Scythrididae: *Areniscythris brachypteris* Powell ♂, lateral aspect, aedeagus below. **75,** Scythrididae: *Areniscythris brachypteris* ♀ [Powell 1976]. **76,** Scythrididae: *Arotrura divaricata* (Braun) ♂, lateral aspect. **77,** Scythrididae: *Landryia matutella* (Clemens) ♂, lateral aspect. **78,** Scythrididae: *Landryia matutella* ♀, detail of sterigma [Landry 1991]. **79,** Cosmopterigidae: *Cosmopterix molybdina* Hodges ♂, ventral aspect. **80,** Cosmopterigidae: *Cosmopterix molybdina* ♀. **81,** Cosmopterigidae: *Periploca nigra* Hodges ♂, ventral aspect. **82,** Cosmopterigidae: *Periploca nigra* ♀. **83,** Cosmopterigidae: *Anoncia conia* (Walsingham) ♂, ventral aspect, aedeagus shaded. **84,** Cosmopterigidae: *Anoncia conia* ♀. **85,** Cosmopterigidae: *Walshia miscecolorella* (Chambers) ♂, ventral aspect. **86,** Cosmopterigidae: *Walshia miscecolorella* ♀ [Hodges 1978].

occur worldwide but are most numerous in arid regions and areas of seasonal drought, such as the southwestern Nearctic; rare or absent in wet tropical habitats. There are more than 40 described species in America north of Mexico, and J.-F. Landry reported having studied by dissections more than 130 undescribed species. Probably there are more than 300 species in collections awaiting study and naming. The western states and provinces harbor a wide diversity of scythridids, especially in desert and Mediterranean climatic areas, but there are montane and boreal species as well, occupying many habitat types, from active sand dunes or grasslands to chaparral, forests, and tundra. A sound phylogeny and framework for the systematics has been developed by Landry, opening the door to anyone interested in devoting a lifetime to the descriptive taxonomy.

The genus on *Arotrura* is primarily western Nearctic and is estimated by Landry to include 60 species, of which only 10 are described. The larval host is known for two undescribed species, one on *Amsinckia* (Boraginaceae) in coastal California, and a widespread species that feeds on greasewood, *Sarcobatus* (Chenopodiaceae). These are generally larger (FW 6–12 mm) with broader FW than most American scythridids. The genus was originally based on *Arotrura eburnea* Walsingham, described from specimens collected in Arizona by Morrison in 1882. It is a large (FW 9–12.5 mm; Plate 6.36), beautiful ivory white species. Females have abdominal terga 1 to 5 ochreous or pale brown. Adults are nocturnal and are taken at lights, flying in May to early July and September to October. It is rare in collections but is recorded from western Texas to California (Owens Valley). *Arotrura divaricata* (Braun)(Plate 6.34, Fig. 76) is widespread in the southwestern United States but was unrecognized for 60 years after it was described from Bahia de los Angeles in Baja California. It ranges through New Mexico, Arizona, Nevada, and California deserts and the Central Valley. There are at least two undescribed species with genitalia similar to *A. divaricata* on the California Channel Islands, one of which feeds on *Eriogonum giganteum* (Polygonaceae) on Santa Barbara Island. The adults are whitish or pale gray, FW (7–11 mm) with a peppered appearance caused by scales darker at the tips; typically there is a distinct, white, zigzag fascia on the basal one-third, which often is accentuated by dark scales on both sides. Female abdominal terga 1 and 2 are white, the others yellowish, progressively darker caudally. The superficially similar *A. longissima* Landry (Plate 6.37) differs markedly in genital morphology, with an enormously long aedeagus, more than two times the rest of the genitalia, almost as long as the abdomen. The FW (7–8.5 mm) is relatively slender, variably suffused with beige and olive brown, often forming an indistinct pattern of longitudinal streaks. The geographic distribution of *A. longissima* is similar to *A. divaricata*, but more southerly, from coastal southern California and the Channel Islands to western Texas, Baja California and its islands, and coastal mainland Mexico to Nayarit. On Santa Barbara Island, the larvae feed on *Lycium californicum* (Solanaceae), during and following the rain season (collections in December and May), and on San Clemente Island adults are active diurnally in early spring, nectaring at flowers of *Senecio lyonii* (Plate 60.1), but are nocturnal during the summer and fall, visiting flowers of *Baccharis pilularis* in September. *Arotrura longissima* has been collected throughout the year on these islands, so another food plant may be used during the dry season and in drought years when *Lycium* is dormant, completely lacking vegetative growth. *Arotrura sponsella* (Busck) differs in phenotype, having a predominantly dark gray brown patterned FW (8.5–9.5 mm; Plate 6.35), coalesced into oblique, jagged transverse striae on basal half and fine longitudinal lines distally, reminiscent of many blastobasids. The species occurs through the desert areas from southern California and northern Baja California to New Mexico. Adults fly in spring, March to May, and are nocturnal.

Rhamphura consists of more than 30 species, of which only four in the southwestern United States are described, and no doubt there are many more to be distinguished. *Rham-*

phura ochristriata (Walsingham)(Plate 6.39) is a member of a complex of closely similar species that differ mainly by the number and distribution of stout pegs on processes of the tegumen in the male genitalia, and until the variation is evaluated, the number and distinguishing characters of species cannot be assessed. The adults are small (FW 4.5–5.5 mm), chocolate brown, FW with an ochreous longitudinal streak in the cell. This species ranges from Mt. Shasta to central California. It was reared from alfalfa (*Medicago sativa*, Fabaceae) in the Sacramento Valley long ago, and at Antioch, on the San Joaquin River delta, larvae of *R. ochristriata* occur in dense webs on *Lupinus albifrons* (Fabaceae) in August and September. The moths are active from early April to October. A similar but larger species, *R. altisierrae* (Keifer)(FW 6.7–7 mm; Plate 6.38), is distinguished by its enormous valvae in the male genitalia. It is known only from the central Sierra Nevada at 8,800′ elevation and at Lebec in the Transverse Ranges of southern California, at 4,000′, by larval collections reared from *Solanum xantii* (Solanaceae), which we have examined at many other sites without finding this scythrid.

"Scythris" inspersella (Hübner) and *"S." noricella* (Zeller) are Holarctic sister species that form an unresolved trichotomy with *Rhamphura* and *Scythris* sensu stricto. Both occupy boreal situations ranging across North America, feeding on *Epilobium* (Onagraceae). In the West, *S. inspersella* (Plate 6.40) is known only from northern British Columbia and at Mt. Shasta, California, where we found larvae on *Epilobium* in June. It is smaller (FW 6–7.5 mm) and not as dark as *"S." noricella*; the scales are beige basally, more conspicuously so distally, and there are scattered white scales in some specimens. *"Scythris" noricella* (FW 8.5–10 mm) is recorded in Oregon, Colorado, Montana, and from many sites in Alberta, British Columbia, and the Yukon. The FW is brownish gray with a purplish hue, distal two-thirds moderately suffused with whitish scales.

The genus *Scythris* in the strict sense is a large group in the Palaearctic but is represented by just five described species in North America, in addition to the type species, *S. limbella* (Fabricius), which is introduced in the northeastern United States and occurs west to Indiana. The most widespread and often collected American *Scythris* is *S. trivinctella* (Zeller) (Plate 6.43), which ranges from New England to Florida through the Great Plains states, Texas, northern Mexico, Arizona, Utah, eastern Oregon, and southern British Columbia. The adults may be primarily diurnal because they are frequently seen at flowers during the daytime, but they also come to lights. They are dark brown, almost black, with three white or whitish, transverse fasciae, the first two sometimes linked in the cell. The larvae have been reared on *Amaranthus* in Kansas and an unidentified amaranth in Illinois. *Scythris immaculatella* (Chambers) is very widespread, from southern Quebec, the Appalachians, Colorado, Wyoming, and southern British Columbia to central Alaska. Excepting two Texas records, *S. immaculatella* is a boreal species. True to its name, the moth has unspotted FW, gray to beige with varying extent of olive or red brown below the R vein, and a brassy luster when fresh (Plate 6.42).

Despite the wide geographic range, the larval host is unknown; but a closely related species in Europe feeds on *Polystichum* ferns. The moths are diurnal and frequently visit flowers, flying from mid-May in coastal British Columbia to late July in the northern Rockies. *Scythris ypsilon* Braun (FW 4–5.5 mm; Plate 6.44) has strongly patterned FW, brown with a metallic golden luster with broad, distinct, cream-colored

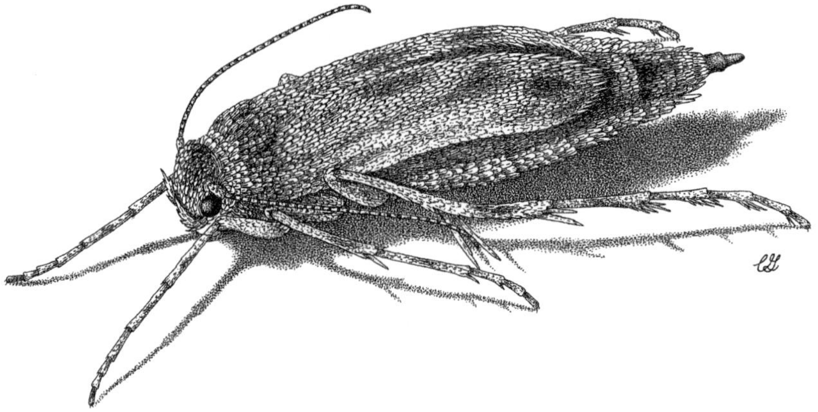

FIGURE 87. Habitus drawing of female *Areniscythris brachypteris* Powell (Scythrididae), an inhabitant of coastal sand dunes in California, the only continental lepidopteran that is flightless in both sexes [Celeste Green, Powell 1976].

bands, the basal and submedial two joined to form an inverted V, a distal band at end of the cell, and in most specimens a subapical cream spot. This is a common species in California, from the Sacramento and Owens Valleys to arid, inland areas of southern California. Spring adults (April to May) are dark olive brown and diurnal, whereas those taken in summer (July to September) are pale golden brown and come to lights as well as diurnally to late season flowering shrubs. *Scythris mixaula* Meyrick (Plate 6.41) is unique in feeding on cacti; several specimens from Wyoming, Arizona, and California are labeled as reared from undetermined cactus or *Opuntia*; feeding in flowers seems likely. This is a moderately large scythridid (FW 6–9.5 mm) with slender, lanceolate wings, color greatly variable, FW gray with white scaling forming diffuse lines along the veins, to primarily white with areas between the veins brownish ochreous, sometimes with a short, oblique brownish streak, lined with white anteriorly. The adults are among the more frequently collected scythridids, with numerous records in desert areas of southern California and Arizona, and along the east side of the Rockies from western Texas to eastern Montana, but oddly, are not recorded in the Great Basin.

Members of the genus **Landryia** (=*Asymmetrura*, a homonym) are mostly quite small (FW 4–6 mm), dark brown moths, diurnal, typically flying in early spring or inhabiting alpine habitats. They are characterized by having grotesquely asymmetrical male genitalia, the structures heavily sclerotized and partly fused (Figs. 77, 78), sometimes severely reduced, such that interpreting their homologies is difficult. There are more than 20 species known, most of them in the West, but only six are named. The few known larval hosts are Asteraceae, and one species in the eastern United States feeds on grasses. *Landryia impositella* (Zeller) was the first named North American scythridid, described in 1855 from New York, and it is one of the most widespread and frequently collected species, ranging from New England to the Gulf states, the northern Rocky Mountains, eastern Oregon and Washington, and southern British Columbia. The adults are brown to blackish brown, typically with a coppery or purplish luster, and usually two white or pale yellowish spots, one in the cell, one above the tornus, the latter sometimes extended as a transverse band. The adults are diurnal and active from mid-May in the south to early September in northern and montane sites and may be bivoltine in some populations. The moths visit flowers, especially Asteraceae, and the larvae create blotch mines in foliage of various *Aster* species. *Landryia*

reducta (Braun)(FW 4.5–5 mm; Plate 6.45), has dark brown FW marked by a variable, longitudinal white streak through the cell and often powdering of whitish scaling distally. This species is poorly documented but occurs in serpentine grasslands of the San Francisco Bay area and southward in the Central Valley, Owens Valley, and Mojave Desert of California. Adults are diurnal and visit flowers in May and June, mostly in habitats that are completely dry for several months following spring.

Neoscythris are small, mostly diurnal species, varying from pale gray desert forms to dark brown, similar in appearance to some *Landryia*. They have asymmetrical but very different genitalia from the latter and are distinguished by features of abdominal sternum 7. The female HW of some desert species have specialized, black scaling. There are four described and at least 20 unnamed species in the western United States, probably many more. The larval host plants are unknown. A complex of species related to **N. confinis** (Braun)(Plate 6.46) occurs in coastal California, from Monterey County northward. The adults are diurnal and nectar at available flowers, such as *Lomatium* in March and *Achillea millefolium* (Asteraceae) in June when most spring wild flowers have dried. Populations occur on serpentine grasslands in Marin, San Mateo, and Santa Clara counties. Available records indicate a single generation. The moths are small (FW 3.5–6 mm) dark brown with a thin ochreous line in the cell, similar to but reduced compared to *Rhamphura ochristriata*. *Neoscythris fissirostris* (Meyrick) is a primarily desert species, ranging from central Texas to the Colorado River in southern California. It is larger than congeners (FW 5.5–8 mm). Adults are nocturnal and come to lights but may shift from primarily diurnal in the spring generation to nocturnal in summer and September.

One of the most remarkably specialized moths in North America is the flightless scythridid, **Areniscythris brachypteris** Powell (Plate 60.2, Plate 60.3, Figs. 74, 75, 87), which lives on unstabilized, coastal sand dunes of central California. This is the only continental moth in the world that is flightless in both sexes, and other modifications include greatly enlarged hind legs that enable leaping 20 times or more the body length, and they bury themselves at night. The sand-colored FW are shorter than the abdomen, thickened, and leathery. The adults scuttle about on the open sand like miniature silverfish, their morphology and behavior evidently having adapted to the prevalent onshore winds. When disturbed they jump and are carried off several meters by the wind, and when winds are too strong the moth creates a pit by several quick movements of the hind legs, to crouch there temporarily. As temperatures decline late in the day, the fore legs are used to flick sand back and cover the moth (Plate 60.3). The sand-burrowing larvae of *Areniscythris* are highly modified, elongate and very slender, with short legs, lack the submental pit and setal rings typical of most scythridids. They feed by mining into buried leaves from silken tubes attached to various plants, including *Phacelia* (Hydrophyllaceae), *Monardella* (Lamiaceae), *Lupinus* (Fabaceae), and *Senecio* (Asteraceae). Numerous other as yet undescribed species of *Areniscythris* occur on isolated sand dune systems, along the coast (Pt. Mugu and Santa Rosa Island) and in interior dunes of the Mojave Desert, eastern Oregon, Columbia River, and southwestern Canada. These species are winged but seem to prefer to run on the sand. They are occasionally found at flowers well above the sand level or are taken in light traps. Each population has evolved coloration to match the sand of its habitat.

FAMILY COSMOPTERIGIDAE

This is a diverse group of tiny to small micromoths (FW 3–13 mm), not defined by any uniquely derived characteristic.

Adult These are small, slender moths with lanceolate or somewhat broad wings, lacking a gnathos and having a strongly hooked aedeagus in the male genitalia; Figs. 79–86.

Larva Morphologically, cosmopterigid larvae are most similar to Gelechiidae. Their habits are diverse; most are internal feeders, leaf miners, bud, stem, bark, or root borers, sometimes causing gall formation by the host plant. They tend to be stout with short legs, without secondary setae, and little integumental pigmentation.

Larval Foods More than 25 families of angiosperms are used. Typical cosmopterigines are mostly leaf miners, *Cosmopterix* often in monocots, and some are seed feeders. Others are scavengers, in and under old bark, dead stems etc., in ferns and palms, especially in the tropics. Chrysopeleines are angiosperm feeders, often as stem or bark miners. Larvae of *Euclemensia* are predaceous on scale insects.

Diversity Cosmopterigidae are worldwide, with 1,650 described species in more than 100 genera, richly represented on oceanic islands. The genus *Hyposmocoma* in Hawaii is the most famous example, with an estimated 450 species, mostly unnamed, that occupy diverse larval niches, including polyphagous on living plants; in dead wood or stems; on lichens; even predaceous on snails, Dan Rubinoff recently has discovered. Some feed from a portable case, and often they live in fresh water and littoral habitats. The radiation is analogous to the Hawaiian *Drosophila* (Diptera) and comparably species rich. In America north of Mexico, there are more than 175 described species of Cosmopterigidae, and the family is relatively well known, based on comprehensive taxonomic studies by R. W. Hodges in the 1960s and 1970s.

SUBFAMILY ANTEQUERINAE

The subfamily Antequerinae is represented in the West by one species, *Antequera acertella* (Busck)(Plate 6.47), a peculiar winter moth that flies primarily in November and December. The adults are gray to brown, (FW 7.5–9 mm), wings elongate, broad, superficially resembling some *Ethmia*, especially the fall and winter active species. Nothing is known about the larval biology; *A. acertella* occurs in coastal southern California and northern Baja California.

SUBFAMILY COSMOPTERIGINAE

Cosmopteriginae are diverse, represented in the West by nine genera and more than 60 species. *Cosmopterix* are small, gracillariid-like moths with extremely slender, broad-fringed wings, and all 30 North American species share a basically similar FW pattern: basal half black, ending abruptly at a white or silvery transverse line near the end of the cell that borders an orange or yellow band, and often the terminal area is black. The basal black area usually and sometimes the terminal area contain longitudinal silvery lines. *Cosmopterix molybdina* Hodges is a small species (FW 4–4.5 mm; Plate 7.1, Plate 7.2, Figs. 79, 80) distinguished by two broad, white bands on the antennae distally, and the FW basal lines are lead colored and fused, transverse fascia pale orange. The geographic range is poorly documented, known only from Maine, Oregon and northern California along the coast. This species has been reared in Oregon from larvae mining leaves of morning glory and moon vine (*Ipomoea*, Convolvulaceae), which are native to tropical America. Adults have been collected in April to June, and larval mines collected in June yielded adults the following January. *Cosmopterix montisella* Chambers is larger (FW 4.5–7 mm; Plate 7.3, Plate 7.4) and shows variation differing from other *Cosmopterix*. Populations in the Rocky Mountains and northern Arizona have the transverse fascia of the FW nearly undifferentiated, black or dark brown, to orange, defined by a broad silver band basally and two silver spots distally, with three separate blue white basal lines. The abdominal terga are ochreous orange. This species occurs from the northeastern United States through the Midwest, Colorado, and the southwestern states to Oregon. There may be two generations in California, with records in July up to 10,000 ft. elevation and September in San Diego County. Specimens taken in September and those from Santa Rosa Island are small. The larval food plant is unknown. Despite its name, *C. opulenta* Braun, is a small (FW 3.2–4.6 mm; Plate 7.5), pallid-colored species. The FW transverse band is yellow or pale orange yellow bordered basally by two silver spots, basal area with four white lines, the antemedial one longest, to three-quarters of the black area. *Cosmopterix opulenta* is a leaf miner in Asteraceae. Braun reared the original series from *Ambrosia psilostachya* collected in Los Angeles in October, adults emerging the following April; we reared it from *Artemisia douglasiana* in July, and adults have been collected in June in northern California, so there appear to be at least two annual generations. This species occurs from northern California to the Mojave Desert and ranges east to Oklahoma.

Eralea albalineella (Chambers)(Plate 7.6) is rare in collections, although it has been recorded from Florida, Texas, Arizona, and southern California. The wings are very slender as in *Cosmopterix*, FW (3.5–3.8 mm) dark gray with a conspicuous white streak along the anterior margin of the Cu crease and with three diffuse white bars from the costa obliquely outward into the cell. There are two flights in Florida, but adults have been taken only in June and July in the mountains of southern Arizona and at San Diego.

The genus *Stagmatophora* includes four North American species, three of which occur in the West, and larvae of all three cause gall-like growth in stems or roots of annual, woody mints of the genus *Trichostema* (Lamiaceae). The tiny but beautiful *S. iridella* (Busck) is the most widespread and frequently collected species. Its FW (3.5–6.5 mm; Plate 7.7) is shining yellow tan to bronze brown with raised, metallic golden spots and three white spots along the costa, the preapical one most prominent. The adults often come to lights and are active diurnally, especially near sundown and sunrise, flying over grassland where vinegar weed, *T. lanceolatum*, grows. The larvae cause irregular swellings in the root crowns of *T. lanceolatum*; our collections at the Big Creek Reserve, California, in September and October produced adults the following spring, and one larva held over in diapause to emerge in May the second year. Adults in the Big Creek population fly from late February to October. This species ranges from eastern Washington to southern California and western Texas. *Stagmatophora enchrysa* Hodges is larger (FW 6–8.5 mm; Plate 7.19) and has dark to rust brown FW with dark golden spots and a large preapical white mark, and lacks ocelli, which *S. iridella* has. This species is known from southern California, where the larvae cause irregular gall-like swellings in above ground stems of bluecurls, *T. lanatum*. Adults emerged in July and early August from stems we collected in late June in Santa Barbara County.

Larvae of *Pyroderces* are primarily detritivores of plant materials in various situations, usually decaying or dried flowers or fruit. The "pink scavenger caterpillar," *P. rileyi* (Walsingham), is widespread in the Neotropics and in eastern North America and occurs in Texas and southern Arizona. The smaller *P. badia* Hodges (Plate 6.48) is superficially similar, distinguished primarily on the basis of differences in genitalia. It ranges along the Atlantic and Gulf coasts and apparently was introduced to the Southwest, having been reported (as *P. rileyi*) of economic concern to seed production in sorghum and milo maize in Arizona by 1923 and in southeastern California in 1925. This *Pyroderces* became widespread in agricultural situations of southern California by 1942. Adults of *P. badia* have slender wings (FW 3.5–5.5 mm), which are rust colored with variable gray and whitish markings, the most consistent being a transverse dark line at basal one-fourth edged outwardly with white. The larvae are pink and have been found in all kinds of seed pods, fruit mummies, flowers of monocots such as coconut, and pine cones. *Pyroderces badia* began to appear regularly at lights in Berkeley in the early 1990s, where it has at least two generations, flying from April to November.

Limnaecia phragmitella Stainton (Plate 7.20) is a familiar insect in Europe and eastern North America as the seed-feeding caterpillar of cattails (*Typha*, Typhaceae), causing the flocculent down to hang in large, conspicuous masses. The adult is slender, tan, with longitudinal markings, like many marsh-inhabiting moths. The wings are attenuate (FW 6–10 mm, females larger than males), tan with a faint to distinct gray, longitudinal streak that encloses a dark spot in the cell and a more distinct one in subterminal area, highlighted by white; fringe pale gray; HW dark gray, fringe broad, whitish. The young larvae feed in the styles of pistillate flowers, moving inward as the cattail matures to feed on the seed. Partially grown larvae overwinter, and feeding resumes in spring. They spin abundant silk that ties up the pappus, preventing it from blowing free, so infested cattails are obvious. Pupation occurs in the fluffy pappus, and adults emerge in July and August. *Limnaecia phragmitella* has been cited as natively Holarctic, but it was not discovered in England until 1851 and in North America until 1899, and it is introduced in Australia and New Zealand. This species occurs in the northeastern United States and

southern Canada west to eastern Wyoming (1965), and eastern New Mexico, where it was first recorded in 2001, so the range may be expanding westward. Inexplicably, one specimen was taken in a light trap at Berkeley, California, in June 2000.

The remaining genera of Cosmopteriginae are broad-winged moths, superficially more similar to depressariines or blastobasids than they are to typical cosmopterigines. Several species of the small, gray, blastobasid-like *Teladoma* occur in southern Arizona, and *T. helianthi* Busck (Plate 7.8) is widespread in the East and Midwest, ranging to Arizona and southern California. The larvae make blister mines in leaves of sunflowers, *Helianthus*, and cocklebur, *Xanthium* (Asteraceae). Adults have been collected mostly in late June to early August, but in March in California. These are nondescript moths that are distinguishable from *Anoncia* species only by inconsistent differences in venation, presence of ocelli, and structure of the aedeagus, but the genera and species are easily differentiated by their genitalia. Species of *Triclonella* are distinctive by having ochreous orange FW with the distal one-third black, with only minor variation between the species. The genus is widespread in Mexico and Central America and is represented by eight species in the southern United States, of which two occur in Arizona. *Triclonella xuthocelis* Hodges is larger than the other U.S. species (FW 6.3–7.7 mm; Plate 7.9) and has paler ochreous FW, terminal dark area with a diffuse white spot followed by rust scaling distally. This species has been collected at lights from the end of June to October in the mountains of southern Arizona and northern Mexico. The immature stages are unknown, but the closely related *T. pergandeella* Busck ties together leaflets of legumes, *Lespedeza* and *Clitoria* (Fabaceae) in the eastern United States. There are two or more generations in *T. pergandeella*, and adults overwinter.

Anoncia is the largest genus of Cosmopterigidae in western North America, with about 30 described species, nearly half of them restricted to California. Probably there are numerous others not yet defined because the moths are mostly exasperatingly similar, gray, blastobasid-like, neglected by collectors, and misplaced in museums. The genitalia and modifications of the seventh and eighth abdominal segments vary markedly between species; the male genitalia are amazingly ornate and often grotesquely asymmetrical and should be examined for accurate identification. However, these structures vary considerably within species, so a single dissection may not closely match a published illustration. The larvae of the few species known are leaf miners or tip tiers that can move from leaf to leaf. Recorded species feed on Lamiaceae, Primulaceae, or Loasaceae. Four quite similar gray species are sympatric in central California. *Anoncia conia* (Walsingham) is small, (FW 6–8 mm; Plate 7.21, Figs. 83, 84), dark gray, with a variable paler area through the cell that includes diffuse darker lines. It occurs in California, from Siskiyou County to San Diego. The dark reddish brown larvae were found mining leaves of pitcher sage, *Lepechinia calycina* (Lamiaceae) in Marin County by Keifer, who described the larva and pupa in detail, and we have reared *A. conia* from coyote mint, *Monardella villosa*, at several sites in the Coast Range. Adults are active from May to August, varying with latitude and elevation. *Anoncia orites* (Walsingham) is superficially similar to *A. conia* but is distinguished by its more distinct black markings, especially an oblique fascia from the end of the cell to the tornus. The species ranges from Mt. Shasta to central California and Mohave County, Arizona, where it was collected in May. Adults fly in June to August in California. The similar but larger *A. sphacelina* (Keifer)(FW 7.5–9.5 mm; Plate 7.22) was discovered feeding on *Lepechinia* (=*Sphacele*) *calycina*, in the foothills of the Sierra Nevada, and we have reared it from *L. calycina* at Mt. Diablo in the Coast Range, whence it ranges south to San Diego. The adult is paler than *A. conia*, with a whitish area in the cell. The larvae form irregular shelters in the new spring growth in April and May, and adults have been collected from April to July. Also at Mt. Diablo, Y.-F. Hsu discovered larvae mining leaves of shooting stars, *Dodecatheon* (Primulaceae) from which another *Anoncia*, apparently a northern population of *A. noscres* Hodges, was reared. *Anoncia noscres* otherwise is known from the Mojave Desert northward along the arid inner Coast Range to San Benito County. This species has slightly narrower FW with a distinct blastobasid-like pattern, which is obscure in the Mt. Diablo specimens. Young larvae in May formed digitate mines outward from a basal blotch and later entire leaves were hollowed out. Species limits among this population, *A. noscres*, and its sister species, *A. loexyla* Hodges (Plate 7.18) in Arizona, need to be evaluated.

Anoncia flegax Hodges is distinctive for an *Anoncia*, pale gray, relatively large (FW 8–10 mm; Plate 7.23) with extensive whitish FW scaling broken by a distinct, broad, black transverse band at one-fourth, two distinct black lines longitudinally beyond the band, and dark gray lines between the veins distally. *Anoncia flegax* occurs at several localities from Mt. Shasta to Mt. Palomar in southern California, May to July, but the immature stages remain unknown. The smaller *A. mosa* Hodges (FW 6–7.7 mm; Plate 7.24) is superficially similar to *A. flegax*, but is darker gray, FW with the similar but even bolder markings. However, *A. mosa* is related to other species by genital structures. *Anoncia mosa* is known from California's Coast Ranges, southern Sierra Nevada, and central Arizona. Adults have been collected at lights from April to early June and in July at Mt. Palomar. The host plant is unknown. *Anoncia leucoritis* (Meyrick) is distinct in having a white FW unmarked or with faint brownish spots at the end of the cell (FW 7–8.8 mm; Plate 7.25). It ranges from southeastern Washington to California, east to Big Bend in western Texas. The larvae feed on the immature ovaries of blazing star, *Mentzelia laevicaulis* (Loasaceae), having been reared in eastern Washington by Clarke and in the Sierra Nevada foothills by Keifer. The largest species in the genus, *A. longa* (Meyrick)(FW 9–13 mm; Plate 7.26), is one of the most widespread, ranging from southeastern Colorado and Utah to southern Arizona and western Texas. The FW varies from white to gray, showing two faint darker spots in and just beyond the end of the cell. Adults have been collected June to early October, but its larval habits are unknown.

SUBFAMILY CHRYSOPELEIINAE

The Chrysopeleiinae contains very slender to moderately stout, tiny to small moths with narrow wings. They are often smooth scaled, shiny, metallic in appearance (e.g., *Periploca*, *Ithome*), while *Walshia*, *Stilbosis*, *Sorhagenia*, and others have strongly upraised FW scale tufts. The labial palpi are upwardly arched but usually short and stout compared to most cosmopterigines. The larvae feed on living plants, mostly as leaf or bark miners, borers in cones, seeds, or stem/root galls they or other organisms cause. The group is worldwide in distribution except in Oceania and is well represented in the Neotropics by a largely undescribed fauna. There are about 90 named species in

America north of Mexico, in 13 genera, and about half of them occur in the West.

Periploca is the dominant western genus, with nearly all of its 27 named species occurring here. These moths are tiny to small, nocturnal, with smooth, shiny, dark bronzy to purplish or black FW that curl over the body in repose. Most species are associated with junipers (Cupressaceae), but larvae of *P. ceanothiella* (Cosens) induce stem galls on *Ceanothus* (Rhamnaceae). This species is relatively large (FW 5–6.2 mm; Plate 7.10), FW broad, HW dark gray including fringe. *Periploca ceanothiella* is transcontinental, occurring from the Northeast to South Dakota, northern Arizona, southern Oregon to southern California. Adults are active May to July. *Periploca atrata* Hodges is similar in size (FW 4–7 mm; Plate 7.11) and color when fresh, but worn specimens become shining pale gray, and the HW is paler than that of *P. ceanothiella*. The larvae feed in cones of *Juniperus californicus*, and adults emerge in April and May via a valve that opens as the moth moves up a silken track, expanding it. This species occurs in the mountains of Arizona, deserts, and inner Coast Ranges of central California. The "juniper twig girdler," *P. nigra* Hodges, is tiny (FW 3–4.5 mm; Fig. 81, 82), shining gray black; HW pale gray, fringe whitish. It occurs in the eastern United States west to Arkansas and in California, where the type locality is Sacramento. Known populations occur only on ornamental junipers in urban situations around Sacramento and the San Francisco Bay area, but *P. nigra* probably is not native in California. The larvae feed under bark of several species of junipers and sometimes cause appreciable dieback as stems are girdled. The biology in California was studied by Koehler and Tauber, who concluded the species is univoltine, with adults emerging mid-May to July. Our records at light in Berkeley indicate a June through August flight, most frequent in August, with occasional records in April and September to October. Another shining black species, *P. juniperi* Hodges, is distinguishable by its genitalia. The adults are small (FW 3.8–4.8), HW pale gray, fringe whitish, and terga of basal abdominal segments pale ochreous. The type series from eastern Wyoming was reared in July from galls caused by *Gymnosporangium* on juniper, and from unspecified feeding site on *J. californicus* at Jacumba, California, near the Mexican border, and we have collected the adults from *J. occidentalis* in the Warner Mountains in northeastern California in June. Also quite similar, *P. mimula* Hodges is dark gray black, with pale HW (FW 3.3–5 mm). The larva forms a cell in the cambium layer of juniper, evidenced by sap and frass exudation, but does not girdle the branch. It has been reared in Tennessee, where there are two generations, with emergences in June and September. This species ranges through Arkansas, Texas, New Mexico, Arizona, and north to Washington.

Stilbosis also are tiny moths, distinguishable by having FW upraised scale tufts, distributed below the Cu crease at one-fourth, on the dorsal margin, above the crease at half and at its end, plus smaller patches along the terminal margin. Some species are nearly uniform dark gray to black and are leaf and stem miners on Betulaceae, Fagaceae, or Fabaceae. There are about 35 species described from North and Central America and northern South America, of which 20 occur in the United States, most of them western, especially Arizona. Several are known from just one locality and/or from one sex, some of which may be mates. *Stilbosis dulcedo* (Hodges)(Plate 7.12) mines leaves of live oaks (*Quercus agrifolia*, *Q. wislizenii*, and *Q. chrysolepis*) in California from Mendocino to San Diego

County. The mine is a full depth, silk-lined, blotch with a hole through which the larva passes the frass and attaches it over the mine surface. Larvae overwinter in the mines, and lab emergences occur from April to July, judged to represent a single generation. The FW is dark gray (2.5–3.5 mm). Also in California, *S. extensa* (Braun)(Plate 7.13) mines the preceding season's green stems of deer weed, *Lotus scoparius* (Fabaceae). Active mines are found in March, producing adults in April and May. The moth is dark gray with diffuse paler gray highlighting the black scale tufts (FW 3.2–3.8 mm). It is known from Antioch on the San Joaquin River delta to coastal areas of central and southern California, including the Channel Islands. *Stilbosis nubila* Hodges is a larger species (FW 4.5–5.5 mm; Plate 7.14) having dark metallic bronzy gray FW with four upraised, dark scale tufts, each surrounded by ochreous tan scaling, faint to well developed and coalescing to form a transverse band at midcell. This species ranges from Utah through Arizona into northern Mexico and has been collected more often than other *Stilbosis*, but the early stages are unknown. At Oak Creek Canyon, *S. nubila* flies in July, but in southern Arizona in August to early October, after the summer rains begin.

Walshia miscecolorella (Chambers) is perhaps the most easily recognizable cosmopterigid in western North America, because it is relatively large (FW 6–9 mm; Plate 7.27, Plate 7.28, Figs. 85, 86). It has pronounced erect FW scale tufts that emphasize the color pattern, a basal brownish black patch ending at an outwardly oblique line at one-fourth, followed by pale to dark ochreous, variably mottled with brownish rust that encloses two large scale tufts, a tan one in the cell and black one at end of the cell. There is considerable variation in hues between and within populations; rarely specimens have almost no darker mottling on the pale distal three-fourths, ranging to darkly mottled with rust and brown. The population in the Owens Valley, California, has all bright rust markings including the basal patch. The larvae are root, bark, and stem borers on various woody legumes (Fabaceae), including lupines, locoweed *(Astragalus)*, and sweetclover *(Melilotus)*. The biology was studied in Nebraska, where there are two generations, larvae overwintering. Collections of adults in coastal California suggest a multivoltine life cycle, with records from April to October, although primarily in April and May and August to October at coastal sites and the Channel Islands. This species ranges from the East Coast to British Columbia and southern California. In the East, specimens average smaller and are easily confused with several similar species, but *Walshia* is represented by just *W. miscecolorella* in the West.

Perimede is a Nearctic genus with 10 species described north of Mexico, of which three occur in western Texas and Arizona. The adults are superficially similar to *Stilbosis* but lack upraised scale patches on the FW. The type species, *P. erransella* Chambers, which is widespread in the eastern United States, is polyphagous, having been reared from foliage of Taxodiaceae, Ulmaceae, Fagaceae, and Juglandaceae, but the immature stages are unknown for the rest of the species. *Perimede battis* Hodges is a small species (FW 3.5–5 mm; Plate 7.15) that occurs in Arizona, recorded at Pine in Gila County, Madera Canyon in the Santa Rita Mountains, and Pena Blanca, Santa Cruz County, flying late July and August. The FW is dark gray when fresh, with poorly defined maculation, a black mark along the Cu crease basally and whitish spots in the cell and preapically; becoming pale gray when worn.

Sorhagenia is a Holarctic genus with three species known in the western United States, two of them from just one or two specimens. The FW (4–5.5 mm) of *S. nimbosa* (Braun)(Plate 7.17) is dark gray with upraised black scale tufts, one below the Cu crease at basal one-third, a smaller one below midcosta, and one at end of the cell, and an oval whitish spot in the cell distally, preceded and followed by whitish powdering caused by white tipped scales. The conspicuous and highly characteristic leaf galls (Plate 60.8) caused by larvae of *S. nimbosa* are familiar on coffeeberry, *Rhamnus californicus* and *R. purshianus* (Rhamnaceae). Larval feeding stimulates abnormal growth of the central portion of the leaf, which closes and bulges between the veins, leaving the ends flared. Silk is not employed to develop the shelter and feeding chamber. Full-sized galls are present in spring, but larvae are too young to complete development if collected before mid-May or June. They leave the galls to form cocoons in leaf litter, and adults have emerged in confinement in June and July and have been taken at lights June to early September. Three European species of *Sorhagenia* are recorded as leaf miners, leaf tiers, or flower feeders on *Rhamnus*. *Sorhagenia nimbosa* occurs from northern Idaho and northwestern Washington to southern California.

The genus *Ithome* is New World in distribution, with most of its species in Central and South America and the Caribbean. There are seven species described in America north of Mexico, the most widespread of which, *I. concolorella* (Chambers), ranges from the midwestern states to Texas and Arizona. This species resembles a tiny *Periploca*, FW (2.5–4.3 mm; Plate 7.16) shining black, worn specimens paler than fresh ones. Several *Ithome* feed on flowers of legumes (Fabaceae) and one on Polygonaceae, and *I. concolor* has been reared from mesquite *(Prosopis)* in Arizona.

FAMILY GELECHIIDAE

One of the major families of micromoths, especially in temperate latitudes, Gelechiidae is defined by the structure of the gnathos in the male genitalia, a pair of articulated sclerites with an articulated mesal hook, but the adults are almost always most easily recognized by the HW shape, with the terminal margin excavated below the acute apex (Plate 7–Plate 10).

Adult These moths are tiny to small (FW 3–12 mm, a few tropical species to 18 mm). Most are nocturnal, somber colored, brown, gray or black, but some are colorfully patterned, especially in tropical regions. Genitalia, Figs. 88–105.

Larva Larvae are concealed feeders, usually leaf tiers, forming shelters in new growing tips of trees and shrubs, but many are leaf miners at least in early instars, or stem, root, or seed borers, and a few live in plant galls they induce. Some feed in dead plant materials.

Larval Foods Most gelechiids are specialists, dependent on one or a few closely related plants. They feed on a broad diversity of gymnosperms and angiosperms, more than 80 families. Some live on ferns or mosses, especially on oceanic islands.

Diversity There are more than 4,500 named species of Gelechiidae placed in more than 500 genera, and unknown numbers of undescribed species. Gelechiids are most diverse in temperate zone areas, including deserts and other seasonally arid habitats, abundant but less diverse than stenomatines and xyloryctids in tropical regions. Several are important agricultural or stored products insects, including the "pink bollworm" *(Pectinophora gossyipiella)*, a threat to cotton growers worldwide; "Angoumois grain moth" *(Sitotroga ceralella),* which feeds in stored grains; "potato tuber moth" *(Phthorimaea operculella);* "peach twig borer" *(Anarsia lineatella);* "lodgepole needleminer" *(Coleotechnites milleri);* and other conifer-needle miners.

Even the North American fauna is too incompletely known to enable an accurate estimate of the diversity. There were 630 species in the 1983 MONA Check List (literature through about 1980), and about 220 have been described subsequently, representing more than half of the Nearctic species in those genera. This dramatic increase in our knowledge during the past quarter century has been primarily through the monographic treatments of Dichomeridinae and the genus *Chionodes* by Ron Hodges and extensive descriptive taxonomy of western Gnorimoschemini by the late Dalibor Povolný. However, rich diversity remains unnamed in other genera (e.g., *Aristotelia, Filatima, Aroga),* based on what we see in collections, and such groups may rival *Chionodes* in the proportion new (60% in 1999). Therefore we estimate not more than 60% of the North American species of Gelechiidae, in collections, are described, which yields an estimated 1,400 or more species, if gelechiid specialists were to get busy and describe all the available species. Unfortunately, there are no American or Canadian gelechiid specialists doing descriptive taxonomy. The family currently includes more than 100 genera in North America, of which 10 are represented by species introduced from other parts of the world.

In Hodges's recent classification (in *Handbook of Zoology),* the world fauna is assigned to five subfamilies, three of which occur in North America—Gelechiinae, Dichomeridinae, and Pexicopiinae—the last represented only by introduced species. Some former groups treated as subfamiles have been subsumed entirely or in part into the Gelechiinae, which now includes the vast majority of Nearctic genera and species. Members of the North American subfamiles possess a female retinaculum consisting of anteriorly directed scales from vein R basally.

SUBAMILY GELECHIINAE

This is the most diverse lineage of the family in North America, with about 90% of the described species, assigned to more than 80 genera, for which there has been no phylogenetic analysis. The subfamily has been divided into four tribes but without precise definition.

TRIBE TELEIODINI

The tribe Teleiodini consists of mostly small gelechiids with narrow FW, often brightly colored, and usually with upraised scale tufts or rows, which rarely are developed in Gelechiini.

The genus **Nealyda** includes tiny (FW 2.8–4.0 mm), broad winged moths characterized by a deep invagination in the distal margin of the HW. **Nealyda bifidella** Dietz (Plate 7.29, Plate 7.30) was described from Glenwood Springs, Colorado. The FW is golden brown with a perpendicular, transverse band of dark brown at 0.4, edged outwardly by a silvery white line and a faint whitish, subterminal line crossing an area of diffuse darker scaling. Y.-F. Hsu reared *Nealyda* specimens matching the description of *N. bifidella* at two sites in New Mexico (Eddy and Guadalupe counties). As noted by Dietz for the type series, the relative intensity of hues varies considerably in these populations. The larvae mine in leaves of *Mirabilis multiflora* (Nyctaginaceae). Specimens with much darker, almost black ground

color were reared by Hsu in coastal San Diego County, California, from *M. californica,* as were similar, smaller *Nealyda* in southern Baja California from *Commicarpus* (Nyctaginaceae). Other *Nealyda* species have been reared from *Pisonia* (Nyctaginaceae) in Florida and Cuba and from *Phytolacca* (Phytolaccaceae) in Florida. Hence, this is one of the few microlepidoptera taxa to specialize on Caryophyllales, sharing the niche with Heliodinidae.

Metzneria lappella (Linnaeus)(Plate 7.31) is a Palaearctic species that was introduced into North America by 1898 and became widespread in the East and Midwest. In June and July 1958 we collected a series at Mt. Shasta City, an unlikely point of introduction, but the species seems not to have been recorded elsewhere in California. We have a recent record (2001) at Santa Fe, New Mexico, but the occurrence of *M. lappella* in western North America remains poorly documented. This species has long, thick but smooth-scaled labial palpi, FW narrow, apex acute, tan with linear brownish streaks, more prevalent on the costal half, a short, dark brown bar on the Cu crease at 0.4 and a dot at the end of the cell. The Mt. Shasta specimens are smaller (FW 6.6–7 mm) than eastern U.S. examples, which range to 8.5 mm in females. The larvae feed in seed heads of burdock, *Arctium lapella* (Asteraceae), and larvae overwinter in the seed heads. *Arctium* has been recorded at scattered, mostly low-elevation places in California as an exotic weed.

Isophrictis is represented by about 20 described species in North America, at least half of which occur in the West. These are narrow-winged moths, FW tan, pale brown, or gray with delicate white lines distally. The head scaling is erect, and the labial palpi have large, spreading scale brushes on the second segment. Species for which larvae are known all feed on Asteraceae. *Isophrictis magnella* (Busck) has pale to moderately dark olive brown FW (6.5–7.75 mm; Plate 7.32), posterior half distally pale straw colored, enclosing three short, black lines, on the crease, in the cell, and beyond the cell, each bordered with white, and there are white lines in the terminal area. It occurs in Colorado, and *I. magnella* or a closely similar species is widespread at low elevations in coastal California, from Monterey southward, including Santa Cruz Island. We have taken the adults from *Hazardia squarrosa* (Asteraceae) at the Big Creek Reserve, Monterey County, in August to early October and reared them from inflorescences of this shrub. *Isophrictis occidentalis* Braun has dull ochreous FW (7.5–8 mm) marked with streaks of white-tipped scales, including a border of white from the costa to the middorsal margin, and there are three longitudinal lines in and below the cell. This species was described from Utah, where the adults were found flying among *Rudbeckia* (Asteraceae), a likely larval host because at least two other *Isophrictis* are recorded feeding in the heads of *Rudbeckia* in the eastern United States. We have reared an *Isophrictis* having pale rust-colored FW (4.5–7.5 mm) from larvae overwintering in stems of *Helianthus annuus* in the San Joaquin Valley, California. The smaller *I. modesta* (Walsingham) was described from Los Angeles and lacks a distinctive FW pattern.

The genus **Monochroa** includes two western species that feed on ferns, a rare adaptation in Lepidoptera. **Monochroa harrisonella** (Busck) ranges from British Columbia to southern California. The moths are slender, *Aristotelia*-like in form, FW (7.5–8.5 mm; Plate 7.33) tan with faint dark brown markings, a dot at the end of the cell. The second and third abdominal terga are covered with modified, ochreous scaling. Adults come to lights and often are found in association with bracken, *Pteridium aquilinum* (Dennstaedtiaceae); and A. Gilbert and T. Eichlin reared *M. harrisonella* from larvae feeding in the rhizomes of bracken in the Sierra Nevada. Larvae collected in September produced adults the following May. Flight records are the end of March to May at the coast and June to July in the mountains. *Monochroa placidella* (Zeller)(Plate 7.34) occupies the same Pacific Coast distribution and the same host fern, but the larvae cause gall-like deformation on the fronds. They feed in the bases of the pinnae, causing the lateral stems to enlarge and curl upward, creating a leafy shelter that does not discolor while the larva is feeding. Later, after the shelter is abandoned, the foliage of the affected area turns reddish. Adults of *M. placidella* are smaller than those of *M. harrisonella* (FW 6–7 mm) and have uniform, bright egg-yolk yellow FW and contrastingly dark gray HW. They are easily detected among bracken during the daytime and come to lights infrequently. We found a similar but smaller species with rust brown FW in the same larval niche on Santa Rosa Island in March and the beginning of May. An unidentified *Monochroa* occurs at Antioch, California, where the larvae bore into main stems and rhizomes of *Polygonum punctatum* (Polygonaceae). Larvae collected in March and early April, produced adults in May and June.

Chrysoesthia lingulacella (Clemens)(Plate 7.35) is a tiny heliodinid-like moth having bright orange FW (3.4–4.6 mm) with a median fascia and irregular, metallic lead-colored spots. The larvae mine leaves of *Atriplex* and *Chenopodium* (Chenopodiaceae). In early literature there was confusion between *Chrysoesthia lingulacella* and *C. drurella* (Fabricius), a European insect believed to have been introduced to eastern North America by the 1870s, and the Old World, *Chrysopora hermannella* (Fabricius). By 1903 Busck believed all the records in the eastern United States referred to *Chrysoesthia lingulacella*, which he assumed to be a native species. In the West, it acts like an immigrant, periodically colonizing weedy places and then dying out. *Chrysoesthia lingulacella* was reported in British Columbia and Washington in the 1920s and 1930s. Colonies have been recorded in California at various times at Berkeley, Palo Alto, San Jose, Davis, and the San Joaquin delta, but the historical and current extent of its distribution in the western United States are poorly documented. Adults have been collected June to October, probably multivoltine.

Aristotelia is a moderately large genus having 35 named species in America north of Mexico and unknown numbers of undescribed species in collections, a group begging for comprehensive taxonomic study. These are distinctive moths, nearly all having a similar, ornate FW pattern consisting of brown to reddish or black ground with a series of parallel pale, often silvery streaks obliquely from the costa and opposite spots along the dorsal margin. The posterior half usually is reddish, pink, lavender, or yellow. The well-known species are food-plant specialists, and diverse hosts are used—Salicaceae, Rosaceae, Fagaceae, Fabaceae, Solanaceae, Asteraceae—typically as foliage feeders, but one eastern species is a leaf miner on *Scirpus* (Cyperaceae). *Aristotelia argentifera* Busck (FW 5–7.5 mm; Plate 7.36) has metallic silvery, slightly raised FW markings on a brown ground. It occurs along the coast in California, the northern Channel Islands, southern California, inland east of the Sierra Nevada (Truckee, Mono Lake), and western Nevada. The larval food plants are Asteraceae, primarily *Ericameria ericoides* on coastal dunes, and coyote brush, *Baccharis pilularis*, inland. There is a pale race in the Panamint Range of the Mojave Desert feeding on *B. sergiloides*. Young larvae burrow into unopened buds and later

form conspicuous webbing at the terminals. This species appears to be double-brooded, with collection records mostly in May to June and September to October, especially the latter, but there are a few July and August records, and larvae can be found almost any time of year at the coast. They sometimes become incredibly abundant in fall, covering branches of *Ericameria* with dense webbing. Keifer found larvae in San Francisco heavily pattered with dark brown, whereas inland *A. argentifera* has pale green larvae without integumental pattern. However, J. W. Tilden reported both color phases in populations at Stanford, and larvae sometimes developed the brown form in the last instar. *Aristotelia adenostomae* Keifer (FW 4.3–6 mm; Plate 7.37) is a dark species, FW dark gray to black, dorsal half tinged with dark reddish. This species occurs in California from the Coast Range foothills around San Francisco Bay and coastal areas to Los Angeles County and Channel Islands. The larvae feed on chamise (*Adenostoma fasciculatum*, Rosaceae) in spring. *Aristotelia adceanotha* Keifer (Plate 7.38) is unusual, feeding on the hard-leafed *Ceanothus cuneatus* in the central Sierra Nevada foothills, and has not been found on the soft-leaved or deciduous *Ceanothus* species used by caterpillars of many other Lepidoptera. Adults of *A. adceanotha* are small (FW 4.5–5 mm) and have the FW basal area blackish brown on the costal half, bright brown on the dorsal area, where the fasciae become orange rose. Keifer found larvae in June, long after spring growth had hardened, and adults emerged in July.

Aristotelia elegantella (Chambers) is a distinctive species, unusual in FW pattern, having rust FW (5.5–6.5 mm; Plate 7.39) with a basal, oblique white fascia and a large white costal patch which is bordered in the cell by a series of upraised, black spots preceded and separated by metallic silver scaling, like a row of jewels. It was described from Texas and later recorded in Mexico from Coahuila to Puebla and across the Southwest to the California Central Valley, north to the Feather River and Santa Rosa. The adults come to lights, but typically only one at a time. Y.-F. Hsu reared a series from *Epilobium* (Onagraceae) in the San Joaquin Valley. *Aristotelia bifasciella* Busck is a pale desert species, FW (6–7.5 mm; Plate 7.40) sordid whitish with two dark brown, transverse fasciae, interrupted at the middle and the dorsal margin by ochreous. The abdominal terga are ochreous. This species occurs widely in southeastern California from Inyo County to San Diego County and through the Mojave into Arizona and Baja California. The early stages are not reported.

Evippe are tiny gelechiids (FW 4.0–4.7 mm) usually with a dark FW having a pale area along the dorsal margin. The larvae mine leaves of various trees and shrubs. *Evippe laudatella* (Walsingham)(Plate 7.41) occurs in the foothills of central and northern California, feeding on blue oak (*Quercus douglasii*) and probably other oaks. The FW is white with a linear black patch through the costal half, projecting to the Cu crease at base and midcell, the most prominent marking. Larvae were collected by Keifer in May, and flight records span June to early October. Other species of *Evippe* feed on several oak species in California. Their FW patterns are mostly brownish gray with just a narrow whitish or pale tan dorsal margin. *Evippe abdita* Braun (Plate 7.42) occurs in northern Utah, Arizona, and the length of California. The FW pattern is characteristic of the genus, brown with whitish dorsal area, evidently more extensively brownish in California populations. We have reared *E. abdita* from *Cercocarpus montanus* (Rosaceae) in coastal counties and from *C. ledifolius* in the Warner Mountains, Modoc

County. Larvae are active March in southern California to June in northeastern California and Utah.

Most of the species assigned to *Recurvaria* in older literature are now included in *Coleotechnites*. True *Recurvaria* are characterized by the males lacking hair pencils and having symmetrical genitalia. Two quite similar species occur in California: *Recurvaria ceanothiella* Braun (Plate 7.43) is known from the Sierra Nevada and Coast Ranges, mines leaves of *Ceanothus leucodermis, C. sorediatus,* and *C. thyrsiflorus.* The adults are small (FW 5–6 mm) FW whitish, densely dusted with dark gray and with darker streaks and a narrow, curved, whitish streak beyond the distal streak. There are three upraised, equally spaced, black scale tufts, two on the fold and one above it in a median dark shade. The larvae begin feeding in early spring, mining the lower side of the leaf next to the midrib, with digitate extensions, then along the leaf margin, with a hole at one end through which frass is ejected. Nearing maturity it is expanded to a blotch. Mines collected in March and April produce adults from April to September. *Recurvaria francisca* Keifer feeds on the same *Ceanothus* in coastal areas. It differs in having a more uniform gray appearance, FW with a black streak from the costal base to the inner raised tuft, a conspicuous black spot surrounded by white at the costal one-third just opposite middle tuft, and distal tuft within the tornal area. The valvae are developed, whereas they are rudimentary in *R. ceanothiella*. The larva differs in having the head and thoracic shield black rather than ochreous in *R. ceanothiella*, and the mine is a full-depth, rust-colored blotch on the outer half of the leaf, with the frass packed in the older portion.

The genus *Coleotechnites* includes about 50 described species in America north of Mexico, and many others are unnamed, particularly in the West. They are small moths characterized by asymmetrical male genitalia and male secondary features, scale brushes on the second segment of the labial palpi, and conspicuous scale tufts at the base of the HW. The larvae are leaf miners, and many *Coleotechnites* mine needles of coniferous trees. The adults have narrow FW, whitish to gray or ochreous mottled with linear black streaks, often arranged as three parallel bars outwardly oblique from the costa. The "lodgepole needleminer," *C. milleri* (Busck)(Plate 7.44), has gained the greatest notoriety among western *Coleotechnites*, historically causing extensive defoliation of pines in the Yosemite National Park region. It has pale gray FW (5.5–6.5 mm) with fine, more or less evenly distributed, short black streaks and incomplete transverse fasciae at middle and beyond the cell in the more heavily maculate specimens. The adults fly in July and August, and young larvae spend the first winter in the mine. Two or more needles are mined the following summer, and the larvae again overwinter before pupating the second year. In Alberta, southeastern British Columbia, and adjacent states, *C. starki* (Freeman)(Fig. 88) has the same biology and is morphologically indistinguishable but has more extensive blackish FW pattern. *Coleotechnites atrupictella* (Dietz) is a widespread species in the northeastern United States and eastern Canada, west to British Columbia. Rearing by the Canadian Forest Service included 40% of 400 records from spruce *(Picea)*, 32% from Douglas-fir *(Pseudotsuga)*, 15% from pines, and 10% from fir *(Abies)*. The same or a closely related species occurred on white fir, *Abies concolor*, in northeastern California during outbreaks of *Choristoneura* (Tortricidae) in the 1970s. Like many Gelechiini, the larvae of these species have linear red stripes, illustrated by Duncan in British Columbia.

By contrast, the larvae of *C. granti* (Freeman) are tan needle miners lacking bands, on grand fir, *Abies grandis*, in British Columbia, and we found this or a similar species mining new foliage of *Abies* abundantly in southern Oregon and California, south to the San Gabriel Mountains. Several species are associated with junipers (*Juniperus*, Cupressaceae), including *C. occidentis* (Freeman) from *J. scopulorum* in southeastern British Columbia. It is very small, FW (4–5 mm) white with ochreous scaling on distal three-fourths and basal, middle, and subterminal, oblique, black fasciae, the last bordered by a white patch containing a small black spot. Other western juniper-feeding *Coleotechnites* have been referred to *C. juniperella* (Kearfott), described from New Jersey, but their taxonomic placement needs reassessment.

Coleotechnites bacchariella (Keifer) is a small species (FW 5.5–6 mm; Plate 7.45) having a nearly uniform gray FW to the unaided eye, with five scale tufts, a white one at the basal one-fourth below crease, two in and below the cell at the middle edged with black, and two at the apical one-third edged with black and connected by a cream white line. There are faint, pale, outwardly oblique fasciae near the middle of the wing and subapically to the tornus. This species occurs in coastal central California, south to Santa Barbara County, and feeds on *Baccharis pilularis* (Asteraceae). It has an unusual larval biology, having a very long period of growth. In early spring the larva mines an unopened foliage bud, forming a shelter tightly closed with silk. The larva continues to burrow into the new twig growth, forming a silken tube, incorporating additional leaves, occupying one terminal. In contrast to other spring-feeding microlepidoptera, *C. bacchariella* development lasts from February to May or June. Tilden observed the larvae to delay feeding for four to 10 days following each molt, and larvae that left their shelters in the lab lived more than 12 weeks without feeding. Adults emerge during the dry season, late June to August, rare for Californian microlepidoptera, and the mode of overwintering is unknown. Two species of *Coleotechnites* feed in Ericaceae in California. Larvae of *C. huntella* (Keifer)(Plate 7.46) mine in the flower buds and sometimes leaf buds of native azalea, *Rhododendron occidentale*. Those collected in April and May in coastal populations yield adults in May and June, and in the Sierra Nevada they are about a month later. This species is relatively large (FW 6.5–7 mm) with white FW uniformly speckled with dark gray and seven or eight black dots around apical margin. There are black-and-white, upraised scale tufts at the base of crease and pairs at basal one-fifth, and at the end of the cell. *Coleotechnites mackiei* (Keifer) is smaller (FW 4–5 mm; Plate 7.47, Plate 7.48, Figs. 89, 90) and darker, with a median longitudinal streak, interrupted at end of the cell and preapically. The larvae mine in fruit of manzanita *(Arctostaphylos)*, maturing in the Sacramento Valley at the end of May and early June, and adults emerge in June, whereas at high elevations in the Sierra Nevada the larvae are much later, July to October, and are appreciably larger. We have reared this or a complex of manzanita-feeding species from *A. patula* in the Mt. Shasta area and central Sierra Nevada, from *A. manzanita* in the North Coast Ranges, *A. virgata* in Marin County, *A. auriculata* in eastern Contra Costa County, and *A. insularis* on Santa Cruz Island.

Exoteleia are tiny to small, mostly brown or brown-and-white moths. The known larvae are pine and oak feeders, including species of economic concern in pine plantations. The type series of *E. burkei* Keifer (Plate 7.49, Fig. 91) was reared from Monterey pine, *Pinus radiata*, on the Stanford University campus, but its occurrence there and the host tree association are not native. The larvae mine in the newly developing shoots in spring, and *E. burkei* has been a pest in Christmas tree plantations and urban plantings of Monterey pine from Mendocino County to southern California and in nursery sugar pine, *P. attenuata*, in the foothills of the Sierra Nevada. However, an extensive survey by Ohmart and Voigt in the three native populations of *P. radiata* failed to discover this moth. Instead, we have found *E. burkei* feeding on gray and Coulter pines, *P. sabiniana* and *P. coulteri*, in the central Coast Ranges, presumably its natural habitat. The adult is tiny (FW 4–5 mm), primarily brown, FW with three whitish, transverse fasciae edged by black scale tufts; active in May and June. The larvae are stout, with a dark head and thoracic shield, body orange to brownish in life, lacking integumental markings. They feed in the staminate cones of *P. sabiniana*, which are present much later in the season in than those of *P. radiata* (April to May vs. January to February), and therefore the latter are not available to *E. burkei* caterpillars. *Exoteleia californica* (Busck) is larger (FW 6.8–8.0 mm; Plate 8.3), narrow, FW white with scattered black dots and faint to distinct, pale ochreous rust pattern on distal one-fourth. This species occurs in north coastal and central California, the Sierra Nevada, and the central Coast Ranges. We reared *E. californica* from canyon live oak, *Quercus chrysolepis* (Fagaceae) at Mt. Diablo. The smaller *E. graphicella* (Busck)(FW 4.5–6.7 mm; Plate 7.50) has white FW with a broad, brown, median transverse band. It ranges from Monterey County to San Diego. Adults have been collected from *Q. berberidifolia* (formerly *Q. dumosa*) several times, and we reared *E. graphicella* from larvae found on that oak and *Q. douglasii*.

The genus *Leucogoniella* has just three species in North America, two of them sympatric in central California. Both are tiny (FW 3.5–4 mm), with shining gray FW patterned with black-and-white. *Leucogoniella distincta* (Keifer)(Plate 7.51), which is known only from Sacramento, has the fascia not as angulate, and the apical area dark gray with a white spot. Additional genitalia dissections are needed to confirm the constancy of FW pattern. *Leucogoniella californica* (Keifer) has an oblique white fascia from the costa at two-thirds to the terminal area then angles back to the tornus, and the subapical area is gray with a white dot subtended by a black dash. This species is widespread in California's Central Valley and coastal counties to southern California and Santa Cruz Island and sporadically comes to lights, with scattered records from late May to early October. These two species are among the few that Keifer described from other than reared specimens. We reared *L. californica* twice from leaf litter under *Holodiscus* (Rosaceae) and suspect that larvae of both are detritivores or fungivores in fallen leaves.

Athrips has no native Nearctic species, but three are introduced from the Old World. *Athrips rancidella* (Herrich-Schaeffer) is entirely black, rather broad winged (FW 4.5–6.2 mm; Plate 7.52). This species was recorded in Oregon by 1929 and became established in coastal British Columbia, Washington, and Oregon. It was discovered in California at Berkeley in 1983 and recorded sporadically in subsequent years. The larvae feed on Rosaceae, including *Prunus* and *Cotoneaster* in the Palaearctic, but the colonist populations seem to be restricted to ornamental cotoneaster. In Berkeley the larvae have been found on *C. congesta*, a low-growing Himalayan plant, which is popular as an evergreen groundcover shrub, but not on more upright species of cotoneaster that are common ornamentals in gardens. In spring the black larvae cover the branches with webbing, from which they forage, sometimes encompassing the entire foliage. There is one generation in the Northwest, and adults have been taken at lights in Berkeley from late May through July.

Telphusa is represented by five species in eastern North America and one in the West. *Telphusa sedulitella* (Busck) (Plate 8.4, Plate 8.5) is one of the most abundant and often collected moths on the Pacific Coast. It ranges from Vancouver Island to Baja California and from sea level to mid-elevations in the mountains of California. The adults are slender moths that curl the wings around the body when perched, FW (5.5–6.8 mm) has rows of strongly upraised scales, giving a roughened appearance. The FW are black with shades of rust brown on anterior half. There is an uncommon form (Plate 8.5) with a sharply distinct white fascia from the basal one-fourth of the costa to the middorsal margin, then along the dorsum except at the tornus and along the termen to just below apex, and there are intermediates. The larvae are green, including the head capsule, without integumental color pattern, and the anal plate has long, conspicuous, black primary setae. The caterpillars are abundant leaf tiers in new foliage of live oaks (*Quercus agrifolia, Q. wislizenii,* and others) in early spring, March and early April at low elevations. The adults emerge in May and June and live in reproductive diapause until the following February, when mating and oviposition occur ahead of oak foliation. There is some question as to host specificity; *T. sedulitella* has been reported from *Salix* and *Corylus* (Betulaceae), and we have reared single individuals twice from *Pseudotsuga* (Pinaceae) and *Arctostaphylos* (Ericaceae) growing with live oak but assumed these were incidental occurrences. During the winter the adults hide in crevices in bark and under debris, often in abundant numbers, sometimes distant from oaks. They are often found inside buildings, and occasionally are attracted to lights.

TRIBE GELECHIINI

Gelechiini is the largest tribe of Gelechiidae in North America, with more than 400 described species, dominated by *Gelechia, Chionodes,* and *Filatima*. These are mostly broad-winged moths, HW much broader than its fringe.

Pseudochelaria are distinctive, large for gelechiids, broad winged, and boldly patterned in purple and brown. There are five North American species, three of which occur on the Pacific Coast, feeding on Ericaceae. *Pseudochelaria manzanitae* (Keifer) is moderately large (FW 7–9 mm; Plate 8.7) with the wings rounded apically, dark gray with a bluish or purple iridescence and a black bar from near the costa at one-fifth, ending at an upraised tuft. This species occurs in California from the foothills of the northern Sierra Nevada to Monterey County. The larvae feed in new foliage terminals of several manzanita species (*Arctostaphylos*), including *A. manzanita* in the Sacramento Valley, *A. virgata* in Marin County, and *A. hooveri* in Monterey County. Adults reared from larvae collected in April emerge in May and June, but light-trap records are July to September. *Pseudochelaria arbutina* (Keifer) is quite similar, averages larger (FW 9–10 mm; Plate 8.8), with a deep black triangle near the dorsal base. Its larvae feed on madrone, *Arbutus menziesii*, in coastal sites. *Pseudochelaria scabrella* (Busck) has narrower FW (7.5–9.5 mm; Plate 8.9) more rectangular distally and is more widespread. We found larvae on *A. insularis* and *Comarostaphylis diversifolia* on Santa Cruz and Santa Rosa islands. At Miramar, San Diego County, *P. scabrella* was taken every month of the year and presumably feeds on *Xylococcus bicolor* (Ericaceae) because the only manzanita there, *A. glandulosa*, is quite rare.

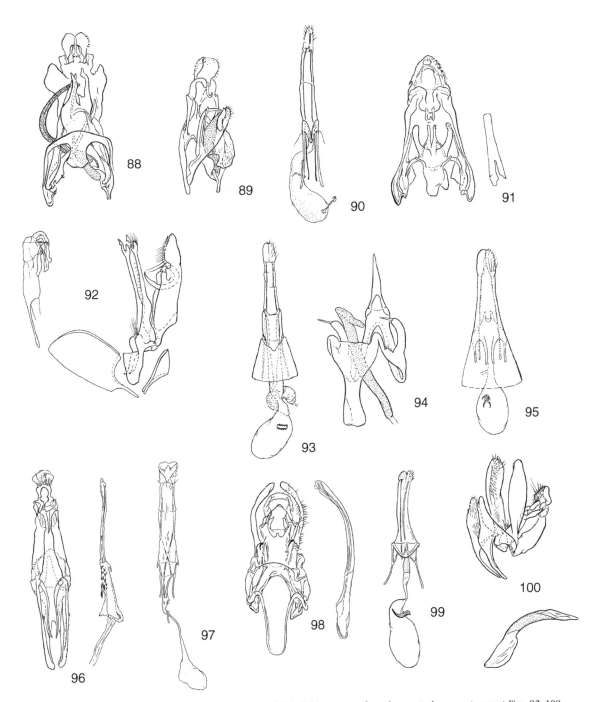

FIGURES 88–100. Genitalia structure of exemplar species of Gelechiidae, venterolateral or ventral or aspect, except Figs. 92, 100 lateral aspect; aedeagus in situ (shaded) in Figs. 88, 89, 94. **88,** *Coleotechnites starki* (Freeman) ♂ [Freeman 1960]. **89,** *Coleotechnites mackiei* (Keifer) ♂. **90,** *Coleotechnites mackiei* ♀. **91,** *Exoteleia burkei* Keifer ♂, aedeagus to right [Keifer 1931]. **92,** *Chionodes trichostola* (Meyrick) ♂, aedeagus to left. **93,** *Chionodes trichostola* ♀. **94,** *Aroga paulella* (Busck) ♂. Fig. **95,** *Aroga paulella* ♀ [Busck 1939]. **96,** *Arla tenuicornis* Clarke ♂, aedeagus to right. **97,** *Arla tenuicornis* ♀ [Hodges 1966]. **98,** *Tuta chiquitella* (Busck) ♂, aedeagus to right. **99,** *Gnorimoschema ericameriae* Keifer ♀ [Powell and Povolny 2000]. **100,** *Anacampsis sacramenta* Keifer ♂, aedeagus below [Keifer 1933].

Teliopsis baldiana (Barnes & Busck)(FW 8.5–10 mm; Plate 8.10) is another large, broad-winged species, resembling a pale *Pseudochelaria*. It is widespread in California but inexplicably rare in collections, considering the abundance of its host plant, poison oak, *Toxicodendron diversilobum* (Anacardiaceae). The larvae are leaf rollers, typically scarce on healthy plants but sometimes numerous on new growth where poison oak has been cut back from trails. In April 1990, the fifth season following a consummate wildfire at the Big Creek Reserve, larvae

of *T. baldiana* appeared in abundance, whereas they were rare before and in later years. Reared specimens have emerged in late May and June; light-trap records are June through August on the mainland, end of April and beginning of May on Santa Rosa and Santa Cruz islands. *Teliopsis baldiana* ranges through much of California at low elevations but apparently does not follow its host north to British Columbia.

North American species of **Xenolechia** are mostly eastern, but there are two described from Arizona and California, along with at least two unnamed species, associated with oaks. **Xenolechia querciphaga** (Keifer) is a small (FW 4.8–5.5 mm; Plate 8.1), dark gray moth having blackish, upraised scales rows on the FW. This species occurs in the foothills on both sides of the California Central Valley. The larvae are green with a black head capsule, body turning pink just before pupation. They feed in the new spring growth of interior live oak, *Quercus wislizenii*, *Q. berberidifolia*, and sometimes deciduous blue oak, *Q. douglasii* (Fagaceae), in April and May. The nocturnal adults are active mid-May to early July.

Species of **Lita** are distinctive in several ways. They are moderately large, FW narrow with rounded apex. The males have slightly to obviously thickened antennae, and the legs are exceptionally long in both sexes. Some species possess strongly produced ridges or spurs on the front and crown. All of the 24 species in North America occur in the West, some boreal, mostly in the Great Basin and other arid regions. Most *Lita* fly in August and September, unusual for univoltine microlepidoptera in the West. The adults are primarily nocturnal, but they often visit flowers during the daytime. Several species have dramatically variable FW patterns, usually dominated by longitudinal lines, with parallel forms among different species. **Lita sexpunctella** (Fabricius)(Plate 8.11), the type species of the genus, is Holarctic-boreal in distribution. It is distinct in having the FW (7–8.5 mm) dark brown with pale tan to cream-colored transverse bands in variable contrast and definition. In the Nearctic it ranges across Canada to Alaska (White Horse) and southward to Colorado, Utah, and the southern Sierra Nevada, California, at elevations of 7,000 to 12,500'. In Europe the larval host plants of *L. sexpunctella* are Ericaceae, *Calluna, Erica, Vaccinium*, but judging from habitats where they are abundant, Ericaceae seem unlikely as hosts for most Nearctic *Lita*. The only species for which larvae are recorded in North America feed on *Ericameria* and *Gutierrezia* (Asteraceae). Hodges reported the larval food plants of several other *Lita* as Asteraceae (*Chrysothamnus* and *Senecio*). It may be that these plants are larval hosts, but the records were based on adults at flowers—label data from the University of California, Berkeley, collection that were misinterpreted by Hodges.

Lita variabilis (Busck) is markedly variable in color and FW (7–10 mm; Plate 8.12) pattern, spanning most of the forms shown by other species; ground color pale tan to ochreous brown clouded with black or dark brown, to having linear bars in the crease, many fine longitudinal lines, or markings reduced to fine strigulation and marginal dots on the termen. This is the most widespread and often collected species of Nearctic *Lita*, with isolated records in New England and the plains provinces in Canada, the Great Basin and desert parts of the western United States to northern Arizona and southeastern California, but the larval biology is unknown. **Lita recens** Hodges and *L. incicur* Hodges, were first collected by Lord Walsingham at Mt. Shasta in 1871, but it was nearly a century before they were named. The FW of *L. recens* (9–10.5 mm; 7.5 mm reared; Plate 8.13) is heavily patterned with longitudinal black lines and three more or less contiguous streaks through the cell, forming a median longitudinal shade. This species ranges from British Columbia to southern California, Montana, Utah, and Colorado. Comstock and Henne found the larvae on goldenbush, *Ericameria cuneata*, in the western Mojave Desert, and we reared *L. recens* from *E. linearifolia* in western Fresno County. Reared specimens emerged in September, and field records are late August to mid-October. **Lita incicur** varies in size (FW 7.5–12.5 mm; Plate 8.14, Plate 8.15) and FW pattern, from

pale tan to orange brown, with variable brown markings, from poorly defined spots near base, middle, and terminal areas to basal two-thirds all dark, terminal area weakly patterned in tan. It is recognizable by its frontal processes, a blunt, oval subtended by a narrow ridge. This species is widespread, mostly at 7,000 to 12,500', in Oregon, California, Wyoming, and Colorado, including alpine fields above timberline in the Sierra Nevada and White Mountains, California, and most collections have been diurnal. The FW of **L. rectistrigella** (Barnes & Busck) has subtle lines of black and white on a gray background, so appear nearly uniform gray at arm's length (FW 8–8.5 mm; Plate 8.16). This species occurs in inland, arid parts of southern California and the deserts of the Southwest to eastern Washington, Montana, Utah, Colorado, and New Mexico. We reared it from *Gutierrezia* in western Fresno County, California. **Lita sironae** Hodges (Plate 8.17) is restricted to southern California and was found in large numbers at San Diego a century ago but has not been recorded there in the past 80 years. It ranges to the desert edge in Los Angeles County and occurs on Santa Cruz and Santa Catalina islands and abundantly on Santa Rosa Island in recent years. Adults are smaller than most *Lita* (FW 6–8 mm) and relatively non variable, FW white with numerous longitudinal streaks of pale tan, pale yellow, brown, and black. They are nocturnal and fly from late September to November.

The genus **Arla** is characterized by having extraordinarily long antennae that are strongly enlarged in the male. *Arla* differs from *Lita* by having a costal process on the valva in the male genitalia and by lack of a signum in the female, but the adults are otherwise similar. **Arla tenuicornis** Clarke (Figs. 96, 97) ranges from southern Washington to central California (Yosemite Valley), whereas *A. diversella* (Busck)(Plate 8.18) occurs in southern California and the northern Channel Islands to the San Joaquin delta (Antioch) and El Dorado County. Thus the two may be narrowly sympatric in central California but have not been taken together. *Arla diversella* (FW 6–8.5 mm) has two color forms in both sexes, most have FW predominately rust colored, but some lack the rust and are pale gray to tan. This species was reared by Keifer from *Lotus* (Fabaceae) and by De Benedictis at San Bruno Mountain and in our inventories at Antioch and Big Creek from *L. scoparius*. The larvae create shelters in the inflorescences, incorporating bits of flower petals into a bulky webbed nest. Larvae have been collected late March to early May and adults late April to June.

Neodactylota has four species in western North America, all known from few specimens, two only in southern Arizona. **Neodactylota snellenella** (Walsingham) is the largest, FW (7–8.5 mm; Plate 8.19), brown with darker red brown spots poorly defined by yellowish to gray scaling, two in the cell before the middle and a larger spot from the dorsal margin extending to the end of the cell. This species is recorded from western Texas and southern Arizona; adults have been collected in May at in Texas and Madera Canyon and in September in the Huachuca Mountains. The larval biology of *Neodactylota* is unknown.

There are seven species of **Friseria** described in Mexico and the southern United States, of which three occur in the Southwest. **Friseria cockerelli** (Busck)(Plate 8.20) is the most widespread and often collected, ranging from the desert regions of California and Nevada to Texas and in eastern Mexico to Hidalgo. The adult is tan, FW (5–7.5 mm) with variable brown and red brown scaling forming poorly defined markings, from just dots along the costa and in the cell to moderate infusion of the dorsal area basally and large spots on the costa at the middle and beyond the end of the cell, and on the veins beyond the cell. The larvae form shelters in foliage of mesquite, *Prosopis glandulosa* (Fabaceae). We reared *F. cockerelli* in May at Cronise Valley, San Bernardino County, and DeLoach studied its biology in central Texas. He found the life cycle was completed in 43 days, and there were at least four peaks of flight numbers, in addition to emergences from overwintered pupae in March to April. The larva has a dark head capsule and thoracic shield, bordered by white and the body is brown with white stripes along the dorsolateral lines. They failed to survive when fed *P. velutina* and would not accept screwbean, *P. pubescens*, and other legumes related to *Prosopis*.

Rifseria fuscotaeniella (Chambers)(Plate 8.6) is distinguished by having FW veins R3 to R5 stalked and terga A2 and A3 with paired, flat patches of stiff setae, visible on intact specimens as rust or orange patches. The head and FW (4.5–7 mm) are white, FW variable, typically with distinct black markings, a basal band, two spots preceding and following midcosta, the latter larger, triangular, and an apical blotch. There is variable gray clouding from the middorsal margin to the subterminal area, so some specimens are much darker. *Rifseria fuscotaeniella* occurs from Manitoba and British Columbia to eastern Colorado, Utah, Nevada, Arizona, and coastal California. Adults are recorded at lights, mostly June to September in the Rocky Mountain and Great Basin regions. There are at least two generations in California, with flight records March to July and September to October. The larvae are leaf miners and feed in growing terminals of everlasting, *Anaphalis*, and cudweed, *Gnaphalium* (Asteraceae), with collections in February to April (San Bruno Mountain) and September (Santa Cruz Island), proceeding to emergence without diapause.

Bryotropha are tiny, dark brown, nondescript, gelechiids, even so the subject of recent systematics study by Rutten and Karsholt. There are seven species in North America, three of which are Holarctic. The larvae of several species feed on mosses in Europe, a most unusual habit for microlepidoptera, overlooked by American lepidopterists. *Bryotropha similis* (Stainton) FW (4–5 mm) dark brown including HW, with indistinct black FW spots. This species is widespread in the Palaearctic and in North America from Greenland and New England, across Canada and southward in the Rocky Mountains to New Mexico, northern Arizona and coastal California. In Europe *B. similis* is a lowland insect in northern latitudes and becomes restricted to higher elevations in the south, but four California records of *similis* are all coastal. *Bryotropha hodgesi* Rutten & Karsholt (Plate 8.2) is similar but has somewhat more visible FW markings, a preapical dark smudge, and the HW is paler. It is known across the continent, from Nova Scotia and Maine to southern British Columbia and Washington, one record in southern Arizona (8,500' in the Chiricahua Mountains), and in California on the coast and inland, to 3,500' at Mt. Shasta City. We found *B. hodgesi* common on Santa Rosa and San Clemente islands, but improbably, there is no record of it on the more mesic and well collected Santa Cruz I. Adults are active from March to October, with most records in California April to June and September.

The genus *Gelechia* is represented in America north of Mexico by about 40 named species, and about 20 occur in the western states. These moths are small to moderate sized gelechiids, generally similar to many *Chionodes* and *Aroga*; but typically they have narrower, often unicolorous FW. *Gelechia desiliens* Meyrick has brown FW (7.5–9.5 mm; Plate 8.21) with a pale subterminal line, often indistinct, and HW hyaline, brownish gray apically. It is widespread in low-elevation parts of Cali-

fornia. The adults overwinter and come to lights primarily in September to November and January to April. The larvae feed in new leaves of western sycamore, *Platanus ramosa* (Platanaceae).Those collected in April to June produce adults in summer, and there may be just a single generation, with adults long-lived. *Gelechia panella* Busck (Plate 8.22) is unusual for a gelechiid and most microlepidoptera in having slender, uniform brick or rust red FW (9–11.5 mm), an apparent adaptation facilitating crypsis on its larval hosts, madrone *(Arbutus)* and manzanita *(Arctostaphylos*, Ericaceae). The HW is dark gray. This species occurs in Arizona and southern California, to north coastal California. Larvae collected in May and June produce adults in June to August, and most blacklight records are those months. There are several records in November and April, so adults may overwinter. *Gelechia lynceella* Zeller (Plate 8.24) was described originally from central Texas but occurs widely in western montane areas, Alberta, Utah, Nevada, and eastern California. The FW (7–11 mm) is blackish gray with variable, poorly defined, white maculation, an oblique, antemedial, transverse band and subterminal fascia. Adults are recorded from late June through August, but the larval biology is not reported.

Gelechia versutella Zeller is a smaller species (FW 6.5–7.5 mm; Plate 8.23) with whitish FW finely speckled with evenly spaced, short, dark gray to black lines, and the abdomen is pale ochreous dorsally. This species is widespread in arid parts of the West, from Montana and Wyoming to Colorado, western Texas, Durango, and the Central Valley of California. The larvae tie leaves of cottonwood, *Populus fremontii*, and aspen, *P. tremuloides* (Salicaceae), sometimes in large numbers. Another small species, *G. ribesella* (Chambers)(FW 6–8 mm; Plate 8.25), has a strong, distinct FW pattern, dark gray tinged with rose pink, markings white, a transverse band beyond end of cell and a round spot in the cell. The species was described from Colorado, reared from currant, *(Ribes*, Grossulariaceae), and it ranges to California, where probably the same species was described as *G. dromiella* Busck from the Sierra Nevada. The phenotype in northeastern California (Plate 8.25) more resembles that of Colorado populations than those to the south. Jim Scott reared *G. ribesella* from *R. inerme* in Colorado, and we have found the larvae on gooseberries, *R. californicum*, *R. menziesii*, and *R. quercetorum*, in the Coast Ranges and on *R. nevadense* in the northern Sierra Nevada.

Chionodes is the most species rich and best-studied genus of Gelechiidae in the Western Hemisphere. This is in large measure due to the monumental systematic treatment by Hodges in 1999 in the MONA series, which was 35 years in gestation. He described and illustrated 187 species in America north of Mexico, of which 115 (61%) were newly named. Some genera that are awaiting comprehensive study (e.g., *Aristotelia, Gnorimoschema, Filatima*) no doubt are represented in collections by greater diversity than their described numbers, but it is not likely any Nearctic genus will match the richness of *Chionodes*, which now includes more than 20% of the described North American species of Gelechiidae. *Chionodes* species are superficially similar in morphology, with smooth head scaling, labial palpi strongly upcurved, second segment with erect scaling but not as broadly spreading as in most *Filatima*, wings usually broader than in *Gelechia* but narrower than in most *Filatima* and many *Aroga*. A convincing synapomorphy for all *Chionodes* is the uniquely derived phallus form, having a well developed caecum (a slender, closed tube extending anteriorly, originating anterior to the ductus ejaculatorius), and with the aedeagus distally cylindrical, asymmetrical, and complexly sculptured.

Many *Chionodes* are somber, more or less unicolorous brown or gray, FWs often with a subterminal paler fascia, and thus are exasperatingly difficult to differentiate, requiring careful examination of male secondary characters and often dissections of the abdomens of both sexes. On the other hand, there are many that are conspicuously patterned in white, yellow, or orange. Without dissection, males of *Chionodes* can be distinguished from species of the closely related and superficially similar genera, *Filatima* and *Aroga*, by lack of long, slender, dorsolaterally curved scales from lateral margin of abdominal tergum 8 and a "curtain" fringe of long scales from the HW Sc vein underside, overlapping the R stem, either or both of which are characteristic of most male *Filatima*. From *Aroga* they differ by lack of a tuft of long scales from the base of the HW dorsally and by lack of slender lobes from the lateral margins of sternum 8. There is considerable variation within many species in color and FW pattern and in structures, so dissection frequently is necessary and more than one because genital characters often vary. Hodges's comprehension and summary of the variation was an amazing accomplishment, based on nearly 2,500 genital preparations. He studied (and databased) more than 19,000 specimens, of which 4,200 representing 77 species originated from California.

Larvae of most *Chionodes* feed externally from conspicuous silken shelters that alter the appearance of vegetation, especially the new growing terminals. As a result, many species have been reared from field collected larvae. Most are specialists, often on one or a few related plants within a genus or on several genera of one family, but a few are polyphagous. A wide array of hosts serve as food plants, including conifers and at least 15 dicot families, usually perennials, and some are grass feeders. In several instances species complexes have evolved along larval host-plant lines (e.g., Pinaceae, Fagaceae, Malvaceae, Polygonaceae, Poaceae). Most Nearctic *Chionodes* feed in spring and proceed through pupation and metamorphosis without diapause, and some hibernate as adults and mate in spring.

Hodges recognized six species groups, two of which are small and occur in the eastern and southeastern United States. About 150 of the 187 species north of Mexico (80%) are represented in the western states and provinces. Many range across the continent, especially in boreal situations, and some are Holarctic in distribution.

FORMOSELLA GROUP

The Formosella Group has 71 species, about 52 western. It is characterized by male secondary features, a pecten of sex scales along the Cu on the HW upper surface, usually a scale pencil arising from the base of A1 on HW, and submesial lobes on the anterior margin of the eighth sternum (shared with the Obscurusella and Phalacrus groups).

Chionodes powelli Hodges (Plate 8.26) is widespread and one of the most often collected species, but it was not differentiated from *C. formosella* (Murtfeldt) of eastern North America for nearly a century. The two are superficially very similar, having dull brown FW (6–8 mm) with darker clouding and pale subapical and weak tornal spots, and both are oak feeders. *Chionodes powelli* occurs from western Texas to the Pacific states. It has been reared from deciduous oaks, *Quercus kelloggii* and *Q. lobata* in California and *Q. garryana* in Oregon. Larvae feed in April and May, and adults emerge in June and July, including in Arizona, ahead of the summer rain season. *Chionodes abitus* Hodges is a slightly larger species (FW 7–9.5 mm; Plate 8.27) having rust brown FW with the three dark spots characteristic of many *Chionodes*, on the Cu crease, in the cell and at the end of the cell, black weakly ringed by pale scaling. The HW scale brush is elongate, conspicuous, orange brown. This species ranges from southern British Columbia to Idaho, Montana, and mountains of southern California. The larvae have been found on willow *(Salix)* in the Warner Mountains of northeastern California and the northern Sierra Nevada in May to June. Flight records are July to August. *Chionodes lophosella* (Busck) is small (FW 3.8–5.2 mm; Plate 8.28) and ranges along the Pacific Coast from northern Washington to San Diego and probably Baja California. The larvae feed on foliage of the beach dune bush lupines, *Lupinus arboreus* and *L. chamissonis*. The moths have black FW speckled with white and usually a subterminal, transverse white fascia, often broken. The species appears to be bi- or multivoltine, with adults active from late March to early June and July, to October in southern California and the northern Channel Islands. *Chionodes nanodella* (Busck) shares the southern part of a similar coastal distribution. Adults are small (FW mostly 4.5–6.5 mm; Plate 8.29, Plate 8.30, Plate 8.31) and average smaller on the southern Channel Islands, particularly in the fall. On the mainland the FW is usually pale brown, fading to an ochreous peach dorsal area, whereas on the southern islands the pattern is wildly variable, from nearly pale yellow tan with tiny dark flecks to heavily black marked, dominated by three outwardly angled, irregular fascia, at times coalescing to nearly all black (Plate 8.30, Plate 8.31). On the northern islands there is a less variable phenotype more like that of the coastal mainland. The larvae seem to be indiscriminate generalists and may be scavengers. We have found them in decaying leaves at the base of *Phacelia distans* (Hydrophyllaceae) and associated with *Lupinus arboreus* (Fabaceae), *Eriogonum latifolium,* and *E. parvifolium* (Polygonaceae) on the mainland and have reared adults from *Achillea millefolium, Artemisia nesiotica,* and *Deinandra clementina* (Asteraceae) on Santa Barbara Island, where *C. nanodella* is very abundant. Flight records are March to June and August to September.

Species related to *Chionodes abella* (Busck)(Plate 8.32, Plate 8.33) are known or suspected to feed on conifers. These are black-and-white moths, and *C. abella* is so variable in FW pattern that its forms resemble those of most of the related species; typical populations in the Rocky Mountains have FW black on the basal two-thirds and white on the distal one-third (Plate 8.33), whereas Pacific Coast *C. abella* are gray densely speckled with white, or black with variable fasciae (FW 4–6 mm). This species ranges from southern British Columbia, northern Idaho, Utah, and Colorado to western Texas, northern Arizona, and the mountains of southern California. Adults have been reared from larvae found on *Abies concolor, A. grandis,* and *A. lasiocarpa,* from *Pseudotsuga, Pinus contorta,* and *P. radiata*. Larvae we collected in June produced adults in July. *Chionodes sabinianae* Powell (Plate 8.34) is a sister species, much less variable, FW black, finely white speckled, without the basal fascia or distal white one-third, usually with a well-developed, thin preapical fascia (FW 5.5–8 mm). It occurs in the Pacific states, from southern British Columbia to southern California, primarily coastal, and feeds on *Pinus sabiniana, P. coulteri,* and *P. ponderosa,* and along the coast on the closed-cone pines, *P. radiata* and *P. muricata*. We reared *C. sabinianae* once at Tehachapi Mountain from *Abies concolor,* where it grows with *P. sabiniana*. *Chionodes nitor* Hodges (Plate 8.35) is sympatric on the California coast and almost indistinguishable from *C. sabinianae* but has broad, symmetrical vinculum lobes at the base of the valvae, unlike any other species of the *C. abella* complex. The larvae feed on *P. radiata* and *P. muricata*, and on Santa Cruz Island *C. nitor* occurs in association with *P. muricata* form *remorata*. Adults fly in May and June, and we collected larvae at Cambria, one of the three native populations of Monterey pine, in March, from which adults emerged in April.

Members of the *Chionodes sistrella* complex differ in phenotype from other *Chionodes* in having brown to black FW with distinct ochreous yellow to whitish markings in various modifications of a basic pattern: dorsal area, connected to a broad transverse band at the end of the cell, and a subapical fascia or blotch to the tornus. In some species the ochreous replaces nearly all the dark ground color. The known larvae (five of 18 species) feed on Chenopodiaceae. *Chionodes kincaidella* (Busck) is the largest in this group (FW 6–11 mm; Plate 8.36) and one of the most widespread, from southern Alberta to southern California, including San Clemente and Santa Barbara islands, Texas, Sonora, and Baja California Sur, Mexico. The wide distribution and variation in FW color, from dark chocolate brown to rust and pattern, especially the extent of ochreous, to nearly obliterating the brown, led to five synonyms, unusual for a western Nearctic gelechiid. This species was reared from *Atriplex acanthocarpa* in Texas by R. Hitchcock. On Santa Barbara Island there are two or more generations, with records in April to June (numerous in late May) and September. The smaller *C. pinguicula* (Meyrick)(FW 5–6.5 mm; Plate 8.38) has dark to pale brown FW ground color and indistinct maculation. This species occurs from southern Colorado to western Texas and southern California. Larvae were reared on *Atriplex canescens* in the low desert of California by C. M. Dammers. *Chionodes xanthophilella* (Barnes & Busck)(Plate 8.37) is a pale desert species having the FW (4.5–9.5 mm) mostly ochreous yellow, the dark pattern reduced to the costal and basal half and a weak transverse fascia beyond the end of the cell. The species ranges from the Mojave Desert in California to southern Nevada, Arizona, and western Texas. The larval host plant is unknown. *C. sistrella* (Busck) is the smallest in the complex (FW 4–6 mm; Plate 8.39) and has black FW with discrete, white to pale orange markings, narrow border at the dorsal edge, and connecting transverse thin line, subterminal fascia broken into two spots. H. H. Keifer reared *C. sistrella* from *Suaeda fruticosa* in the San Joaquin Valley, California, in 1930, and subsequently *S. torreyana* and Australian saltbush, *Atriplex semibaccata*, have been recorded as food plants, according to Hodges. This species occurs from the Gulf states and Texas to western Nevada, the deserts and Central Valley of California, and on the southern Channel Islands, where the species probably is multivoltine, with flights in April to May and August to December.

Members of the *Chionodes dammersi* complex are brown to gray moths, mostly without distinctive FW pattern. Six of eight western species feed on Polygonaceae, either buckwheat *(Eriogonum)* or dock *(Rumex)*, and one on Salicaceae, poplar *(Populus)*. *Chionodes dammersi* (Keifer) has dark to milk-chocolate-colored FW without maculation (7–10 mm; Plate 8.40), and the HW costal scale brush is dark ochreous. This is a coastal Californian species distributed from the San Francisco Bay area to San Diego and on Catalina and the northern Channel Islands. The larvae feed on several species of *Eriogonum* at various times of year, and adults have been recorded in every month. *Chionodes bardus* Hodges (Plate 8.41), which is known only from Anacapa, Santa Barbara, and Catalina islands, and *C. helicostictus* (Meyrick)(Plate 8.42) are superfi-

cially quite similar to *C. dammersi*. *Chionodes bardus* is larger (FW 9–11 mm) and on Santa Barbara Island, the type locality, has mottled FW, with pale gray scaling outlining darker marks, and gray scaling distally on the HW. Its larvae feed on insular endemic buckwheats, *Eriogonum giganteum* and *E. grande*. By contrast, both *C. dammersi* and *C. helicostictus* tend to have more unicolorous FW and white HW. The latter species (FW 6.5–9.5 mm) occurs from western Texas to southern and central California and can be distinguished by the color of its head scales, pale yellowish gray basally, variably darker gray distally. The occipital scales of *C. dammersi* usually are gray with the distal margin narrowly pale gray or yellowish gray. Like *C. dammersi*, *C. helicostictus* has been recorded in all months of the year. We reared *C. helicostictus* from *E. nudum* in central California, and no doubt it feeds on other buckwheats. Unlike related species, *C. notandella* (Busck) is larger (FW 9.5–13.5 mm; Plate 8.43) and has whitish gray FW with strongly contrasting longitudinal, dark gray lines along the veins and forming a streak through the cell. It occurs from Lake County to southern California and Arizona and flies primarily during the winter, October to February, although there are scattered records for other months. We reared *C. notandella* from *E. latifolium* in San Francisco in July. *Chionodes ochreostrigella* (Chambers) has dark brown FW (6.5–9 mm; Plate 8.44, Plate 60.9) fading to ochreous brown posteriorly, streaked with dark brown along the veins. This species has been reared in California on *Rumex crispus* and *R. acetosella* (Polygonaceae), both of which are of Eurasian origin. Presumably *C. ochreostrigella* also feeds on native *Rumex*, unless it is an unrecognized alien moth. If so the colonization was early, as *C. ochreistrigella* was described from San Francisco in 1875. It occurs from northern Oregon to southern California, including Santa Cruz Island, in disturbed habitats in coastal areas and inland valleys, where one might expect an introduced species.

OBSCUROSELLA GROUP

The Obscurosella Group has 42 species in America north of Mexico, 28 of them represented in the West. Many of these are oak feeders, or suspected to be, based on taxonomic relationships, but several species may be general feeders. Males are characterized by having brushes of slender scales in the cell on the HW underside, and the anterior margin of sternum 8 has paired, anteriorly directed, slender lobes. The females have the anterior apophyses fused with lateral margins of the antrum, which has a posteriorly directed extension.

Chionodes thoraceochrella (Chambers)(Plate 8.45) is one of the commonest gelechiids at lights, occurring from eastern North America to the Pacific Coast, Washington to central California, Utah, Colorado, Arizona, and Texas. The FW (5–7 mm) varies in color and maculation, from dark brown with obscure darker and sometimes ochreous brown areas to gray brown with whitish gray costal area and poorly defined black markings. The larvae are leaf tiers on several species of oaks in the East, and there are isolated rearing records from *Vaccinium* (Ericaceae), *Liquidambar* (Hamamelidaceae) in Michigan and Georgia, and L. Crabtree reared *C. thoraceochrella* from *Prunus virginiana* in California, so this species appears to be a general feeder on trees and shrubs. *Chionodes chrysopyla* (Keifer) (Plate 8.46) is a sister species that occurs from Mt. Shasta to southern California, including Santa Cruz and Catalina islands. It is smaller (FW 4.3–6 mm) with a similar range in FW maculation to that of *C. thoraceochrella* but usually paler and strongly mottled. *Chionodes chrysopyla* is strictly an oak feeder, known to use both evergreen *(Quercus agrifolia, Q. berberidifolia, Q. pacifica, Q. wislizenii)* and deciduous species *(Q. kelloggii)*. Larvae we collected in April and May yielded adults in May to June, but adults are recorded in the field from April to May and June to early August on the Channel Islands.

Chionodes occidentella (Chambers) and *C. trichostola* (Meyrick) are commonly seen species on the Pacific Coast, and both appear to be general feeders that fly late in the season. *Chionodes occidentella* has relatively narrow FW (5.5–6.5 mm; Plate 9.1), very dark brown, sometimes with dull orange tan forming a transverse subterminal fascia and defining a dark band at the basal one-fourth on the costal half. The adults emerge in August to September and evidently hibernate in reproductive diapause. They occasionally come to lights during the fall months and become numerous in January and February. Keifer reared *C. occidentella* from *Quercus agrifolia*, *Ceanothus* (Rhamnaceae), and *Cercocarpus* (Rosaceae); and *Arctostaphylos* (Ericaceae) has been recorded in northern California and Oregon. The caterpillars feed in July to August, late in the season compared to other live oak–feeding and *Ceanothus*-feeding microlepidoptera. *Chionodes trichostola* is larger (FW 6–8 mm; Plate 9.2, Figs. 92, 93), broader winged, and distinguished by a band of black scales near base of the FW, and a prominent, black scale brush along the cell on the male HW underside. This species occurs from British Columbia to southern California, east to Utah, southern Colorado, and central Texas. It was reared by Keifer from several inland species of oaks in California, both deciduous and evergreen. This is the most commonly encountered *Chionodes* on the larger California islands, where endemic plant species serve as hosts: oak, *Q. tomentella* and *Q. pacifica*, and manzanita, *Arctostaphylos insularis* and *A. tomentosa* (Ericaceae). According to Hodges, *C. trichostola* also has been reared from *Sorbus* (Rosaceae). The larvae feed in March and April, and adults emerge in May and June, then live through the dry season and winter before mating and oviposition occur in early spring. We have found numerous adults hiding under loose bark of island oak and ironwood, *Lyonothamnus* (Rosaceae), in May and September.

Chionodes lector Hodges is similar superficially to paler forms of *C. chrysopyla*, FW (5–6.7 mm; Plate 9.3, Plate 9.4) ochreous tan when fresh, fading to gray brown on older flown specimens, with poorly defined, black transverse bands, at basal one-fourth angled outward. This species occurs in southern Washington and in the mountains and foothill canyons of California from Siskiyou County to Los Angeles County. The larvae feed in the catkins and developing seeds of bigleaf maple, *Acer macrophyllum* (Aceraceae), in April to May, and although our lab emergences occurred late April to June, flight records are July to September. *Chionodes mediofuscella* (Clemens)(Plate 9.5) is very widespread, eastern North America across to British Columbia and south to Colorado, northern Arizona, and northern California. In contrast to related *Chionodes*, the larvae feed on ragweed, *Ambrosia* (Asteraceae), in the East, but larvae of western populations have not been discovered. Emergence records from dry corn stalks and green cotton bolls are dismissed as incidental or pupation sites. The moths are small (FW 4–7 mm), variable in FW color pattern, in part geographically, from brown with basal and dorsal area ochreous tan, sharply defined from the costal half in the cell, to brown with the basal fascia black or almost unicolorous dark brown. The HW scale brush is pale and weak. Adults are recorded from early spring in southern areas to September, primarily May to July. *Chionodes terminimaculella* (Kearfott)(Plate 9.6) ranges across the continent north from Quebec

to British Columbia and south in the Rocky Mountains to Colorado, in association with aspen, *Populus tremuloides*, and balsam poplar, *P. balsmifera*, the recorded larval hosts. The adult is larger than related species (FW 6.5–8.7 mm) and has dull, gray brown FW marked by blackish, a fascia at the basal one-fourth from the costa to the Cu crease and irregular spotting on the costal half and a row of dots around the termen at the base of the fringe. The flight is mid-May to early July.

A group of four oak-feeding species, including *C. messor* Hodges and *C. optio* Hodges, are very similar phenotypically and reliably distinguished by genitalia and geographical occurrence. These species have brown to faintly rust brown FW (5–8.5 mm) with short, black streaks in the crease near the base and in the cell near the middle. *Chionodes messor* (Plate 9.7) occurs in the Sierra Nevada of extreme western Nevada and California and the Coast Ranges, from Trinity south to San Diego County, mostly at 5,000 to 6,500'. Larvae have been recorded on both deciduous and evergreen oaks (but not on coast live oak), including *Quercus berberidifoliella* (=*dumosa*), *Q. chrysolepis*, *Q. douglasii*, *Q. kelloggii*, *Q. turbinella*, *Q. vaccinifoliella*, and *Q. wislizenii*. *Chionodes optio* (Plate 9.8) is recorded at some of the same localities in California and extensively in Arizona to Colorado and western Texas. It would be interesting to learn if preference for different oaks is shown in sympatry. We reared *C. optio* from *Q. arizonica* in the Chiricahua Mountains and Oak Creek Canyon, Arizona, from larvae collected in June. Both species fly mainly in July to August, but there are records into November, suggesting that adults overwinter. *Chionodes petalumensis* (Clarke)(FW 6.5–9 mm; Plate 9.9) is widespread, British Columbia to southern California and east to Colorado and southern Arizona, but was not recognized as distinct until the 1940s. Opler described the biology in California (as *C. raspyon* Hodges, a manuscript name); the larvae feed on new spring foliage of both live oaks, *Q. agrifolia*, *Q. wislizenii*, *Q. durata*, and *Q. chrysolepis*, and deciduous species, *Q. garryana*, *Q. kelloggii*, and in Utah on *Q. gambellii*. Adults emerge in May and June and evidently live a long time, with flight records at Berkeley from mid-May to early November. Eggs are believed to overwinter. *Chionodes pastor* Hodges is similar and occurs in the mountains of Utah and northern Arizona.

PHALACRUS GROUP

Phalacrus Group comprises eight species, of which five occur in America north of Mexico, and three in the southwestern United States. The larvae feed on Malvaceae. Males lack specialized HW sex scaling.

Chionodes phalacrus (Walsingham)(Plate 9.10) was described from Mexico and occurs in western Texas and Arizona. The head and thoracic dorsal scaling is cream colored, sharply contrasting with the blackish brown FW (3.2–5.5 mm), maculation whitish, variable from nearly absent to well developed; HW yellowish white. The similar but larger *C. popa* Hodges (FW 6–8.5; Plate 9.11) occurs from northern to southern California. Larvae were found in January on *Sphaeralcea* at Mint Canyon by Chris Henne and on *Sidalcea* at San Bruno Mountain by De Benedictis, with adults emerging in February. Flight records range from March near the coast to July at 8,800' in the Sierra Nevada, and the end of August in the Ivanpah Mountains, suggesting a second generation in desert areas following summer rains. *Chionodes petro* Hodges (FW 4.5–6 mm; Plate 9.12) is known from San Diego County from specimens we reared on *Sphaeralcea ambigua* in May more than 40 years ago, and one specimen was taken at light in late September, indicating bivoltine development. The head and thoracic scaling is pale ochreous, and the HW is darker than in the two preceding species.

The Lugubrella Group includes 66 species in America north of Mexico, about 80% of which occur in the West. They are characterized by sternum 8 heavily sclerotized medially and a broad, anterior, invaginated lobe that is free, or nearly free, of the sternal plate.

Chionodes lugubella (Fabricius)(Plate 9.13) is a Holarctic species that occurs widely in the West. In North America it ranges from Newfoundland and northeastern United States to Alaska, south at higher elevations to Colorado, western Texas, and eastern California. Specimens are numerous in collections, in part owing to the distinctive FW pattern, black with two distinct white, transverse lines (5.5–9 mm). The larvae are recorded on legumes, including *Dorycnium*, trefoil *(Lotus)*, clover *(Trifolium)*, and vetch *(Vicia)*. There is a European record from birch *(Betula)*, but *C. lugubrella* was not encountered in the Forest Lepidoptera Survey in Canada. Adults have been recorded June to early August.

Chionodes ceanothiella (Busck)(Plate 9.14) and *C. obelus* Hodges (Plate 9.15) are western species that cannot be distinguished by external features, although *C. ceanothiella* is larger (FW 6.5–9 mm vs. 5.5–7.5 mm) and is more widespread, occurring in Alberta and British Columbia to Utah and southern California, whereas *C. obelus* is known only in California. Both feed on *Ceanothus* (Rhamnaceae), *C. ceanothiella* on several soft-leaf species *(Ceanothus integerrimus, C. sorediatus, C. sanguineus, C. thrysiflorus)*, and *Chionodes obelus* from *Ceanothus cuneatus*, a hard-leaf species. The FW and thorax are dark rust to dark chocolate brown. Adults of both species fly from May to July, with records of the more widespread *Chionodes ceanothiella* to early September. A sister species, *C. kubai* Hodges (Plate 9.16, Plate 9.17), is quite similar in phenotype but is remarkable for its adaptation to *Fremontodendron californicum*, an isolated western representative of the tropical plant family Sterculiaceae, which ranges in California along the foothills of the Sierra Nevada from El Dorado County to Kern County and the desert side of the San Gabriel Mountains.

Chionodes retiniella (Barnes & Busck) differs from all other gelechiine species in North America by having pale rust orange FW (5.3–9 mm; Plate 9.18) crossed by four evenly spaced, white bands, a convergence in pattern to conifer-feeding Tortricidae. The larvae feed on pines, *Pinus sabiniana* and *P. ponderosa,* and spruce, *Tsuga,* in British Columbia and California and likely other Pinaceae over the broad distribution, from North Dakota, Colorado, and New Mexico to the Pacific Coast. The caterpillar is brown with white lateral spots at the spiracles and subventral line, illustrated by Duncan. There are flight records from May through September, mainly June to July. *Chionodes grandis* Clarke (Plate 9.20) usually is recognizable by its fine, yellow or yellowish orange scaling of abdominal terga 1 to 3; the FW (7–11 mm) varies from uniform dark gray or black to having short, linear streaks of paler gray in the cell and along the Cu crease and a weak subterminal fascia. *Chionodes grandis* is widespread, from British Columbia and Manitoba to New Mexico and southern California and from coastal dunes to 10,000' in the White Mountains, California, where the FW is entirely dark and the moths resemble *C. bicolor.* Over the broad range, flight records span late June to early October, presumably representing a univoltine pattern, and probably adults do not overwinter. The early stages are not recorded. *Chionodes figurella* (Busck) is a large species (FW 6.5–12 mm; Plate 9.19)

that has a grass-matching FW pattern, unique among western *Chionodes*, pale tan with brownish longitudinal lines along the veins, like other Poaceae-feeding Lepidoptera. Adults were reared from larvae found on *Elymus* (Poaceae) in Los Angeles by Henne and Martin.

Two widespread species, *Chionodes psilopterus* (Barnes & Busck) and *C. praeclarella* (Herrich-Schaeffer), also are grass feeders but are not superficially similar to *C. figurella*. The moths are somber colored, FW (7.5–12 mm, males larger than females) gray brown usually with orange brown scales posterior to the costa and dorsal area, contrasting with the darker vein scaling and terminal area. *Chionodes psilopterus* has relatively broad wings, FW (6–9 mm; Plate 9.21) dull gray brown with blackish spots in the cell, on the crease, and at the end of the cell. It has a northern range, in the West from Montana and eastern Washington to Ft. Yukon, Alaska, on the Arctic Circle. *Chionodes psilopterus* has been reported as a pest of bluegrass seedling fields *(Poa pratensis)* in Washington. The larvae feed on the lower stems, crown, and roots through summer and fall, then overwinter and resume feeding in spring. There is a single annual generation, and adults fly from May to mid-August, mainly June to July. *Chionodes praeclarella* is a Holarctic species, occurring in the Alps and central Asia and across North America to British Columbia, Oregon, and the mountains of New Mexico and Arizona. *Chionodes bicolor* Clarke (Plate 9.22) flies in the fall and is sometimes numerous at black light in southern California. The larva is unknown, as is true of most members of the species group, possibly because they are all concealed grass feeders. Specimens of *C. bicolor* from the type locality, Petaluma, California, have shining, dark brown FW streaked with orange brown basally and in the dorsal area, but in southern populations the FW (5.5–7.5 mm) usually is uniformly dark brown with a shining, purplish luster, sharply contrasting with bright yellow orange scaling on abdominal terga 1 to 3. This species occurs from eastern Nevada to coastal California, including Santa Rosa Island, where it flies abundantly in open, formerly grazed fields in late September.

Species of the *C. distinctella* complex are primarily conifer feeders, based on what is known for Palaearctic species. *Chionodes continuella* (Zeller), a Holarctic species, and *C. sattleri* Hodges are similar, having dark gray to black FW (6.5–10 mm; Plate 9.23, Plate 9.24) with variable, whitish transverse bands, basal, median, and subterminal, the inner two sometimes reduced to vague indications. Larvae of *C. continuella* have been found feeding on lichen *(Cladonia)* and spruce *(Picea)* in Europe and on white spruce *(P. glauca)* in Alberta and British Columbia. Larvae of *C. sattleri* are unreported, but the two species were confused prior to the recent description of *C. sattleri*, which is one of the most frequently collected *Chionodes*. Presumably *C. sattleri* also feeds on boreal conifers. Both occur across North America in the north and Colorado to Alaska, and adults fly in late June to early September. *Chionodes lictor* Hodges (FW 7–7.5 mm; Plate 9.25) has the same basic FW pattern but the whitish scaling of the FW is primarily shining white, with only partial gray transverse bands on the costal half. This is a local species, occurring from southern Oregon and Mt. Shasta to Sonoma County, California. It was reared from Douglas-fir *(Pseudotsuga menziesii)* by J. Volney's group during studies of the spruce budworm *(Choristoneura*, Tortricidae). *Chionodes praetor* Hodges is smaller (FW 5–8 mm; Plate 9.26) and has a FW pattern similar to that of *C. continuella* and related species. It occurs from Wyoming to Oregon and the mountains of southern California. Adults have been reared from alpine fir *(Abies lasiocarpa)* and Colorado spruce *(Picea pungens*) in British Columbia, and lodgepole pine (*Pinus contorta*) in the Sierra Nevada. Flight records are June to August.

Chionodes braunella (Keifer)(Plate 9.27) is a widespread legume-feeding species, one of the most commonly collected *Chionodes* in the western Nearctic, ranging from Alberta and British Columbia to Colorado, Arizona, and California. The FW (6–10 mm) varies from pale to dark brown, but *C. braunella* is recognizable by having the terminal area, distal to a faint subterminal fascia, darker than the rest of the wing, and by a patch of dark gray scales on the underside of the male FW. The larvae have been found on a variety of Fabaceae, including native pea (*Lathyrus*), deerweed (*Lotus*), several species of bush lupines (*Lupinus*), clover (*Trifolium*), and vetch (*Vicia*). Hodges regarded a population in southern Ontario as conspecific, including a specimen reared from *Gaylussacia* (Ericaceae) by McDunnough.

The genus *Filatima* includes robust species, most of which are larger and broader winged than most *Chionodes*, and usually the labial palpus scaling is thick and more broadly spreading. The abdominal terga 1 to 3 are ochreous-scaled in most species, rarely so in *Chionodes*. Males are distinguished from those of *Chionodes* and *Aroga* by having tufts of long, slender, dorsolaterally curved scales from the lateral margins of abdominal tergum 8 and usually a fringe of elongate scales on the underside of the HW, directed posteriorly from the Sc vein. *Filatima* lack the tuft of long scales from the base of the HW dorsally, characteristic of *Aroga*. The male genitalia are enclosed by a hood formed from the enlarged A8 and often are asymmetrical, with the split valvae differing markedly, and the aedeagus is heavily and complexly sclerotized. Most of the species are somber, brown to gray, with darker spots and a pale subterminal fascia, whereas both *Chionodes* and *Aroga* include many species that are boldly patterned in yellow or white. There are 55 described species of *Filatima* in America north of Mexico, about 40 of which occur in the West, and probably at least that many more undescribed species in collections.

Filatima ornatifimbriella (Clemens) is a widespread species that feeds on legumes. It occurs in the East and Midwest to Colorado, Texas, and Arizona, feeding on lupines (*Lupinus*), false indigo (*Amorpha*), *Thermopsis* in Colorado, and locust (*Robinia*) in Arizona. The moth is entirely dark scaled, FW blackish brown with bright violet iridescence when fresh. There are two conspicuous, black stigmata in middle and at end of the cell, and lesser ones along the Cu crease and in cell, and a row of black dots at base of fringe. The caterpillar is gray with fine reddish stripes and sometimes is numerous, causing damage to locust trees.

Filatima aulaea (Clarke) was described from Mt. Wilson in southern California and occurs north to Siskiyou and Modoc counties. We have found the larvae feeding on gooseberries (*Ribes*, including *R. aureum*, *R. californicum*, *R. menziesii*, and *R. roezelii*, Grossulariaceae) in May and June. Reared adults vary (FW 7–8.5 mm; Plate 9.28), gray or brown with small to conspicuous black stigmata variously highlighted by pale ochreous scaling. Clarke distinguished two *Ribes*-feeding species in Utah, *F. vaniae* Clarke and *F. prognosticata* (Braun), which differ from *F. aulaea* by having paler FW. *Filatima vaniae* has pale gray to whitish FW (6.5–8 mm) with the costal area ochreous tawny, a dark streak in the crease, dark discal spot at the end of the cell, and a whitish transverse subterminal fascia. It was described from specimens reared by P. H. Timberlake. *Filatima prognosticata* is larger and darker and has much longer scales posterior to the HW scale fringe.

Filatima demissae (Keifer) is a moderately large gelechiid, FW (7.5–10 mm; Plate 9.29) uniform dark ash gray with dark stigmata at basal one-third of crease, two discal marks, outwardly angled from the crease, and a larger, transverse, blotch at the end of the cell, all highlighted by pale scales. The HW underside fringe is directed obliquely outward except a row from vein R, which is directed posteriorly. This species is widespread in California, and the gray, white-striped larvae have been collected on choke-cherry (*Prunus virginiana* var. *demissa*, Rosaceae) in Sonoma County (Keifer), the Sierra Nevada (EME), and Lassen-Modoc area (Crabtree). Adults emerge in June and July. The larvae tie up the young foliage, feed within a mass of living and dead leaves, and kill the terminals. The same or a closely similar species feeds on *Purshia* on Mt. Shasta and in the Warner Mountains. *Filatima saliciphaga* (Keifer)(Plate 9.30) has similar larval habits on willows, *Salix sessilifolia*, in the Sacramento Valley, adults emerging in September, and on *S. lasiolepis* and *S. exigua* on Santa Cruz Island, California, where there are two generations. Larvae we collected in April produced adults in June and July, and those taken in September yielded adults in November. The moths are paler than most *Filatima*, FW (7.7–9.5 mm) pale tan, scale tips ochreous brown. There is a dark mark along the dorsal margin at the base, faint, dark stigmata at one-third and two-thirds in the cell. A closely related species, *F. procedes* Clarke, was described from Presidio, Texas, west of Big Bend, also reared from *Salix exigua*.

Filatima depuratella (Busck)(Plate 9.31) is quite distinctive, having cream-colored wings and body scaling, fading to white with age, FW with a bold, blackish mark angled outward from cell at one-third to the Cu crease at middle of the wing. *Filatima depuratella* occurs in desert foothills of southern Arizona, flying in April and September to October. The larval biology is unknown.

Superficially, the genus *Aroga* includes two groups: broadwinged, primarily brown species, FW with a pale subterminal, usually broken fascia, that resemble many *Filatima*, or species having narrower FW, usually dark brown with white to ochreous tan pattern, analogous to members of the *sistrella* complex in *Chionodes*. Males are distinguished from *Chionodes* and *Filatima* by having a tuft of long scales at the base of the HW dorsally (which is hidden from view on spread specimens). About 20 described species occur in North America, at least 15 in the West, and many other unnamed species are in collections.

Aroga eriogonella (Clarke)(Plate 9.32) is one of a group of species that feeds on Polygonaceae. It was described from eastern Washington where Clarke found the larvae on two species of *Eriogonum*. The moths are moderately large (FW 9–11 mm), light to dark brownish gray, FW with a sprinkling of black scales and three stigmata in the cell. Populations of this or a closely related species occur in central California and have the FW tinged with rust brown. Adults fly in August and September. A series of phenotypically similar, brown *Aroga*, including *A. xyloglypta* (Meyrick) and several undescribed species, occurs in coastal habitats along the length of California. We made larval collections from *Eriogonum latifolium* and *E. nudum* in the San Francisco Bay area and *E. parvifolium* in the Santa Maria dunes; we found adults associated with *Polygonum paronychia* at Seaside on Monterey Bay, and S. Baron discovered a population at Sunset Beach on Monterey Bay feeding on *Chorizanthe*. *Aroga eldorada* (Keifer)(FW 6.5–7.5 mm; Plate 9.33) is superficially similar, brown with the dark stigmata indistinct and the pale tan subapical fascia broken. Keifer discovered *A. eldorada* in the foothills of the Sierra Nevada on mugwort, *Artemisia douglasiana* (Asteraceae), on which we have found larvae in northern California, Siskiyou and Plumas counties, and near Clear Lake in the north Coast

Range. Adults of the last population are more gray than brown. Larvae were collected in April and May, and adults emerged from late May to July.

Aroga paulella (Busck) is the largest and most often collected of the black-and-white-patterned species. The FW (8.5–12 mm; Plate 9.34, Figs. 94, 95) has a white fascia angled outward from the costa at one-fourth, a linear blotch along the midcosta, a transverse fascia beyond the cell meeting a bar along the dorsal margin at the tornus. This species is widespread including Colorado, Arizona, and California, in diverse habitats, from desert ranges of the Mojave, coastal serpentine grasslands, Mt. Shasta, mountains of southern California, and the Sierra San Pedro Martir in Baja California. *Aroga paulella* has a single annual generation, adults recorded from the end of March, mostly April to May, to early July at Mt. Shasta. It was reared from *Eriogonum inflatum* by L. Anderson. A complex of species related to *A. unifasciella* (Busck)(Plate 9.36) ranges from Alberta to western Texas and southern California. These have a FW pattern reduced to a white dorsal margin and distal transverse fascia. *Aroga unifasciella* is the largest (FW 7–8.5 mm) and has relatively broad FW lines. It was described from northern Arizona and occurs from Glacier National Park, Montana, to western New Mexico and eastern and southern California. Adults fly in July to August, to early October near the coast. Its larva has not been recorded. *Aroga paraplutella* (Busck)(Plate 9.37) occurs on beach dunes in California from Mendocino to San Diego and on San Miguel Island. The adults are small (FW 5–6.5 mm) with the white to pale orange markings reduced to a thin line along the dorsal margin, and the distal transverse fascia often is broken or angled in the middle. We reared *A. paraplutella* from *Eriogonum parvifolium* at Dune Lakes, San Luis Obispo County; larvae collected in late March and early May produced adults in June. Flight records span May to August. A population at the Antioch Dunes National Wildlife Refuge has broader white FW markings and flies from April to September. *Aroga camptogramma* (Meyrick), described from western Texas, is similar. There are many inland records, from the mountains of southern Arizona, deserts, Sierra Nevada, to Alberta; several species may be involved. *Aroga morenella* (Busck)(Plate 9.35) has broad, longitudinal white bands along the costa and dorsal margin. It is a local species endemic to southern Nevada, southern California, and northern Baja California, in desert and other arid habitats to 4,800 to 5,700′ in the Peninsular Ranges. One specimen from Riverside County is labeled "on *Eriogonum fasciculatum*" (K. Osborne). Adults were recorded from April to July in the Miramar survey.

Aroga trachycosma (Meyrick)(FW 5.5–6.6 mm; Plate 9.38, Plate 9.39) is unique in feeding on alkali heath, *Frankenia salina* (Frankeniaceae). The larvae make sparse silken shelters along the upright stems and sometimes become so abundant that they cover the foliage with webbing. This species occurs at the upper level of tides along the coast of Baja California and California, including Santa Catalina and San Nicolas islands, as far north as San Francisco Bay, but it is not known from inland colonies of *Frankenia* on alkaline soils. The FW in southern populations is gray brown with darker spotting through the cell and dorsal area, and a whitish subapical fascia; rarely the wing is entirely dark grayish. In San Francisco Bay populations, there is a dark form (Plate 9.39), dark gray brown with blackish stigmata in the cell. We have collected *trachycosma* in April and May in southern California and April, September to October at San Francisco Bay; and probably it is multivoltine in all populations. *Aroga elaboratella* (Braun), from Bahia de Los

Angeles, Golfo de Baja California, appears to be the same species. Both were described in 1923.

Faculta has only two described species in the western United States, and although *F. inaequalis* (Busck)(Plate 10.20) is sometimes abundant at lights and is numerous in collections, evidently nothing is known of its larval biology. The adult is small (FW 4.5–6.5 mm), primarily brown, FW pale basally with a well defined blackish fascia from the costa before the middle angled outward, followed by variable rose rust scaling. This species occurs in low desert areas along the Colorado River in southeastern California to montane (7,000–8,500′) sites in the Huachuca and Chiricahua mountains, Arizona, into Sonora and Baja California Sur. Adults are recorded in January and April to August, suggesting a multivoltine life cycle.

TRIBE GNORIMOSCHEMINI

The tribe Gnorimoschemini was proposed by Dalibor Povolný in 1964 to accommodate genera related to *Gnorimoschema*, and subsequently he described numerous genera on the basis of subtle differences in male genitalia without phylogenetic analysis. Contemporary gelechiid specialists do not agree with Povolný's generic concepts, but many of the genera represent apparent radiations on Chenopodiaceae, Solanaceae, or Asteraceae. In America north of Mexico, Gnorimoschemini include the species 1968–2053 in the MONA Check List along with several later-described genera. Descriptive taxonomic work since 1980, carried out by W. E. Miller, P. Huemer, Hodges, and particularly Povolný, has increased the number of described North American species almost 100%, to about 165, with about 75% of them western, the species richness concentrated in the southwestern states.

The genus **Gnorimoschema** has only recently received careful attention, increasing our knowledge and appreciation of its taxonomic and biological diversity. More than 80 species are described in America north of Mexico, at least 60 of them occur in the West, at elevations from sea level to 12,500′. About 60% of the western species were named by Povolný between 1998 and 2004. Although most *Gnorimoschema* are larger than species of related genera, they are slender, streamlined appearing moths that hold the wings tightly to the body in repose. The FW of most species is gray to brown or rust, often with a basal patch of contrasting whitish, yellow, or rust. Species for which larvae are recorded feed on Asteraceae, except for a few sand-dwelling species, which may be general feeders. *Gnorimoschema* includes a remarkable array of larval biology. There are leaf miners, foliage tip tiers, and stem borers, and many induce the host plant to develop galls, in which the larvae feed (Plate 60.11, Plate 60.12, Plate 60.13). Several sibling species in the eastern United States that cause stem galls on goldenrod (*Solidago*) were studied in detail by Miller, but western goldenrod feeders have not been discovered.

Larvae of a few *Gnorimoschema* are leaf miners or live in active sand dunes and feed by mining buried leaves from silken tubes. *Gnorimoschema saphirinellum* (Chambers) is a slender, narrow-winged species with reddish brown to cinnamon brown FW (5.5–6.5 mm; Plate 9.40), varying to grayish and tan in some populations. The source of the type specimen is uncertain; it may have been from San Francisco (Behrens) or from Texas (Belfrage). The name is applied to a species that occurs on the coasts of New England and Florida and in widespread inland areas of the Southwest. In California, *G. saphirinella* occurs in coastal beach dune habitats from Mendocino to the Channel Islands. The larvae feed on buried leaves of beach bur, *Ambrosia chamissonis*, and were recorded by Goeden and Ricker as leaf miners on *A. confertifolia* and *A. psilostachya* inland in southern California. The life cycle is not well defined; adults, which appear to be diurnal, have been observed in the field from March to July and have emerged in the lab from May to August and in December. ***Gnorimoschema vastificum*** Braun is small (FW 5.7–7.5 mm; Plate 9.41) but broader winged, white ventrally with the FW color resembling sand, tan mottled with ochreous, pale brownish, and rust. It occurs in southern Manitoba and Alberta, the Great Basin, and on the Sacramento River delta in California. The adults run about on the riverine sand during the daytime, and the larvae live in silken tubes attached to buried leaves of willow *(Salix)* and may be indiscriminate feeders. Adults are recorded in April and July to August, and there is a record from San Francisco in October, probably from Koebele in the 1880s prior to destruction of the San Francisco sand dunes. *Gnorimoschema vastificum* appears to be closely related to *G. bodillum* Karsholt & Nielsen, which occupies the same niche on beach dunes of coastal Denmark and Germany.

Larvae of **Gnorimoschema debenedictisi** Povolný and Powell (Plate 9.42) mine the new spring leaves of *Erigeron glaucus* in early instars, then make silken shelters in subtending leaves, from which they burrow downward into the meristem. They form cocoons in tightly rolled leaves. De Benedictis recorded this species also on *Solidago*, which grows intermingled with *Erigeron* at the type locality, San Bruno Mountain on the San Francisco Peninsula, but he expressed some doubt about the association of the latter plant. This species has relatively broad FW (5.5–8.5 mm), gray, irregularly black spotted, a triad of stigmata in the cell and submarginal stigmata, more distinctive in females. Larvae were collected in March and April, adults emerged in May and June and were taken at lights May to July.

Soft, ephemeral stem and tip galls of several kinds are caused by *Gnorimoschema*. **Gnorimoschema coquillettella** Busck and *G. ericameriae* Keifer are sister species or races that induce onion dome–shaped tip galls on goldenbush, *Ericameria*. The terminal leaflets become thickened and sealed, forming a chamber at the twig tip (Plate 60.11). At maturity the larva leaves to pupate in litter on the ground, and the gall remnants wither and dehisce. The larvae mature in April and May, and adults emerge in June and July. *Gnorimoschema coquillettella* has a bright yellow basal patch and pale gray FW markings (FW 5.5–7.5 mm; Plate 9.44). It occurs in inland areas of California and feeds on *E. arborescens*, *E. linearifolia*, and *E. pinifolia*, and there are nineteenth-century records for *Isocoma veneta* and *Acamptopappus sphaerocephalus* in the western Mojave, which need confirmation. *Gnorimoschema ericameriae*, which is strictly a coastal species on *E. ericoides*, has dark gray FW (5–7.5 mm; Plate 9.43, Fig. 99) mottled with deep reddish, including the basal patch. Members of a complex related to *G. octomaculellum* (Chambers)(Plate 9.45) induce similar tip galls on *Artemisia* and *Chrysothamnus* in the Great Basin and *Acamptopappus* in the Mojave Desert. The moths are pale gray, FW (5–10 mm) uniformly peppered with fine black dots and crossed by three bright rust orange bands or pairs of spots, the basal one sometimes reduced to a trace.

Gnorimoschema grindeliae Povolný & Powell (Plate 10.3) is host specific on gumplant, *Grindelia hirsutula*, growing on sandstone bluffs above San Francisco Bay. This plant is perennial but dies back in late summer, and the galls grow in the newly developing stems in early spring, causing compression of the stem nodes and clumping of foliage above them (Plate 60.12). Larvae of *G. grindeliae* are mature by late April and leave to pupate on the ground. The stem continues growth, the gall tissue deteriorates, and by summer the galls are evidenced only by the clustering of nodes and twigs they caused in spring. The moths are moderately large (FW 8–11 mm) with rust brown to purplish brown FW when freshly emerged, subtly mottled with ochreous and whitish basally and a subterminal line. Older, flown specimens become pale rust with extensive whitish areas. A sister species, *G. crypticum* Povolný & Powell, is smaller (FW 5.5–8.5 mm; Plate 10.4, Plate 10.5) and paler rust when fresh, with gray ground color in island populations. This species induces similar stem galls to those of *G. grindeliae*, on *Hazardia squarrosa* along the California coast in Monterey County and *Isocoma acradenia* and *I. menziesii* near Morro Bay and on the Channel Islands.

Presumed more-derived galls induced by *Gnorimoschema* are persistent and turn woody after the stem dries. Larvae of some species leave the gall to pupate on the ground, whereas others prepare an emergence window and track from the cocoon, facilitating emergence of the moth from the gall. **Gnorimoschema baccharisella** Busck (Plate 10.2) is the best-known western species, having been studied by J. W. Tilden and others. The FW (6.5–9.5 mm) is mostly streaked longitudinally with gray and whitish, basal patch sharply defined, rust and ochreous. Adults fly in late summer and fall, deposit eggs that overwinter on coyote brush, *Baccharis pilularis*. Newly hatched larvae enter the growing tips in early spring, causing a spindle-shaped gall to develop, which reaches full size (18–36 mm; Plate 60.13) by late February or March. However, young larvae do not survive if galls are removed from the plant before late May or June. At maturity the larva cuts a hole to the outside, drops from the plant, and forms a cocoon in ground litter. Galls collected in May, June, and July produce adults from late July to October, and most collections of adults at lights have been in September and October. *Gnorimoschema baccharisella* occurs throughout the range of its host plant in coastal and foothill regions of California, the Channel Islands, and probably southern Oregon and Baja California. Although heavily parasitized in the larval stage by a complex of wasps, the moths sometimes are numerous at lights. **Gnorimoschema powelli** Povolný is our largest species (FW 8.5–11.5 mm; Plate 10.1), FW dark gray, with the veins outlined by white scaling, lacking a basal patch, the stigma at the end of the cell usually evident. The larvae induce stem galls on *Baccharis sarothroides* in coastal southern California and southern Arizona. Galls collected in early October produced moths in late October and November, and flight records in the Miramar survey spanned November to early February. The geographical distribution of **G. subterraneum** Busck (Plate 10.6, Plate 10.7) is a mystery. It was described from Connecticut, where stem galls on *Aster* occurred below the soil surface level, then were not reported again for 75 years until we found populations at several sites around San Francisco Bay. Here the watermelon-shaped stem galls (17–27 mm long) occur on *Aster (Symphyotrichum) chilensis*, mostly 12 to 15 cm or more above the ground, although one colony had galls buried in leaf litter, 2 to 5 cm above the crown. Larvae are fully grown by April or May and remain in the gall for pupation, and galls collected in April, May, June, and August produced adults from late August to early October. Moths reared in California are larger (FW 7.5–10.5 mm) than those in Connecticut, but structurally are indistinguishable, FW uniform brick red with pale gray fringe. Larvae of a parasitic wasp (*Microdontomerus*, Torymidae), which are gregarious, nine to 12 per gall, wait in the woody gall until the following year to transform and emerge in July, after the new crop of *Gnorimoschema* larvae are fully grown.

Scrobipalpopsis is a primarily Nearctic genus with seven species, larvae of which also are Asteraceae feeders, on foliage or in stem galls. *Scrobipalpopsis arnicella* (Clarke) was described from southeastern Washington from females that were reared from *Arnica*. Supposed conspecific males from northeastern Washington and Siskiyou County, California, differ from one another and from *S. madiae* Povolný & Powell, which we reared from *Madia madioides* at Big Creek, Monterey County. Thus three or more species may be involved in this complex. All have white FW (6–7 mm) with black, somewhat elongated stigmata in the cell and other isolated dots, including a submarginal row around the termen. *Scrobipalpopsis madiae* has an ochreous brown shade subtending the stigmata and in the distal area. Larvae of these species are foliage feeders in spring.

Euscrobipalpa (=*Scrobipalpa* of authors) includes a huge Palaearctic fauna, more than 250 species, whereas only nine occur in the western United States, three of them believed to have been introduced. *Euscrobipalpa artemisiella* (Treitschke) (Plate 10.11) is widespread in Europe and has been reared from growing tips of *Artemisia douglasiana* at five scattered locations in California. Also, it has been reported in Saskatchewan and Nevada, so is assumed to be natively Holarctic. In Europe some populations called *E. artemisiella* feed on *Thymus* (Lamiaceae), and some authors have speculated that nineteenth-century misidentifications led to the notion that this species feeds on *Artemisia*. No Nearctic gnorimoschemine is known to feed on Lamiaceae, and none feeds on plants of two families. There are no appreciable differences in structural features between California and Palaearctic specimens, so there is an unresolved question concerning possible sibling species. In California, *E. artemisiella* has dark gray to rust-colored FW (6–7 mm) with typical triad of stigmata, and there is geographic variation: populations at the coast are darker more gray moths, inland ones pale rust. This species is univoltine in Europe and may be bivoltine in California. Our larval collections were made in April and May, with adults emerging in May and June, and there are flight records in July and September. *Euscrobipalpa obsoletella* (Fischer von Röslerstamm)(Plate 10.10), also a Palaearctic species, has colonized western North America. It was described in Washington by Clarke in 1932 (as *E. miscitatella*, a synonym), and a few years later was reported by Keifer in California, where it occurs in interior valleys and coastal southern California. The larvae are stem borers in the European weed *Chenopodium murale* (Chenopodiaceae), killing the inflorescences. Pupation occurs in the stem. *Euscrobipalpa obsoletella* is multivoltine in England, with adults active from May to September.

Scrobipalpula is a primarily Neotropical genus with more than 40 described species, including several in the Nearctic. The known larvae are vegetative tip borers in Asteraceae. *Scrobipalpula psilella* (Herrich-Schäffer) is a member of a complex of closely similar species over an enormous geographic range in the Palaearctic. After study of the complex, Huemer and Karsholt suggested that Western Hemisphere populations assigned to this name represent phenotypically similar sibling species. This may be true, and we have found seemingly distinct phenotypes in California, where *S. psilella* sensu lato (Plate 10.12) seems to be restricted to cudweeds, *Anaphalis* and *Gnaphalium*, whereas in Europe *S. psilella* feeds on *Artemisia* and *Aster* as well as *Gnaphalium*. The moths are small (FW 4.5–6.5 mm) and not easily distinguished superfi-

cially from several other species of *Scrobipalpula* and other genera, FW pale to dark gray with the characteristic triad of stigmata usually well defined and highlighted by ochreous in the cell. The larvae feed as terminal foliage and bud miners and bore into the meristem or a stem axil. Pupation usually occurs in a silk-lined hollow at the point of feeding. In California larvae are found in November after winter rains start until May, and adults emerge February to May. Collections of adults at lights, with most records March to June and October, suggest two or more generations. None of our larvae expressed a summer diapause. *Scrobipalpula gutierreziae* Povolný & Powell is similar, FW (4.5–6.5 mm; Plate 10.13) dark gray with the stigmata distinct, black, outlined by dark ochreous. The larvae feed on snakeweed, *Gutierrezia californica*, March to May, and adults emerge August to October. *Scrobipalpula gutierreziae* is known from Antioch on the San Joaquin delta and in eastern Alameda County, California. *S. antiochia* Povolný & Powell (Plate 10.14) is similar, with the costal half of the FW ochreous orange. Its larvae fed on *Senecio douglasii* in February–March, and adults emerged in March–April.

Scrobipalpulopsis lutescella (Clarke)(Plate 10.15) is widespread, from Ontario to the Pacific states, from eastern Washington to coastal central California and Santa Rosa Island. Related species are Neotropical, and *S. lutescella* is unique among Nearctic gnorimoschemines in feeding on Scrophulariaceae. Larvae live in the tubular corollas and feed on immature seed of Indian paintbrush, *Castilleja*, including *C. affinis*, *C. foliolosa*, and *C. wrightii* on the California coast. *Scrobipalpulopsis lutescella* is multivoltine, with larvae observed from March to August. There is a facultative summer diapause; larvae collected in April produced some adults in May while sibs aestivated, delaying emergence until September to October. Those collected in June to August eclosed in the fall. Adults are recorded at lights in April, July, and November. They are small (FW 5–6.5 mm), FW ochreous tinged with rose pink, variably obscured by dark gray to blackish through the cell or most of wing other than the dorsal area.

Species of *Tuta* are leaf miners in Chenopodiaceae and occur in saline and maritime habitats of the Atlantic and Pacific coasts in both north and south temperate zones. They are the smallest gnorimoschemine gelechiids in our region (FW 4–5.5 mm). *Tuta chiquitella* (Busck)(Plate 10.16, Fig. 98) is a gray species resembling a diminutive *Scrobipalpula*. It occurs from New Mexico to coastal California, including the Channel Islands and San Francisco Bay area. The larvae mine leaves of *Chenopodium* and *Atriplex*, both native *(A. patula)* and introduced *(A. semibaccata)*. On the last, a semisucculent species, a few leaves are tied together into a shelter and mined at their bases. *Tuta chiquitella* is multivoltine, with larval collections in February, May, July, and August, and adults have been reared or taken at lights in every month from March to December.

Phthorimaea operculella (Zeller)(FW 6.5–8 mm; Plate 10.21), the "potato tuber moth," which is thought to have been native in the Neotropical region, is a serious pest of potato *(Solanum tuberosum)* and tobacco *(Nicotiana)* worldwide, and there is an extensive literature on its biology. The larvae mine in leaves of tobacco, tomato *(Lycopersicon)*, eggplant *(S. melongena),* and other Solanaceae; and they reduce the value of harvested tobacco. In potatoes the larvae enter the stems and tubers, in the field and in storage, destroying their market values. The life cycle can be completed in a month, and there may be four to six generations per year in warmer climates.

Phthorimaea operculella was known in the United States by the mid 1800s and was introduced to Europe, Africa, and Australia by 1900. It occurs across the southern United States and northward into Colorado, Washington, and California. In addition to field crops, the species persists during the winter and early spring as a leaf miner on weedy plants such as Jimson weed *(Datura stramonium)* and on garden plants, including eggplant, nightshade *(Solanum),* and ground cherry *(Physalis).* The adults are small, tan to pale tan with the dark stigmata indistinct, the three in the cell highlighted by white scaling, and dorsal area often black smudged.

Symmetrischema is a large, primarily Neotropical genus with about 10 species in the Nearctic, one of which is introduced worldwide and is a pest of potatoes in some regions. All feed on Solanaceae. *Symmetrischema striatellum* (Murtfeldt)(Plate 10.17) was described in Missouri in 1900 and recorded in California by the 1930s by Keifer, who assumed it had been introduced because he found it only on *Solanum nigrum,* an introduced plant. If true, the introduction was an early one because there are specimens from Alameda and Los Angeles in the Koebele collection, dating to the 1880s. There seem to be no records from other western states, and its occurrence in California is limited to the coastal counties, including the Channel Islands, in urban and other disturbed situations. The adult is quite distinctive, unlike any other Nearctic gnorimoschemine, with FW pattern wood-grain-like (5.5–7 mm), linear striae of various shades of brown and tan. The larvae make shelters by sealing the margins of a leaf together and skeletonizing the enclosed upper surface. In contrast to Keifer's experience, we have found *Symmetrischema striatellum* on native nightshades, *Solanum americanum, S. furcatum, S. douglasii, S. wallacei,* and *S. xantii,* in widely differing situations, acting like a native insect. *Symmetrischema striatellum* is multivoltine, with larvae found in every month of the year and adults recorded March to December. *Symmetrischema tangolias* (Gyen) is a larger, relatively broad winged moth (FW 6–8 mm; Plate 10.8) having gray FW with variable pattern dominated by a blackish patch from the basal one-fifth of the costa, angled outward to the Cu crease. It was first described from Chile, then introduced into Australia and named *S. plaesiosema* by Turner, by which it was known for many years after introduction to North America. The larvae are stem borers in *Solanum,* reared in California on both *S. nigrum* and *S. douglasii,* but *Symmetrischema tangolias* has not adapted to potatoes or tomatoes here.

The notorious "tomato pinworm," *Keiferia lycopersicella* (Walsingham)(Plate 10.18), has a curious nomenclatural history. It was first described in 1897 from the West Indies, where likely it had been introduced. But Walsingham's species was not recognized, and after it began to be noticed in other tomato growing regions, it was described again, this time from Hawaii by Busck, who coincidentally gave it the same species name. Later, when the two were recognized as congeneric, Busck's name became both a synonym and a homonym. The tomato pinworm occurs across the southern United States and was identified in California by 1923. The early-instar larvae create serpentine leaf mines and later fold leaves with silk. They cause economic damage by boring into buds and developing fruit, leaving tiny holes. This insect has been the subject of eradication programs and may not persist as a field pest in the United States except in the warmest regions. There is a closely similar sibling species in California, *Keiferia elmorei* (Keifer), with differences in genitalia, that depends upon native nightshades, including *Solanum umbelliferum, S. clokeyi,* and *S. xantii.*

Both are dark gray, obscuring darker FW markings and ochreous longitudinal streaks, the triad of stigmata hardly distinguished (FW 4–5.5 mm), *K. elmorei* averaging larger than *K. lycopersicella*). Both are multivoltine, with larval collections in all months except November to January.

Exceptia sisterina Povolný & Powell is atypical for Gnorimoschemini, larger than most, with elongate-narrow FW (10–11 mm; Plate 10.9) that are nearly uniform pale tan to brownish, with ochreous streaks between the veins distally. There are two slightly elongate, blackish stigmata in and at the end of the cell, either or both of which are often reduced to a trace or lost. The labial palpi segment 2 has erect, bushy scaling. This species occurs along the coast of California, from Humboldt County to Monterey County, in beach dune habitat, chaparral, and oak-madrone woodland, but nothing is known of the larval biology. Adults are recorded June to August.

Caryocolum is a Holarctic genus with more than 60 described species, most of them in Europe and Central Asia, but several are either Holarctic in distribution or occur only in North America, mostly in montane regions of the West. These are broad-winged moths, superficially like many Gelechiini. *Caryocolum pullatellum* (Tengström)(Plate 10.19) occurs in northern Europe, northwestern Russia, and Japan and is the most commonly collected *Caryocolum* species in North America. It ranges across the northern United States to Utah, Oregon, and northern Arizona. The FW (5–6.5 mm) shows extraordinary variation in color and pattern, nearly entirely dark brown to having two white patches in the cell at one-fifth and middle, and subterminal costal and tornal spots separate or confluent, forming a broad transverse fascia; and sometimes there are orange brown subcostal and dorsal streaks. The larval food plant is unknown, but Palaearctic *Caryocolum* so far as known (33 species) feed exclusively on Caryophyllaceae, such as *Cerastium, Silene,* and *Stellaria.* The larvae feed within shoots, seed capsules, and stems, some causing galls.

TRIBE ANACAMPSIINI

Mirificarma eburnella (Denis & Schiffermüller), a Palaearctic species, is distinctive among western gelechiids, having narrow, apically pointed, rust orange FW (6–7 mm; Plate 10.25) streaked with yellow. It was first recognized in North America when larvae were found defoliating ladino clover (*Trifolium repens,* Fabaceae) in the Sacramento Valley, California, in 1969 (reported as *M. formosella* and later *M. flamella,* which are synonyms). By that time, however, it was already widespread, having been first collected by O. W. Richards at Georgetown (2,650′ in the Sierra Nevada) in 1961 and by A. and G. Keuter near Sacramento in 1965. Later, *M. eburnella* spread, using native and weedy clovers, 200 km along the Sierra foothills and west to the inner Coast Range. Evidently there are two generations, with most flight records late April to June, a few in September to October.

Syncopacma has five North American species, two of which occur in the West, *S. nigrella* (Chambers)(FW 4.3–6 mm; Plate 10.22) and *S. metadesma* (Meyrick). The identity of *S. nigrella* is uncertain because the type locality was not stated, and the alleged type specimen is poor and does not match the original description. Chambers's description follows a series of 14 species of *Gelechia* and *Tinea* he described from California, which he received from Behrens in San Francisco, so coastal California may be the original source of

S. nigrella. Populations assumed to be this species occur widely in the East, to Wyoming, Washington, and California. Chambers described the moth as entirely black, with no mention of a white, transverse fascia, but the supposed type has a thin fascia, as does *S. metadesma*, described from southern California. Keifer reported that one-third of specimens reared in San Francisco lacked or had only traces of the fascia. Specimens from elsewhere in California, Sierra San Pedro Martir in Baja California, and southern Arizona have a complete transverse fascia. The larvae feed on bush lupines (*Lupinus*, Fabaceae), tying the leaflets into a bundle and mining them from within the shelter. In Washington, Clarke reported pupae overwintering, adults emerging in early spring, whereas in California most larval collections have been made in late February and March, and adults emerge in March, but a few were collected in August with adults emerging in September. A smaller species (FW 3.8–5 mm; Plate 10.23), apparently undescribed but possibly true *S. nigrella*, is entirely black and occurs in coastal central California, sympatric with *S. nigrella* at some sites, and on all the Channel Islands. Adults fly in association with Fabaceae, often *Lotus,* in spring (April to May) and fall (August to September). We reared it from *Astragalus didymocarpus* on Santa Rosa Island in March and *A. miguelensis* on San Miguel Island in October.

Batteristis also has five described Nearctic species, two of which occur in Arizona and California. *Batteristis pasadenae* (Keifer)(Plate 10.24) was described from males collected at Pasadena, California, in September, one of the few species Keifer named based on specimens he did not rear. The adults are small (FW 4.2–6 mm) but broad winged, FW gray brown with a thin, pale, transverse line at the end of the cell, strongly angled outwardly at the middle; ground color usually paler in the terminal area than preceding the line. We reared *B. pasadenae* from root crowns of *Gutierrezia* and *Isocoma acradenia* and *I. menziesii* (=*Haplopappus venetus*) infested with larvae of *Sonia* and *Eucosma* (Tortricidae) at three localities in San Diego County. Collections made in September and early October yielded *Battaristis* in October to early January. The larvae may be scavengers in abandoned insect galleries.

Anacampsis includes more than 20 described species in America north of Mexico, most of them in the eastern United States. Genitalia, Figs. 100, 101. *Anacampsis fragariella* (Busck)(Plate 10.29), called the "western strawberry leaf roller" in older literature, was described from an infestation on strawberries (*Fragaria*, Rosaceae) at Pullman, Washington, and was later recorded damaging foliage of strawberry in Idaho and California. The adult is small (FW 6.5–7 mm), broad winged, milk chocolate brown with an indistinctly defined, darker brown, transverse band at distal one-third of the FW. Probably this is a native insect that opportunistically adopted strawberry as a host. We have reared *A. fragariella* in northern California from native Rosaceae, including thimbleberry *(Rubus parviflorus)* at Mt. Shasta and *Potentilla* in Trinity County, and M. Buegler found the larva on *Horkelia* in coastal Mendocino County. *Anacampsis niveopulvella* (Chambers)(Plate 10.30) is perhaps the most frequently recorded Nearctic *Anacampsis*, occurring in boreal forests across the continent to Utah, Idaho, northern Arizona, and along the eastern Sierra Nevada. The larva is a leaf roller, feeding primarily on aspen (*Populus tremuloides*, Salicaceae); for example, 95% of 627 larval rearing records by the Canadian Forest Insect Survey, and most of the remainder were *Salix* and other species of *Populus*. The adult is moderately large, FW (7.5–9 mm) dark gray crossed by a conspicuous, zigzag, white, subterminal fascia and dusted with whitish, more densely beyond the fascia; HW dark brown. Larval, pupal, and adult collections have been made in July and August. *Anacampsis lacteusochrella* (Chambers)(Plate 10.31) is a smaller, pale-colored species that occurs in desert and coastal sandy habitats in Baja California and California, north to the San Joaquin Delta at Antioch. The FW (5.5–7.5 mm) is grayish tan becoming paler with age, with faint, darker stigmata in the cell and a whitish subterminal fascia, often reduced to its costal part. We have reared this species from several localities feeding in concealed shelters on *Croton californicus* (Euphorbiaceae). Adults have been collected from March to October, larvae in February and May to August. On Santa Cruz Island, where *Croton* is absent, *A. lacteusochrella* has adapted to the weedy euphorb, turkey mullein *(Eremocarpus setigerus),* possibly a result of introduction of both the plant and the moth. On the mainland, *Eremocarpus* is widespread and abundant in disturbed places such as grazed land in the California Central Valley, but *A. lacteusochrella* is not recorded on it.

The "peach twig borer," *Anarsia lineatella* Zeller, is a small (FW 5.5–7 mm; Plate 10.26), dark gray moth which is recognized by its peculiar projecting labial palpi, and the FW is gray with darker longitudinal streaks, the most conspicuous beyond the middle on the costa. It has been distributed throughout stone fruit–growing regions of the world by transport of dormant nursery stock. In the western United States, *Anarsia lineatella* was most damaging on the Pacific Coast. It has been suppressed in commercial orchards by winter spray treatments but persists in unsprayed orchards and garden trees. Young larvae overwinter in hibernacula just under the bark, evidenced by tiny turrets of frass; in spring they burrow into newly developing twigs, and each can kill one to several terminals. Later, one or two facultative generations may feed in the fruit. Peach (*Prunus persica*, Rosaceae) is the preferred host plant, but apricot *(P. armeniaca)* and other stone fruits are used. Now *A. lineatella* is found primarily in marginal agricultural and suburban situations and does not seem to have adapted to native *Prunus*. Flight records in California are May to August.

SUBFAMILY DICHOMERIDINAE

Members of this subfamily differ superficially from most other gelechiids in having the labial palpi only weakly upcurved, more porrect rather than evenly arched, and the second segment often has a conspicuous scale tuft, whereas the third segment is erect, slender, and acute tipped. The FW termen is truncated, giving a rectangular appearance to the wing; and the HW termen is only weakly indented below the apex. As a result, species of Dichomeridinae often were described in Oecophoridae (sensu lato) or other taxa, and many tropical species remain incorrectly assigned. This is a relatively well studied group in North America, having been monographed by Hodges in 1986.

Dichomeris is a large genus, diverse in the Neotropical Region, northward into the southeastern United States. There are 84 described species in America north of Mexico, and only about one-fourth of them occur in the West. The larvae are leaf tiers and host-plant specialists on several unrelated plant families, especially Asteraceae. *Dichomeris stipendaria* (Braun) occurs from British Columbia to Oregon, Utah, and northeastern California. It is relatively large (FW 6–9 mm; Plate 10.32, Figs. 102, 103) and has pale rust tan FW with distinctly defined dark brown pattern, a basal bar on the posterior half, curving into the cell at the middle, and the termen broadly. The larvae are leaf rollers on Asteraceae, *Solidago* in northern Utah, *Aster* and

Erigeron in eastern Washington, in June, adults emerging in July. **Dichomeris delotella** Busck is similar but smaller (FW 6–8 mm; Plate 10.33). Its FW color is uniform rust brown, with the dorsal mark restricted to a transverse triangle into the cell at the basal one-fourth. This species ranges from Sonora, Mexico, through southern Arizona, to southeastern California, and on Santa Cruz Island. *Dichomeris delotella* has been collected from March to December in southern Arizona, July to September in California. The larval host is unknown but suspected by Hodges to be *Eupatorium* and *Solidago* (Asteraceae). **Dichomeris baxa** Hodges is a Californian endemic, occurring from Trinity County to San Diego, often in coastal habitats. The FW (6–8.5 mm; Plate 10.34) is tan with the posterior half shaded dark brown, indistinct distally, and the termen is narrowly dark. This species appears to be a southern counterpart of **D. gnoma** Hodges (Plate 10.35), described from British Columbia and Washington, which is larger with the same FW pattern more brightly colored. Larvae of *D. baxa* were found feeding in terminal foliage of *Lessingia* (=*Corethrogyne*) *californica* by Clarke on coastal dunes at Monterey Bay, by us at the Santa Maria dunes, and by Pierce at Playa del Rey, Los Angeles County, with adults emerging in April and May. Adults were collected in February, May, and June at Miramar and June to July at inland, higher elevations. **Dichomeris mulsa** Hodges is distinctively patterned and our largest *Dichomeris* (FW 8–10 mm; Plate 60.10); it is found in central and southern Arizona, yet nothing is known of the larval biology.

Dichomeris leuconotella (Busck)(Plate 10.36) has very dark, almost black FW, with the darker basal bar and termen scarcely discernible and is more variable in size than most *Dichomeris* (FW 5–8.5 mm). It occurs from the Northeast along the Canadian border to eastern Washington, and there is an isolated record from Denver, Colorado. The larvae feed on several Asteraceae, *Solidago, Helianthus,* and *Aster* in southeastern Canada. *Dichomeris leuconotella* is univoltine, with adult flight from June to August, mostly July.

The "juniper webworm," **Dichomeris marginella** (Fabricius)(Plate 10.37), which has been introduced from Europe several times, has a distinctive color pattern contrasted to native species, rust brown FW (5.5–8 mm) with longitudinal, white stripes along the costa and dorsal margin. Several species of *Juniperus* (Cupressaceae) grown as garden ornamentals serve as larval hosts in North America. There have been adventive colonies in the Northeast, Midwest, British Columbia, Oregon, and California, where *D. marginella* was found infesting ornamental juniper at Los Angeles in the 1920s and at Auburn in the Sierra Nevada foothills in 1944 and eradicated.

Helcystogramma is widely distributed, with 75 described species, primarily in the Palaearctic, African, and Neotropical regions. There are nine in North America, five of which occur in the West. **Helcystogramma badium** (Braun) has pale orange or rust tan FW (6–7.5 mm; Plate 10.27) with variable expression of three discal spots and a row of dots along the termen. *Helcystogramma badium* occurs from Vancouver Island to the Sierra Nevada and coastal California south to San Diego, and in the mountains of eastern Nevada to eastern Utah. The larva has purple thoracic and abdominal segments 1 and 2 and is white posteriorly with purple subdorsal line and diagonal slashes. We reared it from scrub oak, *Quercus berberidifolia* (Fagaceae), on Santa Rosa Island in March. **Helcystogramma chambersella** (Murtfeldt) is widespread in eastern North America and ranges west to Arizona and southern California. This is a small dichomeridine (FW 3.5–5 mm; Plate 10.28) with dark brown FW showing only slightly darker basal markings, outlined by pale

scaling. Murtfeldt reported the larva as a case maker on foliage of ragweed, *Ambrosia artemisiella,* in Missouri, while Goeden and Ricker reared *H. chambersana* from *A. confertifolia* and *A. ptilostachya* in California.

Scodes deflecta (Busck)(Plate 10.38) is readily distinguished from other Dichomeridinae by its enormous, porrect labial palpi, second segment with broad scale brush, and third segment tiny, erect, thin. The wing shape is similar to that of *Helcystogramma.* This is a Neotropical species that ranges from Guerrero, Mexico, to southern Arizona. The moth is larger than most gelechiids (FW 9–11 mm), FW gray with a dark gray costal blotch at one-third extending into the cell and with whitish scaling forming longitudinal striae. The larva is a leaf roller on Malvaceae. It was reared from cotton, *Gossypium thurberi* (=*Thurberia thispesioides,* Malvaceae) in Arizona by Dwight Pierce. *Scodes deflecta* flies in July to the end of September in Arizona, coinciding with the summer rain season.

SUBFAMILY PEXICOPIINAE

This subfamily is primarily pantropical, and the three species in America north of Mexico are introduced from other parts of the world, as pests of hibiscus, cotton, or grain. The larvae are internal feeders in seed and developing seed pods. The antennal scape has a pecten, like that of Depressariidae and unlike other Gelechiidae.

The "Angumois grain moth," **Sitotroga cerealella** (Olivier), is a small, iridescent, pale straw-colored moth (FW 6–8 mm; Plate 10.39); wings lanceolate, FW with faint dark brown streaks, along the Cu crease, at apex, and basally in the fringe. The origin of this species is unknown because it was carried in cereals and distributed extensively before recorded times. *Sitotroga cerealella* is said to have been the first insect discussed in an American scientific publication, in 1743. All kinds of grains are infested, including wheat, corn, even dried flower arrangements, but not processed products such as flour and cornmeal. Usually one larva occupies a kernel, and *Sitotroga* can cause great losses of grain in the field and in storage situations. The moths are most abundant in the fall when the grain ripens, but there are continuous generations in storage.

Pectinophora gossypiella (Saunders), the "pink bollworm" of an extensive economic entomological literature, is a small (FW 7–8 mm; Plate 10.40), broad-winged, drab moth, FW dull brown to pale tan with blackish smudges. It has been a serious pest throughout cotton growing regions and has been cited as one of the six most economically destructive insects in the world. This species was originally described in 1844 from India and was imported to Mexico by 1911, whence it spread northward into southern Texas by 1917. During the following decades, *P. gossypiella* became established throughout the southeastern United States and westward to Arizona by 1950. Losses in the United States of several hundred million dollars annually were estimated. The pinkish larvae bore into the flower buds, causing them to fail to open. There are four to six generations per year, and as the season progresses, the bolls are destroyed as the larvae bore into them and feed on the seeds, cause the lint to fail, and greatly reduce the yield of oil from the seed. After discovery of *P. gossypiella* in southeastern California, it has been the subject of constant surveillance through the Cooperative Pink Bollworm Program, an integrated pest management (IPM) project administered by the State Department of Food and Agriculture and funded almost entirely by assessments on the growers. It has been in operation since 1967, possibly the longest running and most successful IPM

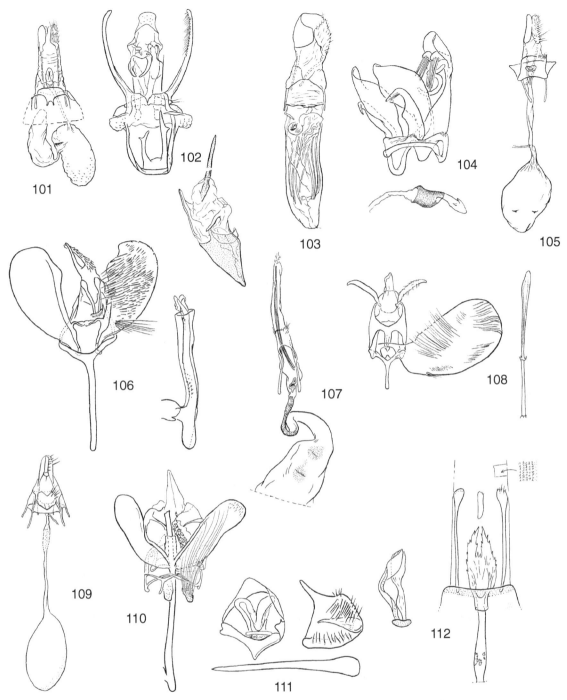

FIGURES 101–112. Genitalia structure of exemplar species of Gelechiidae, Douglasiidae, and Yponomeutoidea; males ventral or venterolateral aspect, females ventral aspect. **101**, Gelechiidae: *Anacampsis sacramenta* ♀ [Keifer 1933]. **102**, *Dichomeris stipendaria* (Braun) ♂, aedeagus below, right, **103**, *Dichomeris stipendaria* ♀ [Hodges 1986]. **104**, *Platyedra subcinerea* (Haworth) ♂, aedeagus below. **105**, *Platyedra subcinerea* ♀ [Powell et al. 2001]. **106**, Plutellidae: *Ypsolopha lyonothamnae* (Powell) ♂, aedeagus to right. **107**, Plutellidae: *Ypsolopha lyonothamnae* ♀ [Powell 1967]. **108**, Yponomeutidae: *Zelleria pyri* Clarke ♂, aedeagus to right. **109**, Yponomeutidae: *Zelleria pyri* ♀ [Clarke 1942]. **110**, Argyresthiidae: *Argyresthia flexilis* Freeman ♂, aedeagus in situ [Freeman 1960]. **111**, Douglasiidae: *Tinagma californicum* Gaedike ♂ (tegumen, valva above aedeagus, anellus to right from another aspect). **112**, Douglasiidae: *Tinagma powelli* Gaedike ♀, detail of A9 [Gaedike 1990].

program in the world. Colonization is monitored weekly by 20,000 pheromone lure traps deployed on 1.2 million acres of cotton. *Pectinophora gossypiella* is not readily attracted to lights. Cultural control, by plowing under the stubble and field trash to destroy overwintering larvae; along with mass releases of sterile males, of which 6 to 7 million are produced in a USDA

lab in Arizona annually; and application of bacterial spray (*Bacillus thuringiensis*, Bt) have produced effective control in lieu of chemical efforts at eradication.

Platyedra subcinerea (Haworth)(Plate 10.41, Figs. 104, 105), which is called the "cotton stem moth" in the Old World, is native to the Palaearctic region but has been adventive in

New England for more than 50 years. It was introduced to California recently; we first found it at Cambria on the central coast in 1990, but this nondescript colonist was not identified until 1997, when it began to appear at lights in urban Berkeley. It was recorded in five counties of the Central Valley, beginning in 1992, where specimens were taken in pheromone lure traps deployed for detection of pink bollworm. We found *P. subcinerea* on Santa Barbara and San Clemente islands in 2001–2002, but it has not yet been detected on mainland southern California. This is a broad-winged, tan moth, superficially resembling an *Agonopterix* (Depressariinae), FW (7.5–9 mm) with poorly defined dark, linear blotches through the cell. The larvae are sluggish, stout, with dark brown head capsule and darkly pigmented thoracic segments 1 and 2, appearing purplish in life, the remainder of the body pale. They bore into flowers, seed, and stems of Malvaceae, including *Lavatera* and *Gossypium* (cotton) in the Middle East, *Malva sylvestris* and *Alcaea rosea* (hollyhock) in England and southern Europe, hollyhock in New England; and we have found them in *M. parviflora* (cheeseweed) in California, where cotton, hollyhock, and *Lavatera* may also serve as hosts. There is a long flight period in California, late March to late July, suggesting two generations, and there are isolated records in February and September, so a third brood may sometimes occur or adults may overwinter.

References for Superfamily Gelechioidea

Adamski, D. A. 2002. A Synopsis of Described Neotropical Blastobasinae (Lepidoptera: Gelechioidea: Coleophoridae). Thomas Say Publications in Entomology: Monographs. Entomological Society of America, Lanham, MD; 150 pages.

Adamski, D., and R. L. Brown 1989. Morphology and systematics of North American Blastobasidae (Lepidoptera: Gelechioidea). Mississippi Agricultural Forest Experiment Station Technical Bulletin 165; 70 pages, 126 figs.

Adamski, D., and R. W. Hodges 1996. An annotated list of North American Blastobasinae (Lepidoptera: Gelechioidea: Coleophorinae). Proceedings of the Entomological Society of Washington 98: 708–740.

Baldizzone, G., H. van der Wolf, and J.-F. Landry 2006. Coleophoridae, Coleophorinae (Lepidoptera). World Catalogue of Insects, Vol. 8. Apollo Books, Stenstrup; 215 pages.

Berenbaum, M. R., and S. Passoa 1999. Generic phylogeny of North American Depressariinae (Lepidoptera: Elachistidae) and hypotheses about coevolution. Annals of the Entomological Society of America 92: 971–986.

Braun, A. F. 1948. Elachistidae of North America (Microlepidoptera). Memoirs of the American Entomological Society 13: 1–110; 26 plates.

Bucheli, S., J.-F. Landry, and J. Wenzel 2002. Larval case architecture and implications of host-plant associations for North American *Coleophora* (Lepidoptera: Coleophoridae). Cladistics 18: 71–93.

Bucheli, S. R., and J. Wenzel 2005. Gelechioidea (Insecta: Lepidoptera) systematics: A reexamination using combined morphology and mitochondrial DNA data. Molecular Phylogenetics and Evolution 35: 380–394.

Busck, A. 1903. A revision of the American moths of the family Gelechiidae, with descriptions of new species. Proceedings of the U.S. National Museum 25: 767–938.

Busck, A. 1939. Restriction of the genus *Gelechia* (Lepidoptera: Gelechiidae), with descriptions of new genera. Proceedings of the U.S. National Museum 86: 563–593, 14 plates.

Clarke, J. F. G. 1941. Revision of the North American moths of the family Oecophoridae, with descriptions of new species. Proceedings of the U.S. National Museum 90: 33–286, 48 plates.

Duckworth, W. D. 1964. North American Stenomidae (Lepidoptera: Gelechioidea). Proceedings of the U.S. National Museum 116: 23–71, 4 plates.

Hodges, R. W. 1966. Revision of Nearctic Gelechiidae, I: The Lita Group (Lepidoptera: Gelechioidea). Proceedings of the U.S. National Museum 119 (3547): 1–66, 31 plates.

Hodges, R. W. 1974. Gelechioidea, Oecophoridae, *in*: Hodges, R. W., et al. (eds.) Moths of America North of Mexico 6.2. Classey and R. B. D. Publications, London; 142 pages, 8 plates.

Hodges, R. W. 1978. Gelechioidea, Cosmopterigidae, *in*: Hodges, R. W., et al. (eds.) Moths of America North of Mexico 6.1. Classey and R. B. D. Publications, London; 166 pages, 6 plates.

Hodges, R. W. 1986. Gelechioidea, Gelechiidae (Part),*in*: Hodges, R. W., et al. (eds.) Moths of America North of Mexico 7.1. Classey and R. B. D. Publications, London; 195 pages, 46 plates.

Hodges, R. W. 1999. Gelechioidea, Gelechiidae (part), *in*: Hodges, R. W., et al. (eds.) Moths of America North of Mexico 7.6. Classey and R. B. D. Publications, London; 257 pages, 51 plates.

Hodges, R. W. 1999. The Gelechioidea, pages 131–158, *in*: Kristensen, N. P. (ed.) Handbook of Zoology. Lepidoptera, Moths, and Butterflies. Vol. 1: Evolution, Systematics, and Biogeography. Handbook of Zoology. W. de Gruyter, Berlin, New York; x + 491 pages.

Kaila, L. 1996. Revision of the Nearctic species of *Elachista*, I: The *tetragonella* group (Lepidoptera: Elachistidae). Entomologica Scandinavica 27: 217–238.

Kaila, L. 1997. Revision of the Nearctic species of *Elachista s. l*, II: The *argentella* group (Lepidoptera: Elachistidae). Acta Zoologica Fennica 206; 93 pages.

Kaila, L. 1999. Phylogeny and classification of the Elachistidae s. s. (Lepidoptera: Gelechioidea). Systematic Entomology 24: 139–169.

Kaila, L. 1999. Revision of the Nearctic species of *Elachista s. l.,* III: The *bifasciella, praelineata, saccharella,* and *freyerella* groups (Lepidoptera: Elachistidae). Acta Zoologica Fennica 211; 235 pages.

Kaila, L. 2004. Phylogeny of the superfamily Gelechioidea (Lepidoptera: Ditrysia): An exemplar approach. Cladistics 20: 303–340.

Kuznetsov, V. I., and A. A. Stekol'nikov 1978. Systematic position and phylogenetic relationships of the superfamily Coleophoroidea (Lepidoptera: Oecophoridae, Coleophoridae, Ethmiidae) as revealed by the functional morphology of the male genitalia [translation]. Entomological Review 57: 91–103.

Landry, J.-F. 1991. Systematics of Nearctic Scythrididae (Lepidoptera: Gelechioidea): Phylogeny and classification of supraspecific taxa, with a review of described species. Memoirs of the Entomological Society of Canada 160; 341 pages.

Minet, J. 1990. Remaniement partiel de al classification des Gelechioidea, essentiellement en fonction de caractères pré-imaginaux (Lepidoptera, Ditrysia). Alexanor 16 (1989): 239–255.

Powell, J. A. 1971. Biological studies on moths of the genus *Ethmia* in California (Gelechioidea). Journal of the Lepidopterists' Society 25(Suppl. 3); 67 pages.

Powell, J. A. 1973. A systematic monograph of New World ethmiid moths (Lepidoptera: Gelechioidea). Smithsonian Contributions to Zoology 120; 302 pages.

Powell, J. A. 1976. A remarkable new genus of brachypterous moth from coastal sand dunes in California (Lepidoptera: Gelechioidea, Scythrididae). Annals of the Entomological Society of America 69(2): 325–339.

Powell, J. A., and P. A. Opler 1996. Moths of Western North America. 4. Distribution of "Oecophoridae" (sense of Hodges 1983) of Western North America. C. P. Gillette Insect Biodiversity Museum, Colorado State University, Ft. Collins; 63 pages.

Powell, J. A., and D. Povolny 2001. Gnorimoschemine moths of coastal strand and dune habitats in California (Gelechiidae). Holarctic Lepidoptera 8 (Suppl. 1): 1–51, 115 figs., 8 color.

Stehr, F. W. 1987. Superfamily Gelechioidea, pages 379–399, *in*: Stehr, F. (ed.) Immature Insects. Vol. 1. Kendall/Hunt, Dubuque, IA.

Superfamily Yponomeutoidea

This superfamily includes a heterogeneous conglomeration of dissimilar microlepidoptera that are grouped in the ditrysian lineages as nonapoditrysian, have nonmotile pupae, and lack the scaled proboscis typical of Gelechioidea. The proposed primary synapomorphy for the included taxa is posterior expansion of abdominal pleura 8, which enclose the male genitalia. The size of the pleural lobes is correlated with that of the valvae, and they are small in Plutellidae, Acrolepiidae, and Heliodinidae. Enlarged lobes are lacking, presumably lost, in Bedellidae and Glyphipterigidae. Hence the monophyly of the Yponomeutoidea is debatable, and there is considerable discrepancy between interpretations of family and subfamily relationships among recent taxonomic treatments. The superfamily as presently conceived is worldwide in distribution, with more than 1,500 described species, of which about 240 occur in America north of Mexico.

FAMILIES YPONOMEUTIDAE AND ARGYRESTHIIDAE

These well-defined groups have been treated by taxonomists as subfamilies or as separate families. We rank them as families.

Adult These are slender moths with elongate FW, ranging from tiny, metallic golden, purple, or gray and white nocturnal *Argyresthia* to larger (FW 9–15 mm) and brightly colored, diurnal moths in *Atteva,* and the white "ermine moths" in *Yponomeuta.* Genitalia, Figs.108–110.

Larva Typical yponomeutines live communally in extensive webs on trees and shrubs, sometimes causing economic damage to fruit trees; those of some *Zelleria* damage growing tips of conifers. Larvae of *Argyresthia* are miners in angiosperm buds or conifer foliage or stems.

Diversity There are about 600 described species of Yponomeutidae sensu lato, occurring in all biotic regions, with both the phylogeny and taxonomy in tropical and south temperate faunas yet to be resolved. More than 140 species are described in America north of Mexico, at least 85 of which occur in the West.

FAMILY YPONOMEUTIDAE

This family is represented in America north of Mexico by about 35 described species, most of which are relatively large, and some are colorful, for microlepidoptera, and therefore the fauna is well known. Genitalia, Figs. 108, 109. The genus *Atteva* is pantropical in distribution and is known in North America by an endemic species in Florida and the conspicuous "Ailanthus webworm," *A. punctella* (Cramer)(Plate 10.42, Fig. 12), which ranges northward in the eastern and midwestern United States, dependent upon the imported Asian tree-of-heaven, *Ailanthus altissima* (Simaroubaceae). A western race, *A. exquisita* Busck, is native in northwestern Mexico, Coahuila to Baja California, and southeastern California, where the larvae feed on crucifixion thorn, *Castela (Holacantha) emoryi* (Simaroubaceae), a native desert shrub. The adults are moderately large (FW 9–14.5 mm), sluggish, and colorful, probably aposematic, warning potential predators of noxious chemical properties. The FW is bright orange with four cream white or pale yellow transverse bands, partitioned and bordered by metallic blue black. The western populations differ by having these bands essentially discrete and each divided into a pair of unbroken fasciae. The moths are attracted to lights, but mating occurs after dawn, and *Atteva* often visit flowers in the daytime. The larvae also are colorful, longitudinally striped with orange, black, and white. They live gregariously, producing copious silk, and feed on foliage, flowers, and seed as available.

Zelleria haimbachi Busck (Plate 10.44, Plate 10.45), the "pine needle sheath miner," is transcontinental, ranging across Canada and through conifer forests of the West. In California, *Z. haimbachi* occurs from coastal areas to 7,000'. The adults are small, FW (7–9.5 mm), white with pale ochreous tan bands along the costa and dorsal margins, the latter broader. In the West, *Pinus ponderosa* and *P. jeffreyi* are favored hosts, but other species of two- and three-needle pines are eaten, and we have reared *Z. haimbachi* from Douglas-fir, *Pseudotsuga menziesii,* in

northern California and *P. macrocarpa* in the San Bernardino Mountains. The first instars mine in needles, overwinter, and in spring the orange larvae migrate to the base of the needle cluster, where needles are severed, and each larva may destroy six to 10 clusters during development. The caterpillars are slender, dull green with indistinct, reddish, dorsolateral stripes, illustrated by Duncan. Adults are recorded June to mid-August. Adults of **Z. gracilariella** Busck are *Caloptilia*-like, with slender, bright rose rust–colored FW (7–9.5 mm; Plate 10.43); some have diffuse subcostal lines of black scales and between the Cu crease and dorsal margin. Specimens from Oregon reared from *Ribes cereum* are paler, rose tan, with at most just a few black scales. This species occurs on the Pacific Coast, from British Columbia to central California, where the larvae feed on currant and gooseberry, *Ribes sanguineum* and *R. menziesii* (Grossulariaceae). The slender, green larvae form frail silken tubes on undersides of the leaves, from late April to early July, at first skeletonizing the leaf surface, later eating the full depth and folding leaves. Adults emerge from July to early September. We have not seen them in the field, diurnally or at lights.

FAMILY ARGYRESTHIIDAE

Argyresthia are tiny to small moths (FW 3.2–6.8 mm), lanceolate winged, shiny, usually white with gray, bronze or golden markings. The front is smooth scaled but the crown roughened, labial palpi porrect, slightly curved, tapering to a point. The moths have the peculiar habit of perching with the head oppressed to the substrate, the body wrapped in the wings raised obliquely. Genitalia, Fig. 110. The larvae are leaf and needle miners, although late instars of some species leave the mines to form shelters in terminal foliage. Larval feeding by most species is specific to one or a few plants; *Argyresthia* feed on at least 13 gymnosperm and dicot families, with more than 40% of recorded species on conifers, including 25% on Cupressaceae, a greater degree of adaptation to conifers than by other lepidopteran lineages. There are more than 50 described species in the Nearctic, of which fewer than half occur in the West, but no doubt many remain to be discovered and named.

Argyresthia goedartella (Linnaeus) occurs in the Palaearctic, and the name has been applied to populations in the Nearctic. Most of these may represent *A. calliphanes* Meyrick, which has a similar FW pattern but a shining white rather than *A. goedartella*'s golden thorax. Western members of this complex are large (FW 5.8–6.8 mm; Plate 11.20), FW white with variable but distinct, coppery golden fasciae: a curved bar basally, strong Y shape at middle, and a double curved band in terminal area. *Argyresthia goedartella* was recorded in British Columbia, north coastal California, New Mexico, and the eastern United States by the late 1800s. Populations in central Alaska differ in having a bolder and darker FW pattern, and there may be several species in the West. In England the larvae of *A. goedartella* feed in catkins and terminal shoots of Betulaceae, birch *(Betula)* and alder *(Alnus)*. *Argyresthia oreasella* Clemens (Plate 11.21) is similar but the FW pattern differs, a strong, outwardly curving bar from the middorsal margin, a curving bar from the tornus with four branches to the costa and termen. This species occurs across the continent, presumably indigenous, having been collected and identified by Walsingham in 1871 from Mt. Shasta, California, and recorded by Busck a century ago from Idaho and New Mexico. *Argyresthia oreasella* has been reported in association with oaks in Missouri and

Colorado, but Prentice recorded four collections reared from choke and pin cherry, *Prunus pennsylvanica* and *P. virginiana*, in Canada.

At least eight western species of *Argyresthia* are Cupressaceae feeders. These are tiny moths (FW 3.5–4.5 mm), including three that mine the terminal twigs of Monterey cypress, *Cupressus macrocarpa*, in coastal California, *A. cupressella* Walsingham (Plate 11.22, Plate 11.23), *A. trifasciae* Braun (Plate 11.24), and *A. franciscella* Busck (Plate 11.25). The first two have silvery white and golden bands. The head scaling is white in *A. cupressella*, which also feeds on ornamental *Thuja* and *Chamaecyparis*, brownish in *A. trifasciae*. By contrast, *A. franciscella* has smoky, pale golden FW crossed by dark brown bars from the dorsal margin, faint toward the costa. Adults of *A. cupressella* are active in spring, as early as January, but mostly April to May, and *A. franciscella* flies late April to June. The pale green larvae move from twig to twig; that of *A. cupressella* is illustrated by Duncan in British Columbia. *Argyresthia thoracella* Busck (Plate 11.27), is similar to *A. cupressella* but the FW is white with paler, more well defined golden markings. It was described from northern Arizona, and populations, possibly a complex of species, occur in Nevada and eastern California in association with junipers *(Juniperus)*. *Argyresthia libocedrella* Busck is larger, FW (6–7 mm; Plate 11.26) bright golden tan with two dark brown spots, at one-third and beyond the middle of the dorsal margin. It lives on incense cedar, *Calocedrus decurrens*, in Oregon and the Sierra Nevada, south to the San Jacinto Mountains in California. The adults fly in June to July. *Argyresthia pseudotsuga* Freeman has pearly gray FW (5–5.5 mm; Plate 11.28) with a golden sheen defining a gray shade at the midcosta and terminal area. It occurs from British Columbia to central coastal California. The larvae mine in the new growing terminals of Douglas-fir, *Pseudotsuga menziesii* (Pinaceae), and induce spindle-shaped, gall-like swellings, ultimately killing the twig. Stems collected in late February to March by D. L. Wood produced adults in April.

FAMILIES YPSOLOPHIDAE AND PLUTELLIDAE

These two families traditionally have been treated as one family, but recent authors have considered the ypsolophids to represent a sister clade to the remaining Yponomeutoidea. This distinction, which needs corroboration, is based on an assumption that *Plutella* is phylogenetically closer to the glyphipterigid-heliodinid complex, supporting evidence for which is contradictory. Moreover, application of that hypothesis to our fauna is difficult because some western Nearctic genera were not included in recent cladistic analyses.

Adult They are small moths (FW 6–13 mm) with distinctive labial palpi, second segment broadly scaled, third slender, smooth scaled, and upcurved from second preapically. FW moderately broad to lanceolate, with a flared terminal fringe. Large pleural lobes enclose the male genitalia in ypsolophids but are small and narrow in plutellids. These are typically nocturnal moths with yellow, brown, or gray FW, often with linear markings. Genitalia, Figs. 106–108.

Larva Larvae are slender, tapered toward both ends, with elongate abdominal prolegs, pale green with unpatterned integument, living in slight webs as external feeders.

Larval Foods Members of this family group use more than 50 families of angiosperms, including Ephedraceae *(Ypsolopha)* and Brassicaceae *(Plutella)*. Nearly 300 described species are assigned to one or the other of these families, but the systematics

relationships and descriptive inventory is incompletely known, especially in Southern Hemisphere faunas.

FAMILY YPSOLOPHIDAE

Our fauna in this family is limited to the genus **Ypsolopha**, which includes about 50 described species in North America, almost all of which occur in the West. These are relatively broad winged moths, mostly larger than plutellids. Several species are markedly variable in FW color pattern, which was not recognized by early descriptive taxonomists. Several species complexes, including *Y. arizonella/schwartzella*, *Y. barberella/flavistrigella*, and *Y. dentiferella/canariella*, likely represent variation within single species. The larvae are very active and react by rapid wriggling when disturbed. They are specialist feeders in new spring foliage on a wide variety of shrubs and trees, typically in a weak silken shelter. Pupation occurs in dense, envelope-like cocoons.

Ypsolopha barberella (Busck) is one of the largest and most widely distributed western species. The adult is gray, FW (9–11 mm; Plate 10.46) usually with a black longitudinal streak from basal one-third to apex and numerous longitudinal black dashes. **Ypsolopha flavistrigella** Busck (Plate 10.47) is distinguished primarily by its median black streak highlighted by white and may be a synonym. *Ypsolopha barberella* occurs in the Great Basin ranges of Utah, Nevada, and northern Arizona to the mountains of eastern and southern California. Adults of this species and the *Y. arizonella/schwarziella* complex apparently are capable of long-distance dispersal because individuals sporadically appear at places far from their normal range, such as on Santa Barbara and San Clemente islands and at Berkeley and Inverness in the San Francisco Bay area, in October to December. *Ypsolopha barbarella* sometimes aggregates under loose bark of fallen trees at high elevations during the summer months, for example at Sonora Pass, 9,600' in the Sierra Nevada on pine in August, and White Mountains, 10,000' in eastern California on aspen. The larval hosts are unknown.

Ypsolopha cervella (Walsingham) occurs from British Columbia to southern California and Santa Cruz Island. The FW (7–9 mm; Plate 10.48) is dark coppery tan with a lavender sheen when fresh, often variously mottled with black along the dorsal margin, becoming pale yellowish tan with age. *Ypsolopha cervella* has a long flight period, May to early October in coastal California, but it is univoltine. The larvae are an abundant component of the spring caterpillar fauna on coast live oak *(Quercus agrifolia)*, and feed on other oaks, including deciduous species (e.g., *Q. garryana*, all 23 rearing records by the Canada Forest Insect Survey in British Columbia). **Ypsolopha sublucella** (Walsingham) is similar but smaller (FW 6–8 mm; Plate 11.1), has white head scaling and well defined white, median stripe through the thoracic notum, and the rose rust FW lacks dark maculation. The larvae feed on scrub oaks, *Q. vaccinifolia*, at higher elevations in the Sierra Nevada and *Q. durata* in the central Coast Ranges in California.

Members of a species complex (subgenus *Trachoma*) have a relatively narrow FW with concave terminal margin and short third segment of the labial palpi, and their larvae feed on Rosaceae in Europe and western North America. **Ypsolopha walsinghamiella** (Busck) is very narrow winged, with bluish gray FW (8–10.5 mm; Plate 11.2) indistinctly mottled with dark shades and streaks. There is a black bar along the dorsal margin and spots along the costa in the most heavily marked examples. This species occurs from southern Idaho to Arizona and in California from Mt. Shasta along the east side of the Sierra Nevada to the Tehachapi Range. The larvae feed on antelope bush, *Purshia tridentata*, and the adults fly from July to September. *Ypsolopha falciferella* (Walsingham) is similar but larger and broader winged, FW (10.5–13.5 mm; Plate 11.3) often black along the dorsal margin and with parallel, oblique, brownish gray fasciae. It ranges from British Columbia and Oregon to southern California and Arizona. The larvae were reared on *Prunus emarginata* and *P. virginiana* in northern California by L. Crabtree, and we found *Y. falciferella* on an unidentified *Prunus* in the Huachuca Mountains, Arizona. Adults have been recorded from early March to June. *Ypsolopha lyonothamnae* (Powell) has pale grayish tan FW (8–10.5 mm; Plate 11.4, Figs. 106, 107) with a narrow cream white band along the dorsal margin. This species is restricted to the larger California islands and feeds on island ironwood, *Lyonothamnus floribundus* var. *aspleniifolius*, a relict, endemic tree, but it has not been discovered on Catalina Island, where the typical form of *Lyonothamnus* occurs. The two forms of this plant had different geographical distributions on the mainland during the Miocene. *Ypsolopha lyonothamnae* may be multivoltine; we have found larvae in March to June and adults April to September and December.

Ypsolopha dentiferella (Walsingham)(subgenus *Harpipteryx*) is a beautiful bright yellow species, FW (8.5–9.5 mm; Plate 11.5, Plate 11.6) with variable dark rust brown shading on the costal half, and a strongly falcate apex. Probably *Y. canariella* Walsingham and *Y. frustrella* Walsingham are synonyms. Walsingham collected specimens of the three forms at different places in northern California, but recent series from single sites show variation between them. *Ypsolopha dentiferella* ranges from Montana to northern Arizona and central California. The larvae are believed to feed on *Lonicera* and *Symphoricarpos mollis* (Caprifoliaceae), based on association of the moths by Walsingham, a series from Utah, and in the Berkeley Hills. *Ypsolopha nella* (Busck) is a narrow-winged species, FW (8.5–9.5 mm; Plate 11.7) mauve rust colored with a thin median, longitudinal line, often accentuated by yellowish shading. It is widespread in montane parts of Colorado, Utah, New Mexico, northern Arizona, to northeastern Oregon. The larvae were found to be frequent associates of western spruce budworm (*Choristoneura occidentalis*, Tortricidae) on white fir, *Abies concolor* (Pinaceae), by R. Stevens and others. *Ypsolopha rubrella* (Dyar) is similar but smaller (FW 7–8 mm in Colorado, only 5.5–6.5 in coastal California; Plate 11.8), with dark rust brown FW. The larvae feed on new foliage of barberry, *Berberis repens*, in Colorado and *B. nervosa* (Berberidaceae) in coastal California. Adults evidently do not come to lights. The FW of *Y. maculatella* (Busck) is variable (Plate 11.9, Plate 11.10), from distinctly and intricately patterned with gray brown to nearly all white; a discrete form, white lightly dusted with gray with a distinct discal dot may represent a second species. We found larvae that produced all these forms feeding on the desert conifer, Mormon tea (*Ephedra viridis* and *E. californica*, Ephedraceae) in the Mojave Desert and western foothills of the San Joaquin Valley in March and Owens Valley, California, in May. The adults are diurnal and sometimes abundant at composite flowers growing near ephedra (Plate 60.14).

FAMILY PLUTELLIDAE

Typical members of this lineage are smaller and relatively slender compared to *Ypsolopha*, and males lack the enlarged pleural lobes of abdominal segment 8 and the corresponding large val-

vae that are characteristic of *Ypsolopha*. Most *Plutella* feed on Brassicaceae, one of the few moth lineages adapted to do so, and species of other genera are specialists on unrelated angiosperms, including Berberidaceae, Ericaceae, and Caprifoliaceae. Pupation occurs in a loose-meshed, less-compact cocoon than that of *Ypsolopha*. About 22 species are known in North America, all of which occur in the West.

The "diamondback moth," *Plutella xylostella* (Linnaeus)(Plate 11.16), a ubiquitous pest of cabbage, cauliflower, and other mustards, is cosmopolitan, having been transported with domestic crucifers to all continents. Its adults are slender, FW (5.5–7.5 mm) gray with a wavy-edged band of yellowish white along the dorsal margin, which is produced to a peak at midwing, forming a narrow diamond-shaped figure when the moth is perched. Specimens vary from pale to dark gray, sometimes obscuring the white pattern. The adults sometimes undergo mass migrations, unusual for microlepidoptera. All kinds of commercial, weedy, and native Brassicaceae serve as larval hosts. Four to seven generations per season may develop in milder climates, and specimens come to lights in every month in coastal California. Among all Lepidoptera, *P. xylostella* probably best lives up to the phrase "occurs throughout the western states"—it ranges from seacoast to high montane and desert habitats. *Plutella porrectella* (Linnaeus)(Plate 11.17) is similar in form and size, but the FW is tan without a differentiated dorsal band, with the terminal fringe dark brown. This seems to be a natively Holarctic species. It occurs in Europe and was known in the eastern United States by the 1860s and collected at several sites in California and Oregon in 1871 by Walsingham. The larvae feed on *Hesperis* (Brassicaceae) in England and Ontario, and adults have been associated with *Erysimum* in California, where *P. porrectella* is much more widespread than the introduced *Hesperis*. *Plutella interrupta* Walsingham is distinctive among North American species; FW (7.5–8.5 mm; Plate 11.11) white with a serpentine, black bar along the crease basally into the cell, and pattered by delicate, black strigulae. It occurs from British Columbia to coastal central California. *Plutella interrupta* seems to be bivoltine, with adults overwintering, recorded in January to March, July to August in Humboldt County and October to November at Inverness, Marin County. The larval host of *P. interrupta* is unknown.

Plutella albidorsella Walsingham (FW 6–7 mm; Plate 11.12) has broader wings, dark brownish gray, having a broad, white, sinuate, variable band along the dorsal margin. *Plutella albidorsella* occurs in California from Mendocino County to San Diego, and on Santa Cruz and San Clemente islands. Adults are recorded from mid-April in desert canyons to July near the coast, mostly June. The larval host is expected to be a crucifer. *Plutella vanella* Walsingham differs markedly from other *Plutella*, having broad wings, FW (7.5–8.5 mm; Plate 11.13) shining bronze colored with bold, whitish ochreous markings lightly dusted with brown. This species is widespread in diverse habitats, near the coast to montane in northern California where it is seen sparingly, one or two at a time, yet it is abundant in deciduous forests in Alberta. We reared *P. vanella* from larvae found on *Vicia gigantea* (Fabaceae), and *Osmorhiza* (Apiaceae) in California, and Dietz and Rude collected a larva on white fir, *Abies concolor*, in Oregon. The latter two hosts might have been incidental cocoon sites, but it appears this species is a generalist. Adults are recorded from May to July. *Plutella dammersi* Busck is a small (FW 5.5–8 mm; Plate 11.18), broad-winged species superficially resembling some *Ypsolopha*, but the antennae in both sexes are thickly scaled, tan dorsally on basal two-thirds, then five segments white with short black scales basally, three segments black, and four pairs of segments alternately white and black. The FW usually is uniform tan, but rare individuals have paler FW, the costal half suffused with gray and two short dark bars in and above the crease. A population on Santa Rosa Island has variations of this and not the typical form. The HW is contrasting gray. This is an endemic species in southern California, known from the Temblor Range in San Luis Obispo County south to Chino Canyon above Palm Springs and on Santa Rosa Island. The larvae feed in the flowers and inflated seed capsules of bladderpod, *Isomeris arborea* (Capparidaceae), in March and April, and adults have been collected in March, June to July, and October, indicating two or more generations. The moths usually occur in close association with the food plant, but we found them numerous in foliage of white fir in the Tehachapi Range and on Mt. Pinos.

Eucalantica polita (Walsingham) is a small (FW 5.7–7.5 mm; Plate 11.19), shining white moth that occurs along the immediate Pacific Coast from British Columbia to central California and on Santa Cruz Island. The FW is white with a black dot at the end of the cell and sometimes black scaling along the dorsal margin, ending in an upcurved spur; HW pale gray. The larvae feed on flowers and foliage of huckleberry, *Vaccinium ovatum* (Ericaceae). The adults often are flushed into flight by day, but evidently they are nocturnal and sometimes are attracted to black light. *Eucalantica polita* appears to be multivoltine, with larvae collected in March, April, and June, and adults recorded in all months except March to April and August. Populations of very similar moths occur in high-elevation forests of Mexico (Durango, Veracruz) and the Cordillera de Talamanca, Costa Rica, at 10,000'.

Euceratia castella Walsingham is superficially similar to *Eucalantica* but is larger, and the FW (7.5–10.5 mm; Plate 11.14) is chalky white, not shining; HW pale to dark gray. Some individuals have the FW lightly peppered with black, and rarely with pale ochreous, transverse strigulae distally. This species occurs from southern Washington to central California, in diverse inland habitats, where the larvae are leaf tiers on honeysuckle, *Lonicera*, and snowberry, *Symphoricarpos* (Caprifoliaceae). We have found larvae in March and April, and adults are recorded from late April to July and August at 6,000' in Mono County. *Eucalantica securella* Walsingham is larger, FW (8.5–11 mm; Plate 11.15) whitish mottled with pale brownish to mostly dark brownish. This species may feed on the same hosts as *E. castella*, and the two have similar geographic ranges, often found together in foothill canyons, but *E. securella* ranges farther south, to San Diego and Santa Cruz Island. It flies April to June through most of its range, June to early September in coastal Humboldt County.

FAMILY ACROLEPIIDAE

Acrolepia and related genera have been included in the Plutellidae or treated as a separate family, and the separation is arguable, based largely on reduction states of genitalia characters. Genitalia, Figs. 113, 114. These are small, secretive moths said to be crepuscular or nocturnal in the Palaearctic, but Pacific Slope *Acrolepiopsis* have not been recorded at lights. The larvae in Europe are leaf miners and borers in stems, buds, and seeds of Liliaceae and several dicot families. There are about 100 named species of Acrolepiinae worldwide, mostly Holarctic, and only *Acrolepiopsis*, with five species, is known in North America. *Acrolepiopsis californica* Gaedike is a small (FW 4.5–7.5 mm; Plate 11.46), dark brown moth, FW with thin,

white dash from the dorsal margin before the middle curving through the crease at midwing. It occurs in Oregon south to coastal Monterey County, California, where the larvae feed on *Disporum hookeri* (Liliaceae) and *Lilium pardalinum*. Early instars may be miners, but later the larvae graze externally in slight webs on flowers, young fruit, and undersides of leaves. We found larvae in April, May, and June in different years, and adults are recorded from April to mid-August.

FAMILY GLYPHIPTERIGIDAE

Adult Sedge moths are tiny to small (FW 3.2–10 mm), diurnal moths with smooth-scaled head, porrect or decumbent, slightly upcurved labial palpi, often metallic gray; FW with iridescent markings and parallel, white chevron marks from the costa and dorsal margin. The last abdominal pleural lobes are greatly enlarged, forming a hood over the genitalia.

Larva They are borers in seeds, flowering stems, terminal buds, or leaves, primarily in sedges and rushes (Cyperaceae, Juncaceae), less commonly grasses, rarely dicots, including Crassulaceae and Urticaceae.

Diversity The family is cosmopolitan and well represented in most faunal regions, Palaearctic, Nearctic, and Australian, including New Zealand. Nearly 400 species are described in over 20 genera. There are 36 species in America north of Mexico, mostly newly described by Heppner in his 1985 monograph, which treats five genera, only two of which occur in the West.

Glyphipterix is distributed worldwide, with 22 species in the western states and provinces. Nearctic species have relatively broad wings, FW characteristically black with one or two broad, white fasciae from the dorsal margin and several thin, whitish to silvery chevrons from the costa. *Glyphipterix urticae* Heppner is distinguished by having the basal half of the FW (4–7 mm; Plate 11.29) black without yellowish overscaling, and the dorsal spot near base quadrate. Larvae of this species have not been discovered, but adults were found in association with nettle, *Urtica* (Urticaceae), at several localities by Heppner. *Glyphipterix urticae* occurs from southern Manitoba and Alberta to Colorado, New Mexico, Utah, and eastern California. Adults fly in late June and July. A sister species, *G. powelli* Heppner (Plate 11.30), occurs along the coast of California, where we found the moths on nettle in the daytime. *Glyphipterix bifasciata* Walsingham is the largest Nearctic species (FW 5–8 mm; Plate 11.31) and is distinguished by having two white fasciae from the dorsal margin complete to the costa and three subapical white chevrons. This elegant species is widespread, from southern British Columbia, Montana, and Colorado to western Nevada and California, in the Cascades south to the San Bernardino Mountains, yet nothing is known about the larval biology. A few adults have been taken at black light, but most were netted diurnally, flying in chaparral and wooded canyons without specific plant association, in June and July. Walsingham collected one male in late September in southern Oregon. *Glyphipterix juncivora* Heppner (Plate 11.32) has not been reared, but Heppner found the moths in association with rushes, *Juncus* (Juncaceae), along creeks in several places. The adult has gray head scaling, a quadrate basal mark on the dorsal margin, and large, distinct costal chevrons bordered with black. *Glyphipterix juncivora* is widespread in the Rocky Mountain region, from Alberta and Wyoming to Colorado, Utah, New Mexico, and the White Mountains of eastern Arizona. *Glyphipterix montisella* Chambers (Plate 11.33) is similarly distributed, from Montana and Wyoming to New Mexico and

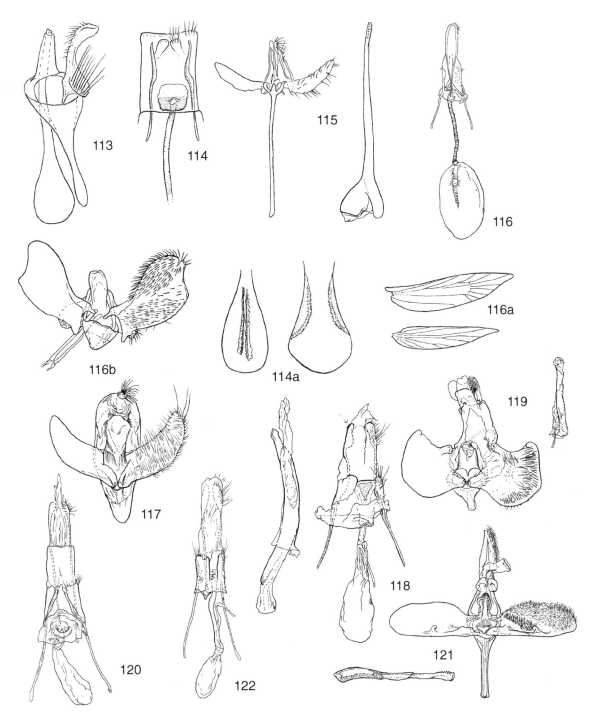

FIGURES 113–122. Genitalia of Yponomeutoidea and Sesiidae, ventral aspect; aedeagus removed and shown to right, except 113, 116b in situ. **113**, Acrolepiidae: *Acrolepiopsis californica* Gaedike ♂, left valva removed. **114**, Acrolepiidae: *Acrolepiopsis californica* ♀, detail of sterigma, a, corpus bursae from two different views [Gaedike 1994]. **115**, Heliodinidae: *Lithariapteryx abroniaeella* (Chambers) ♂, aedeagus to same scale, to right. **116**, Heliodinidae: *Lithariapteryx elegans* Powell ♀; a, Heliodinidae: *Lithariapteryx jubarella* Comstock, wing venation [Powell 1991]; b, Bedelliidae: *Bedellia somnulentella* (Zeller) ♂ [Pierce and Metcalf 1935]. **117**, Sesiidae: *Zenodoxus palmii* (Neumoegen) ♂, aedeagus to right. **118**, Sesiidae: *Zenodoxus maculipes* Grote and Robinson ♀. **119**, Sesiidae: *Sesia tibialis* (Harris) ♂, aedeagus to right. **120**, Sesiidae: *Sesia tibialis* ♀. **121**, Sesiidae: *Penstemonia edwardsii* (Beutenmüller) ♂, aedeagus below, left. **122**, Sesiidae: *Penstemonia edwardsii* ♀ [Eichlin and Duckworth 1988].

northern Arizona, and there is one record from eastern California, mostly at elevations above 7,000′. This species has relatively narrow wings and typically six FW costal chevrons. It occurs in marshy areas and montane meadows, often in association with *Juncus*, the presumed larval host plant. Adults fly in July and August. **Glyphipterix feniseca** Heppner is a small

Californian species (FW 5.5–6.5 mm; Plate 11.34) with large, quadrate dorsal spots and four equally distributed, bright sky blue spots in the tornal area. It occurs in the San Francisco Bay–San Joaquin delta area and the Sierra Nevada in diverse habitats, beach dunes, coastal mountain seepage, riverine marsh, and we found hundreds of *G. feniseca* swarming in a lush, wet meadow among patches of sedges, *Carex* (Cyperaceae), and *Juncus* in Alpine County. The species probably is univoltine, with records in April and May near the coast to July at 7,500′.

Diploschizia impigritella (Clemens) is the most widespread and commonly encountered glyphipterigid in North America. It occurs throughout the eastern half of the United States and Canada and in British Columbia, Oregon, Nevada, and northern California. However, western collections have been rare, only eight specimens taken on seven dates cited by Heppner. The moth is smaller than *Glyphipterix* (FW 2.7–4.5 mm; Plate 11.35), narrow winged, with a slender, strongly curved white chevron from before the middorsal margin to the end of the cell. *Diploschizia impigritella* has been reared from stems and seeds of nutsedge, *Cyperus* (Cyperaceae), in Virginia and Mississippi. Adults of western populations are recorded in May to July, indicating a single generation.

FAMILIES HELIODINIDAE, LYONETIIDAE, AND BEDELLIDAE

The remaining families of Yponomeutoidea (Heliodinidae, Bedellidae, and Lyonetiidae) comprise an unresolved trichotomy of moth lineages having reduced maxillary palpi (not visible in the pupa), and the pupa has fused, nonmovable abdominal segments and strong lateral ridges. The larvae feed primarily on monocots and herbs and have short prolegs, unlike other yponomeutoids, presumably a secondary reversal. Bedellidae are considered to be members of Lyonetiidae by some authors and are so treated here.

FAMILY HELIODINIDAE

Adult Adults are tiny, diurnal moths (FW mostly 3.2–5 mm; rarely to 8 mm), resplendent in shining iridescent body and wing scaling, FW often red or orange with raised silver- or lead-colored spots. Adults of many species hold the hind legs aloft when perched, which has been regarded as characteristic of heliodinids, but not all of them do so, and this behavior is not limited to heliodinids. Wing venation, Fig. 116a. Genitalia, Figs. 115, 116.

Larva Unpigmented and grublike, larvae are host specific as leaf miners or as stem or seed borers; a few feed externally in flowers and fruit. Pupation by most species occurs affixed to a substrate without visible cocoon, like miniature lycaenid chrysalids.

Larval Foods This is the only lepidopteran family to specialize on Caryophyllales (90% of the species for which hosts are known), especially Nyctaginaceae.

Diversity There are about 75 species described worldwide, the majority in the southwestern United States and Mexico. Although overlooked by most lepidopterists, heliodinids are relatively well known, having been recently monographed by Y.-F. Hsu, who proposed a phylogeny based primarily on remarkable variation in genitalia characteristics. He recorded 31 species in America north of Mexico, of which 24 occur in the western United States. Larval biologies are known for about 70% of the Nearctic species, in large measure the result of Hsu's research and studies by Terry Harrison in Illinois.

The genus *Lithariapteryx*, with four species, is widespread in the western United States associated with sand verbena, *Abronia*, and four-o'clock, *Mirabilis* (Nyctaginaceae). The moths perch on the food plant or visit flowers of other nearby plants for nectar. They have prominent bumps of metallic lead-colored

scales, which from a posterior aspect, give an impression of the eyes of a small jumping spider (Salticidae). *Lithariapteryx* do not raise the hind legs in repose. *Lithariapteryx abroniaeella* Chambers (Plate 11.36, Fig. 115) is the most widespread species, occurring in inland sand dune and desert areas of northern Oregon on the Columbia River, Idaho, and Colorado to southern Arizona and coastal California and Baja California, in association with several species of *Abronia* and *Mirabilis*. The moth is shining lead gray, FW (2.8–5.0 mm) with three conspicuous, raised silvery spots bordered by black on the basal half and an orange V-shaped mark from the costa toward the tornus. The HW fringe is white in contrast to dark gray in related species. The larvae mine the semisucculent leaves of the hosts, sometimes from an external web and move from leaf to leaf. Coastal populations are multivoltine, with adults recorded from March to October, but in desert habitats *L. abroniaeella* may be facultatively uni- or bivoltine depending on availability of the host plants. *Lithariapteryx elegans* Powell (Plate 11.37, Figs. 116, 123) is a bulky heliodinid having dark gray FW (3.2–5.0 mm) with rust orange costa and three pairs of large, metallic purplish scale tufts, and the thoracic venter is snow white. This elegant species is endemic to the Monterey Bay and Pismo-Guadalupe dune systems in central California, where it is dependent on *Abronia latifolia* and displaces *L. abronaeella*. Adults are recorded March to October.

Nearctic species formerly treated as the genus **Heliodines**, primarily on the basis of their superficial similarity owing to bright red or orange FW with metallic markings, have been assigned to five genera after phylogenetic study by Hsu, with *Heliodines* restricted to the European type species. **Neoheliodines** has nine species in the United States and Mexico, five of which occur in the southwestern states. The hind legs are not raised during perching. **Neoheliodines hodgesi** Hsu has bright orange FW (3–4 mm; Plate 11.39) with broad patches of black basally on the costal half and along the dorsal margin, crossed by rows of silvery, raised spots. This species is known from central Texas, southern Arizona, and southern Baja California. The adults are recorded from late June to early October, following summer rains. The larvae live in flat webs on scarlet spiderling, *Boerhavia coccinea* (Nyctaginaceae), and graze the upper leaf surface. Those collected in Baja California in November produced moths the following September to October, and some pupae remained in diapause another year. Larvae taken in July at Ft. Worth, Texas, produced adults in late July, so the records indicate a facultative second generation. **Neoheliodines vernius** Hsu (Plate 11.38) is abundant in southern California and northern Baja California, to the Owens Valley and in southern Arizona. The FW (3.0–4.3 mm) is pale orange with costal and dorsal margin silvery spots, without black markings. The larvae feed in spring on *Mirabilis californica* and *M. bigelovii*, boring into unopened flowers and young fruits. *Neoheliodines vernius* is bi- or multivoltine because larvae from a given collection produce nondiapausing pupae and others that enter diapause, from which adults were obtained by Hsu the following spring and the second year following feeding.

Embola includes six species in the Southwest, two known only from the original type specimen. Some species raise the hindlegs when perched, others do not. **Embola ciccella** (Barnes & Busck) is the largest heliodinid in North America (FW 6.5–8 mm; Plate 11.41, Plate 60.15) and is the only species that regularly comes to lights. There have been many collections of the adults, but paradoxically, nothing is known of the larval biology. The FW is pale orange with rather strongly raised lead-colored spots along the costa and dorsal margin, extended as a bar the

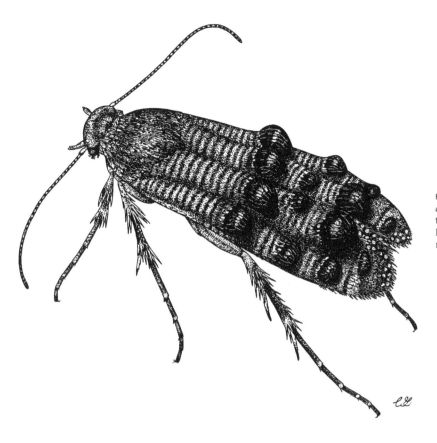

FIGURE 123. Habitus drawing of *Lithari-apteryx elegans* Powell, a diurnal heliodinid that is endemic to the Monterey Bay and San Luis Obispo coastal dune systems in California [Celeste Green, Powell 1991].

length of the termen. *Embola ciccella* occurs in southern Arizona, western New Mexico, to Zacatecas, Mexico. ***Embola powelli*** Hsu (Plate 11.40) is interesting on account of its larval and pupal habits and corresponding morphology. The larvae bore in stems of *Mirabilis californica* and *M. tenuiloba*, pupate there, and the pupa is strongly modified, with reduction of the lateral ridges characteristic of heliodinids, so the body form is cylindrical. The adult has narrow wings, FW (2.8–5.5 mm) dark reddish orange with prominent silvery spots, and the hind legs are raised in repose. This species is widespread, in southern California including Santa Cruz and San Clemente islands, the Mojave Desert, southeastern Arizona, and in Mexico, Chihuahua, Durango, and Baja California Sur. Subsequent to its description, we found a darker race feeding in stems of *Abronia umbellata* on San Clemente Island.

Aetole is the most diverse heliodinid genus, with 17 Western Hemisphere species, including eight in the southwestern United States; others occur in South America and the Galapagos Islands, Central America, and the Antilles. *Aetole* all hold their hindlegs aloft when perched. ***Aetole bella*** Chambers (Plate 11.42) is the most widespread American heliodinid, in part owing to its adaptation to purslane, *Portulaca oleracea* (Portulacaceae), an Old World plant that is established extensively in North and Central America. *Aetole bella* is recorded from Kansas and Colorado through the Southwest, Texas, Florida, and Mexico, to Costa Rica. Larvae have been found on *P. oleracea* in Baja California, Florida, Jamaica, Belize, and Nicaragua, but native hosts have not been identified. The FW (3–4.5.0 mm) is slender, bright vermillion on basal two-thirds, mostly lead gray distally, with silvery spots. The larvae mine in leaves or opportunistically into stems or cecidomyiid (Diptera) galls or feed from an external silk net in later instars, and pupation occurs outside the mine. Specimen records indicate that *A.*

bella has colonized well outside its native range and later became extinct. For example it was collected at San Diego in 1890 and reared in large numbers at Riverside, California, in the 1930s but has not been recorded in California during the past 65 years. Records at other localities may represent historical colonies. ***Aetole prenticei*** Hsu (Plate 11.43) is remarkable for its close association with *Sesuvium verrucosum* (Aizoaceae), distributed in isolated alkaline wetlands, from the California Central Valley and Coachella Valley to New Mexico. The adults are brilliantly colored, FW (3–4 mm) bright vermillion with relatively subdued silvery spots and black distal margin and fringe. This species is multivoltine, with larval collections in March, July, September to October in the San Joaquin delta. ***Aetole extraneella*** (Walsingham)(Plate 11.45) deviates from the Caryophyllales-feeding groundplan and is the only heliodinid known to depend upon Onagraceae. This species is the common heliodinid of central California, occurs in southern British Columbia, and there is an unconfirmed record in northern Utah. The larvae are leaf miners and bore into immature seed pods of *Clarkia* and several *Epilobium* species, particularly *E. (Zauschneria) canum*, and Braun recorded *A. extraneella* from *Gaura* in Utah. The moths are small (FW 2.8–4.2 mm) and similar to *A. prenticei* in color pattern, FW bright vermillion, its silvery spots small and restricted to the costal and dorsal margins. The species is bivoltine in California, more abundant in the fall, when the moths are found at various flowers, especially Asteraceae. ***Aetole cera*** Hsu (Plate 11.44) is unique, an endemic species in the seemingly uninhabitable Death Valley, California, where the larvae feed on *Anulocaulis annulatus* (Nyctaginaceae), which is endemic there. Hsu recognized leaf mines in herbarium specimens of *Anulocaulis*, leading to the discovery of this moth. Adults and larvae were collected in May, and the larvae produced emergences the same season. Probably *Aetole* cera can produce two or more facultative generations and has the ability to delay diapause development over two or more years. The FW (2.8–4.5 mm) is similar to that of related species, but its silvery lead-colored markings are coalesced into longitudinal bars along the distal half of the costa and dorsal margins. The abdomen has two iridescent orange patches.

FAMILY LYONETIIDAE

Members have been treated as a single family including the genus *Bedellia*. The latter is separated as a monobasic family by some recent authors, but the evidence for affinities of heliodinids, lyonetiids, and bedellids is ambiguous. The latter groups differ from Heliodinidae by lack of ocelli and possession of a scale tuft between the antennae, hairy dorsal side of the hind tibia, and a prominent fringe on the underside of the larval labrum. Lyonetiidae sensu stricto, but not *Bedellia*, have the antennal scape slightly flattened and bordered with scales, forming an eye cap, and the larval head is flattened.

Adult These are tiny, slender-winged moths, typically shining white or gray, nocturnal. Genitalia, Fig. 116.b.

Larva The larvae are leaf miners that form linear or irregular blotch mines visible from the upper side, eating the full depth of the leaf in later instars. Pupation occurs outside the mine.

Larval Foods Host plants include at least 27 families of angiosperms, with no marked preference shown for one. A few feed on monocots, and a few are polyphagous.

Diversity About 250 described species are distributed in all faunal regions. There are about 25 species known in North America, half of which occur in the West.

Populations regarded as comprising a single species, sometimes called the "apple leaf miner," **Lyonetia speculella** Clemens (Plate 11.49), are widely distributed across the continent and, contrary to leaf miners generally, feed on a variety of trees and shrubs. The moth has slender FW (4.3–5 mm), shining white with pale brownish gray maculation distally, and there is a black dot preceding a tail-like extension of the fringe at the apex. The FW color pattern varies, and so-called *L. speculella* may represent a complex of species. The larvae mine leaves of *Ceanothus* (Rhamnaceae), maple (*Acer*, Aceraceae), birch (*Betula*, Betulaceae), *Prunus* (Rosaceae), and grape (*Vitis*, Vitaceae). Mines at first are linear, containing the frass, later a blotch, from which the frass is extruded from several holes. Pupation occurs in a hammocklike cocoon formed on a curled leaf. There may be two or more generations, with adults found from March to June and October and, according to Forbes, overwintering under bark. A similar species, *L. candida* Braun, mines the leaves of native azalea, *Rhododendron occidentale* (Ericaceae), in Washington, Oregon, and coastal California.

Moths of the genera **Leucoptera** and **Paraleucoptera** are tiny (FW 3–4 mm), shining snow white with a raised, golden spot at the base of the fringe and delicate, pale ochreous orange lines in the FW apically. **Paraleucoptera heinrichi** Jones (Plate 11.48) was described from Berkeley, California, where the larvae mine leaves of ornamental *Prunus ilicifolia* and Catalina cherry, *P. lyonii* (Rosaceae). The larvae are sometimes abundant on *P. lyonii* in native situations on the larger Channel Islands, and whether *Paraleucoptera heinrichi* was a native species on the mainland is unknown. It occurs on *Prunus ilicifolia* in natural communities in the San Francisco Bay area but rarely elsewhere. Larvae form large blotch mines with digitate

extensions from February to May. A similar species, **Paraleucoptera albella** (Chambers), occurs in Arizona and eastern California north to the Sacramento Valley, on willow and poplar in April to May and August. After colonies of **Leucoptera spartifoliella** (Hübner), a European species, were discovered at Tacoma, Washington, and in Del Norte County, California, this species was introduced as a biological control agent for brooms (*Cytisus* and *Genista*, Fabaceae), several species of which are noxious weeds in the Pacific states. Releases were made at six sites in the Sierra Nevada and San Francisco Bay area in 1960–1962. Recent records include abundant larvae in British Columbia (Vancouver) and northern California (Arcata) in July, but *L. spartifoliella* has not been successful in central California. The moths are similar to *Paraleucoptera*, slightly smaller with faint FW markings.

The "morning glory leaf miner," **Bedellia somnulentella** (Zeller), is a dull gray moth, FW (3.5–5.5 mm; Plate 11.47, Fig. 116.b) sprinkled with darker scales and with a poorly defined pale band along the dorsal margin. In fresh specimens there are three equally spaced smudges of darker scaling extending into the pale posterior band. The labial palpi are quite short and the head scaling roughened like a tineid. *Bedellia somnulentella* was described first in Europe in the midnineteenth century, but because its primary larval hosts, species of *Ipomaea* (Convolvulaceae), are natively pantropical, it may have been a tropical species. Subsequently this species has been spread worldwide as a pest of garden morning glories and sweet potato *(I. batatas)*, and on bindweed, *Convolvulus arvensis*. The larva first excavates a short, linear mine, then exits and creates one or more larger, blotch mines, where it is easily visible, covered only by the nearly transparent leaf cuticle. The frass is ejected and placed in clumps on the underside of the mine. The pupa is exposed, heliodinid-style, anchored with just a few strands of silk. This species is generally distributed in North America. In California in addition to field crop sweet potatoes, gardens, and weedy areas, *B. somnulentella* lives on native morning glories, *Calystegia*, on coastal dunes and on the Channel Islands. Probably there are continuous generations in warmer areas, with collection records in every month, most frequently in September and October.

References for Superfamily Yponomeutoidea

Dugdale, J. S., N. P. Kristensen, G. S. Robinson, and M. J. Scoble 1999. Yponomeutoidea, pages 119–130, *in*: Kristensen, N. P. (ed.), Handbook of Zoology. Lepidoptera, Moths, and Butterflies. Vol. 1: Evolution, Systematics, and Biogeography. W. de Gruyter, Berlin, New York; x + 491 pages.

Freeman, T. N. 1972. The coniferous feeding species of *Argyresthia* in Canada (Lepidoptera: Yponomeutidae). Canadian Entomologist 104: 687–697.

Gaedike, R. 1967. Zur systematischen Stellung einiger Gattungen der Heliodinidae/Schreckensteiniidae sowie Revision der Paläarktischen Arten der Gattung *Pancalia* Curtis, 1830. Beitrag Entomologische, Berlin 17: 363–374.

Gaedike, R. 1984. Revision der nearktischen und neotropischen Acrolepiidae (Lepidoptera). Entomologische Abhandlungen, Dresden 47: 179–194.

Gaedike, R. 1990. Revision der nearktischen Douglasiidae (Lepidoptera). Beitrag Entomologische, Berlin 40: 287–300.

Heppner, J. B. 1985. The sedge moths of North America (Lepidoptera: Glyphipterigidae). Flora and Fauna Handbook 1; 254 pages.

Heppner, J. B. 1987. Glyphipterigidae, Plutellidae, Yponomeutidae, Argyresthiidae, Acrolepiidae, Heliodinidae (Yponomeutoidea), pages 403–411, *in*: Stehr, F. (ed.) Immature Insects. Vol. 1. Kendall/Hunt, Dubuque, IA.

Hsu, Y.-F., and J. A. Powell 2005. Phylogenetic relationships and systematics of moths formerly assigned to *Heliodines* Stainton (Lepidoptera: Yponomeutoidea). University of California Publications in Entomology 124; xii + 158 pages, 172 figs., 3 color plates. Available online at http://repositories.cdlib.org/ucpress/ucpe/vol124.

Kyrki, J. 1983. The Yponomeutoidea: A reassessment of the superfamily and its suprageneric groups (Lepidoptera). Entomologica Scandinavica 19: 437–453.

Kyrki, J. 1990. Tentative reclassification of Holarctic Yponomeutoidea (Lepidoptera). Nota Lepidoptera 13: 28–42.

Landry, J.-F. 2007. Taxonomic review of the leek moth genus *Acrolepiopsis* (Lepidoptera: Acrolepiidae) in North America. Canadian Entomologist 139: 319–353.

Powell, J. A. 1991. A review of *Lithariapteryx* (Heliodinidae), with description of an elegant new species from coastal sand dunes in California. Journal of the Lepidopterists' Society 45(2): 89–104.

MOTHS OF WESTERN
NORTH AMERICA: PLATES

Euproersinus euterpe Henry Edwards: taking nectar from filaree *(Erodium)*, California, San Benito Co., Pinnacles National Monument, March 2008. Photo by Paul Johnson II.

Comprehensive Color Plates

Nearly every species discussed in the text is illustrated on the color plates. For species exhibiting marked sexual dimorphism, genetic polymorphism, or pronounced geographic variation (polytypy), we provide two or more images illustrating the variation. Thus, the nearly 2,350 color images of pinned specimens exceed the number of species discussed. Nearly all of the living moths and caterpillars (Plates 59 to 64) are also represented by pinned specimens of the same species, as are the immature stages (Plates 59 to 64).

Sizing and Scales of Moth Images

We wanted to illustrate the microlepidoptera by images sufficiently enlarged that they would be distinguishable by the naked eye. The great range in size, even among species of the same family (e.g., Tineidae, Prodoxidae, Gelechiidae), made it impractical to group together images of similar-size species and retain a taxonomic sequence that reflects the text. Therefore, every species displayed on plates 1 to 12 and 14 to 26 was photo-imaged and reduced or enlarged individually. We used two basic image frame sizes, ca. 26 mm or 32 mm in width, which allowed us to print images at approximately 1 inch or 1.25 inches in wingspan. We accomplished this by adding a *scale bar* to each image, the length of which is equal to the wingspan of the specimen shown. This urges readers to rely on the scale bar, and not on the specimen, as a measure of size. Hence, very tiny moths such as nepticulids and heliozelids (Plate 1) are shown much larger (3X to 5X) than life size, whereas most larger micros are depicted at about 1.5X natural size, and some of the largest species (e.g., some tortricids and pyraloids) appear at life size or even reduced. In addition, images on these plates were taken against a gray background of varying hues, depending upon the color of the moth, to provide contrast with dark or white or pale gray wings and their fringes.

Specimens on Plate 13 (the largest "micros" and pyralids) and Plates 27 to 58 (the "macros") are all shown at life size without the scale bars, floated onto a uniform background. As a result, images of some of the smallest geometrids and noctuids are actually smaller than those of tiny microlepidoptera.

The Captions

The plate caption for each species provides the scientific name, its author, the sex of the specimen illustrated, and its collec-tion data, including larval rearing data (plant name, larval collection date, and adult emergence date), collector(s), and collection where the specimen is housed. Common names are omitted from the figure legends; these appear in the text and in the index to moth names.

Selection of Specimens and Creation of the Plates

After we had selected the roster of species to be discussed (according to criteria listed on page ix), we sought suitable specimens to represent each species. For the micro moths and pyraloids (Plates 1 to 12 and plates 14 to 26), we selected specimens from the collection at the Essig Museum at the University of California, Berkeley (hence, the emphasis on California localities). We looked for specimens in good condition that were best suited to display (well spread with the wings on a flat plane) and to depict variation. That is, we did not attempt to illustrate specimens from particular regions, localities, or collectors, except when geographic variation was a goal. We tended to favor specimens originating from larval rearing when available because they often were the ones in best condition. Next, Jerry Powell took a list of these to the Smithsonian and borrowed the best specimens of those species not available at Berkeley. Specimens of Zygaenoidea (Plates 20 and 21) were provided by Marc Epstein and borrowed from several institutions, as acknowledged.

Using the same criteria, with emphasis on species for which biological information is available and species of economic importance, Opler selected specimens from the Gillette Museum of Arthropod Diversity at Colorado State University, along with a few from the Essig Museum at Berkeley, and shipped them to Ottawa. Then he took a list of the specimens to be imaged to Ottawa, where he selected examplars of most of them from the Canadian National Collection. Many examples are from the Troubridge collection, parts of which subsequently have been deposited at the University of Alberta. Staff at the Canadian National Collection selected additional specimens to represent variation and fill gaps in the plates. During that visit and subsequently, Jocelyn Gill prepared and manipulated images of them.

We obtained images of the living moths and early stages (Plates 59 to 64) by scanning copies, primarily from 40-year-old to recent Kodachrome slides. Most we provided or borrowed from our friends, as acknowledged on page iv.

MOTH FAMILIES AND PLATES

Acanthopteroctetidae (plate 1)
Acrolepiidae (plate 11)
Acrolophidae (plates 2, 59)
Adelidae (plates 2, 59)
Alucitidae (plate 12)
Apatelodidae (plates 35, 64)
Arctiinae (plates 46–49, 64)
Argyresthiidae (plate 11)
Blastobasidae (plates 4, 60)
Bombycidae (plate 64)
Bucculatricidae (plates 3, 4)
Carposinidae (plates 21, 61)
Choreutidae (plates 12, 60)
Coleophoridae (plates 5, 6, 60)
Copromorphidae (plate 21)
Cosmopterigidae (plates 1, 6, 7, 60)
Cossidae (plate 13)
Crambidae (plates 21–24, 61)
Dalceridae (plates 20, 61)
Deoclonidae (plate 5)
Depressariidae (plates 4, 5)
Doidae (plate 33)
Douglasiidae (plate 4)
Drepanidae (plates 27, 61)
Elachistidae (plates 5, 6, 59)
Epermeniidae (plate 12)
Epipyropidae (plate 21)
Eriocraniidae (plates 1, 59)
Ethmiidae (plates 5, 59)
Gelechiidae (plates 7–10, 60)
Geometridae (plates 27–32, 61, 62)
Glyphipterigidae (plate 11)
Gracillariidae (plates 3, 59)
Heliodinidae (plates 11, 60)
Heliozelidae (plates 1, 59)
Hepialidae (plate 13)
Incurvariidae (plates 1, 59)

Lasiocampidae (plates 34, 35, 37, 38, 62)
Limacodidae (plates 20, 61)
Lithosiinae (plates 46, 64)
Lymantriinae (plates 49, 62, 63)
Lyonetiidae (plate 11)
Megalopygidae (plates 20, 61)
Micropterigidae (plate 1)
Mimallonidae (plate 37)
Momphidae (plate 6)
Nepticulidae (plate 1)
Noctuidae (plates 42–58, 63, 64)
Notodontidae (plates 33, 42, 63, 64)
Oecophoridae (plates 4, 59)
Opostegidae (plate 1)
Phyllocnistidae (plates 3, 59)
Plutellidae (plate 11)
Prodoxidae (plates 1, 2, 59)
Psychidae (plates 2, 59)
Pterophoridae (plates 12, 60)
Pyralidae (plates 13, 24–26, 61)
Saturniidae (plates 35–38, 62, 63)
Schreckensteiniidae (plate 11)
Scythrididae (plates 6, 60)
Sematuridae (plate 35)
Sesiidae (plates 12–14, 60)
Sphingidae (plates 39–41, 63)
Stenomatidae (plates 4, 59)
Symmocidae (plate 4)
Thyrididae (plates 21, 61)
Tineidae (plates 2, 59)
Tischriidae (plates 1, 59)
Tortricidae (plates 14–20, 60, 61)
Uraniidae Epipleminae (plates 27, 61)
Yponomeutidae (plate 10)
Ypsolophidae (plates 10, 11, 60)
Zygaenidae (plates 21, 61)

1.1 *Epimartyria pardella* Walsingham ♂ — California, Humboldt Co., Arcata, VI.11.1969 (Powell) EME

1.2 *Dyseriocrania auricyanea* (Walsingham) ♀ — California, Alameda Co., U.C. Berkeley Campus, II.26.1970 (Opler) EME

1.3 *Eriocraniella aurosparsella* (Walsingham) ♂ — California, Tuolumne Co., NE of Tuolumne City, III.23.1966 (Powell) EME

1.4 *Eriocraniella xanthocara* Davis ♀ — California, Contra Costa Co., NE of Orinda, III.6.1970, adult on *Quercus agrifolia* (Opler) EME

1.5 *Eriocrania semipurpurella* (Stephens) ♀ — California, San Mateo Co., III.15.1984 (Powell) EME

1.6 *Neocrania bifasciella* Davis paratype ♀ — California, Monterey Co., S of Jamesburg, V.5.1975 (J. Chemsak, Powell) EME

1.7 *Acanthopteroctetes unifascia* Davis ♂ — California, Monterey Co., Pfeiffer–Big Sur State Park, III.11.1984 (D. Wagner, JP 84C16, r.f. *Ceanothus thyrsiflorus*, emgd. VIII.1984) EME

1.8 *Ectoedemia platanella* (Clemens) ♀ — California, Contra Costa Co., Antioch Natl. Wildlife Refuge, V.31.1992 (J. Powell 92E247, r.f. *Platanus racemosa*, emgd. VI.28.1992) EME

1.9 *Stigmella rhamnicola* (Braun) ♂ — Ohio, Hamilton Co., Cincinnati, L[arva] V.14.1916, b[red] 676 (A. F. Braun) USNM

1.10 *Stigmella diffasciae* (Braun) ♂ — California, San Luis Obispo Co., SW of Atascadero, II.4.1987 (J. Powell, 87B99, r.f. *Rhamnus californica*, emgd. III.25.1987) EME

1.11 *Stigmella heteromelis* Newton & Wilkinson ♀ — California, Alameda Co., U.C. Berkeley Campus, I.23.1960 (J. Powell 60A1, r.f. *Photinia arbutifolia*, emgd. II.24.1960) EME

1.12 *Stigmella variella* (Braun) ♂ — California, Alameda Co., Berkeley, V.3.1987 (Powell) EME

1.13 *Stigmella variella* (Braun) ♀ — California, Contra Costa Co., Pt. San Pablo, II.20.1989 (J. Powell 89B2, r.f. *Quercus agrifolia*, emgd, III.17.1989) EME

1.14 *Micracalyptris lotella* Wagner, paratype ♀ — California, Contra Costa Co., Antioch Natl. Wildlife Refuge, III.15.1985 (D. Wagner, JP 85C14, r.f. *Lotus scoparius*, emgd. IV.16.1985) EME

1.15 *Opostega bistrigulella* Braun ♂ — California, Santa Barbara Co., Santa Rosa Island, III.15.1997 (P. Jump & Powell) EME

1.16 *Opostega albogaleriella* Clemens ♂ — California, Monterey Co., Big Creek Reserve, IV.12.1990 (Y. F. Hsu & Powell) EME

1.17 *Opostega albogaleriella*, same data, frontal view showing upraised wing tips and eyecaps

1.18. *Tischeria consanguinea* Braun, paratype ♂ — California, El Dorado Co., Cool, III.23.1965 (Powell) EME

1.19 *Tischeria mediostriata* Braun ♀ — California, San Diego Co., W of Guatay, VI.7.1968 (Opler, JP 67F67, r.f. *Quercus dumosa* [=*berberidifolia*], emgd. VII.22.1968) EME

1.20 *Tischeria splendida* Braun paratype ♀ — California, Contra Costa Co., Russellman Park, Mt. Diablo, II.24.1961 (J. Powell 61B12, r.f. *Rubus vitifolia*, emgd. III.7.1961) EME

1.21 *Tischeria omissa* Braun ♂ — California, Solano Co., E of Vacaville, X.10.1956 (J. Powell 56K1, r.f. *Sida hederacea* [=*Malvella leprosa*], emgd. X.19.1956) EME

1.22 *Tischeria ceanothi* Walsingham ♂ — California, Santa Barbara Co., Gaviota Pass, II.3.1987 (Powell & Wagner, JP87B42, r.f. *Ceanothus spinosus*, emgd. II.13.1987) EME

1.23 *Coptodisca powellella* Opler, paratype ♂ — California, Santa Barbara Co., E of Los Prietos, II.13.1968 (Opler, JP 68B76 r,f, Quercus agrifolia, emgd. III.14.1968) EME

1.24 *Coptodisca arbutiella* Busck ♀ — California, Marin Co., Marin Island, IV.13.1989 (Powell & M. Prentice, JP 89D28, r.f. *Arbutus menziesii*, emgd. V.16.1989) EME

1.25 *Vespina quercivora* (Davis) ♀ — California, Los Angeles Co., Malibu Canyon, IV.28.1977 (D. Green & Powell, JP 77D228, r.f. *Quercus agrifolia*, emgd. VIII.17.1977) EME

1. 26 *Tridentaforma fuscoleuca* (Braun) ♂ — California, Siskiyou Co., Ash Creek Ranger Station, VI.7.1974 (Powell) EME

1.27 *Tridentaforma* species ♀ — California, Marin Co., Liberty Gulch near Alpine Lake, IV.19.1975 (Powell) EME.

1.28 *Tridentaforma* species ♀ — California, Siskiyou Co., McBride Springs, VI.23.1974 (Powell) EME

1.29 *Greya politella* (Walsingham) ♀ — California, Yuba Co., N of Smartville, V.7.1980 (Powell) EME

1.30 *Greya variabilis* Davis & Pellmyr, paratype ♂ — Alaska, N of Cantwell, 2,200′, VI.26.1979 (Opler & Powell) EME

1.31 *Greya variabilis* Davis & Pellmyr, paratype ♂ — Alaska, Eagle Summit, VII.2.1979 (Opler & Powell) EME

1.32 *Tegeticula synthetica* (Riley) ♀ — California, Kern Co., NW of Willow Springs, III.7.1991 (Powell) EME

1.33 *Tegeticula maculata* (Riley) ♀ — California, Kern Co., E of Caliente, V.15.1963, *Yucca whipplei* (Powell) EME

1.34 *Tegeticula maculata* (Riley) ♂ — California, Santa Barbara Co., S of Buellton, V.11.1965 (Powell) EME

1.35 *Tegeticula maculata extranea* (H. Edwards) ♀ — California, San Bernardino Co., NE of Highland, IV.16.1965, on *Yucca whipplei* (R. Langston) EME

1.36 *Tegeticula elatella* Pellmyr ♂ — New Mexico, Grant/Luna Co. line, W of Deming, VIII.4.1973, on *Yucca elata* (Powell) EME

1.37 *Tegeticula maderae* Pellmyr ♂ — Arizona, Pima Co., W of Molino Basin, VIII.19.1972 (Powell), *Yucca schottii* (Powell) EME

1.38 *Tegeticula mojavella* Pellmyr, paratype ♂ — California, San Bernardino Co., Mountain Pass, Clark Mtn., V.8.1995, in *Yucca schidigera* flower (Pellmyr & J. Leebens-Mack) UID

1.39 *Parategeticula pollenifera* Davis ♀ — Arizona, Cochise Co., Cave Creek, Chiricahua Mtns., VIII.11.1971 (J. Powell 71H3, r.f. *Yucca schottii*, emgd. VII.28.1973) EME

1.40 *Prodoxus quinquepunctellus* (Chambers) ♀ — Colorado, Weld Co., Roggen, III.8.1992 (J. Powell 92C2, r.f. *Yucca glauca*, emgd. IV.28.1993) EME

1.41 *Prodoxus barberellus* (Busck) ♀ — Arizona, Santa Cruz Co., Madera Canyon, VI.6.1968 (Opler & Powell, JP 68F44, r.f. *Agave palmeri*, emgd. VII.13.1968) EME

1.42 *Prodoxus pallidus* (Davis) ♂ — Mexico, Baja California Norte, Diablito Canyon, Sierra San Pedro Martir, IV.6.1973 (J. Powell 73D1, r.f. *Agave*, emgd. IV.24.1973) EME

1.43 *Prodoxus coloradensis* Riley ♂ — California, Riverside Co., Pinyon Flat, IV.7.1963 (J. Powell 63D7, r.f. *Yuccas schidigera*, emgd. IV.15.1963) EME

1.44 *Prodoxus phylloryctus* Wagner & Powell, paratype ♀ — Colorado, Dolores Co., Dove Creek, II.25.1985 (J. Terry, JP 85C7, r.f. *Yucca baccata*, emgd. IV.3.1985) EME

1.45 *Prodixus y-inversus* Riley ♀ — Nevada, Clark Co., Kyle Canyon, IV.2.1970 (R. Dietz & Powell, JP 70D21, r.f. *Yucca baccata*, emgd. IV.28.1989) EME

1.46 *Prodoxus ochrocarus* Davis ♀ — Arizona, Santa Cruz Co., Madera Canyon, VI.5.1968 (Opler & Powell, JP 68F45, r.f. *Yucca schottii*, emgd. IX.12.1969) EME

1.47 *Prodoxus sordidus* Riley ♀ — California, Los Angeles Co., WNW of Lancaster, III.7.1991 (J. Powell 91C4, r.f. scape *Yucca brevifolia*, emgd. III.15.1993) EME

1.48 *Prodoxus weethumpi* Pellmyr & Balcazar-Lara ♀ — California, Los Angeles Co., WNW of Lancaster, III.7.1991 (J. Powell 91C4, r.f. seed pod *Yucca brevifolia*, emgd. III.8.1994) EME

1.49 *Prodoxus marginatus* Riley ♀ — California, Kern Co., NW of Dove Well, V.1.1964 (J. Powell 64D15, r.f. seed pod *Yucca whipplei*, emgd. V.16.1964) EME

1.50 *Prodoxus marginatus pulverulentus* Riley ♀ — California, San Diego Co., Cardiff, III.23.1967 (J. Powell 67C36, r.f. seed pod *Yucca whipplei*, emgd. III.28.1967) EME

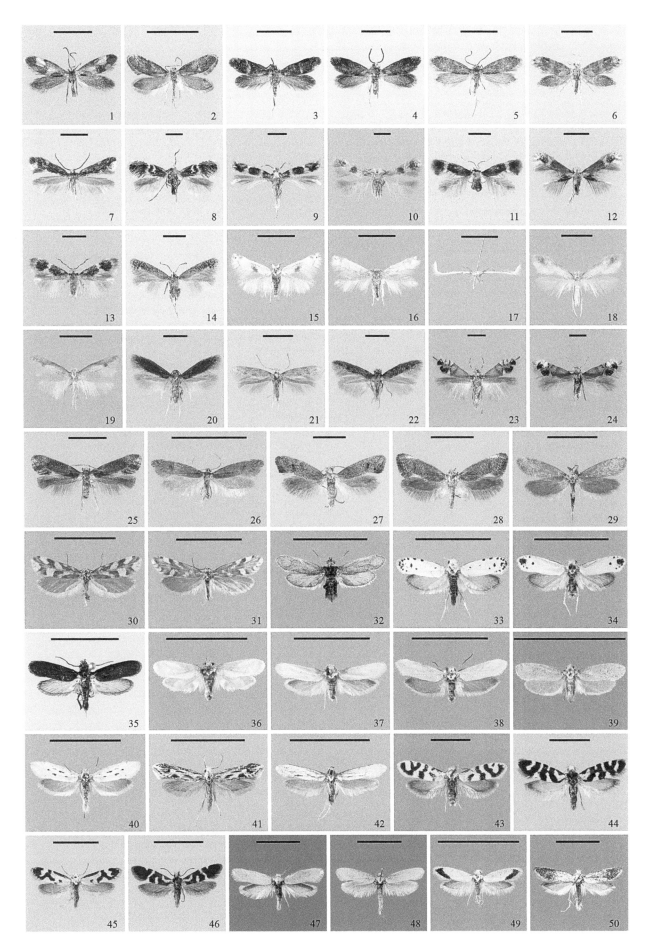

PLATE 1

PLATE 2 Adelidae, Prodoxidae, Tineidae, Acrolophidae, Psychidae

2.1 *Prodoxus cinereus* Riley ♂ — California, Tulare Co., Fairview, IV.27.1964 (J. Powell 64D9, r.f. *Yucca whipplei* stalk, emgd. V.10.1964) EME

2.2 *Prodoxus aenescens* Riley ♀ — California, San Benito Co., New Idria, IV.24.64 (J. Powell 64D4, r.f. *Yucca whipplei* stalk, emgd. V.12.1964) EME

2.3 *Mesepiola specca* Davis ♀ — Arizona, Pima Co., Molino Basin, Santa Catalina Mtns., VIII.2.1970 (Powell & P. Rude) EME

2.4 *Cauchas simpliciella* (Walsingham) ♀ — California, Plumas Co., SE of Quincy, V.15.1982, adult on *Arabis* #1428 (Powell) EME

2.5 *Adela oplerella* Powell ♂ — California, Santa Clara Co., W of San Martin, III.14.1986 (D. Murphy & Powell) EME

2.6 *Adela punctiferella* Walsingham ♂ — California, Santa Clara Co., W of San Martin, III.5.1986 (D. Murphy & Powell) EME

2.7 *Nemophora bellela* (Walker) ♀ — Canada, Yukon, Dawson, VI.30.1949 (W. Judd) CNC

2.8 *Adela eldorada* Powell, paratype ♂ — California, Tulare Co., V.14.1963 (C. Toschi) EME

2.9 *Adela trigrapha* Zeller ♂ — California, San Luis Obispo Co., Nacimiento Dam, IV.14.1967 (Powell) EME

2.10 *Adela thorpella* Powell, paratype ♂ — California, Contra Costa Co., Russellman Park, Mt. Diablo, IV.9.1958 (Powell) EME

2.11 *Adela flammeusella* (Chambers) ♀ — California, Lake Co., Upper Lake, V.11.1961 (Powell) EME

2.12 *Adela septentionella* Walsingham ♂ — California, Marin Co., W of Fairfax, IV.17.1961 (Powell) EME

2.13 *Adela septentionella* Walsingham ♀ — California, Monterey Co., Big Creek Reserve, IV.24.1987 (Powell) EME

2.14 *Adela singulella* Walsingham ♀ — California, Napa Co., E of Conn Dam, VI.5.1964 (J. Chemsak & Powell) EME

2.15 *Adela singulella* Walsingham ♂ — California, Napa Co., E of Conn Dam, V.18.1966 (Powell) EME

2.16 *Phryganeopsis brunnea* Walsingham ♀ — California, Siskiyou Co., Mt. Shasta City, VII.6.1958 (Powell) EME

2.17 *Tinea pallescentella* Stainton ♀ — California, Contra Costa Co., El Cerrito, (P. Rude coll. no. 800302, r.f. blood meal, emgd. III.18.1980) EME

2.18 *Tinea occidentella* Chambers ♀ — California, Monterey Co., W. of Greenfield, V.2.1975 (Powell) EME

2.19 *Tinea niveocapitella* Chambers ♀ — California, San Mateo Co., San Bruno Mtn. III.5.1986 (J. De Benedictis, JP 86C15, r.f. owl pellets, emgd. III.31.1986) EME

2.20 *Morophagoides burkerella* (Busck) ♀ — California, Marin Co., Inverness, III.17.1966 (J. Wolf, JP 66C18, r.f. *Polyporus versicolor*, emgd. V.11.1966) EME

2.21 *Morophagoides burkerella* (Busck) [form *gracilis* Wlsm.] ♂ — California, Alameda Co., Berkeley, VIII.29.1999 (Powell) EME

2.22 *Tineola bisselliella* (Hummel) ♀ — California, Alameda Co., Berkeley (J. Powell, r. f. tennis ball cover, emgd. II.29.1960) EME

2.23 *Tinea pellionella* Linnaeus ♀ — California, Contra Costa Co., Tilden Park, I.1983 (A. Kaplan, JP 83A12, r.f. decayed plant material, emgd. V.17.1983) EME

2.24 *Tinea pallescentella* Stainton ♂ — California, Alameda Co., Berkeley, IX.26.1979 (Powell) EME

2.25 *Elatobia carbonella* Dietz ♂ — California, Yuba Co., N of Smartville, V.2.1980 (J. De Benedictis, JP 80E1, r.f. *Polyporus volvatus*, emgd. VI.29.1980) EME

2.26 *Phereoeca praecox* (Gozmany & Vari) ♀ — California, San Diego Co., Clairemont, IX.26.1999, r.f, case on wall (D. Faulkner) EME

2.27 *Morophagoides berkeleyellus* (Powell) holotype ♂ — California, Contra Costa Co., SE of Canyon, II.5.1967 (J. Powell 67B1, r.f. *Polyporus gilvus* on *Quercus*, emgd. III.31.1967) EME

2.28 *Monopis crocicapitella* (Clemens) ♀ — California, Alameda Co., Oakland, IV.7.1957 (B. Adelson) EME

2.29 *Dryadaula terpsichorella* (Busck) ♀ — California, San Diego Co., San Diego River at Midway, II.23.1999 (Powell) EME

2.30 *Homosetia marginimaculella* (Chambers) ♀ — California, Alameda Co., Berkeley, VIII.24.1989 (Powell) EME

2.31 *Oinophila v-flava* (Haworth) ♀ — California, Alameda Co., Berkeley, I.17.1988 (D. Carver, JP 88A7, r.f. decayed log, emgd. V.5.1988) EME

2.32 *Opogona arizonensis* Davis, paratype ♂ — Arizona, Santa Cruz Co., Madera Canyon, VII.11.1959 (R. Hodges) USNM

2.33 *Nemapogon granellus* (Linnaeus) ♂ — California, Alameda Co., Claremont Canyon, II.10.1967 (J. Powell 67B4, r.f. *Quercus* bark with *Hypoxylon*, emgd. II.28.1967) EME

2.34 *Nemapogon molybdanellus* (Dietz) ♂ — California, Contra Costa Co., S of Clayton, I.22.1961 (J. Lawrence no. 748, r.f. *Polyporus sulfureus* on *Quercus*) EME

2.35 *Nemapogon oregonellus* (Busck) ♀ — Oregon "8877" [no further data] USNM

2.36 *Nemapogon defectellus* (Zeller) ♀ — California, Riverside Co., E of Blythe, I.26.1964 (J. Lawrence lot 253, r.f. *Trametes hispida*, emgd. II.26.1964) EME

2.37 *Acrolophus laticapitanus* (Walsingham) ♀ — California, San Diego Co., McAs Miramar, VII.10.1998 (N. Bloomfield) ENE

2.38 *Amydria arizonella* Dietz ♂ — California, Contra Costa Co., Antioch Natl Wildlife Refuge, IX.20.1990 (Powell) EME

2.39 *Amydria obliquella* Dietz ♂ — Mexico, Baja California Sur, SW of La Paz, VIII.8.1966 (Powell) EME

2.40 *Amorophaga cryptophori* (Clarke) ♀ — California, Trinity Co., S of Scott Mtn Peak, 4,400¢, VI.25.1980 (J. Powell 80F56, r.f. *Polyporus volvatus*, emgd. VII.28.1980) EME

2.41 *Dyotopasta yumaella* (Kearfott) ♂ — California, Riverside Co., Pinyon Flat, IV.13.1963 (R. Langston) EME

2.42 *Daviscardia coloradella* (Dietz) ♀ — Utah, Sanpete Co., Ephraim Canyon, 8,850¢, VII.21.1980 (D. Ferguson) EME

2.43 *Opogona omoscopa* (Meyrick) ♀ — California, Alameda Co., Berkeley, IX.23.1978 (Powell) EME

2.44 *Acrolophus variabilis* (Walsingham) ♂ — Arizona, Cochise Co., Cave Creek SW of Portal, VII.2.1972 (Powell) EME

2.45 *Acrolophus laticapitanus* (Walsingham) ♀ — California, San Diego Co., N of Pine Valley, VI.27.1962 (J. Lawrence) EME [pale form]

2.46 *Acrolophus laticapitanus* (Walsingham) ♂ — California, Monterey Co., Big Creek Reserve, VIII.6.1993 (M. McIntosh & Powell) EME [dark form]

2.47 *Acrolophus pyramellus* (Barnes & McDunnough) ♂ — Arizona, Maricopa Co., White Mtns. Regional Park, IV.1.1976 (R. Wielgus) USNM

2.48 *Hyaloscotes fumosa* Butler ♂ — California, Sierra Co., SE of Sierraville, VII.28.1995 (Powell) EME [abd. dissected JAP 7145]

2.49 *Thyridopteryx meadii* H. Edwards ♂ — Texas, Brewster Co., Big Bend Natl Park, VIII.29.1965 (A. & M. Blanchard) USNM

PLATE 2

PLATE 3 Gracillariidae, Phyllocnistidae, Bucculatricidae

3.1 *Caloptilia acerifoliella* (Chambers) ♀ — California, Monterey Co., Big Creek Reserve, IX.3.1991 (J. Powell 91J1, r.f. *Acer macrophyllum*, emgd. IX.17.1991) EME

3.2 *Caloptilia alnivorella* (Chambers) ♀ — California, Monterey Co., Big Creek Reserve, IX.4.1991 (J. Powell 91J11, r.f. *Alnus rhombifolia*, emgd. X.5.1991) EME

3.3 *Caloptilia alnivorella* (Chambers) ♀ — California, Monterey Co., Big Creek Reserve, IX.4.1991 (J. Powell 91J11, r.f. *Alnus rhombifolia*, emgd. X.12.1991) EME

3.4 *Caloptilia azaleella* (Brants) ♀ — California, Alameda Co., Berkeley, X.28.1987 (Powell) EME

3.5 *Caloptilia reticulata* (Braun) ♂ — California, Alameda Co., Berkeley, X.9.1991 (Powell) EME

3.6 *Caloptilia agrifoliella* Opler ♀ — California, Monterey Co., Big Creek Reserve, X.30.1989 (Powell) EME

3.7 *Caloptilia agrifoliella* Opler ♂ — California, Marin Co., Inverness Ridge, VI.4.1996 (Powell) EME

3.8 *Caloptilia ovatiella* Opler, paratype ♂ — California, Orange Co., S. Laguna Beach (H. Sweet lab colony, r.f. *Rhus integrifolia*, emgd. VIII.11.1968) EME

3.9 *Caloptilia diversilobiella* Opler ♀ — California, Contra Costa Co., Briones Reservoir, IV.22.1984 (J. Powell 84D66, r.f. *Rhus [Toxicodendron] diversilobum*, emgd. V.14.1984) EME

3.10 *Caloptilia nondeterminata* (Braun) ♂ — California, Marin Co., Inverness Ridge, VI.26.1997 (J. Powell 97F16, r.f. *Ribes sanguineum*, emgd. VII.24.1997) EME

3.11 *Caloptilia palustriella* (Braun) ♂ — California, San Mateo Co., Pacifica, IX.1.1980 (J. De Benedictis 80245C, r.f. *Salix*, emgd. IX.21.1980) EME

3.12 *Caloptilia palustriella* (Braun) ♂ — California,—Big Creek Reserve, V.27.1987 (J. Powell 87E16, r.f. *Salix lasiolepis*, emgd. VI.14.1987) EME

3.13 *Neurobathra bohartiella* Opler ♀ — California, Alameda Co., Berkeley, VII.30.1980 (Powell) EME

3.14 *Micrurapteryx salicifoliella* (Chambers) ♀ — California, Contra Costa Co., Bethel Island, VII.27.1969 (Opler, JP 69G41, r.f. *Salix*, emgd. VIII.12.1969) EME

3.15 *Parectopa albicostella* Braun ♀ — Utah, Weber Co., E of Ogden, 6,500′, VII.7.1992 (D. Davis 1147, host: Leguminosae) USNM

3.16 *Acrocercops insulariella* Opler, paratype ♀ — California, Santa Barbara Co., Santa Cruz Island, IV.25.1966 (Powell, A. Slater, J. Wolf, JP 66.D45, r.f. *Quercus agrifolia*, emgd. V.26.1966) EME

3.17 *Acrocercops arbutiella* Braun ♀ — Arizona, Miller Canyon, Huachuca Mtns., 1775m, IV.11.1988 (J. Powell 88D23, r.f. *Arbutus arizonica*, emgd. V.5.1988) EME

3.18 *Leucanthiza dircella* Braun ♀ — California, Contra Costa Co., Wildcat Creek, Berkeley Hills, III.6.1972 (Powell) EME

3.19 *Marmara arbutiella* Busck ♂ — California, Contra Costa Co., NE of Orinda, I.19.1967 (J. Powell 67A7, r.f. *Arbutus menziesii*, emgd. II.6.1967) EME

3.20 *Protolithocolletis lathyri* Braun ♀ — California, Plumas Co., Meadow Valley, IX.13.1992 (Y.-F. Hsu, JP 92J23, r.f. *Lathyrus*, emgd. IX.25.1992) EME

3.21 *Cremastobombycia grindeliella* (Walsingham) ♂ — California, Contra Costa Co., Pt. Isabel, II.29.1992 (J. Powell 92B7, r.f. *Grindelia humilis*, emgd. III.12.1992) EME

3.22 *Cremastobombycia* species ♀ — California, Monterey Co., Big Creek Reserve, IV.12.1985 (De Benedictis & Powell, JP 85D16, r.f. *Baccharis pilularis*, emgd. IV.15.1985) EME

3.23 *Cremastobombycia* species ♀ — California, Monterey Co., Big Creek Reserve, IV.28.1990 (Hsu & Powell, JP 90D66, r.f. *Gnaphalium bicolor*, emgd. V.9.1990) EME

3.24 *Cameraria gaultheriella* (Walsingham) ♂ — California, Sonoma Co., Plantation, V.1.1958 (J. Powell 58E1, r.f. *Gaultheria shallon*, emgd. V.13.1958) EME

3.25 *Cameraria nemoris* (Walsingham) ♀ — Oregon, Lincoln Co., Waldport, VI.22.1975 (Opler & Powell, JP 75F15, r.f. *Vaccinium ovatum*, emgd. VII.6.1975) EME

3.26 *Cameraria agrifolia* (Braun) ♂ — California, Alameda Co., Fairmont Ridge, San Leandro, III.13.1990 (J. Powell 90C11, r.f. *Quercus agrifolia*, emgd. III.31.1990) EME

3.27 *Cameraria wislizeniella* Opler, paratype ♂ — California, Marin Co., Woodacre, III.15.1968 (Opler, JP 68C23, r.f. *Quercus wislizenii*, emgd. IV.8.1968) [abd. dissected, EME 5428]

3.28 *Cameraria lobatiella* Opler & Davis, paratype ♂ — California, Contra Costa Co., Walnut Creek, VIII.27.1962 (Powell) EME

3.29 *Cameraria pentekes* Opler & Davis, paratype ♂ — California, Kern Co., Keene, VI.2.1968 (Opler, JP 68F1, r.f. *Quercus douglasii*, emgd. VI.14.1968) EME

3.30 *Cameraria jacintoensis* Opler & Davis, paratype ♀ — California, Solano Co., Cold Canyon Reserve, III.16.1990 (J. Powell 90C16, r.f. *Quercus dumosa [berberidifolia]*, emgd. III.31.1990) EME

3.31 *Phyllonorycter deserticola* Davis & Deschka, paratype ♀ — California, San Bernardino Co., Yucca Valley, VII.17.1977 (G. Deschka 1270, r.f. *Populus macdougalii [fremontii]*, emgd. VII.19/29.1977) USNM

3.32 *Phyllonorycter erugatus* Davis & Deschka, paratype ♂ — Canada, British Columbia, Hoodoos, Yoho Natl. Park, VIII.8.1982 (G. Deschka 1687, r, f *Populus balsamifera*, emgd. VIII.29/IX.2.1982) USNM

3.33 *Phyllonorycter nipigon* (Freeman) ♂ — Canada, Alberta, E of Red Deer, VIII.27.1982—(G. Deschka 1698, r, f *Populus balsamifera*, emgd. IX.2/9.1982) USNM

3.34 *Phyllonorycter salicifoliella* (Chambers) ♀ — California, Kern Co., S of Monolith, X.2.1967 (J. Powell 67K25, r.f. *Salix*, emgd. X.18.1967) EME

3.35 *Phyllonorycter apicinigrella* (Braun) ♀ — California, Monterey Co., Big Creek Reserve, IX.27.1987 (Powell) EME

3.36 *Phyllonorycter felinelle* Heinrich ♀ — California, Monterey Co., Big Creek Reserve, X.30.1989 (J. Powell 89K5, r.f. *Platanus racemosa*, emgd. I.30.1990) EME

3.37 *Phyllonorycter arbutusella* (Braun) ♀ — California, Marin Co., Alpine Lake, IV.14.1979 (J. Powell 79D12, r.f. *Arbutus menziesii*, emgd. V.23.1979) EME

3.38 *Phyllonorycter arbutusella* (Braun) ♂ — California, Monterey Co., Big Creek Reserve, IV.12.1990 (Hsu & Powell) EME

3.39 *Phyllonorycter arbutusella* (Braun) ♂ — California, Big Creek Reserve, V.4.1991 (Hsu & Powell, JP 91E24, r.f. *Arbutus menziesii*, emgd. V.29.1991) EME

3.40 *Phyllonorycter manzanita* (Braun) ♀ — California, Marin Co., Inverness Ridge, IV.24.1976 (J. Powell 76D26, r.f. *Arctostaphylos virgata*, emgd. VI.1.1976) EME

3.41 *Phyllonorycter ribefoliae* (Braun) ♂ — California, Monterey Co., Big Creek Reserve, I.25.1988 (J. Powell 88A18, r.f. *Ribes sanguineum*, emgd. II.18.1988) EME

3.42 *Phyllonorycter antiochella* Opler paratype ♀ — California, Contra Costa Co., Antioch, III.26.1969, adult on *Quercus agrifolia* (Powell)

3.43 *Phyllonorycter antiochella* Opler paratype ♀ — California, Contra Costa Co., Cowell, II.16.1969 (Opler, JP 69B16, r.f. *Quercus agrifolia*, emgd. II.21.1969) EME

3.44 *Phyllonorycter inusitatella* (Braun) ♀ — California, Tulare Co., Ash Mtn. HQ, IV.28.1979 (J. Powell 79D21, r.f. *Quercus wislizenii*, emgd. IV.28.1979) EME

3.45 *Phyllonorycter inusitatella* (Braun) ♂ — California, Ventura Co., Newbury Park, III.31.1969 (Opler, JP 69C67, r.f. *Quercus agrifolia*, emgd. IV.2.1967) EME

3.46 *Phyllonorycter inusitatella* (Braun) ♂ — California, Contra Costa Co., Mt. Diablo summit, IV.13.1969 (Opler, JP 69D14, r.f. *Quercus wislizenii*, emgd. IV.27.1969) EME

3.47 *Phyllonorycter sandraella* Opler, paratype ♂ — California, Alameda Co., U.C. Berkeley campus, IX.10.1960 (Powell) EME

3.48 *Phyllonorycter sandraella* Opler, paratype ♂ — California, San Diego Co., Descanso Ranger Station, III.31.1961 (J. Powell 61C24, r.f. *Quercus agrifolia*, emgd. IV.5.1961) EME

3.49 *Phyllonorycter sandraella* Opler ♂ — California, Santa Barbara Co., Santa Cruz Island, V.2.1976, r.f *Quercus agrifolia*, emgd. V.10.1976 (D. Green) EME

3.50 *Phyllocnistis populiella* Chambers ♀ — California, Nevada Co., Donner Lake, VIII.30.1993 (D. Davis, DRD 1303, r.f. *Populus tremuloides*, emgd. VIII.31.1993) USNM

3.51 *Phyllocnistis* species ♀ — California, Alameda Co., Albany, IV.8.1989 (Y.-F. Hsu, JP 89C97, r.f. *Prunus ilicifolia*, emgd. IV.15.1989) EME

3.52 *Phyllocnistis vitifolia* Chambers ♀ — California, Yolo Co., Davis, III.18.1989 (De Benedictis, JP 89C16, r.f. *Vitis*, emgd. III.25.1989) EME

3.53 *Phyllocnistis vitifolia* Chambers ♀ — California, Yolo Co., Davis, III.18.1989 (De Benedictis, JP 89C16, r.f. *Vitis*, emgd. III.25.1989) EME

3.54 Cocoons of *Bucculatrix separabilis* Braun, affixed to twig, California, Contra Costa Co., Pleasant Hill, V.19.1958 (Powell) EME

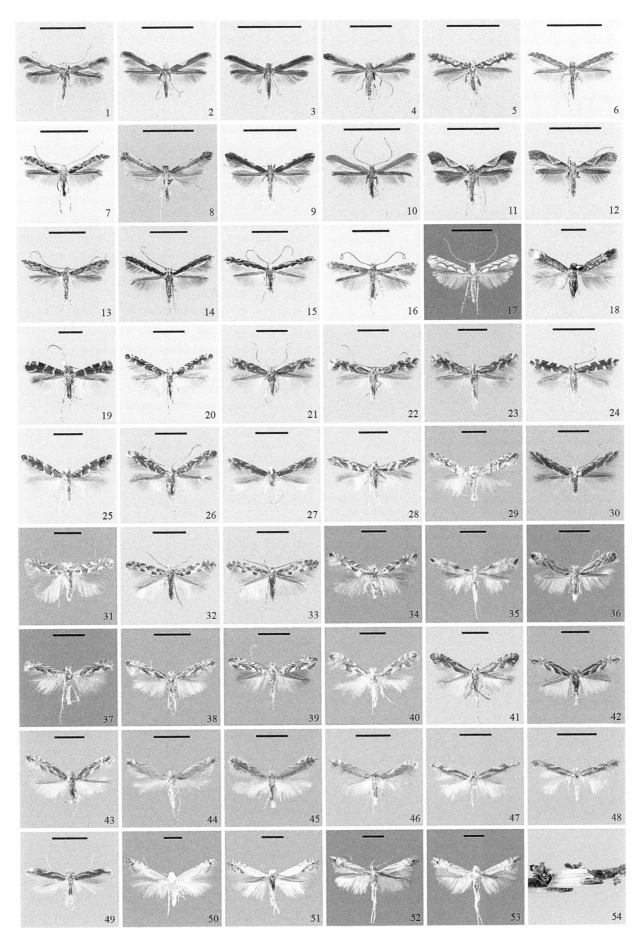

PLATE 3

4.1 *Bucculatrix longula* Braun, paratype ♀ — Washington, Whitman Co., Wilma, IV.29.1934, r.f. *Helianth annuus*, J. Clarke) USNM

4.2 *Bucculatrix angustata* Frey & Boll ♂ — New York, Tompkins Co., Ithaca, V.11.1959 (Hodges) USNM

4.3 *Bucculatrix variabilis* Braun ♂ — California, Monterey Co., Big Creek Reserve, III.22.1989 (Hsu & Powell, JP 89C43, r.f. *Baccharis pilularis*, emgd. IV.3.1989) EME

4.4 *Bucculatrix separabilis* Braun ♂ — California, Marin Co., Alpine Lake, IV.14.1979 (J. Powell 79D4, r.f. *Baccharis pilularis*, emgd. V.13.1979) EME

4.5 *Bucculatrix dominatrix* Rubinoff & Osborne, paratype ♂ — California, Marin Co., Ring Mtn., IV.11.1994 (J. Powell 94D53, r.f. *Baccharis pilularis*, emgd. IV.20.1994) EME

4.6 *Bucculatrix sexnotata* Braun ♂ — California, Placer Co., S. of Tahoe City, VIII.20.1980 (De Benedictis & Powell, JP 80H14, r.f. *Wyethia mollis*, emgd. VIII.29.1980) EME

4.7 *Bucculatrix salutatoria* Braun, paratype ♀ — Wyoming, Teton Co., Grand Teton Natl. Park (A. Braun, B2290, r.f. *Artemisia tridentata*, emgd.VII.15.1959) USNM

4.8 *Bucculatrix tridenticola* Braun ♂ — Nevada, Elko Co., Bear Creek Summit, VIII.4.1981 (Powell) EME

4.9 *Bucculatrix koebelella* Busck ♀ — California, Santa Barbara Co., Santa Rosa Island, IV.27.1995 (Powell) EME

4.10 *Bucculatrix albertiella* Busck ♀ — California, Marin Co., Phoenix Lake, V.18.1927 (H. Keifer, r.f. *Quercus lobata*) CAS

4.11 *Bucculatrix albertiella* Busck ♂ — California, Monterey Co., Big Creek Reserve, III.17.1994 (M. McIntosh & Powell, JP 94C30, r.f. *Quercus agrifolia*, emgd. IV.11.1994) EME

4.12 *Bucculatrix zophasta* Braun ♀ — Oregon, Hood River Co., Hood River, r.f. *Quercus garryana*, emgd. VII.9.1951 (V. W. Olney) USNM

4.13 *Bucculatrix quadrigemina* Braun ♂ — California, Alameda Co., Berkeley, IV.30.1980 (J. Powell 80D24, r.f. *Althea rosea*, emgd. V.17.1980) EME

4.14 *Bucculatrix ceanothiella* Braun ♀ — California, Alameda Co., Berkeley, VI.1.2004 (Powell) EME

4.15 *Tinagma californicum* Gaedicke, paratype ♀ — California, Alameda Co., Berkeley Hills, Oakland, V.10.1962 (Powell) EME

4.16 *Tinagma californicum* Gaedicke, paratype ♀ — California, Modoc Co., E of Pine Creek, VI.10.1970, adult on *Phacelia heterophylla* (Powell) EME

4.17. *Tingma powelli* Gaedicke, paratype, paratype ♂ — California, Marin Co., Ring Mtn., III.30.1985 (Powell) EME

4.18 *Pigritia* species ♂ — California, Monterey Co., Big Creek Reserve, 675m, IX.3.1991 (Powell) EME

4.19 *Symmoca signatella* (Herrich-Schaeffer) ♀ — California, Los Angeles Co., Avalon, Santa Catalina Island, III.31.1968 (J. Powell 68D15, r.f. *Cupressus macrocarpa* cones, emgd. VI.21.1986) EME

4.20 *Oegoconia quadripuncta* (Haworth) ♀ — California, Alameda Co., Berkeley, VII.17.2001 (Powell) EME

4.21 *Glyphidocera septentrionella* Busck ♂ — California, Marin Co., Inverness, VII.13.1994 (Powell) EME

4.22 *Holcocera gigantella* (Chambers) ♂ — California, Monterey Co., Salmon Creek Rgr. Sta., X.25.1984 (J. Powell 84K6, r.f. *Yucca whipplei*, emgd. I.27.1985) EME

4.23 *Holcocera iceryaeella* (Riley) ♂ — California, Alameda Co., Alameda, June [1880s, Koebele] (through C.V. Riley) [abd. dissected] USNM

4.24 *Holcocera iceryaeella* (Riley) ♂ — California, Marin Co., Mt. Vision, IV.24.1982 (J. Powell 82D44, r.f. *Ceanothus thyrsiflorus*, emgd. V.29.1982) EME

4.25 *Calosima munroei* Adamski ♂ — California, Marin Co., Carson Ridge, II.10.1957 (J. Powell 57B2, r.f. *Cupressus sargentii*, emgd. III.19.1957) EME

4.26 *Blastobasis glandulella* Riley ♂ — California, Alameda Co., Berkeley, Strawberry Canyon, VIII.31.1991 (Powell) EME

4.27 *Inga concolorella* (Beutenmuller) ♂ — Arizona, Pima Co., SE of Continental, VI.6.1968 (Opler & Powell, JP 68F58, r.f. *Haplopappus [Isocoma] tenuisectus* roots, emgd. VIII.12.1968) EME

4.28 *Endrosis sarcitrella* (L.) ♀ — California, Alameda Co., Berkeley, IV.21.1995 (Powell)

4.29 *Polix coloradella* (Walsingham) ♀ — California, Alameda Co., Berkeley, Strawberry Canyon, VII.1.1991 (Powell) EME

4.30 *Hoffmanophila pseudospretella* (Stainton) ♀ — California, Alameda Co., Albany, IV.22.1958 (Powell) EME

4.31 *Hoffmanophila pseudospretella* (Stainton) ♂ — California, Alameda Co., Berkeley, IV.28.1988 (Powell) EME

4.32 *Esperia sulphurella* (Fabricius) ♀ — California, Alameda Co., Strawberry Canyon, II.3.1984 (J. Powell 84B9, r.f. under bark, *Quercus agrifolia* log, emgd. II.18.1984) EME

4.33 *Pleurota albastrigulella* (Kearfott) ♂ — California, Monterey Co., Arroyo Seco, V.3/7/1975 (Powell) EME

4.34 *Hypatopa* species— ♂ — California, Monterey Co., Big Creek Reserve, VII.8.1993 (Hsu & Powell) EME

4.35 *Endrosis sarcitrella* (L.) ♂ — California, Alameda Co., Berkeley, I.1996 (D. Rubinoff, JP 96A7, r.f. soy meal, emgd. III.2.1996) EME

4.36 *Borkhausenia nefrax* Hodges ♀ — California, Alameda Co., Berkeley, V.9.1990 (Powell) EME

4.37 *Decantha stonda* Hodges ♀ — California, Monterey Co., Big Creek Reserve, 670m, VIII.17.1988—(Powell) EME

4.38 *Batia lunaris* (Haworth) ♀ — California, Marin Co., Marin Island, V.2.1989 (Powell & M. Prentice) EME

4.39 *Rectiostoma fernaldella* (Riley) ♂ — California, Tulare Co., SW of Three Rivers, V.2.1979 (J. Powell 79E10, r.f. *Quercus wislizenii*, emgd. VI.2.1979) EME

4.40 *Antaeotricha lindseyi* (Barnes & Busck) ♀ — Arizona, Santa Cruz Co., Madera Canyon, VII.13.1977 (Powell) EME

4.41 *Antaeotricha schlaegeri* (Zeller) ♀ — Wisconsin, Oneida Co., Lake Katherine, VI.25.1961 (H. Bower) EME

4.42 *Antaeotricha manzanitae* Keifer ♂ — California, Mendocino Co., Hopland Field Station, IV.27.1975 (J. Powell 75D17, r.f. *Arctostaphylos manzanita*, emgd. VI.6.1975) EME

4.43 *Gonioterma crambitella* (Walsingham) ♂ — Arkansas, Logan Co., Cove Lake, V.23.1991 (Powell) EME

4.44 *Agonopterix alstroemeriana* (Clerck) ♂ — California, Marin Co., Inverness Ridge, X.19.1997 (Powell) EME

4.45 *Agonopterix oregonensis* Clarke ♂ — California, San Mateo Co., San Bruno Mtn., III.5.1986 (J. De Benedictis, JP 86C8.1, r.f. *Sanicula crassicaulis*, emgd. IV.22.1986) EME

4.46 *Agonopterix fusciterminella* Clarke ♀ — California, Marin Co., W of Fairfax, IV.21.1969 (J. Powell 69D37, r.f. *Senecio aronicoides*, emgd. V.23.1969) EME

4.47 *Agonopterix canadensis* (Busck) ♀ — Washington, Whitman Co., Pullman, VI.21.1935, r. f. *Senecio sarra* (Clarke) USNM

4.48 *Agonopterix psoraliella* (Walsingham) ♂ — California, Contra Costa Co., N of Orinda, V.9.1977 (D. Green, JP 77E74, r.f. *Psoralea physodes*, emgd. Vi.2.1977) EME

4.49 *Agonopterix posticella* (Walsingham) ♂ — California, Contra Costa Co., Antioch Natl. Wildlife Ref., IV.22.1992 (J. Powell 92D83, r.f. *Psoralea macrostachya*, emgd. V.18.1992) EME

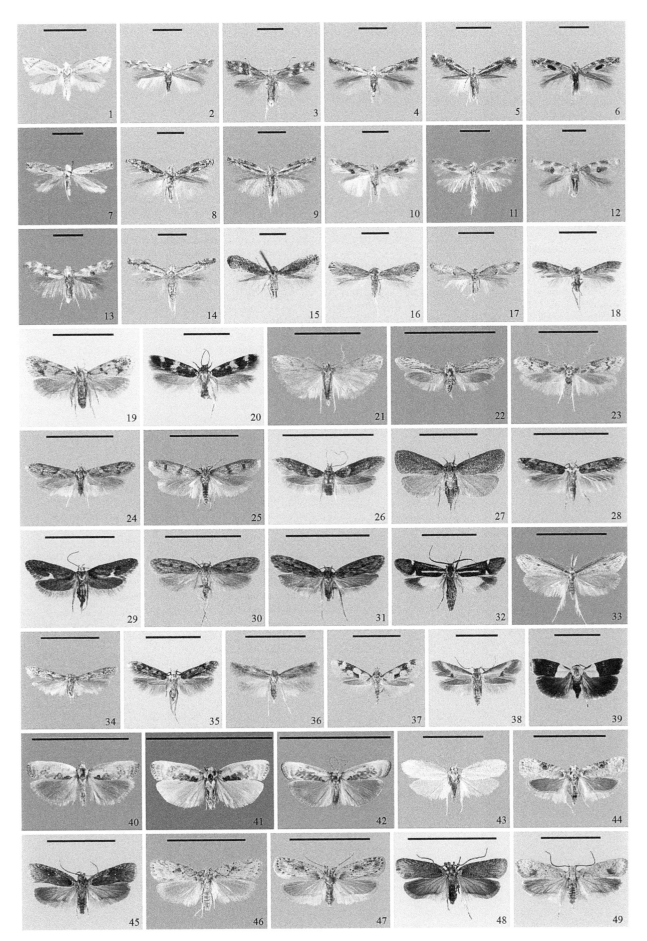

PLATE 4

5.1 *Agonopterix nervosa* (Haworth) ♀ — California, Alameda Co., Berkeley, IX.26.1979 (Powell) EME

5.2 *Exaeretia nechlys* (Hodges) paratype ♂ — California, Orange Co., W of Silverado, III.30.1968 (J. Powell 68C96, r.f. *Sphaeralcea*, emgd IV.21.1968) EME

5.3 *Exaeretia gracilis* (Walsingham) ♂ — California, Los Angeles Co., W Cove, San Clemente Island, X.1.2002 (J. Powell 02J23, r.f. *Heterotheca grandiflora*, emgd. X.24.2002) EME

5.4 *Exaeretia umbraticostella* (Walsingham) ♂ — Arizona, Coconino Co., Walnut Canyon, 6,500', VII.24.1965 (R. Hodges) USNM

5.5 *Semioscopis merricella* Dyar ♀ — New York, Tomkins Co., Dryden, IV.27.1964 (Franclemont) USNM

5.6 *Semioscopis inornata* Walsingham ♀ — California, Plumas Co., Meadow Valley, V.15.1982 (Powell) EME

5.7 *Depressaria artemisiae* Nickerl ♂ — Arizona, Cochise Co., Miller Canyon, Huachuca Mtns., IV.12.1986 (J. Powell 86D26, r.f. *Artemisia dracunculoides*, emgd. IV.22.1986) EME

5.8 *Depressaria pastinacella* (Duponchel) ♂ — Washington, Whitman Co., Almota, VI.28.1975 (J. Powell75F39, r.f. *Pastinaca sativa*, emgd. VII.27.1975) EME

5.9 *Depressaria daucella* (Denis & Schiffermuller) ♀ — Washington, Skagit Co., Hamilton, r.f. *Oenanthe sarmentosa*, emgd. VII.24.1950 (Clarke) USNM

5.10 *Depressaria alieniella* Busck ♂ — California, Siskiyou Co., Ash Creek at McCloud River, VII.8.1974 (J. Powell 74F7, r.f. *Artemisia douglasiana*, emgd. VII.5.1974) EME

5.11 *Apachea barberella* (Busck) ♂ — California, Yuba Co., NW of Smartville, V.4.1980 (J. De Benedictis, JP 80E9, r.f. *Ptelea baldwini*, emgd. VI.14.1980) EME

5.12 *Himmacia huachucella* (Busck) ♀ — Arizona, Cochise Co., Carr Canyon, Huachuca Mtns., VII.31.1986 (D. Wagner & Powell) EME

5.13 *Himmacia stratia* Hodges ♀ — Arizona, Cochise Co., Pinery Canyon, 6'950', Chiricahua Mtns., VI.15.1985 (P. Jump) EME

5.14 *Psilocorsis arguta* Hodges ♀ — Arizona, Santa Cruz Co., Madera Canyon, VI.17.1993 (Y.-F. Hsu, JP 93F35, r.f. *Quercus hypoleucoides*, emgd. VII.26.1993) EME

5.15 *Psilocosis cirrhoptera* Hodges ♀ — Arizona, Cochise Co., Miller Canyon, Huachuica Mtns., VIII.6.1974 (Powell) EME

5.16 *Ethmia minuta* Powell ♂ — California, San Diego Co., SE of El Cajon, III.28.1974 (Powell) EME

5.17 *Ethmia minuta* Powell, paratype ♀ — California, San Diego Co., NE of Lakeside, III.29.1961 (Powell) EME

5.18 *Ethmia brevistriga* Clarke ♂ — California, San Francisco, Laguna Puerca, IV.7.1961 (Powell) EME

5.19 *Ethmia umbrimarginella* Busck ♀ — Arizona, Cochise Co., E foothills Dragoon Mtns., II.21/28.1986 (P. Jump) EME

5.20 *Ethmia lassenella* Busck ♀ — California, Plumas Co., Meadow Valley, VII.7.1982 (J. Powell 82G2, r.f. *Phacelia procera*, emgd. by VII.7.1983) EME

5.21 *Ethmia coquillettella* Busck ♂ — California, Stanislaus Co., Del Puerto Canyon, III.25.1969 (Powell) EME

5.22 *Ethmia scylla* Powell, paratype ♂ — California, Yolo Co., NW of Rumsey, III.8.1964 (Powell) EME

5.23 *Ethmia charybdis* Powell ♂ — California, Kern Co., Red Rock Canyon, XII.21.1995, 31–35(F. (D. Giuliani & Powell) EME

5.24 *Ethmia discostrigella* (Chambers) ♀ — Nevada, Clark Co., Kyle Canyon, Charleston Mtns., VI.10.1959 (J. & S. Burns) EME

5.25 *Ethmia discostrigella* (Chambers) ♀ (race *subcaerulea* Walsingham) — California, Lake Co., Lake Pillsbury, IV.3.1962 (Powell) EME

5.26 *Ethmia semitenebrella* Dyar ♀ — California, Orange Co., Silverado Canyon, Santa Ana Mtns., V.12.1979 (G. Marsh) EME

5.27 *Ethmia semitenebrella* Dyar ♂ (detail of exposed HW) — California, San Diego Co., Pine Valley, IV.17.1950 (E. Johnston) EME

5.28 *Ethmia arctostaphylella* (Walsingham) ♂ (spring form) — California, Stanislaus Co., Del Puerto Canyon, III.4.1969 (Opler) EME

5.29 *Ethmia arctostaphylella* (Walsingham) ♂ (summer form) — California, Stanislaus Co., Del Puerto Canyon, VI.12.1980 (J. De Benedictis, 80164-B, r.f. *Eriodictyon californicum*, emgd. VII.15.1980) EME

5.30 *Ethmia albistrigella* (Walsingham) ♂ — California, Tuolumne Co., Chipmunk Flat, W of Sonora Pass, VI.25.1962 (Powell) EME

5.31 *Ethmia semilugens* (Zeller) ♂ — Mexico, Chihuahua, W of Parrita, VIII.30.1956 (D. Linsdale) EME

5.32 *Ethmia macneilli* Powell ♂ — California, Kern Co., E of Keene, IV.11.1986 (J. Brown & Powell, JP 86D23, r.f. Phacelia, emgd. IX.29.1986) EME

5.33 *Ethmia macneilli* Powell ♀ — California, Kern Co., E of Keene, IV.11.1986 (J. Brown & Powell, JP 86D23, r.f. Phacelia, emgd. X.1.1986) EME

5.34 *Ethmia monticola* (Walsingham) ♂ — California, Modoc Co., NW of Cedarville, VII.4.1962 (J. Buckett) EME

5.35 *Ethmia monticola emmeli* Powell, paratype ♂ — Arizona, Coconino Co., NW of Flagstaff, VI.22.1961 (R. Hodges) USNM

5.36 *Ethmia hagenella josephinella* Dyar ♂ — Texas, Jeff Davis Co., Davis Mtns. 5,800', V.3.1959 (J. & S. Burns) EME [abd. dissected JAP 2596]

5.37 *Ethmia marmorea* (Walsingham) ♂ — California, Los Angeles Co., Table Mtn. 7,200', N of Big Pines, VII.20.1979 (De Benedictis, Powell, Wagner) EME

5.38 *Ethmia albicostella* (Beutenmuller) ♂ — Mexico, Durango, W of Ciudad Durango, 8,400', VII.4.1972 (D. MacNeill, Powell, D. Veirs) EME

5.39 *Ethmia sphenisca* Powell, paratype ♀ — Arizona, Gila Co., Tonto Creek Camp, VI.29.1956 (J. Comstock, L. Martin, W. Rees) EME [abd. dissected, JAP 2828]

5.40 *Pseudethmia protuberans* Clarke ♂ — California, W of Rice, IV.1.1978 (Powell) EME

5.41 *Pyramidobela quinquecristata* Braun ♂ — California, Trinity Co., Hayfork Ranger Sta.,V.20.1973 (J. Powell 73E23, r.f. *Penstemon deustus*, emgd. VI.22.1973) EME

5.42 *Pyramidobela angelarum* Keifer ♀ — California, Alameda Co., Berkeley, XI.8.1983 (J. Powell 83L2, r.f. *Buddleia*, emgd. XI.30.1983) EME

5.43 *Deoclona yuccasella* Busck ♂ — California, Riverside Co., SE of Valle Vista, VII.12.1963 (J. Powell 63D15, r.f. 1962 pod *Yucca whipplei*, emgd. V.16.63) EME

5.44 *Coelopoeta glutinosi* Walsingham ♀ — California, Monterey Co., Arroyo Seco, V.3.1973 (J. Powell 75E83, r.f. *Eriodictyon californicum*, emgd. VI.6.1975) EME

5.45 *Coelopoeta phaceliae* Kaila ♂ — California, Plumas Co., Meadow Valley, VII.7.1982 (J. Powell, 82G2, r.f. *Phacelia procera*, emgd. VIII.18.1982) EME

5.46 *Perittia passula* Kaila ♂ — California, Alameda Co., Strawberry Canyon, II.8.1968 J.(Powell 68B48, r.f. *Lonicera*, emgd. VIII.2.1968, emgd. VIII.2.1968) EME

5.47 *Perittia cygnodiella* (Busck) ♀ — California, Alameda Co., Albany Hill, IV.5.1995 (J. Powell 95D1, r.f. *Symphoricarpos mollis*, emgd. II.3.1996) EME

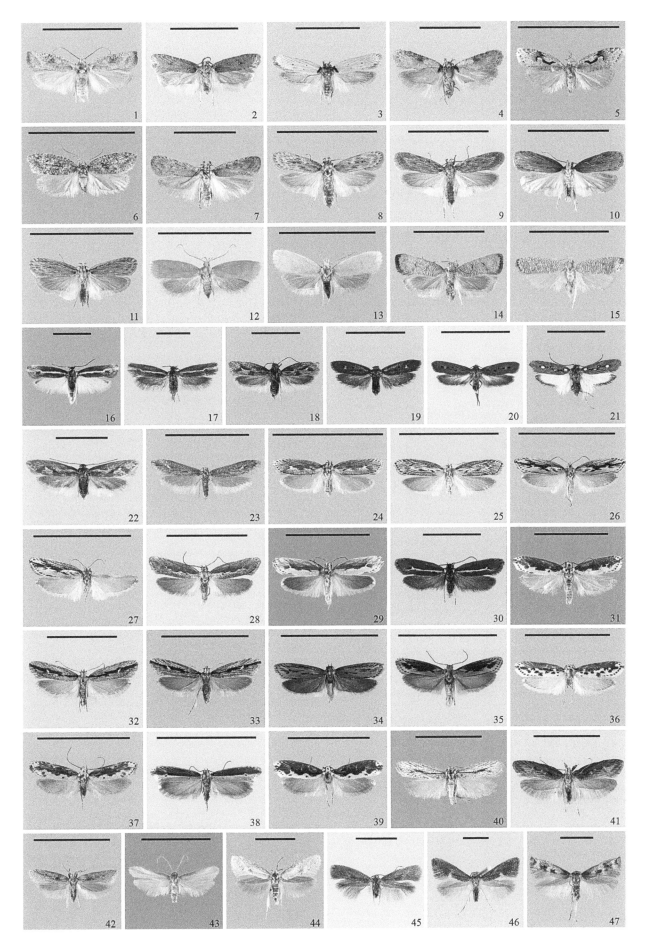

PLATE 5

6.1 *Elachista coniophora* Braun ♂ — California, San Benito/Fresno Co. line, Big Panoche Creek, III.4.1968 (Powell) EME

6.2 *Elachista coniophora* Braun ♂ — California, San Benito/Fresno Co. line, Big Panoche Creek, III.4.1968 (Powell) EME[abd. disseceted, EME 4566]

6.3 *Elachista controversa* Braun ♂ — California, Mendocino Co., Ham Pass, VI.15.1972 (Powell) EME

6.4 *Elachista dagnirella* Kaila ♂ — Alaska, Centennial Park, Anchorage, VI.24.1979 (Opler & Powell) EME

6.5 *Elachista marachella* Kaila, paratype ♂ — California, Alameda Co., Strawberry Canyon, II.23.1985 (J. Powell 85B26, r.f. *Ehrhardta*, emgd, III.10.1985) EME

6.6 *Elachista finarfinella* Kaila, paratype ♂ — California, Humboldt Co., Arcata, VI.24.1969 (Powell) EME [abd. dissected, EME5115]

6.7. *Elachista cucullata* Braun ♀ — California, Monterey Co., Big Creek Reserve, Brunetti Creek, VI.14.1991 (Y.-F. Hsu & Powell) EME

6.8 *Elachista indisella* Kaila ♀ — California, Monterey Co., Big Creek Reserve, Brunetti Creek, V.13.1992 (Powell) EME

6.9 *Elachista gildorella* Kaila, holotype ♂ — California, Alameda Co., Albany—Hill, IV.21.1998 (Powell) [abd. dissected, EME4979]

6.10 *Elachista telcharella* Kaila, paratype ♂ — California, Marin Co., Inverness Ridge, IV.15.1997 (Powell) EME

6.11 *Elachista caranthirella* Kaila, paratype ♂ — California, Los Angeles Co., Avalon Canyon, Santa Catalina Island, III.31.1968 (Powell) EME [abd. dissected, JAP2590]

6.12 *Elachista excelsicola* Braun ♂ — Alaska, N of Cantwell, 2,200′, VI.26.1979 (Opler & Powell) [abd. dissected, EME 5181]

6.13 *Batrachedra striolata* (Zeller) ♂ — California, Los Angeles Co., Carnavon Way, Los Angeles, XI.25.1985 (F. Sala, JP 86A2, r.f. galls on *Salix laevigata*, emgd. II.13.1986) EME

6.14 *Batrachedra enormis* Meyrick ♂ — Arizona, Santa Cruz Co., Madera Canyon, X.23.1959 (Hodges) USNM

6.15 *Batrachedra praeangusta* (Haworth) ♂ — Mexico, Nuevo Leon, W of Iturbide, IX.13.1976 (Chemsak & Powell) EME

6.16 *Coleophora mayrella* (Hübner) ♀ — California, Tulare Co., Ash Mtn. HQ, IV.27.1979 (Chemsak & Powell) EME

6.17 *Coleophora maritella* McDunnough ♀ — Canada, Nova Scotia, Lunenburg, VII.3.1967, "Host Plant Juncus gerardi" (B. Wright) USNM

6.18 *Coleophora sacramenta* Heinrich ♀ — California, Alameda Co., Berkeley, VI.9.1979 (Powell) EME

6.19 *Coleophora annulicola* Braun, paratype ♂ — Utah, Cache Co., [no locality] (A. Braun, B 1140, emgd. VIII.23.1924) ANSP

6.20 *Coleophora wyethiae* Walsingham ♂ — Washington (county unknown), Kamiack Butte, VI.28.1933 (J. Clarke, r.f. *Balsamorrhiza sagitata*) USNM

6.21 *Coleophora accordella* Walsingham ♂ — California, Contra Costa Co., Antioch Natl Wildlife Refuge, III.22.1986 (J. Powell 86D42, r.f. *Lotus scoparius*, emgd. IV.11.1986) EME

6.22 *Coleophora accordella* Walsingham ♂ — California, San Luis Obispo Co., Oso Flaco Lake, III.20.1974 (Powell) EME

6.23 *Coleophora pruniella* Clemens ♀ — California, S.L.O. Co., Dune Lakes S of Oceano, V.1.1974, r.f. Salix (J. Powell 74E4, emgd. V.28.1974) EME

6.24 *Coleophora glaucella* Walsingham ♀ — California, Santa Barbara Co., Mission Canyon, III.3.1987 (J. Powell 87C11, r.f. *Arctostaphylos*, emgd. V.22.1987) EME

6.25 *Coleophora irroratella* Walsingham ♀ — Washington, Whitman Co., Pullman, VII.30.1930 (Clarke) USNM

6.26 *Coleophora lynosiridella* Walsingham ♂ — California, Shasta Co., Old Station, VII.1994 (M. Valenti)(J. Landry 94-15, r.f. *Chrysothamnus viscidiflorus*, emgd. VIII.17.1994) CNC

6.27 *Coleophora kleimeschella* Toll ♂ — California, Kings Co., Kettleman City, XII.30.2003 (B. Villegas)(J. Landry 2004-02, r.f. *Salsoa tragus*, emgd. V.17.2004) CNC

6.28 *Coleophora parthenica* Meyrick ♂ — California, Kings Co., Kettleman City, XII.30.2003 (B. Villegas)(J. Landry 2004-01, r.f. *Salsoa tragus*, emgd. V.3.2004) CNC

6.29 *Mompha eloisella* (Clemens) ♀ — California, Contra Costa Co., Antioch Natl. Wildlife Refuge, III.13.1982 (J. Powell 82C13, r.f. *Oenothera deltoides*, emgd. IV.22.1982) EME

6.30 *Mompha murtfeldtella* (Chambers) ♂ — California, Alameda Co., Berkeley, VIII.22.2002 (Powell) EME

6.31 *Mompha franclemonti* Hodges, paratype ♂ — California, Monterey Co., Big Creek Reserve, III.22.1989 (Y.-F. Hsu & Powell) EME

6.32 *Mompha franclemonti* Hodges, paratype ♀ [lateral aspect] — California, Monterey Co., Big Creek Reserve, III.22.1989 (Y.-F. Hsu & Powell) EME

6.33 *Mompha cephalanthiella* (Chambers) ♂ — California, Sutter Co., Feather River near Nicolaus, X.11.1982 (J. Powell 82K5, r.f. *Cephalanthus occidentalis*, emgd. X.29.1982) EME

6.34 *Arotrura divaricata* (Braun) ♀ — New Mexico, Socorro Co., Gran Quivara Natl. Monument, VII.1.1964 (D. Davis) USNM

6.35 *Arotrura sponsella* (Busck) ♂ — California, San Bernardino Co., Ord Mtn., IV.18.1960 (Powell) EME

6.36 *Arotrura eburnea* (Walsingham) ♀ — California, Inyo Co., Lone Pine, VI.26.1992 (Powell) EME

6.37 *Arotrura longissima* J.-F. Landry paratype ♂ — California, Los Angeles Co., Eel Pt., San Clemente Island, IV.11.1980 (Powell) EME

6.38 *Rhamphura altisierrae* (Keifer) ♂ — California, Kern Co., Lebec, VI.25.1982 (De Benedictis & Powell, JP 82F31, r.f. *Solanum xantii*, emgd. VII.4.1982) EME

6.39 *Rhamphura ochreistriata* (Walsingham) ♂ — California, Contra Costa Co., Antioch Natl. Wildlife Refuge, IX.10.1981 (Powell) EME

6.40 *Scythris inspesella* (Hübner) ♀ — California, Siskiyou Co., McBride Springs, Mt. Shasta, VI.14.1974 (D. Green & Powell, JP 74F36, r.f. *Epilobium*, emgd. VII.17.1974) EME

6.41 *Scythris mixaula* Meyrick ♂ — Colorado, Alamosa Co., Zapata Ranch, 7,900′ VI.26.1982 (Hodges) USNM

6.42 *Scythris immaculatella (Chambers)* [paratype of *pacifica* McDunnough] ♂ — Canada, Alberta, Waterton Lakes, VII.12.1923 (J. McDunnough) USNM

6.43 *Scythris trivinctella* (Zeller) ♂ — Mexico, Coahuila, S of Sabinas, IX.11.1976 (Powell) EME

6.44 *Scythris ypsilon* Braun ♀ — California, Kern Co., Kettleman Jct., III.30.2002, adult assoc. Atriplex polycarpa (Powell) EME

6.45 *Landrya reducta* (Braun) ♀ — California, Kern Co., N of Red Rock Canyon, V.1.1968, adult on *Haplopappus cooperi* (J. Chemsak) EME

6.46 *Neoscythris confinis* (Braun) ♂ — San Mateo Co., Edgewood Park, IV.4.1990 (Y.-F. Hsu & Powell) EME

6.47 *Antequera acertella* (Busck) ♀ — California, San Diego Co., Saratoga Springs, X.11.1999 (N. Bloomfield) EME

6.48 *Pyroderces badia* Hodges ♀ — California, San Diego Co., Cardiff, X.6.1967 Opler, Powell, P. Rude, JP 67K86, r.f. *Yucca whipplei*, emgd. X.11.1967) EME

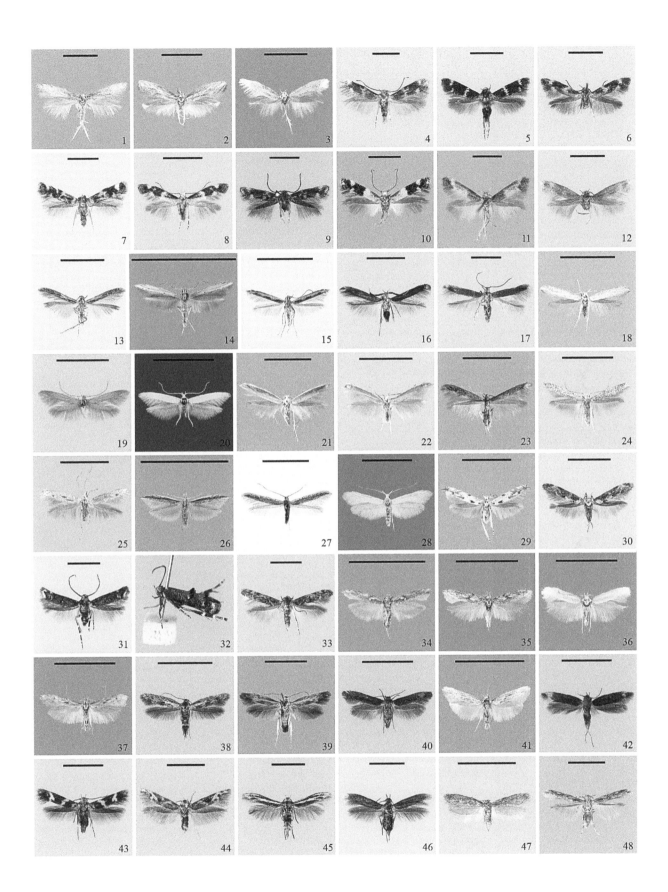

PLATE 6

PLATE 7 Cosmopterigidae, Gelechiidae

7.1 *Cosmopterix molybdina* Hodges ♂ — California, Humboldt Co., Shelter Cove, V.21.1976 (J. Chemsak & R. Dietz) EME

7.2 *Cosmopterix molybdina* Hodges ♂ — California, Marin Co., Ring Mtn., IV.25.1990 (Y.-F. Hsu & Powell) EME

7.3 *Cosmopterix montisella* Chambers ♂ — California, Siskiyou Co., Mt. Shasta City, VII.13.1958 (Powell) EME

7.4 *Cosmopterix montisella* (Chambers) ♀ — Arizona, Coconino Co., Mt. Elen, VII.22.1991 (De Benedictis & Powell) EME

7.5 *Cosmopterix opulenta* Braun ♀ — California, Mendocino Co., E of Eel River Ranger Sta., VI.9.1972, adult on *Artemisia douglasiana* (Powell) EME

7.6 *Eralea albalineella* (Chambers) ♂ — Arizona, Cochise Co., Ash Canyon, Huachuca Mtns., VII.17/31. 1986 (D. Wagner) EME

7.7 *Stagmatophora iridella* (Busck) ♂ — California, Monterey Co., Big Creek Reserve, X.3.1991 (J. Powell 91K7, r.f. *Trichostema lanceolatum*, emgd. V.20.1992) EME

7.8 *Teladoma helianthi* Busck ♀ — Arkansas, Washington Co., Devils Den St. Park, VI.7.1966 (Hodges) USNM

7.9 *Triclonella xuthocelis* Hodges ♀ — Mexico, Tamaulipas, SW of Ciudad Victoria, IX.18.1976 (Chemsak & Powell)

7.10 *Periploca ceanothiella* (Cosens) ♂ — California, Siskiyou Co., McBride Spring, Mt. Shasta, VI.23.1974 (T. Eichlin & Powell) EME

7.11 *Periploca atrata* Hodges ♀ — California, Contra Costa Co., Russellman Park, Mt. Diablo, II.23.1962 (J. Powell 62B8, r.f. *Juniperus californicus*, emgd. III.16.1962) EME

7.12 *Stilbosis dulcedo* (Hodges) ♀ — California, Contra Costa Co., Antioch, IV.17.1969 (Opler, JP 68D57, r.f. *Quercus agrifolia*, emgd. V.16.1967) EME

7.13 *Stilbosis extensa* (Braun) ♀ — California, Contra Costa Co., Antioch Natl. Wildlife Refuge, III.15.1985 (Powell) EME

7.14 *Stilbosis nubila* Hodges ♀ — Arizona, Coconino Co., SW of Flagstaff, VII.13.1995 (Powell & F. Sperling) EME

7.15 *Perimede battis* Hodges ♂ — Arizona, Santa Rita Co., Madera Canyon, VIII.5.1959 (Hodges) USNM

7.16 *Ithome concolorella* (Chambers) ♂ — Arkansas, Washington Co., Devils Den St. Park, VI.23.1966 (Hodges) USNM

7.17 *Sorhagenia nimbosa* (Braun) ♂ — Oregon, Benton Co., NW of Corvallis, VI.20.1975 (Opler & Powell, JP 75F24, r.f. *Rhamnus purshianus*, emgd. VII.24.1975) EME

7.18 *Anoncia loexyla* Hodges ♂ — Arizona, Gila Co., Pine, VI.4.1968 (Opler & Powell) [abd. dissected, EME 759]

7.19 *Stagmatophora enchrysa* Hodges ♀ — California, Santa Barbara Co., E of Los Prietos, VI.25.1965 (J. Powell 65F3, r.f. *Trichostema lanatum*, emgd. VIII.1.1965) EME

7.20 *Limnacia phragmitella* Stainton ♂ — Canada, Ontario, Ottawa, V.25.1986 (J. Powell 86E21, r.f. *Typha*, emgd. VII.5.1986) EME

7. 21 *Anoncia conia* (Walsingham) ♂ — California, Stanislaus Co., Del Puerto Canyon, IV.9.1977 (J. Powell 77D74, r.f. *Monardella*, emgd., V.11.1977) EME

7.22 *Anoncia sphacelina* (Keifer), paratype ♂ — California, El Dorado Co., Shingle Springs, IV.17.1934, r.f. *Sphacele [Lepechinia] calycina*, emgd. IV.28.1934 (Keifer) EME

7.23 *Anoncia flegax* Hodges paratype ♂ — California, Siskiyou Co., Mt. Shasta City, VII.18.1958 (Powell) EME

7.24 *Anoncia mosa* Hodges paratype ♂ — California, San Luis Obispo Co., W of Paso Robles, IV.25.1968 (J. Chemsak & Powell) [abd. dissected EME 737]

7.25 *Anoncia leucoritis* (Meyrick) ♂ — Colorado, Alamosa Co., Great Sand Dunes, VI.28.1982 (Hodges) USNM

7.26 *Anoncia longa* (Meyrick) ♂ — Colorado, Chaffee Co., The Castles, E of Buena Vista, 8,800', VII.4.1982 (D. Ferguson) USNM

7.27 *Walshia miscecolorella* (Chambers) ♀ — California, Marin Co., Pt. Reyes Natl. Seashore, N Beach, V.1.1976 (J. Lawrence & Powell, JP 76E1, r.f. *Lupinus arboreus*, emgd. VI.20.1976) EME

7.28. *Walshia miscecolorella* (Chambers) ♂ — California, S.L.O. Co., Oso Flaco Lake, III.20.1974 (Powell) EME

7.29 *Nealyda bifidella* Dietz ♂ — New Mexico, Eddy Co., SW of Whites City, VI.18.1993 (Y.-F. Hsu, JP 93F42, r.f. *Mirabilis multiflora*, emgd. VII.6.1993) EME

7.30 *Nealyda bifidella* Dietz ♀ — New Mexico, Eddy Co., SW of Whites City, VI.18.1993 (Y.-F. Hsu, JP 93F42, r.f. *Mirabilis multiflora*, emgd. VII.6.1993) EME

7.31 *Metzneria lappella* (L.) ♂ — Washington, Jefferson Co., Quilcene, VII.22.1988 (D. Ferguson) USNM

7.32 *Isophrictis magnella* (Busck) ♀ — California, Monterey Co., Big Creek Reserve, X.3.1991 (Powell) EME

7.33 *Monochroa harrisonella* (Busck) ♀ — California, El Dorado Co., Blodgett Forest, V.29.1970 (Powell) EME

7.34 *Monochroa placidella* (Zeller) ♀ — California, Marin Co., Ring Mtn., IV.30.1986 (J. Powell 86D133, r.f. *Pteridium aquilinum*, emgd. V.13.1986) EME

7.35 *Chrysoesthia lingulacella* (Clemens) ♀ — California, Yolo Co., Davis, X.2.1927, r.f. *Chenopodium murale* (Keifer) EME

7.36 *Aristotelia argentifera* Busck ♀ — California, Monterey Co., Big Creek Reserve, VIII.17.1988 (Powell) EME

7.37 *Aristotelia adenostomae* Keifer ♀ — California, Ventura Co., Sespe Canyon, V.26.1984 (J. Powell 84E110, r.f. *Adenostoma fasciculatum*, emgd. VI.25.1984) EME

7.38 *Aristotelia adceanotha* Keifer ♀ — California, Placer Co., Colfax, VII.28.1932 (Fourness) EME

7.39 *Aristotelia elegantella* (Chambers) ♀ — California, Kern Co., Kern Natl. Wildlife Refuge, VI.3.1993 (Y.-F. Hsu, JP 93F2, r.f. *Epilobium*, emgd. VII.6.1993) EME

7.40 *Aristotelia bifasciella* Busck ♂ — California, San Bernardino Co., Zzyzx Springs, VI.22.1977 (Powell) EME

7.41 *Evippe laudatella* (Walsingham) ♂ — California, Contra Costa Co., Walnut Creek, IV.20.1962 (Powell) EME

7.42 *Evippe abdita* Braun ♂ — Arizona, Santa Cruz Co., Pena Blanca Canyon, IX.1.1959 (Hodges) USNM

7. 43 *Recurvaria ceanothiella* Braun ♂ — California, Contra Costa Co., Brooks Island, IV.6.1993 (Hsu & Powell, JP 93D1.4, r.f. *Ceanothus arboreus*, emgd. VII.31.1993) EME

7.44 *Coleotechnites milleri* (Busck) ♂ — California, Mono Co., S of Lee Vining, VII.1.1964, r.f. *Pinus contorta*, emgd. VII.10.1964 (R. Stevens) USNM

7.45 *Coleotechnites bacchariella* (Keifer) ♀ — California, San Luis Obispo Co., S of Oceano, IV.26.1973 (J. Powell 73D36, r.f. *Baccharis pilularis*, emgd. VII.16.1973) EME

7.46 *Coleotechnites huntella* (Keifer) ♂ — California, Marin Co., Alpine Lake, IV.15.1972 (J. Powell 72D24, r.f. *Rhododendron occidentale*, emgd. V.15.1972) EME

7.47 *Coleotechnites mackiei* (Keifer) ♀ — California, Siskiyou Co., E of McCloud, VII.7.57 (J. Powell 57G1, r.f. *Arctostaphylos*, emgd. VII.9.1957) EME—

7.48 *Coleotechnites* species near *mackiei* (Keifer) ♂ — California, El Dorado Co., E of Georgetown, III.29.1981 (D. Wagner, JP 81C10, r.f. Arctostaphylos, emgd. IV.15.1981) EME

7.49 *Exoteleia burkei* Keifer ♀ — California, Alameda Co., Arroyo Mocho, IV.3.1957 (D. Burdick, JP 57D4, r.f. *Pinus sabiniana* (cones, emgd. V.16.1957) EME

7.50 *Exoteleia graphicella* (Busck) ♀ — California, Monterey Co., W of Greenfield, V.2.1975 (J. Powell 75E6, r.f. *Quercus douglasii*, emgd. V.28.1975) EME

7.51 *Leucogoniella distincta* Keifer, paratype ♂ — California, Sacramento Co., Sacramento, VII.17.1933 (H. Keifer) EME

7.52 *Athrips rancidella* (Herrich-Schaeffer) ♂ — California, Alameda Co., Berkeley, V.13.1983 (J. Powell 83E50, r.f. *Cotoneaster*, emgd. VI.1.1983) EME

PLATE 7

PLATE 8 Gelechiidae

8.1 *Xenolechia querciphaga* (Keifer) ♀ — California, Alameda Co., Del Valle Reservoir, V.15.1975 (J. Powell 75E21, r.f. *Quercus agrifolia*, emgd. VI.7.1975) EME

8.2 *Bryptropha hodgesi* Rutten & Karscholdt, paratype ♂ — California, Santa Barbara Co., Santa Rosa Island, IV.29.1995 (Powell) EME

8.3 *Exoteleia californica* ♀ — California, El Dorado Co., Blodgett Forest, VII.17.1980 (Powell & Wagner) EME

8.4 *Telphusa sedulitella* (Busck) ♀ — California, Alameda Co., Berkeley, IV.26.2004 (Powell) EME

8.5 *Telphusa sedulitella* (Busck) ♀ — California, Santa Barbara Co., Santa Rosa Island., IX.14.2000 (Powell) EME

8.6 *Rifseria fuscotaeniella* (Chambers) ♂ — California, Trinity Co., Hayfork Ranger Station, V.23.1973 (Powell) EME

8.7 *Pseudochelaria manzanitae* (Keifer) ♀ — California, Plumas Co., NW of Quincy (De Benedictis, JP 82E45, r.f. *Arctostaphylos*, emgd. VI.30.1982) EME

8.8 *Pseudochelaria arbutina* (Keifer), paratype ♀ — California, Marin Co., Mill Valley, VI.6.1927, larva on *Arbutus menziesii* (H. Keifer) EME

8.9 *Pseudochelaria scabrella* (Busck) ♀ — California, Marin Co., Inverness Ridge, V.11.1973 (J. Powell 73E9, r.f. *Arctostaphylos virgata*, emgd. VI.11.1973) EME

8.10 *Teliopsis baldiana* (Barnes & Busck) ♂ — California, Monterey Co., Big Creek Reserve, IV.17.1992 (F. Arias, JP 92D111, r.f. *Toxicodendron diversilobum*, emgd. V.14.1992) EME

8.11 *Lita sexpunctella* (Fabricius) ♂ — Colorado, Grand Co., Berthoud Pass, 11,500′ VIII.2.1967 (Powell) EME

8.12 *Lita variabilis* (Busck) ♂ — California, Siskiyou Co., Mt. Shasta City, VIII.22.1958 (Powell) EME

8.13 *Lita recens* Hodges, paratype ♂ — California, Tuolumne Co., Sonora Pass, 9,650′ VIII.21.1959 (Powell) EME

8.14 *Lita incicur* Hodges ♂ — California, Inyo Co., Mono Pass, 12,500′, IX.3.1965 (Powell) EME

8.15 *Lita incicur* Hodges, paratype ♂ — California, Tuolumne Co., Tuolumne Meadows, VIII.27.1960 (W. Ferguson) EME [abd. dissected, JAP 8444]

8.16 *Lita rectistrigella* (Barnes & Busck) ♂ — California, San Benito/ Freno Co. line, Big Panoche Creek, IV.21.1967 (P. Rude, JP 67D93, r.f. *Gutierrezia*, emgd. IX.29.1967) EME [abd. dissected, JAP 3311]

8.17 *Lita sironae* Hodges ♀ — California, Santa Barbara Co., Santa Rosa Island, NPS Station, IX.20.2000 (Powell) EME

8.18 *Arla diversella* (Busck) ♂ — California, Monterey Co., Big Creek Reserve, V.22.1992 (Powell) EME

8.19 *Neodactylota snellenella* (Walsingham) ♀ — Texas, Brewster Co., Panther Pass, 6,000′, VI.4.1973 (R. Hodges) USNM

8.20 *Friseria cockerelli* (Busck) ♀ — Arizona, Cochise Co., Cave Creek SW of Portal, VII.3.1972 (Powell) EME

8.21 *Gelechia desiliens* Meyrick ♀ — California, Alameda Co., Berkeley, IX.9.1985 (Powell) EME

8.22 *Gelechia panella* Busck ♀ — California, Napa Co., Diamond Mtn., S of Calistoga, V.22.1993 (J. Powell 93E34, r.f. *Arctostaphylos manzanita*, emgd, VI.29.1993) EME

8.23 *Gelechia versutella* (Chambers) ♂ — California, S of Davis, Solano Co., X.17.1956 (J. Powell 561D, r.f. *Populus*, emgd. IV.14.1957) EME

8.24 *Gelechia lynceella* Zeller ♀ — Nevada, White Pine Co., Lehman Creek 7,500′, VII.25.1981 (De Benedictis & Powell) EME

8.25 *Gelechia ribesella* (Chambers) ♂ — California, Modoc Co., NW of Ft. Bidwell, VI.12.1970 (J. Powell 70F90, r.f. *Ribes*, emgd. VII.15.1970) EME

8.26 *Chionodes powelli* Hodges, paratype ♂ — California, El Dorado Co., Sly Park Dam, V.19.1968 (J. Powell 68E31, r.f. *Quercus kelloggii*, emgd. VI.21.1968) EME

8.27 *Chionodes abitus* Hodges, paratype ♀ — California, Modoc Co., S of Buck Creek Ranger Station, 6,300′ VI.11.1970 (J. Powell 70F65, r.f. *Salix*, emgd. VII.3.1970) EME

8.28 *Chiondes lophosella* (Busck) ♂ — California, Monterey Co., Big Creek Reserve, IV.24.1987 (Powell) EME

8.29 *Chiondes nanodella* (Busck) ♂ — California, Marin Co., Marin Island, IV.13.1989—(J. Powell 89D19, r.f. *Eriogonum latifolium*, emgd. VI.4.1989) EME

8.30 *Chiondes nanodella* (Busck) ♀ — California, Santa Barbara Co., Santa Barbara Island, V.24.2001 (Powell) EME

8.31 *Chiondes nanodella* (Busck) ♀ — California, Santa Barbara Co., Santa Barbara Island, V.24.2001 (Powell) EME

8.32 *Chionodes abella* (Busck) ♀ — California, Kern Co., Tehachapi Mtn. 5,000′ (De Benedictis & Powell, JP 81F101, r.f. *Abies concolor*, emgd. VII.14.1981) EME

8.33 *Chionodes abella* (Busck) ♂ — Arizona, Coconino Co., Mt. Elden, 8,300′, VII.24.1989 (Hsu, Powell, M. Prentice) EME

8.34 *Chionodes sabinianae* Powell ♀ — California, Alameda Co., U.C. Berkeley Campus, III.15.1961 (J. Chemsak) EME

8.35 *Chionodes nitor* Hodges, holotype ♂ — California, Alameda Co., Berkeley, VI.1.1979 (D. Wagner) [abd. dissected EME 4383]

8.36 *Chionodes kincaidella* (Busck) ♂ — California, Inyo Co., NE of Lone Pine, V.11.1969 (Opler) EME

8.37 *Chionodes xanthophilella* (Barnes & Busck) ♀ — California, San Bernardino Co., Zzyzx Sorings, IV.22.1977 (Powell) EME

8.38 *Chionodes pinguicula* (Meyrick) ♀ — Colorado, Pueblo Co., Pueblo West, VII.14.1989 (B. & J.-L. Landry) USNM

8.39 *Chionodes sistrella* (Busck) ♀ — Arizona, Gila Co., Maricopa, VII.24.1956, r.f. *Sueda torreyana* (F. Bibby) USNM

8.40 *Chionodes dammersi* (Keifer) ♀ — California, San Luis Obispo Co., Dune Lakes, V.2.1974 (J. Powell 74E11, r.f. *Eriogonum parvifolium*, emgd. VI.17.1974) EME

8.41 *Chionodes bardus* Hodges, paratype ♂ — California, Santa Barbara Co., Santa Barbara Island, V.25.2001 (J. Powell 01E29, r.f. *Eriogonum giganteum*, emgd. VII.6.2001) EME

8.42 *Chionodes helicostichus* (Meyrick) ♀ — California, Contra Costa Co., Walnut Creek, X.2.1961 (Powell) EME

8.43 *Chionodes notandella* (Busck) ♂ — California, San Francisco, Baker Beach, IV.13.1977 (J. Powell 77D33, r.f. *Eriogonum latifolium*, emgd. VII.11.1977) EME

8.44 *Chionodes ochreostrigella* (Chambers) ♀ — California, Contra Costa Co., Richmond Field Station, IX.16.1994 (Y.-F. Hsu, JP 94J17, r.f. *Rumex*, emgd. XI.3.1994) EME

8.45 *Chionodes thoraceochrella* (Chambers) ♂ — Utah, Sanpete Co., Ephraim Canyon, 8,800′ (Hodges) USNM

8.46 *Chionodes chrysopyla* (Keifer) ♀ — California, Alameda Co., Del Valle Lake, IV.30.1975 (J. Powell 75D27, r.f. *Quercus* (*dumosa*?), emgd. VI.11.1975) EME

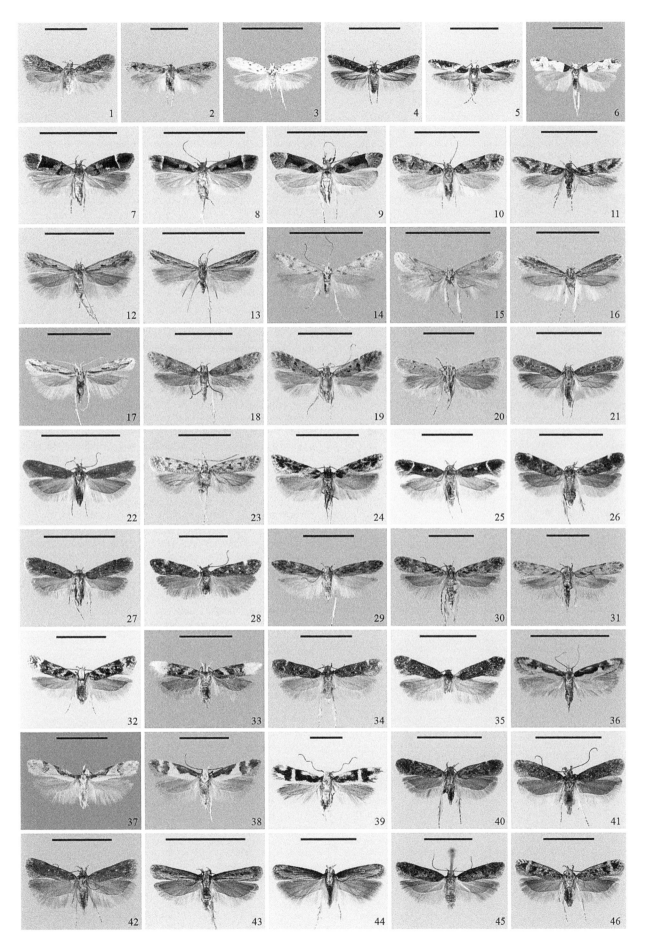

PLATE 8

PLATE 9 Gelechiidae

9.1 *Chionodes occidentella* (Chambers) ♂ — California, Alameda Co, Berkeley, X.12.1991 (Powell)

9.2 *Chionodes trichostola* (Meyrick) ♂ — California, Santa Barbara Co., Santa Cruz Island, Canada Cervada, IV.26.1966 (J. Wolf, JP 66D31, r.f. *Quercus dumosa* [= *berberidifolia*], emgd. V.21.1966) EME

9.3 *Chionodes lector* Hodges, paratype ♀ — California, Siskiyou Co., Shasta Retreat, IX.1/7 [1915, McDunnough] USNM

9.4 *Chionodes lector* Hodges, paratype ♂ — California, Santa Clara Co., W of Saratoga, IV.20.1962 (J. Powell 62D20, r.f. flowers *Acer macrophyllum*, emgd. IV.27.1962) [abd. dissected, EME 0000]

9.5 *Chionodes mediofuscella* (Clemens) ♂ — Colorado, Alamosa Co. Great Sand Dunes, VI.22.1982 (Hodges) USNM

9.6 *Chionodes terminimaculella* (Kearfott) ♂ — Colorado, Alamosa Co., Great Sand Dunes, 8,200', VI.24.1982 (Hodges) USNM

9.7 *Chionodes messor* Hodges, paratype ♂ — California, El Dorado Co., Blodgett Forest, VII.18.1967 (Powell) EME

9.8 *Chionodes optio* Hodges, paratype ♂ — California, El Dorado Co., Blodgett Forest, V.27.1978 (Powell) EME

9.9 *Chionodes petalumensis* Clarke ♀ — California, Marin Co., Marin Island, IV.13.1989 (J. Powell 89D14, r.f. *Quercus agrifolia*, emgd. V.30.1989) EME

9.10 *Chionodes phalacrus* (Walsingham) ♂ — Texas, Presidio Co., Shafter, V.31.1973 (Hodges) USNM

9.11 *Chionodes popa* Hodges, paratype ♂ — California, Los Angeles Co., Mint Canyon, I.24.1942, r.f. *Sphaeralcea fasciculata*, emgd. II.25.1942 (C. Henne) USNM

9.12 *Chionodes petro* Hodges, holotype ♀ — California, San Diego Co., NE of Lakeside, V.16.1962 (J. Powell 62E11, r.f. *Sphaeralcea ambigua*, emgd. VI.5.1962) [abd. dissected. EME 4042)

9.13 *Chionodes lugubrella* (Fabricius) ♀ — Alaska, S of Anderson, VI.28.1979 (Opler & Powell) EME

9.14 *Chionodes ceanothiella* (Busck) ♂ — California, Marin Co, Alpine Lake, IV.22.1973 (J. Powell 73D9, r.f. *Ceanothus thyrsiflorus*, emgd. VI.8.1973) EME

9.15 *Chionodes obelus* Hodges, paratype ♂ — California, Contra Costa Co., Mt. Diablo, I.14.1984, r.f. *C. [eanothus] cuneatus*, emgd. II.12.1984 (D.Wagner 84A24) [abd. dissected EME 4482]

9.16 *Chionodes kubai* Hodges, paratype ♂ — California, Los Angeles Co., Devils Punchbowl, V.1.1968 (Opler & Powell) [abd. dissected EME 4462]

9.17 *Chionodes kubai* Hodges, paratype ♂ — California, El Dorado Co., Pine Hill, VII.4.1979, r.f *Fremontodendron [californicum] decumbens*, emgd. VII.22.1979 (S. Kuba) USNM

9.18 *Chiondes retinana* (Busck) ♂ — California, Yuba Co., NW of Smartville, V.5.1980 (Powell) EME

9.19 *Chionodes figurella* (Busck) ♀ — California, Contra Costa Co., Walnut Creek, VI.20.1970 (Powell) EME

9.20 *Chiondes grandis* Clarke ♂ — Colorado, San Miguel Co., Telluride, 8,750', VII.11.1977 (D. Ferguson) USNM

9.21 *Chiondes psilopterus* (Barnes & Busck), cotype ♂ — Canada, Quebec, Ottawa Co., Meach Lake, May 24/31 (Barnes Collection) USNM

9.22 *Chiondes bicolor* Clarke ♂ — California, Santa Barbara Co., Santa Rosa Island, Lower Windmill Canyon, IX.20.2000 (Powell) EME

9.23 *Chiondes sattleri* Hodges, paratype ♂ — Colorado, Grand Co., SW of Fraser, 9,950', VII.3.1992 (T. Dickel) USNM

9.24 *Chionodes sattleri* Hodges, paratype ♀ — Alaska, S of Anderson, VI.28.1979 (Opler & Powell) EME

9.25 *Chionodes lictor* Hodges, paratype ♂ — Oregon, Jackson Co., S of Mt. Ashland, VII.28.1982 (De Beendictis & Powell) EME

9.26 *Chionodes praetor* Hodges, paratype ♂ — California, Glenn Co., Plaskett Meadows, VII.3.1960 (Powell) EME

9.27 *Chionodes braunella* (Keifer) ♂ — California, Monterey Co., Big Creek Reserve, IV.22.1993 (J. Powell 93D30, r.f. *Lathyrus*, emgd. V.24.1993) EME

9.28 *Filatima aulaea* (Clarke) ♀ — California, Siskiyou Co., McBride Springs, VI.10.1974 (J. Powell 74F12, r.f. *Ribes roezelii*, emgd. VII.22.1974) EME

9.29 *Filatima demissae* (Keifer) ♂ — California, Trinity Co., SE of Hayfork, V.25.1973 (J. Powell 73E50, r.f. *Prunus virginiana* var. *demissa*, emgd. VII.20.1973) EME

9.30 *Filatima saliciphaga* (Keifer), paratype ♀ — California, Sacramento Co., Sacramento, VII.31.1936, r.f. *Salix sessilifolia*, emgd. IX.7.1936 (Keifer) EME

9.31 *Filatima depuratella* (Busck) ♀ — Arizona, Pima Co., SE of Continental, IX.5.1969 (Powell) EME

9.32 *Aroga eriogonella* (Clarke) ♂ — California, Monterey Co., Big Creek Reserve, IX. 3.1991 (Powell) EME

9.33 *Aroga eldorada* (Keifer) ♀ — California, Placer Co., Applegate, IV.24.1961 (J. Powell 61D6, r.f. *Artemisia ?vulgaris* [= *douglasiana*], emgd. V.25.1961) EME

9.34 *Aroga paulella* (Busck) ♂ — Mexico, Baja California Norte, Las Encinas, Sierra San Pedro Martir, 6,000', VI.3.1958 (Powell) EME

9.35 *Aroga moreonella* (Busck) ♂ — California, Los Angeles Co., NW of Valyrmo, V.1.1968 (Powell) EME

9.36 *Aroga unifasciella* (Busck) ♂ — California, Monterey Co., Big Creek Reserve, X.2.1991 (Powell) EME

9.37 *Aroga paraplutella* (Busck) ♂ — California, San Luis Obispo Co., Dune Lakes, V.2.1974 (J. Powell 74E11, r.f. *Eriogonum parvifolium*, emgd. VI.17.1974) EME

9.38 *Aroga trachycosma* (Meyrick) ♀ — California, San Diego Co., Border Field State Park, IV.10.1977 (J. Doyen & P. Rude, JTD lot 77D4, r.f. pupa in sand around *Frankenia*, emgd. IV.16.1977) EME

9.39 *Aroga trachycosma* (Meyrick) ♀ — California, Santa Clara Co., Milpitas, IX.6.1990 (C. Koehler, JP 90J9, r.f. *Frankenia grandifolia* [= *salina*], emgd. X.4.1990) EME

9.40 *Gnorimoschema saphirinella* (Chambers) ♂ — California, Mendocino Co., Inglenook Fen & dunes (r.f. pupa in sand under *Ambrosia*, emgd. II.1.1975 (M. Buegler) EME

9.41 *Gnorimoschema vastificum* Braun ♀ — California, Sacramento Co., W end Grand Island, VII.13.1975 (Opler & Powell) EME

9.42 *Gnorimoschema debenedictisi* Povolny & Powell, paratype ♀ — California, San Mateo Co., San Bruno Mtn., III.26.1988 (J. De Benedictis, JP 88C9, r.f. *Solidago spathulata*, emgd. VI.10.1988) EME

9.43—*Gnorimoschema ericameriae* Keifer, paratype ♂ — California, San Francisco, VII.4.1927, r.f. gall on *Ericameria ericoides* (Keifer) EME abd. dissected JAP 7432]

9.44 *Gnorimoschema coquillettella* Busck ♀ — Arizona, Mohave Co., Hualapai Mtn. Park, VI.2.1968 (Opler & Powell) EME

9.45 *Gnorimoschema octomaculellum* (Chambers) ♂ — California, Modoc Co., Buck Creek Ranger Station, VI.9.1970 (Opler & Powell, JP 70F44, r.f. *Chrysothamnus*, emgd. VII.24.1970) EME

PLATE 9

PLATE 10 Gelechiidae, Yponomeutidae, Ypsolophidae

10.1 *Gnorimschema powelli* Povolny ♀ — California, San Diego Co., Miramar NAS, XII.4.1996 (N. Bloomfield) EME

10.2 *Gnorimoschema baccharisella* Busck ♀ — California, Monterey Co., Big Creek Reserve, VI.7.1993 (J. Powell 93F12, r.f. *Baccharis pilularis*, emgd. VIII.16.1993) EME

10.3 *Gnorimoschema grindeliae* Povolny & Powell, paratype ♀ — California, Contra Costa Co., Pt. Richmond, IV.10.1996 (J. Powell 96D34, r.f. *Grindelia hirsutula*, emgd. VIII.17.1996) EME

10.4 *Gnorimoschema crypticum* Povolny & Powell, paratype ♀ — California, Monterey Co., Big Creek Reserve, III.1.1989 (J. Powell 89C4, r.f. *Haplopappus* [= *Isocoma*] *squarrosus*, emgd. VIII.7.1989) EME

10.5 *Gnorimoschema crypticum* Povolny & Powell, paratype ♀ — California, Santa Barbara Co., San Miguel Island, V.10.1997 (J. Powell 97E39, r.f. *Isocoma menziesii*, emgd. IX.15.1997) EME

10.6 *Gnorimoschema subterraneum* Busck ♂ — California, Contra Costa Co., Pt. Molate, VIII.25.1988 (J. Powell 88H12, r.f. *Aster chilensis*, emgd. IX.23.1988) EME

10.7 *Gnorimoschema subterraneum* Busck ♂ — California, Contra Costa Co., Richmond Field Station, V.20.1992 (J. Powell 92E236, r.f. *Aster chilensis*, emgd. IX.26.1992) EME

10.8 *Symmetrischema tangolias* (Gyen) ♂ — California, Monterey Co., Big Creek Reserve, VI.6.1992 (Powell) EME

10.9 *Exceptia sisterina* Povolny & Powell ♂ — California, Monterey Co., Big Creek Reserve, VI.5.1998 (Powell & Sperling) EME

10.10 *Euscobipalpa obsoletella* (F. von Roeslerstamm) ♀ — California, Contra Costa Co., Antioch Natl. Wildlife Refuge, IX.20.1990 (Powell) EME

10.11 *Euscobipalpa artemisiella* (Treitschke) ♀ — California, Contra Costa Co., Antioch Natl. Wildlife Refuge, IV.3.1990 (Hsu & Powell, JP 90D18, r.f. *Artemisia douglasiana*, emgd. V.3.1990) EME

10.12 *Scrobipalpula psilella* (Herrich-Schaeffer) ♀ — California, Contra Costa Co., Pt. San Pablo, II.20.1989 (J. Powell 89B1, r.f. *Gnaphalium californicum*, emgd. III.18.1989) EME

10.13 *Scrobipalpula gutierreziae* Povolny & Powell, paratype ♀ — California, Contra Costa Co., Antioch Natl. Wildlife Refuge, X.10.1991 (Powell) EME

10.14 *Scrobipalpula antiochia* Povolny & Powell, paratype ♀ — California, Contra Costa Co., Antioch Natl. Wildlife Refuge, II.20.1982 (J. Powell 82B18, r.f. *Senecio douglasii*, emgd. III.18.1982) EME

10.15 *Scrobipalpulopsis lutescella* (Clarke) ♀ — California, San Mateo Co., San Bruno Mtn., III.8.1984 (J. De Benedictis 84068-B, r.f. *Castilleja wrighti*, emgd. IV.4.1984) EME

10.16 *Tuta chiquitella* (Busck) ♀ — California, Contra Costa Co., Martinez Regional Park, VIII.10.1992 (Hsu, JP 92H36, r.f. *Chenopodium*, emgd. IX.30.1992) EME

10.17 *Symmetrischema striatellum* (Murtfeldt) ♂ — California, Contra Costa Co., Brooks Island, XI.13.1995 (J. Powell 95L8, r.f. *Solanum furcatum*, emgd. XII.28.1995) EME

10.18 *Keiferia lycopersicella* (Walsingham) ♂ — California, Kern Co., Shafter, III.26.1934, r.f. potato leaves, emgd. IV.11.1934 (Keifer) EME

10.19 *Caryocolum pullatellum* (Tengstrom) ♀ — Oregon, Clatsop Co., Seaside, VII.4.1955 (Clarke) USNM

10.20 *Faculta inaequalis* (Busck) ♂ — Arizona, Santa Cruz Co., Madera Canyon, VII.10/26.1964 (D. Davis) USNM

10.21 *Phthorimaea operculella* (Zeller) ♂ — California, Monterey Co., Castroville, X.1959, r.f. potatoes, emgd. XII.23.1959 (E. Lindquist) EME

10.22 *Syncopacma nigrella* (Chambers) ♀ — California, Contra Costa Co., Antioch Natl. Wildlife Refuge, IV.9.1982 (J. Powell 82D23, r.f. *Lupinus albifrons*, emgd. V.30.1982) EME

10.23 *Syncopacma* species ♀ — California, Monterey Co., Big Creek Reserve, IX.26.1987 (Powell) EME

10.24 *Batteristis pasadenae* (Keifer) ♀ — California, San Diego Co., Buckman Springs, X.5.1967 (Opler, Powell, Rude, JP 67K65, r.f. *Gutierrezia sarothrae* crown, emgd. XI.21.1967) EME

10.25 *Mirificarma eburnea* (Denis & Schiffermuller) ♀ — California, Yuba Co., NW of Smartville, V.3.1980 (Powell) EME

10.26 *Anarsia lineatella* Zeller ♂ — California, Alameda Co., Berkeley, VI.15.1960 (J. Lawrence) EME

10.27 *Helcystogramma badium* (Braun) ♀ — California, San Luis Obispo Co., Cambria, IX.19.1990 (Powell) EME

10.28 *Helcystogramma chambersella* (Murtfeldt) ♂ — Texas, Brewster Co., Chisos Mtns., 6,000′ VI.2.1973 (Hodges) USNM

10.29 *Anacampsis fragariella* (Busck) ♀ — California, Siskiyou Co., Mt. Shasta City, VI.29.1958 (J. Powell 58F10, r.f. *Rubus parviflorus*, emgd. VII.22.1958) EME

10.30 *Anacampsis niveopulvella* (Chambers) ♀ — California, Alpine Co., W of Monitor Pass, VI.22.1962 (J. Powell 62F5, r.f. *Populus tremuloides*, emgd. VII.15.1962) EME

10.31 *Anacampsis lacteusochrella* (Chambers) ♂ — California, Contra Costa Co., Antioch Natl. Wildlife Refuge, V.28.1982 (J. Powell 82E96, r.f *Croton californicus*, emgd. VI.30.1982) EME

10.32 *Dichomeris stipendaria* (Braun) ♂ — Washington, Whitman Co., Pullman, VII.14.1933, r.f *Aster eatoni* (Clarke) USNM

10.33 *Dichomeris delotella* (Busck) ♂ — Arizona, Santa Cruz Co., Madera Canyon, V.23.1963 (Franclemont) USNM

10.34 *Dichomeris baxa* Hodges, paratype ♂ — California, San Luis Obispo Co., Morro Bay, III.21.1974 (Doyen & Powell, JP 74C13, r.f. *Corethrogyne californica* [= *Lessingia filaginifolia*], emgd. IV.10.1974) EME

10.35 *Dichomeris gnomia* Hodges, paratype ♂ — Washington, Yakima Co., Satus Creek, VIII.19.1949 (Clarke) [abd. dissected USNM 9392]

10.36 *Dichomeris leuconotella* (Busck) ♀ — Michigan, Presque Isle Co., Ocqueoc Lake, VII.25.1970 (Hodges) USNM

10.37 *Dchomeris marginella* (Denis & Schiffermuller) ♂ — Virginia, Fairfax Co., Alexandria, VI.24.1971 (Powell) EME

10.38 *Scodes deflecta* (Busck) ♀ — Arizona, Pima Co., Baboquivari Mtns., VIII.1.1923 (O. Poling) USNM

10.39 *Sitotroga cerealella* (Olivier) ♀ — California, Alameda Co., Berkeley, IX.25.1984 (J. Powell 84J21, r.f. dried flowers, emgd. IX.29.1984) EME

10.40 *Pectinophora gossypiella* (Saunders) ♀ — Texas, Nueces Co., SW of Mathis, VIII.12.1963 (Davis & Duckworth) USNM

10.41 *Platyedra subcinerea* (Haworth) ♂ — California, Alameda Co., Berkeley, IV.24.2004 (Powell) EME

10.42 *Atteva punctella exquisita* Busck ♂ — California, Imperial Co., SE of Coyote Wells, VI23.1966 (J. Powell 66F13, r.f. *Holacantha emoryi*, emgd. VI.29.1966) EME

10.43 *Zelleria gracilariella* Busck ♂ — California, Monterey Co., Big Creek Reserve, VII.8.1986 (J. Powell 86G15, r.f. *Ribes sanguineum*, emgd.VIII.18.1986) EME

10.44 *Zelleria haimbachi* Busck ♂ — California, Contra Costa Co., Lafayette, VI.22.1960 (S. Cook) EME

10.45 *Zelleria haimbachi* Busck ♂ — Virginia, Page Co., NW of Luray, VII.23.1971 (Powell) EME

10.46 *Ypsolopha barberella* (Busck) ♂ — California, Mono Co., Sonora Pass, VIII.13.1960, in crack *Pinus murrayana* log (J. Lawrence) EME

10.47 *Ypsolopha flavistrigella* (Busck) ♂ — California, Mono Co., Sonora Pass, VIII.13.1960, in crack *Pinus murrayana* log (J. Lawrence) EME

10.48 *Ypsolopha cervella* (Walsingham) ♂ — California, Alameda Co., Albany Hill, IV.4.1997 (J. Powell 94D2, r.f. *Quercus agrifolia*, emgd. IV.29.1997) EME

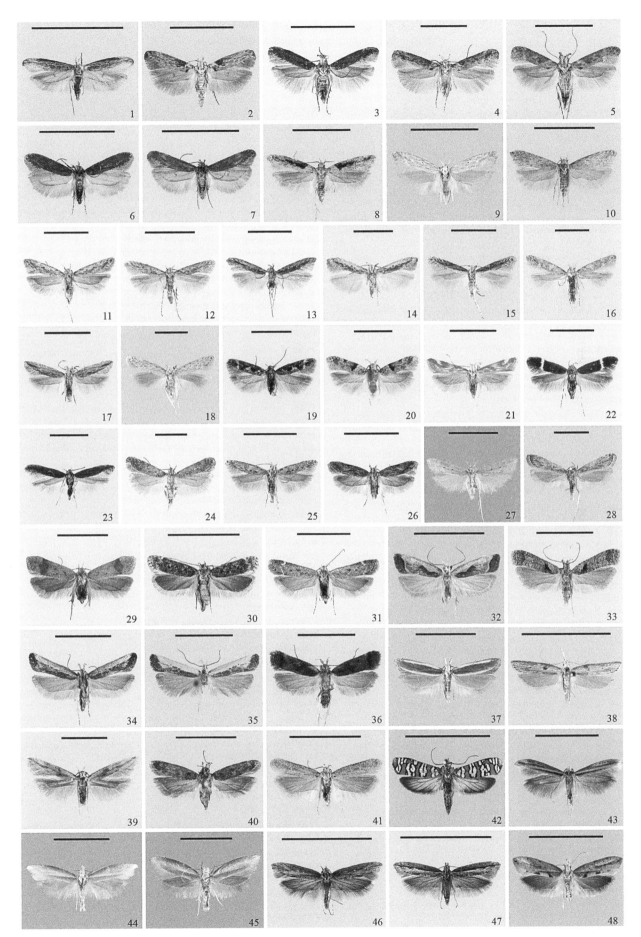

PLATE 10

11.1 *Ypsolopha sublucella* (Walsingham) ♀ — California, El Dorado Co., VI.30.1968 (Opler, JP 68F103, r.f. *Quercus vaccinifolia*, emgd. VII.30.1968) EME

11.2 *Ypsolopha walsinghamiella* (Busck) ♂ — Nevada, Washoe Co., Incline Village, VI.30.1968 (Opler, JP 68F115, r.f. *Purshia tridentata*, emgd. VII.29.1968) EME

11.3 *Ypsolopha falciferella* (Walsingham) ♀ — Arizona, Cochise Co., Miller Canyon, 1775m, (J. Powell 88D37, r.f. *Prunus*, emgd. V.10.1988) EME

11.4 *Ypsolopha lyonothamnae* (Powell) ♂ — California, Santa Barbara Co., Santa Cruz Island, V.23.1984, from shaggy bark of *Lyonothamnus* (Powell) EME

11.5 *Ypsolopha dentiferella* (Walsingham) ♀ — California, Contra Costa Co., Tilden Park, VI.14.1966 (Powell) EME

11.6 *Ypsolopha dentiferella* (Walsingham) ♂ — California, Contra Costa Co., Tilden Park, VI.14.1966 (Powell) EME

11.7 *Ypsolopha nella* (Dyar) ♂ — Arizona, Yavapai Co., Mingus Mtn., 7,600', VII.31.1991 (Hsu & Powell) EME

11.8 *Ypsolopha rubrella* (Dyar) ♀ — California, Monterey Co., Big Creek Reserve, IV.9.1994 (Hsu & M. McIntosh, JP 94D3, r.f. *Berberis nervosa*, emgd. VI.1.1994) EME

11.9 *Ypsolopha maculatana* (Busck) ♂ — California, Fresno Co., Ciervo Hills, III.18.1975 (Powell & P. Rude, JP 75C5, r.f. *Ephedra californica*, emgd. IV.20.1975) EME

11.10 *Ypsolopha maculatana* (Busck) ♂ — California, Fresno Co., Ciervo Hills, III.18.1975 (Powell & P. Rude, JP 75C5, r.f. *Ephedra californica*, emgd. V.2.1975) EME

11.11 *Plutella interrupta* Walsingham ♀ — California, Marin Co., Inverness, X.9.2000 (Powell) EME

11.12 *Plutella albidorsella* Walsingham ♂ — California, Yuba Co., NW of Smartville, V.5.1980 (Powell) EME

11.13 *Plutella vanella* Walsingham ♀ — California, Napa Co., Mt. St. Helena 2,900', VI.5.1980 (Powell) EME

11.14 *Euceratia castella* Walsingham ♂ — California, Stanislaus Co., Del Puerto Canyon, IV.30.1963 (Powell) EME

11.15 *Euceratia securella* Walsingham ♀ — California, Santa Barbara Co., Santa Cruz Island, IV.29.1966 (Powell) EME

11.16 *Plutella xylostella* (L.) ♂ — California, Del Norte Co., Little Grayback Pass, VII.9.1958 (Powell) EME

11.17 *Plutella porrectella* (L.) ♂ — Arizona, Mohave Co., Hualapai Mt. Park, IV.11.1986 (J. Brown & Powell) EME

11.18 *Plutella dammersi* Busck ♂ — California, Kern Co., E of Edison, III.20.1975 (J. Powell 75C9, r.f. *Isomeris arborea*, emgd. V.15.1975) EME

11.19 *Eucalantica polita* (Walsingham) ♀ — California, Marin Co., Inverness Park, X.8.1998 (Powell) EME

11.20. *Argyresthia goedartella* (L.) ♀ — Nevada, White Pine Co., Lehman Creek, 7,500', VII.25.1981 (De Benedictis & Powell) EME

11.21 *Argyresthia oreasella* Clemens ♂ — Utah, Utah Co., Mt. Timpanogos, 7,500' VII.24.1981, black light (De Bebedictis & Powell) EME

11.22 *Argyresthia cupressella* Walsingham ♂ — California, San Francisco Co., Golden Gate Park, IV.30.1958 (Powell) EME

11.23 *Argyresthia cupressella* Walsingham ♂ — California, Alameda Co., Berkeley, IV.2.1985 (Powell) EME

11.24 *Argyresthia trifasciae* Braun ♂ — California, Alameda Co., Berkeley, U.C. Campus, IV.1.1959 (J. Chemsak & Powell) EME

11.25 *Argyresthia franciscella* Busck ♀ — California, San Francisco Co., Lobos Creek, II.16.1961 (J. Powell 61B9, r.f. *Cupressus macrocarpa*, emgd. IV.6.1961) EME

11.26 *Argyresthia libocedrella* Busck ♀ — California, El Dorado Co., Blodgett Forest, VI.30.1967 (Powell) EME

11.27 *Argyresthia thoracella* Busck ♂ — California, Siskiyou Co., Big Springs, V.30.1999, adult ex *Juniperus* (Powell) EME

11.28 *Argyresthia pseudotsuga* Freeman ♀ — California, Santa Cruz Co., Año Nuevo St. Reserve, III.20.2002 (D. Wood, r.f. stem gall *Pseudotsuga menziesii*) EME

11.29 *Glyphipterix urticae* Heppner, paratype ♂ — Colorado, Park Co., SW of Lake George, 8,200', VII.1.1976, assoc. *Urtica* (J. Heppner) EME

11.30 *Glyphipterix powelli* Heppner, paratype ♂ — California, San Luis Obispo Co., Dune Lakes, II.24.1975, on *Urtica* in sunshine (Powell) EME

11.31 *Glyphipterix bifasciata* Walsingham ♂ — Washington, Kittitas Co., NE of Cle Elum, VII.2.1988 (Powell) EME

11.32 *Glyphipterix juncivora* Heppner, paratype ♂ — Utah, Sanpete Co., E of Ephraim, 8,500', VII.10.1976, assoc. *Juncus* (J. Heppner) EME

11.33 *Glyphipterix montisella* Chambers ♂ — Utah, Utah Co., Mt. Timpanogos, 7,400', VII.29.1967 (Powell) EME

11.34 *Glyphipterix feniseca* ♀ — California, Alpine Co., Hope Valley, 7,500', VII.18.1980 (Powell) EME

11.35 *Diploschizia impigritella* (Clemens) ♂ — Canada, British Columbia, Mission City, VI.18. 1953 (W. Mason) CNC

11.36 *Lithariapteryx abroniaeella* Chambers ♀ — California, Mendocino Co., dunes N of Cleome, VII.24.1975 (Powell) EME

11.37 *Lithariapteryx elegans* Powell, paratype ♀ — California, San Luis Obispo Co., Oso Flaco Lake, VI.7.1973 (Powell) EME

11.38 *Neoheliodines vernius* Hsu, paratype ♀ — California, Inyo Co., SW of Lone Pine, V.15.1969, adult on *Mirabilis bigelovii* (Powell) EME

11.39 *Neoheliodines hodgesi* Hsu, paratype ♂ — Mexico, Baja California Sur, SE of El Triunfo, XI.9.1993 (Hsu, JP 93L13, r.f. *Boerhavia coccinea*, emgd. X.21.1994) EME

11.40 *Embola powelli* Hsu, paratype ♀ — California, San Diego Co., NE of Lakeside, III.16.1994 (Hsu & H. Chuah, JP 94C54, r.f. *Mirabilis californica*, emgd. IV.7.1994) EME

11.41 *Embola ciccella* (Barens & McDunnough) ♂ — Arizona, Cochise Co., SW of Portal, VII.11.1972 (Powell) EME

11.42 *Aetole bella* Chambers ♂ — Mexico, Baja California Sur, SE of San Bartolo, XI.10.1993 (Hsu, JP 93L23, r.f. *Portulaca oleracea*, emgd. XII.4.1993) EME

11.43 *Aetole prenticei* Hsu, paratype ♀ — California, Solano Co., Grizzly Island, X.10.1991 (J. Powell 91K9, r.f. *Sesuvium verrucosum*, emgd. X.25.1991) EME

11.44 *Aetole cera* Hsu, paratype ♀ — California, Inyo Co., Stovepipe Wells, Death Valley, V.6.1993 (Hsu, JP 93E18, r.f. *Anulocaulis annulatus*, emgd. V.30.1993) EME

11.45 *Aetole extraneella* (Walsingham) ♀ — California, San Mateo Co., Edgewood Park, IX.18.1993 (Hsu, JP 93J11, r.f. *Epilobium densiflorum*, emgd. IX.19.1993) EME

11.46 *Acrolepiopsis californica* Gaedicke ♀ — California, Monterey Co., Big Creek Reserve, IV.28.1990 (J. Powell 90D67, r.f. *Dispoum hookeri*, emgd. V.16.1990) EME

11.47 *Bedellia somnulentella* (Zeller) ♂ — California, Santa Barbara Co., Santa Barbara Island, II.23.2004 (J. Powell 04B35, r.f. *Calystegia macrostegia*, emgd. III.7.2004) EME

11.48 *Paraleucoptera heinrichi* Jones ♀ — California, San Mateo Co., San Bruno Mtn., II.25.1984 (J. Powell 84B63, r.f. *Prunus ilicifolia*, emgd. III.20.1984) EME

11.49 *Lyonetia speculella* Clemens ♀ — California, Monterey Co., Big Creek Reserve, V.12.1992 (Powell & B. Scaccia) EME

11.50 *Schreckensteinia festaliella* (Hübner) ♀ — California, San Francisco Co., Golden Gate Park, IV.30.1958 (J. Powell 58D15, r.f. *Rubus vitifolius*, emgd. V.22.1958) EME

11.51 *Schreckensteinia felicella* (Walsingham) ♀ — California, Mendocino Co., MacKerricher Beach, V.2.1977, adult on *Castilleja latifolia* (Powell) EME

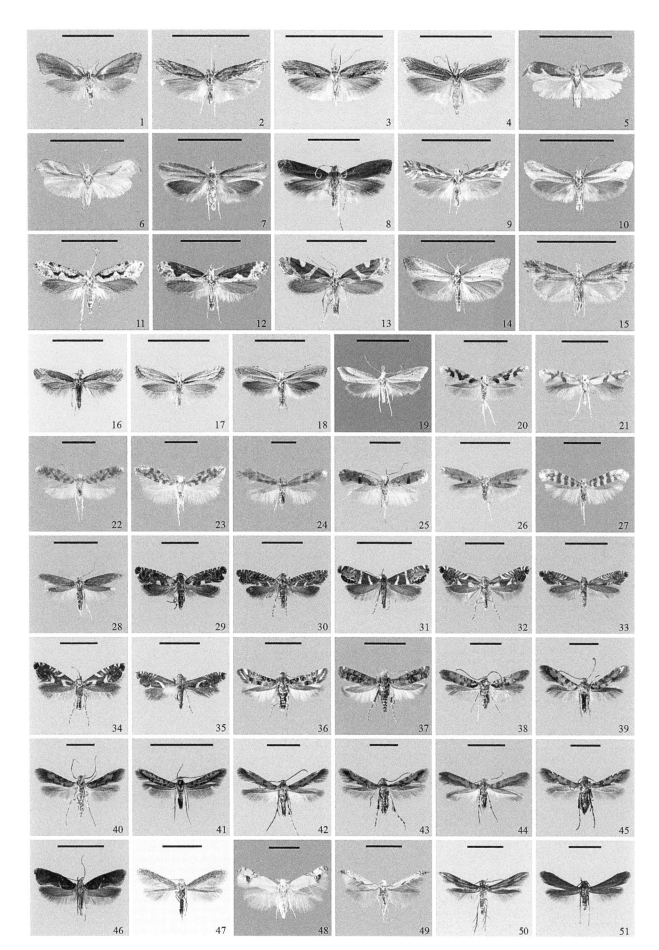

PLATE 11

12.1 *Epermenia californica* Gaedicke ♀ — California, Monterey Co., Big Creek Reserve, XI.7.1988 (J. Powell 88L10, r.f. *Aralia californica*, emgd. XII.9.1988) EME

12.2 *Epermenia californica* Gaedicke ♀ — California, Marin Co., Inverness Park, X.9.1998 (J. Powell 98K12, r.f. *Heracleum lanatum*, emgd. X.24.1998) EME

12.3 *Epermenia cicutaella* Kearfott ♀ — California, Contra Costa Co., Brooks Island, XI.13.1995 (Powell) EME

12.4 *Epermenia cicutaella* Kearfott ♂ — California, Monterey Co., Big Creek Reserve, III.27.1987 (Powell) EME

12.5 *Alucita montana* Barnes & Lindsey ♀ — California, Contra Costa Co., Russell Reserve, I.19.1974 (Powell) EME

12.6. *Alucita adriendenisi* Landry & Landry, paratype ♀ — Arizona, Coconino Co., SW of Flagstaff, VIII.20.1961 (R. Hodges) USNM

12.7 *Geina tenuidactyla* (Walsingham) ♀ — California, Siskiyou Co., SW of Mt. Shasta City, VI.13.1974 (J. Powell 74F27, r.f. *Rubus parviflorus*, emgd. VI.28.1974) EME

12.8 *Trichoptilus pygmaeus* Walsingham ♂ — California, Siskiyou Co., Bartle, VI.11.1974 (Powell) EME

12.9 *Dejongia californica* (Walsingham) ♂ — California, Stanislaus Co., Del Puerto Canyon, IV.4.1992 (J. Powell 92D12, r.f. *Grindelia camporum*, emgd. IV.28.1992) EME

12.10 *Anthophila alpinella* (Busck) ♀ — California, Marin Co., Inverness Ridge, IV.18.2006 (J. Powell 06D4, r.f. *Urtica holosericea*, emgd. V.4.2006) EME

12.11 *Prochoreutis sororculella* (Dyar) ♂ — California, Trinity Co., W. of Hayfork, V.21.1973 (Powell) EME

12.12 *Caloreas melanifera* (Keifer), paratype ♀ — California, Placer Co., Missouri Flat, V.24.1937, r.f. *Artemisia vulgaris* [=*douglasiana*], emgd. VI.7.1937 (Keifer) EME

12.13 *Caloreas melanifera* (Keifer) ♂ — California, Sonoma Co., SE of Plantation, VI.9.1979 (J. Powell 79F13, r.f. *Artemisia douglasiana*, emgd. VI.22.1979) EME

12.14 *Caloreas apocynoglossa* (Heppner) ♀ — California, Monterey Co., Big Creek Reserve, III.23.1989 (Hsu & Powell, JP 89C68, r.f. *Cynoglossum grande*, emgd. III.28.1989) EME

12.15 *Tebenna gemmalis* (Hulst) ♀ — California, Nevada Co., Upper Sagehen Creek, VII.5.1962 (J. Powell 62G6, r.f. *Wyethia mollis*, emgd. VII.12.1962) EME

12.16 *Tebenna gnaphaliella* (Kearfott) ♀ — California, Contra Costa Co., Brooks Island, IV.6.1993 (Hsu & Powell, JP 93D10, r.f. *Gnaphalium bicolor*, emgd. V.6.1993) EME

12.17 *Choreutis diana* (Hübner) ♀ — Nevada, Lander Co., Kingston Canyon, 7,000', VII.22.1968 (J. Powell 68G43, r.f. *Betula occidentalis*, emgd. VII.29.1968) EME

12.18 *Choreutis diana* (Hübner) ♀ — California, Humboldt Co., Kneeland, IX.6.2004 (R. Wielgus) EME

12.19 *Agdistis americana* Barnes & Lindsey ♀ — California, San Diego Co., S. D. River at Highway 5, III.28.2002 (J. Powell 02C17, r.f. *Frankenia*, emgd. IV.25.2002) EME

12.20 *Capperia ningoris* (Walsingham) ♀ — California, Monterey Co., Big Creek Reserve, VI.6.1993 (Powell & Zuniga) EME

12.21 *Oxyptilus delawaricus* (Walsingham) ♀ — California, Siskiyou Co., Bartle, VII.20.1966 (Powell) EME

12.22 *Gillmeria pallidactyla* (Haworth) ♀ — California, Contra Costa Co., Pt. Molate, IV.8.1988 (Powell) EME

12.23 *Platyptilia carduidactyla* (Riley) ♀ — California, San Mateo Co., Half Moon Bay (E. Clark, r.f. artichoke, emgd. III.26.1954) EME

12.24 *Platyptilia williamsi* Grinnell ♀ — California, Contra Costa Co., Pt. Molate, XII.12.1989 (J. Powell 89M2, r.f. *Grindelia humilis*, emgd. I.12.1990) EME

12.25 *Anstenoptilia marmarodactyla* (Dyar) ♂ — California, Monterey Co., Big Creek Reserve, VI.15.1991 (Hsu & Powell, JP 91F66, r.f *Salvia mellifera*, emgd. VII.12.1991) EME

12.26 *Amblyptilia pica* (Walsingham) ♀ — California, Monterey Co., Big Creek Reserve, V.27.1987 (J. Powell 87E10, r.f. *Stachys bullata*, emgd. VI.13.1987) EME

12.27 *Lioptilodes parvus* (Walsingham) ♀ — Hawaii, Maui, Napili Bay, XII.26.1982 (Powell) EME

12.28 *Paraplatyptilia albiciliata* (Walsingham) ♂ — California, San Mateo Co., San Bruno Mtn., VI.18.1985 (De Benedictis, JP 85F17, r.f. *Castilleja wrighti*, emgd. VII.3.1985) EME

12.29 *Paraplatyptilia fragilis* (Walsingham)] ♀ — Arizona, Coconino Co., SW of Flagstaff, VII.13.1961 (R. Hodges) USNM

12.30 *Adaina montana* (Walsingham) ♂ — California [Type ex Walsingham, Fernald Collection] USNM

12.31 *Adaina ambrosiae* (Murtfeldt) ♀ — Texas, Harris Co., Houston, XII.23.1982 (D. Wagner, JP 82M17, r.f. *Ambrosia psilostachya*, emgd. I.18.1983) EME

12.32 *Oidaematophorus occidentalis* (Walsingham) ♂ — California, Modoc Co., N of Davis Creek, VI.24.1999 (L. Crabtree, r.f. *Prunus subcordata*, emgd. VII.8.1999)(abd. dissected, Matthews 1117)

12.33 *Oidaematophorus phaceliae* McDunnough ♀ — California, San Mateo Co., San Bruno Mtn., IV.27.1986 (De Benedictis & Wagner, JP 86D115, r.f. *Phacelia californica*, emgd. V.29.1986) EME

12.34 *Oidaematophorus fieldi* (W. S. Wright) ♂ — California, Santa Barbara Co., Cachuma Camp, V.5.1967 (P. Rude, JP 67E28, r.f. undet. shrub, emgd. V.17.1967) EME

12.35 *Oidaematophorus grandis* (Fish) ♀ — California, Alameda Co., Berkeley, IX.13.1990 (Powell) EME

12.36 *Oidaematophorus glenni* Cashatt ♀ — California, Contra Costa Co., Antioch Natl. Wildlife Refuge, IV.2.1991 (J. Powell 91D10, r.f. *Aster lentus*, emgd. V.6.1991) EME

12.37 *Oidaematophorus sulphureodactylus* (Packard) ♂ — California, San Mateo Co., Edgewood Park, V.11.1990 (J. Powell 90E24, r.f. *Helianthus*, emgd. VI.22.1990) EME

12.38 *Oidaematophorus confusus* Braun ♀ — California, Alameda Co., Albany Hill, IV.9.1997 (J. Powell 97D11, r.f. *Baccharis pilularis*, emgd. V.14.1997) EME

12.39 *Oidaematophorus longifrons* (Walsingham) ♀ — California, Santa Barbara Co., Goleta, VII.18.1969 (Powell) EME

12.40 *Emmelina monodactyla* (Linnaeus) ♀ — California, Contra Costa Co., Walnut Creek, V.21.1972 (J. Powell 72F20, r.f. *Convolvulus arvensis*, emgd. VI.1972) EME

12.41 *Pennisetia marginata* (Harris) ♀ — Washington, Pierce Co., Puyallup, VIII.6.1936, reared (W. Baker) USNM

12.42 *Zenodoxus canescens* H. Edwards ♂ — California, Riverside Co., Blythe, XI.2.1936 (C. Dammers) USNM

12.43 *Zenodoxus canescens* H. Edwards ♀ — California, Riverside Co., Blythe, XI.2.1936 (C. Dammers) USNM

12.44 *Zenodoxus sidalceae* Englehardt ♀ — Washington, Whitman Co., Pullman, V.9.1935, r.f. *Sidalcea nervata* (Clarke) USNM

12.45 *Parathrene robiniae* (H. Edwards) ♀ — Texas, El Paso Co., El Paso, III.7.1950, reared from *Salix* [no collector label] USNM

12.46 *Albuna pyramidialis* (Walker) ♀ — Colorado, Gilpin Co., Tolland, 9,000 to 12,000', VII.15.1922 [no collector label] USNM

12.47 *Euhagena nebraskae* H. Edwards ♂ — Colorado, Jefferson Co., Golden, X.3.1926 (Oslar) USNM

12.48 *Calasesia coccinea* Beutenmuller ♀ — New Mexico, Hidalgo Co., N of Rodeo, VIII.17.1963 (J. Haddock) EME

PLATE 12

PLATE 13 Hepialidae, Cossidae, Sesiidae, Pyralidae

13.1 *Sthenopis quadriguttatus* (Grote) ♀ — Canada, Alberta, NW of Winfield, VII.21.2003 (D. Rubinoff & Powell) EME

13.2 *Paraphymatopus behrensi* Stretch ♂ — California, Marin Co., S. P. Taylor St. Park, VII.14.1895, 2055-2110 PDT (D. Wagner) EME

13.3 *Paraphymatopus behrensi* Stretch ♂ — California, Marin Co., S. P. Taylor St. Park, VII.14.1895, 2055-2110 PDT (D. Wagner) EME

13.4 *Paraphymatopus californicus* (Boisduval) ♂ — California, San Mateo Co., San Bruno Mtn., III.13.1981 (D. Wagner) EME

13.5 *Paraphymatopus californicus* (Boisduval) ♀ — California, Sonoma Co., Bodega Bay, III.15.1968 (Opler & Powell, JP 68C33, r.f. *Lupinus arboreus*, emgd. III.25.1968) EME

13.6 *Paraphymatopus hectoides* (Boisduval) ♂ — California, Mendocino Co., N of Branscomb, V.31.1980, hovering above grass 2050-2100 PDT (Powell) EME

13.7 *Gazoryctra mcglashani* H. Edwards ♂ — California [no data] (W. Schaus collection) USNM

13.8 *Hypopta palmata* (Barnes & McDunnough) ♂ — California, San Bernardino Co., Ivanpah Mtns., VII.18.1947 (C. I. Smith) EME

13.9 *Givira marga* Barnes & McDunnough ♂ — California, Kern Co., Woffard Heights, VI.12.1975 (J. Johnson) EME [greasy]

13.10 *Givira marga* Barnes & McDunnough ♂ — California, Orange Co., Silverado Canyon, IX.13.1979 (G. Marsh) EME

13.11 *Givira theodori* Dyar ♂ — Arizona [county unknown], Mesa Verde River at Coon_s Bluff, VIII.27.1963 (Hasbrouck) USNM

13.12 *Comadia bertholdi* (Grote) ♂ — California, Monterey Co., Big Creek Reserve, VI23.1992 (R. Zuniga & B. Scaccia) EME

13.13 *Acossus populi* (Walker) ♂ — California, Inyo Co., Cottonwood Creek, VI. 10.1972 (L. Orsak) EME

13.14 *Miacora perplexa* (Neumoegen & Dyar) ♂ — California, Monterey Co., Big Creek Reserve, IX.15.1993 (Powell) EME

13.15 *Prionxystus robiniae* (Peck) ♂ — California, Kern Co., Shafter, V.26.1948 (R. van den Bosch) EME

13.16 *Prionxystus robiniae* (Peck) ♀ — California, Monterey Co., Big Creek Reserve, VII.17.19988 (Powell) EME

13.17 *Morpheis* species, *cognatus* group ♂ — Costa Rica, Prov. Heredia, SE of La Virgen, 550m, II.22.2003 (M. Heddle & Powell) EME

13.18 *Melittia grandis* (Strecker) ♀ — Arizona, Cochise Co., Douglas, VII.31.1975, 1 on *Cucurbita foeditissima* 0530-0559 (E. & J. Linsley) EME

13.19 *Melittia snowi* H. Edwards ♀ — New Mexico, Sandoval Co., SW of Algodones, VII.27.1989, adult on *Cucurbita foetidissima* (Powell) EME

13.20 *Melittia gloriosa* H. Edwards ♀ — Oregon, Benton Co., Corvallis, VIII.10.1931 [no collector label] USNM

13.21 *Melittia gloriosa* H. Edwards ♂ — California, Riverside Co., E of Anza, VII.15.1963, adult on *Cucurbita foeditissima* (R. Langston) EME

13.22 *Sarata edwardsialis* (Hulst) ♂ — Arizona, Yavapai Co., II.27.1970 (L. Martin) USNM

13.23 *Melitara dentata* (Grote) ♂ — Nevada, White Pine Co., Lehman Creek, 7,700′, VIII.8.1978, 0430-0500 MDT (Powell) EME

13.24 *Cactobrosis fernaldialis* (Hulst) ♀ — Arizona, Pima Co., Tucson, IV.14.2004 (R. Wielgus) EME

13.25 *Alberada parabates* (Dyar) ♂ — Colorado, Chaffee Co., N of Buena Vista, VII.14.1982 (D. Ferguson) EME

13.26 *Cahela ponderosella* (Barnes & McDunnough) ♀ — California, Riverside Co., S of Palm Desert, IV.6.1963 (R. Langston) EME

PLATE 13

PLATE 14 Sesiidae, Tortricidae, Olethreutinae

14.1 *Sesia tibialis* (Harris) ♀ — Utah, Uintah Co., Green River near Dinosaur Natl Monument, VII.7.1993 (Powell) EME

14.2 *Synanthedon exitiosa* (Say) ♂ — California, Alameda Co., Berkeley, V.3.1977, attracted to ODDA, 1230 PDT (Powell) EME

14.3 *Synanthedon exitiosa* (Say) ♀ — Colorado, Larimer Co., Big Thompson Canyon, VIII.7.1973 (Powell) EME

14.4 *Synanthedon albicornis* (H. Edwards) ♂ — California, Contra Costa Co., Antioch, VIII.20.1979 (Powell) EME

14.5 *Synanthedon navaroensis* H. Edwards ♂ — California, Marin Co., Inverness Ridge, VIII.11.2000 (Powell) EME

14.6 *Synanthedon bibionipennis* (Boisduval) ♂ — California, Contra Costa Co., Mitchell Canyon, VI.27.1989 (Powell) EME

14.7 *Synanthedon bibionipennis* (Boisduval) ♂ — California, Contra Costa Co., Fish Ranch Canyon, VI.20.1987 (Powell) EME

14.8 *Synanthedon polygoni* (H. Edwards) ♂ — California, Marin Co., Pt. Reyes dunes, V.30.1949 (C. I. Smith) EME

14.9 *Synanthedon polygoni* (H. Edwards) ♂ — California, Siskiyou Co., Ash Creek Ranger Station, VI.25.1974 (J. Powell 74F63, r.f. *Eriogonum sulphureum*, emgd. VII.5.1974) EME

14.10 *Podosesia syringiae* (Harris) ♀ — West Virginia, Pendleton Co., Spruce Knob Lake, VII.4.1983 (Opler) EME

14.11 *Carmenta mariona* (Beutenmuller) ♂ — New Mexico, Taos Co., Chimayo, VIII.5.1932 (A. B. Klots) USNM

14.12 *Carmenta verecunda* (H. Edwards) ♂ — California, Mono Co., Sonora Pass, 10,000', VII.10.1957 (J. W. MacSwain) EME

14.13 *Carmenta mimuli* (H. Edwards) ♂ — Arizona, Cochise Co., Pueblo del Sol, Huachuca Mtns., IV.8.1986 (R. Wielgus) USNM

14.14 *Carmenta prosopis* (H. Edwards) ♂ — Texas, Brewster Co., Alpine, IX.14.1927 (G. Englehardt) USNM

14.15 *Penstemonia clarkei* Engelhardt ♂ — California, Alameda Co., Berkeley, VI.18.1959 (Powell) EME

14.16 *Alcathoe verrugo* (Druce) ♀ — California, Orange Co., San Juan Capistrano, "7-8"-1936 (T. W. Hower) EME

14.17 *Bactra verutana* Zeller ♂ — California, San Luis Obispo Co., Pozo, IV.27.1962 (Powell) EME

14.18 *Bactra verutana* Zeller ♀ — California, Marin Co., Inverness Park, X.15/21.1999 (Powell) EME

14.19 *Bactra furfurana* (Haworth) ♂ — Oregon, Lane Co., SW of Florence VI.24.1975 (Opler & Powell) EME

14.20 *Endothenia hebesana* (Walker) ♂ — California, Monterey Co., Carmel St. Beach, VIII.12.1984 (J. Powell 84H44, r.f *Castilleja affinis*, emgd. VIII.23.1984) EME

14.21 *Endothenia rubipunctana* (Kearfott) ♂ — California, Lake Co., near Boggs Lake, VII.1/7.1992 (L. Lawyer, r.f. *Iris macrophyllum*, JP 92G10, emgd. V.5.1993) EME

14.22 *Episimus argutanus* (Clemens) ♂ — California, Tulare Co., Potwisha 2,000', V.4.1979 (Powell) EME

14.23 *Apotomis removana* Kearfott) ♂ — Arizona, Coconino Co., Hochderffer Hill 8,500', VII.17.1964 (J. G. Franclemont) USNM

14.24 *Apotomis capreana* (Hübner) ♂ — Utah, Ephraim Canyon 8,850', VII.20.1981 (R. W. Hodges) USNM

14.25 *Pseudosciaphila duplex* (Walsingham) ♀ — California, Mono Co., Monitor Pass, VI.23.1962 (J. Powell 62F7, r.f. *Populus tremuloides*, emgd. VII.8.1962) EME

14.26 *Olethreutes cespitanus* (Hübner) ♂ — California, Tuolumne Co., Pinecrest, VIII.1.1961 (Powell) EME

14.27 *Olethreutes glacianus* (Moeschler) ♀ — Alaska, SW of Ester, VI.30.1979 (Opler & Powell) EME

14.28 *Olethreutes deprecatorius* Heinrich ♀ — California, Glenn Co., Plaskett Meadows, VI.14.1972 (J. Powell, 72F 4, r.f. *Veratrum californicum*, emgd. VI.20.1972) EME

14.29 *Olethreutes deprecatorius* Heinrich ♀ — California, Glenn Co., Plaskett Meadows, VI.14.1972 (J. Powell 72F4, r.f. *Veratrum Californicum*, emgd. VI.21/27.1972) EME

14.30 *Olethreutes chalybeana* (Walsingham) ♂ — California, Plumas Co., NW of Quincy, 4,000', V.20.1982 (J. A. De Benedictis, JP 82E43, r.f. *Spiraea douglasii*, emgd. VI.10.1982) EME

14.31 *Hedya ochroleucana* Froelich) ♂ — California, Siskiyou Co., Mt. Shasta City, VII.3.1958 (Powell) EME

14.32 *Rhyacionia buoliana* (Denis & Schiffermuller) ♀ — New York, Westbury, Long Island, reared VI.15.1915 (C. Heinrich, Hopkins U.S. 13905a, r.f. Scotch & Austrian pines) USNM

14.33 *Rhyacionia zozana* (Kearfott) ♀ — California, Tuolumne Co., Groveland, VII.19.1962 (R. Stevens, r.f. *Pinus ponderosa*, emgd. II.1963) EME

14.34 *Rhyacionia monophylliana* (Kearfott) ♂ — California, Kern Co., Frazier Mtn., V.7.1959 (Powell) EME

14.35 *Retinia albicapitana* (Busck) ♀ — Idaho, Kootenai Co., Coeur d' Alene (J. C. Evendeen, reared V.13.1916, *Pinus contorta*) USNM

14.36 *Retinia arizonensis* Heinrich ♀ — Arizona, Santa Cruz Co., Madera Canyon, VII.7.1959 (J.G. Franclemont) USNM

14.37 *Retinia metallica* (Busck) ♂ — California, El Dorado Co., China Flat, 4,200' (Kinnee & Eichlin, r.f. *Pinus murrayana*, emgd, II.21/26.1991) EME

14.38 *Retinia metallica* (Busck) ♂ — Oregon, Klamath Co., Chemult, III.7.1963 (D. L. Wood, emgd. IV.1963, r.f. *Pinus contorta*)(abd. dissected, EME slide 498)

14.39 *Barbara colfaxiana* (Kearfott) ♂ — Oregon, Baker Co., Blue Mtns., 4,000', 1970 (J. H. Baker) USNM

14.40 *Spilonota ocellana* (Denis & Schiffermuller) ♂ — California, Alameda Co., Berkeley, VI.10.1959 (Powell) EME

14.41 *Phaneta corculana* (Zeller) ♀ — California, Modoc Co., Fandango Pass, 6,100', VI.7.1970 (Powell) EME

14.42 *Phaneta amphorana* (Walsingham) ♀ (spring form) — California, Contra Costa Co., Pt. Molate VIII.25.1988 (J. Powell 88H13, r.f. *Grindelia humilis*, emgd. XI.13.1988) EME

14.43 *Phaneta amphorana* (Walsingham) ♀ (summer form) — California, Contra Costa Co., Antioch Natl. Wildlife Refuge, V.11.1983 (J. Powell 83E40, r. f. *Grindelia camporum*, emgd. VIII.6.1983) EME

14.44 *Phaneta apacheana* (Walsingham) ♀ — California, Contra Costa Co., Pt. San Pablo, II.20.1989 (J. Powell 89B1, r.f. *Gnaphalium californicum*, emgd. III.20.1988) EME

14.45 *Phaneta apacheana* (Walsingham) ♂ — California, Santa Barbara Co., San Miguel Island, 400', V.6.1997 (J. Powell 97E12, r.f. *Gnaphalium*, emgd. VI.11.1997) EME

14.46 *Phaneta misturana* (Heinrich) ♀ — California, Contra Costa Co., Briones Reservoir, IV.22.1995 (Powell) EME

14.47 *Phaneta pallidarcis* (Heinrich) ♂ — California, Monterey Co., Big Creek Reserve, VIII.17.1998 (Powell) EME

14.48 *Rhyacionia neomexicana* (Dyar) ♂ — Colorado, Larimer Co., Ft. Collins (reared III.1973, Hopkins U.S. 36736) USNM

14.49 *Rhyacionia neomexicana* (Dyar) ♀ — Arizona, Sitgreaves Natl. Forest, Chevelon Road, IV.21.1969 (D. T. Jennings, *Pinus ponderosa*, Hopkins U.S. 37279-y) EME

14.50 *Retinia picicolana* (Dyar) ♂ — California, El Dorado Co., Blodgett Forest, VI.27.1967 (Powell) EME

14.51 *Retinia sabiniana* (Kearfott) ♂ — California, San Luis Obispo Co., La Panza, V.2.1962 (Powell) EME

14.52 *Barbara colfaxiana* (Kearfott) ♂ — Arizona, Cochise Co., Miller Canyon, 5,800', IV.12.1988 (Powell) EME

PLATE 14

PLATE 15 Tortricidae, Olethreutinae

15.1 *Phaneta argenticostana* (Walsingham) ♂ — New Mexico, Eddy Co., near Rattlesnake, III.27.1993 (D. C. Ferguson) USNM

15.2 *Phaneta argenticostana* (Walsingham) ♂ — California, Riverside Co., Chino Canyon, IV.20.1960 (Powell) EME

15.3 *Phaneta subminimana* (Heinrich) ♂ — California, Monterey Co., Big Creek Reserve, VIII.17/19.1988 (Powell) EME

15.4 *Phaneta artemisiana* (Walsingham) ♂ — California, Alameda Co., Redwood Canyon, IV.23.1984 (J. Powell 84D69, r.f. *Artemisia douglasiana*, emgd. VII.16.1984) EME

15.5 *Phaneta scalana* (Walsingham) ♂ — California, Alameda Co., Redwood Canyon, IV.23.1984 (J. Powell 84D69, r.f. *Artemisia douglasiana*, emgd. IX.4.1984) EME

15.6 *Phaneta straminiana* (Walsingham) ♀ — California, Los Angeles Co., San Clemente Island, X.1.2002 (J. De Benedictis & Powell, JP 02J25, r.f. *Isocoma menziesii*, emgd. X.15.2002) EME

15.7 *Phaneta straminiana* (Walsingham) ♂ — California, San Bernardino Co., Zzyzx Springs, IV.27.1977 (Powell) EME

15.8 *Eucosma sonomana* (Kearfott) ♂ — California, Marin Co., Inverness, II.17.1961 (Powell) EME

15.9 *Eucosma bobana* Kearfott ♀ — California, Kern Co., Walker Pass, VII.17.1956 (H. Ruckes jr., r.f. 1956 cone of *Pinus monophylla*, emgd. VII.18.1957) EME

15.10 *Eucosma ponderosa* Powell ♀ — California, Mono Co., near Leevining, IX. 1958 (H. Ruckes jr., r.f. 1958 cone, *Pinus jeffreyi*, emgd. VI.21.1959) EME

15.11 *Eucosma siskiyouana* (Kearfott) ♂ — California, Humboldt Co., Black Lassic Botanic area, VII.8.2003 (R. Wielgus) EME

15.12 *Eucosma avalona* McDunnough ♂ — California, Santa Barbara Co., Santa Cruz Island, IX.28.1978 (Powell) EME

15.13 *Phaneta offectalis* (Hulst) ♀ — Nevada, Nye Co., Currant Creek. VII.20.1968 (Opler & Powell, JP 68G35, r.f. *Chrysothamnus viscidiflorus*, emgd. IX.9.1968) EME

15.14 *Phaneta bucephaloides* (Walsingham) ♀ — California, Lassen Co, Litchfield, VII.24.1968 (Opler & Powell, JP 68G66, r.f. *Chrysothamnus nauseosus*, emgd. VIII.23.1968) EME

15.15 *Eucosma canariana* Kearfott ♂ — Utah, Grand Co., Green River, VII.23.1965 (J. Doyen) EME

15.16 *Eucosma sandiego* Kearfott ♂ — California, San Diego Co., E of Banner, IX.3.1966 (Powell, P. Rude., J. Wolf, JP 66J13, r.f. *Haplopappus [Isocoma] venetus*, emgd. IX.21.1966) EME

15.17 *Eucosma crambitana* (Walsingham) ♂ — California, Mono Co., near Leevining, VII.12.1968 (J. Heppner, Opler & Powell, JP 68G23, r.f. *Chrysothamnus*, emgd IX.9.1968) EME

15.18 *Eucosma crambitana* (Walsingham) ♂ — California, Mono Co., near Leevining, VII.12.1968 (J. Heppner, Opler & Powell, JP 68G23, r.f. *Chrysothamnus*, emgd VIII.23.1968) EME

15.19 *Eucosma optimana* (Dyar) ♀ — California, Mono Co., Walker, VII.8.1968 (Opler & Powell, JP 68G5, r.f. *Artemisia tridentata*, emgd. IX.9.1968) EME

15.20 *Eucosma optimana* (Dyar) ♂ — California, Mono Co., White Mtns., 10,150', VII.21.1961 (Powell) EME

15.21 *Eucosma subflavana* (Walsingham) ♀ — Oregon, Jefferson Co., Pelton Dam, VI.24.1963 (C. Toschi) EME

15.22 Eucosma grandiflavana (Walsingham) ♂ — California, Nevada Co., Upper Sagehen Cr., VIII.5.1962 (Powell & Toschi) EME

15.23 *Eucosma ridingsana* (Robinson) ♂ — Arizona, Yavapai Co., NE of Bridgeport, VI.4.1968 (Opler & Powell, JP 68F33, r.f. *Gutierrezia* sp., emgd. VIII.19.1968) EME

15.24 *Eucosma ridingsana* (Robinson) ♀ — California, San Luis Obispo Co., Oso Flaco Lake, IV.27.1968 (J. Powell 68D194, r.f. *Corethrogyne [Lessingia] filaginifolia*, emgd. V.24.1968) EME

15.25 *Eucosma fernaldana* (Robinson) ♂ — Colorado, Teller Co., Florissant, VIII.14.1960 (T. Emmel) LACM

15.26 *Eucosma fernaldana* (Robinson) ♂ — Colorado, Grand Co., Shadow Mt. Lake, VIII.1.1967 (Powell) EME

15.27 *Eucosma morrisoni* (Walsingham) ♀ — Colorado, Larimer Co., Big Thompson Canyon, VIII.7.1973 (Powell) EME

15.28 *Euscosma agricolana* (Walsingham) ♂ — California, Modoc Co., Buck Creek Ranger Station, 5,150' VI.6.1970 (Powell) EME

15.29 *Eucosma hennei* Clarke ♂ — California, San Luis Obispo Co., Dune Lakes, VII.11.1973 (Powell)—abd. dissected JP 5389, EME

15.30 *Eucosma hasseanthi* Clarke ♂ — California, Riverside Co., Riverside, V.13.1962 (J. Powell 62E9, r.f. *Phacelia ramosissima*, emgd. V.27.1962) EME

15.31 *Eucosma scintillana* (Clemens) ♂ — California, Kern Co., Tehachapi Mtn. Park, 5,000' VI.16.1980 (J. De Benedictis & S. Meredith) EME

15.32 *Eucosma williamsi* Powell ♂ — California, Santa Barbara Co., Cuyama River, IV.12.1967 (Powell & Rude, JP 67D43, r.f. *Baccharis pilularis*, emgd. VI.3.1967) EME

15.33 *Eucosma maculatana* (Walsingham) ♂ — California, San Benito Co., Limekiln Canyon, IV.24.1968 (J. Powell 68D177, r.f. *Eriophyllum lanatum*, emgd. V.30.1968) EME

15.34 *Eucosma juncticiliana* (Walsingham) ♂ — California, Marin Co., North Beach, Pt. Reyes, VI.4.1977 (Powell) EME

15.35 *Pelochrista passerana* (Walsingham) ♂ — California, Santa Barbara Co., Santa Barbara Island, II.20.2004 (J. Powell 04B38, r.f. *Achillea millefolium*, emgd. VI.1.2004) EME

15.36 *Pelochrista expolitana* (Heinrich) ♂ — California, Mendocino Co., Eel River Ranger Station, VI.9.1972, associated with *Artemisia douglasiana* (Powell) EME

15.37 *Pelochrista rorana* (Kearfott) ♂ — California, Yuba Co., Sierra Foothill Field Station, V.7.1980 (Powell) EME

15.38 *Suleima helianthana* (Riley) ♀ — California, Los Angeles Co., SE of Gorman, VIII.31.1966 (J. Powell 66H17, r.f *Helianthis annuus*, emgd. VIII.31.1966) EME

15.39 *Suleima lagopana* (Kearfott) ♀ — California, Riverside Co., La Sierra, VIII.22.1968 (J. Powell 68H26, r.f. *Helianthus annuus*, emgd. VIII.30.1968) EME

15.40 *Sonia vovana* (Kearfott) ♂ — California, San Luis Obispo Co., Upper Cuyama River, X.6.1967 (Opler, Powell, P. Rude, JP 67K92, r.f. *Gutierrezia*, emgd. X.8.1967) EME

15.41 *Sonia vovana* (Kearfott) ♂ — California, San Luis Obispo Co., Morro Strand, VIII.20/22.1990, associated with *Haplopappus [Isocoma acradenia]* (Powell) EME

15.42 *Sonia filiana* Busck ♂ — California, San Diego Co., El Cajon, IX.2.1966 (Powell, Rude & J. Wolf, JP 66J10, r.f. *Haplopappus [Isocoma acradenia]*, emgd. XI.1.1966) EME

15.43 *Sonia comstocki* Clarke ♀ — California, Los Angeles Co., W of Lancaster, X.2.1967 (Opler, Powell & Rude, JP 67K33, r.f. *Haplopappus [Isocoma acradenia]*, emgd. X.7.1967) EME

15.44 *Epiblema strenuana* (Walker) ♀ — California, San Diego Co., E of Banner VII.13.1963 (Powell) EME

15.45 *Epiblema strenuana* (Walker)(pale beach form) ♂ — California, Contra Costa Co., Pt. Molate XII.12.1989 (J. Powell 88M3, r.f. *Ambrosia chamissonis*, emgd. III.10.1989) EME

15.46 *Epiblema sosana* (Kearfott) ♂ — California, Los Angeles Co., S of Valyrmo, V.1.1968 (Powell) EME

15.47 *Epiblema rudei* Powell ♂ — California, Kings Co., Kettleman City, III.17.1975 (J. Powell 75C4, r.f. *Gutierrezia*, emgd. III.30.1975) EME

15.48 *Epiblema macneilli* Powell paratype ♀ — California, Inyo Co., near Mono Pass, VIII.20.1965 (C. D. MacNeill) EME

15.49 *Gypsonoma salicicolana* (Clemens) ♀ — California, San Luis Obispo Co., Dune Lakes, V.2.1974 (J. Powell 74E3, r.f. *Salix*, emgd. V.31.1974) EME

15.50 *Proteoteras aesculana* (Riley) ♂ — California, Alameda Co., Berkeley, III.12.1960 (Powell) EME

15.51 *Proteoteras arizonae* (Walsingham) ♂ — California, San Mateo Co., VI.30.1959, (A. E. Pritchard, JP 59F2, r.f. *Acer negundo*, emgd. VII.30.1959) EME

15.52 *Chimoptesis chrysopyla* Powell ♂ — California, San Francisco Co., Golden Gate Park, IV.5.1968 (Opler, JP 68D5, r.f. *Quercus agrifolia*, emgd. II.5.1969) EME

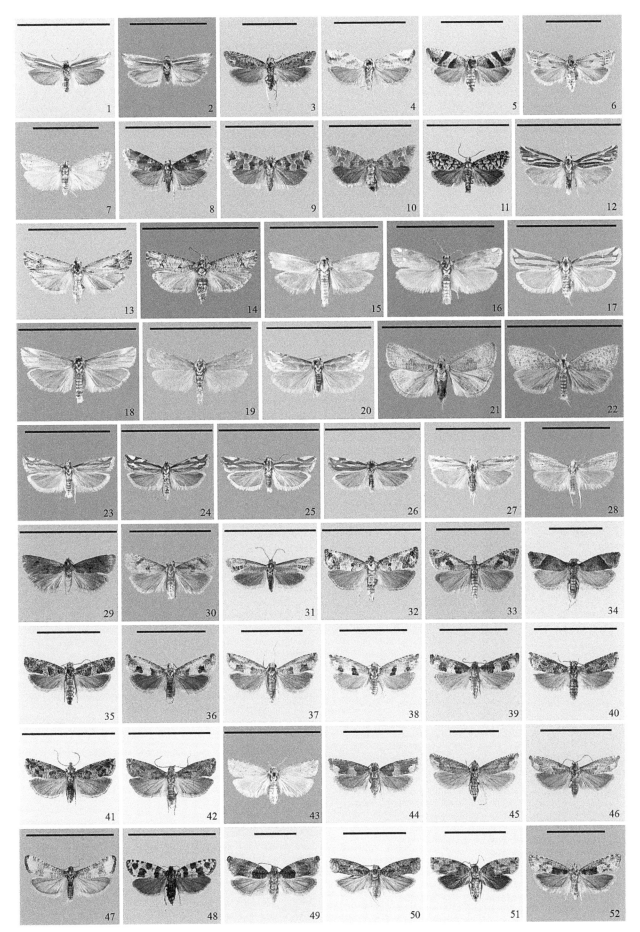

PLATE 15

PLATE 16 Tortricidae, Olethreutinae

16.1 *Zeiraphera canadensis* Mutuura & Freeman ♂— Arizona, Coconino Co., SW of Flagstaff, VII.13.1961 (R. Hodges) USNM

16.2 *Pseudexentera oregonana* (Walsingham) ♂— Canada, British Columbia, Victoria, III.26.1922 (Blackmore) USNM

16.3 *Pseudexentera habrosana* (Heinrich) ♂ — California, Alameda Co., Berkeley, I.21.1990 (Powell) EME

16.4 *Pseudexentera habrosana* (Heinrich) ♀ — California, Tuolumne Co., Don Pedro Dam, II.26.1996 (D. Rubinoff) EME

16.5 *Crocidosema plebejana* Zeller ♂ — California, Alameda Co., Berkeley, XI.16.1990 (Powell) EME

16.6 *Crocidosema plebejana* Zeller ♀ — California, Alameda Co., Berkeley, X.22.1994 (Powell) EME

16.7 *Notocelia culminana* (Walsingham) ♂ — Nevada, Elko Co., Pine Creek 6,600′ VIII.5.1981 (Powell & E. Randal) EME

16.8 *Epinotia radicana* Heinrich (green form) ♂ — California, Siskiyou Co., SW of Gazelle, VI.24.1980 (De Benedictis & Powell, JP—80F77, r.f. *Pseudotsuga menziesii*, emgd. VIII.15.1980) EME

16.9 *Epinotia radicana* Heinrich (brown form) ♂ — California, Marin Co., Inverness Park, 150m, X.8/14.1997 (Powell) EME

16.10 *Epinotia hopkinsana* (Kearfott) ♂ — California, San Luis Obispo Co., Cambria, III.24.1980 (J. Powell 80C3.4, r.f. *Pinus radiata*, emgd. V.17.1980) EME

16.11 *Epinotia columbia* (Kearfott) ♀ — California, Contra Costa Co., Lafayette, VI.12.1960 (S. F. Cook jr.) EME

16.12 *Epinotia columbia* (Kearfott) ♀ — California, Contra Costa Co., Pt. Molate, III.8.1988 (J. Powell 88C6, r.f. *Salix lasiolepis*, emgd. IV.16.1988) EME

16.13 *Epinotia emarginana* (Walsingham) ♂— California, Marin Co., Ring Mtn., III.19.1986 (J. Powell 86C38, r.f. *Q. agrifolia*, emgd. V.2.1986) EME

16.14 *Epinotia emarginana* (Walsingham) ♂— California, Mendocino Co., Eel River Ranger Station, VI.9.1972 (Powell) EME

16.15 *Epinotia emarginana* (Walsingham) ♀ — California, Trinity Co., Hayfork, V.20.1973 (J. Powell 73E28, r.f. *Quercus kelloggii*, emgd. VI.15.1973) EME

16.16 *Epinotia emarginana* (Walsingham) ♀ — California, Mendocino Co., N Calif. Coast Redwood Preserve, V.20/24.1985 (J. Brown, JP 85E109, r.f. *Arctostaphylos*, emgd. VI.20.1985) EME

16.17 *Epinotia johnsonana* (Kearfott) ♂— California, Monterey Co., Big Creek Reserve, V.1.1992 (J. Powell 92E4, r.f. *Holodiscus discolor*, emgd. VI.15.1992) EME

16.18 *Epinotia siskiyouensis* Heinrich ♀ — California, Ventura Co., SE of Pine Mtn. Summit, V.26.1984 (J. Powell 84E98, r.f. *Clematis*, emgd. VII.16.1984) EME

16.19 *Epinotia solandriana* (L.) ♂— California, Marin Co., Inverness Ridge, 1,000′ V.14.1972 (J. Powell 72E8, r.f. *Alnus*, emgd. VI.9/19.1972) EME

16.20 *Epinotia solandriana* (L.) ♂— California, Marin Co., Inverness, V.2.1969 (J. Powell 69D79, r.f. *Alnus rubra*, emgd. V.28.1969) EME

16.21 *Epinotia solandriana* (L.) ♂ — Utah, Sanpete Co., Ephraim Canyon, 8.800′ (Powell & Sperling) EME [abd. dissected JP 7336]

16.22 *Epinotia castaneana* (Walsingham) ♂ — Nevada, Elko Co., Bear Creek Summit, 8,480′ VIII.4.1981 (Powell) EME

16.23 *Epinotia bigemina* Heinrich ♀ — California, Los Angeles Co., N of Castaic, IV.17.1986 (J. Powell 86D102, r.f. *Arctostaphylos*, emgd. XI.1.1986) EME

16.24 *Epinotia arctostaphylana* (Kearfott) ♀ — California, Monterey Co., Big Creek Reserve, V.1/3.1992 (J. Powell 92E27, r.f. *Arctostaphylos hooveri*, emgd. XI.10.1992) EME

16.25 *Epinotia kasloana* McDunnough ♀ — California, Contra Costa Co., Walnut Creek, XI.10.1961 (Powell) EME

16.26 *Epinotia kasloana* McDunnough ♂ — California, Alameda Co., Berkeley, XI.18.1992 (Powell) EME

16.27 *Epinotia kasloana* McDunnough ♀ — California, Monterey Co., Big Creek Reserve, XI.18/20.1990 (Powell) EME

16.28 *Epinotia signiferana* Heinrich ♀ — California, Marin Co., Inverness Ridge, X.22.2000 (Powell) EME

16.29 *Epinotia meritana* Heinrich ♂ — California, Glenn Co., Plaskett Meadows, 6,200′ VII.3.1960 (Powell) EME

16.30 *Epinotia trossulana* (Walsingham) ♂ — California, Modoc Co., Lily Lake, VI.13.1970 (J. Powell 70F97, r.f. *Abies*, emgd. VII.7.1070) EME

16.31 *Epinotia nisella* (Clerck) ♀ — Nevada, White Pine Co., Lehman Creek, 7,500′ VII.13.1981 (De Benedictis & Powell) EME

16.32 *Epinotia nisella* (Clerck) ♂ — California, Humboldt Co., Orick, VIII.7.1976 (Powell) EME [abd. dissected, JAP 9077]

16.33 *Epinotia subplicana* (Walsingham) ♂ — Mexico, Baja California Norte, Sierra San Pedro Martir, 6,500_ VI.1.1958 (J. Powell 58F1, r.f. *Arctostaphylos*, emgd. VI.11.1958) EME

16.34 *Epinotia subplicana* (Walsingham) ♂ — California, Alameda Co., Cedar Mtn., V.14.1958 (Powell) EME

16.35 *Epinotia albangulana* (Walsingham) ♂— California, Ventura Co., Pine Mtn., V.25.1984 (J. Powell 84E97, r.f. *Alnus,* emgd. VI.4.1984) EME

16.36 *Epinotia miscana* (Kearfott) ♀ — California, Mendocino Co., Hopland Field Station, IV.29.1972 (J. Powell 72D53, r.f. *Arctostaphylos manzanita*, emgd. VI.3.1972) EME

16.37 *Epinotia terracoctana* (Walsingham) ♀ — California, Siskiyou Co., Mt. Shasta City, VII.14.1958, associated with *Arctostaphylos* (Powell) EME

16.38 *Epinotia nigralbana* (Walsingham) ♀ — California, Monterey Co., Big Creek Reserve, I.24/26.1988 (J. Powell 88A22,, r.f. *Arctostaphylos hooveri*, emgd. III.3.1988) EME

16.39 *Epinotia digitana* Heinrich ♂ — California, Monterey Co., Big Creek Reserve, II.2.1994 (J. Powell 94B3, r.f. *Heuchera pilosissima*, emgd. IV.7.1994) EME

16.40 *Epinotia lindana* (Fernald) ♀ — California, Trinity Co., Hayfork, V.23.73 (J. Powell 73E40, r.f. *Cornus glabrata*, emgd. IX.11.1973) EME

16.41 *Catastega plicana* R. Brown paratype ♀ — Mexico, Durango, W of Durango, 8,400′ VIII.3/8.1972 (Powell & Veirs) EME

16.42 *Ancylis comptana* (Frölich) ♂ — California, Mono Co., Mono Lake, VII.11.1968 (Opler & Powell) EME [abd. dissected JP 2365]

16.43 *Ancylis mediofasciana* (Clemens) ♂ — California, Los Angeles Co., Calamigos, III.16.1967 (Powell) EM

16.44 *Ancylis simuloides* McDunnough ♂ — California, Monterey Co., Big Creek Reserve, IV.27/29.1990 (Hsu & Powell, JP 90D91, r.f. *Ceanothus papillosus*, emgd. V.24.1990) EME

16.45 *Ancylis simuloides* McDunnough ♂ — California, Monterey Co., Big Creek Reserve, IV.13.1990 (Hsu & Powell, JP 90D25, r.f. *Ceanothus papillosus*, emgd. V.7.1990) EME

16.46 *Ancylis pacificana* (Walsingham) ♂ — California, Plumas Co., SE of Quincy, 3,500′ V.18.1982 (Powell) EME

16.47 *Hystrichophora leonana* Walsingham ♂ — California, Yolo Co., Cold Cyn. Reserve, VI.25.2007 (P. O_boyski) EME

16.48 *Hystrichophora stygiana* (Dyar) ♂ — California, Inyo Co., Wyman Canyon, 10,000′ VI.21.1961 (Powell) EME

16.49 *Hystrichophora roessleri* (Zeller) ♂ — California, Marin Co., SE of Corte Madera, VI.3.1964 (R. Langston) EME

16.50 *Dichrorampha simulana* (Clemens) ♂ — Nevada, White Pine Co., Timber Creek, 8,500′ VII.18.1996 (Powell & Sperling) EME

16.51 *Dichrorampha ?simulana* (Clemens) ♂ — Oregon, Lane Co., N of Roosevelt Beach, VI.22.1975 (Powell) EME

16.52 *Dichrorampha sedatana* (Busck) ♀ — Idaho, Latah Co., NE of Moscow, VI.29.1975 (Powell & W. Turner) EME [abd. dissected, EME 1632]

16.53 *Cydia populana* (Busck) ♂ — Nevada, White Pine Co., Lehman Canyon 7,500′ VII.25.1981 (De Benedictis & Powell) EME

16.54 *Ecdytolopha occidentana* Adamski & J. Brown (paratype) ♀ — Arizona, Cochise Co., Cave Creek Ranch, VIII.2.1974 (Powell & Szerlip) EM

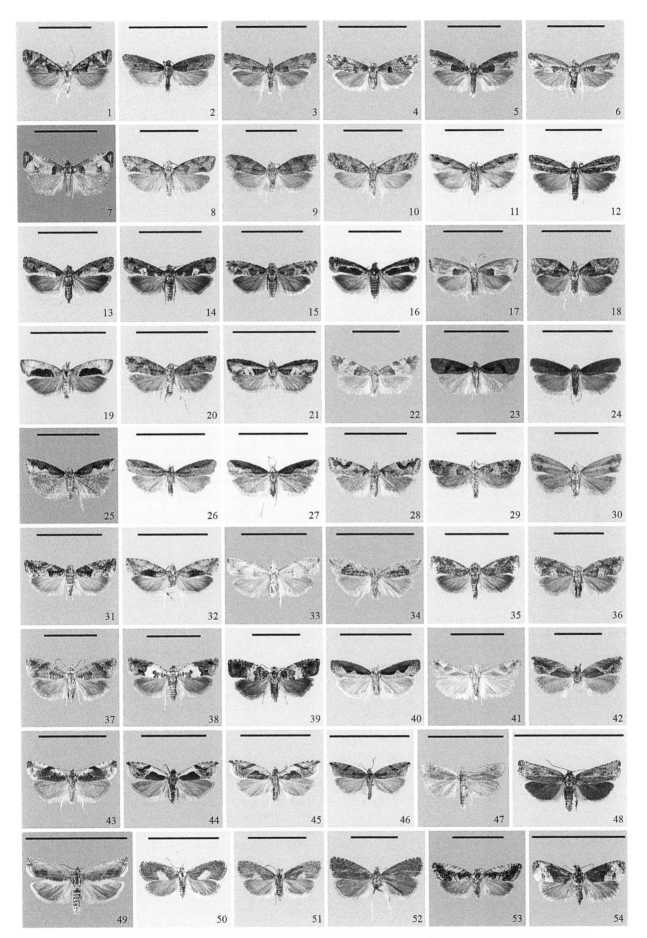

PLATE 16

PLATE 17 Tortricidae, Olethreutinae, Tortricinae

17.1 *Grapholita molesta* (Busck) ♂ — California, Orange Co., Anaheim, IV.26.1944 (R. Dickson) EME

17.2 *Grapholita caeruleana* Walsingham ♂ — California, Tuolumne Co/. NE of Tuolumne City, III.23.1966 (Powell) EME

17.3 *Grapholita conversana* Walsingham ♂ — California, El Dorado Co., Cool, III.23.1965 (Powell) EME

17.4 *Grapholita vitrana* Walsingham ♂ — California, Monterey Co., Big Creek Reserve, IV.11/13.1985 (Powell) EME

17.5 *Grapholita edwardsiana* (Kearfott) ♀ — California, San Francisco Co., Baker Beach, III.18.1977 (J. Powell 77C11, r.f. *Lupinus arboreus*, emgd. IV.14.1977 (EME)

17.6 *Grapholita lunatana* Walsingham ♂ — California, San Francisco Co., Twin Peaks, II.16.1961 (Powell) EME

17.7 *Cydia latiferreana* Walsingham ♂ — California, Contra Costa Co., Walnut Creek, VI.18.1962 (Powell) EME

17.8 *Cydia latiferreana* Walsingham ♂ — California, Contra Costa Co., Pleasant Hill, VII.13.1957 (W. Ferguson, JP 57G5, r.f. *Quercus lobata* gall, emgd. VII.19.1957) EME

17.9 *Cydia deshaisiana* (Lucas) ♂ — Mexico. VIII.1961 (J. Powell 61H11, r.f. purchased Mexican jumping bean [*Euphorbia* seed], emgd. I.00.1962) EME

17.10 *Cydia pomonella* (L.) ♂ — California, Contra Costa Co., Pinole Valley, X.14.2002 (C. Barr, JP 02K3, r.f. cocoon under *Crataegus*, emgd. V.12,2003) EME

17.11 *Cydia piperana* Kearfott ♂ — California Alameda Co., Berkeley, IV.6.1992 (Powell) EME

17.12 *Cydia injectiva* (Heinrich) ♀ — California, San Diego Co., Green Valley, II.14.1957 (H. Ruckes, jr.) EME

17.13 *Cydia cupressana* Kearfott ♀ — California, Marin Co., Carson Ridge, III.15.1968 (J. Powell 68C30, r.f. pitch nodule of cypress cone, emgd. V.16.1968) EME

17.14 *Cydia nigricana* (Stephens) ♂ — Pennsylvania, Allegheny Co., Oak Station, V.24.1914 (F. Marloff) EME

17.15 *Cydia prosperana* (Kearfott) ♂ — California, Monterey Co., Big Creek Reserve, V.3/5.1991 (Hsu & Powell) EME

17.16 *Cydia americana* (Walsingham) ♂ — California, Monterey Co., Big Creek Reserve, IV.10.1989 (Powell) EME

17.17 *Cydia membrosa* (Heinrich) ♀ — Mexico, Baja California Sur, W of La Paz, VIII.13.1966 J. Chemsak, J. Doyen & Powell) EME

17.18 *Ofatulena duodecemstriata* (Walsingham) ♀ — Mexico, Baja California Sur, W of La Paz, VIII.8.1966 (Powell) EME

17.19 *Ofatulena luminosa* Heinrich ♂ — Mexico, Baja California Sur, S of La Paz, VIII.12.1966 (Chemsak, Doyen, & Powell)

17.20 *Acleris hastiana* (L.) ♂ — Oregon, Lane Co., SW of Florence, VI.24.1975 (Opler & Powell, JP 75F26, r.f. *Salix hookeriana*, emgd. VIII.14.1975) EME

17.21 *Acleris hastiana* (L.) ♂ — Oregon, Lane Co., SW of Florence, VI.24.1975 (Opler & Powell, JP 75F26, r.f. *Salix hookeriana*, emgd. VIII.2.1975) EME

17.22 *Acleris hastiana* (L.) ♀ — Oregon, Lane Co., SW of Florence, VI.24.1975 (Opler & Powell, JP 75F26, r.f. *Salix hookeriana*, emgd. VII.31.1975) EME

17.23 *Acleris variegana* (Schiffermüller) ♂ — California, Alameda Co., San Lorenzo, VIII.6.1960 (J. Powell 60H1, r.f. *Pyracantha*, emgd. IX.2.1960) EME

17.24 *Acleris variegana* (Schiffermüller) ♂ — California, Alameda Co., San Lorenzo, VIII.19.1960 (J. Powell 60H4, r.f. *Prunus avium*, emgd. IX.4.1960) EME

17.25 *Acleris variegana* (Schiffermüller) ♂ — California, Alameda Co., San Lorenzo, IX.11.1960 (P. Hurd jr., JP 60J-, r.f. *Prunus avium*, emgd. IX.29.1960) EME

17.26 *Acleris britannia* Kearfott ♂ — California, Humboldt Co., Kneeland, VIII.31.2002 (R. Wielgus) EME

17.27 *Acleris britannia* Kearfott ♀ — California, Humboldt Co., Kneeland, IX.6.2002 (R. Wielgus) EME

17.28 *Acleris keiferi* Powell ♀ — California, Monterey Co., Big Creek Reserve, V.3/5.1991 (Hsu & Powell, JP 91E2, r.f. *Rubus ursinus*, emgd. VI.12.1991) EME

17.29 *Acleris keiferi* Powell ♀ — California, Monterey Co., Big Creek Reserve, IV.27/29.1990 (Hsu & Powell, JP 9D43, r.f. *Rubus ursinus*, emgd. VI.8.1990) EME

17.30 *Acleris cervinana* (Fernald) ♀ — California, Humboldt Co., Kneeland, X.7.2002 (R. Wielgus) EME

17.31 *Acleris cervinana* (Fernald) ♀ — California, Humboldt Co., Kneeland, X.18.2002 (R. Wielgus) EME

17.32 *Acleris santacrucis* Obraztsov ♂ — California, Marin Co., Inverness, X.9/16.2000 (Powell) EME

17.33 *Acleris santacrucis* Obraztsov ♀ — California, Marin Co., Inverness, IX.10.1998 (W. Patterson) EME [abd. dissected, JP 7897]

17.34 *Acleris gloverana* (Walsingham) ♂ — California, El Dorado Co., Blodgett Forest, VI.25.1981 (J. De Benedictis, JP 81F118, r.f. *Abies concolor*, emgd. VIII.9.1981) EME

17.35 *Acleris gloverana* (Walsingham) ♂ — California, Sonoma Co., VI.6.1979 (J. Powell 79F3, r.f. *Pseudotsuga menziesii*, emgd. VI.24.1979) EME

17.36 *Acleris gloverana* (Walsingham) ♂ — California, Placer Co., Kaspian Recr. Area, VII.12.1979 (J. Powell 79G102, r.f. *Abies magnifica*, emgd. VIII.10.1979) EME

17.37 *Acleris gloverana* (Walsingham) ♂ — California, Modoc Co., E of Adin, 5,800',VII.11980 (K. Sheehan, JP 80G19, r.f. *Abies concolor*, emgd. IX.2.1980) EME

17.38 *Acleris foliana* (Walsingham) ♀ — California, Santa Barbara Co., Santa Cruz Island, VI.8.1966 (Powell) EME

17.39 *Acleris foliana* (Walsingham) ♂ — California, Santa Barbara Co., Santa Cruz Island, VI.7.1966 (Powell) EME

17.40 *Acleris foliana* (Walsingham) ♀ — Colorado, Larimer Co., W of Ft. Collins, VI.29.1979 (D. Wagner, r.f. *Cercocarpus montanus*, emgd. VII.20.1979) EME

17.41 *Dorithia semicirculana* (Fernald) ♀ — New Mexico, Bernalillo Co., E of Cedar Crest, 7,000' VII.30.1989 (Hsu, Powell, & M. Prentice) EME

17.42 *Dorithia trigona* Brown & Obraztsov ♂ — Arizona, Cochise Co., Miller Canyon, Huachuca Mtns., 5,800', IV.12.1988 (J. Powell 88D31, r.f. egg,, on *Quercus*, emgd. VI.4.1988) EME

17.43 *Acleris senescens* (Zeller) ♀ — California, Marin Co., Inverness Park, I.25/31.1995 (Powell) EME

17.44 *Acleris senescens* (Zeller) ♀ — California, Marin Co., Inverness Park, I.25/31.1995 (Powell) EME

17.45 *Acleris senescens* (Zeller) ♂ — California, Contra Costa Co., Brooks Island, III.21.1996 (J. Powell 96C33, r.f. *Salix lasiolepis*, emgd. IX.20.1996) EME

17.46 *Acleris nigrolinea* (Robinson) ♂ — Wisconsin, Oneida Co., Lake Katherine, IV.20.1962 (H. M. Bower) EME

17.47 *Acleris maximana* (Barnes & Busck) ♀ — California, Plumas Co., E of Chester, VII.19.1995 (L. Crabtree, r.f. *Prunus emarginata*, emgd. IX.21.1995) EME

17.48 *Croesia albicomana* (Clemens) ♀ — California, Trinity Co., Hayfork, V.20.1973 (J. Powell 73E24, r.f. *Rosa gymnocarpa*, emgd. VI.11.1973) EME

17.49 *Eana argentana* (Clerck) ♀ — California, Sierra Co., E. of Bassetts, VI.29.2001 (Powell) EME

17.50 *Eana georgiella* (Hulst) ♀ — Inyo Co., Ruby Lake, 11,250' VIII.13.1957 (Powell) EME

17.51 *Cnephasia longana* (Haworth) ♂ — California, Humboldt Co., Shelter Cove, V.21.1976 (J. Powell 76E22, r.f. *Erigeron glauca*, emgd. VI.16.1976) EME

17.52 *Cnephasia longana* (Haworth) ♀ — California, Monterey Co., Marina, V.28.1995 (Powell) EME

17.53 *Apotomops wellingtoniana* (Kearfott) ♂ — Oregon, Jackson Co., Mt. Ashland, 6,000', VI.23/24.2000 (Powell) EME

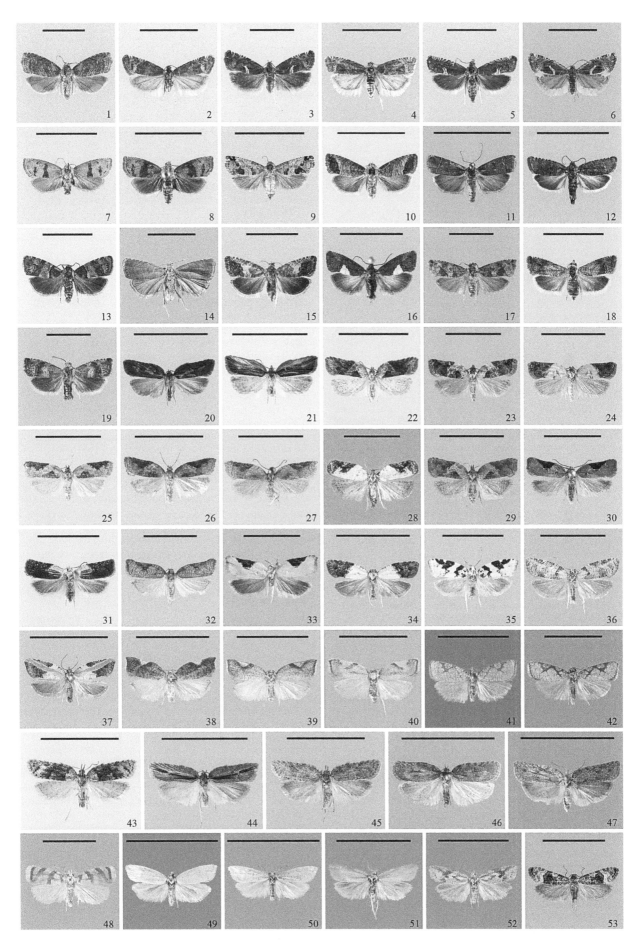

PLATE 17

PLATE 18 Tortricidae, Tortricinae

18.1 *Decodes horarianus* (Walsingham) ♂ — California, Modoc Co., NW of Ft. Bidwell, VI.12.1970 (J. Powell 70F90, r.f. *Ribes*, emgd. after VII.1970) EME

18.2 *Decodes helix* Powell & J. Brown, holotype ♂ — California, Santa Barbara Co., Santa Rosa Island, Torrey Pines area, IV.28.1995 (J. Powell 95D93, r.f. *Arctostaphylos confertifolia*, emgd. I.6.1996) [abd. dissected JAP 7262] EME

18.3 *Decodes aneuretus* Powell ♂ — California, Marin Co., Inverness Ridge, 900′, II.10.2000 (Powell) EME

18.4 *Decodes bicolor* Powell ♀ — California, Napa Co., Pope Creek, IV.6.1957 (Powell), paratype EME

18.5 *Decodes tahoense* Powell ♂ — California, Plumas Co., Meadow Valley, 4,100′ IX.9/16.1983 (De Benedictis & Powell) EME

18.6 *Decodes basiplaganus* (Walsingham) ♀ — California, Monterey Co., Big Creek Reserve, 2,000′, VIII.25.1989 (Powell) EME

18.7 *Decodes fragarianus* (Busck) ♀ — California, Alameda Co., Del Valle Reservoir, V.15.1975 (J. Powell 75E16, r.f. *Quercus agrifolia*, emgd. IX.19.1975) EME

18.8 *Anopina triangulana* (Kearfott) ♂ — California, Alameda Co., Berkeley, VI.6/11.1984 (J. Powell 84F18, r.f. egg, on synthetic diet, emgd. VIII.16.1984) EME

18.9 *Anopina triangulana* (Kearfott) ♀ — California, Alameda Co., Berkeley, VII.11.1982 (Powell) EME

18.10 *Anopina eleonora* Obtraztsov ♀ — Arizona, Cochise Co., Miller Canyon, Huachuca Mtns. IV.16.1986 (Powell) EME [abd. dissected, EME 3913]

18.11 *Anopina internacionana* Brown & Powell ♂ — Mexico, Durango, E. of El Salto, 8,000_ VII.22.1964 (J. Chemsak & Powell), paratype EME

18.12 *Acroplectis haemanthes* Meyrick ♀ — Arizona, Cochise Co., Cave Creek SW of Portal, VII.24/291972 (Powell) EME [abd. dissected, EME 5450]

18.13 *Eulia ministrana* (L.) ♂ — North Carolina, Swain Co., NW of Clingman_s Dome, VI.10.2002 (Powell & Rubinoff) EME

18.14 *Argyrotaenia coloradana* (Fernald) ♂ — Utah, Sanpete Co., Ephraim Canyon, 8,800′ VII.19.1996 (Powell & Sperling) EME

18.15 *Argyrotaenia niscana* (Kearfott) ♀ — California, Los Angeles Co., Bouquet Canyon, V.30.1959 (J. Powell 59E5, r.f. *Adenostoma fasciculatum*, emgd. VI.3.1959) EME

18.16 *Argyrotaenia franciscana* (Walsingham) [typical race] ♂ — California, Sonoma Co., Bodega, IV.18.1992 (Powell) EME

18.17 *Argyrotaenia franciscana* [race *insulana* Powell] ♀ — California, Santa Barbara Co., San Miguel Island, X.15.1995 (J. Powell 95M6, F2, r.f egg, on synthetic diet, emgd. III.16/31.1996) EME

18.18 *Argyrotaenia citrana* (Fernald) [typical race] ♂ — California, Santa Barbara Co., San Marcos Pass, II.4.1987 (J. Powell 87B86, r.f *Ribes amarum*, emgd. II.23.1987) EME

18.19 *Argyrotaenia citrana* (Fernald) ♀ — California, Contra Costa Co., Tilden Park, III.14.1983 (J. De Benedictis 83073-A, r.f *Dirca occidentalis*, emgd. IV.2.1983) EME

18.20 *Argyrotaenia isolatissima* Powell ♂ — California, Santa Barbara Co., Santa Barbara Island, V.28.1939 (L. Martin), paratype LACM

18.21 *Argyrotaenia isolatissima* Powell ♂ — California, Santa Barbara Co., Santa Barbara Island, V.22/28.2001 (J. Powell 01E31, r.f. egg,, on synthetic diet, emgd. VII.15.2001) EME

18.22 *Argyrotaenia franciscana* (Walsingham) San Clemente I. race] ♀ — California, Los Angeles Co., San Clemente Island, III.20.2004 (De Benedictis & Powell, JP 04L3, F4 r.f. egg, on synthetic diet, emgd. I.30.2005) EME

18.23 *Argyrotaenia franciscana* (Walsingham) [San Nicolas I. race] ♀ — California, Ventura Co., San Nicolas Island, IV.7.2005 (De Benedictis & Powell, JP 05D60, r.f. egg, on synthetic diet, emgd. VI.18.2005) EME

18.24 *Argyrotaenia provana* (Kearfott) ♂ — California, Modoc Co., E of Adin, 5,800′ VII.16.1980 (K. Sheehan, JP 80G19, r.f. *Abies concolor*, emgd. VIII.2.1980) EME

18.25 *Argyrotaenia dorsalana* (Dyar) ♀ — Oregon, Jackson Co., Robinson Butte, VII.5.1970 (Dietz & Rude, JP 70G13, r.f. *Abies magnifica*, emgd. VIII.5.1970) EME

18.26 *Argyrotaenia dorsalana* (Dyar) ♂ — California, Modoc Co., Fandango Peak, VII.12.1974 (USFS coll., JP r. f. *Abies*) EME

18.27 *Argyrotaenia dorsalana* (Dyar)(rust form) ♂ — Utah, Sanpete Co., Ephraim Canyon, 8,800′ VII.19.1996 (Powell & Sperling) EME

18.28 *Argyrotaenia lautana* Powell ♂ — California, Kern Co., Tehachapi Mtn. Park, VI.18.1989 (J. Powell 89F51, r.f. *Abies concolor*, emgd. VII.2.1989) EME

18.29 *Aphelia alleniana* (Fernald) ♂ — Wisconsin, Oneida Co., Lake Katherine, VII.28.1961 (H. Bower) EME

18.30 *Archips argyrospilus* (Walker) ♂ — California, Stanislaus Co., Modesto, IV.24.1973 (J. Powell 73D21, r.f. *Quercus coccinea*, emgd. V.21.1973) EME

18.31 *Archips rosanus* (L.) ♂ — Oregon, Yamhill Co., Williamson St. Park, VI.25.1975 (Opler & Powell, JP 75F29, r.f. *Rubus parviflorus*, emgd. VII.8.1975) EME

18.32 *Archips cerasivoranus* (Fitch) ♂ — California, Siskiyou Co., E of McCloud, VI.21.1958 (J. Powell 58F4, r.f. *Prunus subcordata*, emgd. VII.8.1958) EME

18.33 *Archips negundanus* (Dyar) ♂ — Montana, Missoula Co., Missoula, VI.23.1987, (S. J. Gast, host: *Acer negundo*) EME

18.34 *Archips packardianus* (Fernald) ♂ — Minnesota, Clay Co., Moorhead, VI.6.1977 (J. Powers) EME

18.35 *Synemis afflictana* (Walker) ♂ — Canada, Alberta, SW of Sherwood Peak, VI.7.1996 (G. Pohl) NFRC

18.36 *Diedra intermontana* Rubinoff & Powell ♀ — Nevada, Washoe Co., Reno, VI.1.1966 (R. Bechtel, r.f. Pfitzler juniper) paratype EME

18.37 *Diedra calocedrana* Rubinoff & Powell, holotype ♂ — California, El Dorado Co., Blodgett Forest, VII.8.1981 (Powell) EME

18.38 *Choristoneura retiniana* (Walsingham) ♂ — California, Kern Co., Tehachapi Mtn. Park, VI.16.1981 (De Benedictis & Powell, JP 81F101.1, r.f. *Abies concolor*, emgd. VII.7.1981) EME

18.39 *Choristoneura retiniana* (Walsingham)(blond form) ♂ — California, Kern Co., Tehachapi Mtn. Park, VI.16.1981 (De Benedictis & Powell, JP 81F101, r.f. *Abies concolor*, emgd. VII.5.1981) EME

18.40 *Choristoneura lambertiana subretiniana* Obraztsov ♂ — California, Nevada Co., Sagehen Creek, 6,000′ VII.10.1980 (De Benedictis & Meredith, JP 80G13, r.f. *Pinus contorta*, emgd. VII.1.1980) EME

18.41 *Choristoneura lambertiana ponderosana* Obraztsov ♂ — New Mexico, Santa Fe Co., Los Alamos, VI.20.1981 (A. Liebhold, r.f. *Pinus ponderosa*, emgd. VII.8.1981) EME

18.42 *Archips argyrospilus* (Walker) ♀ — California, Lassen Co., Eagle Lake, VI.4.1994 (L. Crabtree, r.f. *Prunus virginiana*) EME

18.43 *Choristoneura rosaceana* (Harris) ♂ — California, Humboldt Co., Kneeland, VIII.7.2002 (R. Wielgus) EME

18.44 *Choristoneura rosaceana* (Harris) ♀ — California, Humboldt Co., Kneeland, VIII.23.2002 (R. Wielgus) EME

18.45 *Choristoneura conflictana* (Walker) ♂ — Arizona, Coconino Co., Little Spring, 8,300′ VII.15.1995 (Powell & Sperling) EME

18.46 *Choristoneura conflictana* (Walker) ♀ — Arizona, Coconino Co., Ski Bowl, S. F. Mtns., 9,200′—VII.15.1995 (J. Powell 95G21, r.f. *Populus tremuloides*, emgd. VII.23.1995) EME

18.47 *Choristoneura carnana* (Barnes & Busck)(typical race) ♂ — California, San Bernardino Co., Mt. Baldy, 6,000′ VI.17.1980 (De Benedictis & Meredith, JP 80F29, r.f. *Pseudotsuga macrocarpa*, emgd. VII.6.1980) EME

18.48 *Choristoneura carnana* (Barnes & Busck)(north coast race, *californica* Powell) ♂ — California, Lake Co., Cobb, VI.5.1980 (De Benedictis & Powell, JP 80F12, r.f. *Pseudotsuga menziesii*, emgd. VII.1.1980) EME

18.49 *Choristoneura occidentalis* Freeman ♂ — Washington, Okanogan Co., Conconully, VII.3.1988 (J. Powell 88G15, r.f. *Pseudotsuga menziesii*, emgd. VII.15.1980) EME

18.50 *Choristoneura occidentalis* Freeman ♂ — Idaho-Montana 1964, twenty-fifth-generation lab colony at Berkeley, California, USFS (R. Lyon) EME

18.51 *Choristoneura occidentalis* Freeman ♀ — Idaho-Montana 1964, twenty-fifth-generation lab colony at Berkeley, California, USFS (R. Lyon) EME

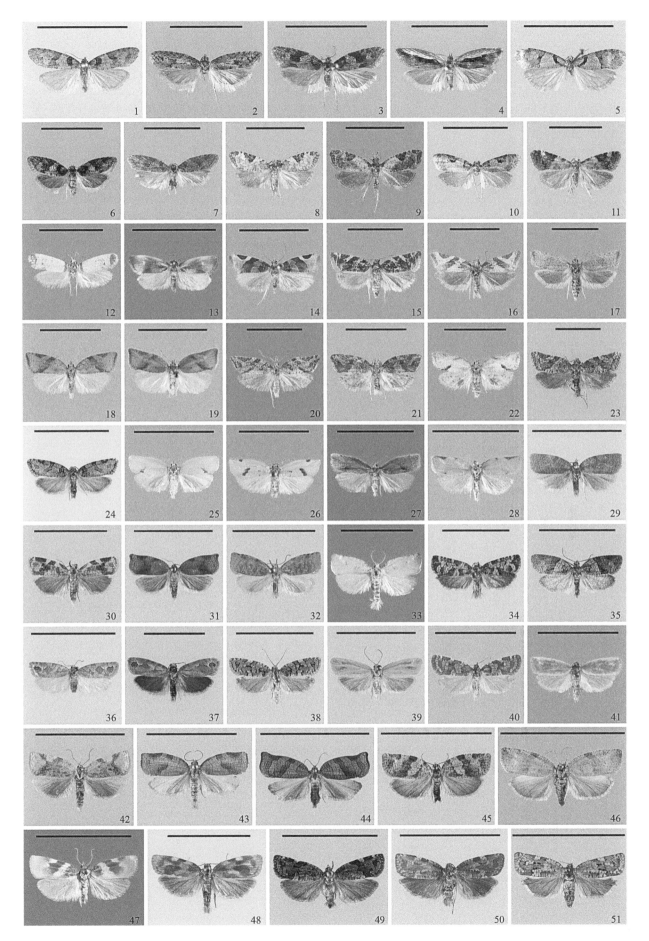

PLATE 19 Tortricidae, Tortricinae, Chilidanotinae

19.1 *Clepsis peritana* (Clemens) ♂ — California, Contra Costa Co., Pleasant Hill, IX.15.1958 (W. Ferguson) EME

19.2 *Clepsis virescana* (Clemens) ♂ — California, Mono Co., E of Monitor Pass, VI.30.1962 (Powell) EME

19.3 *Clepsis penetralis* Razowski ♂ — Utah, Garfield Co., W of Bryce, VI.28.1992 (Powell) EME

19.4 *Clepsis fucana* (Walsingham ♂ — California, Monterey Co., Big Creek Reserve, IV. 23.1987 (J. Powell 87D31, r.f. *Stachys*, emgd. V.16.1987) EME

19.5 *Clepsis persicana forbesi* Obraztsov ♀ — California, Siskiyou Co., McBride Springs, 5,200′ VI.10.1974 (J. Powell 74F15, r.f. *Abies concolor*, emgd. VI.28.1974) EME

19.6 *Clepsis clemensiana* (Fernald) ♂ — Washington, Okanogan Co., Brewster, IV.21.1953 (J. Hopfinger) EME

19.7 *Ditula angustiorana* (Haworth) ♀ — California, Humboldt Co., Arcata, VII.26.1969, on *Taxus* (Powell) EME

19.8 *Epiphyas postvittana* (Walker) ♂ — California, San Francisco— (J. Powell 07F6, r.f. *egg*, on synthetic diet, emgd. VIII.15.2007) EME

19.9 *Epiphyas postvittana* (Walker) ♂ — California, San Francisco, (J. Powell 07H4, reared from egg, on synthetic diet, emgd. IX.30.2007) EME

19.10 *Epiphyas postvittana* (Walker) ♀ — California, San Francisco (J. Powell 07F60, reared from egg, on synthetic diet, emgd. VIII.15/ 18.2007 EME

19.11 *Epiphyas postvittana* (Walker) ♀ — California, San Francisco (J. Powell 07F60, reared from egg on synthetic diet, emgd. VIII.15/ 18.2007 EME

19.12 *Cenopis directana* (Walker) ♂ — California, Siskiyou Co., E of Mc-Cloud, VII.7.1957 (J. Powell 57G2.1, r.f. *Prunus*, emgd. VIII.15.1957) EME

19.13 *Cudonigera houstonana* (Grote) ♀ — New Mexico, Hidalgo Co., Guadalupe Canyon, VIII.30.1971 (J. Doyen) EME

19.14 *Pandemis canadana* Kearfott ♂ — Canada, Alberta, Bilby, VII.12.1924 (O. Bryant) EME

19.15 *Pandemis limitata* (Robinson) ♀ — California, Humboldt Co., Kneeland, VII.15.2003 (R. Wielgus) EME

19.16 *Pandemis pyrusana* Kearfott ♂ — California, Alameda Co., San Lorenzo, VIII.19.1960 (J. Powell 60H4, r.f. *Punus avium*, emgd. VIII.26. 1960) EME

19.17 *Xenotemna pallorana* (Robinson) ♀ — California, Humboldt Co., Samoa dunes, VI.26.1969 (Powell) EME

19.18 *Amorbia cuneana* (Walsingham) ♂ — California, Alameda Co., Berkeley, V.16.1988 (Powell) EME

19.19 *Amorbia cuneana* (Walsingham) ♂ — California, Alameda Co., Berkeley, V.12.1990 (Powell) EME

19.20 *Amorbia cuneana* (Walsingham) ♂ — California, Alameda Co., Berkeley, VII.1991 (D. Rubinoff) EME

19.21 *Amorbia cuneana* (Walsingham) ♀ — California, Siskiyou Co., Mt. Shasta City, VII.5.1958 (J. Powell 58G4, r.f. *Arctostaphylos*, emgd. VIII.5.1958) EME

19.22 *Amorbia cuneana* (Walsingham) ♀ — California, Alameda Co., Huckleberry Preserve, IX.30.1984 (D. Wagner, JP 84K2, r.f. egg, on synthetic diet, emgd. XII.15.1984) EME

19.23 *Amorbia synneurana* Barnes & Busck ♀ — Arizona, Pima Co., Brown Canyon, Baboquivari Mtns., VIII.8.2005 (D. & J. Powell, JP 05J25, r.f. egg, on synthetic diet, emgd. XII.25.05) EME

19.24 *Sparganothis vocaridorsana* Kearfott ♂ — Oregon, Wallowa Co., Wallowa Lake, VII.7.1950 (N. Crickmer) EME

19.25 *Sparganothis unifasciana* (Clemens) ♂ — Arizona, Coconino Co., Hochderffer Hill, 7,700″ VII.24.1989 (J. Powell 89G6, r.f. egg, on synthetic diet, emgd. XII.10.1989) EME

19.26 *Platynota wenzelana* (Haimbach) ♂ — Arizona, Cochise Co., Huachuca Mtns., Carr Canyon 5,500′ VII.31.1986 (D. Wagner, JP 86H2, r.f. egg on *Amorpha*, emgd. X.2.1986) EME

19.27 *Platynota wenzelana* (Hainbach) ♀ — Arizona, Cochise Co., Huachuca Mtns., Carr Canyon 5,500′ VII.31.1986 (D. Wagner, JP 86H2, r.f. egg on *Amorpha*, emgd. IX.30.1986) EME

19.28. *Amorbia synneurana* Barnes & Busck ♂ — Arizona, Pima Co., Brown Canyon, Baboquivari Mtns., VIII.8.2005 (D. & J. Powell, JP 05M1, F2 r.f. egg, on synthetic diet, emgd. III.17.06) EME

19.29 *Sparganothis senecionana* (Walsingham) [typical race] ♂ — California, Mendocino Co., SE of Yorkville, V.18.1966 (Powell) EME

19.30 *Sparganothis senecionana* (Walsingham) [Central Coast Interior race] ♂ — California, Alameda Co., Arroyo Mocho, V.24.1957 (Powell) EME

19.31 *Sparganothis senecionana* (Walsingham) ♂ — California, San Mateo Co., San Bruno Mtn., V.3.1963 (Powell) EME

19.32 *Sparganothis senecionana* (Walsingham) ♂ — California, San Mateo Co., Highway 92 at 280, IV.4.1990 (Powell) EME

19.33 *Sparganothis senecionana* (Walsingham) ♂ — California, Monterey Co., Big Creek Reserve, VI.5/8.1989 (Hsu & Powell, JP 89F4, r.f. egg, on synthetic diet, emgd. II.2.1990) EME

19.34 *Sparganothis senecionana* (Walsingham) [coastal dunes race] ♂ — California, San Luis Obispo Co., Morro Bay, V.3.1974 (Powell) EME

19.35 *Sparganothis senecionana* (Walsingham) ♂ — California, Los Angeles Co., Santa Catalina Island, V.3.1978 (Powell) EME

19.36 *Sparganothis senecionana* (Walsingham) [Sierran race] ♂ — California, Sierra Co., E of Downieville, VI.3.1984 (De Benedictis & Powell, JP 84F8, r.f. *Hypericum*, emgd. VII.11.1984) EME

19.37 *Sparganothis senecionana* (Walsingham) [high Sierran race] ♂ — California, Mono Co., Saddlebag Lake, 9,500′ VIII.25.1983 (De Benedictis & Powell) EME

19.38 *Sparganothis tunicana* (Walsingham) ♀ — California, Modoc Co., NE of Adin, VI.12.1974 (J. Powell 74F24, r.f. *Balsamorhiza saggitata*, emgd. VII.5.1974) EME

19.39 *Sparganothoides machimiana* (Barnes & Busck) ♀ — Arizona, Cochise Co., Huachuca Mtns., VIII.2.1989 (J. Powell 89H5.2 r.f. egg, on synthetic diet, emgd. XII.10.1989) EME

19.40 *Sparganothoides hydeana* (Klots) ♂ — Mexico, Durango, S of El Salto, 8,000′ VIII.10.1986 (J. Brown & Powell, JP 86H23.5, r.f. egg, on synthetic diet, emgd. XII.7.1986) EME

19.41 *Platynota labiosana* (Zeller) ♂ — Arizona, Cochise Co., Huachuca Mtns., IV.15.1986 (J. Powell 86D55, r.f. egg, on synthetic diet, emgd. VII.14.1986) EME

19.42 *Platynota stultana* Walsingham ♂ — California, Contra Costa Co., Pt. Molate, IX.15.1987 (J. Powell 87J8, r.f. egg, on synthetic diet, emgd. XI.18.1987) EME

19.43 *Platynota stultana* Walsingham ♀ — California, Contra Costa Co., Antioch Natl. Wildlife Refuge, IV.2.1991 (J. Powell 91D8, r.f. *Melilotus*, emgd. IV.18.1991) EME

19.44 *Platynota larreana* (Comstock) ♂ — California, Riverside Co., Joshua Tree Natl. Mon., III.31.1970 (J. Powell 70C15, r.f. *Larrea*, emgd. IV.13.1970) EME

19.45. *Synnoma lynosyrana*—Walsingham ♂ — Nevada, Nye Co., W of Clark Station, VII.20.1968 (J. Powell 68G33, r.f. *Chrysothamnus viscidiflorus*, emgd. IX.18.1968) EME

19.46 *Synnoma lynosyrana*—Walsingham ♀ — California, Siskiyou Co., NE of Weed, VIII.23.1958 (J. Powell 58H3-4, r.f. *Chrysothamnus*, emgd. IX.23.1958) EME

19.47 *Synalocha gutierreziae* Powell ♂ — New Mexico, Grant Co., Gold Gulch Rd., VIII.9.1999 (D. Wright) USNM

19.48 *Henricus umbrabasanus* (Kearfott) ♂ — California, Alameda Co., Berkeley, IX.7.1990 (Powell) EME

19.49 *Henricus macrocarpanus* (Walsingham) ♂ — California, San Mateo Co., San Bruno Mtn., I.10.1987 (J. De Benedictis, 87A4.2, r.f. *Cupressus macrocarpa*, emgd. IV.8.1987) EME

19.50 *Henricus fuscodorsanus* (Kearfott) ♂ — California, Monterey Co., Big Creek Reserve, V.2.1992 (S. O_Keefe) EME

19.51 *Henricus infernalis* (Heinrich) ♀ — Texas, Culberson Co., Sierra Diablo, 6,000′ V.29.1973 (R. Hodges) USNM

50% scale

PLATE 19

20.1 Aethdeut *Aethes deutschiana* (Zetterstedt) ♂ — Utah, Wasatch Co., Ephraim Plateau, 10,000', VII.22.1980 (D. Ferguson) USNM

20.2 *Aethes smeathmanniana* (Fabricius) ♂ — South Dakota, Spearfish Co., T3N, R1C, S6, VII.16.1965 (R. Hodges) USNM

20.3 *Cochylis hospes* (Walsingham) ♂ — Utah, Washington Co., Springdale, IX.12.1996 (Powell) EME

20.4 *Cochylis carmelana* (Kearfott) ♀ — California, Monterey Co., Big Creek Reserve, V.4.1991 (Hsu & Powell) EME

20.5 *"Cochylis" pimana* (Busck) ♀ — Arizona, Santa Cruz Co., Madera Canyon, IX.3.1969 (J. Powell 69J18, r.f. *Yucca schottii* pod, emgd. VII.15.1970) EME

20.6 *"Cochylis" yuccatana* (Busck) ♀ — Mexico, Baja California Norte, El Rosario, III.26.1973 (J. Powell 73C4, r.f. *Agave orcuttiana*, emgd. IV.24.1973)(abd. dissected, EME 6006)

20.7 *Platphalonidia felix* (Walsingham) ♀ — California, San Luis Obispo Co., S of Creston, IV.13.1967 (J. Powell 67D45, r.f. *Senecio blochmaniae*, emgd. V.4.1967) EME

20.8 *Platphalonidia plicana* (Walsingham) ♂ — California, San Diego Co., Mason Valley, V.20.1939 (L. Martin) LACM

20.9 *Saphenista latipunctana* (Walsingham) ♂ — California, San Mateo Co., San Bruno Mtn., VI.10.1985 (J. De Benedictis, JP 85F5, r.f. *Eriophyllum stachaedifolium*, emgd. VIII.2.1985) EME

20.10 *Saphenista saxicolana* (Kearfott) ♂ — California, Marin Co., Inverness Ridge, V.7.2006 (J. Powell 06D15, r.f. *Ceanothus thyrsiflorus*, emgd. VI.20.2006) EME

20.11 *Saphenista nomonana* (Kearfott) ♂ — California, San Mateo Co., San Bruno Mtn., IV.27.1986 (De Benedictis & Wagner) EME

20.12 *Saphenista nomonana* (Kearfott) ♂ — California, Marin Co., Inverness Ridge, V.15.1970 (Powell) [abd. dissected EME 2770]

20.13 *Saphenista campicolana* (Walsingham) ♀ — California, San Mateo Co., San Bruno Mtn., V.19.1988 (De Benedictis) EME

20.14 *Eugnosta mexicana* (Busck) ♂ — Colorado, San Miguel Co., Telluride, 8,750' VI.30.1977 (D. Ferguson) USNM

20.15 *Rudenia leguminana* (Busck) ♀ — California, San Bernardino Co., S of Saratoga Springs, III.18.1978 (P. Rude) EME

20.16 *Phtheochroa aegrana* (Walsingham) ♂ — Washington, Klickitat Co., Brooks Memorial Park, V.30.1949 (E. C. Johnston) CNC

20.17 *Phtheochroa aegrana* (Walsingham) ♂ — California, San Mateo Co., San Bruno Mtn., VII.16.1988 (De Benedictis) EME

20.18 *Phtheochroa aureoalbida* (Walsingham) ♂ — Arizona, Coconino Co., Hart Paririe, 8,500' VII.1.1961 (R. Hodges) USNM

20.19 *Phtheochroa fulviplicana* (Walsingham) ♂ — Arizona, Coconino Co., Hart Paririe, 8,500' VII.1.1961 (J. Franclemont) USNM

20.20 *Phtheochroa vulneratana* (Zetterstedt) ♂ — Colorado, Summit Co., Loveland Pass, 12,000' VIII.9.1973 (Powell) EME

20.21 *Tinacrucis noroesta* Powell ♂—Mexico, Durango, W of El Salto, VII.5.1964 (J. Martin) CNC

20.22 *Tinacrucis noroesta* Powell ♀ — Mexico, Durango, W of El Salto, VII.1.1964 (J. Martin) CNC

20.23 *Thaumatographa regalis* (Walsingham) ♂ — California, Kern Co., Kernville, IV.28.1964 (Powell) EME

20.24 *Megalopyge bissesa* Dyar ♂ — Arizona, Cochise Co., Douglas, VI.1.1933 (W. Jones) CSCA

20.25 *Megalopyge lapena* Schaus ♂ — Arizona, Cochise Co., Carr Canyon, Huachuca Mtns., 5600', VIII. 4.1999 (D. Ferguson) USNM

20.26 *Norape tenera* (Druce) ♂ — Arizona, Santa Cruz Co., Pena Blanca Lake Recreation Area, 3,950', VIII.10.1999 (D. Ferguson) USNM

20.27 *N. ovina* (Sepp) ♂ — Texas, Cameron Co., E of Brownsville, X.13.1979 (E.C. Knudson)

20.28 *Trosia obsolescens* Dyar ♂ — Arizona, Santa Cruz Co., Madera Canyon,VII.30.1970 (J.M. Cadiou) USNM

20.29 *Dalcerides ingenita* (H. Edwards) ♂ — Arizona, Cochise Co., Huachuca Mtns., Ash Canyon Rd, W of Hwy 92, 5,100', V.22.1981 (N. McFarland) USNM

20.30 *Euclea incisa* (Harvey) ♂ — Texas, Culberson Co., Guadalupe Mtns. Natl. Park, Ship on the Desert, VI.23.1989 (E.C. Knudson)

20.31 *Euclea obliqua* H. Edwards ♂ — Arizona, Cochise Co., Paradise, VII. (Barnes Collection) USNM

20.32 *Euclea. obliqua* H. Edwards (yellow form) ♀ — Arizona, Santa Cruz Co., Madera Canyon, Santa Rita Mtns., VIII.27.1946 (J. Comstock & L.M. Martin)(abd. dissected, MEE 10, LACM)

20.33 *Monoleuca occidentalis* Barnes & McDunnough ♂ — California, San Diego Co., Del Mar, VIII.6.1944 (J. Comstock) LACM

20.34 *Monoleuca occidentalis* Barnes & McDunnough (light form) ♂ — California, Orange Co., Lower San Juan Campground, VIII.16.1955 (R. Ford) LACM

20.35 *Parasa chloris* (Herrich-Schäffer) ♂ — Arizona, Cochise Co., Ash Canyon Rd., Huachuca Mtns., 5,100', VIII.3.1999 (D. Ferguson) USNM

20.36 *Isa schaefferana* Dyar ♂ — Arizona, Cochise Co., Ash Canyon Rd., Huachuca Mtns., 5,100', VIII.3.1999 (D. Ferguson) USNM

20.37 *Natada nigripuncta* Barnes & McDunnough ♂ — Guatemala, Quirigua, V. (Schaus—& Barnes) USNM

20.38 *Natada ceres* (Druce) ♂ — Guatemala, Cayuga, VIII. (Schaus & Barnes) USNM

20.39 *Perola clara* Dyar ♂ — Arizona, Santa Cruz Co., Pena Blanca Canyon, Oro Blanco Mtns., VIII.8.1960 (L. Martin) LACM

20.40 *Perola clara* Dyar ♀ — Arizona, Santa Cruz Co., Pena Blanca Canyon, Oro Blanco Mtns., VIII. 8. 1960—(L. Martin) LACM

20.41 *Paleophobetron perornata* (Dyar) ♀ — Arizona, Santa Cruz Co., Madera Canyon, Santa Rita Mtns., VIII.24.1946 (J. Comstock & L. Martin) LACM

20.42 *Paleophobetron perornata* (Dyar) ♂ — Arizona, Santa Cruz Co., Madera Canyon, Santa Rita Mtns., VIII.27.1946 (J. Comstock & L. Martin) LACM

20.43 *Prolimacodes trigona* (H. Edwards) ♂ — Arizona, Cochise Co., Huachuca Mtns., Carr Canyon, 5,600', VII.25.1998 (D. Ferguson) USNM

20.44 *Apoda latomia* (Harvey) ♂ — Arizona, Gila Co., North Payson, East Verde River, 5,000' (C. Henne) LACM

20.45 *Cryptophobetron oropeso* (Barnes) ♂ — Texas, Culberson Co., N of Van Horn, IX. 2. 1979 (E. Knudson)

20.46 *Tortricidia testacea* Packard ♂ — Oregon, Clatsop Co., E of Elsie, VII.8.1967 (S. Jewett, Jr.) EME

20.47 *Tortricidia testacea* Packard ♀ — California, Humboldt Co., Hoopa, VI. 22. 1962 (T. Gallion) CSCA

PLATE 20

21.1 *Acolithus novaricus* Barnes and McDunnough ♂ — Texas, Hemphill Co., Lake Marvin, VI. 23.1984 (E. Knudson)

21.2 *Acolithus rectarius* Dyar ♂ — Texas, Hidalgo Co., Bensten State Park, VI.12.1976 (E. Knudson)

21.3 *Acoliathus falsarius* Clemens ♂ — Texas, Walker Co., Huntsville, reared from cocoons under bark, IV.18.1992 (R. Wharton) TAMU

21.4 *Neoilliberis fusca* (H. Edwards) ♂ — Arizona, Santa Cruz Co., Madera Canyon, VII.25.1967 (C. Baker) CSCA

21.5 *Neofelderia rata* (H. Edwards), holotype ♂ — Arizona (H. Morrison) USNM

21.6 *Neoalbertia constans* (H. Edwards) ♂ — Alternate: Arizona, Cochise Co., Ash Canyon Rd., Huachuca Mtns., 5,170', VIII. 20. 2004 (C. Ferris)

21.7 *Pyromorpha dyari* (Jordan) ♂ — Santa Cruz Co., Madera Canyon, Santa Rita Mtns., 5,800', VI.20.1960 (J. Franclemont) USNM

21.8 *Pyromorpha latercula* (H. Edwards), lectotype ♂ — Arizona (H. Morrison, Coll.) (abd. missing [lost?])—USNM

21.9 *Triprocris yampai* Jordan, paratype ♂ — Arizona, Baboquivari Mtns. (O. Poling) USNM

21.10 *Triprocris smithsoniana* (Clemens) ♂ — California, Los Angeles Co., Llano, IV.6.1957,(W. Simonds) CSCA

21.11 *Harrisina metallica* Stretch ♂ — California, San Diego Co., Alpine, VIII.19.1942, from wild grape (S. Lockwood) CSCA

21.12 *Harrisina americana* Guérin-Méneville ♂ — New Mexico, Grant Co., Gila Hot Springs, 5,600', IV.14.2006 (C. Ferris)

21.13 *Fulgoraecia exigua* (H. Edwards) ♂ — California, San Diego Co., NW of Warners Springs, VIII.20.1976 (r.f. *Neathus* on *Quercus*, emgd. IX.12.1976)(P. Rude) EME

21.14 *Lotisma trigonana* (Walsingham) ♂ — California, Marin Co., Inverness Park, I.17/24.1995 (Powell) EME

21.15 *Lotisma trigonana* (Walsingham) ♀ — California, Marin Co., Inverness Park, I.17/24.1995 (Powell) EME

21.16 *Ellabella editha* Busck ♂ — Arizona, Coconino Co., NW of Flagstaff, VIII.17.1961 (Hodges) USNM

21.17 *Ellabella bayensis* Heppner, paratype ♂ — California, San Mateo Co., San Bruno Mtn., IV.15.1981 (De Benedictis 81105-A, r.f. *Mahonia pinnata*, emgd. I.11.1982) EME

21.18 *Bondia comonana* (Kearfott) ♀ — California, Monterey Co., Hastings Reserve, VI.10.2004 (Powell) EME

21.19 *Bondia comonana* (Kearfott) ♀ — Arizona, Cochise Co., Miller Canyon, Huachuca Mtns., IV.12.1988 (Powell) EME

21.20 *Bondia shastana* Davis ♂ — California, El Dorado Co., Blodgett Forest, VI.21.1982 (De Benedictis) EME

21.21 *Bondia shastana* Davis (dark form) ♀ — California, El Dorado Co., Blodgett Forest, V.29.1970 (Powell) EME

21.22 *Bondia fidelis* Davis ♂ — Arizona, Coconino Co., NW of Flagstaff, 7,000', VII.22.1989 (Powell) EME

21.23 *Tesuquea hawleyana* Klots ♀ — New Mexico, Bernalillo Co., N of Cedar Crest, 7,000', VII.30.1989 (Hsu & Powell) EME

21.24 *Dysodia granulata* (Neumoegen) ♂ — Arizona, Cochise Co., SW of Portal, VIII.2.1973 (Powell) EME

21.25 *Thyris maculata* Harris ♂ — West Virginia, Randolph Co., W of Spruce Knob Lake, VII.11.1983 (Powell) EME

21.26 *Gesneria centuriella* (Denis & Schiffermüller) ♂ — Colorado, Garfield Co., Glenwood Springs, July 8–15. USNM

21.27 *Cosipara tricoloralis* (Dyar) ♂ — Washington, Olympic Mtns., Rosemary Inn, VI.29.1939 (G. & J. Sperry) USNM

21.28 *Scoparia palloralis* Dyar ♂ — California, Santa Barbara Co., Santa Cruz Island, IV.26.1966 (Powell) EME

21.29 *Scoparia biplagialis* Walker ♀ — Arkansas, Washington Co., Devil's Den St. Park, V.29.1966 (Hodges) USNM

21.30 *Eudonia rectilinea* (Zeller) ♀ — California, San Diego Co., San Diego, V.20.1953 (Powell) EME

21.31 *Eudonia expallidalis* (Dyar) ♂ — California, Siskiyou Co., Mt. Shasta City, VII.4.1958 (Powell) EME

21.32 *Eudonia echo* (Dyar) ♂ — Colorado, San Miguel Co., Telluride, VI.27.1977 (D. Ferguson) USNM

21.33 *Eudonia torniplagalis* (Dyar) ♂ — Utah, Emery Co., Reeder Canyon, VII.11.1972 (D. Ferguson) USNM

21.34 *Eudonia spenceri* Munroe ♀ — Utah, Sanpete Co., Ephraim Canyon, 7,100', VII.20.1981 (D. Ferguson) USNM

21.35 *Donacaula maximella* (Fernald) ♀ — California, Inyo Co., 1/15.VI.1922.(O. Poling) USNM

21.36 *Microtheoris ophionalis* (Walker) ♂ — California, Alameda Co., U.C. Berkeley campus, VI.23.1960 (Powell) EME

21.37 *Frechinia laetalis* (Barnes & McDunnough) ♀ — California, Imperial Co., NW of Glamis, VI.26.1978 (Rude & Powell) EME

21.38 *Frechinia helianthales* (Murtfeldt) ♂ — Arizona, Cochise Co., Miller Canyon, VIII.5.2005 (D. & J. Powell) EME

21.39 *Edia semiluna* (Smith) ♂ — Arizona, Cochise Co., Pinery Canyon, 7,000', VII.23.1967 (J. Franclemont) USNM

21.40 *Dichozoma parvipicta* (Barnes & McDunnough) ♂ — California, Inyo Co., Surprise Canyon, IV.24.1957 (Powell) EME

21.41 *Dichozoma parvipicta* (Barnes & McDunnough) ♂ — California, San Bernardino Co., Saratoga Springs, III.17.1978 (P. Rude) EME

21.42 *Gyros muiri* (H. Edwards) ♀ — California, Alpine Co., Ebbetts Pass, 8,730', VI.30.1960 (Powell) EME

21.43 *Gyros powelli* Munroe, paratype ♀ — Mexico, Baja California Norte, Sierra San Pedro Martir, 6,500', V.29.1958 (Powell) EME

21.44 *Anatralata versicolor* (Warren) ♂ — California, Contra Costa Co., Richmond Field Station, IV.7.1992 (Powell) EME

21.45 *Metaxmeste nubicola* Munroe ♀ — Colorado, Chaffee Co., Cottonwood Pass, 12,200' VII.17.1982 (R. Hodges) USNM

21.46 *Pogonogenys proximalis* (Fernald) ♂ — California, Inyo Co., Panamint Mtns., 5,680', IV.13.1957 (R. Langston) EME

21.47 *Chrismania pictipennalis* Barnes & McDunnough ♀ — California, Los Angeles Co., Lovejoy Buttes, IV.9.1941 (C. I. Smith) EME

21.48 *Chrismania pictipennalis* Barnes & McDunnough (dark form) ♀ — Mexico, Baja California Norte, W of Cantillas Canyon, III.18.1967 (Powell) EME

21.49 *Nannobotys commortalis* (Grote) ♂ — California, San Benito Co., S of Paicines, IV.24.1968 (D. Veirs) EME

21.50 *Psammobotys alpinalis* Munroe, paratype ♂ — California, Alpine Co., Ebbetts Pass, 8,730', VI.21.1962 (Powell) EME

21.51 *Mimoschinia rufofascialis* (Stephens) ♂ — California, Alameda Co., Berkeley, VII.27.1979 (Powell) EME

21.52 *Mimoschinia rufofascialis* (Stephens)(dark form) ♀ — Arizona, Cochise Co., SW Research Station, VIII.4.1959 (E. Linsley) EME

21.53 *Pseudoschinia elautalis* (Grote) ♀ — Arizona, Pima Co., Tucson, IV.12.2005 (R. Wielgus) EME

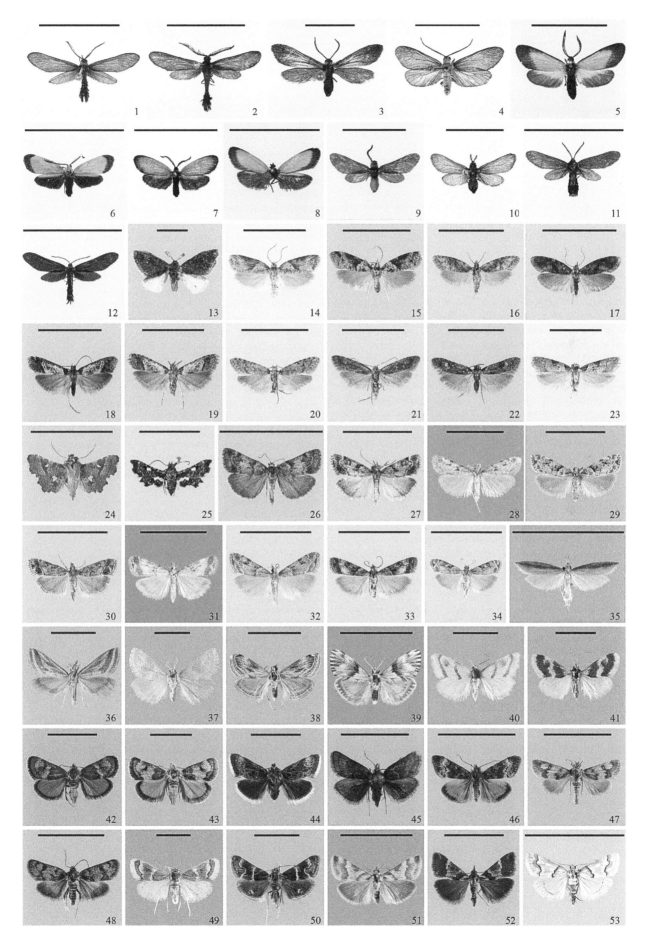

PLATE 21

PLATE 22 **Crambidae**

22.1 *Noctueliopsis palmalis* (Barnes & McDunnough) ♂ — California, San Bernardino Co., Saratoga Springs, III.17.1978 (P. Rude) EME

22.2 *Noctueliopsis aridalis* (Barnes & McDunnough) ♂ — California, San Bernardino Co., Saratoga Springs, IV.22.1977 (J. Doyen & G. Ulrich) EME

22.3 *Noctueliopsis virula* (Barnes & McDunnough) ♂ — California, Riverside Co., Hopkins Well, IV.16.1958 (Powell) EME

22.4 *Mojavia achemonalis* (Barnes & McDunnough) ♀ — Arizona, Cochise Co., Cave Cr., 5,400', Chiricahua Mtns., VIII.9.1966 (Franclemont) USNM

22.5 *Abegesta concha* Munroe ♂ — Arizona, Apache Co., Concho, VI.18.1941 (G. & J. Sperry) USNM

22.6 *Abegesta remellalis* (Druce) ♀ — California, San Diego Co., National City, VIII.13.1956 (A. Lee) EME

22.7 *Stegea salutalis* (Hulst) ♀ — California, Siskiyou Co., Happy Camp, VII.8.1958 (Powell) EME

22.8 *Lipocosma albabasalis* Barnes & McDunnough ♀ — California, Orange Co., Aliso Creek, X.7.1987 (Powell) EME

22.9 *Lipocosma polingi* Munroe, paratype ♀ — Arizona, Pima Co., Baboquivari Mtns., IX.1/15.1923 (O. Poling) USNM

22.10 *Dicymolomia metalliferalis* ♀ — California, Alameda Co., Berkeley, VI.23.1959 (Powell) EME

22.11 *Dicymolomia opuntialis* Dyar ♀ — California, Los Angeles Co., Santa Catalina Island, Middle Ranch, IX.14/18.2004 (De Benedictis & Powell) EME

22.12 *Chalcoela iphitalis* (Walker) ♂ — California, Siskiyou Co., Mt. Shasta City, VI.20.1958 (Powell) EME

22.13 *Hellula rogatalis* (Hulst) ♂ — California, Contra Costa Co., Walnut Creek, X.21.1968 (Powell) EME

22.14 *Cornifrons phasma* Dyar ♀ — California, Inyo Co., Argus Mtns., V.1891 (Koebele) USNM

22.15 *Cylindrifrons succandialis* (Hulst) ♀ — Texas, Hemphill Co., Canadian, VIII.15.1971 (A. & M. Blanchard) USNM

22.16 *Orenaia trivialis* Barnes & McDunnough ♀ — Colorado, Chaffee Co., Cottonwood Pass, 12,500', VII.18.1982 (F. Ferguson) USNM

22.17 *Orenaia sierralis* Munroe, paratype ♀ — California, Inyo Co., Mono Pass, 12,500', IX.3.1965 (Powell) EME

22.18 *Orenaia macneilli* Munroe, paratype ♀ — California, Inyo Co., Mono Pass, 12,600', VIII.31.1967 (Opler) EME

22.19 *Evergestis pallidata* (Hufnagel) ♀ — Canada, Alberta, Fairview, VII.10.1984 (J. Clarke) USNM

22.20 *Evergestis rimosalis* (Guenèe) ♀ — Texas, Presidio Co., Shafter, X.16.1973 (A. & M. Blanchard) USNM

22.21 *Evergestis obliqualis* (Grote) ♀ — California, Inyo Co., Mt. Whitney trail, 9,000' VII.7.1961 (Powell) EME

22.22 *Evergestis angustalis* (Barnes & McDunnough) ♀ — California, Kern Co., Kernville, IV.28.1964 (Powell) EME

22.23 *Psorasea pulveralis* (Warren) ♀ — California, Siskiyou Co., Mt. Shasta City, VIII.3.1958 (Powell) EME

22.24 *Saucrobotys futilalis* (Lederer) ♂ — California, Contra Costa Co., Antioch Natl Wildlife Refuge, VI.11.1991 (J. Powell 91F12, r.f. *Apocynum cannabinum*, emgd. II.9.1962) EME

22.25 *Ostrinia nubilalis* (Hübner) ♀ — Minnesota, Wabash Co., Lake City, VI.19.1964 (J. Powers) EME

22.26 *Ostrinia nubilalis* (Hübner) ♀ — Kentucky, Barren Co., W of Cave City,VIII.3.1971 (Powell) EME

22.27 *Uresiphita reversalis* (Guenée) ♂ — California, Alameda Co., Berkeley, XII.27.1983 (Powell) EME

22.28 *Loxostege sticticalis* (L.) ♂ — California, Kern Co., Shafter, VIII.17.1955 (Powell) EME

22.29 *Phlyctaenia coronata* (Hufnagel) ♂ — California, Humboldt Co., Kneeland, VII.27.2004 (R. Wielgus) EME

22.30 *Anania funebris* (Stromberg ♂ — Idaho, Latah Co., VI.29.1975 (Powell) EME

22.31 *Hahncappsia mancalis* (Lederer) ♂ — Arizona, Santa Cruz Co., Madera Canyon, VII.16.1960 (Franclemont) USNM

22.32 *Hahncappsia coloradensis* (Grote & Robinson) ♀ — Texas, Jeff Davis Co., Old Ft. Davis, VIII.30.1970 (Franclemont) USNM

22.33 *Hahncappsia marculenta*—(Grote & Robinson) ♀ — Arizona, Cochise Co., Cave Creek 4,880', VIII.4.1967 (Franclemont) USNM

22.34 *Achyra occidentalis* (Packard) ♀ — California, Contra Costa Co., Antioch, V.15.1958 (Powell) EME

22.35 *Achyra occidentalis* (Packard) ♀ — California, Kern Co., Jawbone Canyon, X.2.2001 (Powell) EME

22.36 *Achyra rantalis* (Fabricius) ♀ — California, Orange Co., Laguna Beach, VII.6.1932 (D. Meadows) USNM

22.37 *Sitochroa chortalis* (Grote) ♀ — Colorado, Boulder Co., Lefthand Canyon, VI.9.1968 (J. Scott) EME

22.38 *Pyrausta fodinalis* (Lederer) ♀ — California, San Mateo Co., San Bruno Mtn. II.28.1961 (J. Powell 61B17, r.f. *Monardella villosa*, emgd. IV.4.1961) EME

22.39 *Loxotege cereralis* (Zeller) ♀ — California, Humboldt Co., Samoa dunes, VI.25.1969 (Powell) EME

22.40 *Loxostege sierralis* Munroe ♀ — California, Mono Co., Barney Lake, 10,200', VIII.25.1977 (J. Powell 77H3, r.f. wandering larva, emgd. IX.24.1977) EME

22.41 *Loxostege albiceralis* (Grote) ♀ — Arizona, Pima Co., NW of Greenville, X.11.1991 (J. De Benedictis) EME

22.42 *Loxostege lepidalis* (Hulst) ♀ — Colorado, Moffatt Co., E of Craig, VIII.12.1973 (J. Powell 73H6, r.f. *Sarcobatus*?, emgd. VI.9.1974) EME

22.43 *Loxostege annaphilalis* (Grote) ♀ — California, Tulare Co., Johnsondale, IV.27.1964 (W. Turner) EME

22.44 *Pyrausta volupialis* (Grote) ♀ — California, Alameda Co., Berkeley, VII.24.1998 (Powell) EME

22.45 *Pyrausta perrubralis* (Packard) ♀ — California, Alameda Co., Albany, VI.20.1957 (Powell) EME

22.46 *Pyrausta semirubralis* (Packard) ♂ — Nevada, Washoe Co., Mt. Rose, 8,900', VII.9.1979 (Powell) EME

22.47 *Pyrausta unifascialis* (Packard) ♂ — California, Marin Co., Alpine Lake, V.28.1960 (Powell) EME

22.48 *Pyrausta unifascialis* (Packard) ♀ — California, Alpine Co., Ebbetts Pass, 8,750', VI.30.1960 (Powell) EME

22.49 *Pyrausta fodinalis* (Lederer) ♂ — California, San Mateo Co., San Bruno Mtn. III.10.1961 (J. Powell 61C2, r.f. *Monardella villosa*, emgd. IV.10.1961) EME

22.50 *Pyrausta fodinalis* (Lederer) ♀ — California, Tuolumno Co., Sonora Pass, 9,600', VIII.21.1959 (Powell) EME

PLATE 22

PLATE 23 **Crambidae**

23.1 *Loxostege immerens* (Harvey) ♀ — California, San Diego Co., above Vicente Reservoir, III.19.1949 (Powell) EME

23.2 *Loxostege immerens* (Harvey) ♀ — California, San Luis Obispo Co., Dune Lakes S of Oceano, II.15.1974 (Powell) EME

23.3 *Pyrausta nexalis* (Hulst) ♀ — California, Mendocino Co., W of Eel River Ranger Station, VI.12.1972 (Powell) EME

23.4 *Pyrausta signatalis* (Walker) ♂ — Missouri, Benton Co., near Warsaw, VI.3.1967 (J. Heitzman) USNM

23.5 *Pyrausta grotei* Munroe ♂ — Arizona, Cochise Co., Miller Canyon, Huachuca Mtns., VIII.6.1974 (Powell) EME

23.6 *Pyrausta pseudonythesalis* Munroe ♀ — Texas, Brewster Co., Panther Pass, Chisos Mtns., VI.4.1973 (D. Ferguson) USNM

23.7 *Pyrausta laticlavia* (Grote & Robinson), (summer form) ♂ — California, Alameda Co., Berkeley, VIII.25.1992 (Powell) EME

23.8 *Pyrausta laticlavia* (Grote & Robinson), (winter form) ♀ — California, Alameda Co., Berkeley, IV.5.1993 (Powell) EME

23.9 *Pyrausta californicalis* (Packard) ♂ — California, Alameda Co., Berkeley, IX.3.1992 (Powell) EME

23.10 *Pyrausta subsequalis* (Guenée) ♂ — California, Alameda Co., Berkeley, XI.6.1985 (Powell) EME

23.11 *Pyrausta subsequalis* (Guenée) ♀ — California, Alameda Co., Berkeley, VIII.20.1981 (Powell) EME

23.12 *Pyrausta dapalis* (Grote) ♀ — California, San Mateo Co., San Bruno Mtn. II.28.1961 (J. Powell 61B18, r.f. *Monardella villosa*, emgd. III.6.1962) EME

23.13 *Arenochroa flavalis* (Fernald) ♀ — California, Kern Co., Red Rock Canyon, V.2.1968 (Powell) EME

23.14 *Udea rubugalis* (Guenee) ♀ — Minnesota, Clay Co., Moorhead, IX.21.1970 (J. Powers) EME

23.15 *Udea profundalis* (Packard) ♀ — California, Marin Co., Inverness Ridge, II.14.1997 (J. Powell 97B13, r.f. *Urtica holosericea*, emgd. III.4.1997) EME

23.16 *Udea profundalis* (Packard) ♀ — California, Alameda Co., Berkeley, III.11.1959 (Powell) EME

23.17 *Udea vacunalis* (Grote) ♂ — California, Sierra Co., E of Bassetts, VI.28.2001 (Powell) EME

23.18 *Loxostegopsis xanthocrypta* (Dyar) ♂ — California, San Diego Co., San Diego, VI.30.1911 (W. S. Wright) USNM

23.19 *Lineodes integra* (Zeller) ♂ — California, Contra Costa Co., Brooks Island, III.21.1996 (J. Powell 96C32, r.f. *Solanum*, emgd. IV.13.1996) EME

23.20 *Lineodes elcodes* (Dyar) ♂ — California, Alameda Co., Berkeley, XII.11.2005 (Powell) EME

23.21 *Hydropionea fenestralis* (Barnes & McDunnough) ♀ — Arizona, Cochise Co., Onion Saddle, Chiricahua Mtns., VIII.21.1967 (Franclemont) USNM

23.22 *Choristostigma plumbosignalis* (Fernald) ♀ — Arizona, Cochise Co., Pinery Canyon, 7,000', Chiricahua Mtns., VIII.24.1967 (Franclemont) USNM

23.23 *Choristostigma zephyralis* (Barnes & McDunnough) ♀ — California, San Bernardino Co., Angeles Oaks, 5,800', VI.24.1982 (De Benedictis & Powell) EME

23.24 *Choristostigma elegantalis* Warren ♂ — California, Monterey Co., Big Creek Reserve, V.27.1987 (Powell) EME

23.25 *Udea turmalis* (Grote) [paratype of subsp. *tularensis* Munroe] ♂ — California, Inyo Co., Mt. Whitney trail, 9,000', VII.7.1961 (Powell) EME

23.26 *Udea washingtonalis* (Grote) ♀ — California, Humboldt Co., Kneeland, VI.25.2002 (R. Wielgus) EME

23.27 *Udea itysalis* (Walker) [race *marinensis* Munroe] ♀ — California, Marin Co., Alpine Lake, IV.22.1973 (J. Powell 73D12, r.f. *Cynoglossum grande*, emgd. V.7.1971) EME

23.28 *Mecyna mustelinalis* (Packard) ♀ — California, Siskiyou Co., Mt. Shasta City, VII.4.1958 (Powell) EME

23.29 *Mimorista subcostalis* (Hampson) ♂ — Arizona, Cochise Co., Cochise Stronghold, VIII.30.1958 (D. Linsdale) EME

23.30 *Nomophila nearctica* Munroe, paratype ♀ — California, Siskiyou Co., Mt. Shasta City, VI.20.1958 (Powell) EME

23.31 *Terastria meticulosalis* Guenée ♀ — Arizona, Santa Cruz Co., Madera Canyon, VIII.17.1959 (Franclemont) USNM

23.32 *Agathodes designalis* Guenée [race *monstralis* Guenée] ♂ — Arizona, Cochise Co., Pueblo del Sol, Huachuca Mtns., VI.25.1986 (R. Wielgus) USNM

23.33 *Phostria tedea* (Stoll) ♂ — Mexico, Guerrero, S of Ixcateopan, IX.13.1982 (J. Powell 82J46, r.f unidentified tree [Malvaceae?], emgd. X.12.1982) EME

23.34 *Herpetogramma pertextalis* (Lederer) ♂ — California, Santa Cruz Co., NE of Soquel, IV.28.1973 (J. Powell 73D40, r.f. *Asarum caudatum*, emgd. V.18.1973) EME

23.35 *Desmia funeralis* (Hübner) ♀ — California, Tehama Co., NE of Red Bluff, VIII.17.2006 (J. Powell 06H2, r.f. *Vitis californica*, emgd. IX.12.2006) EME

23.36 *Diathrausta harlequinalis* Dyar ♀ — Arizona, Cochise Co., Cochise Stronghold, VIII.20.1958 (Opler) EME

23.37 *Hymenia recurvalis* (Fabricius) ♂ — Virginia, Northampton Co., Kiptopeke, X.4.1986 (W. Steiner) USNM

23.38 *Antigastra catalaunalis* (Duponchel) ♀ — California, Inyo Co., W of Lone Pine, VIII.28.1983 (J. Powell 83H140, r.f. *Penstemon*, emgd. X.5.1983) EME

23.39 *Diaphania hyalinata* (L.) ♂ — Arizona, Santa Cruz Co., Madera Canyon, IX.24.1959 (Franclemont) USNM

23.40 *Lygropia octonalis* (Zeller) ♂ — California, San Diego Co., La Jolla, VI.20.1963 (Powell) EME

23.41 *Palpita gracialis* (Hulst) ♂ — California, Sacramento Co., Sacramento, VIII.21.1956, r.f. privet hedge [*Ligustrum*; no collector listed] EME

23.42 *Palpita quadristigmalis* (Guenée) ♂ — Arizona, Cochise Co., Cave Creek, Chiricahua Mtns., VII.31.1967 (Franclemont) USNM

23.43 *Diastictis sperryorum* Munroe ♀ — California, San Diego Co., San Diego River near Highway 5, VII.15.1998 (J. Powell 98G12, r.f. *Isocoma menziesii*, emgd. VIII.8.1998) EME

23.44 *Diastictis fracturalis* (Zeller) ♀ — California, Santa Barbara Co., Santa Cruz Island, IV.30.1966 (Powell) EME

23.45 *Diastictis fracturalis* (Zeller) ♀ — California, Contra Costa Co., Walnut Creek, IX.26.1963 (Powell) EME

23.46 *Diastictis caecalis* (Warren) ♀ — California, San Mateo Co., junction Highway 280 at 92, IV.4.1990 (Hsu & Powell) EME

23.47 *Synclita occidentalis* Lange ♀ — California, Napa Co., Conn Creek, IX.15.1985 (D. Ferguson) USNM

23.48 *Petrophila confusalis* (Walker) ♂ — Nevada, Washoe Co., Reno, VII.28.1954 (A. Grigarick) EME

23.49 *Petrophila jaliscalis* (Schaus) ♀ — California, Imperial Co., Sidewinder Road, V.21.1999 (N. Bloomfield) EME

23.50 *Petrophila schaefferalis* (Dyar) ♂ — California, Santa Barbara Co., Santa Cruz Island, V.21/24.1984 (D. Wagner)—EME

23.51 *Petrophila schaefferalis* (Dyar) ♀ — California, Santa Barbara Co., Santa Cruz Island, V.22.1984 (Powell) EME

23.52 *Usingeriessa brunnidalis* (Dyar) ♀ — Texas, Presidio Co., Shafter, X.19.1973 (A. & M. Blanchard) USNM

PLATE 23

PLATE 24 Crambidae, Pyralidae

24.1 *Pseudoschoenobius opalescalis* (Hulst) ♂ — California, Imperial Co., N of Glamis, I.27.1977 (P. Rude) EME

24.2 *Crambus pasculellus* (Linnaeus) ♀ — Idaho, Bonner Co., Priest River, VI.16.1979 (D. Ferguson) USNM

24.3 *Crambus whitmerellus* Klots ♀ — Wyoming, Albany Co., NW of Centennial, 8,800′, VII.18.1982 (Powell) EME

24.4 *Crambus occidentellus* Grote ♀ — California, Marin Co, Inverness Ridge, X.8.1996 (Powell) EME

24.5 *Crambus perlellus* (Scopoli) ♂ — California, Siskiyou Co., Mt. Shasta City, VI.21.1958 (Powell) EME

24.6 *Agriphila plumbifimbriella* (Dyar) ♂ — Colorado, San Juan Co., Silverton, VII.8.15 (Barnes collection) USNM

24.7 *Agriphila attenuata* (Grote) ♂ — California, Contra Costa Co., Walnut Creeek, X.4.1961 (Powell) EME

24.8 *Pediasia dorsipunctella* (Kearfott) ♀ — California, Mono Co., Crooked Creek, White Mtns., 10,150′ VI.18.1961 (Powell) EME

24.9 *Pediasia dorsipunctella* (Kearfott) ♀ — Nevada, Nye Co., Currant Creek, VII.20.1968 (Opler & Powell) EME

24.10 *Thaumatopsis repanda* (Grote) ♂ — Nevada, Clark Co., Charleston Mtn., 8,500′, VII.22.1996 (Powell & Sperling) EME

24.11 *Crambus labradoriensis* Christoph ♀ — Canada, Quebec, Mt. Albert, 3,300′, VII.11.1976 (E. Kiel) USNM

24.12 *Crambus guasapalis* Hulst ♀ — California, El Dorado Co., Lake Creek, VI.29.1957 (Powell) EME

24.13 *Crambus cypridellus* Hulst ♀ — Oregon, Marion Co., Aumsville, IX.11.1964 (K. Goeden) USNM

24.14 Crambus sperryellus Klots ♂ — California, Yolo Co., Davis, VIII.8.1956 (Powell) EME

24.15 *Crambus unistriatellus* Packard ♀ — Minnesota, Clay Co., Moorhead, VIII.2.1970 (J. Powers) EME

24.16 *Crambus rickseckerellus* Klots ♀ — California, Santa Barbara Co., Santa Barbara Island, IX.16.2003 (Powell) EME

24.17 *Agriphila anceps* (Grote) ♂ — California, Contra Costa Co., Walnut Creek, IX.25.1963 (Powell) EME

24.18 *Agriphila anceps* (Grote) ♀ — California, Contra Costa Co., Pt. Molate, IX.10.1963 (Powell) EME

24.19 *Agriphila costalipartella* (Dyar) ♀ — Nevada, Elko Co., Angel Creek, 7,000′ VII.18.1971 (D. Ferguson) USNM

24.20 *Parapediasia teterrella* (Zincken) ♀ — California, Alameda Co., Berkeley, V.19.1989 (Powell) EME

24.21 *Catoptria latiradiella* (Walker) ♂ — Colorado, Alamosa Co., Zapata Ranch, 9,600′, VI.27.1982 (D. Ferguson) USNM

24.22 *Catoptria oregonica* (Grote) ♂ — California, Humboldt Co., Kneeland, X.7.2002 (R. Wielgus) EME

24.23 *Chrysoteuchia topiaria* (Zeller) ♂ — California, Siskiyou Co., Mt. Shasta City, VI.23.1958 (Powell) EME

24.24 *Chrysoteuchia topiaria* (Zeller)(dark form) ♂ — California, Sonoma Co., Sebastopol, V.26.1966 (Powell) EME

24.25 *Microcrambus copelandi* Klots ♀ — California, Santa Barbara Co., Santa Cruz Island, V.22.1984 (Powell) EME

24.26 *Microcrambus copelandi* Klots ♀ — Arizona, Cochise Co., Turkey Creek, 5,600′, Chiricahua Mtns., VIII.1.1986 (Powell) EME

24.27 *Tehama bonifatella* (Hulst) ♂ — California, Alameda Co., Berkeley, III.30.1988 (Powell) EME

24.28 *Euchromius ocelleus* (Haaworth) ♂ — California, Kern Co., Shafter, VI.24.1955 (Powell) EME

24.29 *Diptychophora harlequinalis* (Barnes & McDunnough) ♀ — Arizona, Cochise Co., Cave Creek SW of Portal, VIII.20.1972 (D. Veirs & Powell) EME

24.30 *Vaxi* species ♂ — California, Siskiyou Co., Gazelle, VII.26.1966 (Powell) EME

24.31 *Haimbachia arizonensis* Capps, paratype ♂ — Arizona, Pima Co., Baboquivari Mtns., VII.27.1923 (O. Poling) USNM

24.32 *Eoreuma loftini* (Dyar) ♀ — California, Imperial Co., Calexico, II.22.1942 ex rice (H. Osborn) EME

24.33 *Hemiplatytes epia* (Dyar) ♂ — California, Orange Co., Silverado Canyon, VII.6.1979 (G. Marsh) EME

24.34 *Hemiplatytes epia* (Dyar) ♀ — California, San Diego Co., San Diego "VIII.10" (L. Ricksecker) USNM [paratype of *H. damon* B. & McD.]

24.35 *Hemiplatytes prosenes* (Dyar) ♂ — California, Los Angeles Co., Santa Catalina Island, IX.19.2004 (De Benedictis & Powell) EME

24.36 *Hypsopygia costalis* (Fabricius) ♀ — New Mexico, Santa Fe Co., Santa Fe, VII.5.2001 (Powell) EME

24.37 *Aglossa caprealis* (Hübner) ♂ — California, Sacramento Co., Sacramento, V.27.1930 (H. Keifer) CDFA

24.38 *Aglossa acallalis* Dyar ♂ — California, Santa Barbara Co., San Marcos Pass, VII.19.1965 (Powell) EME

24.39 *Herculia phoezalis* Dyar ♀ — California, Contra Costa Co., Richmond, IV.15.1972 (J. Memott, JP 72D29, r.f. sphagnum moss, emgd. V.31.1972) EME

24.40 *Achroia grisella* (Fabricius) ♀ — California, Alameda Co., Berkeley, V.27.1993 (Powell) EME

24.41 *Cacotherapia angulalis* (Barnes & McDunnough) ♀ — California, Los Angeles Co., Avalon, Santa Catalina Island, V.4.1978 (J. Powell 78E24, r.f. *Cupressus macrocarpa,* emgd. V.30.1978) EME

24.42 *Alpheias transferrens* Dyar ♂ — California, Santa Barbara Co., N of Refugio Beach, VI.28.1965 (Powell) EME

24.43 *Decaturia pectinalis* Barnes & McDunnough ♂ — Arizona, Santa Cruz Co., Pena Blanca Lake, VIII.11.1974 (Powell) EME

24.44 *Arta epicoenalis* Ragonot ♂ — California, San Diego Co., San Diego, V.16.1912 (G. Field) USNM

24.45 *Galasa nigrinodis* (Zeller) ♂ — Texas, Bexar Co., W of Leon Valley, VI.21.1972 (Powell) EME

24.46 *Galasa nigripunctalis* Barnes & McDunnough ♂ — Arizona, Santa Cruz Co., Madera Canyon, VIII.13.1974 (Powell) EME

24.47 *Pyralis farinalis* Linnaeus ♂ — California, Alameda Co., Berkeley, VI.20.1960 (J. Burns) EME

24.48 *Pyralis cacamica* Dyar ♂ — California, San Diego Co., San Diego, VI.30.1924 [E. Piazza] USNM

24.49 *Pyralis cacamica* Dyar ♂ — California, Orange Co., Silverado Canyon, VI.19.1979 (G. Marsh) EME

24.50 *Herculia olinalis* (Guenèe) ♂ — Utah, Wasatch Co., SW of Midway, VII 29.1971 (D. Ferguson) USNM

24.51 *Galleria mellonella* (Linnaeus) ♂ — California, Contra Costa Co., Kensington, III.2.1978 (P. Rude, r.f. inactive bee hive) EME

PLATE 24

PLATE 25 **Pyralidae**

25.1 *Caphys arizonensis* Munroe ♀ — Arizona, Pima Co., Madera Canyon, 4,500′, VII.30.1973 (Powell & S. Szerlip) EME

25.2 *Macalla zelleri* (Grote) ♀ — Arizona, Santa Cruz Co., Madera Canyon, VIII.19.1959 (J. Franclemont) USNM

25.3 *Cacozela basiochrealis* Grote ♂ — Arizona, Cochise Co., Ash Canyon, Huachuca Mtns., VIII.2.1989 (Hsu & Powell) EME

25.4 *Cacozela basiochrealis* Grote ♀ — Arizona, Santa Cruz Co., Madera Canyon, VII.12.1959 (J. Franclemont) USNM

25.5 *Oneida luniferella* Hulst ♀ — Arizona, Santa Cruz Co., Madera Canyon, IX.4.1969 (Powell) EME

25.6 *Satole ligniperdalis* Dyar ♀ — California, San Bernardino Co., Kelso dunes, VII.14.1974 (J. Doyen, JP 74G9, r.f. *Chilopsis linearis*, emgd. IX.1.1975) EME

25.7 *Satole ligniperdalis* Dyar ♂ — California, San Bernardino Co., Kelso dunes, VII.14.1974 (J. Doyen, JP 74G9, r.f. *Chilopsis linearis*, emgd. VII 23.1974) EME

25.8 *Satole ligniperdalis* Dyar ♀ — California, San Bernardino Co., Kelso dunes, VII.27.1974 (J. Doyen, JP 74G9, r.f. *Chilopsis linearis*, emgd. IX.1.1975) EME

25.9 *Tallula fieldi* Barnes & McDunnough ♀ — California, Los Angeles Co., Avalon, Santa Catalina Island, III.31.1968 (J. Powell 68D14, r.f. *Cupressus* cones, emgd. VI.31.1968) EME

25.10 *Acrobasis tricolorella* Grote ♀ — California, Napa Co., SE of Angwin, V.20.1980 (De Benedictis & Powell) EME

25.11 *Acrobasis tricolorella* Grote ♂ — California, Siskiyou Co., Mt. Shasta City, VII.2.1958 (Powell) EME

25.12 *Acrobasis comptella* Ragonot (paratype of *kofa* Opler) ♀ — Arizona, Yuma Co., Palm Canyon, Kofa Mtns., IV.8.1963 (J. Powell 63D11, r.f. *Quercus turbinella*, emgd. IV.19.1963) EME

25.13 *Trachycera caliginella* (Hulst) ♂ — Arizona, Santa Cruz Co., Madera Canyon, VI.24.1963 (J. Franclemont) USNM

25.14 *Trachycera caliginoidella* Dyar ♂ — California, Alameda Co., Berkeley, IX.3.1988 (Powell) EME

25.15 *Myelopsis alatella* (Hulst) ♂ — Colorado, Chaffee Co., Buena Vista, VII.11.1988 (R. Hodges) USNM

25.16 *Apomyelois bistriatella* (Hulst) ♀ — California, Santa Barbara Co., Prisoners Harbor, Santa Cruz Island, V.1.1966 (Powell, A. Slater, J. Wolf, JP 66E4, r.f *Hypoxylon occidentale*, emgd. VI.7.1966) EME

25.17 *Amyelois transitella* (Walker) ♂ — California, Alameda Co., Berkeley, V.13.1984 (Powell) EME

25.18 *Catastia bistriatella* (Hulst) ♀ — California, Inyo Co., Mono Pass, 12,500′, IX.3.1965 (Opler & Powell) EME

25.19 *Salebriaria equivoca* Neunzig ♂ — Texas, Culberson Co., Sierra Diablo, 6,000′, VII.11.1971 (A. & M. Blanchard) USNM

25.20 *Qualsisalebria admixta* Heinrich ♂ — Texas, Culberson Co., Guadalupe Mtns., 5,750′, V.22.1973 (D. Ferguson) USNM

25.21 *Meroptera pravella* (Grote) ♀ — Wisconsin, Vilas Co., Northern Highland, VIII.8.1957 (P. Jones) USNM

25.22 *Pyla viridsuffusella* Barnes & McDunnough ♂ — California, Tuolumne Co., Kerrick Meadows, Yosemite Natl. Park, 9,250′, VII.28.1934 (E. O. Essig) EME

25.23 *Tacoma feriella* Hulst ♂ — California, Contra Costa Co., Walnut Creek, VI.5.1962 (Powell) EME

25.24 *Toripalpus trabalis* (Grote) ♂ — California, San Luis Obispo Co., Dune Lakes, III.20.1974 (J. Powell 74C10, r.f. *Eriogonum parvifolium*, emgd. V.5.1974) EME

25.25 *Toripalpus trabalis* (Grote) ♀ — California, San Diego Co., San Diego, III.6.1953 (Powell) EME

25.26 *Dasypyga alternosquamella* Ragonot ♂ — California, Marin Co., Inverness, X.9.2000 (Powell)

25.27 *Etiella zickenella* (Treitschke) ♀ — California, Santa Barbara Co., Canada Cervada, Santa Cruz Island, IV.25.2001 (Powell) EME

25.28 *Pima occidentalis* Heinrich ♀ — California, Contra Costa Co., Pt. Molate, VI.3.1964 (R. Langston) EME

25.29 *Pima albocostalialis* (Hulst) ♀ — California, Santa Barbara Co., Canada Cervada, Santa Cruz Island, IV.25.2001 (Powell) EME

25.30 *Interjectio denticulata* (Ragonot) ♀ — California, Contra Costa Co., Antioch Natl Wildlife Refuge, IV.21.1982 (J. Powell 82D33.1, r.f. *Lupinus albifrons*, emgd. V.22.1982) EME

25.31 *Ambesa laetella* Grote ♀ — Utah, Sanpete Co., E of Ephraim, 7,100′, VII.24.1980 (D. Ferguson) USNM

25.32 *Ambesa walsinghamiella* (Ragonot) ♀ — California, Modoc Co., Fandango Pass, VI.12.1970 (P. Rude, JP 70F88, r.f. *Prunus fremontii*, emgd. VII.16.1970) EME

25.33 *Catastia actualis* (Hulst) ♀ — Colorado, Denver [no date] (Oslar) USNM

25.34 *Sciota basilaris* (Zeller) ♀ — Montana, Sweet Grass Co., N of Big Timber, VII.2.1969 (J. Franclemont) USNM

25.35 *Sciota bifasciella* (Hulst) ♀ — Arizona, Coconino Co., Hart Prairie, 8,500′, VI.29.1964 (J. Franclemont) USNM

25.36 *Telethusia ovalis* (Packard) ♂ — California, Tuolumne Co., Tuolumne Meadows, VII.16/23 [Barnes Collection] USNM

25.37 *Phobus curvatellus* (Ragonot) ♂ — California, Monterey Co., Big Creek Reserve, VI.14.1991 (Hsu & Powell) EME

25.38 *Phobus funerellus* (Dyar) ♀ — California, Monterey Co., Hastings Reservation, VI.1.1958 (D. Linsdale) EME

25.39 *Pyla fusca* (Haworth) ♀ — Colorado, Garfield Co., Glenwood Springs, VIII.16/23 [Barnes Collection] USNM

25.40 *Pyla areneoviridella* Ragonot ♂ — Montana, Glacier [Natl.] Park, VI.27.[19]21 (H. Dyar) USNM

25.41 *Pyla scintillans* (Grote) ♂ — California, Sierra Co., Packer Saddle, VI.29.2001 (Powell) EME

25.42 *Dioryctria abietivorella* (Grote) ♀ — California, El Dorado Co., Badger Hill, VII.29.1988 (J. Volney BH-80-109 ex *Pseudotsuga* cone) EME

25.43 *Dioryctria pseudotsugella* Munroe ♀ — California, Kern Co., Tehachapi Mtn., 5,000′, VI.16.1980 (De Benedictis & S, Meredith, JP 80F26, r.f. *Abies concolor*, emgd. VII.13.1980) EME

25.44 *Dioryctria reniculelloides* Mutuura & Munroe) ♀ — Utah, Utah Co., N of Timpanagos, VII.15.1981 (De Benedictis & Powell) EME

25.45 *Dioryctria muricativorella* Neunzig ♀ — California, Alameda Co., Berkeley, V.2.2004 (Powell) EME

25.46 *Dioryctria auranticella* (Grote) ♂ — California, Contra Costa Co., Walnut Creek, VI.13.1964 (Powell) EME

25.47 *Dioryctria cambiicola* (Dyar) ♀ — Arizona, Mohave Co., Hualapai Mtn. Park, VII.20.1989 (Hsu & Powell) EME

25.48 *Dioryctria ponderosae* Dyar ♂ — Montana, Sweet Grass Co., N of Big Timber, VII.7.1969 (J. Franclemont) USNM

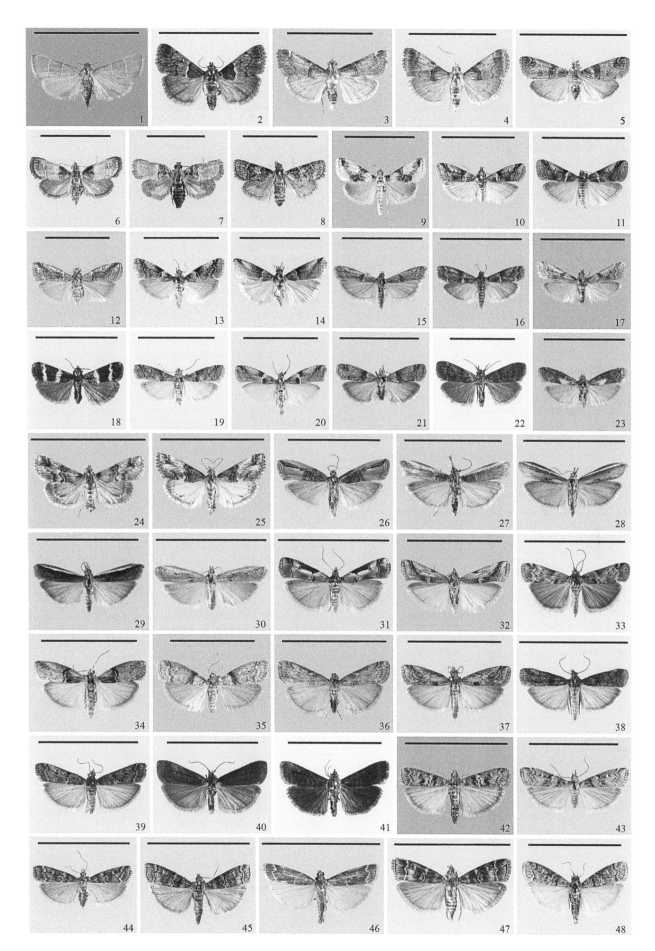

PLATE 25

PLATE 26 **Pyralidae**

26.1 *Pseudadelphia ochripunctella* (Dyar) ♀ — Mexico, Puebla, SW of Tehuacan, 5,300', X.4.1975 (T. Eichlin, T. Friedlander, Powell) EME

26.2 *Elasmopalpus lignosellus* (Zeller) ♂ — California, Santa Barbara Co., Coal Oil Pt., XII.17.1979 (Powell) EME

26.3 *Acroncosa minima* Neunzig ♂ — California, San Bernardino Co., NW of Fenner, VII.30.1991 (Powell) EME

26.4 *Acroncosa albiflavella* Barnes & McDunnough, paratype ♀ — New Mexico, McKinley Co., Ft. Wingate, VIII.1/7 (Barnes collection) USNM

26.5 *Martia arizonenella* Ragonot ♂ — California, Kern Co., Shafter, VIII.17.1955 (Powell) [abd. dissected EME 0139]

26.6 *Martia arizonenella* Ragonot ♂ — Mexico, Baja California Norte, San Felipe Valley, IV.6.1973 (Powell) EME

26.7 *Lipographis fenestrella* (Packard) ♂ — California, San Francisco, Lobos Creek, V.6.1960 (Powell) EME

26.8 *Eumysia pallidipennella* (Hulst) ♂ — California, Ventura Co., San Nicolas Island, V.7.1978 (Powell) EME

26.9 *Homeoesoma striatellum* Dyar ♀ — California, Inyo Co., SW of Lone Pine, V.9.1969 (Opler & P. Welles) EME

26.10 *Heterographis morrisonella* Ragonot ♂ — California, San Bernardino Co., Zzyzx Springs, XII.18.1977 (Powell)

26.11 *Hulstia undulatella* (Clemens) ♂ — California, Contra Costa Co., Pleasant Hill, VIII.18.1960 (W. Ferguson) EME

26.12 *Honora mellinella* Grote ♂ — Texas, Kerr Co., Kerrville, III.10 (H. Lacey) USNM

26.13 *Lipographis fenestrella* (Packard) ♀ — California, Santa Barbara Co., Santa Rosa Island, IV.30.1995 (Powell) EME

26.14 *Lipographis umbrella* (Dyar) ♂ — California, Alameda Co., NE of Livermore, IX.22.1994 (Powell) EME

26.15 *Ragonotia dotalis* (Hulst) ♂ — California, San Diego Co., SE of Ocotillo, III.29.1978 (R. Dietz & Powell) EME

26.16 *Staudingeria albipennella* (Hulst) ♀ — Oregon, Co. unknown, Bone Springs, Blue Mtns., VII.13.1935 (J. Clarke) USNM

26.17 *Staudingeria albipennella* (Hulst) ♀ — Washington, Kittitas Co., Vantage Ferry, V.2.1928 (Clarke) USNM

26.18 *Honora dotella* Dyar ♂ — California, San Luis Obispo Co., IV.25.1968 (Chemsak & Powell) EME

26.19 *Homeoesoma albescentellum* Ragonot ♂ — Arizona, Coconino Co., NW of Flagstaff, VI.29.1961 (Franclemont) USNM

26.20 *Homeoesoma electellum* (Hulst) ♂ — California, Stanislaus Co., Del Puerto Canyon, IX.6.1962 (Powell) EME

26.21 *Laetilia zamacrella* Dyar ♀ — California, Alameda Co., Albany, X.8.1957 (Powell) EME

26.22 *Rhagea stigmella* (Dyar) ♀ — California, Monterey Co., Big Creek Reserve, V.4.1991 (J. Powell 91E17, r.f. *Dudleya caespitosa*, emgd. VII.18.1991) EME

26.23 *Zophodia grossulariella* (Hübner) ♀ — Washington, Whatcom Co., Bellingham, V.4.1923 (Clarke) USNM

26.24 *Ozamia fuscomaculella* (Wright) ♀ — California, Santa Barbara Co., Santa Cruz Island, IX.26.1978 (Powell) EME

26.25 *Rumatha bihinda* (Dyar) ♂ — Texas, Culberson Co., Guadalupe Mtns., V.24.1973 (A. & M. Blanchard) USNM

26.26 *Yosemitia graciella* (Hulst) ♂ — California, San Bernardino Co., SW of Barstow, IV.1989 (R. May, JP 89E7, r.f. *Sclerocactus polyancistrus*, emgd. V.22.1989) EME

26.27 *Euzophera semifuneralis* (Walker) ♀ — Oregon, Benton Co., N of Corvallis, VI.4.1989 (D. Ferguson) USNM

26.28 *Eremberga creabates* (Dyar) ♂ — California, San Diego Co, Borrego, V.2.1956 (Powell) [abd. dissected JAP 864] EME

26.29 *Eremberga leuconips* (Dyar) ♂ — Arizona, Cochise Co., Cave Creek Canyon, Chiricahua Mtns., VII.9.1967 (Franclemont) USNM

26.30 *Vitula edmansae* (Packard) ♂ — California, Alameda Co., Berkeley, XII.10.1988 (Powell) EME

26.31 *Ephestia kuehniella* Zeller ♀ — California, Alameda Co., Berkeley, IX.29.1988 (Powell) EME

26.32 *Ephestia kuehniella* Zeller ♀ — California, Sonoma Co., Petaluma, X.30.1936 (E. Johnston) CNC

26.33 *Bandera virginella* Dyar ♂ — Washington, Whitman Co., Pullman, VII.3.1935 (Clarke) USNM

26.34 *Patagonia peregrina* (Heinrich) ♀ — California, Santa Barbara Co., Santa Cruz Island, V.24.1984 (J. Powell 84E89, r.f. *Gnaphalium*, emgd. VII.11.1984) EME

26.35 *Phycitodes mucidellum* (Ragonot) ♂ — California, Kern Co., SW of Mojave, III.28.1968 (Powell) EME

26.36 *Laetilia dilatifasciella* (Ragonot) ♀ — California, Santa Barbara Co., Santa Cruz Island, VII.13.1977 (D. Green, r.f. scale on *Opuntia*) EME

26.37 *Rostrolaetilia minimella* Blanchard & D. Ferguson ♂ — California, San Bernardino Co., Zzyzx Springs, VI.30.1978 (Powell)

26.38 *Rhagea packardella* (Ragonot) ♂ — California, San Bernardino Co., Yermo, r.f. *Orobanche cooperi*, emgd. V.24.1941 [no collector label] USNM

26.39 *Ephestiodes gilvescentella* Ragonot ♂ — California, Alameda Co., Berkeley, IV.1.1959 (Powell) EME

26.40 *Ephestiodes gilvescentella* Ragonot ♂ — California, Alameda Co., Berkeley, V.23.1979 (Powell) EME

26.41 *Vitula pinei* Heinrich ♂ — Colorado, Costilla Co., Blanca (R. Furniss, r.f. *Pinus edulis*, emgd. VIII.5.1941, Hopkins U.S. 33126-D) (R. Furniss) USNM

26.42 *Sosipatra thurberiae* (Dyar) ♀ — California, Tuolumne Co., Twain Harte, VIII.18.1960 (M. Lundgren) EME

26.43 *Sosipatra insula* Neunzig ♀ — California, Los Angeles Co., Avalon, Santa Catalina Island, III.31.1968 (Opler) EME

26.44 *Sosipatra rileyella* (Ragonot) ♀ — California, Los Angeles Co., E of Gorman, III.11.1963 (Chemsak & Powell, JP 63C8, r.f. *Yucca brevifolia*, emgd. IV.20.1963) EME

26.45 *Plodia interpunctella* (Hübner) ♀ — California, Contra Costa Co., Walnut Creek, X.16.1962 (Powell) EME

26.46 *Plodia interpunctella* (Hübner) ♀ — Canada, Ontario, Ottawa, VII.15.1930 (H. Grey) CNC

26.47 *Ephestia elutella* (Hübner) ♀ — Canada, British Columbia, Wellington,VII (G. Taylor) USNM

26.48. *Ephestia elutella* (Hübner) ♀ — Canada, British Columbia, Salmon Arm, "10.7".1926 (W. Buckell) CNC

26.49 *Cadra cautella* (Walker) ♂ — California, San Francisco, Pier 17, X.19.1971, ex cocoa bean bag (Powell) EME

26.50 *Cadra figulilella* Gregson ♂ — California, Contra Costa Co., Walnut Creek, 1X.11.1962 (Powell) EME

PLATE 26

PLATE 27 Drepanidae, Geometridae

27.1 *Drepana arcuata* Walker ♀ — Oregon, Benton Co., McDonald Forest, V.24.1999 (Opler & E. Buckner-Opler) CSU

27.2 *Drepana bilineata* (Packard) ♀ — Colorado, Larimer Co., Rocky Mtn. Natl. Park, VII.15.1995 (D. Katz) CSU

27.3 *Habrosyne scripta* (Gosse) ♂ — Colorado, Larimer Co., Fort Collins, VII.14.1984 (Opler) CSU

27.4 *Habrosyne gloriosa* (Guenée) ♂ — Colorado, Larimer Co., Rocky Mtn. Natl. Park, VII.14.1990 (Opler) CSU

27.5 *Pseudothyatira cymatophoroides* (Guenée) ♂ — California, Humboldt Co., Arcata, VII.22.1969 (Powell) CSU

27.6 *Thyatira mexicana* Henry Edwards ♂ — Arizona, Cochise Co., Onion Saddle, VII.30/VIII.1.1999 (Opler & E. Buckner-Opler) CSU

27.7 *Euthyatira lorata* (Grote) ♂ — Washington, Klickitat Co., Lyle, IV.24.1998 (J. Troubridge) CNC

27.8 *Euthyatira pudens* (Guenée) ♂ — Pennsylvania, Berks Co., Sinking Spring, IV.26.1936 (H. Moyer) CNC

27.9 *Euthyatira semicircularis* (Grote) ♂ — Oregon, Linn Co., W of Santiam Junction, VI.27.1997 (J. Troubridge) CNC

27.10 *Ceranemota fasciata* (Barnes and McDunnough) ♂ — British Columbia, Quamichan, Vancouver Island, IX.25.1907 (O. Taylor) CNC

27.11 *Ceranemota tearlei* (Henry Edwards) ♂ — Colorado, Larimer Co., Rist Canyon, IX.19.1991 (Opler) CSU

27.12 *Ceranemota improvisa* (Henry Edwards) ♂ — California, Orange Co., Silverado Canyon, V.5.1979 (G. A. Marsh) EME

27.13 *Bycombia verdugoensis* (Hill) ♂ — California, Orange Co., Silverado Canyon, 1,650′, III.10.1979 (G. Marsh) EME

27.14 *Callizzia amorata* Packard, ♀ — Colorado, Chaffee Co., base of Mt. Yale, VII.13.1982 (J. Landry) CNC

27.15 *Archiearis infans* (Möschler) ♂ — Montana, Carbon Co., Custer Natl. Forest, 7,490′, IV.25.1999 (C. Harp) CSU

27.16 *Leucobrephos brephoides* (Walker) ♂ — Alberta, Calgary, IV.7.1915 (F. Wolley-Dod) CNC

27.17 *Leucobrephos brephoides* (Walker) ♀ — Alberta, Calgary, IV.7.1915 (F. Wolley-Dod) CNC

27.18 *Protitame matilda* (Dyar) ♀ — California, Contra Costa Co., Orinda, VII.12.1957 (S. Cook) EME

27.19 *Eumacaria latiferrugata* (Walker) ♂ — Colorado, Larimer Co., Rist Canyon, V.22.1989 (Opler) CSU

27.20 *Digrammia californiaria* (Packard) ♀ — Wyoming, Platte Co., W of Wheatland, V.27.1990 (Opler) CSU

27.21 *Digrammia teucaria* (Strecker) ♀ — Washington, Yakima Co., Toppenish, VIII.23.1960 (D. Hardwick) CNC

27.22 *Digrammia setonana* (McDunnough) ♀ — California, Shasta Co., Hat Creek Insect Lab., VI.7.1991 (M. Valenti) CNC

27.23 *Digrammia setonana* (McDunnough) ♀ — British Columbia, Doctor Creek, II.23.1961, reared on *Juniperus scopulorum* (Forest Insect Survey) CNC

27.24 *Digrammia curvata* (Grote) ♀ — Utah, Iron Co., Cedar Canyon, V.13.1999 (J. Nordin) CSU

27.25 *Digrammia irrorata* (Packard) ♀ — Colorado, Larimer Co., Fort Collins, V.17.1988 (Opler) CSU

27.26 *Digrammia neptaria* (Guenée) ♀ — Colorado, Larimer Co., Rocky Mtn. Natl. Park, VI.17.1990 (Opler) CSU

27.27 *Digrammia muscariata* (Guenée) ♀ — California, San Diego Co., Palomar Observatory Rd., VIII.1/7.1999 (T. Mustelin & N. Bloomfield) CSU

27.28 *Macaria truncataria* (Walker) ♂ — Colorado, Larimer Co., Rocky Mtn. Natl. Park, V.16.1998 (Opler) CSU

27.29 *Macaria ulsterata* (Pearsall) ♀ — Colorado, Larimer Co., Rocky Mtn. Natl. Park, VI.20.1996 (S. Simonson) CSU

27.30 *Macaria adonis* (Barnes and McDunnough) ♀ — Colorado, Larimer Co., Rocky Mtn. Natl. Park, VII.2.1990 (Opler) CSU

27.31 *Macaria sexmaculata* (Packard) ♀ — Alberta, Nordegg, VI.22.1921 (J. McDunnough) CNC

27.32 *Macaria sexmaculata* (Packard) ♀ — British Columbia, Cranbrook, I.23.1943, reared from *Larix occidentalis* (Forest Insect Survey) CNC

27.33 *Macaria signaria* (Walker) ♀ — California, Plumas Co., U.C. Forestry Camp, Meadow Valley, VI.6/7.1997 (Opler & E. Buckner-Opler) CSU

27.34 *Macaria unipunctaria* (Wright) ♀ — Colorado, Larimer Co., Rocky Mtn. Natl. Park, V.25.1995 (Opler & E. Buckner-Opler) CSU

27.35 *Macaria occiduaria* (Packard) ♂ — Alberta, Waterton Lakes, VII.8.1923 (J. McDunnough) CNC

27.36 *Macaria brunneata* (Thunberg) ♂ — Colorado, Grand Co., Rocky Mtn. Natl. Park, VIII.13.1995 (Opler & E. Buckner-Opler) CSU

27.37 *Macaria plumosata* (Barnes and McDunnough) ♂ — Colorado, Larimer Co., SE of Glen Haven, VII.22.1994 (Opler & E. Buckner-Opler) CSU

27.38 *Macaria bitactata* (Walker) ♀ — Colorado, Larimer Co., Rocky Mtn. Natl. Park, VIII.28.1990 (Opler) CSU

27.39 *Macaria colata* (Grote) ♀ — Colorado, Larimer Co., Phantom Canyon Ranch, VI.27.1990 (Opler) CSU

27.40 *Macaria lorquinaria* (Guenée) ♂ — Oregon, Multnomah Co., Forest Park, IX.11.1994 (Opler) CSU

27.41 *Narraga fimetaria* (Grote and Robinson) ♂ — Colorado, Las Animas Co., Pinyon Canyon site, VII.13.1991 (Opler) CSU

27.42 *Aethalura anticaria* (Walker) ♂ — Washington, Okanagan Co., Foggy Dew Trail, V.19.1947 (E. C. Johnston) CNC

27.43 *Glena nigricaria* (Barnes & McDunnough) ♂ — California, Los Angeles Co., Angeles Crest Highway, VII.2.1948 (C. Smith) EME

27.44 *Stenoporpia pulmonaria* (Grote) ♂ — British Columbia, Gabriola Island, VIII.26.1988 (G. Anweiler) CSU

27.45 *Stenoporpia excelsaria* (Strecker) ♂ — Colorado, Larimer Co., Rocky Mtn. Natl. Park, VI.17.1990 (Opler) CSU

27.46 *Tornos erectarius* Grossbeck ♀ — Arizona, Pima Co., Tucson, IV.21.2005 (R. Wielgus) EME

27.47 *Glaucina epiphysaria* Dyar ♂ — California, Los Angeles Co., Mint Canyon, III.14.1968 (F. Sala) EME

27.48 *Glaucina eupitheciaria* Grote ♂ — California, Riverside Co., Coachella Valley, I.24.1947 (C. Smith) EME

PLATE 27

PLATE 28   Geometridae

28.1 *Glaucina interruptaria* (Grote) ♂ — Colorado, Mesa Co., John Brown Canyon, IV.30.1996 (J. Nordin) CSU

28.2 *Glaucina ochrofuscaria* Grote ♀ — California, Riverside Co., Coachella Valley, I.24.1947 (C. Smith) EME

28.3 *Synglochis perumbraria* Hulst ♀ — California, San Diego Co., Borrego, IV.23.1953 (R. Schuster) EME

28.4 *Nepterotaea memoriata* (Pearsall) — no data

28.5. *Chesiadodes cinerea* Rindge ♂ — California, San Diego, VI.13.1920 (W. Wright (CNC) CNC

28.6 *Chesiadodes morosata* Hulst ♂ — California, Inyo Co., NE of Big Pine, I.27.1996 (D. Giuliani) EME

28.7 *Hulstina wrightiaria* (Hulst) ♀ — California, Contra Costa Co., U.C. Russell Farm, VI.5.1972, reared from *Adenostoma fasciculatum* (Powell) EME

28.8 *Pterotaea comstocki* Rindge ♂ — California, San Diego Co., Descanso, VII.13.1947 (G. & J. Sperry) CNC

28.9 *Pterotaea lamiaria* (Strecker) ♀ — California, Contra Costa Co., Walnut Creek, VI.19.1962 (Powell) EME

28.10 *Iridopsis clivinaria* (Guenée) ♂ paratype — California, San Diego Co., Pine Valley, IV.17.1950 (E. Johnston) CNC

28.11 *Iridopsis emasculata* (Dyar) ♀ — Colorado, Grand Co., Radium State Wildlife Area, VII.25.1995 (T. Dickel) CSU

28.12 *Anavitrinella pampinaria* (Guenée) ♂ — British Columbia, Lillooet, VIII.1927 (A. Phair) CNC

28.13 *Protoboarmia porcelaria* Guenée ♂ — British Columbia, Cultus Lake, VII.7.1951, reared from hemlock (Forest Insect Survey) CNC

28.14 *Neoalcis californiaria* (Packard) ♀ — British Columbia, Gabriola Island, IX.6.1988 (G. Anweiler) CSU

28.15 *Orthofidonia exornata* (Walker) ♀ — Alberta, Edmonton, V.10.1918 (D. Mackie) CNC

28.16 *Hesperumia sulphuraria* Packard ♀ — Alberta, Edmonton, VIII.12.1994 (G. Anweiler) CSU

28.17 *Ectropis crepuscularia* (Denis and Schiffermüller) ♀ — California, Humboldt Co., Patricks Point State Park, V.28.1981 (R. Klopshinske) CSU

28.18 *Dasyfidonia avuncularia* (Guenée) ♂ — California, Lassen Co., Eagle Lake, VII.15.1995, *Prunus subcordata* (L. Crabtree) EME

28.19 *Mericisca gracea* Hulst ♀ — Colorado, Larimer Co., Redstone Estates, VII.6.1989 (Opler) CSU

28.20 *Merisma spododea* (Hulst) ♀ — Colorado, Larimer Co., Rocky Mtn. Natl. Park, VII.17.1990 (Opler) CSU

28.21 *Tracheops bolteri* (Hulst) ♂ — Arizona, Cochise Co., Onion Saddle, VII.30/VIII.1.1999 (Opler & E. Buckner-Opler) CSU

28.22 *Melanolophia imitata* (Walker) ♂ — British Columbia, Gabriola Island, IV.20.1988 (G. Anweiler) CSU

28.23 *Carphoides incopriarius* (Hulst) ♀ — Colorado, Las Animas Co., Spool Ranch, V.29.1994 (Opler & E. Buckner-Opler) CSU

28.24 *Galenara lixarioides* McDunnough ♂ — Wyoming, Albany Co., NE of Pole Mtn., VI.20.2003 (J. Nordin) CSU

28.25 *Vinemina opacaria* (Hulst) ♂ — Arizona, Cochise Co., Onion Saddle, VII.30/VIII.1.1999 (Opler & E. Buckner-Opler) CSU

28.26 *Astalotesia hollandi* Rindge ♂ — Arizona, Cochise Co., Ash Canyon Road, IV.20.2004 (C. Ferris) CDF

28.27 *Eufidonia discospilata* (Walker) ♂ — Alberta, Berrymoor Ferry, VI.20.1982 (J. Landry) CNC

28.28 *Biston betularia* (Linnaeus) ♂ — Colorado, Jefferson Co., Lookout Mtn., VII.7.1992 (Opler) CSU

28.29 *Cochisea recisa* Rindge ♂ — California, Los Angeles Co., Caswell's, Gorman, X.22.1946 (C. Smith) EME

28.30 *Cochisea sinuaria* (Barnes and McDunnough) ♂ — California, San Diego Co., 3 miles E of Julian, IX.25/29.1999 (N. Bloomfield) CSU

28.31 *Lycia rachelae* (Hulst) ♂ — Colorado, Larimer Co., Rocky Mtn. Natl. Park, IV.6.1991 (Opler) CSU

28.32 *Lycia ursaria* (Walker) ♂ — British Columbia, Clearwater, IV.4.1992 (G. Anweiler) CSU

28.33 *Phigalia plumogeraria* (Hulst) ♂ — Colorado, Larimer Co., Cache La Poudre River, IV.9.1996 (J. Nordin) CSU

28.34 *Hypagyrtis unipunctata* (Haworth) ♂ — Mississippi, Oktibeha Co., SW of Starkville, IV.19.1987 (R. & B. Brown) CSU

28.35 *Ematurga amitaria* (Guenée) ♂ — Wisconsin, Forest Co., Crandon, V.30.1937 (H. Bowers) CNC

28.36 *Paleacrita vernata* (Peck) ♂ — Colorado, Larimer Co., Fort Collins, IV.19.1990 (Opler) CSU

28.37 *Paleacrita longiciliata* (Hulst) ♂ — California, Contra Costa Co., Walnut Creek, XI.30.1968 (Powell) EME

28.38 *Erannis vancouverensis* Hulst ♂ — Colorado, Larimer Co., Rocky Mtn. Natl. Park, X.7.1997 (Opler) CSU

28.39 *Lomographa semiclarata* Walker ♀ — Colorado, Larimer Co., Roosevelt Natl. Forest, V.21.1994 (Opler & E. Buckner-Opler) CSU

28.40 *Sericosema juturnaria* (Guenée) ♀ — Colorado, Jefferson Co., Coal Creek Canyon, VII.3.1988 (Opler) CSU

28.41 *Cabera erythremaria* Guénee ♂ — Wyoming, Albany Co., Road 712H, 8,200′, VI.29.2000 (J. Nordin) CSU

28.42 *Apodrepanulatrix litaria* (Hulst) ♂ — California, Marin Co., Alpine Lake, IV.20.1974, reared from *Ceanothus thyrsifolius*, emerged II.15.1975 (Powell) EME

28.43 *Apodrepanulatrix litaria* (Hulst) ♂ — California, Humboldt Co., Kneeland, III.29.2003 (R. Wielgus) EME

PLATE 28

PLATE 29 **Geometridae**

29.1 *Apodrepanulatrix litaria* (Hulst) ♂ — California, San Bernardino Co., Lake Arrowhead, IX.28.1946 (C. Smith) EME

29.2 *Drepanulatrix monicaria* (Guenée) ♂ — California, Humboldt Co., Kneeland, II.12.2003 (R. Wielgus) EME

29.3 *Drepanulatrix monicaria* (Guenée) ♀ — California, Humboldt Co., Kneeland, XI.12.2002 (R. Wielgus) EME

29.4 *Drepanulatrix carnearia* (Hulst) ♂ — California, Marin Co., Alpine Lake, IV.25.1958 (Powell) EME

29.5 *Drepanulatrix carnearia* (Hulst) ♀ — California, Humboldt Co., Kneeland, VI.2.2003 (R. Wielgus) EME

29.6. *Eudrepanulatrix rectifascia* Hulst ♂ — California, San Diego Co., Palomar Mtn. Road, VIII.3/7.1999 (T. Mustelin & N. Bloomfield) CSU

29.7 *Ixala desperaria* (Hulst) ♂ — Colorado, Larimer Co., Rocky Mtn. Natl. Park, VI.23.1995 (D. Katz) CSU

29.8 *Chloraspilates bicoloraria* Packard ♀ — Arizona, Cochise Co., Fort Huachuca, VIII.1.1996 (Opler & E. Buckner-Opler) CSU

29.9 *Stergamataea inornata* Hulst ♀ — Colorado, Mesa Co., Colorado Natl. Monument, VI.1.1996 (B. Rodgers family) CSU

29.10 *Euchlaena tigrinaria* (Guenée) ♀ — Colorado, Larimer Co., E of Glen Haven, VII.22.1994 (Opler & E. Buckner-Opler) CSU

29.11 *Xanthotype sospeta* (Drury) ♀ — Colorado, Larimer Co., Fort Collins, VII.3.1989 (Opler) CSU

29.12 *Aspitates orciferarius* (Walker) ♂ — Yukon Territory, Windy Pass, VI.21.1987 (J. Troubridge) CNC

29.13 *Aspitates orciferarius* (Walker) ♀ — Yukon Territory, LaForce Lake, 5,300′, VII.1.1960 (J. Martin) CNC

29.14 *Pero meskaria* (Packard) ♀ — Colorado, Las Animas Co., Pinyon Canyon, VII.13.1991 (Opler) CSU

29.15 *Pero behrensaria* (Packard) ♀ — Colorado, Larimer Co., Rocky Mtn. Natl. Park, V.31.1993 (Opler & E. Buckner-Opler) CSU

29.16 *Pero mizon* Rindge ♀ — British Columbia, Gabriola Island, VII.13.1988 (G. Anweiler) CSU

29.17 *Pero behrensaria* (Packard) ♀ — Wyoming, Albany Co., Laramie Range, VI.17.2002 (J. Nordin) CSU

29.18 *Pero occidentalis* (Hulst) ♂ — California, Humboldt Co., Kneeland, VI.12.2003 (R. Wielgus) EME

29.19 *Ceratonyx permagnaria* (Grossbeck) ♂ — New Mexico., Grant Co., Gila Natl. Forest, VI.22.1988 (C. Covell) CSU

29.20 *Phaeoura mexicanaria* (Grote) ♂ — Colorado, Larimer Co., Rocky Mtn. Natl. Park, VII.21.1991 (Opler) CSU

29.21 *Phaeoura cristifera* Hulst ♀ — Colorado, Larimer Co., Viestenz-Smith City Park, V.22.1996 (Opler & E. Buckner-Opler) CSU

29.22 *Holochroa dissociarius* (Hulst) ♂ — Arizona, Pima Co., Box Canyon, VIII.2.1991 (Opler) CSU

29.23 *Aethaloida packardaria* (Hulst) ♀ — California, Butte Co., 10 miles ESE of Chico, VIII.11.1993 (Opler & E. Buckner-Opler) CSU

29.24 *Gabriola dyari* Taylor ♀ — California, Napa Co., NE of Angwin, V.20.1980, reared from *Pseudotsuga menziesii* (J. Powell & J. DeBenedictis) EME

29.25 *Gabriola dyari* Taylor ♀ — California, San Luis Obispo Co., NE of Cambria, III.24/25.1980, reared from *Pinus radiata* (J. Powell) EME

29.26 *Yermoia perplexata* McDunnough ♀ — Colorado, Mesa Co., Colorado Natl. Monument, IX.21.1995 (B. Rodgers family) CSU

29.27 *Animomyia smithii* (Pearsall), ♂ — Colorado, Washington Co., Akron, IX.19.1990 (Opler) CSU

29.28 *Animomyia smithii* (Pearsall) ♀ — California, San Bernardino Co., SE of Trona, larva collected IV.3.1983, adult emerged VIII.4.1983, reared from *Chrysothamnus* and *Gutierezia* (Powell) EME

29.29 *Campaea perlata* (Guenée) ♂ — Colorado, Routt Co., Ferndale Recreation Area, VII.16.1995 (T. Dickel) CSU

29.30 *Ennomos magnaria* Guenée ♂ — Colorado, Larimer Co., Rocky Mtn. Natl. Park, X.8.1991 (Opler & T. Dickel) CSU

29.31 *Spodolepis substriataria* Hulst ♀ — Colorado, Grand Co., Beaver Creek Road, V.25.1988 (T. Dickel) CSU

29.32 *Philedia punctomacularia* (Hulst) ♀ — British Columbia, Gabriola Island, X.16.1988 (G. Anweiler) CSU

29.33 *Thallophaga hyperborea* (Hulst) ♀ — British Columbia, Gabriola Island, III.12.1987 (G. Anweiler) CSU

29.34 *Thallophaga taylorata* (Hulst) ♀ — Oregon, Clackamas Co., SE of Zigzag, IV.22.1994 (Opler) CSU

29.35 *Slossonia rubrotincta* Hulst ♂ — California, Los Angeles Co., Bouquet Canyon, VII.8.1945 (C. Smith) EME

29.36 *Selenia alciphearia* Walker ♀ — Oregon, Clatsop Co., Elsie, V.19.1971, CSU

29.37 *Metanema inatomaria* Guenée ♀ — Colorado, Larimer Co., Rocky Mtn. Natl. Park, VII.2.1990 (Opler) CSU

29.38 *Metarrhanthis duaria* (Guenée) ♀ — Colorado, Larimer Co., Viestenz-Smith City Park, V.15.1994 (Opler & E. Buckner-Opler) CSU

29.39 *Probole amicaria* Herrich-Schäffer ♂ — Alberta, Edmonton, V.26.1994 (G. Anweiler) CSU

29.40 *Philtraea utahensis* Buckett ♀ — Colorado, Mesa Co., Colorado Natl. Monument, VI.6.2000 (Opler & E. Buckner-Opler) CSU

29.41 *Eriplatymetra coloradaria* (Grote and Robinson) ♀ — Colorado, Larimer Co., Rocky Mtn. Natl. Park, VII.14.1990 (Opler) CSU

PLATE 29

PLATE 30 **Geometridae**

30.1 *Melemaea magdalena* Hulst ♂ — Colorado, Larimer Co., Rocky Mtn. Natl. Park, VIII.27.1994 (Opler) CSU

30.2 *Melemaea virgata* Taylor ♂ — Arizona, Graham Co., 8820,' X.1.2006 (C. Ferris) CDF

30.3 *Lychnosea helveolaria* Hulst ♂ — Colorado, Larimer Co., Rocky Mtn. Natl. Park, VII.29.1994 (R. Muckenthaler) CSU

30.4 *Neoterpes trianguliferata* (Packard) ♀ — Montana, Carbon Co., Cooney Reservoir State Park, V.31.1997 (C. Harp) CSU

30.5 *Neoterpes edwardsata* (Packard) ♀ — California, Sonoma Co., Petaluma, VI.28.1937 (E. Johnston) CNC

30.6 *Neoterpes edwardsata* (Packard) ♂ — California, Sonoma Co., Petaluma, II.1.1938 (E. Johnston) CNC

30.7 *Caripeta aequaliaria* Grote ♀ — Colorado, Larimer Co., Rocky Mtn. Natl. Park, VII.2.1990 (Opler) CSU

30.8 *Caripeta ocellaria* (Grossbeck) ♀ — Arizona, Santa Cruz Co., Madera Canyon, XI.10.1958 (R. Leuschner) CNC

30.9 *Caripeta pulcherrima* Guedet ♂ — Arizona, Cochise Co., Upper Pinery Canyon, VII.30/31.1996 (Opler & E. Buckner-Opler) CSU

30.10 *Snowia montanaria* Neumoegen ♀ — Arizona, Cochise Co., Pinery Canyon, VII.30/31.1996 (Opler & E. Buckner-Opler) CSU

30.11. *Nemeris speciosa* (Hulst) ♂ — Wyoming, Albany Co., Woods Landing, VIII.7.1999 (J. Nordin) CSU

30.12 *Meris alticola* Hulst ♀ — Colorado, Larimer Co., SE of Glen Haven, VIII.11/12.1995 (Opler & E. Buckner-Opler) CSU

30.13 *Destutia excelsa* Strecker ♂ — Utah, San Juan Co., SW of Monticello, V.11.1996 (J. Nordin) CSU

30.14 *Besma quercivoraria* (Guenée) ♀ — British Columbia, Gabriola Island, VI.24.1988 (G. Anweiler) CSU

30.15. *Lambdina fiscellaria* (Guenée) ♂ — Oregon, Clatsop Co., Elsie, IX.10.1984, CSU

30.16 *Nepytia umbrosaria* Packard ♂ — Colorado, La Plata Co., N of Hermosa, VIII.30.1991(Opler) CSU

30.17 *Nepytia phantasmaria* (Strecker) ♀ — California, Humboldt Co., Board Camp Mtn., X.11.2004 (R. Wielgus) EME

30.18 *Sicya macularia* Harris ♂ — Colorado, Larimer Co., Rocky Mtn. Natl. Park, VIII.15.1990 (Opler) CSU

30.19 *Sicya morsicaria* (Hulst) ♀ — Utah, San Juan Co., Canyonlands Natl. Park, VI.6.1994 (Opler, E. Buckner-Opler, & B. Kondratieff) CSU

30.20 *Eucaterva variaria* Grote ♀ — Arizona, Santa Cruz Co., Pena Blanca Lake, VIII.9/11.1999 (Opler & E. Buckner-Opler) CSU

30.21 *Plataea trilinearia* (Packard) ♂ — Colorado, Moffat Co., Dinosaur Natl. Monument, VI.9.1994 (T. Dickel) CSU

30.22 *Plataea personaria* (Henry Edwards) ♂ — California, San Luis Obispo Co., Dune Lakes, VI.1.1972 (Powell) EME

30.23 *Plataea californiaria* Herrich-Schäffer ♀ — California, Orange Co., Silverado Canyon, IV.15.1979 (Powell) EME

30.24 *Eusarca falcata* Packard ♂ — California, Los Angeles Co., La Tuna Canyon, VI.13.1956, reared from *Artemisia californica* (W. Evans) CNC

30.25 *Somatolophia ectrapelaria* Grossbeck ♂ — California, San Bernardino Co., Granite Hills, V.24.1939 (G. & J. Sperry) CNC

30.26 *Somatolophia cuyama* J.A. Comstock, paratype ♂ — California, San Luis Obispo Co., Cuyama Valley, VI.4.1937, reared from *Hymenoclea salsola* (D. Tiemann) CNC

30.27 *Pherne subpunctata* Hulst ♀ — California, Los Angeles Co., Dalton Canyon, IV.11.1927, CNC

30.28 *Synaxis cervinaria* (Packard) ♀ — California, Los Angeles Co., La Tuna Canyon, IV.22.1949 (W. Evans) CNC

30.29 *Synaxis cervinaria* (Packard) ♀ — California, Lake Co., Mt. San Hedron, VI.23.1936 (E. Johnston) CNC

30.30 *Synaxis jubararia* (Hulst) ♀ — Alberta, Lethbridge, IX.19.1922 (H. Seamans) CNC

30.31 *Synaxis formosa* (Hulst) ♂ — Colorado, La Plata Co., Durango, IX.26.1945 (E. Johnston) CNC

30.32 *Tetracis cachexiata* Guenée ♀ — Ontario, Ottawa, VI.13.1949 (T. Freeman) CNC

30.33 *Phyllodonta peccataria* Barnes and McDunnough ♀ — Arizona, Cochise Co., Huachuca Mtns.,V.5.1949 (A. Melander) CNC

PLATE 30

PLATE 31 **Geometridae**

31.1 *Prochoerodes forficaria* (Guenée) ♀ — California, Marin Co., Inverness, I.28.1940 (E. Johnston) CNC

31.2 *"Pityeja"ornata* Rindge ♂ — Arizona, Cochise Co., Chiricahua Mtns., VII.30/VIII.1.1999 (Opler & E. Buckner-Opler) CSU

31.3 *Sabulodes aegrotata* (Guenée) ♀ — California, Sonoma Co., Petluma, VIII.27.1938 (E. Johnston) CNC

31.4 *Sabulodes edwardsata* (Hulst) ♀ — British Columbia, Penticton, VII.7.1954 (Forest Insect Survey) CNC reared from *Pinus ponderosa*

31.5 *Sabulodes niveostriata* (Cockerell) ♀ — Colorado, Clear Creek Co., Mt. Evans, VII.31.1961 (E. Rockburne) CNC

31.6 *Enypia venata* (Grote) ♂ — British Columbia, Lihumpton Park, VIII.10.1927 (C. Young) CNC

31.7 *Nematocampa resistaria limbata* (Haworth) ♂, Washington, Okanogan Co., Black Canyon, VII.1.1949 (E. Johnston) CNC

31.8 *Alsophila pometaria* Harris ♂ — Alberta, Lethbridge, X.17.1922 (H. Seamans) CNC

31.9 *Alsophila pometaria* Harris ♂ — Colorado, Larimer Co., Fort Collins, 5,100′, III.23.1991 (Opler) CSU

31.10 *Alsophila pometaria* Harris ♀ — Alberta, Seven Perse, X.29.1955, reared from *Ulmus* (Forest Insect Survey) CNC

31.11 *Chlorosea banksaria* Sperry ♂ — Washington, Chelan Co., Berne, VII.12.1942 (E. Johnston) CNC

31.12 *Nemoria pulcherrima* (Barnes & McDunnough) ♀, brown form — California, Sacramento Co., Sacramento, III.3.1937 (E. Johnston) CNC

31.13 *Nemoria darwiniata* (Dyar) ♂ — Washington, King Co., Factoria, VI.4.1949 (E. Johnston) CNC

31.14 *Nemoria pulcherrima* (Barnes & McDunnough) ♂, green form — California, Marin Co., Inverness, I.28.1940 (E. Johnston) CNC

31.15 *Nemoria arizonaria* (Grote) ♀, spring form — Arizona, Cochise Co., Chiricahua Mtns., IV.13.2005 (C. Ferris) CDF

31.16 *Nemoria arizonaria* (Grote) ♀, summer form — Arizona, Cochise Co., Chiricahua Mtns., VIII.11.1990 (C. Ferris) CDF

31.17 *Nemoria darwiniata* (Dyar) ♂ — Washington, King Co., Factoria, VI.11.1949 (E. Johnston) CNC Geometroidea

31.18 *Nemoria leptalea* Ferguson ♀ — California, San Diego Co., La Mesa, IV.14.1950 (E. Johnston) CNC

31.19 *Dichorda illustraria* (Hulst) ♀ — California, San Diego Co., La Mesa, IV.14.1950 (E. Johnston) CNC

31.20 *Synchlora aerata* (Fabricius) ♂ — Newfoundland, Doyles, VIII.6.1962 (D. Ferguson) CNC

31.21 *Synchlora bistriaria* (Packard) ♂ — Arizona, Navaho Co., Showlow, VI.25.1950 (E. Johnston) CNC

31.22 *Chlorochlamys appellaria* Pearsall ♂ — New Mexico, Eddy Co., White City, V.14.1950 (E. Johnston) CNC

31.23 *Xerochlora viridipallens* (Hulst) ♂ — New Mexico, Grant Co., Signal Peak, VIII.18.2004 (C. Ferris) CDF

31.24 *Hemithea aestivaria* (Hübner) ♂ — British Columbia, E of Longley, VII.29/VIII.4.1990 (J. Troubridge) CNC

31.25 *Mesothea incertata* (Walker) ♂ — Alberta, Kananaskis Forest Experiment Station, I.21.1963 (W. McGuffin) CNC

31.26 *Lobocleta plemyraria* Guenée ♂ — California, Los Angeles Co., Hawthorne, IX.15.1944 (E. Johnston) CNC

31.27 *Idaea demissaria* (Hübner) ♀ — British Columbia, Seton Lake, VIII.9.1933 (J. McDunnough) CNC

31.28 *Idaea bonifata* (Hulst) ♂ — New Mexico, Sandoval Co., Jemez Springs, VII.25.1925 (J. Woodgate) CNC

31.29 *Cyclophora dataria* (Hulst) ♂ — New Mexico, McKinley Co., McGaffey, VII.24.1962 (E. & I. Munroe) CNC

31.30 *Pigia multilineata* (Hulst) ♀ — Arizona, Cochise Co., Ash Canyon Road, VIII.6.1996 (Opler & E. Buckner-Opler) CSU

31.31 *Haematopis grataria* Fabricius ♀ — Missouri, Cooper Co., Boonville, IX.22.1945 (E. Johnston) CNC

31.32 *Scopula junctaria* (Walker) ♀ — British Columbia, Keremeos, VII.6.1936 (A. Gartell) CNC

31.33 *Scopula sentinaria* (Geyer) ♂ — Alberta, Waterton Lakes, VI.26.1929 (J. Pepper) CNC

31.34 *Leptostales rubromarginaria* (Packard) ♂ — California, Sonoma Co., Spring Mtn., IV.10.1938 (E. Johnston) CNC

31.35 *Dysstroma formosa* (Hulst) ♀ — British Columbia, Keremeos, VI.23.1923 (O. Garrett) CNC

31.36 *Dysstroma formosa* (Hulst) ♀ — British Columbia, Hedley, VI.25.1967 (W. McGuffin) CNC reared from *Ribes cereum*

31.37 *Dysstroma citrata* (Walker) ♀ — British Columbia, Lillooet, IX.18.1938 (J. Jacob) CNC

31.38. *Eulithis xylina* (Hulst) ♂ — California, Plumas Co., Nelson Creek, VIII.11.1940 (W. Bauer) CNC

31.39 *Eustroma semiatratra* (Hulst) ♀ — Washington, Cowlitz Co., Toutle, V.30.1947 (E. Johnston) CNC

31.40 *Ecliptoptera silaceata* (Denis and Schiffermüller) ♂ — Washington, King Co., Seattle, IX.12.1941 (E. Johnston) CNC

31.41 *Plemyria georgii* (Hulst) ♂ — British Columbia, Agassiz, VIII.4.1915 (R. Treherne) CNC

31.42 *Thera otisi* (Dyar) ♀ — Alberta, Snaring River, VIII.10.1953, reared from *Juniperus* (Forest Insect Survey) CNC

31.43 *Ceratodalia gueneata* (Packard) ♀, dorsal — British Columbia, Seton Lake, VI.2.1926 (J. McDunnough) CNC

31.44 *Ceratodalia gueneata* (Packard) ♀ — ventral — British Columbia, Seton Lake, VI.2.1926 (J. McDunnough) CNC

31.45 *Hydriomena furcata* Borgstrom ♀ — Saskatchewan, Rutland, VII.30.1940 (A. Brooks) CNC

31.46 *Hydriomena nubilofasciata* Packard ♂ — California, Napa Co., Spring Mtn., I.26.1940 (E. Johnston) CNC

31.47 *Hydriomena nubilofasciata* Packard ♂ — California, Napa Co., Spring Mtn., I.14.1940 (E. Johnston) CNC

31.48 *Hydriomena renunciata* Walker ♀ — British Columbia, Manning Provincial Park, III.5.1962 (Forest Insert Survey) CNC

31.49 *Hydriomena albimontanata* McDunnough ♂ — Arizona, Coconino Co., Kaibab Lodge, VI.16.1938 (G. & J. Sperry) CNC

31.50 *Hydriomena macdunnoughi* Swett ♂ — Alberta, Nordegg, VI.8.1921 (J. McDunnough) CNC

31.51 *Hydriomena irata* Swett ♂ — California, Napa Co., Spring Mtn., III.13.1940 (E. Johnston) CNC

PLATE 31

PLATE 32 **Geometridae**

32.1 *Hydriomena manzanita* Taylor ♂ — California, Napa Co., Spring Mtn., III.13.1940 (E. Johnston) CNC

32.2 *Hydriomena manzanita* Taylor ♂ — California, Napa Co., Spring Mtn., III.25.1946 (E. Johnston) CNC

32.3 *Hymenodria mediodentata* (Barnes and McDunnough) ♂ — Arizona, Gila Co., Globe, V.1.1915 (S. Cassino) CNC

32.4 *Ersephila grandipennis* Hulst ♂ — Arizona, Yavapai Co., Prescott, III.26.1971 (L. Martin) CNC

32.5 *Triphosa haesitata* (Guenée) ♀ — Alberta, Lloydminster, X.4.1944 (P. Bruggemann) CNC

32.6 *Triphosa californiata* (Packard) ♀ — California, Napa Co., Mt. St. Helena, III.24.1939 (E. Johnston) CNC

32.7 *Coryphista meadi* (Packard) ♂, dorsal — Washington, Snohomish Co., Verlot, V.14.1949 (E. Johnston) CNC

32.8 *Coryphista meadi* (Packard) ♂, ventral — Washington, Snohomish Co., Verlot, V.14.1949 (E. Johnston) CNC

32.9 *Rheumaptera hastata* (Linnaeus) ♀ — Yukon Territory, Dawson, VI.21.1949 (P. Bruggemann) CNC

32.10 *Rheumaptera hastata* (Linnaeus) ♀ — Yukon Territory, Dawson, VI.21.1949 (P. Bruggemann) CNC

32.11 *Archirhoe neomexicana* Hulst ♀, dorsal — Arizona, Apache Co., Ft. Defiance, 1948 (E. Oberg) CNC

32.12 *Archirhoe neomexicana* Hulst ♀, ventral — Arizona, Apache Co., Ft. Defiance, 1948 (E. Oberg) CNC

32.13 *Entephria multivagata* (Hulst) ♂ — Washington, Whatcom Co., Ross Lake, IX.18.1993 (J. Troubridge) CNC

32.14 *Entephria polata* Duponchel ♂ — Yukon Territory, Dempster Highway, VII.8/12.1973 (G. & D. Wood) CNC

32.15 *Mesoleuca ruficillata* (Guenée) ♂ — British Columbia, Mission City, VII.9.1953 (W. Mason) CNC

32.16 *Spargania magnoliata* Guenée ♀ — Alberta, Kananaskis Forestry Station, VIII.11.1963, reared from fireweed (W. McGuffin) CNC

32.17 *Perizoma custodiata* (Guenée) ♀ — Washington, Yakima Co., Toppenish, V.16.1940 (E. Johnston) CNC

32.18 *Perizoma costiguttata* (Hulst) ♀ — British Columbia, Keremeos, VI.9.1935 (A. Gartrell) CNC

32.19 *Anticlea vasiliata* Guenée ♀ — Washington, Snohomish Co., Silverton, IV.21.1936 (E.C. Johnston) CNC

32.20 *Stamnodes morrisata* (Hulst) ♂, dorsal — New Mexico, McKinley Co., McGaffey, CNC

32.21 *Stamnodes morrisata* (Hulst) ♂, ventral — New Mexico, McKinley Co., McGaffey, CNC

32.22 *Stamnodes albiapicata* Grossbeck ♂, dorsal — California, Los Angeles Co., Burbank, XII.4.1961 (F. Sala) CSU

32.23 *Stamnodes albiapicata* Grossbeck ♂, ventral — California, Los Angeles Co., Burbank, XII.4.1961 (F. Sala) CSU

32.24 *Stamndes tessellata* (Packard) ♀, dorsal — Utah, Juab Co., Dividend, VI.22 (T. Spalding) CNC

32.25 *Stamndes tessellata* (Packard) ♀, ventral — Utah, Juab Co., Dividend, VI.22 (T. Spalding) CNC

32.26 *Stamnodes seiferti* (Neumoegen) ♂, dorsal — Arizona, Cochise Co., Fort Huachuca, X.5.1994 (D. Ferguson) CSU

32.27 *Stamnodes seiferti* (Neumoegen) ♂, ventral — Arizona, Cochise Co., Fort Huachuca, X.5.1994 (D. Ferguson) CSU

32.28 *Stamnodes animata* (Pearsall) ♀, dorsal — Nevada, Clark Co., Charleston Mtns., V.13.1934 (G. & J. Sperry) CNC

32.29 *Stamnodes animata* (Pearsall) ♀, ventral — Nevada, Clark Co., Charleston Mtns., V.13.1934 (G. & J. Sperry) CNC

32.30 *Stamnodes marmorata* (Packard) ♀ — California, Siskiyou Co., McCloud, VI.5.1935 (E. Johnston) CNC

32.31 *Stamnodes marmorata* (Packard) ♀ — California, Siskiyou Co., McCloud, VI.5.1935 (E. Johnston) CNC

32.32 *Stamnodes formosata* (Strecker) ♂ — Colorado, Chaffee Co., Buena Vista, VI.22.1961 (M. MacKay) CNC

32.33 *Stamnodes fervifactaria* (Grote) ♂, dorsal — Arizona, Apache Co., W of Eagar, VIII.11.1962 (E. & I. Munroe) CNC

32.34 *Stamnodes fervifactaria* (Grote) ♂, ventral — Arizona, Apache Co., W of Eagar, VIII.11.1962 (E. & I. Munroe) CNC

32.35 *Stamnodes topazata* (Strecker), white form ♂, dorsal — Washington, Mason Co., Stimson Creek, IV.3.1948 (E. Johnston) CNC

32.36 *Stamnodes topazata* (Strecker), white form ♂, ventral — Washington, Mason Co., Stimson Creek, IV.3.1948 (E. Johnston) CNC

32.37 *Stamnodes topazata* (Strecker), orange form ♀, dorsal — Colorado, Park Co., near Leavick, VI.24.1999 (S. Johnson) CSU

32.38 *Stamnodes topazata* (Strecker), orange form ♀, ventral — Colorado, Park Co., near Leavick, VI.24.1999 (S. Johnson) CSU

32.39 *Xanthorhoe munitata* Hübner ♂ — Alberta, Nordegg, VII.4.1921 (J. McDunnough) CNC

32.40 *Xanthorhoe defensaria* (Guenée) ♀ — California, Sonoma Co., Petaluma, III.9.1937 CNC

32.41 *Epirrhoe plebeculata* Guenée ♂ — California, Sonoma Co., Petaluma, III.24.1935 (E. Johnston) CNC

PLATE 32

PLATE 33 Geometridae, Doidae, Notodontidae

33.1 *Euphyia implicata* (Guenée) ♂ — California (J. Grossbeck) CNC

33.2 *Enchoria lacteata* (Packard) ♂ — British Columbia, Victoria, IV.7.1922 (W. Downes) CNC

33.3 *Zenophleps lignicolorata* Packard ♀ — California, Lake Co., Mt. San Hedron, IV.27.1939 (E. Johnston) CNC

33.4 *Orthonama obstipata* (Fabricius) ♀ — California, Sonoma Co., Petaluma III.7.1940 (E. Johnston) CNC

33.5 *Venusia cambrica* Curtis ♂ — Washington, Yakima Co., Mt. Baker, VII.18.1942 (E. Johnston) CNC

33.6 *Trichodezia albovittata* (Guenée) ♂ — Washington, Jefferson Co., Dosewallips River, VIII.4.1947 (D. Frechin) CNC

33.7 *Epirrita autumnata* Borkhaus ♀ — Alberta, Seebe Lake, IX.3.1955 (W. McGuffin) CNC

33.8 *Operophtera bruceata* Hulst ♂ — British Columbia, Langley, XI.29.1990 (J. Troubridge) CNC

33.9 *Operophtera bruceata* Hulst ♂ — California, Contra Costa Co., Walnut Creek, XII.20.1964 (Powell) EME

33.10 *Operophtera danbyi* (Hulst) ♂ — California, Contra Costa Co., Walnut Creek, XII.24.1971 (Powell) EME

33.11 *Operophtera danbyi* (Hulst) ♂ — British Columbia, Victoria, XI.27.1923 (W. Downes) CNC

33.12 *Tescalsia giulianata* Ferguson ♀ — California, Inyo Co., NE of Big Pine, I.13.1996 (D. Giuliani) EME

33.13 *Tescalsia giulianata* Ferguson ♂ — California, Inyo Co., NE of Big Pine, XII.2.1996 (D. Giuliani) EME

33.14 *Eubaphe unicolor* Robinson ♀ — Texas, Brewster Co., Big Bend Natl. Park, V.10.1959 (M. MacKay) CNC

33.15 *Eupithecia pseudotsugata* Mackay ♂, British Columbia, Oliver, I.29.1955 (Forest Insect Survey) CNC

33.16 *Eupithecia bryanti* Taylor ♂ — Washington, King Co., Factoria, VI.11.1949 (E. Johnston) CNC

33.17 *Eupithecia misturata* (Hulst) ♂ — Washington, Mason Co., Stinson Creek, V.7.1949 (E. Johnston) CNC

33.18 *Eupithecia anticaria* Walker ♂ — Colorado, Chaffee Co., E of Buena Vista, VII.17.1982 (J. Landry) CNC

33.19 *Eupithecia nevadata* Packard ♂ — California, Riverside Co., Aguanga, III.6.1976 (R. Leuschner) CNC

33.20 *Eupithecia ravocastaliata* Packard ♀ — Washington, King Co., Factoria, IV.9.1949 (E. Johnston) CNC

33.21 *Eupithecia niphadophilata* (Dyar) ♀ — Montana, Fergus Co., NE of Lewiston, VIII.27.1970 (D. Hardwick) CNC

33.22 *Eupithecia satyrata* (Dyar) ♀ — Prince Edward Island, Alberton, III.31.1941 (J. McDunnough) CNC reared from fireweed

33.23 *Eupithecia nimbicolor* (Hulst) ♂ — Arizona, White Mtns., VI.17.1935 (G. & J. Sperry) CNC

33.24 *Eupithecia cretaceata* (Packard) ♀ — Colorado, Grand Co., St. Louis Creek Road, VIII.15.1995 (T. Dickel) CSU

33.25 *Eupithecia perfusca* (Hulst) ♀ — British Columbia, Kaslo (J. Cockle) CNC

33.26 *Eupithecia multistrigata* (Hulst) ♂ — Nevada, Elko Co., Wells, VIII.31.1969 (D. Hardwick) CNC

33.27 *Eupithecia edna* (Hulst) ♀ — Colorado, Grand Co., Radium State Wildlife Area, VI.14.1993 (T. Dickel) CSU

33.28 *Eupithecia flavigutta* (Hulst) ♂ — Arizona, White Mtns., VI.26.1935 (G. & J. Sperry) CNC,

33.29 *Eupithecia biedermanata* Cassino & Swett ♀ — Arizona, Cochise Co., Huachuca Mtns., III.25.1986 (R. Wielgus) USNM

33.30 *Eupithecia graefii* (Hulst) ♂ — California, Siskiyou Co., Mt. Shasta, VII.24.1965 (E. & I. Munroe) CNC

33.31 *Eupithecia intricata taylorata* Swett ♀ — British Columbia, Kaslo, XI.20.1969 (Forest Insect Survey) CNC reared from *Thuja plicata*

33.32 *Eupithecia lafontaineata* Bolte ♀ — Colorado, Teller Co., Sanborn Ranch, VII.18.1982 (G. Balogh) CNC

33.33 *Eupithecia appendiculata* McDunnough ♂ — California, San Gabriel Mtns., VIII.22.1971 (R. Leuschner) CNC

33.34 *Nasusina vaporata* Pearsall ♂ — California, Santa Cruz Co., Santa Cruz, V.20, CNC

33.35 *Prorella opinata* Barnes and McDunnough ♀ — New Mexico, Sandoval Co., Frijoles Canyon, IX.8.1941 (G. & J. Sperry) CNC

33.36 *Lithostege rotundata* Packard ♂ — California, San Bernardino Co., Apple Valley, VI.13.1955 (D. Hardwick) CNC

33.37 *Cladara limitaria* (Walker) ♀ — Washington, Mason Co., Shelton, IV.16.1949 (E. Johnston) CNC

33.38 *Lobophora magnoliatoidata* (Dyar) ♀ — Yukon Territory, Dawson, VI.10.1940 (W. Judd) CNC

33.39 *Doa ampla* Grote ♂ — Arizona, Cochise Co., Ash Canyon Road, VIII.8.1998 (N. McFarland) CNC

33.40 *Leuculodes lacteolaria* Hulst ♂ — Arizona, Santa Cruz Co., Pena Blanca Lake, VIII.9/11.1999 (Opler & E. Buckner-Opler) CSU

33.41 *Clostera albosigma* Fitch ♂ — Saskatchewan, Christopher Lake, VI.21.1938 (A. Brooks) CNC

33.42 *Clostera apicalis* (Walker) ♂ — California, Alameda Co., Oakland, V.4.1908 (G.R. Pilate) CNC

33.43 *Hyperaeschra georgica tortuosa* Tepper ♂ — Utah, Morgan Co., Porterville, VII.6.1939 (C. Knowlton & F. Harmston) CNC

33.44 *Odontosia elegans* (Strecker) ♂ — Colorado, Boulder Co., Nederland, VI.29.1961 (M. MacKay) CNC

33.45 *Gluphisia septentrionis* (Walker) ♂ — Arizona, Yavapai Co., Prescott, V.6.1950 (E. Johnston) CNC

33.46 *Gluphisia septentrionis* (Walker) [subsp. *albafascia*] ♀ — Utah, Utah Co., Pleasant Grove, VI.24.1927 (G. Knowlton & L. Hansen) CNC

33.47 *Gluphisia septentrionis* (Walker) [subsp. *wrightii*] ♀ — Arizona, Cochise Co., Sunnyside, IX.7.1966 (R. Sternitzky) CNC

33.48 *Gluphisia severa* Henry Edwards ♂ — Washington, Yakima Co., Kusshi Canyon, V.28.1949 (E. Johnston) CNC

33.49 *Furcula cinerea* (Walker) ♂ — California, Sonoma Co., Petaluma, V.29.1938 (E. Johnston) CNC

33.50 *Datana perspicua mesillae* Cockerell ♀ — California, San Bernardino Co., Clark Mtn., VII.27.1935 (C. Dammers) CNC

33.51 *Furcula scolopendrina* (Boisduval) ♀ — British Columbia, Seton Lake, VIII.12.1933 (J. McDunnough) CNC

33.52 *Datana integerrima cochise* Dyar ♂ — New Mexico, Eddy Co., Carlsbad, V.17.1950 (E. Johnston) CNC

33.53 *Datana neomexicana* Doll ♂ — Texas, Culberson Co., Sierra Diablo Wildlife Management Area, VII.14.1971 (A. & M. Blanchard) CNC

33.54 *Datana ministra californica* Dyar ♂ — California, Sonoma Co., Petaluma, VIII.1.1939 (E. Johnston) CNC

33.55 *Nadata gibbosa* (J.E. Smith) ♂ — British Columbia, Fraser River Canyon (L. Crabo) CNC

33.56 *Nadata oregonensis* Butler ♂ — Oregon, Josephine Co., SE of Williams, VI.28.1997 (J. Troubridge) CNC

PLATE 33

PLATE 34 **Lasiocampidae**

34.1 *Gloveria arizonensis* Packard ♂ — California, Riverside Co., Whitewater, IX.12.1930 (C. Dammers) CNC

34.2 *Gloveria arizonensis* Packard ♀ — California, Riverside Co., Whitewater, IX.9.1929 (C. Dammers) CNC

34.3 *Gloveria gargamelle* (Strecker) ♂ — Arizona, Cochise Co., Cave Creek, IX, CNC

34.4 *Gloveria gargamelle* (Strecker) ♀ — Arizona, Cochise Co., Cave Creek, IX, CNC

34.5 *Gloveria medusa* (Strecker) ♂ — California, Los Angeles Co., LaTuna Canyon, VIII.30.1949 (W. Evans) CNC

34.6 *Gloveria medusa* (Strecker) ♀ — California, Los Angeles Co., LaTuna Canyon, IX.2.1949, reared on *Eriogonum fasciculatum* (W. Evans) CNC

34.7 *Gloveria howardi* ♂ — no data, CNC

34.8 *Quadrina diazoma* Grote ♂ — Arizona, Cochise Co., Cave Creek, IX, CNC

34.9 *Quadrina diazoma* Grote ♀ — Arizona, Cochise Co., Cave Creek, IX, CNC

34.10 *Dichogaster coronada* (Barnes) ♂ — Arizona (O. Bryant) CNC

34.11 *Dichogaster coronada* (Barnes) ♀ — Arizona, Cochise Co., Ash Canyon Road, VIII.2.1990(B. Landy & J. Landry) CNC

PLATE 34

PLATE 35 Lasiocampidae, Apatelodidae, Sematuridae, Saturniidae

35.1 *Caloecia juvenalis* (Barnes and McDunnough) ♂ — Arizona, Cochise Co., Onion Saddle, Chirichaua Mtns., VII.15.1966 (J. Franclemont) CNC

35.2 *Caloecia entima* Franclemont ♀ — Mexico, Durango, W of El Salto, VI.8.1964 (J. Martin) CNC

35.3 *Apatelodes pudefacta* Dyar ♂ — Arizona, Cochise Co., Miller Canyon, VII.30.1967 (R. Sternitzky) CNC

35.4. *Anurapteryx crenulata* Barnes & Lindsey ♂ — Arizona, Cochise Co., Huachuca Mts., Ash Canyon Road, 5100′ (N. McFarland) CNC

35.5 *Eacles oslari* Rothschild ♂ — Arizona, Cochise Co., Ash Canyon Road, VIII.2.1988 (B. Landry & J. Landry) CNC

35.6 *Citheronia splendens sinaloaensis* Hoffman ♂ — Arizona, Cochise Co., Peloncillo Mtns., VII.31.1993 (B. Landry & J. Landry) CNC

35.7 *Olceclostera seraphica* (Dyar) ♂ — New Mexico, Eddy Co., White City, V.15.1950 (E. Johnston) CNC

35.8 *Olceclostera seraphica* (Dyar) ♀ — New Mexico, Eddy Co., White City, V.16.1950 (E. Johnston) CNC

35.9 *Anisota oslari* Rothschild ♂ — Arizona, Pima Co., Texas Canyon, August, CNC

35.10 *Anisota oslari* Rothschild ♀ — New Mexico, McKinley Co., Fort Wingate, CNC

PLATE 35

PLATE 36 **Saturniidae**

36.1 *Coloradia pandora* Blake ♂ — California, Plumas Co., Almanor, VIII.3.1965 (E. & I. Munroe) CNC

36.2 *Coloradia pandora* subsp. *davisi* Barnes & Benjamin ♂ — Nevada, Lincoln Co., Panaca, IX.2.1965 (E. & I. Munroe) CNC

36.3 *Hemileuca nevadensis* Stretch ♂ — Alberta, E of Edgerton, IX.8.2001 (G. Anweiler & C. Schmidt) CNC

36.4 *Coloradia doris* Barnes ♂ — Colorado, Boulder Co., Silver Saddle Motel, VI.13.1961 (M. MacKay) CNC

36.5 *Hemileuca nuttalli* (Strecker) ♂ — Washington, Okanogan Co., Brewster, VIII.7.1944 (J. Hopfinger) CNC

36.6 *Hemileuca hera* (Harris) ♂ — Wyoming, Sublette Co., N of Cora, VII.23.1950 (K. Wilson) CNC

36.7 *Hemileuca eglanterina* (Boisduval) ♂ — California, Los Angeles Co., Angeles Crest Highway, VI.20.1949 (W. Evans) CNC

36.8 *Hemileuca electra* W.G. Wright ♀ — California, Riverside Co., Riverside, XI.8.1933 (C. Dammers) CNC

36.9 *Sphingicampa raspa* (Boisduval) ♂ — Arizona, VI.10.1937, CNC

36.10 *Sphingicampa montana* (Packard) ♂ — no data, CNC

36.11 *Sphingicampa hubbardi* (Dyar) ♂ — Arizona, Yavapai Co., Cottonwood, IX.1.1949 (D. Bauer) CNC

36.12 *Hemileuca tricolor* (Packard) ♂ — Arizona, Pima Co., Tucson, I.28.1936 (O. Bryant) CNC

36.13 *Hemileuca hualapai* (Neumoegen) ♂ — Texas, Presidio Co., Presidio, XI.13.1928 (E. Tinkham) CNC

PLATE 36

PLATE 37 **Saturniidae, Lasiocampidae**

37.1 *Hemileuca burnsi* Watson ♂ — California, San Bernardino Co., Oro Grande Wash, X.2.1995 (D. Rubinoff & S. McElfresh) CNC

37.2 *Automeris zephyria* Grote ♂ — New Mexico, Otero Co., High Rolls, CNC

37.3 *Hemileuca neumoegeni* Henry Edwards ♂ — Arizona, Coconino Co., Kaibab Plateau, VIII.20.1960 (E. Munroe) CNC

37.4 *Automeris io* (Fabricius) ♂ — Colorado, Boulder Co., Silver Saddle Motel, 5,300′, VI.18.1961 (M. MacKay) CNC

37.5 *Agapema anona* (Ottolengui) ♂ — Texas (Frey & Boll) CNC

37.6 *Antheraea polyphemus* (Cramer) ♀ — California, Sonoma Co., Santa Rosa, IV.21.1936 (E.C. Johnston) CNC

37.7 *Agapema homogena* Dyar ♂ — Colorado, Boulder Co., Silver Saddle Motel, VI.17.61 (M.R. MacKay) CNC

37.8 *Cincinnus melsheimeri* (Harris) ♂ — South Carolina, Charleston Co., N of McClellanville, V.31.1977 (E. & I. Munroe) CNC

37.9 *Malacosoma californicum* (Packard) ♂ — California, Sonoma Co., Petaluma, V.11.1940 (E. Johnston) CNC

37.10 *Malacosoma disstria* Hübner ♂ — Utah, Utah Co., Logan, VII.12.1955 (S. Wood) CNC

37.11 *Antheraea oculea* (Neumoegen) ♂ — Arizona, Cochise Co., Peloncillo Mtns., VII.30.1999 (B. & J.F. Landry) CNC

PLATE 37

PLATE 38 Saturniidae, Lasiocampidae

38.1 *Eupackardia calleta* (Westwood) ♂ — Texas, Hidalgo Co., San Benito, CNC

38.2 *Saturnia albofasciata* (Johnson) ♂ — California, San Diego Co., Laguna Mtns., X.30.1997 (D. Rubinoff & J. Kruse) CNC

38.3 *Saturnia mendocino* Behrens ♀ — California, Lake Co., Clear Lake, III.1936 (W. Hooten) CNC

38.4 *Apotolype brevicrista* (Dyar) ♂ — Arizona (O. Bryant) CNC

38.5 *Artace colaria* Franclemont ♂ — Arizona, Cochise Co., Ramsey Canyon, XI.17.1967 (R. Sternitzky) CNC

38.6 *Phyllodesma americana* (Harris) ♂ — Colorado, Boulder Co., Nederland, VI.29.1961 (M. MacKay) CNC

38.7 *Hyalophora euryalus* (Boisduval) ♂ — California, San Diego Co., La Mesa, IV.14.1950 (E. Johnston) CNC

38.8 *Hyalophora columbia gloveri* (Strecker) ♂ — Colorado, Boulder Co., Silver Saddle Motel, VII.6.1961 (M. MacKay) CNC

38.9 *Tolype glenwoodi* Barnes ♂ — Colorado, Huerfano Co., SW of Gardner, IX.10.1968 (D. Hardwick) CNC

38.10 *Tolype* species ♂ — California, Riverside Co., N of Anza, IX.5.1996 (D. Hardwick) CNC

38.11 *Tolype* species ♀ — California, Los Angeles Co., Pine Canyon, VIII.8.1941, reared from *Cercocarpus betuloide* (C. Henne) CNC

PLATE 38

PLATE 39 **Sphingidae**

39.1 *Pachysphinx occidentalis* (Edwards) ♂ — Arizona, Cochise Co., Guadalupe Canyon, VII.30.1989 (B. Landry & J. Landry) CNC

39.2 *Pachysphinx modesta* (Harris) ♂ — California, Boulder Co., Nederland, V.1.1961 (M. MacKay) CNC

39.3 *Smerinthus jamaicensis* (Drury) ♂ — British Columbia, Chilcotin River, VI.18.1994 (J. Troubridge) CNC

39.4 *Paonias myops* (J.E. Smith) ♂ — British Columbia, SE of Okanogan Falls, IV.26/30.1992 (J. Troubridge & Gardiner) CNC

39.5 *Smerinthus cerisyi* Kirby ♀ — California, Sonoma Co., Santa Rosa, III.16.1939 (E. Johnston) CNC

39.6 *Sagenosoma elsa* (Strecker) ♀ — Arizona, CNC

39.7 *Smerinthus saliceti* Boisduval ♂ — Arizona, Santa Cruz Co., Atascosa Mtns., Sycamore Canyon, VIII.7.2005 (B. Schmidt) CNC

39.8 *Sphinx chersis* (Hübner) ♂ — New Mexico, Sandoval Co., Frijoles Canyon, VII.18.1962 (E. & I. Munroe) CNC

39.9 *Sphinx asella* (Rothschild and Jordan) ♂ — Colorado, Eagle Co., State Bridge, VI.23.1961 (M. MacKay) CNC

PLATE 39

PLATE 40 **Sphingidae**

40.1 *Sphinx dollii* Neumoegen ♂ — Arizona, Santa Cruz Co., Madera Canyon, VII.21.1947 (J. Comstock & L. Martin) CNC

40.2 *Sphinx sequoiae* Boisduval ♂ — California, Placer Co., Forest Hill, VI.30.1937, CNC,

40.3 *Sphinx perelegans* Edwards ♀ — California, San Diego Co., San Diego, IV.14.1950 (E. Johnston) CNC

40.4 *Sphinx vashti* Strecker ♂ — Colorado, Chaffee Co., Buena Vista, VI.21.1961 (M. MacKay) CNC

40.5 *Manduca sexta* (Linnaeus) ♂ — California, Glenn Co., Orland, IX.30.1970 (D. Hardwick) CNC

40.6 *Manduca quinquemaculata* (Haworth) ♂ — California, Sonoma Co., Petaluma, VIII.15.1934 (E. Johnston) CNC

40.7 *Manduca florestan* (Cramer) ♂ — Arizona, Cochise Co., Chiricahua Mtns., VII.28.1989 (B. Landry & J. Landry) CNC

40.8 *Manduca muscosa* (Rothschild and Jordan) ♀ — Arizona, Cochise Co., Ash Canyon Road, VII.2.1989 (B. Landry & J. Landry) CNC

40.9 *Manduca rustica* (Fabricius) ♂ — North Carolina, Carteret Co., Beaufort, VII.11.1967 (J. Sullivan) CNC

PLATE 40

PLATE 41 **Sphingidae**

41.1 *Hemaris diffinis* (Boisduval) ♂, dorsal — California, San Luis Obispo Co., Atascadero, VII.10.1945 (H. Dickenson) CNC

41.2 *Hemaris diffinis* (Boisduval) ♂, ventral — California, San Luis Obispo Co., Atascadero, VII.10.1945 (H. Dickenson) CNC

41.3 *Hemaris senta* (Strecker) ♂ dorsal — Colorado, Boulder Co., Boulder, VI.17.1961 (C. Mann) CNC

41.4 *Hemaris senta* (Strecker) ♂, ventral — Colorado, Boulder Co., Boulder, VI.17.1961 (C. Mann) CNC

41.5 *Proserpinus flavofasciata* (Walker) ♂ — Manitoba, Churchill, III.15.1998 (H. Hensel) CNC

41.6 *Arctonotus lucidus* Boisduval ♂ — California, Kern Co., Wheeler Ridge, XII.16.1944 (G. Beevor) CNC

41.7 *Euproserpinus wiestii* Sperry ♀ — Colorado, Weld Co., Roggen, V.28.1939 (R. Wiest) CNC

41.8 *Proserpinus clarkiae* (Boisduval) ♂ — Washington, Okanogan Co., Brewster, V.1.1934 (J. Hopfinger) CNC

41.9 *Amphion floridensis* Clark ♂ — Florida, Dade Co., Florida City, VI.14.1940 (H.A. Gibbons) CNC

41.10 *Deilephila elpenor* (Linnaeus), ♀ — British Columbia, Pitt Meadows, VII.23.2002 (J. Troubridge) CNC

41.11 *Euproserpinus phaeton* Grote and Robinson ♂ — California, Kern Co., Antelope Valley, III.25.2001 (T. Dimock) CNC

41.12 *Hyles gallii* (Rottenburg) ♂ — Alaska, Wrangell, VII.23.1942 (E. Johnston) CNC

41.13 *Hyles euphorbiae* (Linnaeus) ♂ — Ontario, Ottawa, VI.24.2004 (J. Troubridge) CNC

41.14 *Aellopus clavipes* (Rothschild & Jordan) ♀ — Arizona, CNC

41.15 *Hyles lineata* (Fabricius) ♂ — Colorado, Moffat Co., Maybell, VIII.17.1965 (D. Hardwick) CNC

41.16 *Xylophanes falco* (Walker) ♂ — Texas, Brewster Co., Big Bend Natl. Park, V.4.1959 (M. MacKay) CNC

41.17 *Eumorpha achemon* (Drury) ♂ — Colorado, Boulder Co., Silver Saddle Motel, VI.10.1961 (M. MacKay) CNC

41.18 *Eumorpha typhon* (Klug.) ♂ — no data, CNC

PLATE 41

PLATE 42 Notodontidae, Noctuidae

42.1 *Pheosia rimosa* Packard ♂ — California, Sonoma Co., Petaluma, VII.21.1936 (E. Johnston) CNC

42.2 *Cargida pyrrha* (Druce) ♂ — Arizona, Santa Cruz Co., Madera Canyon, VII.17.1947 (J. Comstock & L. Martin) CNC

42.3 *Cargida pyrrha* (Druce) ♀ — Arizona, Graham Co., Aravaipa Canyon, VII.24.1989 (B. Landry & J. Landry) CNC

42.4 *Crinodes biedermani* (Skinner) ♂ — Arizona, Cochise Co., Hereford (C. Biederman) CNC

42.5 *Theroa zethus* (Druce) ♀ — Arizona, Cochise Co., Sierra Vista, VIII.28.1966 (R. Sternitzky) CNC

42.6 *Ursia noctuiformis* Barnes and McDunnough ♂ — California, San Bernardino Co., Split Rocks Tank, V.20.1938 (G. & J. Sperry) CNC

42.7 *Litodonta alpina* Benjamin ♂ — Texas, Kerr Co., Kerrville, IV.5.1959 (R. Wigmore) CNC

42.8 *Litodonta gigantea* Barnes and Benjamin ♀ — Arizona, Cochise Co., Sierra Vista, IX.6.1967 (R. Sternitzky) CNC

42.9 *Litodonta wymola* (Barnes) ♂ — Arizona, Pima Co., Box Canyon, VIII.21.1949 (L. Martin) CNC

42.10 *Macrurocampa dorothea* Dyar ♀ — New Mexico, Grant Co., Cherry Creek, VI.30.1977 (C. Ferris) CNC

42.11 *Heterocampa ruficornis* Dyar ♀ — Arizona, Santa Cruz Co., Nogales, VIII.25.1906 (A. Koebele) CNC

42.12 *Heterocampa subrotata* Harvey ♂ — Texas, Williamson Co., Georgetown, IV.19.1937 (L. Milne) CNC

42.13 *Heterocampa lunata* Henry Edwards ♂ — Colorado, Chaffee Co., Maysville, VIII.3.1946 (H. Ramstadt) CNC

42.14 *Hyparpax aurora* (J.E. Smith) ♂ — Texas, Kerr Co., Kerrville, IV.15.1959 (R. Wigmore) CNC

42.15 *Oligocentria alpica* (Benjamin) ♀ — Texas, Jeff Davis Co., Ft. Davis, V.31.1959 (M. MacKay) CNC

42.16 *Oligocentria coloradensis* (Henry Edwards) ♂ — Utah, Utah Co., American Fork, VII.8.1936 (H. Thornley) CNC

42.17 *Oligocentria perangulata* (Henry Edwards) ♀ — Colorado, Garfield Co., Glenwood Springs (W. Barnes) CNC

42.18 *Oligocentria pallida* (Strecker) ♂ — Oregon, Josephine Co., Bolan Mtn., VII.27.2001 (J. Troubridge) CNC

42.19 *Schizura biedermani* Barnes and McDunnough ♂ — Arizona, Santa Cruz Co., Nogales, VIII.16.1906 (A. Koebele) CNC

42.20 *Schizura ipomoeae* Doubleday ♂ — Quebec, Papineau-Labelle PWR, VI.29.2003 (J. Troubridge & J. Lafontaine) CNC

42.21 *Schizura unicornis* (J.E. Smith) ♂ — British Columbia, Lillooet, VI.13.1926 (J. McDunnough) CNC

42.22 *Schizura concinna* (J.E. Smith) ♀ — California, Sonoma Co., Petaluma, VI.14.1937 (E. Johnston) CNC

42.23 *Schizura concinna* (J.E. Smith) ♂ — California, Sonoma Co., Petaluma, VI.14.1937 (E. Johnston) CNC

42.24 *Dasylophia anguina* (J.E. Smith) ♂ — Alberta, Writing-on-stone Provincial Park, VII.20.1982 (J. Landry) CNC

42.25 *Symmerista suavis* (Barnes) ♀ — New Mexico, Colfax Co., Cimarron Canyon, VII.7.1962 (E. & I. Munroe) CNC

42.26 *Phryganidia californica* Packard ♂ — California, Los Angeles Co., La Tuna Canyon, VI.5.1942, reared from *Quercus agrifolia* (W. Evans) CNC

42.27 *Afilia oslari* Dyar ♀ — Arizona, Santa Cruz Co., Madera Canyon, VIII.10.1947 (J.A. Comstock & L. M. Martin) CNC

42.28 *Lirimiris truncata* (Herrich-Schäffer) ♂ — Brazil, Petropolis, CNC

42.29 *Eublemma minima* (Guenée) ♂ — California, Riverside Co., Riverside, VI.30.1940 (G. & J. Sperry) CNC

42.30 *Metalectra quadrisignata* Walker ♂ — New York, Kings Co., Brooklyn, VI.18.1962 (Shoemaker) CNC

42.31 *Schrankia macula* (Druce) ♀ — Florida, Sarasota Co., Siesta Key, II.26.1954 (C. Kimball) CNC

42.32 *Idia americalis* (Guenée) ♀ — California, San Diego Co., Laguna Mtns., V.22.2001 (T. Mustelin) CNC

42.33 *Idia occidentalis* (Smith) ♂ — California, Ventura Co., Cuyama Valley, VIII.20.2000 (T. Dimock) CNC

42.34 *Reabotis immaculalis* (Hulst) ♀ — Oregon, Lake Co., Alkali Lake dunes, VII.19.2001 (J. Lafontaine & J. Troubridge) CNC

PLATE 42

PLATE 43 **Noctuidae**

43.1 *Tetanolita palligera* Smith ♀ — California, Lake Co., Anderson Springs, VIII.13.1950 (W. Bauer) CNC

43.2 *Bleptina caradrinalis* Guenée ♀ — Texas, Jeff Davis Co., Limpia Canyon, V.20.1950 (E. Johnston) CNC

43.3 *Hypenula cacuminalis* (Walker) ♀ — Arizona, Cochise Co., Ash Canyon Road, VII.15.1988 (N. McFarland) CNC

43.4 *Renia hutsoni* Smith ♂ — Colorado, Larimer Co., Viestenz-Smith Mtn. Park, 9.8 miles E of Loveland, 5,700', VIII.5.1996 (T. Dickel) CNC

43.5 *Palthis angulalis* (Hübner) ♂ — British Columbia, Salmon Arm, VI.8.1966, reared from *Pinus monticola* (Forest Insect Survey) CNC

43.6 *Hypena bijugalis* (Walker) ♀ — Alberta, Edmonton, Fulton Ravine, V.25.1998 (G. Anweiler) CNC

43.7 *Hypena abalienalis* (Walker) ♀ — British Columbia, Green Mtn., Agassiz, VIII.10.1999 (J. Troubridge) CNC

43.8 *Hypena humuli* Harris ♀ — Washington, Yakima Co., Bethel Ridge, VIII.22.1997 (J. Troubridge) CNC

43.9 *Hypena humuli* Harris ♂ — British Columbia, Duncans, Vancouver Island (C. Livingston) CNC

43.10 *Hypena californica* Behr ♀ — California, Monterey Co., Carmel, VII.19.1991 (F. Sala) CNC

43.11 *Hypena decorata* Smith ♀ — British Columbia, Wellington (T. Walker) CNC

43.12 *Hypena scabra* (Fabricius) ♀ — Colorado, Prowers Co., Lamar, IX.24.1945 (E. Johnston) CNC

43.13 *Hemeroplanis finitima* (Smith) ♀ — Washington, Sixmile Creek, VIII.6.1999 (L. Crabo) CNC

43.14 *Spargaloma sexpunctata* Grote ♀ — British Columbia, Kirby Flats, VI.21.2001 (J. Troubridge) CNC

43.15 *Nychioptera noctuidalis* (Dyar) ♂ — Arizona, Cochise Co., Ash Canyon Road, Huachuca Mts., 5,100', VII.15.1988 (N. McFarland) CNC

43.16 *Phobolosia anfracta* (Edwards) ♂ — Washington, Yakima Co., Bear Canyon, VIII.15.1998 (J. Troubridge) CNC

43.17 *Ascalapha odorata* (Linnaeus) ♂ — Mexico, Nuevo Leon, Linares, IX.16.1937 (T. Irwin) CNC

43.18 *Anomis erosa* Hübner ♀ — New Jersey, Ocean Co., Lakehurst, IX.21.1930 (F. Lemmer) CNC

43.19 *Alabama argillacea* (Hübner) ♂ — Ontario, Ottawa, IX.2.1938 (E. Lester) CNC

43.20 *Gonodonta pyrgo* (Cramer) ♂ — Mexico, Durango, 10 miles W of El Salto, 9,000', VIII.1.1964 (J. Martin) CNC

43.21 *Hypsoropha hormos* (Hübner) ♀ — Mississippi, Oktibbeha Co., 6 miles SW of Starkville, VII.29.1984 (R. & B. Brown) CNC

43.22 *Plusiodonta compressipalpis* Guenée ♂ — Texas, Uvalde Co., Concan, 1300' (N. McFarland) CNC

43.23 *Scoliopteryx libatrix* (Linnaeus) ♀ — Oregon, Jackson Co., Mt. Ashland, 7,000', VIII.3.1995 (Troubridge & Crabo) CNC

43.24 *Litoprosopus coachella* Hill ♀ — California, Los Angeles Co., IX.7.1930, CNC

43.25 *Tathorhynchus exsiccatus* (Lederer) ♂ — Wyoming, Albany Co., Laramie, 7,500', X.6.2001 (C. Ferris) UASM

43.26 *Phoberia atomaris* Hübner ♀ — Colorado, CNC

43.27 *Cissusa indiscreta* (Henry Edwards) ♀ — California, San Diego Co., Mt. Palomar, 4,800', III.22.1998 (N. Bloomfield) CNC

43.28 *Lygephila victoria* (Grote) ♂ — British Columbia, Kirby Flats Road, VII.7.2000 (J. Troubridge) UASM

43.29 *Litocala sexsignata* (Harvey), ♀ — Oregon, Josephine Co., 8 Dollar Mtn., IV.5.1996 (J. & L. Troubridge) UASM

43.30 *Melipotis jucunda* Hübner ♀ — California, Mono Co., Lee Vining, 7,140', VI.4/5.2002 (Crabo & Troubridge) CNC

43.31 *Melipotis jucunda* Hübner ♀ — CNC

PLATE 43

PLATE 44 **Noctuidae**

44.1 *Melipotis indomita* (Walker) ♂ — Arizona, Pima Co., Baboquivari Mts., Brown Canyon, 4,100′, VIII.20.2006 (T. Mustelin) CNC

44.2 *Melipotis indomita* (Walker) ♀ — Arizona, Santa Cruz Co., Nogales, IX.8.1906 (A. Koebele) CNC

44.3 *Forsebia perlaeta* (Henry Edwards) ♂ — Arizona, Maricopa Co., 12 miles NW of Apache Junction, III.20.1997 (J. Troubridge) CNC

44.4 *Forsebia perlaeta* (Henry Edwards) ♀ — Arizona, Maricopa Co., 12 Miles NW Apache Junction, III.20/27.1997 (J. Troubridge) UASM CNC

44.5 *Bulia deducta* (Morrison) ♂ — California, San Diego Co., Scissors Crossing, VII.16.1996 (T. Mustelin) — CNC

44.6 *Bulia deducta* (Morrison) ♀ — Arizona, Cochise Co., Texas Canyon, 5,460′, IV.19.2001 (C. Ferris) CNC

44.7 *Drasteria pallescens* (Grote and Robinson) ♂ — California, San Diego Co., Jacumba, 3,000′, VII.25.1998 (T. Mustelin) CNC

44.8 *Drasteria pallescens* (Grote and Robinson) ♂ — Colorado, Boulder Co., Silver Saddle Motel, 5,500′, VI.6.1961 (M. MacKay) CNC

44.9 *Drasteria inepta* Henry Edwards ♀ — Arizona, Cochise Co., Ash Canyon Road, Huachuca Mtns., 5,199′, IX.15.1992 (N. McFarland) CNC

44.10 *Drasteria inepta* Henry Edwards ♀ — Colorado, Eagle Co., State Bridge, 7,000′, VI.23.1961 (M. MacKay) CNC

44.11 *Drasteria howlandi* (Grote) ♀ — Washington, Grant Co., 5 miles SE of Vantage, V.22.1994 (J. Troubridge) CNC

44.12 *Drasteria ochracea* (Behr) ♀ — California, Plumas Co., Jackson Creek, 5,400′, V.21.2001 (J. Troubridge) CNC

44.13 *Drasteria petricola* (Walker) ♀ — Alberta, Hailstone Butte, VII.14.1990 (J. Troubridge) UASM

44.14 *Caenurgia togataria* (Walker) ♂ — California, San Diego Co., La Mesa, IV.15.1950 (E. Johnston) CNC

44.15 *Caenurgina caerulea* (Grote) ♂ — California, Sonoma Co., the Geysers, V.9.1939 (W. Bauer) CNC

44.16 *Caenurgina erechtea* (Cramer) ♂ — California, Sonoma Co., Petaluma, IX.13.1934 (E. Johnston) CNC

44.17 *Caenurgina erechtea* (Cramer) ♂ — Oregon, Harney Co., 1 mile N Denio, Nevada, VII.20.2001 (Lafontaine & Troubridge) CNC

44.18 *Caenurgina crassiuscula* (Haworth) ♀ — Alberta, Writing-on-stone Provincial Park, VII.20.1982 (J. Landry) CNC

44.19 *Caenurgina crassiuscula* (Haworth) ♀ — Alberta, Lost River, 3000′, VII.14.1999 (J. Troubridge) UASM

44.20 *Euclidia ardita* Franclemont ♀ — California, Los Angeles Co., La Tuna Canyon, IV.1.1953 (W. Evans) CNC

44.21 *Callistege intercalaris* (Grote) ♀ — Arizona, Yavapai Co., 4 miles N of Prescott, VII.9.1972 (L. Martin) CNC

44.22 *Mocis latipes* (Guenée) ♀ — North Carolina, Macon Co., Highlands, VII.15.1957 (C. Durden) CNC

44.23 *Anticarsia gemmatalis* Hübner ♀ — Arizona, Cochise Co., Ash Canyon Road, 5,100′, IX.22.1992 (N. McFarland) CNC

44.24 *Anticarsia gemmatalis* Hübner ♀ — no data, CNC

44.25 *Heteranassa fraterna* (Smith) ♀ — California, Riverside Co., W of Palm Springs, IV.19.9150 (E. Johnston) CNC

44.26 *Zaleops umbrina* (Grote) ♀ — Arizona, Cochise Co., Huachuca Mtns., VII.5.1988 (N. McFarland) CNC

44.27 *Zale lunata* (Drury) ♂ — Oregon, Jefferson Co., N of Venora, VIII.25.1997 (J. Troubridge) UASM

44.28 *Zale lunata* (Drury) ♂ — California, Los Angeles Co., Pasadena, VII.21.1910 (V. Clemence) CNC

44.29. *Toxonprucha volucris* (Grote) ♀ — California, San Diego Co., Sweeney Pass, III.24.2001 (T. Mustelin) CNC

44.30 *Zale termina* (Grote) ♀ — Oregon, Josephine Co., W of Selma, VIII.24.1997 (J. Troubridge) UASM

44.31 *Zale colorado* (Smith) ♀ — Arizona, Cochise Co., Ash Canyon Road, VII.12.1988 (N. McFarland) CNC

44.32 *Zale colorado* (Smith) ♀ — Arizona, Cochise Co., Huachuca Mts., Ash Canyon Road, 5,100′, VII.15.1988 (N. McFarland) CNC

PLATE 44

PLATE 45 **Noctuidae**

45.1 *Catocala piatrix* Grote ♂ — Arizona, Cochise Co., Cave Creek, IX, CNC

45.2 *Euparthenos nubilis* ♂ — New York, Orange Co., Bear Mtn., VIII.19.1925 (H. Gibbon) CNC

45.3 *Catocala neogama euphemia* Beutenmüller ♀ — New Mexico, Socorro Co., W of Socorro, IX.1.1975 (Lafontaine & Bowen) CNC

45.4 *Catocala aholibah* Strecker ♀ — Utah, Millard Co., S of Kanosh, VIII.27.1965 (D. Hardwick) CNC

45.5 *Catocala ilia* (Cramer) ♀ — California, Lake Co., Clear Lake, VI.7.1935 (E. Johnston) CNC

45.6 *Catocala relicta* Walker ♂ — Alberta, Bashaw, IX.4.1973 (D. Lafontaine) CNC

45.7 *Catocala grotiana* Bailey ♂ — Wyoming, Albany Co., Pole Mtn., VIII.29.2001 (C. Ferris) CNC

45.8 *Catocala junctura* Walker subsp. *arizonae* Grote ♀ — Nevada, Storey Co., Sixmile Creek, VIII.29.2000 (J. Troubridge) CNC

45.9 *Catocala semirelicta* Grote ♂ — British Columbia, Seton Lake, VIII.9.1933 (J. McDunnough) CNC

45.10 *Catocala delilah* Strecker subsp. *desdemona* Henry Edwards ♀ — Arizona, Cochise Co., Garces [Palmerlee] (Biederman) CNC

45.11 *Catocala minuta* Edwards ♂ — New Jersey, Hudson Co., Kearny, VII.27.1920 (T. Mayfield) CNC, ex larva

45.12 *Catocala faustina cleopatra* Henry Edwards ♀ — California, Napa Co., St. Helena, VIII.15.1909 (C. Fuchs) CNC

45.13 *Catocala verrilliana* Grote ♂ — California, San Diego Co., NE of Julian, VIII.15/17.1998 (T. Mustelin & N. Bloomfield) CNC

45.14 *Catocala violenta* Henry Edwards ♂ — Colorado, La Plata Co., Hermosa, VIII.17.1971 (D. Hardwick) CNC

PLATE 45

PLATE 46 **Noctuidae**

46.1 *Marythyssa inficita* (Walker) — Oregon, Baker Co., Brownlee, VI.12.1999 (J. Troubridge) CNC

46.2 *Paectes abrostolella* (Walker) ♀ — Arizona, Yavapai Co., 4 miles N of Prescott, VII.5.1971 (L. Martin) CNC

46.3 *Paectes declinata* (Grote) ♂ — California, Los Angeles Co., La Tuna Canyon, VI.19.1947 (W. Evans) CNC

46.4 *Eutelia pulcherrima* (Grote) — Florida, Escambia Co., W. Pensacola, IV.17.1963 (V. Grant) CNC

46.5 *Meganola fuscula* (Grote) ♀ — Arizona, Coconino Co., Walnut Canyon, 6500', VIII.1.1964 (J. Franclemont) CNC

46.6 *Nola minna* Butler ♀ — California, Sonoma Co., The Geysers, II.4.1940 (E. Johnston) CNC

46.7 *Nola apera* Druce ♀ — Mexico, Vera Cruz, Jalapa, CNC

46.8 *Characoma nilotica* (Rogenhofer) ♀ — Texas, Uvalde Co., Concan, 1,300', III.28.1990 (N. McFarland) CNC

46.9 *Nycteola frigidana* (Walker) ♀ — Washington, Snohomish Co., Verlot, V.14.1949 (E. Johnston) CNC

46.10 *Nycteola cinereana* Neumoegen and Dyar ♀ — California, Lake Co., Anderson Springs, IX.11.1948 (W. Bauer) CNC

46.11 *Iscadia aperta* Walker ♂ — Arizona, Cochise Co., Ash Canyon Road, 5,100', IX.22.1992 (N. McFarland) CNC

46.12 *Acsala anomala* Benjamin ♂ — Yukon Territory, British Mtns., 69 17, 140 03W, 950m, IV.17.84 G. & M. Wood & D. Lafontaine (CNC)

46.13 *Eudesmia arida* (Skinner) ♂ — Arizona, Cochise Co., Ash Canyon Road, VIII.8.1988 (N. McFarland) CNC

46.14 *Cisthene liberomacula* (Dyar) ♂ — California, San Diego Co., San Diego, CNC

46.15 *Cisthene deserta* (Felder) ♂ — California, Sonoma Co., Petaluma, VII.24.1936 (E. Johnston) CNC

46.16 *Cisthene faustinula* (Boisduval) ♂ — California, Sonoma Co., Petaluma, VII.30.1938 (E. Johnston) CNC

46.17 *Cisthene dorsimacula* (Dyar) ♂ — California, San Diego Co., San Diego, V.29.1911 (W. Wright) CNC

46.18 *Cisthene perrosea* (Dyar) ♂ — California, San Diego Co., Escondido, IX.7.1994 (J. Adams) CNC

46.19 *Cisthene tenuifascia schwarziorum* (Dyar) ♂ — Arizona, Pima Co., Madera Canyon, IX.16.1986 (D. & V. Hardwick) CNC

46.20 *Cisthene juanita* Barnes and Benjamin ♂ — Arizona, Cochise Co., Fort Huachuca, X.5.1994 (D. Ferguson) CNC

46.21 *Cisthene martini* C.B. Knowlton ♂ — Arizona, Cochise Co., Fort Huachuca, VIII.3.1999 (J. Landry) CNC

46.22 *Cisthene angelus* (Dyar) ♂ — Arizona, Maricopa Co., Wickenburg, V.7.1980 (D. & V. Hardwick) CNC

46.23 *Cisthene barnesii* (Dyar) ♂ — pink HW form — Utah, Sanpete Co., Mayfield, VIII.25.1965 (D. Hardwick) CNC

46.24 *Cisthene barnesii* (Dyar) ♂ — yellow HW form — New Mexico, McKinley Co., McGaffey, VII.24.1962 (E. & I. Munroe) CNC

46.25 *Ptychoglene phrada* Druce ♂ — Arizona, Cochise Co., Upper Pinery Canyon, VII.3.1956 (L. Martin & J. Comstock) CNC

46.26 *Lycomorpha grotei* (Packard) ♂ — Colorado, Gunnison Co., E of Sapinero, VIII.15.1971 (D. Hardwick) CNC

46.27 *Lycomorpha regulus* (Grinnell) ♂ — California, Los Angeles Co., Camp Baldy, VI.26.1950 (R. Schuster) CNC

46.28 *Lycomorpha splendens* Barnes and McDunnough ♂ — Texas, Jeff Davis Co., Limpia Canyon, V.20.1950 (E. Johnston) CNC

46.29 *Lycomorpha pholus* (Drury) ♂ — Colorado, Douglas Co., Platte Canyon, VII.25.1937, CNC

46.30 *Hypoprepia miniata* (Kirby) ♂ — Alberta, Hanna, VII.16.1960 (D. Hardwick) CNC

46.31 *Hypoprepia miniata cadaverosa* Strecker ♂ Texas, Brewster Co., S of Alpine, VIII.9.1991 (E. Metzler) CNC

46.32 *Hypoprepia inculta* Henry Edwards ♂ — Arizona, Cochise Co., Ash Canyon Road, IX.15.1991 (N. McFarland) CNC

46.33 *Bruceia pulverina* Neumoegen ♂ — Wyoming, Laramie Co., C. Gowdy State Park, VII.7.1993 (J. Landry) CNC

46.34 *Clemensia albata* Packard ♂ — British Columbia, Wellington, VI.19.1923 (W. Downes) CNC

46.35 *Eilema bicolor* (Grote) ♀ — British Columbia, Radium Hot Springs, VII.23.1960 (D. Hardwick) CNC

46.36 *Crambidia dusca* Barnes and McDunnough ♂ — California, San Diego Co., Miramar Naval Air Station, V.7.1997 (N. Bloomfield) CNC

46.37 *Crambidia suffusa* Barnes and McDunnough ♂ — California, Riverside Co., Riverside, X.13.1944 (J. Smith) CNC

46.38 *Crambidia impura* Barnes and McDunnough ♂ — Colorado, Larimer Co., Rocky Mtn. Natl. Park, X.8.1991 (Opler & T. Dickel) CNC

46.39. *Crambidia casta* (Packard) ♂ — Alberta, Waterton Lakes, VIII.20.1923 (J. McDunnough) CNC

46.40 *Crambidia cephalica* (Grote and Robinson) ♂ — Colorado, Eagle Co., State Bridge, VI.23.1961 (M. MacKay) CNC

46.41 *Agylla septentrionalis* Barnes and McDunnough ♀ — Arizona, Cochise Co., Upper Pinery Canyon, VII.4.1956 (J.A. Comstock & W.A. Rees) CNC

46.42 *Gnamptonychia ventralis* Barnes and Lindsey ♂ — Mexico, Durango, W of El Salto, VI.2.1964 (W.McGuffin) CNC

46.43 *Gardinia anopla* Hering ♂ — Arizona, Cochise Co., Hereford (C. Biederman) CNC

46.44 *Virbia ostenta* (Henry Edwards) ♂ — Arizona, Cochise Co., Fort Huachuca, VIII.3.1989 (J. Landry) CNC

46.45 *Virbia fragilis* (Strecker) ♀ — New Mexico, Lincoln Co., Ruidoso, VII.29.1962 (E. & I. Munroe) CNC

46.46 *Neoarctia beanii* (Neumoegen) ♂ — Alberta, Hailstone Butte, 2036m, VIII.12.2004 (C. Schmidt) CNC

46.47 *Neoarctia brucei* (Henry Edwards) ♂ — Washington, Chelan Co., Junior Point, VIII.27.1998 (J. Troubridge) CNC

46.48 *Grammia obliterata* (Stretch) ♂ — Colorado, Moffat Co., Dinosaur Natl. Monument, VIII.13.1994 (T. Dickel) CNC

46.49 *Grammia parthenice* (W. Kirby) ♂ — Ontario, Stillsville, VIII.4.2003 (J. Troubridge) CNC

46.50 *Grammia ornata* (Packard) ♂ — California, Sonoma Co., Petaluma, V.4.1939 (E. Johnston) CNC

46.51 *Grammia ornata* (Packard) ♂ — no data, CNC

46.52 *Grammia cervinoides* (Strecker) ♂ — Colorado, Park Co., Pennsylvania Mtn., VII.14.1980 (D.F.) CNC

46.53 *Grammia williamsii* (Dodge) ♂ — Wyoming, Albany Co., Medicine Bow Natl. Forest, VII.17.1982 (E. Metzler) CNC

46.54 *Grammia nevadensis* (Grote and Robinson) ♂ — Nevada, White Pine Co., SE of Ely, IX.5.1965 (D. Hardwick) CNC

46.55 *Grammia incorrupta* (Henry Edwards) ♂ — New Mexico, McKinley Co., McGaffey, Zuni Mts., 7,500', VII.23.1962 (E. & I. Munroe) CNC

46.56 *Grammia f-pallida* (Strecker) ♂, pink HW form — Colorado, Boulder Co., Silver Saddle Motel, VI.18.1961 (M. MacKay) CNC

46.57 *Grammia f-pallida* (Strecker) ♂, yellow HW form — Colorado, Boulder Co., Silver Saddle Motel, VI.18.1961 (M. MacKay) CNC

46.58 *Notarctia proxima* (Guérin-Méneville) ♂ — California, San Diego Co., La Mesa, IV.14.1950 (E. Johnston) CNC

46.59 *Notarctia proxima* (Guérin-Méneville) ♀ — California, San Diego Co., La Mesa, IV.14.1950 (E. Johnston) CNC

PLATE 46

PLATE 47 **Noctuidae**

47.1 *Parasemia plantaginis* (Linnaeus) ♀ — British Columbia, Summit Lake, Alaska Highway, Milepost 392, 4,500', VII.17/19.1959 (E. MacDougall) CNC

47.2 *Parasemia plantaginis* (Linnaeus) ♂ — Yukon Territory, Burwash Flats, VII.10.1980 (Wood & Lafontaine) CNC

47.3 *Parasemia plantaginis* (Linnaeus) ♂ — Colorado, Boulder Co., Nederland, VII.6.1961 (M. MacKay) CNC

47.4 *Parasemia plantaginis* (Linnaeus) ♀ — CNC

47.5 *Platarctia parthenos* (Harris) ♀ — Alberta, Bragg Creek, VI.11.1972, CNC

47.6 *Platyprepia virginalis* (Boisduval) ♀ — California, Marin Co., Point Reyes, VI.17/21.1952 (D. Hardwick) CNC

47.7 *Platyprepia virginalis* (Boisduval) ♀ — Oregon, Wasco Co., Maupin, VI.21.1942 (E. Johnston) CNC

47.8 *Arctia caja utahensis* (Henry Edwards) ♂ — Nevada, Elko Co., 11 miles SW of Wells, VII.22.2001 (J. Lafontaine & J. Troubridge) CNC

47.9 *Phragmatobia fuliginosa rubricosa* (Harris) ♀ — Alberta, W of Edmonton, V.21.1998 (C. Schmidt) CNC

47.10 *Phragmatobia assimilans* Walker ♀ — British Columbia, Oliver, V.8.1932 (C. Garrett) CNC

47.11 *Sonorarctia fervida* (Walker) ♂ — Arizona, Cochise Co., Huachuca Mtns., Garden Canyon, reared from *Melilotus officinalis*, (N. McFarland) USNM

47.12 *Sonorarctia fervida* (Walker) ♀ — Arizona, Cochise Co., Huachuca Mtns., Garden Canyon, reared from *Melilotus officinalis*, (N. McFarland) USNM

47.13 *Leptarctia californiae* (Walker) ♀ — British Columbia, Victoria, V.12.1922 (G. Taylor) CNC

47.14 *Leptarctia californiae* (Walker) ♂ — British Columbia (G. Taylor) CNC

47.15 *Leptarctia californiae* (Walker) ♂ — Colorado, Jefferson Co., Chimney Gulch, Golden (E. Oslar) CNC

47.16 *Kodiosoma fulvum* Stretch ♂, yellow HW form — California, Riverside Co., Gavilan Hills, IV.8.1950 (S. Nicolay) CNC

47.17 *Kodiosoma fulvum* Stretch ♂, red HW form — Arizona, Santa Maria Valley, IV.3.1980 (D. Hardwick) CNC

47.18 *Pyrrharctia isabella* (J.E. Smith) ♀ — British Columbia, Langley, VI.25.1997 (J. Troubridge) UASM

47.19 *Estigmene acrea* (Drury) ♂ — Oregon, Baker Co., Brownlee, VI.12.1999 (J. Troubridge), UASM

47.20 *Hyphantria cunea* (Drury) ♂ — Colorado, Boulder Co., Silver Saddle Motel, VI.13.1961 (M. MacKay) CNC

47.21 *Hyphantria cunea* (Drury) ♂ — British Columbia, Kelowna, VI.6.1935 (A. Gartrell) CNC

47.22 *Estigmene acrea* (Drury) ♀ — British Columbia, Vernon, V.8.1936 (A. Dannus) CNC

47.23 *Estigmene albida* (Stretch) ♂ — Arizona, Cochise Co., Ramsey Canyon, VIII.17.1967 (R. Sternitsky) CNC

47.24 *Spilosoma virginica* (Fabricius) ♂ — Washington, Grant Co., The Potholes, V.19.2001 (J. Troubridge) UASM

47.25 *Spilosoma vagans* (Boisduval) ♂ — Oregon, Jesephine Co., Eightdollar Mtn. Road, VI.15.1999 (J. Troubridge) UASM

47.26 *Spilosoma vagans* (Boisduval) ♂ — Oregon, Baker Co., Brownlee, VI.12.1999 (J. Troubridge) UASM

47.27 *Hypercompe permaculata* (Packard) ♀ — Alberta, Writing on Stone, VI.28.2000 (J. Troubridge) UASM

47.28 *Spilosoma vestalis* Packard ♂ — California, Plumas Co., Squirrel Creek, VI.14.1999 (J. Troubridge) UASM

47.29 *Arachnis picta* Packard ♀ — California, Monterey Co., Monterey area, X.21.1989 (F. Sala) CNC

47.30 *Arachnis picta* Packard ♂ — Arizona, Pima Co., Mt. Lemmon, emerged X.2006 (R. Nagle) CNC

47.31 *Arachnis zuni* Neumoegen ♂ — Mexico, Durango, W of Durango, VI.19.1964 (J. Martin) CNC

PLATE 47

PLATE 48 **Noctuidae**

48.1 *Tyria jacobaeae* (Linnaeus) ♀ — British Columbia, Naino, V.13.1969, CNC

48.2 *Utetheisa ornatrix* (Linnaeus) ♀ — Arizona, Cochise Co., Sierra Vista, X.30.1966 (R. Sternitzky) CNC

48.3 *Gnophaela clappiana* Holland ♂ — New Mexico, White Mtns., 8,600′, VII.20.1930 (E. Tinkham) CNC

48.4 *Gnophaela latipennis* (Boisduval) ♂ — California, Mendocino Co., Navarro, VI.21.1937 (E. Johnston) CNC

48.5 *Gnophaela vermiculata* (Grote) ♂ — Colorado, Clear Creek Co., Chicago Creek, VIII.5.1961 (E. Rockburne) CNC

48.6 *Cycnia tenera* Hübner ♀ — British Columbia, Oliver, VII.12.1953 (J. Martin) CNC

48.7 *Dysschema howardi* Henry Edwards ♂ — Texas, Jeff Davis Co., Limpia Canyon, VIII.5.1981 (E. Metzler) CNC

48.8 *Euchaetes antica* (Walker) ♀ — Arizona, Graham Co., Aravaipa Canyon, VII.26.1989 (B. Landry & J. Landry) CNC

48.9 *Cycnia oregonensis* (Stretch) ♂ — California, Sonoma Co., Spring Mtn., V.26.1938 (E. Johnston) CNC

48.10 *Euchaetes antica* (Walker) ♂ — Arizona, Santa Cruz Co., Nogales (E. Oslar) CNC 30414

48.11 *Euchaetes elegans* Stretch ♀ — California, San Bernardino Co., Rattlesnake Creek, X.13.1932 (C. Dammers) CNC

48.12 *Pygoctenucha terminalis* (Walker) ♀ — Arizona, Cochise Co., Sierra Vista, IX.4.1966 (R. Sternitzky) CNC

48.13 *Ectypia clio* (Packard) ♀ — California, Riverside Co., Riverside, XI.2.1932 (C. Dammers) CNC

48.14 *Pygarctia murina* (Stretch) ♂ — Arizona, Cochise Co., Chiricahua Mtns., VII.29.1989 (B. Landry & J. Landry) CNC

48.15 *Pygarctia roseicapitis* (Neumoegen & Dyar) ♀ — Arizona, Santa Cruz Co., Riverwash, VII.28.1986 (J. Peacock) CNC

48.16. *Halysidota tessellaris* (J.E. Smith) ♂ — Utah, Cache Co., Logan, VII.14.1955 (S. Wood) CNC

48.17 *Lophocampa ingens* (Henry Edwards) ♀ — Utah, Duchesne Co., W of Duchesne, VIII.20.1965 (D. Hardwick) CNC

48.18 *Lophocampa argentata* (Packard) ♂ — British Columbia, Burns Bog Delta, VIII.20.2002 (J. Troubridge) CNC

48.19 *Lophocampa mixta* (Neumoegen) ♀ — Arizona, Cochise Co., Ramsey Canyon, III.23.1967 (R. Sternitzky) CNC

48.20 *Lophocampa maculata* Harris ♀ — Idaho, Franklin Co., VI.5/6.2000 (J. Troubridge) UASM

48.21 *Lophocampa indistincta* (Barnes and McDunnough) ♀ — California, 1 km SW of Centinela, Santa Cruz Island, emerged VIII.3.2001, reared from *Vaccinium ovatum* (Powell) EME

48.22 *Aemilia ambigua* (Strecker) ♂ — Colorado, Chaffee Co., 8 km W of Buena Vista, VII.16.1982 (J.F. Landry) CNC

48.23 *Hemihyalea edwardsii* (Packard) ♂ — New Mexico, Hidalgo Co., Skeleton Canyon, X.15/16.2001 (C. Ferris) CNC

48.24 *Hemihyalea labecula* (Grote) ♂ — Arizona, Cochise Co., Peloncillo Mtns., VII.31.1989 (B. Landry & J. Landry) CNC

48.25 *Hemihyalea splendens* Barnes and McDunnough ♂ — Arizona, Cochise Co., Garden Canyon, X.1.1967 (R. Sternitzky) CNC

PLATE 48

PLATE 49 Noctuidae

49.1 *Bertholdia trigona* (Grote) ♀ — Arizona, Cochise Co., Ash Canyon Road, 5,100′, VI.28.1982 (N. McFarland) CNC

49.2 *Ctenucha venosa* Walker ♂ — Arizona, Cochise Co., Ash Canyon Road, 5,100′, IX.24.1992 (N. McFarland) CNC

49.3 *Ctenucha cressonana* Grote ♂ — Colorado, Larimer Co., Grey-rock trail, VII.16.1995 (S. Fitzgerald) CNC

49.4 *Ctenucha rubroscapus* (Ménétries) ♂ — Oregon, Lincoln Co., Newport, VII.29.1995 (Troubridge & Crabo) CNC

49.5 *Ctenucha brunnea* Stretch ♂ — California, Los Angeles Co., Zuma Beach, VII.6.1966 (F. Sala) CNC

49.6 *Cisseps fulvicollis* (Hübner) ♂ — British Columbia, Richmond, V.21.2002 (J. Troubridge) CNC

49.7 *Cisseps fulvicollis* (Hübner) ♂ — California, Los Angeles Co., Inglewood, IX.17.1949 (H. Hollingsworth) CNC

49.8 *Horama panthalon texana* (Grote) ♀ — Mexico, Nuevo Leon, Monterrey, VII.11.1963 (H. Howden) CNC

49.9 *Lymantria dispar* (Linnaeus) ♂ — Ontario, Pakenham alvar, VIII.15.2004 (J. Troubridge) CNC

49.10 *Lymantria dispar* (Linnaeus) ♀ — British Columbia, Langley, VI.25.1997 (J. Troubridge) UASM

49.11 *Gynaephora rossii* Curtis ♂ — Yukon Territory, Dempster Highway, VI.11/19.1987 (J. Troubridge) CNC

49.12 *Gynaephora rossii* Curtis ♀ — Northwest Territory, Cameron Bay, IX.10.1937 (T. Freeman) CNC

49.13 *Dasychira vagans grisea* (Barnes and McDunnough) ♂ — British Columbia, Byne Bog Delta, VII.2.2002 (J. Troubridge) CNC

49.14 *Dasychira mescalera* Ferguson, paratype ♂ — New Mexico, McKinley Co., McGaffey, VII.22.1962 (E. & I. Munroe) CNC

49.15 *Dasychira grisefacta* (Dyar) ♂ — Montana, Powder River Co., E of Ashland, IX.5.1964 (D. Hardwick) CNC

49.16 *Orgyia antiqua* (Linnaeus) ♂ — New Brunswick, Edmundston, VIII.31.1987 (H. Hensel) CNC

49.17 *Orgyia antiqua* (Linnaeus) ♀, dorsal and lateral views — Montana, Glacier Natl. Park, VIII.24.1973 (M. Hobson) CNC

49.18 *Orgyia vetusta* (Boisduval) ♂ — California, Monterey Co., Carmel, IX.8.2001 (F. Sala) UASM

49.19 *Orgyia vetusta* (Boisduval) ♀ — California, Monterey Co., Carmel, VIII.19.2001 (F. Sala) UASM

49.20 *Orgyia pseudotsugata* (McDunnough) ♂ — California, Mt. Shasta, VIII.27.2000 (J. Troubridge) CNC

49.21 *Orgyia leucostigma* (J.E. Smith) ♂ — Ontario, Pakenham alvar, VIII.15.2004 (J. Troubridge) CNC

49.22 *Leucoma salicis* (Linnaeus) ♂ — British Columbia, E of Longley, VI.24/30.1990 (J. Troubridge) UASM

49.23 *Abrostola urentis* Guenée, Alberta, Edmonton, VI.30.1942, CNC

49.24 *Mouralia tinctoides* (Guenée) ♀ — California, Los Angeles Co., IX.21.1936 (J. Comstock) CNC

49.25 *Trichoplusia ni* (Hübner) ♀ — Nevada, Humboldt Co., 17 km N of Winnemucca, 1300m, VII.21.2001 (Lafontaine & Troubridge) CNC

49.26 *Chrysodeix includens* (Walker) ♀ — British Columbia, Delta, X.12.2006, greenhouse tomatoes (D. Holden) CNC

49.27 *Autoplusia egena* (Guenée) ♂ — California, Los Angeles Co., Hawthorne, IX.13.1938 (M. Paris) CNC

49.28 *Rachiplusia ou* (Guenée) ♂ — Illinois, Cook Co., Chicago, VIII.11.1951 (Woodcock) CNC

49.29 *Polychrysia morigera* (Henry Edwards) ♀ — Oregon, Jackson Co., Mt. Ashland, 6,400′, VIII.2.1998 (J. Troubridge) CNC

49.30 *Polychrysia esmeralda* (Oberthür) ♀ — Alberta, Edmonton, VII.19.1942 (K. Bowman) CNC

49.31 *Pseudeva purpurigera* (Walker) ♂ — Colorado, Larimer Co., Viestenz-Smith Mtn. Park, 9.8 miles W of Loveland, 5,700′, VII.19.1994 (E. Buckner-Opler & P. Opler) CNC

49.32 *Eosphoropteryx thyatyroides* (Guenée) ♂ — Washington, King Co., Factoria, VII.16.1949 (E. Johnston) CNC

49.33 *Autographa californica* (Speyer) ♂ — Washington, Walla Walla Co., Walla Walla, V.13.1948 (W. Cook) reared, #1948-113, CNC

49.34 *Autographa mappa* (Grote & Robinson) ♂ — Colorado, Grand Co., St. Louis Creek Road, VII.10.1988 (T. Dickel) CNC

49.35 *Autographa pseudogamma* (Grote) ♂ — Colorado, Clear Creek Co., West Chicago Creek campground, 2985m, VII.14.1993 (B. Landry) CNC

49.36 *Autographa corusca* (Strecker) ♀ — British Columbia, Cultus Lake, VII.21.1927 (D. Young) CNC

49.37 *Autographa labrosa* (Grote) ♀ — California, Sonoma Co., Petaluma, V.24.1938 (E. Johnston) CNC

PLATE 49

PLATE 50 **Noctuidae**

50.1 *Megalographa biloba* Stephens ♀ — California, Los Angeles Co., La Tuna Canyon, VI.19.1944, reared from *Phacelia* (W. Evans) CNC

50.2 *Syngrapha viridisigma* (Grote) ♂ — Colorado, Grand Co., St. Louis Creek Road, 8,900′, VIII.10.1988 (T. Dickel) CNC

50.3 *Syngrapha sackenii* (Grote) ♀ — Colorado, Grand Co., Corona Pass, 11,000′, VIII.6.1980 (T. Dickel) CNC

50.4 *Syngrapha ignea* (Grote) ♀ — Alberta, Hailstone Butte, Kananaskis Country, VII.15.1995 (J. Troubridge) CNC

50.5 *Syngrapha alticola* (Walker) ♀ — Alaska, Umiat, VII.21.1959 (J. Martin) CNC

50.6 *Syngrapha celsa* (Henry Edwards) ♀ — Oregon, Jackson Co., Mt. Ashland, 6,400′, VIII.2.1998 (J. Troubridge) CNC

50.7 *Syngrapha angulidens* (Smith) ♂ — Colorado, Grand Co., Fool Creek, 5 miles SW of Fraser, VII.15.1988 (T. Dickel) CNC

50.8 *Anagrapha falcifera* (Kirby) ♀ — British Columbia, Mt. Kobau, VIII.13.1998 (J. Troubridge) CNC

50.9 *Plusia putnami* Grote ♀ — New Brunswick, St.-Basile, VI.21.1986 (H. Hensel) CNC

50.10 *Plusia nichollae* (Hampson) ♀ — British Columbia, Langley, VI.18.2002 (J. Troubridge) CNC

50.11 *Amyna octo* Guenée ♀ — Arizona, Cochise Co., Ash Canyon Road, Huachuca Mtns., 5,100′, X.28.1992 (N. McFarland) CNC

50.12 *Bagisara buxea* (Grote) ♂ — Arizona, Yavapai Co., 4 miles N of Prescott, VII.12.1972 (L. Martin) CNC

50.13 *Pseudeutrotia carneola* (Guenée) ♂ — Colorado, Jefferson Co., Strontia Springs, VII, CNC

50.14 *Cobubatha dividua* (Grote) ♂ — California, San Diego Co., Split Mtn. Canyon, Anza Borrego Desert State Park, 400′, IV.18.2001 (T. Mustelin) CNC

50.15 *Tripudia balteata* (Smith) ♂ — California, Riverside Co., Chino Canyon, Palm Springs, IV.19.1950 (E. Johnston) CNC

50.16 *Tarachidia semiflava* (Guenée) ♂ — Alberta, Highway 41 at Saskatchewan River, VI.25.2000 (J. Troubridge) UASM

50.17 *Tarachidia candefacta* (Hübner) ♂ — Oregon, Baker Co., Brownlee, VI.12.1999 (J. Troubridge) UASM

50.18 *Tarachidia libedis* (Smith) ♂ — Texas, Brewster Co., Panther Junction, Big Bend Natl. Park, IV.29.1959 (M. MacKay) CNC

50.19 *Tarachidia tortricina* (Zeller) ♂ dark form — British Columbia, Fernie, VI.8.1934 (H. Leech) CNC

50.20 *Tarachidia tortricina* (Zeller) ♂ yellow form — New Mexico, Grant Co., 17 miles NE of Santa Rita, 8,000′, IX.5.1975 (Lafontaine & Bowen) CNC

50.21 *Conochares acutus* Smith ♂ — California, San Diego Co., Split Mtn. Canyon, Anza Borrego Desert State Park, 400′, IV.18.2001 (T. Mustelin) CNC

50.22 *Therasea augustipennis* (Grote) ♀ — Alberta, Lost River, Onefour, VII.22.2003 (Troubridge & Lafontaine) CNC

50.23 *Spragueia magnifica* Grote ♂ — New Mexico, Eddy Co., Carlsbad, V.17.1950 (E. Johnston) CNC

50.24 *Ponometia sutrix* (Grote) ♀ — Nevada, Humboldt Co., 17 km N of Winnemucca, 1300m, VI.6.2002 (Troubridge & Crabo) UASM

50.25 *Acontia aprica* (Hübner) ♂ — Texas, San Patricio Co., Welder Wildlife Refuge, IV.6.1989 (D. & V. Hardwick) CNC

50.26 *Acontia major* Smith ♂ — Montana, Ravalli Co., Hamilton, VII.3.1926 (R. Parker) CNC

50.27 *Acontia areli* Strecker ♂ — Texas, Brewster Co., Big Bend Natl. Park, The Basin, V.18.1959 (M. MacKay) CNC

50.28 *Pseudalypia crotchii* Henry Edwards ♂ — California, Los Angeles Co., Lovejoy Buttes, Mojave Desert, emerged IV.18.1945, reared from *Malvastrum exilis* (L. Martin) CNC

50.29 *Panthea virginarius portlandia* Grote ♀ — British Columbia, Burns Bog, Delta, VIII.20.2002 (J. Troubridge) CNC

50.30 *Raphia frater* Grote ♂ — Washington, Okanogan Co., Hart's Pass Road, VII.23.1997 (L. Crabo) CNC

50.31 *Acronicta dactylina* Grote ♂ — British Columbia, Upper Clearwater, 700m, VI.10.1991 (H. Knight) CNC

50.32 *Acronicta funeralis* Grote & Robinson ♂ — Washington, Walla Walla Co., Walla Walla, V.21.1963 (W. Cook) CNC

50.33 *Acronicta grisea* Walker ♂ — Colorado, Boulder Co., Nederland, VI.26.1961 (M. MacKay) CNC

50.34 *Acronicta hesperida* Smith ♂ — British Columbia, 10 miles W of Duncan, 400′, VIII.18.1974 (J. Lafontaine) CNC

50.35 *Acronicta impleta* Walker ♂ — California, Nevada Co., Dog Bar at Lode Star, 2,660′, IV.21.1992 (R. Miller) CNC

50.36 *Acronicta innotata* Guenée ♀ — British Columbia, Castlegar, VI.16.1965 (J. Troubridge) CNC

50.37 *Acronicta lepusculina* Guenée ♂ — Alberta, highway 41 at South Saskatchewan River, VI.25.2000 (J. Troubridge) CNC

50.38 *Acronicta mansueta* Smith ♂ — British Columbia, E end of Seton Lake, VII.7.1995 (J. Troubridge) UASM

50.39 *Acronicta marmorata* Smith ♀ — Oregon, Josephine Co., 3 miles SE of Williams, VIII.24.1997 (J. Troubridge) UASM

50.40 *Acronicta perdita* Grote ♀ — British Columbia, 3 km SE of Okanogan Falls, IV.19/25.1992 (Troubridge & Gardiner) CNC

50.41 *Acronicta quadrata* Grote ♀ — Alberta, Calgary, VI.20.1914 (F. Wolley Dod) CNC

50.42 *Merolonche lupini* (Grote) ♀ — Colorado, Boulder Co., Nederland, 8,300′, V.1.1961 (M. MacKay) CNC

50.43 *Pseudopanthea palata* (Grote) ♀ — Arizona, Yavapai Co., 4 miles N of Prescott, IV.5.1972 (L. Martin) CNC

50.44 *Charadra deridens* (Guenée) ♂ — New Brunswick, Edmundston, VI.14.2001 (H. Hensel) CNC

50.45 *Simyra insularis* (Herrich-Schäffer) — Washington, Walla Walla Co., Walla Walla, VII.15.1958 (W. Cook) CNC

50.46 *Diphthera festiva* (Hübner) ♂ — Arizona, Cochise Co., Ash Canyon Road, Huachuca Mtns., 5,100′, IX.22.1992 (N. McFarland) CNC

50.47 *Agriopodes tybo* Barnes ♂ — Arizona, Cochise Co., Ash Canyon Road, Huachuca Mtns., 1550m, VIII.3.1989 (B. Landry & J. Landry) CNC

50.48 *Miracavira brillians* (Barnes) ♀ — Arizona, Santa Cruz Co., Madera Canyon, Santa Rita Mtns., 5,600′, VI.18.1963 (J. Franclemont) CNC

PLATE 50

PLATE 51 **Noctuidae**

51.1 *Gerra sevorsa* (Grote) ♂ — Arizona, Cochise Co., Sunny Flat campground, Chiricahua Mounains, VII.28.1989 (B. Landry & J. Landry) CNC

51.2 *Euscirrhopterus gloveri* Grote and Robinson ♂ — Arizona, Cochise Co., Huachuca Mts., Ash Canyon Rd., 5,100′, VIII.8.1989 (J. Landry) CNC

51.3 *Euscirrhopterus cosyra* (Druce) ♀ — Texas, Brewster Co., Castolon, Big Bend Natl. Park, V.14.1959 (Howden & Becker) CNC

51.4 *Alypiodes bimaculata* (Herrich-Schäffer) ♂ — Arizona, Cochise Co., Ramsey Canyon, Huachuca Mtns., VII.8.1965 (R. Sternitzky) CNC

51.5 *Alypia octomaculata* (Fabricius) ♀ — Colorado, Boulder, V.19.1936 (T.N. Freeman) CNC

51.6 *Alypia langtoni* Couper ♀ — British Columbia, Steelhead, VI.25.1933 (H.B. Leech) CNC

51.7 *Alypia langtoni* Couper ♀ — California, Monterey Co., Soledad, VII.24.1921 (E.C. Johnston) CNC

51.8 *Alypia mariposa* Grote and Robinson ♀ — California, Stanislaus Co., IV.16.1923 CNC

51.9 *Alypia ridingsi* Grote ♀ — California, Los Angeles Co., Lovejoy Buttes, III.23.1952, reared from *Oenothera* (C. Henne) CNC

51.10 *Androloma maccullochii* (Kirby) ♂ — British Columbia, Mt. Revelstoke, 6,000′, VII.26.1952 (G.P. Holland) CNC

51.11 *Cucullia montanae* Grote ♂ — Arizona, Yavapai Co., Granite Dells, VIII.9.1971 (L. Martin) CNC

51.12 *Cucullia lilacina* Schaus ♂ — Arizona, Cochise Co., Ash Canyon Road, VII.16.1998 (N. McFarland) CNC

51.13 *Cucullia antipoda* Strecker ♂ — Washington, Grant Co., S of Moses Lake, V.21.1994 (J. Troubridge) CNC

51.14 *Cucullia luna* Morrison ♂ — Alberta, Saskatchewan River, VII.13.1999 (J. Troubridge) CNC

51.15 *Cucullia dammersi* (McDunnough) ♀ paratype — California, Riverside Co., Riverside, XII.12.1934 (C. Dammers) CNC

51.16 *Pleromelloida cinerea* (Smith) ♂ — Oregon, Co., John Day, IX.11.1961 (D. Hardwick) CNC

51.17 *Catabena lineolata* Walker ♀ — Arizona, Yavapai Co., 4 miles N of Prescott, VIII.16.1972 (L. Martin) CNC

51.18 *Oxycnemis fusimacula* Smith, ♀ — California, San Diego Co., Anza Borrego Desert State Park, III.24.2001 (T. Mustelin) CNC

51.19 *Homohadena fifia* Dyar, ♀ — British Columbia, Lillooet, VI.27.1926 (J. McDunnough) CNC

51.20 *Adita chionanthi* (J.E. Smith) ♂ — Wyoming, 5 miles W of Newcastle, 4,500′, IX.12.64 (D.F. Hardwick) CNC

51.21 *Homoncocnemis fortis* (Grote) ♂ — California, Los Angeles Co., La Tuna Canyon, IX.18.1943, ex larva on *Fraxinus* (W. Evans) CNC

51.22 *Oncocnemis astrigata* Barnes and McDunnough ♂ — California, San Diego Co., Cottonwood Creek, VII.12.2001 (T. & S. Mustelin) CNC

51.23 *Oncocnemis columbia* McDunnough ♂ — Washington, Yakima Co., Bethel Ridge, VIII.27.1998 (J. Troubridge) CNC

51.24 *Oncocnemis definita* (Barnes and McDunnough) ♀ — Oregon, Harney Co., Catlow Rim, 4,500′, IX.5.1999 (J. Troubridge) UASM

51.25 *Oncocnemis chalybdis* Troubridge and Crabo ♂ — Montana, Flathead Co., Kalispell, VIII.23.1961 (D. Hardwick) CNC

51.26 *Oncocnemis punctilinea* Hampson ♂ — Idaho, Gooding Co., Wendell, VII.18.1963 (R. Miller) CNC

51.27 *Oncocnemis ragani* Barnes ♂ — California, Ventura Co., Valle Vista Campground, VI.23.2000 (T. Dimock) CNC

51.28 *Oncocnemis singularis* Barnes and McDunnough ♂ — California, Riverside Co., Chino Canyon, IV.19.1950 (E. Johnston) CNC

51.29 *Oncocnemis umbrifascia* Smith ♂ — Nevada, Elko Co., Angel Creek Campground, VII.22.2001 (J. Lafontaine & J. Troubridge) CNC

51.30 *Lepipolys perscripta* Guenée ♂ — California, San Diego Co., Penasquitos Canyon, III.19.2000 (T. Mustelin) CNC

51.31 *Sympistis heliophila* Paykull ♀ — Alberta, Caribou Mtns., VI.16.2003 (G. Pohl) CNC

51.32 *Sympistis zetterstedtii* (Staudinger) ♀ — Yukon Territory, S of Carcross, VII.18.1985 (J. Lafontaine) CNC

51.33 *Stylopoda cephalica* Smith ♀ — California, Los Angeles Co., Mint Canyon, III.26.1947, CNC

51.34 *Behrensia conchiformis* Grote ♂ — California, San Diego Co., Torrey Pines State Park, I.14.2006 (N. Bloomfield) CNC

51.35 *Amphipyra pyramidoides* Guenée ♀ — British Columbia, Richmond, VIII.27.2002 (J. Troubridge) CNC

51.36 *Psaphida damalis* (Grote) ♂ — California, Sonoma Co., The Geysers, III.19.1939 (E. Johnston) CNC

51.37 *Pseudocopivaleria anaverta* Buckett & Bauer ♀ — California, Riverside Co., Pinyon Crest, II.20.1971 (R. Leuschner) CNC

51.38 *Pleuromella opter* Dyar ♂ — California, Napa Co., Spring Mtn., IV.2.1940 (E. Johnston) CNC

51.39 *Provia argentata* Barnes & McDunnough ♀ — Arizona, Yavapai Co., N of Prescott, III.10.1972 (L. Martin) CNC

51.40 *Feralia deceptiva* McDunnough ♂ — Oregon, Marion Co., Salem, III.22.1960 (H. Foster) CNC

51.41 *Feralia februalis* Grote ♀ — California, Los Angeles Co., La Tuna Canyon, IV.13.1948 (W. Evans) CNC

51.42 *Aleptina inca* Dyar ♀ — New Mexico, Otero Co., Sacramento Mtns., U.S. 82, VI.13.1975 (D.F. Hardwick) CNC

51.43 *Copibryophila angelica* Smith ♂ — California, Stanislaus Co., Modesto, IX.13.1942 (W. Cook) CNC

51.44 *Metaponpneumata rogenhoferi* Möschler ♀ — Arizona, Santa Cruz Co., Pena Blanca Lake, VII.25.1971 (L. Martin) CNC

PLATE 51

PLATE 52 **Noctuidae**

52.1 *Fala ptychophora* Grote ♀ — California, San Diego Co., Pine Valley, IV.17.1950 (E. Johnston) CNC

52.2 *Plagiomimicus spumosum* (Grote) ♀ — Arizona, Apache Co., Ft. Defiance, VI/IX.1948 (E. Oberg) CNC

52.3 *Plagiomimicus dimidiata* (Grote) ♀ — Arizona, Yavapai Co., 4 miles N of Prescott, IX.16.1971 (L. Martin) CNC

52.4 *Plagiomimicus tepperi* (Morrison) ♀ — Colorado, Platte Canyon, VII.25.1927, CNC

52.5 *Lineostriastiria hutsoni* (Smith) ♂ — California, San Bernardino Co., Needles, IV.11.1952 (D. Hardwick) CNC

52.6 *Xanthothrix ranunculi* Henry Edwards ♂ — California, Los Angeles Co., 1 mile S of Gorman, 3,800', IV.1.1987 (D. & V. Hardwick) CNC

52.7 *Chrysoecia atrolinea* (Barnes & McDunnough) ♀ — Arizona, Cochise Co., Ash Canyon Road, Huachuca Mtns., 5,100', IX.2.1991 (N. McFarland) CNC

52.8 *Chrysoecia scira* (Druce) ♀ — Arizona, Cochise Co., Carr Canyon, Huachuca Mtns., 5,600', VIII.19.1993 (J. O'Hara) CNC

52.9 *Basilodes chrysopis* Grote ♂ — Arizona, Santa Cruz Co., Nogales, VIII.19.1906 (A. Koebele) CNC

52.10 *Cirrhophanus dyari* Cockerell ♀ — New Mexico, Sierra Co., 18 miles NNE of Silver City, 7,900', IX.7.1975 (Lafontaine & Bowen) CNC

52.11 *Eulithosia plesioglauca* (Dyar) ♀ — Arizona (O. Bryant) CNC

52.12 *Stiria intermixta* Dyar ♀ — Arizona, Yavapai Co., Prescott, IX.9.1908 (R. Kunze) CNC

52.13 *Chalcopasta howardi* (Henry Edwards) ♂ — Texas, Jeff Davis Co., near Fort Davis, Davis Mtns., 5,000', VIII.25.1928 (O. Poling) CNC

52.14 *Axenus arvalis* Grote ♂ — California, Napa Co., 6 miles NNE of Pope Valley, IV.21.1978 (D. Hardwick) CNC

52.15 *Annaphila diva* Grote ♂ — Washington, Mason Co., Belfair, V.8.1949 (E. Johnston) CNC

52.16 *Annaphila arvalis* Henry Edwards ♂ — California, Los Angeles Co., Tie Canyon, 4,700', III.17.1950 (W. Evans) CNC

52.17 *Annaphila astrologa* Barnes and McDunnough ♀ — California, Los Angeles Co., Bob's Gap, Llano, IV.1.1948, visiting *Coleogyne* flowers (W. Evans) CNC

52.18 *Annaphila lithosina* Henry Edwards ♀ — California, Plumas Co., Jackson Creek, 5,400', VI.3.2002 (J. Troubridge & L. Crabo) CNC)

52.19 *Annaphila casta* Henry Edwards ♀ — Oregon, Benton Co., MacDonald Forest Reserve, VI.2.1970 (J. Clarke) CNC

52.20 *Annaphila evansi* Rindge and Smith ♂ — California, Los Angeles Co., Mint Canyon, IV.9.1948 (W. Evans) CNC

52.21 *Neumoegenia poetica* Grote ♀ — Arizona, Santa Cruz Co., Peña Blanca Lake, 3,700', VII.25.1971 (L. Martin) CNC

52.22 *Argentostiria koebelei* (Riley) ♂ — California, Riverside Co., Thousand Palms, III.21.1955 (J. Martin) CNC

52.23 *Thurberiphaga diffusa* (Barnes) ♂ — Arizona, Santa Cruz Co., Madera Canyon, Santa Rita Mtns., VIII.29.1946 (J. Comstock & L. Martin) CNC

52.24 *Baptarma felicita* Smith ♂ — Arizona, 9 miles S of Quartzite, 1,000', emerged IV.1.1984, larva on *Phacelia* (D. & V. Hardwick) CNC

52.25 *Eutricopis nexilis* Morrison ♂, dorsal — British Columbia, Keremeos, Twin Lake, 4,700',CNC

52.26 *Eutricopis nexilis* Morrison ♂, ventral — British Columbia, Keremeos, Twin Lake, 4,700',CNC

52.27 *Pyrrhia exprimens* (Walker) ♂ — Quebec, Sheenboro, 1993, reared from egg (D. & V. Hardwick) CNC

52.28 *Helicoverpa zea* (Boddie) ♀ — California, Sonoma Co., Petaluma, X.6.1936, CNC

52.29 *Heliothis phloxiphaga* Grote and Robinson ♀ — California, Kern Co., 9 miles NE of Arvin, 1,800', V.1988 (D. Hardwick) CNC

52.30 *Heliothis oregonica* (Henry Edwards) ♂ — Colorado, Grand Co., 3.9 miles SW of Fraser, St. Louis Creek Road, 8,900', VII.3.1991 (T. Dickel) CNC

52.31 *Heliothis virescens* (Fabricius) ♂ — Arizona, Maricopa Co., Wickenburg, 2,000', emerged X.1990 (D. & V. Hardwick) CNC

52.32 *Heliothis subflexa* (Guenée) ♂ — Colorado, Larimer Co., Vistenz-Smith Mtn. Park, 9.8 miles W of Loveland, 5,700', VI.22.1998 (T. Dickel) CNC

52.33 *Heliocheilus paradoxus* Grote ♀ — Arizona, Cochise Co., Ash Canyon, Huachuca Mtns., 5,100' (D. Hardwick) CNC

52.34 *Heliocheilus julia* (Grote) ♂ — Arizona, Santa Cruz Co., Nogales, VIII.26.1906 (A. Koebele) CNC

52.35 *Schinia nuchalis* (Grote) ♂ — British Columbia, Lillooet, VII.27.1916, CNC

52.36 *Schinia citrinella* (Grote and Robinson) ♀ — Texas, San Patricio Co., Welder Wildlife Refuge, Sinton, IX.23.1984 (D. Hardwick) CNC

52.37 *Schinia aurantiaca* (Henry Edwards) ♀ — California, Riverside Co., Gavilan Hills, IV.19.1948, visiting *Cryptantha* flowers (W. Evans) CNC

52.38 *Schinia jaguarina* (Guenée) ♀ — Arizona, Yavapai Co., 4 miles N of Prescott, VIII.9.1970 (L. Martin) CNC

52.39 *Schinia meadi* (Grote) ♂ — Wyoming, Fremont Co., 35 km W of Jeffrey City, VII.18.1982 (J. Landry) CNC

52.40 *Schinia pulchripennis* (Grote) ♀ — California, Riverside Co., 2 miles N of Aguanga, 2,600', IV.8.1992 (D. & V. Hardwick) CNC

52.41 *Schinia villosa* (Grote) ♀ — Arizona, Coconino Co., De Mothe Park, North Kaibab, VII.25.1975, on *Erigeron* (R. Wielgus) CNC

52.42 *Schinia arcigera* (Guenée) ♀ — Arizona, Yavapai Co., 4 miles N of Prescott, VIII.25.1970 (L. Martin) CNC

52.43 *Schinia miniana* (Grote) ♂ — Arizona, Maricopa Co., Wickenburg, 2,000', IX.26.1980 (D. & V. Hardwick) CNC

52.44 *Schinia gaurae* (J.E. Smith) ♂ — Texas, Jeff Davis Co., Limpia Canyon, VIII.3.1950 (E. Johnston) CNC

52.45 *Schinia masoni* (Smith) ♀ — Colorado, Jefferson Co., Chimney Gulch, VI.17.1927, CNC

52.46 *Schinia niveicosta* (Smith) ♂ — California, Riverside Co., La Quinta, III.4.1955 (D. Hardwick) CNC

52.47 *Schinia acutilinea* Grote ♀ — Alberta, Dinosaur Peak, 2,100', VIII.14.2001 (Troubridge & Anweiler) CNC

52.48 *Schinia walsinghami* (Henry Edwards) ♀ — Washington, Walla Walla Co., Wallula, VIII.27.1957 (W. Cook) CNC

52.49 *Condica discistriga* (Smith) ♀ — British Columbia, above Vasseaux Lake, V.27/31.1990 (J. Troubridge) CNC

52.50 *Ogdoconta cinereola* (Guenée) ♀ — Texas, Uvalde Co., Concan, V.21.1990 (N. McFarland) CNC

52.51 *Crambodes talidiformis* Guenée ♂ — Alberta, Lethbridge, VII.8.1955 (E. Sterns) CNC

52.52 *Cryphia viridata* (Harvey) ♂ — California, Marin Co., Lucas Valley, V.31.1937, CNC

52.53 *Cryphia cuerva* Barnes ♀ — British Columbia, Bute Inlet, 1.9m, VIII.21.2006 (D. Holden) CNC

52.54 *Cryphia olivacea* (Smith) ♀ — California, Mono Co., Lee Vining, VII.20.1938 (E. Johnston) CNC

52.55 *Spodoptera. exigua* (Hübner) ♀ — California, Sonoma Co., Petaluma, X.15.1935 (E. Johnston) CNC

52.56 *Spodoptera praefica* (Grote) ♀ — California, Sierra Co., 6 miles WNW Sierraville, 6,700', VIII.21.1967 (D. Hardwick) CNC

52.57 *Spodoptera ornithogalli* (Guenée) ♂ — California, Los Angeles Co., Hawthorne, IX.15.1944 (E. Johnston) CNC

PLATE 52

PLATE 53 **Noctuidae**

53.1 *Galgula partita* Guenée ♂ — Texas, Uvalde Co., Concan, 1,300', III.21.1990 (N. McFarland) CNC

53.2 *Galgula partita* Guenée ♀ — California, Sonoma Co., Santa Rosa, VI.21.1937 (E. Johnston) CNC

53.3 *Caradrina montana* (Bremer) ♂ — British Columbia, Kirby Flats Road, VI.12.1998 (J. Troubridge) CNC

53.4 *Proxenus miranda* (Grote) ♂ — Washington, Mason Co., Stimson Creek, V.24.1947 (E. Johnston) CNC

53.5 *Nedra stewarti* (Grote) ♂ — California, Lake Co., Anderson Springs, VIII.21.1948 (W. Bauer) CNC

53.6 *Euplexia benesimilis* McDunnough ♂ — British Columbia, Agassiz, Green Mtn., VIII.12.1998 (J. Troubridge) CNC

53.7 *Phlogophora periculosa* Guenée ♀ — British Columbia, Mallory Road, NW of Enderby, VII.25/30.1996 (A. Borkent) CNC

53.8 *Apamea lignicolora* (Guenée) ♀ — South Dakota, Hill City, Black Hills, VII.22.1964 (D. Ferguson) CNC

53.9 *Apamea occidens* (Grote) ♀ — Oregon, Jackson Co., Mt. Ashland, 2050m, VII.30.1991 (L. Crabo) CNC

53.10 *Apamea amputatrix* (Fitch) ♂ — Colorado, Boulder Co., Nederland, 8,300', VII.3.1961 (M. MacKay) CNC

53.11 *Apamea amputatrix* (Fitch) ♂ — Colorado, Grand Co., Radium State Wildlife Area, 7,040', VIII.29.2005 (T. Dickel) CNC

53.12 *Apamea alia* (Guenée) ♂ — Oregon, 8-Dollar Mtn. Rd., V.23.2001 (J. Troubridge) CNC

53.13 *Apamea sordens* (Hufnagel) ♂ — Colorado, Grand Co., Arapaho Natl. Forest, 8,400', VII.2.1999 (T. Dickel) CNC

53.14 *Apamea digitula* (Mustelin & Mikkola) ♂ — Oregon, Marion Co., Salem, V.8.1961 (K. Goeden) CNC

53.15 *Apamea cogitata* (Smith) ♂ — Alberta, 25 miles W Calgary, VII.19.1960 (D. Hardwick) CNC

53.16 *Apamea devastator* (Brace) ♂ — Montana, Gallatin Co., 12 miles E Bozeman, 5,700', VIII.14.1964 (D. Hardwick) CNC

53.17 *Apamea devastator* (Brace) ♂ — Alberta, Calgary, head of Pine Creak, VIII.11.1903 (F. Wolley Dod) CNC

53.18 *Apamea plutonia* (Grote) ♂ — Montana, Sweetgrass Co., Crazy Mtns., 5,743', VI.29.1966 (D. Ferguson) — CNC

53.19 *Resapamea passer* (Guenée) ♂ — Washington, Grant Co., Potholes dunes, V.19.2001 (J. Troubridge) CNC

53.20 *Eremobina hanhami* (Barnes & Benjamin) ♂ — Colorado, Grand Co., Hot Sulphur Springs, 7,670', VIII.22.1991 (T. Dickel) CNC

53.21 *"Oligia" indirecta* (Grote) ♀ — Nevada, Elko Co., Angel Creek campground, 6,700', VII.22.2001 (Lafontaine & Troubridge) CNC

53.22 *"Oligia" rampartensis* Barnes and Benjamin ♀ — Oregon, Baker Co., Cornet Creek, 3,600', VII.12.2002 (J. Troubridge) CNC

53.23 *"Oligia" violacea* (Grote) ♂ — British Columbia, Mt. Kobau, 6,040', VIII.4.1999 (J. Troubridge) CNC

53.24 *"Oligia" mactata* (Guenée) ♂ — British Columbia, Mallory Road, Enderby, IX.1/14.1995 (Borkent & Troubridge) CNC

53.25 *Cobalos angelicus* Smith ♂ — California, San Luis Obispo Co., San Luis Obispo, VI.3.1937 (E. Johnston) CNC

53.26 *Parastichtis suspecta* (Hübner) ♂ — Colorado, Routt Co., Ferndale Recreation Area, Routt Natl. Forest, 8,000', VII.29.1993 (T. Dickel) CNC

53.27 *Photedes inquinata* (Guenée) ♀ — Alberta, Edmonton, VII.17.1918, CNC

53.28 *Capsula oblonga* (Grote) ♂ — Washington, Walla Walla Co., Walla Walla, VII.30.1966 (W. Cook) CNC

53.29 *Amphipoea americana* (Speyer) ♀ — Washington, Chelan Co., Junior Point, 6,100', VIII.27.1998 (J. Troubridge) CNC

53.30 *Amphipoea lunata* (Smith) ♀ — California, Lake Co., Anderson Springs, VII.12.1947 (W. Bauer), CNC

53.31 *Papaipema sauzalitae* (Grote) ♂ — California, Marin Co., Inverness, X.6.1948 (W. Bauer) CNC

53.32 *Papaipema insulidens* (Bird) ♂ — British Columbia, Duncans, Vancouver Island (C. Livingston) CNC

53.33 *Hydraecia perobliqua* Harvey ♂ — Utah, Heepey, VII.27.1937 (D. Hardy) CNC

53.34 *Hydraecia stramentosa* Guenée ♂ — New Mexico, Otero Co., 13 miles SE of Cloudcroft, 7,100', IX.9.1975 (Lafontaine & Brown) CNC

53.35 *Xylena nupera* (Lintner) ♂ — British Columbia, Kamloops, IX.18.1937 (J. Jacob) CNC

53.36 *Xylena brucei* (Smith) ♂ — British Columbia, Kirby Flats Road, IX.20.1999 (J. Troubridge) CNC

53.37 *Lithomoia germana* (Morrison) ♀ — British Columbia, Green Mtn., Agassiz, IX.12.1999 (J. Troubridge) CNC

53.38 *Homoglaea californica* (Smith) ♂ — California, Lake Co., Anderson Springs, II.9.1950 (W. Bauer) CNC

53.39 *Litholomia napaea* (Morrison) ♂ — Oregon, Marion Co., Salem, III.22.1960 (H. Foster) CNC

PLATE 53

PLATE 54 Noctuidae

54.1 *Lithophane pertorrida* (McDunnough) ♂ — British Columbia, Salmon Arm, III.20.1915 (W. Blackmore) CNC

54.2 *Lithophane georgii* Grote ♂ — Colorado, Grand Co., Blacktail Creek, Radium State Wildlife Area, 7,040′, IX.28.1998 (T. Dickel) CNC

54.3 *Lithophane dilatocula* (Smith) ♂ — California, Lake Co., Anderson Springs, II.9.1950 (W. Bauer) CNC

54.4 *Lithophane vanduzeei* (Barnes) ♂ — California, Monterey Co., Carmel, III.1.1991 (F. Sala) CNC

54.5 *Lithophane amanda* (Smith) ♀ — British Columbia, Vanderhoof, VIII.14.1953, reared from *Salix* (Forest Insect Survey) CNC

54.6 *Lithophane contenta* Grote ♂ — California, Sonoma Co., Spring Mtn., X.13.1939 (E. Johnston) CNC

54.7 *Lithophane innominata* (Smith) ♀ — British Columbia, Shingle Creek Road, Keremeos, X.5.1935 (A. Gartell) CNC

54.8 *Lithophane thaxteri* Grote ♂ — Alberta, Sunnyside, Lloydminster, IX.9.1948 (P. Bruggemann) CNC

54.9 *Eupsilia fringata* (Barnes & McDunnough) ♂ — British Columbia, Salmon Arm, V.10.1926 (Blackmore) CNC

54.10 *Mesogona rubra* Crabo and Hammond ♀ — Oregon, Lane Co., Florence, IX.10.1960 (K. Goeden) CNC

54.11 *Agrochola pulchella* (Smith) ♂ — Washington, Clallam Co., Elwa River, IX.13.1947 (E. Johnston) CNC

54.12 *Agrochola decipiens* (Grote) ♀ — Washington, Chelan Co., Derby Canyon, IX.23.2002 (L. Crabo) CNC

54.13 *Anathix puta* (Grote and Robinson) ♀ — British Columbia, 8 miles W of Keremeos, VIII.5.1960 (D. Hardwick) CNC

54.14 *Anathix aggressa* (Smith) ♀ — Colorado, Lincoln Co., 6 miles NW of Hugo, 5,200′, IX.9.1968 (D. Hardwick) CNC

54.15 *Xanthia tatago* Lafontaine and Mikkola ♂ — British Columbia, 7 miles N of Lac La Hache, 2,800′, IX.4.1960 (D. Hardwick) CNC

54.16 *Hillia maida* (Dyar) ♂ — British Columbia, Kaslo, IX.1.1902 (J. Cockle) CNC

54.17 *Aseptis perfumosa* (Hampson) ♂ — California, San Diego Co., Laguna Mtns., Kitchen Creek Road, 5,500′, V.22.2001 (T. Mustelin) CNC

54.18 *Aseptis binotata* (Walker) ♂ — Washington, Mason Co., 2 miles SE of Little Hoquiam, 25m, VII.24.1989 (L. Crabo) CNC

54.19 *Aseptis characta* (Grote) ♂ — California, Ventura Co., Canyon Valley, 3,690′, V.29.2001 (T. Dimock) CNC

54.20 *Aseptis adnixa* (Grote) ♂ — California, Sonoma Co., Spring Mtn., IV.24.1939 (W. Bauer) CNC

54.21 *Aseptis monica* (Barnes & McDunnough) ♂ — California, Riverside Co., Split Rock Tank, V.20.1938 (G. & J. Sperry) CNC

54.22 *Aseptis ethnica* (Smith) ♂ — California, Madera Co., North Fork, 4,440′, VI.29.1987 (E. Metzler) CNC

54.23 *Brachylomia populi* (Strecker) ♂ — Colorado, La Plata Co., 10 miles WSW of Durango, 7,600′, VIII.21.1971 (D. Hardwick) CNC

54.24 *Brachylomia algens* (Grote) ♂ — British Columbia, Lower Post, VIII.16.1948 (W. Mason) CNC

54.25 *Epidemas obscurus* Smith ♂ — British Columbia, 5 km SE of Okanogan Falls, IX.16.1990 (J. Troubridge) CNC

54.26 *Cosmia calami* (Harvey) ♂ — Arizona, Yavapai Co., 4 miles N of Prescott, VII.15.1972 (L. Martin) CNC

54.27 *Cosmia praeacuta* (Smith) ♂ — Washington, Okanogan Co., Harts Pass Road, IX.7.1999 (L. & E. Crabo) CNC

54.28 *Zotheca tranquilla* (Grote) ♂ — Washington, Chelan Co., Leavenworth, VII.3.1949 (E. Johnston) CNC

54.29 *Enargia decolor* (Walker) ♂ — Alberta, 6 miles W of Cowley, 3,900′, IX.8.1973 (J. Lafontaine) CNC

54.30 *Ipimorpha pleonectusa* (Grote) Arizona, Apache Co., Greer, White Mtns., 8,500′, VIII.8.1962 (E. & I. Munroe) CNC

54.31 *Andropolia contacta* (Walker) ♀ — Nevada, Elko Co., Angel Creek campground, 6,700′, VII.22.2001 (Lafontaine & Troubridge) CNC

54.32 *Andropolia theodori* (Grote) ♀ — Washington, Whatcom Co., South Ross Lake, VIII.16/20.2004 (L. Crabo) CNC

54.33 *Rhizagrotis albalis* (Grote) ♀ — California, Alpine Valley, V.24.1955 (W. Richards) CNC

54.34 *Fishia yosemitae* (Grote) ♂ — Colorado, Larimer Co., Viestenz-Smith Mtn. Park, 5,700′, X.9.1998 (T. Dickel) CNC

54.35 *Platypolia contadina* (Smith) ♀ — Oregon, Lane Co., Frissell Point, H.J. Andrews Experimental Forest, 4,800 to 4,960′, IX.9.1996 (L. & E. Crabo) CNC

54.36 *Dryotype opina* (Grote) ♂ — Oregon, Coos Co., Bandon, IX.27.1956 (K. Goeden) CNC

54.37 *Mniotype ducta* (Grote) ♂ — Alberta, Lethbridge, VI.3.1983 (J. Byers) CNC

54.38 *Ufeus satyricus sagittarios* Grote ♂ — Alberta, Pine Creek, Calgary, IX.21.1914 (F. Wolley Dod) CNC

54.39 *Hyppa brunneicrista* Smith ♂ — Oregon, Lane Co., Frissell Point, Forest Road 1056, VI.27.1997 (J. Troubridge) CNC

54.40 *Chytonix divesta* (Grote) ♂ — Oregon, Douglas Co., Thom Prairie, Umpqua River Valley, 1040m, IX.13.1991 (L. Crabo) CNC

PLATE 54

PLATE 55 Noctuidae

55.1 *Properigea albimacula* (Barnes & McDunnough) ♀ — Washington, Columbia Co., Dayton, 1,620′, VIII.3.1955 (R. Miller) CNC

55.2 *Pseudobryomima fallax* (Hampson) ♂ — California, Lake Co., Anderson Springs, XI.14.1949 (E. Johnston) CNC

55.3 *Pseudanarta flavidens* (Grote) ♂ — Colorado, Chaffee Co., Maysville, VIII.17.1946 (H. Ramstadt) CNC

55.4 *Pseudanarta flava* (Grote) ♂ — British Columbia, Oliver, IX.10.1953 (J. Martin) CNC

55.5 *Magusa orbifera* (Walker) ♀ — Arizona, Yavapai Co., 6 miles NE of Prescott, 5,400′, IX.12.1971 (D. Hardwick) CNC

55.6 *Magusa orbifera* (Walker) ♀ — Arizona, Yavapai Co., 6 miles NE of Prescott, 5,400′, IX.12.1971 (D. Hardwick) CNC

55.7 *Perigonica angulata* Smith ♂ — California, Lake Co., Anderson Springs, IV.16.1950 (W. Bauer) CNC

55.8 *Acerra normalis* Grote ♂ — Washington, Yakima Co., Satus Creek, IV.4.1942 (E. Johnston) CNC

55.9 *Stretchia muricina* (Grote) ♂ — British Columbia, Depot Creek, Chilliwack Lake, IV.21.1994 (J. Troubridge) CNC

55.10 *Orthosia pacifica* (Harvey) ♀ — British Columbia, 15 km W of Oyster River, IV.11.2000 (J. Troubridge) CNC

55.11 *Orthosia hibisci* (Guenée) ♂ — California, Marin Co., Inverness, II.7.1940 (E. Johnston) CNC

55.12 *Orthosia praeses* (Grote) ♂ — Washington, Grays Harbor Co., Burrows Road at route 109, Ocean City, III.26.1993 (J. Troubridge) CNC

55.13 *Orthosia praeses* (Grote) ♂ — British Columbia, 5 km E of Langley, IV.1/7.1991 (J. Troubridge) CNC

55.14 *Orthosia mys* (Dyar) ♂ — California, Napa Co., Spring Mtn., I.2.1940 (E. Johnston) CNC

55.15 *Egira simplex* (Walker) ♂ — California, Napa Co., Spring Mtn., III.13.1940 (E. Johnston) CNC

55.16 *Egira perlubens* (Grote) ♀ — California, Napa Co., Mt. St. Helena, III.24.1939 (E. Johnston) CNC

55.17 *Egira dolosa* (Grote) ♂ — Washington, Walla Walla Co., Walla Walla, IV.20.1961 (W. Cook) CNC

55.18 *Anarta trifolii* (Hufnagel) ♂ — Arizona, Yavapai Co., 4 miles N of Prescott, VIII.24.1972 (L. Martin) CNC

55.19 *Nephelodes minians* Guenée ♂ — Alberta, 10 miles W Pincher Creek, 4,500′, VIII.8.1961 (D. Hardwick) CNC

55.20 *Anarta oregonica* (Grote), ♂ — Washington, Chelan Co., Junior Point, 6,100′, VII.12.1998 (J. Troubridge) CNC

55.21 *Anarta farnhami* (Grote), ♂ — Colorado, Larimer Co., Estes Park, 12 miles WNW, VII.27.1967 (D. Hardwick) CNC

55.22 *Anarta crotchi* (Grote), ♀ — Oregon, Harney Co., Catlow Valley Road, VI.4.1994 (J. Troubridge & L. Crabo) CNC

55.23 *Anarta nigrolunata* Packard, ♀ — British Columbia, Lisadele Lake, 4,000′, VIII.6.1960 (W. Moss) CNC

55.24 *Scotogramma fervida* Barnes & McDunnough, ♂ — Oregon, Harney Co., 33 miles S of Burns, VI.13.1999 (J. Troubridge) CNC

55.25 *Polia nimbosa* (Guenée) ♂ — Washington, King Co., Factoria, VII.29.1949 (E. Johnston) CNC

55.26 *Polia purpurissata* (Grote) ♂ — Washington, Okanogan Co., 1 mile NE of Bridgeport, IX.4.1997 (J. Troubridge) CNC

55.27 *Scotogramma submarina* (Grote) ♂ — Alberta, Lost River, Onefour, VII.22.2003 (J. Troubridge) CNC

55.28 *Coranarta luteola* (Grote and Robinson) ♀ — Yukon Territory, Dempster Highway, milepost 155, 950m, VI.29/VII.3.1980 (Wood & Lafontaine) CNC

55.29 *Lacanobia nevadae* (Grote) ♀ — British Columbia, Brent's Lake, Penticton, V.30.1935 (A. Gartrell) CNC

55.30 *Lacanobia subjuncta* (Grote & Robinson) ♀ — California, Plumas Co., Jackson Creek, 5,400′, V.21.2001 (J. Troubridge) CNC

55.31 *Melanchra picta* (Harris) ♂ — Washington, Walla Walla Co., Walla Walla, V.12.1944 (W. Cook) CNC

55.32 *Papestra cristifera* (Walker) ♀ — Wyoming, Washakie Co., Tensleep Preserve, 6,175′, V.27.2000 (C. Ferris) CNC

55.33 *Hada sutrina* (Grote) ♀ — Wyoming, Albany Co., 8,640′, VI.29.99 (C. Ferris) CNC

55.34 *Mamestra configurata* (Walker) ♂ — British Columbia, Mt. Kobau, 6,040′, VII.20.2000 (J. Troubridge) CNC

55.35 *Sideridis rosea* (Harvey) ♂ — Washington, Grant Co., dunes S of Moses Lake, V.21.1994 (J. Troubridge) CNC

55.36 *Admetovis oxymorus* Grote ♀ — Oregon, Josephine Co., Bolan Mtn., 5,200′, VII.27.2001 (J. Troubridge) CNC

55.37 *Hadena variolata* (Smith) ♂ — California, Ventura Co., Pine Mtn. campground, 6,750′, VII.24.2000 (T. Dimock) CNC

55.38 *Hadena capsularis* (Guenée) ♂ — British Columbia, 5 km SE of Okanogan Falls, VI.10.1990 (J. Troubridge) CNC

55.39 *Trichoclea uscripta* (Smith) ♂ — Oregon, Lake Co., Alkali Lake dunes, 4,100′, V.20.2001 (J. Troubridge) CNC

55.40 *Faronta diffusa* (Walker) ♀ — New Mexico, Colfax Co., Cimarron Canyon, 7,900′, VII.10.1962 (E. & I. Munroe) CNC

55.41 *Dargida procincta* (Grote) ♂ — British Columbia, 5 km E of Langley, VIII.19.1990 (J. Troubridge) CNC

55.42 *Mythimna oxygala* (Grote) ♂ — British Columbia, Wickaninnish Beach, VI.12.2001 (N. Page) CNC

PLATE 55

PLATE 56 **Noctuidae**

56.1 *Mythimna unipuncta* (Haworth) ♂ — Oregon, Harney Co., 1 mile W of Frenchglen, IX.26.1997 (J. Troubridge) CNC

56.2 *Leucania multilinea* Walker ♀ — Washington, Grant Co., Potholes dunes, VIII.19.1999 (J. Troubridge) CNC

56.3 *Leucania commoides* Guenée ♂ — Wyoming, Albany Co., Laramie, 7,500′, VIII.5.2001 (C. Ferris) CNC

56.4 *Leucania insueta* Guenée ♂ — British Columbia, Summerland, V.8/14.1994 (M. Gardiner) CNC

56.5 *Leucania farcta* (Grote) ♂ — Washington, Grant Co., 5 miles SE of Vantage, VI.2.1994 (J. Troubridge) CNC

56.6 *Lasionycta impingens* (Walker) ♂ — Colorado, Summit Co., Copper Mtn., 10,000′, VII.25.2000 (T. Mustelin) CNC

56.7 *Lasionycta perplexa* (Smith) ♀ — British Columbia, Watch Peak, VII.16.1995 (J. Troubridge) CNC

56.8 *Lasionycta arietis* (Grote) ♀ — British Columbia, Queen Charlotte Islands, Moresby Island sandspit, VIII.1.1985 (J. Clarke) CNC

56.9 *Lacinipolia anguina* (Grote) ♀ — British Columbia, Salmon Arm, VI.21/VII.9.1994 (A. Borkent) CNC

56.10 *Lacinipolia renigera* (Stephens) ♀ — Ontario, Stittsville, 430′, VII.2.2003 (J. Troubridge) CNC

56.11 *Lacinipolia lorea* (Guenée) ♂ — British Columbia, Salmon Arm, VI.21/VII.9.1994 (A. Borkent) CNC

56.12 *Lacinipolia olivacea* (Morrison) ♂ — Washington, Cowlitz Co., 7 miles N of Castle Rock, 200′, VIII.20.1960 (D. Hardwick) CNC

56.13 *Lacinipolia lustralis* (Grote) ♀ — Washington, Yakima Co., Bethel Ridge, 6,300′, VII.25.1996 (J. Troubridge) CNC

56.14 *Trichocerapoda oblita* (Grote) ♂ — Washington, Grant Co., 2 miles N of Wanapum Dam, IX.17.1994 (J. Troubridge) CNC

56.15 *Anhimella contrahens* (Walker) ♀ — Utah, Provo Co., Provo, VII.25.1909 (T. Spalding) CNC

56.16 *Homorthodes furfurata* (Grote) ♂ — New Mexico, Santa Fe Co., Frijoles Canyon, Bandelier Natl. Monument, 6,080′, VII.17.1962 (E. & I. Munroe) CNC

56.17 *Homorthodes communis* (Dyar) ♂ — California, San Diego Co., Palomar Mtn. Road, 4,800′, IX.13.1998 (Mustelin & Bloomfield) CNC

56.18 *Protorthodes oviduca* (Guenée) ♂ — Colorado, Grand Co., Pioneer Park campground, Hot Sulphur Springs, 7,670′, V.22.1992 (T. Dickel) CNC

56.19 *Protorthodes rufula* (Grote) ♀ — California, Los Angeles Co., La Tuna Canyon, IV.24.1949 (W. Evans) CNC

56.20 *Protorthodes utahensis* (Smith) ♂ — Colorado, Chaffee Co., Poncha Creek, 3 km S of Poncha Springs, VII.14.1982 (J. Landry) CNC

56.21 *Ulolonche disticha* (Morrison) ♂ — Colorado, Chaffee Co., Buena Vista, 7,800′, VI.21.1961 (M. MacKay) CNC

56.22 *Pseudorthodes irrorata* (Smith) ♂ — California, Alameda Co., Berkeley, V.28.1931 (D. Meadows) CNC

56.23 *Orthodes majuscula* Herrich-Schäffer ♀ — Arizona, Cochise Co., Ramsey Canyon, Huachuca Mtns., 6,000′, X.29.1967 (R. Sternitzky) CNC

56.24 *Hexorthodes agrotiformis* (Grote) ♂ — Wyoming, Albany Co., 7,500′, VII.31.1982 (C. Ferris) CNC

56.25 *Zosteropoda hirtipes* Grote ♂ — California, Sonoma Co., Petaluma, VII.12.1936 (E. Johnston) CNC

56.26 *Neleucania bicolorata* (Grote) ♂ — Alberta, Lost River, Onefour, VI.26/27.2000 (J. Troubridge) CNC

56.27 *Miodera stigmata* Smith ♂ — California, San Diego Co., Sorrento Valley marsh, Torrey Pines State Park, I.20/27.2006 (N. Bloomfield) CNC

56.28 *Engelhardtia ursina* (Smith) ♂ — Utah, Juab Co., Eureka, IV.24.1922 (T. Spalding) CNC

56.29 *Tricholita chipeta* Barnes ♂ — California, San Diego Co., McCain Valley, 4,200′, X.2.1997 (T. Mustelin) CNC

56.30 *Hydroeciodes serrata* (Grote) ♂ — Texas, Brewster Co., Green Gulch, Big Bend Natl. Park, X.7.1966 (A. & M. Blanchard) CNC

56.31 *Peridroma saucia* (Hübner) ♂ — Colorado, Larimer Co., Viestenz-Smith Mtn. Park, 9.8 miles W of Loveland, 5,700′, X.3.1998 (T. Dickel) CNC

56.32 *Anicla (Euagrotis) exuberans* (Smith) ♂ — British Columbia, Kirby Flats Road, VI.12.1998 (J. Troubridge) CNC

56.33 *Hemieuxoa rudens* (Harvey) ♀ — Arizona, Yavapai Co., 4 miles N of Prescott, VIII.6.1972 (L. Martin) CNC

56.34 *Hemieuxoa rudens* (Harvey) ♀ — California, Imperial Co., Mountain Springs, 2,500′, IV.28.1998 (T. Mustelin & N. Bloomfield) CNC

56.35 *Striacosta albicosta* (Smith) ♀ — New Mexico, Grant Co., Emory Pass, Gila Natl. Forest, 7,160′, VIII.5.1991 (E. Metzler) CNC

56.36 *Actebia fennica* (Tauscher) ♀ — Alberta, Lethbridge, II.1983, reared from egg, CNC

56.37 *Dichagyris (Loxagrotis) grotei* (Franclemont and Todd) ♀ — Arizona, Yavapai Co., 4 miles N of Prescott, VIII.16.1972 (L. Martin) CNC

56.38 *Dichagyris (Loxagrotis) grandipennis* (Grote) ♀ — Arizona, Santa Cruz Co., Madera Canyon, Santa Rita Mtns., VII.1.1963 (J. Franclemont) CNC

56.39 *Dichagyris (Pseudorthosia) variabilis* (Grote) ♂ — British Columbia, Kirby Flats Road, VIII.23.2000 (J. Troubridge) CNC

56.40 *Dichagyris (Mesembragrotis) longidens* (Smith) ♂ — Arizona, Yavapai Co., 4 miles N of Prescott, VIII.15.1971 (L. Martin) CNC

56.41 *Eucoptocnemis (Eucoptocnemis) rufula* Lafontaine ♂ — Arizona, Cochise Co., Texas Canyon, 5,460′, IV.19.2001 (C. Ferris) CNC

56.42 *Richia parentalis* (Grote) ♀ — Colorado, Larimer Co., Viestenz-Smith Mtn. Park, 9.8 miles E of Loveland, 5,700′, VIII.5.1996 (T. Dickel) CNC

56.43 *Richia parentalis* (Grote) ♂ — Nevada, Douglas Co., 13 miles SE Gardnerville, 5,900′, IX.18.1969 (D. Hardwick) CNC

56.44 *Richia praefixa* (Morrison) ♀ — Wyoming, Albany Co., 7,450′, VIII.27.2003 (C. Ferris) CNC

PLATE 56

PLATE 57 **Noctuidae**

57.1 *Copablepharon pictum* Fauske & Lafontaine ♂ — Colorado, Saguache Co., Indian Spring Natural Area, VIII.11.1999 (P. Pineda & T. Nevins) CNC

57.2 *Copablepharon spiritum* Crabo & Fauske ♂ — Washington, Grant Co., Potholes, 1,095′, VIII.30.2002 (L. Crabo) CNC

57.3 *Copablepharon canariana* McDunnough ♂ — Utah, Cache Co., Logan, VII.22.1936 (G. Knowlton) CNC

57.4 *Copablepharon viridisparsa* Dod ♂ — California, Mono Co., Lee Vining, VI.4.2002 (Crabo & Troubridge) CNC

57.5 *Protogyia postera* Fauske & Lafontaine ♀ — Nevada, Elko Co., 11 miles SW of Wells, 2500m, VII.23.2001 (Lafontaine & Troubridge) CNC

57.6 *Protogyia album* (Harvey) ♂ — California, San Bernardino Co., 2 miles S of Kelso, 2,400′, IV.30.1986 (D. & V. Hardwick) CNC

57.7 *Protogyia milleri* (Grote) ♂ — California, Tuolumne Co., 3 miles ENE of Tuolumne Meadows, 9,200′, VIII.4.1967 (D. Hardwick) CNC

57.8 *Euxoa auxiliaris* (Grote), ♀ — Colorado, Chaffee Co., Buena Vista, 7,800′, VI.21.1961 (M. MacKay) CNC

57.9 *Euxoa bochus* (Morrison) ♀ — Washington, Chelan Co., Derby Canyon, VIII.22.2002 (L. Crabo) CNC

57.10 *Euxoa auxiliaris* (Grote) ♀ — British Columbia, Blowdown Pass, Coast Mtns., VII.12.1992 (J. Troubridge) CNC

57.11 *Euxoa intermontana* Lafontaine ♂ — Washington, Chelan Co., Chumstick Mtn., 5,800′, VIII.24.2002 (J. Troubridge) CNC

57.12 *Euxoa mimallonis* (Grote) ♂ — Nevada, Elko Co., 11 miles SW of Wells, 2500m, VII.22.2001 (Lafontaine & Troubridge) CNC

57.13 *Euxoa messoria* (Harris) ♀ — British Columbia, Apex Mtn., 7,380′, VII.21.2000 (J. Troubridge) CNC

57.14 *Euxoa lewisi* (Grote) ♀ — British Columbia, Watch Peak, 8,070′, VIII.1.2000 (Troubridge & Hensel) CNC

57.15 *Euxoa leuschneri* Lafontaine ♀ [paratype] — California, San Bernardino Co., Barton Flats, 6,300′, VI.14.1985 (R. Leuschner) CNC

57.16 *Euxoa tristicula* (Morrison) ♂ — Oregon, Lake Co., Alkali Lake dunes, 4,100′, VII.19.2001 (Lafontaine & Troubridge) CNC

57.17 *Euxoa vetusta* (Walker) ♀ — Oregon, Lane Co., Frissell Point, 4,900′, VII.26.2001 (J. Troubridge) CNC

57.18 *Euxoa fuscigera* (Grote) ♂ — California, Colusa Co., 6 miles WSW of Stonyford, 1,600′, IX.27.1966 (D. Hardwick) CNC

57.19 *Euxoa olivia* (Morrison) ♂ — British Columbia, E end of Seton Lake, IX.9.1994 (J. Troubridge) CNC

57.20 *Euxoa olivia* (Morrison) ♀ — Washington, Grant Co., 2 miles NW of Wanapum Dam, IX.17.1993 (J. Troubridge) CNC

57.21 *Euxoa atomaris* (Smith), (pale aridland form) ♂ — Oregon, 5 miles SE of Follyfarm, 4,250′, IX.15.1961 (D. Hardwick) CNC

57.22 *Euxoa atomaris* (Smith), (dark forest form) ♀ - Washington, 4 miles N of Lyle, 1,500′, VIII.21.1960 (D. Hardwick) CNC

57.23 *Euxoa simona* McDunnough ♀ — Oregon, Jackson Co., Mt. Ashland, 7,000′, VIII.3.1995 (Troubridge & Crabo) CNC

57.24 *Euxoa bifasciata* (Smith) ♂ — California, San Bernardino Co., Upper Santa Ana River, VII.18.1948 (G. & J. Sperry) CNC

57.25 *Euxoa annulipes* (Smith) ♀ — California, Los Angeles Co., La Tuna Canyon, VI.22.1949 (W. Evans) CNC

57.26 *Euxoa pluralis* (Grote) ♀ — California, Kern Co., Mt. Pinos, Los Padres Natl. Forest, 8,200′, VII.18.1943 (C. Henne) CNC

57.27 *Euxoa cinnabarina* Barnes & McDunnough ♀ — California,Los Angeles Co., Buckhorn Camp, San Gabriel Mtns., VII.18.1953, CNC

57.28 *Euxoa flavidens* (Smith) ♂ — Arizona, Apache Co., Greens Peaak, White Mtns., 9,500′, VIII.6.1962 (E. & I. Munroe) CNC

57.29 *Euxoa silens* (Grote) ♀ — California, Plumas Co., Jackson Creek, 5,400′, V.21.2001 (J. Troubridge) CNC

57.30 *Euxoa pimensis* Barnes & McDunnough ♂ — California, Los Angeles Co., La Tuna Canyon, IV.24.1949 (W. Evans) CNC

57.31 *Euxoa hollemani* (Grote) ♀ — Washington, Kittitas Co., Colockum Road, 1200m, VIII.27.1988 (L. Crabo) CNC

57.32 *Euxoa comosa* (Morrison) ♀ — Nevada, Elko Co., Angel Lake, 2550m, VII.23.2001 (Lafontaine & Troubridge) CNC

57.33 *Euxoa comosa* (Morrison) ♂ — Washington, Yakima Co., Bethel Ridge, VII.29.1989 (L. Crabo) CNC

57.34 *Euxoa piniae* Buckett & Bauer ♀ — California, Plumas Co., 2 miles NW of Almanor, 4,500′, VIII.1.1965 (D. & V. Hardwick) CNC

57.35 *Euxoa obeliscoides* (Guenée) ♀ — California, San Diego Co., Palomar Mtn., 4,800′, VIII.25.1999 (Mustelin & Bloomfield) CNC

57.36 *Euxoa idahoensis* (Grote) ♀ — Washington, Yakima Co., Bethel Ridge, 6,300′, VII.25.1996 (J. Troubridge) CNC

57.37 *Euxoa auripennis* Lafontaine ♀ — British Columbia, Mt. Kobau, 6,040′, VIII.2.2000 (Troubridge & Hensel) CNC

57.38 *Euxoa teleboa* (Smith) ♀ — Nevada, 24 miles E of Lund, 6,000′, IX.4.1965 (D. Hardwick) CNC

57.39 *Euxoa wilsoni* (Grote) ♂ — California, Marin Co., McClure's Beach, Point Reyes, VI.17.1951 (W. Bauer) CNC

57.40 *Euxoa churchillensis* (McDunnough) ♂ — Colorado, Park Co., Pennsylvania Mtn., VII.22.1980 (D. Ford) CNC

57.41 *Feltia* (*Trichosilia*) *mollis* (Walker) ♂ — Colorado, Grand Co., 3.9 miles SW of Fraser, St. Louis Creek Road, 8,900′, VII.11.1995 (T. Dickel) CNC

57.42 *Feltia* (*Feltia*) *jaculifera* (Guenée) ♂ — Colorado, Moffat Co., Pool Creek Canyon, Dinosaur Natl. Monument, 5,200′, VIII.12.1994 (T. Dickel) CNC

57.43 *Feltia* (*Feltia*) *subterranea* (Fabricius) ♂ — California, San Diego Co., S rim of Penasquitos Canyon, V.6.1999 (T. Mustelin) CNC

57.44 *Feltia* (*Feltia*) *subterranea* (Fabricius) ♀ — California, Los Angeles Co., Hawthorne, IX.15.1944 (E. Johnston) CNC

PLATE 57

PLATE 58 Noctuidae

58.1 *Agrotis vetusta* (Walker) ♂ — California, Sonoma Co., Petaluma, X.3.1937 (E. Johnston) CNC

58.2 *Agrotis orthogonia* Morrison ♂ — Nevada, 16 miles NNW of Babbitt, 4,000', IX.20.1969 (D. Hardwick) CNC

58.3 *Agrotis vancouverensis* Grote ♂ — Colorado, Grand Co., St. Louis Creek Road, 4 miles SW of Fraser, 9,000', VI.26.1988 (T. Dickel) CNC

58.4 *Agrotis malefida* Guenée ♂ — Arizona, Cochise Co., 1 mile SW of Portal, 4,900', X.8.1969 (D. Hardwick) CNC

58.5 *Agrotis ipsilon* (Hufnagel) ♂ — California, San Diego Co., Sorrento Valley marsh, Torrey Pines State Park, II.12/14.2006 (N. Bloomfield) CNC

58.6 *Ochropleura implecta* Lafontaine ♂ — Washington, Grays Harbor Co., Ocean Beach Road, Carlisle, 50', VI.20.1990 (L. Crabo) CNC

58.7 *Diarsia rosaria* (Grote) ♀ — British Columbia, Summerland, V.1.1994 (M. Gardiner) CNC

58.8 *Chersotis juncta* (Grote) ♂ — Alberta, Calgary, VII.6.1901 (F. Wolley Dod) CNC

58.9 *Noctua pronuba* (Linnaeus) ♀ — Washington, Ferry Co., 1.5 miles S of Jim Mtn., 2,175', VII.6.2005 (L. & E. Crabo) CNC

58.10 *Cerastis enigmatica* Lafontaine & Crabo ♂ — British Columbia, 15 km W of Oyster River, IV.1.2000 (J. Troubridge) CNC

58.11 *Noctua comes* Hübner ♀ — British Columbia, 5 km E of Langley, VIII.4/10.1991 (J. Troubridge) CNC

58.12 *Cryptocala acadiensis* Bethune ♀ — British Columbia, Revelstoke, VII.15.1931 (A. Gartrell) CNC

58.13 *Spaelotis clandestina* (Harris) ♂ — Colorado, Larimer Co., Viestenz-Smith Mtn. Park, 9.8 miles W of Loveland, 5,700', VI.20.1996 (T. Dickel) CNC

58.14 *Spaelotis bicava* Lafontaine) ♀ — California, San Bernardino Co., Cienega Seca, near Onyx Summit, San Bernardino Natl. Forest, VI.27.1998 (R. Rockwell) CNC

58.15 *Eurois occulta* (Linnaeus) ♀ — Arizona, Apache Co., Greer, 8,500', VIII.3.1962 (E. & I. Munroe) CNC

58.16 *Graphiphora augur* (Fabricius) ♀ — Colorado, Routt Co., Ferndale Recreation Area, Routt Natl. Forest, 8,000', VI.27.1987 (T. Dickel) CNC

58.17 *Anaplectoides prasina* (Denis and Schiffermüller) ♀ — Washington, Yakima Co., Mt. Baker Ski Area, VII.27.1992 (J. Troubridge) CNC

58.18 *Xestia (Xestia) smithii* (Snellen) ♀ — Wyoming, Albany Co., Laramie, 7,500', VIII.15.2000 (C. Ferris) CNC

58.19 *Paradiarsia littoralis* (Packard) ♂ — Colorado, Gunnison Co., Gothic, VII.20.1949 (W. Reinthal) CNC

58.20 *Xestia (Xestia) oblata* (Morrison) ♀ — Quebec, Temiscouata Co., VI.26.1987 (H. Hensel) CNC

58.21 *Xestia (Megasema) c-nigrum* (Linnaeus) ♀ — California, San Diego Co., Mira Mesa At Penasquitos Canyon, III.3.1997 (T. Mustelin) CNC

58.22 *Xestia bolteri* (Smith) ♀ — Arizona Co., Greenlee Co., Hannagan Meadows, VI.25.1966 (R. Sternitzky) CNC

58.23 *Xestia (Pachnobia) speciosa* (Hübner) ♀ — Colorado, Grand Co., 5 miles SW of Fraser, VIII.4.1988 (T. Dickel) CNC

58.24 *Xestia (Pachnobia) perquiritata* (Morrison) ♂ — Alberta, Nordegg, VI.28.1921 (J. McDunnough) CNC

58.25 *Xestia (Pachnobia) fabulosa* (Ferguson) ♂ — Colorado, Grand Co., 5 miles SW of Fraser, 9,440', VII.3.1992 (T. Dickel) CNC

58.26 *Parabarrovia keelei* Gibson ♂ — Yukon Territory, British Mtns., 600m, VI.22.1984 (G. & M. Wood & J. Lafontaine) CNC

58.27 *Agnorisma bugrai* (Koçak) ♂ — Ontario, Marlborough Forest, VIII.30.2003 (J. Troubridge) CNC

58.28 *Setagrotis pallidicollis* (Grote) ♀ — California, San Diego Co., Laguna Mtns., VII.25.1997 (T. Mustelin) CNC

58.29 *Tesagrotis corrodera* (Smith) ♂ — Nevada, Elko Co., Angel Creek campground, 6,700', VII.22.2001 (Lafontaine & Troubridge) CNC

58.30 *Adelphagrotis indeterminata* (Walker) ♂ — Oregon, Lane Co., Frissell Point, H.J. Andrews Experimental Forest, IX.2.1995 (J. Troubridge) CNC

58.31 *Parabagrotis exsertistigma* (Morrison) ♀ — Nevada, Elko Co., Angel Lake, 2550m, VII.23.2001 (Lafontaine & Troubridge) CNC

58.32 *Parabagrotis insularis* (Grote) ♀ — California, San Bernardino Co., Cienega Seca, near Onyx Summit, San Bernardino Natl. Forest, VI.27.1998 (R. Rockwell) CNC

58.33 *Protolampra rufipectus* (Morrison) ♀ — Colorado, Routt Co., Ferndale Recreation Area, Routt Natl. Forest, 7,900', VIII.15.1996 (T. Dickel) CNC

58.34 *Abagrotis trigona* (Smith) ♂ — Oregon, Jackson Co., Mt. Ashland, 7,000', VIII.3.1995 (J. Troubridge & L. Crabo) CNC

58.35 *Abagrotis trigona* (Smith) ♂ — Oregon, Lane Co., Frissel Point, 5,000', VII.26.1996 (J. Troubridge) CNC

58.36 *Abagrotis vittifrons* (Grote) ♂ — Colorado, Larimer Co., Viestenz-Smith Mtn. Park, 5,700', IX.3.1996 (T. Dickel) CNC

58.37 *Abagrotis bimarginalis* (Grote), ♂ — Colorado, Gilpin Co., 4 km NW of Central City, 8,809' m, VII.13.1993 (B. Landry) CNC

58.38 *Abagrotis glenni* Buckett ♀ — Oregon, Deschutes Co., Horse Ridge Summit, IX.3.1995 (J. Troubridge) CNC

58.39 *Abagrotis nanalis* (Grote) ♀ — Oregon, Deschutes Co., Horse Ridge Summit, IX.3.1995 (J. Troubridge) CNC

58.40 *Abagrotis discoidalis* (Grote) ♀ — Nevada, Elko Co., Angel Lake, 2550m, VII.23.2001 (J. Lafontaine & J. Troubridge) CNC

58.41 *Abagrotis variata* (Grote) ♀ — British Columbia, Lilooet, VII/IX.1927 (A. Phair) CNC

58.42 *Abagrotis forbesi* (Benjamin) ♀ — California, San Diego Co., Laguna Mtns., IV.26.1997 (T. Mustelin) CNC

58.43 *Pronoctua peabodyae* (Dyar) ♀ — Washington, Yakima Co., Bethel Ridge, 6,300', VII.25.1996 (J. Troubridge) CNC

PLATE 58

Micropterigidae

59.1 *Epimartyria pardella* (Walsingham) — California, Humboldt Co., June 1982

Eriocraniidae

59.2 *Dyseriocrania aurosparsella* (Walsingham) — California, San Benito Co. [P. Johnson]

Nepticulidae

59.3 *Stigmella variella* (Braun) — California, Berkeley, May 1997, abandoned mine on *Quercus agrifolia*,

59.4 *Stigmella heteromelis* Newton & Wilkinson — California, Alameda Co., April 1976, abandoned mines on *Heteromeles arbutifolia*

Heliozelidae

59.5 *Coptodisca arbutiella* Busck — California, Berkeley, May 1966, ovipositing into leaf underside of *Arbutus menziesii*

59.6 *Coptodisca arbutiella* — California, Berkeley, March 1995, mature mines in *Arctostaphylos bowermaniae*

59.7 *Antispila voraginella* Braun — Utah, Zion Canyon, July 1992, abandoned, communal mine in *Vitis* leaf

Incurvariidae

59.8 *Vespina quercivora* (Davis) — California, Kern Co., March 1968, second-stage larva and case on *Quercus wislizenii*

Adelidae

59.9 *Adela trigrapha* Zeller — California, Santa Clara Co., May 1969, ovipositing into floral ovary of *Linanthus androsaceus*

59.10 *Adela flammeusella* (Chambers) — California, Marin Co., May 1966, on *Ranunculus*

59.11 *Adela oplerella* Powell — California, Santa Clara Co., April 1984, ovipositing into floral ovary of *Platystemon californicus*

59.12 *Adela singulella* Walsingham — California, Napa Co., May 1966, mating pair on *Gilia capitata*

59.13 *Adela septentrionella* Walsingham — California, Berkeley Hills, May 1966, females ovipositing into buds of *Holodiscus discolor*

59.14 *Adela septentrionella* Walsingham — California, Marin Co., March 1970, larva and cases

59.15 *Adela septentrionella* Walsingham — California, Marin Co., March 1970, pupa

Prodoxidae

59.16 *Greya reticulata* (Riley) — California, Alameda Co., April 1995, female on *Sanicula crassicaulis*

59.17 *Tegeticula maculata* (Riley) — California, Tulare Co., May 1979, female ovipositing in floral ovary of *Hesperoyucca whipplei*

59.18 *Tegeticula maderae* Pellmyr — Arizona, Chiricahua Mtns., August 1971, saggital section of *Yucca schottii* fruit, revealing section of seed eaten by larva

59.19 *Parategeticula pollenifera* — Arizona, Madera Canyon, August 1973, row of eggs in open pits on inflorescence branch of *Y. schottii*

59.20 *Parategeticula pollenifera* Davis — Arizona, Madera Canyon, September 1974, larva and cyst it causes in fruit of *Y. schottii*, which aborts growth of several seeds

59.21 *Prodoxus cinereus* Riley — California, Tulare Co., May 1979, herd of females ovipositing into floral scape of *Hesperoyucca whipplei*

Tischeriidae

59.22 *Tischeria splendida* Braun — California, Monterey Co., July 1992, abandoned larval mines on *Rubus ursinus*

Tineidae

59.23 *Tinea pellionella* Linnaeus — California, Marin Co., larva and its portable case

59.24 *Tinea occidentella* Chambers — California San Luis Obispo Co., June 1976, larval tubes below coyote scat, excavated from sand

Acrolephidae

59.25 *Acrolophus variabilis* (Walsingham) — Arizona, Santa Cruz Co., August 2005

Psychidae

59.26 *Thyridopteryx meadii* H. Edwards — California, Darwin Falls, Mojave Desert, May 1969, larval case on *Larrea*

Gracillariidae

59.27 *Caloptilia reticulata* (Braun) — California, Berkeley, March 1984, showing typical perch posture of gracillariines

59.28 *Cameraria nemoris* (Walsingham) — California, Marin Co., April 1976, mature larval mines on *Vaccinium ovatum*

59.29 *Cameraria gaultheriella* (Walsingham) — California, Marin Co., May 1995, larval mine on *Gaultheria shallon*

59.30 *Marmara arbutiella* Busck — California, Marin Co., April 1977, larval mines on *Arbutus menziesii*

59.31 *Marmara opuntiella* Busck — California, San Clemente Island, April 1980, larval mines on *Opuntia littoralis*

Phyllocnistidae

59.32 *Phyllocnistis populiella* Chambers — Alaska, near Anderson, July 1979, larva in mine on *Populus tremuloides*

Stenomatidae

59.33 *Antaeotricha schlaegeri* (Zeller) — Arizona, Huachuca Mtns., August 2005

Elachistidae

59.34 *Perittia passula* Kaila — California, Contra Costa Co., April 1999, larvae in mines in *Lonicera involucrata*

Oecophoridae

59.35 *Esperia sulphurella* (Fabricius) — California, Contra Costa Co., May 1987

Ethmiidae

59.36 *Ethmia arctostaphylella* (Walsingham) — California, Marin Co. April 1972, adult on *Eriodictyon californicum*, resembling bird dropping

PLATE 59

Scythrididae

60.1 *Arotrura longissima* Landry — California, San Clemente Island, March 1972, nectaring on *Senecio lyonii*

60.2 *Areniscythris brachypteris* Powell — California, San Luis Obispo Co., June 1973, female on rootlet avoiding hot sand surface

60.3 *Areniscythris brachypteris* — California, San Luis Obispo Co., July 1973, adult self-buried in pit with just the head and antennae visible

Blastobasidae

60.4 *Holcocera iceryaeella* (Riley) — California, Contra Costa Co., June 1986

Coleopharidae

60.5 *Coleophora pruniella* Clemens — California, Contra Costa Co., April 1991

60.6 *Coleophora glaucella* Walsingham — California, Pinnacles Natl. Monument, March 2005, larval cases on *Arctostaphylos*

60.7 *Coleophora* species — California, Contra Costa Co., April 1990, larval case on *Baccharis pilularis*

Cosmopterigidae

60.8 *Sorhagenia nimbosa* (Braun) — California, Lake Co., September 1973, larval "gall" on *Rhamnus californicus*

Gelechiidae

60.9 *Chionodes ochreostrigella* (Chambers) — California, San Benito Co., Pinnacles Natl. Monument, March 2007

60.10 *Dichomeris mulsa* Hodges — Arizona, Huachuca Mtns., August 2005

60.11 *Gnorimoschema coquillettella* Busck — California, Stanislaus Co., May 1975; tip gall induced by larva on *Ericameria linearifolia*

60.12 *Gnorimoschema grindeliae* Povolny & Powell — California, Contra Costa Co., April 1999, node galls induced by larvae on *Grindelia hirsutula*

60.13 *Gnorimoschema baccharisella* Busck — California, Alameda Co., July 1998, stem galls induced by larvae on *Baccharis pilularis*

Ypsolophidae

60.14 *Ypsolopha maculatella* (Busck) — California, Mojave Desert, L.A. Co., May 1968, nectaring on *Senecio*

Heliodinidae

60.15 *Embola ciccella* (Barnes & Busck) — Arizona, Baboquivari Mtns., August 2005

Schreckensteiniidae

60.16 *Schreckensteinia felicella* Walsingham — California, Mendocino Co., May 1977, cocoon, pupal shell, and newly emerged adult, reared from *Castilleja*

Pterophoridae

60.17 *Anstenoptilia marmarodactyla* (Dyar) — California, Contra Costa Co., September 2002

60.18 *Emmelina monodactyla* (Linnaeus) — California, Marin Co., October 2004

60.19 *Oidaematophorus fieldi* (W. S. Wright) — California, Pinnacles Natl. Monument, October 2005

60.20 *Oidaematophorus* species — California, Santa Cruz Co., April 1973, cryptic larva on *Petasites palmatus*

Sesiidae

60.21 *Synanthedon bibionipennis* (Boisduval) — California, Contra Costa Co., July 2007

60.22 *Synanthedon polygoni* (H. Edwards) — California, Sierra Co., July 2007

60.23 *Synanthedon sequoiae* (H. Edwards) '— California, Berkeley, May 1986

Choreutidae

60.24 *Tebenna gemmalis* (Hulst) — California, Plumas Co., July 2002

Tortricidae

60.25 *Retinia metallica* (Busck) — California, Nevada Co.,1972, pitch nodule created by larva on *Pinus jeffreyi*

60.26 *Epiblema rudei* Powell — California, Fresno Vo., February 1970, stem galls induced by larvae in *Gutierrezia*, pupal shell, and newly emerged adult

60.27 *Epinotia emarginana* (Walsingham) — California, Contra Costa Co., July 2002

60.28 *Epinotia nigralbana* (Walsingham) — California, Monterey Co., November 1990, overwintering mines in *Arbutus menziesii* leaf

60.29 *Epinotia nigrabana* (Walsingham) — California, Mendocino Co., April 1972, fully developed leaf mine in *Arctostaphylos manzanita*

60.30 *Cydia latiferreana* (Walsingham) — California, Pinnacles Natl. Monument, September 2005

60.31 *Grapholita edwardsiana* (Kearfott) — California, San Francisco, January 1978, overwintering larva in its gallery in inflorescence stem of *Lupinus arboreus*

60.32 *Archips cerasivoranus* (Fitch) — California, Siskiyou Co., June 1974, extensive larval tents on *Prunus subcordata*

60.33 *Archips cerasivoranus* (Fitch) — California, Siskiyou Co., June 1974, larval aggregation exposed in tent from *Prunus subcordata*

60.34 *Argyrotaenia cupressae* Powell — California, San Luis Obispo Co., June 1968, cryptically colored larva in foliage of *Cupressus sargenti*

60.35 *Argyrotaenia cupressae* Powell — California, Stanislaus Co., April 1978, newly emerged adult and its pupal shell (lower left) on foliage of *Juniperus californicus*

60.36 *Ditula angustiorana* (Haworth) — California, Humboldt Co., July 1969, egg masses on *Taxus* foliage

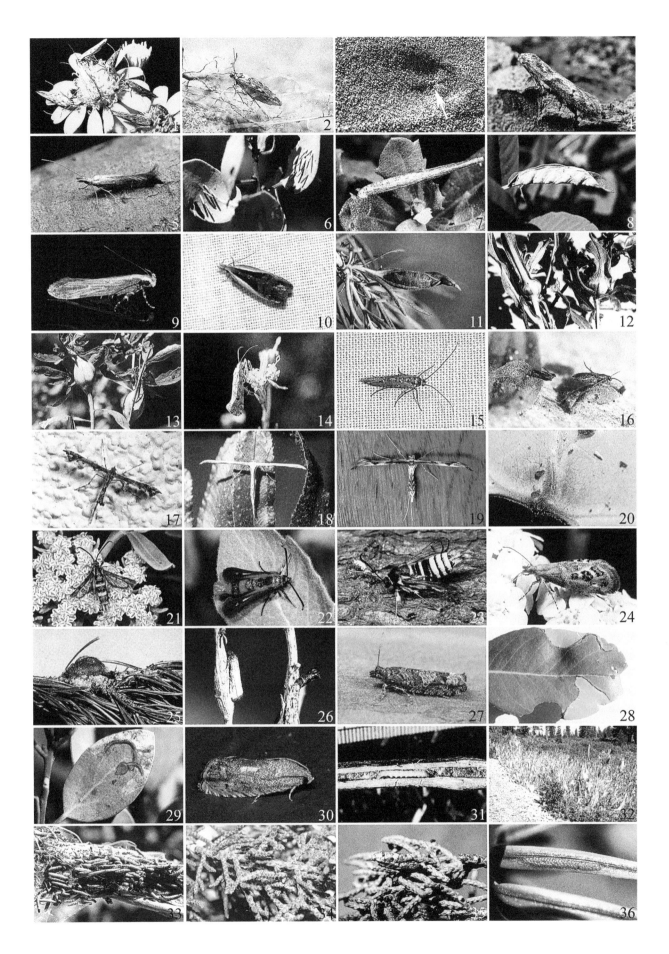

PLATE 60

Tortricidae

61.1 *Choristoneura carnana californica* Powell — California, July 1979, egg masses on *Pseudotsuga* needles

61.2 *Choristoneura lambertiana subretiniana* Obraztsov — California, Sierra Co., July 2007

61.3 *Synnoma lynosyrana* Walsingham — Nevada, Elko Co. August 1973, female perched in calling posture with egg mass on *Chrysothamnus nauseosus*

61.4 *Platynota wenzelana* (Haimbach) — Arizona, Huachuca Mtns., August 2005

61.5 *Irazona comes* (Walsingham) — Arizona, Huachuca Mtns., August 2005, showing the dent in the FW plane distally, characteristic of most cochyline tortricids.

Zygaenidae

61.6 *Harrisina metallica* Stretch larva — California [CDFA]

Limacodidae

61.7 *Cryptophobetron oropeso* (Barnes) larva — Arizona, Chiricahua Mtns.

61.8 *Prolimacodes trigona* (H. Edwards) larva — Arizona, Huachuca Mtns.

61.9 *Isa schaefferana* Dyar larva — Arizona, Huachuca Mtns.

61.10 *Euclea obliqua* H. Edwards larva — Arizona, Huachuca Mtns.

61.11 *Euclea incisa* (Harvey) larva — Texas

Dalceridae

61.12 *Dalcerides ingenita* (H. Edwards) — Arizona, larva, showing its gelatinous coating being teased off

61.13 *Dalcerides ingenita* (H. Edwards) — Arizona, pair *in copulo* [C. Hansen]

Megalopygidae

61.14 *Megalopyge bissesa* Dyar — Arizona, August 2005

Carposinidae

61.15 *Bondia comonana* Kearfott — Arizona, Huachuca Mtns., August 2005

Thyrididae

61.16 *Dysodia granulata* (Neumoegen) — Arizona, Baboquivari Mtns., August 2005

Crambidae

61.17 *Microtheoris ophionalis* (Walker) — Arizona, Santa Cruz Co., August 2005

61.18 *Terastia meticulosalis* Guenée — Arizona, Baboquivari Mtns., August 2005, showing upcurved abdominal posture characteristic of many spilomeline crambids

Pyralidae

61.19 *Cacozelia basiochrealis* Grote — Arizona, Santa Cruz Co., August 2005, showing upcurved abdominal posture characteristic of many chrysaugine pyralids

61.20 *Caphys arizonensis* Munroe — Arizona, Huachuca Mtns., August 2005

61.21 *Galasa nigripunctalis* Barnes & McDunnough — Arizona, Huachuca Mtns., August 2005

61.22 *Laetilia dilutifasciella* (Ragonot) — Arizona, Huachuca Mtns., August 2005

61.23 *Sarata* species — Colorado, Jefferson Co., April 1987, female on snow

61.24 *Plodia interpunctella* (Hubner) — larvae and their silk trackways filling a kitchen cannister of corn meal

Drepanidae

61.25 *Drepana bilineata* (Packard), late instar larva on *Betula papyrifera*, Quebec, Quebec City

61.26 *Habrosyne scripta* (Gosse), late instar larva on *Rubus odoratus*, Quebec, Quebec City

61.27 *Pseudothyatira cymatophoroides* (Guenée), late instar larva on *Betula papyrifera*, Quebec, Quebec City

61.28 *Euthyatira pudens* (Guenée), late instar larva on *Cornus*, Ontario

Uraniidae, Epipleminae

61.29 *Callizzia amorata* Packard, late instar larva on *Lonicera*, Victorria Island, British Columbia

Geometridae

61.30 *Eumacaria latiferrugata* (Walker), late instar larva on *Prunus*, Fort Niobrara, Nebraska

61.31 *Macaria sexmaculata* (Packard), late instar larva on *Pinus*, Quebec, Quebec City

61.32 *Anavitrinella pampinaria* (Guenée), late instar larva on *Solidago*, Ontario

61.33 *Carphoides incopriarius* (Hulst), late instar larva on *Juniperus*, Ephraim, Utah

61.34 *Lycia ursaria* (Walker), late instar larva, Quebec, Quebec City

61.35 *Drepanulatrix unicalcararia* (Guenée), late instar larvae on *Ceanothus cordulatus*, California

61.36 *Phaeoura cristifera* Hulst, late instar larva on *Quercus*, Sierra Vista, Arizona

PLATE 61

PLATE 62 Larvae of Geometridae, Lasiocampidae, and Saturniiidae

Geometridae

62.1 *Aethaloida packardaria* (Hulst), late instar larva on *Adenostoma fasciculatum*, California

62.2 *Campaea perlata* (Guenée), late instar larva on *Rosa*, Colesville, Maryland

62.3 *Spodolepis substriataria* Hulst, late instar larva on *Salix*, Ephraim, Utah

62.4 *Thallophaga taylorata* (Hulst), late instar larva on *Polystichum munitum*, Marin Co., California

62.5 *Metanema inatomaria* Guenée, late instar larva on *Populus*, Fort Niobrara, Nebraska

62.6 *Melemaea virgata* Taylor, late instar larva on *Pinus*, Catalina Mtns., Arizona

62.7 *Caripeta aequaliaria* Grote, late instar larva on *Pinus*

62.8 *Destutia excelsa* Strecker, late instar larva

62.9 *Nepytia umbrosaria* Packard, late instar larva on *Pseudotsuga*, Nevada Co., California

62.10 *Tetracis cachexiata* Guenée, late instar larva on *Quercus rubra*, Quebec, Quebec City

62.11 *Phyllodonta peccataria* Barnes & McDunnough, late instar larva on *Pinus*, Huachuca Mtns., Arizona

62.12 *Prochoerodes forficaria* (Guenée), late instar larva on *Salix*, California

62.13 *Nematocampa resistaria limbata* (Haworth), late instar larva on *Vaccinium*, Gatineau, Quebec

62.14 *Alsophila pometaria* Harris, late instar larva on *Aceer negundo*, Colesville, Maryland

62.15 *Chlorosea banksaria* Sperry, late instar larva on *Ceanothus integerrimus*, El Dorado Co., California

62.16 *Synchlora aerata* (Fabricius), late instar larva on *Rudbeckia*, Colesville, Maryland

62.17 *Dysstroma formosa* (Hulst), late instar larva on *Ribes*, Placer Co., California

62.18 *Dysstroma citrata* (Walker), late instar larva on *Rubus spectabilis*, California

62.19 *Eulithis xylina* (Hulst), late instar larva on *Prunus*, California

62.20 *Hydriomena renunciata* Walker, late instar larva on *Alnus*, Quebec, Quebec City

62.21 *Spargania magnoliata* Guenée, late instar larva on *Clarkia affinis*, California

62.22 *Perizoma custodiata* (Guenée), late instar larva on *Atriplex*, Marin Co., California

62.23 *Operophtera bruceata* Hulst, late instar larva on *Acer*, Quebec, Quebec City

62.24 *Eupithecia ravocostaliata* Packard, late instar larva on *Salix*, Quebec, Quebec City

62.25 *Cladara limitaria* (Walker), late instar larva on *Tsuga*

Lymantriinae

62.26 *Gynaephora rossii* Curtis, late instar larva, Boulder Co., Colorado

Lasiocampidae

62.27 *Gloveria arizonensis* Packard, late instar larva on *Pinus ponderosa*, Platte Co., Wyoming

62.28 *Tolype glenwoodi* Barnes, late instar larva, Mt. Graham, Arizona

62.29 *Gloveria howardi* (Dyar), late instar larva on *Quercus emoryi*, Pena Blanca, Arizona

62.30 *Dichogaster coronada* (Barnes), late instar larva on *Quercus oblongifolia*, Santa Rita Mtns., Arizona

62.31 *Caloecia juvenalis* (Barnes & McDunnough), late instar larva on *Quercus gambelii*. Chiricahua Mtns., Arizona

62.32 *Malacosoma disstria* Hübner, late instar larva on *Populus*, Ontario

62.33 *Malacosoma californicum* (Packard) larval tent on *Quercus agrifolia*, Marin Co., California

62.34 *Malacosoma constrictum* (Henry Edwards), late instar larva on *Quercus agrifolia*, San Benito Co., California

Saturniiidae

62.35 *Citheronia splendens sinaloensis* Hoffman, late instar larva on *Gossypium thurberi*, Santa Rita Mtns., Arizona

62.36 *Eacles oslari* Rothschild, late instar larva on *Quercus oblongifolia*, Pena Blanca, Arizona

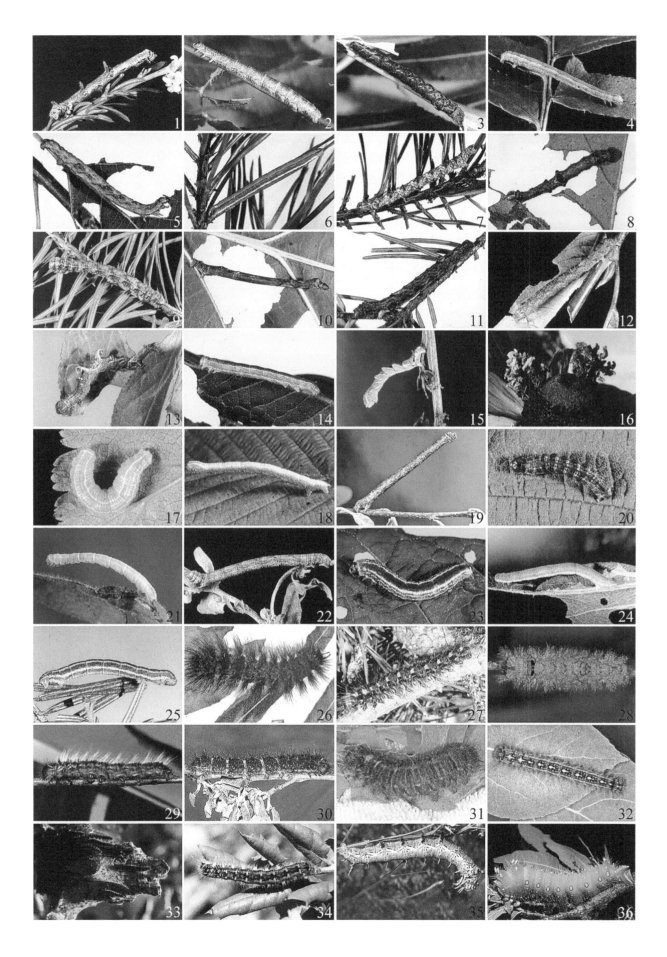

PLATE 62

Saturniidae

63.1 *Sphingicampa hubbardi* (Dyar), late instar larva on *Prosopis juliflora*, Santa Rita Mtns., Arizona

63.2 *Saturnia mendocino* Behrens, late instar larvae on *Arctostaphylos virgata*, Marin Co., California

63.3 *Hemileuca hera* (Harris), late instar larva on *Eriogonum umbellatum*, Sierra Valley, California

63.4 *Hemileuca hualapai* (Neumoegen), late instar larva on grass, Santa Rita Mtns., Arizona

63.5 *Hemileuca eglanterina* (Boisduval)., cluster of early instar larvae on *Purshia tridentata*, Sierra Co., California

63.6 *Rothschildia cincta* (Tepper), late instar larva on *Stylosanthes viscosa*, Pena Blanca, Arizona

63.7 *Eupackardia calleta* (Westwood), late instar larva on *Fouquieria splendens*, Santa Rita Mtns., Arizona

63.8 *Hyalophora columbia gloveri* (Strecker), Late instar larva on *Purshia tridentata*, St. Anthony's, Idaho

63.9 *Antheraea polyphemus* (Cramer), late instar larva on *Cornus stolonifera*, Larimer Co., Colorado

Sphingidae

63.10 *Manduca rustica* (Fabricius), late instar larva on *Chilopsis linearis*, Santa Rita Mtns., Arizona

63.11 *Manduca muscosa* (Rothschild & Jordan), late instar larva on *Viguiera dentata*, Santa Rita Mtns., Arizona

63.12 *Pachysphinx modesta* (Harris), late instar larva on *Populus*, Quebec, Quebec City

63.13 *Pachysphinx occidentalis* (Henry Edwards), late instar larva on *Salix*, Santa Rita Mtns., Arizona

63.14 *Sphinx chersis* (Hübner), late instar larva on *Fraxinus*, Santa Rita Mtns., Arizona

63.15 *Sphinx asella* (Rothschild & Jordan), late instar larva on *Arctostaphylos*, Pena Blanca, Arizona

63.16 *Sagenosoma elsa* (Strecker), late instar larva on *Lycium*, Cochise Co., Arizona

63.17 *Eumorpha achemon* (Drury), late instar larva on *Vitis arizonica*, Santa Rita Mtns., Arizona

63.18 *Eumorpha typhon* (Klug), late instar larva on *Vitis arizonica*, Santa Rita Mtns., Arizona

63.19 *Smerinthus cerisyi* Kirby, late instar larva on *Salix*, Canada

63.20 *Hemaris diffinis* (Boisduval), late instar larva on *Lonicera*, Ontario

63.21 *Hemaris senta* (Strecker), late instar larva on *Symphoricarpos*, Jackson Hole, Teton Co., Wyoming

63.22 *Arctonotus lucidus* Boisduval, late instar larva on *Clarkia*, San Benito Co., California

63.23 *Euproserpinus euterpe* Henry Edwards, late instar larva on *Camissonia campestris*, Carrizo Plain, San Luis Obispo Co., California

63.24 *Euproserpinus euterpe* Henry Edwards, adult, Carrizo Plain, San Luis Obispo Co., California

63.25 *Proserpinus clarkiae* (Boisduval), late instar larva on *Clarkia*, San Benito Co., California

63.26 *Hyles euphorbiae* (Linnaeus), late instar larva

63.27 *Hyles gallii* (Rottenburg), late instar larva. Dark form

63.28 *Hyles lineata* (Fabricius), late instar larva, green form, on *Portulaca*, Cochise Co., Arizona

63.29 *Hyles lineata* (Fabricius), Late instar larva, black form, Larimer Co., Colorado

Lymantriinae

63.30 *Hyles lineata* (Fabricius), Adult nectaring on *Gentiana*, Hinsdale Co., Colorado

63.31 *Leucoma salicis* (Linnaeaus), late instar larva on *Populus*, Quebec, Quebec City

63.32 *Lymantria dispar* (Linnaeus), late instar larva on *Quercus*, Montgomery Co., Maryland

63.33 *Orgyia leucostigma* (J.E. Smith), late instar larva on *Cornus*, Ontario

Notodontidae

63.34 *Clostera albosigma* Fitch, late instar larva on *Populus*, Beltsville, Maryland

63.35 *Pheosia rimosa* Packard, late instar larva on *Populus*, Quebec

63.36 *Odontosia elegans* (Strecker), late instar larva on *Populus*, Quebec, Quebec City

PLATE 63

Notodontidae

64.1 *Datana ministra californica* Dyar, late instar larva, Quebec, Quebec City

64.2 *Schizura biedermanni* Barnes & McDunnough, late instar larva on *Quercus emoryi*, Pena Blanca, Arizona

64.3 *Schizura concinna* (J.E. Smith), late instar larva

Noctuidae

64.4 *Scoliopteryx libatrix* (Linnaeus), late instar larva on *Salix*, Ontario

64.5 *Catocala relicta* (Walker), late instar larva on *Populus*, Ontario

64.6 *Nycteola frigidana* (Walker), late instar larva on *Salix*, Ontario

Lithosiinae

64.7 *Clemensia albata* Packard, late instar larva on algae and lichens, Pocomoke, Maryland

64.8 *Gardinia anopla* Hering, late instar larva on lichens, Catalina Mtns., Arizona

64.9 *Hypoprepia miniata* (Kirby), late instar larva

Arctiinae

64.10 *Utetheisa ornatrix* (Linnaeus), late instar larva on *Crotalaria*, Sinaloa, Mexico

64.11 *Tyria jacobaea* (Linnaeus), late instar larva on *Senecio*, Oregon

64.12 *Ectypia clio jessica* (Barnes), late instar larva on *Asclepias*, Platte Co., Wyoming

64.13 *Spilosoma virginica* (Fabricius), late instar larva on *Melilotus officinalis*, Quebec

64.14 *Hyphantria cunea* (Drury), late instar larva on *Populus*, Sutherland, Nebraska

64.15 *Kodiosoma fulvum* Stretch, late instar larva, Cochise Co., Arizona

64.16 *Halysidota tessellaris* (J.E. Smith), late instar larva on *Quercus*, Montgomery Co., Maryland

64.17 *Lophocampa sobrina* (Stretch), late instar larva on *Alnus*, Marin Co., California

64.18 *Lophocampa maculata* Harris, late instar larva on *Salix*, Mono Co., California

64.19 *Chrysodeix includens* (Walker), late instar larva, Urbana, Illinois

64.20 *Eosphoropteryx thyatyroides* (Guenée), late instar larva

64.21 *Syngrapha ignea* (Grote), late instar larva, Quebec

64.22 *Anagrapha falcifera* (Kirby), late instar larva on *Melilotus officinalis*, Ontario

64.23 *Tarachidia candefacta* (Hübner), late instar larva on *Ambrosia*

64.24 *Acronicta dactylina* Grote, late instar larva on *Alnus*, Quebec, Quebec City

64.25 *Acronicta impleta* Walker, late instar larva on *Corylus*, Quebec, Quebec City

64.26 *Alypia octomaculata* (Fabricius), late instar larva on *Parthenocissus*, Beauport, Quebec

64.27 *Adita chionanthi* (J.E. Smith), late instar larva on *Chionanthus*, Ontario

64.28 *Amphipyra pyramidoides* Guenée, late instar larva on Mothernort-Leonurus

64.29 *Orthosia hibisci* (Guenée), late instar larva, Quebec, Quebec City

64.30 *Helicoverpa zea* (Boddie), late instar larva on *Zinnia*

64.31 *Schinia masoni* J.B. Smith, adult on *Gaillardia* flower, Gilpin Co., Colorado

64.32 *Egira dolosa* (Grote), late instar larva on *Populus*

64.33 *Mythimna unipuncta* (Haworth), late instar larva

64.34 *Xestia c-nigrum* (Linnaeus), late instar larva

64.35 *Abagrotis glenni* Buckett, late instar larva on *Cupressus*, Mendocino Co., California

Bombycidae

64.36 *Apatelodes pudefacta* Dyar, late instar larvae on *Viguiera dentata*, Santa Catalina Mtns., Arizona

PLATE 64

Synanthedon polygoni (Henry Edwards): California, Sierra Co., San Francisco State U., Sierra Nevada Field Campus, July 2007. Photo by Rollin Coville.

Anatralata versicolor (Warren): freshly emerged specimen perched on *Wyethia* leaf, in which the larvae mine, California, Contra Costa Co., El Cerrito Hillside Park, March 2007. Photo by Rollin Coville. Species is also shown in plate 21.44.

Apoditrysia

All of the more-derived Lepidoptera are grouped in the lineage Apoditrysia by their shared possession of shortened apodemes on the second abdominal sternum with enlarged bases, contrasted with the more ancestral state, continuations of longitudinal costae (venulae) of the sternal plate. Several superfamilies, the nonobtectomeran Apoditrysia, retain the ancestral movable and spined abdominal segments in the pupa, which moves forward to protrude from the shelter, enabling emergence of the moth. The more-derived superfamilies of the Apoditrysia, the Obtectomera (the Copromorphoidea, Thyridoidea, Pyraloidea, macromoths, butterflies), have obtect, nonmotile pupae.

Nonobtectomeran Superfamilies

Among the nonobtectomeran ditrysians, the Sesioidea, Choreutoidea, Cossoidea, Tortricoidea, and Pterophoroidea are convincingly defined superfamilies with well-supported membership, whereas the placement and relationships of other taxa (e.g., Schreckensteiniidae, Epermeniidae, Alucitidae) are uncertain.

Superfamily Schreckensteinioidea

FAMILY SCHRECKENSTEINIIDAE

The genus *Schreckensteinia*, with five known species, traditionally has been placed in the Heliodinidae (Yponomeutoidea), but in the Tineoidea by Heppner. The adults resemble heliodinids, small, slender, iridescent moths that hold the hind legs aloft when perched. Male genitalia, Fig. 124. The larvae feed externally on leaves and at maturity construct a cocoon that resembles the open network design of *Plutella*, but the pupal abdominal segments are movable, and the pupa protrudes from the cocoon at eclosion (Plate 60.16). The adults have stiff dorsal spines on the hind tibia and a single-bristle female frenulum. There are three species in North America, two of which occur on the Pacific Coast.

Schreckensteinia festaliella (Hübner)(Plate 11.50) is Holarctic in distribution, occurring in Europe and the Pacific Coast of North America, where it was encountered by Walsingham in 1871–1872, presumably native in northern California. The moth is diurnal, slender, shiny dark bronze colored, FW (4–6 mm) with brassy scaling forming poorly defined longitudinal streaks above the crease and through cell. This species occurs from British Columbia to southern California (Ventura) and might be regarded as introduced, but native host plants are preferred. In California the pale green larvae feed on several species of *Rubus* (Rosaceae) in coastal habitats, most commonly on California blackberry, *R. ursinus*, also on thimbleberry, *R. parviflorus*, and salmonberry, *R. spectabilis*. Several older records from introduced blackberries, *R. discolor* and *R. vitifolius*, need confirmation. In central California, *S. festaliella* is multivoltine, with adults active from March to October, and larvae can be found throughout the year. *Schreckensteinia felicella* (Walsingham) is similar but smaller (FW 2.3-4.5 mm; Plate 11.51, Plate 60.16), gray brown, not as iridescent, and without differentiated pale scaling on the FW; there is a weakly developed dark streak above the crease in some specimens. This species was described from northern Oregon and near Eureka, California, where Walsingham reared it in 1872 from *Orthocarpus*, presumably a species now included in *Castilleja* (Scrophulariaceae). *Schreckensteinia felicella* was not recorded again for a century, until 1977 when we found the larvae abundant on *Castilleja (affinis* or *medocinensis)* at MacKerricher Beach dunes, 50 miles south of Eureka. Later, De Benedictis reared *S. felicella* at San Bruno Mountain near San Francisco from *C. affinis* and *C. wrightii*. The larvae burrow into flower buds and feed externally on inflorescence parts in March to June, possibly bivoltine. *Schreckensteinia felicella* also occurs widely inland (Jefferson County, Oregon; Yosemite Valley; and Mt. Pinos at 7,400' in Kern County, California).

Superfamily Epermenioidea

FAMILY EPERMENIIDAE

Like the Schreckensteiniidae, this taxon has been orphaned from former superfamily placements, Yponomeutoidea or Copromorphoidea.

Adult These are small, sluggish, diurnal moths, not often seen in the field, and most collections have been by rearing. The hind tibia has stiff bristles but not confined to the dorsal surface as in *Schreckensteinia*; FW with upraised scale tufts and fringe groups of lamellar scales as in Alucitoidea and Pterophoroidea. Male genitalia, Fig. 125.

Larva Larvae have the submentum with posterior protuberance as in Alucitoidea and some Copromorphoidea, and perhaps most suggestive of relationships, the prothoracic L group is bisetose, as in Alucitoidea and most Obtectomera. The caterpillars are sluggish like those of pterophorids.

Larval Foods Holarctic species feed on umbells (Apiaceae, Araliaceae), whereas several other families of angiosperms serve as hosts in Australia and India. *Epermenia albapunctella* Busck was reported from conifers in eastern Canada by Prentice but feeds on umbells *(Heracleum, Osmorrhiza)* elsewhere. Pupation occurs under a sparse, loose net cocoon.

Diversity The family has about 100 species described from all faunal regions, of which 11 are Nearctic, mostly western, revised by Gaedike in 1977.

Epermenia cicutaella Kearfott (Plate 12.3, Plate 12.4) was described from New Jersey, reared from *Cicuta maculata*, and it oc-

curs in southeastern Washington and California in the Coast Ranges, Sierra Nevada, and San Bernardino Mountains, recorded almost entirely by single specimen collections. The moth is relatively large (FW 6–7 mm), pale gray, FW with whitish areas, maculation variable, faint to dark and distinct, a weak brownish fascia from the costa beyond the cell and black lamellar scales beyond the tornus form the most conspicuous markings of the FW. According to Gaedike, *E. cicutaella* was collected in southern California by Comstock from caraway *(Carum)*, a nonnative garden herb. Native larval hosts are unknown in the West but may include *Cicuta, Lomatium,* or *Sanicula*. *Epermenia californica* Gaedike is a smaller (FW 5–6 mm; Plate 12.1, Plate 12.2, Fig. 125) and darker moth, FW dark to milk chocolate brown sometimes rust tinged, with basal 0.4 paler, tan to whitish, separated by an oblique line at midcell. The lamellar scale patches of the fringe are black and conspicuous. HW dark gray, fringe pale gray. This species is known only from central coastal California, from Pt. Reyes Peninsula to the Santa Cruz Mountains and coastal Monterey County. In the latter areas, the larvae feed on *Aralia californica* (Araliaceae) and are infrequently abundant on *Sanicula crassicaulis*, whereas at Inverness Ridge, Marin County, they are common on cow parsnip, *Heracleum californicum*, and are not found on *Aralia*, which grows lushly nearby. Early instars mine the leaves, and later the caterpillars feed externally, grazing on the undersurface of the leaves, often in groups, forming large blotches that appear brown from above. We found larvae in mid-March, five months following a consummate wildfire, to the end of October, pupae in November, and adults in January, with no indication of diapause.

Superfamily Alucitoidea

FAMILY ALUCITIDAE

The "many-plume moths" are so called because the wings are deeply cleft to the basal forks of the veins, so each wing has six fringed, plumelike segments. Historically alucitids have been placed with the plume moths (Pterophoridae) in the Pyraloidea, but both families lack abdominal tympana and scaled proboscis that define pyraloids. From the pterophorids, the alucitids differ by having discrete bands of spines on some or all of abdominal terga 2 to 7 and a relatively unspecialized pupa formed in a cocoon. A sister group relationship to Pterophoridae is ambiguous because some traits suggest closer relationship to other taxa of nonobtectomeran Apoditrysia.

Adult Small (FW 3–13 mm), slender moths, they are unmistakable by their wing structure. The wings are divided only a short distance in two tropical genera, including the largest members of the family. The antennae are filiform, proboscis well developed and labial palpi usually fairly long, porrect or upcurved. Most alucitids are gray or brown, delicately banded with tan or white. They are nocturnal and when at rest collapse the wing plumes so as to resemble narrow-winged pyralids, but when active they strut about with the wings fully expanded, like miniature peacocks. Genitalia, Figs. 126, 127.

Larva They are internal feeders in flower buds, shoots, or galls; body stout with short legs, short setae on inconspicuous pinacula. Pupation occurs in a cocoon in the larval shelter or in leaf litter.

Larval Foods At least eight dicot families are used, especially Caprifoliaceae in the Holarctic, Bignoniaceae and Rubiaceae in Australia and tropical regions; one species is a coffee pest in Africa.

Diversity About 130 species have been described, and likely there are many more in tropical regions. There are just three species in America north of Mexico.

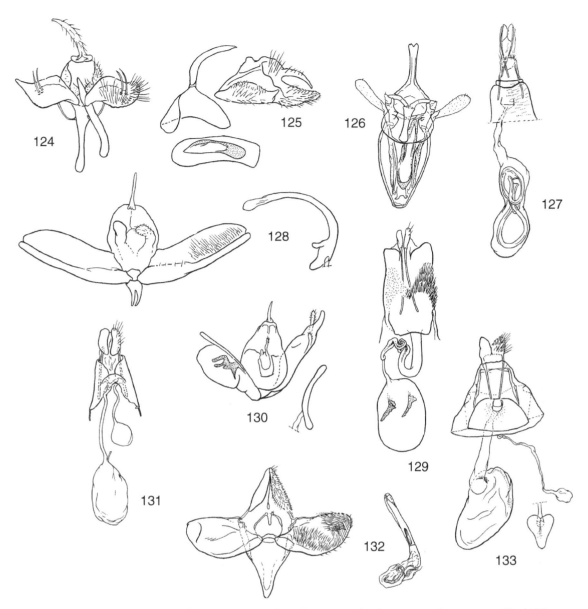

FIGURES 124–133. Genitalia structure of exemplar species of nonobtectomeran Apoditrysia, ventral aspect, except Fig. 125 fragmented. **124,** Schreckensteiniidae: *Schreckensteinia festaliella* (Hübner) ♂, aedeagus in situ [Pierce and Metcalfe 1935]. **125,** Epermeniidae: *Epermenia californica* Gaedike ♂, tegumen and uncus upper left, valva inner aspect, upper right, aedeagus below [Gaedike 1977]. **126,** Alucitidae: *Alucita montana* Barnes & Lindsey ♂, aedeagus in situ. **127,** Alucitidae: *Alucita montana* ♀ [Landry and Landry 2004]. **128,** Pterophoridae: *Anstenoptilia marmarodactyla* (Dyar) ♂, aedeagus to right. **129,** Pterophoridae: *Anstenoptilia marmarodactyla* ♀. **130,** Pterophoridae: *Emmelina monodactyla* (L.) ♂, left valva rotated to show inner aspect, aedeagus to right. **131,** Pterophoridae: *Emmelina monodactyla* ♀ [Gielis 1993]. **132,** Choreutidae: *Caloreas apocynoglossa* (Heppner) ♂, aedeagus to right. **133,** Choreutidae: *Caloreas apocynoglossa* ♀ [Heppner 1976].

Alucita montana Barnes & Lindsey (Plate 12.5, Figs. 126, 127) is widespread and commonly encountered, often in winter when the moths hide under loose bark, in wood piles, and in buildings. The adults are small (FW 5.5–6.5 mm), pale gray with faint, transverse banding of dark brown and tan, more distinct on FW, especially the costal edge. The larvae feed on snowberry (*Symphoricarpos*, Caprifoliaceae), probably as leaf or flower miners. D. Wagner collected cocoons on *S. vaccinioides* at Lake Tahoe in early July, and adults emerged in August. The species may be univoltine, but because adults overwinter and populations occur in a wide range of elevations and latitudes, there are collection records throughout the year. This *Alucita* occurs in southeastern Canada and throughout the western provinces and states, except in desert regions, from central Alberta to western Texas. Pacific Coast populations in northern

California consist of darker and slightly larger specimens. *Alucita adriendenisi* Landry & Landry is similar but appreciably larger (FW 6.5–8 mm; Plate 12.6) and darker colored with more distinct transverse bands. It ranges from New York across Canada to southern Northwest Territory and, according to available records, is disjunct in Arizona and western Texas. A larva was found in flowers of honeysuckle (*Lonicera dioica*, Caprifoliaceae) by G. Balogh in Michigan.

Superfamily Pterophoroidea

FAMILY PTEROPHORIDAE

The "plume moths" are recognizable by their deeply cleft wings in all but the most ancestral genera, lack of scaling on the proboscis, lack of abdominal tympana, hind tibia two or more times the length of the femur, and abdominal terga 2 and 3 elongated.

Adult These moths are small (FW 4–18 mm), with long and slender bodies, legs, and wings; FW cleft for one-fourth to one-third its length and the HW deeply twice cleft in most species, each of the three divisions bearing broad fringes, from which the family common name derives. When at rest, the plumes are overlaid and rolled under the leading edge of the FW to varying degrees, in some genera resembling sticks held out from the body (Plate 60.17, Plate 60.18, Plate 60.19). These moths are usually nocturnal and dull colored, tan, brown, or gray with paler and darker markings; a few species in Central and South America are believed to be members of mimetic complexes, diurnal, black with purplish iridescence, having the abdomen spotted with white to give an effect of the constricted petiole of wasps, and the hind legs have enlarged scale tufts comparable to those of ctenuchine Arctiidae. Genitalia, Figs. 128–131.

Larva Larvae are characteristically elongate, cylindrical with numerous secondary setae and long prolegs. Some species are borers in stems or roots and have short setae, but most are external feeders on foliage and have dense, long setae (Plate 60.20) that may be forked, clubbed or glandular, secreting a sticky fluid. The pupae are strongly spined or setose and either are formed in the larval galleries or affixed to host-plant stems or debris, fully exposed, with the anal cremaster attached to a silk pad, and they can bend and curl over by their movable abdominal segments.

Larval Foods They feed on living angiosperm plants, including about 70 families of dicots recorded, especially Asteraceae and Lamiaceae, usually herbs, but few monocots. The larvae of *Buckleria* are remarkable for feeding on the carnivorous plant sundew (Droseraceae).

Diversity More than 1,000 species have been described worldwide, representing about 92 genera, with probably many undescribed tropical species. There are more than 150 named species in America north of Mexico, of which about 75% occur in the West. There has not been a comprehensive systematic analysis of North American species based on genitalia, so additional new species and synonymies are likely.

SUBFAMILY AGDISTINAE

Agdistinae are presumed ancestral with respect to wing division, lacking the deep clefts in both FW and HW. There are more than 80 described species, primarily in the Oriental and Palaearctic regions, but just one in North America. *Agdistis americana* Barnes & Lindsey (Plate 12.19) occurs in salt marsh situations along the Pacific Ocean in southern California, including its Channel Islands and on both coasts of Baja California, Mexico. The larvae feed on alkali heath, *Frankenia salina* (Frankeniaceae), which is an unusual plant in our flora, but *Frankenia* is widespread in the Old World temperate zones, where several species of *Agdistis* are specialists on it. *Agdistis americana* is a smallish pterophorid (FW 9–10.7 mm, smaller in Baja California, from 8 mm), pale slate gray, FW with a thinly scaled, linear, triangulate area from the basal one-third of the cell to the distal margin, with poorly defined black dots along its lower margin and at the base; remainder of FW lightly peppered with black scales in fresh specimens. The larvae occur in brown and bright green phases unrelated to size and bear scolilike protuberances on the thorax. We found larvae in March and May, with adults emerging in April and June, and adults

have been collected from March to May and September to October, indicating two or more generations. The green larvae, which are remarkably cryptic on *Frankenia* foliage, were rare in March (20%) at San Diego but outnumbered their brown sibs at the same site in May and on San Nicolas Island in March.

SUBFAMILY PTEROPHORINAE

Recent research by Gielis and others treat the Pterophorinae as encompassing all of the remaining plume moths in North America, relegating genera of the former Platyptiliinae to more basal positions among pterophorids and not comprising a monophyletic lineage. *Geina tenuidactyla* (Fitch), the "berry plume moth," is a small (FW 5.5–8 mm; Plate 12.7) but colorful species, bronzy brown with a dark reddish or coppery luster when fresh, wings marked by delicate white lines distally, anal plume of HW with a black tuft distally, and abdomen white at base and around segments 3 and 5. This species occurs from the Atlantic states and Quebec to the Midwest, Colorado, Texas, and on the Pacific Coast from British Columbia to southern California. The larvae feed externally on buds and foliage of wild and cultivated berries (Rosaceae), including strawberry *(Fragaria)*, blackberry, raspberry *(Rubus)*, and thimbleberry *(R. parviflorus)*. *Geina tenuidactyla* is univoltine, with flight records in June and July.

Capperia ningoris (Walsingham) is larger (FW 7.5–10 mm; Plate 12.20), brown with a dull purplish sheen and the wings more strongly banded with white than on *Geina tenuidactyla*, and all abdominal terga spotted with cream white. The adults are occasionally attracted to lights but more often are found flying in association with hedge nettle, *Stachys* (Lamiaceae), and we have reared adults from larvae found on *S. albens, S. bullata,* and *S. rigida* in California. This *Capperia* occurs from the Columbia River, Washington, to southern California and occupies diverse habitats, along the coast south to Morro Bay, in the Sacramento Valley, and in the high Sierra Nevada at Whitney Portal (8,350'). Adults have been recorded from late April to early September, and there may be two generations in maritime habitats where *Stachys* is persistent. Walsingham recalled having seen *C. ningoris* in association with *Teucrium* in California, but according to present concepts that plant is restricted to desert areas in California, and *Stachys* probably was the plant he encountered.

Oxyptilus delawaricus Zeller is distinctive in having rust orange wings, FW (8–10 mm; Plate 12.21) with delicate white lines crossing the costal plume, bordering the apex, and obliquely before the notch; metathorax white, abdomen rust orange dorsally, white ventrally. This pterophorid occurs from the Atlantic states and Tennessee to British Columbia and south to the central Sierra Nevada in California. The larvae feed on hawkweed, *Hieracium* (Asteraceae), along with other species of *Oxyptilus*, adults of *O. delawaricus* having been reared from *H. cymosum* and *H. albertinum* in Canada and northeastern California. Older records of this species feeding on grape are believed to have been based on misidentified specimens of **Sphenarches ontario** (McDunnough) before the latter was described.

Trichoptilus pygmaeus Walsingham, a tiny, *Alucita*-sized plume moth, is the smallest pterophorid in western North America (FW 3.5–5.5 mm, rarely to 6.5 mm; Plate 12.8). It has pale, rust brown FW with white bands, plume fringes with small tufts of black scales; abdomen white basally, rust brown on distal half. This species occurs from Washington to southern California. Clarke reared *T. pygmaeus* from manzanita, *Arctostaphylos* (Ericaceae), in Washington, and we twice reared it from

Ceanothus papillosus (Rhamnaceae) at Big Creek, Monterey County. But *T. pygmaeus* or a closely related species in Florida is reported to feed on *Chrysopsis* (Asteraceae) by Matthews, so it may be a general feeder. In California, *T. pygmaeus* occurs in coastal areas, the Coast Ranges foothills, the Sierra Nevada, and the Peninsular Ranges at elevations below 4,000'. Flight records span June to September and larval collections April to June.

Dejongia californicus (Walsingham) adults are small (FW 5.5–7.75 mm; Plate 12.9), generally pale brown to rust colored, with alternating black-and-white streaks in the FW fringes and a small black tuft distally on the HW anal plume. The larvae feed on woody Asteraceae, including several species of gumplant, *Grindelia*, and goldenbush, *Isocoma*; Matthews reared *T. californicus* from *Heterotheca* in Florida. The report by Matthews and Lott of *Phacelia* (Hydrophyllaceae) as a food plant was based on erroneously labeled specimens from Fresno County, California, which we reared from *Grindelia*. *Trichoptilus californicus* occurs at low elevations in California, including coastal areas, the San Joaquin delta, foothills of the Sierra Nevada, Santa Cruz Island, desert slopes of the Transverse Ranges, and in mountains of southern Arizona, Texas, and in Florida. Adults have been collected in January from the Coachella Valley to November in the San Francisco Bay area, but primarily in April to May and September to October.

Gillmeria pallidactyla (Haworth)(Plate 12.22) is a Holarctic species, widespread in Europe and discovered in North America by early collectors, in New York by 1854, Texas by 1867, and California in 1871, so presumably was native. In the West, *G. pallidactyla* occurs from central Alaska to Idaho, Wyoming, Colorado, British Columbia, and coastal California. Adults differ from Nearctic pterophorids by having blurred, longitudinal brownish streaks rather than distinct markings distally on the FW, densely streaked with brown when fresh, fading to mostly white with age (FW 10–12.5 mm). The larvae feed on the circumboreal yarrow *Achillea millefolium* (Asteraceae), at first in flowers, and later instars tunnel the stems to overwinter, then feed in new shoots and developing inflorescences in spring. *Gillmeria pallidactyla* is univoltine, with flight records from April to August, primarily May to July, depending upon latitude and elevation.

Platyptilia carduidactyla (Riley), the "artichoke plume moth," is a relatively large *Platyptilia* (FW 9–12 mm; Plate 12.23), which is economically the most important pterophorid in North America. The larvae tunnel in stems and flower heads of thistles, including native and weedy species of *Cirsium, Carduus, Centaurea, Arctium, Silybum,* and most notoriously artichoke thistles, *Cynara scolymus* and *C. cardunculus* (Asteraceae). Turner and others found the larvae infesting 17 species of thistles in California, including 15 native species. The larvae start by boring into a stem at a leaf axil or enter young flower heads, and several may mature within a single floral receptacle. Control guidelines have been established for artichoke culture, and populations are monitored by synthetic pheromone lure trapping. This *Platyptilia* is gray brown to rust brown with strongly contrasting and sharply defined, dark FW markings, especially the characteristic triangle preceding the distal cleft. This species occurs across North America and through most of the western states. In coastal California there are three overlapping generations, and all stages can be found at any time of year, according to Lange. *Platyptilia percnodactyla* (Walsingham) is a paler, brownish species that also is widespread in the West but primarily in montane situations to 9,000'. It has been recorded feeding in flower heads of several genera of Asteraceae

(Erigeron, Helenium, Rudbeckia). A record by Goeden from *Carduus* in southern California may refer to *P. carduidactyla*. *Platyptilia williamsii* Grinnell is similar but smaller (FW 7–10 mm; Plate 12.24) and tends to be more gray than brownish, having variable but usually less well defined FW markings, with the costal triangle weak or obsolete, a subapical streak on the costal plume darker, and the FW sometimes nearly unicolorous. This species is abundant along the California coast, where the larvae are active throughout the year, feeding in flower heads of seaside daisy, *Erigeron glaucus*, and numerous other Asteraceae, including *Artemisia, Eriophyllum, Grindelia, Senecio blochmaniae,* artichoke, chrysanthemum, marigolds *(Calendula),* and other commercially grown flowers. *Platyptilia williamsii* occurs from Humboldt Bay south to the Channel Islands and is the only plant-feeding moth species known on the Farallon Islands west of San Francisco, where it feeds on *Erigeron*.

Anstenoptilia marmarodactyla (Dyar)(FW 7–9 mm; Plate 12.25, Plate 60.17, Figs. 128, 129) is generally gray brown, characterized by a tan or peach-colored subapical patch on the FW costal plume following and preceding dark patches. This is perhaps the most often observed western plume moth, occurring widely in Colorado, Utah, New Mexico, Arizona, Nevada, and California and in Mexico, central Baja California, and Sinaloa. Populations occur in diverse habitats, usually associated with mints (Lamiaceae), including rosemary and other garden herbs as well as *Lepechinia, Monardella, Salvia, Trichostema,* and other native genera. Also, Brian Scaccia reared specimens from *Phacelia imbricata* in coastal central California. This species is superficially similar to species of *Lantanophaga, Stenoptilodes,* and *Amblyptilia,* but the genitalia differ markedly. *Anstenoptilia marmarodactyla* also occurs in Hawaii, where apparently it was accidentally introduced prior to 1880 and is recorded by Zimmerman as feeding on *Ageratum* (Asteraceae) and lantana (Verbenaceae) but not on mints. In California, adults are attracted to black light in every month of the year at Berkeley and are recorded from low desert in Imperial County to montane and coastal situations and on most of the Channel Islands.

Amblyptilia pica (Walsingham), the "snapdragon plume moth," is *Platyptilia*-like (FW 7–10.5 mm; Plate 12.26), primarily brown to gray brown, FW with distinct darker markings and quite variable, irregularly mottled by whitish transverse strigulae along the costa, sometimes across the wing. *Amblyptilia pica* is widespread in the West, from Alaska to southern California and east to Wyoming, Colorado, and Illinois. Its range may have increased via garden and greenhouse snapdragon (*Antirrhinum*, Scrophulariaceae) and geranium (*Pelargonium*, Geraniaceae) through commercial shipment of cut flowers. This is one of the very few microlepidoptera named twice by Lord Walsingham, and Lange described several poorly distinguished subspecies. Natural populations are primarily associated with Indian paintbrush *(Castilleja), Penstemon,* and *Scrophularia* (Scrophulariaceae), as are several species of *Paraplatyptilia*. The larvae feed inside the flowers, evidenced by the ejected frass. These plants are characterized by production of the alkaloid rhexifoline, which is concentrated in the seeds, the preferred larval food, and are sequestered in the larvae. Although scrophs are the primary hosts, *A. pica* has been reared from several unrelated plants in California, including *Calendula, Eriophyllum* (Asteraceae), *Phacelia* (Hydrophyllaceae), native honeysuckle (Caprifoliaceae), and *Stachys* (Lamiaceae). Larvae are recorded from February to September and adults from May to December in coastal California, and spring, summer, and fall generations were defined by Lange.

Lioptilodes parvus (Walsingham)(Plate 12.27), which was described originally from one specimen Walsingham collected at Mt. Shasta, California, in 1871, recently has been placed in synonymy of **L. albistriolatus** (Zeller), named in 1877 from Colombia. It has been found to range widely in the southwestern United States, Florida, Mexico, Costa Rica, Peru, Paraguay, and Argentina. This is a distinctive little species, FW (6.5–8.5 mm; 5 mm in Florida) tan to pale gray with weak black speckling above the cell, along the Cu crease, and dots in the fringe. The larvae feed on several species of *Baccharis* in Arizona and New Mexico, northern Mexico, and Australia, and on numerous other Asteraceae (*Aster, Erigeron, Conyza, Senecio, Solidago*, etc.) in Florida, Bermuda, and Hawaii, where *L. parvus* appeared at Pearl Harbor by 1946. Sparse records indicate spring and fall generations in California.

Paraplatyptilia albiciliata (Walsingham)(Plate 12.28) includes a complex of named forms that needs careful study. The typical form, from Mendocino and San Mateo counties, California, is distinctive, FW (8–10 mm) cinnamon brown on the costal half, blending to pale ochreous below the cell, without maculation except fringe white, dark brown at the base. Lange named a race *P. rubricans*, also from San Mateo County and Monterey Bay, which has a FW triangle indistinctly defined, and implausibly treated *P. orthocarpi* (Walsingham) from northern Oregon, which has distinct and ornate FW maculation, as a subspecies. Munroe in the MONA check list included *P. canadensis* (McDunnough), which has well-defined FW markings, as a race from British Columbia, but all of these were relegated to synonymy by Gielis. Larvae of the coastal forms have been collected from seaside painted cups, *Castilleja affinis*, *C. latifolia*, and *C. wrightii* (Scrophulariaceae), whereas Walsingham reared the original series of *P. orthocarpi* from an unidentified species of *Orthocarpus*, now treated as *Castilleja*. Lange collected adults from April to August, and De Benedictis found the larvae at San Bruno Mountain in April, May, and June in different years and reported the early season adults match the typical specimens, while later ones resembled Lange's *P. rubricans*. *Paraplatyptilia fragilis* (Walsingham) is a smaller (FW 7.5–11.5 mm; Plate 12.29) and paler species, grayish brown with the FW triangle well defined, dark brown, and outer pale line faintly indicated. *Paraplatyptilia fragilis* was described from the arid Klamath Lakes area in extreme northern California, and it ranges widely in Great Basin, inland montane, and desert ranges of eastern British Columbia to Utah, Nevada, and New Mexico to southern California. McDunnough in British Columbia and Braun in northern Utah found the larvae boring into seed capsules of *Penstemon cyananthus* (Scrophulariaceae).

Adaina montana (Walsingham) is a small species, FW (7.5–8.5 mm; Plate 12.30), cream white with delicate rust ochreous to brownish streaks on the costal plume and a subterminal shade on the costa. The geographic distribution of this species is poorly known, originally described from Mt. Shasta, California, and later recorded from New York, Ontario, and Colorado. Larvae have been found on flowers of Asteraceae, including *Aster* and *Solidago*, by McDunnough, Forbes, and Ferguson, and there is an unpublished record from *Xanthium* by Comstock and Henne. (Several other plants listed by Robinson and others, citing Forbes, are erroneous.) The life cycle is univoltine, with caterpillars in June, eggs overwintering, according to Forbes. *Adaina ambrosiae* (Murtfeldt)(Plate 12.31) is similar, slightly smaller, FW whitish irregularly dusted with black and with a tan to brown streak through the costal half into the costal plume. Fringe dark gray crossed by three white bars. This species occurs in eastern North America to the Pacific Coast and Central America. It has been reared

from *Ambrosia* in Missouri, Texas, and California, and Hsu found the larvae on *Xanthium* in the San Joaquin Valley, California. *Adaina ambrosiae* is bivoltine, with larvae in June and late August to October.

The large genus *Oidaematophorus* Wallengren, 1852, as formerly conceived (more than 200 species worldwide, 75 in the Nearctic region) was split by Gielis, based on presence versus absence of cornuti in the male and signa in the female. He assumed absence of cornuti and signa to be homologous among species of *Oidaematophorus* sensu stricto, and presence within this lineage represents reversals. Neither assumption is convincing, and lacking a uniquely derived synapomorphy to define either of the sister branches, resurrection of the name *Hellinsia* Tutt, 1905, to accommodate most of the former *Oidaematophorus* species is equivocal. Therefore we prefer to follow a conservative approach and retain *Oidaematophorus* sensu lato. The larvae are specialists on a wide variety of herbaceous and woody plants. Some are stem borers; others feed externally without a shelter, often depending upon cryptic form, color, and behavior to escape detection (Plate 60.20).

Oidaematophorus occidentalis Walsingham (Plate 12.32) and *O. balsamorrhizae* McDunnough make up a complex that needs careful study. They are moderately large species (FW 12–13 mm), pale tan or straw colored with variable FW maculation, from nearly immaculate to having a brownish shade through the cell ending abruptly at a dark crescent. This variation was recognized by Walsingham, who found all degrees of intermediates. McDunnough believed an *Aster*-feeding species in eastern British Columbia was *O. occidentalis* and described *O. balsamorrhizae* based on a series reared from *Balsamorhiza*, which are more heavily patterned specimens than flown ones. McDunnough distinguished that population from *O. occidentalis* by having a sprinkling of black scaling basally in the costal area. Subsequently both species have been reported from various Asteraceae, including *Balsamorhiza, Helianthus*, and *Wyethia*, and *O. occidentalis* from *Eupatorium, Grindelia*, and *Lessingia*. Weakly patterned populations occur at elevations of 6,000 to 8,000' in several mountain ranges of eastern Nevada, the Wasatch Range in Utah, northern Arizona, and mountains of southern California. The life cycle apparently is univoltine, with adults in June to September. *Oidaematophorus helianthi* (Walsingham), which was tentatively placed with *Hellinsia* by Gielis, was described from southern Oregon feeding on *Helianthus*; it is smaller (FW 11 mm), creamy white rather than ochreous or tan, with a row of distinct blackish spots beyond the cell and in fringes of the plumes.

Oidaematophorus phaceliae McDunnough is a moderately large species (FW 9.5–13 mm; Plate 12.33), FW boldly and distinctly patterned with purplish gray bars on the costa distally and the characteristic triangulate spot preceding the cleft darkest. This species was described from Alberta and ranges to southern California, where FW maculation is paler. The larvae feed on various species of *Phacelia* (Hydrophyllaceae) and their hybrids, including *P. heterophylla* in Alberta, *P. californica, P cicutaria, P. imbricata, P. malvifolia*, and *P. ramosissima* in California, and a population judged by McDunnough to be *O. phaceliae* occurred on *Hydrophyllum virginianum* in Ontario, Canada. The larvae feed externally with very little webbing, even on densely bristled species such as *P. malvifolia*, in March, April, and May. Adults have been recorded at lights primarily in June and July but April on Santa Rosa Island. *Oidaematophorus fieldi* (Wright)(Plate 12.34, Plate 60.19) is tan, FW (11–12.7 mm) shaded with rust brown on the costal half, becoming darker distally and edged with whitish. The larvae have been found April to early June and reared on *Eriodictyon californicum* (Hydrophyllaceae), which is toxic to all but a few specialist Lepidoptera, by De Benedictis and Rude and during our survey at Big Creek, Monterey County, where *O. fieldi* flies into September. It ranges from Monterey County, to San Diego, California.

Oidaematophorus grandis (Fish) is our largest pterophorid (FW 14–17 mm; Plate 12.35). Its distinctive larva and biology were described a century ago by F. X. Williams and later by Lange and Tilden. The larvae, which are borers in woody branches of coyote brush, *Baccharis pilularis* (Asteraceae), have a strongly sclerotized anal shield with two stout hornlike processes resembling the urogomphi of coleopteran wood borers. Development requires a year, with pupation and adult emergence in late summer and fall. This species occurs along the Pacific Coast from northern coastal to southern California and the northern Channel Islands, and Barnes and Lindsey reported *O. grandis* from Mexico. We have reared adults from larvae collected as early as February, emerging in April, but in the field the flight period is July to October. The moths are tan, FW immaculate, with faint brownish scaling along the costa distally and preceding the cleft in fresh specimens, and there is a paler, whitish dusting through the dorsal half. *Oidaematophorus glenni* Cashatt is similar, nearly as large (FW 12.5–16 mm; Plate 12.36) but more variable in size than *O. grandis*. The FW is tan, sometimes with a brownish smudge preceding the base of the cleft and usually with the ends of the veins brownish, but worn specimens are difficult to distinguish from *O. grandis* except by slight differences in male genitalia. *Oidaematophorus glenni* had been confused with other *Oidaematophorus* until Cashatt reviewed the group in 1972. This species occurs from the New England states and Virginia across the northern half of the United States to eastern Washington and central California. The larvae are borers in Asteraceae, having been reared from root crowns of goldenrod, *Solidago canadensis*, by M. O. Glenn in Illinois, stems of *Solidago* at Almota, Washington, by Clarke, recorded by Comstock and Henne from *Machaeranthera canescens*. We reared it from stems of the federally endangered plant Suisun marsh aster, *Symphyotrichum lentum*, at Antioch on the San Joaquin River delta, a locality where we also reared *O. grandis* from *Baccharis*. *Oidaematophorus glenni* is univoltine, with flight records in May and June. Larvae are fully fed by fall and hibernate in the stems—Clarke's reared specimens emerged from January to May; our aster stems from the preceding season were collected in early April and produced adults in 4 to 6 weeks.

Oidaematophorus sulphureodactylus (Packard) has distinctive, bright yellow FW (10–13 mm; Plate 12.37) with obscure brownish to pale rust along the costa on the distal half and a dot preceding the cleft; the HW is dark brownish gray; and abdomen yellow, usually with a thin, brownish middorsal line. This species is widespread in the West but poorly documented, occurring in Montana, Colorado, and New Mexico, and in California from Siskiyou County to central coastal and mountains of southern California. The larvae feed in new foliage and flower buds of sunflowers (Asteraceae), *Helianthus pumilis* in Colorado, and in California *H. californicus, H. gracilentus,, Helianthella californica,* and *Heliomeris multiflora*, in March, April, and May, and adults fly in June and July. *Oidaematophorus confusus* Braun is smaller (FW 9–11.5 mm; Plate 12.38) and is dark brownish gray with a darker pattern resembling *O. phaceliae* and related species but appearing rather uniform brown to the naked eye. *Oidaematophorus confusus* is a coastal California species occurring from Humboldt County to Pt.

Conception and the northern Channel Islands. Like *O. grandis*, this species feeds on *Baccharis pilularis* in coastal California but is an external foliage and bud feeder. Larvae were reared by Tilden, De Benedictis, and us in March and April, and adults have been collected at lights in May and June (March to May on the Channel Islands) and August to November, so there are at least two annual generations. *Oidaematophorus longifrons* (Walsingham) is distinctive by its elongate front and unusually long labial palpi, which exceed the head by its length, FW (13–14 mm; Plate 12.39) gray becoming brownish on the dorsal area with scattered black-and-white scales. There are black dots near the middle of the cell and above the base of the cleft. Larval feeding was recorded from *Acourtia microcephala* (Asteraceae) by Comstock and Henne, and we reared *O. longifrons* from unidentified Asteraceae on Catalina and Santa Cruz islands. Flight records on these islands span January to November, indicating a multivoltine life cycle.

Emmelina monodactyla (Linnaeus)(Plate 12.40, Plate 60.18, Figs. 130, 131), the "morning-glory plume moth," is the most widespread and familiar pterophorid, occurring widely in the Palaearctic and North Africa and in North America from the Atlantic to the Pacific and southern Canada to Mexico. The adults are moderately large (FW 12.5–14 mm) and variable in FW color, typically either primarily tawny tan or pale gray, with variable brownish or darker gray streaks. There is a dark dot in the cell, a spot before the base of the cleft sometimes extended into it, and some dark scaling at the tips of the veins. The larvae feed primarily on Convolvulaceae, including several species of morning glory, bindweed, and sweet potato *(Convolvulus and Ipomoea),* but also are confirmed on Solanaceae *(Datura, Hyoscyamus),* and *E. monodactyla* has been recorded on Asteraceae, Chenopodiaceae, and other plants. Convolvulaceae typically are twining vines that cover surrounding vegetation, and some rearing records may have originated from larvae wandering off vines to pupate, or from unreliable identifications because the variable moths resemble species of *Oidaematophorus*. Specialization on both Convolvulaceae and Solanaceae may reflect underlying phytochemical links, particularly tropane alkaloids. The larvae feed externally with minimal webbing, on leaves or flowers. *Emmelina monodactyla* flies from March to November, even in northern areas, and throughout the year in coastal California.

Superfamily Choreutoidea

FAMILY CHOREUTIDAE

The family Choreutidae comprises a small group of phenotypically similar moths that were included in the Glyphipterigidae or Sesioidea in the past, based on wing venation and other ancestral similarities. Choreutids have a scaled proboscis, presumably derived independently from those of the Gelechiidae and Pyraloidea, and minute, one- or two-segmented maxillary palpi. The eggs are deposited upright.

Adult These are small (FW 2–9 mm), diurnal moths with tortricid-like broad wings, which some genera flick in a characteristic fashion as they strut jerkily about on host-plant leaves. The ocelli are large; antennae in males usually with long ventral setae; labial palpi upturned but not strongly curved. These are mostly dark-colored moths with black or brown wings marked by metallic gray, white, or silvery blue;

some tropical species are orange with harlequin patterns. Genitalia, Figs. 132, 133.

Larva   Larvae are slender with elongate abdominal prolegs, living externally in slight webs, from which they graze on leaf surfaces.

Larval Foods   Diverse dicot angiosperms of at least 17 families are used; larvae concentrate on Moraceae in tropical regions, Urticaceae, Fabaceae, and Asteraceae in the Holarctic.

Diversity   There are about 400 described species worldwide, with many undescribed tropical species. There are about 30 species in seven genera described in America north of Mexico, 60% of which occur in the West. In addition, J. B. Heppner studied the Nearctic choreutids in connection with his work on Glyphipterigidae and designated numerous new western species but has not published their descriptions.

Anthophila alpinella (Busck)(FW 5.5–6.7 mm; Plate 12.10) was described from British Columbia and has been treated as a subspecies of the Holarctic *A. fabriciana* (Linnaeus), but recently the two have been regarded as separate species. *Anthophila alpinella* occurs from the northeastern United States and southern Canada to British Columbia, the Rocky Mountains, and south on the Pacific Coast to Marin County, California. This species has differently shaped wings than other choreutids, narrower, somewhat rectangular, FW dark olive brown, sprinkled with white scales that form two obscure, transverse bands, at the basal one-third, and beyond the middle; apical area bronze tinged. HW dark brown with a whitish submarginal band. The larvae feed on nettle, *Urtica* (Urticaceae), crumpling new leaves and inflorescence buds into a shelter. They were numerous on *U. holosericea* in February to April the first and second years following an extensive wildfire at Inverness Ridge, California, when nettle flourished; then *A. alpinella* and its host became scarce. Adults have been collected from March to July and October.

Prochoreutis, *Caloreas*, and *Tebenna* are similar, with broad, rounded wings having maculation patterns that feature slightly raised, metallic lead- or silver-colored spots. The genera have been defined by differences in wing venation and genitalia, without phylogenetic analysis. *Prochoreutis sororculella* (Dyar) is a small species (FW 4.8–6 mm; Plate 12.11) whose FW has slightly raised, silvery spots on black patches between whitish median and subapical fasciae and a row of silvery scales outside a subterminal white line. This species occurs in the mountains of California, from Mt. Shasta and the Trinity Alps to the Sierra Nevada. The moths often are numerous in boggy meadows. The larval biology is not recorded, but a sister species in the eastern United States feeds on skullcap, *Scutellaria* (Lamiaceae). The adults fly May to July.

Caloreas has 10 Neotropical species, eight named in America north of Mexico, and Heppner has identified about 12 undescribed species, mostly from montane areas of western North America. *Caloreas leucobasis* (Fernald) occurs in the northeastern United States and southern Canada and was reported from British Columbia by Dyar, who reared it from *Anaphalis margaritacea* (Asteraceae). The adult has whitish head, thorax, and basal one-fourth of FW, a median transverse, brownish band, containing five metallic silvery spots surrounded by black, and there is a row of metallic spots along the outer dorsal margin to the tornus. Later, a similar species was described, *C. multimarginata* Braun from northern Utah, based on a single female that Braun reared from a mint (Lamiaceae). She noted its similarity to *C. leucobasis* but viewed the different

larval food plant and white-margined, rather than white-tipped, black scaling as diagnostic. *Caloreas melanifera* (Keifer)(FW 4.5–6 mm; Plate 12.12, Plate 12.13) was described from the central Sierra Nevada in California, reared from *Artemisia douglasiana* (="*vulgaris*"). Although he found the genitalia very similar to that of *C. leucobasis*, Keifer contrasted the blackish HW of *C. melanifera* with the reddish brown HW of *C. leucobasis*. The three, along with *C. augustella* (Clarke), described from Washington, may represent populations of a single, widespread species. The Californian race has been reared from artichoke (*Cynara scolymus*) at Half Moon Bay by Lange, and we found larvae on *A. douglasiana* in coastal Sonoma County, confirming specificity to Asteraceae. Coastal populations apparently are multivoltine, larvae having been collected in January and June and an adult in August. *Caloreas apocynoglossa* (Heppner), the type species of the genus, is distinctive in having white HW, averages larger than *C. melanifera* (FW 5.5–7.5 mm; Plate 12.14, Figs. 132, 133) and is univoltine, flying in March and April in central coastal California. The larvae feed in the newly developing foliage of *Cynoglossum grande* (Boraginaceae) in shaded woods in February to April. Pupation occurs in a fluffy white cocoon in a pinch-fold on the underside of a leaf. *Caloreas occidentalis* (Dyar), which flies in May and June in the foothills of Oregon and the Sierra Nevada is similar but has brown HW.

Tebenna has eight described North American species plus several in the western United States identified by Heppner but not yet named. In western montane areas they tend to live in drier, more open habitats than *Prochoreutis* and *Caloreas*, and the larvae specialize on Asteraceae. *Tebenna gemmalis* (Hulst)(FW 6.2–8 mm; Plate 12.15, Plate 60.24) occurs in northeastern California, the Sierra Nevada, the mountains of southern California, and basin ranges of Nevada and Utah. The adults are brightly colored, thorax and basal one-fourth of FW golden to bright orange brown, with metallic bluish streaks. A broad, whitish, transverse fascia is sprinkled with golden scales and is followed by golden greenish and another whitish area. The larvae feed gregariously during June and July on the upperside of the leaves of *Balsamorhiza sagittata* (Asteraceae). They form shelters with copious silk, pulling the margins of the large leaves together. Pupation takes place at the edge of the shelter in a weak cocoon, and adults emerge in July. *Tebenna gnaphaliella* (Kearfott) is the smallest Nearctic choreutid (FW 3.2–4.8 mm; Plate 12.16) but one of the most widespread, occurring in the East and Midwest, and apparently as a disjunct population on the Pacific Coast. In California, *T. gnaphaliella* is a coastal species, ranging from Mendocino County to Monterey County and the Channel Islands. Adults are dark gray with two dark ochreous streaks at base of the FW, black patches on distal half, each surrounding metallic silver blue lines. The larvae mine and later web the leaves of cudweeds, *Gnaphalium* (Asteraceae). Murtfeldt found larvae on *G. polycephalum* and not other cudweeds in Missouri, whereas we have reared *T. gnaphaliella* on the introduced *G. luteo-album* as well as on several native *Gnaphalium* in California.

Choreutis diana (Hübner)(Plate 12.17, Plate 12.18) was described from Europe and is Holarctic, one of the first microlepidoptera to have been recorded from North America. It is transcontinental, from the Northeast to British Columbia, the mountains of Colorado, Utah, and Nevada to above 10,000′, and south on the Pacific Coast to the Santa Cruz Mountains, California. The FW (6–7 mm) is triangulate, dark gray powdered with pale gray, sometimes with a greenish cast, and there are three obscure, broken, transverse bands of whitish. On the West Coast the FW is heavily overlaid with whitish, giving a pale bluish appearance (Plate 12.18). The translucent, greenish yellow larva has a reddish head and black pinacula and feeds on broadleaf trees, primarily Betulaceae, alder *(Alnus)*, and birch *(Betula)*, sometimes in great abundance, also on Salicaceae, willow *(Salix)* and balsam poplar *(Populus balsamifera)*, and Rosaceae, cherry *(Prunus)*. They fold and skeletonize the leaves, which then turn brown. Pupation occurs in a tough cocoon with blunt ends under a sheet of dense silk. *Choreutis diana* is univoltine, adults in April and May on the Colorado Front Range, late May and June in coastal California, and July to early September in the Sierra Nevada and Great Basin.

Superfamily Sesioidea

This superfamily consists of three families, Brachodidae, Sesiidae, and Castniidae, the adults of which are markedly dissimilar in appearance and behavior, but their proposed relationship is based on subtle adult and larval features: the ocular diaphragm is more strongly pigmented anteriorly; they have large patagia; and the larvae have the crochets of the abdominal prolegs arranged in two transverse bands (Fig. 23a) rather than a circle or ellipse typical of most microlepidoptera. Castniidae and Brachodidae are tropical families not represented in America north of Mexico.

FAMILY SESIIDAE

Adult These moths are wasplike, with FW basally narrow, HW relatively short, usually lacking scales except along the veins and distal margin. The coupling mechanism is unique, with the posterior margin of the FW bent down, engaging with the upcurved costal margin of the HW, and both have rows of stiff scales that interlock, in addition to the normal frenulum and retinaculum. Sesiids are small to moderately large and heavy bodied (FW length 5–28 mm, females usually larger and more heavy bodied than males), diurnal or crepuscular, and almost all species resemble wasps or bees, often startlingly so, presumably a nonspecific mimicry that aids vertebrate predator avoidance. This involves not only clear wings and a colorful, banded abdomen, often narrowed basally or colored to create that illusion, but the legs are modified with tufts, even to an extent of having yellow-tipped scales simulating pollen carried by bees. *Alcathoe* are black with bright orange wings and resemble tarantula hawks *(Pepsis)*. Genitalia, Figs. 117–122. Sesiids often visit flowers with quick, darting flights, and until the advent of pheromone baits in recent decades, they were difficult to find and collect.

Larva The larvae are borers in stems, bark, and roots or in galls caused by other insects; stout, with heavily sclerotized head and mandibles and unpigmented integument.

Larval Foods Larvae feed on more than 40 families of flowering plants, with no strong preference for one. Several species feed on conifers, none on monocots.

Diversity There are more than 1,100 described species in 120 genera, speciose in both temperate and tropical regions. In North America, this family perhaps is the most completely known of any microlepidoptera. Many species are agricultural pests, borers in stems of berries, grape and squash vines, trunks of fruit trees, or conifers. This group historically has been a collector favorite owing to the colorful beauty and unusual form of specimens and the challenge of obtaining them. North American sesiids became well known through a beautifully illustrated monograph by Beutenmüller in 1901 and a revision by Engelhardt in 1946 based on many years of larval rearing. Synthetic pheromone baits that are effective for many genera have been developed since the 1970s to monitor pest species. These have greatly enhanced collection of male specimens, led to discovery of new species and recovery of rare ones, and increased the popularity of sesiids among collectors. Eichlin and Duckworth treated the Sesiidae in the MONA series in 1988, illustrating 86 species, of which 70% occur in the western states and provinces.

SUBFAMILY TINTHIINAE

Members of the subfamily Tinthiinae are characterized by lack of a scale tuft at the tip of the antenna and lack of chaetosemata at the posterior margin of the head. The subfamily is represented in our region by 10 species in three genera. *Pennisetia marginata* (Harris)(FW 8–16 mm; Plate 12.41), known as the "raspberry crown borer," occurs widely in the eastern United States, British Columbia, and the Pacific and Rocky Mountain states. This is a robust moth, resembling a yellow jacket *(Vespula, Vespidae)*, thorax black, abdominal segments strongly banded with yellow. Adults have been reared from *Rubus* (Rosaceae) including raspberry and blackberry. The eggs are laid in August and September, and early-instar larvae overwinter, then feed on newly developing canes in spring. Later instars often girdle the stems above ground level and cause considerable damage to berry crops. They may complete development in late summer, but most overwinter a second time.

Zenodoxus canescens Henry Edwards (Plate 12.42, Plate 12.43) is unusual in having the wings almost fully scaled, FW (6–11 mm, males larger than females, unusual among sesiids) gray with mixed brown scaling, red ventrally in females; HW hyaline medially on basal two-thirds in males, fully scaled, red in females. This species occurs from the northwestern states and Wyoming to Colorado, western Texas, Arizona, and extreme southeastern California. Adults have been reared from alkali mallow, *Malvella leprosa* (formerly *Sida hederacea*), and fly in association with globemallow *(Sphaeralcea)* in Colorado, so probably various Malvaceae serve as hosts. Most collections have been made in September to November, but a few specimens from Texas and California were taken in February and March, so a spring generation may develop in southerly, arid areas. *Zenodoxus sidalceae* Engelhardt (Plate 12.44) has similarly colored, almost fully scaled wings, but the abdomen is yellow banded. It has been reared from checkerbloom, *Sidalcea oregona* and *S. nervata* (Malvaceae), and probably feeds on other *Sidalcea*, occurring from Alberta and Idaho to Washington, Oregon, and coastal southern California, but not in arid regions.

SUBFAMILY SESIINAE

The subfamily Sesiinae has been redefined by Edwards et al. to include Paranthreninae of earlier authors, so accommodates the remainder of the North American Sesiidae. *Paranthrene* includes some slender species that have evolved an apparent Batesian mimicry relationship to vespid paper wasps of the genera *Mischocyttarus* and *Polistes*. *Paranthrene robiniae* (Henry Edwards)(FW 11–18 mm; Plate 12.45) is widespread and may be our most often observed species, occurring from sea level to timberline, from the Rocky Mountains to the Pacific Coast, southern Alaska to southern California, including Santa Cruz and Santa Catalina islands, and the southwestern deserts. Adults of the typical form are colored almost exactly like a paper wasp, FW nearly fully scaled, yellow brown, HW hyaline in both sexes, abdomen yellow with segments A1 and A2 black except a narrow band on A2. They often crawl slowly about on willow branches in a manner reminiscent of *Polistes* species' search behavior. A form in the Rocky Mountains has the thoracic dorsum and abdomen rust red except for a narrow, yellow band on A2 and a broad band on A4, and a desert form is all yellow except for rust orange powdering on the thorax, wing bases and anteriorly on A2 and A3. The larvae are borers in limbs of willow, *Salix*, and poplar, *Populus* (Salicaceae), often trees weakened by wood-boring beetles or physical damage. The flight period is May to July in most of the range, but adults are recorded in February to May and November in southern California. The larvae of the closely related, primarily eastern Nearctic *P. dollii* (Neumoegen) have a two-year feeding cycle, but that may not be true for *P. robiniae*, at least in lower-elevation and southern populations.

Albuna pyramidalis (Walker)(FW 8–14 mm; Plate 12.46) occurs across Canada and the northern United States, extending through the Rocky Mountains, Cascades, and Sierra Nevada, ranging from sea level to above timberline. The moth has several color forms but has a diagnostic wide discal band on the mostly hyaline FW. Colors vary from orange red to black in both sexes, and the HW is sometimes mostly scaled in red forms. The abdomen varies, irrespective of wing color, from entirely blue black to yellow segmental bands on all segments except 1 and 3. The larvae of *A. pyramidalis* are borers in deep, perennial roots of Onagraceae, including *Epilobium* and *Oenothera*, feeding from silken tunnels outside the roots. The flight period is from late May to mid-August. Engelhardt believed that *A. pyramidalis* had increased in abundance through enormous spread of fireweed, *Epilobium angustifolium* and *E. latifolium*, following extensive fires.

The species of *Euhagena* also are Onagraceae feeders. *Euhagena nebraskae* Henry Edwards (FW 6–11 mm; Plate 12.47) occurs in sandy situations from South Dakota and western Nebraska, to Colorado, northern New Mexico, and Utah, and there is an apparently disjunct population in the western Mojave Desert. Its color pattern varies remarkably, wings fully scaled and pearly white, semitranslucent, with orange powdering on anal area and veins, to bright orange red with dark gray discal bar and fringes and sometimes veins, and the abdomen brownish black with terga 2, 4, 6, and 7 narrowly edged with whitish posteriorly to having narrow or broad yellow bands on 4, 6, and 7; females are not appreciably larger except the abdomen. The larvae feed in perennial roots of evening primrose, *Oenothera*, probably several species. This species flies in September and October.

The genus *Melittia* includes the largest and most magnificent clearwing moths in North America. Larvae of all six species are borers in squash vine stems or roots (Cucurbitaceae), and all but one occur in the southwestern states. They are characterized by having a tuft of scales below the antenna that extends over the eye, and by enormous scale tufts on the hind tibia and tarsus. The adults have olive green FW, matching the leaves of wild gourd (*Cucurbita foetiditissima*). At rest the FW are closed over the body, completely obscuring the colorful abdomen and orange and black hindleg scale tufts. Stealthy search during cool mornings will reveal the moths perched along the midrib facing outward. Later, after temper-

atures rise, *Melittia* are almost impossible to approach and are very rapid fliers. *Melittia snowi* Henry Edwards is the smallest North American species (FW 8–12 mm; Plate 13.19). It is recognizable by a broad, gray dorsal band on the abdomen. The larvae feed in galls they induce on wild gourds in the arid Southwest, from Kansas and eastern Colorado to Texas, southern Arizona, and Mexico. The eggs are laid near the apex of a shoot, and the newly hatched larva bores into the growing tip, which causes development of an oval gall about 4 to 8 cm long in the stem or on the fruit. At maturity the larva leaves the gall to form a tough cocoon in the sand, to pass the remainder of summer and winter in diapause. Adults emerge from March to September, but in Colorado there is a brief flight, May to June. *Melittia gloriosa* Henry Edwards (Plate 13.20, Plate 13.21) is the largest and most spectacular North American sesiid (FW length 15–28 mm, i.e., the largest females are 6.25 cm in wingspan), with HW moderately to almost completely scaled, bright red orange, the hind-leg tufts brilliant red orange basally and medially, becoming shining brown black posteriorly, and the abdomen is banded in yellow and brown, orange laterally. Males are smaller with HW scaled only at the base, bright orange. This species is the most widespread *Melittia*, occurring from Oklahoma and western Texas to central Mexico, southeastern Colorado, Arizona, and California, including San Clemente, Catalina, and Santa Rosa islands, thence northward to southern Oregon. The larvae bore in the large tubers of wild gourds (*Cucurbita*) and wild cucumber (*Marah fabaceus* and *M. oreganus*), which has enabled occurrence on the Channel Islands, northern California, and Oregon. Larval development may require two years in some localities. The adults are recorded from May to October, varying with locality, but only in July and August in southern California. *Melittia grandis* (Strecker) is similar but smaller (FW 16–21 mm; Plate 13.18) and is distinguished by having dark rather than yellow segmental rings on the abdomen, and olive pale-tipped scales on the FW and thorax. It occurs from southeastern Colorado and western Texas to the Colorado River and northern Mexico.

Calasesia coccinea Beutenmüller is a peculiar little species (FW 6–8 mm; Plate 12.48) having fully scaled wings, FW bright red with a black discal dot and HW gray brown to black, thinly scaled in males. This species occurs from western Kansas and eastern Colorado to western Texas, New Mexico, and southeastern Arizona. The larval biology is unknown, but numerous adults have been taken in association with *Hoffmanseggia* (Fabaceae) in Kansas and New Mexico, providing a clue to the host plant. Adults also visit flowers of composites, and most collection records are from July to September, a few in April and June.

Sesia tibialis (Harris) is large (FW 12–19 mm; Plate 14.1, Figs. 119, 120), is yellow jacket–like, and occurs from New England across southern Canada to British Columbia, through the Rocky Mountains to eastern Arizona and the Pacific states to southern California. The larvae bore under bark in several species of willow and poplar (Salicaceae) in a wide range of habitats. In California, *S. tibialis* occurs on the coast, in the hot Central Valley, and along the eastern Sierra Nevada. This species is conspicuous, more so since the advent of Z-ODDA pheromone baits, to which *S. tibialis* males are strongly attracted, even entering open windows or doors of parked cars containing a lure, scattering uninitiated occupants in alarm. The adults fly from May to August.

Synanthedon is the largest genus of Sesiidae in the Nearctic, including 41 described species, about half of which occur in the West. They are small to medium-sized, mostly slender moths that vary in color, black or black with yellow markings, or blue and red. Most have partially hyaline FW and hyaline HW, but females often have opaque FW and sometimes HW. The "peach tree borer," *S. exitiosa* (Say)(FW 6–15 mm; Plate 14.2, Plate 14.3), is the most notorious and one of the most widespread sesiids in North America. The larvae bore under bark of peach *(Prunus persica),* almond *(P. amygdalus),* cherry *(P. avium),* native chokecherry *(P. virginiana* var. *demissa),* and other *Prunus* species, usually around the crown, and can cause considerable loss of young trees in plantations and orchards. In the West, most moths are shining blue black; males have hyaline FW and HW, females opaque FW. The abdomen varies from all black to black with thin yellow bands on some or all segments dorsally, and in eastern and Colorado populations some individuals have segment A4 or A4 and A5 ringed with red (Plate 14.3). This species occurs in eastern North America, in the Rocky Mountains to New Mexico, and in the Pacific states south to southern California, primarily in agricultural and urban orchards and in montane situations. The adults are recorded from April to November, mostly during the summer months, but *S. exitiosa* is believed to be univoltine. *Synanthedon albicornis* (Henry Edwards)(Plate 14.4) is quite similar, but the antennae have a white preapical band dorsally (inconsistent in coastal populations), the labial palpi are smooth scaled, and the labial palpi and costa of both wings are pale yellow ventrally (FW 9–11 mm). This is a western species, occurring from northern British Columbia through Idaho, Colorado, and the mountains of Nevada and the Pacific states to southern California. The larvae live in willows *(Salix),* boring under bark of larger limbs, in exposed roots, or in smaller canes. Adults are recorded mostly June to September.

Synanthedon bibionipennis (Boisduval)(Plate 14.6, Plate 14.7, Plate 60.21), the "strawberry crown moth," occurs throughout the Rocky Mountain region and British Columbia to the Pacific states, where it is one of the most frequently observed clearwing moths. The FW (8–10 mm) are partially hyaline in both sexes, with black scaling, HW hyaline except the fringe, which is broader in females. The abdomen is black with segments 2 and 4 yellow dorsally and other segments variably banded, hardly an excuse for the roster of nine synonyms by Henry Edwards. The adults are seen hovering about and mating on blackberry *(Rubus),* both native and introduced species, or visiting flowers of other plants for nectar. They also have been reared from strawberry *(Fragaria),* rose *(Rosa),* and cinquefoil *(Potentilla)* and probably feed on other Rosaceae. The larvae bore into the lower stems or root crowns, and after feeding until September or October, overwinter in a silken chamber before resuming feeding in spring. Emergence occurs from April through mid-August, varying with elevation and latitude. *Synanthedon polygoni* (Henry Edwards)(FW 7–12 mm; Plate 14.8, Plate 14.9, Plate 60.22) is a frequently collected western sesiid. It is markedly polytypic, the extent and color of scaling varying among populations, from primarily hyaline wings to fully scaled FW and hyaline HW in males, or fully scaled FW and HW in both sexes, and from almost entirely steel blue to red at base of wings, red HW in females, and variable red abdominal bands. Ten synonyms by Henry Edwards for his own species speak of an era of typological descriptive taxonomy, without regard for variation within species. The larvae tunnel in the basal stems and roots of numerous species of *Eriogonum* and *Polygonum paronychia* (Polygonaceae). Our rearing from prickly phlox, *Leptodactylon* (Polemoniaceae), in Inyo County, California, 40 years ago warrants confirmation. *Synanthedon polygoni* occurs throughout the western states from western

Wyoming and Colorado to the Pacific Coast and north to Alaska. The flight period ranges from April and May on coastal dunes to June to August in northern and montane localities.

Synanthedon novaroensis H. Edwards, the "Douglas-fir pitch moth," is one the largest *Synanthedon* species (FW 10–12 mm; Plate 14.5), steel blue with variable extent of bright orange on tegulae, legs, and abdominal bands. It occurs in conifer-forested regions of the Pacific Northwest, from Montana to Alaska and south to the San Francisco Bay area and mountains of southern California. The larvae bore in the cambium layer and wood of spruce *(Picea),* Douglas-fir *(Pseudotsuga),* and several species of pines (Pinaceae). Emergence occurs from March to September, primarily June to July. We have seen *S. novaroensis* flying in June above 7,000′ on Mt. Shasta, with extensive patches of snow still on the ground.

Podosesia syringiae (Harris)(Plate 14.10) is known as the "ash borer" or "lilac borer," based on the larval food preferences, which include *Fraxinus* and other native Oleaceae, as well as introduced ashes and lilacs. The FW (10–17 mm) is nearly opaque, brownish basally and in western populations orange on the costa and dorsal margin, segment A4 yellow, A5 to A7 and anal tuft mostly red. This species probably was native over eastern North America, west to eastern Alberta and the Rocky Mountains, and in the 1980s *P. syringiae* was discovered in urban ash in Yakima, Washington, and olive *(Olea europaea)* in the Sacramento Valley, California. In Colorado, adults fly from late April to late June.

The genus *Carmenta* has 30 described species in America north of Mexico, of which 22 occur in the West, primarily in arid regions of the Southwest into Mexico. The moths are superficially similar to *Synanthedon,* mostly black with hyaline wings, abdomen often with yellow bands, and genitalic differences separate the two genera. *Carmenta mariona* (Beutenmüller)(Plate 14.11) is a widespread but relatively rarely collected species, larvae of which feed in roots of Boraginaceae. The moth is small (FW 7–10 mm), greenish black with tegulae and a contiguous lateral line on the thorax bright red orange. Females' wings are fully scaled except for HW basally, with a red band along the dorsal margin of FW; male FW partially and HW fully hyaline. This species occurs in Nebraska, Kansas, Colorado, New Mexico, and Arizona. Engelhardt reared it from roots of fiddleneck, *Amsinckia,* in Oak Creek Canyon, and adults were reared from *Lithospermum* in southern Arizona by G. Muenchow. Flight records are June to early August. *Carmenta verecunda* (Henry Edwards)(Plate 14.12) also feeds in roots of Boraginaceae, having been reared from *Lithospermum ruderale* in Washington, and other borages probably are used over the broad range from the western Great Plains, Rocky Mountain region south to western Texas, eastern Arizona, the Cascades and Sierra Nevada, to southern California. The moth is brownish black with whitish or straw-colored markings, including bands on posterior margin of A2, A4, A6, and A7, and the FW is white between the black veins. *Carmenta mimuli* (Henry Edwards)(Plate 14.13) is the most widespread *Carmenta,* extending from North Dakota to Montana, south through the Rocky Mountains to Oaxaca in southern Mexico, largely determined by pheromone trapping (Z,Z-ODDA). The FW (8–11 mm) is mostly hyaline, with broad distal margin scaled black, powdered with orange, and discal spot orange. The female has yellow orange labial palpi and more extensive orange scaling on the wings. F. X. Williams noted adults alighting on *Chamaesaracha coronopus* (Solanaceae), and females were observed ovipositing on the same plant in Arizona by M. Cazier, according to specimen label data. *Carmenta prosopis*

(H. Edwards) is a small (FW 5–8 mm; Plate 14.14), nondescript, black species with hyaline wings but is interesting because the larvae feed as inquilines in galls induced by other insects on leguminous desert shrubs. Engelhardt reported adults reared from stem galls caused by encyrtid wasps on mesquite, *Prosopis glandulosa*, and from stem galls on *Mimosa biuncifera* in Texas and Arizona. *Carmenta prosopis* occurs from western Texas to southern Arizona and south to Durango, Mexico. Adults have been collected from April to November, mostly June to August, when collectors tend to visit those areas. Equally specialized, **C. tecta** (H. Edwards) is a mistletoe stem borer. It occurs in mountainous parts of eastern Arizona and western New Mexico, associated with live oaks supporting colonies of *Phoradendron orbiculatum* (Viscaceae). Engelhardt captured a freshly emerged female still near its exit hole, and in cut mistletoe stems he found sesiid pupae. The moth is small (FW 7–10 mm) with hyaline wings, FW nearly opaque in the female, powdered with orange scaling on the veins and between them distally. The adults have been recorded mostly in May to June and August to September, so *C. tecta* may be double-brooded.

Penstemonia has five described species, larvae of which feed in roots of *Penstemon* (Scrophulariaceae), and all occur in the West. They resemble species of *Synanthedon* and *Carmenta* but are distinguished by having rudimentary, evidently nonfunctional proboscis, hence are not found visiting flowers and may be short-lived. Genitalia, Figs. 121, 122. **Penstemonia clarkei** Engelhardt (FW 5–10 mm; Plate 14.15) occurs from western Montana and southern British Columbia to Monterey County, California, and the Sierra Nevada. Earlier records of this species in the mountains of eastern California refer to the later-described **P. pappi** Eichlin. Both are similar to *Synanthedon bibionipennis*, but *P. clarkei* has more extensive yellow scaling, including the opaque areas of FW between the veins and the abdominal banding, although both are variable. *Penstemon richardsonii* is a larval host of *P. clarkei* in Oregon, and this species was reared from an introduced scroph, Dalmatian toadflax, *Linaria genistifolia*, in northern California.

Species of **Alcathoe** are black with bright orange or black wings, and the males have a long, slender, scaled process from the tip of the abdomen, which in flight resembles the trailing legs of a tarantula hawk *(Pepsis)*. There are five species in America north of Mexico, of which two occur in the Southwest. **Alcathoe verrugo** (Druce) varies greatly in size (FW 8–15 mm; Plate 14.16) and color, body and legs shining bluish black, wings completely scaled, opaque, bright vermillion orange with brownish on veins to all brown black or blue black, and both color forms as well as intermediates originate from the same larval collections, according to Engelhardt. *Alcathoe verrugo* was described originally from Sonora, Mexico, and it ranges to southern California, thence north to the Santa Lucia Mountains. The larvae of *Alcathoe* are borers in roots and stems of Virgin's bower, *Clematis* (Ranunculaceae), and *C. ligusticifolia* is a recorded host of *verrugo*.

Superfamily Cossoidea

FAMILY COSSIDAE

"Goat moths" are small to large and hawk moth–like (FW 0.5–7 cm, rarely to 12.5 cm, females usually much larger than males), belying their membership in "microlepidoptera."

Adult   These are robust, heavy-bodied moths, with short, usually strongly bipectinate antennae, the branches (rami) longer in male; in the subfamily Zeuzerinae bipectinate to

about half their length, then filiform. The proboscis not functional, reduced to a small triangular lobe; labial palpi three-segmented, short, upturned, reaching middle of frons or less. Wing venation more complete than in related superfamilies, with median vein branched within the cell, which with the basic format of male genitalia structure and larval characters, especially the primary setal pattern and arrangement of crochets on the prolegs, is shared by the more ancestral lineages of Tortricidae. Cossids typically are nocturnal, drab, mostly gray with black striae; a few are brown or have orange patches, and some Neotropical species have metallic markings. The eggs are usually laid in crevices or under bark with an extensible ovipositor and may be produced in vast numbers—18,000 counted for one species in Australia.

Larva   First instars bore into branches or trunks of living shrubs or trees, sometimes living gregariously; later instars are stout, cylindrical, with heavily sclerotized head and mandibles, unpigmented integument. Larvae require 1 to 4 years to mature.

Larval Foods   Recorded hosts include at least 17 families of angiosperms, including one monocot, with woody legumes (Fabaceae) accounting for 25% of the records. More than 20% of the species are polyphagous and most may be.

Diversity   Cossids occur worldwide with more than 670 described species placed in 83 genera, greatest numbers in tropical regions (40% Neotropical). There are about 50 species in America north of Mexico, 75% of which are western. Because they are large, nearly all Nearctic species in collections have been described. However, there has been no comprehensive study of Nearctic Cossidae employing genitalic characters or molecular analyses, leading to a sound phylogenetic classification.

SUBFAMLY HYPOPTINAE

The primarily Neotropical subfamily Hypoptinae is represented in the United States, by about 20 species in three genera, occurring in southern Arizona, Texas, and Florida. Its species are characterized by having a cross-vein between the anal veins in the distal one-fourth of the FW. **Hypopta palmata** (Barnes & McDunnough)(Plate 13.8) is gray with antennal scaling, notum, underside; patagia, tegulae, and margins of thorax posteriorly black. FW (male 13–15.5 mm, female to 20 mm) shaded with smoky brown on posterior half, except there are elongate white patches. The veins, especially M and Cu, are sharply outlined in black. This species occurs in the deserts of Sonora, western Arizona, and presumably Nevada, and in California, mountain ranges of the Mojave, west to the San Jacinto Mountains and north to the Owns Valley. The adults fly at the end of June to early September.

Givira includes 15 or more species, possibly several undescribed, most of which occur in the Southwest. These are small, smooth-scaled, almost shining, mostly unicolorous gray to brownish moths. The larval biology evidently is unknown for most species. **Givira theodori** Dyar is small but varies in size (FW 9–14 mm; Plate 13.11) and extent of dark scaling. Head, thorax, and FW white, basal half with minute black dots along margins, becoming more distinct distally; in spaces between veins beyond cell there are diffuse, purplish brown patches separated by white along veins, forming a clouded transverse band from the costa subapically toward the tornus. This maculation varies from dark and broad to pale brown and diffuse. *Givira theodori* occurs from central Texas to Arizona, with collection records May to July. **Givira marga** Barnes and McDunnough is pale brownish gray, FW (11.5–15 mm, females rare in collections, may range larger; Plate 13.9, Plate 13.10) almost uniform velvety gray, shading to whitish beyond cell in fresh specimens, with faint, sparse pale strigulae obscured in the pale area, and there is a distinct, white comma-shaped mark at end of the cell. This species was described originally from Santa Catalina Island, California, and it occurs on Santa Cruz Island, but it is not an insular endemic, ranging on the mainland from the Los Angeles Basin in foothill situations to Santa Clara County and inland in the San Jacinto Mountains and along the Kern River. Flight records are end of June to September, mostly July. *Givira lotta* Barnes & McDunnough, the "pine carpenterworm," is similar and occurs in Arizona and Colorado, probably throughout pine-forested regions of the Southwest. The larvae mine the rough outer bark at the base of ponderosa pine.

SUBFAMILY COSSINAE

The subfamily **Cossinae** includes most of the western goat moths. The "aspen carpenterworm," *Acossus populi* (Walker)(Plate 13.13), and *A. undosus* (Lintner) form a complex that needs resolution. Possibly three or more species are involved. These moths are large (FW male 2.5–3 cm, female to 4 cm) and broad winged, especially the females, generally pale gray but ranging from white ground to heavily mottled with dark gray. Typically there is a black line transversely crossing the cell, more distinct than other transverse strigulae. The larvae are large grubs, 3.5 to 4 cm long, white with a dark head capsule, thoracic shield, and pinacula. They tunnel in the bole of aspen, *Populus tremuloides*, and other *Populus* (Salicaceae), sometimes causing infested trees to break off. *Acossus populi* sensu lato ranges from the northeastern United States and southern Canada though the mountains of the western states to eastern California. Adults are recorded in June to August.

Comadia includes a dozen described species in North America, all in the southwestern states except one introduced from Europe. They are phenotypically similar, medium-sized cossids with triangulate gray to brownish FW and often obscure darker maculation. The males vastly outnumber females at lights. *Comadia bertholdi* (Grote)(Plate 13.12), with three subspecies, is the most widespread, occurring from western Colorado and Utah to New Mexico, Arizona, and California. FW (14.5–16.5 mm in male, to 18 mm in female) pale gray without transverse strigulae, veins in the male are black scaled, especially M in the cell, ending at a black smudge at lower corner of cell; the female is larger with reduced blackish scaling, so markings are obscure. Female antennal pectines are reduced, serrate. The flight occurs in June and July in Colorado. J. J. Rivers found larvae of the species later described as *C. intrusa* Barnes & Benjamin (identified as *C. bertholdi* by Rivers) feeding in the main stem and larger roots of a bush lupine (Fabaceae) in southern California. The larva is yellowish white, gradually becoming red as it matures, which may require more than one season because Rivers found larvae of several ages in the same shrub. A similar species, *Comadia suaedivora* R. M. Brown and Allen, occurs in the San Joaquin Valley, California, where its larvae feed in the crown and roots of *Suaeda fruticosa* (Chenopodiaceae), a woody, halophytic shrub. The bulky larvae attain 3 cm in length and have rose lavender highlights that are lost in preserved specimens. They live gregariously, several hollowing out the crown area, usually killing the host, an unusual habit, as wood-boring larvae of other insects often bite and kill a neighboring larva when contacted during feeding. *Comadia suaedivora* flies in May and June.

The "carpenterworm," *Prionoxystus robiniae* (Peck), is the largest and best-known American goat moth, FW (females 3–3.5 cm; Plate 13.15, Plate 13.16) broader than those of *Acossus*, FW heavily and more or less uniformly mottled with black, with a poorly defined blotch at outer, lower corner of cell and lacking the transverse line across the cell seen in *Acossus*. Males are smaller (FW 2.3–2.8 cm; Plate 13.5) with narrow wings and are easily recognized by their rust orange HW bordered by black. When fully grown, the larvae are enormous grubs up to 7.5 cm long, pale with dark brown head and thoracic shield. They tunnel into the solid wood of a wide variety of hardwoods, including orchard and ornamental trees and are reported to require three years to complete development, with guilds of different age classes sometimes feeding in the same tree. *Prionoxystus robiniae* occurs across the continent and in the West is recorded in Colorado, Utah, Nevada, California, and Arizona. Flight records in the West range from the end of January to July, mostly April to May.

Miacora perplexa (Neumoegen and Dyar) is large (male FW 1.7–2.1 cm, female 2–2.4 cm; Plate 13.14), having gray wings tinged with brown, especially the posterior half of FW basally, and with well-defined transverse, black lines across the cell and beyond it, forked at the middle to form a double-Y-shaped pattern, and there are numerous small, transverse strigulae evenly distributed over the wing. HW pale gray without strigulae. This species is widely distributed in western montane regions, from Colorado to southern Arizona, Oregon, and in the Sierra Nevada, Coast Ranges, and Transverse Ranges of southern California. Adults have been collected from late June to September, mostly in July. The larval biology is unknown.

SUBFAMILY ZEUZERINAE

The subfamily **Zeuzerinae** is primarily tropical and is represented in America north of Mexico by the "leopard moth," *Zeuzera pyrina* (Linnaeus), which is introduced to the East Coast from Europe, and by two species in southern Texas and southern Arizona. *Morpheis clenchi* Donahue flies in July in the Oro Blanco–Pena Blanca area, Arizona, and probably occurs in Sonora and southward in Mexico. Zeuzerine cossids are distinctive in having the male antennae bipectinate on the basal half only, with long, downward curved rami, then abruptly narrowed and serrate to the tip. *Morpheis clenchi* is moderately large, sphingidlike, with narrow wings, male FW (2.8–3.6 cm) whitish with a longitudinal, blackish brown stripe along the costa basally, through the outer half of the cell to the termen below the apex. The female is unknown but is assumed to be larger with uniformly serrate antennae. This species is similar to *M. cognatus* (Walker), which has stronger transverse strigulae on the FW (a *Morpheis* species from Costa Rica is illustrated, Plate 13.17) and ranges from southern Mexico to Brazil. The larvae of *Z. pyrina* bore in the branches and trunks of a huge variety of trees and shrubs and have been recorded feeding on more than 125 plant species; *M. clenchi*, likely has similar habits.

References for the Smaller Nonobtectomeran Superfamilies of Apoditrysia

Barnes, W., and A. W. Lindsey 1921. The Pterophoridae of America, north of Mexico. Contributions to the Natural History of the Lepidoptera of North America 4(4): 280–283.

Barnes, W., and J. McDunnough 1911. A taxonomic revision of the Cossidae of North America. Contributions to the Natural History of the Lepidoptera of North America 1(1): 1–35.

Beutenmüller, W. 1901. Monograph of the Sesiidae of America north of Mexico. Memoir of the American Museum of Natural History 1: 217–315.

Brown, R. M. 1976. A revision of the North American *Comadia* (Cossidae). Journal of Research on the Lepidoptera 14: 189–212 ["1975"].

Dugdale, J. S., N. P. Kristensen, G. S. Robinson, and M. J. Scoble 1999. The smaller Microlepidoptera-grade superfamilies, pages 218–232, *in:* Kristensen, N. P. (ed.), Handbook of Zoology. Lepidoptera, Moths, and Butterflies, Vol. 1: Evolution, Systematics, and Biogeography. W. de Gruyter, Berlin, New York; x + 491 pages.

Edwards, D, P. Gentili, M. Horak, N. Kristensen, and E. Nielsen. 1999. The Cossoid/Sesioid assemblage, pages 181–197, *in:* Kristensen, N. P. (ed.), Handbook of Zoology. Lepidoptera, Moths, and Butterflies. Vol. 1: Evolution, Systematics, and Biogeography. W. de Gruyter, Berlin, New York; x + 491 pages.

Eichlin, T. D., and W. D. Duckworth 1988. Sesioidea, Sesiidae, *in:* Hodges, R. W. et al. (eds.) Moths of America North of Mexico 7.1. Classey and R. B. D. Publications, London; 176 pages, 6 plates.

Engelhardt, G. P. 1946. The North American clearwing moths of the family Aegeriidae. Bulletin of the U.S. National Museum 190; 222 pages.

Gaedike, R. 1977. Revision der nearktischen und neotropischen Epermeniidae (Lepidoptera). Beitrag Entomologische, Berlin 27: 301–312.

Gaedike, R. 1996. Epermeniidae, *in:* Heppner, J. B. (ed.) Lepidopterorum Catalogus. Fascicle 48. Association of Tropical Lepidoptera, Gainesville, FL; 20 pages.

Gielis, C. 1993. Generic revision of the superfamily Pterophoroidea (Lepidoptera). Zoologische Verhandelingen, Leiden 290; 139 pages.

Gielis, C. 2003. Pterophoridae and Alucitoidea. World Catalogue of Insects. Vol. 4. Apollo Books, Stenstrup, Denmark; 2005; 198 pages.

Heppner, J. B. 1977. A new genus and new assignments in the American Choreutidae (Lepidoptera: Sesioidea). Proceedings of the Entomological Society of Washington 79: 631–636.

Heppner, J. B. 1987. Copromorphidae, Alucitidae, Carposinidae, Epermeniidae, Glyphipterigidae (Copromorphoidea), pages 399–404, *in:* Stehr, F. (ed.) Immature Insects. Vol. 1. Kendall/Hunt, Dubuque, IA.

Heppner, J. B., and W. D. Duckworth 1981. Classification of the superfamily Sesioidea (Lepidoptera: Ditrysia). Smithsonian Contributions to Zoology 314; 144 pages.

Kuznetsov, V. I., and A. Stekolnikov 1981. Functional morphology of the male genitalia and phylogenetic relationships of some primitive superfamilies of the primitive infraorder Papiliomorpha (Lepidoptera: Sesioidea, Cossoidea, Zygaenoidea) of the Asiatic part of the U.S.S.R. Zoologicheskii Institut Akademya Nauk SSSR, Leningrad 92: 38–73. [English translation for Smithsonian Institution, 1988.]

Landry, B., and J.-F. Landry 2004. The genus *Alucita* in North America, with description of two new species (Lepidoptera: Alucitidae). Canadian Entomologist 136: 553–579.

Lange, W. H., Jr. 1950. Biology and systematics of plume moths of the genus *Platyptilia* in California. Hilgardia 19(19): 561–668.

Matthews, D. L., and T. A. Lott 2005. Larval hostplants of the Pterophoridae (Lepidoptera: Pterophoridae). Memoir of the American Entomological Society 76; 324 pages.

Neunzig, H. H. 1987. Pterophoridae (Pterophoroidea), pages 497–500, *in*: Stehr, F. (ed.) Immature Insects. Vol. 1. Kendall/Hunt, Dubuque, IA.

Walsingham, Lord T. de Grey 1880. Pterophoridae of California and Oregon. J. Van Voorst, London; xvi + 66 pages.

Superfamily Tortricoidea

FAMILY TORTRICIDAE

This is a large and relatively homogeneous family, with three subfamilies, Chlidanotinae, Tortricinae, and Olethreutinae, each of which, along with several subordinate taxa, at times has been treated as a family. Females possess modified ovipositor lobes (papillae anales) that have been rotated 90° from the ancestral, lateral position to form setose, flat, expanded pads facing ventrally, usually having the outline of shoe soles. The papillae anales are modified in Cnephasiini (Tortricinae) and a few other taxa, for scraping and carrying debris to spread over the eggs.

Adult   These moths are small to moderately large (FW mostly 3–25 mm), generally with rectangular FW and broader, plicate HW. The head scaling is long, tufted, and oriented downward from the crown over the upper half of the frons, very short, smooth on the lower half; the antennae usually are 0.5 to 0.6 FW length, filiform, with sensory setae in males typically short, but long in some groups; labial palpi are porrect or bent upward but not curved as in Gelechioidea. Most species are nocturnal with the FW cryptically colored in gray, brown, rust, or tan, held rooflike over the body when at rest, but some species have colorful markings. A few are spectacularly polymorphic, notably species of *Acleris* (Tortricini). A few tropical groups are diurnal and brightly colored. Genitalia, Figs. 134–154.

Biology   Because many tortricids are economically important as agricultural and forestry pests, there is a vast literature on the biology, ecology, host-plant selection, oviposition behavior, pheromone chemistry, and so on. For example, there are more than 6,000 literature references on spruce budworms (*Choristoneura fumiferana* species complex) in North America. The eggs of most tortricids are flat, scalelike, deposited singly by Chlidanotinae and Olethreutinae and the more ancestral tribes of Tortricinae, but in small to large, shingled masses (100–150 eggs) by females of derived tortricines, including some Cnephasiini and nearly all Archipini, Atteriini, and Sparganothini, most of which are polyphagous foliage feeders. Females of Atteriini, a Neotropical tribe with one species in Arizona, possess thick mats of specialized (corethrogyne) scales of two types on the abdomen, which they deposit on and as upright fences around the egg masses.

Larva   Typical of generalized Ditrysia, the larvae are cylindrical without numerous secondary setae, setal pattern and crochet arrangements similar to those of Cossidae; usually with little or no integumental pigmentation other than the prothoracic shield and setal pinacula. Most external feeders of all three subfamilies possess an anal fork, which is used to flip frass away from the larval shelter. Larvae of most Chlidanotinae and Olethreutinae are internal feeders (endophagous), borers or miners in stems, leaves, roots, buds, or fruits and seeds, which is postulated to be a derived behavior and therefore a reversal of the general trend in Lepidoptera. Larvae of a few species induce plant galls (e.g., Plate 60.26). North American Tortricinae mostly are external feeders that form leaf rolls or other shelters in foliage, but nearly all Cochylini are internal feeders. Pupation usually occurs in the larval shelter or gallery, although larvae of some species drop to the ground to pupate, especially those that diapause over the winter as pupae or pre-pupal larvae.

Larval Foods   Most tortricid species are specialists in food-plant selection, and an enormous array of plants serves as hosts. Several lineages in Olethreutinae and Tortricinae feed primarily on Asteraceae, and members of a few small tribes are specialists (e.g., Bactrini on monocots and Australian Epitymbiini on fallen *Eucalyptus* leaves).

Diversity   The world fauna of this large superfamily recently was cataloged by John Brown and others. There are about 9,000 described species in 1,000 or more genera and incalculable numbers unnamed in tropical regions, for example, 70 to 80% of the species treated in several recent monographs of Neotropical tortricines have been previously unnamed. Rich faunas occur in all

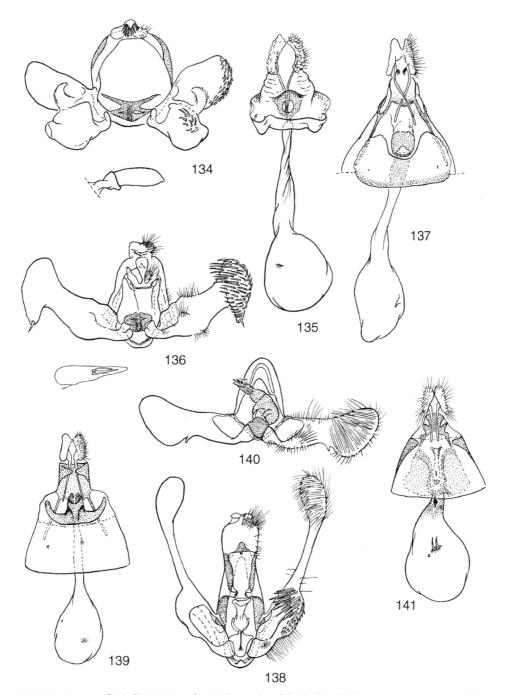

FIGURES 134–141. Genitalia structure of exemplar species of Tortricidae, Olethreutinae, ventral aspect. **134,** *Bactra miwok* Powell ♂, aedeagus below. **135,** *Bactra miwok* ♀ [Powell 1997]. **136,** *Eucosma langstoni* Powell ♂, aedeagus below [Powell 1963]. **137,** *Eucosma monoensis* Powell ♀ [Powell 1968]. **138,** *Pseudosciaphila duplex* (Walsingham) ♂, aedeagus in situ. **139,** *Olethreutes deprecatorius* (Heinrich) ♀. **140,** *Cydia ponomella* (L.) ♂, aedeagus in situ. **141,** *Cydia latiferreana* (Walsingham) ♀ [Heinrich 1926].

biogeographic regions. In addition to the spruce budworms *(Choristoneura),* economically important Tortricinae include the "fruit-tree leaf-roller" *(Archips argyrospilus)* and other orchard pests in North America, and the "light brown apple moth" *(Epiphyas postvittana)* in Australia, which is recently introduced in California. Notorious Olethreutinae include the "codling moth" *(Cydia pomonella,* the ubiquitous worm in apples); "pea moth" *(C. rusticella)* and "Oriental fruit moth" *(Grapholita molesta,* all

of which are introduced to North America); several conifer feeders, larch and spruce budworms *(Zieraphera),* seed and cone borers *(Cydia),* and pine tip borers *(Rhyacionia).*

SUBFAMILY OLETHREUTINAE

This is the major lineage of Tortricidae in North America, with more than 870 described species assigned to about 60 genera.

Western species are grouped into four tribes, Bactrini, Olethreutini, Eucosmini, and Grapholitini, each of which has characteristic morphological and biological features. Genitalia, Figs. 134–141. The vast majority of species are specialists in larval host-plant selection, with most Olethreutini and basal Eucosmini feeding externally, and derived Eucosmini and Grapholitini internal feeders, stem, root, bud, and seed borers. Diverse kinds of gymnosperms and angiosperms are used.

TRIBE BACTRINI

The tribe Bactrini, with more than 100 species worldwide in seven genera, is dominated by the genus **Bactra,** which is most diverse in the Southern Hemisphere, especially in the Indo-Australian and African regions. There are lesser Neotropical and Nearctic radiations. Nine species occur in America north of Mexico, six of them in the western states, of which two or more are introduced to our region. The adults are distinctive, superficially resembling crambine moths, with relatively narrow FW lacking a costal fold in males and having an acute apex and linear markings, usually tan or brown with longitudinal striae, sometimes a darker broad band through the middle of the wing or on the dorsal half ("striate type" of Diakonoff). Some species are polymorphic, having two or more forms, which are parallel among species, so genitalia characters are needed for accurate identification (e.g., Figs. 134, 135). All *Bactra* for which larvae are known are monocot feeders, and their phenotype is an adaptation parallel to grass-feeding crambines that affords crypsis while resting on sedges and rushes. **Bactra verutana** Zeller is a smallish *Bactra* species (FW 6–9.5 mm; Plate 14.17, Plate 14.18) and is relatively nonvariable, having the FW tan with dark discal maculation, strongest at the end of the cell, more obscure in females; there is a rare form with median stripe or dorsal half dark brown. This species is widespread in Central and South America and the southern United States. Heinrich described the Rocky Mountains and Arizona populations as a variety, **B. v. albipuncta,** which Diakonoff later treated as a sibling species, evidently based on slight genitalia differences. *Bactra verutana* has been reared from *Cyperus esculentus* (Cyperaceae) in Florida and Cuba, and in California this native species has been employed as a biological control agent against a Eurasian weed, purple nutsedge, *C. rotundus.* Adults fly from April to October, most numerous in September and October. **Bactra furfurana** (Haworth)(Plate 14.19) is a Holarctic species that occurs in the Atlantic states, Texas, and California and is found in association with pond vegetation. The FW (7–8 mm) differs from that of other *Bactra* by having "fasciate type" markings, transverse, irregular, and variable transverse bands on a paler brown ground. The larvae are reported to feed in stems of rushes (*Juncus,* Juncaceae) and rushes (*Eleocharis* and *Scirpus,* Cyperaceae) in Europe and New York.

Olethreutini are typically broad winged, FW with broad, transverse fasciae of various forms, and males lack a costal fold. Worldwide, this is a major tortricid radiation, with more than 1,400 described species in about 150 genera, richest in Southeast Asia, the Palaearctic, and Neotropical regions. In the Holarctic the larvae feed primarily on Betulaceae, Salicaceae, Ericaceae, and Rosaceae, usually as external leaf tiers, which is interpreted by Richard Brown as an ancestral trait within the subfamily. There are about 175 described species in America north of Mexico, but in marked contrast to the other tortricid lineages and to most Lepidoptera, Olethreutini are poorly developed in the West, with only about 20% of the species represented. Many other species occur across Canada to the northern Rocky Mountains and Alaska but not into British Columbia and the western United States.

Endothenia and a few related genera are sometimes designated as a separate tribe, based on genitalia differences. **Endothenia hebesana** (Walker), the "verbena bud moth," occurs in much of eastern North America, in Texas, and on the Pacific Coast from California to British Columbia. The moths are brown, FW (6.5–9 mm; Plate 14.20) with subtle darker pattern dominated by a black smudge in the costal half of the median fascia, crossing the end of the cell. The larvae bore into inflorescences and eat immature seed of a variety of herbaceous plants, especially Scrophulariaceae, including *Antirrhinum, Castilleja, Orthocarpus, Penstemon, Verbascum,* and *Verbena,* but they also are recorded from unrelated plants, including *Stachys* (Lamiaceae), *Solidago* (Asteraceae), and even pitcher plant (Sarraceniaceae). F. Arias reared *hebesana* from a larva boring in the inflorescence scape of *Scrophularia californica.* Pupation often occurs in the seed capsule, and in coastal California at least two generations develop without diapause, with larval collections from *Castilleja* in May and August and flight records from March to September (mostly June to July). **Endothenia rubipunctana** (Kearfott) also is widespread in the western states but is a larger species (FW 7–10.5 mm; Plate 14.21), FW pale grayish brown with poorly defined, darker transverse fasciae and a bluish lead-colored blotch in the cell, broadly whitish suabpically. The larvae feed in seed pods of native *Iris macrosiphon* (Iridaceae) in California and overwinter as prepupal larvae in the seed; flight occurs in April to July.

Episimus is a tropical and subtropical genus with nine species recorded in the United States, mostly in south Florida. **Episimus argutanus** (Clemens)(Plate 14.22) is widespread from the Atlantic states to the Pacific Coast, recorded in Texas, New Mexico, Colorado, and California. The moth is small (FW 6–8 mm) with rectangular FW, dull reddish brown mottled with dark brown, steel blue, and dull rust in the ocellar spot to the apical area. The larva is a leaf roller or tip tier in a variety of plants, including Asteraceae *(Solidago),* Euphorbiaceae, Hamamelidaceae, Rosaceae *(Crataegus),* and especially Anacardiaceae *(Rhus, Toxicodendron).* Flight records in California span March to July.

Moths of the genus **Apotomis** are broad winged, black-and-white or gray. The larvae are foliage feeders on Betulaceae and Salicaceae There are 17 species in North America, of which seven occur in the West, mostly as extensions of eastern and northern distributions. **Apotomis removana** (Kearfott) is a dark gray species, FW (9.5–11 mm; Plate 14.23) densely mottled blackish, having variable paler costal and subterminal areas. This is the most widely distributed Nearctic *Apotomis,* ranging from Nova Scotia and the Great Lakes region across Canada and through the Rocky Mountain states to northern New Mexico, Arizona, and eastern California. The larvae feed on aspen *(Populus tremuloides),* according to Adamski and Peters. Adults are recorded May to July. **Apotomis capreana** (Hübner) occurs in the Palaearctic and across Canada to British Columbia and Washington, southward to northern New Mexico and Arizona and northern California. The FW (9–10 mm; Plate 14.24) is darkly mottled basally to the end of the cell, blotched with white subapically. Larvae are recorded feeding on foliage of willow, aspen, and birch (Betulaceae), with a single flight in July and August.

Pseudosciaphila duplex (Walsingham) occurs transcontinentally and is one of the largest (FW 10–13 mm; Plate 14.25, Fig. 138) and most prevalent tortricids throughout aspen-forested areas of the West. There are two principal FW patterns, whitish with a broad basal blotch, median fascia, and terminal patch of brownish gray, and a form having the entire FW dark gray or brownish gray, with occasional intermediate forms. The larvae create conspicuous leaf rolls on aspen (*P. tremuloides*) and rarely other Salicaceae, as well as birch (Betulaceae, e.g., 17 records, fewer than 1.5% of 1,128 rearing records reported by Prentice in Canada). *Pseudosciaphila duplex* is univoltine, with flight in June to early August.

The genus **Olethreutes** occurs on all continents and is richly represented in the Nearctic, with about 90 species, especially in deciduous forest areas of eastern North America. About a dozen are recorded in the West. *Olethreutes cespitana* (Hübner)(Plate 14.26) is a Holarctic species that occurs in Europe and is widespread in eastern North America to Utah and the Sierra Nevada through the Northwest. This species is smaller than typical for the genus, FW 7.5 to 9 mm in western specimens, and only 5.5 to 6 mm in north coastal California populations, similar to eastern and midwestern specimens. FW dark brown with irregular, yellowish, transverse lines; HW dark brown. The larvae are leaf tiers on herbaceous plants (e.g., *Trifolium*, Fabaceae; *Fragaria*, Rosaceae, even grass, according to Heinrich), as well as trees (*Aesculus*, Hippocastanaceae; *Populus*). Adults are recorded from May to September, presumably two or more generations. *Olethreutes glaciana* (Moeschler) (Plate 14.27) is a northern species, ranging from Labrador to North Carolina and across Canada to Alaska, southward in the Rocky Mountains at higher elevations to northern Arizona. The FW (7–8 mm) is banded with black and whitish. The larvae roll leaves of *Betula*, *Populus*, and *Acer*, among others, according to W. E. Miller. *Olethreutes deprecatorius* Heinrich flies in wet mountain meadows of the Pacific states and British Columbia. The moth is moderately large (FW 8–10.5 mm; Plate 14.28, Plate 14.29, Fig. 139) with broad, tan or yellowish brown transverse FW bands. Adults fly in association with corn lily (*Veratrum*, Liliaceae) in northern California, and we found the black larvae of *O. deprecatoria* feeding on *Veratrum* leaves in June. Prentice reported one specimen from southeastern Manitoba reared from aspen, an identification that needs confirmation. *Olethreutes chalybeana* (Walsingham) is a small (FW 6.3–7.3 mm; Plate 14.30), dark brown species. It was described from the Siskiyou Mountains on the Oregon–California border and subsequently has been reported from British Columbia and Montana. The larvae feed in May and June on *Spiraea douglasii* (Rosaceae) at Mt. Shasta and in the northern Sierra Nevada. *Olethreutes chalybeana* has been treated as a subspecies of the Palaearctic *O. siderana* (Treitschke).

Hedya is a small group of species separated from *Olethreutes* primarily by having two signa in the female corpus bursae rather than one or none. Superficially the moths resemble various species of *Olethreutes*. *Hedya ochroleucana* (Frölich)(Plate 14.31) is Holarctic in distribution, Europe and across North America, including the western states north of Texas, New Mexico, and Arizona. The FW (8–11 mm) is brownish gray to black on the basal two-thirds, cream white or pale tan distally, divided by a nearly straight line through the end of the cell. The larvae are leaf rollers on Rosaceae, including *Rosa*, *Malus*, and *Sorbus*. Adult captures recorded from May through August.

TRIBE EUCOSMINI

Eucosmini is by far the most diverse and species-rich tortricid tribe in America north of Mexico, with about 600 species in 30 genera. Most are relatively narrow winged moths that perch with the wings tightly appressed to the body. Nearly all Eucosmini are specialists in larval feeding, preferring closely related plants of one genus or family. The majority are internal feeders, borers in roots or stems (a few induce galls), often in Asteraceae, or under confer bark or in cones. However, larvae of the *Epinotia-Catastega* Lineage form shelters in the newly developing foliage of a wide range of plants. Eucosmini have independently adopted conifer-feeding as a mode of radiation several times, including *Rhyacionia*, *Retinia* (=*Petrova*), *Barbara*, the Bobana Group of *Eucosma*, and two or more species groups within *Epinotia* and *Zeiraphera*. *Rhyacionia* and *Retinia* have been associated in check lists, but we lack phylogenetic evidence to indicate they are sister groups with a shared origin as conifer feeders. The moths of both genera, as well as *Eucosma bobana* and related species and some unrelated taxa of Tortricinae, have converged in phenotype, with the FW orange or rust colored, checkered with tan or brown, so as to resemble dead conifer foliage and catkins when the moths are perched, or they have the FW transversely striate with gray on basal two-thirds, appearing bluish, with the distal area rust.

Rhyacionia are the pine tip moths; their larvae feed in the new terminals of pines, causing distortion and dieback of the tips. There are about 40 species of *Rhyacionia* worldwide, primarily Holarctic, Caribbean, and northern Neotropical, with the greatest recorded diversity, 16 species, in the southwestern United States and northern Mexico. Our knowledge of the western species was greatly enhanced by a U.S. Forest Service survey during the 1970s employing synthetic sex pheromone traps extensively in coniferous forested areas. The larval damage stunts growth and is of economic concern, especially in pine plantations and Christmas tree farms, notably by the "European pine shoot moth," **R. buoliana** (Denis & Schiffermüller)(Plate 14.32), a Palaearctic species that was introduced into northeastern North America in the early 1900s. Later *R. buoliana* spread to the Pacific Northwest, where it ranges from southern British Columbia to south coastal Oregon and, in contrast to native species, feeds on many species of the subgenus *Pinus*. The adults are bulkier than native *Rhyacionia*, with a broad FW (8–11 mm, females range larger than males) bright rust with blurred whitish blotches through the subcostal, dorsal, and subterminal areas. *Rhyacionia buoliana* is univoltine in our area, with flight primarily in June. **Rhyacionia zozana** (Kearfott)(male FW 10–11 mm, female FW 8–9 mm; Plate 14.33), the "ponderosa pine tip moth," has bluish gray FW with a rust orange preapical blotch. Males are larger than females, typical for Nearctic *Rhyacionia*. This is the most widely distributed western pine tip moth, occurring in New Mexico, Colorado, Utah, and Montana and from southern British Columbia to the central Sierra Nevada in California. Its larvae feed on lodgepole pine, *P. contorta*, in the northern Rocky Mountains, on *P. ponderosa* and *P. jeffreyi* in California, and on pines of many species in plantations. This is one of eight *Rhyacionia* species known to feed on ponderosa pine. Their flight is mid-March to mid-May in Oregon and California, timed with early new terminal growth of the host pines. The "southwestern pine tip moth," **R. neomexicana** (Dyar), is our largest species (male FW 12.5–13.5 mm, female FW 8.5–10 mm; Plate 14.48, Plate 14.49). It ranges from northern Arizona and New Mexico through the Rockies to Montana and western Nebraska. Adults are similar to *R. zozana* but have pinkish buff covering the distal part of the FW, from the tornus to midtermen. Its habits also mirror those of *R. zozana*: flight from late March in northern Arizona to July in Colorado. *Pinus ponderosa* is the larval host. Whereas most *Rhyacionia* are restricted to pines in the subgenus *Pinus*, a few are known or suspected to feed on pinyon pines (subgenus *Strobus*). One of these, **R. monophylliana** (Kearfott)(Plate 14.34), flies in May in the mountains bordering the Mojave Desert in California. The FW (8–8.5 mm in males) is paler than that of other *Rhyacionia*, the basal area bluish gray with whitish striae, the remainder of the wing pale ochreous with terminal fringe dark reddish.

Species of **Retinia** (=*Petrova* of authors) are called the pitch-nodule or pitch-blister makers because the larvae mine in bark of living conifers, creating a hollow shelter covered by pitch and silk, usually at the base of the current season's growth (Plate 60.25). These moths are larger than *Rhyacionia*, and some species feed on pines, others on Pinaceae other than *Pinus*, or on cypress (Cupressaceae). There are 16 named species in America north of Mexico, of which seven occur in the West, along with several undescribed races or species. **Retinia albicapitana** (Busck), the "northern pitch twig moth" (Plate 14.35), ranges from the Maritime Provinces to eastern British Columbia, Montana, and Idaho. The head scaling is white, FW brick red orange with steel gray transverse striae. The larva is a solitary nodule-maker on various pines, especially jack pine, *Pinus banksiana*, across Canada and *P. contorta* in Montana and Idaho. **Retinia picicolana** (Dyar) is a large species (FW 13–17, rarely 10 mm; Plate 14.50) having pale gray FW with dark gray transverse striae and pretornal and apical markings. *Retinia picicolana* flies in May to early July in the mountains of the Pacific Northwest south to the southern Sierra Nevada. The larvae feed on fir *(Abies)*. **Retinia sabiniana** Kearfott has pale orange FW (9–13.5 mm; Plate 14.51) with broad, whitish, transverse bands. It is associated with lower Transition Zone pines, *P. coulteri* and *P. sabiniana*, from southern Oregon through much of California. The "metallic pitch nodule moth," **R. metallica** (Busck), is a a smaller species, FW (6.5–9 mm; Plate 14.37, Plate 14.38) dark gray crossed by numerous lead-colored striae. The larvae feed at the base of the current season growth on lodgepole and ponderosa pines (Plate 60.25) from the northern Rockies in the Yukon Territory, British Columbia, Alberta, and Montana, to Colorado and the central Sierra Nevada. Larvae collected in the winter yield adults in March and April, but the normal flight period may be May to June.

Barbara colfaxiana (Kearfott), the "Douglas-fir cone moth" (Plate 14.39, Plate 14.52), is widespread in western montane areas, where its larvae feed in cones of Douglas-fir (*Pseudotsuga menziesii*, Pinaceae) and true fir *(Abies)*. The FW (7.5–11 mm) is checkered with pale and dark gray, and the termen is narrowly rust orange. Adults fly from April to September, primarily June to July. *Barbara colfaxiana* varies in size and color pattern, which led to varietal and race names, including erroneous confusion with *Eucosma siskiyouana* (Kearfott) in early literature.

Spilonota ocellana (Denis & Schiffermüller)(Plate 14.40) is the "eyespotted budmoth," a pest of apples. It was introduced into eastern North America from Europe before 1860, then became widely distributed in the Northeast and possibly by separate introduction on the Pacific Coast, in British Columbia before 1923, Montana, Idaho, and the Pacific states. The moth is small (FW 5.5–8 mm), and the male antenna has a deep notch dorsally near its base. The FW is gray with a broad, cream-colored median band and bluish gray–clouded tornal area. *Spilonota ocellana* is recorded feeding on a variety of plants, primarily Rosaceae, but also oak and "laurel."

The genera *Phaneta* and *Eucosma*, along with *Pelochrista*, *Sonia*, *Suleima*, and *Epiblema*, comprise an extremely speciose complex for which no phylogenetic classification has been proposed. All have similar male genitalia, upon which Heinrich based the classification in 1923, and share similar larval habits,

primarily as borers in stems or roots, less commonly inflorescences, usually of Asteraceae. Species of one group of *Eucosma* feed in cones of Pinaceae. This lineage warrants comprehensive study by an ambitious student, including assessment of the female genital characters, which were omitted by Heinrich. The two largest genera are separated primarily on the basis of a costal fold, absent in males of *Phaneta*, present in *Eucosma*, a character state that is unstable among closely related species in other tortricids. Many species have an "ocellar spot"—a pale area above the tornus that contains one or two rows of raised, metallic black or gray dots. More than 280 described species of *Phaneta*, *Eucosma*, and *Pelochrista* occur in America north of Mexico, about 60% of them in the western states, and many other species remain unnamed.

Probably most species of this lineage are specialists in larval hosts, but we lack data for many and lack multiple records for most of those that have been reared. We conducted an extensive survey of tortricid borers in woody Asteraceae in the late 1960s, part of broader investigations on microlepidoptera biologies, funded by the National Science Foundation. We reared about 20 species of *Phaneta-Eucosma-Sonia* root borers from more than 60 collections, about half of them from multiple localities. That study remains the primary source of host data for western members of this guild. We found one widespread species, *E. ridingsana*, feeding on several genera of Asteraceae, whereas other species for which we had multiple records were host specific. A species of *Eucosma* and a *Sonia* or *Phaneta* often occupied the same roots, but two congeneric species did so only once.

Phaneta contains more than 100 described Nearctic species, of which more than half occur in the West. The adults on average are smaller than those of *Eucosma* and tend to be less colorful, with FW tan, rust, brown, or gray, often without distinctive maculation other than the ocellar spot. The recorded larval host plants are Asteraceae. Larvae of only about 20% of described species are known, feeding as vegetative shoot, stem, or root borers. **Phaneta corculana** (Zeller), which has dark brownish gray FW (6.5–9 mm; Plate 14.41) with a sharply contrasted, white ocellar spot, is one of our most widespread species, ranging from Vancouver Island to Colorado and California. It is primarily a boreal species, occurring at Crater Lake and in the high Sierra Nevada at 8,000 to 10,000' and the White Mountains of eastern California to 14,000'. The larval biology is unknown, but we found adults associated with *Wyethia angustifolia* at Richmond on San Francisco Bay. Adult captures are mostly June to July but April near the coast. **Phaneta amphorana** (Walsingham), which occurs from northern Oregon to southern California, has two seasonal forms at Antioch on the San Joaquin River delta (Plate 14.42, Plate 14.43). Early spring moths average larger (FW 7–9.5 mm) than summer ones (to 8 mm), and southern populations larger than north coastal moths (FW 6.5–7.5 mm). Clarke reared *P. amphorana* from *Grindelia* in coastal Washington. At Antioch the larvae feed on *G. camporum*, in the new vegetative terminals in spring, giving rise to summer moths that have lemon yellow FW with grayish brown terminal area. Presumed second-generation larvae feed in the flower heads in fall, from which dark brown adults emerge in early spring. At other sites, however, moths collected in May or July to November are brown to partially pale tan, but the FW is not strongly yellow basally. At San Miguel Island, *P. amphorana* taken in May and September do not differ appreciably. On San Nicolas Island, where the larval host is *Isocoma menziesii*, both forms emerged from spring collections, but only the dark form in the fall.

Phaneta apacheana (Walsingham)(Plate 14.44, Plate 14.45) was described from Arizona but has been recorded along the Pacific Coast from British Columbia to southern California. The adult is small (FW usually 4.3–7.5 mm, larger females rarely to 9.5 mm), FW pale gray with two variable, dark purplish brown chevrons from the dorsal margin before the middle and at the tornus. The larvae feed in flower heads of several native cudweeds, *Gnaphalium*, and on the California Channel Islands on the Eurasian weed *G. luteo-album*. *Phaneta apacheana* is multivoltine in coastal California, with larvae present throughout the year and adults collected from March to June and September to October.

Numerous *Phaneta* are specific to one or more species of *Artemisia*. *Phaneta misturana* (Heinrich) occurs from Saskatchewan to Utah, southern Alaska, and coastal California. It is an early spring flying, dark gray species having the FW (5–7.8 mm; Plate 14.46) mottled dark gray with strongly contrasted silvery white preapical strigulae, ocellar patch, and linear streak through the cell, which is emphasized by black along the Cu crease. This species has been reported feeding on *A. tridentata* (Asteraceae) and *Atriplex confertifolia* (Chenopodiaceae) in Idaho; the latter host needs confirmation. In coastal California adults are found associated with *Artemisia californica*, March to early May, and the larvae feed in the new spring terminals. There are isolated records in July and September, indicating a partial second generation. *Phaneta argenticostana* (Walsingham)(FW 7.5–11.5 mm; Plate 15.1, Plate 15.2) is distinctive in having the costa and a well-defined streak through the cell silvery white on a brownish to yellowish gray ground. This species ranges from the Great Lakes region to the Pacific Coast, south to southern California and southern Arizona. Clarke reared *P. argenticostana* from *A. dracunculus* in southeastern Washington, and specimens reared by E. Boyd from the same area are labeled "galls." We have found *P. argenticostana* in association with *A. douglasiana* in widely distributed localities in California. *Phaneta pallidarcis* (Heinrich) and *P. subminimana* (Heinrich) are small gray species that were both described from San Diego. The former has pale, faintly brownish or grayish FW (5.5–7.5 mm; Plate 14.47) with white costa and a streak through the basal half of the cell, and ocellar spot with two parallel black bars, whereas *P. subminimana* is slightly larger and has darker gray FW (6–7 mm; Plate 15.3) with the ocellar spot distinct. The two are quite distinct in male genitalia. *Phaneta pallidarcis*, which ranges from the coast to the Great Basin, was reared from *A. californica*, in the 1880s at Los Angeles (by Koebele), and a century later in the San Francisco Bay area. Braun in Utah and McDunnough in British Columbia found the adults common on sage brush (*A. tridentata*). Adults of *P. subminimana* fly in the same coastal sage scrub habitats dominated by *A. californica*, in July and August, but this species apparently does not range inland. *Phaneta artemisiana* (Walsingham) and *P. scalana* (Walsingham) also are dependent upon *Artemisia*, but the larvae feed in spring and the adults fly in late summer to fall. *Phaneta artemisiana* is broad winged, FW (7.5–11 mm; Plate 15.4) cream white with pale brownish gray markings: a partial transverse band outwardly angled before the middle and another beyond the cell. It was described from one specimen Walsingham reared from *Artemisia* at Mt. Shasta and since has been recorded from southern Idaho, Nevada, Utah, and diverse situations in California, from the coast to the Sierra Nevada and Owens Valley. We have reared *P. artemisiana* from *A. douglasiana* and *A. dracunculus* at several coastal stations, and De Benedictis reported it from *Gnaphalium* at San Bruno Mountain. Larvae burrow in

the vegetative terminals and form tightly closed shelters in April and May. Adults emerge in July and August. *Phaneta scalana* (Plate 15.5) is slightly smaller, similarly broad winged with a mottled whitish ground color but has a large, distinctly defined basal patch and transverse dark rust to purplish brown band. It occurs at some of the same sites as *P. artemisiana*, once from the same larval collection, but adults of *P. scalana* emerge later, in September, with flight records to mid-October.

Phaneta stramineana (Walsingham) has a pale yellow FW (5.5–11.5 mm; Plate 15.7) over most of its range in the southwestern United States, from Colorado to the deserts of California, whereas populations in coastal southern California and on San Clemente Island have pale tan or rust tan FW (Plate 15.6). The larvae feed in terminals of *Isocoma menziesii* in spring, and adults fly in August and September. *Phaneta offectalis* (Hulst) is large (FW 10–15 mm; Plate 15.13) with a narrow FW, superficially resembling a crambid, at least to Hulst, who described it as a *Crambus*. The FW is rust brown broadly suffused with gray along the dorsal area, with the ocellar patch usually weakly indicated by two to four parallel lines. Paler individuals with washed-out color pattern occur in more arid regions. This species ranges from southern Wyoming and Colorado to eastern California. Heinrich cited "*Artemisia*" as a host, and we found the larvae feeding in roots of *Senecio spartioides* at 10,000′ in the White Mountains, California, in July. *Phaneta bucephaloides* (Walsingham) is a large, *Eucosma*-like, relatively heavy bodied species having a whitish FW (12.5–15 mm; Plate 15.14) peppered with gray, the terminal area broadly pale rust tan, and with white costal strigulae from near the base to the apex. There are remnants of two transverse, dark bars across the lower edge of the cell at the middle and before the tornus. This species occurs in the Great Basin from northeastern California to southern Utah and western Colorado. Flown individuals are rare in museums. Engelhardt reared *P. bucephaloides* from *Chrysothamnus linifolius* in Colorado, and we found the larvae in roots of *C. nauseosus* and *C. viscidiflorus* at six sites in eastern California and Nevada; adults emerged from late July to September. This species was described from a single female from Siskiyou County, California, and was recognized by Heinrich, who suspected it might be a variety of *P. offectalis*. However, the two are quite distinct in morphology and FW color pattern. Specimens of *P. bucephaloides* from Mono Lake and the Mojave Desert are paler.

Eucosma and its sister genus *Pelochrista*, with more than 180 named species in America north of Mexico, comprise the largest generic group lineage of microlepidoptera in our fauna. These are generally larger moths than *Phaneta*, although some are quite small, and many have more colorful FW patterns. Genitalia, Figs. 136, 137. Nearly all evidently are root borers in woody Asteraceae, but larvae of one lineage are Pinaceae specialists, feeding in the growing tips or cones. The Bobana Group of 12 species has adapted to conifers as a mode for speciation, presumably a derived trait from an Asteraceae-feeding groundplan. The "western pine shoot borer," *Eucosma sonomana* Kearfott, is a beautiful moth having deep red to orange FW (8–11 mm; Plate 15.8) with partial transverse, pale gray maculation. The flight period is early—January to May in coastal California, May to June inland—and the larvae bore into the new shoots early in the season. This species was described from coastal populations associated with Bishop pine, *Pinus muricata*, and has become a minor forest pest of *P. ponderosa*, *P. contorta*, and Engelmann spruce (*Picea engelmanni*). *Eucosma sonomana* occurs throughout the range of ponderosa pine in the western states and on Bishop pine south to Monterey County, California. *Eucosma bobana* Kearfott

(FW 8.5–11 mm, females larger than males; Plate 15.9). is the most widespread conifer cone species, ranging through pinyon-juniper woodland of the southwestern states. The larvae feed in cones of *Pinus monophylla* and *P. edulis*, and there is a single annual generation, with flight in June to August. The closely related **E. ponderosa** Powell, which occurs in eastern Oregon and California in association with *P. ponderosa* and *P. jeffreyi*, is larger (FW 8.5–13.5 mm; Plate 15.10) and has rust and tan checkered FW. The two *Eucosma* coexist in a narrow zone of the northern Sierra Nevada. **Eucosma siskiyouana** (Kearfott)(Plate 15.11) was misplaced by Kearfott, and following Heinrich, *E. siskiyouana* was treated as a variety of *Barbara colfaxiana* for many years. This is a broad-winged *Eucosma*, FW (9–11.5 mm) dark brown, strongly checkered with squarish gray spots. The larvae feed in cones of white fir (*Abies concolor*), and likely other firs. Adults were reared along with those of *Barbara*, leading to the confusion. *Eucosma siskiyouana* occurs throughout midelevation zones of the western mountains.

The silver-marked *Eucosma* species comprise a diverse and not necessarily monophyletic group that have bold silver or white, linear or serpentine FW maculation. There are more than 30 described species, 70% of them western, and all species for which larvae are known feed in roots of woody Asteraceae. **Eucosma ridingsana** (Robinson)(Plate 15.23, Plate 15.24) is the most widespread species, occurring in arid parts of all the western states in a wide range of elevations, north on the California coast to San Francisco Bay and east to Illinois and Mississippi. The adults vary considerably in size (FW 7.5–13 mm) and FW color from pale tan to dark brown, in part geographically, with linear silver spots; the median one through the cell is always broken. Clarke reared *E. ridingsana* from *Heterotheca villosa* in eastern Washington, as did we from roots of several species of *Gutierrezia* and *Isocoma*, but not from *Chrysothamnus*, in desert areas of Arizona, Nevada, and California and from California-aster, *Lessingia filaginifolia*, and gumplant, *Grindelia hirsutula*, along the Pacific Coast, where the FW color is much darker (Plate 15.24). There is an old record from "greasewood" [Chenopodiaceae?] in Colorado that needs explanation. Records by Prentice of *E. ridingsana* reared from *Pinus ponderosa* and *P. contorta* in British Columbia are erroneous and possibly should refer to *E. sonomana*. Adults fly late in the season, August to October. **Eucosma fernaldana** (Grote)(Plate 15.25, Plate 15.26) is similar in size and wing maculation, but the FW ground color is bright pink. This species occurs in a high-elevation band in sagebrush habitats from southern Wyoming to central Colorado (Teller County) and Utah, sympatric with *E. ridingsana*. **Eucosma crambitana** (Walsingham) is the largest species of the silver-marked group (FW 13–16, rarely 11 mm; Plate 15.17, Plate 15.18). It is characterized by having broad costal, median, and dorsal silver stripes, the median one unbroken from base to apex. The ground color is less variable than in *E. ridingsana*, tan to milk chocolate brown. This species occurs in the Great Basin from eastern Washington to Colorado and Arizona in association with rabbit brush. We found larvae in roots of *Ericameria nauseosus* and *Chrysothamnus viscidiflorus* at several sites in eastern California and Nevada in July, and adults emerged in August and September. A coastal race or closely related species occurs from southern California to Monterey Bay, California, where larvae feed in roots of *Ericameria linearifolia* and *E. ericoides*, and adults fly in September and October. **Eucosma optimana** (Dyar) is a relatively broad winged species (FW 12.5–15.5 mm; Plate 15.19, Plate 15.20) having pale gray brown FW with thin silvery white lines. This

species also occurs in the Great Basin, from Colorado to eastern California. We reared *E. optimana* from larvae in roots of *Artemisia tridentata* at three sites in Mono County.

Eucosma avalona McDunnough and **E. williamsi** Powell are coastal species in California. *Eucosma avalona* has dull brownish FW (11–15 mm; Plate 15.12) with thin, linear white markings. It was described from Santa Catalina Island and occurs on the northern Channel Islands and on the coastal mainland north to the San Francisco Bay area. The larvae feed in roots of *Artemisia californica*. *Eucosma williamsi* has dull whitish FW (10–13.5, rarely 8.5 mm; Plate 15.32) variably clouded with patches of blue gray, sometimes covering all but a large blotch on the dorsal margin and a preapical costal triangle, and there is a conspicuous black ring in the ocellar area. It ranges from north coastal California to San Diego, associated with coyote brush, *Baccharis pilularis*, from which F. X. Williams reared the type series a century ago in the Oakland Hills. Our later collections in San Luis Obispo and Santa Barbara counties have confirmed the host specificity. Larvae collected in January, February, and April yield adults in July and August, and there are flight records into October.

Eucosma morrisoni (Walsingham) is a small *Eucosma* FW (7.3–10.5 mm; Plate 15.27) having off-white linear streaks along the costa, through the cell, below the Cu crease, and narrowly at the dorsal margin, characteristically with a variable, poorly defined black streak through the cell, subtending the whitish costal streak, and the ocellar patch is reduced to a trace. This species is very widespread, from Michigan to Washington and all the western states, often taken at lights in sagebrush country, yet the larval host remains unknown. Adults fly mostly June to August. **Eucosma agricolana** (Walsingham) is morphologically similar, pale gray (FW 7–11 mm; Plate 15.28), but the FW pattern consists of numerous gray, parallel lines that converge distally above the well-defined ocellar patch. This species is even more widespread than *E. morrisoni*, from the Great Lakes region to the Pacific Coast, from Alaska to southern California, at elevations from sea level to above 6,000' in the Rocky Mountain states. Adults are collected March to May in coastal California to June to July inland. Clarke reared *E. agricolana* from *Artemisia vulgaris* in eastern Washington, we assume as a root borer.

Eucosma subflavana (Walsingham) and **E. grandiflavana** (Walsingham) are two of our largest western *Eucosma* species and both are wide ranging, yet nothing is known of their larval biologies. *Eucosma subflavana* has FW (13–15.5 mm; Plate 15.21) broad, pale rust brown, sometimes with a vague darker brown patch extending from the dorsal margin into the cell, and with some mottling in the terminal area. It was described from southern Oregon and occurs from eastern Washington and Idaho to northeastern California and is disjunct in the Vaca Mountains and adjacent hills near the San Joaquin–Sacramento River strait. Braun collected *E. subflavana* in a sagebrush habitat in northern Utah. Adults are recorded June to September, primarily June and July. **Eucosma grandiflavana** (Plate 15.22) was described from Lake County north of San Francisco Bay and ranges east of the Cascade-Sierran cordillera but has not been recorded again in northwestern California. The FW (13.5–16 mm) is primarily pale yellowish tan, with variable brownish gray speckling. Collection records are July to early August.

Eucosma maculatana (Walsingham) is a broad-winged species, FW (7.5–11 mm; Plate 15.33) dull brownish with an outwardly curved, dark brown bar from the middorsal margin; finely striate over the whole FW but faint in the area surrounding the dorsal mark, appearing whitish. This species occurs from Wyoming and Washington to California, from sea level to 9,600' in the Sierra Nevada; we reared *E. maculatana* from roots of wooly sunflower, *Eriophyllum lanatum*, in Marin and San Benito counties and collected numerous adults from *E. confertifolium* in southern California. Two additional broadwinged species, **Eucosma hasseanthi** Clarke and **E. hennei** Clarke, occur in southern California and are the only rootfeeding *Eucosma* in our region known to feed on plants other than Asteraceae. *Eucosma hennei* (Plate 15.29) was discovered at the El Segundo sand dunes, which are now mostly under the Los Angeles International Airport, in 1940 when Chris Henne reared specimens from larvae feeding in the roots of *Phacelia ramosissima* (Hydrophyllaceae). The specimens vary greatly in size (FW 7.5–14 mm) and in color, FW pale yellowish brown with varying degrees of infuscation and a darker fascia from the costa at the basal one-third through the cell, broadened inwardly to the dorsal margin. *Eucosma hasseanthi* has light ochreous to buff head, thorax, and FW (10–11.5 mm; Plate 15.30), with a variable, inwardly oblique, purplish rust band from middorsal margin to the middle of the cell and a weaker, curved fascia from the costa beyond the cell to the tornus. T. W. Hower reared the original series at Orange, California, from *Hasseanthus* (=*Dudleya*) *variegatus* (Crassulaceae), which probably is *D. blochmaniae*, according to present concepts. Clarke did not compare his two species, but the FW pattern is similar and genitalia are essentially identical. Subsequently we found *E. hasseanthi* feeding in roots of *P. ramosissima* at Riverside, California, and reared *E. hennei* from *P. ramossisima* at the Santa Maria dunes, San Luis Obispo County. Some specimens from the last population resemble *E. hasseanthi*, suggesting the two are not distinct species. The disparate host plants are puzzling, given the pattern of host specificity among *Eucosma* species, yet *Dudleya* and *Phacelia* are so different taxonomically and in growth form, that misidentification seems unlikely. Reared moths have emerged April to June and August to October, and specimens of the *E. hasseanthi* phenotype have been taken at lights in August and September at Palomar Mountain, San Diego County, July and October at Santa Maria dunes. *Eucosma juncticiliana* (Walsingham)(Plate 15.34) was described from Shasta County, California, and has been recorded in Washington and Colorado. The FW (6.5–9 mm, narrower and usually smaller in female) has an oblique, transverse, white line beyond the cell, preceded and followed by rust brown, which is darker in coastal populations. The larval biology is unknown. Host records listed by Robinson et al., specifically, *Chrysocoma* and *Solidago* (Asteraceae), refer to *E. derelecta* Heinrich in eastern North America, recorded before the latter was differentiated by Heinrich in 1929. In California, *E. juncticiliana* occurs along the coast from San Francisco Bay northward, in the Sierra Nevada, and at Mt. Shasta, possibly as a root borer in *Solidago*; adults fly in June and July.

Pelochrista are distinguished from *Eucosma* primarily by having very slender valvae in the male genitalia with a single, strong cucullar seta, in silhouette shaped like the neck and head of a swan. There are 25 named species in North America, all but a few of which occur in the West, but many are poorly known and some doubtfully distinct. A complex of small species related to **Pelochrista passerana** (Walsingham) share similar FW pattern and male genitalia, differing mainly by hue and intensity of dark scaling. The FW (5–9 mm; Plate 15.35) is dark gray, obscuring two dark transverse bars on the posterior half, at the basal one-third and near the tornus. *P. passerana* occurs along the northern California coast south

to San Diego and the Channel Islands, and we reared it from root crowns of yarrow, *Achillea millefolium*, on Santa Barbara Island. **Pelochrista rorana** (Kearfott) (Plate 15.37) and *P. expolitana* (Heinrich)(Plate 15.36) were both described from Spalding's Utah collections, have the FW transverse bars prominent, gray or brown, strongly contrasting with a pale background, varying between populations. *Pelochrista rorana* was reared from *Helianthus annuus* roots by Clarke in eastern Washington. **Pelochrista metariana** (Heinrich) occurs in interior northern California. We found larvae of one of the Utah species or *E. metariana* in roots of *Heliomeris multiflora* in western Nevada. Heinrich questioned the validity of these species, including the two he named, and their status has not been clarified in the ensuing more than 80 years. **Pelochrista scintillana** (Clemens) is a distinctive species, having brown FW (7.5–11.5 mm; Plate 15.31) clouded with paler areas of orange and a large ocellar patch containing three rows of prominent metallic spots. It occurs in the eastern United States and in Colorado, Washington, New Mexico, Arizona, and California. *Pelochrista scintillana* has been reared from *Helianthus* roots in Kansas. Western flight records span May to July.

Epiblema has more than 40 described species in America north of Mexico, most of which occur in the Southeast, and only about a dozen in the West. Many are dark brown or blackish, often with a broad white median spot on the dorsal margin, distinguished from *Eucosma* by having a short, spurlike projection at the interior base of the valva in males. Larval habits of *Epiblema* differ from those of *Eucosma* and *Pelochrista*, being primarily stem rather than root borers, and most induce gall formation or cause irregular swellings in response to their feeding. Consequently, their effects are more obvious, and they have been reared more often than *Eucosma*. **Epiblema strenuana** (Walker)(Plate 15.44, Plate 15.45) is the most widespread and frequently recorded species, ranging from the East Coast across the United States to Texas, Colorado, Utah, and California. The adults are exceptionally variable in color pattern and genitalia morphology, FW (4.5–8.5 mm) typically dark gray or gray brown with a strongly contrasting white ocellar spot and middorsal patch, which is sometimes reduced or absent. There is a smaller, pale gray race that lacks the FW dorsal patch (Plate 15.45) that may be a distinct species. It is characteristic of beach and riverine dune systems in Texas and in California, where the typical form occurs in inland valleys and foothills. Larvae of typical *E. strenuana* are stem borers and induce galls in ragweeds *(Ambrosia)*, cocklebur *(Xanthium)*, and guayule *(Parthenium)* in eastern North America and the Virgin Islands. We have collected larvae of the gray coastal race at several California localities from *A. chamissonis* and *A. psilostachya*, May to September and December. They feed in the lateral, decumbent, rhizomelike stems, causing irregular deformities at the nodes but not stem galls. Adults are recorded April to September. **Epiblema sosana** (Kearfott)(Plate 15.46) is superficially distinguishable from the pale form of *E. strenuana* by having more whitish scaling, especially distal to the FW basal patch, and the dorsal patch is extended through the cell, sometimes to the costa. Typical *E. sosana* has much broader valvae than *E. strenuana*, and darker specimens of *E. sosana* may be masquerading as *E. strenuana* in collections. *Epiblema sosana* is widespread in the Great Basin, Utah to eastern California, and an almost entirely white form with only a faint gray indication of the FW markings occurs in the Mojave and Colorado deserts, associated with sand dune systems, where adults are recorded from late March to June. Goeden and Ricker reared *E. sosana* from *Ambrosia acanthicarpa* and "probably *sosana*" from *A. chamissonis* and *A. confertifolia*, in southern California. **Epiblema rudei** Powell differs from the typical *Epiblema* phenotype, having a whitish gray FW (7.5–10.5 mm; Plate 15.47) with broad basal and postmedial gray bands and a dark termen strongly contrasted with the pale subterminal area. The larvae induce spindle-shaped stem galls on *Gutierrezia* (Plate 60.26), in which they overwinter fully grown, adults emerging in March and April. This species occurs along the dry western side of the Central Valley and interior southern California, but apparently not in desert populations of *Gutierrezia*. We did not find galls of *E. rudei* in eastern California, Nevada, and Arizona. **Epiblema macneilli** Powell is the largest Nearctic *Epiblema* (FW 10.5–14 mm, males larger than females; Plate 15.48), the most distinctive in FW maculation, and probably the species seen by the fewest people. It is confined to Arctic-alpine zone habitats above timberline in the southern Sierra Nevada: at elevations above 12,000' near Mono Pass; Todd Gilligan netted one specimen near the Mt. Whitney Trail Crest Pass at about 13,500'; and one was taken at the summit of Mt. Whitney, 14,500', by an unknown collector. The FW is pale gray with large, black botches that tend to coalesce into antemedial and postmedial transverse bands on the female, which superficially resembles the Siberian *Eucosma victoriana* (Kennel) more so than it does any other Nearctic *Epiblema*. The diurnal adults fly in July, August, and early September, and those we observed seemed to be associated with *Hulsea algida* (Asteraceae) growing on alpine talus slopes. *Epiblema macneilli* has been collected only in odd-numbered years.

The genera **Suleima** and **Sonia** are distinguished from related Eucosmini by lack of an uncus in the male genitalia. *Sonia* differs from *Suleima* by having HW veins M3 and CuA1 stalked, rather than united, and by having a costal fold and a basal projection of the valva. Most of the species of both genera are western. The valva of *Suleima* is simple, lacking a basal projection, and the cucullus is scarcely differentiated. **Suleima helianthana** (Riley)(Plate 15.38) occurs from Michigan and Illinois to Texas and California, feeding in stems and buds of sunflowers, *Helianthus* and *Coreopsis* (Asteraceae). The FW (7.5–11 mm) is pale brownish gray with two dark brown transverse bars, at the middorsal margin into the cell and at the tornus adjacent to the ocellar patch, which is strongly contrasting white with a thin black line and dot below it. In California this species occurs in south coastal areas and the Central Valley. The larvae tunnel in the central stem from axils of side branches of sunflower, *Helianthus annuus*, both the typical roadside weed and large commercial and garden varieties. Adults have been collected in April and July to September. **Suleima lagopana** (Walsingham) differs by having the middle dark band of the FW (7.5–9.5 mm; Plate 15.39) enlarged to cover the whole basal one-third and tornal band complete to the top of the ocellar patch, with the area between them cream white. The larvae tunnel in the slender, peripheral stems of *H. annuus* inflorescences. *Suleima lagopana* occurs in inland valleys of California from Colusa County south to Riverside County.

The three western species of **Sonia** occur in arid areas of the Southwest, where their larvae are root borers. **Sonia vovana** (Kearfott)(Plate 15.40, Plate 15.41) is the most widespread, occurring in the Great Basin, from Utah through Arizona, the Mojave Desert, and the Central Valley of California. The moth is primarily gray, FW (7.5–12 mm) pale gray with a complex pattern of dark gray, dominated by a basal patch highlighted distally by white, and a triangular mark at the tornus. The larvae feed in roots of snakeweed *(Gutierrezia)* and *Isocoma*, both *I. tenuisecta* in southern Arizona and *I. veneta* in

southern California. Adults are recorded from July to October. **Sonia filiana** (Busck) is similar but larger (FW 9.5–13 mm, rarely 8 mm; Plate 15.42) and brown with darker brown pattern. It occurs in coastal southern California north to Morro Bay, sympatric with *S. vovana* in San Diego County, where the larvae feed in roots of *I. acradenia*. **Sonia comstocki** Clarke (FW 10.7–13.3 mm; Plate 15.43) is a pale tan species in the western Mojave Desert, whose larvae also feed in *I. acradenia*.

Gypsonoma is a small genus of small tortricids, with seven Nearctic species, only two of which range into the West. The larvae are foliage feeders primarily on Salicaceae. **Gypsonoma salicicolana** (Clemens) is polymorphic in color within populations, FW (4.3–6.2 mm; Plate 15.49) usually with a dark gray to blackish brown basal patch and partially diffuse postmedial transverse band contrasted with a white to rust brown ground color, but on some specimens the pattern is suffused to a more or less uniform dark to pale gray. This species is widespread in the Northeast and upper Midwest, ranging to southern Texas, British Columbia, and California. It has been reared from willow *(Salix)* at numerous locations; young larvae tie up the new foliage terminals and later form single-leaf shelters, April to June. In California, *G. salicicolana* flies in May to June at the coast to July to August at 4,000' in the Sierra Nevada.

Members of **Proteoteras** are characterized by patches of black wing scaling in the males that differ between species. **Proteoteras aesculana** Riley (Plate 15.50) is the most widely distributed and commonly collected species, with both sexes often attracted to lights. It occurs from the East Coast to the Pacific states. The FW (6–8 mm) scaling is tufted, pale gray with dark olive to blackish green maculation, especially an outwardly angled, transverse band from midcosta curving distally to the terminal area above the ocellar spot. The male HW has a patch of black scaling on the basal two-thirds in the subcostal area on both upper- and undersides, enclosing a silvery gray streak of specialized scales on the upperside. The larvae of *P. aesculana* feed in the seeds and terminal shoots of box elder (*Acer negundo*, Aceraceae) and in eastern North America on the imported horsechestnut (*Aesculus hippocastanum*, Hippocastanaceae), but not on native species of buckeye *(Aesculus)*. There are two or more generations in the West, with adults recorded from March to October in coastal California. **Proteoteras arizonae** Kearfott (Plate 15.51) is similar but lacks greenish on the FW and has the sex scaling on the costal edge and distal half of the FW underside and entire upper and lower surfaces of the HW except at the base. It occurs widely in the southwestern states, and the larvae feed in shoots of *Acer negundo* in California. There are two generations, with flight records in April and August to October.

Pseudexentera are distinctive for their narrow FW, which they wrap around the body when at rest, and for their early-season flight, timed to precede the onset of spring growth of trees. The eggs are deposited around the vegetative buds. There are 17 described Nearctic species, primarily in the deciduous forest areas east of the Great Plains, and only four occur in the West. **Pseudexentera oregonana** (Walsingham), belying its name, ranges from northern Oregon to British Columbia, across Canada and the Great Lakes states to Nova Scotia. The FW (8–9.5 mm; Plate 16.2) varies in both sexes from nearly all dark gray to having weakly defined median and subterminal, coalescing, grayish brown bands. The larvae feed in rolled leaves of Salicaceae, including willow *(Salix)* and aspen (*Populus tremuloides*). Flight records across the broad distribution span March to June. **Pseudexentera habrosana** (Heinrich) occurs along the length of California from sea level to 2,000' on both sides of the Central Valley. The FW (7–10.5 mm; Plate

16.3, Plate 16.4) is gray, distinctively patterned with the dorsal half of the basal patch dark brown to black at the outer margin, and vague bronzy or coppery scaling in the apical area; females are more distinctly patterned than males, basal patch defined inwardly and apical area whitish with bronzy pattern and a partial transverse band through the end of the cell. Adults typically fly January to mid-March, with occasional captures in December and rarely November. The larval hosts are oaks, especially *Quercus agrifolia* and *Q. wislizenii*, and in dry interior sites, *Q. berberidifolia* and *Q. alvordiana*.

Chimoptesis chrysopyla Powell (Plate 15.52) cohabits coast live oak *(Q. agrifolia)* with *Pseudexentera habrosana* in California and has a similar life history. The slightly smaller *C. chrysopyla* (FW 6–8 mm) has primarily whitish FW ground color variably infuscate with blackish and gray, sometimes mostly fuscous. Its flight is confined to January, February, and early March. The larvae feed in March and April, at first in new foliage buds, later instars in developing foliage. *Chimoptesis chrysopyla* occurs at low elevations in the coastal counties, from Marin to southern California and Santa Cruz Island.

The genus *Zeiraphera* is morphologically similar to *Epinotia*, but the uncus is rudimentary and not split or notched, and the female has long, telescoping eight to tenth segments and one or two signa. Eight of nine North American species depend on abietoid Pinaceae and are transcontinental; one eastern species feeds on *Aesculus*. They are generally similar in FW pattern, with dark ante- and postmedial fasciae separated by a broad, pale medial band of ground color that usually is broken, forming a distinct triangle on the dorsal margin. *Zeiraphera canadensis* Mutuura & Freeman, the "spruce bud moth" (Plate 16.1), which was confused with a Palaearctic species, *Z. ratzeburgiana*, in older literature, occurs across Canada and the northern states to Utah and north coastal California. It is smaller than other western species, FW (5–7 mm) dark brown to rust a with well-defined pattern of ochreous tan, its dorsal margin triangle usually nearly connected to the opposing costal patch. The larva forms a shelter in the newly developing foliage buds, and the partially eaten needles die, causing the twigs to appear reddish brown. Severe infestations stunt and deform young trees. *Zeiraphera canadensis* feeds primarily on white spruce, *Picea glauca* (80% of 1,268 rearing records reported by Prentice in Canada), but also uses other spruces, Douglas-fir *(Pseudotsuga)*, fir *(Abies)*, and western hemlock *(Tsuga)*. There is a single annual generation. The dull green larvae of *Z. canadensis* and other *Zeiraphera* are illustrated by Duncan in British Columbia.

Crocidosema plebejana Zeller (Plate 16.5, Plate 16.6) is our sole representative of a large Neotropical genus, which includes several sibling species or races in Hawaii and other Pacific islands, a complex that begs for DNA analysis. *Crocidosema plebejana* was described originally from Sicily, where no doubt it had been introduced, as it had been transported globally with malvaceous plants, including okra and cotton, by the mid 1800s. Ongoing studies by Baixeras and R. Brown suggest the native range was Neotropical, perhaps lowland situations on the west coast of South America and the Juan Fernandez Islands, where its sister species occurs. In North America *C. plebejana* is recorded in the Gulf states, Texas, New Mexico, and California, where it has been resident for nearly a century. Males have dark gray to black FW (5–8 mm; Plate 16.5) with white dorsal and ocellar patches, and a prominent tuft of long scales at the base of the HW, whereas females lack the HW tuft and have tan FW with a basal black patch (Plate 16.6). The larvae feed on various Malvaceae, including cotton *(Gossypium)*, okra and other *Hibiscus*, hollyhock *(Althea)*, and cheeseweed *(Malva parviflora)*. *Crocidosema plebejana* has been a pest of seedling cotton in Australia, where it is called the "cotton tipworm," but not in the United States. Diapause is not known in *C. plebejana*, and its persistence is dependent upon alternate use of weeds such as *Malva*, which develop inflorescences throughout the year. Therefore in the United States it cannot survive on cotton alone. In California adults are recorded throughout the year, and we have found the larvae feeding in the seed heads of cheeseweed.

Notocelia is a Holarctic genus formerly treated as a subgenus of *Epiblema* owing to similarity in male genitalia, but the vesica has thornlike cornuti that are not shed during mating. Moreover, the larvae are external feeders on foliage of Rosaceae, and based on larval morphology MacKay believed this group to be closely related to *Epinotia* and not to *Epiblema*. Four of five species in the Nearctic occur in the Northwest, including two introduced from the Palaearctic. *Notocelia culminana* (Walsingham)(Plate 16.7) was described originally from northern California, yet ranges to the Northwest states, British Columbia, Colorado, and the Atlantic Coast. It has been reared from *Rosa* and apple *(Malus)* in British Columbia and Nova Scotia. The FW (6–9 mm) is mostly whitish lightly sprinkled with gray, with dark basal and subapical patches. Flight records in the West span August to October.

Epinotia provides a marvelous poster child for coexistence of species, contradicting naysayers who disdain a recognition or reproductive isolation species concept and advocate naming lineages, clusters, subspecies, or other vague concepts. There are more than 80 described species of *Epinotia* in America north of Mexico, more than 55 of them in the western states and provinces, and almost all of them are sympatric with at least a few of the others, without any indication of hybridization. *Epinotia* are specialists on larval food plants as foliage feeders, most of them externally in the new terminals, a few in leaf or needle mines, and consequently their biologies are better known than those of other Eucosmini. There are more than 40 species in California alone, and 22 occur at the Big Creek Reserve, Monterey County, all of them larval host specialists, one species each on *Cercocarpus* and *Holodiscus* (Rosaceae) and *Salix* (Salicaceae); two on *Alnus* (Betulaceae); five on *Arbutus* and/or *Arctostaphylos* (Ericaceae); one species on *Heuchera* (Saxifragaceae), one on *Quercus*, and another on *Lithocarpus* (Fagaceae); one each on *Pinus radiata* and *P. ponderosa* (Pinaceae); one on *Lupinus* (Fabaceae); two on *Ceanothus* (Rhamnaceae); one each on *Ribes menziesii* and *R. sanguineum* (Grossulariaceae); and one on *Clematis* (Ranunculaceae).

Moreover, the biologies present a workshop in life-cycle strategies that have evolved to cope with the long dry season in this Mediterranean climate. Although a few species have multiple generations, most are univoltine and feed in spring when woody plants of the region are foliating, then undergo one of several life cycles. For example, larvae of *Epinotia emarginana* feed in March to April, and adults emerge in May to June, then adults aestivate and hibernate in reproductive diapause until early the following spring, when mating and oviposition occur. Other species aestivate in diapause as prepupal larvae and metamorphose in the fall (e.g., *E. kasloana, E. lomonana, E. radicana, E. saggitana*) or late fall to winter, November to January *(E. bigemina)*, while a few are active as adults in summer *(E. johnsonana)* or are bivoltine with late-spring and summer flights *(E. infuscana, E. subplicana)*, presumably overwintering as eggs

Richard Brown completed a phylogenetic analysis of *Epinotia* based on morphological characters in a 1980 Ph.D. thesis at Cornell University and proposed 17 species groups in two lineages for the North American fauna. Heinrich had divided *Epinotia* (including *Catastega* and *Crocidosema*) into two groups based on presence or absence of a FW costal fold in males. This character is almost entirely consistent within species groups but not with the two major lineages in Brown's analysis. Host-plant relationships are partially correlated with his species groups: the conifer feeders comprise independent groups in the two lineages, and adaptation to particular plant families two or more times has been more the rule than the exception (e.g., Salicaceae, Ericaceae, Grossulariaceae, Betulaceae). Brown's lineages are followed in the examples here, rather than grouping species by host-plant preferences.

In the Stroemiana Lineage, the male is characterized by a simple uncus, bifid at tip or fused; socii subquadrate, and cucullus defined by a pronounced emargination in the sacculus; female with a denticulate band (cestum) around the ductus bursae.

Epinotia pulsatillana (Dyar) has broader FW than most *Epinotia* (0.35–0.40 FW length), with a conspicuous costal fold that creates an angled costa shape contrasted to the evenly oval FW of females. This is a gray moth, FW (7–9 mm) with a black basal patch and mottled with dark gray. The larvae, reared by Dyar and Braun, feed on Ranunculaceae, *Pulsatilla* in Colorado and virgin's bower, *Clematis*, in Utah. This species occurs in eastern Washington and Oregon, Nevada, and the Rocky Mountain states to New Mexico and northern Arizona. *Epinotia siskiyouensis* Heinrich (Plate 16.18) differs in having a longer costal fold (0.45–0.5 FW) and by the genitalia, especially the papillae anales, which taper anteriorly in *E. pulsatillana* but are uniform in width along their length in *E. siskiyouensis*. The latter has dark olive gray FW scaling, often with a brownish patch adjacent to the basal patch, and a rare form with brown and tan, wood grain pattern on the FW dorsal half. *Epinotia siskiyouensis* occurs in coastal and foothill parts of California from Mt. Shasta to Ventura County and on Santa Cruz and Santa Rosa islands, ranging to 7,800' in the southern Sierra Nevada. The larvae feed in inflorescences and new foliage of *Clematis lasiantha* and *C. ligusticifolia* February to May, and adults are taken at lights mid-June to early November.

Epinotia emarginana (Walsingham)(Plate 16.13, Plate 16.14, Plate 16.15, Plate 16.16, Plate 60.27) is the most abundant western *Epinotia* in collections and is exceptional in several ways. The adults are polymorphic in both sexes within populations, which remarkably, Lord Walsingham recognized in his original 1871 collections, so no synonyms were proposed. The FW (7–8.5 mm) termen is notched, unlike the termen of most *Epinotia*, and usually is dark brown mottled with black maculation, forming weakly defined transverse bands; sometimes there are two or three white spots along the dorsal half; less commonly there is a well-defined, rectangular bar along the dorsal margin, which may be whitish or rust on a dark ground, or black, rust, or tan on a pale gray ground. In contrast to other *Epinotia* and Tortricidae generally, the larvae are specialists on two unrelated plant groups, Fagaceae and Ericaceae. At any one locality the populations seem to specialize either on oaks (*Quercus*, including both deciduous and evergreen) or on manzanita and madrone (*Arctostaphylos* and *Arbutus*). The adults live for nine to 10 months in a state of reproductive diapause and frequently can be flushed from the bark of oak limbs or found under loose bark They sporadically come to lights and are found in homes and sheds in urban areas, from May until the following March. *Epinotia emarginana* occurs along the Pacific Coast from British Columbia to south-

ern California, including Santa Cruz and Catalina islands, in coastal canyons and foothills to 7,000' in the Trinity Alps and Sierra Nevada. Populations in Arizona were regarded as an undescribed species by Richard Brown. *Epinotia columbia* (Kearfott)(Plate 16.11, Plate 16.12) is superficially quite similar, and Walsingham collected specimens 30 years before Kearfott described *E. columbia* but interpreted them as varieties of *E. emarginana*. Both have the notched termen and are polymorphic, but the ground color in *E. columbia* is gray brown, and the morphs are not as distinctly patterned. This species averages smaller (FW 6–7.5 mm) and varies from uniform gray brown to having black streaks along a line through the FW cell to the apex, or the dorsal half is pale gray or pale rust posterior to a sinuate line along the Cu. The larvae feed on willows *(Salix)*, March to May, and the life cycle is similar to that of *E. emarginana*, with adults living through winter. *Epinotia columbia* occurs across Canada to British Columbia, Utah, Nevada, and northern Arizona and along the Pacific Coast to southern California, including the Channel Islands.

Epinotia solandriana (Linnaeus) is one of three Holarctic species in the Stroemeriana Group. This is a large (FW 8–12 mm; Plate 16.19, Plate 16.20, Plate 16.21), broad-winged, polymorphic species, which ranges across Canada and the western states to Alaska and central California and from sea level to 9,500' in New Mexico. The FW varies from pale gray to brown, dark brown, or olive brown, usually with indistinct darker clouding on the basal one-third and a transverse, median band, which between them define a dorsal edge triangle that may be only slightly paler than the ground color or white. In contrast to the host specificity shown by most species in this genus, *E. solandriana* feeds on Betulaceae, Salicaceae, and a few Rosaceae. Among 836 rearing records by the Canada Forest Insect Survey, primarily in the East and Midwest, 68% were Salicaceae, 31% Betulaceae including 1.5% *Alnus*, but we have found larvae of *E. solandriana* only on alders in coastal California, in April to May and in June in central Alaska. *Epinotia johnsonana* Kearfott is a pretty little species, FW (6–8 mm; Plate 16.17) with a moderately long costal fold (0.35 costa length), inner half of the basal patch and costal area into the cell bright to dull rose pink or rose orange, fading distally into pale peach or tan clouded with dull olive brownish; basal patch dark in outer half. This species ranges from British Columbia to southern California, and the mountain ranges of Nevada and Colorado. The larvae feed in tightly rolled shelters on ocean spray, *Holodiscus discolor* (Rosaceae), in April to May, and adults fly in June to July in California to August to September in the Northwest.

Five of seven species in the Hopkinsana Group have recorded larval hosts, all conifers, three on Cupressaceae, two on Pinaceae. *Epinotia hopkinsana* (Kearfott) occurs along the immediate Pacific Coast from Vancouver Island to central California. The FW (7–9 mm; Plate 16.10) is narrow with a strong costal fold, lime to dull olive green with raised, black scaling on outer margin of basal patch and on margins of subasal and medial fasciae; HW whitish to pale gray. This species feeds on Pinaceae, sitka spruce *(Picea sitchensis)* in British Columbia and Washington, closed cone pines *(Pinus muricata* and *P. radiata)* in California. Larval and adult collections span March to September, suggesting two or more generations. The larvae of this and the following species are dull olive to rust greenish with dark head and thoracic shield, illustrated by Duncan. *Epinotia radicana* (Heinrich) is smaller (FW 5.5–7.5 mm; Plate 16.8, Plate 16.9), greenish gray with distinct rust maculation to all rust brown, basal patch distally and tornal patch dark brown to dark rust. It ranges across Canada to central British Columbia and widely in coniferous areas of the western United States. This species was recorded in more than 2,000 larval collections of the Forest Insect Survey in Canada on various conifers, including hemlock, pine, and junipers, but 66% from spruces *(Picea)*, 22% Douglas-fir *(Pseudotsuga)*, and 16% firs *(Abies)*. In southern Oregon and California *E. radicana* feeds on *Abies* and *Pseudotsuga* in the Cascades, North Coast Ranges, and Sierra Nevada, May to July, and adults fly July to October.

Epinotia castaneana (Walsingham)(Plate 16.22) is a brightly colored species that occurs across Canada and in montane areas of the West, from Wyoming to New Mexico, northern Arizona, and California, to 12,500' in the Sierra Nevada and above 10,000' in Utah. The FW (6–8.5 mm) is pale tan to peach colored with dark rust, distinctly defined basal and apical patches and poorly developed medial fascia. The larvae feed on buds and foliage of *Ribes* (Grossulariaceae), including *R. sanguineum* in Washington. Adults are recorded in July and August.

Members of the Bigemina, Kasloana, and Lindana species groups lack a male costal fold and share a similar life cycle, feeding on new foliage in spring, aestivating as prepupal larvae, and flying in late fall to winter. Most of the species also have a strongly mottled HW underside. *Epinotia bigemina* Heinrich is small (FW 6–8 mm; Plate 16.23), bright rust red to brownish purple, unicolorous or with dark, weakly defined basal and medial fasciae, and flies in winter on the Pacific Coast. The narrow FW has a line of silvery male sex scales near the base of the costa, which is curled upward, not folded, from base to near midwing. This species occurs from southeastern British Columbia to coastal California south to the northern Channel Islands. The larvae are numerous on several manzanita species, *Arctostaphylos*, in California and feed on *A. insularis* and on summer holly, *Comarostaphylis diversifolia* (Ericaceae), on Santa Cruz Island, March to May. Adults emerge late October to January, mostly in November, and have been collected from October to March. The females have modified papillae anales facilitating insertion of the eggs into the vegetative buds of manzanita. Larvae of *E. arctostaphylana* (Kearfott) feed with those of *E. bigemina* on the same hosts, including *Comarostaphylis* on Santa Cruz Island, and on *A. patula* at Mt. Shasta, and on madrone *(Arbutus menziesii)*. Adults average larger (FW 7–10 mm; Plate 16.24), have broader FW, lack the male sex scaling and modified papillae anales of *E. bigemina*, and are polymorphic. The most common form has FW unicolorous reddish brown, others are grayish orange with the basal patch and termen brown, or costal area reddish brown, with a dark brown streak along the Cu. *Epinotia arctostaphylana* occurs from Alberta and British Columbia to southern California. Adults presumably overwinter, recorded in September to October and once in March. *Epinotia kasloana* McDunnough varies in color pattern, FW (7.5–10 mm; Plate 16.25, Plate 16.26, Plate 16.27) nearly uniform dark brown to having the dorsal half rust to tan, separated by a sinuate dark brown line through the cell, but this species does not have discrete polymorphic forms. It ranges from southeastern British Columbia to southern California. The larvae feed in inflorescences of *Ceanothus thyrsiflorus* (Rhamnaceae) in April to May, and adults emerge in late October to November. There are collection records at lights from December to February, and mating occurs soon after emergence rather than being delayed to early spring. A population in southern California studied by Pinto and Frommer has the same life cycle but feeds on leaves and fruit of jojoba, *Simmondsia chinensis* (Buxaceae, Euphorbiales). *Epinotia signiferana* Heinrich is similar, smaller (FW

7–8 mm; Plate 16.28), FW gray with a broad, dark brown, sigmoid bar from midcosta to the terminal area, sometimes obscured by dark gray brown. This species ranges from British Columbia to southwestern Colorado and southern California. In coastal California it feeds on *Ceanothus thyrsiflorus* with *E. kasloana*, and adults are recorded from September to December. Richard Brown found debris packed among the setae of some females' papillae anales, suggesting oviposition and egg covering occurs in the fall. *Epinotia lindana* (Fernald) occurs across North America from New England to British Columbia, Washington, and California. The FW (7–10 mm; Plate 16.40) pattern is divided longitudinally by a strongly sinuate line along the Cu basally and into the cell, dark brown on the costal half and the dorsal half pale brownish orange or brownish gray, sometimes with scattered dark brown scales. The larvae feed in spring on dogwood, *Cornus* (Cornaceae), and adults are active August to October.

In the Nisella Lineage, the male has a deeply bifid uncus, fused only at the base; socii elongate, sickle shaped, tapered to a point; valva simple, cucullus not well defined; female lacking denticulate cestum.

Epinotia meritana Heinrich and *E. trossulana* (Walsingham) are small tortricids that lack a male costal fold, in contrast to other members of the Nisella Lineage. Their larvae mine the needles of firs *(Abies)*. The "white fir needleminer," *E. meritana*, was described from Utah, reportedly from pine, and is widespread in the West. It has whitish FW (4.5–5.5 mm; Plate 16.29) with dark gray forming an angulate basal patch and postmedial dark fascia, outer one-third brown with white costal strigulae. Outbreaks causing defoliation of red fir *(A. magnifica)* and white fir *(A. concolor)* have occurred periodically in Utah, Arizona, and the Sierra Nevada in California. *Epinotia trossulana* is distinctive, having tan FW (5.5–7 mm; Plate 16.30) with a pale streak through the cell, more distinct toward the termen, flanked by variable peppering of black scaling, and there are silvery gray lines in the apical area. It ranges from Vancouver Island to Utah and southern California, generally distributed in association with *A. concolor*, from which we reared it in northern California, as well as from *A. magnifica* at Lake Tahoe. The larvae mine the needles or bundles of needles and create sheets of silk exteriorly, moving from needle to needle. Adults fly in July and August.

Epinotia nisella (Clerck) is a Palaearctic species that occurs across North America from Newfoundland to British Columbia, Colorado, and northern California. The FW (7–9 mm; Plate 16.31, Plate 16.32) is variable, usually pale gray with dark gray basal patch and medial transverse fascia, white between them or with rust basal patch. Less commonly the basal patch may be dark brown and medial fascia bright cinnamon brown, rust distally. The larvae feed on Salicaceae, in catkins or new terminals of balsam poplar *(Populus balsamifera)*, aspen *(P. tremuloides)*, willow *(Salix)*, and rarely on birch *(Betula)* in Canada. In the West adults fly July and August, infrequently into September.

Larvae of the Terracoctana Group differ from other *Epinotia* and most tortricids by mining mature leaves, often even pupating in the mine. *Epinotia miscana* (Kearfott) and *E. terracoctana* (Walsingham) are similar species that feed on Ericaceae. The FW of *E. miscana* (7–8 mm; Plate 16.36) is predominately dark brick red, while that of *E. terracoctana*, a smaller moth (FW 5.5–7.5 mm; Plate 16.37), is pale orange. The FW basal patch is dark brown, distinctly defined on the posterior half, postmedian band darker than ground color, and the terminal area is dusted with bluish gray scaling across the anterior portion of the whitish ocellar patch. *Epinotia terracoctana* occurs from Vancouver Island to southern California, including Santa Cruz Island, from sea level to 7,000' in the San Bernardino Mountains, and in Arizona. The larvae mine leaves of several species of manzanita *(Arctostaphylos)* and madrone *(Arbutus)*, February to April, and adults are recorded from late April in southern Arizona and May in coastal California to September, mostly July to August. *Epinotia miscana* occurs in southern Oregon, California, and western Nevada, at 7,000 to 8,000' at Mt. Shasta, the Sierra Nevada, and the San Bernardino Mountains. This is not a coastal species and is not recorded feeding on madrone. Large-leaved *Arctostaphylos* species, *A. glauca*, *A. manzanita*, and *A. patula* serve as hosts. *Epinotia nigralbana* (Walsingham) and *E. digitana* Heinrich are similar black-and-white species that feed on unrelated food plants, Ericaceae and Saxifragaceae. Both have FW primarily white on the basal half, black distal half, but FW of *E. digitana* have a well-defined dark basal patch, so the median white fascia is distinct (FW 5.5–7 mm; Plate 16.39). Its larvae mine leaves of alumroot, *Heuchera*, early in the season. The known distribution is quite odd, interior British Columbia and Washington, eastern Sierra Nevada to 12,500', and at the immediate coast of California. Adults emerge March to May on the coast, June to July inland. By contrast, *E. nigralbana* (Plate 16.38) mines the leaves of madrone and manzanita during winter, maturing in March to May, and adults fly April to August. On *Arctostaphylos hooveri* in coastal California young larvae of *E. nigralbana* mine a thin loop from the leaf margin, defining a circular area (Plate 60.28), into which they mine digitate extensions after overwintering (Plate 60.29). Pupation occurs in the mine.

Epinotia subplicana (Walsingham) has broad, somewhat falcate-tipped FW (7–11 mm; Plate 16.33, Plate 16.34) with a long costal fold and concave termen, nearly uniform gray, varying to a large, blackish gray dorsal blotch and to whitish with several black, basal dots, the last named *E. basipunctana* by Walsingham, one of the very few times he failed to recognize variation in his pioneer collections of western microlepidoptera. This species occurs from Washington to Nevada and the mountains of northern Baja California. The larvae feed on several species of manzanita *(Arctostaphylos patula* at Mt. Shasta, *A. manzanita*, *A. hooveri*, and *A. virgata* in coastal California, *A. insularis* on Santa Rosa Island), March to April, and adults are recorded from late April to July and September to October, indicating two generations. *Epinotia albangulana* (Walsingham) has the FW (6–8 mm; Plate 16.35) variably mottled with bluish gray and rust brown, with a median fascia and ocellar patch variably defined, usually white, but sometimes the FW is entirely mottled gray except whitish at the distal edge of the ocellar patch. This species rages from Alaska to southern California and Idaho to northern Arizona and is multivoltine, at least in coastal areas. It occurs in low-elevation canyon situations in association with alder, up to 5,000' in the mountains of California. At Big Creek, Monterey County, we have collected the larvae on *Alnus rhombifolia* in March, April, June, August, and October, none of which entered diapause, and adults are active from March to November.

Catastega Clemens was resurrected from synonymy by R. Brown to accommodate *C. timidella* Clemens in eastern North America and *C. marmoreana* (Heinrich) in Utah and Colorado, and he described several new species from the Southwest. *Catastega plicata* R. Brown (FW 5.4–7.8 mm; Plate 16.41) is a small, pale tan species, male with a long fold extending beyond midcosta, enclosing a dense brush of white hair scales from the

base. The FW pattern is brownish basally becoming peach brown to rose brown apically. This species occurs in the Huachuca and Santa Rita mountains of southern Arizona and mountains of Durango and Sinaloa, Mexico, at elevations of 6,400 to 8,800', often in oak woodland. Adults are recorded in July and August.

Members of the Holarctic genus *Ancylis* are recognizable by the FW shape, concave below the acute, almost falcate apex, and usually by the male HW anal margin rolled, enclosing a hair pencil. The genus is distinct, and some authors have placed it in a separate tribe, Enarmoniini, which has its greatest diversity in the Oriental/Australian region. There are about 35 described species of *Ancylis* in America north of Mexico, of which only 20% occur in the West. Most are poorly understood, with considerable geographic variation within species and only slight differences in genitalia. Heinrich suggested there may be too many names, and not much has been resolved in the ensuing 80 years. The "strawberry leaf roller," *A. comptana* (Frölich)(Plate 16.42), is thought to have been a Palaearctic species that was introduced early in American history, but it may have been Holarctic originally. Varieties now treated as synonyms were described in the 1860s and 1870s, from Nova Scotia, Illinois, Ohio, and northern California. Now, *A. comptana* occurs in many parts of the country, including Colorado and the Pacific Northwest. The FW (4.5–7 mm) is brown, darker on the posterior half, with a pale tan costal area on the basal half, rust distally with numerous costal strigulae, and a sigmoid, gray fascia from midcosta into the terminal area. There are numerous accounts of its biology in the economic entomology literature. The primary larval hosts are cultivated strawberry *(Fragaria)*, blackberry, and raspberry *(Rubus)*, but native Rosaceae may be used. There are two or three annual generations, depending upon latitude, with moths active in April to May and again in summer. *Ancylis mediofasciana* (Clemens), a native species, also is transcontinental. The FW (7–11 mm; Plate 16.43) is white on the costal one-third, dark bluish gray below midcell, obscuring darker markings, with a strong, dark transverse bar angled outward from midcosta. Biological information is fragmentary; Ferguson reared it from chokeberry *(Pyrus melanocarpa)* in Nova Scotia, and Bird found the larvae on *Amelianchier* and *Prunus* in Manitoba, so this species may be a Rosaceae specialist. Klots thought *Ancylis mediofasciana* may be a sphagnum bog obligate, but In California it occurs in diverse situations, from coast to 8,000' in the Sierra Nevada. Males were attracted to a synthetic pheromone lure formulated for *Rhyacionia zozana* at numerous montane sites in the western states. *Ancylis simuloides* McDunnough (Plate 16.44, Plate 16.45) was described from British Columbia, along with supposed subspecies in California, based on scant material. The FW (7–9 mm) pattern is distinct and strongly contrasting, purplish brown basal patch and outwardly angled transverse bar from midcosta expanded into the subterminal area, and costal strigulae distally. Populations believed to be *A. simuloides* and/or the closely similar *A. columbiana* McDunnough are prevalent throughout coastal and montane California, where the larvae feed on several species of *Ceanothus* (Rhamnaceae). At Big Creek, Monterey County, we have found the larvae abundant in April to June on *Ceanothus papillosus*, and adults late April to July, evidently univoltine. *Ancylis pacificana* (Walsingham) is a larger species (FW 7–10.5 mm; Plate 16.46) known from montane situations in Oregon and northern California. The FW ground color is pale gray, stippled with brownish gray strigulae, and the HW is dark gray, giving an overall dark appearance to the moth. On at least three occasions we have flushed adults from mats of *Ceanothus prostratus* in May and June, to late July in Oregon, and they are often attracted to lights, but evidently larvae of *A. pacificana* have not been reared.

Hystrichophora species are moderately large tortricids superficially resembling some *Eucosma*, but they lack a male costal fold and are distinguished by their grotesquely asymmetrical male genitalia, which vary within species to a degree that it is almost impossible to identify them with confidence. They are under study by Todd Gilligan, who believes there are 11 valid Nearctic species, six of which occur in the West, mostly in the Pacific Northwest and California. *Hystrichophora leonana* Walsingham, the type species of the genus, is slender and smaller than other *Hystricophora*, FW (7–10 mm; Plate 16.47) tan shaded to brownish in the discal area, with brownish, transverse strigulae distally; the HW is pale gray to brown. It occurs in central California, in the Coast Ranges and Sierra Nevada, flying late May to August. The larva probably is a root borer, and we have found adults associated with *Rupertia* (formerly *Psoralea*, Fabaceae). *Hystrichophora stygiana* (Dyar)(FW 10–14.5 mm; Plate 16.48) has variable phenotypes, nearly black to gray, light tan, or whitish, sometimes with a tinge of ochreous brown in the dorsal area. The variation is marked within and between populations and resulted in several names having been proposed when there were few specimens available for study. This species is widespread in western North America, from Alaska to California and Colorado, primarily at higher elevations, including the Sierra Nevada and White Mountains above 10,000'. The larvae feed in root crowns of bush lupines (*Lupinus*, Fabaceae), reared by Heinrich and by John Bishop at Mt. St. Helens in studies of vegetation succession following its eruption. Adults fly June to September. *Hystrichophora roessleri* (Zeller) is a robust species, FW (usually 13–15 mm, rarely 10 mm; Plate 16.49) whitish on basal half, blending to ochreous brown on posterior half distally, dark brown toward the tornus. The HW is dark brown with a whitish fringe. *Hystrichophora roessleri* occurs in central coastal California to southern Oregon. We found the moths numerous in association with two species of perennial lupines along the immediate coast of San Francisco Bay in June.

TRIBE GRAPHOLITINI

The tribe Grapholitini is characterized by reduction in male genitalia, especially loss of uncus and socii, and is considered by Marianne Horak and Richard Brown to be the most-derived lineage of Tortricidae. The male costal fold is present only in *Dichrorampha*, and the FW often is dark colored with conspicuous, white costal strigulae. Genitalia, Figs. 140, 141.The larvae are nearly all internal feeders, borers in inflorescences, seeds, stems, or roots, primarily in legumes or conifers. The tribe has about 110 described species in America north of Mexico, but only seven of 15 genera and fewer than half the species occur in the West, in addition to numerous undescribed species in western collections.

Dichrorampha are broad winged, often with a pale patch on the FW dorsal margin. *Dichrorampha simulana* (Clemens) and *D. sedatana* (Busck) are the most widespread of five western species, both range across Canada and into the western United States. *Dichrorampha simulana* has brown FW (5.5–7.5 mm; Plate 16.50, Plate 16.51) with an ochreous dorsal patch well developed to absent, distally with iridescent bluish streaks interspersed with dark ochreous, and a row of black dots at base

of the fringe. This species occurs in the northwestern states, south to the central coast of California and Santa Cruz Island. We have taken specimens in association with beach lupine, but the larvae, probably root borers, have not been reared *D. sedatana* (Plate 16.52) is a Holarctic species that occurs from Alaska, to Idaho, Colorado, and the Sierra Nevada in California. It has nearly unicolorous brown FW (5–7 mm), darker than *D. simulana*, distal half with iridescent bluish lines and ochreous scaling. In Europe the larva feeds in rootstocks of *Chrysanthemum vulgare* (Asteraceae), and pupation occurs in spring. Western flight records are June and July.

About a dozen species of **Grapholita** occur in the western states and provinces, including both conifer and legume feeders, each guild with nocturnal and diurnal species. The latter frequently visit flowers for nectar, whereas the nocturnal species apparently do not. The genus has been treated as a subgenus of *Cydia*, but *Grapholita* males have eversible coremata and correlated elaborate courtship behavior that are lacking in *Cydia*. Males display a stereotyped behavioral sequence leading to display of hair pencil organs directed toward the females, accompanied by rapid puffs of air generated by the vibrating wings. Phylogenetic studies are needed to determine if male secondary structures warrant the current generic split of conifer- and legume-feeding lineages in each of the two genera. The "Oriental fruit moth," *G. molesta* (Busck)(Plate 17.1), is the most notorious species; it was imported to the eastern United States with nursery stock prior to 1915—some references suggest it was introduced from Japan with the cherry trees planted around the tidal basin in Washington, D.C.—and by the 1920s this moth became the most important pest to the peach industry. In 1942, *G. molesta* was discovered in California and was the subject of intensive but unsuccessful eradication efforts. This is an extensively studied moth, with more than 300 publications on its biology, pheromone chemistry, monitoring, economic importance, parasites, and control. The adult is smaller (FW 4.5–6.5 mm) than the codling moth, brownish gray, FW with distinct whitish, costal strigulae, bluish scaling in the terminal area, and numerous slightly paler gray, parallel striae, the most conspicuous forming an obscure dorsal margin patch. The larvae burrow into new vegetative shoots of fruit trees in spring and summer and later enter the fruit, causing decay and increased brown rot. There are one to seven generations, varying with latitude and temperature, the last entering diapause as prepupal larvae in tough cocoons on the bark or in litter on the ground. *Grapholita molesta* typically feeds on peach and nectarine *(Prunus persica)* but also damages other stone fruits, including apricot, cherry, and plum, even pear *(Pyrus communis)* and apple *(Malus)*, especially where grown near peach orchards. In the West, the Oriental fruit moth is most destructive in pears. Males are attracted to synthetic pheromone lures formulated for *G. molesta* or for codling moth.

Several western species of *Grapholita* comprise a guild of spring-flying, diurnal, dark moths whose larvae feed in flowers, seeds, or stems of legumes. Males are larger than females, and with one exception, they are univoltine. Two species are the most often collected of these, occurring from British Columbia to Colorado and California: **Grapholita caeruleana** Walsingham is dark brown, FW (4.5–7.5 mm; Plate 17.2) with gray forming a faint dorsal margin patch; ocellar patch tan, offset by vertical, shining lead-colored bars and enclosing three or four short black, transverse lines. This species was reared from *Astragalus* in northern California by Clint Kellner, and adults often are found in association with *Lotus*. They fly from late February to April along the coast, May to early July in the Sierra Nevada and northward. *Grapholita conversana* Walsingham is smaller in both sexes (FW 4.3–5.5 mm; Plate 17.3), FW dark brown to the end of the cell, shaded with ochreous beyond, and parallel white lines forming a conspicuous chevron-shaped dorsal patch. It occurs in the Pacific states and is recorded feeding on several species of clover *(Trifolium)* in Oregon and Idaho. In California the flight period is coincident with that of *G. caeruleana* and often in the same habitats. Males have been attracted in numbers to carbamone traps and synthetic pheromone lures for codling moth. *Grapholita lunatana* Walsingham (Plate 17.6) also is widespread, from Saskatchewan to British Columbia, Colorado, and southern California, including San Clemente Island. FW (males 5.5–7.5 mm, females 4.3–5.3 mm) with a strong, single dorsal half-lunule at the middorsal margin in both sexes, distal area iridescent ochreous with three curving lead-colored lines, fringe pale gray to whitish; female more distinctly marked; HW of male pale gray, of female dark brown. *Grapholita lunatana* is an early spring moth, males appearing in the first warm days in late January and February to early April in the Sierra Nevada. The larvae feed and aestivate/hibernate in folded leaves of wild pea *(Lathyrus)*. *Grapholita vitrana* Walsingham (Plate 17.4) occurs in Oregon, California, and Baja California and is geographically variable, especially in high montane populations and on the Channel Islands. The adults are larger and more robust than other native *Grapholita*, FW (5.5–9.5 mm), dark gray dusted with pale gray on the basal two-thirds, ocellar patch with two perpendicular silver bars. The pale scaling is more extensive on males and on the coast and islands, but reduced in some montane populations where the moths appear much darker. The larvae feed on immature seed in pods of locoweed *(Astragalus)*. This species is univoltine in inland populations where locoweed dries in late spring, but has facultative additional generations along the immediate coast where flowering persists through summer. On the Monterey coast and San Miguel Island, *G. vitrana* flies from March to July and in October, and we found larvae June to October. Prepupal larvae spend winter in tough cocoons and are capable of maintaining the diapause for several years. We obtained multiple emergences after two and three years, synchronized with the normal, spring flight period.

Grapholita edwardsiana (Kearfott)(FW 4.3–6.5 mm, females larger than males; Plate 17.5) was proposed originally for specimens in the Henry Edwards collection thought to have been collected in or near San Francisco during the time Edwards lived there (1860s–1870s); then *G. edwardsiana* was not recorded for nearly a century, until we rediscovered the species in a relictual fragment of the San Francisco sand dunes in 1960. Its apparent restriction to a habitat once shared by the Xerces blue butterfly *(Glaucopsyche xerces)*, which occurred abundantly on the dunes but went extinct during their destruction in the early 1900s, led to designation of *G. edwardsiana* as a federally threatened species, the first such notice of an American tortricid. We found the larvae feed and overwinter in the elongate inflorescence stems of *Lupinus arboreus* (Plate 60.31), which enabled efficient survey of its occurrence. Now we know this species occurs along the immediate coast from Humboldt County to Monterey County, California. *Grapholita edwardsiana* is distinguished by having white fringes on the dark brown HW; the fringes are dark in *G. lana* (Kearfott), which occurs inland, associated with *L. latifolia*. The two are virtually indistinguishable morphologically and probably should be regarded as races of one species.

Cydia is the largest genus of Nearctic Grapholitini, having about 50 named species, more than half of which occur in the western states and northern Mexico. The larvae of this group are seed and bark feeders on diverse plants: conifers and legumes, as well as Rosaceae, Salicaceae, Aceraceae, and Fagaceae. The "codling moth," *C. pomonella* (Linnaeus)(FW 7.5–10 mm; Plate 17.10, Fig. 140), which generates the worms in apples, is one of the most notorious lepidopterans worldwide. If left unchecked, it will destroy 95% of the apples in any given orchard. This species likely was native to southern Eurasia and was distributed throughout Europe centuries before Linnaeus named it. The codling moth was introduced to North America in colonial times and reached the Pacific Coast by 1872. It occurs everywhere apples are grown, even isolated trees in urban yards, and it feeds on native crab apple *(Crataegus)*, pear, stone fruits, and even walnuts. This is one of the most intensively studied moths, with hundreds of literature references detailing the biology, life cycle, pheromone chemistry, monitoring, and control in various parts of the world. The moths are brown, FW with a bluish pattern consisting of parallel striae, and the ocellar patch is dark coppery brown with metallic golden transverse bars edged by black. The adults are mainly crepuscular in spring and nocturnal in summer. Winter is passed by prepupal larvae in tough cocoons affixed to the tree bark or in ground litter, and adults emerge in spring in time to oviposit on new leaves or very young fruit. The larvae enter the fruit to lodge in the seed capsules, and at maturity they tunnel to the exterior, leaving conspicuous holes surrounded by brown rot. Later generations oviposit on green fruit, and there are two to four generations, varying with latitude.

Cydia latiferreana Walsingham (Plate 17.7, Plate 17.8, Plate 60.30, Fig. 141), the so-called "filbertworm," is one of the most widespread and easily recognizable tortricids. It ranges throughout the United States and northern Mexico, wherever oaks grow. The adult is variable in color and genitalia, especially the form of uncus and aedeagus, in part geographically, according to Heinrich, and a detailed study is warranted. The FW (6–9.5 mm) varies from pale tan to dark brown, with two broad, metallic coppery or golden transverse bars. The larvae feed in acorns and the large galls ("oak apples") induced by cynipid wasps on oaks, in beech nuts *(Fagus)*, chestnut burs *(Castanea,* Fagaceae), and filberts or hazel nuts *(Corylus,* Betulaceae). The life cycle is not clearly documented. In coastal California adults are recorded continually from June through October, sporadically in November, March, and April. The "Mexican jumping-bean moth," *C. deshaisiana* (Lucas), has bluish gray FW (7.5–9.5 mm; Plate 17.9) with a band of dark, variably coalesced spots from the costa before the apex to the dorsal margin preceding the tornus, but the adults are rarely seen. Instead, the larvae, which occupy seeds of *Croton* (Euphorbiaceae) in northern Mexico, are sold by the thousands by toy companies and by street vendors in border towns as "jumping beans." When heated, the larvae thrash about, causing the seeds to move. In nature presumably this behavior carries the seed from hot, barren soil surfaces to crevices or other secluded sites; then the larva prepares an emergence window, becomes inactive, and pupates in the seed. In confinement, emergence of the moths occurs at various times of year.

Cydia populana (Busck)(Plate 16.53) was described from Montana, having been reared from larvae feeding in the bark of *Populus trichocarpa*, and it has been recorded from aspen *(P. tremuloides)*. *Cydia populana* is numerous at lights in Alberta and British Columbia to Colorado, Utah, northern Arizona, and eastern California, often in association with aspen. Flight records span July to September. The relative development of black-and-white in the FW pattern is variable.

There appear to be at least two lineages of conifer feeders in *Cydia*, pine and cypress feeders. Species related to *C. piperana* Kearfott, the "ponderosa pine seed moth" (Plate 17.11), have dark brownish gray FW, finely speckled by whitish scale tips, and with raised, transverse bars of metallic lead- or bronze-colored scaling. *Cydia piperana* is the largest western species, FW (8–10.5 mm, males larger than females) dark gray, finely speckled by whitish scale tips, with three metallic lead-colored, transverse, raised lines. The HW is entirely smoky brown. This is a widely distributed species in ponderosa and Jeffrey pine forests *(Pinus ponderosa, P. jeffreyi)*. The eggs are deposited on the surface of green cones, and the larvae burrow in the main axis of the cone and into the seeds. Pupation occurs in fall and winter in the larval galleries, so cones collected for Thanksgiving and Christmas decoration sometimes produce moths in homes far from their origin. Adults are recorded from February to June, varying with elevation and latitude. Prepupal larvae in diapause are capable of holding over until a second or third season prior to development and emergence. *Cydia injectiva* (Heinrich) and *C. miscitata* (Heinrich) are similar, somewhat smaller species, and their larvae sometimes occupy the same cones as *C. piperana*. The HW of *C. miscitata* basally and its fringe are whitish. *Cydia injectiva* (Plate 17.12) is distinguished by having long, dense hairs on the body and legs. The "cypress bark moth," *Cydia cupressana* Kearfott (Plate 17.13), occurs along the California coast on native cypresses *(C. macrocarpa, C. goveniana, C. sargentii)*, planted cypresses, and *Chamaecyparis* (Cupressaceae), *Cryptomeria*, and redwood, *Sequoia* (Taxodiaceae).The moths are smaller than the pine-feeding *Cydia* species (FW 5.5–8.5 mm), black, FW with silvery leaden-colored lines forming a pattern of squares. They are diurnal and rarely collected in the field. The larvae live in the bark and infest persistent cones, especially where tightly clumped. They occur primarily in places damaged by other insects (e.g., regularly associated in cypress cones with the cochyline tortricid *Henricus macrocarpanus* and in branches with *Coryneum*-cankers or cerambycid beetle larvae). *Cydia cupressana* larvae can be found throughout the year, but Frankie and Koehler studied the biology and identified a bivoltine cycle, flights beginning in spring and fall.

The legume feeders include the "pea moth," *C. nigricana* (Stephens) [=*rusticella* (Clerck), an older name that was suppressed by ICZN decision], which is a Palaearctic species that became established in northeastern North America in the 1800s. It reached British Columbia and Washington by the 1920s. This species is similar to the Oriental fruit moth *(Grapholita molesta)*, but the FW (5–7 mm; Plate 17.14) is brown, without a trace of *G. molesta*'s dorsal gray chevron. *Cydia nigricana* has white costal strigulae distally that become bluish below the costa, interspersed with dark brown; the ocellar patch is faint with four short dark bars. The larvae feed in seed pods of commercial and garden peas *(Pisum sativum)*, sweet pea *(Lathyrus odoratus)*, vetch *(Vicia)*, and other legumes. They eat the seeds, partially filling the pods with frass and silk before leaving to form tough cocoons in the soil. Adult emergence occurs in early summer, timed with the flowering of the pea vines. *Cydia prosperana* (Kearfott)(Plate 17.15) is closely related to the Palaearctic *C. succedana* (Schiffermüller). The lead gray FW is crossed by a variable, whitish, medial band, and the ocellar patch is pale gray with longitudinal black lines, divided by a lead gray band. *Cydia prosperana* occurs from Alaska to Alberta, Colorado, Utah, and the Pacific states, in association with bush lupines. The flight occurs in March to April in

coastal California, May to July in the Sierra Nevada, coastal Oregon, and Alaska, and in early August in Wyoming. *Cydia americana* (Walsingham)(Plate 17.16) is a dark brown moth recognized by a large, distinct, white patch on the middorsal edge of the FW (5.5–8.5 mm, females larger than males). This is a Pacific Coast species, ranging from British Columbia to central California (Monterey County). The adults are diurnal and fly in spring, end of February to early May at Big Creek. MacKay recorded the larval host as *Lathyrus*, based on preserved larvae from Seattle in the USNM, and D. Wagner reared a series from seed pods of *Lathyrus* in the Berkeley Hills, California. Larvae collected in early June yielded adults the following May. *Cydia membrosa* (Heinrich)(Plate 17.17) is a native species of the southwestern deserts from southern Texas to southeastern California, southern Nevada, and Baja California. The FW (5.5–9 mm) is dark gray with whitish markings and scale tips, giving an ashy appearance. The costal strigulae and dorsal margin patch are obscure, ocellar patch defined by black vertical bars. The larvae feed in the long pods of mesquite and probably screwbean *(Prosopis)*. Adults are recorded from April to July.

Two species of *Ofatulena* are associated with leguminous shrubs and trees in desert areas. *Ofatulena duodecemstriata* (Walsingham)(Plate 17.18) occurs in Texas, Arizona, Utah, Nevada, southern California, and northern Mexico, where the larvae feed in the seed pods of mesquite and screwbean *(Prosopis)*. The moth is similar to many *Cydia*, except in color, FW (5–8 mm) rectangular, pale gray with numerous dark gray transverse striae and a white ocellar patch. The moths often are attracted to lights or flushed from the host plants during the day. Adults are recorded April to September. *Ofatulena luminosa* Heinrich (Plate 17.19) lacks the transverse striae, and the distal area including the ocellar patch is ochreous. This species is more southerly, occurring along the lower Colorado River and southern Arizona to southern Baja California and Sinaloa, Mexico. We found the adults associated with palo verde *(Cercidium* and *Parkinsonia)* in Baja California.

The genus *Ecdytolopha*, recently monographed by Adamski and J. Brown, contains 10 Nearctic species, all quite similar in appearance (Plate 16.54), and most of them occur in the Southwest and northern Mexico. *Ecdytolopha insiticiana* Zeller, the "locust twig borer" (FW 8–12.5 mm), is mottled dark gray on the basal two-thirds, whitish clouded with gray distally, with a dark tornal bar. The larvae feed in the pods of locust *(Robinia pseudoacacia*, Fabaceae), burrowing into new growth stems and causing a spindle-shaped gall to develop, in which the larva feeds, ejecting the frass from an open hole. Overwintering occurs on the ground in cocoons coated with pieces of fallen leaves, and adults emerge in May and June. This is the most widespread *Ecdytolopha*, occurring through the Northeast and Mississippi Valley to Colorado, Wyoming, Utah, and eastern California. However, west of the Great Plains, where locust is not native, records are sparse, and populations may depend upon planted locust. In Utah and California occurrence of *E. insiticiana* is represented only by single records in 1922. *Ecdytolopha occidentana* Adamski & Brown (Plate 16.54), and *E. coloradana* Adamski & Brown occur in Arizona and Colorado and New Mexico, respectively. They are nearly indistinguishable superficially from *E. insiticiana* but lack the latter's male hair pencil on the HW.

SUBFAMILY TORTRICINAE

Tortricines differ from Olethreutinae in nearly all species by having two rings of scales per antennal segment, absence of the

HW cubital hair pecten, and having the juxta and aedeagus articulated, not strongly fused. Genitalia, Figs. 142–154. This is a worldwide lineage well represented in both tropical and in temperate regions of both hemispheres. There are more than 400 species in America north of Mexico, assigned to about 44 genera. Tortricinae worldwide are classified in 11 tribes, of which seven occur in western North America.

TRIBE TORTRICINI

Tortricini are medium-sized tortricids having rectangular FW, with rows of upraised scales, lacking a costal fold in the male. Three genera occur in the Nearctic, two of which range into western North America.

Acleris is a large, Holarctic genus with about 60 described Nearctic species, of which more than half occur in the West, mostly as extensions of transcontinental, northern distributions. There are only a few *Acleris* endemic to our region, but several species complexes are in need of detailed study. *Acleris* are characterized by remarkable polymorphism in FW pattern and color—two species in England have more than 100 named color forms each—other species have seasonal forms. Males lack the uncus and possess a broad, sclerotized plate derived from the ventral side of the anal tube (subscaphium). Many are generalist feeders on trees, but some host-plant records may be based on misidentifications caused by parallel polymorphism among different species. Accurate determination should rely on genitalic characters, which are diagnostic in both sexes for most species.

Acleris hastiana (Linnaeus)(Plate 17.20, Plate 17.21, Plate 17.22) may comprise a complex of species. In Europe this species is incredibly polymorphic—more than 100 names have been given to varieties by industrious lepidopterists, especially in England. Several FW pattern elements are independently expressed, each may occur in two or more colors, and there is pronounced geographic variation in the genetic makeup, so different localities have subsets of the polymorphism. North American populations are widespread in the West and are less variable, particularly those on the Pacific Coast, where most specimens have brown or dull reddish gray FW (7.5–11 mm) often with a pale gray basal patch or antemedial, transverse band, a form expressed in numerous *Acleris* species. The larvae are leaf tiers, primarily on willows *(Salix)* and other Salicaceae, including poplar and aspen *(Populus)*, or occasionally birch *(Betula*, Betulaceae), and there are a few records from Ericaceae and Rhamnaceae that need confirmation. There are at least two generations in coastal areas; we reared larvae that produced adults without diapause in March, April, May, and September to October on the California Channel Islands. Adults of the fall brood overwinter. *Acleris variegana* (Schiffermüller)(Plate 17.23, Plate 17.24, Plate 17.25) is a Palaearctic species that was introduced to the East and West coasts of North America. Its earliest western records are 1917 at Berkeley, California, and 1920 at Victoria, British Columbia. In Europe there are five discrete color forms, three of which were brought by the founders to the Pacific Coast. In the San Francisco Bay area, the FW (7–9.5 mm) is *a* (Plate 17.23) whitish on the basal half enclosing a bluish gray dorsal margin spot, clouded gray and dark bluish gray on the distal half (ca. 75%); *b* (Plate 17.24) nearly pure white on the basal half, dark bluish gray on the distal half (20%); or *c* (Plate 17.25) tan with bluish gray dorsal margin spot and outer costal triangle (5%). The larvae feed on Rosaceae, primarily *Prunus, Pyracantha*, and *Rubus* in California. Several fruit trees are hosts in Europe, but this species has not been reported as an orchard pest in North America. There

are two annual generations: adults fly from April to July, primarily May to June and August to October, and they overwinter. Resident colonies are known from the Puget Sound area to the Willamette Valley, Oregon, and in the San Francisco Bay area to Russian River and Monterey Bay areas of California, but not inland. *Acleris britannia* Kearfott (Plate 17.26, Plate 17.27) is similar in size and has form *a* of *A. variegana*, but more often the FW (8–10 mm) is pale yellow to rust tan with a dark costal triangular patch. This is a native species on the Pacific Coast, from British Columbia to central California, which feeds on Rosaceae including thimbleberry *(Rubus parviflorus)*, California blackberry and loganberry *(R. ursinus)*, and California wild rose *(Rosa californica)*. Evidently *A. britannia* is univoltine, with flight records in August to September. *Acleris keiferi* Powell (FW 7–9 mm; Plate 17.28, Plate 17.29, Fig. 143) is a similar Pacific Coast species that also feeds on Rosaceae. Its polymorphism differs in having form *a* dominant, with the pattern in some individuals heavily suffused by rust brown, or FW white with a dark bluish brown costal triangle and adjacent apical area. The HW undersides are heavily mottled with brownish. *Acleris keiferi* has been reared from *Rosa californica*, nonnative blackberry *(Rubus vitifolius)*, and strawberry *(Fragaria)*. The moths fly late June to September.

Acleris cervinana (Fernald) is a beautiful little species (FW 7–8 mm; Plate 17.30, Plate 17.31) with jaw-dropping different color forms. FW typically orange brown with obscure markings or tan finely reticulate with brownish; in other forms the ground color is pale fawn with a pale spot in the dark costal triangle, or ground color milk chocolate with a dark chocolate costal triangle; or rarely the thorax and basal half of the FW white, the remainder of the wing deep purplish brown, or entirely whitish with just the costal triangle deep purplish brown. This species ranges across Canada to the Pacific Northwest and north coastal California. The larva is recorded on hazelnut *(Corylus*, Betulaceae). *Acleris santacrucis* Obraztsov (Plate 17.32, Plate 17.33), described from central coastal California, may be a race of *A. cervinana*. We reared specimens on *C. cornuta* from eggs laid by a female taken on *Corylus* at Inverness, Marin County, in July, and adults emerged in September, indicating there are at least two generations. All the reared family were of the typical, orange brown form (Plate 17.32), as are several specimens from this population taken in the winter, but the FW of two are white with a dark costal triangle (Plate 17.33).

There are several larger, primarily gray species of *Acleris*. *Acleris nigrolinea* (Robinson), *A. disputabilis* Obraztsov, and *A. maximana* (Barnes & Busck) form a complex of morphologically similar species having a deeply emarginate saccular margin of the valva. The complex ranges from eastern Canada to the Rocky Mountains and Pacific states. Foreseeing, Obraztsov named *A. disputabilis* from widely scattered parts of the West; its specimens are variable in genitalia and not clearly distinguishable from eastern *A. nigrolinea*. The FW (11–14 mm) of both species (Plate 17.46) is gray, mottled with black or with a black or rust line from base to apex through the cell. Larval host records are mostly Salicaceae, rarely birch (Betulaceae) and cherry (Rosaceae). *Acleris maximana* has broader, usually paler gray FW (12–15 mm; Plate 17.47) sparsely mottled with black, sometimes forming an obliterated dark fascia beyond the cell, but apparently lacking discrete polymorphism in FW pattern. It occurs from Saskatchewan to British Columbia and widespread montane parts of the western states. The larvae feed on Salicaceae, primarily willow, also on aspen *(Populus tremuloides)* and balsam poplar (*P. balsmifera*), and are recorded on bitter cherry, *Prunus emarginata* (Rosaceae),

FIGURES 142–154. Genitalia structure of exemplar species of Tortricidae, Tortricinae, ventral aspect. **142,** *Acleris gloverana* (Walsingham) ♂, aedeagus upper left. **143,** *Acleris keiferi* Powell ♀. **144,** *Decodes fragarianus* (Busck) ♂, aedeagus upper left. **145,** *Decodes bicolor* Powell ♀ [Powell 1964]. **146,** *Dorithia semicirculana* (Fernald) ♂, aedeagus above. **147,** *Dorithia peroneana* (Barnes & Busck) ♀ [J. Brown and Powell 1991]. **148,** *Archips argyrospilus* (Walker) ♂, aedeagus below. **149,** *Clepsis virescana* (Clemens) ♀ [Powell 1964]. **150,** *Synnoma lynosyrana* Walsingham ♂, aedeagus below. **151,** *Synnoma lynosyrana* ♀. **152,** *Synalocha gutierreziae* Powell ♂ antennal cilia [Powell 1985]. **153,** *Platphalonidia felix* (Walsingham) ♂, aedeagus below. **154,** *Platphalonidia felix* ♀ [Powell].

by Keifer at Lake Tahoe and Crabtree in northern California. Spring and fall records of adults suggest two generations. *Acleris senescens* Zeller is superficially similar, dark gray with several polymorph forms (FW 9–12 mm; Plate 17.43, Plate 17.44, Plate 17.45) including pale basal and apical patches or a black or rust longitudinal streak, fine lines radiating along the distal veins, or rose rust in the cell or the whole area distal to the basal patch. But *A. senescens* is distinguished by narrower FW and conspicuously longer labial palpi than the preceding species and differs markedly in genitalia. It ranges from Manitoba to British Columbia and California, where the larvae are found commonly on willow in the spring (March to early June), producing adults in the fall (September to December). *Acleris senescens* also was recorded feeding on apple (*Malus*, Rosaceae) in the Forest Insect Survey in Canada (six of 32 records). The adults fly in January to February in coastal California, after overwintering.

Acleris gloverana (Walsingham), the "western black-headed budworm" (Plate 17.34, Plate 17.35, Plate 17.36, Plate 17.37,

Fig. 142), is a widespread forest insect that sometimes reaches outbreak population levels, causing defoliation of fir *(Abies)*, Douglas-fir *(Pseudotsuga)*, and hemlock *(Tsuga)*, especially in British Columbia and Alaska. This species was described originally from Mt. Shasta, California, but the single specimen was a rare polymorph form, and its identity was not recognized for nearly a century. As a result, all of the early forest entomology literature on *A. gloverana* was recorded under the name *A. variana* (Fernald), an eastern species that differs by having huge abdominal tufts of scales in the female, which are used to cover the eggs, and by male genitalia. Both species are incredibly polymorphic, and *A. gloverana* has remarkably different forms: FW (8.5–10.5 mm) commonly white with variable black basal patch and costal triangle, fading to weak markings; or the same with ochreous orange basal patch; or half white, half black as in form *b* of *A. variegana*, or the same or entirely dark gray with an orange or yellow stripe from base to apex; dark gray with orange ochreous basal patch (the rare form of the holotype), and so on. The larvae are bright green with a black head, hence the common name, until the final instar, which has a brownish head and thoracic shield; Duncan presents a color photo. Larvae reach maturity in summer, and adults fly in August and September.

Acleris foliana (Walsingham) is a western species that differs markedly from other *Acleris* in color pattern and in genitalia. The FW (6.5–10 mm; Plate 17.38, Plate 17.39, Plate 17.40) in most regions is tan to pale rust with a pale yellow costal triangle edged with rust; but in some places, notably around San Diego and on Santa Cruz Island, there is a greater range of ground color, tan to bright orange and dark rust, obscuring the costal triangle; or rarely tan and reticulate with fine brown lines. This species occurs from Montana, Utah, western Colorado, to northern Arizona and California. There is a single generation, adults flying May to July, in close association with mountain mahogany (*Cercocarpus*, Rosaceae), the larval host. The female has modified setae on the papillae anales, which retain dirt particles to be spread over the eggs, analogous to Cnephasiini.

Croesia is a small genus of small tortricids that are characterized by having rectangular FW with transverse lines of upraised, iridescent scales, vein R5 extends to the termen below the apex rather than to the costa as in *Acleris*, and the socii are enormous, well exceeding the tegumen. There are five species in North America, two of them introduced from the Palaearctic, and four occur in the West. *Croesia albicomana* (Clemens) is the most widespread, ranging transcontinentally in the north and widely in the western states. The FW (7–8 mm; Plate 17.48) is bright lemon yellow with four iridescent, rose brown fasciae from the costa, becoming diffuse on the dorsal half; HW white. The larvae roll the new spring leaves of *Rosa californica*, *R. gymnocarpa*, and other *Rosa*, including garden roses where grown near native *Rosa*. In California larvae are present March to April and adults May to June.

TRIBE CNEPHASIINI

Cnephasiini is a Holarctic tribe characterized by uniquely derived traits (highly modified ovipositor pads and correlated behavior, spined gnathos in the male genitalia, 12-carbon-based sex pheromone). The ovipositor lobes (papillae anales) are concave and bear capitate setae, enabling the female to scrape up and carry dirt particles, to be spread over the eggs, which are laid singly or in imbricate masses by different species. Genitalia, Figs. 144, 145. The larvae of some cnephasiines are leaf miners in early instars. North America has a restricted fauna compared to the Palaearctic, with about 25 species; only *Decodes* is endemic and species rich in the West.

Eana is a diverse genus in the Palaearctic, with two of its species Holarctic, and there are two endemic to the Nearctic. *Eana argentana* (Clerck)(Plate 17.49) is a Holarctic-boreal species that ranges across Canada and in the western mountains to Colorado, Idaho, and northern California. The moth is shining cream white, larger (FW 11.5–15 mm), and rare in collections compared to the quite similar *E. georgiella* (Hulst), which has chalk white FW (11–13 mm; Plate 17.50) and pale gray HW. The latter species is more widespread in the West, occurring in mountain meadows from British Columbia to New Mexico and the southern Sierra Nevada. Hulst's name is a classic example of taxonomic incompetence preserved by nomenclatural priority. He described *E. georgiella* from Colorado as a crambid (!), and its identity was overlooked by Obraztsov, who named the same species *E. subargentana* 75 years later. It often flies with the shining white *Crambus perlellus innotatellus*

Walker, and the two sometimes are mixed in collections. In Europe *E. argentana* deposits eggs in imbricate masses and covers them with debris and scales; first-instar larvae construct individual silken hibernacula and overwinter in the sod. In spring the larvae emerge to feed on diverse plants, angiosperms and gymnosperms. We found the same oviposition and hibernation behavior by *E. georgiella*. Both species fly in July and August.

Cnephasia has about 40 species in the Old World, one of which, *C. longana* (Haworth), is introduced to the Pacific Coast of North America. This is the "omnivorous leaf-tier" of American agricultural literature, a dimorphic moth, FW (7.5–11 mm; Plate 17.51, Plate 17.52) of male uniform tan, of female whitish with variable brown, zigzag fasciae. The larvae are general feeders on low-growing herbs, especially in the flowers, and have damaged field crops, including strawberries, flax, and commercial cultivated flowers, in Oregon and California. The eggs are laid singly on upright objects—tree trunks, fence posts, and so forth—and the life cycle is similar to that of *Eana*. In spring the early instars create slender, serpentine mines in herbs and later tie up the flowers with silk. Adult emergence is May and June. This species was first discovered in North America in 1929, when larvae were found on strawberry in Oregon, and within five years *C. longana* had become extensively established in the Willamette Valley in Washington and Oregon. It was reported in California in 1947 on flax and cultivated flowers, and by 1949 had become widespread in the San Francisco Bay area. Later, *C. longana* became a pest of strawberries in the Monterey Bay area and extended its range along the coast, north to the Humboldt Bay area by 1960, and south to Lompoc and Goleta in Santa Barbara County by 1970–1976. There was a massive colony infesting all kinds of herbs on Santa Rosa Island in 1995, apparently having been introduced after 1978, and *C. longana* was numerous at Miramar in San Diego County, by 1996. *Decodes* is the only endemic cnephasiine lineage to have radiated in North America. There are 18 species described, primarily from the Southwest and northern Mexico, 14 of them in California, and perplexing specimens possibly representing several other species are known. There are two life history guilds, larvae of both feed in spring, one group flying in fall, the other, mostly larger species, fly in early spring. Pupae aestivate in tough silken cocoons in the soil. *Decodes basiplaganus* (Walsingham) and *D. fragarianus* (Busck) are superficially similar, widespread, oak-feeding species, which are readily distinguished by genital morphology in both sexes. *Decodes basiplaganus* (Plate 18.6) occurs in the East and Midwest to central Texas, Colorado, Utah, and Arizona to central California. The moth is gray, FW (8–9 mm) with a black transverse band at the basal one-fourth, pale gray medial and darker gray postmedial fasciae, and some brown scaling at the base. Specimens are distinguished by the contrastingly darker basal fascia, whereas the basal and postmedial fasciae are a uniform hue, pale to dark gray, in *D. fragarianus* (FW 6.5–8.5 mm; Plate 18.7, Fig. 144), which lacks the basal brown scaling. These two species feed on various species of oak (*Quercus*). The larvae feed on new foliage in spring, and adults emerge August to October, rarely in June to July or November. *Decodes fragarianus* was described from Vancouver Island, where Blackmore reported the species "breeds commonly in the buds at the base of the crowns of strawberry." Lacking any further implication with *Fragaria* to substantiate the misnomer, one might guess that larvae from adjacent oaks were found pupating in a strawberry field. This species occurs along the Pacific Coast to northern Baja California and separately inland, Utah, Colorado, northern Arizona, and New Mexico. In regions where these two

Decodes are sympatric, *D. basiplaganus* has only scattered colonies and is vastly outnumbered at sites where the two occur together. The data suggest that where they use the same oaks, the more widespread *D. basiplaganus* is at a disadvantage and its abundance is greatly reduced or the species is eliminated by competitive displacement.

Decodes aneuretus Powell and ***D. helix*** Powell & J. Brown are dark gray species on the California coast and northern Channel Islands, which feed on Ericaceae and fly in winter and spring. *Decodes aneuretus* (FW 10–12 mm; Plate 18.3) occurs from Mendocino and Marin counties to Monterey County, feeding on *Arctostaphylos virgata* and *A. hooveri*, and adults fly in April and May. *Decodes helix* is smaller (FW 8–9.5 mm; Plate 18.2), paler gray with black maculation. The larvae feed on *A. confertifolia* and *A. insularis* on Santa Cruz and Santa Rosa islands. At Miramar, San Diego County, the host plant evidently is *Xylococcus bicolor,* and adults fly from November to March. ***Decodes catherinae*** Powell is a spring-flying species having narrow FW (11–13 mm) with variable maculation. The FW is gray, usually with the basal area or antemedial, transverse fascia paler, defined by an outwardly angled line from midcostal toward the tornus, often emphasized distally by black, and there may be a black triangle on the Cu crease into the cell. *Decodes catherinae* is unique among Nearctic Tortricidae in feeding on Sterculiaceae, *Fremontodendron californicum*. We found larvae in the flowers in June, and lab emergence occurred the following January. The host association dictates *D. catherinae*'s peculiar distribution, southern Sierra Nevada to the desert slopes of the Transverse Ranges and inland foothills of southern California. ***Decodes bicolor*** Powell is distinctive, having the FW (10–12 mm; Plate 18.4, Fig. 145) white on the costal half, dark gray on the dorsal half, lacking polymorphic forms. It occurs in southern Oregon and California south to the San Gabriel Mountains. The flight period is early spring, March near the coast to June at Lake Tahoe.

Decodes horarianus (Walsingham), a larger (FW 10.5–12 mm; Plate 18.1), fall-flying species in southern Oregon and northern California, represents a complex of species in California, Colorado, and Canada east to Quebec, larvae of which feed on gooseberries and currants *Ribes* (Grossulariaceae). These species vary from FW having strong, distinct, transverse bands in *D. horarianus* and ***D. tahoense*** Powell (Plate 18.5) to having obscure and broken bands in other species. Owing to their late seasonal flight, members of this complex have been rare in collections, and for many populations the variation and species status has not been adequately assessed.

TRIBE EULIINI

The tribe Euliini is defined primarily on ancestral state (plesiomorphic) characters and may not be monophyletic. There is an expandable hair pencil at the base of the prothoracic femur in males of about half the genera, which may represent a shared, derived feature (synapomorphy). Genitalia, Figs. 146–147. This is a conglomeration of diverse, primarily Neotropical forms, with nearly 600 described species in the New World fauna and a single Holarctic species. The larval biology is unknown for most species, but the few observed deposit the eggs singly or in small groups; and there is lab evidence that leaf-litter-feeding may be a common larval habit, analogous to the Epitymbiini in Australia. There are 20 species assigned to eight genera in Canada and the United States. All except one occur in the western states, and there are many others in northern Mexico.

Eulia ministrana (Linnaeus)(Plate 18.13) is a Holarctic species that ranges from Nova Scotia and New York to British Columbia, Washington, central Alaska, and north coastal California. The adult is unlike any other Nearctic tortricid, FW (9.5–11.5 mm) broad, olivaceous to orange brown, shading into red at the terminal margin, with a diffuse, pale median fascia, and a prominent brown patch on the dorsal margin, the most conspicuous mark. The larvae live in folded leaves of many kinds of trees and shrubs (Betulaceae, Salicaceae, Rosaceae, Rhamnaceae, Onagraceae, etc.) and overwinter, pupating in spring. Flight records are June and July.

Apotomops wellingtoniana (Kearfott)(Plate 17.53) resembles species of *Apotomis* (Olethreutini) in having the basal portion of the FW (6.5–9.5 mm) banded or reticulated with black and gray, and the subterminal area variably whitish. This species occurs from Newfoundland to British Columbia, north coastal California, and the Rocky Mountain states to northern Arizona and New Mexico, south to Veracruz. Larvae of *A. wellingtoniana* were reared from Pinaceae: fir *(Abies amabilis),* hemlocks *(Tsuga heterophylla, T. mertensiana),* and spruce *(Picea)* by the Canadian Forest Insect Survey. The species is univoltine in the north, but records in Arizona indicate spring (April) and summer (July to August) flights.

Dorithia ranges from Wyoming to Costa Rica, and three of its 17 described species occur in the western United States. They are broad winged, FW somewhat triangular with a slightly produced apex, often ochreous to orange with a crisscross pattern of rust lines, and the HW is white. ***Dorithia semicirculana*** (Fernald) is the most often recorded species, occurring from southern Wyoming to western Texas, Arizona, and Chihuahua. The FW (8–12 mm; Plate 17.41, Fig. 146) varies from cream whitish to tan and the pattern from a nearly complete double cross of rust brown, formed by outwardly angled lines from the costa at the basal one-fourth and midcell, inwardly angled lines from midcell and the end of the cell, to having the lines weak or broken below their crossing or rarely obsolete; ground color more or less uniformly reticulated with thin lines. Although the adults have been collected for more than a century, larvae of *D. semicirculana* have not been discovered in the field. We suspect they are leaf-litter feeders, but this species appears to be associated with white oak *(Quercus gambellii)* in much of its range, and we reared both this and the following species from eggs to maturity on green leaves of valley oak *(Q. lobata)* The larvae ignored synthetic diet, suggesting that they are not polyphagous. ***Dorithia trigonana*** J. Brown & Obraztsov is similar, having orange FW (9–12 mm, lab-reared runts from 7 mm; Plate 17.42) and more distinct, darker reticulation. It occurs from northern Arizona to New Mexico and the Sierra Madre Occidental west of Durango, Mexico. ***Dorithia peroneana*** (Barnes & Busck)(Fig. 147) has a similar distribution. It differs by having the FW markings restricted to a triangular patch on the costa, like many *Acleris* (=*Peronea*). In Arizona all three species have a spring flight (March to April) and one or two summer generations (June to October).

Anopina is the largest genus of Euliini, with 65 described species, ranging from southern Canada to Costa Rica. Males of most are gray with a dark costal triangle, and many exhibit appreciable sexual dimorphism in color and FW pattern. Ten species occur in the West, of which *A. triangulana* (Kearfott), a Californian endemic, is the best known. Males average larger than females (FW 5.5–8 mm; Plate 18.8, Plate 18.9) and have pale gray FW mottled with dark gray and a well-defined dark triangle beyond midcosta; females have the distal half of the

FW mostly fuscus clouded, obscuring the costal triangle, and the basal area is pale ochreous or tan. This species occurs from central California to northern Baja California. There are two fairly discrete generations in northern populations, with May to July and August to September flights, but the moths are active April to November in southern California. Adults are frequent at lights, yet larvae have not been discovered in the field. We reared *A. triangulana* from eggs to adults on willow leaves and on synthetic diet, but when given a choice, larvae selected old brown or blackened leaves in favor of fresh, green or yellowing ones, usually *Prunus* (Rosaceae) over various others, even other Rosaceae *(Rosa* and *Pyracantha)* or willow.

Anopina internacionana J. Brown & Powell is distinctive, a small, colorful species, FW (5.5–7.5 mm; Plate 18.11) tan with costal patches of yellow, orange, and dark brown, the largest covers most of the cell and is partially reddish brown. The HW is dark brown, and there is no conspicuous sexual dimorphism. This species occurs from Colorado and Utah through Arizona to the mountains of Durango, mostly at elevations of 5,000 to 9,000'. In Arizona it flies during and just before the rain season, July to August (more than 20 records). By contrast, sympatric *A. eleonora* Obraztsov (Plate 18.10), a gray species with well-defined, dark basal and costal patches, flies in the dry season, April to early July (more than 50 records). Although *A. internacionana* probably has been taken at lights by other collectors, nearly all of our collections have been diurnal, including moths flushed from *Arctostaphylos,* and *Ceanothus.* In tests similar to those with *A. triangulana,* larvae of *A. eleonora* fed on old brown, drying or decaying *Prunus* leaves in preference to fresh ones. Larvae even fed on brown areas but did not move to green areas of the same leaf. Hence, larvae of this genus are suspected to be leaf-litter feeders on the ground.

Anopina silvertonana Obraztsov is a moderately large species (FW 6–8.5 mm), similar to *A. eleonora,* but lacking the basal FW patch, and it has a somewhat bicolored FW in the female. It is restricted to the Rocky Mountain region, from northern Wyoming to eastern Arizona and New Mexico, and is one of the more often collected *Anopina* species. This species occurs mostly at high elevations, 8,000' to above 12,000', and is univoltine, with flight records from late June to August. We reared *A. silvertonana* from eggs on synthetic diet, and three individuals emerged after108 to 140 days, the longest tortricid larval life without an apparent diapause phase that we have observed.

Acroplectis haemanthes Meyrick is a peculiar, slender-winged species, superficially resembling a phycitine pyralid. The FW (6.5–8 mm; Plate 18.12) is whitish with dark reddish pink basal transverse band and apical area. This species occurs from western Texas (Alpine and Palo Duro Canyon) to southeastern California (San Diego County desert, Owens Valley) but has been rarely collected. Nothing is known of the larval biology. The adults are recorded in April and July in both southern Arizona and California.

TRIBE ARCHIPINI

The tribe Archipini is worldwide in distribution, better represented in the Holarctic and south temperate zones than in the tropics. Although there has been no comprehensive phylogenetic analysis of Tortricinae or Archipini, this tribe is thought to be polyphyletic, as it contains taxa based on plesiomorphic features as well as derived lineages. Archipines possess well-developed gnathos arms, joined apically, and most have a strong, awl-shaped signum on or protruding into the female

corpus bursae, often with a clublike capitulum protruding externally, like the handle of a dagger. Genitalia, Figs. 148, 149. Members of this tribe and the following two, Sparganothini and Atteriini, deposit the eggs in imbricate masses, typically a large one first (100–200 or more eggs), followed by successively smaller patches over several days. The larvae then disperse to upper and peripheral vegetative terminals to form shelters in which they feed. About 40% of the of archipine species for which larvae are known, and nearly all Sparganothini and Atteriini, are general feeders, in contrast to the pattern of host specificity prevalent in Tortricidae. Archipini is represented in North America by about120 species placed in 15 genera. Two of the genera and at least 10 species are introduced to this continent from the Palaearctic.

Argyrotaenia is a diverse but seemingly monophyletic group, with about 35 described species in North America and greater numbers in Mexico and Central and South America, many of which await naming. About two-thirds of the Nearctic species occur in the West, including many endemics, and several species complexes are in need of careful study. Taxonomy is complicated by marked sexual dimorphism in some species. Reared families from individual females resolve the problem of matching the sexes when two or more species are sympatric, and many *Argyrotaenia* are polyphagous, readily accept synthetic diet, and are easy to rear after obtaining oviposition by confined females.

Argyrotaenia coloradana (Fernald)(Plate 18.14) occurs in the Rocky Mountain and southwestern states to western Arizona. Freeman listed *A. coloradana* in California, but we have not seen specimens to confirm it. This is a distinctive species having broad FW (8.5–11 mm) pale with a well-defined reddish brown basal patch, median band, and outer costal spot. Posterior half of FW ochreous; HW white. Dyar recorded *Pulsatilla* (=*Anemone*)(Ranunculaceae) as a larval host in Colorado. We reared specimens from eggs on synthetic diet, which suggests that *A. coloradana* is polyphagous. Adults are recorded in late June to August. Eggs deposited in July produced larvae that grew very slowly during fall months, but adults emerged October to January, so in natural situations winter may be passed by young larvae.

Argyrotaenia franciscana (Walsingham), which has been called the "apple skinworm," and *A. citrana* (Fernald), the "orange tortrix," are members of an intriguing complex of semispecies along the Pacific Coast and California Channel Islands. *Argyrotaenia franciscana* sensu stricto occurs on the immediate coast from Washington to central California. Males have narrow FW (5.5–8.5 mm; Plate 18.16, Plate 18.17) with a sinuate costa, gray with whitish antemedial band, dorsal area, and outlining the distal costal spot. Females have broader FW with diffuse markings. The HW is gray with a whitish fringe in both sexes. *Argyrotaenia citrana* (Plate 18.18, Plate 18.19) was described from southern California in the 1880s, where it had become a pest of citrus. The native distribution is unknown, but probably it occurred in California inland from the coast and west of the mountains. It has dark rust brown (winter) to pale orange (summer) FW (6–10 mm) with darker brown pattern similar to *A. franciscana,* but the FW is measurably broader and the costa not as sinuate. The HW are white. Thus, although the genitalia are quite similar, the typical forms of these two appear to be distinct species. However, the larvae of both are polyphagous, and *A. citrana* readily adapted to urban situations, expanding its range to include coastal cities. It hybridized with *A. franciscana,* as evidenced by voucher specimens representing successive changes in phenotype during 1910–1960 in San

Francisco. Reciprocal mating in the lab between various populations produce fully viable progeny, and evidence from mitochondrial DNA failed to distinguish between populations along the coast, including typical *A. franciscana* and on the northern Channel Islands (subspecies *A. f. insulana* Powell), and *A. citrana* from southern California, which interbreed in the lab. *Argyrotaenia isolatissima* Powell (Plate 18.20, Plate 18.21) was described from the smallest island, Santa Barbara, and its distinctness has been confirmed by mitochondrial and nuclear DNA analysis. Phenotypically distinct races also occur on the other southern islands, small and dark brown on San Nicolas (FW 5.5–6.5 mm; Plate 18.23), larger and pale tan on San Clemente (FW 6–9.5 mm; Plate 18.22), which like *A. isolatissima*, show reduced embryonic success when crossed with mainland "*citrana.*" All of these are multivoltine and lack diapause, and all stages overwinter, becoming active as temperatures permit. Their larvae feed on all kinds of herbaceous shrubs, occasionally trees, even conifers, and readily accept synthetic diet. Color forms of the larvae on various host tress in British Columbia are shown by Duncan. *Argyrotaenia niscana* (Kearfott)(FW 6–9 mm; Plate 18.15) is similar morphologically but is dark reddish, and this species is univoltine, a specialist on chamise, *Adenostoma fasciculatum* (Rosaceae), an endemic shrub. It ranges the length of the Californian Province, Monterey County to northern Baja California, including Santa Cruz and Santa Rosa islands.

Argyrotaenia dorsalana (Dyar) occurs from southern British Columbia to eastern California, western Colorado, northern Arizona, and New Mexico. Typically the FW (8–12 mm; Plate 18.25, Plate 18.26) is pale yellow with variable black markings on the dorsal margin and tornal area, but in parts of the Great Basin the ground color is rust orange (Plate 18.27). The type series was recorded from oak in northern Arizona, but all subsequent larval collections have been from conifers (*Abies, Pseudotsuga, Larix,* and *Picea;* even Cupressaceae, *Juniperus* and *Thuja*). This species is more boreal in distribution in California than is Douglas-fir, which is a preferred host in British Columbia. The bright green larva is illustrated by Duncan. We reared the rust-colored Nevada form by starting the larvae on pine needles, then weaning them to synthetic diet, and pinyon pines may be the primary hosts in Great Basin populations. Larvae collected from white fir during our *Choristoneura* survey in the Warner Mountains and Sierra Nevada in June and July produced adults without diapause. *Argyrotaenia lautana* Powell (Plate 18.28) is a larger sister species, judged on similarity of genitalia, that occurs in the Greenhorn Mountains and Transverse Ranges of southern California to Mt. Palomar. The FW (9–13 mm) is tan with weak, brown costal markings. We reared *A. lautana* from *Abies concolor* at Tehachapi Mountain, from *Pseudotsuga macrocarpa* in the San Gabriel Range and from both of these hosts in the San Bernardino Mountains. Both of these *Argyrotaenia* fly in June to July. *Argyrotaenia provana* (Kearfott) also is a western conifer feeder, but it differs by having whitish FW (8.5–11 mm; Plate 18.24) banded and mottled with gray. This species occurs from Vancouver Island to central California on the coast and widely inland, to Colorado and New Mexico. The larvae feed on *Pseudotsuga* and *Abies.* Those we reared were collected in June and July, and adults emerge late July to September. There are flight records in May at the coast to July inland.

In contrast to most *Argyrotaenia,* members of a western complex of species are conifer feeders, and their FW pattern resembles those of unrelated tortricids (e.g., *Eucosma bobana* Group, *Cudonigera, Diedra,* and some *Choristoneura*), being

primarily orange and tan checkered. *Argyrotaenia cupressae* Powell (Plate 60.34, Plate 60.35) feeds on native and cultivated cypresses in coastal California and inland in juniper (Cupressaceae). A sister species, *A. paiuteana* Powell, is paler colored and occurs in pinyon-juniper woodland of eastern California, Nevada, and Utah.

Diedra are similar to *Argyrotaenia* but have much more heavily sclerotized male genitalia, with elongate, spatulate gnathos, and narrow FW that are rust with slightly raised paler scaling in a checkered pattern, characteristic of conifer feeders. *Diedra cockerellana* (Kearfott)(FW 8.5–11 mm) is the most widespread of five species, occurring from southern Ontario to Nebraska, Colorado, Arizona, and the Pacific Northwest. The larvae feed on junipers (*Juniperus,* Cupressaceae). Nearly all collections of adults have been August to September, rarely July and early October, but there is one record in May at 5,600′ in Colorado. The quite similar **D. intermontana** Rubinoff & Powell (Plate 18.36) occurs in the Great Basin from Utah to northern Arizona and eastern California. *Argyrotaenia calocedrana* Rubinoff & Powell is a smaller species (FW 7.5–9.5 mm; Plate 18.37) with darker FW and dark gray HW. It occurs on the west slope of the Sierra Nevada to the San Jacinto Mountains of southern California, in association with the presumed larval host, incense cedar, *Calocedrus decurrens* (Cupressaceae). Adults fly in July and August.

Archips is a large, primarily Holarctic genus richly represented in the Palaearctic. There are 28 described species in America north of Mexico, more than half of which range across the continent, but none is endemic to the West. Four species have been introduced from the Old World. *Archips* are broad winged compared to most Archipini, with the apex somewhat produced in the larger females, and the males have a broad costal fold. Most species are univoltine, and the eggs remain in diapause over summer and winter, protected by a thick, paintlike layer of colleterial secretion. Females of some species also cover the egg masses with scales from large brushes on the underside of the abdomen. Unlike other Nearctic Tortricinae, larvae of some species are gregarious, forming large communal shelters like those of tent moth caterpillars (Lymantriinae).

Archips argyrospilus (Walker)(Plate 18.30, Plate 18.42, Fig. 148), which may comprise a complex of species, occurs across the Nearctic. This is the well-known "fruit-tree leaf-roller" of economic entomology literature. Historically this species developed outbreak population levels as new orchard areas became available, probably representing local movements from native hosts. The adults are sexually dimorphic and geographically variable. Males have dark brown to reddish brown FW (7–12 mm) with distinct white markings preceding and following a median transverse fascia, which becomes diffuse toward the dorsal area. The HW is gray with pale fringe. Females have blurred FW pattern, and in western populations vary to a pale, golden tan form with white HW. Populations in Colorado (*A. a. vividanus* Dyar) tend to have bright reddish brown scaling, whereas those in the Pacific Northwest (*A. a. columbianus* McDunnough) and California are darker, with an olive cast. The larvae are bright green with a black HC until the last instar, which has a dark gray dorsum caused by minute, black spiculae, and a brownish head capsule. They feed on a wide variety of trees and shrubs, including *Salix* (Salicaceae), *Acer* (Aceraceae), *Betula* (Betulaceae), *Cercocarpus* (Rosaceae), *Ceanothus* (Rhamnaceae), *Eriodictyon* (Hydrophyllaceae), even conifers occasionally, but in native habitats in California most frequently on oak (*Quercus,* Fagaceae). Many of the recorded hosts, especially herbs, may have resulted from secondary feeding by dense, defoliating populations. Adaptation to fruit trees seems to have been opportunistic. MacKay found two larval types in California, one in fruit orchards, which suggests introduction of a sibling species, but Kruse and Sperling produced molecular evidence supporting a Pacific Coast clade differentiated from Colorado and eastern *A. argyrospilus*. The adults are recorded from late April to July but fly for only three to four weeks at any given site. *Archips negundanus* (Dyar) is similar but the FW is much paler (8.5–11 mm; Plate 18.33), especially the females, which resemble the pale form of *A. argyrospilus*. Females of *A. negundanus* possess huge brushes of erect scales covering the abdominal venter, which are laid in neat, imbricate rows across the egg masses. The larvae feed on box elder, *Acer negundo* (99% of 190 records by the Canada Forest Insect Survey). Freeman listed this species and the similar *A. semiferanus* (Walker) in California, but we have not seen specimens of either species from the Pacific Coast states. The latter species is polyphagous and widespread, eastern United States to Colorado with dark purplish gray female abdominal scaling. *Archips rosanus* (Linnaeus)(FW 7–10.5 mm; Plate 18.31) is widespread in Europe, where the larvae are general feeders on shrubs and trees, occasionally conifers. This species was discovered in New York by 1890 and later from Nova Scotia to Pennsylvania, then colonized at Vancouver Island in 1920, whence it spread widely in western Washington and Oregon, its ubiquitous larvae feeding on all kinds of native and ornamental trees and shrubs. *Archips rosanus* is univoltine, larvae developing in May and June, adults in late June to August.

Archips cerasivoranus (Fitch)(Plate 18.32), the "cherry-tree ugly-nest caterpillar," is remarkable for its gregarious larval habits, living in large silken nests on native and cultivated cherries (*Prunus*, Rosaceae). Sometimes the tents cover many shrubs and contain hundreds of caterpillars (Plate 60.32, Plate 60.33). Adults are bright to pale ochreous orange, FW (10–12.5 mm) crossed by broken, transverse lines of shining, dark rose rust, prominent only at the costa, and HW ochreous yellow, without much sexual dimorphism in color. The larvae were recorded on a variety of host shrubs and trees by the Canada Forest Insect survey, but more than 90% of the more than 2,300 records were native cherries. *Archips cerasivoranus* occurs from the Northeast to British Columbia, Montana, Colorado, and northern California. Adults fly in July and August.

Archips packardianus (Fernald)(Plate 18.34) is one of several northern conifer-feeding *Archips*, but it differs by having dark gray FW crossed by incomplete whitish fasciae that are finely speckled with brown and is smaller (FW 7.5–9.5 mm). In Canada the larvae are found primarily on spruce *(Tsuga,* more than 90% of 92 Forest Insect Survey records), rarely on balsam fir *(Abies basamifera). Archips packardianus* occurs from Nova Scotia to British Columbia, Wyoming, Montana, and Washington. Duncan illustrated the larva, which is green with black head capsule and thoracic shield.

Syndemis has been perceived as monotypic in North America until recently. Dan Rubinoff has produced molecular evidence revealing two host-plant races on the California coast that represent undescribed species, which are derived from the widespread **S. afflictana** (Walker). These are gray moths, FW (8.5–10.5 mm; Plate 18.35) with a black median band and less distinct dark gray basal and outer costal patches. In Canada, *S. afflictana* is primarily a fir feeder *(Abies)* and occasional larvae are found on willow, alder, and birch, but in New York Chapman and Lienk considered it to be a general feeder, with apple and other Rosaceae primary hosts. By contrast, available evidence indicates that three lineages in California are host specific

on conifers, *Abies*, Monterey pine *(Pinus radiata),* and coast redwood *(Sequoia sempervirens,* Taxodiaceae). Serendipitously, males of *Syndemis* are attracted to pheromone lures developed specifically for *Choristoneura retiniana* (Walsingham). This led us to successfully survey for montane and Pacific coastal populations of *Syndemis,* which had been rare in collections. *Abies*-feeding *Syndemis* fly early in the season in contrast to *Choristoneura* and rarely come to lights, whereas the Monterey pine species flies at various times of year, evidently multivoltine, and is primarily nocturnal; and the redwood species flies June to September and is almost exclusively diurnal. The larvae are bright to dark green including the head, with lateral white lines, variation illustrated by Duncan.

Aphelia alleniana (Fernald)(Plate 18.29) is the most widespread of four similar species of *Aphelia* in North America. It occurs transcontinentally, from the Northeast to British Columbia, Washington, Oregon, and Colorado. The FW (11–15 mm) is broad and varies from reddish to golden ochreous, rarely grayish olive, in males usually with a median band and outer costal spot and ground color crossed by many transverse dark strigulae, whereas the markings are indistinct in females. The larvae are reported to be general feeders on low herbaceous vegetation as well as deciduous and coniferous trees. Adults fly in June and July.

Choristoneura is primarily Nearctic, with 16 described species, and there are four in the Palaearctic. There are two groups; about half the species are typical archipine in facies, with tan to rust FW crossed by a dark median fascia and a preapical spot at the costa, and these are general feeders. The second group consists of conifer-feeding species, mostly having checkered or reticulate orange to brown FW patterns. Females are appreciably larger than males in both groups. *Choristoneura rosaceana* (Harris)(Plate 18.43, Plate 18.44), the "oblique-banded leaf-roller" of agricultural entomology literature, is the most widespread and best known species of the first group. It ranges throughout the continent, from Newfoundland and northern Alberta to Florida, Texas, and all the western states, except at high elevations and in the deserts. Males have tan FW (10–13 mm) with weakly upturned costal fold and brown pattern, white HW, whereas females are larger (FW 14–16 mm), darker rust brown, FW pattern less distinct, and the HW is ochreous with the anal area dark gray. The larvae feed on a wide variety of trees and shrubs, and *C. rosaceana* is a pest of apples and other orchard crops. There are two generations in most parts of the range, adults in May to July and August to September. *Choristoneura conflictana* (Walker), the "large aspen tortrix," is pale gray to brownish gray with the FW (12–17 mm; Plate 18.45, Plate 18.46) pattern indistinct. Males lack a costal fold. This boreal species occurs from Newfoundland to the southern Yukon and New England to interior British Columbia, south in the western mountains to New Mexico, eastern Arizona, Great Basin ranges, and Sierra Nevada. The larvae feed primarily on aspen *(Populus tremuloides)*—90% of 3,380 records by the Canada Forest Insect Survey—and other Salicaceae, sometimes reaching defoliating levels, along with *Pseudosciaphila duplex* (Olethreutini). *Choristoneura conflictana* is univoltine, larvae maturing in June, adults active June to July.

The "spruce budworm" complex comprises several species and semispecies related to **Choristoneura fumiferana** (Clemens) of eastern North America. These are the most destructive foliage-feeding insects of coniferous forests in North America. Researchers and forest entomologists have investigated every aspect of their biology, ecology, physiology, population

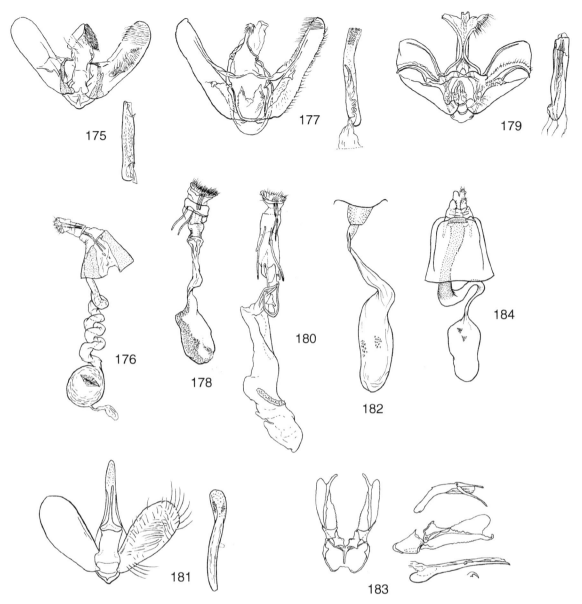

FIGURES 175–184. Genitalia structure of exemplar species of Crambidae, ventral aspect, except Figs. 176, 178 lateral. **175,** Pyraustinae: *Pyrausta subsequalis* (Guenèe) ♂, aedeagus below, right. **176,** Pyraustinae: *Pyrausta fodinalis* (Lederer) ♀ [Munroe 1976]. **177,** Spilomelinae: *Udea profundalis* (Packard) ♂, aedeagus to right. **178,** Spilomelinae: *Udea profundalis* ♀ [Munroe 1966]. **179,** Spilomelinae: *Nomophila nearctica* Munroe ♂, aedeagus to right. **180,** Spilomelinae: *Nomophila nearctica* ♀ [Munore 1973]. **181,** Acentropiinae (=Nymphulinae): *Petrophila longipennis* (Hampson) ♂, aedeagus to right. **182,** Acentropiinae (=Nymphulinae): *Usingeriessa brunnilalis* (Dyar) ♀, bursa copulatrix [Lange 1956b]. **183,** Crambinae: *Crambus sperryellus* Klots ♂, fragmented, tegumen and valvae left, valva inner aspect to right middle, uncus above, aedeagus below. **184,** Crambinae: *Crambus sperryellus* ♀ [Klots 1940].

HW submarginal, yellow streak is better developed. Adults fly throughout the summer, with two or more generations. Immense migrations are recorded in the Palaearctic, and many Nearctic records may represent vagrants. *Loxostege cereralis* (Zeller)(Plate 22.39), the "alfalfa webworm," formerly was not distinguished from *L. commixtalis* (Walker), a boreal species, and other members of its complex. *Loxostege cereralis* is larger than the related species (FW 13–16 mm) and has numerous black streaks forming basally directed wedges on the veins. This species ranges from Quebec to British Columbia, Colorado, and the southwestern states, reaching eastern and north coastal California. Larval habits are similar to those of *L. sticticalis*; adults are active May to August. *Loxostege sierralis* Munroe is smaller (FW 9–11 mm; Plate 22.40) and paler and occurs in the high

Sierra Nevada. The caterpillar is black with red brown pinacula; we found one in a rocky area with sparse grass at 10,200' in August, and the adult emerged in September.

Loxostege albiceralis (Grote) is a large and distinctive species, FW (13–18 mm; Plate 22.41) bluish gray with the costal area broadly pale tan terminated by an oblique dark shade. This species occurs in arid areas from the deserts of southern California to Texas and Mexico. The larva was reared in Texas by Roy Kendall on wolfberry, *Lycium carolinianum* (Solanaceae). Flight records are late March and April in Baja California and the Mojave Desert, but August to October in southern Arizona and New Mexico. The smaller *L. lepidalis* (Hulst)(FW 13–15 mm; Plate 22.42) has the FW costa narrowly ochreous tan, the remainder a patchwork of gray and white with reniform and orbicular spots well defined. We found the larvae forming nests in foliage of a shrub we thought was greasewood, *Sarcobatus* (Chenopodiaceae), in northwestern Colorado in August, and adults emerged the following June. This species occurs from Alberta and eastern Washington and Oregon to southern California, Nevada, and New Mexico. Adults have been recorded June to August. *Loxostege immerens* (Harvey)(Plate 23.1, Plate 23.2) is a small, day-flying species superficially resembling other diurnal spring moths, including **L. annaphilalis** (Grote)(Plate 22.43), *Pogonigenys proximalis*, and some *Annaphila* (Noctuidae), in having dark FW and dark to bright orange HW. *Loxostege immerens* is smaller than most *Loxostege* (FW 10–11 mm). It flies in early spring, February to early April in southern California north to the coastal dunes in San Luis Obispo County and in the southern Sierra Nevada. *Loxostege annaphilalis* (Grote) is larger (FW 12–16 mm), occurs in the same area, flies in April to May, but is rare in collections.

Pyrausta forms the largest generic lineage of Pyraustinae in North America, with about 60 described species. These are small to medium-sized, often colorful moths, although most of them are nocturnal. The frons usually is flat or rounded but not prominent. The FW is distinctly triangular. This genus is defined mainly by male genitalia characteristics: uncus narrow at tip and not deeply cleft, gnathos rudimentary, valva with a well-defined clasper having a single ventrally directed process but lacking a group of erect scales on a rounded prominence. The larvae roll leaves and inflorescences of herbaceous plants, primarily mints (Lamiaceae). About 45 species occur in the western states and provinces.

Pyrausta nexalis (Hulst) is distinctive, having black wings with a crisscross pattern of distinct, white lines and FW (6–9 mm; Plate 23.3) costa tan. This species occurs in Washington and Montana to Texas and southern California. We found the larvae in northern California in tightly closed leaf shelters on *Monardella crispa* (Lamiaceae). Caterpillars collected in May yielded adults in June. According to Munroe, flight records are August to September in the north, June to August in California, and March to October in Texas.

Pyrausta signatalis (Walker) has bright rose pink FW (7–10 mm; Plate 23.4) with the postmedial fascia curving and zigzag on the dorsal half. It is widespread in the eastern United States and ranges to Texas, Arizona, and British Columbia. The larvae feed on horsemint, *Monarda*, and likely other mints, according to Munroe. *Pyrausta volupialis* (Grote) resembles *P. signatalis*, but the FW (8–10.5 mm; Plate 22.44) is deep crimson to rust pink with a zigzag ante- and straight postmedial, white line; HW pale gray, darker distally. Formerly it ranged from southern Mexico to Texas, Colorado, and Arizona. In 1991 *P. volupialis* appeared at lights in coastal Los Angeles County and within a few years was widely established in southern California, then spread northward to the San Francisco Bay area by 1997. The larvae were found on rosemary *(Rosmarinus)* in southern California and probably feed on other garden mints. In California, *P. volupialis* flies year-round, most numerous August to October. *Pyrausta grotei* Munroe is similar, having deep purplish pink FW (10.5–12 mm; Plate 23.5) but lacks the antemedial line, and the postmedial is indicated by a wedge at the costa and sinuate, broken line. This species occurs from eastern Washington and Wyoming to California, mainly montane, to 10,000' in Arizona, but the early stages are unreported.

Pyrausta laticlavia (Grote & Robinson)(Plate 23.7, Plate 23.8), which is similar to *P. volupialis* in color but not genitalia, had a comparable history on the West Coast. This species occurs in the eastern United States to Texas and southern Arizona.

It was first recorded at Los Angeles by one specimen collected by Koebele, presumably in the 1880s, then was taken in San Diego between 1908 and 1912 by Field and Wright, more than a decade after they had begun collecting there. It became widespread in southern California by the 1920s but was not recorded north of Monterey until 1965 and did not become extensively established in the San Francisco Bay area until the 1990s. *Pyrausta laticlavia* has less sexual dimorphism but greater seasonal variation than most *Pyrausta*. Summer moths have bright pink FW with fuzzy yellowish bands, whereas those flying in fall and early spring are much darker, almost uniform dark gray brown in the extreme. This species was reared by Jerry Davidson in Santa Barbara from rosemary, *Rosmarinus* (Lamiaceae), but a native host plant has not been identified.

Pyrausta pseudonythesalis Munroe (Plate 23.6) is similar to *P. onythesalis* (Walker) of eastern North America and was long confused with it. The western species has pale yellowish, semitranslucent FW (9–10 mm) with acute apex and dull reddish brown maculation, most prominent as a broad subterminal band. It occurs from the Owens Valley and deserts of California to western Texas. The larva has not been recorded, but that of *P. onythesalis* was found on *Salvia* by Roy Kendall in Texas. Imponderably, in recent years we trapped single specimens of *P. pseudonythesalis* on the smallest of the California islands, Santa Barbara, where there is no mint, and on the most distant island, San Nicolas, which has the poorest insular fauna, owing to its remoteness and devastation by sheep in the nineteenth century.

Pyrausta californicalis (Packard) is another species that has benefitted from garden mints. The moth is small, FW (7–9 mm; Plate 23.9) reddish brown with obscure paler markings; HW pale orange with black bands. It occurs from southern British Columbia to southern California, commonly in urban situations, and has been reared from garden mints *(Mentha,* including *M. suaveolens)* by Dammers, Lange, Rubinoff, and others, and from *Monardella villosa* at San Bruno Mountain by Wagner. Munroe differentiated a Sierran race, which is larger and has primarily dark rust wings. *Pyrausta subsequalis* (Guenée)(Plate 23.10, Plate 23.11, Fig. 175) is one of the most often observed western *Pyrausta* because it lives in weedy areas over widespread areas and is primarily diurnal. This species occurs throughout North America, from Newfoundland to the Yukon and Alaska, to the Gulf states, Texas, and the Pacific Coast south to central California. It is scarcely distinguishable from the European *P. cespitalis* (Denis & Schiffermüller) but presumably is an indigenous species. In western populations, the moth has rust brown FW (8–14 mm) and bright orange HW with black bands, females (Plate 23.11) more strongly patterned and smaller than males. Winter individuals are somber, gray brown with the orange suppressed. Larvae of *P. cespitalis* feed on plantain, *Plantago* (Plataginaceae), whereas *P. subsequalis* is reported on thistles *(Carduus, Cirsium, Cynara),* including a naive species, *C. occidentalis,* at Monterey Bay by S. Palmisano. In the Berkeley area we have reared it from larvae burrowing in the stems of bull thistle *(C. vulgare)* and feeding at the base of English plantain, *P. lanceolata,* both introduced weeds. Frequently adults are associated with plantain in disturbed habitats in absence of thistles. The disparity in host plants invites speculation that we are dealing with both an introduced and a native moth species. The adults come to lights, mostly February to April and September to November, but are commonly seen active during the daytime throughout the year.

Pyrausta perrubralis (Packard)(FW 9–13 mm; Plate 22.45) is a moderately large, colorful species, yellow with variable pink maculation, especially a broad terminal margin and costal bar at the end of the cell. This species ranges from Vancouver Island to southern California, coastal and inland in the mountains to 7,500'. Adults are common at lights, and larvae were found at San Bruno Mountain by De Benedictis, tying leaves and flowers of *Monardella villosa*. *Pyrausta scurralis* (Hulst) is similar and formerly was considered to be an inland race of *P. perrubralis*; its maculation is less variable, lacking yellow crossed by pink lines along the veins within the pink bands. *Pyrausta semirubralis* (Packard) differs from other *Pyrausta* in having the FW (9–11 mm; Plate 22.46) more or less bicolored, basally grayish ochreous with the distal region beyond the cell dark rust red, varying to broken postmedial and subterminal bands. Females are smaller and darkly colored, HW dark gray brown versus pale brown with weak postmedial and subterminal bands in males. This species also ranges from southern British Columbia to southern California, coastal to 9,000' in the Sierra Nevada, and northern Arizona, but occurrences are scattered, and the early stages evidently are unknown.

Pyrausta unifascialis (Packard)(Plate 22.47, Plate 22.48) is a widespread and highly variable species that ranges across North America and in the West from British Columbia to southern Arizona and California. The males have olive gray FW (10–14 mm), with a well-developed pale postmedial line and discal spot to immaculate; females are smaller and darker, FW (7–11 mm; Fig. 48) with more consistently developed markings. Populations in the north tend to have darker individuals, while those in southern California are paler, HW nearly white in males, with a large, white, postmedial band in females. The larvae are polyphagous, having been reported feeding on a legume (*Phasiolus*), buckwheat (*Eriogonum*, Polygonaceae), *Gayophytum* (Onagraceae), and pussy toes, *Antennaria* (Asteraceae). Adult flight spans April to August, and probably *P. unifascialis* is univoltine. *Pyrausta fodinalis* (Lederer)(Plate 22.38, Plate 22.49, Plate 22.50, Fig. 176) ranges across Canada and south in the Rocky Mountain and Pacific states, but apparently not in the Southwest. Its variation is comparable to that of *P. unifascialis:* males are larger, FW (10–15 mm) pale tan with faint to well developed, dark submarginal band and discal spot; HW white with a dark marginal band. Females are darker and more strongly patterned. We reared this species from coyotemint, *Monardella villosa*, at San Bruno Mountain and found adults on *M. odoratissima* at Sonora Pass, 9,600' near timberline in the Sierra Nevada. Larvae collected in February to March produced adults in late March to April. Adults are active May to August, varying with elevation.

Pyrausta dapalis (Grote) is a distinctive, diurnal species that flies in early spring. It has black FW (7–9.5 mm; Plate 23.12) and bright red, black-bordered HW, similar to the red-HW species of *Annaphila* (Noctuidae). *Pyrausta dapalis* is a California endemic, ranging from Lake County and the Sierra Nevada to San Diego. We reared this species from pitcher sage, *Salvia spathacea*, at several localities, from *Monardella villosa* at San Bruno Mountain, from black sage, *S. mellifera*, at Big Creek, Monterey County, and *S. leucophylla* at Sedgwick Reserve, Santa Barbara County (all Lamiaceae). Larvae fed in new foliage and inflorescences, February to May, and adults emerged January to March, when flight normally occurs.

SPILOMELINE GROUP: SUBFAMILY SPILOMELINAE

The Spilomelinae in North America consists of 86 genera, but more than 70% (50 genera) are represented by either a single species or two, most of them northern extensions of more diverse subtropical and tropical genera. Many such examples are known in the United States only in Florida, southern Texas, or Arizona. There are about 220 species recorded in America north of Mexico, mostly in the Southeast to eastern and southern Texas.

Udea comprises the largest lineage of Spilomelini in North America, with 23 species, nearly all of which are northern in distribution, and several represent Holarctic ranges. All but one occur in the West, some only in the far north. *Udea rubigalis* (Guenée)(Plate 23.14), the "celery leaf-tier," and the closely related "false celery leaf-tier," *U. profundalis* (Packard)(Plate 23.15, Plate 23.16, Figs. 177, 178), are the best-known *Udea* species in North America. Both have broad wings, rust orange with frail, brown linear markings. *Udea profundalis* is larger (FW 8.5–12 mm vs. 6.5–10 mm in *U. rubigalis*) and varies to a brownish form in winter. The HW usually is primarily white in *U. profundalis*, patterned with dull rust in *U. rubigalis*. In male genitalia, *U. rubigalis* has prominent lateral, spinose lobes of the juxta, lacking in *U. profundalis*, and two small, strongly curved spurs in the vesica, whereas there is one larger, slightly curved spur in *U. profundalis*. Older Pacific Coast specimens (before 1950) were called *U. rubigalis*, which ranges transcontinentally, whereas later ones have been identified as *U. profundalis*. Probably the name rather than the species *U. rubigalis* has been replaced. The larvae are polyphagous, primarily on low-growing, herbaceous plants, and are pests of house plants, in greenhouses, and on crops such as celery, cabbage, cauliflower, and a wide variety of commercial flowers, as well as weeds and native plants in disturbed places, including cheeseweed, *Malva parviflora* (Malvaceae), manroot, *Marah fabaceus* (Cucurbitaceae), and stinging nettle, *Urtica holosericea* (Urticaceae). *Udea profundalis* feeds on the noxious tansy ragwort, *Senecio jacobaea*, and was said by Essig to commonly infest hawksbeard, *Crepis* (Asteraceae). The larvae tend to stay on the underside of a leaf, with a slight shelter, and they have nearly transparent integument, so their color matches the foliage eaten. Pupation occurs in a weak cocoon on the ground. The adults are active throughout the year in coastal California.

Many of the remaining *Udea* are widespread and similar in appearance: gray to brownish, with dark costa and orbicular and reniform spots, often with pronounced geographic variation, to which Munroe applied subspecific names in several species. Forms of different species sometimes resemble one another more so than they do other races of their own species. *Udea itysalis* (Walker)(Plate 23.27) is transcontinental, Nova Scotia to the Yukon, and Kodiak Island to British Columbia, the Rocky Mountain states to northern Arizona and New Mexico, and the Pacific Coast to central California. There is considerable polytypy, and Munroe designated 10 subspecies. The FW (10–12 mm) color is variable, dark gray or brown to pale tan or whitish, typically rather uniform, not mottled, orbicular and renal spots usually distinct. McDunnough reared the larvae on bluebells, *Mertensia maritima* (Boraginaceae), in the Northeast, and we have found them on other borages, stickseed, *Hackelia*, and hounds tongue, *Cynoglossum grande*, at several sites in California. They feed early in the season, March and April, and adults fly in April to May, but they are localized and not common at lights. *Udea turmalis* (Grote)(Plate 23.25) is widespread in the West, from British Columbia through the Rocky Mountain states to New Mexico, often at timberline elevations, with subspecies named from New Mexico, thae Sierra Nevada, and the San Bernardino Mountains in southern California. This species is larger (FW 11–14 mm), darker, and more mottled than *U. itysalis*, with which it occurs at many localities, but the two do not show consistent differences in genital morphology. The FW is gray to brownish, usually with a subcostal dark shade obscuring the orbicular and reniform spots, and with a pale streak through the apical area; HW pale to dark gray with darker marginal band. The larvae have been found on thistle (*Cirsium*, Asteraceae) in California in May and June, and adult records are July and August in most of the range, but June to July and October to November in southern California. *Udea washingtonalis* (Grote)(FW 8.5–13 mm; Plate 23.26) occurs from the Bering Coast on the Kamchatka Peninsula and Komandorskie Island on the Siberian side and Nome, Alaska, south along the coast, including Pribolof and Unalaska in the Aleutian Islands, to northern California, with color pattern varying individually and according to population. Compared to most *Udea*, the FW is broad and pale, white, tan or grayish tan, with the orbicular and reniform spots large, dark, strongly contrasting in most populations. Both FW and HW usually have a row of strong dark dots along the base of the fringe. The larval biology evidently is unknown. *Udea vacunalis* (Grote)(Plate 23.17) is unique among American *Udea* in having cream white wings with at most only faint, gray traces of the orbicular spot and evenly curved postmedial band. It occurs from the Siskiyou Mountains at the Oregon border south through the Sierra Nevada to the San Bernardino mountains, at moderate elevations. This species is relatively rare in collections and nothing is known of the early stages.

Lineodes integra (Zeller)(Plate 23.19) and related species, mostly Neotropical, are distinctive among Spilomelinae in having long, narrow wings, very long antennae exceeding the FW, and long, slender legs. They perch in a characteristic manner, with the wings splayed back against the substrate and the abdomen curled upward at a 45° or greater angle (Plate 61.18). *Lineodes integra* is plume moth–like, FW (9.5–12.5 mm) gray with brownish black markings, especially a linear, sinuate streak that borders a white, semicircular patch on the dorsal margin. This moth occurs in the Lower Sonoran and coastal parts of California and its Channel Islands, Baja California, southern Arizona, and northern Mexico. The larvae feed as leaf rollers on native nightshades, *Solanum*, and rarely tree tobacco, *Nicotiana glauca* (Solanaceae), with several broods per year. Adults are active from March to November; we have found the larvae on *S. douglasii* in January, March, and August to October. *Lineodes elcodes* (Dyar)(Plate 23.20), which was described from Puebla, Mexico, has recently become established in the Los Angeles, Santa Barbara, and San Francisco Bay areas of California. The larvae were found by Jerry Davidson feeding on night jessamine, *Cestrum nocturnum* (Solanaceae), an ornamental plant from the West Indies. This species is larger (FW 12.5–13.5 mm) and darker than *L. integra*, FW dark brown with delicate white and yellow lines angled outward from the middorsal margin, followed by a hyaline patch.

Loxostegopsis xanthocrypta (Dyar)(Plate 23.18) was described from Mexico City and San Diego, so this may be a name applied to two species. These and related species are dark brown with pale ochreous labial palpi, head scaling, and obscure FW markings. *Loxostegopsis xanthocrypta* is small, FW (8–10.5 mm) dull brownish gray with indistinct dark, transverse fasciae and a pale discal dot and outer costal spot, subtended by a row of indistinct dots bordering a curved subterminal line. The HW is darker in females. It occurs in southern California including San Clemente, Santa Cruz, and Santa Rosa islands, where adults are recorded in April to May and August to September, sometimes abundant in the fall. *Loxostegopsis curialis* Barnes & McDunnough was described

from Utah, southern Arizona, and Camp Baldy in the San Gabriel Mountains of southern California and occurs in the arid Yreka area north of Mt. Shasta. It is larger (FW 10–12.5 mm) and darker than *L. xanthocrypta*, with longer palpi and rudimentary proboscis; FW markings obscure. The larval biology is not recorded.

Hydropionea fenestralis (Barnes & McDunnough) is a curious species unlike any other in our area. The moth has more oval than triangular FW (12–15 mm; Plate 23.21), scaling rust tan with a reticulate pattern of whitish and hyaline spots covering all but the costa and termen, which is sinuate, concave above and below the middle. The HW is shining white with a dark rust line at the base of the fringe. This species occurs in the mountains of southern Arizona south to the pine forests above 8,500' in Durango, Mexico. Adults are recorded in April and August at several sites in Arizona, but the larval biology is unknown.

Species of *Choristostigma* are distinguished primarily by the structure of the male antenna, thickened 1.5 to more than two times the diameter of female's, with bands of darker scaling, and the FW is acute apically. *Choristostigma plumbosignalis* (Fernald)(Plate 23.22) is perhaps the most commonly observed pyraloid in the mountains in Colorado and Utah, ranging to southern Arizona. The FW (9.5–11 mm) is sulfur yellow, orbicular and reniform spots dark rust are overlaid with lead-colored scales, and there is a rust lead–colored stripe from the tornus, branching at the middle to form a Y-shaped pattern, with rust filling the terminal area in heavily marked specimens; HW white. The adults fly in July and August. *Choristostigma zephyralis* (Barnes & McDunnough) is a similar, larger, bright yellow moth, FW (10–12.5 mm, rarely 9 mm; Plate 23.23) having brownish markings tinged with rose, appearing lavender to the unaided eye; costa and the orbicular and reniform spots faintly to distinctly shaded in the subterminal area. The HW is white basally, yellow toward the anal margin. This species is common in inland chaparral areas of central California from the foothills of the Sacramento Valley south to 6,000' in the San Bernardino Mountains and Sierra San Pedro Martir in Baja California Norte. Flight records are March to June. Most collections have been at lights, and the larvae are not known, but we found adults associated with coyote-mint, *Monardella villosa* (Lamiaceae). *Choristostigma elegantalis* Warren is a small, heavily patterned species having pale yellowish FW (9–11.5 mm; Plate 23.24) variably tinged with rust, concentrated along the veins. The orbicular and reniform spots are whitish, bordered by rust, the outer one sometimes gray. This crambid is widespread in the West, from Wyoming and Colorado at high elevations to California, where it occurs from the coast to 10,000' and on Santa Cruz and Catalina islands, and it ranges in the mountains west of Durango, Mexico. Flight records are April to June at low elevations and mostly July to August in the mountains. Nothing is known of the larval biology.

Mecyna mustelinalis (Packard) is moderately large (FW 12.5–16.5 mm; Plate 23.28), brown with obscure pattern, varying to distinct, whitish spots in the cell and forming a distal transverse row from the costa, curving inward above the tornus. This species is widespread in lower-elevation chaparral zones the length of California, to the Sierra San Pedro Martir at 7,000' in Baja California Norte, and is one of the most commonly collected species of Spilomelinae, yet nothing seems to have been recorded on its larval biology. Adults fly from April in the low desert foothills to August at Mt. Shasta, evidently univoltine. *Mecyna luscitialis* (Barnes & McDunnough) is similar but smaller (FW 11.5–13 mm) and paler, gray brown with the same pattern, but the markings are white and much more prominent, including two on the HW distally. It occurs in desert areas of southern Arizona and California.

Mimorista subcostalis (Hampson), a large (FW 15.5–19.5 mm; Plate 23.29) and distinctive spilomeline, is widespread and commonly collected in Mexico and southern Arizona .It ranges to southern Utah and central California, although northern records may represent late season migrants. The wings are yellowish, hyaline, with broad brown margins. Specimens from Arizona and Mexico are bright yellow, with unscaled areas of the wings covered with yellowish hairs. Those from southern California average smaller and are less colorful, whitish. Flight records in Arizona are mostly August to September but April to June in southern California and July, September to October in central California, so there may be at least two generations. The larval biology is unrecorded.

Nomophila nearctica Munroe is a moderately large spilomeline with narrow FW (14.5–16 mm; Plate 23.30, Figs. 179, 180), one of the most ubiquitous moths in North America and with its sister species, *N. noctuella* (Denis & Schiffermüller), throughout the world. They fly throughout the year in warmer areas. The FW are pale to dark gray brown with linear streaks, in a wood-grain-like pattern. The larvae are general feeders on low herbs, including legumes (Fabaceae) and Polygonaceae. Easton in Australia found the larvae of *Nomophila* fed gregariously, forming in later instars communal nests beneath the plants and ultimately silk-lined burrows in the soil, feeding at night. *Nomophila nearctica* is multivoltine, with flight records April to November in California.

Desmia funeralis (Hübner)(Plate 23.35), the "grape leaffolder," is a distinctive, black, white-spotted moth that occurs throughout much of the United States, including the Southwest and Pacific Coast from British Columbia to southern California, to 5,500' in southern Arizona and 6,000' in Chihuahua. Males of this diurnal moth have two large white spots on the FW (11–14 mm), and a broad white band on the HW, and the antennal scape greatly elongated and flared apically with an enlarged scale tuft. Females lack antennal modifications and have smaller spots, the HW band usually broken into two offset spots. The larvae roll the leaves of wild and cultivated grapes and Virginia creeper, *Parthenocissus* (Vitaceae), also occasionally *Oenothera* (Onagraceae) and *Cercis* (Fabaceae) according to Forbes. There are two or three generations, with adults active from April to October in warmer regions.

Spoladea recurvalis (Fabricius)(Plate 23.37), placed in the genus *Hymenia* by some authors, is the "Hawaiian beet webworm," one of the world's most widespread moth species, occurring in Africa, the Indo-Australian region, Australia, Pacific Islands, and throughout the Neotropical region, into the southern United States. It disperses far beyond regions in which it normally overwinters, regularly in the Atlantic states north to New York in the fall, and rarely to central California. The adults are small (FW 10–11.5, rarely 9 mm), wings black to brownish black with a broad, white band crossing the HW from anal angle to midcosta, and on the FW, forming, when the moth is perched, a continuous stripe nearly to the costa, bordered by black on paler varieties. The larvae feed on many kinds of low plants, especially Amaranthaceae, and field crops, including beets, Swiss chard, and spinach (*Beta*, *Spinacia*, Chenopodiaceae), and a variety of weeds. The moths fly all year in warmer regions.

Diathrausta harlequinalis Dyar is superficially similar to *S. recurvalis*, but smaller, FW (6–9.5 mm; Plate 23.36) black with

two basal, orange ochreous, transverse lines and outer costal and pretornal bars forming a broken fascia, and there are white spots in the disc. This species is frequent at lights and sometimes visits flowers diurnally, in southern Arizona, late July to early September, and it occurs north to Colorado.

Antigastra catalaunalis (Duponchel) is a distinctive species having a narrow FW (10–11.5 mm, rarely 8 mm; Plate 23.38) striate longitudinally with rust orange on the veins and reticulated by cross bars between them. There is a dark brown line at the base of the ochreous fringe. This species occurs in Mexico, southern Arizona, mountains of the eastern Mojave Desert, and the Owens Valley, California, to northern Utah, flying in July to August. The larva is a leaf-tier on *Tecoma stans* (Bignoniaceae), collected in October in southern Baja California by Ev Schlinger, and adults were reared from *Penstemon* (Scophulariaceae) in the San Joaquin Valley, California, in April, presumably an introduction.

Diaphania hyalinata (Linnaeus)(Plate 23.39), the "melonworm moth," is a widespread species in Mexico and Central America, a pest of pumpkins and other cucurbit vines in the Gulf and Midwest states, primarily feeding on the foliage. It occurs in the West in Colorado, Texas, and rarely in southern Arizona and California at the Colorado River. The wings, thoracic venter, and abdomen are shining pearly white; FW (13–14.5 mm) and HW are broadly bordered with dark brown. The expandable anal tuft of the male is dark brown to mostly ochreous. Probably *D. hyalinata* is active year-round in warmer areas.

Palpita gracilalis (Hulst)(Plate 23.41) occurs in Mexico and from Texas to southern California, sometimes as a pest of ornamental privet (*Ligustrum*, Oleaceae), which is not a native plant in northern parts of the moth's range, including Arizona and California. This moth has translucent, white wings with pearly iridescence, FW (10–12.5 mm) costa brown, with dark gray, large reniform and small orbicular spots, a broken basal and diffuse postmedial bands, and variable gray dusting. *Palpita quadristigmalis* (Guenée) is larger (FW 13.5–14.5 mm; Plate 23.42) with nearly uniform, cream white translucent wings; FW costa rust brown, and there are four black dots, which give the species its name, two along the Sc vein in the cell, two at outer corners of the cell. *Palpita quadristigmalis* occurs in the West Indies and Central America, Gulf and Atlantic states, Colorado, New Mexico, and southern Arizona. It was called the "privet moth" by Holland after it became a pest of garden *Ligustrum* in the 1880s.

Terastia meticulosalis Guenée (Plate 23.31, Plate 61.18) is one of the largest spilomelines in the Southwest. It is a tropical species that ranges from South America through Mexico and reaches southern Arizona, perhaps only sporadically. The wings are elongate, the narrow FW (16.5–23 mm, i.e., to 2 inches wingspan) with a sinuate distal margin and acute apex, tan irregularly patterned with brown and ochreous brown scaling over a semitranslucent pale ground. This species flies from May to August in Central America and has been taken in Arizona from June to August. The early stages are recorded feeding on *Erythrina* (Fabaceae) in the pods, in Florida.

Agathodes designalis Guenée is a beautiful tropical moth with long, narrow wings, FW (13–18.5 mm, rarely 12.5 mm; Plate 23.32) olive brown, costa white, with an outwardly curved, rose purple stripe from the middorsal margin toward the apex, blending to rust orange above the tornus and interrupted by a silvery patch at the apex. It occurs from northern South America to Mexico and southern Arizona. The larva feeds on *Erythrina flabelliformis* (Fabaceae) in Arizona, making a shelter by folding over leaves, on *Erythrina*, *Inga* (Fabaceae), and *Citharexylum* (Verbenaceae) in Florida and Puerto Rico. Flight records in Arizona are July to August, whereas *A. designalis* is multivoltine in Florida, Costa Rica, and French Guiana.

Lygropia octonalis (Zeller) is a pretty little species contrasted to the large, tropical genera with which it is placed in our lists. The moth is pale yellow, FW (7.5–8.5 mm; Plate 23.40) with four partial transverse fasciae consisting of pale to dark brown spots, diminishing to a wavy line in the cell. This species is widespread in the Southwest at low-elevation, often arid habitats, from central California to southern Baja California, Arizona, and Texas. *Lygropia octonalis* was reared by Y.-F. Hsu in the San Joaquin Valley, from *Heliotropium curassivicum* (Boraginaceae), an inhabitant of saline soils. In California, It flies from late March to October, and larvae collected in September produced adults the following May.

Most members of the genus *Diastictis* are recognizable by their silvery white FW markings, but a few lack them, having obscurely patterned brown wings. There are nine species of *Diastictis*, four in the Southeast and Cuba, and the remainder in the Southwest. *Diastictis fracturalis* (Zeller)(Plate 23.44, Plate 23.45) is the most widespread species, ranging from Arkansas and Texas to California, and Mexico to Guatemala. The moth is pale tan to dark brown with highly variable silvery markings on the FW (9.5–14.5 mm), from small, isolated spots to elongated, connected ones. Adults are active from February through November in coastal California. *Diastictis sperryorum* Munroe (Plate 23.43) differs by having more or less round or quadrate spots, smaller than those of *D. fracturalis*, the spot in the distal part of the cell always separate from the one below the cell, and there are differences in genitalia. *Diastictis sperryorum* was described from numerous localities, Oklahoma to southern California, sympatric with *D. fracturalis*, and the two are confused in collections because specimens available to Munroe 50 years ago were insufficient to express the variation in FW pattern. Collection records in California indicate two generations, April to June and September to October. We reared specimens from larvae found on Asteraceae, inflorescences of *Gnaphalium* in central coastal California in August, and from *Isocoma menziesii* at San Diego, where larvae webbed the lower stems and fed on the foliage in July. The genitalia of these and Channel Islands specimens of this complex do not match Munroe's figures of *D. fracturalis* or *D. sperryorum*, indicating *Diastictis* is in need of further taxonomic research. *Diastictis caecalis* (Warren) lacks the white FW maculation; FW (9–12.5 mm; Plate 23.46) chocolate brown with an irregular, darker postmedial, transverse line; HW milk chocolate brown to ochreous brown with a postmedial line extended to anal margin. *Diastictis caecalis* occurs in the foothills of the Sierra Nevada and Coast Ranges from Shasta County to the Transverse Ranges in southern California. Adults fly in spring, March to May, and appear to be diurnal, often associated with serpentine soil grasslands.

Phostria tedea (Stoll) is a large (FW 17–19 mm; Plate 23.33), spectacular spilomeline, wings blood red with large white, semitranslucent spots. It ranges from northern South America to Mexico, and occasionally into southern Arizona. We found its larvae on an unidentified tree in Guerrero and on morning glory, *Ipomoea* (Convolvulaceae), in coastal Jalisco, Mexico, in September and October. Flight records are July to September.

Herpetogramma pertextalis (Lederer)(Plate 23.34) is a moderately large, cream white to pale tan species, which occurs across North America and widely in the West. The wings are patterned with linear, dark lines of variable intensity along the veins and dark smudges at the orbicular and reniform spot positions (FW 12–16 mm; rarely 10 mm in British Columbia). In Washington, Oregon, and California this species occurs primarily in shaded habitats along the coast. The larvae roll the leaves of unrelated herbaceous plants including several notorious for their defensive chemicals, *Asarum* (Aristolochiaceae), *Urtica* (Urticaceae), *Rubus* (Rosaceae), and *Viola* (Violaceae), April to June, and flight records are mostly July to August.

SPILOMELINE GROUP: SUBFAMILY ACENTROPINAE

Acentropinae (= Nymphulinae) are unique among Lepidoptera in having adapted to aquatic larval life. The adults are small, slender moths with long legs and narrow wings, usually with transverse markings, and often with metallic spots along the posterior margin of the HW. Genitalia, Figs. 181, 182. The larvae of nearly all species are aquatic; some live in ponds and feed on vascular plants, others live as borers or in cases and breath air trapped in the cases via the spiracles. Alternatively, many species live in rapid streams under webs on rocks, feeding on algae, and breathe dissolved oxygen through external gills, with their spiracles closed. More than 700 species are described, and there is a rich, mostly unstudied tropical fauna. About 50 species in 14 genera are known in America north of Mexico, mostly in the Southeast, and only a dozen are described in the West.

Synclita occidentalis Lange is somewhat broader winged than most acentropines, brown with a reddish hue, FW (8–11 mm; Plate 23.47) crossed by an indistinct, dark, median band and whitish lines, with the reniform spot white encircled by dark scaling. The larva lacks gills and lives under a piece of leaf that it cuts out and attaches to the lower surface of floating leaves. According to Lange, the preferred food plant is yellow water-weed, *Echinodorus* (Alismataceae), and hosts include both dicots and monocots, pondweed, *Potamogeton* (Potamogetonaceae), water-hyssop, *Bacopa* (Scrophulariaceae), and water primrose, *Ludwigia* (= *Jussiaea*, Onagraceae). There are at least two generations per season in the Sacramento Valley, with flight records from April to October, mostly July to August. *Synclita occidentalis* occurs in Texas and Colorado to northern Arizona and central and southern California. An eastern U.S. *Synclita* is a minor pest of water lilies, occurs in greenhouses, and is adventive in Hawaii.

Petrophila (= *Parargyractis*) is a moderately diverse genus in North America, and more than half the 15 Nearctic species occur in the West. These are slender-winged moths that occur along rapid streams, where their larvae live on alga-covered rocks. Genitalia, Fig. 181. The female enters the water to oviposit, carrying a plastronlike layer of air as a source of oxygen. Munroe synonymized **Petrophila confusalis** (Walker)(Plate 23.48), which was originally described from eastern North America, with *P. truckeealis* (Dyar), the most commonly collected species of the northwestern and Pacific states. The adults vary considerably in size, FW (5–11 mm) brownish with ante- and postmedial transverse, whitish fasciae, followed by a dark band, which is white in related species; HW with a strongly developed row of metallic marginal spots, preceded by a white patch. The biology was studied in detail by Lange and Tuskes. There are two or three generations per season in northern California, but *P. confusalis* is univoltine in Montana. The larvae overwinter. According to Tuskes, females can remain alive under water for 4 to 12 hours, and they oviposit in two distinct methods: either they enter the water and deposit all their eggs (200–300) in one or two bouts, as deep as 4 m, then die without leaving the water, or they submerge only partially, prevented by surface tension, and lay eggs shallowly several times over a few days. *Petrophila jaliscalis* (Schaus) has similar wing maculation but is larger (FW 7–11 mm; Plate 23.49) and the colors are brighter, more contrasting, featuring orange brown, and the HW marginal spots are bold, subtended by white. This species occurs from Alberta to Texas and Mexico, to Arizona and California, including Santa Rosa Island. The larval biology is similar to that of *P. confusalis*, but *P. jaliscalis* is adapted to slower, warmer, less well oxygenated streams than is *P. confusalis*. *Petrophila schaefferalis* (Dyar) is larger (FW 10.5–14 mm; Plate 23.50, Plate 23.51) and darker than other *Petrophila*, with the maculation subdued by brownish gray; HW marginal spots smaller than on related species. This species occurs sparingly in southern Arizona, Mexico, and southern California, including Catalina and Santa Cruz islands, where adults fly late April to September.

Usingeriessa species are similar to *Petrophila*, but the HW differs, having a shorter discal cell, to about half the length of the wing, and the distal margin is weakly invaginated, with a row of black and metallic lead-colored spots. *Usingeriessa brunnildalis* (Dyar) is small, FW (7.5–11 mm; Plate 23.52, Fig. 182) with dark brown maculation, moderately well defined whitish median fascia, and dark basal patch. This species occurs from western Texas to central and southern California. The moths are found near permanent streams, and the larval biology is similar to that of *Petrophila*.

SPILOMELINE GROUP: SUBFAMILY CRAMBINAE

Crambinae are characterized by having long, porrect labial palpi in most genera ("snout moths"), usually narrow FW with grass-matching, longitudinal patterns, rendering the moths cryptic as they perch on monocots, their larval hosts. The HW stem of CuA usually has a pecten of erect scales. The ovipositor lobes are small. Genitalia, Figs. 183, 184. Crambine larvae are of two general types, root and leaf feeders living on the ground, or stem borers in grasses, sedges, or rushes. Both types cause economic damage, as lawn turf and sod pests or in grain crops, including sugar cane, corn, and rice. This large lineage has nearly 2,000 described species occurring on all continents and many oceanic islands, and in all life zones from Arctic-alpine to tropical. Root- and grass-feeding types are prevalent in temperate zones, stem borers in tropical and subtropical zones. There are about 35 genera and more than 180 species recognized in America north of Mexico (including those formerly distinguished as Ancylolomiinae), and numerous undescribed species in collections. Crambines are especially rich in the Southeast, but more than 70 species occur in the western states and provinces.

Pseudoschoenobius opalescis (Hulst)(Plate 24.1) is a peculiar species with very slender, long labial palpi, body, FW, and legs. The FW (12.5–16 mm, 4–5 X the width) is pale gray with delicate dark gray markings. This species was described in 1886 from Arizona, locality and collection date unknown, and it occurs in western Nebraska, western Texas, Colorado, and southern Utah. These populations vary slightly in morphological details, and both sexes are fully winged. Adults are recorded from April to September. On the other hand, *Pseudoschoenobius* in southeastern California, the Algodones dunes near the Colorado River, Kelso dunes in the Mojave, and Eureka dunes in northern Inyo County are active in winter, January to March; their spidery females are brachypterous and flightless, FW about three-fourths and HW one-third the

abdomen length. The males of these populations seem to be morphologically indistinguishable from Great Basin *P. opalescis*, but examples of geographic variation to this degree are rare in Lepidoptera. The complex begs for detailed study.

Crambus is the largest genus of the subfamily and the one most familiar to nonspecialists. There are about 50 Nearctic species, many of them occurring at high latitudes and elevations and ranging across the continent. More than 30 species occur in the western states and provinces, most of them characterized by having a longitudinal silver stripe though the FW cell. *Crambus pascuellus* (Linnaeus) is a Holarctic species, its Nearctic populations are known as *C. floridus* Zeller. The FW (11–12.5 mm, smaller in Alaska, 8–9.5 mm; Plate 24.2) is pale rust ochreous, weakly bordered by dark gray, median stripe divided beyond the cell. The terminal area has four or five parallel silvery lines bordered by dark borders. This species is transcontinental in boreal North America and occurs southward in the Rocky Mountains and Cascades to the central Sierra Nevada, flying in June to July. *Crambus whitmerellus* Klots is similar to *C. pascuellus*, but the FW (10.5–13.5 mm; Plate 24.3) is darker, ochreous brown, with the costa narrowly pearly white along the basal half, and the silvery stripe has a notch in the midposterior margin, creating a strong spur. This is one of the commonest *Crambus* in the Rocky Mountain region, occurring from Alberta to Utah, Wyoming, and Colorado, flying in July to August. *Crambus gausapalis* Hulst has the FW (9–12.5 mm, males much larger than females; Plate 24.12) pattern washed out, with just a trace of the linear white stripe, bordered by weak rows of dark dots. It occurs from southern Oregon to central California at middle elevations in the Sierra Nevada. Adults are recorded in July. *Crambus cypridellus* Hulst (Plate 24.13) is widespread in the Great Basin and Oregon south to the San Bernardino Mountains in southern California. The FW is pale brownish with darker streaks between the veins, with a prominent notch and spur in the posterior margin of the silver stripe. The terminal area beyond the stripe is yellowish. It flies in June to August. A Californian species, *C. occidentalis* Grote (Plate 24.4), occurs in natural and weedy grasslands but not in urban situations and flies in fall, late September to November. It resembles *C. cypridalis* but tends to have dark gray brown FW (10.5–14 mm) and a strongly developed, distinct silver stripe, with its notch in the posterior margin accentuated by black scaling. The stripe is more or less filled with brownish scaling in occasional specimens. The termen is brownish, usually with four or five elongated spots or lines, but sometimes absent. This species occurs from the San Francisco Bay area to San Diego and on Santa Cruz, Santa Rosa, and San Nicolas islands. *Crambus rickseckerellus* Klots (Plate 24.16) occurs around San Diego and on the southern Channel Islands and also flies in the fall. This species has very long labial palpi, exceeding the eye by 4.5 times its diameter, contrasted to about three times in *C. occidentalis* and *C. sperryellus*. The FW of *C. rickseckerellus* (10.5–15 mm) is gray brown with the white stripe reduced to half the cell width and broken distally by a black streak extending toward the apex. The terminal patch is long and whitish with its six black dots more widely spaced than on related species.

Crambus sperryellus Klots has pale to dark brown FW (9–12 mm; Plate 24.14, Figs. 183, 184) with a distinct silvery stripe that lacks a notch on the posterior margin. There is a gray patch above the tornus with five parallel black lines. In urban areas *C. sperryellus* flew from April to early October and caused economic damage to lawns in cities, including bluegrass *(Poa)* in the Central Valley. Historical records indicate this species occurred natively in diverse habitats, desert margins to montane. Subsequently it adapted to urban situations and became the most widespread and commonly collected *Crambus* in California, occurring from Mt. Shasta to the San Bernardino Mountains at 6,000', even in the Mojave Desert (China Lake, Victorville) and around Phoenix and Tucson, Arizona, perhaps introduced there. It was abundant where lawns were grown until another crambid, *Parapediasia teterrella*, was introduced and became widely established in California cities; then *C. sperryellus* abruptly disappeared at Berkeley within five years. *Crambus sperryellus* was recorded on Santa Cruz Island on numerous dates from 1966 to 1984, April to September, but not in April, May, or September in 1999–2001. Its competitor, *P. teterrella*, is not known to occur there, but major increase in weedy, annual grasses occurred following removal of feral sheep in the 1980s.

Crambus unistriatellus Packard (Plate 24.15) is a northern species that ranges from Labrador to British Columbia, coastal Washington, and Oregon and was reported in California by Forbes. This is an easily identified species having brown FW (10.5–11.5 mm) with a silvery white stripe that continues uninterrupted to the distal margin. It flies from July to September in the Pacific Northwest. *Crambus perlellus* (Scopoli)(Plate 24.5) is Holarctic, and its Nearctic populations, which are known as *C. innotatellus* Walker, have shining pure white wings (FW 11.5–12.5 mm). This species flies in July in mountain meadows in the northern Rocky Mountains to 9,000' in Colorado and the Pacific states to northern and eastern California (Mono Lake). *Crambus labradoriensis* Christoph, as the name suggests, is a northern species, having dark brown FW (9–10 mm; Plate 24.11) with diffuse white patches, and a thin, short, indistinct line in the cell interrupted by a black zigzag line. There are five parallel black lines in the paler brown terminal area. *Crambus labradorensis*, which is believed to be a sister species to *C. alienellus* Zincken in the Palaearctic, occurs across Canada and the northern states, south to Oregon.

Chrysoteuchia topiaria (Zeller) is a distinctive species having a shining golden cast with metallic silvery gray or lead gray FW fringes. The FW (9–12.5 mm; Plate 24.23, Plate 24.24) is pale tan to ochreous, orange tinged toward the dorsal margin. There are thin to broad, dark brownish gray lines between the veins, and there is a strong chevron of shining lead-colored scaling crossing the cell distally. Brown suffusion almost completely obscures the ground color in some populations. The terminal area usually is paler, with three black dots at base of the fringe. According to Forbes, the larvae feed on various plants, sometimes injurious to cranberry (*Vaccinium*, Ericaceae). This crambine is transcontinental and occurs in the Pacific Northwest, south to central California in the Coast Ranges and both sides of the Sierra Nevada, often in bogs. Adults fly in late May to July.

Agriphila plumbifimbriella (Dyar)(Plate 24.6) is superficially like *Chrysoteuchia topiaria* in having many parallel dark lines on a pale tan or whitish FW, but the lines consist of numerous isolated dark scales. This species lacks strong chevron marks and is ochreous orange in the terminal area with six or seven black dots along the base of the shining lead-colored fringe (FW 9–10.5 mm). *Agriphila plumbifimbriella* occurs from Washington and the Rocky Mountain states to northern and eastern California and flies in June to July. *Agriphila costalipartella* (Dyar) is similar, but the FW (10–10.5 mm; Plate 24.19) is whitish gray, and the dotted lines usually are densely aggregated to form a dark, subcostal stripe. When the dark scaling is sparse, specimens resemble *A. plumbifimbriella*. *Agriphila*

costalipartella occurs in the mountains of Utah, Nevada, southwestern Idaho, Oregon, and eastern California, with flight records in July. *Agriphila anceps* (Grote)(Plate 24.17, Plate 24.18) is one of three *Agriphila* that fly in late September to October in foothill California. The FW (8.5–11 mm, males larger than females) varies from a rich, almost uniform chocolate brown to mostly pale tan, usually with two parallel blackish brown chevrons crossing at midcell and just beyond it. There is a row of black dots at the base of the fringe, which is shining dark lead gray. This species occurs along the coast from Humboldt Bay to San Luis Obispo County and on Santa Cruz and Santa Rosa islands; populations on the latter have mostly pale gray brown FW with a paler streak through the cell. *Agriphila attenuata* (Grote) is larger (FW 11–14.5 mm; Plate 24.7) with narrower FW, which are less variable in color and maculation than those of other *Agriphila* species. The FW is tan to yellowish tan with a dark dot at the end of the cell, preceded by variable brownish streaks. The HW is pale gray. This species occurs along the coast of California and inland in the Central Valley, where individuals are paler. They fly in late September to early November in grasslands, even grazed fields.

Catoptria latiradiella (Walker)(Plate 24.21) is distinctive, having rust brown to pale brownish FW (11–14 mm, larger than in the East), with a well-defined white stripe from base along the Cu crease, expanding to midcell distally, crossed by two inwardly angled, dark brown fasciae near the end of the cell and in the terminal area. The labial palpi are short for a crambine, exceeding the eye by approximately 1.5 times its diameter. This species occurs from the Great Lakes region to the northern Rocky Mountain states, Washington, and eastern Nevada in montane situations, to 8,800' in Utah. Western records are July to August. *Catoptria oregonica* (Grote)(Plate 24.22) has broader wings, FW dark gray, with a narrow white stripe to midcell, diffuse distally, and thin, black streaks below and distal to the cell. The labial palpus is longer than that of *C. latiradiella*, exceeding the eye by 2.5 times its diameter. *Catoptria oregonicus* ranges from Montana to British Columbia, Oregon and California, on the north coast and in the Sierra Nevada to 12,500', with flight records from July to early September.

Species of *Pediasia* are relatively large, broad-winged, nearly monochromatic gray to tan moths. *Pediasia dorsipunctella* (Kearfott)(Plate 24.8, Plate 24.9) occurs in the West from Manitoba and Colorado to coastal Oregon, Nevada, and California. It ranges to 10,000' in the Sierra Nevada and 12,500' in the White Mountains of eastern California. The FW (12–14 mm) is tan with the posterior half to the tornus pale gray speckled with dark gray brown scales. Some individuals have weak indications of two transverse, brownish fasciae near the end of the cell and distally, usually represented only on the posterior half. Adults are recorded late June through August.

Microcrambus was proposed by Bleszynski for some small, mostly eastern United States and Mexican species formerly assigned to *Crambus*. *Microcrambus copelandi* Klots (Plate 24.25, Plate 24.26) was described from Santa Catalina Island and also occurs on Santa Cruz Island and widely in California, Arizona, and northern Mexico, ranging from lowland situations to 5,600' in the Chiricahua Mountains, and 6,000' in Sinaloa. The FW (6–8.5 mm) is shining white with variable rust maculation, usually two chevrons at the end of the cell and beyond, which angle acutely back to the middorsal margin and before the tornus, where they are darker, the most distinct markings of the wing. Flight records in California indicate two generations, May to June and August to early October.

Thaumatopsis species are large, sometimes rather husky crambids, characterized by strongly pectinate antennae in the males. *Thaumatopsis fernaldella* Kearfott occurs across the United States to Wyoming, Colorado, Utah, and Nevada; and there is a race in California and northern Baja California (*T. f. lagunella* Dyar). The labial palpi are very long, exceeding the eye by 3.5 times its diameter, and the male antennal setae are long and curled. The FW (11–15.5 mm) is tan with a thin white streak in the cell, bordered on its costal side by black, which interrupts the white stripe at the end of the cell and extends to the apex. The Californian race is more brightly colored, and individual moths average smaller than in the Rocky Mountains. They are recorded June to September. *Thaumatopsis repanda* (Grote)(Plate 24.10) is a large western species that has serrate male antennae, its setae modified, short, thick, saw-toothlike. The FW (14–15 mm) is gray streaked with whitish and dark brownish gray. This species occurs in Colorado, Texas, and Arizona to California, ranging to 8,500' at Charleston Mountain in southern Nevada, and flies from late June to early August.

Parapediasia teterrella (Zincken), described in 1821 from Georgia, was one of the first pyraloid moths recorded in North America. It is small (FW 7–10 mm; Plate 24.20), tan with the head scaling white, widespread and often very abundant at lights in the eastern United States. The geographical distribution no doubt was altered by human colonization. By the late nineteenth century *P. teterrella* encompassed the midwestern states, and Murtfeldt reported that it had become more abundant around St. Louis during the early 1890s. In 1935–1940 it was recorded at Albuquerque, New Mexico, and Tucson, Arizona. *Parapediasia teterrella* first appeared in California at Los Angeles in 1954, then gradually spread through the Central Valley in the 1960s but did not reach the San Francisco Bay area until 1988. During the following several years this species replaced its lawn moth competitors, *Tehama bonifatella* and *Crambus sperryellus*, in urban situations. Flight records from April through November in California indicate this species is multivoltine.

Tehama bonifatella (Hulst)(Plate 24.27) was described from Colorado and occurs natively in western montane regions, such as the White Mountains of eastern California. In addition, *T. bonifatella* found its way to urban situations and became the most abundant lawn moth in cities of the West Coast. The adults are similar to *Parapediasia teterrella* but have variable black pattern on the FW (9–13 mm) and lack the white head scaling of the latter species. Adults of *T. bonifatella* were active continuously from February to November in Berkeley but became scarce (one record per year) within four years after *Parapediasia* colonized in 1988, then disappeared (none seen since 1995).

Euchromius includes about 50 species worldwide, primarily in the Mediterranean region and North Africa, and most of them are similar in appearance, with tan to brownish FW crossed by a pair of yellow stripes. *Euchromius ocelleus* (Haworth)(FW 8–11 mm; Plate 24.28), which is called the "necklace veneer" in England, is a widespread Palaearctic species thought to have been native in the Mediterranean region. It has been imported worldwide and reached western North America in the nineteenth century, even Hawaii by 1882. This is one of the most widespread moths in the western United States, often seen at lights, and sometimes incredibly abundant. For example more than 2,000 were caught in an 8-watt blacklight trap in weedy grassland on tiny Santa Barbara Island, 25 miles from the nearest other land. The larvae are dead-leaf feeders, including Asteraceae and dry thatch, and must do well on Mediterranean annual grasses in the Pacific states. The FW (8–10.5 mm) is elongate, tan sprinkled with black scales, and has two parallel, ochreous yellow, transverse bands beyond midwing, enclosing a whitish streak and bordered by gray. There is a row of golden metallic spots preceded by black bars along the termen from below the apex to the tornus, like the jewels of a necklace. The HW is white. *Euchromius ocelleus* occurs from coastal Oregon through Utah, Colorado, California, the southwestern states, and much of Mexico. Adults are collected mostly April to October in California, but their abundance is wildly erratic, as though they sporadically emigrate in large numbers. *Euchromius californicalis* (Packard) was described in 1873 and is considered to be a native species. It averages larger (FW 9.5–12 mm) and tends to more brightly colored, with more distinct markings, the inner transverse band curved, whereas it is straight in *E. ocelleus*. The HW is pale gray, and there are differences in genitalia of both sexes. *Euchromius californicalis* is broadly sympatric with *E. ocelleus*, but more northern in distribution, from southern British Columbia through the Pacific states to coastal southern California and Rocky Mountain states to northern New Mexico, and from near sea level to 8,000'. Adults are recorded from March to September, most commonly June to August.

Diptychophora harlequinalis (Barnes & McDunnough) is a curious little species quite unlike any other western Nearctic moth. The FW (5–6.5 mm; Plate 24.29) are triangular, bright egg-yolk yellow with black maculation, and three dots on the terminal margin. This species occurs in the mountains of southern Arizona at elevations of 4,000 to 6,000' and Durango and Sinaloa, Mexico, at 6,400 to 8,700'. Adults are recorded in July to August. The larval habits are unknown.

Most species of the genera *Argyria*, *Urola*, and *Vaxi* are small with broadly triangular, silver or white FW. Nearly all the species occur in Florida and the Atlantic states, but there is one in California, a beautiful silvery moth with bright rust markings on the FW, a median, transverse band connected along the dorsal margin to a band along the termen. This moth had been identified as **Vaxi auratella** (Clemens)(FW 7.5–10.5 mm; Plate 24.30), occurs in the eastern United States, but it is undescribed in the opinion of the late A. B. Klots. It occurs from the Shasta Valley north of Mt. Shasta to the central Sierra Nevada at middle elevations, where adults fly in June to July, and in the Vaca Mountains, Napa County, in May.

Chilo and related genera (Chilini) are represented by more than 40 species in North America, mostly broad winged, members of tropical genera, including the following three. **Haimbachia** includes 10 species in the United States, all of them similar in appearance but with marked differences in frontal structure and genitalia. They occur in the East and Midwest, Florida, and southern Texas, and there are three species in Arizona. *Haimbachia arizonensis* Capps (Plate 24.31) was described from the Baboquivari Mountains. The frons is conical. The FW (7–7.5 mm) is grayish white with typical *Haimbachia* pattern faint: thin, yellow brown, postmedial and subterminal transverse lines, which extend outward then angle back to the dorsal margin beyond the middle. The pattern is further obscured by evenly scattered, dark brown scales over the basal half. Evidently the life cycle and larval biology remain unknown for all the *Haimbachia*.

Eoreuma loftini (Dyar)(FW 10–13 mm; Plate 24.32) was described from single specimens reared from "Mexican cane" (presumably *Saccharum*, Poaceae) and wheat stems (*Triticum*) in Arizona, and probably *E. loftini* was not native there. Subsequently it has been recorded in Mexico south to Yucatan. This species was found in California along the Colorado River at Calexico, Bard, and Blythe in 1944–1945 infesting stems of various Poaceae, rice *(Oryza)*, corn and milo maize *(Zea)*, sugar cane *(Saccharum)*, barley *(Hordeum)*, grasses, and *Canna* (Cannaceae). The frons is conical and the FW is tan, usually with a faint, dark discal dot and whitish along the veins, sometimes a sprinkling of dark scales, and the apex is acute. Larvae were collected in February and July to November.

Hemiplatytes species have broad, triangular FW and marked sexual dimorphism in color pattern but not in antennal pectination. *Hemiplatytes epia* (Dyar)(Plate 24.33, Plate 24.34) and *H. prosenes* (Dyar)(Plate 24.35) were described on the basis of two females from Laguna Beach, California, collected a century ago when natural vegetation existed there, and C. F. Baker camped on the beach. The two names may refer to worn and fresh specimens of the same species, but later specimens from Great Basin/Mojave sites in Idaho, southern Utah, Nevada, and eastern California were identified as *H. prosenes* and seem to represent a distinct species. *Hemiplatytes epia* has small males (FW 6.5–9 mm), fresh specimens with dark golden brown FW having white veins and variable blackish scales between the veins. Females are larger (FW 8.5–10 mm; Plate 24.34) and when fresh have white FW with pale ochreous tan between the veins, and there are lines of scattered black scales distally. The HW is dark gray. Flown specimens become nearly all white, and partially worn specimens match the description of *H. prosenes. Hemiplatytes epia* occurs in coastal southern California to 5,000′ in the San Jacinto Mountains, and on Santa Catalina Island, with flight records at Miramar, June to October. The inland populations of *H. "prosenes"* consist of larger individuals (FW 11–13.5 mm), whereas the original type specimen was smaller than *H. epia*. The FW is silvery white broadly along the veins with distinct ochreous patches between them and scattered black scales along the vein edges, sometimes very few. These moths apparently do not fade or wear to white. Collection records are late June to July.

FAMILY PYRALIDAE

These are small to relatively large moths (FW 5–75 mm, mostly under 30 mm) that have the tympanal bullae invaginated into the abdominal cavity and almost completely closed, with their connecting membranes (conjunctiva) and tympana in the same plane. They lack a sclerotized flap (praecinctorium) between the tympanal lobes and secondary venulae.

Adult FW vein R5 is stalked or fused with R3 and R4. The male genitalia possess lateral arms at the base of the uncus. Genitalia, Figs. 185–200. Other features, such as form of the labial palpi, FW shape, colors, and male secondary features are extremely variable.

Larva The larvae of almost all species have a sclerotized ring around the base of seta SD1 on abdominal segment 8. They are usually stout, cylindrical, with relatively short legs and setae; characterized by sclerotized rings around the setal bases of abdominal segment 8 and often the meso- and metathorax; lateral setal group of T1 bisetose, of T2 and T3 trisetose. The body is typically unpigmented, although some species are well patterned, even brightly colored. Almost all are concealed feeders, most often borers in seed, fruit, or stems, or they live in tunnels in the soil beneath plants. Many others construct nestlike shelters among tied leaves, sometimes of quite tough silk.

Larval Foods Diverse living plants are consumed, including wood rot fungi (Xylariaceae), as well as dry vegetable matter including seeds, the papery structure of social Hymenoptera nests, and wax in bee hives (*Galleria, Achroia, Aphomia*, Galleriinae). Many are household and granary pests that have been transported worldwide by human activities (*Corcyra*, Galleriinae; *Pyralis*, Pyralinae; *Plodia, Ephestia, Cadra, Ectomyelois*, Phycitinae). A few species are predaceous on scale insects (Phycitinae) or live in ant nests (some Chrysauginae); three species of Chrysauginae feed in the dung of sloths, and the adults live in their fur. A few species of Phycitinae (e.g., *Cactoblastis cactorum*) have been used for biological control of cactus in Australia, Hawaii, and elsewhere. The majority of species feed on flowering plants, angiosperms and conifers (*Dioryctria*, Phycitinae), monocots, including pests of coconut and other palms (*Tirathaba*, Galleriinae), and corn (Epipaschiinae), but mostly dicots, and a wide range of forest trees, ornamental shrubs, and crops are damaged (e.g., pecans and walnuts, *Acrobasis*, Phycitinae), especially in tropical regions.

Diversity More than 6,000 species are described worldwide, and the tropical faunas are not thoroughly studied. More than 600 species occur in America north of Mexico, about 70% of them Phycitinae, with several subfamilies in need of modern study.

SUBFAMILY PYRALINAE

Moths of the typical subfamily are small to medium sized, mostly red, brown, or yellow, rather than the gray or black-and-white characteristic of Phycitinae, and typically they have a glossy appearance. The third segment of the labial palpus is upturned in both sexes. The female frenulum has two bristles. The male genitalia have uncus arms extending at 90°, and the phallobase is not curved ventrad. Genitalia, Figs. 185, 186. The larvae of many species feed on dried, fungi-permeated, or decaying plant matter, while some feed on living vegetation. Several are cosmopolitan stored food products pests. There are more than 900 described species, mostly in Asia and Africa, and the North American fauna has only 27 species, including several introduced from other parts of the world.

Pyralis farinalis (Linnaeus), known as the "meal moth" or "meal snout moth," is a cosmopolitan species, thought to have originated in Asia. The adults are rather narrow winged, FW (9–14 mm; Plate 24.47) rust brown basally and distally, pale ochreous brown in the middle, separated by white transverse lines. They are secretive, hiding in dark places where grains are stored and infrequently come to lights. This species occurs in most of the United States except at high elevations, primarily associated with human activities. The larvae are scavengers in all kinds of damp vegetable matter and cause serious damage to alfalfa hay and stored grain when damp or infested by other insects. They form tunnels of silk plastered with frass, concealed in corners of storage bins, granary elevators, and the like. There may be four or more generations per year in storage situations, but in adverse circumstances, larvae may live for two years. Out of doors, *P. farinalis* has been reared from bumblebee and bird nests. *Pyralis cacamica* Dyar (Plate 24.48, Plate 24.49), an endemic species in the Southwest, occurs from southern Oregon through California, Nevada, southern Utah, and northern Arizona. It has broad, oval FW (9.5–12.5 mm), like an *Aglossa*, tan clouded with dark brown defining a curved, transverse, postmedial line, parallel to the termen, sometimes all tan beyond it. Specimens from Nevada and Arizona have a rust tinge. The larval biology is not well known, but likely they depend upon vegetative detritus in situations such as wood rat nests. *Pyralis cacamica* lives in diverse habitats, coastal to 6,000′

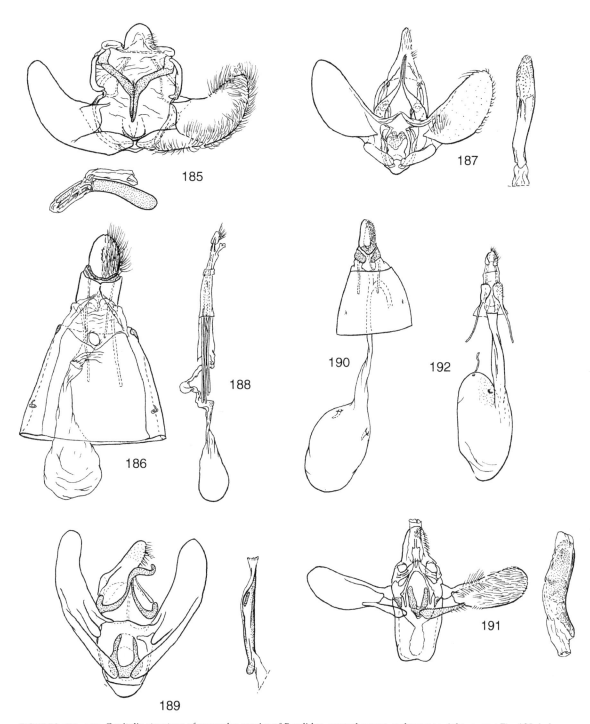

FIGURES 185–192. Genitalia structure of exemplar species of Pyralidae, ventral aspect, aedeagus to right, except Fig. 185. below. **185,** Pyralinae: *Pyralis farinalis* Linnaeus ♂. **186,** Pyralinae: *Pyralis farinalis* ♀ [Solis and Shaffer 1999]. **187,** Chrysauginae: *Caphys arizonensis* Munroe ♂. **188,** Chrysauginae: *Caphys arizonensis* ♀ [Munroe 1970]. **189,** Epipaschiinae: *Oneida luniferella* Hulst ♂. **190,** Epipaschiinae: *Oneida luniferella* ♀ [Solis 1991]. **191,** Phycitinae: *Acrobasis tricolorella* Grote ♂. **192,** Phycitinae: *Acrobasis tricolorella* ♀ [Heinrich 1956].

elevation in southern California; records of adults span June to September.

Aglossa caprealis (Hübner)(FW 6.5–10 mm; Plate 24.37), which is called the "small tabby" in England, apparently was introduced from Europe by the nineteenth century, became widespread in the eastern United States by the early 1900s, and subsequently in the West. The larvae feed from silken tubes among thatch and hay refuse in stables and barns and often live for two years. They usually occur in damp, moldy situa-

tions, sometimes on corn or other foods. *Aglossa caprealis* rarely damages stored products in good condition but sometimes infests houses and cellars, associated with fungus and damp wood or books, even wine bottle corks. It has been reared from a mouse nest and gallery of the wood wasp, *Xiphidria*. In California, this *Aglossa* evidently lives outdoors because adults are taken at lights distant from human habitation, from May to September, and larvae are active throughout the year. ***Aglossa acallalis*** Dyar is a small, pale tan species, FW (7.5–9 mm; Plate

24.38) with indistinct brown scaling defining ante- and post-medial transverse fasciae, dark brown concentrated preceding and following the pale lines. Fresh specimens have a pinkish or peach tinge. This species occurs in Arizona and southern California, in foothills to 6,000' at Mt. Baldy. Specimens from the Kofa Mountains in western Arizona are more rust colored. Flight records are June to August.

Hypsopygia costalis (Fabricius), the "clover hayworm," has rust red wings broadly bordered by yellow, and the FW (8–9.5 mm; Plate 24.36) has two broad yellow bars from the costa extended through the cell. This species was introduced from Europe in early settler times, then spread to the Rocky Mountains by the late 1800s. It occurs in Colorado, Utah, Washington, and eastern California (Owens Valley). The biology is similar to that of *P. farinalis*, larvae feeding on decaying vegetable matter and infesting stacked alfalfa, clover, and timothy hay, especially where moist. *Hypsopygia costalis* has been reared from nests of birds and squirrels in Europe and bumblebee nests in Michigan. Two generations per year have been reported in the midwestern United States, but flight records in the West are in July.

Herculia phoezalis Dyar (Plate 24.39) was described from Los Angeles based on specimens reared by Coquillett and Koebele in the 1880s. Most subsequent records are from urban situations, and this species may be introduced. The wings are dark reddish brown with curving, transverse, dark brown lines outwardly edged with yellowish (FW 7.5–12 mm). Koebele's specimens were labeled from old branches of cypress and fresh bark of citrus; we reared one *H. phoezalis* from cones of Monterey cypress at Avalon, Santa Catalina Island; and larvae infested "Spanish moss" (presumably *Tillandsia usneoides*, Bromeliaceae) in a planter box at Richmond, California. Probably they feed primarily as scavengers. Adults are active June to September, rarely late April and early October. *Herculia olinalis* (Guenée)(FW 12–14 mm; Plate 24.50) is similar to *Hypsopygia costalis*, dark, uniform rose brown to purplish brown, with similar yellow markings. *Herculia olinalis* occurs across North America through the Rocky Mountain states, Texas, Arizona, and Nevada to eastern California. Covell cites hosts of *H. olinalis* as oaks *(Quercus)*, but probably the larvae are scavengers. *Herculia infimbrialis* Dyar is similar and is widespread in the East and Midwest, to Texas and Colorado.

SUBFAMILY GALLERIINAE

These are small, slender to relatively large, stout moths, with narrow to broad wings. The frenulum of females has three bristles, of males one. The tympanal venulae are well developed; male genitalia lack the uncus arms. In females, the ductus seminalis originates on the ductus bursae. Larval habits include feeding on dry vegetable matter, in honeycombs of bees, in root crowns of Asteraceae, and in cones of Cupressaceae, probably mostly as scavengers.

Achroia grisella (Fabricius), the "lesser wax moth," is a small, mouse gray pyralid (FW 7.5–11.5 mm; Plate 24.40) with pale yellow head scaling. The larvae feed almost exclusively on honeycombs and associated detritus in bee hives. They sometimes damage stored honeycombs by eating away the wax cell caps and are often associated with *Galleria* in abandoned beehives, but *A. grisella* is not an important pest of healthy hives. There are two to five generations per year, depending upon temperature, and in mild climates larvae are active through winter. This species does not seem to have been reared from bumblebee or wasp nests. *Achroia grisella* probably was intro-

duced to the Pacific Coast soon after honeybees in the mid-nineteenth century, but adults rarely come to lights, and it was not collected until 1908, at San Diego, spreading northward into the Sacramento Valley and San Jose area by the 1950s.

Galleria mellonella (Linnaeus)(Plate 24.51), the "wax-worm" or "greater wax moth," is a common inhabitant of beehives and sometimes a serious pest. The adult is bulky, moderately large, with broad wings, FW (10–14 mm in male, 13–16 mm in female) with a concave termen, gray suffused with rust on the costal half and the dorsal area with dark rust and brown to black streaks, bordered by upraised scale ridges. The male labial palpi are very reduced and hidden in the scaling. The larvae burrow through the combs, forming silk-lined tunnels, and feed on wax, frass, and exuviae of the bees. In time they can destroy a hive. In addition, *G. mellonella* is reported to feed on dried fruits, dead insects, but not stored grain or flour. There are three generations in a season, and adults emerge from April to October.

In contrast to the typical galleriines, members of the tribe **Macrothecini** are slender, gray, phycitine-like moths, which are recognizable by broad, upraised, black scales on the terminal abdominal segment and genital valvae of males. They live in desert and semiarid regions. *Cacotherapia* (formerly *Macrotheca*) *angulalis* (Barnes & McDunnough) is a gray species, FW (7.5–11.5 mm; Plate 24.41) with two black, transverse lines, bordered inwardly with white, and there is a faint to pronounced dark brown shade in the median area between the black lines. This species occurs in southern California, coastal to montane at 6,000'. We found the larvae under sparse webs among cones of Monterey cypress (*Cupressus macrocarpa*), at Avalon, Santa Catalina Island, in March and May. They occurred at the base of cones of various ages, feeding on sterile tissue, along with larvae of another pyralid, *Tallula fieldi*, and the gelechioid scavenger *Symmoca signatella*. *Cacotherapia angulalis* also occurs on Santa Cruz Island, with adult captures in May and September. *Cupressus* is not native or is absent from most mainland and island sites where *C. angulalis* occurs, so it is associated with habitats other than conifer cones, perhaps as a scavenger. Based on the variation in reared series, this species, *Cacotherapia ponda* (Dyar), also described from southern California, and *C. bilinealis* Barnes & McDunnough from Arizona, may all refer to a single species.

Species of the closely related genus *Alpheias* are gray, variously mottled with black, usually with the inner transverse line white, weakly zigzag. We reared specimens of *A. transferrens* Dyar (Plate 24.42) and a smaller, less distinctly patterned *Alpheias* species, from root crowns of woody *Chrysothamnus* (Asteraceae) in Nevada and eastern California, collected in July. They may have been scavengers in larval galleries of eucosmine tortricids or beetles. *Alpheias transferrens* occurs in southern California east to Utah. *Decaturia pectenalis* Barnes & McDunnough (Plate 24.43) is a distinctly patterned, small species with strongly pectinate male antennae. The FW (5.5–6.5 mm) is pale gray, having well-defined, pale brown markings bordered with black and a less distinct smudge at midcosta terminated by an oblong black line in the cell. This species occurs in several mountain ranges of southern Arizona. Its larva is unknown.

SUBFAMILY CHRYSAUGINAE

This group is primarily Neotropical, with small to moderately large species, most of which are colorful, red, yellow, greenish, or brown, FW often with two transverse pale lines. The FW morphology is sexually dimorphic in most genera, males with the costa, shape of the cell, stalking of the R veins, and frenulum modified, sometimes bizarrely so, with a conspicuous open pouch below the costa basally. Males have a single frenulum bristle, sometimes thickened, females three. The male genitalia have lateral arms of the uncus extending at 90°. Genitalia, Figs. 187–188. The larval habits are diverse, usually they are borers in roots, stems, fruits, or seeds, but some are leaf rollers. Members of three genera live in the fur of sloths, and their larvae feed on sloth dung. A few chrysaugines develop in ant or wasp nests. There are more than 400 described species, of which about 50, mostly token representatives of widespread genera, extend into the southern United States, Atlantic states, and Southwest.

Caphys arizonensis Munroe is a relatively large pyralid, FW (11–14 mm; Plate 25.1, Plate 61.20, Figs. 187, 188) triangular, rust brown or rose brown with two conspicuous yellowish, transverse lines. The HW is white, tinged with yellowish and peach. The ovipositor is very long, slender, and extensible. *Caphys arizonensis* occurs in northern Mexico (Sinaloa, Jalisco) and southern Arizona to western Texas, the northernmost of a dozen Neotropical *Caphys*. The adults are frequently attracted to lights in July to August, but the larval biology is unknown. *Parachma ochracealis* (Walker) is similar but the FW (6.5–8 mm) is broader, ochreous, striate with rust. The two transverse lines are indistinct, the distal one nearly straight. This species occurs in eastern North America to the Rocky Mountains. Adults fly from March to September in Florida, July to August in the West.

Arta epicoenalis Ragonot is similar but smaller and more variable in size (FW 6.5–10 mm; Plate 24.44), color, and the ochreous FW transverse lines are distinct to obsolete. This species is widespread and commonly collected from Oregon to northern Baja California and Arizona, from low-elevation chaparral habitats to 6,000' in the Sierra Nevada and southern Arizona. Adults are recorded from April to early September in California, April and July to August in Arizona. This species may be a western component of the smaller, darker red *A. statalis* Grote, which is widespread in the East and Midwest.

Species of *Galasa* are recognized by their peculiar FW shape, oval with two concave indentations in the costa in males, one in females, and large scale tufts on the middle and hind tibiae and tarsi in both sexes. This too is a primarily Neotropical genus, with about two dozen described species, of which two occur north of Mexico. *Galasa nigrinodis* (Zeller) is widespread in the eastern United States, to western Texas and the Rocky Mountains. The FW (8.5–9.5 mm; Plate 24.45) is reddish orange basally, dark rust crimson beyond, with pale gray shading along the costal indentation. The larvae tie up and eat dead leaves of boxwood (*Buxus*, Buxaceae), according to Covell, and are recorded on devilwood (*Osmanthus*, Oleaceae). In Florida this species flies year-round; western records are July to September. *Galasa nigripunctalis* Barnes and McDunnough (Plate 24.46, Plate 61.21) is similar but has scaling of the head, palpi, and tegulae primarily ochreous, the FW (8–10 mm) pale brick red sprinkled with ochreous, and the costal area suffused with black. This species occurs from southern Arizona to Colorado.

Satole ligniperdalis Dyar (FW 7–9 mm, females larger than males; Plate 25.6, Plate 25.7, Plate 25.8), like many tropical chrysaugines, has pronounced dimorphism, males having the base of the costal area curled, forming a pocket sealed by a tympanum-like membrane, which is recessed under a hood of long scales, while the female FW is unmodified. There are two primary color forms in both sexes: FW dark gray with whitish transverse lines or black basally, pale peach tan beyond, with a dark gray fringe. This species occurs in Arizona and southern California desert areas in association with desert willow (*Chilopsis linearis*, Bignoniaceae). Dyar reported larvae of *S. ligniperdalis* boring in "solid wood" of *Chilopsis*, whereas we found them in the green, living stems. Larvae collected in July 1974 produced adults sporadically over two years: several days following collection, after 14 and 23 moths, and in late fall 1976, 26 months after larvae completed feeding.

SUBFAMILY EPIPASCHIINAE

These are medium sized, rather robust pyralids, FW elongate-triangular, often with upraised scale tufts, mostly patterned in gray, black, and white. The labial palpi have the third segment upturned and pointed, and the antennal scape is greatly modified in some genera. The frenulum of males has one bristle, of females two. The male genitalia possess elongate lateral arms of the uncus extended at an angle of 110° or more, and the phallobase usually is curved downward. Genitalia, Figs. 189, 190. This subfamily is distributed in tropical and temperate regions, exclusive of Europe, and has nearly 600 described species, of which about 50 occur in America north of Mexico, primarily in Florida and the Atlantic states. There are only about 15 species recorded in the West.

Males of **Macalla** have a long, heavily scaled extension of the antennal scape, which curves beyond the head and dwarfs the short antenna. *Macalla zelleri* (Grote)(Plate 25.2) has thick scape projections that extend nearly the length of the thorax and bear flared scale brushes distally. The FW (11–13.5 mm) is dark brown basally, black at the antemedial line, followed by a white fascia with brownish, upraised scaling at the postmedial fascia, and the terminal area is brown. This species ranges transcontinentally but has been only sporadically recorded in the West. It occurs in interior California (Mt. Shasta to Mt. Palomar), Utah, southern Arizona, and probably throughout the Southwest, with flight records from May to July.

In *Cacozelia* the male antennae have the peculiar extensions of the scape, but shorter than in *Macalla*. *Cacozelia basiochrealis* Grote (Plate 25.3, Plate 25.4, Plate 61.19) is widespread, from the Midwest to Texas into northern Mexico and southern Arizona. It is commonly collected from late July to early September, and there are records in June and October in Texas. The scape projections extend beyond the collar and have flared scaling dorsally, ochreous tinged with rust, like the basal area of the FW (9–11.5 mm), which is gray beyond the antemedial line, with a rust preapical patch.

Toripalpus trabalis (Grote) is widespread in the West, whereas its congeners all occur in southern Texas and Florida. This is a large, broad-winged epipaschiine (FW 11–15 mm; Plate 25.24, Plate 25.25), having short antennal scape projections and long labial palpi, exceeding the crown. The FW color pattern is variable: pale gray with transverse, dark gray fasciae, shaded with rose red, giving a bluish appearance to the unaided eye; or basal area bordered near midwing by a distinct, sinuate line, followed by a broad white band that is well developed in some populations to nearly obscured by gray or reddish fasciae in others. *Toripalpus trabalis* occurs from eastern Oregon through the Southwest to western Texas. We found the foliage-feeding larvae forming extensive webbing on the lower stems of buckwheat shrubs (*Eriogonum grande*,

Polygonaceae) on Santa Cruz Island, and *E. parvifolium* and *E. nudum* at coastal and inland sites on mainland California. Larvae were active in March, May, and June, and adults are recorded from April to November.

Male **Oneida**, like *Toripalpus*, have elongate labial palpi and short extensions of the antennal scapes. **Oneida luniferella** Hulst (Plate 25.5, Figs. 189, 190) is widespread in the mountains of northern Mexico (Nuevo León, Durango, Sinaloa) and Arizona north to Mingus Mountain in Yavapai County. The antennal projections are short, not exceeding the back of the head, with broad scale tufts. The FW (14–16 mm) is dark gray with a black and rust, strongly upraised scale ridge at the inner border of an antemedial band and a rust brown preapical patch enclosing a white crescentic line. Adults fly in August to September.

Tallula fieldi Barnes & McDunnough is a distinctive little species having triangular FW (7.5–10 mm; Plate 25.9), which are white, variably patterned with brownish gray, on basal half; a darker patch with upraised scale tufts on the dorsal margin beyond a white submedial line; and a subapical fascia weakly connected to the dorsal patch is overlaid with whitish, giving a bluish appearance to the middle portion of the fascia. The larvae live in sparse webs externally on densely clumped cones of cypress (*Cupressus*), *C. sargenti* in coastal San Luis Obispo County and Monterey cypress (*C. macrocarpa*) on Santa Catalina Island, where cypress is not native. *Tallula fieldi* shared its larval niche with native pyralids, *Cacotherapia angulalis* and *Herculia phoezalis*, and an introduced scavenger, *Symmoca signatella*, none of which has been collected in native habitats on the island. Larvae we found in March and May produced adults in May and June. Adults are recorded June to August.

SUBFAMILY PHYCITINAE

This subfamily comprises the largest, most diverse lineage in the Pyralidae, with an estimated 4,000 described species worldwide. These are mostly slender moths with narrow FW and broad, fanlike HW, which are held around the body when the moth is perched. The FW usually is gray with white or dark gray maculation, but a few species are colorful. Phycitines are characterized by lacking secondary venulae associated with the tympanal structures, and both sexes have a single-bristle frenulum. Genitalia, Figs.191–200. The larvae mostly feed on living plants as external feeders on foliage, often spinning a tough silken shelter, or in flower heads of Asteraceae, while others are borers in shoots, cactus stems, or cones and cambium of conifers. A few are leaf miners in early instars, inquilines in galls or nests of other insects, or are predaceous on scale insects. Several are notorious as stored products pests, feeding in flour, meal, nuts, and dried fruit. There are more than 450 described species in America north of Mexico, assigned to 118 genera, and we treat western species representing half of the genera (Table 1, see p. 199). About 90% of the species were treated by H. H. Neunzig in a series of MONA volumes during the past 20 years. He and others have raised the number substantially through descriptions of new species and by treating named geographic races as species. More than 260 species occur in the West. This group suffers from extreme generic splitting after Heinrich in his 1956 monograph followed earlier taxonomists in emphasizing male secondary features in proposing relationships. As a result, females of many genera cannot be distinguished on superficial bases, and this is the only major group of North American moths that approaches the butterflies in degree of generic restriction, with about 3.8 species per

genus. Neunzig followed this practice and has not attempted to contribute to our understanding of phylogenetic relationships, having placed genera in the individual fascicles "mainly with ease of identification in mind and deferred a phylogenetic classification." Heinrich divided the subfamily into three groups based on presence and absence of HW veins, but the groups do not correlate well with genitalia anatomy. The Phycitinae beg for a sound phylogenetic analysis leading to tribal level classification and generic reassessment. We treat the genera in Heinrich's sequence and give an index to Neunzig's treatment (Table 1).

Acrobasis includes a large group of relatively broad winged species that are characterized by usually having a transverse, antemedial scale ridge on the FW in both sexes, and males of most species have the antennal scape enlarged or angulate. The larvae feed on foliage, buds, or shoots of trees and shrubs, especially Juglandaceae, including notorious pests of walnut (*Juglans*) and pecan (*Carya*), but some species feed on various other plants, including Betulaceae, Fagaceae, Rosaceae, or Ericaceae. There are about 40 Nearctic species, largely in eastern deciduous forests, and only nine range into the West, mostly widespread eastern species that reach western Texas or New Mexico or have been introduced locally in British Columbia or southern California. Incongruously, numerous morphologically similar, allopatric host-plant specialists in the East have been treated as species, whereas western host races have been placed in synonymy. **Acrobasis tricolorella** Grote (Plate 25.10, Plate 25.11, Figs. 191, 192) occurs across southern Canada and the northern United States and in the West through the Rocky Mountain states to New Mexico, Arizona, and California. The adult is small (FW 8–11 mm), distinguished by lack of a FW scale ridge and is colorful compared to most *Acrobasis*, having a white transverse line at the basal one-fourth, usually followed by a rust orange patch, and the postmedial line is sinuate, well defined, and often followed by rust scaling. The red larvae feed in flowers and fruit of Rosaceae, including *Prunus*, both native and orchard, and *A. tricolorella* is known as the "destructive prune worm" in the Pacific Northwest. Hosts include cherries and plums, mountain ash (*Sorbus*), rose, and Christmas berry (*Heteromeles*), and larvae are reported to feed in sporophores of *Dibotryon* (Ascomycetes) on cherry. Larvae we collected fed in the growing terminals of *Heteromeles* in April and produced adults in May. Capture records of adults in California span June through October, possibly two generations. We found one larva in old foliage of Douglas-fir, presumably a vagrant from adjacent vegetation. **Acrobasis comptella** Ragonot averages slightly smaller (FW 8–9.5 mm; Plate 25.12) and is nondescript, with ante- and postmedial transverse lines black, not raised, bordered basally by white and followed by a dull rust brown or orange brown patch that varies from a trace to moderately broad. There is geographic variation as well, including distinctive races in the Kofa Mountains, southwestern Arizona (*A. kofa* Opler)(Plate 25.12), associated with the hybrid desert oak, *Quercus turbinella* x *Q. ajoensis*, and in the Carson Range at 7,300' in western Nevada (*A. neva* Opler), feeding on chinquapin (*Chrysolepis sempervirens*, Fagaceae). This species occurs from Washington to Utah and western Texas to the Pacific Coast. The larvae feed on deciduous oaks, *Q. douglasii* and *Q. garryana*, in addition to evergreen desert oak and chinquapin.

Trachycera species are quite similar to *Acrobasis* species but are distinguished by having the basal segment of the male antenna unmodified, whereas usually it is enlarged and angulate in *Acrobasis*. **Trachycera caliginella** (Hulst) occurs in Arizona

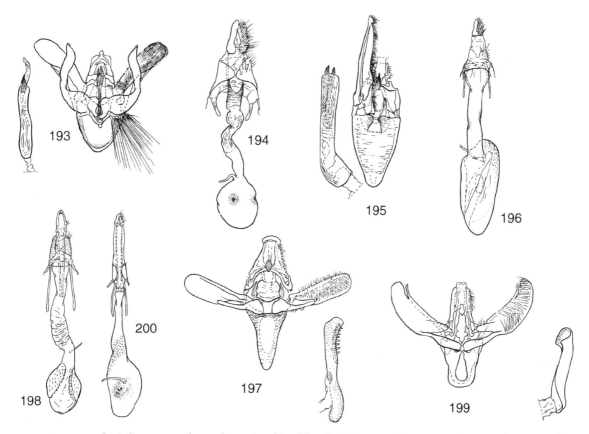

FIGURES 193–200. Genitalia structure of exemplar species of Pyralidae, Phycitinae, ventral aspect. **193**, *Dasypyga alternosquamella* Ragonot ♂, aedeagus to left. **194**, *Dasypyga alternosquamella* ♀. **195**, *Pima albiplagiatella occidentalis* Heinrich ♂, aedeagus to left. **196**, *Pima occidentella* ♀. **197**, *Phycitodes mucidellum* (Ragonot) ♂, aedeagus below, right. **198**, *Phycitodes mucidellum* ♀. **199**, *Ephestia kühniella* (Zeller) ♂, aedeagus below, right. **200**, *Ephestia kühniella* ♀ [Heinrich 1956].

and in California from sea level to 7,000′ in the Sierra Nevada and includes several host races. Adults are similar to *A. comptella* but are appreciably larger, FW (15–21 mm; Plate 25.13) with the color pattern similarly variable, and the two species sometimes occur together. The larvae form tough silken tubes among leaves of evergreen oaks, feeding after the foliage has hardened, overwintering and maturing in spring but not emerging until the new foliage has developed. Larvae taken in April produce moths from June to August. *Trachycera caliginella* was described from Arizona, locality and larval host not specified. In California, *Quercus agrifolia* and *Q. wislizenii* are hosts in coastal areas, where the name *T. caliginoidella* Dyar applies (Plate 25.14), and there are isolated populations on huckleberry oak, *Q. vaccinifolia*, in the northern Sierra Nevada, *T. yuba* (Opler); on a chaparral species, leather oak, *Q. durata*, *T. durata* (Opler); and on scrub oak, *Q. berberidifolia*, on Santa Cruz and Catalina islands, *T. cruza* (Opler).

Myelopsis alatella (Hulst) has slender, smooth-scaled FW (8–13.5 mm; Plate 25.15), gray with a more or less complete transverse, dark fascia at basal one-fourth, a median spur outward, usually preceded by whitish. The early stages are unknown. A European species of *Myelopsis* feeds on Salicaceae, but in California the collection sites tend to be chaparral or pine forest habitats lacking willow and poplar. *Myelopsis alatella* is widespread in the West, from Montana to Texas, Oregon, and the northern half of California and Santa Cruz Island. Adults come to lights, late March to August.

Apomyelois bistriatella (Hulst) has simple antennae, in the male thickened with longer, dense cilia; wings smooth scaled, broader than related genera. The FW (7.5–12.5 mm; Plate 25.16) is charcoal gray with a variably complete, antemedial, transverse, white fascia from the dorsal margin followed by a darker shade. This species is Holarctic, occurring across Canada and the eastern and midwestern United States, and on the Pacific Coast, ranging from coastal foothills to the Sierra Nevada and desert margins. In California the larvae feed in the charcoal-briquette-like sporophores of the ascomycete fungus, *Hypoxylon occidentale* (Xylariaceae) growing on recently fallen live oak *(Quercus agrifolia)* and poplar *(Populus)*. In Europe Beirne reported this species feeding on *Daldinia* (Xylariaceae) on hardwoods, including gorse and birch. In Great Britain *A. bistriatella* is univoltine, but in California we have collected larvae February to June and October, and none entered diapause.

Amyelois transitella (Walker)(Plate 25.17), the "naval orangeworm" of agricultural literature, occurs throughout the southern half of the United States. It is distinguished from related genera by male genitalia features, an abbreviated uncus, incomplete transtilla, long lateral elements of the juxta, and narrowed distal half of the valva. The moth has heavily scaled labial palpi, and pale gray FW (11.5–24 mm) with the outer border of the antemedial line enlarged below the costa into a conspicuous dark spot, and there is variable ochreous to reddish orange scaling toward the dorsal margin. *Amyelois transitella* gained much of its notoriety infesting stored walnuts and almonds, and the common name may have originated as much from the color of the larva, which is yellowish, often tinged with a pink to orange, as from its damage to citrus. Larvae of this species also feed on many other plants, primarily in the fruit and seed, including apple, pistachio (*Pistacia,* Anacardiaceae), date palm (*Phoenix,* Arecaceae), and acacia (Fabaceae).

Tacoma feriella Hulst has unmodified antennae in both sexes and slender, upturned labial palpi. The FW (7–10 mm; females larger than males; Plate 25.23) are dark gray with intermixed powdering of black-and-white scales, giving a bluish gray appearance, and there is a prominent white blotch on the middorsal margin. This species is widespread, from central Texas to southern California, north inland to the Sacramento Valley. *Tacoma feriella* was reared from mistletoe (presumably *Phoradendron,* Viscaceae) by Dammers at Riverside, California. Adults fly July to September in California, and there are May and June records from Texas and Arizona.

Dasypyga alternosquamella Ragonot is one of the most distinctive phycitine moths, having salmon pink to dark reddish salmon FW (8.5–14 mm; Plate 25.26, Figs. 193, 194) with basal one-third dark gray crossed by a transverse rust fascia, termen gray, and the salmon distally divided longitudinally by a white line, which often is partially replaced by dark rust. The male genitalia are unique, with the sacculus separated and heavily sclerotized. The larvae feed on dwarf mistletoe, *Arceuthobium* (recorded by Heinrich from Oregon and Colorado as *Razamofskya,* a synonym) on conifers, primarily pines. This pyralid is associated with various pines from southern British Columbia and coastal Washington to California, Colorado, at 8,500' in the Clark Mountains of southern Nevada and Chiricahua Mountains, Arizona, and above 9,000' in Durango, Mexico. Coastal and high-elevation specimens are darker, whereas those from pinyon pine areas (eastern California, midelevations in Arizona) are pale colored, the FW gray nearly obsolete. The adults fly in May and July through October in coastal California, indicating two generations.

Etiella zinckenella (Trietschke), the "lima-bean pod borer" or "bean pod borer" (FW 8–12.5 mm; Plate 25.27), was described originally from Sicily in 1832, probably introduced there. It occurs on all continents and is widespread in western North America, especially inland and in desert areas. The larvae feed on the immature seed of legumes, including native *Lupinus, Astragalus,* and *Robinia,* as well as various commercial beans (*Phaseolus*) and soybean (*Glycine*). The moth has very long, porrect labial palpi (extending three times the length of the head beyond it), and the male has a slender scale brush at the base that fits into a groove along the inner surface of the palpus, which is formed by smooth scaling. The FW is gray with the costal edge white and crossed basally by a line of tufted, brassy ochreous scales, edged inwardly with rust. There is a paler race in the deserts and on the southern California islands, where there are two generations, March to May and August to December. Larvae of *E. zinckenella* sometimes occur with *Pima* in the same legume pods.

Pima are relatively narrow winged species, FW often with a longitudinal white, subcostal streak, and the larvae feed in legume seed pods. There are nine described species in America north of Mexico, most of which occur in the West. *Pima occidentalis* Heinrich (FW 10.5–14.5 mm; Plate 25.28, Figs. 195, 196) and *P. albocostalialis* (Hulst)(FW 11–14 mm; Plate 25.29) are similar and widespread, from British Columbia to southern California and the Rocky Mountains, and both feed primarily on locoweed, *Astragalus,* but they rarely occur together. *Pima*

occidentalis tends to occupy more arid regions, across the deserts to western Texas, whereas *P. albocostalis* is northern and ranges southward only to northern Arizona and New Mexico. The latter's FW is dark with a narrow to broad black shade bordering the white subcostal streak, often suffusing most of the wing. The FW of *P. occidentalis* is mostly pale brown, darker adjacent to the subcostal streak, with a dark dot at the end of the cell. The fully grown larvae of other *Pima* species are reported to overwinter in a silken hibernaculum in the seed pod, then leave to form a cocoon at a new location. Flight records for these two species are primarily April to June, but we also collected specimens of *P. albocostalialis* in August and September at Big Creek, suggesting a partial second generation near the seacoast.

Interjectio denticulella (Ragonot) is morphologically similar to *Pima* but is larger and the FW (11.5–16 mm; Plate 25.30) is finely streaked with black, forming a smudge on the dorsal half of the antemedial fascia. This species occurs from British Columbia to Wyoming and central California, with adult records from April near the coast to July and August in northeastern California. At Antioch on the San Joaquin River, the larvae feed on inflorescences of *Lupinus albifrons,* discovered by Clint Kellner. We found larvae late March to late May, and adults emerged in May to June, so the life cycle differs from that of *Pima.*

Ambesa are larger and more colorful than most phycitines, particularly the larvae, and the five species all occur in western North America. *Ambesa laetella* Grote (FW 12–15 mm; Plate 25.31) is distinctive, having white FW with bold, well-defined, purplish brown maculation. The larvae, which are black and orange striped, feed on native roses (*Rosa californica* and *R. gymnocarpa*). We collected them in May, adults emerged in July, and flight records span June to August. This species occurs from Alberta through the Rocky Mountains to New Mexico and southern California. *Ambesa walsinghami* (Ragonot)(Plate 25.32), including *A. mirabella* Dyar, a southern California race, is smaller, FW (9–12 mm) gray with black streaks. The two were treated as a separate species by Neunzig; *A. mirabella* has ochreous scaling on the abdomen and HW, whereas the scaling is gray or brownish on specimens referred to *A. walsinghami,* and differences in genitalia intergrade. This species ranges from Washington to Utah, Nevada, and California. The larva is gray and black with an orange dorsal stripe; it feeds in a webbed shelter of rolled leaves on *Prunus* (Rosaceae), including chokecherry, *P. virginiana,* in Washington reared by Clarke, bitter cherry, *P. emarginata,* and Sierra plum, *P. subcordata,* reared by Crabtree in northern California, and prune, *P. domestica,* at Hopland, California, reared by Essig.

Catastia is Holarctic with three species in the Palaearctic and four in the western Nearctic, all northern and/or at high elevations. Larval hosts have not been discovered in North America, but a European species feeds on *Alchemilla* and *Potentilla* (Rosaceae). The adults are dark colored and some, possibly all species in our fauna, are diurnal. *Catastia actualis* (Hulst)(Plate 25.33) is the largest and most often collected, occurring from Saskatchewan and Alberta to Colorado and northeastern California to Lake Tahoe. The wings are brown, FW (12–14 mm) with partially developed, sinuate ante- and postmedial, whitish lines. Flight records are June to July. *Catastia bistriatella* (Hulst) is smaller, FW (10–11 mm; Plate 25.18) dark metallic blue with distinct, nearly straight, white, transverse lines. This is a true child of the cloudlands, occurring in the Arctic-alpine zone above timberline (11,000–14,000') in the southern Sierra Nevada and White

Mountains of eastern California, where adults at times are numerous, July to early September.

Salebriaria is a moderately diverse genus with 25 described species in America north of Mexico, most of which are local host races in the eastern and southeastern United States. Only four are named in the West. The male antennal scape has a large scale tuft arising from a deep concavity. The larvae are oak feeders and have the peculiar habit of skeletonizing the lower surface, usually without a shelter, even in later instars. The FW of many eastern species has a white band or dorsal blotch basally, but not the western species. *Salebriaria equivoca* Neunzig (FW 7–10 mm; Plate 25.19) has gray FW with white transverse lines bordered outwardly in the costal area by black and inwardly in the dorsal area with brown. This species occurs from western Texas to southern Arizona, recorded from late June to August. The early stages are unknown, but presumably the hosts are oaks. *Salebriaria maximella* Neunzig is quite similar, with paler FW maculation, and averages larger (FW 10–11 mm). It is known only from the mountains of Utah and Colorado.

Quasisalebria is distinguished from *Salebriaria* by differences in an enlarged scale tuft on the male antennal scape. *Quasisalebria* have a discrete group of small, thin scales at the distal end of the sinus from which the large scale tuft arises. *Quasisalebria admixta* Heinrich is distinct among species of both genera by having white HW and the FW (7.5–9 mm; Plate 25.20) shaded with olive brown or rust brown, with a strong white antemedial line, emphasized by dark scale patches. This species ranges from Utah and Colorado to western Texas and southern California. Flight records span April to October, including April and July at Madera Canyon, Arizona, so *Q. admixta* may be bivoltine. *Quasisalebria occidentalis* (Neunzig) is almost indistinguishable superficially, differing by genitalic details including short lateral extensions of the juxta rather than long, curved ones, and longitudinal ridges or folds on the corpus bursae, which are absent from *Q. admixta* females. *Quasisalebria occidentalis* occurs in central California and flies in April to June. The early stages are unknown for both these species; in the eastern United States the larvae of *Q. atratella* (Blanchard & Ferguson) feed on oaks.

Meroptera is one of a group of mostly eastern Nearctic genera that are distinguished by having the scale tuft of the male scape developed only on the anterior side of the antenna, whereas males of related genera have anterior and posterior opposing tufts, or they are reduced. *Meroptera pravella* (Grote) is a nondescript species resembling many *Salebriaria*, *Sciota*, *Quasisalebria*, and related phycitines, having gray FW (9–10 mm; Plate 25.21) with the antemedial line preceded by pale gray and followed by blackish shades. This is a northern species, occurring across Canada and the adjacent states to British Columbia, Idaho, Utah, Colorado, and eastern Oregon. The larvae feed primarily on aspen, *Populus tremuloides* (91% of 332 rearing records by the Forest Lepidoptera Survey in Canada), and other Salicaceae, rarely on birch and alder. Adults fly in May to July.

Sciota is Holarctic, occurring in Europe and Asia and in North America from Alaska to Mexico, with 24 described Nearctic species, about half of which occur in the West. The male antennal shaft is modified basally with clusters of scales forming a tuft on the anterior side. The FW of most of the species has a basal pale sector, often tinged with rust or brownish, or the wing is pale with a dark patch following the antemedial line. The larvae feed between two appressed leaves, skeletonizing the opposed surfaces, and in later instars usually form a shelter by folding a leaf or tying leaves together. Diverse host plants are used by *Sciota*, but individual species are specialists. *Sciota basilaris* (Zeller) ranges across southern Canada to British Columbia, Colorado, and Utah. The FW (9.5–12 mm; Plate 25.34) is brownish with the white antemedial line enclosed in a black patch. The larvae are recorded on Salicaceae (*Populus, Salix*) in Canada. Adults fly in June to July. *Sciota bifasciella* (Hulst) occurs in Oregon, California, and Arizona. The FW (10–11 mm; Plate 25.35) is gray speckled with whitish. In southern California, the larvae feed on Anacardiaceae; McFarland reared *S. bifasciella* from squaw bush, *Rhus trilobata*, and we found the larva on laurel sumac, *Malosma laurina*. Moth captures are July to August. *Sciota dammersi* (Heinrich) also occurs in southern Arizona and southern California, where Dammers reared it from *Amorpha californica* (Fabaceae). The FW (10–11 mm) is pale gray, lightly tinged with rust on the basal half. Collection records in April and July indicate a bivoltine life cycle.

Telethusia ovalis (Packard)(Plate 25.36) is larger than *Sciota*, and the male antenna is strongly bowed basally, encasing a large, triangular scale tuft. The ovipositor lobes are hardened and terminated in a nipplelike tip, and its subtending membrane is peculiarly spiculate. The FW (10–12.5 mm) is variable, dark to pale gray streaked with dark gray, rust, and white. This species occurs from Ontario to British Columbia and Montana to Texas, northern Arizona, and central California. Clarke reared *T. ovalis* from Asteraceae, pussy paws (*Antennaria*), and woolly sunflower (*Eriophyllum*) in Washington. There are June to July and September to October flights in coastal California.

Phobus are similar to *Telethusia*, with the same antennal structure, but the ovipositor is unmodified. The four species are all western and widespread, although poorly documented, according to Neunzig, and nothing is recorded on their larval biology. *Phobus curvatellus* (Ragonot) has dark gray FW (10–12 mm; Plate 25.37) with a whitish antemedial line on the posterior half and a weak indication of the postmedial and subterminal lines. The HW are white, in contrast to congeners. This species occurs in Colorado, Utah, Arizona, and southern California. There are collection records in May through early September at Santa Cruz Island and coastal Monterey County, California. *Phobus funerellus* (Dyar) has dark brown or brownish black FW (11.5–14.5 mm; Plate 25.38) with the transverse lines indicated by darker shades following them, and the HW is brownish gray. It occurs from British Columbia to northern Arizona and Baja California Norte, where *P. funerellus* ranges above 8,000' in the Sierra San Pedro Martir. Flight records at the extensively sampled Big Creek Reserve are limited to June and July.

Species of *Pyla* comprise a rich lineage in the Nearctic, with 21 described species including the Holarctic type species. There are two guilds of *Pyla*, nocturnal species having mostly gray FW with whitish transverse markings, and diurnal species with dark brown or black FW, usually with metallic greenish, coppery, or bronze sheens. About 15 species occur in the West, including most of the latter group. *Pyla fusca* (Haworth) is a gray moth, FW (10.5–12.5 mm; Plate 25.39) darker than typical for nocturnal species of this genus, with the transverse lines weak to nearly obsolete. It occurs across the Palaearctic and is a northern species in the Nearctic, ranging from Newfoundland to Alaska and British Columbia, Colorado, and northern California. Its larvae are recorded feeding on Ericaceae (*Calluna, Erica, Vaccinium*) in Europe; and *P. arenaeola* Balogh & Wilterding, a nocturnal species in the Great Lakes region, feeds on bearberry, *Arctostaphylos uva-ursi*, so various Ericaceae may be hosts of other *Pyla*. Balogh discovered larvae of *P. arenaeola* on shoreline dunes, living in silken tubes just under the sand surface, where they cut pieces of leaves and drag them into their shelters. *Pyla areneoviridella* Ragonot is similar to several other diurnal species, FW (10–11 mm; Plate 25.40) dark brown with greenish reflections. It is the most widespread of the montane species, ranging from the Olympic Mountains in Washington to Montana, Colorado, Utah, and eastern Oregon, flying mostly in July. *Pyla scintillans* (Grote) has dark brown FW (9.5–11.5 mm; Plate 25.41) with a metallic bronzy sheen and a darker median band when fresh. It is the most commonly seen diurnal *Pyla* in California, occurring at mid- to high elevations (4,000–10,000') in the Cascade Range and Sierra Nevada. *Pyla viridisuffusella* Barnes & McDunnough has ruffled, bright green scaling on the FW (8–8.5 mm; Plate 25.22) and body. It is one of several species restricted to high elevations, occurring in Arctic-alpine habitats in the southern Sierra Nevada at 9,000 to 12,000'. E. O. Essig collected a series at Kennick Meadows in Yosemite National Park in late July 1934, and flight records are mostly in August.

Dioryctria is the most diverse genus of Phycitinae in the western Nearctic, with many colorful species, several of which are important in forest entomology. The larvae feed in cones, foliage shoots, or under bark of conifers, primarily Pinaceae, and cause economic losses, especially in seed production and in nursery plantations. Most of the species are univoltine with a summer flight. There are 40 described species in America north of Mexico, 25 of which are recorded in the West, and likely numerous species remain to be distinguished and clearly defined in the United States and Mexico. Structurally and in external appearance many *Dioryctria* are very similar to one another, and more refined approaches to differentiation, including DNA analyses, are in progress.

Dioryctria abietivorella (Grote), the "fir coneworm," is transcontinental, and its larvae bore in cones of many conifers, particularly firs *(Abies)* and Douglas-fir *(Pseudotsuga)*. They are a serious concern in fir, spruce, and pine seed orchards and also mine the buds, shoots, and under bark. The adults have a dark gray, nearly black FW (10–12.5 mm, Plate 25.42) crossed by transverse, zigzag, white, antemedial and subterminal lines, the former preceded by a faint, gray tan patch, and there is a white spot at the end of the cell. This species ranges across southern Canada and the Pacific Northwest to the central Sierra Nevada and San Francisco Bay area. Adults fly from May to August. The "spruce coneworm," **D. reniculelloides** Mutuura & Munroe (Plate 25.44), and **D. pseudotsugella** Munroe (Plate 25.43), which has nomenclatural priority, comprise a widespread complex in the West, abundantly represented in collections. The former species has a dark gray FW (9.5–13 mm) with distinct, white, ante- and postmedial lines, sometimes inconspicuous pinkish brown between them. It occurs across southern Canada, the northern United States, and the Rocky Mountain and Pacific states. By contrast, *D. pseudotsugella* FW (10–12 mm) has weak ante- and postmedial lines and a conspicuous pinkish brown patch between them, and it is limited to the western states. The two are not clearly distinguished and require further study. The larvae of both are rust colored with darker longitudinal bands bordered by yellowish lines, illustrated by Duncan. Caterpillars of *D. reniculelloides* feed primarily on spruce *(Picea)*, those of *D. pseudotsugella* mainly on Douglas-fir, but other Abietoideae serve as hosts, including western hemlock *(Tsuga heterophylla)*, firs, and rarely lodgepole pine *(Pinus contorta)*. The prevalent form in California is

D. pseudotsugella, although reared series sometimes include both. Powell and De Benedictis found the larvae to be the most abundant foliage-feeding species other than *Choristoneura* (Tortricidae) during outbreaks of the latter on white fir *(A. concolor)*. **Dioryctria ponderosae** Dyar (FW 10–14 mm; Plate 25.48) is superficially similar to *D. reniculelloides*, primarily dark gray or black, with white transverse lines, but there is a conspicuous patch of pale rust brown scales following the postmedial line, and the head and thorax have some reddish brown scaling. This species ranges from eastern Washington to Colorado, western Texas, Arizona, and California. The larvae bore in the cambium under bark of *Pinus ponderosa*. **Dioryctria muricativorella** Neunzig (FW11–14 mm; Plate 25.45) is similar but has only inconspicuous patches of tan scales preceding the antemedial line and following the postmedial and lacks rust scaling on the head and thorax. This species was reared by Don Burdick from cones of Bishop pine, *P. muricata*, a coastal, closed-cone pine in coastal central California. Neunzig designated paratypes from Berkeley and Walnut Creek, where another closed-cone pine, *P. radiata*, is extensively planted and presumably serves as the host tree. Flight records in May to June and August to October indicate two generations. A collection of similar males at black light on Santa Rosa Island, the only known association of *Dioryctria* with the extremely restricted Torrey pine *(P. torreyana)*, presumably refer to *D. muricativorella*. **Dioryctria cambiicola** (Dyar) is a large species (FW 13.5–15.5 mm; Plate 25.47) that differs from the above *Dioryctria* by having the FW color pattern dominated by dark rust brown. The HW is dark brown. This species is widespread, from British Columbia and Montana to the Sierra Nevada in California, northern Arizona, and New Mexico. The larvae bore in the cambium of pines, especially *P. ponderosa*. Copious pitch and frass exudations accumulate at the burrow sites. *Dioryctria cambiicola* flies in July and August.

Dioryctria auranticella (Grote)(Plate 25.46), the "ponderosa pine coneworm," is the most widespread of a group of species having primarily reddish orange FW, occurring from South Dakota to New Mexico, and British Columbia to southern California. It has paler orange FW (10.5–14 mm) than related species and ochreous white HW, rather than pale gray. In addition to *Pinus ponderosa*, *D. auranticella* has been reared from various pines in California, including knobcone *(P. attenuata)*, Jeffrey *(P. jeffreyi)*, gray *(P. sabiniana)*, Parry pinyon *(P. quadrifolia)*, and Monterey *(P. radiata)*, and presumably *P. remorata*, a derivative of *P. muricata*, serves as the host on Santa Cruz Island. This species flies in late May to June near the coast to July and August in the mountains. **Dioryctria erythropasa** (Dyar) has beautifully colored FW (11–14.5 mm), dark reddish brown to pale orange brown with distinct white transverse lines, a transverse bar at the end of the cell, and a yellow patch following the antemedial line. There is a subcostal streak of reddish brown on the FW underside of the male. This species occurs in the mountains of southern Arizona and widely in Mexico. The larvae feed in cones of Chihuahua pine *(P. leiophylla)* in Arizona, reported by Dyar, and on several other pines, in the mountains of Durango, Sinaloa, and central Mexico, recorded by Cibrián-Tovar, feeding in cones and shoot galls infected by *Cronartium*.

Sarata includes an interesting group of species because the sexual dimorphism is extreme, such that it is difficult to associate males with females with confidence. The immature stages and larval foods are unknown, so reared series that might provide direct evidence are not available. Moreover, the adults fly in winter or early spring, in advance of typical collector efforts, so most collections consist of one or a few specimens netted diurnally. As a result, association of the sexes is speculative, and Heinrich treated the taxonomy provisionally, providing "temporary" names for the females *(alpha, beta, gamma, delta, epsilon, iota, kappa,* and *phi),* and suggested that some or most of these are destined for synonymy when the sexes are associated (e.g., which might be accomplished by DNA analysis). Neunzig ignored those names, although of course they are available, and treated just the males. All 12 species known from males occur in the western United States. They have gray FW with obscure or no transverse pattern, whereas the much smaller females have well-defined basal, medial, and postmedial, dark, transverse fasciae, often defined by whitish (Plate 61.23). *Sarata perfuscalis* (Hulst), which was described originally from females, is the only species for which Neunzig associated the sexes, evidently based on specimens collected in absence of other species. This species occurs in Washington, Utah, and coastal central California. Adults fly in March and April, and both sexes have been collected diurnally at several sites in the San Francisco Bay area and Lake County, California. Males have gray FW (12–12.5 mm) without any transverse pattern and the veins lined in black. In females (FW 9–11 mm) the basal area, medial, and postmedial fasciae are dark gray. *Sarata edwardsialis* (Hulst) is the largest *Sarata* and the most widespread in the West, occurring in Colorado, Utah, Washington, and California. Males have brownish gray FW (17–20 mm; Plate 13.22) with faint indications of ante- and postmedial, pale gray lines, and the veins are tinged with black, terminating in a subterminal row of black dots. *Sarata dnopherella* Ragonot, the type of the genus, was described from California without locality data and was recognized by Neunzig from males we collected on Mt. Shasta above 7,000' in June, which were collected in fir woods, associated with a female taken from a patch of snow. The male has dark gray FW (14–15.5 mm) with the ante- and postmedial lines weakly indicated by dark scaling, following and preceding them, respectively. The female is smaller, darker with distinct transverse lines.

Lipographis, with three species, is restricted to western North America. These are primarily brownish moths. The male antenna has a basal curve, where the apices of its segments are produced into black, spinelike processes, and its scale tuft is strongly developed; the labial palpi are conspicuously longer than in related genera; FW with only weakly upraised scales in fresh specimens. *Lipographis fenestrella* (Packard) has pale rust brown to ochreous brown FW (10–11 mm; Plate 26.7, Plate 26.13), antemedial line indicated by fragmented spots, postmedial line distinct, white. There is seasonal variation, spring individuals darker, summer specimens faded, particularly a gray shade preceding the subterminal band. This species occurs in Manitoba, the Great Basin from Utah to eastern and coastal California, and on the Channel Islands, where there are discrete April to May and August to October generations. J. Ruygt found larvae of *L. fenestrella* feeding on bird's-beak, *Cordylanthus mollis* (Scrophulariaceae), a salt marsh succulent, at Napa, California, but we have been unable to detect an association with *Cordylanthus* or other scrophs at several localities where the adults were numerous. They occasionally come to lights, but most collections have been by flushing the moths from barren sand or alkaline flats, often near coastal, bay, or riverine beaches, sometimes associated with sand-spurrey, *Spergularia macrotheca* (Caryophyllaceae), and saltgrass, *Distichlis* (Poaceae). *Lipographis umbrella* (Dyar) is larger and more colorful, FW (13–14.5 mm; Plate 26.14) orange tan with

brownish streaks. This species is known only from alkaline and salt flats in coastal areas in the San Francisco Bay area, San Diego, and Baja California. Neunzig stated that our specimens from the San Joaquin delta and Livermore Valley, Alameda County, were light trapped, but they were not. The moths were netted diurnally, in September, associated with alkali heath, *Frankenia salina* (Frankeniaceae), and *Cordylanthus palmatus.*

Elasmopalpus lignosellus (Zeller)(Plate 26.2), the "lesser cornstalk borer," was described in 1848 from Brazil and is known in South America, the Caribbean Antilles, Mexico, and eastern North America. Apparently disjunct populations, presumably introduced, occur in the Phoenix, Arizona, area and coastal southern California, including Santa Cruz Island. The adults are dimorphic: the FW (6.5–10 mm) of males is primarily tan, with black streaks along the costal and dorsal margins and a dot at the end of the cell; females have more extensive dark scaling, tan reduced to a longitudinal streak, or the FW is all black with some reddish scaling at the base. The larvae of *E. lignosellus* feed mainly on grasses (Poaceae) but occasionally on dicots, especially Fabaceae. They form silken tunnels in the soil at the base of the plant and at first feed on the stem surface. Later instars burrow into the stems. Adults are recorded from May to September.

Pseudadelphia ochripunctella (Dyar)(Plate 26.1) was considered by Neunzig to warrant a monotypic genus based on distinctive features of the genitalia. Superficially the moth resembles the female of *Elasmopalpus lignocellus,* having dark brownish black FW (8–10 mm), but *P. ochripunctella* has a weakly developed, pale antemedial line and a pale reddish or ochreous spot at the end of the cell. This species is known only from coastal southern California, from Santa Barbara to San Diego, mostly from beach habitats that are long gone. The larvae were found feeding on turkey mullein, *Eremocarpus setigerus* (Euphorbiaceae), at the El Segundo dunes by W. D. Pierce in October the 1930s, and we reared *P. ochripunctella* from another euphorb, *Croton californicus,* at Border Field adjacent to the Mexican border. Both these plants are more widespread than are records for this moth, especially *Eremocarpus,* which occupies dry, often highly disturbed areas, inland to eastern Washington. Larvae we collected at the end of March produced adults in May, and there are flight records March to November.

Acroncosa consists of four mostly white species in the western United States. The male vesica has a large cluster of cornuti, and the female bursa has two opposed plates with numerous broad spines, as in *Passadena* and related genera, but taxonomic relationships to other phycitines are uncertain. *Acroncosa albiflavella* Barnes & McDunnough has white FW (10–11 mm; Plate 26.4) with an orange antemedial band and weak subterminal fascia. This species occurs from the foothills east of Lake Tahoe in western Nevada to the San Bernardino, California, area but apparently not in the deserts. Adults fly June to August. *Acroncosa minima* Neunzig is similar, smaller, FW (7–8 mm; Plate 26.3), with relatively broader orange bands, and the labial palpi are mostly orange. This species was abundant one incredibly hot, midsummer night in the Woods Mountains, central Mojave Desert, but it has not been collected elsewhere.

The remaining several genera of this group feed on Asteraceae, primarily in the flowers and seed. *Homoeosoma* is the largest, with 20 described species in North America, of which 12 occur in the West. *Homoeosoma electellum* (Hulst), the "sunflower moth," occurs across the continent. It has two color phases in both sexes, FW (7.5–12 mm; Plate 26.20) either gray with a white subcostal streak or ochreous tan with or without the subcostal streak. The larvae, which are yellowish to bluish green with broad, rust brown to purplish longitudinal streaks, feed in the flower heads of composites. They infest commercial sunflowers (Helianthus annuus), reducing seed production and oil quality, as well as weedy colonies of this plant and a wide variety of other indigenous and cultivated Asteraceae. Recorded hosts include desert marigold (Baileya), Spanish needles (Bidens), thistles (Carduus and Cirsium), safflower (Carthamus tinctorius), tickseed (Coreopsis), gumplant (Grindelia), snakeweed (Gutierrezia), goldenbush (Isocoma), coneflower (Rudbeckia), groundsel (Senecio), and garden Zinnia. The sunflower moth occurs all over the West from British Columbia to Mexico, is multivoltine in southern parts of its range, and does not overwinter in northern areas, immigrating annually. Most of the other Homoeosma species are gray, typically with a variably developed dark band following the antemedial line and two dots representing the postmedial. The FW of **H. albescentellum** Ragonot (8–12 mm; Plate 26.19) is whitish with a strong, black transverse band and usually distinct discal dots. This species is widespread in the Southwest from Utah and Colorado to New Mexico, Arizona, and the desert mountains of California. The larval host is unknown but assumed to be thistle or other native Asteraceae. Adults are recorded from April to September, presumably multivoltine. **Homoeosoma striatellum** Dyar (FW 8.5–11.5 mm; Plate 26.9) is distinctive in having black lines along the veins, with the antemedial transverse band and discal dots weak to obsolete. This species occurs from southeastern Utah to Arizona and California (Greenhorn Mountains, Owens Valley, deserts, Santa Catalina Island) and Baja California. We reared H. striatellum from an unidentified composite shrub in the Mojave Desert. Adults have been taken at lights from January to August.

Patagonia peregrina (Heinrich) is completely nondescript, FW (5.5–8 mm; Plate 26.34) slate gray without maculation, the costal area dusted with white. Structurally Patagonia differs from Homoeosoma by minor details in the male genitalia and absence of a signum in the female corpus bursae. The genus is Neotropical, especially in South America, and P. peregrina is its only representative in our fauna. The larvae feed on seed of cudweeds (Gnaphalium), less commonly on everlasting (Anaphalis), and are ubiquitous on native or introduced cudweeds in coastal California and its islands throughout the year. This species ranges from California to Mexico and Central America.

Phycitodes mucidellus (Ragonot) is a small species similar to Patagonia and Ephestiodes but averages larger, FW (6.5–10 mm; Plate 26.35, Figs. 197, 198) white anteriorly, its antemedial fascia is broken and postmedial represented by two, usually distinct, black spots. This species is widespread in the western United States and southwestern Canada. Phycitodes mucidellus has been reared from flower heads of several species of thistle (Cirsium) by Goeden and Ricker and from Senecio and Gnaphalium in California, as well as Grindelia and Senecio in Texas and Arizona. Adults are recorded from March to October in California and April to July, October, and December in Arizona.

Ragonotia dotalis (Hulst) is sometimes abundant at lights in the southwestern deserts from Nevada and southeastern California to New Mexico. This is a moderately large phycitine (FW 10.5–13 mm; Plate 26.15), with oval, whitish FW that are peppered with gray, giving a bluish appearance, and crossed by indistinct, antemedial and subterminal, ochreous fasciae, each emphasized by black scaling distally. In California, R. dotalis is restricted to the Colorado Desert and lower elevations of the Mojave. Collection records are February to August, evidently two or more generations.

Martia arizonella Ragonot also ranges through arid parts of the Southwest, from western Texas to Utah and California. This is a smaller species, with triangular FW (7.5–9.5 mm; Plate 26.5, Plate 26.6), which have mixed gray and pale rust scaling to mostly dark gray. The costa and two transverse lines are white, at midwing and parallel to the termen. Adults are active from April along the lower Colorado River and in Baja California east of the mountains to September but only June to September in southern Arizona.

Eumysia are slender moths having elongate antennal cilia in the male (about three times the width of the segment) and porrect labial palpi projecting 2 to 2.5 times the eye diameter beyond the head. There are six named species, all occurring in the West. **Eumysia pallidipennella** (Hulst)(Plate 26.8) was described from Colorado and occurs in the Great Basin, Utah, New Mexico, and deserts and coastal southern California, including most of the Channel Islands, where adults are recorded March to June and October to December. The FW (8–10 mm) is dull brownish gray, nearly immaculate to having an indistinct antemedial fascia of dull ochreous. The larval hosts of most Eumysia are unknown, but larvae of **E. idahoensis** Mackie defoliate shadscale (Atriplex confertifolia, Chenopodiaceae), a valuable desert foliage plant, and feed on saltsage (A. nuttalli) and four-winged saltbush (A. canescens) in southern Idaho.

Heterographis morrisonella Ragonot (Plate 26.10) is similar, but the labial palpi are upturned, not elongate and porrect; the FW is not as slender as that of Eumysia; and the antennae have short cilia in both sexes, that of the male is curved basally. The size (FW 5.5–10.5 mm) and FW maculation are extremely variable, from cream whitish with only faint darker dusting to forms with flecks and streaks of brown and reddish, or the median area or whole wing is mostly dark reddish gray. This species occurs from Utah through the deserts to coastal California, Arizona, New Mexico, Texas, and Durango, Mexico, reaching 6,500' in the Sierra San Pedro Martir, Baja California Norte. It was reared from Ambrosia chamissonis (=Franseria bipinnatifida, a synonym) by Dwight Pierce at the El Segundo dunes prior to construction of the Los Angeles International Airport, and we reared H. morrisonella from the same plant at Coal Oil Point, University of California, Santa Barbara, where the larvae occupied silken tubes in the sand, from which they fed on buried leaves in unstabilized dunes. Owing to its wide geographic range, other Asteraceae likely serve as host plants of Heterographis. The adults come to lights but appear to be active in the daytime on coastal dunes, darting by quick flights, then virtually disappearing when they suddenly alight on the sand, which the FW pattern resembles.

Staudingeria albipennella (Hulst)(Plate 26.16, Plate 26.17) is similar to Heterographis, differing by having the third segment of the labial palpus very short and the second segment in the male grooved to hold the maxillary palpus. Staudingeria is larger (FW 8–12.5 mm), and the FW color varies from pale ochreous tan to dark gray or reddish with the costal edge white. This species is widespread in the West, primarily in arid regions, from eastern Washington and Oregon, Idaho, Wyoming, and Colorado to Arizona and southern California. The larval host is locoweed (Astragalus, Fabaceae), according to Heinrich. Flight records are June to September.

Hulstia undulatella (Clemens), the "sugar-beet crown borer," has brownish gray FW (7–10 mm; Plate 26.11) with zigzag, whitish ante- and postmedial fasciae, preceded by darker gray shades, sometimes with a reddish brown streak basally. This species was described from Pennsylvania, but it occurs throughout the western states. Essig reported the larvae lodged just below the base of the leaves on sugar beet (*Beta vulgaris*, Chenopodiaceae) and bored into the crown and down the center of the root, whereas Heinrich recorded them eating the foliage. *Hulstia undulatella* occurs in urban and rural disturbed places, as well as in habitats with primarily native vegetation, and it may be a general feeder. Flight records in California span April to September.

Honora are slender-winged phycitines having unmodified antennae, and the labial palpus is oblique with the third segment porrect, deflected forward. The eighth abdominal segment of the male has a pair of elongate, ventrolateral scale tufts. Five species occur in the West. *Honora mellinella* Grote is recorded in the southeastern states, west through Texas to Arizona, southeastern California, and eastern Washington. The FW (7–11 mm; Plate 26.12) is dark with the costal area whitish into the cell; antemedial line nearly vertical, white, preceded by blackish scaling and a yellowish patch, and there is a similar yellowish patch at the base of the wing. *Palafoxia* (Asteraceae) was recorded as a larval host, reared by C. M. Dammers, and we found larvae of *P. mellinella* feeding in the flower heads of *P. arida* in March, also in the Mojave Desert. *Honora dotella* Dyar is quite similar but averages larger (FW 8.5–12.5 mm; Plate 26.18) and has the rust red FW basal area preceding the antemedial line. This species occurs in the coastal mountains of California from Monterey County southward, with adults recorded February to July.

Laetilia is an American genus with an extensive Neotropical fauna and nine described species north of Mexico, of which four occur in the West. Morphologically, this appears to be a composite group, some of its species sharing features with different related genera, but the larvae so far as known are all predaceous, feeding on scale insects or mealybugs (Homoptera) or on flowers of plants infested with them. The name *L. dilatifasciella* (Ragonot)(Plate 26.36, Plate 61.22) is applied to western populations as a species closely related to the widespread eastern *L. coccidivora* (Comstock). They differ slightly in color and genital structures. These are small phycitines, FW (6–8 mm) pale gray with variable dark gray patches beyond the ante- and postmedial lines, and the posterior area is tinged with ochreous tan. The larvae are predaceous on scale insects of several families (Cerococcidae, Coccidae, Dactylopiidae, and Kermesidae). *Laetilia dilatifasciella* ranges from southern New Mexico to southern California. It was inadvertently introduced to Santa Cruz Island, where it feeds on the cochineal scale, *Dactylopius*, which had been successfully introduced in the 1950s to control prickly pear cactus *(Opuntia)*. Larvae are recorded in May and September and adults at lights in May and July through October. Larvae we collected in September spun silken sheets covered with waxy debris, under which they fed and pupated in tough cocoons, producing adults in October to November and the following spring. *Laetilia zamacrella* Dyar is the largest species of *Laetilia* (FW 7–13 mm; Plate 26.21), and it has dark gray FW with well-defined, sinuate, white ante- and postmedial lines. This species occurs in coastal California, where it is associated primarily with scale insects on Monterey pine *(Pinus radiata)*, a coastal endemic, which is planted widely in cities and parks. We found larvae feeding on *Toumeyella pinicola* (Coccidae) on *P. radiata* in March in the native grove of *P. radiata* at Cambria. *Laetilia zamacrella* also ranges into the Sierra Nevada. Adults at lights appear to represent two generations, April to June and August to November.

Rostrolaetilia has 10 species, all western. These moths appear to be related to *Laetilia*, but *Rostrolaetilia* lack a functional proboscis, possess much longer labial palpi, and the ductus seminalis originates from the corpus bursae posterior to the signum, rather than at the anterior end of the bursa as in *Laetilia*. Each species occurs locally, western Texas, southern Colorado, central Utah, Arizona, or southern California, and most are known from few specimens. In size and FW maculation they resemble *L. dilatifasciella*. *Rostrolaetilia minimella* Blanchard & Ferguson (Plate 26.37) is appreciably smaller than its congeners (FW 4.5–6 mm vs. 6–11 mm) and has a pale, washed-out FW pattern. It occurs in southeastern California and adjacent Nevada, and there are female specimens judged by Blanchard and Ferguson to be conspecific from Salt Lake County, Utah, and southern New Mexico, the latter reared from scale insects of the genus *Orthezia*.

Larvae of **Rhagea** species are borers in succulent and parasitic plants but not cactus; *R. packardella* (Ragonot) has pale gray FW (7.5–10 mm; Plate 26.38) heavily streaked with black along the veins, concentrated into a spot at the end of the discal cell. This species ranges from western Texas and Colorado to eastern Washington and California, where it occurs in the deserts, and on Santa Cruz Island. Dyar reared *R. packardella* from flowers of broomrape, *Orobanche* (Orobanchaceae), in Washington, and we found the larvae in buried stems of another parasitic plant, *Pholisma arenarium* (Lennoaceae), in the Mojave Desert. *Rhagea stigmella* (Dyar) is larger with more uniform gray FW (11–17 mm; Plate 26.22) having darkened veins and two dots in the cell. It occurs in the California deserts and on the coast. The larvae feed in the cortex and roots of several species of live-forever, *Dudleya* (Crassulaceae), and cause problems for succulent gardeners. This and probably both species have at least two generations, with adult records from May to November at Big Creek, Monterey County and on Santa Cruz Island.

Species of **Zophodia** most resemble *Ozamia* and *Cactobrosis*, particularly in sharing clusters of scalelike sensilla near the inner base of the male antenna, but are not cactus feeders as are members of those genera. *Zophodia grossulariella* (Hübner) [=*convolutella* (Hübner)], the "gooseberry fruitworm," has pale gray FW (11–16 mm; Plate 26.23) with longitudinal black lines and a prominent black smudge following the antemedial line, resembling the maculation of *Myelopsis*. This species was first described in Europe in 1809, but it has no close relatives there, and it may have been introduced via nursery stock from North America, where it occurs over much of southern Canada and the northern United States. In the West, *Z. grossulariella* occurs from southern British Columbia to central Utah, the Rocky Mountain states, and northern California. The larvae feed first in the developing fruit, later tying up several berries, and on leaves of Grossulariaceae, including European garden currant, *Ribes grossularia*, and black currant, *R. nigrum*. The species is univoltine, with flight records from April to June.

THE CACTUS-FEEDING PHYCITINAE

Members of the following nine genera, along with others outside our region, are cactus-feeding species believed to comprise a monophyletic lineage. Nearly all are larger, more bulky moths than other Phycitinae, and their larvae are borers in flowers, fruit, or the cortex and fleshy stems of cacti. Larval hosts were studied by Mann, who compiled a comprehensive summary of them. *Cactoblastis cactorum* (Berg), the "cactus moth," is the most famous, having won worldwide acclaim owing to its remarkable success as a biological control agent against prickly pear cacti. In a span of 12 years, *C. cactorum*, a native of South America, effectively eradicated *Opuntia* from 30 million acres of rangeland in Australia that had been completely occupied by this weed. *Cactoblastis cactorum* also was used effectively in the West Indies, but it has moved into Florida, posing a threat to native cacti.

Cactobrosis includes four species in Mexico, one of which ranges into the southwestern United States. *Cactobrosis fernaldialis* (Hulst)(Plate 13.24) is a large phycitine (FW 14.5–20.5 mm) having a narrow, gray FW, usually with a black streak along the Cu crease in males that is lacking in the much larger females. Both sexes have a whitish antemedial fascia bordered by dark gray distally, and sometimes the whole basal area is black. This species occurs in the southwestern United States from Texas to southern California. The larvae feed in barrel cactus, *Ferocactus wislizenii*, and presumably related cacti, including *F. cylindraceus*. There are collections records for late March to April and July to November, indicating two or more generations.

Ozamia includes four Nearctic species and several in Mexico and the Antilles. They are similar to *Zophodia* and are distinguished by having larger, squamous maxillary palpi, and the male antennae are serrate basally. *Ozamia fuscomaculella* (Wright) has pale gray FW (10.5–12 mm; Plate 26.24) with the dark pattern typical of *Rostrolaetilia* and related genera, quite similar to *O. clarefacta* Dyar in Texas, and superficially resembling *Amyelois transitella*. *Ozamia fuscomaculella* occurs in coastal southern California and is abundant on the Channel Islands. This species feeds on the flowers and fruit of prickly pear, *Opuntia*, according to Heinrich, and is multivoltine, with flight records from late March to November on the islands.

Melitara has seven species in America north of Mexico, three of which occur in the West. They can be recognized by their bipectinate antennae in both sexes, and HW veins M 2 and M3 and CuA1 connate or very short stalked. These are mostly large pyralids (wing span 3–5 cm or more), whose larvae burrow into the stems of prickly pear (*Opuntia*). *Melitara dentata* (Grote)(Plate 13.23) is widespread in the West, from Alberta through the Rocky Mountain states to northern Arizona and western Texas. The FW (1.5–2.3 cm) is pale gray with thin, strongly zigzag ante- and postmedial lines (which are weak to obsolete in related species). Recorded larval hosts are *O. fragilis*, *O. macrorhiza*, and *O. polyacantha*, and flight records are July to August.

Alberada has five species that are similar to *Melitara* but the female antennae are simple, not bipectinate, and the HW veins M2 and M 3 and CuA1 are stalked. Their larvae feed on cholla cactus rather than prickly pear. *Alberada parabates* (Dyar) has gray FW (1.8–2.3 cm; Plate 13.25) with longitudinal black lines and faint, zigzag ante- and postmedial whitish lines; the HW is white. This species occurs from Mexico and western Texas to southern Arizona. The larvae are bluish and feed on tree cholla, *Opuntia (Cylindropuntia) imbricata*, in Mexico, Texas, and New Mexico, and on *O. fulgida* in Arizona. Several populations formerly treated as this species have been described as new species by Neunzig, including *A. californiensis*, which feeds on *O. prolifera* in coastal and desert areas of southern California. Its FW are paler.

Cahela ponderosella (Barnes & McDunnough)(Plate 13.26) is a monobasic genus in the southwestern United States and adjacent Mexico. The antennae are simple in both sexes, but those of the male are thickened. The labial palpi are upturned in the male, porrect in females, and the proboscis is well developed. *Cahela* is not as large as *Alberada*, with which it shares larval hosts, and has dark gray FW (11–20 mm) with a strong,

black line along the R stem nearly to the termen, and a shorter one along the Cu. This species occurs in the deserts from Mexico and western Texas to southern Colorado, Utah, and southern California. The larvae are recorded by Mann from a variety of chollas, especially *O. imbricata*.

The four species comprising *Rumatha* occur in the southwestern United States and adjacent Mexico and are closely related to *Cahela* but are smaller. The two genera share a uniquely derived feature, with the gnathos fused apically into a single structure. *Rumatha* have a short uncus and a subbasal, sclerotized concavity on the valva in the male genitalia. *Rumatha bihinda* (Dyar) has the FW (14.5–16 mm; Plate 26.25) streaked with black but lacks the strong lines of *Cahela*, and the postmedial line is strongly zigzag. This species occurs from western Texas to southern California and in Mexico. The larval hosts are unknown, but other *Rumatha* species feed on *Cylindropuntia* chollas. Flight records span May to August.

Yosemitia graciella (Hulst) resembles a small version of *Rumatha bihinda*, having the posterior half of the FW (9–14 mm; Plate 26.26) dusted with rust brown. The longitudinal lines are rust to black. According to Mann, this species occurs from western Texas to southern California, Nevada, and Colorado. The larvae are recorded from species of hedgehog cactus, *Echinocereus*, and beehive cactus, *Escobaria* (=*Coryphantha*), and we reared a series from infested pineapple cactus, *Sclerocactus polyancistrus*, collected near Barstow, California, by R. May. Neunzig treated *Y. longipennella* (Hulst) as a synonym, but Mann reported different larval hosts in southern Texas, *Y. longipennella* feeding on *Homalocephala* and *Neomamillaria*. *Yosemitia* was among the last names representing a peculiar fad launched by Ragonot in 1887, and enthusiastically endorsed by Hulst, to propose generic names for Nearctic pyraloids using place names, often having little or nothing to do with the species included.

Eremberga has three species in the Southwest. *Eremberga leuconips* (Dyar) has dark gray FW (15–17 mm; Plate 26.29) with black scaling along the veins and a strong line through the cell, resembling that of *Cahela ponderosella*. *Eremberga leuconips* occurs in Mexico and southern Arizona and has been reared from *Echinocereus triglochisiatus*; it flies in July to September. *Eremberga creabates* (Dyar)(Plate 26.28) has very pale gray FW with only faint black lines. It is more localized, coastal and desert San Diego County, California, and presumably also feeds on hedgehog cactus.

DETRITIVORE AND STORED FOOD PRODUCTS–FEEDING PHYCITINES

Larvae of the remaining genera of Phycitinae feed as scavengers on dried fruit, seed pods, and grain products, and as inqilines in galls caused by other insects, or on dry or decayed plant matter, rather than living plants. The group includes our most notorious stored food products pests. *Euzophera* are characterized by having the FW rust brown or magenta brown with the area between the transverse lines forming a broad, dark band. This is a Holarctic genus, with about 50 species, of which seven occur in America north of Mexico, four of them in the West. *Euzophera semifuneralis* (Walker), the "American plum borer" (FW 16–26 mm; Plate 26.27), is a widespread, polyphagous species that sometimes causes economic damage boring in and weakening trunks and limbs of ornamental and orchard trees. It is of minor importance in healthy trees because feeding is initiated primarily at damaged or diseased spots. All kinds of woody hosts are used, including apple *(Malus)*, stone fruits

(*Prunus*), walnut (*Juglans*), pecan (*Carya*), ginkgo (Ginkgoaceae), olive (*Olea*), persimmon (*Diospyros*, Ebenaceae), and sweetgum (*Liquidiambar*, Hamamelidaceae). There are two generations with spring and summer flights.

Ephestiodes are small phycitines, gray with a paler basal area followed by ante- and postmedial, dark transverse lines; the dorsal area posterior to the Cu crease is often pale peach gray. The male has a costal fold, which is sometimes opened on spread specimens, giving a flared shape to the costa. There are complex, heavy, black dorsal scale tufts on A8 in the male. Four of the seven North American species occur in the West, and all are quite similar. *Ephestiodes gilvescentella* Ragonot is variable, FW (4.5–8.5 mm; Plate 26.39, Plate 26.40) ground color pale gray to reddish. This species may be the most often collected Nearctic phycitine, occurring throughout the West, sometimes abundant at lights; but the life history and larval habits are poorly known. Larvae have been reared from raisins and cotton but are not reported infesting field or stored products such as dried fruit. We have reared this species from galls caused by cynipid wasps ("oak apples") that also had larvae of *Cydia latiferreana* (Tortricidae) and from root crowns of woody composite shrubs (Asteraceae) infested by eucosmine tortricid larvae, where *Ephestiodes* larvae may be scavengers. We have not found *E. gilvescentella* feeding on foliage or inflorescences of plants independent of other insect infestation. In coastal California this species is active throughout the year and occurs in native as well as urban habitats.

Vitula includes eight species in America north of Mexico, of which five are western. These are mostly small, pale gray moths, which along with *Ephestiodes* are distinguished from *Sosipatra* and the following genera by having 10 FW veins and by genital characters, *Vitula* lack specializations of the uncus, gnathos, or cornuti. *Vitula edmansae* (Packard)(FW 8.5–11 mm; Plate 26.30) is the most widespread and often collected species, ranging throughout the United States and southern Canada (including the scarcely differentiated western race, *V. serratilineella* Ragonot, which has been called the "dried fruit moth"). The FW is pale gray with a zigzag ante- and sometimes postmedial line, the latter often reduced to two short dash marks in the cell. This species typically infests nests of bumblebees and other aculeate Hymenoptera, particularly in weak colonies, feeding on pollen and honey, and hosts include yellow jackets (*Vespula*) and the leafcutter bee (*Megachile rotundata*), which is cultivated in portable nest boards for pollination of alfalfa. *Vitula edmansae* also is reported to infest dried fruit, including apples, figs, raisins, and prunes. The moths come to lights, March to October, and there are several generations per season. This species has been inadvertently introduced into Europe. (The name *edmandsii* was proposed by Packard, named for Miss A. Edmands, and the spelling was correctly emended to *edmandsae* by Heinrich, in accordance with the ICZN [Latin genitive ending]. But that decision was not accepted by Neunzig.) *Vitula pinei* Heinrich is larger (FW 9–10 mm; Plate 26.41) and differs from other *Vitula*, having distinct FW ante- and postmedial lines, each preceded by a white and a partial dark gray line. This species occurs from northern California and the Great Basin, to northern Arizona and New Mexico. One larva was found feeding in a cone of pinyon pine (*Pinus monophylla*) in Nevada, but as surmised by Neunzig, probably it was a secondary detritivore, based on what is known of other *Vitula*. Flight records are June to July.

The *Sosipatra*, *Plodia*, and *Ephestia* group of genera are distinguished by having eight or nine FW veins versus 10 in Vitula and Ephestiodes. All six species of Sosipatra occur in western North America. *Sosipatra thurberiae* (Dyar) has a broad, dark gray, transverse band between the ante- and postmedial lines on the FW (5–9 mm; Plate 26.42), at times darkly shaded to the termen. The dark band is also expressed in *V. insula* Neunzig (Plate 26.43) from the California islands, but not consistently. *Sosipatra thurberiae* occurs in southern Oregon, California, northern Arizona, and Mexico. Its larvae have been found in cynipid wasp galls on oak and seed pods of cotton and redbud (*Cercis occidentalis*, Fabaceae), probably as scavengers. Adults recorded June to September. *Sosipatra rileyella* (Ragonot) is distinctive, having cream white FW (6.5 mm; Plate 26.44) with two pairs of black dots in place of the ante- and postmedial lines, and this species is a specialist associated with Agavaceae. The larvae feed in the debris and dry seed after yucca moth larvae (Prodoxidae) have abandoned the seed capsules, with several species of Yucca and *Nolina* recorded as hosts. *Sosipatra rileyella* occurs in arid areas of most of the western United States from Wyoming to Mexico.

The "Indian meal moth," *Plodia interpunctella* (Hübner)(Plate 26.45, Plate 26.46), may be the moth seen by more people in North America than any other because it infests dry foods in homes, and its presence usually is detected when the moths are seen in kitchens (Plate 61.24). They are distinctive, with the FW (5–8.5 mm) sordid whitish basally, dark rust red beyond the antemedial line, ornamented with patches of metallic gray scaling. The larvae feed in all kinds of seeds, cereal, ground meal, nuts, dried fruit, chocolate, and spices, in both commercial and household storage. Once an infestation is noticed, all opened containers of such foods need to be inspected and discarded if necessary, and the remainder stored in the refrigerator or in airtight containers, because cocoons may be secreted in corners of the cupboards, and emergence of new moths may occur for some weeks. Life cycle development can occur within 18 days in optimum conditions, producing as many as eight generations annually. This species in believed to be native to the American tropics, based on its relatives, but was first described in the early nineteenth century in Europe, having already been transported there via infested meal or corn. Subsequently it has been distributed worldwide, and it was recognized in New York by the 1850s. In most of North America, *P. interpunctella* is found primarily in buildings, but it occurs out of doors in warmer areas, feeding on waste fruit in orchards, in seed heads—we have reared it from old flower heads of gum plant (*Grindelia*, Asteraceae)—and in nests of bumblebees and solitary bees (*Anthophora*, *Osmia*). Adults occasionally come to lights.

Ephestia and the closely related *Cadra* contain the other Phycitinae that are stored products pests. These are gray or tan moths, easily distinguished from the colorful *Plodia* but not from one another. *Cadra* is considered to be a separate genus on the basis of the female genitalia, having reduced ovipositor lobes (papillae anales), scale tufts associated with the lamella postvaginalis, and a longitudinal row of sclerotized ridges forming a spiral band on the ductus bursae. *Ephestia kuehniella* Zeller, the "Mediterranean flour moth," is the largest of these species. The FW (7.5–13 mm; Plate 26.31, Figs. 199, 200) has a more contrasted, darker shade following the antemedial line than its close relatives, and the male lacks a costal fold. In addition to its notoriety as a processed foods pest, *E. kuehniella* became the lab animal of choice for a great number of European studies on development, physiology, genetics, and so on. This species has been distributed worldwide by human activities during recorded times, but its origin is unknown. It was not recognized in Europe until 1879. Almost simultaneously, it was found to be widespread in North America, appearing in eastern Canada by 1884 and California by 1889, where every mill in the state was said to be infested by 1895. Neunzig speculated that *E. kueniella* was indigenous to North America because it occurs in natural habitats and because a closely related species, *E. columbiella* Neunzig, occurs in the southeastern United States. Although *E. kuehniella* is an indoor insect in most of North America, Keifer provided the first of several larval records from acorns stored by woodpeckers in California. If not native on the Pacific Coast, this species has adapted well, with flight records in every month at the Big Creek Reserve, Monterey County, and Santa Cruz Island, remote from any human habitation. The Mediterranean flour moth is an economic pest in flour mills but infests households only infrequently. It is recorded from a wide variety of seeds and/or milled flour of wheat, barley, oats, rice, and corn, and in chocolate, cakes, and nuts in candy. *Ephestia elutella* (Hübner)(FW 8.5–14 mm; Plate 26.47, Plate 26.48), the "tobacco moth," is cosmopolitan in distribution but not as ubiquitous as *Plodia* or *Cadra*. *Ephestia elutella* was described in Europe in 1775, where it is considered to be native and occurs in haystacks and feeds on seed of wild grasses. This species can be a serious pest in a variety of stored vegetable products, including tobacco, cocoa, cereals, and nuts, but not in milled flour. The larvae also feed out of doors in tropical regions on decaying wood, as scavengers in orchards, including citrus, and on ripe fruits. The adult is superficially similar to *Cadra cautella*, dark to pale gray with a darker transverse antemedial fascia, and infestations by the two have been mistaken for one another. *Ephestia elutella* shows a greater tendency to diapause than the other phycitine storage pests, allowing it to overwinter in unheated situations, but in North America it is exclusively an indoors insect, according to Neunzig.

Cadra cautella (Walker)(Plate 26.49), the "almond moth" or "fig moth," also is worldwide in distribution, thanks to human commerce, and is more often encountered as a stored products pest than related species. Superficially the adult is very similar to *E. elutella* but is smaller (FW 5–9 mm) and the genitalia structures of *C. cautella* are diagnostic and should be examined for positive identification. This species is cosmopol-itan in warmer climates but apparently cannot survive cold winters. It was described from Ceylon (Sri Lanka) and Australia in the 1860s and was recorded from North America by the 1880s, having been imported from the Middle East to the eastern United States with figs, but it was not recognized as established until the late 1890s. *Cadra cautella* was recorded in California by 1913 and was a common moth in almond warehouses by 1930. Like the related species, *C. cautella* larvae are destructive to a wide variety of dry fruit, grains, nuts, and other stored vegetable products. *Cadra figulilella* (Gregson)(Plate 26.50), the "raisin moth," is similar, but the FW (5.5–8.5) tends to be tan rather than gray. This species was first described in 1871 from Liverpool, England, where it occurred in warehouses containing imported raisins and figs, and subsequently it has been found widely distributed in the Mediterranean region and places with similar climate in Australia and the Western Hemisphere. It occurs in Arizona and California, where the larvae feed outdoors on fallen fruits and infest fruit such as raisins and apricots during the drying process. *Cadra figulilella* undergoes several generations from April to December.

Bandera has three species in western North America, all similar, recognizable by their nearly uniform tan or pale gray FW, which are rounded distally, and by their elongate, porrect labial palpi, so the moths resemble peorine pyralids, with which they were classified based on lack of a functional proboscis. **Bandera virginella** Dyar has pale gray FW (9.5–11.5 mm; Plate 26.33) with faint, pale yellow, longitudinal lines between the veins. It occurs from Washington and Oregon to Colorado, New Mexico, Arizona, and California. The moths fly in fall. This species occurs on Santa Cruz and was abundant on Santa Rosa Island in late September in formerly grazed grassland. **Bandera cupidinella** Hulst has a patch of pale yellow scales near the base of the FW costa. It occurs in Colorado, western Texas, and New Mexico. The early stages of *Bandera* are unknown, but they are suspected to be thatch feeders in grasslands.

Eurythmia yavapaella Dyar has gray, obscurely patterned FW (7.5–9 mm) shaded with rust in the posterior half. It occurs in Arizona and southern California, north to Contra Costa County, with flight records May to July. The larval habits are unknown.

TABLE 1

Index of North American Genera of Phycitinae

Treated by H. H. Neunzig in Four MONA Fascicles, 1986–2003

Genus, Fascicle

Acrobasis*, 1	Dasypyga*, 0	Martia*, 0	Quasisarata, 4
Acroncosa*, 4	Davara, 0	Melitara*, 3	Ragonotia*, 0
Actrix, 4	Dioryctria*, 4	Meroptera*, 4	Rhagea*, 3
Adanarsa, 2	Diviana, 0	Mescinia, 3	Rostrolaetilia*, 3
Adelphia, 4	Divitiaca, syn., 4	Metephestia, 4	Rumatha*, 3
Alberada*, 3	Ectomyelois, 2	Monoptilota, 0	Salebriacus, 4
Ambesa*, 4	Elasmopalpus*, 4	Moodna, 2	Salebriaria*, 4
Amyelois*, 2	Ephestia*, 2	Moodnodes, 2	Sarasota, 0
Anabasis, 1	Ephestiodes*, 2	Myelopsis*, 2	Sarata*, 4
Anadelosemia, 0	Erelieva, 2	Ocala, syn., 4	Sciota*, 4
Anagasta, syn., 4	Eremberga*, 3	Olybria, 4	Selga, 2
Ancylostomia, 4	Etiella*, 0	Olycella, 3	Sosipatra*, 2
Anderida, 3	Eulogia, 2	Oreana, 4	Staudingeria*, 0
Anerastia, 0	Eumysia*, 0	Ortholepis, 4	Stylopalpia, 4
Apomyelois*, 2	Eurythmia*, 2	Ozamia*, 3	Tacoma*, 0
Athaloca, 0	Eurythmidia, 0	Palatka, 4	Tampa, 2
Australephestiodes, 2	Euzophera*, 2	Passadena, 4	Telethusia*, 4
Bandera*, 2	Fundella, 0	Passadenoides*, 4	Tlascala, 4
Baphala, 3	Glyphocystis, 0	Patagonia*, 3	Trachycera*, 1
Barberia, 0	Glyptocera, 4	Philocrotona , 4	Tulsa, 4
Bertelia, 2	Homoeosoma*, 3	Philodema, 4	Ufa, 4
Cabnia, 2	Heinrichessa, 2	Phobus*, 4	Ulophora, 4
Cactobrosis*, 3	Heterographis*, 0	Phycitodes*, 3	Unadilla, 3
Cadra*, 2	Honora*, 0	Pima*, 4	Valdivia, 3
Cahela*, 3	Hulstia*, 0	Pimodes, 4	Varneria, 2
Canarsia, 4	Hypsipyla, 2	Plodia*, 2	Vitula*, 2
Caristanius, 4	Immyrla, 4	Polopseustis, 4	Wakulla, 2
Catastia*, 4	Interjectio*, 4	Promylea , 0	Welderella, 3
Caudellia, 2	Laetilia*, 3	Protasia, 0	Wunderia, 0
Chararica, 2	Lascelina, 3	Pseudadelphia*, 4	Yosemitia*, 3
Coenochroa, 0	Lipographis*, 4	Psorosina, 4	Zamagiria*, 4
Crocidomera, 2	Macrorrhinia, 4	Pyla*, 4	Zophodia*, 3
Cuniberta, 2	Manhatta, 2	Quasisalebria*, 4	

1 = fasc. 15.2, 1986; 2 = 15.3, 1990; 3 = 15.4, 1997; 4 = 15.5, 2003; 0 = not treated.
*, genera represented in this book.

References for Superfamily Pyraloidea

Capps, H. W. 1967. Review of some species of *Loxostege* Hübner and descriptions of new species (Lepidoptera, Pyraustidae: Pyraustinae). Proceedings of the U.S. National Museum, Washington DC; 120: 1–83.

Goodson, R. L., and H. H. Neunzig 1993. Taxonomic Revision of the Genera *Homoeosoma* Curtis and *Patagonia* Ragonot (Lepidoptera: Pyralidae: Phycitinae) in America North of Mexico. Technical Bulletin 303. North Carolina Agricultural Research Service, North Carolina State University; v + 105 pages.

Heinrich, C. 1956. American moths of the subfamily Phycitinae. Bulletin of the U.S. National Museum, Washington DC; 207; 581 pages.

Klots, A. 1940. North American Crambus, I: The silvery-striped species of California (Pyralididae). Bulletin of the Southern California Academy of Sciences 39: 53–70.

Landry, B. 1995. A phylogenetic analysis of the major lineages of the Crambinae and of the genera of Crambini of North America (Lepidoptera: Pyralidae). Memoirs on Entomology, International. Vol. 1. Associated Publishers, Gainsville, FL; 242 pages.

Lange, W. H., Jr. 1956. A generic revision of the aquatic moths of North America: (Lepidoptera: Pyralidae, Nymphulinae). Wasmann Journal of Biology 14: 59–144.

Lange, W. H., Jr. 1956. Aquatic Lepidoptera, pages 271–288, *in:* Usinger, R. L. Aquatic Insects of California. University of California Press, Berkeley, Los Angeles, London; ix + 508 pages.

Munroe, E. 1961. Synopsis of North American Odontiinae, with descriptions of new genera and species (Lepidoptera: Pyralidae) Canadian Entomologist, Supplement 24; 93 pages.

Munroe, E. 1966. Revision of North American species of *Udea* Guenée (Lepidoptera: Pyralidae). Memoirs of the Entomological Society of Canada, 49; 57 pages.

Munroe, E. 1970. A new genus and three new species of Chrysauginae (Lepidoptera: Pyralidae) Canadian Entomology 102: 414–420.

Munroe, E. 1972–1973. Pyraloidea. Pyralidae (part), subfamilies Scopariinae, Nymphulinae, Odontiinae, Glaphyriinae, and Evergestiinae, Fascicle 13.1A, B, C; xx + 304 pages, 24 plates, *in:* Dominick, R. et al. (eds.) The Moths of America North of Mexico. E. W. Classey, Ltd., and Wedge Entomological Research Foundation, London.

Munroe, E. 1973. A supposedly cosmopolitan insect: The celery webworm and allies, genus *Nomophila* Hubner)(Lepidoptera: Pyralidae: Pyraustinae). Canadian Entomology 105: 177–216.

Munroe, E. 1976. Pyraloidea. Pyralidae (part), subfamily Pyraustinae (part), Fascicle 13.2A, B; xvii + 150 pages, 29 plates, *in:* Dominick, R. et al. (eds.) The Moths of America North of Mexico. E. W. Classey, Ltd., and Wedge Entomological Research Foundation, London.

Munroe, E., and M. A. Solis, 1999. The Pyraloidea, pages 233–256, *in:* Kristensen, N. (ed.) Lepidoptera, Moths, and Butterflies. Vol. 1: Evolution, Systematics, and Biogeography. Handbook of Zoology. W. de Gruyter, Berlin, New York; x + 491 pages.

Neunzig, H. H. 1986–2003. Pyraloidea. Pyralidae (part), subfamily Phycitinae (part), Fascicle 15.2, 1986, xii + 112 pages; 15.3, 1990, 165 pages; 15.4, 1997, 157 pages; 15.5, 2003, 335 pages, *in:* Hodges, R. W. et al. (eds.) The Moths of America North of Mexico. Wedge Entomological Research Foundation, Washington, DC.

Neunzig, H. H. 1988. A taxonomic study of the genus *Salebriaria* (Lepidoptera: Pyralidae: Phycitinae) in America north of Mexico. Technical Bulletin 287. North Carolina Agricultural Research Service, North Carolina State University; iii + 95 pages.

Solis, M. A. 2008. Aquatic Lepidoptera, *in:* Merritt, R. W., and K. W. Cummins (eds.) An Introduction to the Aquatic Insects of North America. Kendall/Hunt, Dubuque, IA; 1–1214 pages.

Solis, M. A., and K. V. N. Maes 2002. Preliminary phylogenetic analysis of the subfamilies of Crambidae (Pyraloidea Lepidoptera). Belgian Journal of Entomology 4: 53–95.

Solis, M. A., and C. Mitter 1992. Review and preliminary phylogenetic analysis of the subfamilies of the Pyralidae (sensu stricto)(Lepidoptera: Pyraloidea). Systematic Entomology 17: 79–90.

Solis, M. A., and M. Shaffer 1999. Contribution towards the study of the Pyralinae (Pyralidae): Historical review, morphology, and nomenclature. Journal of the Lepidopterists' Society 53: 1–10.

Zimmerman, E. C. 1958. Insects of Hawaii. Vol. 8: Lepidoptera: Pyraloidea. University of Hawaii Press, Honolulu; xi + 456 pages.

Macrolepidoptera

Superfamily Drepanoidea

Larvae of the two western families have at least some of their abdominal prolegs either vestigial or absent. In this superfamily, adults of most species have broad wings accompanied by slender bodies; furthermore, they lack scales on their proboscises. Their abdominal tympana are sufficiently different from those of other Lepidoptera that they may have separately evolved. There are about 700 species worldwide with by far the greatest species richness found in the Oriental region. About 20 species are reported for North America with most in the Drepanidae. The included family Epicopeiidae is a small Asian group of mostly brightly colored diurnal moths that often mimic butterflies such as pierids and papilionids.

FAMILY DREPANIDAE

Adult   Adults possess abdominal tympana that are unlike those of other Lepidoptera; these connect the dorsum of the first abdominal segment and the venter of the second abdominal segment. The tympanal opening is dorsal. Adults are medium to large but mostly medium and are relatively broad winged. The moths are nocturnal and are attracted to light. The eggs, which are ribbed and flat, are laid in small groups, rarely singly.

Larva   Larvae have the head globular or occasionally bifid dorsally. There are at least two pair of secondary setae on the second thoracic through eighth abdominal segments. Larvae have few secondary setae or, rarely, numerous short setae. Prolegs are usually vestigial, but those of the third through sixth abdominal segments are well developed. The anal shield is conspicuously elongate.

Larval Food   Most species feed on leaves of broad-leaved trees and shrubs. The larvae may be gregarious in early instars. The fully developed larvae usually spin a cocoon and pupate between leaves, but occasionally almost without silk. The pupae of some (e.g., *Drepana bilineata*) have a powdery waxy coating.

Diversity   This is the largest family of Drepanoidea, with 21 North American species, 18 of which occur in western North America.

SUBFAMILY DREPANINAE

The larvae have an eversible vesicle just dorsal to the prothoracic coxa. The medium adults are very similar to typical geometrids, with broad wings, and they often have curved or falcate FW apices. The moths rest with their wings appressed to the substrate, as in most Geometridae. There are five North American drepanines, but only two are found in the West—both *Drepana* Schrank species.

Drepana arcuata Walker, "arched hooktip" (FW 1.5–2.1 cm; Plate 27.1), is widespread across temperate North America from southern Alaska east across Canada to Newfoundland and south to northern California, Colorado, Texas, and the Southeast. The adults are either yellow tan or pale tan with thin curved and wavy lines. They are easily distinguished by their sharp falcate FW apexes. Adults fly from May to August (mainly June to July) and are attracted to light. The larvae feed on alders, *Alnus rubra* and *A. viridis* var. *sinuata* (Betulaceae), during July and August; they live in silk-lined shelters on leaves that have the edges folded up. McFarland found that when disturbed they can make a faint tapping noise that may serve to space out competing individuals. Larvae are green with mottled brown, purple, and black on the dorsum. There are paired prominent subdorsal bumps or verrucae on the second thoracic through second abdominal segments. The head is whitish with two brown purple bands. The last abdominal segment is acuminate with a short dorsal horn and is held in an elevated position at rest. The head is cream yellow with brown semicircular bands. Pupation is within the folded leaf shelter in a tough, pale brown cocoon. *Drepana bilineata* (Packard), "two-lined hooktip" (FW 1.6–1.7 cm; Plate 27.2, Plate 61.25 larva), is also widespread in temperate North America from Alaska east to Newfoundland. In the West it ranges south to Oregon and Colorado. The adults are either orange yellow with faint lines or whitish tan with stronger dark lines and vermiculations; they can be distinguished by their moderately hooked FW apex and scalloped outer margin. In the East, there are two flights, and McFarland reports that they fly during May and August in western Oregon. The adults are nocturnal and are attracted to light. Host plants include alders, *Alnus,* and birches, *Betula* (both Betulaceae), as well as elms, *Ulmus* (Ulmaceae). Young larvae skeletonize the host leaves, and older larvae rest on top of the leaves in a curled posture with anterior and posterior portions raised above the leaf. The last instars are mottled brown with warty integument. The larvae have the same head-tapping behavior as that of *D. arcuata.* Pupation is within a cocoon of yellow yarnlike silk within a curled leaf shelter. The pupa acquires a heavy coat of white material a few days after pupation.

SUBFAMILY THYATIRINAE

Larvae have dorsal protuberances as in Notodontidae. Adults are usually medium to large and appear like noctuids and are relatively stout bodied. When resting, the wings are held rooflike, also as in noctuids. Antennae are short, bipectinate, or lamellate to the tip but sometimes are filiform. There are 16 North American thyatirines, and most are endemic to the West, although several also occur in eastern North America.

Habrosyne scripta (Gosse). "lettered habrosyne" (FW 1.8–2.0 cm; Plate 27.3, Plate 61.26 larva), is widespread across much of temperate North America. The species has been found from Alaska east to Labrador and southward in the mountains. In the West, *H. scripta* is found south to the northern Sierra Nevada and central coast of California, Montana, and Colorado. The adults are nocturnal and are attracted to light. In the East, there are two flights, but the moths have only a single flight in the West from mid-May to early September, mainly June to July. In California, flight dates from May to early November indicate that there may be two flights, at least along the coast. Larvae feed on several native *Rubus* species (Rosaceae), including salmonberry, *R. spectabilis*. Reports of *Betula,* birch, are most unlikely and require confirmation. The cutwormlike larvae are mottled golden orange brown with a thin, black middorsal line. *Habrosyne gloriosa* (Guenée)(FW 1.6–1.8 cm; Plate 27.4) ranges from Washington and South Dakota (Black Hills) south along the Rocky Mountain front to New Mexico and southeastern Arizona. It is extremely similar to *H. scripta* but can be separated by the more extensive scripting on the postmedian of the FW and the white antemedial line on the FW that is bent at right angles—not at a shallow angle as in *H. scripta*. The adults fly in July and August. The larval host of *H. gloriosa* is reported as ninebark, *Physocarpus opulifolius* (Rosaceae).

Pseudothyatira cymatophoroides (Guenée), "tufted thyatirid" (FW 1.8–2.1 cm; Plate 27.5, Plate 61.27 larva), ranges across temperate North America, perhaps discontinuously, from British Columbia east to Newfoundland. In the West, it may be found primarily along the Pacific Coast from British Columbia south to northern California. The adults may be either of two forms: one is more or less uniformly gray, whereas the second has the FW with a pink basal band and a postbasal black band. Adults are nocturnal and are attracted to light. There is a single flight from June to early July. Larvae are light brown, green, and yellowish with brown blotches and lines. The moth seems relatively polyphagous on deciduous trees and shrubs. The larval hosts include salmonberry, *Rubus spectabilis*, thimbleberry, *R. parviflorus* (Rosaceae), *Alnus rubra* (Betulaceae), and *Corylus* (Betulaceae) in the Pacific Northwest. Elsewhere, *Acer* (Aceraceae), *Quercus* (Fagaceae), *Populus*, and *Salix* (Salicaceae) are reported hosts. The host range thus includes as many as seven plant families. *Thyatira mexicana* Henry Edwards (FW 1.5–2.1 cm; Plate 27.6) is relatively widespread in Mexico, Central America, and northern South America; it reaches the United States only in southeastern Arizona, where it may be found in small numbers. There it flies in the Chiricahua Mountains during July and August.

Euthyatira lorata (Grote)(FW 1.9–2.0 cm; Plate 27.7) is found coast to coast in mesic temperate forests. In the West it occurs from Washington south to California (Marin and Tulare counties). Adults are nocturnal and are attracted to light. They have a single flight from late March to mid-June, mainly April to May. The larval hosts are dogwoods, *Cornus* species (Cornaceae), especially *C. sericea*, red-osier dogwood. *Euthyatira pudens* (Guenée), "dogwood thyatirid" (FW 1.9–2.2 cm; Plate 27.8, Plate 61.28), is an eastern species that also occurs in the Rocky Mountains south to Colorado. Its larval hosts are limited to various dogwoods, *Cornus* species. There is a single adult flight during April and May. *Euthyatira semicircularis* (Grote)(FW 1.9–2.2 cm; Plate 27.9) ranges from southern British Columbia south to southern California and extends inland to western Idaho and western Nevada. The moths are nocturnal and are attracted to light. There is a single flight from late March to early August, rarely in early September. The life history is unreported.

Ceranemota Clarke was revised by Clarke and Benjamin who included eight species. As far as is known all species feed on various Rosaceae. *Ceranemota fasciata* (Barnes & McDunnough)(FW 1.8–2.0 cm; Plate 27.10) is found in coastal wet forests from British Columbia south to northern California (Mendocino County). Adults fly in late September. The larval hosts are serviceberry, *Amelanchier alnifolia, Prunus ilicifolia lyonii*, and *P. virginiana* (all Rosaceae). Larvae are tan with white, yellow, and black mottling. The posterior end of the larva is held slightly elevated above the substrate. *Ceranemota tearlei* (Henry Edwards)(FW 1.5–2.0 cm; Plate 27.11) is the most widespread *Ceranemota* in temperate western North America; it is found in more arid habitats than *C. fasciata*. It has been documented from Oregon south to northern California, Utah, and Colorado. The moth has a single flight from September to mid-October. A July record requires confirmation. Its larvae have been recorded feeding on *A. alnifolia* and *Sorbus scopulina*. *Ceranemota improvisa* (Henry Edwards)(FW 1.6–1.8 cm; Plate 27.12) ranges from southwestern British Columbia south to northwestern California. Adults are nocturnal and have a single flight from mid-October to mid-November. McFarland described the life history based on observations from west of Corvallis, Oregon. The larval hosts are *Crataegus douglasii* and *Prunus* (Rosaceae). Females lay 100 or more eggs singly or in short chains. Second- to fourth-instar larvae are reddish brown laterally and are white and yellow dorsally. The larvae rest on the host with the posterior portion of the body tilted abruptly upward. *Ceranemota crumbi* Benjamin (FW 1.5–1.9 cm; not illustrated), known from Oregon and Washington, has a single flight from mid-September to October. Its larvae are reported to feed on *S. sitchensis*.

Bycombia verdugoensis (Hill)(FW 1.5–1.7 cm; Plate 27.13) has two forms, a rather plain morph and a strongly marked one. The moth is reported only from southern California (Los Angeles south to San Diego County), where it is rare and little known. Adults have been collected from February to early April. The life history is unreported.

Superfamily Geometroidea

This superfamily, called geometers, comprises the Geometridae, Sematuridae, and Uraniidae. Stehr characterizes the larvae by several setal characters including having a fourth subventral seta on the sixth abdominal segment, having setae L1 (lateral) and L2 widely separated on the on the first eight abdominal segments (only on the fourth through eighth in Epiplemidae), and having the L3 seta more anterior than its placement in most other lepidopterous larvae.

FAMILY SEMATURIDAE

According to Holloway and coauthors, the moths of this family lack tympanal organs and have distally swollen, sometimes clubbed antennae that have a swollen scape and elongate basal flagellomere; moreover the chaetosemata have long setae that overhang the eyes. They are primarily tropical moths.

SUBFAMILY SEMATURINAE

Anurapteryx crenulata Barnes & Lindsey (FW 2.0–2.1 cm; Plate 35.4) has antennae that are superficially similar to those of a pyrgine skipper; the species is known only from southeastern Arizona and, presumably, ranges south into Mexico. Adults fly during early August. The life history is unreported.

FAMILY URANIIDAE

The Uraniidae is characterized by a marked sexual difference in the structure and position of the tympanal organs. The HW are usually tailed and angled. These are primarily tropical moths with strong representation in the Neotropical, Oriental, and Australasian regions.

SUBFAMILY EPIPLEMINAE

The Epipleminae is a pantropical group that includes the smaller uraniids; they often have doubly tailed HW with a concave margin between the tails. Holloway and coauthors point out that the larvae are covered with fine spines but lack secondary setae. The crochets are in a strongly curved mesoseries. Young larvae may be gregarious and live in webs. *Callizzia amorata* Packard, "the gray scoopwing" (FW 1.0–1.1 cm; Plate

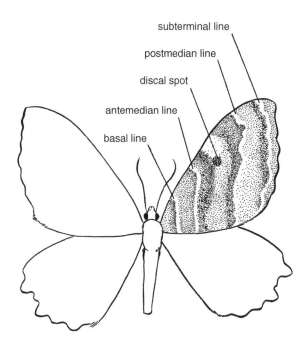

FIGURE 201. Schematic representation of Geometridae FW with color pattern elements used in diagnoses.

27.14, Plate 61.29), our sole representative, ranges from Washington across the continent to Quebec and south to Oregon, California (Marin County), Arizona, New Mexico, and western Texas. We have found the species primarily in wooded hilly or mountainous terrain. The adults are nocturnal and are readily attracted to light. In the West, the single annual flight is from late June through August, while in the East they appear as early as April and probably have several broods. The larval hosts are honeysuckles, *Lonicera* species (Caprifoliaceae).

FAMILY GEOMETRIDAE

Among the Lepidoptera, the Geometridae, called inchworm moths or geometrids, is one of the most species-rich families; among the macrolepidoptera, only the Noctuidae comprises more species. Adults have unique tympanal structures at the base of the abdomen in deep ventrolateral cavities. Additionally, in the FW one or two areoles (accessory cells) are usually formed by the first R and Sc **veins;** while in the HW vein Sc is strongly bent (Fig. 14). Larvae, or "inchworms," usually lack prolegs on the third through fifth abdominal segments—having them only on the sixth and tenth abdominal segments (Fig. 20). As a result, they walk with a measured looping progression.

Adult Adults are small to large and most often broad winged with a slender body. Male antennae are fasciculate or bipectinate, while those of females are simple. The majority are dull colored, often gray, but a number, especially in the tropics, are brightly colored members of mimicry rings. The disposition, color, and presence of various FW transverse lines and a discal dot (. 201) are used to describe various geometrids; other wing features are typical of those found in other Lepidoptera, but a fovea may or may not be present at the base of the FW. Many geometrid species are dimorphic or extremely variable in color pattern. In many genera, species identification is difficult without genitalic dissection (e.g., Figs. 202–217). Most species rest with their wings appressed against the substrate and often appear to blend into their background, commonly in shades of gray, green, or tan. These often rest on tree trunks or branches, and foliage (live or dead). Some others hold their wings above their back in a butterfly-like manner, and these often have cryptically marked ventral HW. Most adults are nocturnal and are variably attracted to lights, however some species are diurnal, especially spring and Arctic-alpine taxa. These species are more brightly or contrastingly colored. Some winter and spring species have flightless females with vestigial or reduced mouthparts and tympana.

Larva Most geometrid larvae are typified by having only two pair of prolegs on the sixth and tenth abdominal segments. A few exceptions among North American taxa, *Archiearis* and *Leucobrephos* (Archiearinae), *Alsophila* (Oenochrominae), and *Campaea* (Ennominae), have three or more pairs of prolegs, but these are always smaller than those on the sixth abdominal segment, and the number and size of the crochets is always reduced. Larvae are naked but are normally cryptically colored, often in shades of green or brown. They may have varied body forms or protuberances, and most appear like or blend in with twigs, foliage, or flowers. Their appearance is interpreted as a way to avoid predators. The larvae of almost all species feed externally, although some of these live in folded or silked leaf shelters, and some cover their bodies with plant parts. The larvae of many *Eupithecia* bore in buds, flowers, and sometimes even cones. The larvae of *Perizoma incultraria* (Herrich-Schäffer) are unique as they are leaf miners of *Primula* (Primulaceae).

Larval Foods Most species feed on the leaves, flowers, or fruits of woody dicotyledonous trees and shrubs, although some feed on conifers, ferns, or herbaceous plants. A few species, notably among the Sterrhinae, are detritivores. As a notable exception, several Hawaiian *Eupithecia*, normally a plant-feeding genus, are predaceous on small flies, which they capture using modified prothoracic legs. The vast majority of species are not of concern, but a few species are sometimes serious defoliators of shade or forest trees, for example, spring cankerworm, *Paleacrita vernata* (Peck), fall cankerworm, *Alsophila pometaria* (Harris), and hemlock looper, *Lambdina fiscellaria* (Guenée).

Pupa The palpi are not exposed, and the cremaster is well developed.

Diversity Worldwide, Scoble et al. estimate there are 21,000 described species. In America north of Mexico, Covell and Ferguson list almost 1,400 species. In western North America there are more than 960 species representing 176 genera. In addition, there may be more than 40 undescribed species, most of which are in the Larentiinae. The majority of species are found in forested or shrubby habitats, and grasslands are generally relatively depauperate.

The names used in this section follow those given by Scoble in his *Geometrid Moths of the World* catalogue. Since the last published checklist of North American Geoemtridae by Ferguson in Hodges's checklist, many changes in the generic names have been published, although not always specifically applied to North American species. Scoble has done his best to apply these changes to the appropriate species. The type localities, where known, are given by Scoble for most species, making it possible to tally the total species for any given zoogeographic region.

SUBFAMILY ARCHIEARINAE

Scoble and Minet point out that this small group is composed of only 12 species that occur disjunctly in the Holarctic, southern South America, and Tasmania. The adults have reduced tympanal organs, the frons is covered with hairlike setae, and they have small oval compound eyes. The moths are all diurnal and tend to be brightly colored. The larvae are unusual in having four pairs of midabdominal prolegs, a feature of all species in the subfamily. During the day, the larvae rest in a leaf shelter. The pupa overwinters. Three species are found in North America.

Archiearis Hübner comprises four species, which are restricted to the Holarctic region in boreal habitats. Two species occur in North America; both have distributions that include the West. *Archiearis infans* (Möschler), "the infant" or "the first-born" (FW 1.4–1.5 cm; Plate 27.15), one of our more colorful geometrids, is the sole North American species of its genus. It ranges from Alaska and Northwest Territories east to Labrador, Newfoundland, and Nova Scotia and south to southern California, Montana, and the Colorado Front Range, and in the East, south to Wisconsin, Pennsylvania, and New Jersey. It is usually found in mixed broad leaf–evergreen forests. The colorful adults are diurnal and have a rapid erratic flight. In most of the West, there is a single generation from late March through May, but in California, there are records from late February to early July. The larval hosts are alders and birches (Betulaceae), as well as poplars and willows (Salicaceae). According to Miller and Hammond the larvae are green with yellow subdorsal spots and a strong yellow subspiracular line.

Leucobrephos Grote, with four species in the Holarctic, is represented in North America by *L. brephoides* (Walker)(FW 1.3–1.4 cm; Plate 27.16 male, Plate 27.17 female), which has been found from Alaska and Yukon east across Arctic Canada to Labrador, thence south to southern British Columbia, Manitoba, Wisconsin, and New York. The adults have a single flight from mid-March to mid-May and are diurnal with flight activity between 11:00 and 15:00. They will visit damp places near aspen woods. Prentice reports that the larval hosts are alders and birches (Betulaceae), as well as poplars and willows (Salicaceae). The larvae make shelters that comprise several leaves. Last instars are green with yellowish intersegmental folds. There is a conspicuous, wide, yellow, spiracular stripe and less obvious yellowish longitudinal lines. The head is pale green. Pupation is in the soil.

SUBFAMILY ENNOMINAE

The subfamily Ennominae with roughly 9,700 species worldwide comprises 45% of the family. Scoble and Minet discuss the group in some detail. The unity of the subfamily is in question, since the unifying character, absence of M2 as a tubular HW vein, may have been evolved more than once. The males of many but not all genera have a band of posteriorly directed setae on the third abdominal segment. Most species are slender bodied and medium to large.

TRIBE ABRAXINI

This tribe was thought by Holloway and coauthors to be restricted to *Abaxas* and several small genera of Asian and Palaearctic species. *Protitame* McDunnough, included in this tribe by Ferguson, is a North America genus of seven species, four of which are western. *Protitame matilda* (Dyar)(FW 1.1–1.2 cm; Plate 27.18) ranges from British Columbia and Alberta south to California, Utah, and Montana. The adult flight is from May to September. Larvae feed on willows and quaking aspen, Salicaceae. The larvae are yellow green with a geminate purple middorsal line or a general reddish tinge dorsally and a yellow lateral line. Related western species are **P. hulstiaria** Taylor and **P. virginalis** (Hulst)

TRIBE MACARIINI

The tribe Macariini is found in most world regions; the North American taxa are revised in a soon-to-be-published, posthumous treatment by Ferguson. Many North American species, formerly referred to *Semiothisa* Hübner, are now included in *Digrammia* Gumppenberg or *Macaria* Curtis in Scoble's catalogue. Another cluster of species comprises those formerly included in the genus *Itame* Hübner, but now classified by Scoble as a group within *Macaria*.

Eumacaria latiferrugata (Walker), "brown-bordered geometer" (FW 0.9–1.2 cm; Plate 27. 19, Plate 61.30 larva), is the only species in its genus. It ranges from central British Columbia east across southern Canada to Nova Scotia, thence south to eastern Washington, northern Idaho, Colorado, Texas, and Florida. The adults are nocturnal and have one or two flights from late April to early August. The larval hosts are various species of *Prunus* (Rosaceae). The larvae are brown with white patches before and after each spiracle on the first through fifth abdominal segments. The pupa overwinters.

Digrammia has a Holarctic distribution with a number of Neotropical species as well. Scoble lists 55 species of which 43 are North American. The majority of these, 37, includes all or a portion of their range in the West. Thus, the majority of the species richness occurs in our area. Some species are best separated by reference to the ventral surfaces. **Digrammia californiaria** (Packard)(FW 1.2–1.4 cm; Plate 27.20) ranges from southeastern British Columbia and southern Alberta south to southern California and New Mexico. Adults have been found from early March to mid-October. **Digrammia teucaria** (Strecker)(FW 1.3–1.5 cm; Plate 27.21) is found from southern Vancouver Island, British Columbia, south to Washington. The adults have a single flight from May to July. The larval host is Oregon white oak, *Quercus garryana* (Fagaceae), but since this tree is not found throughout the moth's range, other hosts are used. Eggs are laid singly on oak leaves, and larval development takes about 40 days. Winter is passed in the pupal stage. As reported by McGuffin, the larvae resemble oak twigs and are gray green with a touch of cinnamon in the intersegmental areas. There are closely spaced fine brown lines on the dorsal and lateral areas. Some larvae have black or yellow marks along the sides. The prolegs are marked with crimson and the head is milky green and is mottled with light brown. **Digrammia setonana** (McDunnough)(FW 1.6–1.8 cm; Plate 27.22, Plate 27.23) occurs from southern British Columbia and southwestern Alberta south to California and Utah. This species forms a complex with **D. excurvata** (Packard) and **D. continuata** (Walker), and adults are best identified by dissection and examination of their genitalia. The adults have two flights from mid-May to mid-August. The reported larval hosts are Rocky Mountain juniper, *Juniperus scopulorum,* and western red-cedar, *Thuja plicata* (Cupressaceae). The larvae vary from green to gray green with a complex pattern of lines and oblique marks. There are six instars instead of the usual five. **Digrammia curvata** (Grote)(FW 1.0–1.3 cm; Plate 27.24) ranges from southern British Columbia and southern Alberta south to eastern California, Nevada, Utah, and Colorado. The moth is found in arid intermountain shrubland habitats in association with its larval host rubber rabbitbrush, *Ericameria nauseosa* (Asteraceae). The adults are in flight from late March

to early September. There are two annual generations, at least in the southern portion of its range. Larvae are white with geminate green, dark, or black longitudinal lines. The head is light gray with a light green herringbone pattern on the lateral lobes. **Digrammia irrorata** (Packard)(FW 1.0–1.5 cm; Plate 27.25) ranges from southern British Columbia south to southern California and Colorado. Adults have one or two flights from late April to September. In California, Richers reports the species as early as February. The larval hosts are poplars and willows (Salicaceae). The ultimate-instar larvae are either green or brown with obscure lines. **Digrammia neptaria** (Guenée)(FW 1.0–1.5 cm; Plate 27.26) ranges from British Columbia and Yukon east across Canada to Labrador and Newfoundland, thence south to California, Arizona, New Mexico, South Dakota, and, in the East, New Hampshire. The adults fly from late April through August in much of the West but have an extended flight from mid-March to September in California, according to records compiled by Richers; there are two generations, at least in the southern portion of its range. The larval hosts, as for *D. irrorata*, are poplars and willows (Salicaceae). The larvae are quite variable with green, yellow green, or gray brown forms. As he does for many other geometrids, McGuffin describes the larvae in great detail. **Digrammia muscariata** (Guenée)(FW 1.5–1.6 cm; Plate 27.27) ranges from southern coastal British Columbia south to southern California. In California, adult flight is from mid-May to mid-September, probably as two generations. The larval hosts are various oaks, *Quercus*, including *Q. garryana*.

Macaria Curtis has 175 described species, which are distributed in the Holarctic and Neotropical regions. There are 78 species known from North America with 40 that have at least a portion of their range in the West. Probably, there are a number of undescribed species, especially in the Neotropics. The species utilize either conifers or broad-leaved plants as their larval hosts. Most, but not all, of the species are grayish, and all hold their wings directly above their back in butterfly-like fashion. Many of the species are widespread, and there are very few warm nights in the montane and interior portions of the West when at least some individuals of this genus are not found at attraction lights. Eggs are laid either singly or in small clusters, often near the growing point of new leaves.

Macaria truncataria (Walker), "variegated orange" (FW 0.9–1.1 cm; Plate 27.28), ranges from Alaska east across Canada to Labrador and Newfoundland south to British Columbia, northern Arizona, and Colorado in the West; the moth is also found south to Wisconsin, Michigan, and New Jersey in the East. The diurnal adults have a single flight from late May to early July. We have seen them flying in and out of light gaps in lodgepole pine forests that have a low understory layer of *Vaccinium*. Larvae feed on several genera of Ericaceae, including *Arctostaphylos* and *Vaccinium*. Eggs are laid singly on the underside of host leaves. The full-grown larva is green with light green or white lines margined in gray. There is a prominent whiter lateral line. The thoracic legs are russet, and the head is russet to green with light brown markings. Winter is passed by the pupal stage.

ULSTERATA SPECIES GROUP

The *Ulsterata* Species Group includes three western moths that are sometimes difficult to distinguish without dissection of their genitalia. Additional species occur in the East and southward into the New World tropics. **Macaria ulsterata** (Pearsall), "birch angle" (FW 1.4–1.5 cm; Plate 27.29), ranges from the Northwest Territories and British Columbia east to Nova Scotia, from which it occurs south to Oregon, northern Idaho, and Colorado in the West and the Appalachians of North Carolina in the East. The adults are strictly nocturnal and have a single flight from late May to July. McGuffin reports that *M. ulsterata*'s larval hosts are alder and birch (Betulaceae) as well as quaking aspen (Salicaceae). The glossy larvae are reddish brown or green banded with gray or brown. There is often a light spot forward of each spiracle. Two related western species are **M. perplexata** (Pearsall) and **M. aemulataria** (Walker).

BISIGNATA SPECIES GROUP

The *Bisignata* Species Group includes several species. **Macaria adonis** (Barnes & McDunnough)(FW 1.3–1.6 cm; Plate 27.30) ranges from southern British Columbia south to southern California, Utah, Colorado, and western South Dakota. In much of its range, adults have a single flight during June and July, but apparently they have a second flight in coastal lowland California, as Richers compiled records of adults from early April to mid-September. The larvae feed on young needles of several pines, Pinaceae, including ponderosa pine, *Pinus ponderosa*, and lodgepole pine, *P. contorta*. They are pruinose green with geminate gray green dorsal and longitudinal lines as well as subdorsal and lateral cream or white lateral lines. **Macaria sexmaculata** (Packard), "green larch looper" (FW 0.9–1.1 cm; Plate 27.31, Plate 27.32, Plate 61.31 larva), ranges from eastern British Columbia and the Northwest Territories east to Newfoundland, and thence south to eastern Oregon, northern Idaho, the Great Lakes region, and Maryland. There may be two adult flights spanning late May through September. The larval host is tamarack, *Larix* species, and, secondarily, Douglas-fir, *Pseudotsuga menziesii*. Larvae are variably green, brown, maroon, or light gray with variable patterns. **Macaria signaria** (Walker), "spruce-fir looper" (FW 1.4–1.8 cm; Plate 27.33), ranges from Alaska and the Queen Charlotte Islands of British Columbia east across Canada to Labrador and Newfoundland, and thence south to northern California, northern Idaho, and northern Wyoming. The adults fly from mid-May to early September. The larval hosts are a wide variety of Pinaceae including Douglas-fir, true firs, larch, hemlock, and spruce. The ultimate-instar larvae are light green to blue green with a complex pattern of geminate gray and black lines together with various white and green stripes. The thoracic legs are brown, and the prolegs are concolorous with the body. **Macaria unipunctaria** (Wright)(FW 1.4–1.7 cm; Plate 27.34) occurs from southern British Columbia and southwestern Alberta south to the central Sierra Nevada of California, southeastern Arizona, and northern Colorado. There is a single flight of the nocturnal adults from late May to early August. Douglas-fir, *P. menziesii* (Pinaceae), is the only reported larval host. Full-grown larvae are light green with various gray and black lines. The thoracic legs are brown and the prolegs are green. The head is russet brown with a chestnut brown streak over each lobe.

"ITAME" SPECIES GROUP

The "Itame" Species Group includes species that occur in the Holarctic realm; most are limited to either North America or Eurasia, but two species, **Macaria loricaria** (Eversmann) and *M. brunneata* (Thunberg), are Holarctic. While the pupae of most *Macaria* overwinter, the eggs of species in this group overwinter, with larval feeding and pupation during the following spring. **Macaria occiduaria** (Packard)(FW 1.4–1.5 cm; Plate 27.35) ranges from southern British Columbia east to southern Ontario and south to Oregon and Colorado. The

adults have a single flight from mid-June through August. Larval hosts are a variety of broad-leaved woody trees and shrubs in several families. The ultimate-instar larva is white with subdorsal rows of small black spots and a wide black subventral line. The thoracic legs are black, and the prolegs are pale with black spots. As mentioned above, **M. brunneata** (Thunberg)(FW 1.1–1.4 cm; Plate 27.36) is Holarctic. In North America it is found from Alaska and British Columbia east across Canada to Newfoundland and Nova Scotia and south to Oregon, Wyoming, Colorado, Michigan, and Massachusetts. The adults are diurnal and fly from late June through mid-August. The hosts are broad-leaved woody trees and shrubs in several unrelated families. The larvae are predominantly green with reddish brown subdorsal lines; they are yellowish laterally and dirty white ventrally. **Macaria plumosata** (Barnes & McDunnough)(FW 1.1–1.3 cm; Plate 27.37) ranges from the southern interior of British Columbia south to eastern California, Arizona, and New Mexico. The adults have a single flight during July and August. **Macaria bitactata** (Walker)(FW 1.1–1.5 cm; Plate 27.38) is one of the most abundant species, and it ranges from Alaska east to Saskatchewan and south to the east side of California's Sierra Nevada, Nevada, Utah, Colorado, and western South Dakota. Adults fly from late June to late September and may have two flights along the Colorado Front Range. According to Miller and Hammond, the larval hosts are currants and gooseberries, *Ribes* (Grossulariaceae). The larvae vary from green to gray brown and have dorsal triangular white patches and lateral oblique white patches on each segment. **Macaria colata** (Grote)(FW 1.1–1.3 cm; Plate 27.39) ranges mainly in the intermountain West from eastern Oregon and eastern California east to eastern Colorado. The adult flight is from mid-May to late October, probably as two generations. The larval hosts include big sagebrush, *Artemisia tridentata* (Asteraceae), and bitterbrush, *Purshia tridentata* (Rosaceae). The larvae are mottled silver, gray, and white with an irregular dark gray patch surrounding each spiracle. In addition, Miller and Hammond illustrate the larvae and adults of **Macaria guenearia** (Packard) and **M. quadrilinearia** (Packard), "the western currant spanworm."

ELPISTE SPECIES GROUP

Formerly included in the genus *Elpiste* Gumppenberg, the following species, as well as *M. marcesaria* (Packard), was transferred by Scoble to *Macaria* as a separate group. **Macaria lorquinaria** (Guenée)(FW 1.2–1.4 cm; Plate 27.40) is a western species that ranges from southern British Columbia south to southern California and Utah. Adults have been collected from late April to mid-October, and there are probably two flights. The larval hosts are alders, birches (Betulaceae), and willows (Salicaceae). Larvae are light yellow or light green with brown lines and a brown patch surrounding each spiracle. Alternatively, in the Pacific Northwest, they are described as also having a green form that has a broad yellow spiracular band.

Narraga Walker, with eight species distributed in the Holarctic, has three North American species, two of which include portions of the West in their range. **Narraga fimetaria** (Grote & Robinson)(FW 0.8–1.0 cm; Plate 27.41) ranges from eastern California and Utah south to Arizona east to southern Wyoming, New Mexico, Oklahoma, and central Texas. There are one or two flights from early May to late August (as early as March in central Texas). In the Mojave Desert, California, McFarland found that *Gutierrezia microcephala* (Asteraceae) is the larval host for *N. fimetaria*. Eggs are laid in ropelike rows. **Narraga stalachtaria** (Strecker) is similar but is mainly diurnal and is orange dorsally. Ferris notes that it is attracted to light on occasion.

TRIBE BOARMIINI

The tribe Boarmiini is one of the largest groups within the Ennominae and may be paraphyletic with respect to several other tribes, according to Holloway and coauthors who present a good discussion of the problem. According to most recent treatments this tribe includes the former Melanolophiini and Bistonini.

Aethalura anticaria (Walker)(FW 1.2–1.4 cm; Plate 27.42) ranges from southern British Columbia east across southern Canada to Newfoundland and Nova Scotia. McGuffin notes that it is common in Nova Scotia, eastern Ontario, and portions of British Columbia. Its distribution extends south to California, Idaho, and Florida. The adults are nocturnal but are easily disturbed during daylight. There is a single flight from late April to early August. Females lay eggs singly or in small groups in expanding terminals of alder or birch, Betulaceae. Fifth-instar larvae are green but assume a dark pattern prior to pupation. The larvae are extremely similar to those of *Macaria lorquinaria*, which have a single group of crochets on their prolegs, whereas those of *A. anticaria* have two groups. The body color usually is light yellow suffused with light green and various longitudinal geminate green lines. Alternatively, the body may be mauve, yellow, or white with geminate maroon or gray longitudinal lines. The larva rests while stretched out along a leaf. The pupa, which overwinters, lies near the soil surface in a cell of silk and soil particles.

Glena Hulst has 44 species, and all are native to the New World; 10 of these are North American with six found in the West. *Glena nigricaria* (Barnes & McDunnough)(FW 1.5–2.0 cm; Plate 27.43, Figs. 204, 205 genitalia) ranges from southern British Columbia to eastern California, Arizona, and Mexico. There is a single flight of nocturnal adults from mid-May to mid-August. The larval hosts are Pinaceae, including ponderosa pine, lodgepole pine, and Douglas-fir. Prentice reported that about 90% of larval samples were found on ponderosa pine. Ultimate-instar larvae are light green with a blue or gray line within a broad green middorsal stripe. There are several other gray or white longitudinal lines as well as a brown subspiracular patch on the thoracic segment and the first abdominal segment. Larval feeding is from mid-July to early September. Full-grown larvae drop from the tree and pupate within a flimsy shelter near the soil surface. *Glena grisearia* (Grote) is a common species in the mountains of southeastern Arizona. McFarland found that its host plant is Emory oak, *Quercus emoryi* (Fagaceae).

Stenoporpia McDunnough comprises 28 species of which 22 are western, one eastern, and the remainder Mexican. Specific identity of species is usually based on dissection and examination of the genitalia. *Stenoporpia pulmonaria* (Grote) (FW 1.7–1.9 cm; Plate 27.44) ranges from coastal southern Alaska, British Columbia, southwestern Alberta, and northeastern Wyoming south to southern California and Arizona. The nocturnal adults have two flights from June to September. The larvae feed on various Pinaceae including hemlock, Douglas-fir, true firs, Sitka spruce, and several species of pines. The pupae overwinter. The larvae are gray with broken black lines, obliques, and markings on the dorsum and along the sides. The head is light gray with brown markings laterally. *Stenoporpia excelsaria* (Strecker)(FW 1.9–2.2 cm; Plate 27.45) ranges from southern British Columbia south through the

northern California, northern Arizona, Colorado, and, discontinuously, to western Durango, Mexico. The adults have a single flight from June to early August. The larval hosts are several Pinaceae and include ponderosa and lodgepole pines, as well as Douglas-fir. Another widespread western *Stenoporpia* is *S. separataria* (Grote).

Tornos Morrison includes 17 New World species; there are seven North American species of which three are eastern and two are western. The moths of this genus have a raised tuft of hairs in the FW discal cell. *Tornos erectarius* Grossbeck (FW 1.4–1.6 cm; Plate 27.46, Figs. 206, 207 genitalia) ranges from coastal southern California, and, discontinuously, in southeastern Arizona and, likely, to Sonora, Mexico. The adults have been found in every month except January. McFarland observed two abundance peaks during July to August and October. The life history is unreported.

Glaucina Hulst, with 39 described species, is limited to the Nearctic and Neotropical regions. There are 32 North American species, with 29 that include the West as at least a portion of their distributions. Dissection and examination of genitalia is often necessary for specific identification. *Glaucina epiphysaria* Dyar (FW 1.3–1.5 cm; Plate 27.47) is found in California from San Benito County south to San Diego County. The moths fly discontinuously from February to December in southern California, and McFarland noted peaks of abundance during May and October that indicate two or more generations. *Eriogonum fasciculatum* (Polygonaceae) is its larval host. *Glaucina eupitheciaria* Grote (FW 1.1–1.3 cm; Plate 27.48) occurs in the deserts of southeastern California and much of lowland Arizona. There appear to be two or more flights, from March to May and again in August and September. A record from December seems odd. Its larval host has been reported as *Olneya* (Fabaceae). *Glaucina interruptaria* (Grote)(FW 1.2–1.9 cm; Plate 28.1) is found throughout much of eastern Utah and Colorado. The adult flight is from late April to late July, probably in two flights. *Glaucina ochrofuscaria* Grote (FW 1.2–1.5 cm; Plate 28.2) is found in the deserts of southern California and Arizona. Adults are reported by Richers from March to June and again in September. Its larvae feed on *Atriplex* (Chenopodiaceae), and *Ambrosia* (Asteraceae).

Synglochis Hulst is monotypic; *S. perumbraria* Hulst (FW 1.0–1.4 cm; Plate 28.3) is endemic to southeastern California and Arizona. The adults have two flights in California, one in March then again in July and August. Hurd reported the host as creosote bush, *Larrea tridentata* (Zygophyllaceae).

Nepterotaea McDunnough was revised by Rindge. There are seven species; six are endemic to the Southwest and one is found in the Ozarks. *Nepterotaea memoriata* (Pearsall)(FW 1.0–1.3 cm; Plate 28.4) ranges from southeastern Arizona east to southwestern Colorado and western Texas. Adults have been found from late April through October. The life history is unreported.

Chesiadodes Hulst was revised by Rindge and comprises 13 species, all of which are endemic to the Southwest and adjacent Baja California. *Chesiadodes cinerea* (FW 1.1–1.5 cm; Plate 28.5) ranges through intermountain sagebrush and juniper habitats in Nevada and Utah. Males are fully winged and capable of flight, whereas females are brachypterous and do not fly. D. Giuliani found females in pitfall traps during December. The adults are univoltine from December to March. The larval hosts are rubber rabbitbrush, *Ericameria nauseosa*, and *Tetradymia* (Asteraceae). The larvae are olive green with a broad white dorsal band and a broken white lateral line.

Chesiadodes morosata Hulst (FW 1.5–1.8 cm; Plate 28.6) is endemic to the arid foothills of interior southern California. It may be separated from the nearly identical *Chesiadodes aridata* Barnes & Benjamin by its possession of a proboscis, which is lacking in the latter species. The adults fly from mid-December until early April. The life history is unreported.

Rindge revised *Hulstina* Dyar, which consists of eight species, all western. *Hulstina wrightiaria* (Hulst)(FW 1.0–1.3 cm; Plate 28.7) ranges from central California, primarily in the coast ranges, south to the mountains of coastal southern California, including Santa Cruz Island, and into Baja California. The adults have been found from late April to August, with occasional individuals found in October and December. The larval hosts are chamise, *Adenostoma fasciculatum,* and mountain-mahogany, *Cercocarpus* (both Rosaceae). J. A. Comstock pointed out that the last instars were extremely variable in color and pattern, and they vary from almost clear yellow green through darker green or brown mottled individuals to a brightly colored yellow and red brown form. The spiracles are yellow brown with black outer rims.

Pterotaea Hulst comprises 26 described species, all of which are endemic to western North America including the Southwest and adjacent Mexico. Twenty-three of the species are found in the southwestern United States, and 21 of these are endemic to California or include California in their range. *Pterotaea comstocki* Rindge (FW 1.3–1.7 cm; Plate 28.8) occurs in the mountains and associated foothills of southern California. The adults have a single flight from late May to early July. The larval host is scrub oak, *Quercus berberidifolia*. Larvae of three other species were collected from California oaks or chamise, *Adenostoma fasciculatum* (Rosaceae). *Pterotaea lamiaria* (Strecker)(FW 1.4–1.6 cm; Plate 28.9) is a well-marked common species in the mountains of southern California. In the Santa Monica Mountains, California, McFarland found that the larvae feed on flowers of chamise, *A. fasciculatum* (Rosaceae).

Iridopsis Warren includes 86 species limited to North, Central, and South America, including the Caribbean. Eighteen of the species are North American, and 11 of these are found in the West, primarily the Southwest. This genus includes species that were formerly treated as *Anacamptodes* McDunnough. *Iridopsis clivinaria* (Guenée), the "mountain-mahogany looper" (FW 2.2–2.5 cm; Plate 28.10), is found from the southern interior of British Columbia east to Idaho, thence south to southern California, Utah, and Colorado. The nocturnal adults have a single flight from March to July. Eggs are placed in bark crevices. Larvae feed on several shrubs including *Arctostaphylos* (Ericaceae) and *Cercocarpus, Philadelphus, Purshia,* and *Rosa* (Rosaceae). The first two larval instars are leaf skeletonizers, whereas the older larvae feed on the full depth of their host leaves. Furniss and Barr found four larval color morphs: black, gray, reddish, and yellow. There are slight dorsal bumps on the second and eighth abdominal segments. The larvae sometimes cause serious defoliation and loss of *C. ledifolius* and *Purshia tridentata* populations. Pupation is in the soil. Another common species is *I. emasculata* (Dyar)(FW 1.5–1.8 cm; Plate 28.11), which is distributed from British Columbia south to northern California, Arizona, New Mexico, and western Texas. The adults have a single flight from the end of May until mid-August. The larvae feed on several woody plants in unrelated families: alder, maple, willow, and blueberry. Larvae are green, yellow, or orange with an irregular red middorsal band that is much narrowed on the third through sixth abdominal segments.

Anavitrinella McDunnough comprises four North American species, two of which are western. *Anavitrinella pampinaria* (Guenée), the "common gray" (FW 1.2–1.6 cm; Plate 28.12, Plate 61.32), ranges from Vancouver Island, British Columbia, east across southern Canada to Nova Scotia and south through much of the United States to Mexico. The adults have one or two flights from April to September. Adults are quite variable, and doubtful specimens should be identified by preparation of genitalia dissections. Females lay their eggs singly under host leaves. Many larval hosts have been reported, but since the larvae of several other species are easily confused with those of *A. pampinaria*, it is best to accept only those plants specifically reported by McGuffin: *Shepherdia canadensis* (Elaeagnaceae), *Viburnum* (Caprifoliaceae), *Spiraea, Crataegus* (Rosaceae), *Salix* (Salicaceae), *Betula papyrifera* (Betulaceae), *Cornus sericea* (Cornaceae), and *Ceanothus* (Rhamnaceae). The larvae, which rest in a twiglike posture, vary from light green to light brown with longitudinal gray or brown geminate lines. There may be black diamond marks on the anterior edges of the first through fourth abdominal segments. There is a swelling of the second abdominal segment as well as a dorsal pair of closely spaced tubercles on the eighth abdominal segment.

Protoboarmia McDunnough includes four species, three of which are restricted to Asia. The fourth species, *P. porcelaria* Guenée, the "dashlined looper" or "porcelain gray" (FW 1.3–1.7 cm; Plate 28.13), is widely distributed in North America, from Northwest Territories and British Columbia east across Canada to Newfoundland and Nova Scotia, whereas the species extends southward to Texas and Florida. The nocturnal adults fly from late June to mid-August. Larval hosts are primarily Pinaceae including balsam fir, Douglas-fir, white spruce, larch, western hemlock, and black spruce, in order of apparent preference as noted by Prentice. Woody, broad-leaved angiosperms used on occasion include birch, aspen, willow, and elm. The larvae assume a twiglike posture at rest and overwinter in the penultimate instar. There may be seven to eight instars. The larvae are brown with a dark brown geminate middorsal band. The lateral area is pinkish. There is a prominent subspiracular swelling on the second abdominal segment and prominent creases that encircle each segment. Late-instar larvae overwinter. McGuffin, as usual, provides a more detailed technical description.

Neoalcis californiaria (Packard)(FW 1.5–1.8 cm; Plate 28.14), the only species in a genus described by McDunnough, ranges from coastal British Columbia south to southern California. Adults are mainly brown but vary from reddish to light gray and have a single flight from July to September. In California, Richers lists collection dates from March through October, with a single record in early December. The larval hosts include Douglas-fir, hemlock, western red-cedar, red fir, coast redwood, and, occasionally, other conifers. We reared it from *Ceanothus* in Marin County, California. Larvae are brown or reddish brown and overwinter. The head is brown and rugulose, whereas the larval cuticle is granular.

Orthofidonia Packard comprises three moth species found in eastern North America, one of which, *O. exornata* (Walker), the "pale viburnum geometer" (FW 1.4–1.5 cm; Plate 28.15), extends its distribution into the West. This moth ranges from British Columbia east to Nova Scotia. The nocturnal adults fly from early May to mid-July. The larvae feed on *Viburnum* (Caprifoliaceae). The larvae are light gray to green with a rust red middorsal stripe that broadens on the last few abdominal segments. The larval head is light green to dark green with rust red marks. The larvae drop to the

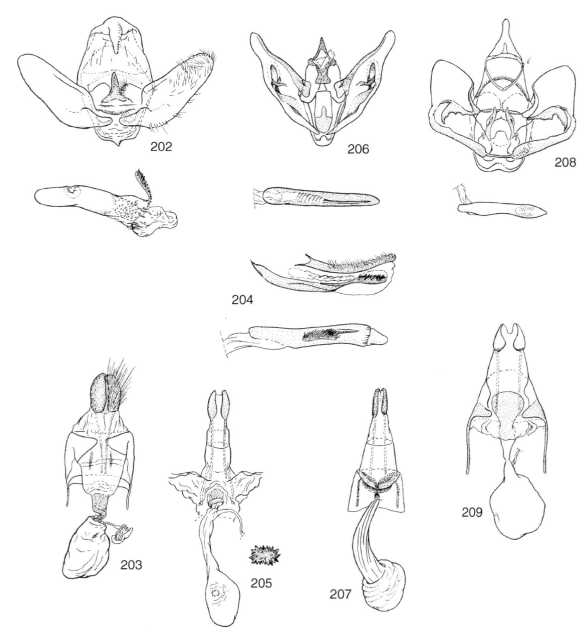

FIGURES 202–209. Genitalia structure of exemplar species of Geometridae, ventral aspect, aedeagus below. **202,** *Tescalsia giuliani-ata* Ferguson ♂. **203,** *Tescalsia giulianiata* ♀ [Powell and Ferguson 1994]. **204,** *Glena nigricaria* (Barnes & McDunnough) ♂, valva inner aspect above, aedeagus below. **205,** *G. nigricaria* ♀, signum enlarged to right [Rindge 1965]. **206,** *Tornos erectarius* Grossbeck ♂. **207,** *Tornos erectarius* ♀ [Rindge 1954]. **208,** *Phaeoura mexicanaria* (Grote) ♂. **209,** *Phaeoura mexicanaria* ♀ [Rindge 1961].

ground and overwinter in a light silk cocoon with incorporated soil particles.

Scoble lists six species in **Hesperumia** Packard. Two of these are endemic to Japan, whereas the remainder are North American, primarily western. **Hesperumia sulphuraria** Packard, known as the "sulphur moth" (FW 1.5–1.8 cm; Plate 28.16), is polyphenic. Most adults have primarily yellow FW, but a small proportion has a large brown patch or markedly dark ante- and postmedial bands. Rare individuals have the FW completely infuscated with brown scaling. There seems to be most variability in southeastern Utah. The moth occurs from Northwest Territories and British Columbia east across Canada to Nova Scotia and south to California, Arizona, New

Mexico, western South Dakota, Missouri, and Virginia. The moths are uncommon in the East but are commonly encountered in the West. The adults are both nocturnal and diurnal and are in flight from April to mid-August, but primarily in June and July. There may be two generations in southern portions of its range. Although the females select and the larvae feed on a wide variety of woody trees and shrubs, including both broad-leaved angiosperms and conifers, McGuffin points out that this species has a decided preference for rosaceous plants. The genera of Rosaceae utilized by this species include *Adenostoma, Amelanchier, Cercocarpus, Holodiscus, Potentilla,* and *Prunus;* whereas other plants fed on with some frequency are *Salix* (Salicaceae), *Arctostaphylos*

(Ericaceae), *Ceanothus* (Rhamnaceae), *Pseudotsuga* (Pinaceae), and *Betula* (Betulaceae). The larvae are variable and may be light green, light gray, or brown. The markings are variable and may be red, brown, or black. There is a pair of distinctive subdorsal bumps on the second abdominal segment. The moth passes the winter in the egg stage. The larva and adult of *H. latipennis* (Hulst), a Pacific Coast moth, were figured and described by Miller and Hammond.

Ectropis Hübner is a worldwide genus with 100 species. There are only a few representative species in Central and South America, whereas only a single Holarctic species occurs in North America. *Ectropis crepuscularia* (Denis & Schiffermüller), the "saddleback looper, small engrailed" (FW 1.4–1.7 cm; Plate 28.17), ranges from southern Alaska east to Newfoundland and south to northern California, Idaho, and, in the East, Florida. The adults are nocturnal and fly from late March to mid-July. The females lay batches of eggs in bark crevices and under moss on tree trunks. The young larvae feed on forest understory and groundcover plants and move later to feed on the foliage of larger trees. More than 30 host genera have been listed by Prentice as hosts, but the species prefers conifers, willows, birches, and alders. Among conifers, western hemlock is preferred, and Douglas-fir, western red-cedar, true firs, and spruces are also among the principal hosts. Severe injury levels have been reached in Alaska and British Columbia. The larvae are light gray to light brown with a geminate gray or black middorsal line. Other markings are light gray or brown. Hammond and Miller add further description, including white and golden with a pair of subdorsal black longitudinal lines extending to the eighth abdominal segment, and a pair of dorsal bumps on the eighth abdominal segment. The pupa overwinters in leaf litter. The larvae have caused noticeable defoliation in parts of British Columbia.

Dasyfidonia avuncularia (Guenée)(FW 1.3–1.6 cm; Plate 28.18) ranges from southern British Columbia and southwestern Alberta south to the southern Sierra Nevada (Tulare County), western Nevada, and western Montana. It is found in open conifer forest and sagebrush habitats. This is a diurnal moth with a single flight from early April through early July. Adults nectar at flowers including those of *Ceanothus* and *Prunus*. Miller and Hammond documented the host as bitter cherry in the Pacific Northwest, and various other *Prunus* species, Rosaceae, are probably the normal hosts since we have found adults in close association with that genus in California and Nevada. Hammond and Miller described the larvae as red brown with shades of pink and a broken yellow spiracular band. This genus has a second species, *D. macdunnoughi* Guedet, which is endemic to the Cascades and Sierra Nevada of California.

Mericisca Hulst comprises eight species endemic to the Southwest and Mexico. Four are southwestern, including *M. gracea* Hulst (FW 1.6–2.2 cm; Plate 28.19) ranges from southern Utah and Colorado south to Durango, Mexico. The adults are double-brooded with adults from February to April and again from June to August. A single individual is known from October. The life history is unreported.

Merisma McDunnough comprises seven species, two of which are found in the Southwest, including *M. spododea* (Hulst)(FW 1.3–1.7 cm; Plate 28.20), which ranges from Utah and northwestern Wyoming south to Coahuila, Chihuahua, and Durango, Mexico. The adults are in flight from June through September. The life history is unreported.

Tracheops bolteri (Hulst)(FW 1.2–1.7 cm; Plate 28.21) is monotypic and ranges from southeastern Arizona and central Colorado south through much of Mexico to Distrito Federal.

There are two flights from April and May and again from late June to early September. The life history is unreported.

Melanolophia Hulst with 104 species is limited to the New World, and the vast majority of species occur in South America. Five occur in North America, and only two of these are western. *Melanolophia imitata* (Walker), "green striped forest looper" (FW 1.7–2.3 cm; Plate 28.22) occurs from southern Alaska, British Columbia, and southwestern Alberta south to southern California, Idaho, and Montana. The adults have a single flight from mid-March to mid-June. The larval hosts are various needled conifers, Pinaceae; the most preferred are Douglas-fir, western hemlock, and western red-cedar, with firs, spruces, larch, and pines fed on with less frequency. Alders and willows are eaten infrequently. Severe defoliation of western hemlock has occurred in coastal portions of British Columbia. Females lay eggs singly on host trunks and branches, but on hatching the larvae move out to outer branches and the crown where they prefer to feed on the previous year's foliage. Larvae are light green to green with a darker green geminate middorsal line. The lateral line is yellow, white, or light green. Full-fed larvae drop to the ground and burrow in the litter where they pupate and overwinter.

Carphoides McDunnough was treated by Rindge who included five species, three of which occur in the West. The other two species occur in Mexico. *Carphoides incopriarius* (Hulst)(FW 1.4–1.6 cm; Plate 28.23, Plate 61.33 larva) occurs from Nevada, Utah, and Colorado south to northern Arizona, New Mexico, and western Texas. It is in flight from late April until early July.

Galenara McDunnough comprises 19 species; seven are western and the remainder is found in Mexico and Guatemala. *Galenara lixarioides* McDunnough (FW 1.6–1.7 cm; Plate 28.24) ranges from southern Utah and central Wyoming south to eastern Arizona, New Mexico, and western Texas. The adults are in flight from April to September. The life history is unreported.

Vinemina McDunnough comprises seven species; three occur in the Southwest, with the remainder found in Mexico and Guatemala. *V. opacaria* (Hulst)(FW 1.3–1.8 cm; Plate 28.25) ranges from Utah and Colorado south to Arizona, New Mexico, and western Texas. There is a single flight from the end of April through August. The life history is unreported.

Astalotesia Ferguson comprises two species: *A. bicurvata* Blanchard & Knudson from western Texas, and *A. hollandi* Rindge (FW 1.5–1.6 cm; Plate 28.26), which was described from Sonora, Mexico, but was collected during April 2004 from Ash Canyon in the Huachuca Mountains of southeastern Arizona and identified by Ferris.

Eufidonia Packard comprises three North American species, two of which occur in the West. *Eufidonia discospilata* (Walker)(FW 1.1–1.3 cm; Plate 28.27) occurs across Canada from coastal British Columbia east to Labrador and Newfoundland. Adults are diurnal and fly from late May to mid-August. The species is most often found in bogs or barrens and the eggs are laid on host stems or under leaves, usually where there is most "woolliness." Hosts are various shrubs and small trees found in bogs or barrens and include several heath and rose family plants as well as viburnum, birch, alder, and willow. The larvae are green with a gray green middorsal line with a yellow spiracular line. Larvae feed over the warmer months, and the species overwinters as pupae.

Biston betularia (Linnaeus), the "pepper-and-salt moth" (FW 2.0–2.8 cm; Plate 28.28), is one of two North American species of *Biston*, a genus that is found in most forested portions of the Old World, excepting Australia and New Zealand. There are 50 described species worldwide with the greatest diversity in Asia and the Indo-Pacific regions. *Biston betularia* was the subject of investigations by Kettlewell on "industrial melanism." The black morph of this normally gray moth had become more abundant in Britain in concert with the blackening of trees and loss of lichens due to coal smoke associated with increasing industry. Kettlewell claimed that it was differential predation by birds that had effected the increased survival of the black forms by their resemblance to the then-blackened tree trunks. In recent years, some have questioned Kettlewell's experimental design, but, despite experimental flaws, the hypothesis of differential predation seems the best explanation. The species ranges across much of North America and Eurasia including the Himalayas. The moth is found east across southern Canada (one record from Northwest Territories) from coastal British Columbia to Newfoundland and Nova Scotia, thence south in the West to central California, Arizona, and Chihuahua, Mexico. In the East, it ranges south in the Appalachians to Tennessee and Georgia. McGuffin gives a full account for the species. The nocturnal adults appear from late March to August, but there may be only a single brood at any single location. The larvae have been reported to feed on a wide variety of broad-leaved plants, and even *Gingko* (Gingkoales) was recorded. Most feeding is found on a limited number of broad-leaved trees. The species occasionally becomes numerous enough to cause local but severe defoliation. The larvae are stout twig mimics and have the head extended dorsally into paired, rounded projections. Like other Geometridae, there are usually five instars, but up to seven have been documented. The body is covered with coarse granules and is variably colored gray, green, or brown without lines; they are sometimes mottled light and dark. The fifth abdominal segment has a pair of conical bumps just above the spiracle, and there is a pair of small bumps on the dorsum of the eighth abdominal segment. The head is brown and deeply bilobed with reddish dorsal projections. Pupation is in an underground cell within which it overwinters.

Cochisea Barnes & McDunnough has nine species, all endemic to the Southwest. *Cochisea recisa* Rindge (FW 1.8–2.1 cm; Plate 28.29) occurs in eastern and southern California and western Nevada. The adults have a single flight from late August to early November. The larval host is single-leafed pinyon, *Pinus monophylla* (Pinaceae). The ultimate-instar larvae are of two forms: predominantly gray with light brown markings, or, more commonly, green with dull white spotting. The green form is highly protected when resting on needles of the host. *Cochisea sinuaria* (Barnes & McDunnough)(FW 1.8–2.8 cm; Plate 28.30) ranges from central California south to southern California west of the Sierra Nevada crest and deserts; it is also found at several isolated locations in Arizona. Adults fly from late August to December. McFarland summarized the life history based on observations in the Santa Monica Mountains. Eggs are laid singly or in flat masses in bark crevices. The larval hosts are several woody shrubs and small trees typical of chaparral, including *Adenostoma* and *Cercocarpus* (Rosaceae), *Arctostaphylos* (Ericaceae), *Quercus* (Fagaceae), and *Malosma laurina* (Anacardiaceae). The ultimate-instar larvae are sticklike and are either rusty brown with scattered gray white dots or with the lateral areas olive green. Pupation is underground near suitable host plants.

Lycia Hübner, with 17 species, extends across the Holarctic region. Three are North American species; two include the West in their range. *Lycia rachelae* (Hulst), the "twilight moth" (male FW 1.2–1.6 cm; Plate 28.31), is primarily western and extends from interior Alaska, southern Yukon, and British Columbia east to Manitoba and south to Utah, Colorado, Pennsylvania, and Massachusetts. Males are fully winged with translucent wings, whereas females are essentially wingless. The female has an exceptionally long ovipositor, and the eggs are laid in bark crevices. First-instar larvae disperse via ballooning in the spring. The adults emerge and are active from March to May, rarely early June. They are most active an hour before sunset and are occasionally attracted to light, even when temperatures are below freezing. A fairly wide variety of broad-leaved deciduous trees and shrubs serves as hosts, but Betulaceae, Rosaceae, and Salicaceae are used most often. The body is gray to gray mauve with a number of longitudinal geminate black lines. There is a white to yellow lateral line that is edged with black ventrally. The legs are yellowish with black markings, and the head is whitish with black or brown markings. The species overwinters as a pupa. *Lycia ursaria* (Walker)(FW 1.8–2.5 cm; Plate 28.32, Plate 61.34 larva) extends the breadth of the continent from Alaska and central British Columbia across Canada to New Brunswick. It extends south to Colorado and to Missouri and New Jersey. Both male and female are fully winged, and the species has a single flight from March to June. Larvae feed on a wide variety of broad-leaved deciduous trees and shrubs. Plants from 11 families were compiled by Robinson, but, as for the previous species, species of Betulaceae, Rosaceae, and Salicaceae are selected most frequently.

Phigalia Duponchel has 14 species that range across the Holarctic region and four of these are North American. Only one, *P. plumogeraria* (Hulst), the "walnut spanworm" (male FW 1.8–2.4 cm, female FW 0.1–0.2 cm; Plate 28.33), is western, whereas a second, *P. titea* (Cramer), is predominantly eastern. *Phigalia plumogeraria* occurs from southern British Columbia south to California, Utah, and Colorado. In California, the adults may be found from December through April and during May in Colorado and Utah. The females are essentially wingless and lay masses of eggs on host twigs and bark. Hosts include oak, maple, willow, walnut, apple, and plum. In California, Essig considered it a minor pest. There, the usual host is coast live oak, *Quercus agrifolia* (Fagaceae), but the larvae rarely cause defoliation of walnut and other orchard trees. Scaccia and Powell reared it several times from larvae found on wild lilac, *Ceanothus* (Rhamnaceae), and *Arbutus* (Ericaceae) in Monterey County, California. Larvae are variable and may be light gray to light brown with several dark lines and patches or, alternatively, black or yellow. In addition, there is a pair of small dorsal pinacula on the first three as well as the eighth abdominal segments. The full-grown larva drops to the ground and enters the soil to a depth of 5–10 cm, where it pupates in a cell without silk.

Hypagyrtis Hübner comprises seven species, five of which are North American. *Hypagyrtis unipunctata* (Haworth), known as the "one-spotted variant" (FW 1.1–1.5 cm; Plate 28.34), occurs from southeastern British Columbia east across southern Canada to Nova Scotia and south to Oregon, Idaho, Colorado, western South Dakota, Mississippi, and Florida. The primarily nocturnal adults have a single flight from late June to late August. The larvae have been found to feed on a wide variety of broad-leaved wood trees and shrubs including primarily Salicaceae, Betulaceae, Rosaceae, Aceraceae, and Cornaceae. The larvae are tan and brown with a diffuse white patch; they overwinter and complete feeding the following spring.

Ematurga amitaria (Guenée), the "cranberry spanworm" (FW 1.3–1.5 cm; Plate 28.35), is found from Alaska east to Nova Scotia and south to Colorado and Pennsylvania. The diurnal adults fly from mid-May to late June. The larval hosts are many species of dicot trees, shrubs, and herbaceous plants; conifers, such as hemlock, are sometimes fed on. In the East, the larvae are serious pests of cranberry plantations. The larvae are dark brown above and light brown below with a black line running through the spiracles.

Paleacrita Riley has three described species noted by Scoble—all from North America. Two species occur in the West. *Paleacrita vernata* (Peck), the "spring cankerworm" (male FW 1.1–1.8 cm; Plate 28.36), occurs from Alberta east to Nova Scotia and Texas. The adults are nocturnal and are active from late February to late May. Females are wingless and lay large batches of eggs, up to 250, under loose bark or in crevices. A wide variety of broad-leaved trees and shrubs are selected as hosts; maple, elm, birch, and plum are among the more frequently selected. The larvae are slender and are light gray to light brown with a variable gray or brown lines and patches. Early-instar larvae may be dispersed to adjacent trees by being ballooned on silken threads. The larvae feed on leaves and young fruit. Full-grown larvae drop from the host and enter the soil where they pupate in a cell without the benefit of silk. Winter is passed by the pupa. *Paleacrita longiciliata* (Hulst)(male FW 1.2–1.7 cm; Plate 28.37) is widespread in California west of the Sierra Nevada crest; it was previously reported as *P. vernata*, but all California reports represent the present species. The adults are found primarily from December through February but may be found as early as November and as late as May. Like those of *P. vernata*, its females are wingless. *Adenostoma fasciculatum* (Rosaceae) has been reported by McFarland as a larval host, and we also reared it from this host in Monterey County. *Paleacrita longiciliata* occurs at many sites where this plant is absent, and therefore, other hosts must be used.

Erannis Hübner comprises 10 species distributed across the Holarctic region, two of which occur in North America. *Erannis vancouverensis* Hulst (male FW 1.7–2.5 cm; Plate 28.38) has been considered a subspecies of *E. tiliaria* (Harris), the "linden looper," but separate status is supported by morphological evidence and DNA barcode data. The western species ranges from British Columbia and Alberta as far south as California, Utah, and New Mexico. McGuffin points out that our moths are extremely close to the Palaearctic *E. defoliaria* (Linnaeus). In most areas there is a single annual flight during October and early November, but adults fly in December in western Oregon and have been found until early January in California. The wingless female lays oval eggs in small clusters of three to 12 on bark or twigs, and these overwinter. Favored hosts in the West are maple, oak, apple, willow, and birch, but many others, including Pinaceae such as western hemlock and western white pine, are utilized on occasion. The full-grown larva is light yellow with brown mottling or reddish brown without mottling with geminate longitudinal black or gray lines, a yellow spiracular line, and yellow or white venter. The larvae feed on leaves and developing fruit; Essig reports that they may cause serious defoliation to several hosts. Pupation is in a loose leaf shelter on the ground.

TRIBE BAPTINI

The tribe Baptini is limited primarily to the Holarctic genus *Lomographa* and several Oriental and Indo-Australian genera that all share features of the male genitalia. *Lomographa* Hübner has 68 species worldwide with four North American species, three of which have at least a portion of their range in the West. *Lomographa semiclarata* Walker, the "wild cherry looper" or "bluish spring moth" (FW 0.9–1.1 cm; Plate 28.39), is a patchy gray-and-white diurnal moth that ranges from coastal British Columbia east across southern Canada to Newfoundland and Nova Scotia. In the West, the moth extends south to Oregon, Idaho, Colorado, and western South Dakota. The strictly diurnal adults fly from mid-May to late July. The larvae feed on several small rosaceous trees or shrubs including *Amelanchier*, *Crataegus*, *Prunus*, and *Sorbus*. When not feeding, the larvae lie flat on top of the host leaf. Along the Rocky Mountain front from Alberta south to Colorado, larvae of this moth may be found feeding together with larvae of *L. vestaliata* (Guenée). Wagner et al. describe the caterpillar as bright green with yellow middorsal spots or a stripe. The yellow is sometimes edged with scarlet. The head is green with a broad red or green line along each side of the ecdysial suture.

TRIBE CABERINI

Sericosema Warren comprises five species, four of which occur in western North America, whereas the fifth occurs in northwestern Mexico. *Sericosema juturnaria* (Guenée)(FW 1.6–2.0 cm; Plate 28.40) is found from Alaska, southern British Columbia, and southwestern Alberta south to California, Utah, and Colorado. Adults have a single flight from May to mid-August and are primarily nocturnal. During the day adults are easily disturbed and fly short distances before alighting with wings held upright over the back in butterfly-like posture. Larvae feed on foliage of *Ceanothus* and *Frangula* (Rhamnaceae). The moth may be most abundant in postfire succession chaparral habitats where the host plants can be dominant. The larval color pattern is gray with alternating dark gray bands. Additionally, there are minute subdorsal tubercles on the second through fifth abdominal segments. The spiracles are yellow. The eggs are laid on the ground, and this is the overwintering stage.

Cabera Treitschke has 30 species worldwide, and most of these are found in Eurasia and Africa, but four species are found in North America, all of which include at least a portion of the West in their ranges. *Cabera erythremaria* Guénee, the "yellow-dusted cream moth" (FW 1.4–1.5 cm; Plate 28.41), ranges from British Columbia east across Canada to Labrador, Newfoundland, and Nova Scotia south to northwestern California, Idaho, Montana, Colorado, and Georgia. Adults are nocturnal but are readily flushed during daylight. The flight period extends from mid-May to mid-August, but records are primarily in June and July; the moth is probably univoltine (single-brooded). Larvae feed on willows and occasionally poplar, Salicaceae. Larvae are green and have a yellow middorsal line that sometimes has flanking red or brown oblique lines. The larvae are green with purple brown or red dash at the anterior edge of the first through eighth abdominal segments.

Apodrepanulatrix Rindge has two North American species, one of which is western. *Apodrepanulatrix litaria* (Hulst)(FW 1.4–1.7 cm; Plate 28.42, Plate 28.43, Plate 29.1) ranges from southern British Columbia and southwestern Alberta south to California, Arizona, and Colorado. Adults are found from April to October, possibly in two broods. The larval hosts are various species of *Ceanothus*, Rhamnaceae. The larvae are light gray with much brown suffusion and a geminate black middorsal line or green with a wide white middorsal band. There also may be various brown, yellow, and white longitudinal lines. The head is light brown with a brown herringbone pattern on each lateral lobe or yellow green.

Drepanulatrix Gumppenberg is a small genus with 12 species endemic or found primarily in western North America. The color pattern of adults of all species is somewhat to extremely variable. Eggs are laid on leaves of wild lilacs, *Ceanothus* (Rhamnaceae), and the larvae feed on *Ceanothus*. Caterpillars of more than a single species may be found feeding on the same plant simultaneously, and adults of several *Drepanulatrix* species may be seen on the same night. *Drepanulatrix falcataria* (Packard)(FW 1.4–1.7 cm; not illustrated) ranges from southern British Columbia and southwestern Alberta south to California, Utah, and Colorado. The adults have a slightly falcate FW apex and are extremely variable, varying from yellow to tan to red brown with or without dark bands or dusting. Adults fly from December to May. McFarland reported an abundance peak during January to February in the Santa Monica Mountains. Larval hosts are several species of *Ceanothus*. In southern California, the larvae are yellow green with various white or green longitudinal lines; whereas in the Pacific Northwest, Miller and Hammond describe the larvae as having a patchwork of white, silver, gray, tan, and black with a very thin yellow spiracular line. Along the Pacific Coast and coast ranges, *D. monicaria* (Guenée)(FW 1.4–1.6 cm; Plate 29.2, Plate 29.3) is most abundant. Its larval hosts are wild lilacs, *Ceanothus* (Rhamnaceaae), whereas *D. carnearia* (Hulst)(FW 1.2–1.4 cm; Plate 29.4, Plate 29.5) seems most abundant in the Sierra Nevada. *Drepanulatrix unicalcaria* (Guenée)(Plate 61.35 larvae) is a related Pacific Coast species whose larvae also feed on *Ceanothus*. *Drepanulatrix bifilata* (Hulst) is extremely abundant in the intermountain West.

Eudrepanulatrix rectifascia Hulst (FW 1.3–1.5 cm; Plate 29.6) is the sole representative of its genus and occurs from southern British Columbia to California, Arizona, and Colorado. There are two flights of adults between late February and mid-August. A late-emerging adult was found in mid-October in San Diego County, California. The hosts are several species of *Ceanothus* (Rhamnaceae). Larvae are light gray with a geminate brown middorsal line; Miller and Hammond report that in the Pacific Northwest larvae are green with faint white lines. The other body areas have brown or gray longitudinal lines.

Ixala Hulst is limited to four species, all of which are endemic to western North America. *Ixala desperaria* (Hulst)(FW 1.4–1.6 cm; Plate 29.7) is the most wide ranging, being distributed from British Columbia south to Arizona, Colorado, and western South Dakota. The adults are easy to identify with a single, round white spot in the middle of the HW. There is a single flight from late May to late August. Larvae feed on *Ceanothus fendleri*, Rhamnaceae, in southeastern Arizona. The larvae are polymorphic, appearing in both brown and green forms. The brown form has a white head with black spotting, and the gray body has dotted darker lines. The green form has the dorsum faintly lined in white and has a yellow spiracular line. Pupation is in the ground over winter.

Chloraspilates Packard comprises four species, three of which are North American, whereas the fourth is endemic to Mexico. *Chloraspilates bicoloraria* Packard (FW 0.9–1.1 cm; Plate 29.8) ranges from southeastern California east through southern Arizona to southern Texas. Adults have been found during April to November, possibly in three flights. The life history is unreported.

There are only two *Stergamataea* Hulst; both are western. One of these, *S. inornata* Hulst (FW 1.4–1.8 cm; Plate 29.9), is

found in arid juniper woodland from southeastern Utah and western Colorado south to southern Arizona and western Texas. The adults are in flight from late April to mid-August, primarily in late May and June. McFarland found that the larvae feed on *Brickellia californica* (Asteraceae) in southeastern Arizona.

TRIBE ANGERONINI

Euchlaena Hübner comprises 19 species: three are Neotropical, whereas the remainder are North American. Six species occur in the West. Ferris points out that each species is polyphenic to at least some degree; some are extremely so, and their identification may be problematic. *Euchlaena tigrinaria* (Guenée), the "mottled euchlaena" (FW 1.8–2.1 cm; Plate 29.10), ranges from southern British Columbia east to Maine and south in the West to Oregon, Utah, and Colorado. In the West, there is a single flight from late June to early August. The larval hosts are a variety of broad-leaved woody trees and shrubs that include oak (Fagaceae), hazel and paper birch (Betulaceae), serviceberry (Rosaceae), and quaking aspen (Salicaceae). Miller and Hammond describe the larvae as silver and gray with dorsal red brown patches. There are dorsal pinacula on the first through fifth abdominal segments. As summarized by Robinson et al., a large proportion of western *Euchlaena* species have had their hosts and larval descriptions reported, including *E. mollisaria* (Hulst), *E. johnsonaria* (Fitch), and *E. madusaria* (Walker)

Xanthotype Warren comprises five species. Two are limited to the East, and two have transcontinental distributions. The fifth species, *Xanthotype barnesi* Swett, is limited to the West. The species are most reliably identified by dissection and examination of the genitalia. *Xanthotype sospeta* (Drury), the "crocus geometer" (FW 1.7–2.5 cm; Plate 29.11), ranges from British Columbia east across Canada to Nova Scotia and south in the Rockies to Colorado. Although there is speculation that there are two flights in the East, only a single flight occurs in the West, from July to September. The adults are primarily diurnal in more northern areas but are more prevalent at light in the South and West. Rindge discussed how the problems of distinguishing species in this genus have led to a confusing collection of reports on the immature stages. Larvae of either this species or *X. urticaria* Swett feed on various woody and herbaceous angiosperms in about 15 plant families. There are seven larval instars, and the penultimate-instar larvae are light brown with the dorsum and sides gray. The pattern of the final instar is suffused with green and is subdued. The species pass the winter as a pupa.

Aspitates Treitschke, with 23 species, has a primarily Holarctic distribution with a few species in South Africa and a single species in South America. In North America, five species occur in the Far North. Three of these range into the western Arctic. *Aspitates orciferarius* (Walker)(FW 1.3–1.5 cm; Plate 29.12 male, Plate 29.13 female) occurs from Alaska east to the western shore of Hudson Bay and Baffin Island. Adults have a single flight from mid-June to late July. The life history is unreported.

TRIBE ANZELININI

There are more than 250 species of *Pero* Herrich-Schäffer, most of which occur in the New World, primarily in South America. An extensive revision was published by Poole. In North America, there are 21 species, 16 of which occur in the West. All of our species are strictly nocturnal and are attracted to light. **Pero behrensaria** (Packard)(FW 1.6–1.8 cm; Plate 29.15, Plate 29.17) occurs from British Columbia and western Alberta south to California, Arizona, and Colorado. There is a single spring to early summer flight from mid-May to early August; in California, occasional individuals may be collected during March. Larval hosts are Pinaceae, primarily Douglas-fir (*Pseudotsuga menziesii*). The larvae are brown with light gray middorsal and subdorsal lines. Winter is passed in the pupal stage. **Pero meskaria** (Packard)(FW 1.3–1.6 cm; Plate 29.14) ranges from California, Utah, and western Colorado south to Arizona, New Mexico, and Texas. Taking the range as a whole, adults are in flight from late March to mid-November, probably as several generations, but in southern Utah and western Colorado, only from late April to July in two flights. The hosts and life history are unreported. **Pero mizon** Rindge (FW 2.1–2.5 cm; Plate 28.16) ranges from southern British Columbia south to central California. There is a single flight from March to November. Eggs are laid in rows on the leaves of host plants, which include Grossulariaceae, Pinaceae, and Rosaceae. Several herbaceous plants have been mentioned, but they may have been those accepted by captive females. The sixth-instar larvae are light gray and mottled with dark gray. There are black bars on the dorsum of the seventh and eighth abdominal segments. The head is light gray with a brown frame on the frons, and the lobes have a brown herringbone pattern. The larvae overwinter in the fourth instar and complete feeding in the late winter or spring. The larva and adult of **P. occidentalis** (Hulst)(Plate 29.18), a widespread intermountain species, were illustrated by Miller and Hammond.

TRIBE NACOPHORINI

Phaeoura Hulst, recently treated as *Nacophora* in North America, has 15 species limited to the New World. There are nine North American species, eight of which are western. **Phaeoura mexicanaria** (Grote)(FW 2.2–3.0 cm; Plate 29.20, Figs. 208, 209 genitalia) has perhaps the greatest range of the western species. It is found from the southern interior of British Columbia, Colorado, and western South Dakota south to California, Arizona, and New Mexico. The adults are nocturnal and have a single flight from mid-May to early August. The larval host is ponderosa pine. The larvae may cause extensive defoliation in rare instances; on one occasion, Furniss and Carolin report that a large area of pines in southeastern Montana were defoliated by this moth and subsequently killed by *Ips* beetles. Females lay clusters of eggs on the needles, twigs, and branches. The larvae have a bifid head, and the body is patterned brown with seta-bearing tubercles. The full-grown larvae leave the host tree and burrow under the duff to mineral soil, where they pupate in a lightly silk-lined chamber and therein overwinter. **Phaeoura cristifera** Hulst (FW 1.7–2.9 cm; Plate 29.21, Plate 61.36 larva) occurs from Colorado south to Arizona and western Texas. The adults are bivoltine and fly from April until August. Comstock reported their larval host is willow, *Salix* (Salicaceae), and McFarland notes that the larvae readily feed on *Quercus* (Fagaceae) and *Arctostaphylos* (Ericaceae) in southern Arizona; all of these hosts were offered to larvae and may not be natural hosts. The last instar is mottled gray, brown, and black with many raised seta-bearing chalazae.

There are four species of **Holochroa** Hulst. Three are found in western Mexico and the fourth is western. **Holochroa dissociarius** (Hulst)(FW 1.5–2.1 cm; Plate 29.22) occurs from Ari-zona, western Colorado, New Mexico, and Texas south to Durango, Mexico. The adults are in flight from March to October and are double-brooded. The life history is unreported.

Aethaloida packardaria (Hulst)(FW 1.3–1.5 cm; Plate 29.23, Plate 62.1 larva), the only species in its genus, ranges along from the Pacific Coast from southwestern Oregon to California. The adults are double-brooded, according to Miller and Hammond, with flights from March to June and August and September. In the Santa Monica Mountains, McFarland reported it from March to early December with abundance peaks during May, August, and November, indicating three generations. The larval hosts are a variety broad-leaved wood plants including *Adenostoma* (Rosaceae), *Arctostaphylos* (Ericaceae), *Ceanothus* (Rhamnaceae), *Myrica* (Myricaceae), and *Quercus* (Fagaceae). The larvae are good twig mimics and are plain gray with dorsal tubercles on the second, fourth, and eighth abdominal segments. There are ventral tubercles on the third abdominal segment, and the third pair of true legs is much enlarged.

Ceratonyx Guenée is limited to the New World with most species in the Neotropics; Fred Rindge's revisionary treatment includes 12 species, three of which are found in North America. Two species are found in the West. **Ceratonyx permagnaria** (Grossbeck)(FW 1.7–2.4 cm; Plate 29.19) is the more widespread and includes southeastern Arizona and southwestern New Mexico in its range. Adults have been collected between February and October, and there are likely at least two flights. Very little is known of the life history. The other western species, **C. arizonensis** (Capps), is known only from Cochise and Santa Cruz counties in southeastern Arizona. Its larvae feed on *Vigueira dentata* (Asteraceae); the early stages were described by J. Franclemont.

Gabriola dyari Taylor (FW 1.3–1.7 cm; Plate 29.24) is one of five western species in the genus; the remaining species are limited to western Mexico. The present species is found from southern Alaska, British Columbia, and western Montana south to central California (San Luis Obispo County). The adults are nocturnal and fly from April to late October in a single flight. The larvae feed on a number of conifers with the vast majority of records from western hemlock and Douglas-fir. The larvae are mottled brown or, rarely, mottled orange. Both forms have transverse white patches. At rest, the larvae resemble bird droppings. Winter is passed in the egg stage.

Yermoia perplexata McDunnough (FW 1.4–1.5 cm; Plate 29.26), one of two species in this small genus, is found from the east side of the Sierra Nevada south to San Diego County. Additionally, there is a seemingly disjunct population in western Colorado. Further sampling may well fill the apparent distributional gap. According to records compiled by Richers, the adults fly from early January to mid-May. McFarland has found that the larvae feed on *Gilia sinuata* (Polemoniaceae) in the Huachuca Mountains of southeastern Arizona.

Animomyia Dyar comprises nine species; seven are known from western North America, and two are apparently endemic to Sonora, Mexico. The known females are flightless. **Animomyia smithii** (Pearsall)(male FW 1.1–1.8 cm; Plate 29.27 male, Plate 29.28 female) ranges from eastern California east to eastern Colorado and south to Arizona. The adults have been found in all months in California and Arizona but seem to be single-brooded in Utah and Colorado during late June and July. The reported larval host is *Ambrosia dumosa* (Asteraceae), whereas larvae were reported by extension agents defoliating extensive stands of sand sagebrush, *Artemisia filifolia* (Asteraceae), in eastern Colorado. Female is flightless.

TRIBE CAMPAEINI

Campaea Lamarck has five species in the Holarctic, four of which are entirely Palaearctic; only one species is North American. *Campaea perlata* (Guenée), known as the "fringed looper" or "pale beauty" (FW 1.7–2.4 cm; Plate 29.29, Plate 62.2 larva), ranges from Alaska across Canada to Labrador, Newfoundland, and Nova Scotia. Its distribution extends south in boreal habitats to central California, Arizona, Colorado, and, in the East, to North Carolina. In the south, the adults are nocturnal and have a single flight from May to August, while in the Yukon they are diurnal and have only a single flight from mid-July to early August. At Inverness in lowland coastal California, we found that *C. perlata* is bivoltine with flights from late April through July and again during September and October. The species is polyphagous with larvae having been reported from more than 65 broad-leaved woody angiosperms and conifers, although Prentice reports that Salicaceae and Betulaceae comprise the majority of larval collections. The females lay flat masses of 75 to 100 eggs. The larvae are light brown with various gray brown lines. The venter is almost white with a fringe of long hairs along each side. McFarland observed that the larvae rested tightly appressed to a stem, much in the same manner as *Catocala* larvae.

TRIBE ENNOMINI

Ennomos Treitschke comprises 17 species that occur primarily in the Palaearctic region. Only a single species occurs in North America. *Ennomos magnaria* Guenée (FW 1.8–2.5 cm; Plate 29.30), the "maple spanworm," is distributed from British Columbia east across southern Canada to Newfoundland and Nova Scotia, thence south to northern California, Utah, Colorado, Arkansas, and Florida. The moth is bivoltine with flights in late June and July then again in September and early October. The eggs, which overwinter, are laid in masses on the undersides of host branches. Larval hosts include a variety of broad-leaved angiosperms in several families. Severe defoliation of elms, hickories, and oaks has been reported in the eastern United States. The larvae stand out rigidly from the twigs on which they rest and are thus excellent twig mimics and may be either olive green or brown. There are banded swellings around the second, third, and fifth abdominal segments. Pupation occurs in a two-layered cocoon of tough white silk.

TRIBE EPIRRANTHINI

Spodolepis substriataria Hulst (FW 1.9–2.4 cm; Plate 29.31, Plate 62.3 larva) is the only member of its genus. The species ranges from Alaska and Yukon Territory discontinuously across Canada to Nova Scotia and south to California, Utah, Colorado, and, in the East, to New Jersey. The adults are diurnal in the north but are nocturnal in the south. The flight period extends from March to early July. Larval hosts include Salicaceae (*Populus* and *Salix*) and Pinaceae (*Larix*, *Picea*, *Pinus*, and *Pseudotsuga*). McGuffin states that larvae are difficult to rear, even on *Salix*, and he believes that the species could not pass through all instars on any conifer. The larvae are twig mimics and, when disturbed, may jump or snap in the manner of Lithinini species. This is a method of predator avoidance. The full-grown larvae are light gray to light brown with darker brown suffusion; there is also a dark brown middorsal line and various other geminate longitudinal brown lines.

TRIBE LITHININI

Philedia punctomacularia (Hulst)(FW 1.6–1.8 cm; Plate 29.32), representing a monotypic genus, is found on the Pacific Coast from Vancouver Island, British Columbia, south along the Pacific Coast to central coastal California. The adult flight is from mid-October to November. A specimen collected in early June is apparently aseasonal. Larval hosts are ferns, Pteridaceae, including bracken, *Pteridium* (Dennstaedtiaceae), and brake fern, *Pteris* (Pteridaceae). McFarland reared this moth and found the female laid about 30 eggs, which overwintered. The larvae are translucent green with closely spaced red brown longitudinal lines.

Thallophaga Hulst includes three species, all of which are restricted to the West. *Thallophaga hyperborea* (Hulst)(FW 1.8–1.9 cm; Plate 29.33) is found from Alaska and British Columbia south to California. The moth has spring (March to May) and summer (July to August) flights with the respective adults of different forms. Records for late December to early February compiled by Richers suggest a third flight. The preferred larval host is western hemlock, but other Pinaceae and, occasionally alder, are utilized. The larvae are light brown with brown middorsal, addorsal, and subventral longitudinal lines. The pupae overwinter. *Thallophaga taylorata* (Hulst)(FW 1.5–1.8 cm; Plate 29.34, Plate 62.4) is found in wet forests along the Pacific Coast from British Columbia and Idaho south to southern California and western Nevada. The adult flight period is from December to early September, and two generations are likely. In the Santa Monica Mountains, California, McFarland observed two abundance peaks during January and May. The larval host is sword fern, *Polystichum munitum* (Dryopteridaceae). The larvae are orange, brown, and tan with faint longitudinal lines; the midabdominal intersegmental areas are orange.

TRIBE ANAGOGINI

Slossonia rubrotincta Hulst (FW 1.2–1.3 cm; Plate 29.35), representing a monotypic genus, is found in California west of the Sierra Nevada crest from Napa and Placer counties south to San Diego County. Adults always rest with their wings over the back in butterfly-like fashion. There is a single flight April to July. McFarland reports the host to be scrub oak, *Quercus berberidifolia* (Fagaceae).

Selenia Hübner comprises 25 species, and the genus is represented across the Holarctic region, with many that occur in Mexico. Two species were described from eastern North America, and one of these ranges across North America. *Selenia alciphearia* Walker (FW 2.0–2.4 cm; Plate 29.36) ranges from Yukon and British Columbia east across southern Canada to Labrador, thence south to northwestern California and Idaho. In British Columbia, two flights are reported—April to May and June to July—with two corresponding forms of adults. In California, there is a report from August, but the flight period there may be incompletely known. Larval hosts in the laboratory are alder, birch, maple, and willow, but the hosts in nature are unreported. The larvae are bizarrely patterned twig mimics; their pattern is broken into four abruptly distinct areas: the head and thorax are gray streaked with white, then the first four abdominal segments are golden yellow, the fifth and sixth abdominal segments are gray and white streaked, and, finally, the sixth through tenth abdominal segments are brown. In addition, there are dorsal humps on the fourth and fifth abdominal segments. The winter is passed in the larval stage. Pupation is in a rolled leaf.

Metanema Guenée has nine species, which are limited to the New World including two transcontinental species in North America. One of these is *M. inatomaria* Guenée, the "pale metanema" (FW 1.3–1.6 cm; Plate 29.37, Plate 62.5 larva), which occurs from Yukon and central British Columbia east across Canada to Newfoundland and Nova Scotia and south to eastern California, Arizona, Colorado, and western South Dakota. In British Columbia, the adults have a single flight from mid-June to mid-July, while in Colorado they fly as early as late May. According to Prentice, the larval hosts are *Populus* and *Salix*, Salicaceae. The eggs are laid singly or in small clusters. The last-instar larva is light gray and covered with seta-bearing tubercles. There is a wavy gray middorsal line. The head is gray with a black herringbone pattern on the lateral lobes and has a black bar on the clypeus. The pupa overwinters. *Metanema determinata* Walker has a similar distribution and life history.

Metarrhanthis Warren comprises 13 species, all of which were described from eastern North America. *Metarrhanthis duaria* (Guenée), the "ruddy metarrhanthis" (FW 1.5–1.8 cm; Plate 29.38), has a transcontinental distribution from southern British Columbia east across southern Canada to southern Newfoundland and Nova Scotia south to Washington, Idaho, Colorado, and, in the East, to Georgia. Adults are variable and several forms have been described in error as separate species. In British Columbia, there is a single flight from April to June. The adults are diurnal, and, according to McGuffin, the female scatters eggs from the air over appropriate habitat. Larval hosts are primarily shrubby plants in the Rosaceae, but several other woody plants are used on occasion. Larvae are medium to dark brown with variable markings, but often with a diamond-shaped mark on the dorsum of each anterior abdominal segment. The pupae overwinter.

Probole Herrich-Schäffer includes four species, three described from the eastern United States and one from Chile. *Probole amicaria* Herrich-Schäffer, the "friendly probole" (FW 1.2–1.6 cm; Plate 29.39), has a transcontinental distribution from Alaska, Northwest Territories, and British Columbia east across Canada to Newfoundland and Nova Scotia and south to northern California, Idaho, Montana, and western South Dakota, and, in the East, to Florida. The adults are variably colored and are both diurnal and nocturnal; they fly from June to August. Prentice reports that larvae were found primarily on birch, alder, and hop hornbeam, *Ostrya virginiana* (Betulaceae), willow (Salicaceae), true fir (Pinaceae), and dogwood (Cornaceae), but a number of other woody broad-leaved angiosperms and conifers are also mentioned. The larvae are petiole mimics or twig mimics and are green, brown, or reddish. A slight transverse hump on the dorsum of the fifth abdominal segment may be white with a pair of brown spots. The head is variably marked in reddish.

TRIBE OURAPTERYGINI

Philtraea Hulst comprises eight New World species. Six are western, whereas one is found in the East and is widespread in northern Mexico. A revision replete with distributional and biological information was provided by Buckett. *Philtraea latifoliae* Buckett is known only from Solano and Yolo counties, California, in the coast ranges just east of San Francisco Bay. Only a single flight of nocturnal adults is known—two individuals were collected during late June and mid-July. The only reported larval host is Oregon ash, *Fraxinus latifolia* (Oleaceae). McFarland found the larvae common in early May, and, when

disturbed, the larvae readily dropped from the host on tough silken threads. The larva, illustrated by Buckett and described by McFarland, is glossy and light gray dorsally. Ventrally, the larva is a paler whitish gray. The body has numerous black setigerous tubercles and has fine narrow transverse lines on the dorsum and lateral areas. There is a subdorsal series of yellow dots and a subspiracular series of yellow orange dots. The posterior portion of the larva, abdominal; segments 4 to 6 are thicker than the more anterior portion. The head is pale gray speckled with black setigerous tubercles. The pupa is quite colorful, being yellow or yellow gray and black with black speckling and a black crescent over each compound eye. *Philtraea utahensis* Buckett (FW 1.4–1.6 cm; Plate 29.40) ranges from southeastern Nevada east across southern Utah to western Colorado. The adults are found in pinyon-juniper woodland habitat. Adults are nocturnal and are in flight from early June to late July. The host plants and life history are unreported.

Eriplatymetra Grote comprises four species, three western and one endemic to Mexico. *Eriplatymetra coloradaria* (Grote & Robinson)(FW 1.7–1.9 cm; Plate 29.41) is distributed across Utah and Colorado south to southern Arizona. There is a single flight of adults from late May to early August, primarily mid-June to mid-July. McFarland reports that larvae feed on *Calystegia* (Convolvulaceae) and *Mimosa* (Mimosaceae) in the Huachuca Mountains of southeastern Arizona. Other western species are *E. lentifluata* Barnes & McDunnough and *E. grotearia* (Packard).

Melemaea Hulst comprises four species, three of which are western and one described from Mexico. These moths are found above 7,000' in coniferous forests mostly during late summer and early fall. *Melemaea magdalena* Hulst (FW 1.4–1.6 cm; Plate 30.1) is found along the Colorado Front Range in the spruce-fir zone. The adults are attracted to light during late August and early September. There is no available life history information. There is a similar undescribed sympatric taxon that flies during October that is paler and has a different maculation pattern. *Melemaea virgata* Taylor (FW 1.3–1.6 cm; Plate 30.2, Plate 62.6 larva) of southeastern Arizona is bivoltine with flights during late June and early July and September to October. Adults of the summer flight are nocturnal, while those of the fall flight are diurnal. Ferguson found that the larval host is pine, *Pinus* (Pinaceae). An image of the larva is shown.

Lychnosea Grote comprises four species, two of which are endemic to western North America, whereas the other two are found in adjacent Mexico. *Lychnosea helveolaria* Hulst (FW 1.4–1.8 cm; Plate 30.3) ranges from Colorado south to southern Arizona and New Mexico. The univoltine adults are in flight from late July through August. The life history is unreported.

Neoterpes Hulst comprises four species, all of which are endemic to western North America. *Neoterpes trianguliferata* (Packard)(FW 1.4–2.0 cm; Plate 30.4) is found from southern British Columbia east to southwestern Alberta, thence south to California, Utah, and Colorado. Adults fly from April to June in British Columbia and from late March through July in California. Eggs are laid in small groups on the undersides of currant or gooseberry leaves, *Ribes* species (Grossulariaceae). The twig-mimic larvae feed during summer and are light gray or light brown with many seta-bearing tubercles on the abdominal segments. There are several longitudinal geminate brown lines. The head is pale brown with a darker brown herringbone pattern on the lateral lobes. Pupation is in a cluster of host leaves within which the winter is passed. *Neoterpes edwardsata* (Packard)(FW 1.7–2.0 cm; Plate 30.5, Plate 30.6) ranges from central coastal California and is in flight from February to December. In the Santa Monica Mountains, McFarland notes three flight periods, possibly corresponding to emergences during May, August, and October. We have found adults at light during every month except August. On San Miguel Island, we found that the larvae feed on California poppy, *Eschscholzia californica* (Papaveraceae), and we believe that this is the probable host in most native California habitats. McFarland reports that in suburban situations, the larvae feed on bush poppy, *Dendromecon rigida,* and Matilijah poppy, *Romneya coulteri* (both Papaveraceae). The life history was described by J. A. Comstock who pointed out that the last instars were extremely variable in color and pattern and are similar to those of *N. ephelidaria* Hulst. Caterpillars varied from light green with faint longitudinal lines to yellow tan with light brown, light tan, and gray brown stripes and bands. The spiracles are black with narrow yellow lines, and the legs are almost white basally. Pupation is on the ground or slightly under in silk webbing that has debris and soil particles intermixed. The pupa is covered with a whitish powdery bloom.

Scoble reports that *Caripeta* Hulst comprises 15 species, three of which are Mexican endemics, whereas the remainder is North American. Seven *Caripeta* species are known from the West. The most widespread and most frequently encountered is *C. aequaliaria* Grote, the "Indian blanket moth" (FW 1.7–2.0 cm; Plate 30.7, Plate 62.7 larva); it ranges from southern British Columbia and southwestern Alberta south to California, Arizona, New Mexico, and western Texas. The moth is strictly nocturnal and flies from late April through August. The larvae feed on Pinaceae, primarily on Douglas-fir, several pines, and, more infrequently, on western hemlock. McFarland observed that the last-instar larvae feed on older needles and resemble old twigs, being drab-mottled gray and brown. *Caripeta ocellaria* (Grossbeck)(FW 2.5 cm; Plate 30.8) occurs in the mountains of southeastern Arizona and southwestern New Mexico and is common from late July to September. The larval host plant reported by Walsh is Arizona white oak, *Quercus arizonica* (Fagaceae). Its life history is unreported. *Caripeta pulcherrima* Guedet (FW 1.6–1.7 cm; Plate 30.9) is known from the mountains of southeastern Arizona and flies during July and August. Its life history is unreported.

Snowia montanaria Neumoegen (FW 1.6–1.8 cm; Plate 30.10), representing a monotypic genus, ranges from eastern Oregon east to southern Wyoming and Colorado south to Arizona and New Mexico. Its usual habitats are montane coniferous or mixed oak-coniferous forests from about 6,000′ to almost 10,000′ elevation. This attractive moth is univoltine and flies from early June to early August. Its life history is unreported.

Nemeris Rindge comprises four species, three of which are western. The fourth occurs in Mexico. *Nemeris speciosa* (Hulst)(FW 1.4–2.0 cm; Plate 30.11) ranges from Utah, southern Wyoming, and Colorado south to Arizona, New Mexico, and western Texas. In more northern or high-elevation sites, the species has a single flight from late June to early August, but in more southern locations has two flights from March to October. The life history is unreported.

Meris Hulst was revised by Rindge who included five species, all endemic to the West. *Meris alticola* Hulst (FW 1.6–2.2 cm; Plate 30.12) ranges from Utah and central Wyoming south to southern Nevada, Arizona, and New Mexico, usually in forested montane habitats. The adults have a single flight from late June to mid-August. The only reported larval host is *Penstemon virgatus,* Scrophulariaceae. Poole de-

scribed the ultimate-instar larva as having alternating black-and-white bands for its entire length. The ultimate-instar larvae overwinter and complete their feeding by grazing on the rosettes of the host; this requires moving from plant to plant to find sufficient food. This behavior is thought to be unique among the Geometridae.

Destutia Grossbeck includes five species, four of which are found in the West, whereas the remaining species is known from Costa Rica. *Destutia excelsa* Strecker (FW 1.2–1.6 cm; Plate 30.13, Plate 62.8) ranges from Utah and Colorado south to southern Arizona, New Mexico, and western Texas. The adults are nocturnal and bivoltine, flying from mid-May through August. McFarland states that the larval host, at least in southeastern Arizona, is oak, *Quercus* (Fagaceae).

Besma Capps comprises seven species–two western, two eastern, and three limited to the Neotropics. *Besma quercivoraria* (Guenée), the "oak besma" (FW 1.5–1.8 cm; Plate 30.14), occurs across Canada from British Columbia across Canada to Newfoundland and south to Oregon, Idaho, Texas, and Florida. In British Columbia, there is a single flight from May to July. McFarland has reared the species in the Huachuca Mountains of southeastern Arizona, where the species has two flights, April and July. According to Ferris, the April adults are larger, darker, and more strongly marked than those that fly in July; without rearing one brood from the other, one might suspect that there were two species. Eggs are laid singly or in small groups on the undersides of host leaves. Prentice reported that the larvae feed on birch, maple, willows, oaks, or alder. The larva at rest is a twig mimic. The last-instar larvae are quite variable, being brown or yellow with various markings; dorsal ridges may be present as well as lateral swellings of the mesothorax and eighth abdominal segment. Winter is passed by the pupal stage.

Lambdina Capps has 12 North American and Mexican species, five of which are found in the West. The species limits of taxa in southern Arizona and New Mexico are unclear, and Ferris points out that there is a need for serious study and revision. *Lambdina fiscellaria* (Guenée), the "hemlock looper" or "western oak looper" (FW 1.6–1.8 cm; Plate 30.15), ranges from British Columbia east across Canada to Newfoundland and Nova Scotia, thence south to California, Idaho, Montana, and northern Colorado, and, in the East, to Wisconsin and Pennsylvania. The adults are nocturnal but occasionally diurnal; they have a single flight during September and October. This moth is one of the most important defoliators of Canadian forests. The larval hosts are a wide variety of conifers and broad-leaved trees. In British Columbia, Prentice found that several conifers are the primary hosts; these include western hemlock, Douglas-fir, spruces, western red-cedar, and various true firs. By contrast, in Oregon, Miller reports that Oregon white oak, *Quercus garryana,* Fagaceae, is a principal host. Eggs are laid on mosses and lichens on understory vegetation and tree limbs. Eggs overwinter, and the first-instar larvae feed on understory vegetation before moving up into the canopy. The larvae are variable–either greenish with a brown geminate middorsal line with the remaining lines black or white, gray, and light brown with fine longitudinal lines of various shades. The setal bases are black. The head is light gray with a black herringbone pattern on the lateral lobes and black epicranial lines. Pupation is on the tree trunk, branches, or in leaf litter below the tree.

Nepytia Hulst has 13 species, 12 described from North America, whereas the remaining species was described from Guatemala. Nine are found in the West. *Nepytia umbrosaria*

Packard (FW 1.6–2.0 cm; Plate 30.16, Plate 62.9 larva) ranges from British Columbia south to central coastal California. Adults have a single flight from June through October, probably as two generations. Larvae feed on several Pinaceae including Monterey pine, Douglas-fir, and western hemlock, but no extensive injury has been reported. The larvae are twig mimics when at rest; they are striped—cream on the dorsum and reddish brown laterally. Pupation is among the host needles with a few strands of silk. *Nepytia phantasmaria* (Strecker), the "phantom hemlock looper" (FW 1.6–1.9 cm; Plate 30.17), occurs from southwestern British Columbia south to California. The adults have a single flight from September to mid-October. Larvae feed primarily on western hemlock and Douglas-fir, Pinaceae, according to Prentice, but several other conifers have also been reported. Furniss and Carolin state that the larvae have had local outbreaks in British Columbia. The final-instar larvae are green or yellow brown with a broad white or yellow band margined in purplish brown and a narrow white spiracular band. At rest, the larvae stretch out along the axis of a needle or twig; if suddenly disturbed they drop on silk threads. Pupation is amidst the foliage among a few silken strands. *Nepytia swetti* Barnes & Benjamin is common in western Colorado south to southeastern Arizona and southwestern New Mexico. Ferris points out the species must have staggered emergence, since it can be taken from early spring to late fall. Furniss and Carolin report *N. janetae* Rindge is a serious forest pest in some southern Arizona mountains (e.g., Mt. Graham).

Sicya Guenée, with 22 described species, is limited to the New World. Although six species are known from the West, the majority are found in Mexico. *Sicya macularia* Harris, the "sharp-lined yellow" (FW 1.5–1.9 cm; Plate 30.18), ranges from Yukon and British Columbia east across Canada to Newfoundland and Nova Scotia and south to California, Utah, and Colorado, and, in the East, to New Jersey. The adults are nocturnal and have a single flight from June to early September— as early as May in southern California. Eggs are laid in small groups on host twigs, and overwinter. The larval hosts are a fairly wide variety of broad-leaved shrubs and trees. The larvae are twig mimics and are variably colored, usually brown and sometimes with a yellow spiracular stripe. There are prominent middorsal pronged hornlike projections from the third and fifth abdominal segments that are about equal to the body width, as well as a much shorter dorsal hornlike projection on the eighth abdominal segment. The head may be light green or brown. The pupa has a brilliant silvery pearl luster. Pupation is in a loose cocoon among the host foliage. *Sicya morsicaria* (Hulst)(FW 1.4–1.6 cm; Plate 30.19) ranges from eastern Oregon and California east to Utah and south to southern California, Arizona, New Mexico, and western Texas. Adults have two flights from mid-May through August in California. McFarland reported the hosts as mistletoe *Phoradendron* (Viscaceae), and oak, *Quercus* (Fagaceae). We found that the larvae are mistletoe green with silver spots.

Eucaterva Grote comprises two species, both endemic to the West. *Eucaterva variaria* Grote (FW 1.5–1.6 cm; Plate 30.20) ranges from southern California east across southern Arizona and New Mexico to western Texas. The adults, as the specific epithet implies, are extremely variable and range in color from mostly black with a white FW costa to primarily white with a sprinkling of black scales. The adults appear to have three flights during March, June, and late July to early August. The larval host is desert willow, *Chilopsis* (Bignoniaceae). McFarland reports that larvae make open-mesh cocoons on the host plant.

Plataea Herrich-Schäffer is a small genus with 10 species, seven of which are western, whereas the remainder is limited to Mexico. The genus was revised by Rindge, who reported that *P. trilinearia* (Packard)(FW 1.5–2.0 cm; Plate 30.21) is by far the most widespread species and ranges from south-central British Columbia east to southern Saskatchewan and south through the arid intermountain western United States and high plains to Mexico. The adults are primarily nocturnal but will readily take flight during the day. In Canada, the single flight extends from May to July, whereas in California, Richers reports them as early as March and April. In Colorado, there are records from late April to mid-July. McFarland found larvae fed on rubber rabbitbrush, *Ericameria nauseosa* (Asteraceae), in Wyoming. *Plataea californiaria* FW 1.5–1.7 cm; Plate 30.22, Plate 30.23) is known from coastal California from San Francisco south to Los Angeles. Adults are found sporadically through the year and, likely, have several flights. It is usually found in association with coastal sage habitats and feeds on *Artemisia californica* (Asteraceae). *Plataea personaria* (Henry Edwards)(Plate 30.22) is similar and is known along the Pacific Coast from central California south to northern Baja California.

Eusarca Hübner is a large genus with 106 species limited to the New World. In North America, 10 species are western and three are eastern. *Eusarca falcata* Packard (FW 2 cm; Plate 30.24) ranges in lowland western California from the San Francisco Bay area south to San Diego County. McFarland indicates flight from May to October, with abundance peaks during May to June and October that indicate two generations. The life history is unreported.

Somatolophia Hulst is known from 13 species found in the Nearctic and Neotropical regions. Ten species are known from the West. *Somatolophia ectrapelaria* Grossbeck (FW 2.0–2.4 cm; Plate 30.25) ranges from eastern Oregon and southern Idaho, and northeastern Colorado south to southern California, central Arizona, and central New Mexico. Adults are in flight from May through October, although the majority is found in June and July. The life history is unreported, although Henne raised the species from eggs laid by a captured female. The reported host, *Ericameria nauseosa*, is most likely artifactual and was probably based on a laboratory rearing. *Somatolophia cuyama* J. A. Comstock (male FW 1.3–1.8 cm; Plate 30.26) is known only from southeastern San Luis Obispo County, California. The female is unique in this genus as it is brachypterous and, presumably, flightless. Adults have never been found in nature, and a single collection of larvae collected on *Hymenoclea salsola* (Asteraceae) produced adults that emerged in early June. The final instar was described by Comstock as having alternating bands of soiled yellow and white; there is a suprastigmatal row of square black spots and a number of small black spots around each spiracle. The spiracles are black or dark gray, and the head is soiled yellow with orange laterally.

Pherne Hulst, with four species, is endemic to the West. *Pherne subpunctata* Hulst (FW 1.3–2.0 cm; Plate 30.27) ranges from the Los Angeles area south to San Diego County. Adults are in flight from February through October, probably in several broods. The larval host is *Salvia mellifera* (Lamiaceae), but the details of the life history are unreported.

Synaxis Hulst has 11 described species with 10 of these endemic to the West. The eleventh species is known from Chile. *Synaxis cervinaria* (Packard)(FW 1.9–2.1 cm; Plate 30.28, Plate 30.29) ranges from British Columbia east to Montana and Utah and south to southern California and northeastern Nevada. Adults are polyphenic and vary in color from tan to orange to brown and in pattern with ante- and postmedial lines from pale yellow to black. The adults are in flight from the end of March to early July; reports from October and November represent a late emergence. Larval hosts include various species of *Populus* (Salicaceae), *Quercus* (Fagaceae), and *Arbutus menziesii* (Ericaceae). The larvae are light brown with the venter even lighter brown. The head is brown with a darker herringbone pattern on the lateral lobes; the prolegs have brown markings. *Synaxis jubararia* (Hulst)(FW 1.8–2.1 cm; Plate 30.30) is very similar to *S. cervinaria* and is found from southern British Columbia east to Idaho and Utah south to central California and Wyoming. Richers reports that in California adults fly from March to May and again from September to November. The larval hosts include many broadleaved shrubs and trees such as alder, ash, maple, and snowberry. The larvae are variably mottled gray to red brown and have paired dorsal bumps on the first thoracic and the third through seventh abdominal segments. *Synaxis formosa* (Hulst)(FW 1.7–2.1 cm; Plate 30.31) is found in arid juniper and pinyon-juniper woodlands in much of the intermountain West from southeastern California east to western Colorado and western New Mexico. Unlike the two *Synaxis* species discussed above, *S. formosa* is relatively monomorphic. The adults have a single flight during September and October. The larval host is rubber rabbitbrush, *Ericameria nauseosa* (Asteraceae). The larvae are gray brown with thin longitudinal lines above and along the sides, whereas the venter is gray yellow. There are minute dorsal bumps on the second through fifth abdominal segments and a slight hump on the dorsum of the eighth abdominal segment.

Tetracis Guenée comprises 10 species, including two described from eastern North America and the remainder from the Neotropics. *Tetracis cachexiata* Guenée, the "white slantline" (FW 1.7–2.0 cm; Plate 30.32, Plate 62.10 larva), occurs from southern British Columbia east to Manitoba and south to Wyoming and Colorado. In Colorado, adults have a single flight from late May to mid-July. The species is polyphagous on a wide variety of woody trees and shrubs; occasionally a nonwoody plant is selected; Pinaceae, Betulaceae, Rosaceae, and Salicaceae are favored.

Phyllodonta Warren is a Neotropical genus comprising 23 species, of which *P. peccataria* Barnes & McDunnough (FW 1.7–2.0 cm; Plate 30.33, Plate 62.11 larva) is found in southeastern Arizona and southwestern New Mexico. There is a single flight during late July and August. Larvae were raised on pine *(Pinus)* by Ferguson.

Prochoerodes Grote has its greatest richness in the Neotropics, where 18 species were described. There are five North American species with four found in the West. Additionally, there is at least one undescribed species. *Prochoerodes forficaria* (Guenée), the "large maple spanworm" (FW 1.8–2.1 cm; Plate 31.1, Plate 62.12 larva), is found from southern British Columbia south to southern California, Arizona, and northern New Mexico. The single adult flight is during June and July in British Columbia, according to McGuffin, but we found adults nearly year-round at Inverness, Marin County, California. McFarland reports adults in the Santa Monica Mountains from January to July with peaks during March to April and June to July, an indication of bivoltinism; we have similar records from Big Creek, Monterey County. Robinson et al. catalogue known hosts including plants in Aceraceae, Juglandaceae, Rosaceae, Rhamnaceae, and Salicaceae. McFarland has raised the species on ash, *Fraxinus* (Oleaceae), in southeastern Arizona, and we have reared it at Big Creek from *Salix* (Salicaceae), *Frangula cal-*

ifornica (Rhamnaceae), and *Rubus ursinus* (Rosaceae). Eggs are laid singly or in small groups. The larvae are light brown or gray and are spotted with black. The head is light brown or gray marked with darker brown. Winter is passed with the species in the pupal stage. *Prochoerodes truxaliata* (Guenée) is another widespread western species found in the intermountain region.

Pityeja Walker has two Neotropical species. "*Pityeja*"*ornata* Rindge (FW 1.5–1.7 cm; Plate 31.2) is one of the more attractive western moths. It is found in southeastern Arizona and ranges south in Mexico's Sierra Madre Occidentale to at least central Sonora. The nocturnal adults may be found during August and September. That the genus name appears within quotes by Scoble indicates that the assignment to this genus is uncertain. The life history is unreported.

Sabulodes Guenée comprises 51 species, most of which occur in the Neotropics. Ten are found in western North America. *Sabulodes aegrotata* (Guenée), the "omnivorous looper" (FW 1.7–2.6 cm; Plate 31.3), is found from northwestern Oregon south along the Pacific Coast to the west of the Cascade–Sierra Nevada crest to northern Baja California. The range includes a number of the California islands. As for color and pattern, this is one of our most variable western moths. Most individuals are unpatterned pale tan, with darker submedian and subterminal lines, or having contrasting rust to dark brown bands. Dark individuals, form "cottlei," are chocolate brown above with blonde head and thorax and are found primarily in urban and suburban areas, whereas they are extremely rare or absent in more natural populations. We have determined through breeding experiments that the maintenance of this form is due to a dominant gene that may have pleiotropic effects. According to an examination of historical collections, the chocolate brown form first appeared in 1926 in San Francisco and reappeared again in 1946 and has occurred sporadically since then in urban populations on both sides of San Francisco Bay, remaining rare (16% of observed individuals at Berkeley). The species may be found year-round, and the number of broods is indeterminate. McFarland points out that the species is often more abundant in suburban cultivations than in adjacent native habitats. Eggs are laid on leaves in small clusters. The larvae feed on a large variety of herbaceous and woody broad-leaved angiosperms. Plants in 27 families have been reported as hosts, and the species' feeding occasionally causes damage to orchard trees such as avocado, citrus, and walnuts. English ivy, *Hedera* (Araliaceae), is favored in urban and suburban situations, and its use may be largely responsible for inland records away from the coast ranges, for example, Shasta Lake, Mariposa County, and urban Palm Springs, California. The larvae rest in a small silken shelter in a leaf fold or between several leaves. The larvae have most of the body with a harlequin pattern of alternating black, white, orange, and gray stripes and lines. The head is orange with a pair of black spots on top. Pupation is within one of the silked leaf shelters. *Sabulodes edwardsata* (Hulst)(FW 1.8–2.3 cm; Plate 31.4) ranges from south-central British Columbia south to western Nevada and central California, mainly to the west of the Sierra Nevada–Cascade crest, and east to western Montana. The adults have a single flight from May to early October. Larval hosts are Pinaceae, primarily pines and Douglas-fir, but Prentice also reports some use of western hemlock and Sitka spruce. Larvae are green and lined with yellowish white. *Sabulodes niveostriata* (Cockerell)(FW 1.8–2.3 cm; Plate 31.5) is found in montane habitats from

Utah and Colorado south to Chihuahua and Durango, Mexico. The adults have a single flight from June to September. This moth is found almost always in association with stands of ponderosa pines *(Pinus ponderosa)*, and pines are the presumed hosts. The adults share a general similarity to *Cabera aequaliaria*, another pine feeder, and this pattern is presumed to lend background resemblance to the moths when resting on pine branches.

Enypia Hulst includes four species all of which are western endemics. Adults are both nocturnal and crepuscular. *Enypia venata* (Grote)(FW 1.5–1.8 cm; Plate 31.6) is the most widespread, being found from southern Alaska and British Columbia east to western Alberta and south to Arizona and New Mexico. There is a single flight from early June to early September. Larval hosts reported by Prentice are all Pinaceae; western hemlock, Douglas-fir, and true firs are preferred. The larvae are golden brown with a geminate brown middorsal line that is filled with black at the anterior portion of each abdominal segment. There are also additional brown or black longitudinal lines. The head is brown with a darker brown herringbone pattern on the lateral lobes. Larvae overwinter, most often in the fourth or fifth instar. *Enypia griseata* Grossbeck is a similar western moth with a comparably large western distribution.

Nematocampa Guenée comprises 16 species, 13 of which are Neotropical, whereas three are North American. Two are known from the West. *Nematocampa resistaria* Herrich-Schäffer, the "filament bearer," is Holarctic, with the subspecies *N. r. limbata* (Haworth)(FW 1.0–1.3 cm; Plate 31.7, Plate 62.13 larva) the North American representative, which ranges from British Columbia east to Nova Scotia and south to northwestern California, Colorado, and, in the East, to Florida. In British Columbia, Jones reports that the adults are found from July to September. The eggs are laid singly or in small groups and overwinter. The larval hosts are a wide variety of broad-leaved and coniferous trees, shrubs, and herbaceous plants. Plants in 11 families are compiled by Robinson et al. The larvae are mottled with brown or yellow brown, including dark brown transverse bands on the venter of each segment and a brown patch surrounding each spiracle. Most notably, there is a prominent forked filament on both the second and third abdominal segments; each fork is white tipped.

SUBFAMILY ALSOPHILINAE

The subfamily Alsophilinae, which includes only one genus, was previously included in the Oenochrominae but was given separate status by Holloway. *Alsophila* Hübner includes 10 species, most of which are distributed across Eurasia, whereas a single species is found in North America. *Alsophila pometaria* Harris, the "fall cankerworm" (male FW 1.3–1.5 cm; Plate 31.8 male, Plate 31.9 male, Plate 31.10 female, Plate 62.14 larva), ranges from western Alberta east to Nova Scotia and south to California, Colorado, Kansas, and North Carolina. Females are wingless. In California, the adults are active from October to December. In Colorado, adult males have flight dates from March to early April. The flightless females lay clusters of 50 to 200 upright eggs on branches and bark of deciduous trees and shrubs. Scales from the female's abdomen are attached to the egg masses. The first-instar larvae let themselves down on a strand of silk and are dispersed by wind to other suitable host trees in the forest. The larvae are serious defoliators of the leaves and young fruit of many kinds of deciduous trees. Preferred host genera reported by Prentice are *Acer, Quercus,* and *Ulmus.* Essig and others report that fruit trees, including apple, apricot, cherry, and plum, are often seriously damaged. The larvae have an extra pair of midabdominal prolegs on the fifth abdominal segment; they are variably gray to light green with or without fine longitudinal lines. Several workers have found that the larvae become much darker green under crowded conditions.

SUBFAMILY GEOMETRINAE

The adults of all North American Geometrinae (greens, emeralds, geometrines) may be best characterized by having at least some green pigmentation. A few species have alternate brown or pinkish adult forms. The antennae of males are usually short and bipectinate, whereas those of females are either simple or bipectinate. The adults usually rest with spread wings on vegetation, where their green coloration makes them difficult to distinguish from their background. The larvae of species in the tribes Nemoriini and Synchlorini have dorsolateral protuberances. In the latter group the protuberance bears hooks to which they attach bits of vegetation. The North American genera and species have been described and discussed in great detail by Ferguson.

TRIBE NEMORIINI

Chlorosea Packard comprises four species, all endemic to western North America. *Chlorosea banksaria* Sperry (FW 1.5 to 2.1 cm; Plate 31.11, Plate 62.15 larva) is found from Vancouver Island south to southern California. Ferguson reports that the adults are nocturnal and fly from May to September, primarily during June and July. The life history was reported by Rindge. Larval hosts are *Ceanothus thyrsiflorus* (Rhamnaceae) and *Holodiscus discolor* (Rosaceae), and other plants may be reported in the future. The larvae are green with oblique yellow white marks or with reddish brown markings and have prominent lateral processes on the abdominal segments. The head is slightly bilobed and is coarsely rugose and reddish brown. The legs are reddish brown. *Chlorosea nevadaria* Packard is closely related and has a more inland distribution. At least one of its hosts is bitterbrush, *Purshia tridentata* (Rosaceae).

Nemoria Hübner comprises 134 species, all of which are either Nearctic or Neotropical. The vast majority of species are found in the Neotropics, but 32 are North American of which 21 are western. *Nemoria pulcherrima* (Barnes & McDunnough)(FW 1.2–1.6 cm; Plate 31.12, Plate 31.13, Plate 31.14) ranges down the Pacific Coast west of the Sierra Nevada–Cascade crest from Oregon south to San Diego County, California. Adults are dimorphic with both a usual green form as well as a brown pink form. In the 1960s we found the pink form dominant at Walnut Creek, Contra Costa County, California; this is a reversal of the usual case. There is a single late winter flight from mid-January to mid-March, but some individuals have appeared as early as late December and some as late as mid-May. Comstock and Dammers have described the larva and pupa and reported that the larvae feed on oak catkins. *Nemoria arizonaria* (Grote)(FW 1.4–1.7 cm; Plate 31.15, Plate 31.16) is found from central Arizona west across southern Arizona and southwestern New Mexico to the Davis Mountains of western Texas. There are two distinct seasonal forms, which were originally described as separate species. Moreover, the genitalia of the two seasonal forms differ from each other, and their true relationship would have remained undiscovered were it not for the rearing of one from the other by Noel McFarland. The flight periods are for the winter–early spring form (middle of February to mid-May) and for the late spring–summer form (mid-June to mid-August). Larvae resulting from a late-winter female fed on catkins and new oak leaves. *Nemoria darwini-ata* (Dyar)(FW 1.3–1.8 cm; Plate 31.17) ranges from southern British Columbia west to southwestern Alberta and south to Baja California, Arizona, and, doubtfully, Colorado. Adults fly from mid-June to mid-August on Vancouver Island. In northern California there seem to be three broods—mid-April to mid-September—while in southern California, there are almost certainly three or more broods—mid-April to mid-October with scattered records during the winter. Larvae have been reported to feed on a wide variety of broad-leaved trees and shrubs. The larvae are variably yellow, tan, brown, or cream colored. Most strikingly, there are prominent lateral processes on most segments reduced to small tubercles. The integument is almost velvety. *Nemoria leptalea* Ferguson (FW 1.2–1.7 cm; Plate 31.18, Figs. 210, 211 genitalia) ranges from coastal central California south into Baja California. The species is also found on Santa Catalina, Santa Cruz, and Santa Rosa islands. Ferguson reports records of adults from every month. The reported natural larval hosts are California buckwheat, *Eriogonum fasciculatum* (Polygonaceae), and toyon, *Heteromeles arbutifolia* (Rosaceae). When disturbed or moving the larva quivers or jerks its body. The last instar is light tan with russet brown on the lateral protuberances of the second to fourth abdominal segments, which are equal in length. The skin is covered with minute cream spicules that give the larva an encrusted appearance. Pupation is on the host plant in a fragile cocoon to which small leaf fragments are attached.

Dichorda Warren includes four North American and four Neotropical species. Three of the North American species are endemic to the West. *Dichorda illustraria* (Hulst)(FW 1.4–2.2 cm; Plate 31.19) ranges from coastal Oregon south through most of California west of the Sierra Nevada and Transverse Ranges, including most of the Channel Islands, to northern Baja California Norte. There is a single flight in Oregon and northern California (June to September) and two in southern California and northern Baja California (March through November with scattered records in the other months). McFarland reported an abundance peak during July and August in southern California. The reported larval host plants are *Malosma laurina*, *Rhus trilobata* (Anacardiaceae), and *Ribes* (Grossulariaceae), although the last plant seems unlikely. The larvae are dark brown with large quadrate dorsolateral plates on the thoracic and first five abdominal segments. Abdominal segments 6 to 8 each bear small lateral humps, and abdominal segment 8 bears a dorsal pair of bifurcate hornlike processes.

TRIBE SYNCHLORINI

Synchlora Guenée comprises 42 species that are limited to the New World. Nine species are found in North America, of which seven are found in the West; the remainder occur in the Neotropics. *Synchlora aerata* (Fabricius), the "wavy-lined emerald" (FW 0.7–1.2 cm; Plate 31.20, Plate 62.16 larva), is found from British Columbia east across southern Canada to Newfoundland and south to California, Arizona, and New Mexico. It is found throughout most of the eastern United States south to Missouri and North Carolina. In southern California, the moths are found from March to November, and it is difficult to determine the number of generations. In contrast, in British Columbia, Washington, and Oregon, Ferguson describes two distinct flight periods: April to June and July to September. In the central Rockies, there seems to be only a single extended flight from June to August. In general, larvae feed on buds and flowers of a wide variety of plants but are found most often on composites, Asteraceae. In California, Comstock and Dammers reported larvae on three composite genera as well as desert willow, *Chilopsis* (Bignoniaceae), and *Eriogonum fasciculatum* (Polygonaceae). Larvae feed on flower heads and usually cover themselves with bits of flowers so that they are well camouflaged. *Synchlora bistriaria* (Packard)(FW 1.1–1.7 cm; Plate 31.21) ranges widely in the West, mainly in interior regions, from the interior of southern British Columbia east to southern Saskatchewan and south to southern California, central Arizona, New Mexico, and western Texas. The adults fly from April to September in central and southern California, where there are two or more broods, and only during June and July more northward is there only a single annual flight. The larvae feed primarily on the flowers and buds of composites such as *Helianthus* and *Solidago*, but occasionally on other plants such as *Rosa* (Rosaceae). The larvae are rather stout and rugose; they vary from whitish to brownish and have dorsolateral extensions on the abdominal segments. Chewed pieces of vegetation are attached to the small spines on the abdominal protuberances, thus lending them to concealment.

TRIBE HEMITHEINI

Chlorochlamys Hulst comprises four species, all of which are endemic to the West. *Chlorochlamys appellaria* Pearsall (FW 0.7–1.0 cm; Plate 31.22) ranges from California, Nevada, and Utah to western Colorado. The species also occurs in northeastern and central Mexico, where it seems to represent a disjunct population. Adults are usually green, but may also be brown, rust, or, rarely, cream. Adults may be found from mid-March to early October, and there seem to be three flights in San Diego County, California. The larval hosts include two species of *Eriogonum* (Polygonaceae), as well as *Baccharis glutinosa* and *Chrysothamnus* (Asteraceae). Miller and Hammond report that the larvae are yellow green with vague lighter lines. The head is yellow.

Xerochlora Ferguson comprises five species, four of which are western; the fifth is endemic to Mexico. *Xerochlora viridipallens* (Hulst)(FW 0.9–1.2 cm; Plate 31.23, Plate 31.24) ranges from southeastern California east across Arizona, southern Colorado, and New Mexico to western Texas. The adults have a single flight in July and early August, rarely to early September. The life history has not been reported.

Hemithea Duponchel has 31 species that are found mainly in Asia and the Indo-Pacific, *H. aestivaria* (Hübner), the "common emerald" (FW 1.3–1.9 cm; Plate 31.24), a common Palaearctic species, was reported by Dôganlar and Beirne to have been accidentally introduced to southern British Columbia in 1979. Ferguson suggested that, because it fed on a wide array of hosts, it might spread widely in North America. The adults have a single flight in July. The species is polyphagous on a wide variety of broad-leaved deciduous trees and shrubs. Larvae vary from green to brown to purplish with a dark dorsal line that widens posteriorly into triangles that are edged with white. Bits of vegetation are attached to the larvae during early instars, but this habit is lost in later instars. Winter is passed in the larval stage.

Mesothea Warren contains only a single species, *M. incertata* (Walker)(FW 0.9–1.1 cm; Plate 31.25), which ranges from central Alaska across Canada to Labrador and south in the boreal habitats to northern California, northeastern Nevada, Utah, and southern Colorado. In the East, it has been found south to Illinois and New York. It is the only member of the Geometrinae that is found in Arctic locations and is the only diurnal representative in North America. It is strictly diurnal and flies in openings with grass or herbs. There is a single flight, and Ferguson cites records from mid-April to very early July. Larval hosts are a variety of woody and herbaceous broad-leaved angiosperms with genera of Betulaceae, Rosaceae, and Salicaceae being the most frequently cited. The larvae, which are twig mimics, were described by Dyar as being slender and green marked with brownish. There are dorsal conelike projections on the mesothorax and anal plate. The head is granular and has two pointed lobes on the vertex; it is whitish green with brown laterally. The pupae overwinter.

SUBFAMILY STERRHINAE

The subfamily Sterrhinae includes approximately 2,800 described species worldwide, with most diversity present in the tropics. The adults are mostly small and have thin wavy lines extending from FW to HW as well as a prominent dark discal spot on all wings. The larvae of most species feed on low herbs, but the larvae of some tropical groups feed on trees.

TRIBE STERRHINI

Lobocleta Warren comprises 54 species found primarily in the Neotropical region, seven are North American with five of these found in the West. *Lobocleta plemyraria* Guenée, the "straight-lined wave" (FW 0.7–0.9 cm; Plate 31.26), ranges from California to southern Arizona. Adults have been found in California from April to September, probably in several generations. The larval host was reported by McFarland as *Chamaesyce* (Euphorbiaceae).

Idaea Treitschke is a worldwide genus with about 680 described species. Twenty-six species are found in North America, and 10 of these are western. *Idaea demissaria* (Hübner), the "red-bordered wave" (FW 0.8–0.9 cm; Plate 31.27), ranges from British Columbia east to Ontario and south to central California, Colorado, and much of the eastern United States. The adults are nocturnal and fly in July and August. *Idaea bonifata* (Hulst)(FW 0.6–0.7 cm; Plate 31.28) is found in California and Colorado. This may be the smallest North American geometrid. In southern California, McFarland observed that the adults fly all year but are most abundant in summer. Adults may be found indoors resting on walls, curtains, ceilings, and so forth. The larvae are detriphagous and are found in habitats with suitable decaying material such as leaf piles. They also feed on stored food products such as oatmeal and raisins and are also occasional pests of herbarium specimens.

TRIBE COSYMBIINI

Cyclophora Hübner is found worldwide, and includes 64 species. Two species are found in the West (one ranges to the East), and five are found only in eastern North America. *Cyclophora dataria* (Hulst)(FW 0.9–1.0 cm; Plate 31.29) is found from southern British Columbia south to California, Utah, and Colorado. The adults fly from late March to early December, indicating several broods, but McFarland found a distinct peak in southern California during July. The larval hosts are catkins of Oregon white oak, *Quercus garryana* (Fagaceae), and flowers of *Hemizonia* (Asteraceae). Larvae are variably colored tan, brown, or gray, are mottled, and have a discontinuous white lateral line. The head is a paler yellow tan.

Pigia multilineata (Hulst)(FW 1.0–1.2 cm; Plate 31.30) occurs in southeastern Arizona (Huachuca Mountains). Adults are found in June and early August, and there are presumably two generations. The life history is unreported.

TRIBE TIMANDRINI

Haematopis grataria (Fabricius), the "chickweed geometer" (FW 1.1–1.2 cm; Plate 31.31), the only representative of a monotypic genus, is found from Northwest Territories east to Quebec and south to Colorado and Texas in the West. The adults have a unique appearance and fly from July to October, possibly in two flights, in disturbed weedy habitats, wetlands, and grasslands. The larval hosts are thought to be various species of *Polygonum* (Polygonaceae). The larvae are yellowish white, but this is largely obscured by blackish brown mottling. The head is blackish brown. The pupa is light brown and light yellow green with a purplish dorsal line.

TRIBE SCOPULINI

Scopula Schrank comprises approximately 640 species that are found on several continents worldwide. Twenty-four species are North American, 11 of which have all or some part of their range in the West. The North American species were studied by Covell. Adults may be found during the day in a wide variety of open habitats, but some species are nocturnal and are attracted to light, for example, *S. junctaria* (Walker). The larval hosts for most species seem to be herbaceous plants. *Scopula junctaria*, the "simple wave" (FW 1.2–1.5 cm; Plate 31.32), is found from southern British Columbia east across southern Canada to Quebec and south in the West to California, Utah, and New Mexico, and in the East, to New Jersey. According to Covell's research we include here the subspecies *S. j. quinquelinearia* (Packard) and *S. j. johnstonaria* McDunnough. Adult flight is from late May to August. McGuffin reports that the larvae feed on chickweed, *Stellaria* (Caryophyllaceae), clover, *Trifolium* (Fabaceae), and elm, *Ulmus* (Ulmaceae). *Scopula sentinaria* (Geyer)(FW 1.1–1.2 cm; Plate 31.33) ranges from Alaska and boreal portions of western Canada east to northern Labrador and ranges south to Alberta and, disjunctly, Colorado, where it is restricted to alpine habitats. The adults have a single flight from mid-June to early August. Larvae hibernate in the fourth or fifth instar.

Leptostales Möschler comprises 55 species that are endemic to the New World. Eight species are North American. The sole western species, *L. rubromarginaria* (Packard)(FW 0.8–0.9 cm; Plate 31.34), ranges from southern British Columbia south to central California, Arizona, and Colorado. Adults are mostly diurnal and have a single flight from April to July. The life history is unreported, but, along the Colorado Front Range, adults are seen in close association with *Prunus americana*, Rosaceae.

SUBFAMILY LARENTIINAE

The subfamily Larentiinae comprises 5,700 described species, and, although most species occur in temperate regions or high-altitude tropics, the group is widely distributed. The moths are small to medium and many have a small dark discal spot on both FW and HW. The FW often have fasciae of closely juxtapositoned lines. The females of some species are flightless winter moths. Most species are external feeders, but the larvae of some bore into buds and flowers, even cones, whereas the larvae of *Perizoma incultraria* (Herrich-Schäffer) mine the leaves of *Primula* (Primulaceae) and, as such, are unique in the family.

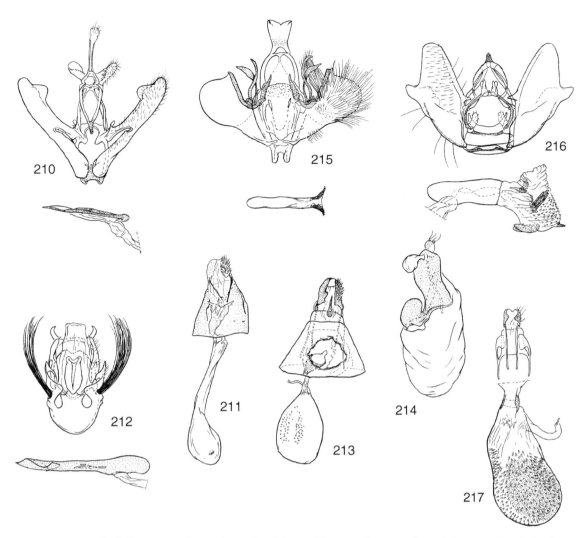

FIGURES 210–217. Genitalia structure of exemplar species of Geometridae, ventral aspect, aedeagus below. **210,** *Nemoria leptalea* Ferguson ♂. **211,** *Nemoria leptalea* ♀ [Ferguson 1985]. **212,** *Scopula sideraria* (Guenèe) ♂. **213,** *Scopula sideraria* ♀ [Covell 1970]. **214,** *Hydriomena renunciata* (Walker) ♀, bursa copulatrix. **215,** *Hydriomena furcata* (Thunberg) ♂ [McDunnough 1954]. **216,** *Eupithecia nevadata* Packard ♂ [Bolte 1990]. **217,** *Eupithecia cretaceata* (Packard) ♀ [McDunnough 1949].

TRIBE HYDRIOMENINI

Dysstroma Hübner includes 50 described species found in the Holarctic and Indian subregion. There are 14 North American species, 12 of which are western or have some portion of their range in the West. Most North American species are polyphenic, and Ferris notes that some species can be identified only by dissection and examination of the genitalia. *Dysstroma citrata* (Walker)(FW 1.5–1.7 cm; Plate 31.37, Plate 62.18 larva) is Holarctic. In North America, it occurs from Alaska and British Columbia east to western Saskatchewan and south to California, Arizona, and Colorado. In northern and interior locations, the adults have a single nocturnal flight from June to August, primarily during July and August, but probably have several broods in coastal central California (February to November). Larvae have been recorded on a wide array of broad-leaved woody shrubs and herbaceous plants. The larvae are green and have a gray green or blue green middorsal line. The head is unmarked pale green. Pupation is in a white cocoon among curled leaves. Miller and Hammond described the larvae of *D. formosa* (Hulst)(FW 1.5–1.6 cm; Plate 31.35, Plate 31.36, Plate 62.17 larva) and report that they represent the most common species found on currant and goose-

berry, *Ribes* (Grossulariaceae), in the Pacific Northwest. The moth ranges from southern British Columbia and western Alberta south to California, eastern Arizona, and Colorado. In California, the adults are in flight during July and early August.

Eulithis Hübner has 27 described species, five of which are western, at least in part. *Eulithis xylina* (Hulst)(FW 1.5–1.8 cm; Plate 31.38, Plate 62.19) ranges across North America from Alaska and British Columbia east to western Alberta south to central California, Utah, and northern New Mexico. Adults are nocturnal and fly from July to mid-September. The usual hosts reported by Prentice are *Alnus* (Betulaceae), *Salix*, (Salicaceae), and *Dasiphora fruticosa* (Rosaceae). Many other trees and shrubs have also been recorded. Larvae are pale gray with the dorsum having each abdominal segment with the anterior portion black and the posterior portion white. The lateral areas and venter are gray. The head is pale gray green with brown marks in a herringbone pattern along the epicranial stem and parietals. Several other *Eulithis* species are similarly patterned but are yellow.

Eustroma Hübner includes three western species. *Eustroma semiatratra* (Hulst), the "black-banded carpet" (FW 1.3–1.7 cm; Plate 31.39), has a Nearctic distribution from Alaska and

British Columbia east across Canada to Nova Scotia and south to southern California, Utah, Colorado, and western South Dakota. Adults fly from April to October in British Columbia and through much of the year in coastal California. The larval hosts are various herbaceous plants, especially fireweed, *Chamerion angustatum* (Onagraceae). The larvae are slightly rugose and are gray mottled with brown or red brown mottled with black. The head is reticulate, white, and has black or dark brown marks in a herringbone arrangement along the epicranial stem and parietals.

Ecliptoptera Warren comprises 45 species that occur in various parts of the Holarctic region. *Ecliptoptera silaceata* (Denis & Schiffermüller), the "small phoenix" (FW 1.4–1.6 cm; Plate 31.40), is Holarctic. In North America, it ranges from Alaska and British Columbia across Canada to Labrador, Newfoundland, and Nova Scotia, and, in the West, south to northwestern California, Idaho, and northern Colorado. Away from the coast, *E. silaceata* has a single flight from June to August but flies as early as May in coastal Oregon. The reported larval hosts are *Epilobium* (Onagraceae) and *Impatiens* (Balsaminiaceae). The larvae are pale yellow or pink and have a gray green or dark brown middorsal line that becomes broader on the posterior abdominal segments. In addition, there are additional pink, yellow green, or dark brown lines. The thoracic legs are pink, whereas the prolegs are pink or green. The head is variably colored pale russet green or brown with a reddish brown or dark brown patch on each side.

Plemyria Hübner includes two species, one in the Palaearctic, and one in the Nearctic. *Plemyria georgii* (Hulst)(FW 1.3–1.6 cm; Plate 31.41) ranges from British Columbia and Northwest Territories east across Canada to Nova Scotia and south to central California, Colorado, and Maine. Adults fly as a single brood from July to early September. The larval hosts are *Alnus* and *Betula* (Betulaceae) and *Salix* (Salicaceae). The larvae are rugose and covered with minute convex spicules; they are yellow green with a pink subspiracular line and yellow intersegmental bands.

Thera Stephen includes 23 species limited to various portions of the Holarctic region. In North America, there are four species, two of which are found in the West. *Thera otisi* (Dyar)(FW 1.2–1.4 cm; Plate 31.42) ranges from British Columbia and western Alberta south to Oregon, eastern Arizona, and southern Colorado. Adults have a single flight in August and September. The recorded larval host is *Juniperus communis* (Cupressaceae). Larvae are rugose and pale green with a darker green subdorsal stripe outlined on both sides with white. A subspiracular white line almost always has a red line above the white. The head is reticulate and russet.

Ceratodalia gueneata (Packard)(FW 1.1–1.3 cm; Plate 31.43 dorsal, Plate 31.44 ventral), the only species of this monotypic genus, occurs from southern Alaska, British Columbia, and southwestern Alberta south to central California and eastern Nevada. Adults are in flight from May to September. The recorded larval host is *Polygonum* (Polygonaceae), and there is a rearing record in the Smithsonian files from western hemlock, *Tsuga heterophylla* (Pinaceae). The larvae are reticulate pale brown with darker brown mottling. The head is similarly pale brown with a darker brown herringbone pattern on each of the parietals. Larvae hibernate.

Hydriomena Hübner comprises 135 species worldwide, and 56 of these are North American, most of which, 51, are western. In addition, there are perhaps 10 undescribed western species. Most of the remaining species are Neotropical, mostly associated with higher-elevation boreal forests, in-

cluding in Mexico and Costa Rica. There are six or seven *Hydriomena* that occur disjunctly in New Zealand, and five or six in Europe. The adults of many species are polyphenic, at least quite variable, and dissection of genitalia is often required for certain identification of several species. *Hydriomena furcata* Borgstrom (FW 1.4–1.7 cm; Plate 31.45, Fig. 215 genitalia) is Holarctic and ranges in North America from Alaska across Canada to Nova Scotia and south to central California, eastern Arizona, and Colorado. Adults fly from March to August in most areas. Larval hosts are primarily *Salix* (Salicaceae) but also include other broad-leaved woody trees and shrubs including *Populus* (Salicaceae), *Alnus* and *Betula* (Betulaceae), and *Prunus* (Rosaceae). Larvae feed in small groups, at least on occasion, on new host leaves. Larvae are red with a black or brown middorsal stripe and a white subdorsal stripe. Spiracular stripe is dark brown or black above and underlined in pale brown. The prothoracic and anal shields are brown. The head is brown and slightly rugose with a darker brown herringbone pattern on the parietals. *Hydriomena nubilofasciata* Packard (FW 1.3–1.5 cm; Plate 31.46, Plate 31.47) ranges from Vancouver Island south to California and Arizona. The adults exhibit an incredible amount of variation in color and markings. There is a single flight from January to May. Adults are nocturnal but fly during the day, especially in shady ravines and under cloudy conditions. Larval hosts include several oaks, among them *Quercus agrifolia* and *Q. garryana* (Fagaceae). We feel the primary hosts are oaks, and reports of feeding on a variety of other woody plants such as elderberry, *Sambucus* (Caprifoliaceae), and Douglas-fir, *Pseudotsuga menziesii* (Pinaceae), are rare, incidental, or erroneous. Larvae feed from March to May and are pale green and have interrupted dusky dorsal and lateral lines. The head is slightly rugose and is light brown. McFarland observed that the larvae rest in a half-curled posture within leaf nests. The pupae overwinter. *Hydriomena renunciata* Walker, the "renounced hydriomena" (FW 1.3–1.5 cm; Plate 31.48, Plate 62.20 larva, Fig. 214 genitalia), ranges across North America south to northern California and northern Idaho. There is a single adult flight from March to May, rarely to early July. The larval hosts are alders, *Alnus* (Betulaceae). The larvae are pale brown with gray brown middorsal lines that may be widened into rectangles. Pupae overwinter. *Hydriomena albimontanata* McDunnough (FW 1.7–1.8 cm; Plate 31.49) ranges from central British Columbia and Alberta south to eastern Arizona. The adults fly from mid-May to June. The larval host is Douglas-fir and occasionally spruces (Pinaceae). The larvae are pink or yellow with pink suffusion. The middorsal line is dark brown, and, on some individuals, may broaden out to form rectangles on the anterior abdominal segments. There is a broken supraspiracular line that is brown in the intersegmental areas. The prothoracic shield is brown with pale brown laterally, and the anal shield is pink anteriorly and pale brown posteriorly. The head is slightly rugose and is pale brown with a darker brown herringbone pattern on the parietals. Pupae overwinter. *Hydriomena macdunnoughi* Swett (FW 1.4–1.6 cm; Plate 31.50) is found in the Rocky Mountains from Yukon Territory south through western Alberta to Colorado. Adults fly from June to mid-July. The larval hosts are willows, *Salix* (Salicaceae). Larvae are pale gray with a darker gray middorsal stripe and darker gray along the sides. The head is light gray with a darker gray herringbone pattern on the parietals. *Hydriomena irata* Swett (FW 1.5–1.7 cm; Plate 31.51) ranges along the Pacific Coast from Alaska south to California. Adult flight is from mid-May to late August. The larval hosts are selected Pinaceae, including spruce, hemlock, and Douglas-fir. The larvae are banded with broken dark brown subdorsal and spiracular lines. *Hydriomena manzanita* Taylor (FW 1.6–1.9 cm; Plate 32.1, Plate 32.2) ranges from Vancouver Island, British Columbia, south along the Pacific Coast to central California (Monterey and Placer counties). The adults fly from mid-February to early June. The larval hosts include madrone, *Arbutus menziesii*, and manzanitas, *Arctostaphylos* (Ericaceae). The larvae are light green with a black middorsal band and a white-mottled pattern. The head is light brown.

Hymenodria McDunnough is monotypic. *Hymenodria mediodentata* (Barnes & McDunnough)(FW 1.3–1.5 cm; Plate 32.3) is found in southeastern Arizona and western Texas. This is a univoltine moth with adult flight from March to May. The life history is unreported.

Ersephila Hulst comprises four species, two of which are western and two others that occur further south in the Neotropical region. *Ersephila grandipennis* Hulst (FW 2.5–2.8; Plate 32.4) ranges from the Colorado Front Range (Douglas County) south to southeastern Arizona. Adults have a single flight during April in Colorado. McFarland reports that the larvae feed on new leaves of oaks, *Quercus* (Fagaceae), in southeastern Arizona.

Triphosa Stephen comprises 52 species distributed in the Holarctic and Neotropical regions. Four species have North American distributions, and three of these are western. *Triphosa haesitata* (Guenée), the "tissue moth" (FW 1.9–2.1 cm; Plate 32.5), ranges from British Columbia east to Labrador and south to central coastal California, Idaho, and Colorado. In the East, the species extends south along the Appalachians to North Carolina. Adults fly through much of the year and overwinter. The preferred larval host in British Columbia is *Frangula* (Rhamnaceae), and we reared the species at Big Creek, Monterey County, seven times from *F. californica*. There are reports by McGuffin of other plants including *Prunus* (Rosaceae), *Pseudotsuga* (Pinaceae), and *Quercus* (Fagaceae), but we believe these must be incidental. The larvae are translucent green with two fine lines on each side of the dorsum and have a yellow white supraspiracular stripe. The venter is white or pale gray. The head is pale russet green. *Triphosa californiata* (Packard)(FW 1.7–2.0 cm; Plate 32.6) is found from Oregon south to California. The adults have one or more flights from January to July, with one aseasonal report from September. The larval host is hollyleaf redberry, *F. ilicifolia* (Rhamnaceae). Mc-Farland observed that the larvae feed at night and rest in a two-leaf shelter during the day. The brightly colored larvae are lined with bands of pink, black, yellow, and light brown and have a distinct lateral yellow band that is edged with black along its upper edge.

Coryphista meadi (Packard), the "barberry geometer" (FW 1.8–2.1 cm; Plate 32.7 dorsal, Plate 32.8 ventral), representing a monotypic genus, has a discontinuous Pacific and eastern distribution. In the West, it occurs from southern British Columbia and southwestern Alberta south to central California, Utah, New Mexico, and western Texas. Adults are nocturnal and fly from March to September. It is likely that there may be two generations. Larval hosts are *Berberis* (Berberidaceae). The larvae are pale with five brown dorsal stripes and white or lightened spiracular area that is broken by brown or black intersegmental bands. Prothoracic and anal plates are black or almost so. The head is unmarked orange to pale yellow orange.

Rheumaptera Hübner comprises 66 species that are found in the Holarctic and Neotropical regions. Scoble included

Hydria in his concept of the genus. Four species occur in North America, two of which include the West in their ranges, and both are Holarctic in distribution. *Rheumaptera hastata* (Linnaeus), the "spearmarked black moth" (FW 1.4–1.6 cm; Plate 32.9, Plate 32.10), ranges across North America in boreal habitats from Alaska to Newfoundland and Nova Scotia and south in the West to Idaho, Colorado, and western South Dakota. The adults are diurnal and fly during June and July. Larval hosts are primarily birches and alders (Betulaceae), and, secondarily, willows (Salicaceae). An extensive area of birch was once defoliated in central Alaska. The larvae live in folded leaf shelters and are very slightly rugose and are dark brown to black and may have four pale dorsal lines. There are dull orange patches around the spiracles. The head is orange brown and is slightly rugose. Pupae overwinter. The closely related *R. subhastata* (Nolcken) has much the same distribution, but its larvae prefer alder.

Archirhoe Herbulot includes four species, all of which occur in the West, but one of these extends to eastern North America. *Archirhoe neomexicana* Hulst (FW 1.7–2.0 cm; Plate 32.11 dorsal, Plate 32.12 ventral) ranges from California, including several of the California Channel Islands, east to Colorado, New Mexico, and western Texas. McFarland reported the species flies throughout the year in southern California and has peaks of abundance in June and October, whereas on the California islands, we found it in most months from February to December. In the Rocky Mountain states it flies from April to September. In southern California, Henne found that the larval host is *Penstemon* (Scrophulariaceae); whereas McFarland reported spiderling, *Boerhavia diffusa* (Nyctaginaceae), in southeastern Arizona, and DeBenedictis reared it from larvae found on *Mirabilis laevis* (Nyctaginaceae) on San Clemente Island.

Entephria Hübner is a Holarctic genus with 47 included species. Eight of these are North American, and seven of these are either endemic to the West, or include the West in their ranges. *Entephria polata* Duponchel (FW 1.1–1.3 cm; Plate 32.14) is Holarctic and, in North America, ranges from Alaska across boreal Canada to northern Manitoba (Churchill) and, disjunctly, to the alpine of the Colorado Rockies. There Opler and Slater have recorded the diurnal adults during July. Larvae feed on dwarf birch, *Betula nana* (Betulaceae), and very likely other plants. *Entephria multivagata* (Hulst)(FW 1.7–2.0 cm; Plate 32.13) ranges from southern British Columbia and western Alberta south to Oregon, Arizona, Colorado, and New Mexico. Adults fly from July to early October. Unlike *E. polata* they are nocturnal and are attracted to light. The larval host is white spruce, *Picea glauca* (Pinaceae), and, since this spruce does not occur throughout the range of the moth, other spruces, and possibly other conifers, must be eaten.

Mesoleuca Hübner includes nine species of striking black-and-white-patterned diurnal moths all found in the Holarctic region. They are usually found in broad-leaved or mixed forest. Three species are North American, and two of these include the West in their range. *Mesoleuca ruficillata* (Guenée), the "white-ribboned carpet" (FW 1.3–1.4 cm; Plate 32.15), ranges from British Columbia east to Quebec and south in the mountains to northwestern California, Idaho, and northern New Mexico. In the East, the moth has been found south to North Carolina. Adults have a single flight from May to July. The larval host is *Rubus* (Rosaceae). The larvae are green with a darker green middorsal line and a red underlined white subspiracular line. There is a red or brown triangular spot at the posterior end of the first through seventh abdominal segments. The prothoracic plate is green and the anal plate is rose. The head is pale green with brown marks in a herringbone pattern along the epicranial stem and lateral lobes. Pupae overwinter.

Spargania Guenée, with 60 species, occurs in the Holarctic and Neotropical regions, according to Scoble. Four of the five North American species are western or include the West in their range. *Spargania magnoliata* Guenée, the "double-banded carpet" (FW 1.3–1.5 cm; Plate 32.16, Plate 62.21 larva), is found in coniferous or mixed forest habitats from Alaska and British Columbia east to Newfoundland and south to central coastal California, Utah, and Colorado. Adults may have several flights in coastal California, where the species is found through much of the year, but primarily from June to August. In Rocky Mountain National Park in the Colorado Front Range, *S. magnoliata* is univoltine and flies only from mid-June to early August. The larval hosts are various herbaceous and woody Onagraceae including *Epilobium*, *Oenothera*, and *Fuchsia*. The slightly rugose ultimate-instar larvae are variably green, yellow green, or pale gray. The middorsal area is gray or green anteriorly and brownish posteriorly. The longitudinal lines are variable. Pupae overwinter.

Perizoma Hübner comprises 168 species, most of which are Neotropical, but some species are found in the Palearctic, Africa, and Asia. Thirteen species are North American, and 11 of these occur in the West. *Perizoma custodiata* (Guenée)(FW 1.2–1.5 cm; Plate 32.17, Plate 62.22 larva) ranges from eastern Washington south to northern Baja California, Utah, New Mexico, and western Texas. There are several flights in coastal central California through much of the year, and it flies year-round on the Channel Islands. The larvae feed on various Chenopodiaceae including *Atriplex* and *Grayia spinosa*. Comstock and Dammers studied the life history and described the last instar as pale blue green with many faint white longitudinal lines. The intersegmental areas are edged with white and the spiracles are yellow. The head is pale blue gray with numerous small brown spots. Pupation is probably in litter below the host in a light silk cocoon. The larvae rest at a 45° angle to host branches and are, thus, excellent green twig mimics. The larva and adult of *P. costiguttata* (Hulst)(FW 1.0–1.2 cm; Plate 32.18) were illustrated by Miller and Hammond. This moth ranges from southern British Columbia and Idaho south to central California and southern Arizona. In most inland locations there is a single flight in March and April, but on the coast at Big Creek, Monterey County, we found it from March to May, as well as during July and September. The larval host is ocean spray, *Holodiscus discolor* (Rosaceae). The larvae are unmarked red brown.

Anticlea Stephen comprises 17 species that are scattered around much of the World. Four species are North American, and all occur in the West. *Anticlea vasiliata* Guenée (FW 1.3–1.6 cm; Plate 32.19) occurs from southern Alaska and British Columbia east across Canada to Quebec and Nova Scotia, thence south to northwestern California and New Jersey. Adults have a single spring flight from March to June. The larval host is *Rubus* (Rosaceae). The larvae are green with a slightly rugose cuticle and longitudinal faint white lines. The head is green and slightly rugose. The pupae overwinter.

TRIBE STAMNODINI

Stamnodes Guenée comprises 62 species that occur in many world regions but primarily in China, Japan, and eastern Eurasia. There are 34 North American species, all of which are western except *S. gibbicostata* (Walker). Ferguson synonymizes under *Stamnodes* the species formerly treated as *Marmopteryx* Packard and *Stamnoctenis* Warren. Adult *Stamnodes* perch with wings closed above their back in butterfly-like posture, and, often, the most diagnostic and colorful pattern is on the underside of the HW. For this reason, we have illustrated the ventral aspects of these species. *Stamnodes morrisata* (Hulst)(FW 1.4–1.7 cm; Plate 32.20 dorsal, Plate 32.21 ventral) is found from southern British Columbia east to Montana and south to eastern California, Arizona, New Mexico, and western Texas. Adults fly in June to August. The larval host is *Juniperus scopulorum* (Cupressaceae). Larvae are green with roughened cuticle and a fine gray middorsal line. The prothoracic plate is pale with brown marks, and the anal plate is brown with a longitudinal middorsal groove. The head is pale brown and roughened with a few darker brown flecks on the lateral lobes. Pupae hibernate. *Stamnodes albiapicata* Grossbeck (FW 1.5–1.7 cm; Plate 32.22 dorsal, Plate 32.23 ventral) occurs in southern California (Los Angeles, Riverside, and San Diego counties), including Santa Catalina Island, and southern Arizona. In Arizona, adults fly in January, in southern California during February. Larval hosts are various Hydrophyllaceae including *Nemophila*, *Phacelia cicutaria*, and *Pholistoma*. *Stamnodes tessellata* (Packard)(FW 1.5–1.8 cm; Plate 32.24 dorsal, Plate 32.25 ventral) ranges in pinyon-juniper woodland from southeastern California and western Arizona east across the intermountain west to western Colorado and western New Mexico. Adults have a single flight from mid-April to early July. *Stamnodes seiferti* (Neumoegen)(FW 1.7–2.0 cm; Plate 32.26 dorsal, Plate 32.27 ventral) ranges from Arizona east across southern New Mexico to western Texas. It has a single flight in the fall, during September, and uses mountain-mahogany, *Cercocarpus* (Rosaceae), as its larval food plant. *Stamnodes animata* (Pearsall)(FW 1.5–1.7 cm; Plate 32.28 dorsal, Plate 32.29 ventral) ranges from eastern Oregon and eastern California east across Nevada to Utah, western Wyoming, and western Colorado. Adults have a single flight from mid-May to mid-July, with occasional records in early September. The larval host is mountain-mahogany, *Cercocarpus ledifolius* (Rosaceae), and its distribution agrees well with the distribution of the moth. The life history and ecology were described by Furniss et al. Eggs are laid in groups on the underside of host leaves. Severe mass defoliations were described by Furniss et al. in northwestern Nevada. The presence of *Formica* ants that tended scale insects was correlated with stands that escaped defoliation by *Stamnodes* larvae; predation of the eggs and larvae by the ants is the presumed cause. Full-grown larvae were polyphenic; most were green or gray, but rare individuals were black. *Stamnodes marmorata* (Packard)(FW 1.5–1.7 cm; Plate 32.30 dorsal, Plate 32.31 ventral) occurs from eastern Washington east to Montana and south to eastern California, Nevada, Utah, and Colorado. There is a single flight from May to mid-August. Adults are usually found in abundance in stands of mountain-mahogany, *Cercocarpus* (Rosaceae). *Stamnodes formosata* (Strecker)(FW 1.3–1.5 cm; Plate 32.32) ranges from Utah and Colorado south to Arizona and New Mexico. *Stamnodes fervifactaria* (Grote)(FW 1.2–1.5 cm; Plate 32.33 dorsal, Plate 33.34 ventral) extends from southern Colorado to southern Arizona. Adults have a single flight during August. *Stamnodes topazata* (Strecker)(FW 1.1–1.2 cm; Plate 32.35 dorsal, white form, Plate 32.36 ventral, white form, Plate 32.37 dorsal, orange form, Plate 32.38 ventral, orange form) occurs from Alaska south along the Pacific Coast to central Oregon and along the Rockies to Colorado. There are two geographically limited color forms, white and orange. The white form extends from Alaska to Oregon, whereas all from Colorado are orange. Adults have been found from late March (coastal Oregon) to late July (alpine of Colorado south to Park County).

Xanthorhoe Hübner comprises approximately 280 species worldwide including 26 North American species of which at least 22 occur in the West. *Xanthorhoe munitata* Hübner (FW 1.4–1.5 cm; Plate 32.39) is Holarctic. In North America, the moth occurs from Alaska and British Columbia east to Manitoba and south to central California, Utah, and Colorado. Adults are nocturnal and fly from June through August, mainly July. Larval hosts are several broad-leaved herbaceous plants including *Senecio* (Asteraceae) and *Stellaria* (Caryophyllaceae). The larvae are pale and mottled with purplish on the first through fifth abdominal segments. There is a black patch anterior to seta D1. The head is pale with a brown herringbone pattern on the lateral lobes. Larvae overwinter, at least in the British Isles. *Xanthorhoe defensaria* (Guenée)(FW 1.2–1.5 cm; Plate 32.40) is found from Alaska, British Columbia, and east to western Manitoba and south to southern California, Arizona, and Colorado. Adults may have several flights along the Pacific coast throughout the year but are most abundant from March to June, and they fly only from June to August at more interior high-elevation sites. The larval hosts are several broad-leaved shrubs and trees including alder (Betulaceae), maple (Aceraceae), currant or gooseberry (Grossulariaceae), and willow (Salicaceae). The larvae have the thorax and last abdominal segments green, whereas the remainder of the body is pale cinnamon. The dorsum has small black dots arranged in the pattern of an X. There is a wavy dark brown lateral line and numerous fine white longitudinal lines on the dorsum and venter.

Epirrhoe Hübner includes 26 species that are more or less scattered worldwide. Four species are found in North America with at least three in the West. *Epirrhoe plebeculata* Guenée (FW 1.1–1.2 cm; Plate 32.41) occurs from British Columbia east to Alberta and south to southern California, Utah, and Colorado. Adults are diurnal and have a single spring flight from January to early April in central California and from March to June in British Columbia. The larval hosts are *Galium* species (Rubiaceae). The larvae are red brown to gray brown with a black middorsal line on the thorax and first four abdominal segments. The first five abdominal segments each have an inverted black or dark brown triangular mark tipped with white. The spiracular line is white. The venter is gray with four longitudinal white lines. The head is pale tan and spotted with black. McFarland describes the cocoon as compact with several layers of silk and interwoven soil particles.

Euphyia Hübner comprises 175 species worldwide; four are found in North American species, all of which are found in the West. *Euphyia implicata* (Guenée)(FW 1.4–1.6 cm; Plate 33.1) is found in loose sand habitats from central California, including several sites on the offshore Channel Islands, south to Baja California Norte. Adults are found throughout much of the year and probably have several generations. The larval hosts are various sand verbenas, *Abronia* (Verbenaceae); larvae are leaf miners in the early instars but feed externally later on. We reared the species several times on *A. maritima*. During the day, the larvae hide under the sand or under buried plant parts. The ultimate-instar larva is rose pink with many scattered straw yellow, light brown, and olive dots. Spiracles are black. The head is rose pink with many scattered dots.

Enchoria Hulst, with three described species, is endemic to the Pacific Coast. *Enchoria lacteata* (Packard)(FW 0.9–1.1 cm; Plate 33.2) is found from British Columbia south to central California. The adults are diurnal and fly from late February to early May in grassy openings or along woodland edges. The known larval hosts are species of miner's lettuce, *Claytonia* (Portulacaceae). The larvae are variably dark fuscous, light gray brown, or buff with a broad black middorsal stripe that is edged with yellow; in addition, there are two thin black lines between the dorsal and spiracular line. The head is pale brown with small brown dots and an inverted V-shaped brown mark on the upper portion.

Zenophleps Hulst has four species, all endemic to western North America. *Zenophleps lignicolorata* Packard (FW 1.2–1.5 cm; Plate 33.3) is found from Alaska, British Columbia, and southwestern Alberta south to California, Arizona, New Mexico, and western Texas. There is one flight of adults during July and August in British Columbia, but the species is seemingly bivoltine (double-brooded) in lowland California, January to May, September to December. The larval hosts include conifers such as Engelmann spruce, *Picea engelmannii*, and limber pine, *Pinus flexilis* (Pinaceae), but both McFarland and our surveys found larvae feeding on *Galium* (Rubiaceae). The larvae are gray with darker gray diamond marks on the dorsum of abdominal segments 2 to 5. The head is gray with darker gray herringbone pattern on the epicranial lines and lateral lobes.

Orthonama Hübner includes 28 species, all Neotropical except one species from eastern North America, and *O. obstipata* (Fabricius), the "gem" (FW 0.8–1.0 cm; Plate 33.4), which is Holarctic and, in North America, ranges from Oregon and California east to western Nebraska, western Oklahoma, and Texas. In coastal California, the moths are found throughout the year from January to December. McFarland notes peaks in the Santa Monica Mountains during October to November. In Colorado, we have records from the end of May through July. Larval hosts include *Polygonum* and *Rumex* (Polygonaceae). Larvae vary from green to reddish brown with longitudinal indistinct geminate stripes and five purplish intersegmental bands crossing the abdomen. The head is either pale brown dotted with dark brown or white dotted and striped with red.

TRIBE ASTHENINI

Venusia Curtis comprises 43 species, mostly Indo-Oriental. In North America, all five species have at least a portion of their range in West. *Venusia cambrica* Curtis, the "Welsh wave" (FW 1.1–1.3 cm; Plate 33.5), ranges from Alaska east across Canada and south to northern California. There is a single flight of adults from March to mid-August. Eggs are laid on emerging leaves and twigs. Larvae have been reported to feed on new leaves and buds of *Alnus* and *Sorbus*, Betulaceae. Larvae are green or yellow green and may have a red triangular mark on the dorsum of each abdominal segment. Winter is passed in the pupal stage.

Trichodezia Warren encompasses four species, one in eastern Eurasia, with the other three in western North American. *Trichodezia albovittata* (Guenée), the "white-striped black" (FW 0.9–1.1 cm; Plate 33.6), is transcontinental being found from Alaska and British Columbia east across southern Canada to Quebec and south to Oregon, western South Dakota, and, in the East, to North Carolina. Adults fly from May to September. The reported larval hosts are *Epilobium* (Onagraceae), *Thalictrum* (Ranunculaceae), and *Impatiens* (Balsaminiaceae). Reports by McGugan that conifers serve as hosts require confirmation. Larvae are green and have a black bar along each side of the thorax at the base of the legs. The head is pale green with a black bar across the upper portion of the clypeus—thus meeting the anterior ends of the two thoracic bars. *Trichodezia californiata* (Packard) is diurnal and flies from late February to August. We have reared the species twice from larvae found on *Oxalis oregana* (Oxalidaceae) in coastal redwood forest. This is unusual because Lepidoptera very rarely use any Oxalidaceae as their hosts.

TRIBE OPEROPHTERINI

Epirrita Hübner comprises nine described species found in the Holarctic region, including three species in North America, two western endemics and a third ranging across the Holarctic, including the West. *Epirrita autumnata* Borkhaus, the "green velvet looper" (FW 1.7–1.9 cm; Plate 33.7), is Holarctic. In North America, it occurs from British Columbia east across Canada to Labrador and Newfoundland and south to central California, and, in the East, to Pennsylvania and New York. Adults have a single fall flight from October to December. The eggs overwinter. The larval hosts are a variety of conifers and broad-leaved trees and shrubs. The larvae are green to yellow green and faintly reticulate with a gray green to dark green middorsal line and a white lateral subspiracular line. The head is faintly reticulate pale green.

Operophtera Hübner comprises 14 species limited to the Holarctic, mainly in eastern Eurasia. There are four North American species, three of which are found in the West, but the application of species names has been confused. As a result, the application of biological information from the literature is difficult. *Operophtera bruceata* Hulst, the "Bruce spanworm" (FW 1.6–1.8 cm; Plate 33.8, Plate 33.9, Plate 62.23 larva), ranges from coast to coast. In the West, the moth is found from Alaska south to British Columbia, Alberta, Manitoba, and western North Dakota. The adults fly in the fall af-

ter a few frosts, usually in October. Eggs are laid singly in bark crevices, overwinter, and hatch in the first warm days of spring. Many first-instar larvae spin down on silk threads and are dispersed by the wind. Larval hosts include *Acer* (Aceraceae), *Populus tremuloides* and *Salix* (Salicaceae), and *Malus* (Rosaceae). The larvae are pale green or green with gray or darker green middorsal and ad-dorsal lines. Prominent white subdorsal lines may be present, according to Miller and Hammond. The head is finely reticulate gray, green, or russet green. In Alberta, there are occasional population outbreaks that have defoliated large extents of aspen; they also are capable of causing damage to apple orchards. The larva and adult of *Operophtera danbyi* (Hulst)(FW 1.6–1.8 cm; Plate 33.10), which ranges from southern British Columbia south to central California, are illustrated by Miller and Hammond. One larval host is ocean spray, *Holodiscus discolor* (Rosaceae).

Tescalsia Ferguson comprises two little-known species with flightless females found in sandy areas of the Deep Springs Valley and Owens Valley of Inyo County, eastern California, and an adjacent part of Mineral County, Nevada. Both sexes lack a proboscis and the tympana, the latter being characteristic of most other geometrids. Females have straplike FW that bear long setae and have exceptionally long legs that are apparently used for climbing (Fig. 218). The males are fully winged and have HW veins Sc and R fused for half of their length resulting in an exceptionally long cell. *Tescalsia giulianiata* Ferguson (FW 1.2–1.8 cm; Plate 33.12 female, Plate 33.13 male, Fig. 14 venation, Fig. 218 adult female, Figs. 203, 204 genitalia) is known from three California localities and is active during frigid evenings during November and December. After D. Giuliani discovered the first female, we collected others in pitfall traps. Cooperating entomologists including DeBenedictis, Giuliani, and Powell investigated the systematics and behavior of this bizarre species. Although females have been found resting on

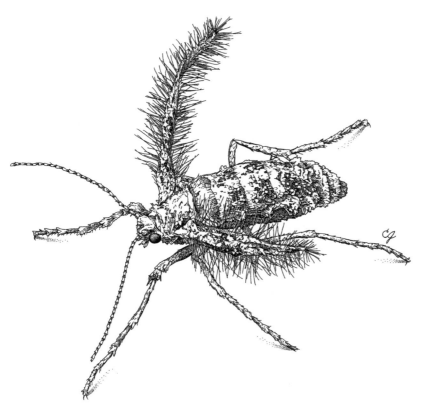

FIGURE 218. *Tescalsia giulianiata* Ferguson, adult female [Tina Jordan, Powell and Ferguson 1994].

sand and caught in pitfall traps set in sandy areas, their long legs seem best suited for climbing about on the twigs and branches of shrubs. Adults are active at temperatures of 0 to –3°C. Eggs were obtained from a captive female, but the first instars did not feed on any offered plant, although several *Atriplex* species (Chenopodiaceae) are the most likely possibility.

TRIBE EUDULINI

Eubaphe Hübner, with 26 species, has a Neotropical distribution south to Argentina. There are five species reported for North America, three in the East and two in West. *Eubaphe unicolor* Robinson (FW 0.9–1.2 cm; Plate 33.14) ranges from Colorado and Kansas south to Arizona, New Mexico, and Texas. The adults have multiple generations where the climate permits. We collected adults from late April to September in the Rocky Mountain states. The larvae are reported to feed on violets, *Viola* (Violaceae), and McFarland found that the larvae feed on *Hybanthus* (Violaceae) in the Huachuca Mountains of Arizona.

TRIBE EUPITHECIINI

Eupithecia Curtis is by far the largest genus with approximately 1,400 species worldwide listed by Scoble. The species are found mainly in more temperate portions of Eurasia, Africa, and the Americas. The moths are generally absent from Australia, the Indo-Pacific, and the more warm to hot tropical portions of world. There are relatively few *Eupithecia* species that may safely be identified without dissection; and there are a number of undescribed species, especially in western North America. Adults of the vast majority of species are nocturnal and are easily attracted to light. They usually rest with their wings fully outstretched. Some species, especially in the far north, are diurnal or crepuscular, and some have been seen nectaring at flowers. Most species are single-brooded. Larvae of all species feed solitarily, and most are polyphagous. Most feed on leaves, but some feed on flowers, and a very few feed on fruits or seeds. Several Hawaiian species are exceptional by being predaceous on small insects such as flies. Most species overwinter as pupae, but a very few overwinter as either larvae or eggs. In southeastern Arizona, the adults of some species have their only annual flight during the winter months! The American fauna was treated in detail by McDunnough, and the Canadian fauna was revised in great detail by Bolte. Miller and Hammond illustrate the larvae of a number of species, including several that are not included in the discussion below. There are about 160 described species in North America, only about a dozen of which do not extend into the West. Bolte describes several species groups within the genus and we have organized the species by these groups.

PALPATA SPECIES GROUP

Eupithecia pseudotsugata Mackay (FW 0.9–1.1 cm; Plate 33.15) occurs in southern British Columbia and western Alberta south in the Rocky Mountains to Colorado. In British Columbia adults have a single flight from late May to late June; while in Rocky Mountain National Park their flight is from June to September. The usual larval host is Douglas-fir, *Pseudotsuga menziesii* (Pinaceae); ponderosa pine and Engelmann spruce are used on rare occasions. The ultimate-instar larvae are rugose and pale green with a fine gray middorsal line. There are two forms, one with red on the dorsum, and the other without such

color. The head is rugose and is either pale green or russet with red reticulation on the parietal lobes. *Eupithecia edna* (Hulst)(FW 1.0–1.1 cm; not illustrated) ranges from Colorado and Utah south to Arizona and New Mexico. A record from New Jersey mentioned by McDunnough seems unlikely and may be due to mislabeled material. There are two flights of adults from May to early September. The life history is unreported.

STELLATA SPECIES GROUP

Eupithecia bryanti Taylor (FW 0.9–1.0 cm; Plate 33.16) is found from Yukon and British Columbia east to western Alberta and south to California. There is a single adult flight from late June to early August in British Columbia and Colorado, but collection dates in central California include April, May, and September. The larval hosts are willows, Salix species (Salicaceae). *Eupithecia misturata* (Hulst)(FW 0.8–1.1 cm; Plate 33.17) ranges from Yukon and British Columbia east to Newfoundland and Nova Scotia, thence south to southern California, Utah, and New Mexico in the West, and the Appalachians of North Carolina in the East. The adult flight is from late March to early September; McDunnough stated the species was double-brooded, and this seems likely in coastal California. In Colorado, at a range of elevations, adults were collected from late May to September, and two generations might be possible. Larval hosts are a wide variety of conifers and woody broad-leaved trees and shrubs including *Larix* and *Picea glauca* (Pinaceae), *Salix* (Salicaceae), *Frangula* and *Ceanothus* flowers (Rhamnaceae), *Spiraea* flowers and *Malus* (Rosaceae), and *Baccharis* (Asteraceae). The larvae are white, pale green, or rose brown with a broad dorsal smoky band and a subdorsal series of brown or yellow triangular marks. The thoracic legs are ringed with brown. The head is white or tan with brown dots on the parietal lobes. *Eupithecia lafontaineata* Bolte (FW 0.9–1.0 cm; Plate 33.32) ranges from British Columbia east to eastern Alberta and south to northern California and Colorado. The species is univoltine with flight records from early June to early August. The life history is unreported.

MUTATA SPECIES GROUP

Eupithecia anticaria Walker (FW 0.9–1.1 cm; Plate 33.18) ranges from British Columbia east across Canada to Newfoundland and south to central Arizona and northern New Mexico. The adult flight is from mid-May to late July. Larval hosts are a variety of herbaceous angiosperms, especially composites, and occasionally shrubs. Flowers are usually selected for feeding. *Eupithecia graefii* (Hulst)(FW 1.0–1.3 cm; Plate 33.30) ranges from coastal Alaska south along the Pacific Coast to California; it is most often found coastally but may occur inland. The adults are distinguished by their large size and black or ruddy oblique discal dash. The adults are variable in color and pattern, and there are two described geographic subspecies. Adults may be found year-round, and have multiple broods, but are most abundant from April to August. Larval hosts are manzanitas, *Arctostaphylos*, or Pacific madrone, *Arbutus menziesii* (Ericaceae). Miller and Hammond described the last instars as variably colored green, pink, to red pink, usually matching the color of their host-plant leaves or twigs.

NEVADATA SPECIES GROUP

Eupithecia nevadata Packard (FW 1.1–1.3 cm; Plate 33.19, Fig. 216 genitalia) occurs from southern British Columbia south to

California and Colorado. In British Columbia and Colorado, there is a single spring flight from early April to mid-June; while in central California, it takes place from March to May. In the Pacific Northwest, larvae are common feeders on *Ceanothus* (Rhamnaceae) and *Purshia* (Rosaceae), in arid brush land or pinyon-juniper forests. We raised it twice from larvae found feeding on *Frangula californica* (Rhamnaceae). In the Santa Monica Mountains McFarland found the larvae fed on deer weed, *Lotus scoparius* (Fabaceae). Miller and Hammond describe the larvae as dark green above and lighter green below, with a white spiracular line that is edged below with irregular red streaks. ***Eupithecia ravocastaliata*** Packard (FW 1.1–1.2 cm; Plate 33.20, Plate 62.24 larva) ranges from British Columbia east to Nova Scotia and south to central California, northern Colorado, and western South Dakota. Adults from coastal Oregon and California are paler with a more darkly marked FW costa than those from Alberta and Colorado. Adults have a single flight from late April to late June (as early as February in coastal California). Larval hosts are usually willows, *Salix* (Salicaceae); but other hosts less frequently selected include poplars, *Populus* (Salicaceae); alders, *Alnus*, and birches, *Betula* (Betulaceae); plums, *Prunus* (Rosaceae); cascara, *Frangula* (Rhamnaceae); and Viburnum, *Viburnum* (Caprifoliaceae). Larvae are rugose and green with blue green or olive green dorsum; there is a yellow or whitish subspiracular line that is red edged below. The head is rugose and is unmarked bright green or russet green.

INTERRUPTOFASCIATA SPECIES GROUP

Eupithecia niphadophilata (Dyar)(FW 1.0–1.2 cm; Plate 33.21) ranges from British Columbia east to eastern Alberta and south to Utah and New Mexico. Adults cannot be separated from those of *E. interruptofasciata* Packard except by dissection and examination of the genitalia. The flight is from early July to early October. Larval hosts are junipers, *Juniperus communis* and *J. scopulorum* (Cupressaceae). The larvae are rugose with either a brown or green phase. The two phases also have differently colored lines and stripes. The head is rugose and, paralleling the two phases, respectively, is either pale brown with darker brown herringbone pattern or unmarked russet green.

SATYRATA SPECIES GROUP

Eupithecia satyrata (Dyar)(FW 1.0–1.2 cm; Plate 33.22) is Holarctic; in North America, it ranges from southern Yukon and British Columbia east across Canada to Labrador and Newfoundland and south to the Sierra Nevada of California. Adults may be found from late May to late August. Larval hosts include a wide variety of herbaceous plants and, occasionally, shrubs of broad-leaved angiosperms; the flowering structures are usually eaten. The ultimate-instar larvae are rugose and are quite variable with several color forms: yellow, pale cream, or light brown. Moreover, they may have a variety of brown dorsal markings on the first five or six abdominal segments. The head is rugose, variably colored, and brown dotted. ***Eupithecia nimbicolor*** (Hulst)(FW 1.0–1.2 cm; Plate 33.23) ranges from Alaska and British Columbia east across Canada to Newfoundland and south to California and the White Mountains of Arizona. The flight is from mid-May to early August. The larval diet includes flowers or leaves of a variety of angiosperm shrubs and perennial herbs. McGuffin found that single females, when given a choice of several suitable hosts, would lay their eggs on only one. The larvae are rugose and have color forms that are either pale green or yellow with variable lines and marks. The head is also rugose and varies from pale brown to light yellow green. ***Eupithecia intricata taylorata*** Swett (FW 1.0–1.2 cm; Plate 33.31) ranges from Yukon Territory and coastal British Columbia (including the Queen Charlotte Islands) east across southern Canada to Newfoundland and New Brunswick south to northern California and the Colorado Front Range. Flight records from early May to mid-July suggest a single flight. On the California coast the species may appear as early as April.

CRETACEATA SPECIES GROUP

Eupithecia cretaceata (Packard)(FW 1.3–1.7 cm; Plate 33.24, Fig. 217 genitalia) ranges from Anchorage, Alaska, and western British Columbia east across Canada to western New Brunswick and south to California, Nevada, Utah, and Colorado; in the East it occurs south to New England. Adults can be found from late May to late August. The larval host is false hellebore, *Veratrum viride* (Liliaceae); flowers, seeds, and leaves are eaten. The larvae are rugose and yellow green with black spots; the plate, thoracic legs, and head are black.

INNOTATA SPECIES GROUP

Eupithecia perfusca (Hulst)(FW 1.0–1.2 cm; Plate 33.25) ranges from Yukon Territory and western British Columbia east across Canada and south to the northern Sierra Nevada of California, eastern Arizona, and Colorado. Adults are in flight from mid-June to early August (mid-September in Colorado). The usual larval hosts are *Salix* (Salicaceae), and *Alnus* and *Betula* (Betulaceae). *Malus* and *Prunus* (Rosaceae) are fed on occasionally. Larvae are rugose and polyphenic with either yellow, pale green yellow, or pale brown. The larval markings, as described by McGuffin, are also variable. ***Eupithecia multistrigata*** (Hulst)(FW 1.0–1.2 cm; Plate 33.26) occurs from western British Columbia east to Saskatchewan and south to the San Bernardino Mountains of California and to Colorado. The adults fly from early June to late September. The usual larval hosts are alders, *Alnus* (Betulaceae).

SPECIES GROUP UNDETERMINED

Eupithecia flavigutta (Hulst)(FW 0.9–1.0 cm; Plate 33.28) is distinguished in our fauna by the smoky violaceous FW with two ochreous patches in the terminal area and the ochreous scaling on the dorsum of the thorax. The species ranges from Colorado south to the mountains of Arizona: Chiricahuas, Huachucas, Santa Catalinas, and White mountains. The species seems to be univoltine, as adults have been collected from July to early August. ***Eupithecia biedermanata*** Cassino & Swett sFW 0.9 cm; Plate 33.29) is distinguished by its ferruginous FW. It is found in southeastern Arizona, where it is known from the Huachuca Mountains. Ferris had investigated the taxonomic identity and affinity of this species. The univoltine adults fly from February to April. The larval host is *Arbutus arizonica* (Ericaceae), and when perched on host branches, the adults are a perfect match. ***Eupithecia appendiculata*** McDunnough (FW 0.8–1.0 cm; Plate 33.33) ranges from northern California east to Colorado and south to southern California and southern Arizona. There are probably two broods as there are collection records for adults from early May to early September, including records for the entire span from Montezuma County, Colorado.

Nasusina Pearsall includes five species in North America—all western; four were described from California. *Nasusina vaporata* Pearsall (FW 0.5–0.6 cm; Plate 33.34) is found from central to southern coastal California, where adults fly during April and May. The reported host is chamise, *Adenostoma fasciculatum* (Rosaceae).

Prorella Barnes & McDunnough is restricted to North America and has 15 species; all but one species are western. None has reported hosts. *Prorella opinata* (Pearsall)(FW 0.7–0.9 cm; Plate 33.35) is the most commonly collected and widespread species in Colorado, where it is found abundantly in western Colorado in Mesa and Moffat counties from late June to early September. It is also found sparingly along the Colorado Front Range in Douglas and Larimer counties.

TRIBE LOBOPHORINI

Lithostege Hübner has 50 described species, mainly from Eurasia. Six species, all western, are reported for North America. *Lithostege rotundata* Packard (FW 1.1–1.2 cm; Plate 33.36) ranges from eastern Oregon and Utah south to California, Arizona, and western New Mexico. Adults fly from March to April. McFarland found that the larvae feed on *Gilia sinuata* (Polemoniaceae), especially its flowers, in southeastern Arizona.

Cladara Hulst comprises three North American species, including two from the West. *Cladara limitaria* (Walker), the "yellow-lined forest looper," or "mottled gray carpet" (FW 1.4–1.7 cm; Plate 33.37, Plate 62.25 larva) is transcontinental across southern Canada and south to California (Placer County), eastern Nevada, and central Colorado. Adults are nocturnal and have a single flight from April through June. Eggs are laid on new host foliage in the spring. Larval hosts are primarily conifers, especially spruce, true firs, larch, western hemlock, and Douglas-fir, all Pinaceae. Other reported hosts are among the Cupressaceae, Taxodiaceae, Ericaceae, Rosaceae, and Salicaceae. Last instars are green and granulose with a yellow green subdorsal stripe and yellow spiracular line along each side. The short setae are pale or dark and are blunt. The head is entirely green. Pupae overwinter.

Lobophora Curtis includes six species found in the Holarctic. In North America, one species is Holarctic but not found in the West, whereas four or five occur in the West. *Lobophora magnoliatoidata* (Dyar)(FW 1.1–1.2 cm; Plate 33.38) occurs from British Columbia and Alberta south to northern California, Nevada, and Utah. Adults fly from May to July. The larval host is quaking aspen, *Populus tremuloides* (Salicaceae). The last instar is blue green with the epidermis slightly rugose and covered with a white bloom; blue green. The abdomen is moniliform, tapering to a point; there are both subdorsal and lateral yellow lines. Pupae overwinter.

References for Superfamilies Drepanoidea and Geometroidea

Barnes, W., and J. H. McDunnough 1917. A revision of the genus *Hydriomena* based on the male genitalia. Contributions to the Natural History of the Lepidoptera of North America 4(1): 6–60.

Bolte, K. B. 1988. Guide to the Geometridae of Canada (Lepidoptera). VI: Subfamily Larentiinae. 1: Revision of the genus *Eupithecia*. Memoirs of the Entomological Society of Canada 151: 1–253.

Buckett, J. S. 1971. Revision of the Nearctic genus *Philtraea*. Journal of Research on the Lepidoptera 9(1): 29–64.

Clarke, J. F. G., and F. H. Benjamin 1938. A study of some North American moths allied to the thyatirid genus *Bombycia* Hübner. Bulletin of the southern California Academy of Sciences 37: 55–73.

Covell, C. V., Jr. 1970. A revision of the North American species of the genus *Scopula* (Lepidoptera, Geometridae). Transactions of the American Entomological Society 96: 101–221.

Ferguson, D. C. 1985. The Moths of America North of Mexico, Fascicle 18.1, Geometroidea, Geometridae (part). Wedge Entomological Research Foundation, Washington, DC; 131 pages, 4 color plates.

Furniss, M. M., D. C. Ferguson, K. W. Voget, J. W. Burkhardt, A. R. Tiedemann, and J. L. Oldemeyer 1988. Taxonomy, life history, and ecology of a mountain-mahogany defoliator, *Stamnodes animata* (Pearsall), in Nevada. Fish and Wildlife Research 3: 1–26.

Kettlewell, H. B. D. 1955. Selection experiments on industrial melanism in the Lepidoptera. Heredity 9: 323–342.

McDunnough, J. H. 1949. Revision of the North American species of the genus *Eupithecia* (Lepidoptera: Geometridae). Bulletin of the American Museum of Natural History 93(8): 537–729, plates 26–32.

McDunnough, J. H. 1954. The species of the genus *Hydriomena* occurring in America north of Mexico (Geometridae, Larentiinae). Bulletin of the American Museum of Natural History 104(3): 237–358, 3 plates.

McGuffin, W. C. 1958. Larvae of the Nearctic Larentiinae (Lepidoptera: Geometridae). Canadian Entomologist Supplement 8; 104 pages.

McGuffin, W. C. 1967. Guide to the Geometridae of Canada (Lepidoptera). I: Subfamily Sterrhinae. Memoirs of the Entomological Society of Canada 50; 67 pages.

McGuffin, W. C. 1972. Guide to the Geometridae of Canada (Lepidoptera). II: Subfamily Ennominae 1. Memoirs of the Entomological Society of Canada 86; 159 pages.

McGuffin, W. C. 1977. Guide to the Geometridae of Canada (Lepidoptera). II: Subfamily Ennominae 2. Memoirs of the Entomological Society of Canada 101; 191 pages.

McGuffin, W. C. 1981. Guide to the Geometridae of Canada (Lepidoptera). II: Subfamily Ennominae 3. Memoirs of the Entomological Society of Canada 117; 153 pages.

McGuffin, W. C. 1987. Guide to the Geometridae of Canada (Lepidoptera). II: Subfamily Ennominae 4. Memoirs of the Entomological Society of Canada 138; 182 pages.

McGuffin, W. C. 1988. Guide to the Geometridae of Canada (Lepidoptera). III, IV, and V: Subfamilies Archiearinae, Ennochrominae, and Geometrinae. Memoirs of the Entomological Society of Canada 145; 56 pages.

Packard, A. S. 1876. Monograph of the geometrid moths or Phalaenidae of the United States. Report of the U.S. Geological Survey of the Territories, Vol. 10. U.S. Department of the Interior, Washington, DC; 607 pages.

Poole, R. W. 1987. A taxonomic revision of the New World moth genus *Pero* (Lepidoptera, Geometridae). U.S.D.A., Agricultural Research Service, Technical Bulletin 1698: 1–257.

Rindge, F. H. 1954. A revision of the genus *Tornos* (Lepidoptera, Geometridae). Bulletin of the American Museum of Natural History 104(2): 177–236, 8 plates.

Rindge, F. H. 1959. A revision of *Glaucina*, *Synglochis*, and *Eubarnesia* (Lepidoptera, Geometridae). Bulletin of the American Museum of Natural History 118: 263–365, plates 23–27.

Rindge, F. H. 1961. A revision of the Nacophorini (Lepidoptera, Geometridae). Bulletin of the American Museum of Natural History 123(2): 87–154, 6 plates.

Rindge, F. H. 1964. A revision of the genera *Melanolophia*, *Pherotesia*, and *Melanotesia* (Lepidoptera, Geometridae). Bulletin of the American Museum of Natural History 126(3): 241–434, 8 plates.

Rindge, F. H. 1965. A revision of the Nearctic species of the genus *Glena* (Lepidoptera, Geometridae). Bulletin of the American Museum of Natural History 129(3): 265–306.

Rindge, F. H. 1966. A revision of the moth genus *Anacamptodes* (Lepidoptera, Geometridae). Bulletin of the American Museum of Natural History 132(3): 175–243, 8 plates.

Rindge, F. H. 1970. A revision of the moth genera *Hulstina* and *Pterotaea* (Lepidoptera, Geometridae). Bulletin of the American Museum of Natural History 142(4): 255–342, 8 plates.

Rindge, F. H. 1972. A revision of the moth genus *Mericisca* (Lepidoptera, Geometridae). Bulletin of the American Museum of Natural History 149(4): 341–406.

Rindge, F. H. 1973. A revision of the moth genera *Nepterotaea* and *Chesiadodes* (Lepidoptera, Geometridae). Bulletin of the American Museum of Natural History 152(4): 205–252.

Rindge, F. H. 1974. A revision of the moth genus *Animomyia* (Lepidoptera, Geometridae). American Museum Novitates 2554: 1–23.

Rindge, F. H. 1975. A revision of the New World Bistonini (Lepidoptera, Geometridae). Bulletin of the American Museum of Natural History 156(2): 69–156.

Rindge, F. H. 1976. A revision of the moth genus *Plataea* (Lepidoptera, Geometridae). American Museum Novitates 2595: 1–27, 42 figs.

Rindge, F. H. 1978. A revision of the genus *Sabulodes* (Lepidoptera, Geometridae). Bulletin of the American Museum of Natural History 160(4): 193–292.

Rindge, F. H. 1978. A revision of the moth genus *Xanthotype* (Lepidoptera, Geometridae). American Museum Novitates 2659: 1–24.

Rindge, F. H. 1979. A revision of the North American moths of the genus *Lomographa* (Lepidoptera, Geometridae). American Museum Novitates 2673: 1–18.

Rindge, F. H. 1980. A revision of the moth genus *Somatolophia* (Lepidoptera, Geometridae). Bulletin of the American Museum of Natural History 165(3): 291–334.

Rindge, F. H. 1981. A revision of the moth genera *Meris* and *Nemeris* (Lepidoptera, Geometridae). American Museum Novitates 2710: 1–28.

Rindge, F. H. 1983. A generic revision of the New World Nacophorini (Lepidoptera, Geometridae). Bulletin of the American Museum of Natural History 175(2): 147–262.

Rindge, F. H. 1990. A revision of the Melanolophiini (Lepidoptera, Geometridae). Bulletin of the American Museum of Natural History 199: 1–147.

Scoble, M. J. 1999. Geometrid Moths of the World: A Catalogue (Lepidoptera: Geometridae). 2 vols. CSIRO Publishing, Canberra, Australia, and Apollo Books, Stenstrup, Denmark; 1015 pages, 258 index pages.

Troubridge, J. T., and J.D. Lafontaine 2007. The Geometroidea of Canada. Canadian Biodiversity Information Facility, Government of Canada. Available at www.gbif.gc.ca/spp_pages/geometroidea.phps/geoimage_e.php.

Wagner, D. L., D. C. Ferguson, T.L. McCabe, and R. C. Reardon 2001. Geometroid Caterpillars of Northeastern and Appalachian Forests. Forest Health Technology Enterprise Team FHTET-2001-10. U.S. Department of Agriculture, Forest Service, Morgantown, WV; 239 pages.

Superfamily Mimallonoidea

FAMILY MIMALLONIDAE

Mimallonidae is the only family in this distinctive superfamily. Mimallonids are characterized by the short rami (sometimes absent altogether) on the distal part of the male antennae—contrasted with longer rami basally—proboscis short or absent, and the second segment of the labial palpus shorter than the first segment; hind tibiae are without medial spurs, and the HW has the Sc and R veins strongly arched beyond the wing base. On the dorsum of the pupal segments the second through seventh abdominal segments there are grooves with dentate edges. The adults are medium to large, are stout bodied, and have broad wings. The family is primarily Neotropical with about 200 species and roughly 30 genera. Four species have been found in North America, and two of these are found in southeastern Arizona, according to Franclemont.

Lacosoma arizonicum Dyar (FW 1.2–1.6 cm; not illustrated) is a medium-sized moth with the sexes similar, unlike those of its eastern North American relative *Lacosoma chiridota* Grote. The species is known only from southeastern Arizona, but is likely to be found in adjacent Sonora, Mexico. The moths are pale pinkish brown with slightly falcate FW apex and serrate wing margins. There is a blurred blackish median line on both wings. Both sexes are nocturnal and are attracted to light; the adults do not fly until late at night (2–4 a.m.) and, as a result, are seldom seen. There seems to be one flight from June to August. A larva was found in an elongate, oval case on oak in southeastern Arizona, as reported by Franclemont.

Cincinnus melsheimeri (Harris), the "Melsheimer's sackbearer" (FW 1.8–2.0 cm; Plate 37.8, Fig. 219 genitalia), is widespread in the eastern two-thirds of the United States, including Colorado and Arizona, and is also found in southern Ontario and northern Mexico, according to Franclemont. The sexes are similar and are gray throughout with scattered black scales. The FW is distinctly falcate. There is an even, thin black median line bent slightly near the FW costa. In Colorado, the moth flies from late May through June. The species is nocturnal, and both sexes are attracted to light. Larvae have been reared only from oaks, *Quercus* (Fagaceae), but the specific host in Arizona is not reported.

Superfamily Lasiocampoidea

This superfamily comprises the Anthelidae, which are restricted to Australia and New Guinea together with the cosmopolitan Lasiocampidae. Lemaire and Minet describe several apomorphic adult characters that unite the superfamily. The adults range from small to very large.

FAMILY LASIOCAMPIDAE

The Lasiocampidae are small to large stout-bodied moths.

Adult The adults lack a thoracic or abdominal tympanum, and mouthparts are lacking or vestigial. The labial palpi are porrect. The compound eyes usually have fine hairs between the facets. The antennae are bipectinate to the tip and have a bare patch with sensory setae on the most-basal segment. The wings are coupled by tiny interlocking setae, and there is no retinaculum or frenulum as in other advanced Lepidoptera. Lasiocampids most often are tan, brown, or blackish, with the females often much larger than males and with extremely large abdomens. The wing shape, color, and pattern of the two sexes are often so different that one might not guess they belong to the same species! The eggs are fully developed on emergence, and the flight ability of females is reduced. Both sexes are attracted to light, except for a few species where the males fly in search of females in the afternoon.

Larva Larvae are often gregarious, as in *Malacosoma* species (tent caterpillars), whose tents are frequent on certain shrubs and trees throughout temperate North America. Larvae often rest in silken shelters or in patches on tree trunks by day and forage after dark. Movements between resting and feeding sites are often accomplished in head-to-tail fashion, thus the name "processionary caterpillars." *Gloveria howardi* (Dyar) of southeastern Arizona is an excellent example of this behavior.

Larval Foods Several species of lasiocampids, notably several *Malacosoma*, are serious forest pests, including defoliation of ornamental trees in urban and suburban settings. Lasiocampids use a wide array of plants on a worldwide basis, primarily broad-leaved woody trees and shrubs, although some species feed on conifers or mistletoes. In western North America, Betulaceae, Fagaceae, Rosaceae, and Salicaceae are most often selected as larval hosts, but four western species, *Tolype laricis* (Fitch), *T. dayi* Blackmore, *T. lowriei* Barnes & McDunnough, and *Gloveria arizonensis* Packard, feed only on conifers (Cupressaceae and Pinaceae). Late instars sometimes wander and may be found on hosts unrelated to those selected by ovipositing females. The caterpillars often wander and form their pupae in tightly woven cocoons, often with urticating larval hairs attached. The silk from these cocoons has been historically used to make silk fabric and is still used today in some parts of the world, according to Franclemont. The original silk

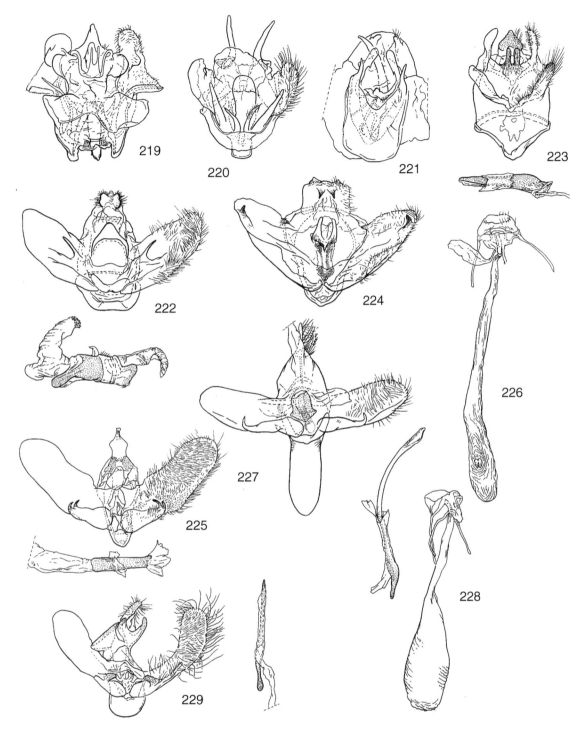

FIGURES 219-229. Genitalia structure of exemplar species of Mimallonoidea, Lasiocampoidea, and Bombycoidea, ventral aspect, aedeagus in situ in Figs. 219–221, removed and shown below in Figs. 222, 223, 225, 227, 229. **219,** Mimallonidae: *Cincinnus melsheimeri* Barnes ♂. **220,** Lasiocampidae: *Tolype glenwoodi* Barnes ♂. **221,** Lasiocampidae: *Malacosoma californicum* (Packard) ♂ [Franclemont 1973]. **222,** Saturniidae: *Sphingicampa heiligbrodti* (Harvey) ♂, aedeagus below. **223,** Saturniidae: *Hemileuca eglanterina* (Boisduval) ♂, aedeagus below. **224,** Saturniidae: *Saturnia mendocino* Behrens ♂, lacks aedeagus [Ferguson 1971–1972]. **225,** Sphingidae: *Agrius cingulata* (Fabricius) ♂, aedeagus below. **226,** Sphingidae: *Agrius cingulata* ♀, A10 removed. **227,** Sphingidae: *Eumoirpha achemon* (Drury) ♂, aedeagus below, right. **228,** Sphingidae: *Eumorpha achemon* ♀, A10 removed. **229,** Sphingidae: *Hemaris diffinis* (Boisduval) ♂, aedeagus to right [Hodges 1970].

made by the ancient Greeks was probably from lasiocampid cocoons. The larval hairs can be irritating and are known to cause a mild dermatitis.

Diversity The family is cosmopolitan with roughly 1,500 species among approximately 50 genera; it is best represented in the tropics and is absent from New Zealand. In North America, Franclemont detailed 10 genera with 36 species. Nine genera and 29 species are either western or include some western areas in their ranges. The Lasiocampidae are divided into five subfamilies by Lemaire and Minet. Three subfamilies are found only in Africa and Eurasia, while two are found in the Americas.

SUBFAMILY MACROMPHALINAE

The Macromphalinae includes three of our genera: *Tolype, Apotolype*, and *Artace*; the remainder are distributed in the Lasiocampinae—a diverse lot parceled out among several tribes.

North American Macromphalinae are medium-sized moths with a combination of gray, white, and black. Adult females are larger than males and have thicker, heavier abdomens. Females of most species have tufts of long scales extending from the distal tip of the abdomen. Larvae are hairy and strongly flattened—"lappet caterpillars." Most western species (nine) are in *Tolype*, and there are single species each in *Apotolype* and *Artace*.

All **Tolype** Hübner, a North American genus with some species existing south into Mexico, are various shades of gray and white, with a median thoracic crest or tuft of curly, metallic black scales. Females usually are paler than males. **Tolype glenwoodi** Barnes (FW 1.4–1.8 cm; Plate 38.9, Plate 62.28 larva, Fig. 220 genitalia) ranges from western Colorado south to southeastern Arizona and likely further south into the Sierra Madre Occidentale of Mexico. Adults are nocturnal and are attracted to light, especially in the early morning. There is a single flight from August to October in the northern portion of the range, but two flights are documented by Franclemont in southeastern Arizona—June to early July, and September to early November. The larvae are external feeders on Gambel oak, *Quercus gambelii* (Fagaceae), and possibly other oaks. The larva was described by Franclemont as brownish gray, much flattened, with lateroventral lappets (extensions). **Tolype lowriei** Barnes & McDunnough (FW 1.7–2.2 cm; not illustrated) is known only from the San Francisco Bay region of central California. Larvae have been reared from pines, *Pinus* species (Pinaceae), and were described by Barnes and McDunnough as gray brown with dorsal black diamond-shaped patches and tinges of yellow. The adults have a single flight in July and August. Other western *Tolype* include **T. velleda** (Stoll), which is widespread in eastern North America and extends into the eastern Rocky Mountain foothills, and **T. distincta** French, which appears to be widespread in the West.

Apotolype brevicrista (Dyar)(FW 1.2–1.4 cm; Plate 38.4) ranges from western Texas to central Arizona, and thence south to central Mexico. Adults are found from March to September, and there are several broods annually. The adults are quite variable and seem to show seasonality with spring and some late-summer individuals that are darker gray than those found in late summer. The adults are similar to those of *Tolype,* but the metallic thoracic patch is much smaller and inconspicuous. McFarland has reared this species from mesquite, *Prosopis glandulosa* var. *juliflora* (Fabaceae), in southeastern Arizona.

Artace colaria Franclemont (male FW 1.5 cm, female FW 2.3 cm; Plate 38.5) is one of two North American species of a genus whose species extend south into South America. Larvae are similar to those of *Tolype,* according to Franclemont, but have a prominent verruca on the dorsum of the eighth abdominal segment. The present species is known from southeastern Arizona and may occur in adjacent portions of New Mexico and south into the Sierra Madre Occidentale of Mexico. Adults are predominantly white with tiny black dots on both wings. The HW has a black adterminal line. The nocturnal adults are attracted to light and have been found from June to October. There may be two broods. The species feeds on manzanita, *Arctostaphylos pungens* (Ericaceae), according to McFarland, and he found that the larvae are red brown and closely match the color of the host-plant stems.

SUBFAMILY LASIOCAMPINAE

Lasiocampinae includes 16 western species in six genera; two other species are restricted to eastern North America. The adults are small to very large brown, tan, or blackish moths, often with medial FW bands or lines. Eggs are most often laid in clusters, and the early instars are gregarious in both resting and feeding sites. Larvae are either less flattened than those of the macromphalines *(Phyllodesma)* or are cylindrical and densely clothed with secondary setae.

The Holarctic genus **Phyllodesma** Hübner is represented in North America by three species, according to Franclemont. One of these, *Phyllodesma carpinifolia* (Boisduval), is limited to the southeastern United States, while *P. americana* (Harris), the "American lappet moth" (FW 1.4–1.8 cm; Plate 38.6), is found throughout in forested temperate and subalpine boreal habitats in the remainder of the United States and southern Canada. *Phyllodesma* species are distinguished by their distinctive appearance at rest or as museum specimens. The outer margin of its FW is slightly scalloped, and there are significant excavations of the FW inner margin and the HW costal margin. The adults are variably colored, ranging from gray to red orange brown; they are excellent dead-leaf mimics. The species is double-brooded in much its range but is single-brooded in cooler montane situations. Spring individuals tend to be grayer, while late summer individuals are more of a unicolorous red orange. The moths are quite variable within and between geographic areas, and past attempts to divide the species into named subspecies and forms was likely not a productive venture. Attempts to apply infraspecific names are more or less futile. Adults are nocturnal and readily attracted to light. There are one or two generations each year depending on the length of the growing season. The adults have a single generation with flight from mid-March to early July. Larval hosts include many broad-leaved trees and shrubs including oaks, *Quercus* (Fagaceae), as well as poplars, *Populus*, and willows, *Salix* (Salicaceae). The larvae rest by day, closely appressed to a host twig or branch. Later-instar larvae are silver and gray with faint black maculations, according to Miller and Hammond. There are faint thin yellow transverse bands and long silver ventrolateral hairs. There are two spectacular orange transverse bands hidden between skin folds between the first two abdominal segments; these may be displayed when the larva is disturbed by arching and uncurling the forward portion of its body. The dorsum of the eighth abdominal segment has a slightly raised area, and there are ventrolateral extensions (lappets). **Phyllodesma coturnix** Lajonquière is found widely in California and, presumably, western Nevada; it may be separated from sympatric *P. americana* only by examination of the male genitalia.

Malacosoma Hübner, known as "tent caterpillars," are best known for the silken communal tents that are so obvious in the landscape during their late-spring feeding periods. The genus is Holarctic, with six North American species of which five occur in the West. The systematic relationships of the North American species have been studied in great detail by Stehr and Cook. These are medium, brown, or tan moths with a rounded FW. The females are larger and heavier bodied than males and have relatively poor flight ability. The antemedian and postmedian FW lines form a median band in all species—sometimes with a different ground color than the remainder of the wing. The adults all have a single late-spring or summer flight, are nocturnal, and are readily attracted to light. Females deposit clusters of 150 to 250 eggs either as bands encircling twigs or as masses on limbs. Furniss and Carolin relate that the eggs clusters, which overwinter, are usually covered by the females with a yellow to dark brown frothy substance. Larvae are not flattened but of typical cylindrical caterpillar form. They are often brightly colored with dense tufts of long hair, especially so laterally. Reddish brown is a recurring color but black, white, tan, and blue gray are frequent. Patterns are quite variable and include lines, bands, spots, and blotches. The early instars of several species live communally in dense silken tents on their host plant and feed gregariously. The larvae spin a silken mat for molting, where many individuals gather and molt more or less simultaneously. On completion of feeding, the larvae become solitary and wander in search of suitable pupation sites. The cocoons are not covered with larval hair but are filled with a pale yellow or white powder, which will erupt outwardly if the cocoon is torn open. McFarland notes that this is diagnostic for *Malacosoma* cocoons. Adults of several species are difficult to separate, especially without reference to host information or larval color patterns, but male genitalia may be used to reliably distinguish the species. *Malacosoma disstria* Hübner, the "forest tent caterpillar" (FW 1.1–1.6 cm; Plate 37.10, Plate 62.32 larva), is the most widespread North American species occurring in boreal forests across southern Canada and all of the United States except Nevada and Arizona. The larvae often occur in outbreaks and strip quaking aspens, *Populus tremuloides* (Salicaceae), of their foliage over large areas. The species will accept other hosts such as paper birch, *Betula papyrifera* (Betulaceae), and other *Populus* species. The larvae congregate in patches on branches or trunks of the host tree (the larvae do not form tents). Although not as widespread, *M. californicum* (Packard), the "western tent caterpillar" (FW 1.2–1.8 cm; Plate 37.9, Plate 62.33 larvae on tent, Fig. 221 genitalia), is more abundant throughout much of the West in temperate and boreal middle-elevation sites, and its silken communal larval tents are familiar on many hosts. There are quite a number of subspecies and local populations that are described by Stehr and Cook. The range extends from southeastern Canada, New York, and the western edge of the high plains west to the Pacific Coast. On a local basis the species may be fairly host specific, but when considered throughout its range, it is more catholic in its choice of hosts. Hosts include various species of *Prunus, Rosa, Purshia tridentata*, all Rosaceae; willows, *Salix,* and aspen and cottonwoods, *Populus* (Salicaceae); oaks, *Quercus* (Fagaceae); and wild lilacs, *Ceanothus* (Rhamnaceae). The other western species are *M. constrictum* (Henry Edwards)(Plate 62.34 larva), the "Pacific tent caterpillar," *M. incurvum* (Henry Edwards), the "southwestern tent caterpillar," and *M. tigris* (Dyar), the "Sonoran tent caterpillar." Two of these species, *M. constrictum* and *M. tigris,* use oaks as a larval host, and their larvae do not construct silken tents. *Malacosoma constrictum* is an oak-feeding Californian species that ranges south into Baja California, Mexico, while *M. tigris* feeds on shrubby oaks and occurs from Colorado in the southern Rocky Mountains and southern Great Plains to the Southwest and Mexico. The third of these, *M. incurvum,* makes tents on Fremont cottonwood, *Populus fremontii,* and other *Populus* species (Salicaceae). It occurs in the desert Southwest and ranges from there to southern Mexico.

There are five western *Gloveria* Packard. These are all dark chocolate brown to tan large moths with densely hairy eyes. The species are strikingly dimorphic. The males are diurnal and have swift flight in the afternoon as they search for receptive females. The heavy-bodied larger females are nocturnal and are found at lights. The males are smaller with more narrow, pointed wings. All species have a single annual flight.

The most widespread is *G. arizonensis* Packard (male FW 2.5–2.7 cm, female FW 3.1–4.0 cm; Plate 34.1 male, Plate 34.2 female, Plate 62.27 larva), a dark brown moth that ranges from eastern Montana and Colorado west to California and south to Arizona, New Mexico, and western Texas. Most likely, its range extends south into Mexico. It is often common at lights. There is a single flight from late June to August. Males fly rapidly in the afternoon, while females are nocturnal and are attracted to light. The larvae feed on junipers, *Juniperus,* and cypresses, *Cupressus* (Cupressaceae), as well as pines, *Pinus* (Pinaceae). Clusters of ivory-colored eggs are laid on the outer limbs of their host during August. Larvae feed in groups for brief periods in the evening but remain inactive at other times. The larva is sooty black with a dark gray middorsal band that bears gray white, long hairlike setae. Laterally, there are diagonal patches of red brown hair, and one or two light spots per segment. Subdorsally, the hairlike setae are gray white. Pupation is within a loosely woven translucent cocoon. The pupae are brown black, strongly pitted, and covered with waxy yellow pile. *Gloveria gargamelle* (Strecker)(male FW 2.8–2.9 cm, female FW 3.5–4.3 cm; Plate 34.3 male, Plate 34.4 female) ranges from central Arizona to western Texas and, presumably, south into Mexico. The adults are found in July and August. Males fly in late afternoon, while females are nocturnal. The larvae feed on leaves of several oaks, *Quercus* species (Fagaceae). *Gloveria medusa* (Strecker)(male FW 2.8 cm , female FW 3.5–3.6 cm; Plate 34.5 male, Plate 34.6 female) is endemic to coastal southern California (San Luis Obispo County south to San Diego County) with a flight during May and June. Females are readily attracted to light, and the males are not found at light and are diurnal. Host plants include California buckwheat, *Eriogonum fasciculatum* (Polygonaceae), and wild lilac, *Ceanothus verrucosus* (Rhamnaceae); and we reared *G. medusa* from eggs to adults on laurel sumac, *Malosma laurina* (Anacardiaceae). J. A. Comstock reports that the eggs are laid in single-layered patches encircling the host stems. Larvae feed gregariously. The larvae are gray speckled with black and have a white middorsal band containing a thin black longitudinal stripe. The body is covered with long gray, white, and tan hairlike setae. The spiracles are yellow and ringed with black. The pupa, formed in spring, is black, pitted, and covered with short orange brown hairs. There is a pad of short recurved spicules at the cremaster. The larva spins a thin loose cocoon. *Gloveria howardi* (Dyar)(FW 2.7–3.1 cm; Plate 34.7, Plate 62.29 larva) is a smaller tan species found in southeastern Arizona, where it is apparently a periodic colonist from the neighboring Sierra Madre Occidentale of Mexico. It can be absent from Arizona for many years at a time. The larvae feed gregariously on oaks and, according to McFarland, live in globular silken nests.

Quadrina diazoma Grote (male FW 2.7–3.0 cm, female FW 3.5–3.8 cm; Plate 34.8 male, Plate 34.9 female) is the only species in this genus that has sparsely hairy eyes. The adults are nocturnal and are attracted to light. There is a single flight in July and August. Comstock described the life stages but mistakenly attributed them to another genus. The full-grown larvae are mottled black with speckled narrow gray transverse bands. There are fleshy tubercles on the dorsum of the prothorax; and there is also a middorsal brush of long yellow hairs emanating from the thorax. The remainder of the body is covered with short black setae and long black hairs. The spiracles are orange with black rims. The head is light yellow tan and is covered with raised brown dots and dense yellow brown hairs.

Caloecia juvenalis (Barnes & McDunnough)(male FW 3.1–3.9 cm, female FW 3.5–4.3 cm; Plate 35.1 male, Plate 62.31

larva) and *C. entima* Franclemont (FW 3.4–4.0 cm; Plate 35.2 female) are both known from the Chiricahua Mountains of southeastern Arizona and extend south into the Sierra Madre Occidentale to at least Durango. *Caloecia juvenalis* is much more common and has its single flight in July and August, while *C. entima* is much rarer and flies during June and July. Adults of both are attracted to light. Little is known of their biology or hosts.

Dichogaster coronada (Barnes)(male FW 3.1–4.5 cm, female FW 4.5–5.0 cm; Plate 34.10 male, Plate 34.11 female, Plate 62.30 larva) is the only described species in its genus, which was characterized by Franclemont by the absence of hairlike setae on its eyes. The FW color is quite variable and ranges from pale tan to deep brown to reddish brown. Both sexes are nocturnal and are readily attracted to lights. The males do not fly until the early morning. Adults may be found from June to August. The life history has been studied in the Huachuca Mountains of southeastern Arizona by McFarland who found that the larval host is Emory oak, *Quercus emoryi* (Fagaceae). The young larvae are gregarious and become solitary in later instars; they develop from September to the following April and pupate in May.

Superfamily Bombycoidea

These are the macromoths that have no thoracic or abdominal tympana. The group is distinguished by deep clefts between the prescutum and mesoscutum of the mesothorax. The larvae have abdominal prolegs on the third through sixth and tenth abdominal segments fully developed, reduced, or absent; in other features, such as secondary setae, they vary greatly. There are 12 families that include Bombycidae, Saturniidae, and Sphingidae. Other bombycoid families are Eupterotidae, worldwide with about 300 species; Endromidae, Mirinidae (Palaearctic), and Carthaeidae (Australian), each with one or two species; and two small Eurasian and African families related to Sphingidae: Lemoniidae and Brahmaeidae. As presently considered these are medium to large moths with primarily nocturnal habits, although some Saturniidae and Sphingidae are strictly diurnal. Forecoxae are anteriorly fused in the last-instar larvae, the eighth abdominal larval segment has the D1 seta arising from a mid-dorsal protuberance, and the adult FW has veins Rs1 and Rs2 closely parallel to stem Rs3 + Rs4. The Bombycoidea includes the largest and best-known moths and is worldwide in occurrence. The silk spun in cocoons is the source of various silk fabrics, including the well-known commercial silk spun by the presumed Eurasian "silkworm" *Bombyx mori* (Linnaeus). The adults range from those with the longest proboscies in Lepidoptera capable of pollination and long-distance flight to those with nonfunctioning adult proboscis. Eggs are laid singly or in masses. The larvae range from polyphagous to monophagous. Some feed communally, at least in the early instars, and the well-known processionary caterpillars are included here. Larvae of some species and the later instars of others feed solitarily. The larvae of most feed or rest on stems or foliage without the benefit of shelters. Most have cryptic coloration and are presumably quite palatable. Stinging or urticaceous spines are possessed by larvae of some Saturniidae. Pupation is in the ground with or without a cocoon or aboveground with a tightly spun cocoon.

FAMILY BOMBYCIDAE

This is a pantropical family with four subfamilies, only one of which, Apatelodinae, occurs in North America.

SUBFAMILY APATELODINAE

Of the six species found north of Mexico, only two occur in the West. Adults of Apatelodinae are medium, with retracted head and irregularly shaped FW outer margin. At rest, the wings are held relatively remote from the body in a flat, rooflike manner. Larvae are densely hairy. The larvae are external feeders and feed on a wide variety of broad-leaved plants, especially trees.

Apatelodes pudefacta Dyar (FW 1.7–1.8 cm; Plate 35.3, Plate 64.36 larvae) occurs only in southeastern Arizona but will likely also be found in adjacent New Mexico and south into Sonora, Mexico, according to Franclemont. This distinctive moth may be found during the late summer in sites close to the border, especially in southern Cochise County and Santa Cruz County. Adults are nocturnal and are attracted to lights. In southeastern Arizona, McFarland has found that *A. pudefacta* larvae feed on *Baccharis bigelovii* (Asteraceae). He has noted that the larvae are densely covered with yellow, white, brown, and black hairlike setae. The full-grown larva pupates in a cell in the soil. Tropical species usually spin a cocoon, but those that occur in North America do not. The second western apatelodid, *Olceclostera seraphica* (Dyar)(FW 2.0–2.5 cm; Plate 35.7 male, Plate 35.8 female), described from Brownsville, Texas, has been reared by Kendall from larvae found in Culberson County, western Texas. The species is also known from Big Bend National Park and southern Arizona (Pima and Santa Cruz counties). There is a single flight from mid-July to early August. The life history is unreported.

FAMILY SATURNIIDAE

Adult The wild silk moths include small to extremely large moths (3–28 cm). Some of the world's more spectacular moths such as the Atlas moth (*Attacus* spp.) and the luna moths and relatives (*Actias* spp.) are included. In our area, species are medium to large and most are colorful. A few genera are primarily diurnal, while most are nocturnal and are attracted to light. The proboscis and maxillary palpi are nonfunctional, being either absent or poorly developed. As a result, the adults do not feed and are relatively short-lived. Antennae are relatively short; they are bi- or quadripectinate in males and filiform to quadripectinate in females. Adults are broad winged with the wings densely scaled. Many species have a central eyespot, often transparent, on each HW, which is thought to deter vertebrate predators. It is common for many species to feign death when closely approached by presumed predators. This is done either with the wings spread, displaying eyespots that may deter vertebrate predators, or with the wings closed above the body and the exposed abdomen in a curled position, sometimes revealing brightly colored intersegmental membranes. Courtship and mating occur in the early morning hours for nocturnal species and from late morning to late afternoon for most diurnal species.

Larva Eggs are usually laid in large masses, and larvae of early instars often rest and feed gregariously. The larvae some species are "processionary" in that they file in head-to-tail lines to and from resting and feeding sites. The larvae of all species are stout and eruciform and are covered with a fairly dense pattern of verrucae or scoli that bear clusters of long hairs, often urticating. In our species these may sting and cause water blisters in humans. Feeding may be either diurnal or nocturnal. The later instars feed solitarily.

Larval Foods The majority of species feed on woody, broad-leaved angiosperms, and most western species are relatively specific in their choice of hosts—one or a few species in a genus or just a few genera. Exceptions include the larvae of a few western species of *Hemileuca* and *Automeris* that are grass feeders. The larvae of *Hyalophora euryalus* will eat Douglas-fir, a conifer, among broad-leaved hosts, and all *Coloradia* feed on only one or a few pine species. Three subfamilies are represented in the West. According to Tuskes et al. the larvae of our western Ceratocampinae (*Citheronia, Eacles, Anisota, Sphingicampa*, and *Adeloneivaia*) feed on leaves of dicotyledonous trees and shrubs. Each species has a relatively narrow choice of hosts that is most often limited to a single plant family—sometimes a single genus—but may span a few plant families. Where more than a single plant is used, local populations may be more specific. Tuskes et al. also report that the larvae of Hemileucinae (*Coloradia, Hemileuca*, and *Automeris* species) range from being relatively specific to a single host genus or even to a single host species. Some species of *Automeris*, especially *A. io*, are relatively polyphagous in their choice and utilization of host plants. Larvae of western Saturniinae (*Saturnia, Agapema, Antheraea, Rothschildia, Eupackardia*, and *Hyalophora*) range from quite specific in *Saturnia* and *Agapema* to polyphagous in *Antheraea, Rothschildia, Eupackardia*, and *Hyalophora*.

Pupa Larvae of saturniid moths most often pupate on their larval hosts aboveground. The larvae make tightly spun silken cocoons that are sometimes covered with leaf or twig fragments. In the genus *Hemileuca* most larvae leave their host and make a silken cell in the leaf litter or duff, within which they pupate. Pupae of some species, especially those of *Hemileuca* and *Coloradia*, may diapause several years before eclosing. Peigler and Maldanado relate how the cocoons of *Eupackardia calleta* and *Rothschildia cincta* have been used for centuries by several Native American tribes as ankle rattles for ceremonial dances.

Diversity There are about 1,500 species and 60 genera worldwide, with most occurring at low to middle latitudes. The Neotropical region has the most species and is most diverse in the South American Andes. In North America, about 70 species in 17 genera are documented, and about 47 species in 15 genera have been found in the West. Although the family is divided into nine subfamilies, only three occur in North America, including the West. These are the Ceratocampinae, Hemileucinae, and Saturniinae. In the Hemileucinae, almost all *Coloradia* and *Hemileuca* are found in the West. In fact, between them 20 species are primarily western—comprising more than 40% of the family's western representation.

SUBFAMILY CERATOCAMPINAE

The western Ceratocampinae comprises six genera. Five of them are represented by only a single western species, but *Sphingicampa* contains three.

Citheronia splendens sinaloaensis Hoffman (FW 4.8–7.0 cm; Plate 35.6, Plate 62.35 larva) occurs from southeastern Arizona south into Mexico. This is an attractive moth similar to the eastern *C. regalis*, the "Imperial moth," but differs in having much darker gray black ground and a more contrasting pattern of dark red vein outlining and creamy maculations. This species has a single flight in July and August. Adults are attracted to light in the early morning; mating takes place between 01:00 and 03:30. Eggs are laid in groups of one to four on host leaves. According to Tuskes et al. the larvae feed on wild cotton, *Gossypium thurberi* (Malvaceae), manzanita, *Arctostaphylos pungens* (Ericaceae), and New Mexico wild sumac,

Rhus virens, and squawbush, *Rhus trilobata* (both Anacardiaceae). Often only a single host is used by each population. Last-instar larvae are purplish brown with the thoracic scoli light brown at the base with black tips. The full-fed larva burrows into the ground and forms a pupal chamber.

Eacles oslari Rothschild (FW 5.1–6.8 cm; Plate 35.5, Plate 62.36 larva) occurs from southeastern Arizona south to Sinaloa, Mexico. The attractive moth is similar to the eastern *E. imperialis* (Drury) and has four striking color forms: yellow, red brown, chocolate, and pale lavender. Like the previous species, adults are attracted to light in the early morning when eclosion and mating often take place. Its larvae feed on Mexican blue oak, *Quercus oblongifolia*, and Emory oak, *Q. emoryi*, both Fagaceae; western soapberry, *Sapindus saponaria* variety *drummondii* (Sapindaceae); and *Pinus discolor* (Pinaceae). It is likely that other hosts are eaten, at least on occasion. Last-instar larvae are either brown or green with turquoise spiracles that are ringed with black. The full-fed larvae dig an underground pupal chamber.

Anisota Hübner has most of its species in eastern North America and only *A. oslari* Rothschild (FW 2.2–4.0 cm; Plate 35.9 male, Plate 35.10 female) occurs in the West, where it ranges from southwestern Colorado south through western Texas, southern New Mexico, and southeastern Arizona into Chihuahua, Mexico. There is a single brood in July and August; males fly by day and females by night. Courtship and mating occur in the morning (08:30–11:30). The larvae feed on leaves of oaks, most often Mexican blue oak, *Quercus oblongifolia*, scrub oak, *Q. turbinella*, and Emory oak, *Q. emoryi*—all Fagaceae. The eggs are laid in clusters, and the young larvae feed gregariously. Last instars are brick red with a variably broken black spiracular stripe.

Three species of ***Sphingicampa*** Walsh are found in the West, and all three extend south into western Mexico. ***Sphingicampa hubbardi*** (Dyar)(FW 2.2–3.3 cm; Plate 36.11, Plate 63.1 larva) is most commonly encountered and ranges from central Texas west across southern New Mexico and Arizona to eastern San Bernardino County, California, according to Tuskes et al. Adults have two flights in spring and midsummer, and males are readily attracted to light. The larvae feed on Catclaw acacia, *Acacia greggii* var. *wrightii*, honey mesquite, *Prosopis glandulosa*, and little-leaf palo verde, *Cercidium microphyllum*, all Fabaceae. Last-instar larvae are green and are often blue green ventrally. The enlarged thoracic scoli are yellow green, and the abdominal scoli, when present, are silver. There is a lateral white line edged with red or pink along its upper edge. Two other *Sphingicampa* are western. ***Sphingicampa montana*** (Packard)(FW 2.8–3.9 cm; Plate 36.10) ranges from southeastern Arizona south in western Mexico to Jalisco. There is a single flight from mid-July to early August. *Mimosa dysocarpa* (Mimosaceae) is its larval host, and last instars are green with a lateral white line edged dorsally with red. The body is covered with minute yellow points and the thoracic and posterior enlarged scoli are often red; the remainder of the scoli, when present, are silver. ***Sphingicampa raspa*** (Boisduval)(FW 2.6–3.1 cm; Plate 36.9) is found in southeastern Arizona, southern New Mexico, and western Texas, and thence southward through much of Mexico to at least Oaxaca. There is a single annual generation with adults during July and early August; there are a few mid- to late-September records of adults that suggest there might be a second brood. Adults fly late at night or early in the morning. Tuskes et al. found that the larval host is *Acacia angustissima* (Mimosaceae); the last-instar larvae are similar to those of *S. montana* but have blue and yellow dorsal scoli and orange spiracles.

SUBFAMILY HEMILEUCINAE

In the subfamily Hemileucinae, four of eight species of *Coloradia* Blake are western, while the others are Mexican endemics. All *Coloradia* species use pines as their larval hosts. The larvae of all species have urticating spines. The most widespread is *C. pandora* Blake, the "Pandora moth" (FW 3.2–4.4 cm; Plate 36.1); *C. p. davisi* Barnes & Benjamin (Plate 36.2) is found from western Texas and the Black Hills west to California and south into northern Mexico. Pandora is biennial and flies in mid- to late summer, June to July. Pandora's larval hosts are various pines, Pinaceae, especially ponderosa pine, *Pinus ponderosa*, Jeffrey pine, *P. jeffreyi*, lodgepole pine, *P. contorta*, and pinyon pine, *P. edulis*. Coulter pine, *P. coulteri*, and sugar pine, *P. lambertiana*, are sometimes eaten in California. The last-instar larvae of *C. pandora* may be black, brown, or green; there is a series of segmental white lines or bars along each side as well as a subspiracular white stripe. *Coloradia pandora* is an important defoliator of pines in the West, especially where soils are loose enough for the larvae to bury themselves prior to pupation; this happens primarily in forests on pumice and decomposed granitic soils. Furniss and Carolin called attention to an outbreak of several thousand acres in southern Oregon; they also state that outbreaks begin about every 20 to 30 years and continue for six to eight years after their onset. Significant losses of pine stands occur only during the heaviest of population outbreaks. The larvae complete development in two years. Eggs are laid in small batches near the base of host trees, and larvae feed gregariously. Young larvae overwinter in small groups on pine needles for the first winter. Pupation occurs in a silken chamber just below the soil surface, and the second winter is passed by this stage. Pupal diapause for up to six years has been reported by Carolin. Native Americans utilized larvae as a food source in eastern California. The three other *Coloradia* found in the West are all smaller and fly earlier in the year. *Coloradia doris* Barnes, the "Black Hills Pandora moth" (FW 2.6–3.4 cm; Plate 36.4), and *C. luski* Barnes & Benjamin (FW 2.5–3.1 cm; not illustrated) are mainly Rocky Mountain and Great Basin species that range south to southeastern Arizona, while *C. velda* Johnson & Walter (FW 2.9–3.0 cm; not illustrated) is endemic to the San Bernardino Mountains. Like *C. pandora*, all three are pine feeders. *Coloradia doris* has had large outbreaks on ponderosa pine in northeastern Wyoming and the Black Hills of South Dakota. Its larvae have longer and more conspicuously branched spines than those of *C. pandora*.

Perhaps the most intriguing hemileucines are members of the genus *Hemileuca* Walker, where 16 of the 20 described species are western. These species represent a diversity of adult phenotypes, host-plant utilizations, and daily flight schedules. Most species are brightly colored large moths. Perhaps the two most widespread species are *H. eglanterina* and *H. nevadensis*. Most *Hemileuca* species select woody broadleaf plants as their hosts, but two species, *H. oliviae* and *H. hualapai,* feed on grasses. All *Hemileuca* species deposit their eggs in ringlike bands around host twigs. The eggs overwinter. First-instar larvae are usually all black, and feed in tight clusters until the fourth instar, when they begin feeding solitarily. Feeding and metamorphosis to the adult occurs in the same year except in some populations of the *H. eglanterina* Group that require two years to complete development (biennial). Moreover, some species of desert regions may remain in diapause for as many as four years. McFarland has shown that daily sunlight is necessary for *Hemileuca* larvae to feed and grow. The spines of all *Hemileuca* species are urticaceous and can raise blisters that last up to several days. All species have a single annual flight in spring, summer, or fall, and most species are diurnal with the exception of some grassland and desert species that fly on mild nights in the autumn or early spring. *Hemileuca oliviae* Cockerell, the "range caterpillar" (FW 2.1–3.5 cm; not illustrated), occurs in northeastern New Mexico and southeastern Colorado. The moths fly in the late fall, mid-September to December, often at fairly cold temperatures; eggs are laid in small clusters, and these overwinter until the following wet season. The caterpillars feed on a variety of native and introduced grasses, Poaceae. The last-instar larvae are light yellow with a scattering of small gray marks. The caterpillars can cause damage to rangelands during years of large populations when caterpillars reduce available grass forage and may cause sores on feeding cattle. The moth is unicolorous but quite variable, ranging from pinkish to tan, usually without distinct markings. The closely related *H. hualapai* (Neumoegen)(FW 2.7–3.1 cm; Plate 36.13, Plate 63.4 larva) is found in southeastern Arizona, where it also feeds on grasses (Poaceae). This species is more pinkish than *H. oliviae* and has a yellow FW costa. Some females have gray median FW bands. Unlike the similar *H. oliviae*, this species overwinters in the pupal stage. Adults fly from March to early May, mainly in April; the species is most abundant in April. Eggs hatch with the onset of the July to August monsoon rains, and larvae feed during this period. Last-instar larvae vary from light brown to gray brown, the scoli are black and gold with white tips, and there is a white spiracular band. *Hemileuca tricolor* (Packard)(FW 2.1–3.4 cm; Plate 36.12) ranges from southern Arizona and New Mexico south into Sonora, Sinaloa, and Baja California. The species flies in the spring from late February through March. Larval hosts are several mimosaceous and fabaceous legumes including palo verde, *Parkinsonia microphylla*, mesquite, *Prosopis juliflora*, and catclaw acacia, *Acacia greggii*. Full-grown larvae are gray to gray brown with yellow pinacula and scoli. *Hemileuca nevadensis* Stretch, the "Nevada buck moth" (FW 2.7–3.8 cm; Plate 36.3), and several of its relatives feed on willows and cottonwoods (Salicaceae), oaks (Fagaceae), or woody rosaceous plants. All are primarily black with a broad central white band and black FW discal spot. *Hemileuca nevadensis* ranges over most of the western and northern plains states from southeastern Washington, Montana, southern Manitoba, and Saskatchewan south to southern California, central Arizona, and southern New Mexico. There is a single generation and the diurnal adults fly primarily in October and early November. Its larvae feed almost exclusively on cottonwoods and willows (Salicaceae). Tuskes et al. report an instance of the larvae feeding on coast live oak, *Quercus agrifolia* (Fagaceae), in southern California. Last instars are yellow and black with golden yellow dorsal scoli. A second species in this group is *H. slosseri* Peigler & Stone, which occurs in southwestern Oklahoma, northern Texas, and southeastern New Mexico, is intermediate between the widespread eastern *H. maia* (Drury) and *H. nevadensis* in adult appearance and uses the same larval host genus as *H. maia*—*Quercus* species, Fagaceae. As such, this species is thought by some to represent a stabilized hybrid population. The larvae likely feed on Havard oak, *Q. shumardii* throughout most of its range. *Hemileuca electra* W.G. Wright, the "Electra buck moth" (FW 2.3–3.1 cm; Plate 36.8), is a brightly colored red, charcoal, and white moth that ranges from the

Mojave and Colorado Deserts of southern California south into Baja California, including Cedros Island. There is also a population in the Sonoran Desert of central Arizona; five isolated populations each with distinctive adults were delineated by Tuskes et al. Several of the populations comprise named subspecies *Hemileuca e. clio* Barnes & McDunnough and *H. e. electra mojavensis* Tuskes & McElfresh, but, according to Rubinoff, the unnamed Colorado Desert populations of southeastern California have the most distinctive DNA and are as distinct genetically as a separate species. Adults fly in the fall, mid-September through early December. Larvae feed on flat top or California buckwheat, *Eriogonum fasciculatum* (Polygonaceae), in California, in Arizona, and on Cedros Island but feed on boojum, *Fouquieria columnaris* (Fouquieriaceae), in mainland Baja California. Despite the disparate relations of the two plant host families, Rubinoff found no difference that would support separate species status for Baja's *H. e. rubra* McElfresh & Tuskes. The last-instar larvae are gray brown to black and have black scoli and three white to off-white lateral lines; the intersegmental areas vary from red in coastal populations to gray or black for inland populations. *Hemileuca grotei* Grote & Robinson (FW 2.2–2.9 cm; not illustrated) is a white-banded black moth found in mountain ranges of the Southwest from central Arizona and Colorado east through New Mexico to central Texas. Adult *H. grotei* are diurnal and fly in the fall, September to November. Their larval hosts are oaks including *Q. gambelii*, *Q. fusiformis*, and *Q. marilandica* (all Fagaceae). Last instars vary from gray to light yellow and have yellow seta-bearing pinacula. Intersegmental area range from brown to brick red in different populations, and the head is shiny black. *Hemileuca neumoegeni* Henry Edwards (FW 2.0–2.8 cm; Plate 37.3) ranges from the Mojave Desert of southeastern California through arid lands of the Colorado River drainage to southern Utah, western Colorado, and central Arizona. The adults are nocturnal, and there is a single flight mid-August to early October. Larval hosts are squawbush, *Rhus trilobata* (Anacardiaceae), and desert almond, *Prunus fasciculata* (Rosaceae). Apache plume, *Fallugia paradoxa* (Rosaceae), has also been reported as a possible larval host. The early-instar larvae feed gregariously and the larger larvae feed solitarily at night. The full-grown larvae vary from pale to dark gray and have several cream white lines along the sides and the intersegmental areas red orange. *Hemileuca burnsi* J. H. Watson (FW 2.1–2.9 cm; Plate 37.1) ranges from southern California, central Nevada, and southern Utah south to central Arizona where it is usually found in dry foothills in arid mountain ranges. It is an off-white moth with a white or white-and-black abdomen. It is primarily a Mojave Desert species whose range overlaps with that of *H. neumoegeni* and extends north to Lassen County, California. The species is univoltine with the adults in flight during September and October. *Hemileuca burnsi* males fly by day with females flying at night; courtship and mating take place between 15:30 and dusk. Larval hosts are cotton-thorns, *Tetradymia axillaris* and *T. glabrata* (Asteraceae), indigo-bush, *Psorothamnus arborescens* (Fabaceae), and desert almond, *Prunus fasciculata* (Rosaceae). Older larvae may complete development on other plants such as buckwheat, *Eriogonum* (Polygonaceae), but first-instar larvae cannot survive on these plants. The full-grown larvae are sooty black and covered with black-centered white pinacula; on the first through sixth abdominal segments there is a broken yellow or yellow subspiracular stripe. There are long white hairs and both short white and yellow spines. The spiracles are orange and are rimmed with black. *Hemileuca eglanterina*

(Boisduval), the "common sheep moth" or "brown day moth" (FW 3.1–4.1 cm; Plate 36.7, Plate 63.5 larval cluster, Fig. 223 genitalia), is perhaps the most frequently encountered western *Hemileuca* due to its wide range and abundance. The moths are found from southern British Columbia and southwestern Alberta south through the western United States to California, northern Arizona, and Colorado. There are a number of distinctively marked populations whose taxonomic status is uncertain. The adults are diurnal and fly in summer and fall. Individuals from more montane habitats tend to fly during summer, July to August, while those of more lowland areas fly in the fall, September and October. Furniss and Carolin report that a British Columbia population is biennial. The larvae feed on a broad variety of primarily shrubby Rosaceae and Rhamnaceae, but also on plants in at least four other families. Last instars are variable between and within populations but are usually black with the dorsal rosette scoli yellow to gold; there may be four longitudinal off-white or pink lines, but these may be absent. Two other closely related species are *H. nuttalli* (Strecker), "Nuttall's sheep moth" (FW 3.2–3.9 cm; Plate 36.5), which is yellow orange and occurs mainly in the Great Basin and interior Columbia River drainage. It flies from July to early September. Peak flight of adults is from 13:30 to 16:30, after the peak of the very similar *H. eglanterina*. Calling females of *H. nuttalli* attract only conspecific males in courtship. Larval hosts are primarily bitterbrush, *Purshia tridentata* (Rosaceae), and snowberry, *Symphoricarpos* (Caprifoliaceae), but other plants are used on occasion. Last instars are black and have numerous secondary setae and either black or yellow dorsal rosette setae. There may be a broken white supraspiracular line. In some cases, larvae of *H. nuttalli* may be found feeding together on the same plants as *H. eglanterina* larvae. *Hemileuca hera* (Harris), the "Hera sheep moth" (FW 3.2–4.7 cm; Plate 36.6, Plate 63.3 larva), which is white, ranges over much of the intermountain West but occurs further east and south than does *H. nuttalli*. There is a single adult flight ranging from mid-July to late September, depending on population and subspecies. *Hemileuca hera* larvae feed on big sagebrush, *Artemisia tridentata*, and sand sagebrush, *A. filifolia* (both Asteraceae). Last instars are black or purplish with three lateral white longitudinal lines along each side.

Automeris Hübner is most species rich in the lowland and middle elevations of the American tropics. Males have quadripectinate antennae, and those of the female are bipectinate. All species have a prominent bright, round eyespot in the center of the HW. At rest the HW is covered by the FW, but, if disturbed, the moth is likely to raise the FW, revealing the eyespots. This is thought to be likely to have a startle affect and to deter would-be vertebrate predators. Adults are nocturnal and are readily attracted to light. Eggs are laid in clusters on leaves or thin twigs. The larvae usually feed gregariously in the early instars and are solitary in later instars. The spines of larvae are urticaceous and can raise blisters on tender skin areas. Seven species occur in North America; six species are found in the West and all except *A. zephyria* Grote and possibly *A. patagoniensis* Lemaire, Smith, & Wolfe range south into Mexico. *Automeris io* (Fabricius), the "Io moth" (FW 2.4–4.3 cm; Plate 37.4), is the most widespread member of this primarily Neotropical genus. The Io moth is found in most of eastern temperate North America and ranges west to New Mexico and Utah. Females have reddish brown FW and are larger than the males, which have yellow FW. There is only a single May to June flight in the West. Larvae are polyphagous, and host plants include a wide variety of dicotyledonous trees and shrubs in widely unrelated plant families; even grasses are eaten on occasion. Commonly eaten host genera include *Celtis* (Ulmaceae), *Prunus* and *Rubus* (Rosaceae), *Ribes* (Grossulariaceae), and *Salix* (Salicaceae). Early instars are dull orange, while later instars are bright green with narrow white and red spiracular lines. The larvae pupate in litter or crevices, and the cocoons are often covered with dead leaves and other debris. *Automeris zephyria* Grote (FW 3.5–4.7 cm; Plate 37.2) is found in the high mountains of New Mexico and the Guadalupe Mountains of western Texas. The moths have brown FW with a prominent diagonal white line running from near the apex to the inner margin. The abdomen is primarily red. As is usual in this family females are slightly larger than the males but otherwise are similar. Larvae have been found only on willow in nature, although McFarland found they would readily feed on *Quercus emoryi*, Emory oak (Fagaceae), in captivity. Adults are nocturnal and are attracted to light. There is a single annual flight in mid-May to mid-June. Full-grown larvae are mainly yellow dorsally with a light blue middorsal line. Laterally, they are black with two white spiracular lines. Four other *Automeris* are found in southeastern Arizona: *A. patagoniensis* Lemaire, Smith, & Wolfe, *A. cecrops* (Boisduval), *A. iris* (Walker), and *A. randa* Druce. The first of these species is unusual in being a grass feeder. Larvae of the other three feed on oaks and plants in a few other families.

SUBFAMILY SATURNIINAE

The subfamily Saturniinae occurs on all of the larger continents except Antarctica. This group includes the best silk-makers aside from the domesticated bombycid *Bombyx mori*. All species have eyespots or distinctive discal marks on all four wings. Larvae of all species have scoli that bear urticating spines. Pupae are enclosed in silk cocoons that range from loosely to tightly woven. In North America, eight genera with 15 species are known, but in the West, the saturniine fauna is limited to six genera—*Saturnia, Agapema, Antheraea, Rothschildia, Eupackardia,* and *Hyalophora*—and nine species.

All three North American **Saturnia** Schrank are restricted to the West and are primarily Californian, with one species ranging north to coastal Oregon and the other two extending south to northern Baja California. **Saturnia mendocino** Behrens (FW 2.6–3.3 cm; Plate 38.3, Plate 63.2 larvae, Fig. 224 genitalia) is found in the Cascades, Sierra Nevada foothills, and coast ranges of coastal Oregon and northern California north. The preferred habitats are chaparral and forests with manzanita up to 6,000′. The adults are diurnal and fly during warm spring days primarily between mid-February to mid-April, into June at higher elevations and latitudes. The eggs are laid singly or in small groups on host leaves. The later-instar larvae are variable in color with or without black saddlelike markings. There are orange scoli that bear short black urticating spines. The larvae feed on *Arctostaphylos* species and *Arbutus menziesii* (both Ericaceae) in spring and pupate in a meshlike brown cocoon attached to a stem or branch of the host plant. The closely related **S. walterorum** Hogue & Johnson ranges from southern California south into Baja California. The third species, **S. albofasciata** (Johnson)(FW 1.9–2.7 cm; Plate 38.2), ranges from northern California (Trinity County) south to the Sierra Juarez of Baja California Norte. This species is not found in any areas where there is winter or summer fog. As a generalization, Wells found them primarily above 3,000′ near the point where foothill chaparral gives way to coniferous forests. In contrast

to *S. mendocino, S. albofasciata* flies in the fall from mid-October to early November. McFarland found that the adults have narrow flight times—males flew only between 2 and 4 p.m. under sunny conditions, while females flew in the early evening. The species occurs in montane chaparral habitats where the larvae feed on several species of *Ceanothus* (Rhamnaceae) and *Cercocarpus montanus* (Rosaceae), in spring and early summer. The final instars are green with transverse white lines and white dorsal scoli on the abdominal segments.

Sometimes considered congeneric with *Saturnia*, **Agapema** Neumoegen & Dyar is most often considered a separate genus. There are seven species of *Agapema*, three of which occur in the West, while the others are found in Texas and adjacent Mexico. All three are black, gray, brown, and white and are nocturnal. There are eyespots on all four wings. The most widespread western species, **A. homogena** Dyar (FW 3.2–4.2 cm; Plate 37.7), occurs in montane Arizona, Colorado, New Mexico, and the Guadalupe Mountains of western Texas. The species is known to range south to Mexico City. There is no size difference between the sexes. Adults are nocturnal and come to lights. There is a single flight from May to July, and the species is found in montane habitats. Known larval hosts include plants in three unrelated families—*Frangula californica* (Rhamnaceae), *Ribes cereum* (Grossulariaceae), and *Salix exigua* (Salicaceae). Eggs are laid in clusters, and the early instars feed in large groups. The late-instar larva is bright yellow with mottled black markings, short white secondary setae, and long white setae emanating from the dorsal and subdorsal scoli. Tuskes et al. postulate that the larvae are distasteful and their yellow color and a sweet aromatic odor are warnings to would-be predators. The other western species, **A. anona** (Ottolengui)(FW 2.7–3.3 cm; Plate 37.5), has disjunct populations ranging from southern Arizona east to south-central Texas and south into central Mexico. Unlike *A. homogena*, this species flies during the fall and occurs in arid low-lying environments. Its larval hosts in southeastern Arizona are *Ziziphus zizyphus* and *C. spathulata* (Rhamnaceae), and the final instars are black with numerous white setae. Middorsal setae are bright yellow, and the dorsal scoli are red with black setae. There is a prominent broken white subdorsal line as well as a thinner white spiracular line along each side.

Antheraea Hübner is a cosmopolitan genus that has two species in North America and several additional species in tropical America. The most widespread is **Antheraea polyphemus** (Cramer), the "Polyphemus moth" (FW 4.7–7.4 cm; Plate 37.6, Plate 63.9 larva). Both sexes are very similar except that males have slightly more falcate FW and broadly quadripectinate antennae, and the females have much narrower quadripectinate antennae. The adults are most often an even tan with blue-ringed transparent eyespots on each wing. Western moths may be olive greenish to reddish. The moth ranges from southern Canada throughout the United States, except Nevada and southern Arizona. The moth is often found in woodlands, but it readily adapts to disturbed areas including cities and suburban gardens. This is one of our most familiar moths. In the West, the species has one, two, or more annual flights. In more northern and cooler areas, the species' single annual flight is from May to July. The species is polyphagous, and its host range includes woody plants in several families, but *Betula* (Betulaceae), *Acer* (Aceraceae), *Salix* (Salicaceae), and *Quercus* (Fagaceae) are favored. The species may use a narrow range in any region; for example, in the San Joaquin River delta region of central California, we found that *Rosa* (Rosaceae) is used exclusively even though *Salix* is locally abundant. Only one to three eggs are laid at a time, and the larvae are solitary in all instars. Larvae are green with six longitudinal rows of golden or orange tubercles, each of which emanates from a red spot. Dorsally, there are accordion-like indentations between each segment. The pupa overwinters inside a large oval cocoon that is suspended from a branch or falls to the ground. In southern Arizona and south into adjacent Mexico, one finds **Antheraea oculea** (Neumoegen)(Plate 37.11), which is extremely similar to the former species except that it is more consistently larger (FW 7.0–8.0 cm) and often has extensive black dorsal wing scaling. A narrow hybrid zone between the two species is found in New Mexico and southwestern Colorado. Laboratory hybridization reported by Tuskes et al. demonstrated reduced fertility between the two species. The adults fly in late summer and *Quercus* species (Fagaceae) are most often selected as larval hosts.

Rothschildia cincta (Tepper)(FW 4.8–6.4 cm; Plate 63.6 larva) is the only western representative of a widespread Neotropical genus that comprises 23 species, according to Lemaire. The genus is characterized by the presence of transparent triangular discal patches on each wing, a jagged or irregular postmedian line on the dorsal wing surfaces, and more or less falcate FW, especially in males. The moths are nocturnal and are attracted to light just before and after midnight. There is a single flight during the monsoons in July and August. *Rothschildia cincta* is found rarely in southern Arizona, especially the Baboquivari Mountains, Pima County, and near Pena Blanca, Santa Cruz County; *R. cincta* extends south in the Sierra Madre Occidentale of Mexico to Guererro. Limberbush, *Jatropha*, and hopbush, *Stylosanthes viscosa* (both Euphorbiaceae), are documented hosts, and cocoons have been found on two genera of woody legumes. The last-instar larvae are green with five transverse white stripes, each of which is dark edged anteriorly. The scoli are minute and variable in their presence. Cocoons are suspended by a strong silk peduncle (stem) from the branch of a host or adjacent woody plant.

Eupackardia calleta (Westwood), the "Calleta silkmoth" (FW 4.1–5.4 cm; Plate 38.1, Plate 63.7 larva), is the only member of its genus and has distinctive black adults. In North America, the moth ranges from southern Texas west to extreme southern Arizona. According to Lemaire, it occurs widely in Mexico as far south as Chiapas. The moth has a single annual flight during the summer monsoons (July and August) in the mountains of southeastern Arizona and from late October to mid-January in the low deserts of south-central Arizona. Males are diurnal with most of their activity between sunrise and midday, while females are nocturnal and are attracted to light. In southeastern Arizona, ocotillo, *Fouquieria splendens* (Fouquieriaceae), is the principal host, while for the western Pima County, Arizona, population *Sapium biloculare* (Euphorbiaceae) is preferred. The southern Texas populations have larvae that feed primarily on Texas Ranger, *Leucophyllum frutescens* (Scrophulariaceae). The larvae are quite differently patterned in the three North American populations. In southeastern Arizona, they are blue with variably extensive black mottling and red surrounding the scoli, while in western Pima County, they are light green with white bands and much reduced black and red markings. The larvae of the southern Texas population are blue green with white dorsal bands and orange or yellow at the bases of the scoli. The cocoons are attached to the host plant by a short peduncle and are usually spun in the shade close to the ground.

The temperate North American genus **Hyalophora** Duncan is one of the most familiar to North Americans, especially because of the widespread familiarity with the predominantly eastern *H. cecropia* (Linnaeus). This name is used by many people for any large moth including the western *H. euryalus* (Boisduval). These moths are well known for their use in pioneering research such as that by Carroll Williams on the role of juvenile hormone in metamorphosis. Moreover, the species are easy to rear and hybridize, and the species, especially *H. cecropia*, are popular in classrooms. Most of the species form blend zones with each other, and reproductive isolation is only partially genetic since some reciprocal male-female crosses are not viable. The sexes are quite similar except that females are larger and have much more slender antennae than males. Adults of all species have large opaque white discal comma marks on each wing. The most widespread western species is **H. columbia** (S. I. Smith), including its western subspecies **H. c. gloveri** (Strecker), "Glover's silk moth" (FW 4.8–7.4 cm; Plate 38.8, Plate 63.8 larva). This moth is widespread in the Canadian Rockies and south through the southern Rockies, intermountain ranges, and into northern Mexico. In the interior of British Columbia, this moth forms a broad hybrid blend zone with *H. euryalus*. McFarland points out that in southeastern Arizona adult *H. c. gloveri* are so red as to be very close in appearance to *H. euryalus*. Adults are nocturnal, being most active after midnight when mating and oviposition occur. They are readily attracted to lights, and males are easily attracted to caged females. Adults have a single flight throughout the species' range, usually during late spring (May to June), but in southeastern Arizona, the adults fly during the summer monsoon (July to August). In its eastern range, nominate *H. columbia columbia* is virtually monophagous on tamarack, *Larix laricina* (Pinaceae), while western populations are relatively polyphagous on broad-leaved trees and shrubs of several unrelated plant families, especially Rosaceae, Salicaceae, Rhamnaceae, Ericaceae, and Elaeagnaceae. The moth is more selective in different geographic regions where only one or a few host families are eaten. According to Tuskes et al. eggs are laid in small groups (one to 10) on host leaves. The larvae may be gregarious in early instars but feed solitarily later. Larvae of *H. c. gloveri* are variably green to blue green with enlarged yellow dorsal scoli on the meso- and postthoracic segments as well as the first abdominal segment. The remaining dorsal and all lateral scoli are white and peglike. Cocoons are tapered at both ends, are tied with silk to host twigs or branches, and are cryptically colored. Like those of other *Hyalophora*, the cocoons have inner and outer silk layers. **Hyalophora euryalus** (Boisduval), the "ceanothus silk moth" (FW 5.0–6.7 cm; Plate 38.7), ranges along the Pacific Coast and Cascade–Sierra Nevada axis from British Columbia south to Baja California. In addition, the range extends eastward across Washington and Oregon to extreme western Idaho. The adults are predominantly reddish but otherwise are similar to those of other *Hyalophora*. There is one flight from January to early July, rarely to August, the specific timing depending on local climate. For example, we found the flight extends longer and later along the central California coast (February 22 to August 5 at Big Creek Reserve, Monterey County). Eggs are laid singly or in small groups on host leaves. At one time thought to be a specialist on *Ceanothus*, *H. euryalus* has proven to be catholic in its host-plant choice, according to Tuskes et al. who cite plants in eight families, notably *Ceanothus* and *Frangula californica* (Rhamnaceae), *Salix* (Salicaceae), *Prunus* (Rosaceae), *Arctostaphylos* (Ericaceae), *Alnus* (Betulaceae), *Ribes* (Grossulariaceae), *Rhus* (Anacardiaceae), and *Pseudotsuga menziesii* (Pinaceae), among others. The larvae are green with the dorsal scoli yellow, and the lateral ones are light blue tipped with white. The cocoons are compact, rounded to tear shaped, and are firmly attached to host twigs or branches.

FAMILY SPHINGIDAE

Sphinx or hawk moths occur worldwide, except in Antarctica, and are well known because of their large size, rapid flight, and importance as pollinators of many nocturnal flowers. Many species are strongly attracted to lights, while others are diurnal and may be seen hovering, hummingbird-like, at flowers. Larvae are external feeders and usually may be found on stems without accompanying shelters.

Adult These moths are medium to large (sometimes very large!)(FW 1.5–9.0 cm) streamlined stout-bodied moths with narrow triangular FW and relatively small HW (Fig. 15). The proboscis is long and coiled, and the globular eyes are large and prominent. The antennae are relatively short and are filiform, or lamellate ventrally and either clavate or pectinate dorsally. For almost all taxa, the apex of each antenna is clublike, upturned, or hooked. The legs are strong and heavily spined. The wings are heavily scaled and may be brightly colored or quite dull.

Larva The larvae are moderate to very large (up to 100 mm), cylindrical, naked with or without terminal spike on abdominal segment 8. Spiracles are oval. The head is hypognathous, rounded or angular. Thoracic legs and prolegs are prominent. Crochets are in biordinal mesoseries. Larvae may be cryptically colored green or brown or brightly colored. Larvae of the same species may often have both green and brown morphs. The eggs are upright but relatively flat; they are laid singly by females but may sometimes be placed in loose groups. Larvae feed externally without shelters but often rest inconspicuously during daylight. Pupation is in an earthen cell belowground or in a cocoon in leaf litter at the soil surface. Several species are of importance, notably *Manduca quinquemaculata* and *M. sexta*, which feed on the leaves of Solanaceae including tomato, potato, and tobacco. Home gardeners may see their tomato plants rendered leafless, seemingly overnight, by a few of these larvae.

Larval Foods The host range is often relatively host specific, limited to single plant families, or broad in the cases of *Erinnyis ello* and *Hyles lineata*. Host plants are often, but not always, woody and include conifers and especially many broadleaved angiosperms. Sphinginae larvae feed on Convolvulaceae (*Agrius cingulata*), Solanaceae (*M. quinquemaculata* and *M. sexta*), Oleaceae, Ulmaceae (*Ceratomia amyntor*), Ericaceae, Rosaceae, Annonaceae, Bignoniaceae, and Pinaceae (*Sphinx dollii*, *S. sequioiae*), among others. Smerinthinae larvae feed on Salicaceae (*Smerinthus* and *Pachysphinx*), Rosaceae (*Paonias*), and Juglandaceae (*Cressonia juglandis*). Macroglossinae larvae feed on a wide variety of plants, partly due to the polyphagous *Hyles lineata*, but other species' hosts include Apocynaceae, Caprifoliaceae, Vitaceae, Onagraceae, and Rubiaceae. *Hyles euphorbiae* was intentionally introduced as a biological control agent for leafy spurge (Euphorbiaceae).

Pupa Stout and fusiform (cigar shaped) with head narrower than thorax, proboscis is usually long. In some Sphinginae, the proboscis takes the form of a tubular loop. The cremaster is often slightly forked. The pupa may be formed belowground in an earthen cell or in a cocoon among litter at the soil surface.

Diversity More than a thousand species are known worldwide on all continents except Antarctica. The family is richest in the tropics, but 121 have been reported for North America

north of Mexico, and 75 of these, some one-time strays from Mexico, have been recorded from western North America. There are three subfamilies of Sphingidae: Smerinthinae, Sphinginae, and Macroglossinae.

SUBFAMILY SMERINTHINAE

Smerinthinae is probably a paraphyletic mixture of several or more basal and primitive lineages within the family. The plant families used as hosts are primarily those used by other non-sphingid bombycoids, such as Anacardiaceae, Bombacaceae, Fabaceae, Fagaceae, Rosaceae, Salicaceae, and Sterculiaceae.

Pachysphinx occidentalis (Edwards), the "big poplar sphinx," is the one of the largest (FW 6.5–7.1 cm; Plate 39.1, Plate 63.13 larva) common sphingids in our area. This heavy-bodied moth has pale gray to tan FW with rounded projections at each vein ending on the outer margin. This western sphinx ranges through most of the western United States and into northern Mexico, especially at lower elevations. The species has but a single midsummer generation in much of the inter-mountain West, but Rubinoff found that it has up to three generations in more southern lowland habitats such as southern California and southern Arizona (February to September). Its larvae feed on the leaves of cottonwoods, especially *Populus fremontii* and *P. sargentii* (Salicaceae). The full-fed larva of *Pachysphinx occidentalis* is similar to that of *P. modesta*, but the integument is smoother and the caudal horn is two to three times longer. The more widespread congener, **P. modesta** (Harris), the "modest sphinx" (FW 4.4–6.0 cm; Plate 39.2, Plate 63.12 larva), tends to be more montane in the West. This moth ranges from southern interior British Columbia east across Canada, and south through the montane portion of western North America and eastward through much of the eastern United States. It is absent from California, Nevada, and Arizona. *Pachysphinx occidentalis* is larger, paler, and has a disklike black mark near the inner margin of the HW (*P. modesta* has a thicker triangular mark). Although Hodges reports hybrids with *P. occidentalis*, we have not seen such specimens. The larvae are whitish green and have seven oblique lateral white lines. The integument is quite granulose and the caudal horn is quite short. The usual caterpillar hosts are probably quaking aspen, *Populus tremuloides*, and narrowleaf cottonwood, *P. angustifolia* (Salicaceae).

Smerinthus cerisyi Kirby, the "one-eyed sphinx" (FW 3.0–4.1 cm; Plate 39.5, Plate 63.19 larva), ranges across southern Canada and then south through most of the western United States into northern Mexico. Although found primarily in montane habitats, in the Pacific Coast states it occurs in many lowland areas and is perhaps the commonest western sphingid. It is a medium sphingid with a pointed FW, slightly irregular outer margin and a largely pink HW with a blue eyespot with a centered black pupil. To complicate matters, *S. cerisyi* is geographically variable and has a number of named geographic and seasonal color forms. In most of the interior and montane West *S. cerisyi* has but a single spring to midsummer generation, but Rubinoff found it has three generations (March to August) in lowland California. The larvae feed on leaves of various willows, *Salix*, and cottonwoods, *Populus* (Salicaceae). Caterpillars range from pale yellow green to blue green and have a pair of pale yellow subdorsal lines that extend from the head almost to the caudal horn; there are six oblique white bands on each side (reduced in some California larvae). The extremely similar **S. saliceti** Boisduval (FW 3.1–4.2 cm; Plate 39.7) occurs in more arid habitats from southern California and southern Arizona south into Mexico. *Smerinthus saliceti* is best separated by examination of the male genitalia, but Hodges states that the black mark in the blue HW spot crosses it as a V-shaped line. The host plants are likely willows and cottonwood. The moth has two seasonal generations in Arizona. Walsh points out that in southeastern Arizona *S. cerisyi* is found at higher elevations and *S. saliceti* is found at lower elevations. **Smerinthus jamaicensis** (Drury), the "twin-spotted sphinx," is a smallish sphingid (FW 2.8–3.1 cm; Plate 39.3) that ranges across southern Canada from Newfoundland west to British Columbia and in most of the eastern United States. In the West, its range is limited to Colorado, northern montane New Mexico, and central eastern Arizona (White Mountains). A record from Yavapai County, Arizona, should be confirmed. *Smerinthus jamaicensis* is like a smaller version of *S. cerisyi* but is much smaller, has a squared-off FW apex, and has a black line dividing the HW blue eyespot. Adults have rudimentary mouthparts and probably do not feed. When resting, adults have all four wings spread, unlike those of *S. cerisyi*. There is a single late-spring generation in the West. Larvae have been reported to feed on a variety of trees including poplar, willow, birch, elm, ash, apple, and plum. They are somewhat variable and are generally blue green with six or seven oblique pale green white lateral bands and a pair of pale white subdorsal bands running from the first to fourth segments. Moreover, the integument is covered with pale greenish white nodules.

Paonias myops (J. E. Smith), the "small-eyed sphinx," is a small orange brown sphingid (FW 2.7–3.8 cm; Plate 39.4) with irregular outer margin on the FW, in combination with a large orange yellow area on the HW containing a blurry black-outlined blue eyespot. The species ranges across southern Canada from Nova Scotia to British Columbia, in much of the eastern Untied States, and is widely distributed in the western Untied States south into northern Mexico. *Paonias myops* is largely absent from the Pacific Coast west of the Cascade–Sierra Nevada axis. This moth has been treated as a species complex by Eitschberger. The species has a single midsummer generation in the West. Larvae have been raised from a number of woody plants, primary among which are several species of *Prunus*, but *Amelanchier* (Rosaceae) and *Vitis* (Vitaceae) are also reported. They are medium green with six yellow oblique lateral lines and two brown to reddish brown blotches on most body segments—one just above the oblique lines and the other surrounding the spiracle. There are also two yellow lines on the side of the head that meet at the vertex.

SUBFAMILY SPHINGINAE

Most species in this group have an inordinately long proboscis and a tongue case separated from the main pupa. These are preeminent long-distance pollinators and are especially rich in tropical habitats. Most species have barklike gray or brown FW and a swept-back resting posture that shows the FW, but not the HW.

Sagenosoma elsa (Strecker), the "Elsa sphinx" (FW 2.8–3.7 cm; Plate 39.6, Plate 63.16 larva), is a striking gray white and black hawkmoth with somewhat rounded FW. The species is geographically restricted to a small range from southern Colorado and northern New Mexico south through Arizona and New Mexico. It has a single spring generation from May to mid-July. Walsh reports that the larval host is wolfberry or box thorn, *Lycium* (Solanaceae). The larvae are green with six red brown oblique lateral stripes along each side; the stripes are bordered with red along their posterior margins. The integument is slightly granulose.

Ceratomia sonorensis Hodges, the "Sonoran sphinx," is limited to southeastern Arizona and neighboring Sonora, Mexico. The adult (FW 4.3–4.4 cm; not illustrated) has a gray FW and blackish HW. It is most similar to *Manduca florestan* (Cramer) but can be separated by its smaller size and two narrow black FW streaks. There is a single generation in July and August during the rainy season. Its larval hosts are ashes, *Fraxinus* (Oleaceae). It is closely related to *C. undulosa* (Walker), a widespread abundant moth of eastern North America that is rare in the West, although common in the low foothill canyons of the Colorado Front Range.

The genus **Sphinx** Linnaeus has a large number of species, most of which occur in the New World, but a few are found in Eurasia. The species are unified by their black-and-white-banded abdomens. **Sphinx chersis** (Hübner), the "great ash sphinx" (FW 5.0–5.7 cm; Plate 39.8, Plate 63.14 larva), is widespread, ranging from British Columbia to Nova Scotia and through most of the United States, and south into northwestern Mexico. In the West, it occurs in most areas except for the northern intermountain territory where, presumably, suitable hosts are absent. The moth has a gray FW with three or four thin black streaks, a black-and-white-striped HW, and alternating black-and-white marks along each side of its abdomen. There is but a single summer generation in the West, but two generations occur in the Southeast. Its larvae have been reared on a variety of woody hosts including ashes, *Fraxinus* (Oleaceae), *Prunus* species (Rosaceae), and quaking aspen, *Populus tremuloides* (Salicaceae). The larvae have also been reared on lilac, *Syringa vulgaris*, and privet, *Ligustrum* (both Oleaceae), although perhaps these represent laboratory rearings. The caterpillars are pale blue green, but darker on the first three segments; there is also a rare pale pink form. The lateral oblique stripes are pale and edged darker green anteriorly. There is a yellow lateral band on each side of the head that meet at the vertex. The closely related **S. asella** (Rothschild & Jordan) averages smaller (FW 4.1–5.5 cm; Plate 39.9, Plate 63.15 larva) and is extremely similar to *S. chersis* but may be separated by a faint black medial line on each tegula. This species is almost entirely restricted to the intermountain West from the Colorado Front Range west to central Nevada and south to southeastern Arizona and western New Mexico. *Sphinx asella* has a single generation from May to July. Its larval host in the Huachuca Mountains of southeastern Arizona is manzanita, *Arctostaphylos pungens* (Ericaceae), but the species occurs in some locations where manzanitas are absent. **Sphinx dollii** Neumoegen, "Doll's sphinx," is a very small sphingid (FW 2.0–2.6 cm; Plate 40.1) that ranges from Nevada east to western Colorado and south to southeastern California (eastern San Bernardino County), Arizona, New Mexico, and western Texas. It is also known from Sonora, Mexico. Although similar to *S. sequoiae*, in its range there is no other tiny *Sphinx* with muted markings. The adults are somewhat variable in color pattern, perhaps geographically. There is a single brood of adults from late June to early August, although there may be two flights in more southern portions of its range. The larval hosts are various junipers, *Juniperus* (Cupressaceae), and the larvae are spectacular juniper foliage mimics. The color pattern renders the larvae almost invisible when resting on juniper foliage; they are bright green with white bands and rows of white dots as well as orange or white lunules below the spiracles. The caudal horn is blunt, and its base and apex may be either white or pale orange. **Sphinx sequoiae** Boisduval, the "sequoia sphinx" (FW 2.2–3.2 cm; Plate 40.2), ranges from Oregon south through the mountains of California to Baja California Norte. Additionally, a paler form *S. s. engelhardti* extends from southeastern California, southern Nevada, and southwestern Utah to Baja California Norte. There is a single annual generation with adults in flight from May to August. The larval hosts are California juniper, *Juniperus californica*, and Utah juniper, *Juniperus osteosperma* (Cupressaceae). The larvae are bright green and have rows of red brown, white, and yellow brown spots; the spiracles are black and rimmed with white. The caudal horn is short and blunt and is yellow tipped. *Sphinx perelegans* Edwards, the "elegant sphinx," is a large (FW 4.0–4.9 cm; Plate 40.3) sphingid with a black thoracic dorsum, and dark gray FW with a central pale gray area, black-and-white-striped HW, and black-and-white checks along both edges of the abdomen. Its FW lacks the submarginal slightly curved black line of *S. vashti* (see below). It ranges most abundantly along the Pacific Coast from Vancouver Island, British Columbia, to northern Baja California. There are scattered records in other western states, but some of these may represent misidentifications. There is a single generation of adults in spring and early summer. Records in different years range from March to August. On the coast, at Big Creek, Monterey County, we found the moths mainly in May to June, but mostly during April on Santa Cruz Island. In the Sierra Nevada, the moths are found mainly in June. The larval hosts are various manzanitas, *Arctostaphylos* (Ericaceae), as well as hollyleaf cherry, *Prunus ilicifolia* (Rosaceae). *Sphinx vashti* Strecker, the "Vashti sphinx" (FW 3.0–4.1 cm; Plate 40.4, Fig. 15 venation), has contrasting dark and light gray FW with a slightly curved submarginal black line (see *S. perelegans* above). It is quite variable in the intensity of its markings and is often abundant where found. It ranges from British Columbia east to Manitoba and the Great Plains south to California, Nevada, Utah, and Colorado. It is mainly absent west of the montane Cascade–Sierra Nevada axis. There is a single generation from late spring to summer, April to August, depending on latitude and elevation. Its larval hosts are various snowberries, including *Symphoricarpos albus* (Caprifoliaceae). The larvae are green with seven oblique white lines along each side each of which is edged anteriorly with black; the caudal horn is black.

Manduca quinquemaculata (Haworth), the "tomato hornworm" or "five-spotted hawkmoth," is a large (FW 4.6–5.6 cm; Plate 40.6), heavy-bodied gray sphingid with four to six, usually five, yellow orange, black outlined spots along each side of the abdomen. The five-spotted hawkmoth differs from the similar and sympatric Carolina sphinx *(Manduca sexta)*(see below) by its slightly sinuous dark brown to black subterminal line and paler HW with each HW having two centered dark jagged line enclosing a clear gray area. The tomato hornworm has a wide range in the Americas from southern British Columbia east to Nova Scotia, although occurring sporadically in Canada, south through most of the United States, Mexico, and Central America, at least to northern South America. In the West, *M. quinquemaculata* is found from southern British Columbia south to southern California and east to Idaho, Utah, and Colorado. This hawkmoth may not be capable of overwintering in areas with harsh winters but may colonize these areas in some years. There may be several flights in Florida and lowland Mexico and Central America, but only a single mid- to late-summer generation in the West. The larval hosts include tomato, potato, and a variety of other Solanaceae such as *Datura, Nicotiana,* and *Physalis.* The larvae are very similar to those of *M. sexta* in having seven white oblique lateral bands

on each side but are completely glabrous and have a black or blue caudal horn. Home gardeners may see their tomato plants rendered leafless, seemingly overnight, by a few of these cigar-shaped green larvae. Pupation is without a cocoon in the soil. The proboscis case of this species and *M. sexta* extend anterad of the pupa in a loop or "handle." *Manduca sexta* (Linnaeus), the "tobacco hornworm" or "Carolina sphinx," like *M. quinquemaculata* (see above), is a large (FW 4.3–5.6 cm; Plate 40.5), heavy-bodied gray sphingid that differs from the similar and sympatric *M. quinquemaculata* by the mottled unremarkable brown gray FW and its darker HW with two centered blurred dark lines enclosing a dark smudged area. The tobacco hornworm has a wide range in the Americas from the United States, Mexico, and Central America, at least to northern South America. The moth rarely reaches southern Canada as a vagrant. In the West, records seem to occur in two large regions: (1) the Front Range of Colorado from western Texas west through southern New Mexico, southern Arizona, and southwestern Utah, and (2) southern Nevada to California west of the Sierra Nevada crest. It may have several flights in Florida and lowland Mexico and Central America but has only a single mid- to late-summer generation in the West. Larval hosts are a variety of Solanaceae, including tobacco, potato, tomato, and others. Reports of plants in other families are almost certainly spurious. The larvae are very much like those of *M. quinquemaculata* but have a minute soft pubescence and the caudal horn is usually reddish. *Manduca rustica* (Fabricius), the "rustic sphinx" (FW 4.8–6.0 cm; Plate 40.9, Plate 63.10 larva), occurs in the southernmost portion of the country and may stray northward in the summer or fall. In the West, it is resident in southern California and southern Arizona and may also be found in southern New Mexico and western Texas. According to Hodges, it resides north to Virginia and the lower Mississippi Valley. The moth is also resident south into Mexico and Central America. The adults are strictly nocturnal and fly during the summer and early fall. Adults nectar at flowers such as petunia and moonflower and are also readily attracted to lights. The larval hosts are primarily Bignoniaceae, Boraginaceae, and Verbenaceae. In the West, desert willow, *Chilopsis* (Bignoniaceae), is the favored host. The larvae are distinctive and have white nodules on the dorsum of the thorax, and diagonal white or blue gray segmental slashes beginning on the first abdominal segment. The caudal horn is whitish at the base and blue gray at the tip. Another southwestern relative is *M. muscosa* (Rothschild & Jordan)(FW 4.7–6.0 cm; Plate 40.8, Plate 63.11 larva), which ranges from southern Arizona south to Costa Rica. The moths fly during July and August in southeastern Arizona. The larval host plants include *Viguiera dentata* and other composites (Asteraceae). *Manduca florestan* (Cramer)(FW 4.3–5.0 cm; Plate 40.7) ranges from southeastern Arizona south to Brazil. Adults are univoltine from late June to early August. Larval hosts are fiddlewood, *Citharexylum* (Verbenaceae), and trumpetbush, *Tecoma stans* (Bignoniaceae). The larvae are bright yellow green and have seven pairs of oblique yellow bands. The thoracic segments are covered with yellow excrescences, and the posterior horn is roughened.

SUBFAMILY MACROGLOSSINAE

This is the largest subfamily worldwide whose monophyly is defined by a patch of microtrichia (small sensory setae) on the first segment of the labial palpus. Favored host-plant families include Caprifoliaceae, Euphorbiaceae, Onagraceae, Rubiaceae, and Vitaceae.

Aellopus clavipes (Rothschild & Jordan)(FW 2.3–2.7 cm; Plate 41.14) ranges from southeastern Arizona and Texas south to central Mexico. The species is univoltine with adults in flight during August. The life history has not been reported.

Hemaris diffinis (Boisduval), the "snowberry clearwing" (FW 1.7–2.2 cm; Plate 41.1 dorsal, Plate 41.2 ventral, Plate 63.20 larva, Fig. 229 genitalia), is a bumblebee-like sphingid that ecloses with fully scaled wings, but a few shudders cause most of the scales to fall away; the wings are then transparent with narrow dark brown or black margins. The moths have a brownish cast and have the abdomen and thorax black scaled below. This moth ranges from coast to coast in Canada and as far north as Northwest Territories. The moth is found, sometimes commonly, in most of the United States and northwestern Mexico. *Hemaris diffinis* is reported to be double-brooded in the East, and two flights have been documented by Rubinoff in lowland California, but the species seems to have but a single generation in the higher-elevation portions of the West, flying from May to August. McFarland observed adults nectaring at various nectar-rich flowers including gilia and thistles. The female, while hovering, lays single blue green eggs on the underside of host leaves. The caterpillars feed mainly on honeysuckle, *Lonicera*, and snowberry, *Symphoricarpos* (Caprifoliaceae). Reports of its larvae feeding on dogbane, *Apocynum* (Apocynaceae), seem unlikely, although we have observed adults nectaring on these plants. McFarland has noted that the larvae rest on the underside of host stems and are difficult to spot from above. The larvae are primarily blue green to yellow green dorsally and laterally but are red brown to a rich purplish brown ventrally. The caudal horn is green. Pupation is in the leaf litter. *Hemaris senta* (Strecker), the "Rocky Mountain clearwing" (FW 1.6–1.9 cm; Plate 41.3 dorsal, Plate 41.4 ventral, Plate 63.21 larva), is extremely similar but has the ventral surface of the abdomen and thorax covered with yellow scales, and the adults have a greenish cast. *Hemaris senta* has a restricted range, mainly in the Rocky Mountains of Colorado, Wyoming, Utah, Idaho, and Montana. Adults fly in the daytime and nectar on flowers of plants such as *Arctostaphylos uva-ursi* (Ericaceae) and *Purshia tridentata* (Rosaceae). There is only a single annual flight from May to August, mainly June. We have found the larvae feeding on *Symphoricarpos* (Caprifoliaceae) in July.

Eumorpha achemon (Drury), the "achemon sphinx," is a medium (FW 3.8–4.4 cm; Plate 41.17, Plate 63.17 larva, Fig. 227 genitalia) sphingid with pale brown FW, each with dark brown blotches at the apex and on the central inner margin. The HW are pink with a row of black dashes near the outer margin. The species is widely distributed in the United States, often in suburban areas, from southwestern Oregon and California east to Massachusetts and south into northern Mexico. *Eumorpha achemon* is seemingly absent from Washington, Idaho, and Montana, with few records for Wyoming and Oregon; in other states it seems well distributed. Although reported to be double-brooded in the East, there seems to be but a single late-spring to summer generation in the West, for example, late May to late August along the Colorado Front Range. Its caterpillars feed mainly on wild and cultivated grape, *Vitis*, and *Parthenocissus* and *Ampelopsis* (all Vitaceae). The larvae are purplish brown to yellow brown and have six oblique cream white blotches along each side. Instead of a caudal horn there is a raised button. *Eumorpha typhon* (Klug)(FW 5.1–5.9 cm; Plate 41.18, Plate 63.18 larva) ranges from southeastern Arizona south into the mountains to Honduras. There is a single generation with adults coming to light

from late June through August. The larval hosts are wild grapes, *Vitis* (Vitaceae). The larvae are either pale green or brown and are covered with tiny black pinacula. There are oblique pale subdorsal bands on the first six abdominal segments, each containing an oval yellow spot. *Proserpinus clarkiae* (Boisduval), "Clark's sphinx" (FW 1.3–1.8 cm; Plate 41.8, Plate 63.25 larva), is a diurnal sphinx that may be seen hovering at low flowers including blue dicks, *Dichelostemma capitatum*. It has green FW and orange, black-margined HW. The moth ranges from montane southern California northward with an increasingly broad range through California (mainly east of the Coast Ranges crest), northern Nevada, northern Utah, Idaho, Oregon, and Washington to British Columbia. Most likely, it will also be found in northern Baja California. There is a single record for Teton County, Wyoming, but reported occurrences from Colorado and Manitoba more likely refer to *P. juanita* (Strecker). The moths fly in spring and, rarely, early summer, from late March and April to early July. The hosts are various species of *Clarkia*, *Camissonia*, and *Oenothera* (Onagraceae). There are six species in North America and one species in Eurasia, *P. proserpina* Pallas, the type of the genus. Four species occur in the West, and all but *P. clarkiae* are seldom seen by collectors. *Proserpinus flavofasciata* (Walker), the "yellow-banded day sphinx" (FW 1.8–2.3 cm; Plate 41.5), is a boreal moth that ranges from Alaska and British Columbia east across Canada to Nova Scotia and, discontinuously, south to Wyoming and Colorado, where it is rare. There is a single generation with adults in flight from April to early June. Adults are diurnal and nectar at flowers of dandelion, *Taraxacum* (Asteraceae), and cherry, *Prunus* (Rosaceae). The larval hosts are fireweed, *Chamerion angustifolium*, or willowherb, *Epilobium* (Onagraceae), and the report of thimbleberry, *Rubus parviflorus* (Rosaceae), probably refers to flower visitation not a larval host. The final instar is brown covered with many black dots and has a pale middorsal line and a dark subdorsal line along each side. There is a sclerotized black caudal button that is surrounded by a white band.

Amphion floridensis Clark, the "Nessus sphinx," is a medium (FW 2.4–2.5 cm; Plate 41.9) sphingid often seen visiting nectar-rich flowers such as lilac and milkweed by day and at dusk. The adults have wavy-edged reddish brown FW and orange brown HW with black outer margins. The abdomen has two prominent yellow intersegmental bands and flared black scale tufts at the tip. The species occurs from Nova Scotia south to Florida and westward to Alberta, Kansas, and the Colorado Front Range. Although double-brooded in the Southeast, in the West the species has a single generation from late spring to early summer. Larvae feed on grapes and *Ampelopsis* (Vitaceae). A record of feeding on pepper, *Capsicum* (Solanaceae), is almost certainly erroneous. The final instar is pale brown dusted with yellow and has nine or 10 oblique lateral dark lines. The caudal horn is short and black.

Arctonotus lucidus Boisduval, the "bear sphinx" or "Pacific green sphinx" (FW 2.2–2.3 cm; Plate 41.6, Plate 63.22 larva), is a furry winter sphingid with bright green FW and dull rose pink HW. The moth is found from eastern Washington and western Idaho through Oregon and California west of the Cascade–Sierra Nevada axis to coastal southern California. The moths have a mid- to late-winter generation from December, rarely November, to mid-March. They fly on relatively mild nights, including those with light precipitation. Adults have been seen at flowers during the day. The larval hosts are evening primrose, *Oenothera*, and *Clarkia* (both Onagraceae). The final-instar larva may be either brown or green. The green form is dark greenish black dorsally and laterally but is lighter gray green below; additionally, there is a series of transverse segmental white bands each with a medial black patch. Further, there are additional white lines and dots. The full-grown larvae lack a prominent caudal horn but, instead, have a sclerotized caudal button ringed with black, which has a tiny central black horn (0.5 mm).

Euproserpinus euterpe Edwards, the "Kern primrose sphinx" (FW 1.6–1.7 cm; Plate 63.23 larva, Plate 63.24 adult), has a mottled gray black FW and cream white black-edged HW. It is listed by the U.S. Fish and Wildlife Service as a threatened species and is found only in the Kern River basin and Temblor Range in Kern and San Luis Obispo counties, California. The life history is described in detail by Jump and coauthors. In the Temblor Range it flies in the winter from late January to late February, while in the Kern River Valley, it is found much later, from late February to early April. The adults rest on the sand in dry washes, where they bask to maintain flight temperature. Larvae feed on primroses, *Camissonia strigulosa* and *C. campestris* (Onagraceae). Final instars are variable from brown and green to brown, orange, black, and green; there are two dorsal white or yellow longitudinal stripes. The similar but more distinctly marked *E. phaeton* Grote & Robinson (FW 1.6–1.7 cm; Plate 41.11) occurs in arid sand dunes and drainages from western Nevada and northeastern California south to interior southern California and northwestern Mexico. It has a single flight in March and early April. The larval hosts are *Camissonia* and *Oenothera* (Onagraceae), and a report of larval use of *Coreopsis* (Asteraceae) must be in error and, more probably, may represent a nectar plant. The ultimate instar varies from green to pink and is covered with transverse rows of white points; there are eight black slightly oblique lateral bands extending from the fourth through eighth abdominal segments. The caudal horn is short and stubby. The spiracles are orange and black-rimmed. The pupa is formed in a soil depression under any available object. McFarland found that some sunlight is not only necessary for larval feeding and daily adult flight, but also for their proper emergence from the pupae. Newly eclosed adults hold their expanded wings directly overhead until they are dry and flight is possible. This is most unusual for a lepidopteran, as most species of macros direct their wings downward during expansion and hardening. *Euproserpinus wiesti* Sperry (FW 2.1–2.4 cm; Plate 41.7) has a discontinuous distribution from eastern Colorado, and thence south to northern Texas, north-central New Mexico, and northern Arizona. This moth is always found in association with sand dunes or sandy streambeds. There is a single spring flight from late March to late May. The larval hosts are evening primroses including *Camissonia latifolia*. Larvae are variable and similar to those of other *Euproserpinus*. *Xylophanes falco* (Walker)(FW 3.0–3.7 cm; Plate 41.16) ranges from southern Arizona east to western Texas and south to Guatemala and Honduras. The species is univoltine, with adults from June through September. Larvae feed on *Bouvardia ternifolia* (Rubiaceae). The larvae are cream white and black with a pair of C-shaped blue and yellow eyespots in a black field on the first abdominal segment.. The caudal horn is short and stubby.

Deilephila elpenor (Linnaeus), the "elephant hawkmoth" (FW 2.6–3.0 cm; Plate 41.10), is a widespread Eurasian species that was accidentally introduced to western British Columbia. It was first discovered by Troubridge in 2002 in a small area of the Fraser River delta. In the British Isles, there is a single brood with adults in flight during June. The moth nectars at flowers such as honeysuckle. Larvae usually feed on bedstraws, *Galium* (Rubiaceae), but occasionally on willowherb, *Epilobium* (Onagraceae), or ornamental plants such as fuchsia or impatiens. Larvae are either green or brown and are finely speckled with black except on the thoracic segments, which have a pale subdorsal line; there are black, lateral kidney-shaped ocelli with lilac centers on the second and third abdominal segments. When disturbed, the larva may retract its head and thoracic segments, which results in the snakelike eyespots being displayed on the swollen anterior portion of the abdomen.

Hyles gallii (Rottenburg), the "bedstraw hawkmoth" (FW 2.4–2.9 cm; Plate 41.12, Plate 63.27 larva), occurs in both North America and Eurasia. It ranges across the Arctic and subarctic areas of Canada and Alaska and down the montane cordillera to northeastern California, northern Utah, and northern Colorado. There is a single generation, and the adults fly from early June to late July. Adults are both diurnal and nocturnal. They nectar at flowers such as Canada thistle, *Cirsium arvense* (Asteraceae), and may be seen at light. The larvae feed on a variety of herbaceous plants, particularly willowherbs, including fireweed, *Chamerion angustifolium*, and clarkias, *Clarkia* (both Onagraceae); bedstraw, *Galium*, and woodruff, *Asperula* (both Rubiaceae). The final instars are variably colored red brown, brown, black, or green and have subdorsal yellow spots on each segment; there are also transverse rows of yellow dots along the sides and the spiracles are pale yellow. The caudal horn is usually black. *Hyles euphorbiae* (Linnaeus), the "spurge hawkmoth" (FW 2.8–3.5 cm; Plate 41.13, Plate 63.26 larva), is a Palaearctic sphingid that was intentionally introduced into western Canada for the biological control of leafy spurge, *Euphorbia esula* (Euphorbiaceae), a widespread invasive weed of western meadows and rangeland. At present, the moth has spread south into Montana and northern Wyoming. Adults have a single generation and are in flight from mid-June to mid-September. The final instars are either yellow or dark red to black, and there are 12 subdorsal yellow dots along each side. The dorsum and venter are marked with transverse rows of yellow or red dots. *Hyles lineata* (Fabricius), the "whitelined sphinx" (FW 2.4–3.7 cm; Plate 41.15, Plate 63.28, Plate 63.29 larvae, Plate 63.30 adult), is without doubt the most widespread and abundant sphingid in North America, including the West. Although it is active by night and attracted to lights, it is often seen by day and at dusk nectaring at flowers. The moth is found in North America, Eurasia, and Africa. In the West, it is continuously resident only in the southernmost areas such as southern California across to western Texas. It ranges south well into Mexico. It regularly emigrates and colonizes many areas to the north, except perhaps the northernmost areas of Canada and Alaska. There are several generations to the south, where the species is resident, and one or two to the north, where the species is an annual immigrant. The species may be found from February to late fall in southern residential areas and for shorter periods to the north. The polymorphic larvae occur in yellow, green, and black forms. They feed on a wide diversity of plants, including portulaca, *Portulaca* (Portulacaceae), four-o'clocks, *Mirabilis* (Nyctaginaceae), willowherb, *Epilobium* (Onagraceae), and others.

References for Superfamilies Mimallonoidea, Lasiocampoidea, and Bombycoidea

Barnes, W., and J. H. McDunnough. 1911. The lasiocampid genus *Gloveria* and its allies. Contributions to the Natural History of Lepidoptera of North America 1(2): 1–17, 4 plates.

D'Abrera, B. 1986. Sphingidae Mundi, Hawk Moths of the World. E. W. Classey, Ltd., Faringdon, U.K.; ix + 226 pages.

Ferguson, D. C. 1971. The Moths of America North of Mexico, Fascicle 20.2A. Bombycoidea, Saturniidae (Part). E. W. Classey, Ltd., and R. B. D. Publications, Inc., Faringdon, U.K., and McClellanville, SC; 1–155, color plates 1–11.

Ferguson, D. C. 1972. The Moths of America North of Mexico, Fascicle 20.2B. Bombycoidea, Saturniidae (Part). E. W. Classey, Ltd., Limited and R. B. D. Publications, Inc., Faringdon, U.K., and McClellanville, SC; 156–275, color plates 12–22.

Franclemont, J. G. 1973. Mimallonoidea. Mimallonidae and Bombycoidea. Apatelodidae, Bombycidae, Lasiocampidae, Fascicle 20.1, in: Dominick, R. B. et al. (ed.) The Moths of America North of Mexico. E. W. Classey, Ltd., and R. B. D. Publications, London; viii + 86 pages, 11 plates.

Hodges, R. W. 1971. The Moths of America North of Mexico, Fascicle 21. Sphingoidea. E. W. Classey, Ltd., Limited and R. B. D. Publications, Inc., Faringdon, U.K., and McClellanville, SC; 164 pages, 14 color plates.

Jump, P. M., T. Longcore, and C. Rich 2006. Ecology and distribution of a newly discovered population of the federally threatened Euproserpinus euterpe (Sphingidae). Journal of the Lepidopterists' Society 60: 41–50.

Kitching, I. J., and J.-M. Cadiou 2000. Hawkmoths of the World: An Annotated and Revisionary Checklist (Lepidoptera: Sphingidae). The Natural History Museum, London, and Cornell University Press, London and Ithaca, NY; viii + 227 pages.

Lemaire, C., and J. Minet 1999. The Bombycoidea and their relatives, pages 321–352, in: Kristensen, N.P. (ed.) Lepidoptera, Moths, and Butterflies. Vol. 1: Evolution, Systematics, and Biogeography. Handbook of Zoology. Vol. 4: Arthropoda: Insecta, Part 35. W. de Gruyter, Berlin, New York; 487 pages.

Peigler, R. S., and R. O. Kendall 1993. A review of the genus Agapema (Lepidoptera: Saturniidae). Proceedings of the Denver Museum of Natural History 3(3); 22 pages.

Peigler, R. S., and M. Maldonado 2005. Uses of cocoons of Eupackardia calleta and Rothschildia cincta (Lepidoptera: Saturniidae) by Yaqui Indians in Arizona and Mexico. Nachrichte entomologische Verein Apollo, N.F. 26(3): 111–119.

Peigler, R. S., and P. A. Opler 1993. Moths of Western North America. 1: Distribution of Saturniidae of Western North America. Maps and References. Contributions of C. P. Gillette Museum of Arthropod Biodiversity, Colorado State University, Fort Collins; 23 pages.

Rubinoff, D., and F. A. H. Sperling 2002. Evolution of ecological traits and wing morphology in Hemileuca (Saturniidae) based on a two-gene phylogeny. Molecular Phylogenetics and Evolution 25(2002): 70–86.

Smith, M. J. 1995. Moths of Western North America. 2: Distribution of Sphingidae of Western North America. Revised edition. Contributions of C. P. Gillette Museum of Arthropod Diversity, Colorado State University, Fort Collins; 26 pages.

Stehr, F. W., and E. F. Cook 1968. A revision of the genus Malacosoma Hübner in North America (Lepidoptera: Lasiocampidae): Systematics, biology, immatures, and parasites. U.S. National Museum Bulletin 276: 1–319.

Stone, S. E. 1991. Foodplants of World Saturniidae. Lepidopterists' Society Memoir 4. Lepidopterist's Society; 186 pages.

Stone, S. E., and M. J. Smith 1990. Buckmoths (Lepidoptera: Saturniidae: Hemileuca) in relation to southwestern vegetation and foodplants. Desert Plants 10(1): 13–30.

Tuskes, P. M., J. P. Tuttle, and M. M. Collins 1996. The Wild Silk Moths of North America. Cornell University Press, Ithaca, NY, and London; 250 pages, 30 color plates.

Tuttle, J. P. 2007. The Hawk Moths of North America. Wedge Entomological Foundation, Washington, DC; xvii + 253 pages.

Superfamily Noctuoidea

The moths of this group are unified by the complex metathoracic tympanal organs and associated abdominal structures. These have been shown by Roeder to function for avoidance and escape of bat predation and may also function in communication during courtship and mating by some groups such as in the Arctiinae. This superfamily is represented by the largest number of species worldwide (70,000) and in

North America (3,700). The number of families believed to compose this group varies, but, since we follow the inclusive definition of the Noctuidae as proposed by Lafontaine and Fibiger, the superfamily comprises the following families: Doidae, Notodontidae, Noctuidae, Micronoctuidae, and Oenosandridae. As such, the size, form, and color of the adults and larvae are widely variable, as is the nature of their life history and behavior. With a small percentage of notable exceptions, the adults of most species are nocturnal and not brightly colored.

FAMILY DOIDAE

The Doidae (Euphorbia moths) is a small family limited to the New World. Its placement within the superfamily has been uncertain. Often placed within the Arctiinae (Franclemont in Hodges et al. 1983), more recently, it has been given separate family status by Donahue and Brown. It seems to have no single unique character, but it shares features of both the trifid Noctuidae and the Notodontidae.

Adult The apices of the adult tibial spurs are smooth, in contrast to the serrate condition in Notodontidae. The wings of doid adults are fragile and relatively thin scaled with a satiny sheen on their FW. The M2 vein in the FW arises from the middle of the cell so that the cubital vein at the bottom of the cell appears to have three branches, as in the Notodontidae.

Larva Doid larvae superficially appear like sawflies because of their swollen thoracic segments; they have biordinal crochets in a homoideous mesoseries, and Brown found that they feed communally in sparse webs on leaves of Euphorbiaceae.

Diversity There are three species found in western North America, and all three range south into Mexico.

Doa ampla Grote (FW 1.4–2.0 cm; Plate 33.39) ranges from central Colorado south through southeastern Arizona to Mexico (Durango and Nuevo León). The adults have a satiny pointed FW and a dark gray HW. In western Texas, the moth flies from May to October and likely has two broods, while in Arizona *D. ampla* flies from April to September (two flights). Further north in Colorado, there are only July records, and the moth must have only a single flight. The larval hosts are *Euphorbia schizoloba* and *E. brachycera*, as well as *Stillingia texana* (Euphorbiaceae). Eggs are laid in neat parallel rows in clusters of 15 to 35. Young larvae feed gregariously, and the last instar is black with broad yellow longitudinal stripes and white spots. The integument is covered with minute spicules, which are especially dense in pigmented areas. The spiracles are elliptical. The head is small and blackish. Pupation is in debris at the base of the host plant in a single-layered open mesh brown cocoon. Larvae of ***Doa dora*** Neumoegen & Dyar feed on *E. misera* (Euphorbiaceae) and have been recorded by J. W. Brown from San Diego County, California. ***Leuculodes lacteolaria*** Hulst (FW 1.3–2.1 cm; Plate 33.40) has pale greenish white translucent wings crossed with parallel wavy lines. The species ranges from Pena Blanca, Santa Cruz County, Arizona, south through western Mexico to Central America. The single annual flight is from June to September. The host and life history details are unreported.

FAMILY NOTODONTIDAE

The Notodontidae (prominents) occur worldwide except on New Zealand and Pacific islands. According to studies by Miller the family is united by a number of characters, most of which are not universal in the family or are shared with some other families.

Adult The adults of most species are nocturnal and are strongly attracted to lights, whereas the Dioptinae are primarily diurnal and often involved in mimicry complexes with distasteful diurnal Arctiinae and other insects. Adults are not known to visit flowers and probably feed opportunistically on moisture. Adults are medium to large (FW 1.1–4.3 cm) relatively thick bodied moths with either a broad or elongate FW and HW. The proboscis is usually well developed and coiled. In males, the antennae are usually bipectinate, while in females the antennae are usually filiform, but sometimes bipectinate. The wings are well scaled and relatively dull colored gray, brown, or tan, although the wings of the Dioptinae are often translucent or brightly colored, for example, the wings of *Phryganidia californica* are translucent and pale sooty gray. The FW has trifid venation (see under Doidae). The abdomen is usually covered with long slender scales, and some genera (e.g., *Astylis*) have bushy tufts at the apex. Abdomens of females in some groups (e.g., *Heterocampa*) may be distended with eggs and relatively weak; their flight may be limited as a result. Adults of almost all species are nocturnal and are attracted to light, but *P. californica* is primarily diurnal and perhaps crepuscular.

Larva These are medium, cylindrical larvae with hypognathous heads. The prolegs have uniordinal, rarely biordinal, crochets that are arranged in homoideous mesoseries. Many notodontid larvae have the anal prolegs highly modified as stemapods, often with distal eversible glands; if not so modified the prolegs of the tenth abdominal segment are much smaller than those on the third through sixth abdominal segments. Notodontid caterpillars have additional secondary setae on the third through sixth abdominal prolegs. All are external feeders, primarily on broad-leaved woody plants, and may be found on leaves without accompanying shelters. The larvae are often highly cryptic in color, pattern, and body profile, superbly matching their food plant leaves. The larvae of some species are rather gaudy and colorful; these are presumably aposematic and are very likely distasteful and emetic to would-be vertebrate predators. The appearance and patterns of many genera are widely variable between instars, and those of many North American species were beautifully illustrated by Packard. The larvae are often positioned in a seemingly awkward posture with their posterior elevated above the substrate. Larvae of some species feed gregariously, for example, *Schizura*, *Datana*, and *Phryganidia*, while those of others feed solitarily. Those that feed in groups respond to disturbance by raising their front and hind portions in unison to a presumably defensive posture. The larvae of many solitary notodontids are cryptically shaped and patterned. The eggs are almost uniformly hemispherical with smooth chorions that lack obvious sculpturing; they are often laid in groups but may sometimes be placed singly. Larvae feed externally without shelters. The pupae are not distinguishable from those of Noctuidae. Several species are of minor importance, notably *Schizura* and *P. californica*.

Larval Foods Oviposition is often genus specific, limited to single plant families, or general in the cases of *Datana ministra* (the "yellow-necked caterpillar") and *Schizura concinna* (the "red-humped caterpillar"). Host plants are usually woody and are primarily broad-leaved angiosperms. Presently, the family is broken into at least 10 subfamilies, seven of which occur in western North America. Pygaerinae larvae feed on leaves of woody plants including Fabaceae, Proteaceae, and Salicaceae.

Notodontinae larvae are known to feed on woody plants including Fagaceae and Salicaceae, but also likely include Flacourtiaceae. Phalerinae larvae feed on grasses and woody plants among the Anacardiaceae, Fabaceae, Fagaceae, and Juglandaceae. Dudusiinae larvae feed primarily on woody plants in the Aceraceae, Rhamnaceae, and Sapindaceae. Heterocampinae larvae feed on a wide variety of woody plants, especially Fagaceae, Juglandaceae, Rosaceae, and Ulmaceae. Nystaleinae larvae feed on woody trees and shrubs including Fabaceae, Fagaceae, and Sapindales. Dioptinae larvae feed on distasteful plants and are presumably distasteful themselves. Known hostplant families include Aristolochiaceae, Fagaceae, Passifloraceae, and Violaceae.

Diversity Notodontidae are richest in the tropics, but 137 species have been reported for North America north of Mexico, and 62 of these, almost all resident, have been recorded from western North America. They are most often associated with wooded habitats, including suburban plantings. Miller felt that the classification within the family is tentative and several genera are not placed within subfamilies. There are a number of undescribed North American species, and the generic placement of described species is likely to change.

SUBFAMILY PYGAERINAE

Pygaerine notodontids are poorly defined and have a notable lack of characteristic derived characters (apomorphies). Larvae are hairy, and adult males have terminal abdominal hair tufts, but these characters are also shared with species in other notodontid subfamilies.

The only North American pygaerines are various *Clostera* Samouelle, a temperate Holarctic genus. There are seven North American species, five of which occur in the West. The most widespread and abundant species is *Clostera albosigma* Fitch, the "sigmoid prominent" (FW 1.4–1.7 cm; Plate 33.41, Plate 63.34 larva), which is found from coast to coast and is ubiquitous in temperate eastern North America. In the West, it ranges from Washington and Alberta south to northern California, Utah, and Colorado. There is a prominent sigmoid line at the FW costa and a clear-cut dark brown or reddish patch on the outer portion of the FW. In the East there are two broods, and the adults can be found from March to August, but in the West the species has but a single flight from June to August. The larvae feed solitarily within leaf shelters on various willows and poplars (Salicaceae). *Clostera apicalis* (Walker), the "willow nestmaker" or "red-marked tentmaker" (FW 1.1–1.5 cm; Plate 33.42), occurs from British Columbia and Alberta east to Newfoundland and south to California and Colorado in the West. There are two flights of adults (June to September), and larvae may be found in August and September. Like those of *C. albostigma*, the larvae feed on various *Populus* and *Salix* (Salicaceae). Caterpillars are gray or pale brown with vague darker stripes.

SUBFAMILY NOTODONTINAE

The Notodontinae are unified by their simple tarsal claws and four L-setae on larval third through sixth abdominal segments. In this subfamily, there are nine western species placed in five genera. As is typical for notodontids, several of the genera require taxonomic research, as the current placement of species is uncertain.

Hyperaeschra georgica tortuosa Tepper, **new status** (FW 1.4–1.9 cm; Plate 33.43) blends with typical *H. georgica* (Herrich-

Schäffer) in the Great Plains (e.g., Nebraska), and we consider it a subspecies of that species. There is a black extension of the FW along the inner margin. Males have a white HW, while that of the larger females is gray. The range is from Utah, Colorado, and Nebraska south to Arizona, New Mexico, and western Texas. Typical *H. georgica* is widespread in eastern North America. Adults fly from late May to early August (late March in Oklahoma); in the East there are two flights. The larvae have been reported to feed primarily on oaks, *Quercus* (Fagaceae), and reports of other woody plants may be in error. Larvae are predominantly green with a pale yellow spiracular stripe that may be narrowly edged in red; there are many tiny white pinacula. A blunt orange red knob is found on the dorsum of the eighth abdominal segment. The head is green with many tiny white pinacula and a continuation of the spiracular stripe.

Pheosia rimosa Packard, the "false-sphinx," "mirrorback caterpillar," or "black-rimmed prominent" (FW 2.0–2.9 cm; Plate 42.1, Plate 63.35 larva), ranges from coast to coast and from southern Canada south in the West to near the Mexican border. In the Pacific Coast states, intermediates with *Pheosia portlandia* Henry Edwards demonstrate that the latter name should be treated as a synonym. In the East, the moths are found from May to August with two flights, but primarily in June and July in the West, where there is probably only a single flight. Host plants are willows and cottonwoods (Salicaceae). Larvae are sphingidlike, glossy green, gray, or dark brown with white-rimmed black spiracles, and a black dorsal horn on the eighth abdominal segment. Larvae rest on the exact leaf edge and thrash about if disturbed. McFarland noted that the shiny black pupae overwinter in a silk-lined underground cell. *Odontosia elegans* (Strecker), the "elegant prominent" (FW 2.2–2.7 cm; Plate 33.44, Plate 63.36 larva), is a large moth with broad uniformly gray FW and extensions that extend from the inner margin. The species ranges from coast to coast across southern Canada and, in the West, is found south to Arizona and New Mexico. There is a single spring to early-summer flight (late May to early August). Larvae feed on aspen and cottonwood, *Populus* (Salicaceae).

Gluphisia septentrionis (Walker), the "common gluphisia" (FW 1.2–1.5 cm; Plate 33.45, Plate 33.46 subsp. *albafascia*, Plate 33.47 subsp. *wrightii*), is one of the more abundant western prominents. Adults are heavy bodied, and the FW is gray with scattered yellow and black scales. There is quite a bit of variation within and between populations, and more than a single species may be found in the West. The species ranges from coast to coast. In the West it is found from Alaska and Yukon Territory south throughout our area to northern Mexico. Adults have two flights—spring and summer. The larval hosts are willows and cottonwoods (Salicaceae). Larvae are pale green with yellow subdorsal longitudinal lines along each side. The intervening dorsum may be filled with wine red and pale yellow quadrate patches. *Gluphisia severa* Henry Edwards (FW 1.7–1.9 cm; Plate 33.48) is a larger, darker species and more limited to montane forests. It ranges from southern British Columbia south to the central Sierra Nevada and to Colorado in the southern Rocky Mountains. There seems to be one flight in the West from June to August. Like its more common widespread congener, its larvae feed on willows and poplars (Salicaeae).

The species now placed in *Furcula* Lamarck were formerly known as *Cerura*, and the species are sometimes referred to as "puss moths." Several of the species seem to blend into each other, and the practical answers to such problems remain to be resolved. *Furcula cinerea* (Walker), the "gray furcula" (FW 1.5–1.8 cm; Plate 33.49), has a pale gray FW with obscure

darker bands and a white HW. The outer margins of both wings have marginal black marks between vein endings. The species ranges from coast to coast and from southern Canada south to northern Mexico. Adults have two flights from May to August. Its larvae feed on willows and cottonwoods (Salicaceae) and often rest with their anal prolegs above the substrate. The full-grown larvae are green with irregular brown saddle marks. Long, pointed anal prolegs ringed with green and brown give a forked-tail appearance. McFarland reports that the anal prolegs are eversible under stress. The head is brown. *Furcula scolopendrina* (Boisduval)(Walker)(FW 1.7–2.0 cm; Plate 33.51) has a cream white FW crossed by an irregular black antemedial band and also has a black postmedian costal band closer to the apex. This striking moth is widespread in the West and ranges from British Columbia to northern Mexico. It is uncommon in the East. The adults are found from April to August (two flights), but most appear in June to July. Its larvae feed on willows and cottonwoods (Salicaceae) as well as birch (Betulaceae). The larva is yellow with dark brown black dorsal saddle marks. The anal prolegs are long and pointed as with *F. cinerea*.

SUBFAMILY PHALERINAE

The subfamily Phalerinae includes six western species. On the whole Miller found the limits of the subfamily to be vague. *Datana integerrima* Grote & Robinson subsp. *cochise* Dyar, the "walnut caterpillar" (FW 2.1–2.6 cm; Plate 33.52), occurs in southeastern Arizona and western Texas. The typical subspecies is found in the East from southern Canada south to Florida and southern Texas. Adults are found from June to early August. A variety of woody host trees and shrubs are used, but Juglandaceae including *Carya* and *Juglans* are selected in the majority of cases. In southeastern Arizona, Arizona walnut, *Juglans major*, is the primary larval host. Like other *Datana* species, the larvae of this species feed gregariously. Larvae are charcoal gray to black with several thin longitudinal yellow stripes and are densely covered with long white hair. *Datana perspicua* Grote & Robinson subsp. *mesillae* Cockerell, the "yellow-striped caterpillar" or "spotted datana" (FW 1.9–2.4 cm; Plate 33.50), is found from coast to coast, including southern Ontario, and in the West occurs from Idaho, Montana, and Wyoming south to southern California, southern Arizona, and southern Texas. Adults fly from late May to mid-August in the Rocky Mountain states. Robinson et al. compiled reports that indicate the larval hosts are various species of *Rhus* (Anacardiaceae). In California, the larvae preferred *R. trilobata* (Anacardiaceae). Eggs are laid in a single-layered cluster. Young larvae are gregarious and feed only on leaf surface tissue. The larvae of nominate *D. perspicua* are similar to those of *D. ministra*, but the yellow stripes are broader than the intervening black areas and the cervical shield is brown. The full-grown larva may, on occasion, retain the red-and-yellow striped color pattern of the preceding instar. Comstock and Dammers noted two larval color forms from a population in the eastern Mojave Desert of California. One form has a black head and is olive black with yellow stripes, and the second form has a maroon head and body and orange stripes. *Datana neomexicana* Doll (FW 1.7–2.5 cm; Plate 33.53) is found from southwestern Colorado south to southeastern Arizona, New Mexico, and western Texas. There is one flight from May to early August. *Datana ministra* (Drury) subsp. *californica* Dyar, the "yellownecked caterpillar" (FW 2.0–2.5 cm; Plate 33.54, Plate 64.1 larva), occurs discontinuously in the West from British Columbia south to central California and Colorado. The species ranges east to Nova Scotia. There is one annual flight from early June to mid-July. Larval hosts are a wide variety of woody trees and shrubs including Fagaceae, Juglandaceae, Betulaceae, and Rosaceae. The white, ovate eggs are laid in masses of 25 to 50 on the undersides of host leaves. Early-instar larvae feed gregariously and skeletonize young host leaves. Final stage larvae are black with narrow yellow longitudinal stripes. The body is covered with fine long white hairs. The head and terminal segments are black or dark red, and there is an orange cervical shield. The species is a common pest of orchard trees, especially apple.

Nadata gibbosa (J. E. Smith), the "white-dotted prominent," "tawny prominent," or "green oak caterpillar" (FW 2.0–2.6 cm; Plate 33.55), is one of the more common prominents and is found coast to coast from southern Canada south through the Rocky Mountains to western Texas. Along the Pacific Coast it occurs to southern California. There are two flights (April to September) in more southern areas. The larvae have been reported to feed on leaves of broad-leaved trees and shrubs in seven different families, primarily Aceraceae, Betulaceae, Fagaceae, and Rosaceae. The larva is light green with a narrow pale yellow subdorsal line, yellow mandibles, and yellow line marking the anal plate. Caterpillars may be found from June to October. *Nadata oregonensis* Butler (FW 2.0–2.4 cm; Plate 33.56) is found in the Pacific coastal states from Washington south to southern California (San Bernardino County). It is similar to *N. gibbosa* but has a blotchy yellow brown FW with a yellow patch in the FW discal cell and a postmedian line that curves slightly basal where it meets the inner margin. There is a single flight from late May through August. The hosts and life history are unreported.

SUBFAMILY DUDUSINAE

Dudusiine adults have large ocelli, which are lacking or vestigial in other Notodontidae. Their larvae lack seta 10 on the tenth abdominal segment. The subfamily includes species from the New World, but primarily species native to tropical Africa and Asia. *Cargida pyrrha* (Druce)(male FW 1.7 cm, female FW 2.4 cm; Plate 42.2 male, Plate 42.3 female) is dimorphic in size and color. Males are smaller with a charcoal black FW, while the larger females have a gray FW. The species is known from central Arizona south to Guatemala. In Arizona, the moths have been found from July to September, and we found them abundant during September in Sonora, Mexico. Both sexes are strongly attracted to light. Females lay masses of eggs in a single layer around twigs of the host plant, wolfberry, *Lycium torreyi* (Solanaceae), in at least one instance. In southeastern Arizona, the larvae feed on *Zizyphus obtusifolia* and other *Zizyphus* species (Rhamnaceae). The larvae are gregarious in the early instars but are solitary later. The full-grown larvae are predominantly black with longitudinal blue gray and yellow stripes and lines. The body has black verrucae, each of which bears short black hairlike setae. The posterior portion has a black dorsal hump with two white spots. Spiracles are black ringed with white. The prolegs are red orange. Larvae rest with both the anterior and posterior ends raised above the substrate.

Crinodes biedermani (Skinner)(FW 3.5–4.3 cm; Plate 42.4) is known from southeastern Arizona, especially the Huachuca Mountains where there is one flight from late July to early August. According to McFarland, the hostplant is *Ceanothus fendleri* (Rhamnaceae). The spectacular full-grown larvae are dull gray black with pale yellow lines and blotches with extensions running down to each proleg. In addition, there are smaller red smudges and lines. The larva feeds with its posterior portion hanging off the substrate. McFarland notes that when disturbed the larvae will regurgitate copiously.

SUBFAMILY HETEROCAMPINAE

The limits of the Heterocampinae are uncertain, according to Miller, and the subfamily is defined largely on setal patterns of the larvae. First-instar larvae of North American species often have the bases of dorsal seta 1 greatly enlarged and extended into antlerlike structures on the prothorax and occasionally other segments. Larvae of many species have elongate anal prolegs. The larvae of this subfamily produce defensive secretions (formic acid and ketones) from the adenosma. Larvae feed on leaves of a wide variety of woody plants; especially favored are Fagaceae, Juglandaceae, Rosaceae, and Ulmaceae. According to McFarland, in southeastern Arizona, the larvae feed on an annual species of *Euphorbia* (Euphorbiaceae) that is present only from July to September during the monsoons. *Ursia noctuiformis* Barnes & McDunnough (FW 1.0–1.2 cm; Plate 42.6) is found in the deserts of southeastern California, Nevada, southwestern Utah, and western Arizona where it flies from March to early September (two flights). Its life history is unreported.

Litodonta alpina Benjamin (FW 1.7–2.0 cm; Plate 42.7) ranges from southeastern Arizona, southern New Mexico, and western Texas. There it is found from April to July and may have two flights. *Litodonta gigantea* Barnes & Benjamin (FW 1.8–2.3 cm; Plate 42.8) is distributed from southeastern Arizona eastward to western Texas. *Litodonta wymola* (Barnes)(FW 1.4–1.6 cm; Plate 42.9) has been recorded from southern California and southern Nevada east to western Texas. The moth has been found from April to October and may have as many as three flights. The hosts and life history for all of our *Litodonta* are unreported.

Macrurocampa dorothea Dyar (FW 1.8–2.4 cm; Plate 42.10) ranges from Utah and central Colorado south through Arizona, New Mexico, and western Texas into western Mexico. It is often found abundantly in the mountains of southeastern Arizona during July and August but flies as early as May. The larval hosts are several oaks, *Quercus* (Fagaceae). McFarland reports that the larvae also feed on Arizona madrone, *Arbutus arizonica* (Ericaceae). As illustrated by Wagner, the full-grown larva is mottled blue green with a red brown and yellow middorsal stripe. There are also lateral pale green diagonal lines. The spiracles are red brown, and there are additional small irregular red brown and yellow spots and blotches. The head is green with lateral red brown and yellow bands.

Heterocampa ruficornis Dyar (FW 1.9–2.2 cm; Plate 42.11) has a subtly mottled gray FW. The female's HW has a gray marginal band. This species is best known from southern Arizona and is known in the central part of the state as well. It flies from July to mid-August.

Heterocampa averna Barnes & McDunnough (FW 2.0–2.7 cm; Fig. 232 genitalia) ranges from southwestern Utah and Arizona east across southern New Mexico to western Texas. The moth flies from February through November in Arizona and in western Texas it flies in March and April then again from July to October, presumably in two flights. The larval host plant is Emory oak, *Quercus emoryi* (Fagaceae). *Heterocampa subrotata* Harvey, the "small heterocampa" (FW 1.3–1.7 cm; Plate 42.12), ranges widely across the eastern United States from New Jersey west to Arizona. Adults may be found from March to October in two flights. The larval hostplant is hackberry, *Celtis* (Ulmaceae). *Heterocampa lunata* Henry Edwards (FW 2.0–2.2 cm; Plate 42.13) is found commonly in the Rocky Mountain region from Wyoming, Utah, and Colorado south to Arizona, New Mexico, and western Texas. It flies mainly in June and July and occasionally is found in May. The life history is unreported.

Hyparpax aurora (J. E. Smith), the "pink prominent" (FW 1.5–1.8 cm; Plate 42.14), which includes *H. a. aurostriata* Graef,

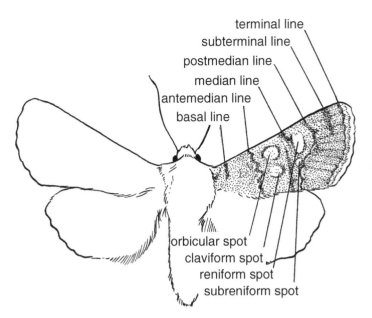

terminal line
subterminal line
postmedian line
median line
antemedian line
basal line

orbicular spot
claviform spot
reniform spot
subreniform spot

FIGURE 230. Schematic representation of Noctuidae FW with color pattern elements used in diagnoses.

is found from southern Ontario and the central Atlantic states west to Utah and Arizona. The adults are polyphenic and vary from the typical *H. aurora* form in the East to the tan western form *(H. a. aurostriata)*, with mixed populations in Oklahoma and Texas. Adults may be found from April to August in two flights, but mainly July and August as one flight in the northern parts of the range. Larvae, which feed on various oaks (Fagaceae), are complexly patterned pale green and darker green and are covered with tiny white speckles. There are white blotches and red lines dividing the boundary between green shades. There are slight dorsal humps on the first and eighth abdominal segments. Each hump is surmounted by a pair of red tubercles.

Oligocentria alpica (Benjamin)(1.4–1.8 cm; Plate 42.15) is found from southeastern Utah and western Colorado south to Arizona, New Mexico, and western Texaas. There are two flights of adults from late March to mid-September. *Oligocentria coloradensis* (Henry Edwards)(FW 1.6–2.0 cm; Plate 42.16) is gray streaked with short thin black lines. There is almost always a thin black streak extending from the thorax onto the basal area of the FW. Females have a smokier gray HW. The species is found from Utah and Colorado south to Arizona, New Mexico, and western Texas. The moths fly from July to September (one flight) but may have two flights in western Texas, where they fly as early as May. *Oligocentria perangulata* (Henry Edwards)(FW 1.6–2.4 cm; Plate 42.17) is very similar to *O. pallida* (Strecker)(Plate 42.18) but has a dark cell spot and two white streaks just outside the discal cell. The inner margin is orange tan, and there is usually a thin dark basal streak. The species has been recorded in the Rocky Mountain region from Wyoming, Utah, and Colorado south to Arizona, New Mexico, and western Texas. The moths are found in two flights from mid-May to September. The flight season is longer in the southern portion of its range.

Schizura biedermani Barnes & McDunnough (FW 2.0–2.8 cm; Plate 42.19, Plate 64.2 larva) is found in southeastern Arizona, where it flies during July and August. McFarland found the larval hosts include Emory oak, *Quercus emoryi*, and Mexican blue oak, *Q. oblongifolia* (Fagaceae). The full-grown larvae, as photographed by Wagner, are purplish with zebra-striped black-and-white patches. Several black plumose setae emanate from each black verruca. The spiracles are red brown. The head

is purplish blue. *Schizura ipomoeae* Doubleday, the "morning-glory prominent" (FW 1.6–2.2 cm; Plate 42.20), is quite variable, ranging from light to dark. Females have a dusky gray HW. The species is wide ranging and is found commonly from coast to coast and from southern Canada south into Mexico. On the Pacific Coast it is found south to the extreme northwestern corner of California (Del Norte County) but has not been reported from most of California, Nevada, and Arizona. There is one flight in the northern part of the range, but two flights to the south. Larvae feed on many woody trees and shrubs including Cornaceae, Ericaceae, Fagaceae, and Rosaceae. The caterpillars are rather similar to those of the unicorn caterpillar but have the head with narrow bands. *Schizura unicornis* (J. E. Smith), the "unicorn caterpillar" (FW 1.5–1.7 cm; Plate 42.21), occurs from coast to coast and is probably the most common species of the genus. There is one flight from June to August in most of the West, but the species has been found from March to September in Texas, where there are likely two flights. Larvae feed on a wide variety of trees and large shrubs, especially alder and birch (Betulaceae) and willow (Salicaceae). The larvae are bicolored light brown and green with notable humps surmounted by spines on the first, fifth, and eighth abdominal segments. The second and third thoracic segments are green, and the pale brown abdomen also has white dorsal patches and red lines. The shape and pattern gives the larva excellent blending protection when resting on host vegetation. *Schizura concinna* (J. E. Smith), the "redhumped caterpillar" (FW 1.2–1.6 cm; Plate 43.22 female, Plate 43.23 male, Plate 64.3 larva), occurs in both eastern and western North America from southern Canada southward but is rare or absent in the Great Plains. There is a single flight in June and July. Hosts include many woody trees and shrubs, among them Betulaceae, Juglandaceae, Rosaceae, and Salicaceae. Many orchard and ornamental trees are fed upon, and the species is considered a minor pest. The white eggs are laid in masses on the underside of host leaves, and the larvae feed gregariously. The larvae rest with last pair of prolegs elevated above the substrate and are yellow with black or dark red longitudinal lines and verrucae. There are prominent white lateral bands and a red dorsal thoracic hump surmounted by black spines. The head is also red. They pupate in loose silken cocoons in the ground litter under the host.

SUBFAMILY NYSTALEINAE

This subfamily is limited to the New World, with most species found in the tropics. There Weller found that there are a number of apomorphies that characterize the subfamily. Many species are characterized by large scale tufts on the antennal scape, and by the resemblance of resting adults to sticks or twigs. Larvae are glossy with medial processes on the eighth abdominal segment. They are often brightly colored and are capable of producing defensive secretions. According to Kitching and Rawlins, host plants include many woody trees, shrubs, and vines distributed primarily in the Fabaceae, Fagaceae, and Sapindales. There are 15 North American nystaleines in six genera, while seven species in four genera have been reported from the West. The most familiar North American species are several eastern *Symmerista*, but in the West, members of the subfamily are most easily found close to the Mexican boundary, especially southeastern Arizona.

Dasylophia anguina (J. E. Smith), the "black-spotted prominent" (FW 1.4–1.8 cm; Plate 42.24), is the most widespread North American nystaleine and ranges from southern Quebec and Nova Scotia west to the Rocky Mountain region from Montana south to New Mexico. There is a single adult flight from June to August. In southeastern Arizona, McFarland has found that the larvae feed on several species of *Desmodium* (Fabaceae). Its larvae are banded red orange, lavender, and yellow with black lines or spots. Larvae may be found from June to October.

Symmerista suavis (Barnes)(FW 1.5–2.0 cm; Plate 42.25, Figs. 233, 234 genitalia) is the most frequently encountered western representative of the subfamily and ranges from southern Utah and western Colorado south through Arizona, New Mexico, and western Texas to Durango, Mexico. There is a single flight during July and August, occasionally as late as September. Its early stages are unreported.

SUBFAMILY DIOPTINAE

Dioptinae is a distinctive group with many larval and adult synapomorphies. Both adults and larvae of many species are brightly colored and are presumably distasteful. The larvae feed in an exposed manner, and the pupae of most species are similar to those of nymphalid butterflies in that they hang head downward from the substrate attached by the pupal cremaster. This group is especially rich in the Neotropical region. Adults are apt to be most frequently seen during the day.

Phryganidia californica Packard, the "California oakworm" (1.4–1.8 cm; Plate 42.26), ranges along the Pacific Coast from central coastal Oregon south into north Baja California, where it is most abundant in areas bathed by coastal fog. On occasion, it can be so abundant that its larvae completely defoliate large expanses of oak forest, especially those of *Quercus agrifolia*, coast live oak. There are two flights each year, during June and July and again in October and November. Furniss and Carolin report three generations in southern California. The adults fly weakly during the day but are also attracted to lights at night. Hosts include any Fagaceae within the moth's range, including chinquapin, *Chrysolepis*, and tan oak, *Lithocarpus*. Unlike many tropical American dioptines, adult *Phryganidia* are not brightly colored but are translucent gray brown with darker wing veins. Males have a small, ill-defined yellowish FW patch. Eggs are laid singly or in patches on leaves or branches. The larvae are smooth and olive green with longitudinal yellow and black stripes and bands. The head is red brown or pale brown. Larvae feed, sometimes in groups, on leaves of any age. The early-instar larvae skeletonize patches of leaf epidermis, while older larvae eat all except the principal veins.

SUBFAMILY UNASSIGNED

The remaining taxa are unplaced to subfamily, since the higher level classification of the Notodontidae remains incomplete. *Afilia oslari* Dyar (FW 1.5–1.7 cm; Plate 42.27) flies in July and August and is common in southeastern Arizona, where, according to McFarland's unpublished research, the larvae feed on the leaves of *Mimosa grahamii* (Mimosaceae). **Lirimiris truncata** (Herrich-Schaeffer)(FW 2.6–3.5 cm; Plate 42.28) is a tropical American species that barely enters our area in Santa Cruz County, southeastern Arizona. The adults, which fly from August to October, are nocturnal and are attracted to light. Larvae feed on *Gossypium thurberi* (Malvaceae). The full-fed larva, photographed by Singer, is yellow with irregular black-and-white-patterned blotches and stripes. There is a large dorsal black hump at the posterior, and the head is shiny black.

FAMILY NOCTUIDAE

The Noctuidae (owlets, cutworms, and millers) are now considered a monophyletic group as reconstituted by Lafontaine and Fibiger. The Noctuidae were paraphyletic under previous schemes that kept Arctiidae, Nolidae, and Pantheidae as separate families. The rationale behind the present classification and problems with previous schemes are explained by Lafontaine and Fibiger. Based on venation of the HW, the family is divided into trifine species—those in which vein M2 is reduced or absent so the cubital vein appears to have three branches—and quadrifine species—those in which vein M2 is normal so that the cubital vein appears to have four branches.

Adult These moths range from very small to very large (*Thysania agrippina* of Central America has the largest wing expanse—in excess of 25 cm). There are several FW pattern elements (Fig. 230) that are present in many Noctuidae and are used to discriminate and describe species and genera. There are several FW spots, referred to as the orbicular, reniform, subreniform, and claviform. In addition, there are FW lines that appear in many taxa at particular points on the FW, and these may be abbreviated by their position from the wing base: basal line or dash, antemedian (am.) line, median line, postmedian (pm.) line, and subterminal (st.) line. Noctuid moths may be diurnal or nocturnal or both; but the majority of these insects are nocturnal and many are attracted to light.

Larval Foods Larval hosts are virtually all higher plants, but most prefer conifers and broad-leaved plants. These are among the most important consumers of terrestrial green plant matter on Earth! Adults may feed on flower nectar, rotting fruit, sap, and other fluids. They do not have a reputation for any potential to act as pollinators, but such has not been tested. Many species are important pests, especially to row crops and pastures. Moreover, many species are emigratory and may colonize regions hundreds of miles from their permanent haunts.

Diversity With more than 39,000 described species in 48 subfamilies, this is the most species rich lepidopteran family. In North America: there are about 3,300 described species, about half of which occur in western North America. The western North American species are divided among 34 subfamilies.

SUBFAMILY BOLETOBIINAE

This subfamily, formerly included in the Rivulinae, has 19 North American species of which nine are known from the West. Included are one or more species each of *Oxycilla*, *Zelicodes*, *Prosoparia*, *Mycterophora*, and *Dasyblemma*. **Dasyblemma straminea** Dyar (FW 0.9–1.0 cm; not illustrated) is found in

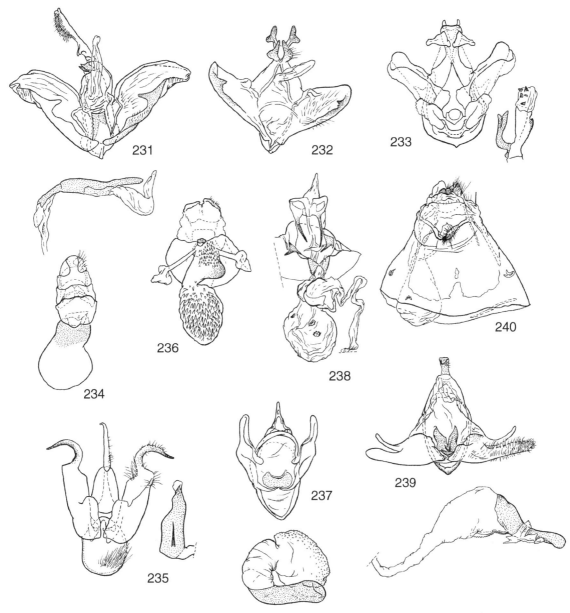

FIGURES 231–240. Genitalia structure of exemplar species of Noctuoidea, ventral aspect. **231**, Notodontidae: *Crinodes beskei* Hübner ♂, aedeagus below. **232**, Notodontidae: *Heterocampa averna* Barnes & McDunnough ♂, aedeagus removed [J. Miller 1991]. **233**, Notodontidae: *Symmerista suavis* (Barnes) ♂, aedeagus to right. **234**, Notodontidae: *Symmerista sauvis* ♀ [Franclemont 1946]. **235**, Arctiinae: *Cisthene martini* Knowlton ♂, aedeagus to right. **236**, Arctiinae: *Cisthene martini* ♀ [Knowlton 1967]. **237**, Arctiinae: *Grammia nevadensis* (Grote & Robinson) ♂, aedeagus below. **238**, Arctiinae: *Grammia nevadensis* ♀ [Ferguson 1985]. **239**, Lymantriidae: *Orgyia cana* H. Edwards ♂, aedeagus below. **240**, Lymantriidae: *Dasychira mescalera* Ferguson ♀, position of bursa copulatrix within A8 shown by dotted lines [Ferguson 1978].

the mountains of southern California. It is attracted to light throughout the year. Its life history is unreported.

SUBFAMILY HYPENODINAE

The Hypenodinae are a small subfamily that includes the smallest noctuids, none of which is important. This subfamily was briefly referred to as the Strepsimaninae. The monophyly of this group is uncertain, and it may represent Hypeninae or Catocalinae with some reduced structures; the prolegs on third and fourth abdominal segments are often absent. The larvae feed on algae, lichens, and fungi. Pupation is in a dense silk cocoon covered with plant or wood fragments. There are 21 species reported for North America, but

only one includes the West in its range. ***Schrankia macula*** (Druce)(FW 0.9–1.0 cm; Plate 42.31) ranges from southern California east across the plains and Gulf states to the Atlantic Coast, where it occurs from Massachusetts south to Florida. It also extends south to Centrtal America and occurs on Bermuda. The species is holodynamic and may be found throughout the year in favorable locations. The larvae feed on bracket fungus (Polyporaceae).

SUBFAMILY EUBLEMMINAE

This is a small group of small moths whose larvae feed on flowers, seeds, fungi, and coccid scales. The larvae are semiloopers with only two pairs of abdominal prolegs.

Eublemma minima (Guenée), the "everlasting bud moth" (FW 0.7–0.8 cm; Plate 42.29), ranges from California east to Texas; the species is distributed widely through Central America and South America to Argentina. The adults have been collected at light throughout the year in southern California, and it is expected that there are several flights. McFarland has found that the larvae burrow and feed in the flower heads of cudweed, *Gnaphalium* (Asteraceae), and we also reared the species from this host.

Metalectra Hübner has 65 species that are restricted to the New World; 10 are found in North America, and five of these occur or range into the West. *Metalectra quadrisignata* Walker, the "four-spotted fungus moth" (FW 1.0–1.2 cm; Plate 42.30), ranges from southeastern Arizona and Oklahoma east through southeastern Canada and most of the eastern United States. Larvae feed primarily on fungi but have been reported to be detritivorous and to occasionally feed on spruce. The larva is blackish and minutely granulose. The anal shield is produced and upturned medially, forming a short, blunt horn. The head is pale with dark blotches.

SUBFAMILY HERMINIINAE

The Herminiinae are unified by a number of autapomorphic features, many of which are highly technical and need not be mentioned here. It has been suggested by some authors that this subfamily be excluded from the Noctuidae because of the position of the countertympanal hood, but Kitching and Rawlins explain the derived nature of this character. Most Herminiinae lack an orbicular spot on the FW, and many males have distinctive secondary features such as antennal knots, costal wing folds, or hair brushes on the legs. Most genera have larvae with a full complement of prolegs, but in some there is a reduction similar to that of other quadrifine noctuids. In one Asian group the first instars have prolegs only on the sixth and tenth abdominal segments—the only such instance in the Noctuoidea. The adults are nocturnal and seem equally attracted to lights or fermented sugar baits. The larvae of most herminiines feed on dead and decaying leaves, but there is actually a full range of feeding habits in various species, from decaying leaves to live leaves. The larvae of some species feed on dead insects, dried fruits, or vertebrate dung. Eggs are laid singly or in masses. The larvae pupate inside silken cocoons covered with bits or leaves or debris. According to Franclemont and Todd, there are 82 species in 19 genera recorded for North America north of Mexico. In the West, there are at least 32 species in 10 genera. The richest diversity seems to be in southeastern Arizona.

Idia Hübner is a widespread New World genus with about 17 North American species of which nine occur in the West. Larvae of *Idia* feed primarily on dead leaves, but some feed on lichens or fungi, and, in exceptional cases, on the leaves of broad-leaved plants or conifers (e.g., compilation by Robinson et al.). Perhaps the most widespread and frequently encountered species is *I. americalis* (Guenée), the "American idia" (FW 1.1–1.3 cm; Plate 42.32). This species ranges from British Columbia east to Nova Scotia and south through most of the West to southern California, western Texas, and northern Mexico. The adults are nocturnal and readily attracted to light. In areas with mild climates, such as lowland California, adults are found from April to October (there may be several flights), but the flight season is more restricted in interior regions with longer cold seasons (e.g., June to September in Colorado). The larvae are detritivorous on dead leaves, fir needles, and twigs; McFarland found them as inquilines in *Formica* ant nests.

Larvae have their heads and bodies brown to gray with pink suffusion. The caterpillar is finely granulose with scattered coarser granules. There are larger tubercles bearing yellow setae. Winter is passed in the larval stage. *Idia occidentalis* (Smith)(FW 1.3–1.7 cm; Plate 42.33) ranges from British Columbia east across southern Canada to Nova Scotia and south to southern California, Arizona, and western Texas. The adults are nocturnal and have one or two flights each year from May to October. The larva is dark gray black with rusty peglike setae on head and body. There are two pairs of raised knobs on the dorsum of each segment—the anterior pair is black and closer to the midline, while the posterior pair is rusty orange. The prolegs on the third abdominal segment are strongly reduced. Larvae feed on lichens, grasses, rotting wood, and other organic matter. Winter is passed by the larval stage. Other western *Idia* include *I. aemula* Hübner, the "common idia," *I. parvulalis* (Barnes & McDunnough), *I. suffusalis* Smith, and *I. dimunendis* (Barnes & McDunnough), the "orange-spotted idia."

Reabotis Smith is monotypic, the only species being *R. immaculalis* (Hulst)(FW 1.6–1.7 cm; Plate 42.34), a broad-winged moth that could be mistaken superficially for a geometrid. The wings are pale tan and completely devoid of markings. The moth is found on the short-grass prairies and arid habitats in intermountain states from eastern California, Nevada, Utah, and southern Wyoming south to Arizona and western Texas. The species flies from June to August as a single brood. Adults are nocturnal and are attracted to light. There are no reports of the species' life history.

There are four North American *Tetanolita* Grote, with the remaining five species restricted to the Neotropics; three species have been found in the West. Adults have long, sickle-shaped upturned labial palpi. *Tetanolita palligera* Smith (FW 1.0–1.3 cm; Plate 43.1) is the most frequently encountered. The species ranges from British Columbia south through the West to the southernmost part of California and Arizona to northern Colorado. The adults are nocturnal and are attracted to light. Mustelin reports that in San Diego County, California, they fly from February to November. Larvae are known to feed on dry leaves. The larva is light gray with reddish intersegmental bands, and its body is covered with fine granules and interspersed larger granules, and the verrucae bear long yellowish hairlike setae.

Bleptina Guenée, with almost 70 New World species, has six found in North America; three of these are found in the West. *Bleptina caradrinalis* Guenée, the "bent-winged owlet" (FW 1.1–1.3 cm; Plate 43.2), is the most widespread, ranging from British Columbia east across Canada and south throughout the United States to Florida, Texas, and Colorado. Adults are nocturnal and are attracted to light. They may be encountered from late April to August. Larvae feed on dead leaves of deciduous trees, according to Crumb. The larva varies from almost black to dark brown and is covered with a mixture of fine and pale coarse spicules giving the larva a speckled appearance. *Bleptina flaviguttalis* Barnes & McDunnough (Arizona) and *B. minimalis* Barnes & McDunnough (Arizona) have also been encountered in the West.

Hypenula Grote has two North American species that include the West in their ranges; three others are found in the Greater Antilles. *Hypenula cacuminalis* (Walker), the "long-horned owlet" (FW 1.3–1.4 cm; Plate 43.3), is a southeastern U.S. species that enters our area in western Texas and southeastern Arizona. The adults are attracted to light from May to October in western Texas. Nothing has been reported concerning its life history.

Renia Guenée comprises about 30 New World species; 13 of these are found in North America, and six have ranges that include portions of the West. *Renia hutsoni* Smith (FW 1.2–1.5 cm; Plate 43.4) is an orange brown species that ranges from Oregon west to Colorado and south to southern Arizona and western Texas. The adults are nocturnal and are attracted to light from June to September. The life history has not been reported, but larvae of other *Renia* species feed on dead leaves.

Palthis angulalis (Hübner), the "dark-spotted palthis" or "spruce harlequin" (FW 1.1–1.2 cm; Plate 43.5), is one of two North American representatives of a New World genus that includes 23 species; it ranges widely in the temperate portions of eastern Canada and the eastern United States. In the West, there are records in Oregon and northern Colorado. There is one flight of adults in more northern areas, and two flights southward. In Colorado, the adults were found in June. Larvae are polyphagous on both broad-leaved trees and shrubs and conifers of at least 11 different families. Moreover, the larvae are detritiphagous and also feed on dead dried plant material. The larva is a dark blotchy brown with arched eighth abdominal segment. There are vague oblique pale stripes on the anterior abdominal segments. Larvae feed from June to November.

SUBFAMILY HYPENINAE

The Hypeninae are poorly defined, according to Kitching and Rawlins. The larval setae are usually borne on chalazae, but these are not conspicuous in later instars. HW vein MA2 is strong and arises well above the lower angle of the cell, running more or less parallel to vein MP1. The *Hypena* group is unified by possession of a spinose "cuff" on the aedeagus. The adults usually have long porrect labial palpi and have been called "snout moths," and those of many species are sexually dimorphic with the males being larger. Kitching and Rawlins report that most species are restricted in larval diet, especially to Fabaceae, Malvales, Polygonaceae, and Urticales. Three subventral setae are present on the first abdominal segment of known larvae. Larvae are often green, without dark markings, and are "semiloopers," as they have reduced prolegs on the third abdominal segment. The anal prolegs are directed posteriorly and offset outward at an angle. The larval head is usually wider than the prothorax. There are 49 North American species of hypenines in nine genera, while at least 27 species in six genera occur in the West.

Hypena Schrank is a worldwide genus that has well over 400 species; more than 20 of these, including those formerly assigned to *Bomolocha* Hübner, occur in North America. *Hypena bijugalis* (Walker), the "dimorphic bomolocha" (FW 1.2–1.3 cm; Plate 43.6), is found from Washington south to California and east through eastern North America. Adults are found from April to September. Larvae feed on various dogwoods, *Cornus* species (Cornaceae). *Hypena abalienalis* (Walker), the "white-lined bomolocha" (FW 1.4–1.5 cm; Plate 43.7), is found from British Columbia and Manitoba east to Quebec and New Brunswick south to California, Texas, and northern Florida. Adults are found from April to August, according to Covell. The larval hosts are dogwoods (Cornaceae) and elms (Ulmaceae). *Hypena humuli* Harris, the "hop looper" (FW 1.3–1.6 cm; Plate 43.8, Plate 43.9), ranges from southern British Columbia south to southern California and east to Alberta, Wyoming, Colorado, and South Dakota. It has been found as far east as Nova Scotia. There are several flights from late April through September. The slender green larvae have

been reported to feed on hops, *Humulus,* and stinging nettle, *Urtica* (Urticaceae). *Hypena californica* Behr (FW 1.4–1.5 cm; Plate 43.10) is found from Alaska south along the Pacific Coast to southern California (San Diego County). Adults are nocturnal and are attracted to light. There is a single flight from spring to fall. The larvae feed on stinging nettle, *Urtica* (Urticaceae), during spring. Larvae are dark green with a faint subdorsal white line. *Hypena decorata* Smith (FW 1.3–1.4 cm; Plate 43.11) is found from British Columbia south to southern California. The larval hosts are nettles, *Urtica* (Urticaceae). *Hypena scabra* (Fabricius), the "green cloverworm" (FW 1.2–1.6 cm; Plate 43.12), ranges from western Texas and throughout eastern North America. Adults are most common in late summer but can be found in at least small numbers during any warm period. The species is emigratory, as it colonizes areas in the northeastern United States and southern Canada where it cannot successfully overwinter. Larvae feed on clovers, beans, alfalfa, other Fabaceae as well as diverse species of herbaceous and woody broad-leaved plants in several families. The moth is considered a minor pest. According to Wagner, larvae are lime green with faint subdorsal and strong lateral white stripes. The anal prolegs are directed posteriorly at an angle. When disturbed, larvae begin thrashing and drop to the ground.

SUBFAMILY PHYTOMETRINAE

Hemeroplanis Hübner has 29 New World species, 11 of which occur in North America. *Hemeroplanis finitima* (Smith)(FW 1.1–1.2 cm; Plate 43.13) is one of six western species and is quite variable, according to McFarland. Its range is from Washington, Utah, and New Mexico south to Baja California, Arizona, New Mexico, and mainland Mexico. In southern California, Mustelin found the adults from March to October. McFarland suspects that certain legumes may be the larval hosts in Arizona. *Spargaloma sexpunctata* Grote (FW 1.1–1.2 cm; Plate 43.14) is the only species in its genus and is found from coastal Oregon east to Nova Scotia and south in eastern North America to Mississippi and Virginia. Adults have been recorded from May to September. The larvae feed on dogbanes, *Apocynum* (Apocynaceae).

SUBFAMILY SCOLECOCAMPINAE

Nychioptera noctuidalis (Dyar)(FW 0.8–0.9 cm; Plate 43.15) is found in southeastern Arizona in the Baboquivaris, Chiricahuas, and Huachuca mountains. In the Baboquivaris it flies at least during August and September. The genus includes two other species, *N. accola* Franclemont and *N. opada* Franclemont, both described from southeastern Arizona. *Phobolosia anfracta* (Edwards)(FW 1.0–1.4 cm; Plate 43.16) is a member of a small genus with nine New World species; *P. anfracta* ranges from Northwest Territories south to southern California, Baja California Norte, southern Arizona, and east to western Texas, where adults are found from March to September. Mustelin reports the species from all months except January in southern California. The life history is unreported.

SUBFAMILY EREBINAE

Ascalapha odorata (Linnaeus), the "black witch" (FW 5.7–8.1 cm; Plate 43.17), is one of the largest moths regularly found in North America and is the only member of its genus. In the older literature it was usually referred to in the genus *Erebus*. It ranges widely throughout the lowlands of tropical America, where it is resident. The species breeds regularly in the southernmost portions of the United States. In the West, this may include only coastal southern California (San Diego County) and possibly southeastern Arizona, where the species might be found during much of the year. The moth is regularly brought northward during the summer monsoons and appears far north of areas where it might breed. It could and does appear over much of the western United States and has been as far north as Anchorage, Alaska. Strays are usually found during summer, often in garages, under eaves, or under bridges. The moth is primarily nocturnal and is attracted to light and fermenting fruit, although it readily takes flight during the day, when it might be mistaken for a bat. Larvae feed on a variety of woody mimosoid (Mimosaceae) and cesalpinoid (Cesalpiniaceae) legumes, including *Acacia dealbata* in southern California. Reports of *Ficus* are very likely in error. The larvae feed at night and rest during the day on bark and branches, where their protective colors make them difficult to spot. The life history stages were described by J. A. Comstock. The fully developed larva is dark gray tinged with brown, covered with dark spots that anastomose in an apparent network of pale reticulation. There is a large dorsal black patch on the first abdominal segment that has a round pale eyespot in the center. There several longitudinal pale and black stripes. The spiracles are black at the center with pale margins and black rim. The larval head has the front pale and most of posterior black. Pupation is in ground litter within a fragile cocoon.

SUBFAMILY CALPINAE

This subfamily, as discussed by Kitching and Rawlins, is characterized by the specialized apical structure of the proboscis, which is modified to pierce thick-skinned fruits or mammalian skin. Three diverse tribes are included: Calpini, Gonopterini, and Anomiini. The FW costal margin of Calpini is often sinuate and often produced as a median lobe and tornal angle. The postmedian line is often oblique—running to the FW apex. Larvae of Calpini feed primarily on Menispermaceae, those of Gonopterini feed on Salicaceae, Annonaceae, and Piperaceae, while those of the Anomiini feed primarily on Malvales (Malvaceae, Sterculiaceae, and Tiliaceae). Pupation often occurs between leaves on the host with little or no silk formation, or in a very slight cocoon.

TRIBE ANOMIINI

Anomis Hübner is a pantropical genus with about 130 species. Nine species are North American, and two of these include the West in their ranges. *Anomis erosa* Hübner, the "yellow scallop moth" (FW 1.2–1.3 cm; Plate 43.18), ranges from southern Arizona east across the southern states to the south Atlantic Coast. In the East, the moths may emigrate as far north as Wisconsin and Quebec. Adults are found during May and October in western Texas. Robinson et al. compiled reports that the larvae feed primarily on various genera of Malvaceae, including cultivated and wild cotton, *Gossypium*, hollyhock, *Alcea*, and hibiscus, *Hibiscus*. There are also reports that they may also feed on *Peperomia* (Piperaceae), but these are suspect. The larva is green with subdorsal, supraspiracular, and subspiracular white lines. Dorsal seta-bearing tubercles are distinctly elevated on posterior segments and rimmed with black. The spiracles are pale yellow and rimmed with brown. The head is green.

Alabama Grote comprises a single North American species, *A. argillacea* (Hübner), the "cotton leafworm" or "cotton moth" (FW 1.6–1.8 cm; Plate 43.19), whose adults are brown with a pointed FW. The reniform is a small white dot and the orbicular is grayish. *Alabama argillacea* is widespread in Central and South America and is an annual immigrant to the southern United States, in our area ranging from southern California as a rare stray east to western Texas. In late summer, the species may reach southern Canada from the Prairie Provinces eastward. Adults are nocturnal and are attracted to light. The species occurs year-round in the tropics but usually appears in our area during the summer and fall. The larvae have been reported to feed on Malvaceae, primarily wild and cultivated cotton, *Gossypium*, but also rarely on Solanaceae. The species is a serious pest of cotton in the Southeast and tropical countries. The larvae are quite variable ranging from unicolorous green with only a white middorsal line and black points at the setal insertions or with varying degrees of black. This may amount to bands of black adjacent to the middorsal white line, ranging to entirely black other than for the venter. The spiracles are black. The head is yellowish with black points marking the setal insertions.

Gonodonta Hübner, the "fruit-piercers," is a tropical American genus with about 40 species, six of which have been recorded in North America. One of these, *G. pyrgo* (Cramer) (FW 2.1–2.5 cm; Plate 43.20) occurs in western Texas and southeastern Arizona. The moths have been found by Knudson and Bordelon during August in western Texas. The adults feed by piercing ripe fruits and imbibing their juices.

Hypsoropha Hübner is a New World genus with four species; two are North American and one of these ranges into the West. *Hypsoropha hormos*, the "small necklace moth" (FW 1.2–1.3 cm; Plate 43.21), has been found from southern Arizona east across southern Texas to North Carolina, Georgia, and Florida. According to Forbes, the adults are found from May to August as two broods. Reports compiled by Robinson et al. show that larvae are defoliators of plants in three families: persimmon, *Diospyros* (Ebenaceae), *Sassafras* (Lauraceae), and cotton, *Gossypium* (Malvaceae). Larvae are green with dark markings or are of a darker form. There are small purplish subdorsal spots on the abdominal segments, fused across dorsum of the fifth and eighth abdominal segments. Spiracles are orange with brown rims. The larval head is reticulated.

Plusiodonta Guenée is a pantropical genus with 35 described species. Two are North American, and both occur in the West. *Plusiodonta compressipalpis* Guenée, the "moonseed moth" (FW 1.1–1.5 cm; Plate 43.22), ranges from southern Arizona east across Oklahoma to North Carolina and north to southern Ontario. The adults have several annual flights from August to November. Several authors report that the larvae feed on *Menispermum*, moonseed vine (Menispermaceae), an eastern plant, and it seems certain that the moth's larvae feed on different plants in the West. Larvae sever leaves of the plant but feed on them only after they have dried. The larva is mixed white and black with white saddle bands on abdomen, giving it a "bird dropping" appearance. Black subdorsal spots are present from mesothorax to the seventh abdominal segment. Abdominal prolegs are absent from the third and fourth abdominal segments. The head is pale with broad black markings. The larvae crawl in a looping fashion, like those of geometrids.

TRIBE SCOLIOPTERYGINI

Scoliopteryx libatrix (Linnaeus), the "Herald moth" (FW 1.8–2.0 cm; Plate 43.23, Plate 64.4 larva), may be characterized by the hooked FW apex and scalloped FW outer margin combined with the mottled red ground on the basal half of the FW, and strong double white postmedial line. This peculiar moth is Holarctic, ranging across the temperate zones of North America, Eurasia, and North Africa; one other species of the genus occurs in China. In North America, the moth ranges from Alaska (Chatanika) south to southern California, Arizona, New Mexico, and east in temperate habitats through much of southern Canada and the eastern United States. The adults are nocturnal and are attracted to both light and fermented fruit bait. There is a single flight of the long-lived adults from July to November, then overwintering (often in caves, according to Covell) in that stage until the following spring. They feed strictly on willows, Salix, and cottonwoods and aspens, *Populus* (Salicaceae). The long, slender larvae are green with two thin, longitudinal yellow lines. Pupation takes place in a whitish cocoon between host leaves that are drawn together with silk.

Litoprosopus Grote has had four species recorded in North America, one of which is western; three other species are restricted to the Neotropics. *Litoprosopus coachella* Hill, the "palm flower moth" (FW 1.6–3.0 cm; Plate 43.24), is found where palms are native or planted from the Central Valley of lowland northern California south to southern California and southeastern Arizona. The adults have two flights in southern California during May to June and August to September. The larvae have been reported by McFarland to feed on the native southern California desert fan-palm, *Washingtonia filifera* (Arecaceae).

SUBFAMILY CATOCALINAE

The Catocalinae are an assemblage of genera whose affinity is established on the similar structure of female genitalia.
Tribe Toxocampini

Lygephila victoria (Grote)(FW 1.8–2.2 cm; Plate 43.28) is the only North American representative of a worldwide genus that has more than 30 species. It ranges from British Columbia south to southern California and Arizona and east sporadically over much of the West to Montana, western Colorado, New Mexico, and western Texas; it also extends south to at least Nayarit, Mexico. The adults are nocturnal and have a single flight from late June through September. Larvae feed on snowberry, *Symphoricarpos* (Caprifoliaceae). The larva is pale gray tinged with brown; there is a brown middorsal stripe of variable width and a few dark spots and pale lines. The setae are borne on flat black tubercles. The spiracles are golden brown. The head is white with black submedian stripes and dark reticulation.

Tathorhynchus exsiccatus (Lederer)(FW 1.3–1.5 cm; Plate 43.25) is a member of a genus that has nine species; this Holarctic moth is found from southern California, Baja California Norte, and eastern Colorado east to western Texas and Oklahoma. Because of the distribution of the remaining species, it seems possible that moth may have been accidentally introduced to North America. In Colorado, the moth has been found from mid-July to late October, and there are likely two flights. The larvae are reported to feed on alfalfa, *Medicago sativa* (Fabaceae), but we reared an adult from the roots of *Isocoma* (Asteraceae).

TRIBE MELIPOTINI

Phoberia atomaris Hübner, the "common oak moth" (FW 1.7–1.8 cm; Plate 43.26), is very likely a complex of several species, according to Metlevski. In the West, this species ranges

from Colorado and California south to Texas. The eastern population ranges from Massachusetts and central New York south to Florida. There is a single spring flight from March to May. The larvae feed on *Quercus* (Fagaceae). Larvae have the first prolegs slightly reduced. According to Forbes the larva is striate and finely mottled with brown, giving it a striped effect. The head is coarsely reticulate but the front and adfrontals are unmarked.

All North American *Cissusa* Walker are western. The FW of *C. indiscreta* (Henry Edwards)(FW 1.5–1.7 cm; Plate 43.27) varies from light to dark brown with fine wavy ante- and postmedian lines, as well as a rectangular black reniform. The moth occurs in oak woodlands along the Pacific Coast from British Columbia to southern California, and, in the interior West, to southern Arizona, and central Colorado. Adults are nocturnal and fly during early spring (February to May in southern California). The larvae feed nocturnally on new oak foliage during spring and are mottled silver, gray, and black.

Litocala sexsignata (Harvey)(FW 1.3–1.5 cm; Plate 43.29) is the only member of its genus, and it ranges from Washington, Montana, Utah, and Colorado south to southern California, Baja California Norte, Sonora, and Chihuahua. The moths have a blackish FW with a distinct white reniform and pale orbicular smudge. The HW is black, each with three round white spots. The adults are strictly diurnal and have small eyes; they are often seen nectaring at flowers or sipping moisture at muddy spots. There is a single annual flight during spring (March to June, mainly April to May). The larvae feed on new foliage of oaks, *Quercus*, or chinquapin, *Chrysolepis* (Fagaceae), and, possibly manzanita, *Arctostaphylos* (Ericaceae). Crumb describes them as gray with a sinuous black stripe along each side. In addition, there is a white subspiracular stripe margined ventrally with purple. Spiracles are dark brown. The head is pale with black reticulation and pale lines. The description by Comstock and Dammers, based on material from southern California, is quite different.

The Neotropical genus *Melipotis* Hübner comprises about 60 described species, and is represented by 13 North American species, eight of which are reported from the West. Adults of *M. indomita* (Walker), the "indomitable melipotis" (FW 2.0–2.3 cm; Plate 44.1 male, Plate 44.2 female), are sexually dimorphic, and, according to Richards, the species is distinguished by the straight oblique transverse anterior line, black interrupted transverse posterior line, and cream, dusky, or fuscous base of the HW. The species ranges from the central California, Utah, Arizona, and New Mexico across the South to Florida. It also ranges south into Mexico. It occurs northward irregularly as an emigrant, sometimes in large numbers. It has been found northward to Colorado, Missouri, and Maine. The moths fly from May to October in coastal southern California and May to July in Colorado, where the species is likely a nonreproductive migrant. Larvae feed on several legume genera, *Albizzia*, *Calliandra*, and *Prosopis* (Mimosaceae). The caterpillar has two pairs of prolegs much reduced and is termed a "semilooper." The larvae are finely striate with dark spots on the tubercles. The venter is pale. In *M. jucunda* Hübner, the "merry melipotis" (FW 1.6–2.0 cm; Plate 43.30, Plate 43.31), the sexes are dimorphic in pattern with males varying greatly in FW pattern and intensity. Females have their FW almost immaculate. This moth is an adventive tropical species that migrates northward, sometimes in large numbers, during late summer and fall. The area of residency in the West ranges from California east to western Texas and south to central Mexico, but the moths appear north to British Columbia and east to Manitoba and through much of our area. On the Atlantic Coast, the moth is found from New Jersey south to Florida. There are likely several ill-defined flights. In western Texas, for example, the adults are found in several flights from March to October, while in southern California, Mustelin reports them from at least January to October. In Colorado, they appear from May to September and may be reproductive since potential hosts are frequent. Larvae have been reported to feed on *Acacia* and *Calliandra* (Mimosaceae), but reports of such hosts as various *Quercus* and *Salix* may be apocryphal. The larvae are dark gray with a purplish brown tinge. The body is marked with various dark and paler lines, many subdued. The ventral area is white with three purple lines on the first and second abdominal segments. Spiracles are dark brown. The head is pale brown with black submedian arcs and sparse reticulation.

Forsebia perlaeta (Henry Edwards)(FW 1.4–1.6 cm; Plate 44.3 male, Plate 44.4 female) is the only North American representative of its genus; its only other member is known from Argentina. The species ranges from southern California east through southern Arizona to western Texas and south into Baja California and Sonora, Mexico. One individual was purportedly collected at Glenwood Springs, Colorado, but this was likely mislabeled. Adults fly from late February to October in Arizona. In southern California, the larvae feed on *Parkinsonia* and other woody legumes

Bulia deducta (Morrison)(FW 1.4–1.7 cm; Plate 44.5 male, Plate 44.6 female) is a widespread Neotropical moth whose range extends northward into the Southeast and Southwest from southern California east to western Texas. The adults are nocturnal but are readily flushed during the day. There are more or less continuous flights in the tropics, but the moth is found more seasonally to the north. In western Texas, for example, the flight is from March to October. Adults have been found north to southeastern Colorado. Larvae feed on mesquite, *Prosopis* (Mimosaceae). The larva is gray with pale and black longitudinal stripes. The spiracles are black. The head is brown with black reticulations. The genus has been treated by Pogue and McLaughlin who describe how to separate this species from the generally scarcer and very "similar" ***Bulia similaris*** Richards (Figs. 241, 242 genitalia).

There are 27 North American ***Drasteria*** Hübner, formerly *Synedoida*, with about 70 species ranging across the Holarctic region. Twenty-six are North American and 23 of these occur in the West. The group is currently under study by Metlevski; the species are quite varied and were studied in detail previously by Richards. ***Drasteria pallescens*** (Grote & Robinson)(FW 1.5–1.8 cm; Plate 44.7, Plate 44.8) has the basal portion of the HW white and the outer portion dusky or black. The moth occurs mainly on the high plains and intermountain region from Idaho and Montana south to Baja California, Utah, New Mexico, and western Texas. The moths have two flights from mid-April through September. The life history is unreported. ***Drasteria inepta*** Henry Edwards (FW 1.6–2.0 cm; Plate 44.9, Plate 44.10) has the FW color quite variable between populations and sometimes within populations; it ranges from red gray to brown to dark gray. This "species" may actually represent a complex, according to Metlevski. The species ranges from Utah, Wyoming, and Colorado south to Arizona, New Mexico, and western Texas; it is found from March to October in western Texas. The moth has one primary flight in May and June and a second, smaller flight in late summer. The adults rest on the ground in dry rocky areas during the day and are active at night, being readily attracted to light. The early stages are unknown. ***Drasteria howlandi*** (Grote)(FW

1.6–1.7 cm; Plate 44.11) is found through much of the intermountain West and high plains from eastern Washington east to Nebraska and south to northern Arizona and western Texas. The moth is nocturnal but is readily flushed during the day, as are most species in this group. There appear to be several broods in western Texas, with adults collected at light from March to September. To the north, the species flies from late April to August. Larvae feed on wild buckwheat, *Eriogonum* (Polygonaceae). The larva is light gray with longitudinal pale and dark stripes. The lateral area is white edged with reddish. The head is pale white with dark stripes. In the desert Southwest, *D. howlandi* is replaced by ***D. tejonica*** (Behr). ***Drasteria ochracea*** (Behr)(FW 2.0–2.1 cm; Plate 44.12) ranges from British Columbia east to Montana and south to southern California, Arizona, New Mexico, and western Colorado. Adults are nocturnal and are attracted to light. The species may be indefintely brooded southward (March to October in southern California), but there is only a single late-spring to early-summer (March to July) flight in the northern portion of its range. Larvae feed on blue and red elderberries, *Sambucus* (Caprifoliaceae). The larva has longitudinal bands of yellow and tan bordered with darker shades. The head has six distinct bands. ***Drasteria petricola*** (FW 1.2–1.4 cm; Plate 44.13) is the smallest *Drasteria* and the only species to have the radial and cubital veins on the ventral portion of the FW streaked with black. The species ranges from Alaska and western Canada south to Utah and Colorado. The adults are diurnal and readily visit flowers such as those of mints. The species has a single annual flight from early May to late July. The moths may be found in more or less boreal habitats from the plains up to subalpine slopes. Larvae feed on *Hedysarum* (Fabaceae).

TRIBE EUCLIDIINI

Caenurgia togataria (Walker)(FW 1.7–1.9 cm; Plate 44.14) is one of two North American species. Three additional species compose the remainder of the genus; two of these are Neotropical, while the other is Eurasian. The species' range includes all of low-elevation California west of the Cascade–Sierra Nevada crest. The adults are nocturnal and found at light through much of the year (e.g., in San Diego County). Mustelin has found adults in all months.

Caenurgina caerulea (Grote)(FW 1.4–1.6 cm; Plate 44.15), one of five species in the genus, all Nearctic, is a strictly diurnal moth of the Pacific Coast from western Oregon south through much of California west of the Cascade–Sierra Nevada divide. Boisduval's blue (*Plebejus icarioides*), a widespread western butterfly, closely resembles this moth in flight. There is a single annual flight from April to mid-June. The larval hosts are wild vetches, *Lathyrus* and *Vicia* (Fabaceae). ***Caenurgina erechtea*** (Cramer), "forage looper" (FW 1.5–2.2 cm; Plate 44.16, Plate 44.17) is one of five North American species which closely resembles ***C. crassiuscula*** (Haworth), the "clover looper" (FW 1.4–1.9 cm; Plate 44.18, Plate 44.19), and may be confidently separated only by close examination of the abdominal structures. The adults are quite variable within and between seasons. The species ranges across much of temperate North America, being absent in most southern locations as well as from most desert and boreal regions. These moths are most abundant in open grasslands of all types. The adults do not seem to have long-distance migrations but do have altitudinal movements, and some individuals appear well above timberline in Colorado. Adults are nocturnal but readily take flight during the day. There are usually at least two flights each year,

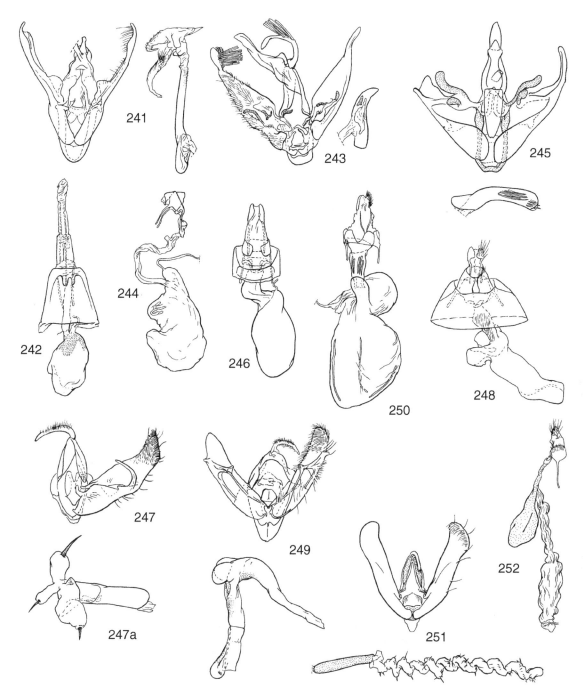

FIGURES 241–252. Genitalia structure of exemplar species of Noctuidae, ventral aspect, except Figs. 244, 252 lateral. **241,** *Bulia similaris* Richards ♂, aedeagus to right. **242,** *Bulia similaris* ♀ [Pogue and Laughlin 2002]. **243,** *Mouralia tinctoides* (Guenèe) ♂, aedeagus to right. **244,** *Autographa labrosa* (Grote) ♀, A10 removed [Eichlin and Cunningham 1978]. **245,** *Annaphila diva* Grote ♂, aedeagus below. **246,** *Annaphila diva* ♀ [Rindge and Smith 1952]. **247,** *Cucullia styx* Poole ♂, left valva removed, aedeagus inflated, below. **248,** *Cucullia styx* ♀ [Poole 1995]. **249,** *Xestia c-nigrum* (L.) ♂, aedeagus below. **250,** *Xestia c-nigrum* ♀ [Lafontaine 1998]. **251,** *Heliocoverpa zea* (Boddie) ♂, edeagus inflated, below. **252,** *Heliocoverpa zea* ♀ [Hardwick 1965].

in spring and summer. In western Texas, adults were found by Knudson and Bordelon from May to October. Primary hosts of both species are native and low weedy Fabaceae, especially clovers *(Trifolium)*. Larvae vary from pale gray to infuscated brown to almost black. There are pale brown or pinkish longitudinal stripes middorsally, subdorsally, and subventrally. The pale stripes are obscured in the darker-colored larvae. Spiracles are yellow or white. The larval head is dark dorsally and lighter frontally and ventrally.

Euclidia Ochsenheimer is a primarily Palaearctic genus with 10 species, but there are disjunct species in Senegal, Sri Lanka, Chile, and North America (two). ***Euclidia ardita*** Franclemont (FW 1.4–1.8 cm; Plate 44.20) is found in coastal Oregon and California west of the Cascade–Sierra Nevada divide. The single spring flight of this strictly diurnal moth is from March to early June. In flight, the adults could be confused with adult duskywing skipper butterflies (*Erynnis* spp.). The slender larvae are yellow tan and feed on deerweed, *Lotus* *scoparius* (Fabaceae), according to McFarland, who observed wild females fluttering near the plants.

Callistege Hübner comprises three North American species, all western, and three Eurasian species. ***Callistege intercalaris*** (Grote)(FW 1.2–1.4 cm; Plate 44.21) is widespread in Arizona, New Mexico, and western Texas. In western Texas, Knudson and Bordelon report the species in flight from June to September, while in New Mexico they are found primarily in July. The adults are nocturnal and readily attracted to light.

Mocis latipes (Guenée), the "small mocis" (FW 1.7–1.8 cm; Plate 44.22), is one of four North American species of a worldwide genus comprising almost 40 species. The species ranges from Texas east to Florida and south at least to Panama. Covell records the species as a vagrant north to southern Ontario. In western Texas, the species flies from August to October. The larvae are polyphagous on a wide variety of herbaceous plants and, especially, grasses (Poaceae).

TRIBE PANOPODINI

Anticarsia gemmatalis Hübner, the "velvetbean caterpillar" (FW 1.6–1.9 cm; Plate 44.23, Plate 44.24), is the only North American member of a worldwide genus with 13 species. The adults are extremely variable in wing color and pattern. This is a Neotropical moth that ranges from southern Arizona east to western Texas and, in the East, is prevalent in the southeastern states but occurs as a vagrant north to Maine and New York. The species has a single flight from August to October in western Texas. Its larvae have been reported to feed on many genera of legumes (Fabaceae), including beans, alfalfa, and black locust. Reports of their use of Linaceae and Malvaceae are almost certainly erroneous. The first two pairs of larval prolegs are reduced. The larvae range from green to dusky olive gray to brown. There are distinct middorsal and subdorsal pale lines both margined with a dusky shade that may be tinged with purplish. Spiracles are light yellow and rimmed with brown. The head may be white or green with dark submedian arcs and reticulation.

TRIBE OPHIUSINI

Heteranassa fraterna (Smith)(FW 1.1–1.4 cm; Plate 44.25) is resident in southern California, southern Arizona, and possibly western Texas, south into the tropics. In western Texas, the moth appears from March to October, according to Knudson and Bordelon. Larvae feed on woody Mimosaceae, usually mesquite, *Prosopis*, and occasionally catclaw, *Acacia greggi* (Mimosaceae). Larvae are extremely variable in color and range from gray to green to black. The larvae are so variable that Crumb segregated what appeared to him to be eight species. The diagnostic features are lateral fleshy filaments on the venter coupled with the absence of a seta on the eighth tubercle of the first two abdominal segments.

There are seven North American ***Toxonprucha*** Möschler, and six of these occur in the West; the remaining five species are solely Neotropical. ***Toxonprucha volucris*** (Grote)(FW 1.2–1.4 cm; Plate 44.29) occurs in southern California, Baja California Norte, southern Arizona, New Mexico, and western Texas. The adults are nocturnal and fly from April to September. Larvae of all North American *Toxonprucha*, including *T. volucris*, feed on catclaw, *Acacia greggi* (Mimosaceae). The larvae are slender and taper strongly posteriorly; they vary from pale gray to dark gray. The white middorsal stripe includes a purplish line and often a black dot on each abdominal segment. There

is a black dorsal saddlemark on the fourth abdominal segment extending ventrally to the spiracle. The venter is white with a central purplish line. The head is also white marked with black and purplish.

Two North American species comprise ***Zaleops*** Hampson, and both occur in the West. ***Zaleops umbrina*** (Grote)(FW 1.4–1.6 cm; Plate 44.26) ranges from Arizona and western Texas and south into northern Mexico. The adults are nocturnal and are attracted to light from April to September. Larvae feed on catclaw, *Acacia greggi* (Mimosaceae). The larva is broad in the middle tapering strongly posteriorly with a middorsal hump on the eighth abdominal segment and is generally blackish with longitudinal pale and black stripes. The venter is black edged with white. The head is pale with dense black reticulation.

Zale Hübner comprises about 60 New World species. There are 36 described North American species, most of which occur in eastern deciduous forests; 14 species occur in the West. ***Zale lunata*** (Drury), the "lunate zale" (FW 2.0–2.4 cm; Plate 44.27, Plate 44.28), is quite variable. The adult has both wings dark brown with shades of yellow, red brown, and black. There may or may not be lunate marginal patches of white or silver on each wing. On the FW, the postmedian line is dentate anteriorly. The moth ranges broadly across North America, and, in the West, is found from Washington south to southern California and east to the Colorado Front Range and New Mexico. The species is nocturnal and is attracted to light. There are one or two flights from May to November. In southern California, the adults appear as early as January. The larvae are polyphagous and feed on a wide variety of broad-leaved trees, shrubs, and perennial herbs; they are variable and may be gray, with brown shades or tan. There is a light middorsal band bordered with darker shades. The first and eighth abdominal segments are swollen dorsally, and each bears two fine points. ***Zale termina*** (Grote)(FW 1.5–1.8 cm; Plate 44.30) is dark gray with fine lines and strong ante- and postmedian dark bands on the FW. This slightly smaller species ranges from the Pacific Coast east to southern Colorado, New Mexico, and western Texas. Adults are nocturnal and are attracted to light. The species has one or two flights depending on latitude. These moths fly from March to September in western Texas and only during July and August in the Pacific Northwest. The larvae feed on various oaks, *Quercus*, and chinquapin, *Chrysolepis* (Fagaceae). The larva is dark gray with darker gray longitudinal lines and a subdorsal dark scalloped line along each side. ***Z. colorado*** (Smith)(FW 1.7–2.0 cm; Plate 44.31, Plate 44.32) is found in Colorado, Utah, New Mexico, and Arizona, primarily near stands of *Quercus*, especially Gambel oak, *Q. gambelii*. In New Mexico, R. Holland has found the species between 4,000 and 9,200'. The adults are nocturnal and are attracted to light. They have two flights from April to early August, but the species is much more abundant during the May to June flight.

Euparthenos nubilis (Hübner), the "locust underwing" (FW 2.2–3.0 cm; Plate 45.2), is the only species in its genus. This species is often mistaken for a species of *Catocala* by its size and general habitus, but it has a more complexly black-marked yellow orange HW. In the West, this handsome moth is found in Arizona and New Mexico. The adults are commonly found at fermented fruit bait but are also attracted to light. In New Mexico there is a single flight from the end of May to early August. The vast majority of larval feeding associations implicate *Robinia* species (Fabaceae), especially black locust, *R. pseudacacia*, as the larval hosts. In the West, adults have been collected on several occasions in association with New Mexican locust, *R. neomexicana.* The larvae have a reticulate brown pattern on a pale gray or yellowish ground. There is a pale middorsal band that is expanded laterally on each abdominal segment forming large ovoid pale areas. The spiracles are large and brown. The head has dark submedian arcs and dark reticulation.

TRIBE CATOCALINI

The Holarctic genus **Catocala** Schrank has about 230 worldwide species with 110 described species in North American, about 47 of which occur in the West. Most of the remaining species are Palaearctic, and only one other occurs in Mexico. Several of our *Catocala* have their nearest relatives in Eurasia, but some groupings seem to be strictly North American. The richest array of species occurs in the deciduous forests of eastern North America, but taken altogether there are also many western species, although not as many as in the East can be found in any single western region or locality. The moths are strictly nocturnal and are variably attracted to lights or fermented fruit baits. The adults usually have a brightly colored banded HW which is covered by their more cryptic FW at rest. When disturbed adult *Catocala* suddenly take flight and, after a quick irregular dash, come to rest and quickly conceal their HW. This "flash" coloration and its apparent disappearance have been shown to be confusing to would-be avian predators. Their FW color and pattern often match the substrate upon which they normally rest—usually tree trunks or rock surfaces. A lengthy treatise on the ecology and behavior of this group, emphasizing eastern species, was prepared by Sargent. There have been intensive recent studies of the group by Gall, Hawks, and Peacock, but the only available treatment is by Barnes and McDunnough as well as an improved later version by Forbes. All *Catocala* species have but a single flight each year, but emergences within the flight period may be protracted and the adults are relatively long-lived. All *Catocala* species are host specific to one degree or another, with each species selecting no more than a few genera within the same plant family, and usually within the same plant genus. The larvae feed nocturnally and are extremely well camouflaged during the day; they are thick and twiglike, resting on host twigs, branches, or in bark crevices.

The tentative treatment of the North American *Catocala* proposed by Barnes and McDunnough, and later modified by Forbes, is followed here. As much as is possible our selection of species is based on a sampling of the different groups set forth by these authors and available descriptions of life history stages. It should be noted that the groups that are not specifically mentioned here do not contain western species.

SECTION I

Section I of Barnes and McDunnough includes those species whose adults have all tibiae spined and whose tarsi have a fourth irregular row of spines. If necessary, the subgeneric name *Mormonia* Hübner could be applied to this group, which encompasses all of the Juglandaceae feeders, *Myrica* feeders, and one aberrant *Gleditsia* feeder. Most species have ribbed eggs, larvae devoid of dorsal protuberances or lateral filaments, and male genitalia usually with the valvae asymmetrical.

Species Group 2 includes only **Catocala piatrix** Grote, "the penitent" (FW 3.1–3.4 cm; Plate 45.1), which ranges through much of temperate eastern North America from southern Ontario southward and west from southern California, southeastern Arizona, southern New Mexico, and western Texas and south into Mexico. There is a single adult flight from July to October. Although there are a few reports of other hosts, the vast majority of rearings are from hickories, *Carya,* and walnuts and butternut, *Juglans* (both Juglandaceae). In southern California, they use California walnut, *J. californica.* Larvae are greenish or brown and are striate rather than mottled. Tubercles are slightly enlarged and are white. The venter has several black spots, which may be partially fused. The head is largely black, with lower front light.

Species Group 6 includes only Juglandaceae feeders, with flat disk-shaped eggs and larvae with lateral filaments but lacking dorsal protuberances. The group includes a number of species from which we discuss **Catocala neogama euphemia** Beutenmüller, "the bride" (FW 3.3–3.7 cm; Plate 45.3). The species is found throughout most of temperate eastern North America from Ontario and Quebec southward and extends westward to the Colorado Front Range, New Mexico, western Texas, and southeastern Arizona. Adults have a single flight during June and July. Larvae feed primarily on *Juglans* and occasionally on *Carya* (both Juglandaceae). A report of *Quercus* may refer to another species of underwing. J. A. Comstock described the last instar and pupa based on larvae found in Gila County, Arizona. The last instar is dull smoky white with innumerable small black and gray dots and dashes. There is a middorsal series of fused diamond-shaped marks that gives the impression of a wide serrated band; the diamonds are alternating light and dark gray. Subdorsally, there is a wide crenulated zigzag, dark longitudinal band heavily sprinkled with black dots; below this there is a wider light band, the lower portion of which is white. The venter is rose pink with a prominent velvety black spot on each segment. Pupation is in a thin fragile cocoon into which leaf fragments and debris are incorporated.

SECTION II

Section II species have the fore tibiae unspined and the tarsi with three or four rows of spines. Species Group 7 includes only **Catocala aholibah** Strecker (FW 2.7–3.8 cm; Plate 45.4), which is characterized by adults with a fourth row of tarsal spines and symmetrical male valvae. Eggs are hemispherical, minutely granulose, and without ribbing. Larvae have a prominence on the fifth abdominal segment and possess lateral filaments. The species ranges over much of the West where oaks occur from the Rocky Mountain Front and northwestern Great Plains westward to the Pacific Coast. Adults are nocturnal and are attracted to both lights and fermented fruit bait. There is a single flight during July to early October. Larvae feed on several oaks, *Quercus* (Fagaceae), including *Q. garryana, Q. gambelii,* and *Q. macrocarpa.* The larva is described by Miller and Hammond as gray tan with a rosy pink overtone and minute black speckles throughout, and small middorsal tubercles on the fifth and eighth abdominal segments. For shadow reduction when at rest on branches there is a ventral row of pale short whiskerlike hairs, and the true legs are red pink. The head is cream with reticulate lines.

Species group 8 consist only of one species, **Catocala ilia** (Cramer), the "Ilia underwing" (FW 3.2–3.6 cm; Plate 45.5). The species has large hemispherical, heavily ribbed eggs, larvae with short lateral rootlike fringes, and the adult male valvae slightly asymmetrical. The species ranges from the Pacific Northwest south to California and east to western Colorado,

New Mexico, and western Texas. The species occurs south into northern Mexico and also includes most of eastern North America from Minnesota, Ontario, Quebec, and Newfoundland southward. The adults have a single flight from June to September. The larvae feed on a variety of oaks, *Quercus* (Fagaceae); the larvae are mottled green and black and, according to Miller and Hammond, match the lichens found on the host Oregon white oak, *Q. garryana*, branches, at least in the Pacific Northwest. Dorsally, the fifth abdominal segment is slightly humped. The head has a broad black band across the vertex.

Species Group 10 includes most western Salicaceae feeders, characterized by ribbed eggs, larvae with lateral filaments, and a transverse wart on the dorsum of the fifth abdominal segment, as well as the adult male having slightly asymmetrical valvae. *Catocala relicta* Walker, the "white underwing" (FW 3.3–3.7 cm; Plate 45.6, Plate 64.5 larva), belongs in this group but is a bit aberrant. This species has a boreal, mainly montane distribution in the West and occurs south to central California, Arizona, and New Mexico. As a whole it ranges across Canada in woodlands and taiga and south in the Appalachians to Pennsylvania and Missouri. The Eurasian *C. fraxini* Linnaeus is a close relative. The adults of *C. relicta* are readily attracted to light and have a single flight from July to early October. Larvae feed on cottonwoods, aspens, and poplars, *Populus* species (Salicaceae) but are also reported to use *Salix*. Records of *Betula*, and *Quercus* are dubious and require confirmation. On the basis of adult collections, it is most likely that quaking aspen, *P. tremuloides*, is the primary host in western montane regions. The larvae are pale with a greenish yellow tinge. The fifth abdominal segment has a middorsal hump that is followed by a contrasting transverse brown band that ends ventrally between the last prolegs. Tubercles on the eighth abdominal segment are only slightly enlarged with an oblique brown band between them. The head is large, squarish and with black reticulation. The front is pale with black lateral bands joined over the vertex. *Catocala junctura* Walker subsp. *arizonae* Grote, the "Arizona underwing" (FW 3.3–3.8 cm; Plate 45.8), ranges from southern California, southeastern Arizona, and western Colorado east to the Midwest. There is a single adult flight during late June to early October. Larvae feed on poplars, *Populus*, and willows, *Salix*. *Catocala grotiana* Bailey (FW 3.0–3.6 cm; Plate 45.7) occurs primarily in boreal forests from northern Colorado to southern New Mexico. The adults are nocturnal and are attracted to either light or fermented fruit bait. The single annual flight is from late August to early October. *Populus* and *Salix* (Salicaceae) are the larval hosts.

Species Group 11 species are Salicaceae feeders with ribbed hemispherical eggs with the rib branches arising regularly from the equatorial region of each egg. Larvae have lateral filaments and a transverse dorsal wart on the fifth abdominal segment. Valvae of the male genitalia are symmetrical. *Catocala semirelicta* Grote, the "semirelict underwing" (FW 2.8–3.4 cm; Plate 45.9), is a widespread western species that ranges from Colorado and New Mexico east to Ontario and Maine. The adults have a single flight from late July to September. Larval hosts are cottonwoods and poplars, *Populus* (Salicaceae). The larva is pale green gray with normal striping usually inconspicuous. A dorsal hump on the fifth abdominal segment is followed by oblique dark patches. The hump and dorsal tubercles are orange and often conspicuous. The head is pale with the continuous black stripe across the vertex.

Species Group 12 species have the valvae of their male genitalia less asymmetrical and more rounded than those of

Species Group 10. Most included species are *Quercus* feeders, and the larvae often feed on buds and catkins. A typical species is *C. delilah* Strecker subsp. *desdemona* Henry Edwards (FW 2.3–2.7 cm; Plate 45.10), which ranges from southern Colorado south through New Mexico and western Texas to northern Mexico, where R. Holland reports that it is found more abundantly. The adults are nocturnal and have an annual flight from late May to mid-August in most areas but fly until October in southeastern Arizona. Larval hosts are oaks, *Quercus* (Fagaceae); reports of *Salix* are likely unreliable. The larva is gray with darker banding and orange bumps. The dorsal humps are fairly pronounced. The head has vertical dark reticulation. The lateral black stripes are thin and partially edged with yellow. Other western species in the group include *C. chelidonia* Grote, *C. frederici* Grote, the "Frederic's underwing," and *C. andromache* Henry Edwards—all oak feeders. The last species is the most common yellow underwing in southern California.

Species Group 16 are all *Quercus* feeders and have hemispherical prominently ribbed eggs. The larvae have lateral filaments and a dorsal conical, pointed, posteriorly directed wart on the fifth abdominal segment. Male valvae are symmetrical and rounded apically. *Catocala verrilliana* Grote, the "Verrill's underwing" (FW 1.8–2.3 cm; Plate 45.13), is found from southern California east to southern Colorado, New Mexico, and western Texas. The species has been found in Sonora, Mexico, according to R. Holland. The adults are found from May to September (mainly August to September) in what is assumed to be one extended flight. Larvae feed on various oaks (Fagaceae). The larva is gray white with a pale pinkish overtone. There are small middorsal tubercles on the fifth and eighth abdominal segments and a ventral row of sparse whiskerlike setae. The head is also gray white with darker fine reticulations. Other western species in this group are *C. violenta* Henry Edwards (FW 2.5–2.6 cm; Plate 45.14), *C. faustina cleopatra* Henry Edwards (FW 3.0–3.5 cm; Plate 45.12)—the most abundant red underwing in coastal southern California—*C. coccinata* Grote, and *C. ophelia* Henry Edwards.

Adults of Species Group 19 have asymmetrical male valvae. Most of the included species are small. *Catocala minuta* Edwards, the "little underwing" (FW 1.7–1.9 cm; Plate 45.11), ranges from the East, where it is rare, west to Nebraska, Colorado, and Texas. The species may occur in suburban developments where their larval host has been cultivated. The adults have a single flight during July and August. The only reported larval host is honey locust, *Gleditsia* (Fabaceae). According to Forbes the larva is mottled dark gray brown with whitish patches. The larval head has heavy black reticulation except for the front, which is pale.

SUBFAMILY EUTELIINAE

The Euteliinae is a subfamily whose monophyly is supported by several autapomorphies, the most consistent of which is the structure of the basal abdominal sternite—the central membranous zone of which is flanked by two internal longitudinal flanges. The adult has a characteristic resting posture with wings folded together and held at a 45 to 90° angle to the body together with its upwardly curled abdomen. This posture allows the moth to resemble a dried leaf or cluster of twigs. Larvae have either homoideous or heteroideous crochets. The pupae lack a cremaster, a feature independently derived by the Nolidae. The most frequent larval hosts are among the Anacardiaceae, but as a whole, the moths in this group utilize an extremely broad taxonomic array. There are 18 euteliines in four genera reported for North America; nine species in all four genera occur in the West.

There are eight species of **Marythyssa** Walker scattered around the world, two of which are found in North America north of Mexico, and one of these occurs in the West. *Marythyssa inficita* (Walker), the "dark marathyssa" (FW 1.2–1.3 cm; Plate 46.1), ranges from eastern Oregon and Idaho to southern California, Arizona, and Texas. Adults are attracted to light and may be found during April and September in western Texas. Larvae are reported by several authors to feed on *Rhus trilobata* (Anacardiaceae). The larva is yellow green with a slightly swollen thorax, according to Wagner. There is a white or yellow white subdorsal longitudinal stripe along each side and many scattered small white spots below these lines. The cocoons have soil particles or other debris incorporated and superficially look like woodrat droppings.

There are 54 species of **Paectes** Hübner distributed worldwide; 11 species of *Paectes* occur in North America, but only four of these seem to occur in the West. *Paectes declinata* (Grote)(FW 1.3–1.6 cm; Plate 46.3) ranges from Los Angeles County, California, south to Baja California Norte. Adults fly from March to October. Larvae feed on mature leaves of laurel sumac, *Malosma laurina* (Anacardiaceae). *Paectes abrostolella* (Walker)(FW 1.1–1.3 cm; Plate 46.2) ranges from Santa Barbara County south to San Diego County in California, thence east, discontinuously, to southeastern Arizona. Adults are found in April and October in southern California and during July in Arizona. The larvae feed on *Rhus* (Anacardiaceae).

Eutelia Hübner comprises about 120 species distributed in most worldwide regions except the Australian. *Eutelia pulcherrima* (Grote), the "beautiful eutelia" (FW 1.2–1.6 cm; Plate 46.4), ranges from eastern North America west to southeastern Arizona (Huachuca Mountains). Adults are nocturnal and fly in August. The larvae feed on both poison sumac, *Toxicodendron vernix*, and poison ivy, *T. radicans* (both Anacardiaceae). *Eutelia furcata* (Walker) is found in Big Bend National Park in western Texas, while an undescribed *Eutelia* occurs in the Huachuca Mountains of southeastern Arizona during August. The larva of *E. adulatrix* Hübner, found in England, may be similar to that of our species.

SUBFAMILY NOLINAE

The Nolinae (tuft moths) have about 1,400 described species worldwide. They are mostly small with dull coloration, the main distinguishing feature being tufts of raised scales on the FW. The larvae also tend to have muted colors and tufts of short hairs.

TRIBE NOLINI

Meganola Dyar is represented by about 200 species worldwide; six species are North American, and all of these are described from the West. *Meganola fuscula* (Grote)(FW 1.0–1.2 cm; Plate 46.5) ranges from central California south to southern California and east through Arizona to Colorado and western Texas. There is a single flight in California from late February to early April, but there may be several flights in Arizona and western Texas. Nothing has been reported about its hosts or life history.

Worldwide, there almost 400 species of **Nola** Leach; 10 are known from North America, and two or three of these are western. *Nola minna* Butler (FW 1.0–1.1 cm; Plate 46.6) ranges along the Pacific Coast west of the Cascade–Sierra Nevada divide and the southern California Transverse and Peninsular Ranges from southern British Columbia south to southern California (San Diego County). There is a single flight from January to early June depending on elevation and exposure. The larval hosts include broad-leaved woody trees and shrubs including *Alnus* (Betulaceae), *Quercus* (Fagaceae), *Ceanothus* and *Frangula* (Rhamnaceae), *Salix* (Salicaceae), and *Nicotiana* (Solanaceae). *Nola apera* Druce (FW 0.8–0.9 cm; Plate 46.7) ranges in southern California from Santa Barbara County south, including Santa Catalina and Santa Cruz islands, through Mexico and Central America to western South America. At the Miramar Marine Base near San Diego, California, adults were found in every month except June, and, by implication, there would seem to be multiple generations. The larval hosts are *Laguncularia* (Combretaceae) and *Salix*, willow (Salicaceae).

TRIBE SARROTHRIPINI

There are 32 worldwide species of **Characoma** Walker, and only one species is North American. *Characoma nilotica* (Rogenhofer), the "black olive caterpiller" (FW 0.7–0.8 cm; Plate 46.8), ranges from southern California, Arizona, and Colorado east to Quebec and Nova Scotia and south to Texas, Missouri, and Florida. The species also occurs on Bermuda. In the West, adults have been found during May and August. Ferguson et al. list trees and shrubs in five families of broad-leaved plants including *Rhododendron* (Ericaceae), *Olea* (Oleaceae), *Prunus* (Rosaceae), *Salix* (Salicaceae), and *Tamarix* (Tamaricaceae). The larvae are pale green with the cervical shield showing a middorsal line broadly margined with black and a transverse black bar across the posterior. The spiracles are pale brown with brown rims. The head is pale with a smoky infuscation with or without black spotting showing through.

Nycteola Hübner has 46 described species worldwide; five are North American including three from the West. *Nycteola frigidana* (Walker), the "frigid owlet" (FW 1.1–1.3 cm; Plate 46.9, Plate 64.6 larva), ranges from British Columbia east to Nova Scotia and south to eastern Washington, Colorado, Tennessee, and Pennsylvania. The species may have two flights in the East (April to September), but only a single flight in Colorado (August). The larval host is willow, *Salix* (Salicaceae). McFarland observed larvae of a related species, **N. cinereana** Neumoegen & Dyar (FW1.1–1.3 cm; Plate 46.10) in Bonner County, Idaho. This closely related species ranges from British Columbia east to Quebec and south to California, Nevada, and Colorado. Adults fly in fall and spring and may overwinter. These larvae made folded leaf edge web nests on poplar, *Populus* (Salicaceaae) leaves. The larvae are dark olive green without markings and have pale spiracles with pale brown rims. The setae are pale and slender. The head is unmarked pale brownish green. The cocoon as observed by McFarland is white pointed at one end and truncate at the other.

TRIBE COLLOMELINI

Iscadia Walker comprises 42 species, and the genus has a primarily Pantropical distribution. *Iscadia aperta* Walker (FW 1.2–1.3 cm; Plate 46.11) is our only North American species and is our largest noline. It was described from the Caribbean but is found from March to September in southeastern Arizona.

SUBFAMILY ARCTIINAE

The Arctiinae are well known as the tiger moths and lichen moths. Some members of the subfamily are among the best known and most colorful North American moths. The larvae of many Arctiini, known as woolly bears, are familiar sights as they usually wander long distances before pupation and often may be seen crossing open areas including sidewalks and highways. The subfamily is unified by the universal presence of a pair of dorsal eversible pheromone glands that are associated with the anal papillae of females. Adults of almost all arctiines have metathoracic microtymbals that consist of a series of grooves on the katepisternum. These tymbals may be reduced or absent in some Ctenuchina. Arctiine larvae are unified by the presence of two setal characters. First, setae D1 and D2 on both the second and third thoracic segments are fused into a single verruca, while, second, seta L3 is bisetose on abdominal segments 3 to 6. These two characters, of course, may be seen only on larvae of species that have only primary setae (e.g., *Tyria* and *Utetheisa*). Moreover, all arctiine larvae lack an adenosma. Most arctiine larvae have heteroideous crochets, but some have homoideous crochets, and, thus, this cannot be a family character.

Adult Adult Arctiinae are small to large moths and, with some exceptions, tend to be brightly colored—primarily red, orange, black, and white. Almost all species produce distasteful chemicals, and, thus, the brightly colored insects are assumed to be distasteful and aposematic. To enhance this aspect many can feign death, especially Arctiini, and produce droplets of distasteful fluid (hematolymph) at their prothorax and leg joints (reflex bleeding) when threatened. These fluids contain acetycholines and histamines and, probably, pyrrolozines. In the tropics, many species, including at least some members of all subfamilies, belong to groups of species that mimic each other or especially unpleasant models. These are referred to as Müllerian mimicry rings. In most of North America, the mimetic aspects of arctiine adults are not always so clear, but mimicry becomes more obvious in groups with more southern ranges. For example, most Ctenuchina are clearly mimetic, and there are many taxa of these groups in the southern portion of the United States. Arctiine genera with more obvious mimetic colors and forms include *Euchaetes* and *Pygarctia*, both of which include a number of species in the southwestern states adjacent to Mexico. Adults of many species are nocturnal and are attracted to light, while a significant number of species are strictly diurnal. A few species (e.g., *Lycomorpha grotei*), are both diurnal and nocturnal. Some genera have both diurnal and nocturnal species (e.g., *Grammia* and *Phragmatobia*), and this habit may be summarized by calculating their relative eye diameters. "Large-eyed" species are usually nocturnal and "small-eyed" species are diurnal. Virtually all alpine and Arctic arctiines are diurnal. As for many Arctic-alpine Lepidoptera, larval development may require more than a single summer. The females of alpine arctiines are relatively heavy bodied and presumably are poor fliers. Moreover, in several arctic species (e.g., *Dodia*), the females are brachypterous or apterous and flightless.

Larva The larvae are hypognathous and are most often cylindrical with the normal noctuoid complement of prolegs. A primary feature of many taxa is the presence of dense clumps of long secondary setae, but these are absent in some Arctiinae and all Lithosiinae. The crochets are heteroideous in most taxa but are homoideous in some. Larvae can crawl rapidly and often incorporate their long setae into the sparsely spun cocoons, but the larvae of many other species form dense silken cocoons packed with long setae ("hairs").

Larval Foods Larvae of arctiines are either general feeders on a variety of dicotyledonous plants or may specialize on particular plant genera or families; a few are specialists on Pinaceae or Cycadaceae. The syntomines feed on lichens, fungi, detritus, flowers of angiosperms, or grasses. These specialists often feed on plants that contain toxins or produce latex. Lithosiines, the lichen moths, feed on algae or the algal portion of lichens.

Pupa Pupae are stout and have either a weak cremaster or none at all. Veruccal scars are surrounded by fine setae.

Diversity More than 11,000 species in about 750 genera are known worldwide. In North America, 260 species in 85 genera have been described. In the West, there are 194 species of 71 genera including 149 Arctiini of 54 genera and 45 Lithosiini of 17 genera. These figures do not include approximately 15 undescribed species. In addition, the systematic status of several arctiine subtribes is unsettled, and there is at least one undescribed genus in the West. We follow the current treatments of the family and include the Pericopina as a subtribe of Arctiinae instead of a separate family or subfamily as previously considered. The group formerly referred to as Ctenuchinae is here considered as the subtribe Ctenuchina, following current treatments.

Economics This family includes no serious pests, but several are considered minor shade tree pests. *Hyphantria cunea* (Drury), the "fall webworm," makes extensive communal silken nests on ornamental shade and orchard trees, especially pecans *(Carya illinoinensis)*. Larvae of *Estigmene acrea* (Drury) (the "salt marsh caterpillar moth"), *Spilosoma virginica* (Fabricius), and others cause minor damage to field crops and garden plantings. *Tyria jacobaeae* (Linnaeus), the "cinnabar moth," was intentionally introduced from Europe to the Pacific Northwest and Nova Scotia for control of *Senecio jacobeae* (tansy ragwort). It is considered to be one of the better examples of the successful use of biological control of exotic weeds.

TRIBE LITHOSIINI

Where known, larvae of all lithosiines (lichen moths) feed either on algae or, more often, the algal component of lichens. The character that unites the tribe is the presence of an enlarged molar area at the base of each larval mandible. Ocelli seem to be absent but are very small and difficult to locate. Adult structures are extremely variable and larvae exhibit a wide range of setal positioning. Larval crochets may be either heteroideous or homoideous. The adults range from small to medium in size (FW 0.7–2.5 cm) and are mostly nocturnal and are attracted to light, but some are diurnal. Adults of many taxa are brightly colored and seemingly aposematic, but others (e.g., *Acsala*, *Eilema*, *Crambidia*, and *Clemensia*), have dull white, tan, or black coloration. Females of *Acsala* have greatly reduced wings and are flightless.

SUBTRIBE ACSALINA

Acsala anomala Benjamin (male FW 1.2–1.3 cm; Plate 46.12) is a small black moth of nonglaciated Arctic mountains of the Yukon Territory and Alaska, where it is found on gravelly scree and talus slopes during June and July. These unusual moths were once placed in the family Lymantriidae, but the discovery of their larvae demonstrated their affinity with the Lithosiinae. Because of their uniqueness, they are placed in a separate tribe Acsalini by Bendib and Minet, and they may be among the more primitive members of the subfamily. Unlike most other arctiines the adults lack ocelli. In addition, the crochets of the prolegs are heteroideous. Larvae are primarily black and have clusters of barbed bristles atop verrucae. Like most arctiines, the pupae have short hairs surrounding verrucal scars. Males fly slowly over the slopes but are easily carried away under windy conditions. Females have much reduced wings, are flightless, and are most easily found under rocks. Eggs are laid in batches of six to 30 in a single layer on the underside of rocks. The larvae feed during late afternoon or evening on black foliose lichens and crustose lichens *(Parmelia, Umbilicaria, Orphniospora, Sphaeraria, Tephromela, Buellia,* and *Lecidia)* found on rocks. Larvae seem to rest in clusters and require an unknown number of years, possibly many, to complete development.

SUBTRIBE EUDESMIINA
Eudesmia arida (Skinner)(FW 1.2–1.5 cm; Plate 46.13) is a member of a group given separate tribal status (Eudesmiini) by Bendib and Minet. The species is a member of a Neotropical genus with many species. The adults have alternating gray and yellow orange bands. In the West, this moth ranges from central Arizona east to western Texas, but its limits in Mexico and, possibly, Central America are unclear because of the difficulty of separating this moth from its close relatives. The moths fly from mid-July to early September with a peak in August.

SUBTRIBE CISTHENINA
Cisthene Walker comprises small New World lithosiines, often with a pink or red HW. There are 20 North American species, including 13 in the West, but the center of diversity for the genus lies in the Neotropics, where there are a myriad of species, mostly undescribed or poorly known. These species are nocturnal and are attracted to light, but, in addition to being nocturnal, *Cisthene juanita* has been found during the day on seep-willow, *Baccharis salicifolia,* flowers (Asteraceae). All are presumed to feed on lichens, but the life history and hosts of most are unknown. These have been placed in the tribe Cisthenini by Bendib and Minet. *Cisthene liberomacula* (Dyar)(FW 0.8–1.0 cm; Plate 46.14) occurs from coastal California (Lake County south to San Diego County) to Baja California Norte. The moths are almost always found in the vicinity of oaks, especially coast live oak *(Quercus agrifolia).* Adults have two flights during July and October in Santa Clara County and occur from March to December (possibly three broods) in San Diego County. To distinguish this species, there is usually a faint basal line extending out for about half the FW. The HW is dusky. The larvae presumably feed on lichens, especially those found on oaks. There are several other gray *Cisthene* species found is association with oaks and associated woody plants in California: *C. deserta* (Felder)(FW 0.9–1.0 cm; Plate 46.15), adults from April to July, *C. faustinula* (Boisduval)(FW 0.9–1.1 cm; Plate 46.16), adults July to September, and *C. dorsimacula* (Dyar)(FW 0.9 cm; Plate 46.17), adults from May to August and October. All three occur in coastal California, but *C. dorsimacula* is found only in southern California, including Santa Catalina Island. The last species is dimorphic and may have either a light pink or light yellow HW. J. A. Comstock described the early stages of *C. deserta,* whose larvae were found feeding on a lichen, *Vermilacinia combeoides,* that was found growing on *Lycium brevipes* (Solanaceae) in northern Baja California. The full-grown larva is yellow green with black spots and blotches. There is a middorsal pale band, and the body is covered densely with short yellow hairs. Most of the body has small black papillae or verrucae each bearing a black hair.

Pupation was in a loose cocoon. In addition, the early stages of *C. faustinula* were partially recorded by F. X. Williams. He reported the host lichen as *Ramalina menziesii.* Another southern California endemic lithosiine is *C. perrosea* (Dyar)(FW 0.9–1.1 cm; Plate 46.18), which differs from the preceding four in having a narrow pink FW band and a distinctly pink HW. There are two flights during June to July and September to October. *Cisthene tenuifascia schwarziorum* (Dyar)(FW 0.7–0.9 cm; Plate 46.19) ranges from southern Arizona to western Texas and, presumably, south into adjacent northwestern Mexico. There it flies from June to October. The typical subspecies is found in the Southeast and southern Texas and may range into northeastern Mexico (February to November). The larva was illustrated by Habeck and is mottled gray, green, and brown. It lacks verruca and is clothed only with long primary setae. Three other *Cisthene* species are found in southeastern Arizona and presumably range into northwestern Mexico, but the knowledge of distribution and species identity in that country is fragmentary at best. These are *C. juanita* Barnes & Benjamin (FW 0.9–1.1 cm; Plate 46.20), a species identified by its primarily black FW, with adults in flight from August to mid-October; *C. martini* C. B. Knowlton (FW 0.9–1.1 cm; Plate 46.21, Figs. 235, 236 genitalia), characterized by a red pink FW band; and the rare *C. coronado* C. B. Knowlton, separable from *C. martini* only by genital examination and the presence of androconial scales at the male FW tornus. *Cisthene angelus* (Dyar)(FW 1.0–1.2 cm; Plate 20.22) ranges more widely in the Southwest from southern Nevada and Utah south to southern Arizona and east to western Texas. Adults are in flight from April to August, possibly as two flights. It has not been reported from adjacent Mexico. It is easily distinguished by its broad yellow FW band. *Cisthene barnesii* (Dyar)(FW 0.8–1.2 cm; Plate 46.23 pink HW form, Plate 46.24 yellow HW form) is a Rocky Mountain species that occurs from eastern Montana and western North Dakota south to Arizona, New Mexico, and western Texas. It flies from May to early September in Texas, possibly as two broods, but only July and August (one flight) to the north. Like *C. angelus,* it is unreported from northern Mexico. In most areas adults have a pink HW, but in western Colorado, New Mexico, and Utah it is dimorphic, as both yellow HW and pink HW forms may be found in the same populations.

Ptychoglene Felder is a tropical American genus with most of its diversity to the south of our area. These species are mimics, likely Müllerian, in a group of species that includes lycid beetles and zygaenid moths (see Zygaenidae chapter). Most species are diurnal and are most easily found at flowers such as those of *Melilotus officinalis* (white sweet clover) or *Baccharis glutinosa* (seep-willow). We have three species found in southeastern Arizona, one of which is relatively frequent. *Ptychoglene phrada* Druce (FW 1.0–1.4 cm; Plate 46.25) is found during July and August in several ranges in southeastern Arizona and adjacent southwestern New Mexico.

Lycomorpha Harris is a compact genus of six western species with some poorly known Neotropical relatives. Several species are diurnal but most are nocturnal and are attracted to light. *Lycomorpha grotei* (Packard)(FW 1.3–1.5 cm; Plate 46.26) is sometimes found on flowers but is primarily nocturnal and is attracted to light. The single adult flight is from June to August. The taxonomy of this species and several likely relatives requires study. All have a red FW and a primarily black HW. Included are *L. grotei pulchra* Dyar, *L. fulgens* (Henry Edwards), *L. desertus* Henry Edwards, and *L. regulus* (Grinnell). J. A. Comstock and Henne described the life history of *L. regulus* (FW 1.2–1.4 cm; Plate 46.27) in its home, the mountains of cismontane southern California. The species is found from June to October and may have two or more broods. The adults are diurnal and visit flowers of scale-broom, *Lepidospartum squamatum*, and *Senecio flaccidus* var. *douglasii* (both Asteraceae). Adults laid eggs in crevices and larvae ate *Parmelia* lichen in captivity. Full-grown larvae were blue green to olive green and mottled with dark brown to charcoal black. Long primary setae arise out of small verrucae. The pupa is covered with sheer white viscous webbing and a larger cocoon into which particles of the substrate have been incorporated encloses the above structure. *Lycomorpha splendens* Barnes & McDunnough (FW 1.2–1.5 cm; Plate 46.28) is a striking moth with a glossy black FW and red HW, has two distributional areas, one in the Great Basin and another in southwestern New Mexico and western Texas. As far as is known it is strictly nocturnal and is attracted to light. Adults are in flight from May to October as two flights. In the Big Bend of Texas, they may be encountered earlier than May. *Lycomorpha pholus* (Drury)(FW 1.2–1.5 cm; Plate 46.29) is a common moth of eastern North America that ranges west to the Rocky Mountain Front from southern Alberta and Montana south to New Mexico. It is wholly diurnal and often found at flowers such as *Apocynum* (Apocynaceae), *Ericameria nauseosa*, or *Solidago* (Asteraceae). In the East and Midwest, populations comprise individuals with a yellow orange FW, while in the Rockies, all individuals have a bright red FW (*L. p. miniata* Packard). Some population samples from northern Texas have an orange red FW. In the West, the moths fly from late June through August.

Hypoprepia Hübner is a small genus of less than 10 species that ranges south into Mexico. All five described North American species occur in the West, although *H. fucosa* (Hübner) ranges west only to northeastern Wyoming (Crook County). Adults of all species are nocturnal and are attracted to light. *Hypoprepia miniata* (Kirby), the "scarlet-winged lichen moth" (FW 1.4–1.7 cm; Plate 46.30, Plate 46.31 *cadaverosa*, Plate 64.9 larva), ranges from British Columbia along the Rocky Mountains from southern Canada south to eastern Arizona, New Mexico, and western Texas. From there it occurs east throughout eastern North America. In the West, this lovely moth flies in July and August, but eastward it has an extended flight from April to September. *Hypoprepia cadaverosa* Strecker appears to represent the same species, as intermediate populations with *H. miniata* are found and the ranges of the two species are complementary. We treat the populations in the combined ranges of those taxa as that of a single species. The larvae are sooty and mottled with yellow. There are one to three long barbed setae emanating from black verrucae. Larvae overwinter and complete development the following spring. There is one annual flight in June or July. *Hypoprepia inculta* Henry Edwards (FW 1.0–1.5 cm; Plate 46.32), with a gray black FW and a pale pink HW, has an entirely western distribution from the Black Hills of South Dakota and Wyoming south through the Rocky Mountains, Colorado Plateau, and southeastern Arizona to central Mexico. The adults are found in June and early July. The moth is sometimes abundant, but nothing is known of its early stages.

Bruceia Neumoegen is a small genus with two described species, both western. The moths are nocturnal and readily come to light. *Bruceia pulverina* Neumoegen (FW 1.2–1.6 cm; Plate 46.33) ranges from the interior of British Columbia and the Rocky Mountains from Wyoming south to southern Arizona and western Texas. This species may be found from June to early August. The life history is unknown. *Bruceia hubbardi* Dyar, found in southern Arizona and the deserts of southern California, is very much like *B. pulverina*. The smaller but similar *Clemensia albata* Packard, the "little white lichen moth" (FW 1.0–1.2 cm; Plate 46.34, Plate 64.7 larva), is found from coast to coast across Canada but in most of the West is found only along the Pacific Coast south to central California (Santa Cruz County). Adults are nocturnal and fly from June through September, rarely early October. The caterpillar is green or white mottled with black and shows a clear middorsal line. Larvae feed in spring on lichens that grow on trees and shrubs, especially Oregon white oak, *Quercus garryana* (Fagaceae); their growth is slow. McFarland found that captive larvae readily accepted the foliose *Lobaria pulmonaria* (Stictaceae), but only feed when the lichens are moistened. The larvae rasp away the upper green algal layer of the lichen, leaving the lower white layer.

SUBTRIBE LITHOSIINA

Eilema bicolor (Grote), the "smoky moth" or "bicolored moth" (FW 1.1–1.6 cm; Plate 46.35), is the only North American representative of a Holarctic genus. The moth ranges from eastern Alaska across southern Canada to the Maritime Provinces and, in the West, south in the Rocky Mountains to Colorado. The single adult flight is during July and early August. The larva is smoky black with a weak pale orange subdorsal stripe and shiny black head capsule. There are rows of black verrucae each bearing small clumps of black setae. Schmidt reared larvae that he found feeding on lichens that grew on the bark of large birches, *Betula* (Betulaceae). The larvae have also been reported to feed on the foliage of several conifers in captivity, but it is more likely that they were eating lichens that grew on the conifers.

There are at least seven western *Crambidia* Packard. Additional taxa are found in eastern North America and south into Mexico. There seem to be a number of undescribed species, and the genus would benefit from further taxonomic research. *Crambidia dusca* Barnes & McDunnough (FW 1.2–1.4 cm; Plate 46.36) is found in Orange and San Diego counties, California, and south along the Pacific Coast of Baja California Norte. Adults fly from April to June and occasionally as a partial second brood in September. The adults are dark brown with a lighter HW. *Crambidia suffusa* Barnes & McDunnough (FW 1.2–1.5 cm; Plate 46.37) is another southern California endemic. The distribution extends from extreme southwestern San Bernardino and Orange counties south to Baja California Norte. The moths are gray with a lighter FW. The vertex has yellow orange scaling. There are two flights during April to June and September to October. *Crambidia impura* Barnes & McDunnough (FW 1.3–1.6 cm; Plate 46.38) ranges from Utah and Colorado south to Arizona, New Mexico, and western Texas. Adults are nocturnal and may be found in montane habitats from late July to October. *Crambidia casta* (Packard), the "pearly-winged lichen moth" (FW 1.3–1.7 cm; Plate 46.39), has a shining silky white FW and lacks any hint of orange on the vertex. The species ranges across southern Canada from coast to coast and, in the West, south to Oregon, northern Arizona, and northern New Mexico. The moths are in flight from June to mid-September. *Crambidia cephalica* (Grote & Robinson)(FW 0.9–1.5 cm; Plate 46.40) ranges from central Nevada, Utah, and Wyoming south to Arizona, New Mexico, and western Texas. It probably occurs in northern Mexico. The adults have several flights, usually two, from March to September.

Agylla septentrionalis Barnes & McDunnough (FW 1.7–1.8 cm; Plate 46.41) is our sole representative of a tropical American genus. There are many species in Central and South America. Our species is found only in the Chiricahua Mountains of southeastern Arizona, where it is nocturnal and is attracted to light. There is a single annual flight during July and August.

Several other rare species of Lithosiines are found in southeastern Arizona: *Gnamptonychia ventralis* Barnes & Lindsey (FW 1.9–2.0 cm; Plate 46.42) ranges from eastern Arizona east to western Texas. Adults are found in June and July. *Gardinia anopla* Hering (FW 2.1–2.5 cm; Plate 46.43, Plate 64.8 larva), the largest North American lithosiine, has an iridescent green black FW, a black HW and an iridescent blue black abdomen. Adults have been found in southeastern Arizona from late June to early August. The caterpillars feed on green tree algae.

TRIBE ARCTIINI

The Arctiini, as presently viewed, contains all groups except the Lithosiini and the Old World Syntomini. The tribe is considered monophyletic and is united by two apomorphic features: the presence of an elongate male retinaculum and heteroideous crochets (larvae of *Virbia* are exceptions, as they have homoideous crochets). Larvae have barbed or plumose setae all arising from verrucae, except on prolegs. This tribe includes the derived Pericopina, Ctenuchina, and Euchromiina as well as the more basal Arctiina, Callimorphina, and Phaegopterina. Relationships within the tribe are poorly understood, as most knowledge is based on temperate groups that are derived from a vast but poorly known assemblage of tropical taxa.

Virbia Walker, now known to include the group formerly referred to as *Holomelina*, comprises 14 North American species including seven in the West. Intensive research by Zaspel has clarified the relationships between *Holomelina* and *Virbia*. There are additional species extending southward into Mexico and Central America. Some may be wholly diurnal [*V. lamae* (Freeman)], others are solely nocturnal *(V. ostenta),* and still others *(V. aurantiaca, V. fragilis)* are both diurnal and nocturnal. The larvae of several species reared from eggs have been fed *Plantago, Melilotus,* and *Taraxacum* and presumably feed on a variety of low herbaceous plants in nature. *Virbia ostenta* (Henry Edwards)(FW 1.2–1.4 cm; Plate 46.44) ranges from Arizona and southwestern New Mexico south into western Mexico. This is our largest *Virbia,* with a tan FW and the HW equally divided between red and black. In the Huachuca Mountains McFarland has found the univoltine adults fly from mid-June to early September. *Virbia fragilis* (Strecker)(FW 1.2–1.3 cm; Plate 46.45) is a widespread, common Rocky Mountain moth found from southern Canada, Idaho, and Montana south to Arizona and New Mexico. Adults are both diurnal and nocturnal. The adults are quite variable; most often they are unmarked orange with a somewhat paler HW, but either the FW or HW may be variably maculated. The larvae will feed on *Taraxacum* in the laboratory. Other western species include *V. costata* (Stretch), *V. ferruginosa* (Walker), and *V. lamae* (Freeman).

Neoarctia Neumoegen & Dyar has only three Nearctic species, all of which are western. *Neoarctia beanii* (Neumoegen)(FW 1.4–1.6 cm; Plate 46.46) occurs in the Rockies of western Canada and Montana. The moth is nocturnal and is attracted to light. There is a single flight from mid-July to early August. The primary habitat seems to be in meadows near timberline. The adults are quite variable in color and markings.

The larvae are dull green mottled with brown. The verrucae are shiny black. There are long, hairlike setae emanating from the verrucae, and these are entirely black on the first two thoracic segments, but a few white setae are included on the last thoracic and abdominal segments. Below the spiracular line the setae are all rust red. The spiracles are black and the head is shiny black. *Neoarctia brucei* (Henry Edwards)(FW 1.3–1.5 cm; Plate 46.47) has a more contrasting wing pattern. It is best known as a diurnal alpine moth above timberline in Colorado but is nocturnal and is attracted to light in British Columbia, Canada. Its single flight is during July.

Grammia Rambür, formerly *Apantesis* (in part), is the most species rich genus of North American tiger moths with about 30 species including several yet to be described. Within species variation and geographic variation continue to make the systematic relationships difficult to untangle, but C. Schmidt has clarified the *G. nevadensis* complex, one of the more difficult challenges. Antennae of males are bipectinate and those of females are biserrate and threadlike. Most species have a black FW with cream lines of variably width, pattern, and extent. The HW are often black and pink but may have the pink substituted by orange or red. In a few cases the HW are entirely or primarily black. Most species are nocturnal and are attracted to light, while a few montane and alpine species are strictly diurnal. In several cases, likely due to their heavy bodies and poor flight capabilities, females are extremely rare or unknown. Larvae of all species are either known or thought to feed more or less indiscriminately on a variety of low herbaceous plants and then overwinter and complete feeding during the following spring. When adequately sun-warmed, the larvae may run more rapidly than those of any other North American macromoth! *Grammia obliterata* (Stretch)(FW 1.3–1.4 cm; Plate 46.48) is a Holarctic species that has a Rocky Mountain distribution in North America. Its range extends from Alaska and the Northwest Territories south to northeastern Utah and northwestern Colorado. Its habitats include taiga, burns, and mountain slopes. Larvae hibernate. This species is nocturnal, attracted to light, and has a single adult flight in the fall. *Grammia parthenice* (W. Kirby), the "Parthenice tiger moth" (FW 1.9–2.4 cm; Plate 46.49), ranges in North America from southern Canada south through most of North America east of the Great Basin. In the West, it is most prevalent in the Rocky Mountains and Great Plains from British Columbia and Alberta south to Arizona and New Mexico. Adults are nocturnal and are attracted to light. There is a single flight in July. According to Gibson, final-instar larvae have the chalazae variably yellow, reddish, or black, and the stiff hairlike setae are yellowish, reddish, or black. The spiracles are orange and are black rimmed. *Grammia ornata* (Packard), the "ornate tiger moth" (FW 1.7–2.1 cm; Plate 46.50, Plate 46.51), ranges from southern British Columbia and western Montana south through the Pacific Coast states, Utah, and northern Nevada to southern California including Santa Cruz Island. Adult males are nocturnal and are attracted to light abundantly, while the heavy-bodied females are found in the day and fly only slowly. McFarland observed females in McDonald Forest, Oregon, fluttering along the ground. There is a single spring flight from March to early July, mainly March to April. The larvae have dense clumps of long hairlike setae—black with a few white setae dorsally and laterally and orange ventrally—and feed during late spring and summer. There is a yellowish, whitish, or orange middorsal line. McFarland found pupation is in a slight cocoon devoid of larval hairs and is under leaf litter. *Grammia cervinoides* (Strecker)(FW 1.1–1.2 cm; Plate

46.52) is a small, black diurnal species found only in alpine habitats in the Colorado Rockies. The FW is black with cream lines, and the HW is black with a few white marks. Adults have a single flight in July, and it is likely that the larvae require more than a single year to complete development. *Grammia williamsii* (Dodge)(FW 1.3–1.5 cm; Plate 46.53) is a black moth with reduced FW lines and a pink HW with broad black margins. The moth is widespread in the West from the Canadian Prairie Provinces south through the Rocky Mountains and eastern Great Basin ranges to Arizona and New Mexico. The single flight extends from late May through the end of August. The *G. nevadensis* (Grote and Robinson)(FW 1.5–1.9 cm; Plate 46.54, Figs. 237, 238 genitalia) complex is a taxonomically difficult group studied by Schmidt who found that it includes typical *G. nevadensis* in eastern California, Nevada, and Utah, *G. n. geneura* (Strecker) in Colorado and Wyoming, *G. n. superba* (Stretch) in the Pacific Northwest, and *G. n. gibsoni* (Mc-Dunnough) in the northern Great Plains from Montana to Manitoba. As a whole, the species is found from Yukon Territory and southwestern Canada south through much of the western United States, south to southern California, Utah, and Colorado. Adults are univoltine from mid-July through September. In southern California, including the Channel Islands, we found the adults fly only in the fall. Females have fully developed wings but are apparently flightless as they are not found at light. Full-grown larvae have black bodies and heads, while the body is covered with black verrucae bearing black, hairlike setae dorsally and subdorsally. Below the stigmata the hairs are red brown. Larval hosts are a wide variety of herbaceous and woody angiosperms, including *Amsinckia* (Boraginaceae), *Lupinus* (Fabaceae), *Purshia* (Rosaceae), *Artemisia* (Asteraceae), and many others. Larvae will drop to the ground when the plant on which they are feeding is only slightly disturbed. In southern California the larvae complete feeding by May and remain dormant during the summer; they pupate shortly before the fall adult emergence. The final-instar larvae from Inyo County, California, are of the *Grammia* type, with black hairs above the spiracular line and reddish brown hairs below. The closely related *G. incorrupta* (Henry Edwards)(Plate 46.55) is found from Arizona and western Texas south to Durango, Mexico. *Grammia f-pallida* (Strecker)(FW 1.3–1.5 cm; Plate 46.56 pink HW form, Plate 46.57 yellow HW form) ranges from southern Utah and Colorado south to Arizona, New Mexico, and western Texas. The related *G. figurata* (Drury), which is difficult to separate, occurs to the east in the adjoining territory of the Plains states, and thence further east. A population of *G. f-pallida* with mixed phenotypes, including some individuals that appear like *G. celia* (Saunders), is found in Jefferson County, Colorado.

Notarctia proxima (Guérin-Méneville), the "Mexican tiger moth" (FW 1.4–2.0 cm; Plate 46.58 male, Plate 46.59 female), is sexually dimorphic with males having a white HW and females having a red pink HW. The abdomen is red with black terminal segments. The moths have a number of flights each year from April to October in many areas, but they fly all year in southern California with no specific indication of broods; only freezing weather seems to halt their appearance. The species ranges from southeastern Oregon and southern Idaho south through all but northern coastal California as well as Nevada and western Utah; it occurs south through Mexico and Central America to at least Costa Rica. The related *N. arizoniensis* (Stretch) is similar but smaller and paler; it occurs in Utah, Arizona, New Mexico, and western Texas.

Parasemia plantaginis (Linnaeus), the "wood tiger" (FW 1.5–1.8 cm; Plate 47.1, Plate 47.2, Plate 47.3, Plate 47.4), is a polyphenic Holarctic moth found throughout much of both boreal North America and Eurasia. In North America, it ranges from Alaska southeasterly to Manitoba and south through the Rocky Mountain region to southern New Mexico. There are isolated populations in the White Mountains of Arizona and the Sierra Nevada of California and Nevada. Most often the moths' FW is black with several broad cream white bands and spots combined with either a primarily white, orange, or black HW. Occasionally, the markings on the FW may be reduced. Many of the color forms have been named. It is a strictly diurnal species with a single flight from late June through July. Adults are most often found in the vicinity of lupine (*Lupinus*) stands in relatively moist habitats such as meadows and adjacent to streamcourses. The adults have a low, quick darting flight. In North America, larvae have been reported to feed on *Aster* (Asteraceae) and *Plantago* (Plantaginaceae). Most likely, however, the larvae are polyphagous on various woody and herbaceous broad-leaved plants as reported for the species in Great Britain and Russia. The larvae have black heads and are covered with relatively short, black hairlike setae emanating from black verrucae except for the first through third abdominal segments, which have red orange verrucae that bear likecolored setae. There are longer gray setae projecting from the last abdominal segments.

Platarctia parthenos (Harris), the "giant tiger moth" or "St. Lawrence tiger moth" (FW 2.5–2.9 cm; Plate 47.5), ranges from Alaska east to Yukon and Northwest Territories, thence south through the Rockies to eastern Arizona (White Mountains) and northern New Mexico. It also ranges east across southern Canada to the Maritime Provinces. There is a single flight in late June through July. Males are readily attracted to light. The larvae feed on foliage of several broad-leaved trees, especially willows, *Salix* (Salicaceae), and alder, *Alnus* (Betulaceae). Other hosts, including shrubs and low herbs, have been reported by la Plante.

Platyprepia virginalis (Boisduval), the "ranchland tiger moth" or "ranchman's tiger moth" (FW 2.4–2.9 cm; Plate 47.6, Plate 47.7) ranges from southern British Columbia south along the Pacific Coast states, Great Basin ranges, and Rocky Mountains to coastal southern California, Nevada, Utah, and Colorado. They are most often found in or near meadows, marshy areas, and dunes. The moths are polyphenic, varying greatly in the relative amount of black or yellow orange on the HW. Adults are diurnal, fly fast and erratically, and have a single flight during June and early July. The larvae usually feed on various low-growing herbaceous plants but prefer perennial bush lupines on the Point Reyes peninsula, California. In western Oregon, McFarland found older instars preferred basal leaves of *Cirsium* (Asteraceae) and *Galium* (Rubiaceae) in wet grassy sites. The full-grown larva has a shining black head and a black body densely covered with long hairlike setae. The setae on the first thoracic segments are rust red, while the segments have very long silky white hairs mixed with shorter black hairlike setae. There are again rust red hairs on the posterior segments. McFarland observed that pupation is in a sticky, weblike cocoon that has few if any incorporated larval setae.

Arctia caja (Linnaeus), the "garden tiger moth" (FW 2.4–3.1 cm; Plate 47.8), is a Holarctic moth that ranges widely in temperate and boreal North America as several described subspecies, the exact taxonomic status of which remains to be determined. As a whole, the species extends from southern British Columbia and Alberta east across southern Canada to the Maritime Provinces. In the West, it ranges south along the Pacific Coast to northwestern California as subspecies **waroi** Barnes & Benjamin and from southeastern British Columbia and Alberta south through the intermountain West to the Sierra Nevada of California, central Nevada, Utah, and western Colorado as subspecies **utahensis** (Henry Edwards). The moths in the East are represented by subspecies **americana** Harris. In the West there is a single flight during July and August. Larvae are polyphagous on low herbaceous plants, but McFarland found that larvae in western Oregon preferred to feed on *Pteridium aquilinum* bracken (Dennstaedtiaceae). The larva is black, covered with long hairlike setae emanating from verrucae. Above the spiracular line, the setae are rust red together with long sweeping silvery white hairs posterior to the thoracic segments. Verrucae posterad to the spiracles bear rusty red hairlike setae. All verrucae ventral to the spiracles also bear rust red setae. The spiracles are white and the head is shiny black. Pupation is in a loosely woven cocoon covered with long setae from the last-instar larva. *Arctia opulenta* (Henry Edwards), now considered a distinct species, ranges from Alaska east across southern Yukon and northern British Columbia to Northwest Territories, while *Arctia brachyptera* Lafontaine & Troubridge is found only on alpine slopes in extreme southwestern Yukon Territory.

Phragmatobia Stephens has a Holarctic distribution with almost 30 species. Three described species are found in North America, two of which occur in the West. *Phragmatobia fuliginosa rubricosa* (Harris), the "ruby tiger moth" (FW 1.3–1.4 cm; Plate 47.9), is the North American subspecies of a widespread, geographically variable Eurasian species. In North America, the moth is relatively common and ranges from coast to coast. In the West, it is found from Alaska, Yukon, and western Northwest Territories south to Oregon, Colorado, and South Dakota. In the East, the moths range south to New Jersey and Pennsylvania. Although Newman and Donahue report two flights in the East, there is only a single adult flight in the West, during June and July. The adults are nocturnal and are attracted to light. Larvae are polyphagous and feed on a taxonomically diverse array of low herbaceous plants. *Phragmatobia assimilans* Walker, the "large ruby tiger moth" (FW 1.3–1.6 cm; Plate 47.10), has a boreal distribution from British Columbia east across southern Canada and south, in the West, to the Rockies of northern Colorado. There is only a single flight from late May to mid-June. A small batch of eggs is laid in a closely abutted single layer. Larvae are reported to feed on low herbaceous plants as well as paper birch, *Betula papyrifera* (Betulaceae), balsam poplar, *Populus balsamifera* (Salicaceae), and blackberry, *Rubus* (Rosaceae). The last-stage larvae overwinter, probably under rocks and debris, and wander in spring in search of pupation sites.

Sonorarctia fervida (Walker)(FW 1.3–1.5 cm; Plate 47.11 male, Plate 47.12 female) occurs from southeastern Arizona south to Costa Rica. Larvae were found by McFarland on white sweet clover, *Melilotus officinalis* (Fabaceae), in the Huachuca Mountains (Cochise County, Arizona). Adults have emerged from May to July.

Leptarctia californiae (Walker)(FW 1.3–1.5 cm; Plate 47.13 female, Plate 47.14 and Plate 47.15 males) is a relatively small geographically polyphenic diurnal moth. The adults are geographically variable as well as sexually dimorphic in the frequency of color morphs. In California, males usually have a red or orange HW with a black marginal band. Only about one in five females in California has red or orange HW, and most have primarily black HW with or without a white band. In

Colorado, almost all males have an orange HW. Males follow ridgelines or fly to hilltops in their search for receptive females. *Leptarctia californiae* ranges from southern British Columbia south through most of the western states to southern California, southern Arizona, and northern New Mexico. It is absent from the arid deserts and much of the Great Basin. It is found in lowland habitats as along the Pacific Coast but has been found by Walsh as high as 8000' in the Santa Catalina Mountains of southeastern Arizona. There is a single annual spring flight. Larvae feed on a variety of trees and low plants, especially broad-leaved plants but also including bracken, *Pteridium aquilinum* (Dennstaedtiaceae). In northern Arizona, McFarland found larvae feeding on white sweet clover, *Melilotus officinalis* (Fabaceae); they are dark brown with a long caudal tuft and an orange middorsal line. The larvae hide by daylight in surface litter and feed at night.

Kodiosoma fulvum Stretch (male FW 0.8–0.9 cm, female FW 1.1–1.3 cm; Plate 47.16 male, yellow HW form, Plate 47.17 female, red pink HW form, Plate 64.15 larva) is a small uncommon tiger moth with several color phases having different HW colors: black, red pink, or yellow. These are each somewhat differently distributed geographically, and several named forms have been treated as distinct species. Yellow HW forms are found mainly to the west of the Sierra Nevada crest, and red HW forms are found mainly to the east. In some localities such as Marin County, California, we found different color forms flying together. The species ranges from northern California, southern Nevada, southwestern Utah, and western Colorado south to southern California, southern Arizona, and southwestern New Mexico. The adults are mainly diurnal and have a single flight during February and March in the Mojave Desert, and during April and May in the intermountain region. In wet years on the Mojave Desert, McFarland found the larvae feed on *Stephanomeria* (Asteraceae) from December to mid-January; they are gray tan and hairy.

Pyrrharctia isabella (J. E. Smith), the "banded woollybear" or "Isabella tiger moth" (FW 1.8–2.3 cm; Plate 47.18), ranges across southern Canada from coast to coast and extends southward through much of the more temperate portions of the conterminous United States. The moths are nocturnal and are attracted to light. There seems to be a single flight in midsummer from late June through much of July, although two flights have been recorded in the eastern United States. The larvae are densely covered with stiff hairlike setae. The "hairs" are red orange in the central portion of the abdomen but black at both the anterior and posterior. The band is of variable width and has been reputed to reflect the likely severity of the upcoming winter. Efforts to support this claim have not been met with success. Larval foods include grasses (Poaceae) and various broad-leaved herbs and woody plants in many plant families. Larvae are mainly found close to the ground.

Estigmene acrea (Drury), the "saltmarsh caterpillar moth," or "acrea moth" (FW 2.0–2.6 cm; Plate 47.19 male, Plate 47.22 female), occurs across southern Canada from coast to coast, and thence southward through the conterminous United States and Mexico to Central America. The adults are nocturnal and are attracted to light. In portions of the West, this moth has been confused with *E. albida* (Stretch) that differs primarily by having black cross bands instead of dots on the abdominal segments and the male's white HW. There are two or more generations each year, with year-round presence in more southerly areas such as coastal southern California. The white eggs are laid in masses on low vegetation. The larvae are extremely variable, ranging from blonde to almost black, but brown or gray are most normal. The hairlike setae are soft and are longer at both the anterior and posterior. The setae arise from prominent black or orange verrucae. The ventrolateral verrucae usually bear orange setae. Spiracles are white. Winter is passed inside a loose hair-covered cocoon. This species is especially prevalent in open disturbed habitats including agricultural fields, where it may be an occasional pest to field crops, and is found more sparingly in undisturbed native habitats. **Estigmene albida** (Stretch)(FW 1.9–2.6 cm; Plate 47.23 male) occurs from Idaho, Montana, and western South Dakota south through Colorado, southeastern California, Arizona, and New Mexico as well as through much of Mexico. *Estigmene albida* occurs most often in more natural and arid habitats than those favored by *E. acrea*, but at times the two may be found together. In some regions, such as southern Arizona, this moth may be by far the most abundant *Estigmene*. In the West, adults may be found from late May through August, while in Mexico they may be found during much of the year.

Hyphantria cunea (Drury), the "fall webworm" (FW 1.4–1.6 cm; Plate 47.20, Plate 47.21, Plate 64.14 larva), is most noticeable due to its large, loose communal webs usually seen during the late summer and fall. The moth ranges across southern Canada from coast to coast, and thence southward to central California, the Rocky Mountains, eastern United States, and much of Mexico. It is absent from the deserts, much of the Great Basin, and most boreal situations. The adults are extremely variable and range from pure white to heavily spotted with black. The coxae and femora of the front legs are distinctively yellow orange. There is one flight in the northern parts of the range, including most of the West; the univoltine adults are in flight from late May until early August. Masses of white or yellow eggs are deposited on the undersides of leaves in June or July and the resultant larvae feed gregariously during August and September. A wide variety of woody plants, especially trees, is selected. Those favored include but are not limited to alders (Betulaceae), cottonwoods and willows (Salicaceae), and walnuts (Juglandaceae). The full-grown larvae are variably cream to black and are marked with yellow. Long and soft white and/or black hairlike setae of several lengths emanate from several rows of red orange or black verrucae. The moth is considered an occasional pest to street and orchard plantings. Essig reported that even cotton and hops were fed on in Arizona and California. The larvae pupate in late fall and overwinter in cocoons, usually located in bark crevices.

Spilosoma Curtis occurs in both Eurasia and North America, but Dubatolov has found that American species are not equivalent to Old World species, and that a new genus name should be assigned. **Spilosoma virginica** (Fabricius), the "yellow woollybear" or "Virginian tigermoth" (FW 1.5–2.1 cm; Plate 47.24, Plate 64.13 larva), ranges across southern Canada and southward to variable extents in the West. Along the Pacific Coast it extends to central California and south through the Rockies to Utah, eastern Arizona (White Mountains), and New Mexico. It is widespread in the East and extends south into northern Mexico. The species is double-brooded in most areas—April to May and July to August—but in the East, the species may fly earlier and later as well. The nocturnal adults have white wings with one or two black spots in the anal corner of the HW. The prolegs are yellow orange, and the abdomen is distinctive, with a dashed central black line subtended by white then yellow orange more laterally. The pupa overwinters in a cocoon and produces an adult during the following spring. The early instars feed gregariouly, but older larvae are solitary. Full-grown larvae are

extremely variable and reported by Wagner as beige or yellow to red brown or black. The body is covered with extremely long, soft hairlike setae that can range from white or yellow to brown or red orange, according to Wagner. Spiracles are white. Larval hosts are included among more than 43 families of woody and herbaceous dicotyledonous and monocotyledonous angiosperms. Larval use of fungi and pines has also been reported. **Spilosoma vestalis** Packard, the "vestal tiger moth" (FW 1.6–2.4 cm; Plate 47.28), ranges from Washington south along the Pacific Coast to Baja California Norte. Within its range, it is the most common, almost ubiquitous *Spilosoma*. The adult is distinguished by its red forelegs and black intersegmental areas on the white abdomen. The wings are white with a variable amount of black spotting. The adults are in flight from late May (rarely April) through June. The larvae are black clothed in relatively short black or blackish brown (dorsally) and white (laterally) stiff hairlike setae that arise from black verrucae. There is a strip of rust brown setae along each side. Larvae have been reported to feed on a variety of both woody and herbaceous broad-leaved plants. At one chaparral locality in southern California, McFarland found that they prefer to feed on young leaves of wild cucumber, *Marah* (Cucurbitaceae), but when that plant's leaves shrivel they wander widely and feed on many other plants. **Spilosoma vagans** (Boisduval), the "wandering tiger moth" (FW 1.3–1.6 cm; Plate 47.25, Plate 47.26), ranges from southern British Columbia south along the Pacific Coast through much of California and the Rocky Mountains south to Utah and northern Colorado. Males are primarily pale tan to brown, with variable amounts of black on the HW, while females are rust colored. Males are strongly attracted to light, but females may be mostly diurnal as they are rarely found at light. Adults are in flight from the end of May to early July. In northern California, larvae were found by McFarland resting under a perennial lupine (Fabaceae) by day and feeding on its leaves at night.

Hypercompe permaculata (Packard)(FW 1.6–1.9 cm; Plate 47.27) ranges from southern Alberta and extreme southeastern British Columbia south through the Rocky Mountains, eastern California, and central Nevada to Arizona, New Mexico, and western Texas to the Sierra Madre Occidentale of Sonora, Mexico. Adults are commonly attracted to light and have a single flight from mid-June to August. The larvae feed on various low herbaceous plants including *Stellaria* (Caryophyllaceae) and *Polygonum* (Polygonaceae).

Arachnis Geyer has seven described North American species, all of which have major portions of their ranges in the West. **Arachnis picta** Packard, the "painted tiger moth" (FW 1.9–2.5 cm; Plate 47.29 female, Plate 47.30 male), is found in two separate groups of populations, one in central and southern California, including several of the Channel Islands, and the Colorado Front Range south to much of Arizona, New Mexico, and western Texas. *Arachnis picta* has also been found in the Sierra Madre Occidentale of Sonora, Mexico. In most areas the adults have a single flight from June to mid-October, the exact timing varying geographically. For example, in coastal California they fly in late summer and fall, while in southern California, they fly from late October to December. Larvae feed on the flowers of a wide variety of broad-leaved herbaceous plants, but perennial legumes such as *Lotus scoparius* and *Lupinus* (Fabaceae) may be preferred. McFarland found they preferred *Lupinus succulentis* in the Santa Monica Mountains of southern California, and we found them on a shrubby lupine at Antioch, California. **Arachnis zuni** Neumoegen (FW 2.3–2.5 cm; Plate 47.30) ranges from eastern Arizona and New Mexico south to southeastern Arizona and western Texas at least to Durango,

Mexico. There is a single flight of adults from May through August. The larvae feed on various low-growing annuals such as *Melilotus* (Fabaceae) and *Chenopodium* (Chenopodiaceae) as well as Siberian elm, *Ulmus pumila* (Ulmaceae), according to McFarland.

SUBTRIBE CALLIMORPHINA

Tyria jacobaeae (Linnaeus), the "cinnabar moth" (FW 1.6–1.9 cm; Plate 48.1, Plate 64.11 larva), was intentionally introduced from Eurasia to the Pacific Northwest and Nova Scotia as a biological control for tansy ragwort, *Senecio jacobaea* (Asteraceae), which is toxic to cattle and horses. Naturalized populations persist in both regions. Currently in the West, the cinnabar moth is found close to the coast from southwestern British Columbia south to northern California (Mendocino County). The FW is gray black with a red costal stripe and two small marginal patches. The HW is predominantly red. The adults are strictly diurnal and have one flight during May and June. The larvae feed almost entirely on *Senecio* (Asteraceae) and have alternating bands of black and yellow orange. There are only relatively few scattered setae. The larvae are poisonous to vertebrates as they sequester pyrrolizadine alkaloids from their host plants.

Utetheisa ornatrix (Linnaeus), the "bella moth" or "rattlebox moth" (FW 1.8–2.1 cm; Plate 48.2, Plate 64.10 larva), ranges from southeastern Arizona east to western Texas in the West, and thence south through Mexico and Central America at least as far south as Costa Rica. The moths are strictly diurnal and are found in close association with their host plants. It is found commonly in eastern North America and blends with *U. bella* (Linnaeus), here considered a synonym, in eastern Texas. Adults seem to fly throughout much of the year whenever the hosts are in leaf. Its larvae feed primarily on various species of *Crotalaria*, rattlebox (Fabaceae), but may also feed occasionally on other legumes such as *Lespedeza* and *Lupinus*. Reports of its use of plants in other families are suspect.

SUBTRIBE PERICOPINA

Adults of most of the North American taxa are black-and-white, but those of the American tropics are often brightly colored and several belong to presumed Müllerian mimicry complexes. Eggs are laid singly or in small groups. Recorded host-plant families are Agavaceae, Apocynaceae, Asclepiadaceae, Asteraceae, Boraginaceae, and Fabaceae. Larvae are usually brightly colored and are presumably aposematic. The larvae are distinguished by having only three verrucae above the coxa on the second thoracic segment and by having dark shiny verrucae on certain thoracic and abdominal segments. All setae are barbed and are located on verrucae except those on prolegs. Spiracles are elliptical. Pupation is in a light, flimsy cocoon.

Gnophaela Walker has five described species. Four of these are endemic to western North America, but one, *G. aequinoctialis* (Walker), is found in the Sierra Madre Orientale of northeastern Mexico. *Gnophaela clappiana* Holland (FW 2.1–2.5 cm; Plate 48.3) is unlike its close relatives with a much reduced FW pattern and a blue iridescent abdomen. The species is distributed from southern Colorado and eastern Arizona (White Mountains) south through the Sacramento Mountains of southern New Mexico. Adults fly during late July to early August. The identity of the larval host has not been reported. Full-grown larvae are black with yellow spots and a lateral bright yellow band. Tufts of black hairs emanate from iridescent blue scoli. The distal portion of the prolegs is maroon. The head is shining dark maroon and slightly lighter along the sutures. The pupa is black with irregular yellow spots and is formed inside a loose cocoon spun by the final instar. There is a bundle of small hooks at the cremaster but also several bundles of similar hooks on the anterior. *Gnophaela latipennis* (Boisduval), the "Sierran pericopid" (FW 2.4–2.6 cm; Plate 48.4), ranges from western Oregon to central California. The adults are strictly diurnal and have a single flight from mid-May to early July. Larvae usually feed on hound's-tongue, *Cynoglossum* (Boraginaceae), during fall and spring, but we raised adults from larvae found on *Horkelia californica* found near Hayfork, Trinity County, California. The larvae are yellow with dorsal and lateral black patches and lines. Relatively sparse, white hairlike setae emanate from black verrucae with iridescent blue spots. The head is red brown. *Gnophaela vermiculata* (Grote)(FW 2.0–2.4 cm; Plate 48.5) is the most widespread member of its genus. It ranges from southern British Columbia south through the Cascades and Rocky Mountains to Oregon, northeastern Nevada, Utah, Colorado, and northern New Mexico. The moths are diurnal, fly during late July and August, and are often found nectaring at flowers including *Rudbeckia* and *Chrysanthemum* (Asteraceae). Its striking yellow and black larvae feed on *Mertensia* (Boraginaceae).

Dysschema howardi Henry Edwards (FW 3.7–4.6 cm; Plate 48.7) belongs to a genus found through much of the New World tropics. This is our largest North American tiger moth. This species is found from southern Arizona, New Mexico, and western Texas south into Mexico. The sexes are dimorphic with males having a white HW and females an orange one. The moths are nocturnal and are attracted to light. There is a single flight during late July and August. Its larvae are known to feed on *Brickellia* and *Viguiera* (Asteraceae) in southeastern Arizona and were reported by G. Forbes to feed on sotol, *Dasylirion* (Agavaceae), in southern New Mexico. The last-instar larva was described by McFarland as transversely banded black and cream yellow with long, soft hairlike setae emanating from large chalazae. The chalazae are orange yellow where found on light areas and are blue black where found on black areas. The lateral shields on the prolegs are iridescent blue black and the head is shining black. The basketlike cocoon was suspended from the container lid, probably under a branch or similar object in nature, and was constructed with 13 strands of tough silk.

SUBTRIBE PHAEGOPTERINA

Cycnia Hübner has five described North American species. The larvae of all species feed on plants in the Apocynaceae. Adults are closely associated and are usually found resting on these plants. *Cycnia tenera* Hübner, the "dogbane tiger moth" or "delicate cycnia" (FW 1.8–1.9 cm; Plate 48.6), ranges in the West from southeastern British Columbia southeasterly across the West including Montana, Wyoming, and Colorado. There are a few isolated records from central California, possibly representing an accidental introduction. In the East, it is found south to Arkansas, Mississippi, and Florida. The moths are nocturnal as they are found resting on their dogbane host plants during the day and flying to lights at night. According to Covell, in the East there are two flights from May to October, but in the West they have only a single flight during June and July. The larvae feed exclusively on dogbanes, *Apocynum* (Apocynaceae), as is usually evidenced by webbing and skeletonized leaves. The larvae feed at night and rest near the base of the plant during the day. Larvae have short gray hairs intermixed with much longer white-tipped setae. *Cycnia oregonensis* (Stretch), the "Oregon cycnia" (FW 1.6–1.8 cm; Plate 48.9), is distinguished by its yellow costal area on the FW. It ranges from southern British Columbia east across southern Canada and south to central California, Arizona, and New Mexico. In the East, it is found south to Texas, Arkansas, and South Carolina. In the West, this species flies from May to mid-July but has two flights in the East from April to August. Like *C. tenera* it also uses only dogbanes as its larval host plants. According to Wagner, the larvae are similar to those of *C. tenera* but have blonder setae.

Euchaetes Harris is a moderately large genus of sexually dimorphic tiger moths found primarily in the states bordering Mexico and south into the tropics. There are 10 species in the West and another four or five in the eastern portion of the United States; additional species are found only south of the border in Mexico and Central America. Males are always much smaller than females and often more wasplike in appearance. Both sexes are usually nocturnal and are attracted to light. The larvae of most species feed in clusters on various plants in the milkweed family (Asclepiadaceae). *Euchaetes antica* (Walker)(male FW 1.3–1.5 cm, female FW 1.6–2.0 cm; Plate 48.8 female, Plate 48.10 male) ranges from central to southeastern Arizona. They have a single flight during July and August. *Euchaetes elegans* Stretch (FW 1.6–1.8 cm; Plate 48.11 female) is found from southeastern California, where it is most common, eastward across southern Arizona to western Texas and south into Mexico. There is a single annual flight during October. Eggs are laid in a mass on a milkweed leaf and covered with a mat of hairlike scales from the posterior portion of the female's abdomen. The eggs hatch in 15 days. The earlier instars were described by J. A. Comstock and Dammers. They noted that the full-grown larva is pale green with dense dorsal black hair tufts (tussocks) on the first through eighth abdominal segments. There is a tuft of very long white hairs on the dorsum of the prothorax that arches anteriorly. The dorsal tufts on the other thoracic segments are a mixture of long black-and-white hairs. Pupation is on the host milkweed in a silken cocoon into which larval hairs are incorporated.

There are only two North American species of *Pygoctenucha*—both western. *Pygoctenucha terminalis* (Walker)(FW 1.6–1.8 cm; Plate 48.12) is much the more common and widespread of the two, being found from eastern Colorado and diagonally across New Mexico to southeastern Arizona. There seem to be no records from Mexico, but the species must certainly occur there. The flight period is from late May to early August—the exact timing possibly dependant on rainfall events and the resultant foliation of the hosts. The moths are diurnal and found in close association with their larval hosts, various species of milkweeds *(Asclepias)*. The second species is *Pygoctenucha pyrrhoura* (Hulst), which is known only from western Texas.

Ectypia clio (Packard)(FW 1.7–1.9 cm; Plate 48.13, Plate 64.12 larva) is a widespread western moth in a small genus that has two other western species: *E. bivittata* Clemens and *E. mexicana* (Dognin). *Ectypia clio* is found from western Oregon south through California to Baja California and from Montana, eastern Wyoming, and western Nebraska south through Colorado, Utah, and southern Nevada to Arizona, New Mexico, and western Texas then south into Mexico. There are spring and fall flights during June and August to September. The life-history details are unreported.

There are seven described North American species of *Pygarctia* Grote. Six of these species are endemic to or range into the West. Three additional species are known from Mexico but do not range north into the United States. *Pygarctia murina* (Stretch)(FW 1.3–1.6 cm; Plate 48.14) ranges from southeastern California and southwestern Utah east across Arizona, New Mexico, southern Colorado to southern Texas, and thence south into Mexico. There is a single flight from late July to September. *Pygarctia roseicapitis* (Neumoegen & Dyar)(FW 1.4–1.7 cm; Plate 48.15) is found from southeastern Arizona across New Mexico to central Texas and south in Mexico to at least Durango. The adults are nocturnal and have a single flight from late July through August. Larvae feed on a variety of annual Euphorbiaceae that appear only during the summer monsoons (July to August).

Halysidota Hübner is an extensive New World genus with 30 species found mostly in tropical America; the genus was the subject of an extensive revision by Watson. *Halysidota tessellaris* (Smith), the "banded tussock moth," or "pale tussock moth" (FW 2.0–2.2 cm; Plate 48.16, Plate 64.16 larva), ranges from Utah and Arizona east through Colorado and southeastern Wyoming and through much of eastern North America including southeastern Canada south to Texas and central Florida. A very similar species, *H. harrisii* (Walsh), the "sycamore tussock moth," is limited to eastern North America and uses American sycamore, *Platanus occidentalis* (Platanaceae), as its only larval food. Adults can be separated from those of *H. tessellaris* only by dissection of the genitalia. In the West, the adults have a single flight from mid-June to mid-July, but two flights are reported for the Southeast. Eggs are laid in masses on the undersides of leaves, and the larvae feed on a wide array of broad-leaved deciduous trees and shrubs. The larvae vary from gray to yellow and have long pencils (tussocks) of black-and-white hairlike setae. The cocoons have larval hairs incorporated.

Lophocampa Harris is restricted to the New World and has about 70 described species, most of which occur in the Neotropics. Many of these are found in cool highland forests. There are 11 North American *Lophocampa*, nine of which are endemic to or extend their ranges into the West. *Lophocampa ingens* (Henry Edwards)(FW 2.1–2.7 cm; Plate 48.17) is found in the southern Rocky Mountains of Utah, Colorado, Arizona, and New Mexico. The adults are nocturnal and fly in a single flight from late July through mid-August. The larvae feed on foliage of pinyon pine, *Pinus edulis*, and ponderosa pine, *P. ponderosa* (Pinaceae). *Lophocampa sobrina* (Stretch)(Plate 64.17 larva) of the Pacific Coast is very similar in appearance to *L. ingens*. Essig mentions that *L. argenta* larvae have caused significant defoliation to native stands of the Monterey pine, *P. radiata*, in California, but these were probably larvae of *L. sobrina*. Larvae sometimes feed on broadleaf plants below the host pine, such as alder.

Lophocampa argentata (Packard), the "silverspotted tiger moth" (FW 1.6–2.2 cm; Plate 48.18), ranges from British Columbia south through the Pacific Coast states to southern California, and, in the Rockies, to Arizona, New Mexico, and western Texas. The species is also reported from the Sierra Madre Occidentale of Sonora, Mexico. The univoltine adults are in flight from mid-June to early August. The hosts are primarily conifers, especially Pinaceae. Pines *(Pinus)*, firs *(Abies)*, and Douglas-fir *(Pseudotsuga menziesii)* are preferred hosts. Junipers and some broad-leaved trees have been reported as hosts but these reports may be based on larvae of related species. The green eggs are laid on loose clusters on host twigs and needles. The larvae feed in groups and overwinter in their webs. In spring the larvae continue to feed communally but disperse and become solitary when about two-thirds grown. The damage referred by Furniss and Carolin to this species may actually have been caused by *L. sobrina*. The larvae are black with black verrucae bearing longitudinal rows of short stiff setae—yellow subdorsally and red orange dorsally. There are batches of shorter black setae middorsally. ***Lophocampa mixta*** (Neumoegen)(FW 1.6–1.8 cm; Plate 48.19) ranges from southeastern Arizona south into northern Mexico. The bivoltine adults fly from mid-February to mid-April and from mid-June to early September. The life history is unreported. ***Lophocampa maculata*** Harris, the "spotted tussock moth" or "yellow-spotted tiger moth" (FW 1.5–2.1 cm; Plate 48.20, Plate 64.18 larva), ranges across southern Canada from coast to coast and south in the conterminous United States to southern California, eastern Arizona, southern New Mexico, and, in the East, North Carolina. There is one flight from late May through July. The adults are nocturnal and are attracted readily to light. Early-instar larvae feed gregariously, but older larvae are solitary. Host plants are various broad-leaved trees, especially willows, *Salix* (Salicaceae), alder, *Alnus* (Betulaceae), and maples, *Acer* (Aceraceae). McFarland reports that the larvae also feed on madrone, *Arbutus menziesii* (Ericaceae). Larvae are densely covered with short hairlike setae. The central four or five segments are covered with a red orange to yellow band, while either end is black. There may also be short tufts of black hairs on the dorsal light-colored segments. There are also scattered tufts of very long white hair as well as light setae scattered along the body. ***Lophocampa indistincta*** (Barnes & McDunnough)(FW 1.5–2.1 cm; Plate 48.21) is endemic to Anacapa, Santa Catalina, Santa Cruz, and Santa Rosa islands; it is one of the most distinctive Channel Island endemics. This species had generally been forgotten since its 1910 description until larvae were found by a University of California expedition in 2001 and were raised to adults. Adults collected by L. Martin on Catalina in 1939 were not recognized until the 1980s, but this discovery was generally unreported. The adults are nocturnal, and we found them in flight during August and September. Larvae have been found in nature feeding on *Rumex* (Polygonaceae) and *Vaccinium* (Ericaceae).

Aemilia ambigua (Strecker)(FW 1.9–2.4 cm; Plate 48.22) is an attractive moth with a rust red FW having evenly dispersed silver white stripes along the wing veins. The species has been found from southernmost Montana south through Utah and Colorado throughout much of Arizona and New Mexico, especially in montane areas where its host pines occur. It appears that this species as well as a larger relative occur to the south in the Sierra Madre Occidentale of Mexico. Adults are nocturnal and are attracted to light, although males are by far the predominant sex at light. There is a single flight from mid-June through August. These moths are always found in close proximity to pines, and they are the likely larval host. Ferguson pointed out that these moths are unrelated to the type species of *Aemilia* Kirby and that a new genus name should be proposed. Schmidt is currently preparing a new genus for this moth and its relatives found to the south of our area.

Hemihyalea Hampson is a tropical American genus with many species, only four of which range north of the Mexican border. These species were placed in *Pseudohemihyalea* Reg-Barros, but this genus was treated as a synonym by Ferguson and Opler. ***Hemihyalea edwardsii*** (Packard), "Edwards' glassywing" (FW 2.6–2.9 cm; Plate 48.23), ranges from western Oregon south to southern California (including several Channel Islands), Arizona, and southwestern New Mexico as well as the Sierra Madre of Sonora and Chihuahua, Mexico. Differentiation between this moth and the next species is not always clear-cut, and systematic studies would be helpful. The adults have translucent cream tan wings and an orange or red abdomen. There is a single August to October flight. Irregular clusters of pale whitish eggs are deposited on bark, frequently in crevices. The larvae rest by day in holes in branches or under bark, then come out to feed at night on mature leaves of various oaks, *Quercus* (Fagaceae), including *Q. agrifolia*, *Q. chrysolepis*, and *Q. emoryi*. The caterpillars are able to feed on old hardened leaves. Full-grown larvae are brown black with tufts of moderately long brown hairlike setae. From each tuft emanates one or two long chestnut setae. The spiracles are orange and the head is large and glistening dark brown. The cocoon is loosely woven and yellow with included larval setae. ***Hemihyalea labecula*** (Grote)(FW 2.1–2.9 cm; Plate 48.24) is wide-ranging and occurs from southern Nevada, Utah, and Colorado south through Arizona, New Mexico, and western Texas south into the Sierra Madre Occidentale of Mexico. Adults are nocturnal and have a single flight from July to early September. The hosts have not been reported but are almost certainly oaks in many circumstances, although the species is found in habitats where no oaks occur (e.g., northern Colorado). ***Hemihyalea splendens*** Barnes & McDunnough (FW 2.6–2.8 cm; Plate 48.25) is known from southeastern Arizona south into the Sierra Madre Occidentale of Sonora, Mexico. The adults fly as a single flight from late July to mid-November with the peak from mid- to late September. The life history is unreported.

Bertholdia Grote is a large genus with many species occurring in the Neotropics. ***Bertholdia trigona*** (Grote)(FW 1.6–1.8 cm; Plate 49.1) is an attractive moth that ranges from eastern Wyoming and western South Dakota south through the Rocky Mountains of Colorado, southwestern Utah, Arizona, New Mexico, and western Texas to the Sierra Madre Occidentale of northern Mexico. Adults fly during mid-July and August, occasionally in May. The female lays single layers of green eggs. McFarland had young larvae feed on lichens that grew on the stems of *Cercocarpus*. J. A. Comstock attempted to rear the species in captivity, and a variety of foods was provided with minimal success. One larva reached the sixth instar but later died. The head was black with the labrum white. The body was a rich yellow orange with a well-defined middorsal stripe. The larva has yellow, brown, and red brown longitudinal stripes and lines. There are prominent black papillae, from each of which projects one long black or white hairlike seta and several shorter white hairlike setae. Those on the dorsum of the first and second thoracic segments are longest and arch forward over the head.

SUBTRIBE CTENUCHINA

Adults of most groups are wasp mimics to one degree or another; some are stunningly exact in their wasp resemblance. Many belong to Müllerian mimicry complexes and are assumed to be poisonous and distasteful; their appearance is therefore considered to be aposematic. Larval host-plant families include Apocynaceae, Moraceae, and Poaceae. The larvae are more or less densely covered with long barbed setae that arise from verrucae except those on the prolegs. Ctenuchine larvae have only three verrucae above the coxa on the second thoracic segment, and their verrucae are similarly colored on all segments.

Ctenucha W. Kirby has six North American species, five of which are western endemics or range into the West. All species are diurnal, most of them exclusively so. All species are presumed to use grasses (Poaceae) as their larval hosts. *Ctenucha venosa* Walker (FW 1.4–1.6 cm; Plate 49.2) ranges from southern Nevada, Arizona, and New Mexico to Oklahoma and Texas, and thence south well into Mexico. Individuals found to the north in Colorado and Kansas are likely vagrants. This is an abundant moth that may be found during the day or night. It is often abundant at lights in southern Arizona and Mexico. *Ctenucha venosa* has several generations, and McFarland has documented that adults fly from April to mid-November in southeastern Arizona. The larvae feed on grasses (Poaceae), including side-oats gramma, *Bouteloua curtipendula*. *Ctenucha cressonana* Grote (FW 2.0–2.3 cm; Plate 49.3) is a Rocky Mountain moth that is found from Colorado south through western New Mexico, eastern Arizona, and, rarely, western Texas. The moths are diurnal and attracted to flowers such as spreading dogbane, *Apocynum androsaemifolium* (Apocynaceae). The single flight is from the end of May through July. *Ctenucha rubroscapus* (Ménétries)(FW 1.8–2.0 cm; Plate 49.4) has a narrowly white tipped smoky black FW and HW, red pink head and tegulae, and a glossy blue black abdomen. The species ranges from southern coastal Washington south along the coast and Sierra Nevada to central California. There is a single flight during July. Adults are diurnal and feed at flowers. The larvae feed on grasses such as *Dactylis* or *Elymus* (Poaceae). When disturbed while feeding, McFarland observed that the larvae snap into a curled position and fall off the host, then rather quickly uncurl and walk rapidly away. The larva is black narrowly marked with white and covered with tufts of moderately long gray tan hairlike setae that emanate from black verrucae. The head and prolegs are shiny golden brown to light brown. *Ctenucha brunnea* Stretch (FW 1.8–2.0 cm; Plate 49.5) occurs along the coast of California from Marin County south to San Diego County. There are rare reports from the southern Sierra Nevada and Arizona that may represent misidentifications. McFarland noted striking declines in this species' abundance in southern California after the 1950s. The adults have been found from mid-May to mid-July; they are especially attracted to flowers of toyon, *Heteromeles arbutifolia* (Rosaceae), in the Santa Monica Mountains of southern California. The larvae feed on grasses (Poaceae), especially giant ryegrass, *Leymus condensatus*.

Cisseps fulvicollis (Hübner)(FW 1.3–1.9 cm; Plate 49.6, Plate 49.7), the "yellow-collared scape moth," ranges from southeastern British Columbia east across the Prairie Provinces (one record from Northwest Territories) to eastern Canada. In the West, it is found sparingly throughout but seems absent from much of Nevada, southern Utah, and Arizona. Overall, the moth has a wasplike appearance and flight. Adults have dull black FW with an orangeish costal edge and a yellow orange collar. The HW and abdomen are glossy blue black. It is possible that *C. packardii* (Grote) and *C. wrightii* (Stretch) are conspecific with *C. fulvicollis,* but conclusive studies have not been undertaken. There is a single flight from late July to mid-October. Adults are both diurnal and nocturnal. During the day they are often seen nectaring at flowers such as goldenrod, *Solidago* (Asteraceae). Larval hosts are primarily grasses (Poaceae), but sedges (Cyperaceae) may be used on occasion. Reports of *Eupatorium* and *Solidago* refer to nectar plants, not larval hosts, and one report of lichen use is likely in error. The larvae are black with longitudinal orange stripes. The verrucae bear tufts of gray white setae with tufts of longer black setae laterally. The head is orange.

Horama panthalon texana (Grote), the "Texas wasp moth" (FW 1.3–1.7 cm; Plate 49.8), a northern representative of a Neotropical genus, occurs from southern Arizona east to Texas and south through the lowlands of Mexico and Central America. It can be an abundant diurnal moth and is found at flowers, especially composites. The adults are attracted to light at low frequency. The moths seem to have a number of flights through much of the year.

SUBFAMILY LYMANTRIINAE

These, together with several genera of Arctiinae, are the so-called tussock moths. The integrity of this subfamily as a monophyletic unit will be in doubt until extensive phylogenetic research is undertaken on all or most of the group's genera. Many genera share several characteristics: (1) Larvae have a single red or yellow eversible middorsal gland on the center of the sixth abdominal segment and often on the seventh abdominal segment. (2) In many genera, males have paired pockets with corrugated tymbal organs on the third abdominal sternite. (3) Adults rest with wings held against the substrate in a triangular outline with the densely hairy forelegs projecting forward of the head. Adults are small to large with generally dull colors of grays, browns, and black maculations. The sexes are often dimorphic with the females large and heavier bodied than the males. Females have long tufts of dehiscent hairlike setae at the tip of the abdomen, which are used to cover their egg masses. In some genera the females have only vestigial wings and are flightless. The western species are all single-brooded. Some species are nocturnal, but several are diurnal. Eggs are often laid in clusters, and the early instars feed gregariously. Young larvae are the primary dispersal stage for many species as they can drift long distances on long silk threads (ballooning). The larvae are usually more brightly colored than the adults, with shades of red, tan, brown, and black. Moreover, the larvae of most of our species have verrucae that bear sparser tufts of longer hairs on the dorsum and along the sides. The larvae of many species have dense toothbrushlike tufts (tussocks) of long erect setae along the dorsum of the abdomen. With the exception of *Gynaephora rossii*, the larvae of our species feed on broad-leaved or coniferous trees, and several are serious pests. The relationships of the lymantriine genera are poorly understood. Worldwide, the subfamily has about 2,500 species in about 360 genera with the greatest diversity in the Old World tropics. There are only about 15 genera and 200 species in the New World. Thirty-three species in six genera are detailed by Ferguson for North America. Three of the genera are introductions from Eurasia—*Lymantria dispar* (Linnaeus), the "Gypsy moth," *Leucoma salicis* (Linnaeus), the "satin moth," and *Euproctis*, "brown-tail moths"—all of which can be of concern to forestry. Native North American lymantriines are limited to *Gynaephora* Hübner, *Dasychira* Hübner, and *Orgyia* Ochsenheimer. The richest genera are *Dasychira* with 16 species, found primarily in eastern North America, and *Orgyia,* which has 11 species. Sixteen species are reported for the West.

TRIBE LYMANTRIINI

Larvae lack dorsal abdominal hair tufts as well as the long anterior and posterior hair pencils.

Lymantria dispar (Linnaeus), the "Gypsy moth" (male FW 1.5–1.9 cm, female FW 2.3–2.9 cm; Plate 49.9 male, Plate 49.10

female, Plate 63.32 larva), accidentally escaped from a laboratory culture in Massachusetts in 1868 or 1869 and is now a serious and widespread forest defoliator in eastern North America. Because it is so abundant, and because its egg masses and cocoons, which may be attached to and readily dispersed long distances by vehicles, local outbreaks may occur almost anywhere in the West. Whenever the moths are found, usually with sentry pheromone traps, intense efforts at eradication soon ensue, and, as a result, the moth is not yet resident in the West. Introduced colonies have appeared in several states including California, Colorado, New Mexico, and Oregon. More recently, adults of an Asian population of Gypsy moths have been found in the Pacific Northwest. These are viewed more seriously than the eastern population because they seem to specialize on conifers and females are capable of flight. Although not the most serious North American forest pest, more funds have been spent to combat this moth and prevent its spread because it causes more damage to eastern deciduous forests, including suburbs with dense human populations. In some years more than a million acres of eastern deciduous forests are defoliated by Gypsy moth larvae. The adults are strikingly dimorphic with brown males and predominantly white females. The females are larger than the males with large expanded abdomens that contain their large complements of eggs. There is a single flight from mid-July through August. The males are diurnal. The eggs overwinter, and the larvae feed in spring. Egg masses of 100 to 800 eggs are laid on tree trunks or branches. The larvae are mottled blue gray with long tan subdorsal hairlike setae, dorsal blue verrucae on the first thoracic through the second abdominal segment, and red verrucae on the third through seventh abdominal segments. Host plants include a wide variety of broad-leaved trees as well as a few conifers such as Douglas-fir, *Pseudotsuga menziesii,* and Colorado blue spruce, *Picea pungens.* First-instar larvae may favor particular hosts, but the larvae later become less discriminating. The range of favored trees includes at least 10 families. The host range of the introduced European race is described and discussed in great detail by Burgess and Baker.

TRIBE ORGYIINI

Gynaephora rossii Curtis (male FW 1.1–1.7 cm, female FW 1.6–19 cm; Plate 49.11 male, Plate 49.12 female, Plate 62.26 larva) is a sexually dimorphic Arctic-alpine moth. The species is North American and ranges across the Arctic and south, discontinuously, through the Rocky Mountain cordillera to Colorado and Wyoming; *G. rossii* is among the most cold-hardy, occurring on some high Arctic islands where few if any other lepidopterans occur. The species is notable for its long life cycle, documented to be seven to 11 years in northern Canada. With the exception of some long-diapausing prepupal yucca moths (see Incurvariidae), this moth is the longest-lived lepidopteran. The females are winged but are essentially flightless, while the males are capable of rapid flight and are diurnal. The adults are found in late June and July on Arctic or alpine tundra. The life history is reported in great detail by Morefield and Lange. Eggs are often laid in masses on the cocoons from which females emerge. Reported hosts include *Dryas* (Rosaceae), *Salix* (Salicaceae), *Saxifraga* (Saxifragaceae), and *Picea* (Pinaceae), although Morefield and Lange showed that the first two are preferred. Thus, they are most likely generalized feeders. The larvae are densely hairy and are notable for having seven dorsal dark hair tufts, instead of five as is usual for the family. Ferguson concludes that the larvae live in one of the most severe environments faced by Arctic insects, and that they must be frozen each winter. The hair tufts are composed of yellow hairs laterally. There are six larval instars. The pupa is densely hairy, more so than that of any other North American lepidopteran. The hairs have been shown to be urticating.

Among the 16 North American ***Dasychira***, only ***D. vagans grisea*** (Barnes & McDunnough), the "variable tussock moth" (male FW 1.4–1.8 cm, female FW 2.2–2.4 cm; Plate 49.13 male), *D. mescalera* Ferguson (male FW 1.5–1.7 cm, female FW 2.1–2.4 cm; Plate 49.14 male, Fig. 240 genitalia), adults from mid-July to early October, and *D. grisefacta* (Dyar)(male FW 1.6–1.9 cm, female FW 2.1–2.2 cm; Plate 49.15 male) are found in the West. For all three species, males are smaller and rounder winged than their associated females. The colors and degree of shading are variable in all species. Identification is often more certain by comparison of larvae and examination of adult genitalia. *Dasychira vagans* ranges in appropriate habitat from British Columbia across southern Canada to Quebec and south to Colorado, and, in the East, to North Carolina. Males of *D. v. grisea* are easily confused with those of *D. mescalera,* which occurs sympatrically in southern Colorado and, most likely, New Mexico as well; while the females have more pale patterning and lack brownish FW scaling. Larvae of *D. v. grisea* are distinctive in having primarily white hairlike setae, with the exception of the black dorsal tussocks and hair pencils. Hosts through the range include a variety of broad-leaved trees including cottonwoods *(Populus)* and willows *(Salix,* Salicaceae), and oaks (e.g., *Quercus garryana,* Fagaceae) in the West. Adults of *D. vagans* fly from mid-June to mid-August in a single brood. *Dasychira grisefacta* is a western conifer-feeding moth that ranges from British Columbia east to Alberta and south in the mountainous West to northwestern California, South Dakota (Black Hills), Arizona, and New Mexico. There is a single flight in July and early August. Ferguson suggested that fresh individuals found in Arizona during late August implied the possibility of a second brood in that region. The adults can be separated easily from *D. vagans* by their uniformly gray appearance (melanistic in Pacific Northwest). Host plants span a wide range of conifers including Douglas-fir *(Pseudotsuga),* hemlock *(Tsuga),* spruces *(Picea),* larch *(Larix),* pines *(Pinus),* and true firs *(Abies)*—all Pinaceae. In rare instances, the species can cause extensive defoliations. For example, Furniss and Carolin describe an episode of extensive defoliation of young ponderosa pine stands in eastern Montana. The larvae are covered with a mixture of gray, brown, black, and white hairs. The dorsal tussocks are gray, and the larvae have the typical anterior and posterior projecting hair pencils long of black, clubbed hairlike setae.

Orgyia Ochsenheimer is cosmopolitan with about 60 species, 11 of which are North American, with another eight found only in Mexico. Nine of the species range into the West or are endemics. The species are strikingly dimorphic with winged males and almost wingless females—the wings being reduced to tiny pads. Females have bipectinate antennae with short side branches, and very densely hairy, large abdomens. Adult males are our smallest lymantriines (FW 1.1–1.5 cm). The males are broad winged and geometroid in gross appearance; the wings are usually dark gray or brownish with variable darker maculations. There are dorsal tufts of lustrous metallic scales on the mesoscutellum and second abdominal segment. They have bipectinate antennae with long side branches. Most species are nocturnal with one or more annual flights. Larvae are similar to those of *Dasychira* but have narrower dorsal tussocks, lack the dorsal tuft on the eighth abdominal segment (often present on *Dasychira* larvae), and have delicately barbed sparse body hairs on verrucae—as opposed to longer rigid barbed spines on *Dasychira* larvae. *Orgyia* species are the only North American lymantriids to overwinter in the egg stage. The flightless females emerge, are mated, and lay a single egg mass on the cocoon from which they emerged. The egg mass is often covered with a frothy coating to which female body hairs often adhere, although, as pointed out by McFarland, the eggs may be packed inside a bundle of female body hairs. Recently emerged larvae are frequently picked up and carried on their silken threads in wind currents; this is the principal means of long-distance dispersal for these species with flightless females. Hosts are primarily trees and shrubs but may also include herbaceous plants on rare occasions. Host specificity ranges from widely polyphagous in *O. antiqua* and *O. leucostigma* to apparently host specific in *O. leuschneri*. *Orgyia antiqua* (Linnaeus), the "rusty tussock moth" (male FW 1.2–1.5 cm; Plate 49.16 male, Plate 49.17 female), is Holarctic and ranges from Alaska and British Columbia across Canada south to the Mid-Atlantic states and northern California. The males are distinguished by their bright rusty brown wings, absence of spurs on the middle of the hind tibia, and reduced eye diameter (most often an indicator of diurnal activity). In most of the range, there is a single flight in August and September. The possibility of two broods on the Pacific Coast was mentioned by Ferguson. The host range is wide and includes many conifers and a broad range of broad-leaved trees. Furniss and Carolin report that the larvae sometimes cause conspicuous defoliation of western hemlock, *Tsuga heterophylla* (Pinaceae), in coastal British Columbia. The larvae differ from those of other North American *Orgyia* by the presence of black lateral hair pencils arising from verrucae on the second abdominal segment. The long, sparse body hairs are pale yellow, and the middorsal tufts are off-white. *Orgyia vetusta* (Boisduval), the "western tussock moth" or "oak tussock moth" (male FW 1.1–1.3 cm; Plate 49.18 male, Plate 49.19 female), ranges from the San Francisco Bay region south to San Diego County and also occurs on several of the Channel Islands. Adults are active during May and June. Ferguson treated *O. vetusta*, *O. magna* Ferguson, and *O. cana* Henry Edwards (Fig. 239 genitalia) as a complex of closely related species. Previous literature reports from California are often unreliable as to their biological information, because earlier workers were unaware of the species differences. The larvae of the species complex are all similar and may be characterized as having mostly long white setae arising from the verrucae and light brown to gray dorsal tufts. Definite host records for *O. vetusta* include coast live oak, *Quercus agrifolia* (Fagaceae), *Ambrosia* (Asteraceae), *Lupinus* (Fabaceae), and *Atriplex* (Chenopodiaceae). Earlier reports of damage to fruit trees and utilization of other hosts are uncertain as to species. *Orgyia pseudotsugata* (McDunnough), the "Douglas-fir tussock moth" (male FW 1.2–1.5 cm; Plate 49.20 male), is a western montane species ranging from British Columbia south to Colorado, Arizona, and the Transverse Ranges of southern California. The adult males may be distinguished by their gray FW with almost straight black antemedial line in combination with a rust brown HW. The adults are active from late July to early September. The larvae are typical for the genus and have long, white hairlike setae arising from the verrucae and brown- or golden-tipped white dorsal tufts. The hosts are restricted to conifers—mainly *Pseudotsuga*, *Abies*, *Picea*, and *Tsuga*. *Orgyia pseudotsugata* may cause severe defoliation of Douglas-fir, *Pseudotsuga menziesii*, and grand fir, *Abies grandis*, in the Pacific Northwest. In southern Oregon, California, and Arizona it may cause serious defoliation and loss of white fir, *A. concolor*. In

some years, it can be the most serious western forest defoliator. In British Columbia, outbreaks seem to occur every 10 to 20 years. According to Furniss and Carolin, when found in mixed stands with firs, the larvae may feed on other conifers including ponderosa pine, *Pinus ponderosa*, and adjacent shrubs such as bitterbrush, *Purshia tridentata* (Rosaceae). *Orgyia leucostigma* (J.E. Smith), the "whitemarked tussock moth" (male FW 1.1–1.4 cm; Plate 49.21 male, Plate 63.33 larva), occurs primarily in eastern North America but ranges to Alberta, Colorado, and New Mexico, where it might occur with or in proximity to populations of *O. pseudotsugata*. There, adults of *O. leucostigma* would be more or less uniformly dark brown with a curved black antemedial line [subsp. *oslari* (Barnes)]. Adults are active during July in Colorado and New Mexico (one brood) but may be found from April to November in Texas (two broods). Larval hosts include a wide variety of conifers and broad-leaved deciduous trees; it occasionally causes damage to shade trees, especially oaks, *Quercus* (Fagaceae). The egg masses are covered with a white frothy material that does not include hairs from the female adult. The larvae have a black dorsal abdominal band with subtending yellow subdorsal stripes. The dorsal tufts are white to gray, and the head is bright red. Two additional species, *O. leuschneri* Riotte and *O. falcata* Schaus occur in the West.

TRIBE LEUCOMINI

Leucoma salicis (Linnaeus), the "satin moth" (male FW 1.7–2.2 cm, female FW 2.1–2.7 cm; Plate 49.22 male, Plate 63.31 larva), was accidentally introduced into North America from Eurasia before 1920 when two widely separate incursions were discovered in British Columbia and New England. The moth is rarely reported as causing damage. In western North America, the satin moth has spread from British Columbia south to the northern Sierra Nevada of California. The large adults are uniformly satiny white with black banding on the legs. Similar all-white arctiines (e.g., *Spilosoma virginica* or *S. vestalis*) possess at least some yellow or red on their legs or abdomen. There is a single flight from June to early August, and the nocturnal adults are attracted to light. Hosts are limited to *Populus* and *Salix* (Salicaceae). Quaking aspen, *P. tremuloides*, and black cottonwood, *P. balsamifera* var. *trichocarpa*, are favored in the West. After its introduction the satin moth was considered of importance, but outbreaks were controlled by the introduction of biological control agents from Europe to British Columbia in 1929–1934. Now, a combination of both native and imported parasites keeps the species under control. Eggs are laid on tree trunks in ovoid patches and are covered with a frothy white substance. In North America, McFarland observed that larvae overwinter in the third instar within round, white weblike cocoons. In spring, the fully developed larvae spin a loose cocoon, which may be tucked between leaves or various other sites. Larvae are distinctive with a series of paired red orange dorsal verrucae and a lateral row of smaller red orange verrucae. The dorsal verrucae bear sparse brown hairlike setae, and the lateral verrucae bear white setae. There are merged intersegmental dorsal cream yellow patches surrounded by black.

SUBFAMILY PLUSIINAE

The subfamily is a primarily a north temperate group with about 400 species found worldwide. There are 77 species in North America north of Mexico of which 60 occur at least occasionally within the confines of western North America. Plusiine adults are defined by their quadrifid HW venation and lashed eyes. Almost all species have an iridescent metallic white or silver stigmatal mark in the center of the FW. Other features distinguishing the Plusiinae include specialized hair pencils on a modified eighth abdominal sternite of males, a double countertympanal hood, and, in the larvae, biordinal crochets and a sclerotized comblike structure on the hypopharynx. The larval epidermis is covered with fine spines (spicules). Larvae are semiloopers. Several species are agricultural pests, especially to row crops.

TRIBE ABROSTOLINI

Abrostola Ochsenheimer has 35 species, which are found mainly in Eurasia and Africa; the only four New World species are found in North America, three of which are western or include the West as a portion of their distribution. *Abrostola urentis* Guenée (FW 1.3–1.5 cm; Plate 49.23) ranges from coast to coast across southern Canada (British Columbia east to the Maritime Provinces) and south to Oregon on the Pacific Coast and south to northeastern Colorado on the Great Plains. In the East, it is found south to South Carolina. Adults are single-brooded in the north (May to July) and double-brooded in the southern portion of their range with the second flight during August and September. Its larvae feed on stinging nettle, *Urtica dioica* (Urticaceae), and possibly other related plants as well. The larvae may be identified by being green and by having a large dark brown diamond-shaped or semicircular dorsal spot on each of the first two abdominal segments.

Mouralia tinctoides (Guenée)(FW 1.7–2.0 cm; Plate 49.24, Fig. 243 genitalia), the only member of its genus, is a distinctive tropical moth that ranges from the southeastern United States south through tropical America, including the Antilles, to Argentina. There is an apparently introduced population in southern California that was first noted by J. A. Comstock in 1935. In North America, adults have been found throughout the year except for the period of May to July. The number of flights has not been defined. The larvae feed on *Tradescantia fluminensis* and *T. zebrina* (both Commelinaceae).

TRIBE ARGYROGRAMMATINI

Trichoplusia McDunnough has 45 described species; all except two are native to Africa, India, and Madagascar, while one is found in China, and *T. ni* (Hübner), the "cabbage looper" (FW 1.4–1.6 cm; Plate 49.25), is widely distributed. This is a nondescript, slightly grizzled gray brown plusiine that is much plainer than other western argyrogrammatines. There is a double white stigmatal mark that should separate this moth from all but a few *Autographa* species. The latter would have more distinctive FW patterns, however. The adults are distributed throughout much of the world in continental areas where beneficent climates permit. Each spring or early summer migratory adults establish temporary populations in more northern areas where there are freezing winter temperatures. In North America, the species may be found throughout the 48 conterminous states and across southern Canada from British Columbia to Newfoundland. The adults have an indeterminate number of generations and may be found through much of the year. In fact, Mustelin has records of *T. ni* from every month in southern California. The larvae feed on the widest possible variety of herbaceous plants but have special predilections for the mustard family (Brassicaceae), particularly species of *Brassica*,

including cabbage, and the moth is a serious pest of cabbage crops. The larvae are green with three pairs of wavy white lines on the back and have the peglike prolegs on the third and fourth abdominal segments.

Chrysodeix includens (Walker), the "soybean looper" (FW 1.3–1.7 cm; Plate 49.26, Plate 64.19 larva), like other economically important plusiines, is a widespread tropical moth that may migrate northward in the spring and summer, thereby establishing temporary breeding populations that die out with the onset of freezing temperatures. The species ranges throughout the Americas including the Antilles, save for those areas with freezing winters. In North America, it is most frequent in the eastern half of the conterminous United States and is resident in the Southeast. In the West, it has been found in southern California, southern Arizona, New Mexico, and southern Colorado. The adults can be easily confused with those of *Autographa precationis* (Guenée), but there is only one western record for that species. The adults may be found throughout the year where lack of freezing temperatures permits, and through the growing season in more northern or inland areas. The number of flights is indeterminate. Larvae are green with two pairs of dorsal and lateral white lines—the upper of which is twice as wide as the lower. The larvae feed on a number of mostly dicotyledonous and monocotyledonous herbaceous plants including a number of crops. It is one of the most important pests of soybean (*Glycine max*) crops in North America.

TRIBE PLUSIINI

SUBTRIBE AUTOPLUSIINA

Autoplusia egena (Guenée), the "bean-leaf skeletonizer" (FW 1.7–2.1 cm; Plate 49.27), is widespread in the Americas from the southern United States south to Argentina. In the West, it occurs in coastal California north to the San Francisco Bay Area. It may be separated from the very similar *A. egenoides* Franclemont & Todd and *A. olivacea* (Skinner) by a rusty tint on the FW and by a pale crescentic patch with even margins on the FW outer margin. The adults fly primarily in summer and fall. The green larvae may be separated from those of related species by minor technical points. The larvae feed on a wide variety of dicotyledonous and a few monocotyledonous herbaceous plants. They are minor pests of snapbeans, *Phaseolus vulgaris*, and soybeans, *Glycine max*, both Fabaceae.

Rachiplusia ou (Guenée), the "gray looper" (FW 1.7–2.1 cm; Plate 49.28), has a dull gray FW with a slightly glossy sheen and is nearly identical to *Autographa californica*. It can be reliably separated from the latter by the presence of tibial spines on all legs and a hair tuft on the frons. The species is found in southern Canada from Manitoba eastward, and in the West from central California, Montana, and Colorado southward. The moth ranges south through the Antilles, Mexico, and Central America south to Ecuador. Most likely, more northern records are the result of the establishment of temporary populations by immigrants from further south. The number of flights is indeterminate. The larvae have been reported to feed on a variety of herbaceous plants, including broad-leaved species and grasses.

SUBTRIBE EUCHACIINA

Polychrysia morigera (Henry Edwards)(FW 1.3–1.5 cm; Plate 49.29) is a distinctive gray brown mottled plusiine with long upturned labial palpi and irregular long nonmetallic stigma. The moths are spottily distributed from Oregon south to northern California, from Montana south to Colorado, and, in the East, from Pennsylvania south to Arkansas and Tennessee. The adults are not very attracted to light, although they occasionally show at light in good numbers; they have a single flight from May to early August. During the day, the adults may be found resting on native larkspurs. Based on records in Colorado and Oregon, the larvae feed from cut-leaf nests only on species of wild larkspurs, *Delphinium* (Ranunculaceae). The cocoon is most distinctive, as it is oval and composed of yellow silk.

Polychrysia esmeralda (Oberthür)(FW 1.5–1.8 cm; Plate 49.30) is a widespread boreal species ranging through temperate Eurasia and, in North America, from Alaska south and west across western Canada and the Prairie Provinces to Saskatchewan. In North America, adults have been found in July and August. Like those of *P. morigera*, larvae feed on wild larkspurs, *Delphinium* (Ranunculaceae).

Pseudeva purpurigera (Walker), the "straight-lined looper moth" (FW 1.4–1.5 cm; Plate 49.31), is distinctive with its light rust brown FW with a straight postmedial line. The FW is slightly falcate with a convex outer margin. Adults are very similar to the more western *P. palligera* (Grote), the only other species in the genus, which is paler, lacks the metallic bronze cast, and has the postmedial line followed outwardly by gray. *Pseudeva purpurigera* ranges from Alberta east to the Maritime Provinces and south, discontinuously, to Colorado, New Mexico, and eastern Arizona, and, in the East, to South Carolina. There is a single flight of adults from late June to early August. The larva was described by Dyar and has been found to feed on meadow-rue, *Thalictrum* (Ranunculaceae). The larvae are pale green with white lines and diagonal slashes. The hump-backed appearance is due to dorsally enlarged first four and eighth abdominal segments. The outer surface of the cocoon has patches and knots of silk.

Eosphoropteryx thyatyroides (Guenée), the "pink-patched looper moth" (FW 1.7–1.9 cm; Plate 49.32, Plate 64.20 larva), is unique with its elongate FW with its pink base and concave inner margin; it is the only species in its genus. The moth has a bimodal distribution in North American Canadian zone forests. In the West, it is found from southern British Columbia and southern Alberta south to western Oregon and Jackson Hole, Wyoming. In the East, the species ranges more widely from Minnesota and Ontario east to Nova Scotia and south through New England and the Appalachians to North Carolina. The adults have a single flight during July to September. Larvae are similar to those of *Pseudeva purpurigera* and, like them, feed on meadow-rues, *Thalictrum*, and wild columbine, *Aquilegia* (both Ranunculaceae). The cocoon differs from that of *Pseudeva* by lacking the silken knots and patches.

SUBTRIBE PLUSIINA

Autographa Hübner has 41 species limited to the Holarctic region; 19 are found in North America with 17 found in the West. *Autographa californica* (Speyer), the "alfalfa looper" (FW 1.4–2.0 cm; Plate 49.33), is a dark gray species with a slight purplish cast to the FW in recently emerged individuals. The reniform spot is a fish-hooked shape and there is a thin black streak extending from the subterminal line to the postmedian line about one-fourth length of the postmedian line from the apex. The species is almost exclusively western with populations in Alaska and from central British Columbia east to southwestern Saskatchewan and south through the western states to southern California, Arizona, and the northwestern

quadrant of New Mexico. The adults are both diurnal and nocturnal, as they are often seen hovering while nectaring at flowers during the day, especially late afternoon, and at lights nocturnally. There are several flights in the south throughout the growing season, and one flight northward during mid- to late summer. The larvae are green with three pairs of white lines on the back. The host range includes a wide variety of woody and herbaceous plants, including diverse families of dicotyledons but also a few gymnosperms and monocotyledonous plants. *Autographa mappa* (Grote and Robinson)(FW 1.5–1.7 cm; Plate 49.34) has a red brown complexly marked FW with contrasting shades and wavy lines. This attractive species ranges from southern British Columbia east across southern Canada to the Maritime Provinces and Newfoundland. In the West, it ranges south to Oregon and Idaho with an apparently isolated population in southern Wyoming and Colorado. The stigma comprises a silver V-shaped mark and a dot. The related Holarctic *A. buraetica* (Staudinger) occurs in Alaska and adjacent northwestern Canada. Adults are nocturnal and are attracted to light. There is a single flight of *A. mappa* in June and July. The larvae are green with three pairs of white wavy lines on the dorsum. On the head there is a black ring around each setal socket. Crumb reports that the larvae feed on nettles, *Urtica* (Urticaceae), but blueberry, *Vaccinium* (Ericaceae), has also been reported. *Autographa pseudogamma* (Grote)(FW 1.7–1.8 cm; Plate 49.35) is similar to *A. californica* but is paler, has a violet tint to the FW, and lacks the black FW streak that runs from the antemedian line to the postmedian line. It ranges from Alaska, Yukon, and eastern British Columbia east to Labrador, Newfoundland, and New England. In the West, it ranges south in the mountains to the central Sierra Nevada, northeastern Nevada, Arizona, New Mexico, and the Black Hills of South Dakota. Adults are nocturnal and are attracted to light. There is a single flight from late June to early August. The early stages are unreported. *Autographa corusca* (Strecker)(FW 1.5–1.6 cm; Plate 49.36) and *A. labrosa* (Grote)(FW 1.5–1.6 cm; Plate 49.37, Fig. 244 genitalia) are closely related, morphologically very similar species. Both are characterized by a mottled and dark rusty red FW. The range of *A. corusca* is from the panhandle of Alaska and British Columbia south along the coast to Mendocino County, California, and the range of *A. labrosa* is from Mendocino County, California, south to the San Francisco Bay area. There is a single strange, though plausible, record of *A. corusca* from Quebec. The hosts of *A. corusca*, and presumably of *A. labrosa*, are alders, *Alnus* (Betulaceae). The adults of both species are nocturnal and are attracted to light. There is a single flight of *A. corusca* from late June to August, while *A. labrosa* has been found from March to September and may have two flights. The larvae of *A. corusca* have been described by Crumb as similar to those of *A. mappa*, but the head lacks black rings around the setal sockets.

Megalographa biloba Stephens, the "bilobed looper moth" (FW 1.5–1.8 cm; Plate 50.1), is a wide-ranging plusiine with a brown FW and a large silver orbicular spot that appears like two merged hemispheres with a notch in between. The species is found in Hawaii and most of the New World from the western United States and eastern Canada south to northern Patagonia. There are two records from Great Britain that likely represent trans-Atlantic migrants. In the West, the species ranges from Washington south and in the Rockies from Montana southward. The species is migratory, and areas with freezing winters are recolonized each summer and early fall from the south. The adults are nocturnal and are attracted to light. In lowland California, the moths have been found year-round, and there are a number of flights. The larvae are polyphagous and reported by Eichlin and Cunningham to feed on a wide variety of herbaceous plants and grasses. The larva is light green with three pairs of white lines, and a series of white dashes above the lateral solid lines. The spiracles are white with brown rims; there is a segmental series of black nodules with one above each spiracle. The head is glistening green with a lateral black margin and black mandibles.

Syngrapha Hübner has 34 species, 24 of which are North American; 21 of these have at least a portion of their ranges in the West. *Syngrapha viridisigma* (Grote)(FW 1.7–1.9 cm; Plate 50.2) is large for a member of its genus, with a light gray FW with darker wavy lines. The stigma is green. The species ranges in boreal coniferous habitats from Alaska east to Labrador and south to the northern conterminous states. In the West, it occurs south to Oregon and, in the Rockies, south to northern New Mexico and the White Mountains of eastern Arizona. The single flight is from mid-July to late August. Larvae feed on the foliage of conifers, especially white spruce *(Picea glauca)*, balsam fir *(Abies balsamea)*, and Douglas-fir *(Pseudotsuga menziesii)*. The larva is green with two pairs of white lines on the back and a pale yellow lateral line on each side. The larval head is yellow brown (save for the green frons) with black flecks. *Syngrapha sackenii* (Grote)(FW 1.5–1.8 cm; Plate 50.3) has a gray brown FW with a black patch at the posterior portion bounded by the narrow white ante- and postmedian lines. The HW is orange yellow with black borders. The species ranges in the southern Rocky Mountains from southern Montana south to central New Mexico. The nocturnal adults have a single flight from mid-July to early August. The early stages are not reported. *Syngrapha ignea* (Grote)(FW 1.3–1.6 cm; Plate 50.4, Plate 64.21 larva) is another species with a black-margined orange yellow HW but is quite different in appearance from *S. sackenii*. It lacks the black FW patch and has an oblique stigma. The ventral surface of the FW is primarily yellow with a black outer margin and, as such, is very similar to *S. alticola* (Walker)(FW 1.0–1.3 cm; Plate 50.5), but *S. alticola* is smaller, with a wavier submarginal line on the FW, and distinctive genitalia. *Syngrapha ignea* has a primarily western North American distribution extending from southern Alaska and northwestern Canada south to the Sierra Nevada of California and in the Rocky Mountains to New Mexico. There is a widely disjunct population in Labrador and Quebec. *Syngrapha alticola* has a similar range to *S. ignea* in the West but is more widely distributed in northern Canada. Both species are primarily diurnal and crepuscular and are most often seen at flowers in alpine or subalpine habitats. *Syngrapha ignea* is attracted to lights nocturnally. There is a single flight from late June to early August. The larval food plant is unreported. The larvae are red brown with the dorsal lines running so close together that they appear fused into a single middorsal stripe. *Syngrapha celsa* (Henry Edwards)(FW 1.5–1.9 cm; Plate 50.6) is a small gray *Syngrapha* that is very closely related to *S. angulidens* (Smith)(FW 1.4–1.6 cm; Plate 50.7); the two species have largely complementary ranges in the montane Canadian and Hudsonian zones of western North America but are sympatric in southern British Columbia and Washington. In addition to genital features, *S. celsa* may be distinguished from *S. angulidens* by having its stigma wider than high and having the satellite spot usually connected. The range of *S. celsa* is from central British Columbia, southern Idaho, Nevada, and the Pacific Coast states to southern California. There are additional records in the White Mountains of Arizona and adjacent New Mexico.

Syngrapha angulidens has a more or less complementary range south through the Rockies to southern New Mexico with a single northern record along the Gulf of Alaska. The adults of both species have single flights, mainly during July and August. The larvae of *S. celsa* are green with three pairs of pale longitudinal lines—the dorsal pairs being white and the lateral being pale yellow—and feed on a variety of needled conifers (Pinaceae) including spruces, *Picea*, true firs, *Abies*, white pine, *Pinus monticola*, and western hemlock, *Tsuga heterophylla*.

Anagrapha falcifera (Kirby), the "celery looper" (FW 1.4–1.7 cm; Plate 50.8, Plate 64.22 larva), is the only member of its genus and has a unique appearance, particularly by the stigma, which has an extension running down to the inner margin of the FW. The species is variable in FW ground color running from mostly brown to gray with restricted brown areas. The variation seems determined, at least in part, by local conditions, with grayer individuals appearing in more arid habitats. The species is often abundant and ranges from southern British Columbia east across Canada, with records extending to northern Labrador. It has been found over most of the conterminous United States but is scarce or absent in California, Nevada, Arizona, and New Mexico. Adults are both diurnal and nocturnal but are primarily seen at lights. The species is very adaptable and invasive. It is often abundant in weedy or cultivated habitats but also appears in natural habitats and, according to records compiled by Slater and Opler, can be found above timberline. In the West, adults have been found from mid-May to September, and there are usually two flights, although the species has three broods in the East, where adults have been found as late as November. Larvae are green with three pairs of wavy white longitudinal lines. The head is brownish green with a short black lateral line running through the stemmata on each side. Host plants are a wide variety of herbaceous plants and sometimes include shrubs. Both native and cultivated plants are selected, and the species has been noted as a sporadically important defoliator of beet, celery, lettuce, and cabbage.

Plusia Ochsenheimer has seven species distributed in the Holarctic. *Plusia putnami* Grote, "Putnam's looper moth" (FW 1.3–1.7 cm; Plate 50.9), is the most widespread of five North American representatives, four of which occur in the West. *Plusia putnami* is Holarctic, ranging across Eurasia and widely in North America from central Alaska east across Canada to Labrador and south, in the West, to northeastern California, Utah, and Colorado. It also occurs in the East south to Wisconsin, Michigan, and Pennsylvania. *Plusia magnimacula* Handfield & Handfield is a closely related recently described northeastern species that has an isolated population in northwestern Nebraska. *Plusia nichollae* (Hampson)(FW 1.4–1.9 cm; Plate 50.10) is endemic to the Pacific Coast from the base of the Aleutians in Alaska south to the San Francisco Bay area, California. The two species have a blotchy golden brown FW with prominent stigmata. The stigma of *P. putnami* appears more triangular as the basal spot appears above the cubital vein, while in *P. nichollae* the basal spot tends to be more oval and does not normally extend above the cubital vein. Both species are nocturnal and are attracted to light. The flight season of *P. putnami* extends from June to early October, and two flights are likely, but only one may occur in more inland montane regions such as Colorado. Similarly, adults of *P. nichollae* are found from late May to early September and may have two broods. Larvae of *P. putnami* feed on a variety of grasses (Poaceae) and sedges (Cyperaceae) as well as bur-reed, *Sparganium* (Sparganiaceae).

SUBFAMILY EUSTROTIINAE

Tripudia Grote has 42 described species limited to the New World; seven are described from North America, and six of these are western. *Tripudia balteata* (Smith)(FW 0.7–0.8 cm; Plate 50.15) ranges from southern California east to western Texas. In southern California, adults have a single flight from March to July, but in adjacent Baja California Norte, Mustelin found them during December and February. The life history is unreported.

Cobubatha Walker comprises 17 New World species, with seven described from North America, all of which are western or nearly so. *Cobubatha dividua* (Grote)(FW 0.8–0.9 cm; Plate 50.14) ranges from southern California east to western Texas. Knudson and Bordelon report that in western Texas the adults are found from March to October, while, in southern California, Mustelin points out that the species is found from January to November. This span likely represents two or more generations. The life history was described by J. A. Comstock and Dammers who found that the larvae feed on *Justicia (Beloperone) californica* (Acanthaceae). The general color of the full-grown caterpillars varies from green to red mauve above with green below to completely red mauve. There is a broad white supraspiracular line. The body is covered with minute white points. Each segment bears six white large raised points on each side of the middorsal line; these are arranged in alternating diagonals. Each of the points is topped with black and bears a short white seta. The prolegs on the fourth abdominal segment are much reduced and almost nonfunctional. The head is olive green and spotted with brown.

Amyna octo Guenée, the "eight spot" (FW 1.0–1.1 cm; Plate 50.11), is a tropical species that is found from southeastern Arizona, Texas, and Missouri east to southern Ontario and south to Florida. In western Texas, the adults are found in May and from September to November. The adults straggle north from August to October. The larvae feed on *Chenopodium* (Chenopodiaceae), but reports of larval usage of Amaranthaceae, Sapindaceae, and Ulmaceae require confirmation.

SUBFAMILY BAGISARINAE

The larvae are all semiloopers with vestigial or missing prolegs on the third and fourth abdominal segments. Larvae of all taxa feed on Malvaceae and other malvalean families.

Bagisara buxea (Grote)(FW 1.3–1.4 cm; Plate 50.12) is found from central California and southern California west to western Texas and then east through the southern United States. The moth is also found in the West Indies. In California, the moth is in flight from April to early September probably as two flights, while Knudson and Bordelon report the adults were found from April to October in western Texas. Larvae feed on *Sphaeralcea* (Malvaceae). The larvae are green mottled with white and have a diffuse subspiracular band.

SUBFAMILY ACONTIINAE

The Acontiinae are characterized by Kitching and Rawlins as having an enlarged, heavily sclerotized alula overlying the tympanum and a reduced countertympanal hood. The male genitalia are often asymmetrical, and the adults frequently have the appearance of bird-droppings. In many cases, the larvae are obligate feeders on plants in the Malvales or Asteraceae. None of the species is of economic importance.

TRIBE ACONTIINI

Tarachidia Hampson has 33 New World species with 20 of these described from North America. At least 12 species are found in or described from the West. *Tarachidia semiflava* (Guenée), the "half-yellow" (FW 1.0–1.2 cm; Plate 50.16), ranges from British Columbia south to Arizona, Colorado, and Texas and thence east to Manitoba, New Jersey and Florida. The moths fly from April to September, likely in two broods. The larvae are reported to feed on pitcher-plant, *Sarracenia* (Sarraceniaceae), but this report no doubt results due to confusion with another moth, *Exyra semicrocea*, which is a pitcher-plant feeder. *Tarachidia candefacta* (Hübner), the "olive-shaded bird-dropping moth" (FW 1.0–1.2 cm; Plate 50.17, Plate 64.23 larva), ranges from Montana and California eastward through much of the United States, southern Canada, and northern Mexico. Knudson and Bordelon report the adults fly from April to September in western Texas. The larval host is reported to be ragweed, *Ambrosia artemisiifolia* var. *elatior* (Asteraceae), by Crumb; the larvae are also reported to feed on *Arctium* and *Aster* (Asteraceae). Wagner reports the larvae to feed until the first frosts and that the species overwinters in the pupal stage. The larvae are green with a distinct middorsal hump on the eighth abdominal segment. The body and head have many fine white longitudinal striations, and there is also a more prominent white spiracular line. The larvae are semiloopers as they lack prolegs on the fourth and fifth abdominal segments. *Tarachidia libedis* (Smith)(FW 0.9–1.1 cm; Plate 50.18) ranges from Colorado south to southern Arizona, east to western Texas, and farther south into northern Mexico. This species is in flight from April to September in western Texas. The larval host is *Iva ambrosiaefolia* (Asteraceae). *Tarachidia tortricina* (Zeller)(FW 0.8–1.0 cm; Plate 50.19, Plate 50.20) varies from pale yellow to dark gray. The species ranges from British Columbia south to southern California and east to Ontario, Illinois, and Texas. This small noctuid is in flight from June to September. The life history is unreported.

Conochares acutus Smith (FW 0.9–1.2 cm; Plate 50.21) ranges from central California south to Baja California and east to Arizona and northwestern mainland Mexico. This moth is in flight from February to early November. The larvae feed on *Ambrosia* (Asteraceae).

Therasea augustipennis (Grote)(FW 1.1–1.3 cm; Plate 50.22) is one of the more abundant moths on the high plains, especially in disturbed habitats. The moth has been found from British Columbia east to Colorado and south to Arizona and western Texas. There are several flights from as early as April to as late as October. McFarland found that a related species feeds on *Sida* (Malvaceae) in southern Arizona, and it is possible that *T. augustipennis* feeds on *Sphaeralcea*, which is often found in its habitat.

Ponometia sutrix (Grote)(FW 0.8–1.0 cm; Plate 50.24) is found from California east to southern Arizona and western Texas. The adults fly from April to September with at least two broods.

Spragueia magnifica Grote (FW 0.7–0.8 cm; Plate 50.23) is found from southern California through southern Arizona to western Texas. The moths are attracted to light from May to October. It is surmised that there at least two and possibly three generations. Related species feed on Malvaceae.

Acontia Ochsenheimer has about 175 species with a primarily Pantropical distribution; 32 species are found in North America, and about 25 of these are found in the West. *Acontia aprica* (Hübner), the "exposed bird-dropping moth" (FW 1.2–1.3 cm; Plate 50.25), is found from Colorado and western

Texas east to Illinois and south to central Mexico and Florida. There have been strays north to Montreal, Quebec. There are two broods from April to September. The larval host is hollyhock, *Alcea rosea* (Malvaceae), and almost certainly native Malvaceae as well. The full-grown larva is green with a dark middorsal strip broken by white spots on the anterior abdominal segments. There is a bold white spiracular stripe bordered above and below by black on the first four abdominal segments. The cervical and anal shields and spiracles are black. The head is white with black reticulation. *Acontia major* Smith (FW 1.3–1.5 cm; Plate 50.26) ranges from British Columbia east to Utah and Colorado and then south to southern California and Baja California Norte. The adults are nocturnal and are attracted to light. Richers summarizes data that show that adults have been found from late March to early August in California. The larvae feed on *Sphaeralcea ambigua* (Malvaceae) and *Fremontodendron californicum* (Sterculiaceae). *Acontia areli* Strecker (FW 1.1–1.2 cm; Plate 50.27) has been found from British Columbia south through the arid interior to southern California, Baja California, and northern mainland Mexico east to Arizona, Colorado, and western Texas. In California, adults are in flight from mid-June to late August (one flight). The species is reported from April to October in western Texas. McFarland found that the larvae feed on *Sphaeralcea* (Malvaceae) in southeastern Arizona. Reports of larval feeding on mesquite, *Prosopis* (Mimosaceae), are almost certainly in error.

Pseudalypia crotchii Henry Edwards (FW 0.9–1.2 cm; Plate 50.28) is endemic to California west of the Sierra Nevada crest and Transverse Ranges from Amador and San Benito counties south to San Diego County, mainly in low-lying arid foothills, deserts, and arid areas. In San Diego County, Mustelin has found the species at 5,800' in the Laguna Mountains. There is a single spring flight of this diurnal moth from early February to May—rarely to early June in coastal habitats. Adults are often found resting in mallow flowers. The life history was described by J. A. Comstock who found that the larvae ate *Eremalche exilis* and *Malva parviflora* (Malvaceae). The larvae are yellow with a dark middorsal line and dark longitudinal crenulations throughout. There are dark subdorsal spots on the first three segments. There is a yellow spiracular line with extensions down each proleg. The spiracles are yellow each with a brown rim. The head is yellow with sparse dark reticulation.

SUBFAMILY PANTHEINAE

The Pantheinae were previously classified as quadrifine noctuids with hairy eyes, an untenable assignment, according to Kitching and Rawlins who treated them as a separate family. Although these moths lack some typical noctuid features including a cylindrical galeal lobe in larvae and features of the tympana and genital musculature, Lafontaine and Fibiger include them in their concept of the family. The adults are nocturnal and are attracted to light. First-instar larvae have secondary setae and an enlarged prothoracic prespiracular verruca, characters otherwise found in the Lymantriinae. Larval foods primarily consist of the foliage of woody trees including Pinaceae, Podocarpaceae, Fagaceae, Salicaceae, Aceraceae, and Rosaceae. The pupae of Pantheinae are in dense cocoons and have well-developed often multisetose cremasters. The prothoracic femora are concealed in the pupae of most pantheines. The family is Holarctic, and its exact composition remains uncertain but does include at least *Panthea, Colocasia, Pseudopanthea, Lichnoptera, Charadra,* and *Meleneta.* In the West, we have 13 species in five genera.

Panthea Hübner has 11 species found across the Holarctic region; seven are limited to North America, with five of these found in the West. *Panthea virginarius portlandia* Grote (FW 1.6–2.0 cm; Plate 50.29) is found in coniferous forests from British Columbia east to New England and south to central coastal California. In California, the moth flies from May to September; although newly emerged individuals appear over a long period, there is believed to be just one protracted generation, although in coastal California there seem to be two flights (April to June and August to September). There are occasional melanic individuals of this species as well as other *Panthea* species. McFarland reported a captive female that laid 150 glossy yellow eggs in elongate masses glued to Douglas-fir needles. Larvae feed on the mature foliage of several conifers such as Douglas-fir, *Pseudotsuga menziesii* (Pinaceae). The full-grown larva is black with red orange verrucae, each bearing a splayed tuft of light brown hairlike setae. The head is black.

Pseudopanthea palata (Grote)(FW 1.4–1.7 cm; Plate 50.43) is found from Arizona and Utah east to southern Colorado and western Texas. The moths are nocturnal and are attracted to light. In western Texas they fly from March to September, and there are likely two broods. The life history details are unreported.

Charadra deridens (Guenée), "the laugher" (FW 1.7–2.0 cm; Plate 50.44), ranges from British Columbia east across southern Canada to Quebec and Nova Scotia and south to Colorado, Texas, and, in the East, Florida. The larvae feed on *Quercus* (Fagaceae), *Betula* (Betulaceae), and *Ulmus* (Ulmaceae), according to Crumb. The full-grown larvae are pale gray green without markings. Verrucae are distinct and each bears a cluster of fine, long whitish hairlike setae. The spiracles are white. The head is usually black but sometimes whitish tinged with green. A related species of the Southwest, *C. pata* (Druce), has a dark yellow HW base.

SUBFAMILY DIPHTHERINAE

Diphthera festiva (Hübner), the "hieroglyphic moth" (FW 1.8–2.1 cm; Plate 50.46), is the only North American representative of its genus; the moth ranges from southern Arizona and southern Texas south through the lower elevations of tropical America to at least Brazil. The adults are nocturnal and are attracted to light. They fly through much of the year in tropical environments. Larval hosts include *Waltheria indica* and *Morongia* (Sterculiaceae). The larvae have bright red orange heads and yellow bodies with black bands.

SUBFAMILY RAPHIINAE

The subfamily Raphiinae, as presently considered, comprises the genus *Raphia* Hübner, which has 12 species worldwide. All except one native to Nigeria occur in the Holarctic region. Seven species occur in North America, and five occur in the West. These moths have been considered as members of the Pantheinae or Acronictinae by other authors. The subfamily is distinguishable as the larvae lack secondary setae but have two subventral setae on the first abdominal segment, a feature shared with Acronictinae and trifid noctuids. Seta D1 on the second thoracic segment is borne subapically on a conical process. The larval prolegs have enlarged plantae with many crochets. The larvae feed primarily on willows, poplars, and relatives (Salicaceae), but alders and birches (Betulaceae) are also known hosts for at least one species. Eggs are laid singly. The pupae are rugulose with two stout spread spines on the cremaster.

Raphia frater Grote , "the brother" or "yellow-marked caterpillar" (FW 1.4–1.9 cm; Plate 50.30), ranges from British Columbia east to Labrador and, in the West, south to central California, Idaho, and Colorado. The adults are nocturnal and have a single flight from April to August. The larvae feed from June to September on several species of *Alnus* and *Betula* (Betulaceae) as well as *Populus* and *Salix* (Salicaceae). The caterpillars are bright translucent green with thin yellow dorsal crescent moon marks on the first, fifth, and eighth abdominal segments. There are short red tubercles on the dorsum of the second thoracic segment, and the body is dappled with transverse rows of small yellow spots. The larvae rest flatly appressed to their host leaf, and, as noted by McFarland, when disturbed will rapidly vibrate the fore part of the body in all directions. The larvae pupate by cutting a depression in bark and cementing the resultant chips in their tough waterproof cocoons. Other western *Raphia* species, more or less similar, include **R. cinderella** Smith (California west of the Sierra Nevada), **R. coloradensis** Putnam-Cramer (British Columbia east to Alberta and south to northern California [Alpine County], Idaho, and Colorado), **R. elbea** Smith (southern California east to southeastern Arizona), and **R. pallula** Edwards (northern California east to southeastern Arizona).

SUBFAMILY ACRONICTINAE

The Acronictinae is largely restricted to the genus *Acronicta*, known as the "dagger moths," and its close relatives. In addition to *Acronicta* species there are also five *Merolonche* species, two *Simyra*, and three *Agriopodes*, of which four *Merolonche*, both *Simyra*, and one of the *Agriopodes* are western. The adults of all species are nocturnal and are attracted to light. The larvae of *Acronicta* are variously specialized on plants within the Fagaceae, Betulaceae, Ulmaceae, Rosaceae and Salicaceae, while a few of the species are polyphagous.

Acronicta Ochsenheimer comprises roughly 175 species that are found mainly in the Holarctic region, with a few ranging farther to the Indo-Pacific and a few occur south into Mexico. There are 76 species in North America, and 36 of these occur in the West. *Acronicta dactylina* Grote, the "alder dagger moth" (FW 2.1–2.4 cm; Plate 50.31, Plate 64.24 larva), ranges from southwestern Canada east to Newfoundland, and thence, in the West, south to northern California (Humboldt and Plumas counties). There is a single adult flight from May to August. Favored hosts include birches and alders (Betulaceae) as well as aspen, poplars, and willows (Salicaceae). The larva is much like that of *A. hesperida* Smith (see below), another alder feeder, except that there are middorsal tufts of long black hairs on the first, third, and eighth abdominal segments.

Acronicta funeralis Grote & Robinson, the "funerary dagger moth" (FW 1.7–1.9 cm; Plate 50.32), has a mottled gray FW with a costal black mark and a black band along the inner margin. This moth is found from British Columbia east to Nova Scotia. In the West, it is found south through Oregon to central California (Monterey, Tulare, and Mono counties). In the East, the adults appear in two flights from May to August, but there seems to be only a single flight in the West. During August the larvae feed on a variety of woody plants including blueberry, *Vaccinium* (Ericaceae), maple, *Acer* (Aceraceae), hazel, *Corylus* (Betulaceae), and Oregon white oak, *Quercus garryana* (Fagaceae). The full-grown larvae are black with pale yellow patches along the dorsum. There are a few scattered long black setae with thickened flat tips. *Acronicta grisea* Walker, the "gray dagger moth" (FW 1.7–1.8 cm; Plate 50.33), has a

mottled gray FW with thin basal and anal marginal dashes together with an irregular thin black postmedian line. The species ranges from southwestern Canada east to Newfoundland and the Maritimes south in the East to Minnesota and Connecticut. In the West, it occurs south to California (Plumas County). The adults have a single flight from April to August. According to Prentice and other authors the larvae feed primarily on birches, *Betula*, and alders, *Alnus* (Betulaceae), as well as willows, *Salix* (Salicaceae), from early July to early October but especially during August and September. Rarely, other broadleaf trees may be fed upon. The larvae have mottled brown heads and are green with an irregular longitudinal brown or purplish patch running along the dorsum. *Acronicta hesperida* Smith (FW 2.1–2.4 cm; Plate 50.34) is gray with faint mottling. The moth ranges from western Washington south to central California (El Dorado and Marin counties). The reniform and orbicular spots are narrowly outlined with black. There is a single flight during July. The larvae feed on alders (Betulaceae) during August and September. The larvae have the dorsum covered with orange or golden brown hairs of medium length with white hairs laterally. There are sparse long white hairs anteriorly and posteriorly. The head is shiny black. Pupation is is in a tough silken cocoon on tree boles or under logs. *Acronicta impleta* Walker, the "yellow-haired dagger moth" (FW 1.6–2.2 cm; Plate 50.35, Plate 64.25 larva), has a dark gray mottled FW without obvious black dashes ("daggers"). The species ranges widely in both eastern North America and the West from southwestern Canada south to southern California (Los Angeles County), Louisiana, and Florida. In the East, according to Covell, there are two flights from April to August, and in the West from March to August. Larvae feed on a wide variety of woody broad-leaved trees, but especially alders and birches (Betulaceae); elms (Ulmaceae); and poplars and willows (Salicaceae) during the spring. Larvae are black with sparse tufts of tan setae emanating from rows of red tubercles and yellow tan hair tufts as well as a few black brushlike tussocks of long setae. *Acronicta impressa* Walker (FW 1.5–1.6 cm; not illustrated) has a mottled gray FW with a dark outlined discal spot. The HW is gray. The species ranges from Yukon Territory and British Columbia east to Newfoundland and Nova Scotia, and through western North America to California (Mendocino and Placer counties). The adults fly from May to August. The larvae feed on wild roses, *Rosa*, and bitterbrush, *Purshia tridentata*, both Rosaceae, during July and August. The larvae are black with tufts of long white hairs emanating from longitudinal rows of orange scoli. There are two dense tufts of gray hairs on the dorsum of the thorax and two dense tufts of longer gray hairs on the dorsum of the eighth abdominal segment. The thoracic tufts are partially encircled by tufts of shorter white hairs. *Acronicta innotata* Guenée, the "birch dagger moth" (FW 1.9–2.1 cm; Plate 50.36), is found from central British Columbia east to Nova Scotia and south to Oregon. The larvae feed primarily on birch, but to a lesser degree on alder, willow, and aspen. During the Forest Lepidoptera Survey in Canada, larvae were found from June to mid-September. The full-grown larvae have purplish black heads with two orange spots on the vertex. Their bodies are dark purplish with tiny white flecks and gray lateral spots. In addition, there are ventrolateral tufts of white hairs. *Acronicta lepusculina* Guenée, the "cottonwood dagger moth" (FW 1.8–2.0 cm; Plate 50.37), has the FW with basal and subanal black dashes. The species ranges widely from British Columbia east across southern Canada to Newfoundland. In the West, it is found south to California (El Dorado and Santa Clara counties) and Colorado. The adults have been found in California during June and July. According to Prentice, the larvae are found mainly from June to September when they feed on aspen, cottonwoods, willows (Salicaceae), and, rarely, birches (Betulaceae). The larva has a black head, sometimes marked with brown, and a pale or black body covered with long yellow hairs. There are pencillike middorsal tufts of long black hairs on most abdominal segments. *Acronicta mansueta* Smith (FW 1.3–1.5 cm; Plate 50.38) ranges from British Columbia east to Saskatchewan and south to southern California (San Bernardino Mountains), Oregon, and Idaho. According to information compiled by Richers, adults were collected in California during June and August. The host and life history details are unreported. *Acronicta marmorata* Smith (FW 1.6–1.9 cm; Plate 50.39) has the FW strongly mottled gray, black, and white with black basal and subanal dashes. A melanic form is frequent in western Oregon. The moth ranges from western British Columbia south to southern California (San Diego County). The adults fly from April to September. The larvae feed on oaks, including *Quercus garryana* (Fagaceae), during July and August. The sparsely hairy larvae are variably light brown with yellow and white spots and blotches to light brown with faint green blotches. There is a brown middorsal line, and the head is brown. *Acronicta perdita* Grote (FW 1.9–2.1 cm; Plate 50.40) has a gray FW that is heavily suffused with black. This species ranges from British Columbia south along the Pacific Coast to southern California (San Bernardino and Ventura counties). Adults are found from April to mid-July. In the mountains of southern California, the larvae feed on bitterbrush, *Purshia tridentata*, and toyon, *Heteromeles arbutifolia* (both Rosaceae), as well as various species of *Ceanothus* (Rhamnaceae) including snowbrush, *Ceanothus velutinus*, and buckbrush, *Ceanothus cuneatus*, during July and August. The larvae are brownish black and have tufts of black hairs emanating from pale orange scoli. There is a prominent white longitudinal lateral band along each side that narrows before and after each spiracle. Pupation is in a cocoon that combines its setae and host leaves with silk. *Acronicta quadrata* Grote (FW 1.6–1.9 cm; Plate 50.41) ranges from British Columbia east across southern Canada to Quebec. The larvae feed on pin cherry, choke-cherry, and Saskatoon (Rosaceae). The larva has a red head, and the black body is sparsely covered with tufts of white hairs. There is a middorsal black band containing an hourglass-shaped red mark on each abdominal segment and a prominent white subdorsal band to each side. It is worth mentioning that *A. othello* Smith is the most abundant member of the genus in southern California, although its life history is unreported, while the recently described *A. browni* Mustelin & Leuschner is a southern California endemic.

Merolonche lupini (Grote)(FW 1.5–1.7 cm; Plate 50.42) ranges from British Columbia and Alberta south to central California, Utah, and Colorado. Adults have a single flight from May to July. Larvae feed on various low plants including *Valerianella* (Valerianaceae) and mountain-ash, *Sorbus* (Rosaceae). The larva is black with brownish areas, and the head is black.

Simyra insularis (Herrich-Schäffer), "Henry's marsh moth," formerly called *S. henrici* (Grote), (FW 1.7–2.0 cm; Plate 50.45), occurs from southeastern Oregon (Harney County) south to at least western Texas and southern California, thence east to Nova Scotia and Maine. The adults are nocturnal and fly from March to October, likely in several broods. The larvae feed on various low plants, especially grasses (Poaceae), in wet places, but McFarland found that they also feed on smartweed, *Polygonum* (Polygonaceae), and cat-tail, *Typha* (Typhaceae), in southeastern Arizona. The tough silk cocoon is often formed in a rounded triangular area in a broad folded grass or sedge blade.

Agriopodes Hampson is represented in North America by three species, but only *A. tybo* Barnes (FW 1.5–1.7 cm; Plate 50.47) is found in western North America. It is pale blue green with various fine wavy thin black lines and some small patches of white scaling. The reniform and orbicular spots are both outlined in black. The moths are reminiscent of *Feralia* species but are unrelated. It is found in southeastern Arizona during August. The adults are nocturnal and readily attracted to light. Its life history is unreported, but larvae of the related eastern *A. fallax* (Herrich-Schäffer) feed on various species of *Viburnum* (Caprifoliaceae). J. A. Comstock described and illustrated the larva and pupa. The last instar is mottled gray green and brownish gray; its surface is overlain with black dots, blotches, and discontinuous wavy lines. The integument is roughened with numerous folds, bumps, and papillae.

SUBFAMILY AGARISTINAE

The Agaristinae, although treated by some past workers as comprising a separate family, are indeed only a distinctive subfamily of Noctuidae. The countertympanal membrane is greatly enlarged and associated with reduced size of the countertympanal hood. There is a paired vesicle at the base of the abdomen. The adults of agaristines, often but not always diurnal, are usually brightly colored, strikingly patterned, or both. The antennae are clavate, and McFarland reports that males of at least some genera patrol for receptive females along ridge tops. We have observed males of *Androloma* awaiting receptive females in gully bottoms, while Scott has observed males of several species patrolling hilltops or ridges for receptive females. The larvae of many species are brightly colored. Larval hosts are often among the Onagraceae or Vitaceae, but some western species feed on Cactaceae and Portulacaceae. Larvae have a bisetose group of setae on the seventh through ninth abdominal segments. Pupation usually occurs in a chamber carved out within soft or rotting wood. The pupa has a truncate cremaster and is densely spinulose. In the West, there are 15 species in eight genera.

Gerra sevorsa (Grote)(FW 1.5–1.9 cm; Plate 51.1) ranges from southern Arizona south to Costa Rica. The moths are strictly nocturnal and are attracted to light. There is a single annual flight during August and September. The pale larvae feed on both Arizona wild grape, *Vitis arizonica*, and Virginia creeper, *Parthenocissus* (both Vitaceae), according to McFarland.

Euscirrhopterus gloveri Grote & Robinson, "Glover's purslane moth" (FW 1.5–2.0 cm; Plate 51.2), ranges from southern Arizona and western Texas south to western Mexico (Sonora and Sinaloa). There is a single annual flight during August. The attractive adults are nocturnal and come readily to light. The larvae are reported to feed on *Beta vulgaris* (Chenopodiaceae), *Talinum*, *Portulaca grandiflora*, and *P. oleracea* (Portulacaceae). The full-grown larva is gray white with a slight middorsal hump on the eighth abdominal segment. On the middle of each segment there is an incomplete black transverse band and a coalesced irregular red orange middorsal patch. Additionally, there is a very thin transverse black line at the anterior of each segment. The head and cervical shield are shiny red orange with black spots mostly bearing white hairlike setae. The related *E. cosyra* (Druce), the "staghorn cholla moth" (FW 1.7–1.9 cm; Plate 51.3), found in southern Arizona, is unusual in its larval specialization on Cactaceae.

Its caterpillars feed externally on the fleshy canes of various chollas *(Opuntia)*.

Alypiodes bimaculata (Herrich-Schäffer)(FW 1.7–2.1 cm; Plate 51.4) ranges from southern Arizona, southern New Mexico, and western Texas south through much of Mexico. The adults are diurnal and fly during July and August; they have a definite predilection for patrolling ridges. They visit flowers such as *Melilotus officinalis*, white sweet clover. The larvae feed on *Boerhavia*, Nyctaginaceae. The larvae are velvety black with a few small orange subdorsal spots. The head is black with orange bands posteriorly.

Alypia Hübner comprises 12 New World species, two of which are endemic to Argentina, with the remainder found in North America. *Alypia octomaculata* (Fabricius), the "eightspotted forester" (FW 1.4–1.8 cm; Plate 51.5, Plate 64.26 larva), ranges from Idaho east to Quebec and south to central California, Colorado, Missouri, and Florida. Western populations are the subspecies **matuta** Henry Edwards, which has a much larger basal HW white patch. Adults are diurnal and fly rapidly about host plants and nectar sources. There is only a single flight in most of the West, but Covell reports a second brood in the Southeast. Adults are often seen nectaring at flowers of various plants such as dogbanes, *Apocynum*, honeysuckles, *Lonicera*, and snowberry, *Symphoricarpos* (all Caprifoliaceae). Larvae feed on various Vitaceae including *Ampelopsis*, *Parthenocissus*, and *Vitis* and sometimes are abundant enough to be considered pests. Reports of larval feeding on Berberidaceae and Rosaceae are almost certainly erroneous. The larvae are white with many transverse black lines and several transverse orange bands. There is a series of raised black verrucae, each bearing a long white hairlike seta, on each segment. Those bands on the fourth to fifth abdominal segments are incomplete. The eighth abdominal segment has a middorsal hump. The head as well as the cervical and anal shields are orange with several seta-bearing black spots. There are five other western *Alypia*; three are most likely to be met with in montane habitats. *Alypia langtoni* Couper (FW 1.0–1.3 cm; Plate 51.6) is found south to the central Sierra Nevada of California during June and July; *A. mariposa* Grote & Robinson (FW 1.0–1.3 cm; Plate 51.8) is found in the California coast ranges and southern Sierra Nevada foothills during April and May; we found larvae feeding on *Clarkia unguiculata* (Onagraceaae); and *A. ridingsi* Grote (FW 1.0–1.3 cm; Plate 51.9) range south in the California coast ranges and west slope of the Sierra Nevada to San Diego County and, discontinuously, to southern Arizona. The adults are univoltine and fly from mid-March to early July. Comstock and Dammers give a thorough description of the life history stages. The larvae of some or most of these species feed on *Oenothera* (Onagraceae), according to McFarland.

Androloma maccullochii (Kirby)(FW 1.0–1.3 cm; Plate 51.10) is a boreal species that ranges from Alaska, western Canada, and east across the forested areas of Canada to Labrador and Quebec. In the West, the species occurs south to California, Idaho, and Colorado. Adults are smaller than similar *Alypia* species and have three spots on the FW with black veins crossing all spots. Males have the costal area of the FW modified as a stridulating structure. There is a single annual flight from May to August. Larvae feed on fireweed *Chamerion angustifolium* or willow-herb, *Epilobium* (Onagraceae). The full-grown larvae are white with a slight middorsal hump on the eighth abdominal segment. There is a faint white middorsal line bordered by black and much fine black vermiculation, and also a single subdorsal orange spot on each side of the eighth abdominal segment. The spiracles

are black. The head is white with black spots, most of which bear setae.

SUBFAMILY CUCULLINAE

The subfamily Cuculliinae, as redefined by Poole, is restricted to the genus *Cucullia* Schrank and a few genera—*Dolocucullia* Poole, *Opsigalea* Hampson, and *Emarriana* Benjamin—that possess similar morphological features. The genus **Cucullia** Schrank comprises roughly 250 species found in the Northern Hemisphere with a few also found in Africa. Although about 40 species are found in North America, with most (more than 36) in the West, the greatest taxonomic diversity is in the Palaearctic region. All species have a long, extended, and somewhat pointed FW apex. The patagia are capable of being raised into a pointed hat-shaped hood. The male genitalia have one to three long spines on the vesica of the aedeagus, each borne on a diverticulum and, except for a few Palaearctic species, the female genitalia by the emergence of the ductus seminalis from the anterior end of the bursa. The species of North American **Cucullia** Schrank are found from southern Canada, the conterminous United States, Mexico, and Central America. Poole mentions that some species of *Dolocucullia* and *Opsigalea* occur in South America. Adults of all species are nocturnal and are attracted to light. Poole found pollen on the proboscises of many individuals and suggested that the species are frequent flower visitors and may be pollinators. The larvae of all known North American species feed on the flowers, seeds, and leaves of composites (Asteraceae), although one group of Palaearctic species specializes on mints (Lamiaceae). The larvae are often brightly colored and pupate in earthen cells. Adults are long-lived and most have single, long flights in the summer. About 250 species of *Cucullia* are found worldwide, and these are distributed in most of the biogeographic regions. Forty species are recorded from North America, and most of these occur in the West. **Cucullia montanae** Grote (FW 1.9–2.3 cm; Plate 51.11) has a gray FW with brown smudges. The FW costa is black. The reniform and orbicular spots of *C. montanae* are distinct. The HW is diffusely bordered with dark gray. The species ranges from southern British Columbia to Manitoba south, mainly through the intermountain West and Great Plains to southeastern California, Arizona, New Mexico, and western Texas. There is a single flight from June to August. The larvae feed on gumplant *(Grindelia)* and rubber rabbitbrush *(Ericameria nauseosa)*, both in the Asteraceae. The larva is described by Crumb as green or yellow green with dark brown to black longitudinal lines. The pair of lines above and below the spiracles is filled with black at the point of the spiracles. The head is green with brown speckles. **Cucullia lilacina** Schaus (FW 1.8–2.0 cm; Plate 51.12) may be identified by the violaceous color of the FW combined with the distinct brown patch along the FW costa extending from the reniform mark to the outer margin. The species ranges from Arizona east through New Mexico to western Texas and from there south through Mexico to Costa Rica. Adults have a single flight period from June to October in the U.S. portion of the species' range. Larvae feed on several species of fleabanes, *Erigeron* (Asteraceae), in southeastern Arizona. The larvae are pale violet or pink with a black transverse band at each thoracic and abdominal segment. The anterior margin of each segment has small black bands and the prolegs are solid black. The eighth abdominal segment is humped dorsally. There is a hint of a dorsal orange longitudinal band. **Cucullia antipoda** Strecker (FW 1.5–1.8 cm; Plate 51.13) is extremely variable in adult appearance. Poole defines two primary morphs. One morph has a more or less broad, uniform pale gray FW combined with a light brown HW. The second morph has a darker and narrower FW with a distinct blue tint. There is a distinct contrast between the lighter HW base and its outer margin. The species has a broad range in the western interior from British Columbia east to Manitoba and south to southern California, central Arizona, and southern New Mexico. The species has been found from May to August in most of the range, but from April to June in southern California. The larval host is unreported but is likely a shrubby composite. *Cucullia luna* Morrison (FW 1.7–2.0 cm; Plate 51.14) cannot be confused with any other cuculline. Its FW is a satiny white punctuated with several black dots. The HW is less lustrous but is still pure white. It is most likely to be mistaken for all-white species of *Copablepharon*, all of which lack the black dots and the pointed FW. The species ranges from the Rocky Mountains and plains beginning with southern Alberta east to southern Manitoba and south to northern Arizona and central Colorado. The species is found in dunes and sandy grassland habitats. The adults have a single flight during June and July. The early stages are unreported. *Cucullia dammersi* (McDunnough)(FW 1.9–2.1 cm; Plate 51.15) is a member of the *Cucullia pulla* Species Group, and, as such, has a late-winter to early-spring flight period that overlaps the beginning of the calendar year. The moths are distinguished by their dark brown FW with scattered smudges of rusty brown in the subterminal area, not black as in *C. pulla* (Grote). The FW costa is paler than the rest of the wing. This is a species of California's Mojave Desert. The early stages were described by Comstock and Dammers. The larval hosts were reported as several species of shrubby Asteraceae including *Acamptopappus sphaerocephalus*, *Ericameria linearifolia*, *E. palmeri*, *E. pinifolia*, and *Gutierrezia sarothrae*. The larva is described as blue gray with a complex pattern of black patches. There is an interrupted dorsal yellow line with two yellow patches per abdominal segment and a yellow band below the spiracles. The prolegs are green each with a central black spot. The head is blue gray with numerous darker blue gray patches.

SUBFAMILY ONCOCNEMIDINAE

Pleromelloida cinerea (Smith)(FW 1.3–1.4 cm; Plate 51.16) ranges from British Columbia south to southern California. The adults are found from late August to April, but one should be aware that there are several very similar species. The larvae have been reported by McFarland to feed on the leaves of *Lonicera subspicata* and *Symphoricarpos* (Caprifoliaceae).

Catabena lineolata Walker, the "fine-lined sallow" (FW 1.1–1.4 cm; Plate 51.17), ranges from British Columbia east to Quebec and south to California, Colorado, and, eastward, to Texas and Tennessee. In California, there is single flight from March to early August with occasional records until November. The larvae feed during spring and summer on several species of *Verbena* (Verbenaceae) and have also been reported to eat *Solidago* (Asteraceae). The larvae are greenish with intermingled dark flecks and lines resulting in an overall dark gray green appearance. The spiracles are white. The head is white or greenish white with longitudinal reddish or brownish stripes giving the appearance of three white lines dorsal to the ocelli on each side.

Homohadena fifia Dyar (FW 1.1–1.5 cm; Plate 51.19) ranges from British Columbia to California, Utah, and Colorado. The adults have a single flight from late June to early

August. The larvae feed on snowberry, *Symphoricarpos* (Caprifoliaceae). The larvae are dusky brown and are broad in the middle and taper posteriorly but have a pronounced dorsal hump on the eighth abdominal segment. There is a subdued pale middorsal stripe that includes a dark line. There is a supraspiracular white stripe edged by thin black lines, and a spiracular line comprised of blackish reticulation. There is also a narrow white subspiracular stripe enclosing a median lavender line. There is a purplish stripe just above the prolegs that encloses a thin yellow line. The venter is purplish and is reticulated with white. The spiracles are pale yellow. The head is coarsely granulose and white with scattered black and brown splotches. The adfrontal areas are black margined with white that forms an inverted V shape.

Adita chionanthi (J. E. Smith), the "fringe-tree sallow" (FW 1.5–1.7 cm; Plate 51.20, Plate 64.27 larva), ranges from British Columbia east across temperate Canada to New Brunswick south to Montana, Colorado, Kansas, and Georgia. There is one flight each year from late August to early October. The larvae feed during June and July on *Triosteum* (Caprifoliaceae) as well as fringe-tree, *Chionanthus*, and ashes, *Fraxinus* (Oleaceae). The larvae are variably colored gray, green, or red, are characteristically striped, and have a somewhat humped eighth abdominal segment. The pale middorsal stripe is broad and broken into spots on the anterior abdominal, and then whitened on the seventh and eighth abdominal segments. The remainder of the larva has alternating black, dark, and pale stripes and lines. There are purplish spots between the prolegs. The spiracles are brown. The head is granulose and is brown with black spotting; it also has a continuation of the white lateral body stripe.

Homoncocnemis fortis (Grote)(FW 1.6–1.7 cm; Plate 51.21) is found in Oregon and California east to Arizona and Colorado. In southeastern Arizona, the adults have a single flight from late August to early September but in California are found from January to early April, according to Richers and Powell, with a late record during October. The larvae feed on *Keckiella antirrhinoides* (Scrophulariaceae) and *Fraxinus latifolia* (Oleaceae). The larvae have a slight dorsal hump on the eighth abdominal segment and are black with a number of yellow and white lines and spots. The spiracles are gray. The head is coarsely granulose and is bluish white marked with black freckles and stripes.

Oncocnemis Lederer is a large Holarctic group (99 described species) whose members are characterized by the presence of a long, curved apical spine on the foretibiae. The adults present a diverse array of wing patterns and colors, and the larvae may or may not have a hump on the dorsum of the seventh and eighth abdominal segments. Larval hosts are found in a diverse array of perennial plants in broad-leaved dicot families. Each species seems to be limited in its host choice to plants in a single family. A few have been reported to feed on a short list of diverse plants, but some of these reports may be in error. Favored host plants are in the Ericaceae, Scrophulariaceae, and Rosaceae. More than 90 North American species are listed by Franclemont and Todd, and several additional species have been described. Troubridge is revising the group and indicates that many species remain to be described. There are roughly 75 described western species with about 50 awaiting description. Adults are nocturnal and are attracted to light, but some species are diurnal and may be found in the afternoon nectaring at flowers such as rubber rabbitbrush, *Ericameria nauseosa* (Asteraceae). *Oncocnemis astrigata* Barnes & McDunnough (FW 1.7–1.9 cm; Plate 51.22) ranges in California from Napa and Lassen counties south to San Diego County. Adults have a single flight from early May to mid-August. Larvae were found in Los Angeles County, California, by McFarland during May on mountain-mahogany, *Cercocarpus montanus glaber* (Rosaceae). **Oncocnemis columbia** McDunnough (FW 1.4–1.5 cm; Plate 51.23) ranges from British Columbia south to eastern California (Mono County). The adults fly in August. Jeff Miller reports that the larvae feed on oceanspray, *Holodiscus discolor* (Rosaceae), during June and July. The larvae have a slight dorsal hump on the eighth abdominal segment and are light brown to gray with discontinuous and irregular light longitudinal lines. **Oncocnemis definita** (Barnes & McDunnough)(FW 1.2–1.6 cm; Plate 51.24) ranges from British Columbia south to California and Colorado. The adults fly from September to mid-October in California. J. Miller found that the larvae feed only on oceanspray, *H. discolor* (Rosaceae), from June to August. **Oncocnemis chalybdis** Troubridge & Crabo (FW 1.4–1.6 cm; Plate 51.25), Oregon east to Manitoba, is a close relative of the eastern *O. piffardi* (Walker), the "three-striped oncocnemis" (not illustrated), which ranges from Alberta east to New Brunswick and south to Michigan and New York. *Oncocnemis chalybdis* has the black thoracic collar edged with white scales, while the collar is solid black in *O. piffardi*. Adults of *O. chalybdis* have a single generation with adults during late summer. **Oncocnemis punctilinea** Hampson (FW 1.3–1.7 cm; Plate 51.26) ranges from California and Utah south to central Arizona. The adults fly in May. The larvae feed on *Fraxinus*. Larvae were described by Crumb and are light gray with alternating dark and pale stripes. There is a subventral white stripe edged with black on its lower edge. The head is granulose and pale gray with black spotting. **Oncocnemis ragani** Barnes (FW 1.3–1.5 cm; Plate 51.27) ranges from southwestern Oregon south to southern California. It is distinguished from the similar *O. semicollaris* Smith by having a brown head, whereas the second species has a gray head. In coastal southern California, McFarland found adults from late March to late June with scattered records to October. He supposed there were three broods there. Larvae feed on chaparral honeysuckle, *Lonicera subspicata* (Caprifoliaceae). Related species with a similar FW pattern are *O. figurata* (Harvey), British Columbia south to California and Colorado, *O. greyi* Troubridge & Crabo, British Columbia south to Oregon with larvae on *Symphoricarpos*, and *O. semicollaris*, British Columbia south to California and Arizona with larvae also on *Symphoricarpos*. **Oncocnemis singularis** Barnes & McDunnough (FW 1.3–1.4 cm; Plate 51.28), Adults are found from February to July in coastal southern California. McFarland found larvae during June feeding on the buds and flowers of *Keckiella cordifolia* (Scrophulariaceae). **Oncocnemis umbrifascia** Smith (FW 1.2 cm; Plate 51.29) ranges from British Columbia and Montana south to central California, Utah, and Colorado. The adults fly during July to August. The larval host is *Penstemon* (Scrophulariaceae); larvae feed from May to mid-June in eastern Washington. The larvae are green with several indistinct pale wavy lines, and there is a distinct white subventral stripe. The spiracles are white and the head is green, granulose, and with dark freckles.

Lepipolys perscripta Guenée (FW 1.1–1.3 cm; Plate 51.30) ranges from California east to New England and south to Arizona, Colorado, and Florida. Richers reports that the adults fly from late September to May in southern California, and flight dates elsewhere in the West fit this pattern. The larvae feed on *Antirrhinum*, *Linaria*, and *Nuttallanthus* (Scrophulariaceae). A report of larval feeding on *Citrus* is likely apocryphal.

Sympistis heliophila Paykull (FW 1.0–1.1 cm; Plate 51.31) is Holarctic and occurs in boreal North America as well as Eurasia, ranging from Alaska and Yukon east across northern Canada to Quebec and south to Alberta. Opler and Slater report a disjunct population in alpine habitats in the Colorado Rockies. Elsewhere the moths are found in grassy bogs. The adults are diurnal and have a single flight in June and July. The larvae feed on *Empetrum* (Empetraceae) and *Vaccinium* (Ericaceae) in Europe and likely feed on these plants in North America. **Sympistis zetterstedtii** (Staudinger)(FW 1.0–1.1 cm; Plate 51.32) is found from Alaska east across Arctic and boreal Canada to Labrador, and thence south to northern Washington, Alberta, and, disjunctly, to the alpine zone of the Colorado Rockies. The moth is Holarctic as it is also native to Greenland and Eurasia.

Stylopoda cephalica Smith (FW 0.7–0.9 cm; Plate 51.33) is native to the inner coast range of central California south to the valleys and foothills of Los Angeles and Riverside counties. There is a single flight from mid-March to mid-April. The life history has not been reported.

Behrensia conchiformis Grote (FW 1.1–1.2 cm; Plate 51.34) ranges from British Columbia south along the Pacific Coast states to San Diego County, California. This striking moth has a single flight from late December to late May, mainly February to early April. In southern California, some individuals have been collected during September and October, indicating the possible occurrence of a second flight. The adults are nocturnal and are attracted to light. Larvae feed at night on the young leaves of chaparral honeysuckle, *Lonicera subspicata*, and snowberry, *Symphoricarpos albus* (Caprifoliaceae). McFarland describes the larvae as gray and brown with a pattern of undulate longitudinal lines; there are, in addition, whitish middorsal, subdorsal, and subspiracular lines. Pupation is in a tough whitish cocoon into which chewed particles of wood or other dry material are incorporated. **Behrensia bicolor** McDunnough is similar to the previous species but lacks the greenish FW scales and has a sharply contrasting black-and-white-banded HW. Its range is in montane habitats extending from Shasta County south to Placer County, California. The adults have a single flight during June. Its life history is unknown.

SUBFAMILY AMPHIPYRINAE

The Amphipyrinae in North America comprises *Amphipyra*, *Cropia*, *Paratrachea*, *Apsaphida*, and *Viridemas*. Other genera formerly included in this subfamily are now placed in the Xyleninae. **Amphipyra** Ochsenheimer adult males have simple genitalia; the larvae have uniordinal crochets. The pupae do not have the extended wings and proboscis of Cuculliinae. **Amphipyra pyramidoides** Guenée, the "copper underwing" (FW 1.8–2.0 cm; Plate 51.35, Plate 64.28 larva), ranges from British Columbia east to the Atlantic Coast and south to central California, southeastern Arizona, and northeastern Colorado. The adults are nocturnal and have a single flight from August to early October. The moths are readily attracted to fermented fruit bait but are also attracted to light. The larvae feed on a wide variety of broad-leaved woody plants and rest along the midrib on the underside of leaves. The larvae are light translucent yellow green with a slight dorsal hump on the first abdominal segment and a distinct middorsal hump on the eighth abdominal segment. There are fine white dorsal, subdorsal, and stigmatal lines. Three other *Amphipyra* species occur in the West. **Amphipyra glabella** Morrison is an eastern species that occurs west to Alberta and Colorado, while *A. brunneoatra* Strand is a southwestern species. ***Amphipyra tragopogonis*** Linnaeus was thought by Forbes to be a European introduction found in the Northeast but also in British Columbia.

SUBFAMILY PSAPHIDINAE

The Psaphidinae, as defined by Poole, includes the tribes Psaphidini, Nocloini, Triocnemidini, and Feraliini. The subfamily is unified by the following apomorphic features: (1) heavy and irregular placing of spines on ventral surface of tarsal segments, (2) elongate vesica covered with stout long cornuti not placed in distinct patterns, (3) male abdomen with tergum heavily sclerotized and enlarged (absent in Feraliini), (4) tympanic region characterized by a large external hood followed by a large bulla in the intersegmental membrane of the first two abdominal segments, (5) known larvae of the "larvae liberae" class, and (6) prothoracic tibial claws present (absent in Feraliini). There are almost 90 North American species, with more than 70 in the West.

TRIBE PSAPHIDINI

Psaphidine noctuids are characterized by the adult's early annual flight period, heavy spination of the tarsal segments, and occurrence of pectinate or strongly serrate male antennae. In North America, there are 13 species in six genera, but only five western species in four genera.

Psaphida damalis (Grote)(FW 1.4–1.6 cm; Plate 51.36) is a subtly marked moth with a plain gray FW and a white HW having pronounced darker veins. The species is known from central lowland California. The species is nocturnal and is attracted to light. There is a single annual flight in late winter and early spring (February). The larvae are reported to feed on oak, *Quercus* (Fagaceae).

Pseudocopivaleria Buckett & Bauer is endemic to California, southern Oregon, and northern Baja California. There are only two species. Adults have a small patch of tan scales at the base of the FW along the inner margin. ***Pseudocopivaleria anaverta*** Buckett & Bauer (FW 1.6–1.9 cm; Plate 51.37) has paler wings than the otherwise nearly identical, but darker, ***P. sonoma*** (McDunnough). *Pseudocopivaleria anaverta* occurs in southern California and northern Baja California, Mexico, mostly at higher elevations. There is a single spring flight (April to early May). The larvae feed during spring on young foliage of canyon live oak, *Quercus chrysolepis* (Fagaceae).

Pleuromella opter Dyar (FW 1.3–1.7 cm; Plate 51.38) is the only representative of the genus and ranges from extreme southern Oregon south through California, mainly west of the Sierra Nevada crest and the Transverse Ranges, to Baja California. The adults are pale gray with a vague diagonal line running from the FW apex to the midpoint of the inner margin. Usually the wing veins are darker than the surrounding ground color. The univoltine adults have been found from February through June, primarily in April. McFarland reports that captive larvae readily ate young leaves of manzanita, *Arctostaphylos glauca* (Ericaceae), in the San Gabriel Mountains of southern California. The life history is unreported.

Provia argentata Barnes & McDunnough (FW 1.2–1.3 cm; Plate 51.39) is the sole member of its genus and is characterized by its mottled gray and brown FW with enlarged white V-shaped mark extending from the reniform and claviform. The moth is distributed from central Utah south and west through northwestern Arizona to southern California. Records are mainly during spring (March to April). The larvae are reported by McFarland to feed on bitterbrush, *Purshia glandulosa* (Rosaceae), in southern California. The larvae are translucent green with vivid red and white marks.

Miracavira brillians (Barnes)(FW 1.7–1.9 cm; Plate 50.48) is an attractive green moth that is very similar to *Feralia* species. It occurs in Arizona and New Mexico. There is a single flight of adults from July to September. The life history is unreported. ***Miracavira sylvia*** (Dyar) ranges from Arizona south into Mexico, and a third member of the genus is endemic to Mexico.

TRIBE FERALIINI

The tribe Feraliini comprises seven species of **Feralia** Grote, five of which occur in the West. One is entirely eastern North American, and a single species, *F. sauberi* (Graeser), occurs in Japan and eastern Russia. The moths in this group are characterized by their early seasonal flight, green wings, hairy bodies, external tympanic structure, and massive irregular spination of tarsal segments. ***Feralia deceptiva*** McDunnough (FW 1.7–1.8 cm; Plate 51.40) is a sister species of the mainly eastern *F. jocosa* (Guenée), which ranges westward to southeastern British Columbia. *Feralia deceptiva* ranges from southern British Columbia south to central California and occurs primarily west of the Cascade–Sierra Nevada divide. A report from Alberta is suspect. The adults of *F. deceptiva* are relatively invariable and have dark apple green FW. The FW postmedial line is strongly accented with black scaling. The moths are attracted to light and fly primarily in spring (late February to May). The larvae feed on young foliage of Douglas-fir, *Pseudotsuga menziesii* (Pinaceae). The larvae are translucent green with conspicuous longitudinal white dorsal lines. The spiracles are white, their upper margins edged with red. There are yellow spots on the prolegs that are reddish-tipped, and the true legs are red purple. The head is green sparsely flecked with black. ***Feralia februalis*** Grote (FW 1.5–1.6 cm; Plate 51.41) is another green early-spring noctuid, found in lowland and foothill locations from coastal Oregon south to southern California. This species and *F. meadowsi* Buckett, a channel island endemic limited to Santa Catalina Island, are distinguished by the presence of quill-like spines along the abdominal dorsum. The postmedian line and discal dot, as well as any dark HW suffusions, are lacking in both species. The green color of adult *Feralia* specimens changes to yellow or orange when exposed to any moisture. The two species may be separated by characters of the male genitalia. The adults fly in late winter and early spring (December to mid-April). The larvae of *F. februalis* have been reported to feed on mountain-mahogany, *Cercocarpus montanus* (Rosaceae), blue oak, *Quercus douglasii* (Fagaceae), and various other woody trees and shrubs. The larvae are similar to those of *F. deceptiva* but have a conical hump on the eighth abdominal segment and the pedicel of the antenna is yellow.

TRIBE TRIOCNEMIDINI

Oxycnemis fusimacula Smith (FW 0.9–1.3 cm; Plate 51.18) is found in California, Baja California, and Arizona. There are several very similar species in this group. Adults fly from March to May, occasionally July. Larvae feed on *Krameria* (Krameriaceae). The larvae are broad in the middle and become narrow posteriorly with a pronounced dorsal hump on the eighth abdominal segment. They are infuscated purple with a white subventral stripe that is sinuate along its upper margin. The spiracles are white. The venter is blue green and is strongly suffused with purple. The base of the prolegs is white. The head is greenish white with purple suffusion and clusters of small black spots.

Aleptina inca Dyar (FW 1.0–1.2 cm; Plate 51.42) ranges from the deserts of southeastern California south to Baja California and east through southern Arizona to western Texas. The moth has been found from May to October in western Texas and likely has several flights. The life history is unreported.

Copibryophila angelica Smith (FW 1.1–1.3 cm; Plate 51.43) ranges from California east to western Texas. The adults are in flight from June to September in California and during August in western Texas. The life history is unreported.

Metaponpneumata rogenhoferi Möschler (FW 1.0–1.2 cm; Plate 51.44) is found from southern California south into Baja California and east to southern Arizona and western Texas. The species has been found during May and June, and again in August to September. It likely has two annual flights. The life history is unreported.

SUBFAMILY STIRIINAE

The Stiriinae is a group of entirely American taxa that are at their most diverse and richest in arid regions of the southwestern United States and Mexico, although some taxa occur beyond these areas in more northern and eastern areas of the United States and Canada. The most complete treatments of the subfamily have been by Hogue and Poole. The adults are extremely diverse and often colorful, but the limits of the subfamily are based on only a few characters, some of which are shared by remotely related Noctuidae. The shape and spination of the vesica (inflated interior of male aedeagus) and the reduction of the larval spinneret to a small scalelike object are the two primary distinctive synapomorphic characters. Other features include a large frontal process on the adult head, terminal claw on the prothoracic tibia, male genitalia of the primitive reduced trifid noctuid type, and female genitalia with elongate, pointed, heavily sclerotized ovipositor lobes (presumably for oviposition into the flowering heads of their hosts). Poole reports that there are 61 species in North America, 50 of which are found in the West. The life histories have been reported for relatively few species, but those known are often associated with the flowers and seeds of Asteraceae. Adults are most often seen in late summer and fall (July to September) and may be either diurnal or nocturnal. The adults often have coloration that matches the flowers of their hosts.

Fala ptychophora Grote (FW 1.4–1.7 cm; Plate 52.1) is a monotypic clay brown moth found in the Mojave and Colorado deserts of southern California, where there is a single spring flight during March and April. The adults are somewhat mottled with white reniform and orbicular spots. The life history is unreported.

Plagiomimicus Grote is the largest genus of the subfamily with a diverse array of small to medium species, mostly united by their size and pointed FW apex. *Plagiomimicus spumosum* (Grote)(FW 1.4–1.7 cm; Plate 52.2) has a brown FW with a fine granular appearance and thin cream ante- and postmedian lines. The species has a broad irregular range from eastern Washington and southern Alberta east to Iowa, Missouri, South Carolina, and Florida. In the West, the moth ranges south to southern California, Arizona, New Mexico, and western Texas. There is a single flight from July to September. The early stages are unreported. *Plagiomimicus dimidiata* (Grote)(FW 1.3–1.4 cm; Plate 52.3) is a small moth with a pointed FW and a silvery white basal area. The outer half is gray brown with a white postmedian line. The species occurs from western Wyoming, Utah, Colorado, and Kansas south through Arizona, New Mexico, and western Texas to Chihuahua. The adults are nocturnal and are attracted to light. There is a single flight of adults during September and October. The host and early stages are unreported. *Plagiomimicus tepperi* (Morrison)(FW 1.3 cm; Plate 52.4) is a variable gray green moth with a pointed FW that has white lines dividing varying-colored darker and lighter patches. The moth ranges from eastern Washington, southern Nevada, and northeastern Colorado south through the Southwest to southern Mexico. The adults have been found from late July to mid-September, and some individuals are labeled from April and May. The life history is unreported. *Plagiomimicus mimica* Poole is a similar species with a broadly overlapping distribution in the Southwest but is so far unreported for Mexico.

There are five western *Lineostriastiria* Poole. *Lineostriastiria hutsoni* (Smith)(FW 0.8–1.1 cm; Plate 52.5) is the smallest and dullest of the subfamily. It has an olive green to gray green FW with a faint yellowish postmedian line. It occurs in the Sonoran Desert of southern California, southern Nevada, and western Arizona. There is a single flight of adults in March and April. The life history is unknown.

Xanthothrix Henry Edwards contains only two small species of diurnal spring moths, both limited to the Mojave Desert of southern California from eastern Ventura and Los Angeles counties south to Riverside County. They are often mistaken for heliothines. Both species have larvae that feed in the flower heads of desert composites (Asteraceae). *Xanthothrix ranunculi* Henry Edwards (FW 0.8–1.1 cm; Plate 52.6) is a small yellow to yellow green diurnal moth with an unmarked FW. There is a single flight in March and April. Larvae feed on the flower heads of *Coreopsis douglasii* (Asteraceae). The larvae were described by Comstock and Dammers as having alternating red brown, cream, and dull brown bands and lines. Pupation is under litter or in sandy soil without benefit of a cocoon. *Chrysoecia* Hampson, which has eight described species from the southwestern United States and Mexico, is defined primarily on the basis of genital features. The adults of four western North American species range from cream to yellow or dark colored and have a more or less ovate FW. *Chrysoecia atrolinea* (Barnes & McDunnough)(FW 1.2–1.4 cm; Plate 52.7) is a small yellow moth that ranges from southeastern Arizona east to western Texas. The adults are nocturnal and have been found during late summer and early autumn. *Chrysoecia scira* (Druce)(FW 1.4–1.6 cm; Plate 52.8) has a distinctive appearance among North American moths. The ovate FW has a broad cream basal area while much of the remainder being a dark purplish brown. This striking moth ranges from southeastern Arizona to southern Mexico. The adults are seen at light in July and August. The life history is unreported.

Basilodes chrysopis Grote (FW 1.7–1.9 cm; Plate 52.9) has the most widespread western distribution of three North American species (an additional species is known from Mexico). The moth ranges from eastern Colorado and western Kansas south to Arizona, New Mexico, and southern Texas. It undoubtedly occurs in Mexico as well. The moths have a bronzy golden FW with thin ante- and postmedian lines, as well as weakly contrasting pale reniform and orbicular spots. The univoltine adults fly in July and August. The host plant is *Verbesina encelioides* (Asteraceae). The larva was described by Crumb as being characterized by alternating transverse orange and black stripes. There is a distinctive pair of thin black lines below the spiracle and the head is orange brown speckled with black.

Cirrhophanus dyari Cockerell (FW 1.6–1.7 cm; Plate 52.10) is one of six species, all of which are unified by their orange yellow wings with darker orange pattern of lines and smudges. The adults often perch on flowers of their host plants, where their color matches that of the flowers. For example, the eastern *C. triangulifer* Grote is referred to as the "goldenrod stowaway." *Cirrhophanus dyari* is the only western species as two have more eastern distributions and three others are restricted to Mexico. This species ranges from southeastern Arizona east to western Texas and south into northern Mexico (Chihuahua). There is one record for southern Colorado (Chaffee County). The adults have a single flight in August and September.

Eulithosia plesioglauca (Dyar)(FW 1.2–1.3 cm; Plate 52.11) is one of four species in this small genus. All three North American species have quite different appearances but are unified by their genitalic features. The present moth is one of the more unique appearing species in the West with its three-toned violaceous FW traversed by thin lines that run from the base to outer portion. The species ranges from southeastern Arizona to southern Mexico. Adults are nocturnal and are attracted to light. The single flight is during August and September. The life history is unreported.

Stiria Grote comprises seven extremely similar medium moths characterized by their yellow FW with several dark brown patches. Each species is geographically restricted and usually only one to three species might occur in a particular region. The most widespread western species is **S. intermixta** Dyar (FW 1.4–1.7 cm; Plate 52.12), which ranges from southern Idaho, Utah, and western Colorado south to southern Mexico, is nearly identical to **S. rugifrons** Grote, whose range does not overlap with that of the present species. The adults are nocturnal and are attracted to light. There is a single flight during August and September. The larval host will likely prove to be a species of *Helianthus* (Asteraceae). **Chalcopasta** Hampson comprises small moths with metallic green areas on their FW. All species occur in the Southwest or Mexico. The most widespread western species is **C. howardi** (Henry Edwards)(FW 1.5–1.7 cm; Plate 52.13), which is a spectacular moth; each FW is iridescent yellow green with a brown triangular area at the base. The species occurs from central Arizona, central New Mexico, and Texas south to southern Mexico. The adults are nocturnal and are attracted to light. There are two flights from April to May and July through October. The life history is unreported.

TRIBE ANNAPHILINI

Axenus arvalis Grote (FW 0.8–0.9 cm; Plate 52.14) is endemic to California west of the Cascade–Sierra Nevada crest and ranges from Humboldt County south to San Diego County. There are specimens labeled from Colorado, but these may be in error. The diurnal adults have a single flight from late March to mid-June with the later dates at higher elevations. The life history is unreported, but we observed males patrolling over stands of a native clover, *Trifolium* (Fabaceae).

Annaphila Grote is a genus of small colorful moths most of which are found in California, where many are endemic. The species were revised by Rindge and Smith and were illustrated and synopsized by Leuschner. Life histories and ecology have been reported by Buckett, Comstock, Henne, and Sala. Twenty-four species have been described, but further taxonomic research would be helpful, because some of these may

be synonyms of each other and it is possible that some undescribed species await discovery. All the species are single-brooded and fly in spring (occasionally to early July), although some **A. pustulata** Henry Edwards specimens are labeled with fall collection dates. The adults of all species are rapid diurnal flyers, and all species for whom life histories are known select herbaceous plants as their hosts. Thus far, according to Leuschner, known hosts are limited to *Emmenanthe*, *Nemophila*, and *Phacelia* (Hydrophyllaceae), *Mimulus* (Scrophulariaceae), *Gilia* and *Linanthus* (Polemoniaceae), and *Claytonia* (Portulacaceae). Larvae usually feed on the reproductive structures of their hosts. The adults often seek nectar at such plants as *Ceanothus* (Rhamnaceae), *Nemophila* (Hydrophyllaceae), or *Salix* (Salicaceae), among others. **Annaphila diva** Grote (FW 0.9–1.0 cm; Plate 52.15, Figs. 245, 246 genitalia) ranges from southern British Columbia south to southern California. The basic wing pattern and genitalic features, rather than its cream white HW, indicate the species' apparent relationship to **A. miona** Smith and **A. casta** Henry Edwards, both of which have a yellow orange HW. Adults are in flight from mid-March to early June. Adults nectar on flowers of *Ceanothus* (Rhamnaceae). Males circle evergreen oaks in the early afternoon, and females have been seen hovering about low vegetation in midmorning. The early-instar larvae feed on the flowers and buds of their host *Claytonia perfoliata* and probably other species of *Claytonia* (Portulacaceae), and the later instars feed on the leaves as well. The larvae are brownish gray with light and blackish lines and spots as described by Rindge and Smith, although the description by J. A. Comstock and Henne is somewhat different. The middorsal stripe comprises two blackish lines consisting of small irregularly shaped spots on the thoracic and abdominal segments. There is a subdorsal black-edged whitish stripe mottled with brown on the thoracic segments that is interrupted by a round black spot on the anterior portion of the first seven abdominal segments. This results in a whitish mottled spot on the posterior third of each abdominal segment. A whitish bar extends laterally across the posterior portion of the eighth abdominal segment. A cream triangular spot then extends to the anal extremity. The head is dirty yellow and has black blotches especially on the vertex. The larva pupates near the soil surface in a cocoon made with soil particles adhered together by silk. **Annaphila arvalis** Henry Edwards (FW 1.0–1.2 cm; Plate 52.16) ranges from Washington south to southern California. The single flight extends from late January to April but is primarily during March. The larvae feed on buds and flowers of their host *Claytonia perfoliata* (Portulacaceae). The fifth-instar larvae are gray black and have a longitudinal cream lateral stripe running the length of the body. There are a few scattered subdorsal white points. The larvae pupate inside hollow stems. **Annaphila astrologa** Barnes & McDunnough (FW 0.9–1.1 cm; Plate 52.17) ranges from central coastal California south to southern California and discontinuously to southeastern Arizona, the apparent origin of the type specimens. The adults have been found from late February to mid-May, rarely to July. Adults nectar at flowers including *Coleogyne ramosissima* (Asteraceae). The larval host is *Emmenanthe penduliflora* (Hydrophyllaceae). The last instar is green with paired middorsal green lines comprised of dots and dashes. There is a spiracular band of purple spots with white that is bordered at its lower edge by white. The spiracles are dark and rimmed with white. The head is dull yellow heavily freckled with brown and with white ocelli. Pupation is in a pupal chamber cut into pithy wood and is inside a cocoon made up of soil particles held together with silk. All members of the Astrologa Complex including *A. pseudastrologa* Sala and *A. vivianae* Sala make a pupal chamber inside pithy wood. These latter species are endemic to southern California, also fly in spring, and use various species of *Phacelia* (Hydrophyllaceae) as their larval hosts. *Annaphila lithosina* Henry Edwards (FW 1.0 cm; Plate 52.18) is restricted to California west of the Cascade–Sierra Nevada divide from Shasta County south to Kern County. Its preferred habitat is near watercourses in steep foothill canyons; here the larval host *Mimulus guttatus* (Scrophulariaceae) abounds. The adults fly from mid-February to mid-July. *Annaphila casta* Henry Edwards (FW 1.1–1.2 cm; Plate 52.19) is endemic to openings in moist coastal forests in western Oregon and northern California. The adults have a single flight during April and May. Adult females fly during midday and select *M. moschatus* (Scrophulariaceae) as their larval host. *Annaphila evansi* Rindge and Smith (FW 0.8–0.9 cm; Plate 52.20) is endemic to central California and southern California and flies from March to early May. This species is very close to *A. spila* Rindge & Smith but differs from that species by having the front black instead of brown. The adults nectar at flowers including *Lasthenia gracilis* (Asteraceae), and *Salix* (Salicaceae). The larvae feed on buds, flowers, and the bases of floral bracts of *Gilia* (Polemoniaceae). The last-instar larvae is pale whitish green with a pale middorsal stripe bordered by two dark green stripes each having an included white dot on each abdominal segment. The head is ivory with a tinge of yellow. The dark green subdorsal stripe is split by a pale line on the thoracic segments. The spiracular stripe is white with the dark spiracles along its upper edge. The ventral area is green with a single white spot on each segment just above the prolegs or in line with them. Pupation is in the ground in a cocoon covered with sand grains or soil.

UNASSIGNED TRIBE

The following three genera and their included species have not yet been assigned to any specific tribe. *Neumoegenia poetica* Grote (FW 1.0–1.2 cm; Plate 52.21) looks superficially like it might be a geometrid but was provisionally placed by Poole with the Stiriinae. The leading edge of the FW above is white while the remainder is a shiny metallic bronze green. The adults are found in southeastern Arizona during July and August. A sister species is known from Mexico. McFarland and Walsh point out that the larval host is *Bidens leptocephala* (Asteraceae). *Argentostiria koebelei* (Riley)(FW 1.0–1.2 cm; Plate 52.22) is the only species in its genus and is only provisionally placed with the Stiriinae. The adult has a pointed FW with four silver white patches each separated by orange brown bands coinciding with the ante- and postmedian, discal, and marginal lines. The species ranges through southern California, Nevada, southern Utah, and western Arizona. Adults are nocturnal and are attracted to light. The adults have a single flight from late March to early June. The life history is unreported. *Thurberiphaga diffusa* (Barnes)(FW 1.3–1.5 cm; Plate 52.23) is another moth that is only provisionally placed in this group. The adults are pale yellow cream with blotches of pink. They are known to occur only in southeastern Arizona. Adults are nocturnal and are attracted to light. There is a single flight from July to mid-September. The larvae are pink and covered with coarse granules. The larvae bore in the stems of wild cotton, *Gossypium thurberi* (Malvaceae), but have never been found on cultivated cotton.

SUBFAMILY HELIOTHINAE

The subfamily has two autapomorphies that unite the group. The larvae have the skin covered with conical granules, each bearing a minute spine. Other noctuids may have granular skin, but these conditions seem not to be homologous. In addition, in larvae beyond the first instar, seta L1 on the prothoracic segment is placed directly anterior to seta L2. There are no universal adult apomorphies, but the male genitalia include elongate straplike valvae and a coiled vesica. The female genitalia often have complementary coiled appendix bursae.

Many of the species are brightly colored flower visitors with larvae that feed on the reproductive parts of various broad-leaved dicots, especially nonwoody annuals and perennials of many families. Monocotyledonous plants are rarely included as hosts, but one, *Helicoverpa zea*, is a major pest of corn. Some of our more serious crop pests are included in this subfamily, especially *H. zea* (Boddie) and the *Heliothis virescens* (Fabricius) complex. In North America, there are almost 170 species of Heliothinae, 140 of which are included in the genus *Schinia* Hübner. The genus is especially rich in the southern Great Plains and Southeast, but about 80 of the species range into or are endemic to the West.

Baptarma felicita Smith (FW 0.9–1.1 cm; Plate 52.24), the only member of its genus, ranges from southern Arizona and the Colorado Desert of southern California north to eastern California (Mono County). The usual flight period is in March and April, but in the White Mountains of California and Nevada, the species has been seen in mid-June above 11,000'. The larvae feed on the seed pods of *Phacelia crenulata* (Hydrophyllaceae). Like a great many noctuids, the full-grown larvae are either green or brown with lighter diagonal marks. Pupation is in the soil.

Eutricopis nexilis Morrison (FW 0.8–0.9 cm; Plate 52.25 dorsal, Plate 52.26 ventral) is a small diurnal species found from Yukon Territory and British Columbia east to Saskatchewan and south to California and Colorado. This species may be found from the Canadian Life Zone forests up to alpine slopes. The adults have a single flight from mid-May to early September depending on latitude and elevation; they are strictly diurnal and usually remain in close association with stands of the host plant, on which they rest and copulate. The larvae feed in the flower heads of pussytoes, *Antennaria* (Asteraceae), and are gray with the dorsal portion brownish to pink. Spiracles are dark. Pupation is in a silken or debris-covered cocoon at or just below soil level.

Pyrrhia exprimens (Walker)(FW 1.5–1.6 cm; Plate 52.27) ranges from central British Columbia east to Ontario, Quebec, and Newfoundland. In the West, it is found south to central California, Arizona, and central Colorado. Adults are attracted to light and may be found from late May to early September. Most individuals fly during June and July, and there are only occasional late-flying individuals of a putative second brood. The larvae feed on reproductive parts and foliage of a wide variety of broad-leaved plants but seem to prefer Ranunculaceae. McFarland found larvae feeding on columbines, *Aquilegia*, in Idaho, while they have caused damage to delphiniums, *Delphinium*, in Canada. The larvae are extremely variable ranging from white to yellow to green, sometimes marked with black and blotches of yellow. There may also be thin longitudinal white or yellow lines.

Helicoverpa zea (Boddie), the "bollworm," "corn earworm," or "tomato fruitworm" (FW 1.5–1.8 cm; Plate 52.28, Plate 64.30 larva, Figs. 251, 252 genitalia), occurs throughout the New World but is resident only in those regions without freezing winters. The adults recolonize more northern areas, usually by midsummer. The species is resident in warmer parts of Latin America and is found south to Chile and Argentina. In

North America, it is resident across the southern United States from southern California east to Florida. Each summer, the adults emigrate northward and become common in many northern areas as far north as southern Canada. In warmer tropical and subtropical regions, the adults may be found year-round, and the seasonal appearance of adults becomes more restricted in more northern localities. This is a major pest to cotton (Malvaceae), corn (Poaceae), tomatoes, and tobacco (Solanaceae); the larvae are polyphagous and feed on a seemingly limitless variety of broad-leaved plants. This is the most serious agricultural pest in the Western Hemisphere. The larvae bore into and feed on the reproductive parts of their host plants. Ultimate instars are extremely variable and range from more or less plain green to brown and may be variegated with longitudinal bands and blotches of various colors.

Heliothis phloxiphaga Grote & Robinson, the "spotted straw" (FW 1.4–1.7 cm; Plate 52.29), ranges from British Columbia east to Maine and southern Ontario south through most of the United States. Adults are found from June to August and are both nocturnal and diurnal. Larvae feed on the flowering parts of many low-lying annual and perennial broad-leaved plants. The ultimate instar is quite variable in color and pattern. The base colors range from shades of green, yellow, red, or brown, but all forms have many fine longitudinal lines. In addition, there can be dark blotches and distinctive longitudinal bands. Larvae complete feeding in five or six instars in 18 days, after which they make a cell in the soil and pupate in a tough soil-covered cocoon. *Heliothis oregonica* (Henry Edwards)(FW 1.1–1.3 cm; Plate 52.30) is found from British Columbia east to Saskatchewan and south in the mountains to the Sierra Nevada of California and White Mountains of Arizona. There is a single flight from early June to early August with the timing depending on latitude and elevation. In most instances, the females select *Castilleja* species, Scrophulariaceae, for oviposition, but, on occasion, the larvae were found by Hardwick feeding on other genera such as *Geranium* (Geraniaceae) and *Artemisia* (Asteraceae). The larvae, which feed on the reproductive parts of their hosts, are variable in color including shades of green, brown, reddish, and even purple; they have a white spiracular line. *Heliothis virescens* (Fabricius), the "tobacco budworm" (FW 1.3–1.5 cm; Plate 52.31), is found from Maine and southern Ontario south to Florida. The moth ranges throughout lowland tropical America and the West Indies in suitable habitats. In the West it is a periodic vagrant and colonist in agricultural and urban portions of Arizona, California, Colorado, and, rarely, to Nevada and New Mexico. In the West, it does not survive the winter north of Mexico. Adults may be easily separated from those of *H. subflexa* by the presence of rust-colored scales on the labial palpi and prothoracic legs. The flight period is from May to October. There are two flights with adults most numerous and wide-ranging in the second flight. The larvae feed on reproductive parts of many plants, especially nightshades (Solanaceae). According to Hardwick they are serious pests of cotton, *Gossypium*, and Covell points out that they also cause serious damage to tobacco, *Nicotiana tabacum*. The first brood immatures feed on the flower buds, and those of the second brood eat the flowers and seed pods. After *Helicoverpa zea*, this moth causess the most economic damage to agriculture in the Western Hemisphere. Ultimate instars are shades of green, pink, or purple, with many irregular and broken fine longitudinal lines. A closely related species *Heliothis subflexa* (Guenée)(FW 1.3–1.5 cm; Plate 52.32) is extremely similar in appearance to *H. virescens* but lacks rust-colored scales on the labial palpi and prothoracic

legs. It is resident in lowland tropical America, and migrant adults may establish temporary breeding populations north to Arizona, southern California, and Colorado. In the East it has been found as far north as Nebraska, Kentucky, thence south to Arkansas, Mississippi, and Florida. Covell records the adults from May to October, but in the tropics the species may be found at any season. This species prefers *Physalis* and *Solanum* (Solanaceae) for its larval foods.

Heliocheilus paradoxus Grote (FW 1.1–1.3 cm; Plate 52.33) ranges from British Columbia east to Minnesota and south to southern California and central Mexico. According to Hardwick, more northerly occurrences may represent vagrants from southern populations. In areas of residency to the south, the species may be found from April to October, but more northern occurrences are mainly from midsummer. In the southern parts of the range, there are likely two or more generations each year. McFarland found that the male may make faint clicking sounds during flight. Although larvae have not been found in nature, they likely feed on grasses, according to both Forbes and Hardwick. The larvae are strongly prognathous and are green or straw colored with longitudinal fine lines. *Heliocheilus julia* (Grote)(FW 1.1–1.4 cm; Plate 52.34) is a distinctive vivid pink moth that ranges from eastern Arizona and central New Mexico south to Jalisco, Mexico. There is a single flight between late July and mid-September. The moths are attracted to light and are sometimes abundant in southern Arizona. The larval hosts and life history have not been reported.

Hardwick discusses the North American species of **Schinia** Hübner in depth and illustrates adults and larvae of most species. Subsequently, Pogue and Harp have been gradually revising the genus one species group at a time. They have described a number of new species and elevated several taxa that had been treated by Hardwick as synonyms. Virtually all *Schinia* species are specific in their use of larval hosts, either to single host species or to single host genera. The larvae of most species feed in the flower heads of various Asteraceae, but Fabaceae is another frequently used family. *Schinia* moths may be either diurnal or nocturnal with about a 50-50 division between the two habits. A very few species are active both by day and night. When inactive, the adults almost always rest on the host plants, usually on or adjacent to the flower structures, and very often the adult colors and maculation match the colors of the plant parts upon which they rest. Although perhaps seeming to be brightly colored obvious insects when seen alone, they are actually well camouflaged when at rest on the appropriate parts of their hosts, usually flowers (Plate 64.31). *Schinia nuchalis* (Grote)(FW 1.2–1.4 cm; Plate 52.35) has a close Palaearctic relative, *S. scutosa* Denis & Schiffermüller. On the basis of its genitalia, Hardwick put it in the genus *Protoschinia* as being intermediate between *Heliothis* and *Schinia*, but Matthews combined it with *Schinia*. The range is from southern British Columbia and southern Saskatchewan south through the high plains and arid intermountain West to Colorado and southern Arizona. There are two flights of adults in spring and late summer. Larvae have been reared from native tarragon, *Artemisia dracunculus* (Asteraceae), which flowers in the late summer and fall, but, according to Hardwick, another host is likely for early summer feeding larvae. *Schinia citrinella* (Grote & Robinson)(FW 0.9–1.1 cm; Plate 52.36) ranges from southern California east to the Texas Gulf Coast and north along the high plains to northern Colorado and southern Nebraska. There are two flight periods, April to June and September to October, in southern California and coastal Texas, but only one during

July and August further north on the Great Plains. The adults are strictly nocturnal and are attracted to light. The larval hosts are *Croton* species (Euphorbiaceae), often where these plants grow on consolidated sand dunes or along the edges of active windblown dunes. The larvae feed on buds, flowers, young fruits, and leaves. Ultimate-instar larvae are gray green heavily mottled with white. The prothoracic and suranal shields are white or pale cream mottled with green. The spiracles are rimmed with light brown. The head is pale cream mottled dorsally with gray or fawn. *Schinia aurantiaca* (Henry Edwards)(FW 0.8–0.9 cm; Plate 52.37), an attractive little moth, ranges from northern California (Sonoma County) south through California, including the Mojave and Colorado deserts, to southern Baja California Norte and east to eastern Arizona. The diurnal adults fly from late March to early June. The larval host is *Eriastrum sapphirinum* var. *dasyanthum* (Polemoniaceae). The ultimate-instar larva is green and brown marked with longitudinal bands and lines of white. The spiracles are light brown with paler centers. The head is grayish fawn mottled with light brown. *Schinia jaguarina* (Guenée)(FW 1.1–1.5 cm; Plate 52.38) is quite variable, with the FW ranging from green to purplish and the maculation of the orange yellow HW ranging from pronounced to obsolete. The moth ranges from the plains of southern Canada (Alberta east to Saskatchewan) south to Texas and Arizona and to central Mexico. It ranges west of the continental divide in Arizona and occurs sparingly east to Florida. The adults are known from March to September in various parts of their ranges but seem to have but a single flight everywhere except in southeastern Arizona where there may be two. Adults are nocturnal and are attracted to light. The larvae feed on the reproductive portions of *Psoralea* including *P. rhombifolia* and *P. tenuiflora* (Fabaceae). The full-grown larvae are green with a number of fine white longitudinal lines above the spiracles and a pair of dorsal black spots on each segment. The spiracles are black. *Schinia meadi* (Grote)(FW 1.0–1.2 cm; Plate 52.39) is distributed from Washington, Alberta, and Saskatchewan south to northern Arizona and New Mexico. *Schinia meadi* is especially common in open habitats with sandy soils. The adults fly from May to mid-August, are nocturnal, and are attracted to lights. The host plants and life history are not reported. *Schinia pulchripennis* (Grote)(FW 0.7–1.0 cm; Plate 52.40) ranges from Trinity County, California, south through the Central Valley, Coast Ranges, and peninsular ranges to Baja California Norte as well as east to southern Arizona. The diurnal adults are in flight from late February to early May. The larval host plant is purple owl's-clover, *Castilleja exserta* (Scrophulariaceae). Adults rest on the flowering portion of the plant, where they blend in well. According to Hardwick, the females deposit the eggs rather carelessly between the flowers and surrounding bracts. Young larvae bore into and feed on developing seed capsules, but later instars feed externally on both flowers and developing seeds. The ultimate-instar larvae vary from light green to dark brown marked with light gray or bright yellow, respectively. *Schinia villosa* (Grote)(FW 0.8–1.0 cm; Plate 52.41) is a montane species that occurs from British Columbia east, in appropriate habitats, across the Prairie Provinces to Saskatchewan and south in the West to northern Arizona and Colorado. The diurnal adults have a single flight from early July to mid-August. The larval hosts are fleabanes, *Erigeron* (Asteraceae), and possibly other related plants. The adults rest on the flower heads and fly about stands of these plants. Females deposit the eggs between the florets in the head and the larvae feed by burrowing underneath the seed

layer. The larvae are cannibalistic, and only one larva per flower head will survive. The larvae will move to a second head partway through development and feed externally during the last instar. The ultimate-instar larvae are light green with darker green longitudinal lines and bands. The head is green. *Schinia arcigera* (Guenée), the "arcigera flower moth" (FW 0.9–1.2 cm; Plate 52.42), is sexually dimorphic with males having a yellow black-margined HW and females with a solid black HW. The moth ranges from southern Manitoba east to Nova Scotia and south through the East to Florida and plains to central Mexico. It occurs to the west of the Rockies from Idaho and western Colorado south to New Mexico and southern Arizona. It has also been found in highlands in the Sierra Madre Occidentale of Sonora, Mexico. The adults are nocturnal and are attracted to light; they fly from mid-July to September in most sites but are found during October in southern Texas. The larval hosts are species of *Aster* and *Machaeranthera* (Asteraceae). The ultimate instar is olive green to brown with many darker fine longitudinal lines. The spiracular line is white. The spiracles are black. *Schinia miniana* (Grote)(FW 1.0–1.2 cm; Plate 52.43) is variable, with the HW ranging from red to brown. The moth ranges from the deserts of southern California and southern Utah to western Texas, and thence south to Chihuahua and southern Arizona. *Schinia miniana* may be in flight from March to September and may have more than one flight per year. Adult *S. miniana* become active in late afternoon but are primarily nocturnal and are attracted readily to light. The species is very flexible in its emergence due to the fact that one or more of its host plants respond to precipitation by flowering almost any time during the warmer months. The larval hosts are three species of *Baileya* (Asteraceae), and the larvae feed strictly on the flowering parts. The female moths lay eggs between florets in opened heads of the host. The resultant larvae feed on the florets and spin a silken feeding and molting shelter over the disk florets. The ultimate instar is greenish yellow, grayish yellow, or pinkish yellow with longitudinal gray green lines. The spiracles are black, and the head is light brown suffused with dark brown. *Schinia gaurae* (J. E. Smith), the "clouded crimson" (FW 1.2–1.4 cm; Plate 52.44), ranges from Colorado east to Kentucky and south to southern Arizona, Texas, and Florida. In most of its western range, the species is univoltine from mid-July through August, but in Oklahoma and Texas, it is bivoltine with extreme collection dates from June through October. Adults are attracted to light and may also be found in the late afternoon flying about flowering inflorescences of *Gaura* (Onagraceae), their larval host. Larvae feed and rest among the host inflorescences. Larvae are white with transverse bands of black and longitudinal middorsal and lateral series of yellow blotches. Thus, the larvae are superficially very similar to those of the monarch, *Danaus plexippus* (Linnaeus), but no connection between the two species is known. *Schinia masoni* (Smith)(FW 1.0–1.2 cm; Plate 52.45, Plate 64.31) has wine red wings and a yellow orange head and thorax that match up extremely well with the flowers of *Gaillardia aristata* (Asteraceae), and the adults usually rest on the flowers so the colors of the moth match up almost exactly with those of the host flowers (Plate 64.31). Byers reported that this species, which ranges along the Rocky Mountain Front from southeastern Wyoming south into central Colorado, actually blends with *S. volupia* (Fitch) in southeastern Colorado. The adults rest on the flowers by day and are more active at dusk. The host *Gaillardia* is a plant that responds positively to fire and, as a direct result, the populations of adults are most abundant within a few years after fires in the ponderosa pine belt. The flight extends from mid-June to mid-July. The life history was described by Byers. Females insert their eggs between florets in the flower head, and the resulting larvae feed on flowers and developing seeds. The larvae are ivory or very pale yellow with dark brown, purple brown, or dark red middorsal, subdorsal, and supraspiracular lines. The prothoracic and suranal shields are black or very dark brown and divided into bands by three ivory or very pale yellow lines. The head is light brown mottled with dark brown or black. *Schinia niveicosta* (Smith)(FW 1.4–1.7 cm; Plate 52.46) ranges from the Mojave and Colorado deserts of southern California and western Arizona south to Baja California Norte and Sonora, Mexico. *Schinia niveicosta* is somewhat variable in FW color, magenta with a white costal margin and pale outer margin. When at rest, they almost perfectly match the flowers of their host. When inactive, the adults rest head downward on flower pedicels of their host plant, *Palafoxia linearis* (Asteraceae), which often grows in sandy habitats such as are found near dune systems. The adults fly from February to late April and occasionally during September and October in response to late-summer rains; they are nocturnal and become active in late afternoon. Young larvae bore in *Palafoxia* flowers, while older larvae feed from the stems at night and rest at the base of the plant during the day. The full-grown larvae are pale gray green with faint longitudinal lines. The head is orange brown mottled with darker brown. *Schinia acutilinea* Grote (FW 1.0–1.3 cm; Plate 52.47) ranges through the intermountain West and high plains from the interior of British Columbia east to Saskatchewan and south to southern California and Arizona. The adults are nocturnal and have a single flight from mid-August to mid-October. The larval host is big sagebrush, *Artemisia tridentata*, as well as other species of *Artemisia* (Asteraceae). The females insert their eggs in the still tightly closed flowering panicles of the host. The larvae feed on the reproductive parts of the plant. The ultimate-instar larvae are pale gray with fine darker lines or bands. The prothoracic shield is white with a median mauve brown band and two narrower submarginal mauve brown bands. The head is creamy white mottled with mauve brown. *Schinia walsinghami* (Henry Edwards)(FW 1.0–1.4 cm; Plate 52.48) is found from southern British Columbia east to Montana and south to southern California, Arizona, and New Mexico. *Schinia walsinghami* has a single flight from mid-August to early October, and is nocturnal and is attracted to light. Females oviposit in the unopened flower buds of rubber rabbitbrush, *Ericameria nauseosa*, and probably *Chrysothamnus* (Asteraceae). Ultimate-instar larvae are green with faint longitudinal shades.

SUBFAMILY CONDICINAE

This subfamily comprises *Condica*, *Ogdoconta*, and relatives, including many species described as belonging to *Perigea* and *Platysenta*. The group is divided into two tribes: Condicini and Leuconyctini, after restriction by Berio and Poole. The larvae have reduced spinnerets. Larval host plants are mainly Asteraceae, especially those with pyrrolizidine alkaloids.

TRIBE CONDICINI

Condica discistriga (Smith)(FW 1.4–1.6 cm; Plate 52.49) ranges from British Columbia east to Saskatchewan and south to California, southeastern Arizona, and Colorado. There is a single flight from May through August. *Ogdoconta cinereola* (Guenée),

the "common pinkband" (FW 1.1–1.3 cm; Plate 52.50), occurs from British Columbia east across southern Canada to Quebec and Nova Scotia, thence south to Colorado and Kansas. The adults are probably bivoltine as they fly from late May to early September in Colorado.

TRIBE LEUCONYCTINI

Crambodes talidiformis Guenée, the "verbena moth" (FW 1.2–1.5 cm; Plate 52.51), ranges from Alberta east across Canada to Quebec and south to Arizona, Colorado, Texas, and North Carolina. Covell reports the moths fly from May to October in at least two broods. This seems to be confirmed for the West, as in eastern Colorado there are records from May to June and again in August. The larvae feed on several vervains, *Verbena* (Verbenaceae). The larvae are green with a white middorsal line that is margined in red. There is a subdorsal white line, and a red spiracular line that is subtended by white. The spiracles are brown. The head is green with darker reticulation.

SUBFAMILY BRYOPHILINAE

The Bryophilinae were split off from the Acronictinae by Kitching and Rawlins. The subfamily is characterized by small lichencolored moths whose first-instar larvae have the seventh abdominal segment unmodified and have the pupal cremaster elongate with the usual noctuid complement of hooks that are dimorphic in size and length. The eggs are laid on rocks or walls near the host lichens, and larvae overwinter in silk-lined burrows. The subfamily is represented by the Holarctic genus *Cryphia* Hübner, which has 14 described species in North America, 11 of which occur in the West. *Cryphia viridata* (Harvey)(FW 1.2–1.3 cm; Plate 52.52) is found in southern California (San Diego County) and adjacent Baja California. The moths are nocturnal and are attracted to light. There seem to be several flights (January to February, July, and September). *Cryphia cuerva* Barnes (FW 1.2–1.4 cm; Plate 52.53) is a dusky gray species found on lava flows from southern British Columbia and Alberta south to eastern Oregon. The adults are attracted to light and fly from July to early September. The life history is unreported. *Cryphia olivacea* (Smith)(FW 1.2–1.5 cm; Plate 52.54) is olive green with mottled black lines. The blackcentered orbicular is pronounced. The moth ranges from British Columbia south through the Rockies (Colorado and Wyoming) to southeastern Arizona. The adults are nocturnal and are found from the end of June through August. They appear earlier in Arizona and later in the Rockies. The life history is unreported.

SUBFAMILY XYLENINAE

TRIBE PSEUDEUSTROTIINI

Pseudeustrotia carneola (Guenée), the "pink-barred lithacodia" (FW 1.0–1.1 cm; Plate 50.13), ranges from Alberta and eastern Colorado east to Newfoundland and south to Georgia. The adults are nocturnal and may be found at light from May to September. There are likely two annual generations. Larvae feed on *Polygonum* and *Rumex* (Polygonaceae).

TRIBE PRODENIINI

Spodoptera Guenée includes 10 North American species, all of which are polyphagous on low, generally herbaceous plants, all of which are multiple-brooded, and all of which may be destructive to crops. The taxonomy of this genus was treated in great detail by Pogue. *Spodoptera exigua* (Hübner), the "beet armyworm" (FW 1.1–1.2 cm; Plate 52.55), was introduced to North American from Southeast Asia about 1876. It was discovered in Oregon and reached the Atlantic Coast by 1924. The species is found throughout the world's tropical and subtropical regions including many oceanic islands. In North America, it is resident in areas where winter freezes are absent or minimal, including southern California, southern Arizona, southern Texas, and southern Florida. Each summer the moths expand their range by emigration to more northern locales, and the species has been recorded in British Columbia, Manitoba, Colorado, Nebraska, and, in the East, Ontario and Maine. The moth regularly reaches northern California, Colorado, and Maryland, but the species can reach destructive levels anywhere to the south. *Spodoptera exigua* can breed during any warm period in the south, including lowland Arizona and California, where the species is most destructive. Northward, the moths may appear during September and October. Hosts include plants in at least 33 families of dicotyledonous and monocotyledonous angiosperms and are listed by Pogue. Most of the hosts are herbaceous and include many crops such as apple, *Malus* (Rosaceae), beans, *Phaseolus vulgaris* (Fabaceae), sugarbeets *Beta vulgaris* (Chenopodiaceae), corn, *Zea mays* (Poaceae), cotton, *Gossypium hirsutum* (Malvaceae), as well as peppers, *Capsicum*, potatoes, *Solanum tuberosum*, and tomatoes, *Solanum lycopersicon* var. *cerasiforme* (Solanaceae). The eggs are laid in clusters of 50 to 150, and the young larvae feed gregariously and skeletonize foliage. Later, the larvae feed solitarily and may chew large holes in foliage or burrow into the buds of their hosts. The larvae vary in color from green to gray or brown. Dorsally and subdorsally there may be dark spots or dashes. There is a white lateral stripe and the venter is pink or yellow. The spiracles are white or yellow with narrow black rims. The head is pale and has black reticulation and coronal stripes. The larvae pupate in the soil within a chamber hardened by an oral secretion. *Spodoptera praefica* (Grote), the "western yellowstriped armyworm" (FW 1.6–1.7 cm; Plate 52.56), is a native western moth that ranges from British Columbia and Alberta south to California, Arizona, and Colorado. The moths fly from late January to early September. *Spodoptera ornithogalli* (Guenée) "yellow-striped armyworm moth, cotton cutworm" (FW 1.4–1.8 cm, Plate 52.57) is native to Mexico, Central America, and South America as well as many Caribbean islands. The species ranges northward in the warmer months, and, though most prevalent in the Southeast, regularly reaches California, Arizona, southern Colorado, and southeastern Canada. The adults may be found from February to November in three or four generations. Twenty families of monocot and dicot angiosperms are listed as hosts by Robinson et al. Composites (Asteraceae), legumes (Fabaceae), and nightshades (Solanaceae) are especially prominent in the host list, and Covell reports the species is especially destructive to cotton, *Gossypium*, clovers, *Trifolium*, grasses, Poaceae, and potatoes, *Solanum tuberosum*, and tobacco, *Nicotiana tabacum* (Solanaceae). Eggs are laid in masses of 200 to 500 on the underside of leaves, and young larvae feed gregariously. The larvae vary from pale gray to deep black. Various markings may be present or absent on any particular larva and include black subdorsal series of triangles, a yellow supraspiracular line, a black spiracular band, and a subventral white band that may be suffused with orange or pink. The spiracles are brown, and the head is brown with the dorsal portion black and the adfrontal areas white. Larvae pupate in the soil within a chamber sparsely lined with silk.

TRIBE ELAPHRIINI

Galgula partita Guenée, the "wedgling" (FW 0.9–1.0 cm; Plate 53.1 male, Plate 53.2 female), is sexually dimorphic with males gray to reddish brown and females dark brown maroon to blackish. This is one of our more widely distributed moths occurring from southern Canada south in sufficiently moist habitats through most of the United States and Mexico to Honduras. The adults are found through much of the year in several generations. For example, in Berkeley, Powell has found adults from January to early November. The larval hosts are wood sorrels, *Oxalis* (Oxalidaceae).

TRIBE CARADRININI

SUBTRIBE CARADRINIA

Caradrina montana (Bremer)(formerly *Platyperigea extima*)(FW 1.3–1.5 cm; Plate 53.3) is widely distributed across the West, mainly in arid habitats. This is a Holarctic moth that is also widespread in Eurasia. The North American range extends from Yukon and Northwest Territories south along the Pacific Coast and Rocky Mountains to California, Arizona, and Colorado. In California, Richers reports that the adults fly from mid-February through November and very likely they have three to four generations. Larval hosts have been reported to be several low herbaceous plants including *Polygonum* (Polygonaceae) and *Taraxacum* (Asteraceae). The larvae vary from light gray to dark gray. The dorsum is darkest and becomes paler ventrally. The spiracles are black. The venter is densely flecked with white, and the head is dark brown with underlying black splotches.

SUBTRIBE ATHETINA

Proxenus miranda (Grote), the "Miranda moth" (FW 1.2–1.3 cm; Plate 53.4), is found from British Columbia east across southern Canada to Nova Scotia and south to California, Arizona, Colorado, and Texas and, in the East, to South Carolina. The adults are found from March to November, and, according to Covell, there are two flights. The larvae have been reported to feed on a variety of low herbaceous plants in several families. The larvae are dusky gray. The dorsal area has some blackish reticulation, and the supraspiracular area is blackish. The spiracles are black. The head is brown with sparse black reticulation and coronal stripes.

TRIBE ACTINOTIINI

Nedra stewarti Grote (FW 1.2–1.5 cm; Plate 53.5) is found primarily along the Pacific Coast from Washington south to the Coast Ranges (Lake County) and Sierra Nevada (Sierra County), California. The adults fly from early May to mid-September, possibly in two generations. Although the life history is unreported, all other species of *Nedra* use various species of *Hypericum* (Guttiferae) as the larval hosts. It is possible that this moth has become scarcer because of the introduction of *Chrysolina* beetles for the biological control of Klamath weed, *Hypericum perforatum*.

TRIBE PHLOGOPHORINI

Euplexia benesimilis McDunnough, the "American angle shades" (FW 1.3–1.4 cm; Plate 53.6), is unusual in that the FW may be pleated when the moth is at rest, a trait it shares with *Phlogophora* species. The species is distributed from British Columbia east across Canada to Newfoundland and south to central California, Colorado, and, in the East, to Arkansas and North Carolina. The adults are attracted to light and may be found in two broods from early February to August. The larvae feed on a diverse array of herbs, shrubs, and trees that encompasses ferns, conifers, and broad-leaved spermatophytes. The larvae occur in both brown and green phases. The brown phase has dark markings that are tinged with purplish brown. The green phase has dark flecks and reticulation as well as minute white spots at the base of each seta. The head is green with dark dorsal reticulation.

Phlogophora periculosa Guenée, the "brown angle shades" (FW 1.8–2.1 cm; Plate 53.7), ranges from British Columbia east to Labrador, south to northern California, and, in the East, to Mississippi and South Carolina. The adults are bivoltine, flying in April and again in late July and early August. The larvae feed on a wide variety of herbaceous plants, shrubs, and trees including ferns, conifers, and broad-leaved spermatophytes. The larvae are variable and range from light brown to dark green. The dorsum may be yellow, gray, or green with a faint reddish or fuscous reticulate pattern. The middorsal line is yellow and is represented on each segment by an anterior dash and two posterior dots. Additionally, there is a series of V-shaped marks on the abdominal segments. The head is light brown to green gray with red, brown, or green reticulation. The setae are inserted in minute black spots.

TRIBE APAMEINI

SUBTRIBE APAMEINA

This group contains about 260 western species in 45 genera. The larger genera are *Apamea*, *Papaipema*, and *Hydroecia*. *Apamea* Ochsenheimer comprises about 125 described species distributed across the Holarctic region including North Africa. About 72 described species occur in North America, with most endemic to or ranging into the West. *Apamea occidens* (Grote)(FW 1.8–2.1 cm; Plate 53.9) ranges from British Columbia and Alberta south to California, Arizona, New Mexico, and Nebraska. In California, adults of the single generation fly from June to mid-August. Reports compiled by Robinson et al. indicate that the larvae feed on grasses (Poaceae). The larvae are a dirty white and are unmarked. The spiracles are orange and the cervical shield is dusky brown. The head is brown with dark reticulation and narrow black coronal stripes. *Apamea amputatrix* (Fitch), the "yellowheaded cutworm" (FW 1.8–2.2 cm; Plate 53.10, Plate 53.11), ranges from British Columbia and Northwest Territories east to Newfoundland and south to California, Arizona, New Mexico, and, in the East, Missouri and Georgia. Although this moth is often found in disturbed habitats, including yards and agricultural environments, it is also found frequently in undisturbed native habitats. Adults are dimorphic and can be either with or without a distinct pattern. Adults have two or three flights from early March to early September. The larvae are polyphagous and feed on a variety of herbaceous plants, grasses, and young trees. They are considered to be destructive to some vegetable crops and shrubs. The larvae are smoky gray, devoid of markings, and have a yellow orange head. *Apamea maxima* (Dyar)(FW 1.6–3.0 cm; not illustrated) is found from southern coastal British Columbia, mainly on beaches of Vancouver Island, south to northern California (Marin County). The species is found on coastal beach dunes, fens, and woodland along the more or less immediate coast. Adults are in flight from late May to early August. *Apamea alia* (Guenée)(FW 1.7–1.9 cm; Plate 53.12) ranges from Alaska to Newfoundland south to California (Mono County), Utah, New Mexico, and New Jersey. The adults have a single flight from late June through July. Larvae feed on grasses (Poaceae), although Robinson et al. compiled reports of their feeding on a variety of broad-leaved trees. The larvae are blackish with faint middorsal and subdorsal pale stripes. The

spiracles are brown. The head is pale brown with black reticulation subdued by infuscation and black coronal stripes. *Apamea sordens* (Hufnagel), the "bordered apamea" (FW 1.1–1.8 cm; Plate 53.13), ranges from British Columbia east to Labrador and south to northeastern Oregon, Colorado, and, in the East, to South Carolina. The species also occurs in Eurasia. This moth has both a pale gray form and a maculate reddish brown one. We illustrate an example of the gray form and a maculate reddish brown form of *A. digitula* (both species have both forms). *Apamea digitula* Mustelin & Mikkola (Plate 53.14) replaces *A. sordens* in western Oregon and all of California; the two species can be separated only by DNA sequence, male genitalia, and range. There is a single flight from May through July. The larvae feed on a variety of meadow grasses (Poaceae), rarely on sedges (Cyperaceae). The larvae are purplish gray with a strong middorsal white stripe and a much fainter pale subdorsal line. The venter is reticulated pink and white. The spiracles are dark gray and are within a whitish area. The head is brown with infuscated brown reticulation and coronal stripes. *Apamea cogitata* (Smith)(FW 1.7–1.9 cm; Plate 53.15) is found in boreal habitats from Alaska to Newfoundland south to southern California, Utah, Colorado, Wisconsin, and New York. It occurs south in the Rockies to the Colorado Front Range. It flies from during July and the first half of August. *Apamea devastator* (Brace), the "glassy cutworm" (FW 1.6–1.8 cm; Plate 53.16, Plate 53.17), ranges from Alaska to Newfoundland south to California, Arizona, Texas, and, in the East, Virginia. The moths have several flights from late March to September. The larvae feed on grasses (Poaceae) and a wide variety of herbaceous plants. The species is found in a wide range of habitats from native to highly disturbed and from mesic to rather xeric. The larvae are translucent greenish white with a dark subcutaneous middorsal line and are without markings. The cervical shield is infuscated brown, and the anal shield is infuscated blackish. The head is bright red brown with only slight evidence of dark reticulation and coronal stripes. It is considered a serious pest of sod grasses, grains, and some vegetable crops. *Apamea plutonia* (Grote)(FW 1.5–1.7 cm; Plate 53.18) ranges from British Columbia east to Nova Scotia and south to central Washington, northern New Mexico, and, in the East, to Virginia. The flight period is during May. The larvae are reported to feed on grasses (Poaceae) and are gray with a purplish tinge. There is a distinct pale middorsal line. The spiracles are brown. The cervical and anal shields are black with three white lines. The head is infuscated brown with black reticulation and coronal stripes.

Resapamea passer (Guenée), the "dock rustic" (FW 1.7–1.9 cm; Plate 53.19), ranges from Yukon Territory east to Newfoundland south to Oregon, Arizona, Colorado, and, in the East, to New Jersey. According to Covell, the adults have two flights from May to July and in August to September. The larvae feed externally as well as by boring into the roots of docks, *Rumex*, including *R. obtusifolius* (Polygonaceae). The larvae are whitish and tinged dorsally with purplish. They are unmarked and have brown spiracles. The cervical and anal shields are brown, and the head is red brown.

Eremobina hanhami (Barnes & Benjamin)(FW 1.4–1.6 cm; Plate 53.20) ranges from British Columbia south to California (Monterey County) and Colorado. Adults have a single flight from late August to late October. The larvae feed on sedges (Cyperaceae). The larvae are pale olive green with pale middorsal and subdorsal stripes that cross the anal shield. The spiracles are white. The head is white with pale green reticulation and coronal stripes.

Oligia Hübner comprises about 50 species distributed through the Eurasian and Ethiopian regions, with two species introduced to eastern North America. More than 20 North American species, mostly western, belong in other genera but are referred to here as "*Oligia*." "*Oligia*" *indirecta* (Grote)(FW 1.2–1.3 cm; Plate 53.20) ranges from British Columbia east to New Brunswick and south to California and Colorado. The adults appear in mid-July and fly until early September. The larvae feed on grasses (Poaceae) and sedges (Cyperaceae); they are pale brown with yellowish middorsal and subdorsal lines. The cervical shield is pale brown with a black anterior margin, and the anal shield is brown with the posterior margin raised and crenulate. The venter is white. The head is pale brown and is darkened at the ocelli. The spiracles are brown. "*Oligia*" *fractilinea* (Grote), the "lined stalk borer" (FW 1.1–1.3 cm; not illustrated), ranges from Alberta east to Saskatchewan and south to Arizona, Colorado, and, in the East, to Tennessee and Florida. The adults are univoltine with their flight during late July and August. Larvae feed on grasses (Poaceae) including Timothy, *Phleum pratense*, and occasionally damage corn crops, according to Rings. The larvae are pale with dark stripes. The spiracles are brown. The cervical shield is pale brown, and the anal shield is brown and is bilobed and raised posteriorly. The head is brown. A pair of attractive related species, "*O.*" *rampartensis* Barnes & Benjamin (FW 1.4–1.5 cm; Plate 53.22) and "*O.*" *violacea* (Grote)(FW 1.3–1.4 cm; Plate 53.23), range from British Columbia and Alberta south along the Pacific Coast to California and in the Rocky Mountains to Utah and Colorado. In general, "*O.*" *rampartensis* has the more eastern range in the Great Basin, whereas "*O.*" *violacea* has a more western distribution along the Pacific Coast. According to Troubridge, "*O.*" *violacea* has a pure white hyaline HW and creamy gray thorax, whereas "*O.*" *rampartensis* has its HW with a smudged reddish submarginal band and red in the center of the thorax. There is a single generation with adults from mid-June through August. The larvae feed on *Juncus filiformis* and other rushes (Juncaceae). The larva is green with a lateral stripe bounded in black. "*Oligia*" *mactata* (Guenée), the "prairie Quaker" (FW 1.4–1.6 cm; Plate 53.24), ranges from British Columbia east to Quebec and New York, and south to southern California and southern Arizona. There may be two generations as adults have been found from mid-June to mid-September. The larvae feed on birch, *Betula* (Betulaceae), and dogwood, *Cornus* (Cornaceae), and likely other broad-leaved woody plants.

Cobalos angelicus Smith (FW 1.3–1.6 cm; Plate 53.25) ranges from central California to southern California and has been found on several of the California islands off the southern California coast. The adults have a single flight from mid-March to mid-July. The life history is unreported.

Photedes inquinata (Guenée), the "tufted sedge moth" (FW 1.0–1.4 cm; Plate 53.27), ranges from British Columbia east across Canada to Newfoundland, and thence south to California, Colorado, and, in the East, to Ohio and Virginia. The univoltine adults are found in July. Sedges, *Carex* (Cyperaceae), are the reported larval hosts.

Capsula oblonga (Grote), the "oblong sedge borer moth" (FW 1.7–1.9 cm; Plate 53.28), ranges from British Columbia across southern Canada to Nova Scotia and south to southern California, and, in the East, to Oklahoma and Florida. It also occurs in the West Indies. This is a species of wetland habitats. According to Covell the adults have a single brood, with adults from July to September in the north and most of the West, but have a second brood in the Southeast, with adults from November to May. The larvae feed primarily on cat-tails, *Typha* (Typhaceae), and occasionally on sedges, *Cyperus* (Cyperaceae). The larvae feed near the water surface or from within the stem of their host cat-tails. Prior to pupation the larva prepares a hole in the cat-tail stem through which it later emerges as an adult. The larvae are pale gray with pale brown longitudinal stripes. The spiracles are pale yellow white and are very narrow. The cervical and anal shields are pale brown. The head is pale gray brown with dark freckles.

Amphipoea americana (Speyer), the "American ear moth" (FW 1.4–1.5 cm; Plate 53.29), ranges from British Columbia east across Canada to Newfoundland and south to central California, Arizona, and, in the East, to North Carolina. The adults are nocturnal and fly from mid-April to September in California and may have two generations, while in Colorado there is a single flight in July and August. The larvae bore into and feed in the roots of grasses (Poaceae), sedges (Cyperaceae), and irises (Iridaceae). The larvae are an occasional pest of corn and irises. *Amphipoea lunata* (Smith)(FW 1.2–1.6 cm; Plate 53.30) has a red FW with a narrow reniform spot and occurs in dry forests of southwestern Oregon and California. In California, the adults fly from May to mid-August, according to data compiled by Richers.

Papaipema Smith is distinctive in that the larvae of all species feed by boring in the stems of their host plants, and adults fly little and are only rarely attracted to light. There are reports, however, that some species in the Northeast are more readily attracted to light. The genus is endemic to North America, with 45 described species of which 10 or more occur in the West. Quinter has been studying the group for many years and has found a number of undescribed taxa. The early-larval instars have pale middorsal, subdorsal, and subventral stripes on a pinkish or purplish ground. The cervical shield is always a paler shade than the head and almost always has black lateral margins, while the anal shield varies in color between species but usually has black lateral margins. The head varies between species from yellow to yellow brown to brown and may have a dark line behind the ocelli. There are two grouping of species based on the ratio of the spiracle width to its height. In one group the spiracles are narrow, and their width is only 45 to 55% of their height, whereas in the other group the width is 60 to 65% of the height. Most *Papaipema* species select only a single plant or group of related plants as their host, although *P. sauzalitae* is broader in its selection of host plants. *Papaipema sauzalitae* (Grote)(FW 1.4–1.7 cm; Plate 53.31) is found in coastal counties from the San Francisco Bay area south to Los Angeles County. Powell found it common on Brooks Island in San Francisco Bay, with larvae in *Scrophularia californica* (Scrophulariaceae) during the spring and adults in September. The flight period is from August through October. The larvae have been found feeding a variety of herbaceous herbs including columbine, *Aquilegia* (Ranunculaceae), thistles, *Cirsium* (Asteraceae), and lily, *Lilium* (Liliaceae) in addition to the plant reported by Powell. *Papaipema insulidens* (Bird)(FW 1.4–1.7 cm; Plate 53.32) ranges from southern British Columbia (Vancouver Island) south along the Pacific Coast to California. Moths fly during August. Caterpillars bore in stems of ragworts, *Senecio hydrophilus*, and *S. serra* (Asteraceae).

Hydraecia perobliqua Harvey (FW 1.5–1.7 cm; Plate 53.33) ranges from British Columbia east to Ontario and south to northern coastal California. Adults have a single flight in mid-July to early September. The life history is unreported. *Hydraecia stramentosa* Guenée (FW 1.7–2.0 cm; Plate 53.34) ranges from Manitoba east to Quebec and south to Arizona, Colorado, and Pennsylvania. Adults fly during August in Colorado. The larval host is *Scrophularia lanceolata* (Scrophulariaceae).

TRIBE XYLENINI

SUBTRIBE XYLENINA

Xylena Ochsenheimer has 11 species, five of which are North American; three of these have all or part of their range in the West. *Xylena nupera* (Lintner), the "American swordgrass moth" or "false armyworm" (FW 2.2–2.5 cm; Plate 53.35), ranges from British Columbia east across Canada to Nova Scotia and south to northern California, Utah, Colorado, and, in the East, to Illinois and New Jersey. The adults emerge in late summer or fall and overwinter. They become active and fly again from late March to May when the females lay their eggs. The larvae have been reported to feed on a wide variety of trees, shrubs, and herbaceous plants including grasses and sedges. Of 12 plant families catalogued by Robinson et al., the widest variety of genera was cited for Rosaceae. The larvae vary from green to dark brown. The middorsal and subdorsal lines may or may not be present, and the dorsum tends to be darker than the lateral areas. There is a dark subventral band including the spiracles, which are orange or red. The head is unmarked and has dark green or pink reticulation. The full-grown larvae rest in the ground for 2 to 6 weeks prior to pupation and presumably undergo a summer diapause. *Xylena brucei* (Smith)(FW 1.9–2.4 cm; Plate 53.36) occurs in western ponderosa pine forests. It ranges from British Columbia south to California, Arizona, and Colorado. The adults emerge in late summer or early fall and, like *X. nupera*, overwinter as adults and resume activity during the following spring. The larval host is bitterbrush, *Purshia tridentata* (Rosaceae). Larvae vary from light brown to orange brown and have a pale middorsal line that is bordered by dark lines on both sides. There is a segmental series of dark shield-shaped marks on the abdominal segments, and a dull white subventral band that contains the orange spiracles. The head is pale brown with some dark marginal reticulation and black coronal stripes.

Lithomoia germana (Morrison), the "American brindle moth" (FW 1.9–2.2 cm; Plate 53.37), ranges from Alaska to Newfoundland and south to California, Colorado, and, in the East, Pennsylvania. There is a single flight of adults from August to early September; and, following winter, the moths continue to fly until April. Larvae feed on a variety of broad-leaved trees and shrubs including Betulaceae, Ericaceae, Rosaceae, and Salicaceae. The larvae are dark gray with white middorsal and subdorsal lines as well as a subdorsal segmental series of black spots. There is a black spiracular line that includes the brown spiracles and a dull white subventral stripe. The head is pale brown with dark brown reticulation and black areas to each side of the front.

Homoglaea californica (Smith)(FW 1.4–1.7 cm; Plate 53.38) ranges from British Columbia south to California. Adults are nocturnal and, in California, fly during winter from late November through March, rarely April. The larvae are reported to feed on *Populus* (Salicaceae). The larvae are light gray suffused with purplish. There is a strong pale yellow middorsal stripe that is broadened at the posterior end of each abdominal segment but does not reach the end of the abdomen. This line is edged with dark lines, and the subdorsal line is inconspicuous. The spiracles are black. The venter is paler than the dorsum and is flecked with white. The head is pale brown and covered with closely placed black spots. The coronal area is unmarked.

Litholomia napaea (Morrison), the "false pinion" (FW 1.3–1.4 cm; Plate 53.39), ranges from British Columbia east across Canada to Newfoundland and south in suitable boreal habitats to the Sierra Nevada of California, and the Rockies of Colorado and New Mexico. The adults emerge in late September and October and then overwinter. They emerge from hibernation and continue their activity in spring from April to June. The larvae were reported to feed on several conifers (Pinaceae) as well as *Alnus* (Betulaceae), *Amelanchier* and *Spiraea* (Rosaceae), and *Populus* and *Salix* (Salicaceae). The larvae are brown with a series of dark subdorsal dashes. The lateral areas are darker and the spiracles are brown. The head is brown with black reticulation, adfrontal area, and coronal stripes. There is also a black spot by the antennae.

Lithophane Hübner has 65 species distributed across the Holarctic region, with 44 that are limited to or include North America in their range. *Lithophane pertorrida* (McDunnough)(FW 2.0–2.3 cm; Plate 54.1) ranges along the Pacific Coast from British Columbia south to southern California. Adults emerge in September and October; subsequently they may reappear from December to May when flight conditions permit. Larvae feed on plum, *Prunus* (Rosaceae), and willow, *Salix* (Salicaceae). The larvae are green flecked with white. There is a continuous white middorsal line and a discontinuous subdorsal series of white spots. The spiracles are white, and there is a yellowish subventral stripe. The head is whitish with green reticulation. *Lithophane georgii* Grote (FW 1.6–2.0 cm; Plate 54.2) ranges from British Columbia east to Newfoundland and south to California, Arizona, Colorado, and, in the East, to North Carolina. Adults emerge in September, then enter hibernation, and subsequently become active from January to May. The larvae feed on a wide variety of woody trees and shrubs, especially Betulaceae and Rosaceae. The larvae were illustrated by J. Miller and described by Crumb. They are pale green with many tiny white flecks and a solid white middorsal line. There is also a white or yellow spiracular stripe. The head is also green with perceptibly darker reticulation. *Lithophane dilatocula* (Smith)(FW 2.2–2.4 cm; Plate 54.3) occurs west of the Cascade–Sierra Nevada crest and Transverse Ranges from British Columbia south to southern California (Mt. Palomar, San Diego County). The adults emerge during October and November, become inactive, and then resume activity from January to March. The larvae feed on alder, *Alnus* (Betulaceae). The last instar is heavily mottled with shades of brown. Middorsal and subdorsal cream bands are discontinuous, and gaps are filled with black patches. The body is covered with white verruca–bearing setae, and the legs and prolegs are translucent ivory. The head is ivory mottled with dark gray. *Lithophane vanduzeei* (Barnes)(FW 1.7–1.9 cm; Plate 54.4) is extremely rare and is known from only Monterey County, California. The moths are found in Monterey pine, *Pinus radiata*, forest. A related species, *L. ponderosa* Troubridge & Lafontaine, occurs in the Sierra Nevada of California and Rocky Mountains of Colorado; the larvae feed on ponderosa pine, *P. ponderosa* (Pinaceae). Adults of both species have been found from October through April. The life history is unreported. *Lithophane amanda* (Smith), the "Amanda pinion" (FW 1.7–1.8 cm; Plate 54.5), ranges from Northwest Territories and British Columbia east to Newfoundland and south to California and Wyoming. Adults have been collected in late April and May. Larvae are reported to feed on a variety of woody trees and shrubs in the Betulaceae, Fagaceae, Rosaceae, and Salicaceae. The larvae are a pale whitish green with faint middorsal and subdorsal lines. The body is densely covered with small white points. There is a thin yellow spiracular line. The head is pale whitish green with many white dots. *Lithophane contra* (Barnes & Benjamin), found in Arizona and New Mexico, looks like a pale form of *L. amanda*. *Lithophane contenta* Grote (FW 1.7–1.8 cm; Plate 54.6) occurs from British Columbia south along the Pacific Coast to California. Adults emerge in late September and October, overwinter, and resume activity from January to early May, according to DeBenedictis and W. Patterson. Larvae feed on Oregon white oak, *Quercus garryana*, and possibly other oaks (Fagaceae). The larvae are dark green with a dense covering of tiny white spots. There is a strong white or pale yellow spiracular band with red or pink areas surrounding the white spiracles. The head is green and is densely spotted with white. *Lithophane innominata* (Smith), the "nameless pinion" (FW 1.5–1.6 cm; Plate 54.7), ranges from British Columbia east to Nova Scotia and south to northern California, and, in the East, to Wisconsin and North Carolina. The adults are in flight during late April and May. Adults usually overwinter in debris on the ground. The larvae feed during June and July on a variety of trees and shrubs in at least nine plant families including broad-leaved trees in Aceraceae, Betulaceae, Rosaceae, and Salicaceae, as well as several coniferous genera (Pinaceae). The larvae vary from light gray to brownish gray with a yellowish or pinkish middorsal line. The dorsum is densely covered with tiny yellow points. There is a black spiracular line that includes the white spiracles, as well as a yellowish or pinkish subspiracular stripe. The head is white with black reticulation and coronal stripes. *Lithophane thaxteri* Grote, "Thaxter's pinion moth" (FW 1.7–1.8 cm; Plate 54.8), ranges from British Columbia east to New Brunswick and south to Oregon, Michigan, and New York. The moth occurs in wet coastal forests along the Pacific Coast and in riparian forests east of the Cascades. Host plants are scattered among several families of broad-leaved trees and shrubs including *Alnus* (Betulaceae), *Chamaedaphne* (Ericaceae), *Myrica* (Myricaceae), *Ceanothus* (Rhamnaceae), *Larix* (Pinaceae), and *Salix* (Salicaceae).

Eupsilia fringata (Barnes & McDunnough)(FW 1.6–1.9 cm; Plate 54.9) ranges from British Columbia south to central California and Utah. The adults emerge in the fall, overwinter, and appear again in late winter or spring (December to May). The larvae live in silked leaf nest shelters and feed on several rosaceous plants including *Rosa nutkana* and *Rubus*, as well as willows, *Salix* (Salicaceae), and asters, *Aster* and goldenrods, *Solidago* (Asteraceae). The larvae are velvety black dorsally and laterally. There is a subventral white or yellowish stripe, and the venter is mixed brown and maroon. The spiracles are black. The anterior two-thirds of the head is intense black, while the posterior third is brown.

Mesogona rubra Crabo & Hammond (FW 2.0–2.1 cm; Plate 54.10) ranges from south coastal Washington south to central California. It is found in dry chaparral and forest habitats to the west of the Cascades and Sierra Nevada. The adults have a single flight from late August to mid-October. Caterpillars feed on the foliage of manzanitas, *Arctostaphylos columbiana*, and probably also *A. nevadensis* (Ericaceae).

Agrochola pulchella (Smith)(FW 1.3–1.5 cm; Plate 54.11) ranges from British Columbia south to California. The adults have a single flight from July to mid-October. The larvae feed on a wide variety of broad-leaved and coniferous trees and shrubs, as well as at least one herbaceous annual. The larvae are brown with a reticulate pattern and have the middorsal and subdorsal lines faint and discontinuous. There is a subdorsal series of dark shield-shaped marks. The spiracles are white and are included in a broad, sharply defined pale white subventral stripe. The head is pale with black reticulation and coronal stripes; the front and adfrontal areas are dark. *Agrochola decipiens* (Grote), the "shield-backed cutworm" (FW 1.3–1.6 cm; Plate 54.12), ranges from British Columbia south to southern California, Utah, and Colorado. It is closely related to *A. bicolorago* (Guenée), which has a complementary more eastern range. The adults have a single flight from August to mid-October. The larvae of *A. bicolorago* feed on seeds of aspen and cottonwoods, *Populus,* and willows, *Salix* (Salicaceae), as well as maples, *Acer* (Aceraceae). The larvae of *A. decipiens* are probably much like those of *A. bicolorago*, which are dark brown tinged with purplish and flecked with white. The middorsal and subdorsal lines are pale but obscure. There is a dorsal segmental series of dark shield-shaped marks on the abdominal segments. The spiracles are black, and there is a yellowish subventral stripe. The head is deep brown with black reticulation and coronal stripes.

Anathix puta (Grote & Robinson), the "puta sallow" or "poplar catkin moth" (FW 1.1–1.3 cm; Plate 54.13), is found from Alaska to Nova Scotia south to California, Utah, and, in the East, Pennsylvania. The adult moths have a single flight during August and September. Larvae feed on quaking aspen, *Populus tremuloides* (Salicaceae). The larvae are dark with a dorsal segmental series of dark shield-shaped marks through which passes a bold yellow middorsal stripe. On some larvae the dorsal dark marks are present only on the seventh and eighth abdominal segments. The lateral area and venter are much paler than the dorsum and have dense white reticulation. The spiracles are brown. The head is brown, is marked with black reticulation, and has coronal stripes. In addition, the front, adfrontal area, and a large spot by the antennal insertions are black. *Anathix aggressa* (Smith), the "pale sallow" (FW 1.2–1.3 cm; Plate 54.14), ranges from British Columbia east to Ontario and south to California, Utah, and New Mexico. The adults have a single flight from August to October. The larvae feed on *Populus* (Salicaceae). Crumb describes the larvae as very similar to those of *A. puta*.

Xanthia tatago Lafontaine & Mikkola (FW 1.3–1.4 cm; Plate 54.15) is nearly identical to the Eurasian *Xanthia togata* (Esper) but may be distinguished by features of the male and female genitalia. The North American moth ranges from Alaska across temperate Canada to Newfoundland and south to Oregon, Montana, Colorado, and, in the East, to Wisconsin and Pennsylvania. Adults have a single flight from late August to early October. Young larvae feed on willow and poplar catkins, *Salix* and *Populus* (Salicaceae), and later on they feed on foliage. According to Lafontaine and Mikkola the life history is likely similar to that of the Eurasian *X. togata*, which deposits its eggs on willow buds that then overwinter. In early spring the larvae bore into the catkins, and these are likely to fall onto the ground. There, the larvae feed on willow leaves as well as herbaceous plants. The larvae are blackish tinged with purple with a short pale middorsal line present only on the thorax and with a dark middorsal spot present on the anterior portion of each abdominal segment. The spiracles are black and the venter is pale and tinged with purplish. The head is pale brown with black reticulation and coronal stripes, as well as a black spot above each antennal insertion.

Hillia maida (Dyar)(FW 1.5–1.7 cm; Plate 54.16) ranges from British Columbia and Alberta south to northeastern California and Colorado. The adults have a single flight during August. The larvae feed on *Salix* (Salicaceae). The larvae are brown with pale middorsal and subdorsal stripes outlined in darker brown. There is a pale subventral stripe tinged with orange. The spiracles are dark brown and each is surrounded by a pale area. The head is yellowish with inconspicuous brown reticulation and coronal stripes.

Aseptis McDunnough comprises 18 species that are endemic to North America; most are found in the West. *Aseptis perfumosa* (Hampson)(FW 1.4–1.7 cm; Plate 54.17) ranges from Oregon south to the Sierra San Pedro Martir of Baja California and to northern Arizona. The adults are found from April to early August and may have two flights. The larvae feed on manzanita, *Arctostaphylos glauca* (Ericaceae). The larvae are green with white longitudinal middorsal and subdorsal lines. In addition, there are angled pale lines emanating from the anterior of each segment, forming a series of V-shaped lateral figures. The spiracles are white. There is also a subventral white stripe edged dorsally with red or purplish. The head is green with darker green reticulation. *Aseptis binotata* (Walker)(FW 1.3–1.5 cm; Plate 54.18) ranges from British Columbia and Alberta south to Baja California and Colorado. The adults are found from mid-March to September, probably as two generations. The larvae feed on a wide variety of broad-leaved and coniferous shrubs and trees in several families. The larvae are green, often with a dorsal whitish suffusion. There are white middorsal and subdorsal lines as well as a subventral white line edged dorsally by a red line. The spiracles are white. The head is green with darker green reticulation and coronal stripes. Pupation is in a light silk cocoon among host-plant leaves. *Aseptis characta* (Grote)(FW 1.3–1.4 cm; Plate 54.19) ranges from southeastern British Columbia east to Saskatchewan and south through interior aridlands to southern California, Nevada, Utah, and Colorado. The preferred habitat is juniper woodland. The univoltine adults are in flight from April to August, mainly early June. The larval host is *Artemisia* (Asteraceae), and the last instar is gray green with a white lateral stripe. *Aseptis adnixa* (Grote)(FW 1.5–1.9 cm; Plate 54.20) ranges from British Columbia south to southern California along the Pacific Coast. The adults have a single flight from March to July, according to Richers. The larvae feed on several species of rosaceous shrubs including Indian plum, *Oemleria cerasiformis* and *Prunus serotina*. The larvae feed among the young terminal leaves of the host during March, and, when not feeding, the young larvae hide in bark crevices. The larvae pupate in the ground. The fully developed larvae are dark gray with darker subdorsal markings. There is a strong white middorsal line and an obscure subdorsal line on each side. The lines are all margined by dark lines. There may be a segmental series of shield-shaped marks. The spiracles are white and lie within a lateral black line. The head is brown with dark reticulation and coronal stripes. *Aseptis monica* (Barnes & McDunnough)(FW 1.5–1.7 cm; Plate 54.21) is found in California, Nevada, and Arizona. The adults fly from March to mid-June. The larval host is the woody shrub *Ericameria laricifolia* (Asteraceae). The larvae are dark green with prominent white stripes. The middorsal and subdorsal stripes are constricted at the intersegmental areas. The spiracles are pale. The head is pale green with black spots along the posterior margin. *Aseptis ethnica* (Smith)(FW 1.9–2.2 cm; Plate 54.22) ranges along the Pacific Coast east of the Cascade–Sierra Nevada divide from eastern Oregon south to southern California. The adults have two flights from mid-April to mid-September. The larvae feed on madrone, *Arbutus menziesii,* and manzanitas, *Arctostaphylos* (Ericaceae). The larvae are green with minute yellow white flecking and an incomplete yellow white spiracular line that is apparent only on the last few abdominal segments.

Brachylomia populi (Strecker)(FW 1.3–1.5 cm; Plate 54.23) ranges from British Columbia and Alberta south to California, Utah, and east-central Arizona. The adults have a single flight during August and September. The larvae usually feed on balsam poplar, *Populus balsamifera,* and quaking aspen, *P. tremuloides* (Salicaceae), but also were reported by Crumb to feed on hazel, oak, and chokecherry. The larvae are variably colored from whitish with dark markings to obsolete to dark gray with distinct markings. The spiracles are brown, and the head is variably marked and patterned depending on the basic larval color and pattern. The 12 North American *Brachylomia* Hampson, all found in the West, were revised by Troubridge and Lafontaine. *Brachylomia algens* (Grote)(FW 1.2–1.4 cm; Plate 54.24) ranges from Yukon to Newfoundland south to Colorado and New England. The adults fly from late August through September. Larvae feed mainly on *Populus* and *Salix* (Salicaceae) but also feed occasionally on *Betula* (Betulaceae), *Rosa* (Rosaceae), *Shepherdia* (Elaeagnaceae), and *Picea* (Pinaceae).

Parastichtis suspecta (Hübner), "the suspected" (FW 1.4–1.7 cm; Plate 53.26), is found from Alaska to Newfoundland south to southern British Columbia, Colorado, and, in the East, to North Carolina. This moth also occurs widely in temperate Eurasia. The adults fly in August. Larvae feed on *Salix* (Salicaceae) and *Betula* (Betulaceae).

Epidemas obscurus Smith (FW 1.6–1.8 cm; Plate 54.25) ranges from British Columbia and Alberta south to central California, Idaho, and Colorado. The adults have a single flight from August to early October. The life history has not been reported.

SUBTRIBE COSMIINA

Cosmia praeacuta (Smith)(FW 1.4–1.6 cm; Plate 54.27) ranges from Manitoba and Saskatchewan south to Oregon, California, Arizona, and Colorado. The adults have a single flight in July to August. The larval hosts are white fir, *Abies concolor,* and Douglas-fir, *Pseudotsuga menziesii* (Pinaceae). In spring, the young larvae feed on opening buds then switch to expanding foliage. The larvae are green with prominent middorsal and subdorsal white lines. The spiracles are brown and the subventral stripe is yellow. The head is green with a white spot by the ocelli. Pupation is in a silk cocoon among fir foliage. Closely related to *C. praeacuta* (Smith) is *C. elisae* Lafontaine & Troubridge, which is found in the same range but has the FW reniform spot constricted and narrow. *Cosmia calami* (Harvey), the "American dun-bar" (FW 1.3–1.4 cm; Plate 54.26), ranges from Manitoba and Saskatchewan east to Nova Scotia and south to southern California, southern Arizona, Texas, and, in the East, to Florida. The moths are found most commonly in oak woodlands and fly in from April to mid-September, mainly June and July. McFarland reports *C. calami* is mainly predaceous by feeding on other caterpillars, especially inchworms; they also feed on several species of oaks, *Quercus* (Fagaceae), including coast live oak, *Q. agrifolia,* and black oak, *Q. kelloggii,* in the West. A report of feeding on *Vaccinium* is probably incorrect. The larvae are green with a narrow white or pale yellow middorsal line and much pale yellow or white flecking. There are faint discontinuous subdorsal and subventral lines. The spiracles are light brown with darker rims. The head is completely green or with darker green reticulation.

Zotheca tranquilla (Grote), the "elder moth" (FW 1.6–1.8 cm; Plate 54.28), ranges from British Columbia south to California and Utah. In California, adults have been found from late April to September, but primarily during June and July. The adults are polyphenic and are usually of a pale green phase; a brown phase is less frequent. The larvae feed on several *Sambucus* species, elderberries (Caprifoliaceae), and a record of *Vaccinium* is likely erroneous. The larvae are described by Crumb and by Miller. The larva lives in a shelter of skeletonized leaves within which the frass is collected. The larva has a dorsal series of squarish pale yellow patches surrounded by ashy black. This feature is subtended by a violet gray band that has irregular margins. A bright yellow lateral band contains many black setal pinacula. The spiracles are black, and the venter and prolegs are green. The head is granulose and black. Pupation takes place on the host plant in a loose silk cocoon.

Enargia decolor (Walker), the "aspen twoleaf tier" (FW 1.6–2.0 cm; Plate 54.29), ranges from southern Alaska and British Columbia east to Quebec and south to California, Nevada, Arizona, New Mexico, and Kansas, and, in the East, to Minnesota and Vermont. Adults fly from late July to mid-October in Colorado. Larvae feed primarily on aspen, *Populus tremuloides,* and occasionally on willow (Salicaceae); they also feed to a degree on alder, *Alnus,* and birch, *Betula* (Betulaceae). First-instar larvae feed on the inside of a rolled up leaf bound with silk and then move to feed on the outside of the leaf. Older larvae bind two leaves together. The species is considered a minor pest of aspen, and when abundant, the larvae may defoliate entire trees, but the trees quickly recover the following season. The larvae are a uniform pale yellow green with a white middorsal line and pale subdorsal lines. The spiracles are dull white. The head is pale and unmarked.

Ipimorpha pleonectusa (Grote), the "even-lined sallow" or "blackcheeked aspen caterpillar" (FW 1.3–1.5 cm; Plate 54.30), ranges from Yukon to Newfoundland south to Washington, Arizona, Colorado, and, in the East, to Texas, Kentucky, and New Jersey. The adults are univoltine with a single flight from July to early September. They lay eggs that overwinter and hatch the following spring. Larvae feed on many species of *Populus* (Salicaceae), and a report of *Larix* may be apocryphal. Larvae feed within a shelter of two leaves sewn together. The larvae are green with white middorsal and subdorsal lines. The spiracles are white. The head is yellow white with the front margined by a black band.

SUBTRIBE ANTITYPINA

Andropolia Grote, with 14 species, is found only in North America. *Andropolia contacta* (Walker)(FW 1.7–2.2 cm; Plate 54.31) is found from Alaska east to Newfoundland south to southern British Columbia, northern California, Nevada, Colorado, and, in the East, to New York and New England. Adults have a single annual flight during July and August. Larvae feed primarily on Betulaceae *(Alnus* and *Betula)* and Salicaceae *(Populus* and *Salix),* but reports compiled by Robinson et al. suggest they might feed on plants in several other families. *Andropolia theodori* (Grote)(FW 1.9–2.1 cm; Plate 54.32) ranges from British Columbia south in moist coastal or boreal habitats to California, southern Arizona, and Colorado. Coastal individuals are relatively dark brown, while those in Nevada are pale gray, and those in Arizona are orangeish. Adults have a single flight from July to October, occasionally as early as mid-May in southern California. The larvae feed on willows, *Salix* (Salicaceae), and *Ceanothus* (Rhamnaceae), as well as ocean spray, *Holodiscus,* and plum, *Prunus* (Rosaceae).

Rhizagrotis albalis (Grote)(FW 1.5–1.6 cm; Plate 54.33) ranges from Alberta south to eastern California, Nevada, Utah, and Colorado. The species is found in pinyon-juniper woodland and prairie habitats. The adults fly from May to mid-July. The life history is unreported.

Fishia yosemitae (Grote)(FW 1.7–2.1 cm; Plate 54.34) ranges from British Columbia and Alberta south in montane habitats to California, Wyoming, and Colorado. There is a single flight of adults during late September to early November. The larvae feed on an unlikely combination of five plants: Douglas-fir, *Pseudotsuga menziesii* (Pinaceae), wild buckwheat, *Eriogonum* (Polygonaceae), Welsh poppy, *Meconopsis* (Papaveraceae), goldenrod, *Solidago* (Asteraceae), and mullein, *Verbascum* (Scrophulariaceae). The larvae are quite variable in ground color from greenish gray, purplish green, to pale brown. Middorsal and subdorsal pale lines are inconspicuous and are margined by dark lines. The spiracles are yellow or pale brown. The head is pale brown and either is unmarked or has green or brown reticulation and coronal stripes.

Platypolia contadina (Smith)(FW 1.2–1.8 cm; Plate 54.35) ranges from British Columbia and Alberta south to California, Utah, and Colorado. The adults fly during October and November. Larvae feed on a variety of broad-leaved shrubs, according to Poole. The larvae are quite variable from green, light brown gray, to deep purplish brown. There are pale inconspicuous middorsal and subdorsal lines that are bordered by dark. There is a straight dark spiracular line. The spiracles are white, and the subventral stripe is also white. The venter is green and tinged with purplish. The head is pale green or brown and may or may not have greenish or brownish reticulation and coronal stripes.

Dryotype opina (Grote)(FW 1.4–1.5 cm; Plate 54.36) ranges from British Columbia south to central California. There is a single flight during October and November. The larvae are reported to feed on *Lathyrus* (Fabaceae). The larvae are green with inconspicuous middorsal, subdorsal, and subventral white lines. The spiracles are white. The head is green and is slightly tinged with brown.

Mniotype ducta (Grote)(FW 1.7–2.0 cm; Plate 54.37) is found from British Columbia east across temperate Canada to Newfoundland and south to northern California and Colorado. Adults have a single flight during June and July. The larvae feed on *Myrica gale* (Myricaceae) and *Prunus serotina* (Rosaceae) according to reports compiled by Robinson et al. The larvae are described by Crumb as brown with pale middorsal and subdorsal lines that are edged with dark. There are dark middorsal spots at the junctures of the abdominal segments. The subventral pale stripes are white edged on its upper side, and the spiracles are pale brown. The venter is green and is spotted with white. The head is green with pale brown reticulation.

SUBTRIBE UFEINA

Ufeus Grote is limited to six species, four of which are North American, one species is a Mexican endemic, and the last species is described from India. Three of the North American species are limited to or range into the West. The larvae have a third L3 seta on the ninth abdominal segment, and the proleg plantae are exceptionally large and bear an unusually large number of crochets. Larval hosts are poplars, *Populus,* and willows, *Salix* (Salicaceae). *Ufeus satyricus* Grote (FW 1.6–2.2 cm; Plate 54.38) ranges from British Columbia east across southern Canada to New Brunswick and Nova Scotia and south to southern California east to Colorado and New Mexico. The species' range is incompletely known, perhaps because of its late flight in September and October. We illustrate the subspecies *sagittarius* Grote, which replaces typical *U. satyricus* throughout the West and is an even dark reddish brown. Lafontaine feels that it may well prove to be a distinct species, although the genitalia are identical. The life history details are unreported.

The subtribal placement is uncertain for **Hyppa** Duponchel and the next five genera. For *Hyppa*, four of five described species are North American, and the fifth is Eurasian. **Hyppa brunneicrista** Smith (FW 1.4–1.5 cm; Plate 54.39) ranges from British Columbia east across southern Canada to Newfoundland and Nova Scotia, and thence southward in montane conifer forests to California, Utah, Colorado, and the Great Lakes region. There are several closely related species of *Hyppa*, which have been reviewed by Troubridge and Lafontaine. The adults have a single flight from mid-June through July. The larval hosts have been reported as alder, *Alnus* (Betulaceae), and fireweed, *Chamerion angustatum* (Onagraceae). More of an indication of the actual host range is suggested by that of the related **H. xylinoides** (Guenée), which feeds on a variety of herbaceous and woody broad-leaved plant families. The larvae of *H. brunneicrista* are pale yellow and tinged with brown, while the more ventral portion is darker. There is middorsal brown band that contains a white line. There is a segmental series of subdorsal dark marks and pale oblique lines. The spiracles are pale and tinged with orange. The head has a brown infuscation underlain with closely placed black spots.

Chytonix divesta (Grote)(FW 1.5–1.6 cm; Plate 54.40) ranges from British Columbia south to southern California. Adults have a single flight from late June to early September. The larvae are reported to feed on grasses (Poaceae). The larvae are a pale black tinged with purple, with a yellow middorsal line and yellow subdorsal stripes. The spiracles are brown and the head is pale brown with dark brown flecks.

Properigea albimacula (Barnes & McDunnough)(FW 1.2–1.7 cm; Plate 55.1) ranges from British Columbia south to central California. The adults fly from May through September, possibly in two flights. The life history is unreported.

Pseudobryomima fallax (Hampson)(FW 1.2–1.5 cm; Plate 55.2) is endemic to California west of the Cascade–Sierra Nevada crest. The adults fly throughout the year. The first-instar larvae are reported by McCabe and colleagues to form blister mines on the leaves of their host ferns but later to feed within shelters. The larvae were reared from eggs obtained from a captured female and were fed the fern *Pellaea andromedifolia* (Pteridaceae). The larvae are green, and the setal pattern is described in detail by Crumb.

Pseudanarta flavidens (Grote)(FW 1.0–1.3 cm; Plate 55.3) ranges from Colorado south to southern California and southeastern Arizona. The adults have a single flight from August to November. The life history is unreported. **Pseudanarta flava** (Grote)(FW 1.0–1.1 cm; Plate 55.4) ranges from British Columbia south to California, Arizona, and Colorado. The adults are both crepuscular and nocturnal; their flight is from August to mid-October at which time they may be found nectaring at yellow-flowered asteraceous shrubs such as rubber rabbitbrush, *Ericameria nauseosa*. The larvae eat various grasses (Poaceae) and are light gray sometimes tinged with brown. There is a faint pale middorsal line broken by a small black spot at the anterior of each abdominal segment. There is also a series of faint dorsal rhomboidal patches that are most distinct on the posterior abdominal segments. The spiracles are brown, and

each is found within a pale area. The head is pale brown with black reticulation and coronal stripes.

Magusa orbifera (Walker), the "orbed narrow-wing" (FW 1.5–1.9 cm; Plate 55.5, Plate 55.6), has a very narrow elongate FW and is polyphenic. The FW may be unicolorous in various colors with or without a pale rounded apical patch. In addition, the FW may have a long pale band long the inner margin. The moth ranges from southern California, Arizona, and southern Texas east to southern Florida and south into the American tropics. The adults are excellent dispersers and have been found as strays north to Colorado, Manitoba, southern Ontario, and Maine. This species is attracted to light and can appear almost anywhere. In southern areas of residence, *M. orbifera* may be found throughout the year but, according to Covell, appear during late July to November in more northern areas. The larval hosts are various shrubby Rhamnaceae including *Condalia*, *Frangula betulifolia*, and *Karwinskia*. According to McFarland, the larvae may be sufficiently abundant to completely defoliate the host plant. The setal pattern was described by Crumb from a sample of preserved larvae from which the color pattern was not distinguishable.

SUBFAMILY HADENINAE

The subfamily Hadeninae is composed of hairy-eye trifid noctuids with larvae that have long, apically fringed spinnerets. Adults of almost all taxa are nocturnal and are attracted to light. The larvae of many species are described in detail by Godrey. The chaetotaxy and detailed comparison of head capsule setation and mandibular structure are included in that treatment.

TRIBE ORTHOSIINI

Perigonica angulata Smith (FW 1.3–1.6 cm; Plate 55.7) ranges from Oregon south to California, Arizona, and New Mexico. Adults fly from March to May. Larvae feed on chinquapin, *Chrysolepis chrysophylla*, and oaks, *Quercus* (Fagaceae). The larvae vary from green to yellow and have faint whitish or yellow longitudinal subdorsal and lateral lines. The head is pale brown with white setal bases and thin black coronal stripes.

Acerra normalis Grote (FW 1.6–2.1 cm; Plate 55.8) ranges along the Pacific Coast mainly west of the Cascade–Sierra Nevada crest from British Columbia south to southern California. Adults have a single flight from late December through April. Larvae feed on a variety of broad-leaved trees and shrubs in six families. Rosaceae, including *Amelanchier*, *Cercocarpus*, *Holodiscus*, *Prunus*, and *Purshia*, seems to be favored. In western Oregon, McFarland found that the larvae rest in leaf shelters or crevices by day and feed at night. The larvae are light gray to dark gray sometimes tinged with brown or purple. There are two black dorsal spots on the eighth abdominal segment that are joined by a transverse black line. The spiracles are black and placed within an intense white subventral stripe. The head is brown with faint pinkish or brownish reticulation and a black coronal area.

Stretchia muricina (Grote)(FW 1.3–1.4 cm; Plate 55.9) occurs from British Columbia and Alberta south to northern California, Wyoming, and Colorado. Adults fly in March and early April. The larvae feed on *Ribes* (Grossulariaceae). The overall larval color is a faint purplish, and the middorsal line and subdorsal lines are white and may be distinct or obscure. The spiracles are white with black rims. The cervical shield is pale brown bordered by a black line, and the head is brown and unmarked. **Stretchia plusiaeformis** Edwards is found from

British Columbia east to Saskatchewan and south to California, Wyoming, and Colorado. The larvae feed on and may cause damage to gooseberry, *Ribes* (Grossulariaceae); they are polyphenic and are either of a brown form or a green form. Both were described in detail by Comstock and Dammers.

Orthosia Ochsenheimer has about 60 described species, all of which are found in various regions of the temperate Northern Hemisphere. There are 23 species in North America with 16 species described from the West. Virtually all species have a single spring flight of nocturnal adults with the resultant larvae feeding on the spring growth of woody trees and shrubs. *Orthosia pacifica* (Harvey)(FW 1.6–1.9 cm; Plate 55.10) ranges from British Columbia south to California with additional records in southeastern Arizona. There is a single annual flight that occurs mainly during winter in California (January to April) but during spring at higher-elevation, interior regions such as Colorado (April to May). The larvae feed at night on the young leaves of trees and shrubs in several families, but *Quercus* species (Fagaceae) and Ericaceae are preferred. The larvae are pale green with distinct white middorsal and subdorsal lines and are flecked with white and have a white midventral suffusion. There is a transverse black mark on the dorsum of the eighth abdominal segment. The spiracles are pale yellow. The head is dark green with a yellow patch by the ocelli and black setal bases. *Orthosia hibisci* (Guenée), the "speckled green fruitworm" (FW 1.5–1.8 cm; Plate 55.11, Plate 64.29 larva), ranges from British Columbia east across southern Canada to Newfoundland and south in the West to California, Arizona, New Mexico, and Texas. *Orthosia hibisci* is found through much of the northeastern United States but is absent from most of the Great Plains and Southeast. There is a single late-winter to spring flight. In California, this occurs from late January to early May but is later in more interior regions such as Colorado (March to April). The larvae feed on a wide variety of trees and shrubs including both conifers and broad-leaved species. The larvae are considered to be pests of rosaceous orchard trees including apple, crab apple, cherries, and plums. The larvae vary from whitish to blackish green and are covered with small white flecks. There are thin middorsal and subdorsal lines, which are most distinct on the dark green phase individuals. The head is pale green and may or may not have pale brown reticulation. *Orthosia praeses* (Grote)(FW 1.4–1.5 cm; Plate 55.12, Plate 55.13) ranges along the Pacific Coast from British Columbia south to California. The moths are polyphenic from plain brown to well marked with pale orbital, reniform, and outer margin contrasting with dark ground. The adults have a single spring flight in California from late December to early May. Its larvae feed on a variety of deciduous broad-leaved shrubs in several families, particularly Rosaceae. The larvae are brown with black subdorsal markings and obscure pale middorsal and subdorsal lines. The spiracles are pale and are rimmed with black. The head is pale brown with black setal insertions and faint black reticulation and coronal stripes. *Orthosia mys* (Dyar)(FW 1.4–1.9 cm; Plate 55.14) ranges south along the Pacific Coast from British Columbia to California and southern Arizona. The FW color varies from brown to reddish orange. The adults have a single winter flight from November through January. We found larvae on manzanita, *Arctostaphylos,* and madrone, *Arbutus menziesii* (Ericaceae).

Like *Orthosia,* **Egira** Duponchel is Holarctic with 22 species found in the Holarctic region, seven of these are found only in Eurasia. There are 15 species in North America, 12 of which range into or are endemic to the West. Most species are single-brooded with a spring flight and the resultant larvae feeding on the new foliage of various woody trees and shrubs. *Egira simplex* (Walker)(FW 1.6–1.7 cm; Plate 55.15) ranges from southern British Columbia south along the Pacific Coast to northern California, and in the interior south through the Rocky Mountains to Arizona and New Mexico. Adults have a single late-winter to spring flight from February to May. Larval hosts include both broad-leaved trees such as *Alnus* (Betulaceae), *Prunus* (Rosaceae), and *Salix* (Salicaceae), as well as conifers (Pinaceae). Coniferous hosts include Sitka spruce, *Picea sitchensis,* Douglas-fir, *Pseudotsuga menziesii,* and grand fir, *Abies grandis.* The early-instar larvae feed in the opening buds, later switch to new foliage. Although not normally considered to be of importance, a large population once caused significant defoliation of Sitka spruce in British Columbia, and in 1964 about 8,000 acres of Douglas-fir and associated conifers were defoliated in western Oregon. Eggs are laid in a group on foliage, and the young larvae feed in opening buds and expanding foliage. Full-grown larvae are brown with faint white mottling, distinct paired black addorsal spots, middorsal and subdorsal white lines, and a broad orange brown spiracular stripe. The head is black and unmarked. The pupa estivates and overwinters in organic matter under the host. *Egira perlubens* (Grote), the "brown woodling" (FW 1.5–1.8 cm; Plate 55.16), ranges from British Columbia south to California, Arizona, and New Mexico. There is a single winter to spring flight in California from January through April. In more inland locations such as the Colorado Rockies the species is found as late as May. The larvae feed primarily on coniferous trees (Pinaceae) but also include several families of broad-leaved trees in their diet. *Egira dolosa* (Grote)(FW 1.5–1.6 cm; Plate 55.17, Plate 64.32 larva) ranges from interior British Columbia and Northwest Territories across Canada and the northern United States to eastern Canada and New England. In the West it is found south to California and Colorado. There is a single annual flight during the second half of April and early May. The larvae feed primarily on Salicaceae but also frequently select birches, *Betula* (Betulaceae), and oaks, *Quercus* (Fagaceae), as hosts. The larvae vary from whitish to gray with heavy black reticulation that is arranged in transverse dorsal bands. There are also obscure thin white middorsal and subdorsal lines. The head is blackish brown with brown reticulation and coronal stripes.

TRIBE THOLERINI

Nephelodes minians Guenée, the "bronzed cutworm" (FW 1.4–1.9 cm; Plate 55.19), ranges from British Columbia and Northwest Territories east across Canada to Newfoundland and south to California, Arizona, New Mexico, Texas, and, in the East, to Georgia. Ferris has shown that western adults are much smaller than those found in the Mid-Atlantic region. There is a single flight of adults from August to October. Larvae feed on various grasses including corn and cereal crops (Poaceae), although there are reports of larval use of several broad-leaved plants in other families. These reports may be erroneous. The larvae are dark brown with a bronzy sheen and are distinctly paler ventrally. There is a broad white waxy band below the spiracles which are black. The cervical shield and anal shield are both shiny black with three pale lines. The head is dark brown with darker reticulation.

TRIBE HADENINI

Anarta trifolii (Hufnagel), the "clover cutworm," or "nutmeg" (FW 1.4–1.6 cm; Plate 55.18), is found from Alaska and Yukon

Territory east to Newfoundland and south to Mexico and Virginia. It is also found widely in Eurasia. There are several flights from May to October. Adults are nocturnal and are attracted to light. Larval hosts include a wide variety of broad-leaved herbaceous plants including crop, ornamental, and weedy species. goosefoot, *Chenopodium*, sugarbeet, *Beta vulgaris*, and Russian thistle (a.k.a. tumbleweed), *Salsola* (Chenopodiaceae), seem to be preferred. The larvae are variable in overall color from the usual dusky green to brown, yellowish, or gray. The integument has a reticulate pattern with black-and-white flecks on the green ground, and there may or may not be various white stripes and black dashes. The spiracles are white. Godfrey provides a detailed technical description of the setal pattern. *Anarta oregonica* (Grote)(FW 1.3–1.5 cm; Plate 55.20) ranges from British Columbia east to Montana and south to California, Utah, and Colorado. This dark species is found in montane habitats usually in coniferous woodlands, but Slater and Opler have found it as high as the lower alpine zone. The adults have a single flight from mid-May through July. The life history has not been reported. *Anarta farnhami* (Grote)(FW 1.4–1.6 cm; Plate 55.21) is a boreal species that ranges from Yukon Territory east to Saskatchewan and south to California, Utah, and Colorado. This is a species of boreal habitats being found in spruce-fir habitats and even in alpine situations on occasion. The adults have a single flight from late May through July. The early stages have not been reported. *Anarta crotchi* (Grote)(FW 1.4–1.7 cm; Plate 55.22) ranges from British Columbia east to Saskatchewan and south to California, Utah, and Colorado. The adults fly from mid-April to mid-June. *Anarta nigrolunata* Packard (FW 1.0–1.1 cm; Plate 55.23) occurs at high elevations, usually at or above timberline in the Rocky Mountains from Washington and Montana south to Utah, Colorado, and northern New Mexico. The adults are diurnal and have a single flight in July and early August. They may be found nectaring on flowers of bistort, *Polygonum*. The larvae feed on *Vaccinium* (Ericaceae). A closely related and nearly identical species, *A. melanopa* (Thunberg), the "broad-bordered white underwing," occurs widely in Eurasia; its larvae feed on *Arctostaphylos uva-ursi* and *Empetrum* (both Ericaceae).

Scotogramma fervida Barnes & McDunnough (FW 1.3–1.5 cm; Plate 55.24) is resident from Alberta and Saskatchewan in western Canada south to California and New Mexico. The species is normally associated with arid desertlike habitats. Adults are found from April through July, and there are possibly two generations. *Scotogramma submarina* (Grote)(FW 1.3–1.4 cm; Plate 55.27) is found from Alberta and Saskatchewan south to the high plains and intermountain regions of Utah and western Colorado. There are two generations with adults in April to May and again in September.

Coranarta luteola (Grote & Robinson)(FW 1.0–1.1 cm; Plate 55.28) ranges from Alaska, Yukon, and Northwest Territories east to Labrador then south in the Rocky Mountains to British Columbia and, discontinuously, to Colorado. The species is found in subarctic and subalpine peat bogs and meadows. There is a single diurnal flight from mid-May to mid-July. Adults have been reared from larvae collected on *Kalmia polifolia* (Ericaceae).

Polia Ochsenheimer comprises 30 species that are limited to the Holarctic region; 11 of these are North American, and most are found in the West. *Polia nimbosa* (Guenée), the "stormy arches" (FW 2.0–3.2 cm; Plate 55.25), is found from British Columbia east to Newfoundland south to California and Colorado. In the East it can be found south to Kentucky

and North Carolina. The adults have a single flight in July. Larvae have been found feeding on trees and shrubs in five plant families but seem to favor *Alnus* and *Betula* (Betulaceae). The larvae are brown with an orange tinge on paler areas. On the dorsum of each abdominal segment there is a dark brown diamond. There are thin, indistinct white middorsal and subdorsal lines as well as dark brown subdorsal oblique lines. Lateral areas are whitish to pinkish brown, and the spiracles are orange and rimmed with black. *Polia purpurissata* (Grote), the "purple arches" (FW 1.8–2.2 cm; Plate 55.26), ranges from Yukon and Northwest Territories east to Nova Scotia and south to California, Arizona, New Mexico, Missouri, and New Jersey. Adults have their primary flight during July and early August. Larvae feed on a variety of broad-leaved trees and shrubs, especially *Vaccinium* (Ericaceae) and various Rosaceae. The larvae are yellow gray or pale violet and are mottled with diffuse gray. The cervical shield is shiny brown. Some individuals have a pale gray diamond mark on each segment.

Lacanobia Billberg has 13 species, and nine of these are restricted to temperate Eurasia; the remainder are North American. *Lacanobia nevadae* (Grote)(FW 1.5–1.8 cm; Plate 55.29) ranges from British Columbia east across southern Canada to Quebec, where it is a rare find, and south in the West to California, Utah, Wyoming, and Colorado. The adults have a single flight from June through early September. The larvae, as described by Godfrey, are violet brown with a slight pink tinge. There is a dorsal series of black chevrons with one on each segment. The chevrons on the seventh and eighth abdominal segments are edged with reddish laterally. The lateral area is pink with numerous white flecks. The head is dark brown with darker brown reticulation and coronal stripes. *Lacanobia subjuncta* (Grote & Robinson), the "speckled cutworm moth" (FW 1.8–2.0 cm; Plate 55.30), occurs from British Columbia east across Canada to Nova Scotia and south in the West to California, Arizona, and New Mexico. In the East, the moth ranges south to Missouri and Virginia. The adults fly during June and July. Robinson et al. compiled a number of reports that the larvae feed on plants in 12 families that include trees, shrubs, and herbaceous plants. A number of cultivated crops such as asparagus, corn, and strawberries are among its hosts. Crumb described the larvae as brown or yellowish with oblique dark marks that meet middorsally and form a small distinct black spot at the junctures of the segments. The spiracles are white, and the head is pale with black reticulation and coronal stripes.

Melanchra picta (Harris), the "zebra caterpillar" (FW 1.4–1.7 cm; Plate 55.31), ranges from British Columbia east to Newfoundland and south to California, Utah, Wyoming, and Colorado. Further east it ranges south to the Great Plains and Mid-Atlantic states. The adults fly from May to September in what must represent two generations. The larvae, as described by Godfrey, are boldly marked white, black, and yellow. The dorsal area is solid velvety black extending from the yellow subdorsal band and interrupting the lower edges of the dorsal pattern. Below the yellow band there is another velvety black band that is interrupted by thin vertical white lines. The lateral area is yellow with black spotting, and the venter is black with white reticulation. The prolegs are pinkish orange. The head is orange or reddish.

Papestra cristifera (Walker)(FW 1.5–1.8 cm; Plate 55.32) occurs from Alaska and Northwest Territories to Newfoundland south to California, Utah, Colorado, and, in the East, to central Quebec. Adult *cristifera* fly from May to early August. Robinson et al. catalogued host plants in nine families, mostly broad-leaved woody trees and shrubs. Several conifers are included among the selected plants. The larvae are pale gray with a dorsal series of dark shield-shaped marks on each segment. Laterally, there is an oblique dark mark on each segment. The spiracles are white, and the head is white with brown reticulation and submedian arcs.

Hada sutrina (Grote)(FW 1.2–1.4 cm; Plate 55.33) ranges in boreal coniferous forests from British Columbia and Yukon Territory east across Canada to Labrador, Newfoundland, and New Brunswick, thence south to Oregon and Colorado. In Colorado, it favors spruce-fir and lodgepole pine forests and their edges. The single annual flight is from mid-June until mid-August. The life history is unreported.

Mamestra configurata (Walker), the "bertha armyworm" (FW 1.5–1.8 cm; Plate 55.34), ranges from British Columbia east to Saskatchewan and south to central California, Arizona, Texas, and Mexico. There are two flights of adults from April to August. The larvae have been reported to feed on herbaceous plants and grasses in 17 plant families, but they favor Chenopodiaceae and Fabaceae. The larvae were described by both Crumb and Godfrey. The larval body color varies from olive green, olive gray, or brown, to black. Conspicuous black dorsal marks may be present or absent. There are inconspicuous pale middorsal and subdorsal lines, and the lateral area is yellow or orange. The spiracles are white. The head is pale brown with or without reticulation and coronal stripes.

Sideridis rosea (Harvey), the "rosewing" (FW 1.5–1.8 cm; Plate 55.35), is found from British Columbia east to New Brunswick and south to Nevada, Utah, and Colorado. The adults fly from mid-May to mid-July, although the specific flight period may be more restricted at any given site. The larval hosts include *Elaeagnus*, *Ribes*, *Salix*, and *Shepherdia*. Larvae are translucent gray brown with a slight pinkish tinge, according to Godfrey. The spiracles are whitish yellow. There are faint thin middorsal and subdorsal lines. The cervical shield is shiny yellow brown, and the head is yellow brown with brown reticulation and coronal stripes.

Admetovis oxymorus Grote (FW 1.6–1.8 cm; Plate 55.36) ranges from British Columbia and Montana south to California, Arizona, and Colorado. There is a single adult flight during June and July. The larvae feed on elderberry, *Sambucus nigra* (Caprifoliaceae). Godfrey describes the setal pattern from preserved larvae, and no color description is available. A related moth, *A. similaris* Barnes, ranges from British Columbia and Idaho south to southern California and southwestern New Mexico. The adults of *A. similaris* fly from mid-March to June and rarely during October.

Hadena variolata (Smith)(FW 1.4–1.7 cm; Plate 55.37) ranges from British Columbia and Montana south to eastern California, Arizona, and New Mexico. The species has a single flight during July. Larvae feed on *Lychnis* (Caryophyllaceae) and are yellow brown with dark markings. A dorsal series of black diamonds on each abdominal segment may or may not be present. There is a diffuse dark oblique line on each segment. The spiracles are blackish brown, and the head is pale with dark posterad reticulation and black coronal stripes. *Hadena capsularis* (Guenée), the "capsule moth" (FW 1.2–1.3 cm; Plate 55.38), ranges from British Columbia and Manitoba east to New England and south to northern California, Arizona, Colorado, and most of the eastern United States. The adults have a single flight from late April to early July. The larvae feed in seed capsules of pinks, *Dianthus* and *Silene* (Caryophyllaceae). Godfrey's description of the larvae is based on preserved material, and the color pattern was not clear, although the larvae did have a series of dark diamonds on the dorsum of the abdominal segments.

Trichoclea uscripta (Smith)(FW 1.4–1.6 cm; Plate 55.39) ranges from Alberta and Manitoba south to the central Sierra Nevada of California and southeastern Arizona. There is a single flight of adults during June and early July. The life history is unreported.

Faronta Smith has 13 species limited to the New World; six are North American, and four of these are western or range into the West. *Faronta diffusa* (Walker), the "wheat head armyworm" (FW 1.2–1.6 cm; Plate 55.40), ranges through much of temperate North America from Northwest Territories and British Columbia eastward across Canada to Newfoundland and most of the United States except for the Southeast. The adults fly from March to October in two to three generations. The larvae feed on various grasses (Poaceae), and the species is considered a pest of corn and wheat. The larvae may be green, olive green, yellow, or pink to brown. The dorsum is uniformly dark with a middorsal pale line. There is a supraspiracular black band. The spiracles are pale yellow with black rims. The head is concolorous with the general color of the body, and there may or may not be darker reticulation and submedian stripes.

TRIBE LEUCANIINI

Dargida procincta (Grote), the "girdler moth" (FW 1.8–2.0 cm; Plate 55.41), ranges from British Columbia east to Saskatchewan and south to southern California, Nevada, Utah, and Colorado. The adults fly through most of the year (except December and most of January) in several flights, are nocturnal, and are attracted to both fermented bait and light. The larvae feed on grasses, especially reed canary grass, *Phalaris arundinacea* (Poaceae). The larvae vary from green to black with pale longitudinal lines that may be tinged with salmon or pink. The spiracles are white with black rims. The head varies in color and may or may not have reticulation. Pupation is in a slight cocoon.

Mythimna oxygala (Grote), the "lesser wainscot" (FW 1.3–1.5 cm; Plate 55.42), ranges from British Columbia east to Saskatchewan and south to northern California and Colorado. In the West, adults fly during July and August. The larvae feed on *Cichorium* (Asteraceae), *Carex*, (Cyperaceae), and *Dactylis* (Poaceae). The larval dorsum is yellow brown with brown mottling, and the lateral area is pale yellow with yellow brown mottling. There is a middorsal pale orange brown stripe with two included thin yellow white lines. The head is yellow brown with brown reticulation and coronal stripes. *Mythimna unipuncta* (Haworth), the "armyworm" (FW 1.6–1.9 cm; Plate 56.1, Plate 64.33 larva), was previously referred to as *Pseudaletia unipuncta*. The species is wide-ranging from southern Canada south through the United States, Mexico, and Central America to northern South America. There are also scattered records from Europe. The adults fly from late January to early December in two or three generations. The larvae prefer grasses (Poaceae), but when preferred hosts are depleted feed on a wide variety of wild and cultivated plants and can reach destructive levels on alfalfa, corn, other grains, vegetables, and young fruit trees. According to Covell the larvae feed in masses at night then migrate to other fields when their food resources are depleted, hence their common name. The larval dorsum is gray brown with mottled black with obscure thin whitish dorsal and subdorsal lines. The subdorsal stripe is brownish with numerous white flecks. The lateral area is reddish brown with numerous white flecks. The head is yellow brown with brown reticulation and dark brown coronal stripes.

Leucania Ochsenheimer comprises roughly 350 species found through much of the temperate and tropical regions of the World; all species for which the larval host is known feed on grasses. *Leucania multilinea* Walker, the "many-lined wainscot" (FW 1.6–1.8 cm; Plate 56.2), ranges from British Columbia and Montana east to New England and south to Arizona, Colorado, and Florida. The adults have a single flight from June to early August. The larvae feed on various grasses (Poaceae) including smooth brome *(Bromus inermis)*, orchard grass, and quackgrasses. Larvae are pale yellow brown with a black-edged dorsal band, and white dorsal stripe with two pinkish longitudinal lines. The lateral area is whitish yellow, and the venter is translucent gray with white flecking. The head is yellow brown with brown reticulation and dark brown coronal stripes. *Leucania commoides* Guenée (FW 1.5–1.7 cm; Plate 56.3) occurs from British Columbia and Montana east to Nova Scotia and south to Arizona, New Mexico, and Florida. There is a single flight in the West during July and August. Larvae feed on various grasses (Poaceae) including *Dactylis* and *Glyceria*. The larvae have the dorsal area dark yellow to lavender brown with a thin middorsal line and blackish margins. The dorsal stripe is white with a thin pink brown longitudinal line, and the lateral area is pink brown with a white dorsal margin and many white flecks. *Leucania insueta* Guenée (FW 1.4–1.6 cm; Plate 56.4) ranges from Yukon Territory and British Columbia east to New Brunswick and New England and south to California, Arizona, Colorado, and the Mid-Atlantic coast. Adults are found from May through August in California; very likely, there are two broods. In the Rockies, they fly from June to early August, and, probably, there is only a single generation. Larvae feed on grasses (Poaceae). Larvae are pale yellow to gray brown. The dorsal area is whitish yellow with black setal bases. The head is yellow brown with light brown reticulation and coronal stripes. *Leucania farcta* (Grote)(FW 1.7–2.1 cm; Plate 56.5) has a distinctive pink overtone to the FW that distinguishes it from similar species. The moth ranges from British Columbia, Idaho, and Montana south to California, including Santa Cruz Island, southern Arizona, and Colorado. Adults fly from April to August in two flights. The larvae feed on broad-leaved grasses (Poaceae) such as *Dactylis* and *Elymus*. The larvae vary from gray to pale brown. The spiracles are black. The head is pale brown with darker reticulation and coronal stripes.

TRIBE ERIOPYGINI

Lasionycta Aurivillius with 51 described species is distributed throughout temperate and boreal habitats in the Holarctic region; 33 species are North American of which 29 are western. *Lasionycta impingens* (Walker)(FW 1.2–1.4 cm; Plate 56.6) ranges from Yukon south to British Columbia, Colorado, and northern Manitoba. The moths are found in high-elevation spruce-fir or alpine habitats. The adults fly during late July and August. Pogue reports a disjunct population in New Hampshire. The adults are diurnal and are often seen nectaring at flowers such as alpine bistort, *Polygonum* (Polygonaceae), and yellow composites (Asteraceae). *Lasionycta perplexa* (Smith)(FW 1.3–1.6 cm; Plate 56.7) ranges from Yukon Territory south to Washington and Alberta and discontinuously to the boreal forest zone of Utah, Wyoming, and Colorado. This seems to be a complex of similar species with one or more that are undescribed. The adults are nocturnal, are attracted to light, and

have a single flight from late June to mid-August. The larvae are light dusky brown dorsally with the sides blackish. The spiracles are deep dusky brown. The cervical shield is black with three white lines. The head is pale laterally, suffused with fuscous dorsally, and with faint dark freckles. *Lasionycta arietis* (Grote)(FW 1.7–2.1 cm; Plate 56.8) ranges from British Columbia south to northern California. Adults have been found during September. Larvae feed on grasses (Poaceae), *Abronia* (Nyctaginaceae), *Lathyrus* (Fabaceae), and *Polygonum* (Polygonaceae) on beach dunes. The larvae are pinkish with a faint middorsal line and a pronounced white subspiracular line. The venter is gray. The spiracles are brown. The head is pale with light brown freckles.

Lacinipolia McDunnough comprises about 95 species; 21 species are endemic to Mexico or the Antilles, and the remainder is North American, mostly western. *Lacinipolia anguina* (Grote), the "snaky arches, " (FW 1.1–1.2 cm; Plate 56.9), ranges from British Columbia east across southern Canada to Newfoundland and south to Colorado, Nebraska, Arkansas, and Kentucky. The adults have a single flight from April to early July. The host and early stages are unreported. *Lacinipolia renigera* (Stephens), the "bristly cutworm" (FW 1.1–1.4 cm; Plate 56.10), is a common moth that ranges from British Columbia and Northwest Territories east across southern Canada through most of the eastern United States and south to California, Arizona, Texas, Mississippi, and Georgia. The adults may be found from May to October. In the West, there are two flights in most regions. The larvae feed on both wild and cultivated broad-leaved plants, particularly herbaceous species, in many families. The larvae are dark gray brown with pinkish brown middorsal and subdorsal lines. There is a diffuse blackish diamond mark on the dorsum of each segment. The lateral areas are pinkish brown. The head is brown with blackish reticulation and coronal stripes. *Lacinipolia lorea* (Guenée), the "bridled arches" (FW 1.2–1.4 cm; Plate 56.11), ranges from British Columbia east across southern Canada and the northern United States to Newfoundland south to Utah, Colorado, and, in the East, to Missouri and Virginia. There is a single adult flight from May to August, the exact timing depending on latitude and elevation. The larvae feed on a wide variety of broad-leaved plants in many families. The larvae are pale violet or brownish gray with a segmental series of gray diamonds. The subdorsal area has a series of blackish oblique lines, and the lateral area is brown with many yellow white flecks. The head is brown with dark brown to blackish reticulation and coronal stripes. *Lacinipolia olivacea* (Morrison), the "olive arches" (FW 1.0–1.1 cm; Plate 56.12), is found from Alaska to Newfoundland south to northern California, Arizona, New Mexico, and North Carolina. The adult coloration and maculation is quite variable over the species' range, but there does seem to be only a single species, based on analyses by Ferris. The adults have a single flight from June to September. Larvae feed on herbaceous broad-leaved plants in a variety of families. The larvae vary from brown to blackish brown. The dorsum is yellow gray with a segmental series of diffuse gray diamonds. The spiracles are black. The head is brown with dark brown to blackish reticulation and coronal stripes. *Lacinipolia lustralis* (Grote)(FW 1.1–1.3 cm; Plate 56.13) ranges from British Columbia east across southern Canada to Newfoundland and south to Arizona, Colorado, Nebraska, and Virginia. There is a single flight in late June and the first half of July. Larvae were reared from eggs obtained from captive females. The hosts used by such larvae do not necessarily reflect the hosts selected by females in nature. The larvae are brown and have the dorsum pink with white flecks and a segmental series of brown diamonds. The cervical shield is dark brown with distinct middorsal and subdorsal pink lines. The lateral area is brownish with white flecks. The spiracles are black. The head is yellow brown with dark brown to blackish reticulation and coronal stripes.

Trichocerapoda oblita (Grote)(FW 1.2–1.3 cm; Plate 56.14) occurs from British Columbia south to southern California and Colorado. The species is found in short-grass prairie and pinyon-juniper woodland. The adults are bivoltine and fly during June and again in October and November.

Anhimella contrahens (Walker)(FW 1.2–1.3 cm; Plate 56.15) occurs from Northwest Territories and British Columbia east across Canada to Nova Scotia and south to California, Arizona, Colorado, and, in the East, to New Jersey. The flight period is from late June to early August. Captive larvae will feed on dandelion, *Taraxacum officinale* (Asteraceae), but the host selected by females in natural populations is unknown. Larvae are yellowish brown with a pink tinge and tiny yellow white spots. The middorsal and subdorsal lines comprise an interrupted series of small yellowish white dots and dashes. The cervical shield and anal plate are gray brown. The spiracles are black. The head is pale yellow brown with light brown reticulation and coronal stripes.

Homorthodes furfurata (Grote), the "scurfy Quaker" (FW 1.0–1.1 cm; Plate 56.16), ranges from British Columbia and Northwest Territories east across Canada and south to California, Arizona, New Mexico, Texas, and, in the East, to Mississippi. In California, adults have been collected from June through August. The larvae are reported to feed on plants in several broad-leaved families. Godfrey states that the larvae are variably dark gray to blackish with middorsal and subdorsal lines comprised of white dots and dashes. The spiracles are brown and the head is dark orange brown and may or may not have brown speckling. *Homorthodes communis* (Dyar)(FW 1.0–1.1 cm; Plate 56.17) ranges from British Columbia south to southern California, southern Arizona, and southwestern New Mexico. Adults have several flights from April to October in California. The larvae feed on alders, *Alnus* (Betulaceae). Crumb describes the larvae as dark brown with pale broken middorsal and subdorsal lines that are bordered by dark lines. There is a segmental series of dorsal dark diamonds as well as obscure pale oblique marks. The dorsum of the eighth abdominal segment has a large yellow or orange patch crossed transversely by a curved black line. The spiracles are dark brown, and the head is brown with a pattern of small black freckles.

Protorthodes McDunnough has 21 species in North America, 17 of which occur in the West. *Protorthodes oviduca* (Guenée), the "ruddy Quaker" (FW 1.1–1.3 cm; Plate 56.18), ranges from British Columbia east to Newfoundland south to California, Arizona, Colorado, and, in the East, to Mississippi and Florida. The bivoltine adults fly during May and again in August and September. The larval foods are a variety of low herbaceous broad-leaved plants and grasses. The larvae are various shades of gray. The cervical shield is cream white with black lateral and anterior edges, while the anal shield is white laterally edged with black. The spiracles are black and the head is infuscated with black. *Protorthodes rufula* (Grote)(FW 1.3–1.6 cm; Plate 56.19) ranges from British Columbia south to California and possibly to southeastern Arizona. In California, adults may be found from March to October, and there are likely two or more broods. The larvae feed primarily on

various rosaceous trees and shrubs including *Malus, Prunus,* and *Pyrus*. Reports of their use of other plants may be erroneous. The larvae are dark gray with a pink or red tinge dorsally and sometimes with a bronze sheen. There are pale middorsal and subdorsal lines and a segmental series of dark dorsal ovoid markings. The spiracles are brown with black rims. The cervical and anal shields are white with black lateral edging. The head is pale with brown or black reticulation and coronal stripes. The larvae pupate in the soil. *Protorthodes utahensis* (Smith)(FW 1.1–1.2 cm; Plate 56.20) ranges across the intermountain West and Canadian prairies from Alberta east to Saskatchewan and from Nevada and Montana south to Arizona and New Mexico. The larvae are reported by Godfrey to feed on filaree, *Erodium cicutarium* (Geraniaceae), and possibly other low herbaceous plants. The larvae are dark gray with a segmental series of dark ovoid marks on a paler ground color. The cervical shield is pinkish white with black anterior and lateral edges. The anal shield is pale with marked laterally with black. The head is pale brown with darker brown coronal stripes and partially obscured dark brown reticulation.

Ulolonche disticha (Morrison)(FW 1.2–1.4 cm; Plate 56.21) is actually a complex of at least three western species based on preliminary studies by Ferris and Pogue. The complex ranges from Alberta south to southern California, Arizona, New Mexico, and Texas. Adults have been found from late August through September. The life history stages are not reported.

Pseudorthodes irrorata (Smith)(FW 1.2–1.6 cm; Plate 56.22) ranges from British Columbia south along the Pacific Coast to California. Adults have been found in California from March to early August. The larvae are reported to be relatively polyphagous on a variety of broad-leaved and coniferous trees and shrubs. The last instars are light brown and tan but vary appreciably between individuals. The dorsum of the first through eighth abdominal segments has longitudinal white streaks. There is a wavy black-and-white lateral line below which the larva is paler. The head is dull red orange.

Orthodes majuscula Herrich-Schäffer, the "rustic Quaker" (FW 1.2–1.4 cm; Plate 56.23), ranges from Alberta to Nova Scotia and southward in the West to Utah and Arizona and in the East to Florida and the West Indies. The adults are in flight from June to mid-September. The larvae feed on several unrelated broad-leaved plants in different families. The larvae are soft gray brown with a complex pattern of markings. The dorsum of the abdomen has a segmental series of dark brown chevrons and dorsal black spots. The black spots of the first two abdominal segments are large, and those on the eighth segment are small. The intervening segments have diffuse spots. The first abdominal segment also has two thin yellow crescents immediately behind the middorsal black spot. The lateral area is gray with heavy white flecking. The spiracles are orange brown, and the head is yellow brown with pale brown coronal stripes and black lateral stripes.

Hexorthodes agrotiformis (Grote)(FW 1.4–1.5 cm; Plate 56.24) occurs from Utah east to Wyoming and Colorado and south to Arizona, and western New Mexico. There is some confusion about this species' identity and, according to Ferris and Pogue, there may be a complex of several species. The adults are found during July. Larvae were obtained from eggs laid by captive females and fed dandelion, *Taraxacum officinale* (Asteraceae). The larvae are brown with the thin white middorsal line interrupted by a series of blackish spots at the anterior margin of each segment. Otherwise, the dorsum is yellow brown with series of brown diamond marks on the first through seventh abdominal segments. The subdorsal areas have heavy dark brown mottling with white flecks. The spiracles are black. The head is yellow brown with distinct brown spots.

Zosteropoda hirtipes Grote (FW 1.2–1.3 cm; Plate 56.25) ranges from British Columbia south along the Pacific Coast states to southern California and south in the Rocky Mountains to Colorado. The adults fly from January to mid-September in California and may have two or three flights. The larvae are reported to feed on plants in several unrelated broad-leaved families as well as grasses (Poaceae). The larvae are gray with the dorsum somewhat paler and flecked with black. There are several alternating longitudinal dark and white lines and stripes. The spiracles are white. The head is pale with black coronal stripes that are each bordered laterally with a white line. A related species, *Z. clementei* Meadows, is endemic to the California Channel Islands and is in flight from March through May.

Neleucania bicolorata (Grote)(FW 1.2–1.3 cm; Plate 56.26) ranges from Alberta and Montana south through the western United States to California, Arizona, New Mexico, and western Texas. The single adult flight is from late May through July. Larvae feed on a variety of broad-leaved grasses (Poaceae) including *Bromus tectorum*, *Dactylis glomerata*, and *Digitaria*. The larvae vary from whitish to yellow brown with brown mottling and black setal bases thus giving them a peppered appearance. There is a dorsal series of yellow brown rhomboidal figures on the abdominal segments. The spiracles are white with black peritremes. The head is yellow brown with dark brown reticulation and coronal stripes. There is a lateral whitish stripe that appears continuous with that of the body.

Miodera stigmata Smith (FW 1.2–1.5 cm; Plate 56.27) is endemic to California. The adults have a single flight from November to January. The larvae feed on *Artemisia californica* (Asteraceae). J. A. Comstock and Dammers described the life history. The larvae are dark green with thin longitudinal dark lines and spots above the spiracular line. There are also white longitudinal middorsal, subdorsal, supraspiracular, and subspiracular white bands or lines. The spiracles are white with black rims. The head is green with inconspicuous brown freckles.

Engelhardtia ursina (Smith)(FW 1.3–1.9 cm; Plate 56.28) ranges from the mountains of Kern and Los Angeles counties east across southern Utah to western Colorado. The species favors pinyon-juniper woodland. The single flight is from January to mid-April. The life history is unreported.

Tricholita chipeta Barnes (FW 1.3–1.4 cm; Plate 56.29) occurs in the Southwest from California, Utah, and Colorado south to Arizona, New Mexico, and western Texas. Adults have a single generation and fly from late August to October. The larvae feed on *Gutierrezia wrightii* (Asteraceae). The larvae vary from dull gray to brown. The dorsum is paler and has an obscure pale middorsal line that is margined on each side by a darker area. The supraspiracular area is blackish. The spiracles are dark brown, and the cervical shield is pale brown flecked with black. The head is pale brown and is densely covered with black spotting.

Hydroeciodes serrata (Grote)(FW 1.2–1.5 cm; Plate 56.30) looks much like a *Papaipema* at first glance. The species ranges from Alberta south to Utah, Colorado, and northern Texas in a wide range of habitats. Adults fly from late July to early September. The host and life history are unreported.

SUBFAMILY NOCTUINAE

The subfamily Noctuinae has traditionally included those trifine noctuids with spined tibiae and unlashed, hairless eyes,

but these characters are considered untenable by Kitching and Rawlins. Nevertheless, there are some potentially apomorphic characters that seem to unite the subfamily. The wings of adults are held in a parallel (not rooflike) posture, and abdominal brush organs are absent in males of almost all taxa. The primary setae of first-instar larvae are borne on clubbed pinacula. The adults of several species are emigratory and may establish pestiferous population levels during the warmer months or may form aestivating masses at high elevations during the summer. The larvae are generally polyphagous and hide in the soil or between leaves by day; they ("cutworms") feed at night by cutting pieces of vegetation and then removing them to their shelters for feeding. A number of species are serious agricultural pests to crops and rangeland. In North America there are 60 genera with roughly 490 species divided among the Agrotini and Noctuini. In the West, Lafontaine reports 43 genera and about 422 species.

TRIBE AGROTINI

In North America, the tribe Agrotini comprises 14 genera and at least 327 species. The majority of these species, 190, belong to the genus *Euxoa* Hübner. All 14 genera have at least one species that includes part of the West in its distribution. Lafontaine reports 284 species that are endemic to or include part of the West in their distribution. Most agrotines occur in association with well-drained soils in arid-land habitats, and the vast majority of species seem to be attuned to particular habitats rather than to one or more specific host plants. The adults of the vast majority of species are nocturnal and are attracted to light, but they are also attracted to flowers, especially those of rubber rabbitbrush, *Ericameria nauseosa* (Asteraceae), during late afternoon and evening. Two exceptions to the nocturnal rule are (1) a group of four far-northern boreal *Feltia* that are diurnal and found only on scree slopes, and (2) a number of alpine *Euxoa* of the southern Rockies that are diurnal and include *E. churchillensis* and *E. intermontana*. These diurnal exceptions are all comparatively small and have relatively small ellipsoid eyes.

SUBTRIBE AUSTRANDESIINA

Peridroma Hübner includes 10 species of which three are endemic to South America, and six are endemic to the Hawaiian archipelago. The tenth species, ***Peridroma saucia*** (Hübner), the "variegated cutworm" (FW 1.7–2.0 cm; Plate 56.31), has a worldwide distribution in part due to its long-distance dispersal ability, and in part due to its opportunistic dispersal on oceanic vessels. The species may be found throughout North America, Central America, and South America below tree line, but the species may be incapable of overwintering where there are long hard winters. The adults are nocturnal and may be found in several generations throughout much of the year. In the South, the species overwinters as pupae from which adults emerge in early spring, which then emigrate and colonize more northern regions. There are four generations in the southern United States, three generations in the central United States, and two generations in more northern or boreal habitats in the northern United States and southern Canada. The species is found most often in disturbed weedy habitats that host a wide variety of plants. The larvae feed on a wide variety of broad-leaved herbaceous plants, including a number of agricultural crops on which it may be a serious pest to flowering and fruiting parts. Reports compiled by Robinson et al. indicate that the larvae feed on plants in at least 40 families, including grasses (Poaceae), conifers (Pinaceae), and bracken fern (Dennstaedtiaceae). According to Lafontaine, most serious crop depredations are to beets (Amaranthaceae), globe artichoke and lettuce (Asteraceae), cabbage and cauliflower (Brassicaceae), and alfalfa (Fabaceae), as well as peppers, potatoes, tobacco, and tomatoes (Solanaceae). The last-instar larvae vary from pale yellow gray to pale reddish brown with dark brown speckling. There may or may not be a dorsal series of four to eight elongate yellow spots. When present, these spots help identify the species and are largest on the first abdominal segments and smaller to the fore and aft. A dorsal black W-shaped or triangular mark is often present along the posterior portion of the eighth abdominal segment. A thin orange subdorsal line may be present and is margined dorsally by a series of black dashes. There is a narrow black lateral line just above the black spiracles. The head is yellow brown with dark brown coronal stripes and reticulation.

Anicla (Euagrotis) exuberans (Smith)(FW 1.5–1.7 cm; Plate 56.32) is a western representative of a new World genus that has 19 species in Mexico and North America, but only nine of which are western. This moth ranges from southern British Columbia and southern Alberta south through the Rocky Mountain cordillera and the Sierra Madre of Mexico to Durango and Nuevo León. The adults have a single flight between mid-June and mid-August with extreme dates from May to early October. The adults are geographically variable in FW color and size; FW color may be gray, gray brown, yellow brown, or orangeish and may or may not be speckled with black. The black band on the prothoracic collar is an even width throughout.

Hemieuxoa rudens (Harvey)(FW 1.3–1.5 cm; Plate 56.33) is the only one of six species of a Neotropical group that ranges north into Mexico and the West. It ranges from central California, Utah, Colorado, and Nebraska south through the Southwest, Mexico, Central America, and South America to Ecuador and Bolivia. The adults have two quite different forms. In its typical form the FW has a black basal streak, a black streak between the reniform and orbicular, plain gray costal margin, and a black prothoracic collar. In the alternate "pellucida" form the two black streaks are absent, the costa is darker than the ground, the FW has a speckled appearance, and the extent of the black prothoracic collar is reduced. The species is bivoltine with spring and fall flights from April to June and then again in September and October. The larval hosts include snakeweed, *Gutierrezia wrightii* (Asteraceae), grasses (Poaceae), and possibly other plants. The larvae were described by Crumb as varying from light gray to dark gray with an obscure pale middorsal line that is interrupted by black intersegmental spots. There are vague blackish subdorsal spots on each segment. The spiracles are black. The head is white with black submedian arcs and reticulation.

SUBTRIBE AGROTINA

Striacosta albicosta (Smith)(FW 1.5–1.8 cm; Plate 56.35), the "western bean cutworm," ranges from southern Idaho, Colorado, Nebraska, and northwestern Iowa south through the arid interior West, Mexico, and Central America to Colombia. The species is most often found in habitats with loose sandy soils, where the larvae may seek shelter by day. The adults have been found from early July to October. The recorded larval hosts include beans, *Phaseolus vulgaris* (Fabaceae), tomato, *Solanum lycopersicon* var. *cerasiforme*, and ground cherry, *Physalis* (Solanaceae), and corn, *Zea mays* (Poaceae). The larvae may cause damage to bean crops by feeding on the reproductive structures. In later instars the larvae rest below the soil surface

by day and feed at night. Larvae were described by Crumb as gray with a dorsal series of ovoid black intersegmental spots. The spiracles are brown with black rims, and the cervical shield is brown with three longitudinal black lines. The head is brown with black submedian arcs and reticulation.

Actebia fennica (Tauscher), the "black army cutworm" (FW 1.8–2.3 cm; Plate 56.36), belongs to a Eurasian genus with 29 species, only two of which occur in North America. This species ranges from Alaska and Yukon Territory east across Canada below tree line to Labrador and Newfoundland, and thence south to northern Oregon, Montana, North Dakota, the Great Lakes region, and New England. The adults fly from early June to mid-September. The larvae feed on a wide variety of plants but prefer the young growth of blueberry, *Vaccinium* (Ericaceae); Robinson et al.'s compilation enumerates plants, including grasses and ferns, in 24 families as hosts. The larvae have caused serious damage to blueberry crops in eastern Canada and New England and may also cause damage to young conifer seedlings where they grow associated with stands of *Vaccinium*. The larvae are primarily black with two irregular subdorsal lines along each side as well as a spiracular band of white speckling. The spiracles together with the cervical and anal shields are black, and the legs are orange brown and the prolegs translucent yellow brown. The head is orange brown with black shading on the frons, adfrontal area, and along the median line.

Dichagyris Lederer is a diverse Holarctic genus whose North American species belong to endemic subgenera. *Dichagyris (Loxagrotis) grotei* (Franclemont and Todd)(FW 1.4–1.6 cm; Plate 56.37) is a pale brown noctuid with sharply demarked apical areas and thin black antemedian and postmedian lines. The species has a wide range, but its area of greatest concentration and abundance is in the West. Overall, its range extends from eastern Oregon east to Pennsylvania and south to Sonora, Mexico, New Mexico, Texas, Missouri, and the Florida panhandle. East of the Great Plains the species occurs in relict prairie habitats. The adults fly from late July to early October in most of the species' range but occur as late as November in Arizona. The larval host is false boneset, *Brickellia eupatorioides*, and likely other related Asteraceae. Crumb's larval description is based on alcohol-preserved examples and the colors are probably not accurate. *Dichagyris (Loxagrotis) grandipennis* (Grote)(FW 2.1–2.5 cm; Plate 56.38) ranges from central Colorado, southeastern Arizona, and New Mexico south to Durango and Vera Cruz, Mexico. There is a single adult flight during June and July. The host and life history are unreported. *Dichagyris (Pseudorthosia) variabilis* (Grote)(FW 1.5–1.9 cm; Plate 56.39) ranges from southern British Columbia east to southwestern Saskatchewan south to southern California, Nevada, Colorado, and New Mexico. The adults have a single flight with records extending from late July to early October. The larval hosts are not reported, but larvae have been reared on at least two occasions. The larva is yellow gray with dark longitudinal bands with a dorsal series of dark V-shaped marks at the posterior portion of each segment. The spiracles are black. The prothoracic shield is black with three prominent longitudinal yellow lines. The head is orange with two black submedian arcs. *Dichagyris (Mesembragrotis) longidens* (Smith)(FW 1.4–1.7 cm; Plate 56.40) occurs in central Arizona and from eastern Colorado south through New Mexico to western Texas. The adult flight is from mid-August to late September, although a series of older specimens from western Texas has a much broader range of dates.

Eucoptocnemis (Eucoptocnemis) rufula Lafontaine (FW 1.4–1.7 cm; Plate 56.41) ranges from central Arizona and southern Arizona east to the western half of Texas. The adults vary geographically from brownish gray to reddish brown. There is a single flight from late March to mid-May. Life history information is unreported.

Richia parentalis (Grote)(FW 1.6–1.9 cm; Plate 56.42, Plate 56.43) has extremely variable FW color and pattern—both in the basic ground color from light gray to dark gray and in the amount of black that ranges from none to extensive in the median two-thirds. The species is found in ponderosa pine forests as well as pinyon-juniper forests in two apparent geographic clusters: one from Montana and the western Dakotas to eastern Wyoming and western Nebraska, and the second from northeastern California east to the mountainous portion of Colorado and south to southern Arizona, New Mexico, and western Texas. The adults have a single flight from early August to mid-October. *Richia praefixa* (Morrison)(FW 1.7–2.1 cm; Plate 56.44) has a relatively narrow north-south distribution from northeastern Wyoming south through the mountainous coniferous forests of Colorado, New Mexico, and east-central Arizona (White Mountains). The species is found in coniferous forests. There is a single annual flight from late August to mid-October.

Copablepharon Harvey comprises 23 species limited to North America, with 21 of these found in the West. Most of the species are characterized by relatively smooth or silky yellow or white FW. Most of the species are limited to sandy soils and many are found in association with windblown (aeolian) sand dunes. There is much endemism among the species, and several are candidates for conservation efforts. Although the adults are nocturnal and are attracted to light, those of some species may easily be seen during the day either resting on vegetation or nectaring on flowers. The abdomens of females have telescoping segments that form an ovipositor that is apparently used for inserting their eggs below the sand surface. Larvae rest during the day under the dune surface and either feed under the surface or above ground at night. *Copablepharon pictum* Fauske and Lafontaine (FW 1.4–1.8 cm; Plate 57.1) is a rather attractive species endemic to Great Sand Dunes National Park in south-central Colorado. The adults have a single flight from late June to mid-August. Although F. Martin Brown, in Lafontaine, reported a long series as taken resting on *Psoralea* (Fabaceae), field observations by Harp and Opler indicate that the adults rest on any available broad-leaved plant, including *Helianthus petiolaris* (Asteraceae). The adults are abundantly attracted to light, according to observations by Pineda Bovin. *Copablepharon spiritum* Crabo & Fauske (FW 1.8–2.2 cm; Plate 57.2) ranges from central Washington east to northwestern Colorado and south to northern Baja California, Nevada, Utah, and northwestern New Mexico. Its distribution is discontinuous as populations are limited to sand dune ecosystems and associated "sand sheets." There is some geographic variation, and two additional subspecies have been described. The species has a single flight during August, September, and, in southern California, the first half of October. *Copablepharon canariana* McDunnough (FW 1.8–2.1 cm; Plate 57.3) occurs in arid interior habitats from eastern Oregon, southern Idaho, and western Montana south to southern California, northern Arizona, and central New Mexico. This species is exceptional in not being a denizen of sand dunes; rather it is found associated with stream and lake margins as well as some upland habitats. The adults have a single flight from mid-June to mid-September with most records from mid-July to mid-August.

Copablepharon viridisparsa Dod (FW 1.6–2.1 cm; Plate 57.4) ranges in interior arid land basins from southern British Columbia east to southwestern Saskatchewan and south to east-central California (Mono County), southern Utah, and northwestern Colorado. There is a long single flight period from early June until late September with flight earlier in the south than in the north. The species is most often associated with sand dunes and associated habitats. The adults are both diurnal and nocturnal, although most activity is probably nocturnal. The adults have been found in numbers during the day resting on vegetation and nectaring at flowers including a white-flowered *Cirsium* in northern Colorado. The larvae were found resting beneath the sand surface and then came above ground to feed on *Brassica nigra* and *Sinapis alba* (Brassicaceae). The larva is cream with reddish tan subdorsal stripes and a darker line at the spiracles. The venter is bluish white. The cervical shield is gray brown with two darker marks at the posterior edge. The head is also gray brown. Eggs are laid soon after adult emergence and mating. The larvae overwinter and pupate during the following year.

Protogyia McDunnough comprises 15 species, all of which are endemic to western North America including northwestern Mexico. Lafontaine and Fauske divide the species into three species groups. *Protogyia postera* Fauske & Lafontaine (FW 1.5–1.8 cm; Plate 57.5) is a member of the Lagena Species Group and ranges from eastern Washington south to southern California and east to southwestern Wyoming, Colorado, and northern New Mexico. *Protogyia postera* is found most often in arid interior habitats. There is a single flight period from mid-June to late September, and most adults fly between mid-July and early September. *Protogyia album* (Harvey)(FW 1.4–1.6 cm; Plate 57.6) is very similar to *Copablepharon spiritum* but, in addition to diagnostic structural features, may be separated on the basis of its smaller size. It ranges from the Columbia River basin of eastern Washington south along the east side of the Cascade–Sierra Nevada–Transverse Ranges axis through eastern Oregon and eastern California to Baja California. It is known further to the east in southwestern Utah and southern Arizona. The larval host plant is *Mentzelia laevicaulis* (Loasaceae). The larvae were noted by Crumb as varying from pale brown gray, whitish, or pinkish with a white subdorsal line subtended by black. The spiracles are dark brown. The head is suffused with dark and has dark brown or black reticulation. *Protogyia milleri* (Grote)(FW 1.6–1.8 cm; Plate 57.7) is found in high-elevation conifer forests (4,000 to 10,000') in the Cascades, Sierra Nevada, and Transverse Ranges from northern Oregon south to southern California. An Arizona record is almost certainly based on a mislabeled or misidentified specimen. This is the only species in the Album Species Group with pronounced transverse lines. The species has a single flight from late May to mid-September. The larvae have been reared on artificial diet and have been described by Lafontaine and Fauske as white with slightly darker brownish gray shading so that white middorsal and subdorsal lines are clearly evident. The setae have a black ring around each socket that is in turn surrounded by a gray pinaculum. The head is pale yellow brown with large dark brown freckles.

Euxoa Hübner is a rich Holarctic genus with about 400 species found across Eurasia, the Indo-Pacific, and Africa as well as North America. This grroup has 171 North American species, 156 are endemic to or range into the West. Our discussion of *Euxoa* relies heavily on the monograph by Lafontaine. Virtually all species are single-brooded and nocturnal, although a few Arctic and alpine species are diurnal and have eyes that are small relative to their overall size. The larvae are cutworms and few have distinctive species-diagnostic markings or colors. *Euxoa (Crassivesica) bochus* Morrison (FW 1.4–1.7 cm; Plate 57.9) is a distinctive species that was recently transferred to *Euxoa* but is the only species in its subgenus. Its black prothoracic collar (not visible in plate) is a trait shared only with selected species of *Anicla* and *Dichagyris (Loxagrotis)*. The species has a primarily interior arid land distribution from southern British Columbia east to southwestern Manitoba and south to central California, Nevada, Utah, and central Colorado. A record from coastal southern California seems out of place. Extreme dates for the single annual flight period are mid-July to mid-October, but most records are from mid-July to mid-September. The larvae were reared in captivity on several broad-leaved herbaceous plants, but Lafontaine believes the larval color pattern is indicative of a grass feeder. The larvae are pale buff to pale greenish buff with pale flecks form pale streaks and dark flecks arranged to form dark streaks. The head is pale orange brown with black submedian arcs and reticulation. *Euxoa auxiliaris* (Grote), the "army cutworm" (FW 1.7–2.2 cm; Plate 57.8, Plate 57.10), is the commonest of seven North American members of the subgenus *Chorizagrotis* Smith, which is characterized by genitalic features described by Lafontaine. The present species occurs throughout western North America from southern Northwest Territories south to northern Mexico and east to Michigan, Missouri, and Texas. The adult FW maculation and color are extremely variable, but can usually be sorted into one of three patterns. The most common has a black subbasal dash, black between the reniform and orbicular spots, a contrasting pale gray or straw yellow costa, and a pale streak beyond the claviform. The second form is much less frequent and has the FW more or less uniformly pale gray or pale brown, and the third form, also much less frequent, has the FW dark brown with the orbicular and reniform spots narrowly rimmed with white. The adults may be found from spring (April and May) when they emerge, until late summer and fall (late August to early October) when they return to their breeding areas. The adults emerge during spring in plains, prairie, and high desert habitats and undertake seasonal mass migrations to higher altitudes. During the summer, the adults favor alpine habitats where they shelter under rocks during day and come out to nectar at flowers nocturnally. As far as is known this phenomenon is most impressive along the Rocky Mountain Front where the adults have their most impressive movements from the eastern plains westward to alpine habitats in the Rockies. Under favorable conditions, the number of moths is incredibly large, and the resting masses have been known to be relied on by grizzly bears *(Ursus arctos horribilis)* in some years, particularly in the Yellowstone ecosystem. In late summer, the adults, which may have survived the summer, have a return migration to their traditional breeding areas. Along the Rocky Mountain Front, these seasonal movements are from east to west and then back eastward, but it is unknown whether all such movements in other parts of the West follow the same cardinal directions. According to the compilation by Robinson et al. larvae feed on plants in 16 families, primarily low broad-leaved herbs, but they favor cereal grasses such as *Avena*, *Elymus*, *Hordeum*, *Poa*, and *Triticum* (Poaceae). This species can cause extensive damage to cereal plantings and other crops in the Great Plains and western North America. In the Southwest, it can be confused with *E. concinna* (Harvey), which usually resembles a smaller washed-out version of the pale form of *E. auxiliaris* but can only be reliably separated by the examination of its genitalia.

The subgenus *Palaeoeuxoa* Lafontaine comprises four western species including *Euxoa intermontana* Lafontaine and *E. mimallonis* (Grote) and is characterized by genitalic features as well as by the epidermal structure of the last-instar larvae, which has the polygons each having a central large raised bump, giving the skin a coarse texture. *Euxoa intermontana* Lafontaine (FW 1.6–1.8 cm; Plate 57.11) ranges across the intermountain West from central Washington eastward to Montana and southward to eastern California, northern Arizona, and northern New Mexico. There is a single flight from late July to late September. *Euxoa mimallonis* (Grote)(FW 1.5–1.7 cm; Plate 57.12) ranges from British Columbia east across southern Canada to Nova Scotia and south to east-central California, central Arizona, central New Mexico, western Nebraska, and the eastern states on the Canadian border. Adults have a single flight in August and September.

The subgenus *Heteroeuxoa* Lafontaine contains three western species including *Euxoa olivia* (Morrison). The three species are unified by genitalic features discussed by Lafontaine. *Euxoa olivia* (Morrison)(FW 1.4–1.5 cm; Plate 57.19 male, Plate 57.20 female) ranges from southern British Columbia east to southern Manitoba and south to southern California, southern Arizona, southern New Mexico, and northern Texas. The sexes are dimorphic; the male is orange brown and streaked, while the female is uniformly gray. The flight period is from late August to late October. Larvae have been reported to feed on *Fragaria*, strawberry (Rosaceae), and *Zea mays*, corn (Poaceae).

The subgenus *Longivesica* Hardwick comprises three species that are characterized by their extremely long vesicas. These must be everted for identification. *Euxoa messoria* (Harris), the "dark-sided cutworm" (FW 1.3–1.8 cm; Plate 57.13), ranges from Yukon Territory east to Newfoundland and south to California, southern Arizona, New Mexico, Missouri, and Virginia. The larvae prefer broad-leaved herbaceous plants and small trees but occasionally feed on several kinds of grasses including corn seedlings. The larva has longitudinal pale brown and dark brown stripes.

The subgenus *Pleonectopoda* Grote is a Holarctic group with 23 North American species, all but one of which occur in the West. *Euxoa churchillensis* (McDunnough)(FW 1.1–1.4 cm; Plate 57.40) has two disjunct areas of occurrence. One area is from the mouth of the McKenzie River in Northwest Territories east to northern Ontario, while the second is in the alpine zone of Colorado and northern New Mexico. The adults are diurnal, and those of the Colorado population have a single flight during July and early August. *Euxoa lewisi* (Grote)(FW 1.4–1.9 cm; Plate 57.14) ranges from southern British Columbia and southern Alberta south to northern Oregon, northern Utah, and Colorado and has a disjunct population in the Sierra Nevada of California. This moth is most abundant in Hudsonian spruce-fir forests but may be found occasionally in either lower-elevation forests or in higher-elevation alpine habitats. The adults are nocturnal and fly during July and August. *Euxoa leuschneri* Lafontaine (FW 1.5–1.6 cm; Plate 57.15) is endemic to pine forests of the high San Bernardino Mountains and Santa Rosa Mountains of southern California. The nocturnal adults have a single flight from mid-June to early August. *Euxoa tristicula* (Morrison)(FW 1.5–2.0 cm; Plate 57.16) may be found from the interior of British Columbia east to Manitoba and Minnesota and south to southern California and Arizona. A disjunct population is found in coastal dunes from eastern Quebec to coastal Maine. There is a single flight from late May to late August, although most individuals fly during June and July. The larvae feed on low herbaceous plants during late fall and early spring, with an intervening winter dormancy. The species' host range is wide and includes *Beta*, beets (Chenopodiaceae), *Medicago*, alfalfa (Fabaceae), and *Linum*, flax (Linaceae). Because the species feeds when commercial fields are fallow, the species is not considered to be of importance. *Euxoa vetusta* (Walker)(FW 1.6–1.9 cm; Plate 57.17) ranges along the Pacific Coast east of the Cascade–Sierra Nevada divide from southern British Columbia to central California. The species has an extended flight period from late April to early September. The life history is unreported. *Euxoa fuscigera* (Grote)(FW 1.4–1.8 cm; Plate 57.18) is a Californian endemic that is found from Lake County south to San Diego County and as far inland as the Mojave Desert. It is usually a moth of relatively arid regions but has been found in moist coastal habitats as well. It has a single fall flight from early September to early December. *Euxoa atomaris* (Smith)(FW 1.3–1.7 cm; Plate 57.21 pale aridland form, Plate 57.22 dark forest form) occurs throughout much of the West from southern British Columbia east to southern Alberta and western North Dakota south to southern California, central Arizona, and central New Mexico. The moths from moister coastal populations are darker than those from more interior arid regions. The primary flight in the northern parts of its range is during August and September and during September and October in the more southern portions. The species may be partially double-brooded as some individuals have been found during March and April. *Euxoa simona* McDunnough (FW 1.8–2.2 cm; Plate 57.23) is found in montane regions from southern British Columbia and southern Alberta south to southwestern California, southern Utah, and southern Colorado. The species may hybridize with *E. pleuritica* (Grote) and *E. pestula* (Smith) in some portions of its range. The adults have a single flight from mid-June to mid-September. *Euxoa bifasciata* (Smith)(FW 1.5–1.7 cm; Plate 57.24) ranges from north-central Washington east to western Montana south to southern California, central Arizona, and central Colorado. The pale gray brown FW is distinctively marked with thin black antemedial and postmedial lines. The species is uncommonly collected and is found primarily in pine-fir habitats. There is a single flight from early July to early September. *Euxoa annulipes* (Smith)(FW 1.4–1.7 cm; Plate 57.25) extends from southeastern Oregon east to western Colorado and south to southwestern California. The adults range from pale gray to creamy white and usually have a black medial line on the FW. The adults have a single flight from early June to mid-August.

Euxoa pluralis (Grote)(FW 1.4–1.8 cm; Plate 57.26) is a member of the Pluralis Species Group and is found from southern British Columbia south to southern California, central Utah, and northern New Mexico. The adults fly from mid-March to mid-August. The latest flight dates are from higher elevations in the Sierra Nevada. *Euxoa cinnabarina* Barnes & McDunnough (FW 1.5–1.8 cm; Plate 57.27) is a distinctive member of the Pluralis Species Group that is found in the southern Sierra Nevada and in the Transverse Ranges and Peninsular Ranges of southern California. The adults are in flight from mid-July to early August. *Euxoa flavidens* (Smith)(FW 1.4–2.0 cm; Plate 57.28) is a member of the Declarata Species Group that ranges from northern Utah and central Colorado south through Arizona and New Mexico to central Mexico. The adults are distinguished by the streaked FW, yellow buff claviform, and the very large yellow buff claviform spot. The adults have a single flight from late July to late September. *Euxoa silens* (Grote)(FW 1.6–1.9 cm; Plate 57.29) is found primarily in arid intermountain desert habitats from southeastern Washington east to southern Alberta, southwestern Montana, and western Colorado south to southern California and central Arizona. Adults are in flight from late April to early August. Three species including *E. pimensis* Barnes & McDunnough (FW 1.5–1.9 cm; Plate 57.30), together with the previous species and the eastern *E. immixta* (Grote), form the Silens Species Group. The range of *E. pimensis* extends from the deserts of southern California east to central Colorado and south at least to the vicinity of the Mexican border in southern Arizona and southern New Mexico. The adults fly primarily in spring from April to early July. *Euxoa hollemani* (Grote)(FW 1.0–1.6 cm; Plate 57.31) ranges from south-central British Columbia east, perhaps discontinuously, to eastern Montana and south to southern California, southeastern Arizona, and central New Mexico. The FW has the fused orbicular and reniform spots in the shape of an hourglass, which runs to the base and is subtended by a long black streak. The adults fly in arid open habitats including plains and pinyon-juniper woodlands from mid-August to mid-October.

Euxoa comosa (Morrison)(FW 1.2–1.8 cm; Plate 57.32, Plate 57.33) is a member of the Comosa Species Group, which has seven North American species and possibly some from temperate Asia as well. The species can be best recognized by genitalic characters, but the combination of a drab gray or brown FW with a pale median line and pale-outlined orbicular and reniform spots is a good guide to the group. *Euxoa comosa* itself is widespread and variable; it has several named geographic subspecies. The species ranges from western Alaska east to Newfoundland and south to southern California, southern Arizona, central New Mexico, Nebraska, Minnesota, Michigan, and New England. It may be found in a wide variety of habitats ranging from deserts to moist coniferous forests. Over this wide range of environments its flight period extends from late July to late September.

Euxoa piniae Buckett & Bauer (FW 1.3–1.5 cm; Plate 57.34) is a member of the Infausta Species Group, which has eight difficult to identify North American species. *Euxoa piniae* has a bright orange brown FW with thin black transverse lines. The present species is a California endemic that is found in coniferous forests to the west of the Cascade–Sierra Nevada crest from Siskiyou and Plumas County south to Tuolumne County. The annual flight is from mid-July to mid-September.

Euxoa obeliscoides (Guenée)(FW 1.3–1.9 cm; Plate 57.35), together with *E. oberfoelli* Hardwick, forms the Obeliscoides Species Group. In the Great Plains region, where the two species are sympatric, the two species may be separated with certainty only by examination of the male vesica. Generally, the present species may be recognized by its distinctive FW pattern and the presence of a dark marginal band on the HW. The species ranges from British Columbia east to Manitoba and south to California, Arizona, New Mexico, and central Kansas. There is a single flight from late July through September. The species overwinters in the egg stage, and it has been reared in the laboratory.

Euxoa idahoensis (Grote)(FW 1.4–1.9 cm; Plate 57.36) is a member of the Detersa Species Group, which has 32 species; this is the largest North America species group of *Euxoa*. The species ranges from central Alaska south and east to southeastern Saskatchewan and south to central California, central Arizona, northern New Mexico, and southwestern South Dakota. The species is most often found in dry coniferous forests. The adults fly from mid-June through late August. This species is extremely close to *E. castanea* Lafontaine, *E. clausa* McDunnough, and *E. foeminalis* (Smith), and the species may be difficult to separate and may even hybridize in some locations.

Three additional species, *E. auripennis*, *E. teleboa*, and *E. cicatricosa* (Grote & Robinson), are also members of the Detersa Species Group. *Euxoa auripennis* Lafontaine (FW 1.1–1.6 cm; Plate 57.37) ranges from central British Columbia east to south-central Manitoba and south to southern California, central Arizona, central New Mexico, and North Dakota. The adults have a single flight from late July to late September. *Euxoa teleboa* (Smith)(FW 1.2–1.6 cm; Plate 57.38) is a distinctive moth with a cream, pale yellow, or orange FW that has a black median line fused to a black filled reniform. The species ranges from southern Alberta and southern Saskatchewan south to northern Arizona, central New Mexico, and northern Texas. The preferred habitat is sagebrush plains or pinyon-juniper woodland. The adults fly from mid-August to late September.

Euxoa wilsoni (Grote)(FW 1.5–1.9 cm; Plate 57.39) is a member of the subgenus *Orosagrotis,* which includes 16 species most of which are Nearctic, but three are entirely Palaearctic. This species occurs in coastal sand dunes from the Queen Charlotte Islands of central British Columbia south along the Pacific Coast to San Luis Obispo County, California. The adults have a single extended flight period from early June to early October. The larvae have been reported to feed on *Lupinus* (Fabaceae), and, on Santa Rosa Island, we found the larvae feeding on *Ambrosia chamissonis* (Asteraceae).

Feltia Walker is limited to North America, where there are 11 species. *Feltia (Trichosilia) mollis* (Walker)(FW 1.7–2.1 cm; Plate 57.41) is a boreal species that ranges from Yukon Territory and British Columbia east across Canada to Labrador and Newfoundland and extends southward to the vicinity of the Great Lakes and New England. There is an apparently isolated set of populations in the Black Hills, South Dakota, and in the Rocky Mountains from southern Wyoming to northern New Mexico. The adults have a single flight from late June to early August. The life history is unreported. *Feltia (Feltia) jaculifera* (Guenée), the "dingy cutworm" (FW 1.3–2.0 cm; Plate 57.42), is most likely a species group of six or more separate but currently inseparable species. Lafontaine reports that males attracted to quite different pheromone blends most likely represent more than a single reproductively isolated species. At least two pheromone species occur in western Canada alone. The distribution of the species complex as a whole extends from Alaska east across Canada south of the Canadian shield to Nova Scotia and south throughout most of the United States except much of California, southern Nevada, and much of the Mississippi Valley, southeastern coastal plains, and peninsular Florida. The "species" flight period extends from late June to early November. Adults are nocturnal but may also be seen nectaring at flowers such as *Helianthus*, *Grindelia*, and *Solidago* (Asteraceae) during the late afternoon and evening. Although not thought of as a pest, the immense numbers of larvae required to support the adult populations on the Great Plains must remove an astronomical amount of foliage from both native and human-maintained ecosystems. The listing of the species' larval hosts is confused because the identity of this "species" has often been comingled with that of *F. subgothica* (Haworth) and *F. tricosa* (Lintner). Lafontaine conservatively lists alfalfa and clover (Fabaceae), flax (Linaceae), tobacco (Solanaceae), raspberry (Rosaceae), and oats and wheat (Poaceae) as definite hosts but states that the actual variety of plants eaten must be much more expansive. There are both light and dark forms of the larvae. The dark form is dull gray brown with darker shading along the sides. The dorsum is usually reddish brown with dark speckling that forms a dorsal series of diamond-shaped

marks on the abdominal segments. The head is pale orange with dark submedian arcs and reticulation. The larva overwinters as a third or fourth instar and completes development in the spring. *Feltia (Feltia) subterranea* (Fabricius), the "granulated cutworm" (FW 1.2–2.0 cm; Plate 57.43, Plate 57.44), ranges eastward from central California extending to more northern latitudes including southern Nevada, northwestern Colorado, southern Wyoming, Nebraska, Michigan, New York, and Nova Scotia. The moth is found in South America to Peru and Brazil. In the South, the species flies throughout the year in five to six generations and has peaks of adult abundance in spring and fall. In the central areas it flies from March to October and has two to three flights. It is likely a seasonal migrant in the West to the north of southern California and southern Arizona. Strangely enough, Lafontaine does not report it from western Texas or New Mexico. According to reports compiled by Robinson et al., larvae will feed on a variety of broad-leaved herbaceous plants in 13 families, especially Brassicaceae, Fabaceae, and Solanaceae. The larvae are cutworms, and their skin has scattered raised conical granules. Otherwise the larvae are gray with black spiracles. The head is pale brown or yellowish with black submedian arcs and black or ferruginous reticulation.

Agrotis Ochsenheimer occurs virtually worldwide except for Antarctica and the northern boreal regions. There are several hundred species. Our North American species, except for *A. ipsilon* and *A. malefida*, belong to Holarctic species groups. Seventeen of 22 North American *Agrotis* have at least a portion of their distribution in the West. *Agrotis vetusta* (Walker), the "old man dart" (FW 1.5–2.1 cm; Plate 58.1), ranges from Alaska and Yukon Territory, where disjunct populations occur in relatively xeric areas, east to Northwest Territories. Further south, where more continuous populations are found, the species ranges from southern British Columbia east across southern Canada to Quebec and Nova Scotia. The species occurs south to southern Baja California and east, including virtually all of the western United States to the Atlantic Coast. The species is present but extremely uncommon in the Southeast. Adults fly from late July to mid-October (early November in the Southwest). The species is found most often in habitats with sandy soils, and Lafontaine posits that the larvae feed on whatever broad-leaved herbaceous plants are present. Plants in 13 families have been recorded as hosts. The host list compiled by Robinson et al. also includes corn, *Zea mays* (Poaceae), and a few woody plants. The larvae are gray with several longitudinal dark gray or gray brown subdorsal lines. The head is pale orange with number dark dorsal spots. This is more similar to the state found in many *Euxoa* larvae and differs from the pattern found in most *Agrotis* larvae. *Agrotis orthogonia* Morrison, the "pale western cutworm" (FW 1.3–1.7 cm; Plate 58.2), occurs throughout the more arid intermountain and high plains sagebrush and desert scrub habitats from southern Alberta east to southwestern Saskatchewan and south to southern California, central Arizona, New Mexico, and western Texas. The species has a single flight from mid-July to early October. Robinson et al. list herbaceous plants and grasses in 11 families. The larvae may cause damage to cereal crops in southern Alberta and Montana. The larvae are pale yellow gray and are darker on the sides than dorsally. The pale middorsal line is bounded on both sides by diffuse darker lines. There are also diffuse thin dark subdorsal lines. The head is unusual by being yellow brown with or without dark shading. When present, the dark shading on the frons and vertex may fuse and form a distinctive H- or X-shaped pattern. *Agrotis vancouverensis* Grote (FW 1.3–1.7 cm; Plate 58.3) is the commonest western *Agrotis* and ranges from Yukon Territory south through western boreal habitats to southern California, eastern Arizona (White Mountains), and northern New Mexico. The species normally flies during May and June but has its flight during July and August at higher elevation in the Sierra Nevada and Colorado's Rocky Mountains, where it may be found as high as the alpine zone (above 12,000′). The adults are often found with those of *A. obliqua* (Smith), and the two species are often confused. Adults of *A. vancouverensis* can be recognized by their paler brown FW, paler HW, and solid black claviform spot and basal dash. The larvae have been reported to feed on low herbaceous plants such as *Trifolium* (Fabaceae), *Fragaria* (Rosaceae), and grass (Poaceae), but the full host range is probably much broader. The larvae are gray with a diffuse dark oval or diamond-shaped patch on the dorsum of each segment. There is a dark gray subdorsal line with dark gray shading speckled with white along the sides. The spiracles and cervical shield are black. The head is closely infuscated with black, including both submedian arcs and a close reticulate pattern. *Agrotis malefida* Guenée, the "palesided cutworm" (FW 1.8–2.3 cm; Plate 58.4), ranges from the southernmost portion of the United States south through the Caribbean, Mexico, Central America, and South America to Chile and Argentina. It is probably resident only in areas adjacent to the Mexican boundary but migrates northward to central Arizona, northern New Mexico, and southern Kansas in the West. In its most northern areas of residence, pupae overwinter and adults emerge in January, but the species flies year-round to the south. The larvae feed on a wide variety of broad-leaved herbaceous plants; plants in 10 families are listed by Robinson et al., and occasionally grasses or woody plants are eaten. The larvae build underground tunnels from which they forage and return to feed. The larval skin has flattened polygons. Pale color extends from the venter up to at least the spiracular line. Above that the larvae have the dorsal and subdorsal areas concolorous in either dark or pale forms. The head is pale gray or light brown with darker submedian arcs and reticulation. *Agrotis ipsilon* (Hufnagel), the "black cutworm" (FW 1.8–2.4 cm; Plate 58.5). is one of the most widespread moths in the World, being found virtually everywhere, including oceanic islands, with the exception of Antarctica and the Arctic and subarctic portions of North America and Eurasia. In North America, the species is resident in the more southern portions of the United States, but the species regularly emigrates northward in spring and may establish temporary populations that may produce up to two generations. The species has been found in most of the Canadian provinces bordering the United States and is occasionally found as far north as Hudson Bay and Labrador. In its southern areas of residence, adults are found throughout the year and the species has four or more generations. The species prefers habitats with rich moist, even mucky, soils in contrast to the preferences of most other *Agrotis*. The larvae build tunnels from which they forage at night. The larvae feed on an impressive list of broad-leaved herbaceous plants and grasses that includes plants in 20 families, according to reports compiled by Robinson et al. The larval skin has raised rounded granules of various sizes. The larvae are unicolorous above the spiracular line and vary from gray to black. The spiracles are black. The head is pale brown with black submedian arcs and reticulation.

TRIBE NOCTUINI

This tribe includes 31 genera that comprise 166 species in North America, while the majority is endemic to or includes the West

in its ranges—28 genera and 138 species. The most species rich western genera are *Xestia* (42 species) and *Abagrotis* (35 species), and these account for more than half of all western Noctuini.

SUBTRIBE AXYLIINA

Ochropleura implecta Lafontaine (FW 1.2–1.5 cm; Plate 58.6) is a North American member of a genus of 15 species, 12 of which are confined to sub-Saharan Africa and two of which occur in Eurasia. The present species ranges from southern coastal Alaska and southern British Columbia east across Canada to southern Labrador and Newfoundland, thence south to southern Oregon and the eastern United States. There is a wide scattering of records in other western states including Arizona, Colorado, and New Mexico. The moth has two flights, from mid-May to September in the northern part of its range and from late April to mid-October in the southern portion. The larval hosts are a small variety of broad-leaved herbaceous and woody plants including (Fabaceae), *Cichorium* (Asteraceae), and *Salix* (Salicaceae). The larva was described by Crumb and, unlike most other noctuine larvae, lacks subdorsal dark wedge-shaped marks and has a slight hump on the dorsal portion of the eighth abdominal segment. The last-instar larva is pale red brown dorsally and light brown or greenish brown laterally. There are narrow broken white middorsal and subdorsal lines edged in black. The setal insertions are comprised of small black rings. The spiracles are white. There are faint oblique subdorsal dark slashes that produce an apparent dorsal herringbone pattern. The head is pale yellow brown with darker brown coronal stripes and reticulation.

SUBTRIBE NOCTUINA

Diarsia rosaria (Grote)(FW 1.3–1.5 cm; Plate 58.7) is a member of a genus with about 75 species, most of which occur in Eurasia, especially in China. In North America, there are 12 species, five or six of which have at least a portion of their distribution in the West. The North American occurrence of *D. rosaria* comprises disjunct eastern and western range segments. In the West, the species occurs from Alaska and Yukon Territory south to central California, Utah, northern New Mexico, and Manitoba. The population in Utah, Colorado, and northern New Mexico is disjunct, as there are no apparent occurrences in neighboring Arizona and Wyoming. In the East, the moth occurs discontinuously from Labrador south to the northern Great Lakes region and Nova Scotia. Likely, there are several flights each year, since flight dates extend from late April to mid-October in most of the range and, in California, from January until early June, then again in November. According to reports compiled by Robinson et al. the larvae feed primarily on grasses (Poaceae) but also on *Geum* and *Fragaria* (Rosaceae). Crumb describes the larva as varying from warm brown to dusky gray with a subdorsal series of broken dark dashes. The spiracles are brown. The head is black with white lateral and coronal areas. Partially grown larvae overwinter and resume feeding in the spring.

Cerastis enigmatica Lafontaine & Crabo (FW 1.2–1.3 cm; Plate 58.10) belongs to a genus with 12 species, seven of which are North American and five of which are restricted to Eurasia. There are two distinctive subgenera; all Palaearctic species belong to the subgenus *Cerastis*, but one occurs in eastern North America. The remaining six species are in the subgenus *Metalepsis*, and all of these occur in the West. *Cerastis enigmatica* ranges along the Pacific Northwest coast in coniferous habitats from the Alaska panhandle south to Humboldt County, California. The moth flies from late March to late April. The larvae feed on salmonberry, *Rubus spectabilis* (Rosaceae). The last-instar larva is pale brown with dark brown mottling and a series of V-shaped dorsal spots on the dorsum of the abdominal segments. The spiracles are white. The head is pale brown to pale yellow brown with darker brown or reddish brown coronal stripes and reticulation. Lafontaine gives a detailed larval diagnosis for the genus. The closely related *C. gloriosa* Crabo & Lafontaine is an attractive moth that ranges from northwestern Washington south to central California. In the northern part of its range it is restricted to bogs but is not so restricted in California. There is a single flight from late February to late April.

Paradiarsia littoralis Packard (FW 1.4–1.7 cm; Plate 58.19) is a close relative to the Asian *P. coturnicola* (Graeser). It is the only North American member of its genus, and it occurs in boreal habitats from Alaska east to Labrador and south to Oregon, northeastern California, Utah, New Mexico, South Dakota, and, in the East, to Wisconsin and New York. These moths are usually found in either alpine or subalpine habitats. The adults are both diurnal and nocturnal and have a single flight from mid-June to early August. The larvae have been reared in captivity on several hosts that were provided, but these may not be the same as those plants used in nature. Among Noctuini the larvae of *Paradiarsia* are distinctive in their lack of the dorsal black triangular marks found on most other noctuine larvae and have black spiracles. There is a broad pale middorsal stripe as well as black subdorsal and supraspiracular lines. The head is light brown with darker brown coronal stripes and reticulation. A more detailed technical description may be seen in Lafontaine's generic characterization.

Chersotis juncta (Grote)(FW 1.0–1.6 cm; Plate 58.8) is the sole North American representative of a Palaearctic genus. *Chersotis juncta* is Holarctic and ranges from northeastern Siberia and Alaska east and south across boreal North America to Newfoundland and south to California, eastern Arizona (White Mountains), Colorado, Ontario, and Maine. There is a single flight from late June to mid-August. The host and early stages are unreported.

Noctua Linnaeus has 12 Eurasian species and one endemic to the Azores Islands. Two of these species have been introduced accidentally into North America. *Noctua pronuba* (Linnaeus), the "large yellow underwing" (FW 2.1–2.6 cm; Plate 58.9), is characterized by its yellow HW and black marginal band. In addition, the HW lacks a dark discal spot (present in *N. comes*). The FW color and pattern are quite variable; in some forms the FW is almost immaculate, but it is well marked in others. In 1979, the species was accidentally introduced to Nova Scotia, and it had spread and become abundant in much of Atlantic Canada, the Northeast, and the Great Lakes region by the late 1980s, according to Passoa and Hollingsworth. Since then this weedy species has spread to California, British Columbia, and Alaska; it likely now occurs in many intervening areas. Lafontaine feels it will soon occur throughout the nondesert habitats of the West. The native range is in Eurasia and Africa. The adults are nocturnal and fly from mid-June to late September. When abundant, the adults may fly short distances during the day upon being disturbed from their resting places. The larvae feed on a wide variety of both weedy and cultivated herbaceous plants and grasses including *Fragaria* (Rosaceae), tomato, *Solanum lycopersicon* var. *cerasiforme* (Solanaceae), *Beta* (Chenopodiaceae), and *Vitis* (Vitaceae). The last-instar larva is pale reddish brown dorsally and laterally; it

is pale pinkish brown below the spiracles. The middorsal line is narrow, broken, and bordered by black speckling. There is a subdorsal white line that is most prominent on the anterior half of each abdominal segment. The abdominal segments have black dashes, and the subdorsal area has faint oblique dashes formed by dark speckling that comprises a herringbone pattern. The spiracles are yellow brown. The head is pale yellow brown with darker brown coronal stripes and reticulation. The winter is passed by nearly full-grown larvae. *Noctua comes* Hübner, the "lesser yellow underwing" (FW 1.6–2.1; Plate 58.11), was introduced accidentally to Vancouver, British Columbia, around 1982 and has since spread to the Fraser River delta and south to Seattle, Washington. It is similar to *N. pronuba* but may be recognized by its smaller size and the presence of a dark discal spot on each HW.

Cryptocala acadiensis Bethune, the "catocaline dart" (FW 1.2–1.5 cm; Plate 58.12), is the only North American species of *Cryptocala*; there is one Eurasian relative. The adults are distinctive and relatively invariable. The species ranges from British Columbia across southern Canada to Newfoundland and Nova Scotia. The species is also found in Montana, the Great Lakes states, and New England. The adults fly in July and August. The larvae have been reared from eggs laid by captive females, but their host utilization may or may not match well with those used in nature. The larvae are light brown flecked with darker brown and with pale longitudinal middorsal, subdorsal, and subventral lines. A distinctive feature is the presence of rather large black pinacula that bear the setae. The spiracles are pale brown. The head is pale brown with darker brown coronal stripes and reticulation.

Spaelotis clandestina (Harris), the "W-marked cutworm" (FW 1.6–1.9 cm; Plate 58.13), is the most widespread of six North American *Spaelotis* Boisduval. Worldwide, there are 25 species, of which 16 occur in Eurasia. *Spaelotis clandestina* is the only member of the genus that is found in eastern North America, and the only species lacking longitudinal FW streaks. The species ranges from southern Alaska and Yukon Territory east across Canada south of the Canadian Shield to Newfoundland and south in the montane habitats to central California (one record), southeastern Arizona, New Mexico, Nebraska, Great Lakes states, and Appalachians (to North Carolina). The species is entirely absent from the Great Basin. There are two flights of the nocturnal adults from mid-May to late September, usually with a distinct gap in early summer. According to reports compiled by Robinson et al. the larvae are polyphagous and feed on a wide variety of broad-leaved trees, shrubs, and herbaceous plants including plants in 16 families. The species also has been reported to feed on conifers and grasses (Poaceae). Extensive damage is attributed to feeding by the larvae. The larva is pale brown with darker brown and black speckles. There is a series of black subdorsal wedge-shaped marks on the abdominal segments that, when viewed from above, have a vague W-shaped appearance thus lending the species its common name. There is a narrow broken pale middorsal line. The spiracles are pale yellow. The head is pale brown with darker brown coronal stripes and reticulation. Partially grown larvae overwinter. *Spaelotis bicava* Lafontaine), the "western W-marked cutworm" (FW 1.5–1.7 cm; Plate 58.14), is the most widespread of the streaked FW western endemic *Spaelotis*. The other species are *S. unicava* Lafontaine, *S. velicava* Lafontaine, *S. quadricava* Lafontaine, and *S. havilae* (Grote), and these all have smaller ranges than *S. bicava*. The other species are most reliably distinguished by the configuration of the male clasper and the sculpturing of the pockets near the end of the females' abdomens. The species recently named by Lafontaine were previously treated under the name *havilae* (Grote), and Crumb described the larva of *S. bicava* under the name *havilae. Spaelotis bicava* ranges from British Columbia east to southwestern Manitoba and south to southern California, southern Arizona, and northern New Mexico. The adults fly from late April to late September in most of its range (as late as November in southern California), and there are probably two generations. The larvae have been found in the vicinity of stands of *Eriogonum* (Polygonaceae). The larvae are similar in most aspects to those of *S. clandestina* but differ primarily by having a broader continuous white middorsal stripe.

Eurois occulta (Linnaeus)(FW 2.4–3.2 cm; Plate 58.15) belongs to a genus with only three species. *Eurois occulta* is widespread in boreal habitats in much of North America as well as across Eurasia and Greenland, while the other two species are restricted to North America. *Eurois occulta*'s North American range is from Alaska east across Canada to Newfoundland and Labrador and south in boreal habitats to California, east-central Arizona (White Mountains), northern New Mexico, the Great Lakes states, and Maryland. There is a single generation with a wide range of dates from mid-June to late September. The larvae feed primarily on broad-leaved trees and shrubs as well as conifers (Pinaceae). Robinson et al. list Aceraceae, Betulaceae, Caprifoliaceae, Ericaceae, Rosaceae, and Salicaceae, as well as one herbaceous plant, *Delphinium* (Ranunculaceae). The last-instar larva is pale gray or gray brown and is mottled with black. Subdorsal black wedge-shaped spots may be prominent or faint but are always evident on the seventh and eighth abdominal segments. In addition, the eighth abdominal segment is slightly humped dorsally. The spiracular line is white but changes to yellow along its ventral margin. The head is pale brown or pinkish brown with darker coronal stripes and reticulation. *Eurois astricta* Morrison and *E. nigra* (Smith) are somewhat similar to *E. occulta* but have more restricted distributions. *Eurois astricta* is widespread in boreal sites from Alaska east to Labrador and south to Oregon, Utah, northern New Mexico, the Great Lakes states, and New England. *Eurois nigra* ranges down the western cordillera from Yukon Territory south to California, Arizona, and New Mexico.

Graphiphora augur (Fabricius)(FW 1.7–1.9 cm; Plate 58.16) is the only species in its genus, and it ranges across boreal Eurasia and, in North America, from Alaska east to Newfoundland and south to central California, southeastern Arizona (Chiricahua Mountains), northern New Mexico, South Dakota, the Great Lakes states, and Maryland. The adults have a single flight from late June through August, but most records are from mid-July to mid-August. The larvae prefer to feed on the opening young leaves and buds of a wide array of broad-leaved trees and shrubs but also feed on a variety of perennial herbaceous plants. The plant families eaten include Rosaceae, Salicaceae, and Urticaceae. The last-instar larvae are pinkish gray or pale gray brown with a diffuse dorsal herringbone pattern, a prominent black spiracular line, and a black transverse line across the dorsum of the eighth abdominal segment. The dorsum of eighth abdominal segment is slightly humped. The spiracles are yellow brown. The head is pale yellow brown with darker brown coronal stripes and reticulation.

Anaplectoides prasina (Denis & Schiffermüller), the "green arches" (FW 2.1–3.0 cm; Plate 58.17), is one of three North American *Anaplectoides*, two of which occur in the West. *Anaplectoides prasina* ranges across boreal Eurasia and, in North America, occurs from central British Columbia east to

Newfoundland, but the species is largely absent from the Canadian prairies and the Great Plains. In the West, the moth ranges south to central Oregon, northern Utah, and southern New Mexico (Sacramento Mountains). The adults have a single flight from mid-June to late September. Robinson et al.'s compilation suggests that the larval hosts are a small number of broad-leaved herbs and shrubs including Ericaceae, Rosaceae, and Scrophulariaceae. The last-instar larva is pale brown to brownish gray with a series of darker brown wedge-shaped spots on the abdominal segments. There is a dark brown transverse line across the posterior margin of the eighth abdominal segment. In addition, the eighth abdominal segment has a slight dorsal hump. The spiracles are white or pale gray. The head is yellow brown with darker brown reticulate pattern that merges with the dark brown coronal stripes. The related *A. pressus* (Grote) also ranges across North America but is smaller and has a less contrasting pattern.

Xestia Hübner, as delineated by Lafontaine, the largest genus in the Noctuini, is worldwide and amazingly diverse. There are 52 species in America north of Mexico and about 200 species worldwide. There are 42 species that are endemic to the West or include it as part of their distributions. *Xestia (Xestia) smithii* (Snellen), "Smith's dart" (FW 1.4–1.7 cm; Plate 58.18), ranges from Alaska across boreal portions of Canada to Newfoundland and south in the West to central California, eastern Arizona (White Mountains), and northern New Mexico. In the East, it extends south in the Appalachians to southern North Carolina. There appears to be a single generation with adults having been captured from mid-July to mid-September, rarely June and October. The larvae feed on a wide variety of broad-leaved trees, shrubs, and perennial herbs. The larvae have been reported by Crumb to cause serious damage to strawberry crops in the Pacific Northwest. The last-instar larva is gray brown to brown with paler interrupted middorsal and subdorsal lines. There is a series of dark brown dorsal X-shaped markings on the abdominal segments. The spiracles are pale yellow to brown. The head is bright brown with black coronal stripes and pinkish to reddish reticulation. *Xestia (X.) oblata* (Morrison)(FW 1.3–1.5 cm; Plate 58.20) is a boreal forest species that ranges from Yukon Territory east to Newfoundland and south to central California, southern Utah, and northern New Mexico. There is a single flight from late June to mid-August. The reported larval hosts are *Vaccinium* (Ericaceae), *Spiraea* (Rosaceae), and *Salix* (Salicaceae). The late-instar larva has not been described. *Xestia (Megasema) c-nigrum* (Linnaeus), the "spotted cutworm" (FW 1.5–1.8 cm; Plate 58.21, Plate 64.34 larva, Figs. 249, 250 genitalia), is the most widespread North American *Xestia* and also occurs widely in Eurasia. In the East, it could be confused with *X. dolosa* Franclemont, but there is no similar western species. In North America, it ranges from Alaska east to Newfoundland and south through Mexico to El Salvador. This species has been expanding its range in the East, mainly in disturbed habitats. There are two flights of adults from May to early July and from late July to October. The larval hosts are an extremely wide range of broad-leaved trees, shrubs, and herbaceous plants in at least 26 families. It is likely that this moth causes some damage to crops, but its confusion with *X. dolosa* prior to 1980 make earlier reports from the East indefinite. The last-instar larva is pale brown to gray with a pair of darker brown wedge-shaped marks on each abdominal segment. The spiracles are white to yellowish. The head is white or pale brown with black coronal stripes and brownish reticulation. The related *X. bolteri* (Smith)(FW 1.6–1.7 cm; Plate 58.22) ranges from southeastern Wyoming south in montane coniferous forests to Durango, Mexico. *Xestia (Pachnobia) speciosa* (Hübner)(FW 1.8–2.0 cm; Plate 58.23) is a Holarctic species more or less restricted to open subalpine coniferous forests. In North America, it ranges from Alaska east to the west coast of Hudson Bay and south in the Cascades to northern Washington and in the Rockies, perhaps discontinuously, to Colorado. There is a single flight during July and August. The larvae have been found on several unrelated broad-leaved plants including *Betula nana* (Betulaceae), *Vaccinium* (Ericaceae), as well as *Solidago* and *Hieracium* (Asteraceae). Because of the dissimilarity of host families, the species is considered polyphagous, and a wide array of hosts in suspected. The last-instar larva is dark reddish brown to mottled gray brown above the spiracular line and a paler gray brown below it. There is a diffuse darker herringbone pattern on the dorsum. *Xestia (P.) perquiritata* (Morrison)(FW 1.7–2.1 cm; Plate 58.24) can be recognized from the similar but more simply marked *X. fabulosa* by its narrow boomerang-shaped reniform spot. It ranges from Yukon Territory east to Labrador and Newfoundland, thence south, disjunctly, to central coastal Oregon, Colorado, and, in the East, to North Carolina. It is a species of subalpine boreal forests. The adults are in flight from mid-July to late August. The adults were reared from *Abies* and *Picea* (Pinaceae), but the larvae were not preserved and described. *Xestia (P.) fabulosa* (Ferguson)(FW 1.5–2.0 cm; Plate 58.25) is a subalpine and alpine moth that ranges from Yukon, British Columbia, and the Cascades of northern Washington east to Labrador, Newfoundland, and New England. In addition, there is an isolated population in the Colorado Rockies. Throughout this range the species is found rather sparingly. The adults are nocturnal and have a single flight from late June to mid-July. The larvae are believed to feed on *Vaccinium* (Ericaceae), as does their close Eurasian relative *X. gelida* (Sparre-Schneider). The larvae are light pinkish cream dorsally with broken white middorsal and subdorsal lines. There is a series of paired black subdorsal spots. The lateral areas are speckled with black and gray.

Parabarrovia Gibson is an enigmatic genus of three species with uncertain affinity. The adult genitalia are reminiscent of *Xestia*, but the larvae combine features of Acronictinae, Agaristinae, and Apameini. Placement in the Noctuini is uncertain. *Parabarrovia keelei* Gibson (FW 1.1–1.5 cm; Plate 58.26) is a small, weak diurnal flier that ranges from Alaska east to Yukon Territory in remote montane habitats. The presence of erect hairlike scales on the wings and lack of any markings should serve to identify the species. The adults have a single flight from late June to mid-July. The larvae were reared in the laboratory but probably feed on low tundra plants such as *Dryas* (Rosaceae) and *Saxifraga* (Saxifragaceae). The larvae are pale gray or pale yellow brown with large prominent dark brown pinacula and pale middorsal and subdorsal lines. The spiracles are black. The head is pale yellow brown with darker brown reticulation and coronal stripes.

Agnorisma bugrai (Koçak)(FW 1.3–1.6 cm; Plate 58.27) is one of three species in this North American endemic genus. This species ranges from central British Columbia east to southern Northwest Territories and Nova Scotia, and thence south to central Oregon, southern Utah, southern Colorado, Nebraska, Ohio, and Pennsylvania. There is a single flight from late July to late September. The larval host plants have not been reported. The larva is pale gray to brown with series of black subdorsal and lateral dashes. The spiracles are yellow brown. The head is pale brown with dark brown or reddish brown coronal stripes and reticulation.

Setagrotis pallidicollis (Grote)(FW 1.5–1.6 cm; Plate 58.28) is one of three species in the genus, the females of which have extended telescoping ovipositors. This species ranges from southern British Columbia and western Alberta south to southern California, northern Idaho, and western Montana. It ranges to the west of the more intermountain *S. vocalis* (Grote). The adults have a single flight from mid-July to early September. Crumb reports the larval hosts as alder, *Alnus* (Betulaceae) and serviceberry, *Amelanchier* (Rosaceae). The last-instar larva is pale brown with an apparent black middorsal stripe. In addition, there is a segmental series of black subdorsal dashes. The spiracles are white. The head is pale brown with black coronal stripes and reticulation.

Tesagrotis corrodera (Smith)(FW 1.5–1.9 cm; Plate 58.29) is one of three species of this western North American genus. The species has a range that extends from southern British Columbia south along the eastern base of the Cascade–Sierra Nevada axis to eastern central California. Additionally, the species occurs in the intermountain region extending from northeastern Nevada east to southern Wyoming and western Colorado. The adults have a single flight from early August to early October. The larval host is bitterbrush, *Purshia tridentata* (Rosaceae). The last-instar larva is dark with a prominent white middorsal line and somewhat less prominent subdorsal lines. Each segment has a dark shield-shaped patch that is black laterally and shading to brown medially. There is an extremely sinuous dark brown spiracular stripe. The spiracles are brown. The head is brown with obscure brown coronal stripes and pinkish reticulation.

Adelphagrotis Smith is a small genus with three western species. The genus seems to be most closely allied to *Parabagrotis* Lafontaine. *Adelphagrotis indeterminata* (Walker)(FW 1.4–1.5 cm; Plate 58.30) is the most widespread species and ranges from southern British Columbia, northern Idaho, and western Montana south to southern California. The adults fly from early July to early September (October in southern California). The larval hosts reported by Crumb are three rosaceous shrubs or vines *(Holodiscus, Rubus,* and *Spiraea)* as well as *Salix* (Salicaceae). The last-instar larva is described by Crumb as light mottled gray to mottled pale brownish gray. There are slight protuberances on the first and second abdominal segments as well as a more pronounced dorsal hump on the eighth abdominal segment. There is a black dash on the dorsum of the seventh abdominal segment and a wedge-shaped black mark on the dorsum of the eighth abdominal segment. The spiracles are white. The head is dark with black coronal stripes and reticulation.

Parabagrotis exsertistigma (Morrison)(FW 1.4–1.8 cm; Plate 58.31) is the commonest, most widespread member of a small western genus with five more or less related species. The present species is the most widespread and ranges from southern British Columbia east to southeastern Manitoba and south to southern California, Nevada, northern Arizona, and central Colorado. Records for Saskatchewan and eastern North Dakota may represent vagrants. The adults fly from early May to late September and might have two generations. The hosts reported by Crumb are grasses (Poaceae). The last-instar larva was described by Crumb as brown to pale gray with a broad white middorsal stripe and less conspicuous subdorsal white lines. There is a series of segmental black subdorsal dashes that are wedge-shaped on the last abdominal segments. The spiracles are white. The head is brown with darker brown coronal stripes and reticulation. The other species are more or less limited to the Pacific Coast region, and, among these, **P. insularis** (Grote)(FW 1.3–1.6; Plate 58.32) is the most common and widespread. It is extremely variable both in basic ground of the FW and in the presence or absence of black costal markings.

Protolampra rufipectus (Morrison)(FW 1.4–1.9 cm; Plate 58.33) represents a small genus with one Eurasian species and two North American species. *Protolampra rufipectus* is widespread in North America and ranges from Alaska and Yukon Territory east to Newfoundland and south in montane habitats to east-central California, Utah, and northern New Mexico, with an outlying population in the White Mountains of east-central Arizona. In the East, it occurs south to the Great Lakes states and New England. The adults have been collected from mid-July to early October. The larvae have been found on *Spiraea* (Rosaceae). The larva is almost unpatterned, unlike other noctuines.

Abagrotis Smith is endemic to the New World and, after *Xestia*, is the second largest genus of Noctuini. There are 41 species, most of which occur in open dry forested habitats in western North America. All *Abagrotis* adults are nocturnal and are attracted to light. **Abagrotis trigona** (Smith)(FW 1.2–1.5 cm; Plate 58.34, Plate 58.35) is a variable moth whose females have abdomens not broad and flattened as in other *Abagrotis*. The species ranges from southern British Columbia east to southern Manitoba and south to California, Arizona, New Mexico, and western Texas. Additionally, there is a disjunct prairie population in Ohio. Adults have been collected from March to October, and there may be two generations. In southern Washington, the larval host is *Salix scouleriana* (Salicaceae), and other species of willows are expected to serve as hosts elsewhere. Crumb describes the last-instar larva as light brown with a segmental series of vague dark shield-shaped marks along the back. There is a transverse black line across the dorsum of the eighth abdominal segment. The spiracles are white. The head is brown with black coronal stripes and dark reticulation that becomes reddish posterior to the ocelli. **Abagrotis vittifrons** (Grote)(FW 1.3–1.5 cm; Plate 58.36) is a distinctive moth that ranges from south-central British Columbia, Montana, and southwestern Manitoba south to southern California, central Arizona, and central New Mexico. The species is typical of relatively arid interior habitats and is not found along the coast. The adults have been recorded from April to early November, mainly mid-July to October, and there may be two or more generations. The life history is unreported. **Abagrotis bimarginalis** (Grote)(FW 1.5–1.8 cm; Plate 58.37) has a chestnut FW with an orange costa and gray marginal band. It is a species of moister montane forests and has a Rocky Mountain distribution from the Black Hills south through Colorado, Arizona, and New Mexico to the Sierra Madre Occidentale of Durango, Mexico.

Abagrotis glenni Buckett (FW 1.4–1.5 cm; Plate 58.38, Plate 64.35 larva) is a member of the Mirabilis Species Group whose larvae feed on coniferous foliage. The adults are relatively similar with contrasting light-colored orbicular and reniform spots. The species ranges from southern British Columbia, Montana, and Wyoming south through the interior West to southern California, northern Arizona, and central New Mexico. Adults have been collected from mid-July to late September. Larvae have been found feeding on *Cupressus sargentii* and *Juniperus scopulorum* (Cupressaceae). The last-instar larva was illustrated by Miller; the larvae vary from green to pinkish brown and have a subdorsal segmental series of black spots. **Abagrotis nanalis** (Grote)(FW 1.2–1.3 cm; Plate 58.39) is the smallest *Abagrotis* and has a very distinctive FW. *Nanalis* ranges through much of the interior West in sagebrush-dominated regions from southern British Columbia east to southwestern Manitoba and south to central eastern California, Nevada, Utah, and northwestern New Mexico. There is a single flight from late July to early September. The larval host is big sagebrush, *Artemisia tridentata* (Asteraceae), and possibly other species of *Artemisia*. The larva is black with strong white middorsal and subdorsal lines. The supraspiracular area is pale dorsally with reddish reticulation. The spiracles are black. The head is pale with black coronal stripes and reddish reticulation posterior to the ocelli. **Abagrotis discoidalis** (Grote)(FW 1.3–1.5 cm; Plate 58.40) ranges from eastern Washington, Montana, and southern Alberta south through much of the interior West to central eastern California, Nevada, Utah, northeastern Arizona, and northwestern New Mexico. The adults are in flight from late May to early October; whether there is more than one generation has not been determined. The larval host is greasewood, *Sarcobatus vermiculatus* (Amaranthaceae). The last-instar larva was described by Crumb as light gray above the spiracular line with paler middorsal and subdorsal lines. There is a series of dark diffuse segmental subdorsal markings that become black along their lower edges. The spiracles are blackish brown with brown rims. The head is pale with darker coronal stripes and reticulation. **Abagrotis variata** (Grote)(FW 1.3–1.8 cm; Plate 58.41) ranges from southern British Columbia and southwestern Alberta south to southern California, Arizona, western New Mexico, and southeastern Wyoming. The species is found primarily in montane habitats and is notably absent from the Great Basin. Adult *A. variata* are in flight from mid-May to late September and have one or more generations. The last-instar larva is dark gray with a dorsal segmental series of black wedge-shaped marks margined laterally and posteriorly by a series of pale V-shaped marks. The spiracles are pale brown. The head is pale brown with darker coronal stripes and reticulation. The larvae have been found feeding on *Amelanchier* and *Prunus* (Rosaceae), *Salix* (Salicaceae), *Alnus* (Betulaceae), as well as *Mimulus* and *Verbascum* (Scrophulariaceae). **Abagrotis forbesi** (Benjamin)(FW 1.4–1.6 cm; Plate 58.42) ranges from southern British Columbia, Idaho, Montana, and western South Dakota south to southern California, Arizona, New Mexico, and northern Texas. The adults fly from late May to early October. The larvae feed on serviceberry, *Amelanchier* (Rosaceae). The last-instar larvae are pale to dark gray with darker reticulation. The spiracles are dark brown. The venter is distinctive in being dusky with dark reticulation and, unlike all other known *Abagrotis* larvae, is as dark as the dorsal coloration.

Pronoctua peabodyae (Dyar)(FW 1.6–1.9 cm; Plate 58.43) is one of four species in the genus, all found in western North America. The genus is closely related to *Abagrotis* but differs in the form of the labial palpi, adult appearance, and larval characters. The present species ranges from southern British Columbia and southern Alberta south through the interior West to Nevada, Arizona, and northern New Mexico. In general, its range is complementary to and lies to the east of **P. pyrophiloides** (Harvey), from which it may be distinguished only by dissection of the male genitalia. A larva was found in the vicinity of *Trifolium* (Fabaceae) and was reared by Crumb. The dorsal portion of the last-instar larva is a mottled gray comprising light and dark mottling. The white spiracles are inside inconspicuous dark patches. The head is brown with black coronal stripes and darker brown reticulation.

References for Superfamily Noctuoidea

Barnes, W., and F. H. Benjamin 1922. A revision of the noctuid moths heretofore referred to the genus *Grotella* Harvey. Contributions to the Natural History of the Lepidoptera of North America 5(1): 8–27.

Barnes, W., and J. McDunnough 1918. Illustrations of the North American species of the genus *Catocala*. Memoirs of the American Museum of Natural History, New Series 3(1): 3–47, 22 plates.

Bendib, A., and J. Minet 1999. Lithosiine main lineages and their possible interrelationships. 1: Definition of new or resurrected tribes (Lepidoptera: Arctiidae). Annales de Société Entomologique de France 35: 241–263.

Brown, J. W. 1990. The early stages of *Doa dora* Neumoegen and Dyar (Lepidoptera: Noctuoidea: Doidae) in Baja California, Mexico. Journal of Research on the Lepidoptera 28(1–2): 26–36.

Burgess, A. F., and W. L. Baker 1938. The Gypsy and brown-tail moths and their control. U.S. Department of Agriculture Bulletin 1469: 1–38.

Clarke, J. F. G. 1941. The North American moths of the genus *Arachnis*, with one new species. Proceedings of the U.S. National Museum 91: 59–70, 9 figs.

Comstock, J. A., and C. Henne 1965. Studies in the life histories of North American Lepidoptera. California *Annaphila*s. Journal of Research on the Lepidoptera 3(3): 175–191.

Comstock, J. A., and C. Henne 1966. Studies in the life histories of North American Lepidoptera. California *Annaphila* II. Journal of Research on the Lepidoptera 5(1): 15–26.

Comstock, J. A., and C. Henne 1967. Studies in the life histories of North American Lepidoptera. California *Annaphila* III. Journal of Research on the Lepidoptera 6(4): 257–262.

Crabo, L., and P. C. Hammond 1997. A revision of *Mesogona* Boisduval (Lepidoptera: Noctuidae) for North America with descriptions of two new species. Journal of Research on the Lepidoptera 34: 83–98.

Crumb, S. E. 1956. The larvae of the Phalaenidae. U.S. Department of Agriculture Technical Bulletin 1135: 1–356, plates 1–11.

Donahue, J. P., and J. H. Newman 1966. The genus *Phragmatobia* in North America, with the description of a new species (Lepidoptera: Arctiidae). Michigan Entomologist 1: 35–74.

Dyar, H. G. 1899. On the larvae of North American Nolidae, with descriptions of new species Canadian Entomologist 31: 61–64.

Eichlin, T. D., and H. B. Cunningham 1978. The Plusiinae (Lepidoptera: Noctuidae) of America north of Mexico, emphasizing genitalic and larval morphology. U.S. Department of Agriculture Technical Bulletin 1567: 1–222.

Ferguson, D. C. 1978. Noctuoidea: Lymantriidae. Fascicle 22.2, *in:* Dominick, R. B. et al. (eds.) The Moths of America North of Mexico. E. W. Classey, Ltd., and the Wedge Entomological Research Foundation, Washington, DC; x + 110 pages, 8 plates.

Ferguson, D. C. 1984. Two new generic names for groups of Holarctic and Palaearctic Arctiini (Lepidoptera: Arctiidae). Proceedings of the Entomological Society of Washington 86: 452–459.

Ferguson, D. C. 1985. Contributions toward reclassification of the world genera of the tribe Arctiini. Part 1: Introduction and a revision of the *Neoarctia-Grammia* group (Lepidoptera: Arctiidae: Arctiinae). Entomography 3:181–275.

Ferguson, D. C., D.J. Hillburn, and B. Wright 1991. The Lepidoptera of Bermuda: Their food plants, biogeography, and means of dispersal. Memoirs of the Entomological Society of Canada 158: 2–100.

Ferguson, D. C., P. A. Opler, M.J. Smith, and J. P. Donahue 2000. Moths of western North America. 3: Distribution of Arctiidae of Western North America. Part 1: Texts, maps, and references. C. P. Gillette Museum of Arthropod Biodiversity, Colorado State University, Fort Collins; 171 pages.

Fibiger, M., and J. D. Lafontaine 2005. A review of the higher classification of the Noctuoidea (Lepidoptera) with special reference to the Holarctic fauna. Esperiana (Buchreihe zur Entomologie) 11: 7–92.

Forbes, W. T. M. 1954. Lepidoptera of New York and Neighboring States. Part III: Noctuidae. Cornell University Agricultural Experiment Station Memoir 329. Cornell University, Ithaca, NY; 433 pages, 290 figs.

Forbes, W. T. M. 1960. Lepidoptera of New York and Neighboring States. Part IV: Agaristidae through Nymphalidae. Cornell University Agricultural Experiment Station Memoir 371. Cornell University, Ithaca, NY; 188 pages.

Gibson, A. 1903. Notes on the earlier stages of some Canadian tiger moths of the genus *Apantesis*. Canadian Entomologist 35: 111–123.

Godfrey, G. L. 1972. A review and reclassification of larvae of the subfamily Hadeninae (Lepidoptera, Noctuidae) of America north of Mexico. USDA Technical Bulletin 1450. U.S. Department of Agriculture, Agricultural Research Service, Washington, DC; 265 pages.

Hardwick, D. F. 1958. Taxonomy, life history, and habits of the elliptoid-eyed species of *Schinia* (Lepidoptera: Noctuidae), with notes on the Heliothidinae. Canadian Entomologist 90(Suppl. 6); 104 pages.

Hardwick, D. F. 1996. A Monograph to the North American Heliothentinae. Centre for Land and Biological Resources Research, Ottawa; 281 pages, 25 color plates.

Harp, C. 2004. Genus *Schinia* (Noctuidae: Lepidoptera). Moths of North America Web Site. Northern Prairie Wildlife Research Center, U.S. Geological Survey, Jamestown, ND. www.butterfliesandmoths.org.

Henne, C. 1967. Life history studies on the *lithosina-miona-casta* complex of the genus *Annaphila*. Journal of Research on the Lepidoptera 6(4): 249–256.

Hogue, C. L. 1963. A definition and classification of the tribe Stiriini (Lepidoptera: Noctuidae). Contributions in Science 64: 1–129, 32 plates.

Holloway, J. D. 2003. The Moths of Borneo. Part 18: Family Nolidae. Southdene Sdn. Bhd., Kuala Lumpur, Malaysia; 455 pages, 611 figs., 10 plates.

Johnson, J. W. 1984 (1985). The immature stages of six California *Catocala* (Lepidoptera: Noctuidae). Journal of Research on the Lepidoptera 23(4): 303–327.

Johnson, J. W., and E. Walter 1978 (1980). Similarities and differences in forewing shape of six California *Catocala* species (Lepidoptera: Noctuidae). Journal of Research on the Lepidoptera 17(4): 231–239.

Kitching, I. J., and J. E. Rawlins 1999. Noctuoidea, pages 355–401, *in:* Kristensen, N. P. (ed.) Lepidoptera, Moths, and Butterflies. Vol. 1: Evolution, Systematics, and Biogeography. Handbook of Zoology. Vol. 4 Arthropoda: Insecta, Part 37. W. de Gruyter, Berlin, New York; 487 pages.

Knowlton, C. B., Jr. 1961. A revision of the species of *Halysidota* Huebner known to occur north of the Isthmus of Tehuantepec (Lepidoptera, Arctiidae, Arctiinae). Ph.D. Thesis, Cornell University, Ithaca, NY; v + 205 pages, 94 figs., 12 plates.

Knowlton, C. B. 1967. A revision of the species of *Cisthene* known to occur north of the Mexican border (Lepidoptera: Arctiidae: Lithosiinae). Transactions of the American Entomological Society 93:41–100, 33 figs.

Lafontaine, J. D. 1987. Noctuoidea: Noctuidae (part), Noctuinae (part–*Euxoa*), Fascicle 27.2, *in:* Dominick, R. B. et al. (eds.) The Moths of America North of Mexico. Wedge Entomological Research Foundation, Washington, DC.

Lafontaine, J. D. 1998. Noctuoidea: Noctuidae (part), Noctuinae (part–Noctuini), Fascicle 27.3, *in:* Dominick, R. B. et al. (eds.) The Moths of America North of Mexico. Wedge Entomological Research Foundation, Washington, DC.

Lafontaine, J. D. 2004. Noctuoidea: Noctuidae (part), Noctuinae (part–Agrotini), Fascicle 27.1, *in:* Dominick, R. B. et al. (eds.) The Moths of America North of Mexico. Wedge Entomological Research Foundation, Washington, DC; 385 pages.

Lafontaine, J. D., and M. Fibiger 2006. Revised higher classification of the Noctuoidea (Lepidoptera). Canadian Entomologist 138: 610–635.

Lafontaine, J. D., J. G. Franclemont, and D.C. Ferguson 1982. Classification and life history of *Acsala anomala* (Arctiidae: Lithosiinae). Journal of the Lepidopterists' Society 36: 218–226.

Lafontaine, J. D., and R. W. Poole 1991. Noctuoidea: Noctuidae (part–Plusiinae), Fascicle 25.1, *in:* Dominick, R. B. et al. (eds.) The Moths of America North of Mexico. Wedge Entomological Research Foundation, Washington, DC; 182 pages.

Leuschner, R. H. 1997. Day-flying moths of the genus *Annaphila* Grote (Noctuidae). News of the Lepidopterists' Society 39(3): 33–34, 39.

McCabe, T. L. 1980. A reclassification of the *Polia* complex for North America (Lepidoptera: Noctuidae). New York State Museum Bulletin 432; vi + 141 pages.

Miller, J. S. 1991. Cladistics and classification of the Notodontidae (Lepidoptera, Noctuoidea) based on larval and adult morphology. Bulletin of the American Museum of Natural History 204: 1–230.

Morefield, W. D., and P. Lange 1997. Immature stages of high Arctic *Gynaephora* species (Lymantriidae) and notes on their biology at Alexandra Fiord, Ellesmere Island, Canada. Journal of Research on the Lepidoptera 34: 119–141.

Murzin, V. S. 2003. The Tiger Moths of the Former Soviet Union (Insecta: Lepidoptera: Arctiidae). Pensoft Publications, Sofia, Bulgaria; 243 pages.

Mustelin, T. 2009. Owlet moths (Family Noctuidae) of Southern California. Wedge Entomological Research Foundation, Washington, DC. In press.

Packard, A. S. 1895. Monograph on the bombycine moths of America north of Mexico. Part 1: Family Notodontidae. Memoirs of the National Academy of Sciences, Washington, DC 7; 293 pages, 44 plates.

Pogue, M. G. 2002. A world revision of the genus *Spodoptera* Guenée (Lepidoptera: Noctuidae). Memoirs of the American Entomological Society 43; 202 pages.

Pogue, M. G., and C. E. Harp 2003. Revised status of *Schinia unimacula* Smith including morphological comparisons with *Schinia obliqua* Smith (Lepidoptera: Noctuidae: Heliothinae). Zootaxa 226: 1–8.

Pogue, M. G., and C. E. Harp 2003. Systematics of *Schinia cupes* (Grote) complex: Revised status of *Schinia crotchii* (Hy. Edwards)(Lepidoptera: Noctuidae: Heliothinae). Zootaxa 294:1–16.

Pogue, M. G., and A. C. Laughlin 2002. A revision of the genus *Bulia* (Walker Lepidoptera: Noctuidae). Journal of the Lepidopterists' Society 56: 129–151.

Poole, R. W. 1989. Lepidopterorum Catalogus (New Series), Fascicle 118: Noctuidae, in: Heppner, J. B. (ed.) E. J. Brill, Leiden, New York; Parts 1–3, 1314 pages.

Poole, R. W. 1991. Noctuoidea: Noctuidae (part). Cuculliinae, Stiriinae, Psaphidinae (part), Fascicle 26.1, *in:* Dominick, R. B. et al. (eds.) The Moths of America North of Mexico. Wedge Entomological Research Foundation, Washington, DC; 250 pages.

Poole, R. W., C. Mitter, and M. Huettal 1993. A revision and cladistic analysis of the *Heliothis virescens* species group (Lepidoptera: Noctuidae) with a preliminary morphometric analysis of *Heliothis virescens*. Mississippi Agricultural and Forestry Experiment Station Technical Bulletin. Mississippi State University, 185.; v + 51 pages.

Richards, A. G., Jr. 1939. A revision of the North American species of the *Phoberia-Melipotis-Drasteria* group of moths. Entomologica Americana 29(1): 1–98.

Rindge, F. H., and C. I. Smith 1952. A revision of the genus *Annaphila* Grote (Lepidoptera, Phalaenidae). Bulletin of the American Museum of Natural History 98(3): 187–256.

Rings, R. W., E. H. Metzler, F. J. Arnold, and D. H. Harris 1992. The owlet moths of Ohio. Order Lepidoptera, Family Noctuidae. Bulletin of the Ohio Biological Survey (New Series) 9(2); vi + 219 pages.

Rockburne, E. W., and J. D. Lafontaine 1976. The Cutworm Moths of Ontario and Quebec. Canada Department of Agriculture Publication 1593. Canada Department of Agriculture, Ottawa; 164 pages.

Sargent, T. D. 1976. Legion of the Night: The Underwing Moths. University of Massachusetts Press, Amherst; 222 pages, 8 color plates.

Schmidt, B. C. 2000. The Tiger Moths (Arctiidae) of Alberta. Published by author, Edmonton, Alberta; 36 pages, 4 plates.

Smith, J. B. 1895. Contribution toward a monograph of the insects of the lepidopterous family Noctuidae of boreal America. A revision of the deltoid moths. Bulletin of the United States National Museum 48; vi + 129 pages, 14 plates.

Smith, M. E. 1938. A revision of the genus *Apantesis* Walker (Lepidoptera, Arctiidae). Ph.D. Thesis, University of Illinois, Urbana; iv + 183 pages, 14 plates.

Sotavalta, O. 1963. The generic position of *Hyphoraia alpina* Quens. (Lep., Arctiidae).Annales Entomologica Fennica 29(4): 257–267, 6 figs.

Sotavalta, O. 1965. A revision of the genus *Hyphoraia* Huebner s. lat. (Lepidoptera, Arctiidae). Annales Entomologica Fennica 31(3): 159–197, 31 figs.

Troubridge, J. T., and J. D. Lafontaine 2002. A review of the North American species of the *"Oligia" semicana* group with descriptions of a new genus and twelve new species (Lepidoptera: Noctuidae). The Canadian Entomologist 134:157–191.

Troubridge, J. T., and J. D. Lafontaine 2006. The Noctuoidea of Western Canada. Canadian Biodiversity Information Facility, Government of Canada. Available at www.cbif.gc.ca/noctuoidea/imagelibrary/westindex_e.php.

Tshistjakov, Y. A., and J. D. Lafontaine 1984. A review of the genus *Dodia* Dyar (Lepidoptera: Arctiidae) with description of a new species from eastern Siberia and northern Canada. Canadian Entomologist 116: 1549–1556.

Watson, A. 1971. An illustrated catalog of the Neotropic Arctiinae types in the United States National Museum (Lepidoptera: Arctiidae). Part I. Smithsonian Contributions to Zoology 50: 1–160.

Watson, A. 1973. An illustrated catalog of the Neotropic Arctiinae types in the United States National Museum (Lepidoptera: Arctiidae). Part II. Smithsonian Contributions to Zoology 128: 1–160.

Watson, A. 1980. A revision of the *Halysidota tessellaris* species-group (*Halysidota* sensu stricto)(Lepidoptera: Arctiidae). Bulletin of the British Museum of Natural History (Entomology) 40(1): 1–65, 106 figs.

Watson, A., and D. T. Goodger 1986. Catalogue of the Neotropical tigermoths. Department of Entomology, British Museum (Natural History), Occasional Papers on Systematic Entomology 1; 71 pages, 4 color plates.

Weller, S. J. 1989. Phylogeny of the Nystaleini (Lepidoptera: Noctuoidea: Notodontidae). Ph.D. Thesis. University of Texas, Austin; xi + 396 pages.

Zaspel, J. M., and S. J. Weller 2006. Review of generic limits of the tiger moth genera *Virbia* Walker and *Holomelina* Herrich-Schäffer (Lepidoptera: Arctiidae: Arctiinae). Zootaxa 1159: 1–68.

Suggestions for Collecting
and Observing Moths

We strongly urge any lepidopterist, beginner to professional, to obtain a copy of the inexpensive book, *Basic Techniques for Observing and Studying Moths and Butterflies*, by William D. Winter, Jr. (2000, edited by W. E. Miller and published by The Lepidopterists' Society). This volume summarizes a great deal of experience by numerous contributors and serves as the standard reference for answering questions on any aspect of Lepidoptera techniques. It includes a reprint of Landry and Landry's fine paper that details techniques for handling microlepidoptera. Here we briefly outline materials and methods, with reference to those aspects we have found to be efficient. Winter's manual includes an appendix listing dealers and suppliers of entomological books, equipment, and chemicals, with addresses and telephone numbers.

We also recommend any serious student or collector join the Lepidopterists' Society, which is the only organization that facilitates communication among moth and butterfly enthusiasts throughout North America and internationally. The LS publishes a bimonthly *News*, quarterly *Journal*, and an updated membership list every other year, cross-indexed by state, town, and special interests. The society sponsors an annual meeting, which is informal and fun, yet informative. In addition, each year that the society meeting is not held in the West, the western states members hold a field station–based meeting.

Collecting Moths

Biologists, including biogeographers, ecologists, and people interested in biodiversity or development of local or regional lists of species, as well as taxonomists and systematists, need specimens and specimen-vouchered collection data to carry out their research. Therefore specimens are collected, preserved, dissected, and studied to understand morphological and geographical variation within and between species and to discover and record where each occurs, historically and now, what times of year their immature stages and the adults are active, their larval foods, and so on.

Making a collection of local species is the easiest and most satisfying way for a student to learn the kinds of moths and their relationships, their larval habits, seasonal flight periods, and life history patterns. There is no evidence that normal collecting endangers or has any lasting effect on population numbers. Each female moth lays hundreds or thousands of eggs, and on average, all but two of the progeny must be eliminated by climatic events, parasites, and predators before they reach reproductive age, in order for stable populations to be maintained. A perception that collecting is harmful to insect species as though they are comparable to vertebrates is deeply misguided. Young naturalists should be encouraged to make collections, for their enjoyment and education and appreciation of insect diversity. Ultimately, their specimens, if well prepared and maintained, will contribute to our baseline inventory and knowledge.

Almost any page of this book reinforces the challenge—there are enormous gaps in our knowledge of the life histories and basic taxonomy, with countless species all around us awaiting description and naming. Specimens are required to provide this needed knowledge. Unfortunately, these days little emphasis is placed on training professionals to carry out this kind of biodiversity study.

Collection Methods and Equipment

Field Notebook

We recommend carrying a pocket notebook in the field to record brief notes on weather, especially temperature, mileages, and plant associations. These can be elaborated in more legible form to a more permanent notebook or an electronic file and augmented by details later, such as the following morning if evening collecting is carried out. Surprisingly, many lepidopterists fail to record field data other than by specimen labels. Younger people think they will be able to remember details of collection circumstances, as we did. Alas, they are wrong; we urge any collector to maintain a record of at least basic aspects, including detail of locality site, date, host-plant associations, and so forth. Number of hours afield can be a measure of relative abundance. Such records have been essential to refresh our own memories or respond to inquiries about the circumstances of past collections, often 30 or 40 years later. Ideally, one can record exact coordinates with a geopositioning instrument (GPS). Elevation above mean sea level is useful and is especially important in mountainous western North America, where many moth species are limited to certain elevations.

Net

Various insect nets are available commercially, from small, pocket-sized ones that fold when not in use, to enormous ones favored by butterfly collectors, even models with several extension segments that enable sweeping high in the forest canopy. We prefer a smaller net with standard 12 inch diameter hoop and a 24 to 28 inch handle, which allows carrying in a suitcase or carry-on case during airline travel and is efficient for sweeping at one's feet and collecting in dense vegetation. Although more expensive, dacron chiffon nets are more durable than cotton or nylon—resisting snagging and tearing—and the fine mesh is essential for collecting tiny moths. The net bag is sewn to a folded canvas or muslin strip, which forms a loop through which the hoop is threaded, and the strip should have a length about twice its diameter. A novice net may be made from a wooden dowel and heavy wire (not easily bent by hand) to form the hoop, available from a hardware store or building supply dealer.

Net Collecting

The great majority of moths fly at night, of course, but we regard diurnal netting as an essential part of moth inventory. Many species, especially those that fly during early spring when nightly temperatures are low, are strictly diurnal, for example Adelidae, Heliozelidae, Heliodinidae, Sesiidae, as well as many prodoxids, scythridids, some species of yponomeutids, plutellids, and tortricids such as *Grapholita*, most of which can be found feeding at flowers. Many crambids, geometrids, noctuids, and other moths that fly in spring, in deserts, or at high elevations are day fliers. Elachistids are best taken at dusk and just after dawn by lightly sweeping. Moreover, while still in our student days, we came across "Directions for Collecting Microlepidoptera," written by Lord Walsingham and published in the *American Naturalist* in 1872. He was 29 years old then but had just completed a 13-month collecting expedition in California and Oregon, so he had more experience than many of us achieve in a lifetime. There is no mention of collecting at light in the trip journals; indeed the expedition camped using a sea lion oil lamp. For us, a lasting impression was his emphasis on the importance of netting late in the day: "I watch the tops of grass, the stems of the flowers, the twigs of the trees; I disturb leaves and low-growing plants with a short switch and secure each little moth that moves." He mentions setting (spreading) 400 specimens after a one-night stop. We have taken the hint and have augmented light collecting whenever possible, collecting during the "dusk flight," almost invariably securing species that did not come to lights at the same place, and sometimes obtaining more specimens than later came to lights, particularly if temperatures fell rapidly after sundown.

Walsingham used the field collection technique prevalent in England at the time, in Australia a century later, and by several American microlepidopterists today. That is to carry a large number of small plastic vials in a knapsack or bag, Walsingham's glass topped pillboxes being no longer available. Upon capture, each moth is introduced into a separate vial, which is then held in a bag to reduce activity. That evening or the next day each is transferred into a suitable receptacle, such as a freezer box with snug-fitting lid, containing a piece of cloth with a few drops of liquid ammonia. The moth quickly dies and can be spread after a few minutes. This approach requires one to carry a large number of containers in the field if hiking for prolonged periods, but major advantages are prevention of damage to the specimens during transit in the field, and that the specimens are relaxed after dying. Specimens killed in cyanide vials become stiff for several hours, so need to be relaxed, which in turn can cause breakdown of DNA sequences. When carrying killing vials in the field, which has been our primary experience, you need to provide loose tissue in which lodge the specimens, and to avoid placing too many in one vial (e.g., six to 10, depending on size), then place the vials in snug pockets in the collecting bag, rather than in pants pockets where they will be jarred with every step.

Beating Sheet

A beating sheet provides the most efficient method for collecting caterpillars of many species. Simply hold the sheet under shrubs or trees and sharply rap the vegetation above it with a stick. A beating sheet can be obtained commercially or be inexpensively constructed by sewing flaps to the corners of a canvas or muslin sheet about 3 feet square, then cutting light weight wood slats to fit diagonally, secured in the middle by a wing nut bolt. When not in use, one of the slats is rotated 90° from its position, and the sheet rolled into a canvas bolt. Or, simply carry a light-weight, collapsible umbrella for this purpose, as was a common habit in the nineteenth century.

Collection Vials

For live specimens to be caged for oviposition or transferred later to ammonia, we use 15, 20, and 40 dram snap-top plastic vials (32–50 × 85 mm). In the field, cyanide, ammonium bicarbonate (not carbonate), and ethyl acetate are commonly used killing agents. Using open-mouth vials with snap-top or cork stoppers, or cylindrical olive jars, effective collection vials can be made by placing crystalline ammonium bicarbonate in the bottom, wedged under a tight-fitting disk of polystyrene foam or blotter paper made by cutting a strip about 3/8 inch wide, which is then tightly rolled into a disk that will wedge into the vial. This is a much safer method than using cyanide as the killing agent and more efficient than ethyl acetate, which is used in liquid form and must be replenished frequently. Depending upon climatic conditions, the ammonium bicarbonate needs to be replenished after a few weeks, whereas cyanide will last two or three seasons and is much more effective in traps, which, however, need to be deployed in controlled circumstances.

Light Attraction

Most moth species are nocturnal and are attracted to light. Whether the intent is to collect or to observe and photograph moths, there are several techniques that involve differing amounts of equipment. Technology continues to provide new options for the use of lights of various wave lengths and strengths. In most situations we use a sheet or white canvas background upon which the attracted moths may rest. These sheets may be deployed upright or flat on the ground or both. In windy conditions, a sheet on the ground may work best. People may wish to construct their own sheet frameworks in concave design to maximize the amount of sheet that is lit by the light, and others prefer to use several light sources, including auxiliary sun lamps!

MV Light

More species and individuals of moths are attracted by mercury vapor (MV) lights than other light sources. MV bulbs and associated ballast may be obtained from hardware or building supply stores. We use a 175 watt bulb in a homemade holder fashioned from a 5 gallon plastic bucket that contains the light's ballast. These lights may be arranged in various ways, but a gasoline generator is required for their operation in habitats away from an electrical source. We use a generator that can accommodate either one or two 175 watt bulbs. It is necessary to ensure the generator has sufficient amperage to light the bulb initially. Alternatively, if alternating current (AC) is available, one may use a self-ballasted MV bulb, for example, 160 watts and 120 volts. Such lights may be plugged directly into an outside fixture or outlet. With MV light, individuals of some species may remain around the periphery and not come all the way in to the sheet.

UV Light

If a 15 watt ultraviolet light, also known as a black light (BL), and a 175 to 250 watt MV light are run near one another, the MV light will attract many more moths. However, disadvantages of MV bulbs are the initial and replacement costs, the amount of gear required, and the electrical source, which usually involves a generator that requires fuel replacement every few hours (and cannot be taken on a commercial airline), or sampling is confined to an electrical source, Moreover, in our experience many smaller moths tend to settle distantly from the light more so than is true with UV light. A 15 watt or even 8 watt BL run in a habitat distant from other lights will attract large numbers of moths when temperature conditions are favorable, enough to keep two people busy at the sheet.

A simple set up, consisting of a sheet, a rope, several clothespins, a 15 watt BL, and a 4 pound lead acid battery, which will run a 15 watt fluorescent light four or five hours, easily can be carried to a different habitat and augment collections at the camp base.

Light Traps

Light traps operated all night provide collections from longer periods than evening sheet collecting, enable sampling at two or more sites simultaneously, and are better for collecting species that tend to fly later at night and early morning, including many noctuids and bombycoids. In our experience, sheet collecting for a few hours yields more species, often twice as many, of microlepidoptera and geometrids, which tend to settle on and in the vicinity of traps rather than in them, whereas, traps tend to capture many more species and individuals of larger moths than one sees at a sheet in the same night. Hence both methods used in tandem enable more sites to be sampled for a more complete array of species than either does alone. Effectiveness of both methods is greatly affected by climatic conditions, particularly temperature, and moon phase. On warm nights, above, 60 to 64°F (15 to 17°C), both are much more productive than on cold nights. In tropical regions where temperatures are above 70°F every night and therefore not a factor, light attraction is dramatically increased during overcast and rainy nights compared to clear nights, irrespective of moon phase.

The basic design of a light trap includes a light and its power source; and the light is suspended over a funnel, which conveys the specimens by gravity into a subtending container (bucket, jar, coffee can, styrofoam cooler box, etc.). Erect baffles, preferably made of clear plastic, may be placed above the funnel and can be fitted to hold a blacklight tube. Incoming larger moths, such as noctuids, hit the baffles and drop, enhancing trap efficacy, but smaller moths and geometrids often settle on the trap and its baffles, and the overall effectiveness of baffles has not been assessed. A smaller funnel with screen cap, leading to a hole in the bottom of the bucket, can serve as a rain drain. The hole is plugged with a cork when there is no threat of rain. We have found small, light weight traps (10 inch diameter plastic funnel and bucket, funnel 6 inches deep) equipped with an 8 watt BL to be about 75% as productive as large traps (18 inch sheet metal funnel 18 inches deep) with a 15 watt BL suspended above, in multiple pairwise tests. The disadvantage is greatly outweighed by the mobility smaller traps provide.

For efficient collecting by overnight traps in areas without electrical source, we use a bucket, baffles, and a 4 pound rechargeable lead acid battery, which will power an 8 watt BL all night (10 hours or more). Two setups of a BL sheet or trap can be carried easily as far as one cares to walk (and in airline luggage). On several of the California Channel Islands, we have trapped 120 or more species in a single night, which is as many as we might expect in adjacent coastal mainland habitats that have much greater diversity. A single 8 watt trap on San Clemente Island captured 75 species, about 35% of the moths known on an island of 56 mi² (145 km²).

Baits

Pheromone Lures

Use of synthetic pheromone lures for detecting economic species, such as the codling moth and recently discovered light brown apple moth in California, is a widespread practice. Many specific sex pheromone lures are available for a wide variety of agricultural and household pests. Among lepidopterists, the most popular lures attract sesiids, sometimes several genera. Among those used for tortricids, a blend (98:2 E:Z-11-tdal) for *Choristoneura retiniana* also attracts *Syndemis*, western specimens of which had been quite rare in collections, and the lure fortuitously provided us with an effective survey tool. There are species known only from individuals collected in pheromone traps, for example, *Decodes stevensi* Powell in Colorado, specimens of which were attracted to a lure for *Eucosma sonomana* (Z-9 duodecynly-acetate). Much knowledge has been gained from cooperative pheromone sampling by forest and agricultural entomologists and systematists (e.g., pine tip moths [*Rhyacionia*] surveys, recovery of *Platyedra* during pink bollworm monitoring).

Sugaring

Attracting moths by fermenting baits, commonly called "sugaring," historically has been a favorite collection method in Europe and eastern North America, particularly for underwings (*Catocala*). The basic ingredients include molasses or brown sugar, stale beer, and rotten fruit, and every underwing enthusiast has a better recipe than any other. Who among us can forget the first reading of Holland's riveting account of a sultry evening with thunderstorms building and legions of *Catocala* attending the baits? We read it in winter as boys and could hardly wait for summer to pester our mothers for the ingredients. Alas, we rarely have warm, humid evenings in California, and many lepidopterists have tried sugaring in various parts of California with little or no success. We guess low humidity affects the condition of sap flows and other natural fermentation sources to the extent that moths here have adopted other strategies for feeding. Considering the effort in preparation and the degree of expectation, failed ventures soon discourage most collectors, even though our *Catocala* come to lights only sporadically. Elsewhere in the West, John Peacock and others have used hanging bait traps successfully for attracting *Catocala*, and Terry Deckel and others have taken many *Lithophane* and other noctuids in Colorado. Other cuculline noctuids are readily attracted in late fall and early spring. The classical method has been to paint the fermenting mixture on tree trunks along the edges of woods at dusk and walk a route checking them every half hour or so. In absence of trees, a cotton rope soaked in the bait can be strung along a fence or bushes. Several methods, bait mixtures, and trap designs are described by Winter in his chapter on collecting adult Lepidoptera. We urge others to try sugaring in the West and report their success in order to encourage those of us who have been skeptical.

Insect Pins

Pins specifically for insect specimens are made of thin noncorrosive steel, and they are longer and have sharper points than sewing pins. Insect pins may be obtained from biological supply houses or directly from their European manufacturers. They are marketed in various sizes determined by the diameter of the pin and are sold in packs of 100, or boxes of 10 packs. For most moths, size 2 or 3 is best, the latter for larger moths. Some people use pins size 0 or 1, but specimens mounted on these must be handled carefully because any hasty movement of the specimen may cause vibration and loss of the abdomen. Smaller sizes, 1, 0, or 00, may be used to move the wings of specimens during the spreading process (see below).

Minutin nadeln (or just "minutins") are very thin, short pins manufactured specifically for double mounts that are preferred for small micros. They are sold by biological supply houses in packs of 500 and sizes ranging in diameter from 0.10 to 0.20 mm. The 0.10 mm size can accommodate the tiniest leaf miners but are more difficult to work with, as they bend easily and the tips may hook.

Spreading Moth Specimens

By tradition, lepidopterists commonly spread specimens in an abnormal posture such that the trailing edge of the FW forms a right angle with the body axis, and the wings are fully visible. This is not usually done with winged insects of other orders, but owing to the varied wing patterns shown in Lepidoptera, most lepidopterists feel spreading is necessary. This antiquated fixation almost certainly has been the single most important factor in discouraging students from specializing on Lepidoptera. Pinned, unspread moths are specimens just like those of other insect orders, with all the same values, data, and morphological features. Yet without spreading, often we can't see some important features (e.g., hind wing, distinguishing Gelechiidae; tympana, defining Noctuidae), so at least some representatives of population samples should be spread.

Spreading Boards

Spreading boards are designed for preparing spread specimens of moths or butterflies. These consist of two parallel boards for the wings with a central groove to hold the body. These vary in size and construction, but the usual lengths are 12 to 18 inches. The parallel boards usually are tilted up at a slight angle and vary in width, dependent on the size of specimens intended for preparation. Similarly, the space between the boards (groove) also varies in proportion to the specimen size. For example, parallel boards of 1 5/8 inch with a 3/16 inch groove work well for large micros and geometrids. White pine is ideal for the parallel boards and plastizote foam lining the groove, into which the pinned specimen is seated. The parallel boards are spaced about 3/4 inch above a hard bottom board by two or three hardwood struts. Here are the steps in the process:

1. Select an insect pin of size appropriate to the specimen.

2. Place moth to be pinned in left hand or help by using forceps.

3. With right hand, insert point of pin into exact center of prothorax at a 90° angle to the dorsal axis of the body.

4. Keep wings slightly spread; they should be loose and flexible.

5. Insert pin into central groove of board at 90° angle.

6. Pin a thin strip of stiff paper or card stock onto inner edge of left board with glass-headed pin. Hold left wings flat on board with strip and tease wings into spread position with a small-diameter (00 or 0) pin or setting needle.

7. Pin another thin strip on right board and repeat procedure used for left wings.

8. Hold strips into position over wings by using glass-headed pins or insect pins. Do not leave pins in wings except momentarily.

9. Additional pins may be used to position antennae and to prevent the abdomen from sagging.

10. Proceed to next specimen and so on, until board is full, proceeding from the upper portion to the lower portion. After moths are spread, cover the exposed wing portions with wider strips and add pins to hold them firmly in place. This will prevent the wings from curling.

11. Place boards in a dry place where they will be undisturbed by people, pets, mice, or insect pests. The time to allow for drying depends on the aridity of the climate where you reside, usually about a week for larger specimens.

Single Specimen Spreading Blocks

Instead of spreading boards, some lepidopterists prefer single specimen blocks, which are less cumbersome to store and transport during the drying period. Typically, the method preferred by any given person is the one he or she learned and became proficient with initially. The blocks are made of soft wood such as pine in a similar range of sizes to spreading boards (e.g., from very large 5 or 6 inches across to tiny, $1 \times 1 \times 1$ inch with a 2 mm wide slot). Two favored sizes are $2.5 \times 2.5 \times 1.25$ inches deep with a $\frac{5}{16}$ inch slot and $1.5 \times 2 \times 1\frac{1}{8}$ inches deep with a $\frac{3}{16}$ inch deep slot (Fig. 253a). As with spreading boards, a slight upward slope of the spreading surfaces from the slot is recommended. A hole equal to the width of the slot is drilled through the block at the middle and filled with a plug of plastazote foam or other suitable material that will hold the pin securely (yucca stalk pith is a favorite). Make three or four equally spaced razor blade cuts along the leading and trailing edges of the spreading surfaces. Here are the steps for the spreading procedure:

1. Pin the specimen through the mesothorax, using a size 2 insect pin, or size 3 for larger species (we avoid using size 1 and 0 pins for tiny moths because vibration is more likely to dislodge the abdomen from dry specimens). If the specimen is too small for a size 2 pin, we recommend a double mount using a minutin, described below.

2. Run the pin into the foam to a depth that aligns the underside base of the wings with the spreading surfaces, making sure they are even on both sides and the legs are tucked into the slot.

3. Using a length of soft thread that will wrap around the block more than six times, tie a knot in one end and lodge it in one of the inner razor cuts, then gently pull the wings down to the spreading surface with the thread and run it snugly but not tightly through the opposing razor cut. Pull the thread around to the opposite (right or left) side and repeat the process (Fig. 253b). If the thread is not placed in a razor cut it will be too tight.

4. Using a size 2 pin, gently pressed behind a major wing vein (avoid poking a hole in the wing), pull the wings of one side forward until they resist the position held by the first wrap of thread and secure by a second wrap of thread, also not tightly. For fresh specimens of smaller moths, this usually will complete the spreading position, with the posterior margin of the fore wing perpendicular to the body). Repeat the process for the opposite side.

5. If necessary, complete the spreading process by further nudging the wings forward. Add a third wrap of thread on each side. If the wings exceed the thread wraps, a small piece of paper or soft card, such as the locality label, can be pinned over the wing tips to prevent curling during drying (Fig. 253c).

6. Arrange antennae and, if desired, for larger moths the abdomen can be supported by two pins placed at 45° under it.

Spreading Microlepidoptera

We have used a method learned from H. H. Keifer, using smooth-surfaced cork boards (Fig. 254). This is an efficient system; it takes about 60 to 90 seconds per specimen, once one becomes proficient. If cork board with smooth paper covering is unavailable, balsa wood from a crafts shop will suffice but needs to be smoothed with fine sandpaper frequently. Perform the following steps (see diagram):

1. Using scissors, cut off the minutin at about two-thirds its length from the tip; discard the blunt one-third.

2. Holding the minutin with flat-surfaced eyebrow tweezers (not serrated or fine-tip forceps), pick up the moth by gently stabbing its venter. Place the moth in the "V" between your

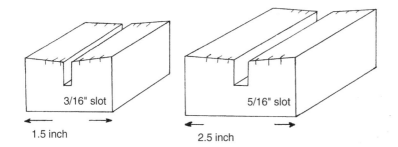

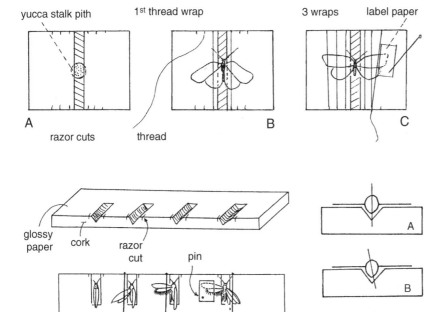

FIGURE 253. Single specimen spreading blocks.

FIGURE 254. The Keifer cork board method of spreading small Lepidoptera.

thumb and forefinger and pull the minutin downward. Do not grasp or pick up the specimen with fingers (legs easily break off). Run the minutin through the pronotum and out the venter about 3 to 4 mm. It is important to place the minutin as near to the center of the thorax and as perpendicular as possible.

3. Pin the minutin into a slot of appropriate width and depth for the size of the moth in the cork board, making sure the thorax is oriented so that both pairs of wings lie flat on the board surface (Fig. 254A); if not quite pinned perpendicularly, compensate by the angle of the minutin (Fig. 254B), to avoid having the wings at different heights on each side (Fig. 254C). Be sure no legs extend under the wings.

4. Place a short length of thread with a knot near the end into the razor blade cut at the left edge of the slot (right-handers). Using your left hand thumb and forefinger at the * shown in the figure to control the thread pressure, gently press the wings to the board. Regulating pressure of the thread, nudge the wings forward by gently teasing with a size 1 insect pin held in your right hand, and blowing lightly if necessary. When in place, hold the wings while pulling the wings forward until the hind margin of the fore wing is perpendicular to the body. Do not puncture the wing membrane.

5. Fluff out the fringes by drawing the pin through them from the apex toward the body. Make final adjustments to orientation of antennae, wings, body, and legs.

6. Holding the wings in place with the thread snug but not tight (avoid scratches or creases in the scaling), secure the wings by placing a small rectangle of stiff card stock over them (soft card stock, such as index cards, will bend and prevent the card from lying flat). Away from the moth, puncture the card off center and run the pin through it about 1 mm, then place firmly over the wings with the pin peripheral to them, making sure the card is not placed at an angle, which will permanently crease the wings.

7. Repeat the process for the right side, using the same piece of thread and another size 1 pin and small card.

8. After drying (three or four days indoors), transfer the specimen to a double mount made from a short strip of plastazote foam or other suitable substance (strips of polyporus fungus were long favored but are no longer available commercially) mounted on a size 2 pin. Again, try to position the wings on a horizontal plane.

Shallow boxes can be constructed with removable slats to hold the cork boards securely (ours are about 8¼ × 11½ inches and about 2 inches in height externally, and hold seven boards with seven or eight specimens each), so specimens can be easily transported. They can be dried in a few hours if a box is subjected to direct sunlight. Lepidopterists at the Smithsonian use a similar system, employing long, narrow spreading boards, such that 300 specimens can be spread and carried in one box. A similar but more elaborate technique has been published in detail by Landry and Landry and is reprinted as an appendix in Winter's manual.

Storage Boxes

Storage Boxes are used in the field for pinned specimens, and their design and expense depends on the purpose to be served. We prefer wooden boxes with pinning bottoms of plastazote foam, which readily accept pins but hold them firmly. Such boxes should be constructed with tight-fitting lids, which help prevent the entrance of museum pests, notably dermestid beetles. Cardboard boxes may be used for temporary storage and are suitable for shipping because of their lighter weight. Permanent collections are stored in glass-topped museum drawers that are housed in steel cabinets or compactor units. Boxes and drawers may be either purchased or constructed in the home workshop.

Handling Specimens after Delay

Ideally, one would like to spread specimens immediately after killing with ammonia, but field conditions or professional and domestic obligations often dictate otherwise. When scheduling in the field does not permit sufficient time for spreading all specimens, several alternatives are available.

If returning to the lab or home the same day, specimens can be placed in a refrigerator freezer in plastic vials or boxes with tight-fitted lids overnight or longer.

If two or three days will elapse until return to home base, place specimens in small containers with snug-fitting lids that have been provided with damp tissue held in place by a nonabsorbent retainer. To hold the specimens, we use cardboard pill boxes containing crumpled tissue, in which specimens lodge, preventing movement during travel. The small relaxing chambers thus created should be carried in a camp ice box to prevent mold development, especially if traveling in warm climates. Upon return home, these containers can be placed in a refrigerator freezer until time is available for spreading.

On longer trips, specimens can be pinned without spreading and allowed to dry. Microlepidoptera can be pinned in shallow boxes (7/8 inch deep) with a tightly fitted lid and polystyrene foam floor. Place the minutin at an angle, with the abdomen appressed to the substrate; the wings can be nudged forward and fringes fluffed out, held in position by friction with the plastazote. These will be much easier to spread properly after relaxing than those pinned without semispreading the wings.

Relaxing Dry Specimens

Some lepidopterists loathe the idea of dealing with dry, unspread moths and even state that it is not possible to obtain good specimens, especially of tiny species. This is not true, but it is difficult, more time-consuming, and variable in results than handling fresh or frozen specimens. The equipment needed is simple, essentially a damp container and mold deterrent. We use a wide-mouth jar lined on the bottom with wet paper toweling. Pinned specimens are placed on a Styrofoam platform or unpinned ones in a pillbox on a glass or plastic coaster on the paper toweling. A pinch of chlorocresol should be added occasionally to prevent mold growth. The length of time required for relaxing varies with how long specimens have been dry, how big they are—small micros may need only a day or two—temperature, and the moisture conditions of the relaxer. One has to learn to judge the time for any particular set of conditions by trial and error. When relaxed properly, most

moths can be spread as effectively and nearly as easily as fresh ones. The most common error is relaxing small specimens too long, so they become wet, which makes them very difficult to spread properly, the fringes often sticking to the body. If left in the relaxer too long, species that were borers as larvae often become greasy, which is caused by breakdown of fat in the body.

Rearing Methods

Host- and Nectar-Plant Documentation

Almost any reader of this book can make contributions to the field of moth biology by documenting the plants on which larvae or adults feed. A critical element of this process is the accurate identification of these plants. Although photographic images may suffice for identification, we recommend that new host associations be documented by preparing a plant specimen with a homemade or standard herbarium press and placing specimens in a university or other museum herbarium for a permanent record. Plants may be identified by using key characters in state or regional floras, then having the identifications verified by a botanist. These floras usually list *all* plant species known in a specific area, and such sources are much preferred over picture books that illustrate only an unknown proportion of plants in the area covered.

We have used primarily two approaches in rearing moths: collecting females, which then deposit eggs, or collecting larvae. Collect females at light or by net in the field, which are retained alive, individually in 15, 20, or 40 dram, snap-top plastic vials (microlepidoptera), provided with a damp cotton wick. They are allowed to oviposit in the vial (mostly polyphagous species) or are transferred to a larger container with suspected host-plant foliage (specialists). Larvae of many species will accept synthetic diet, which is available commercially. Specialist feeders often need their host to induce oviposition and at least initial larval feeding, and sometimes larvae can by weaned to synthetic diet in later instars. Diet is added to the vial containing eggs, and after several days, using a fine brush, we transfer second or third instars to 1 ounce plastic cups containing the diet. About five larvae of tortricid-sized species can be reared per cup without excess crowding.

We collect larvae in the field by a visual search for affected foliage or stems (tied leaves, mines, galls), by digging up suspect plants for root borers, or by beating, which is more productive at night, in our experience. Beating is especially useful for detecting the presence of a larval species initially, facilitating visual search of the foliage, and for collections of macromoth larvae that come out from hiding to feed at night. Typically, we carry polyethylene plastic bags (8 × 18 inches and heavier than a market vegetable bag) and plastic vials, in which batches of larvae are carried, usually held in a camp cooler during travel. Later, we often hold the rearing lots in the plastic bags, lined with paper toweling, and secured by a clothespin. These need to be tended frequently, daily if time permits, often reversed to prevent excess moisture or spritzed with water to reduce desiccation.

Dan Janzen and Brian Scaccia employ systems that are more precise, by assigning a code to each individual larva. This enables a more definite association of the moth, host, parasitoid, and larval exuvium than if multiple larvae are reared in one container in the belief they are all one species. It still requires a subjective assumption that the larva is the same species as others that appear to be identical, to associate parasitoids or preserved larvae. Also, of course, such systems are enormously more labor intensive. Janzen has a team of gusaneros (caterpillar hunters) to monitor hundreds of collections simultaneously and conducts the largest moth-rearing program in the Neotropical region.

Larval Specimens

Study of moth larvae lags well behind that of the adults, with properly associated larval specimens available for study representing fewer than half of the described species in virtually all families. This is an area of research that begs for increased attention. Whereas photos and images may be sufficient for generic or species identification in many macromoths and a few microlepidopteran taxa, well-preserved larvae are important for scientific study of macrolepidoptera, as well as micros. Habitus paintings and photos, however beautiful, including those in this volume, are not adequate substitute for careful morphological studies. Crumb, Godfrey, MacKay, McGuffin, Stehr, and others have contributed fine basic taxonomic treatments, but there are many lepidopterous families for which we have only limited knowledge of the larval taxonomy.

Well-preserved specimens may be produced by killing larvae in just-boiled water for a suitable period to allow expansion of the larva, until its prolegs are fully protruded (often 20 or 30 seconds for tiny larvae, 1 to 5 minutes for larger species). Then transfer to 70% ethanol for permanent storage. If specimens are placed directly into ethanol or isopropyl alcohol, bacterial action continues in the gut, and they turn black. Moreover, the integument shrivels, which pulls it and the prolegs inward, and structures important in identification and classification are difficult to examine.

Rearing Records

We strongly urge any person planning to or already engaged in rearing moths to adopt a consistent number code system that correlates the association of reared adults with preserved larvae, pupal shells, parasitoids reared, and observations recorded during the rearing, including the plant identification and its source. Our numbering system, which was adopted from H. H. Keifer, in 1956, is based on a year-month-collection number code, for example, 07A1 was the first collection in January 2007, and 69D76 was the seventy-sixth collection we made in April 1969. Each collection of one or a group of larvae believed to be a single species taken on the same host on the same day at one locality is assigned a rearing lot number, which is entered on a standardized rearing lot sheet or computer file page, along with the collection data. The lot number is placed in or on the container with the larvae, in vials with preserved larvae, with pressed larval mines, photographs, and later as temporary and ultimately permanent labels on any moth or parasitoid specimens that are reared. If it turns out the lot needs to be subdivided because more than one moth species was included, or two or more host plants are tested, or sublots are exposed to differing conditions during diapause, and so forth, a decimal designation can be added (e.g., 07E4, E4.1, E4.2). Any other system of coding can be developed, so long as it remains consistent. Annette Braun had a basic method; she simply numbered all rearing lots consecutively, irrespective of year, host plant, or locality, such as "B.1103" (Bred number 1103). Brian Scaccia's codes were acronyms based on the host-plant names. Our code, placing the year first followed by code for the month enables immediate recognition of the approximate date of collection and allows chronological sorting in a database, in addition to state, county, family, host-plant name, and so on. A large number of rearing lots can be managed by one person using this system (e.g., a few hundred per year). Our database has more than 15,000 records.

Genitalia Study

Preparation of Genitalic Dissections and Slides

For most moths, study of the external male and internal female genitalia is essential for accurately differentiating genera and species, using available literature. Lepidopterists use various methods to prepare dissections. Various factors, including the nature of structures in different taxa and their orientation, whether slides or glycerine storage is desired, and whether photographs are a goal, affect personal choices of procedures employed. We use a method geared to storage of slide-mounted dissections that are not overly flattened and distorted, which has been developed with modifications from various sources over the years. We work primarily with smaller moths, but the technique is applicable to macromoths. Follow these steps:

1. First, preferably on colored card stock, prepare a small, permanent label to affix to the specimen pin that will unequivocally link it to the microscope slide. A series of these bearing the preparator's name or initials and a sequence of numbers can be generated by computer. Make a temporary label with the same number to retain with the dissected parts during processing.

2. Remove the abdomen by gently pressing upward on the venter, and place it in a solution of approximately 10% KOH or NaOH. After about 12 to 14 hours, or longer for larger specimens with more heavily sclerotized structures, transfer the abdomen to water, and remove the macerated viscera and musculature by flattening the abdomen dorsoventrally with a spatulate tool. For faster maceration, warming under a desk lamp for a few hours is effective.

3. For female specimens, use fine dissecting scissors to cut the abdominal pelt longitudinally along the pleuron, and remove the genital apparatus by cutting the segment 7 and 8 intersegmental membrane. Take care to avoid severing the ductus bursae at its origin; sometimes the ductus bursae and ostium are inextricably connected to the eighth sternum, and it is necessary to remove the latter along with the genitalia. Wash out any remaining body contents.

4. For males, cut or tear the intersegmental membrane just anterior to the genital capsule. The aedeagus can be removed by pulling it free from the interior side; gently tease the vesica out with fine forceps and distend it sufficiently to reveal the form and distribution of the cornuti. For larger moths such as many noctuids, this can be accomplished by water pressure through a fine pipette and may be necessary for identification in some taxa. For tiny specimens that lack cornuti, it may be advisable to retain the aedeagus in situ through the staining process to avoid losing it. A syringe can be used to assist cleaning the abdominal pelt.

5. For both sexes, place the genital apparatus and abdominal pelt in Cellosolve (ethylene glycol monoethyl ether). This dehydrates the preparation while retaining its flexibility and can be used for storage for several hours to several days, during study and illustrating. Finish descaling the pelt and male genitalia, which will be easier after a day or two in Cellosolve.

6. Except for heavily sclerotized males, genital preparations should be stained. We prefer chlorozol black, which when used properly, quickly stains membranes pale greenish blue and leaves sclerotized structure unstained. It is sold in powder form and when mixed with dilute ethanol forms a black-appearing solution. Use two drops of the solution in a small dish of 50% ethanol; the dilution enables one to observe the staining process. The time needed varies with different taxa, but often only 10 or 20 seconds is needed for females, especially if unmated. If neglected and overstained, membranous and sclerotized parts become unicolorous. They can be partially destained by soaking in caustic solution again, but the differential coloring of the membranous and sclerotized parts cannot be retrieved. Acid fuchsin, used in concert with chlorozol black or separately, stains sclerotized parts pink and yields excellent results, especially for tiny specimens. When sufficiently stained, transfer all parts to 90% ethanol (100% if acid fuchsin used because it is highly soluble in water) to wash off excess stain and any remaining bits of debris or scales. After several minutes, transfer back to Cellosolve.

7. Transfer the dissection to essence of euparal or 100% iso-propyl alcohol, and spread the males' valvae with chips of glass for 30 minutes or more, which fixes their position for slide mounting.

8. Place about three drops of somewhat thickened euparal (about the consistency of honey) at the middle of a clean microscope slide. Add all parts of the preparation, including abdominal pelt, which is filleted in females. By custom, the genitalia are placed venter up, with the caudal end away from the observer. For males the aedeagus is aligned to the right, with the abdomen horizontal, below. The female abdomen normally is placed alongside to the right in the same orientation, with the exterior surface up. If the genital capsule or delicate female structures are bulky, tiny squares of plastic can be used as retainers to support the cover slip and minimize distortion.

9. Add the cover slip by gently leaning it onto one side of the euparal and slowly lowering it to prevent trapping bubbles.

10. Inscribe the preparation number assigned to the specimen in step 1 on the slide with a glass etcher, wax pencil, or felt-tip marker, temporarily until a permanent label is added. The label should include slide number, preparator's name or initials, date, genus and species name if known, and name of the mounting medium.

For some taxa, such as Prodoxidae, Tineidae, many phycitine Pyralidae, where the tegumen is bulky and the valvae short, it may not be feasible to mount males in the conventional manner. If a slide mount is desired, one valva can be removed and placed to one side with its inner face up. Alternatively, and with females of many taxa that have intricate, membranous structures, some taxonomists prefer to store the dissected genitalia in glycerine, and tiny neo-prene-stoppered vials can be purchased for this purpose. This method has the advantage of comparing structures from a lateral or any other aspect. The disadvantages are maintenance of the glycerine and difficulty of handling and comparing specimens as contrasted with simply stacking two or three slides.

Photography

Taking images of adult or immature moths has become relatively easy with new techniques of digital photography, and excellent results are achieved. Moreover, old 35 mm slides can be scanned, and the digital images may be enhanced by the manipulation by such computer software as Adobe Photoshop. A bewildering diversity of camera models and styles is changing at an ever-increasingly rapid rate. Point and shoot cameras and single lens reflex cameras are both available in digital format. There is a wealth of published information about photographic techniques; readers might begin by referring to chapter 2 in Winter's techniques manual. Photographers should record detailed information explaining the circumstances when and where each images was taken. Plant associations should be documented by pressed specimens where possible.

There are no permanent repositories or archives for photographic images of moths. There are several Web sites and electronic databases, but none of these has the same intent as a permanent record. For bird photographic images, for example, most states have records committees that maintain records, including images, to document special records. There are no similar efforts and

processes for moth images and records. Records including images and associated data are being gathered in a database by the U.S. Geological Survey in their Butterflies and Moths of North America database, but there is no guarantee this will comprise a permanent record in the way research or voucher specimens in museum collections are intended.

Labels

Labeling every specimen collected or photographed is of critical importance to the value of the specimen (or image). At a minimum, the label information should include the country, state or province, county or other subunit, specific location or GPS coordinates, date, and collector or photographer. Additional information can include elevation, host plant, emergence date, and management unit, for example, national park, national forest, and so on.

Printing Labels on a Computer

Labels may be printed on a computer using processing software such as Microsoft Word. Set the computer for 4.5 points in a standard font such as Century Gothic or New Times Roman and to four or five columns. Keyboard the data in for one label in three to five rows of information, preferably not exceeding 18 digits (including spaces) per line. Then, use the copy and paste function to make enough labels for that information. Go on until the entire page is filled with labels. Alternatively, without special software, labels can be produced easily by typing a row of different labels across the top of a page (one row all the way across at a time), spaced by tabs. Then highlight and duplicate to the end of the page, which yields about 33 copies of each label, using 4.5 point type for four-line labels. If fewer are needed, just stop duplicating at the number needed and change to another label in that or those columns for the rest of the page.

Print the labels on archival paper with as high a quality as possible. We prefer to use acid-free 65 pound card stock that has high rag content.

Legal and Other Considerations

Certainly, no one wishes to disregard laws or trespass on the property of others. There is only a single moth species, *Euproserpinus euterpe* (Sphingidae), that is protected under the specifications of the U.S. Endangered Species Act. This moth is found only in relatively remote areas of Kern and San Luis Obispo counties, California, and it is illegal to collect adults or any of its life stages. We are aware of no other western moth species that are specifically protected by any federal or state statutes.

Recreational or incidental collecting is allowed on most federal lands, except National Park Service units, without a specific permit, but a few national forests are closed to insect collecting. The rules relating to state, provincial, and county parks are quite variable, and one should check first in each case. Most parks, refuges, and preserves will readily provide permission for sampling and research on their lands and usually are anxious to receive information about natural resources under their aegis. Arrangements for study on public lands should always be made well in advance of your planned visits. Always ask private landowners before entering their lands. For further information refer to the Lepidopterists' Society policy on collecting.

GLOSSARY

For anatomical terms, see Figs. 2–24.

ADFRONTAL SUTURES, SCLERITES   On the lepidopteran head, a pair of sutures defining narrow sclerites that border the frons.

AEDEAGUS   Sclerotized, distal part of the phallus, containing the vesica.

AEOLIAN   Windborne, e.g., sand deposits.

ANAL VEINS   The posterior-most of the five principal wing vein systems.

ANAL FORK (COMB)   Sclerotized, musculated prong below the anal plate, usually with four to six tines, used to flip frass away from the larva.

ANAL PLATE (SHIELD)   Sclerotized, dorsal shield on the caudal segment of larvae.

ANAL PROLEGS   Pair of prolegs on the last abdominal segment.

APODEME   Interior projection of integument serving as muscle attachment.

APOMORPHIC   Derived, later development in evolutionary change.

APOPHYSES, ANTERIOR, POSTERIOR   Pair of anteriorly projecting apodemes originating from the ninth (anterior) and tenth (posterior) segment in females, facilitating abdominal movement during oviposition.

AROLIUM   Lobe between the tarsal claws.

BI-, TRI-, UNIORDINAL   Reference to crochets, two, three, or all one size.

BI-, UNISERIAL   Reference to crochets, two rows or one.

BIVOLTINE   Two generations per year.

BRACHYPTEROUS   Having reduced wings, incapable of sustained flight.

BULLA SEMINALIS   Blind sac located along the ductus seminalis, between the ductus bursae and oviduct, where sperm is stored.

BURSA COPULATRIX   Internal female copulatory duct, consisting of the ductus bursae and corpus bursae.

CALLI, AMBULATORY   Enlarged bumps that function as thoracic legs and abdominal prolegs on some leaf- and stem-mining larvae.

CHAETOSEMA (PL. CHAETOSEMATA)   Enlarged, setate, sensory organs located between the antennae and ocelli, unique to Lepidoptera.

CHAETOTAXY   Arrangement of primary setae on larval head, thorax, and abdomen.

CHORION   Egg shell.

CLAVIFORM   Thickened or broadened distally, club shaped.

COLLETERIAL GLANDS, FLUID   Glands that open into the oviduct and produce an adhesive substance that glues the eggs to the substrate and sometimes forms a protective covering over them.

CONJECTIVUM (PL. CONJECTIVA)   A pair of opaque membranes separating the tympanal cavities; the conjectiva meet the tympanal cavities at a distinct angle in Crambidae, on the same plane in Pyralidae.

COREMATA   Eversible, membranous tubes with glandular cells, bearing scales or hairs that are flared during dissemination of chemical signals.

CORETHROGYNE   Usually pertains to specialized scaling of the female, deployed during oviposition.

CORNUTUS (PL. CORNUTI)   One or more sclerotized structures attached to the vesica, sometimes deciduous during mating.

CORPUS BURSAE   Enlarged, distal portion of the bursa copulatrix, where the spermatophore is deposited.

COSTA   Anterior vein of the wing; leading edge of the male valva.

COXA   Basal segment of the thoracic leg.

CREMASTER On the pupa, a group of modified setae, usually hooklike, at the abdominal tip, used to anchor the pupal shell during emergence of the adult.

CROCHETS Tiny, hooklike, sclerotized setae on the abdominal and anal prolegs of larvae.

CUBITUS, CUBITAL VEIN Wing vein between the radial and anal veins.

CUCULLUS Enlarged, distal portion of the male valva.

DIAPAUSE State of metabolic arrest enabling insects, usually an immature stage, to delay development during seasons when necessary resources are not available.

DISCAL CELL An open area formed between the radial and cubital veins by loss of the basal portion of the medial and cross veins.

DITRYSIAN, DITRYSIA Refers to a female genital system in which there are separate mating and egg-laying orifices; defines the derived lineage Ditrysia, which includes 98% of lepidopteran species.

DIVERTICULUM A blind tube or sac branched off from a canal or cavity.

DORSAL MARGIN ("DORSUM") Antiquated term, refers to the posterior margin of the wing; the margins of the two FW meet along the dorsal midline when most moths are at rest.

DORSUM, DORSAL Morphologically, the upper side or back side of animals standing erect.

DUCTUS BURSAE The tubular, proximal portion of the bursa copulatrix.

DUCTUS SEMINALIS A tube connecting the ductus bursae with the oviduct, through which the sperm passes.

ENDOPHAGOUS Feeding internally, in Lepidoptera referring to leaf-mining larvae and stem, seed, and root borers.

EPIPHYSIS Articulated process on the prothoracic tibia, used to clean the antenna, unique to Lepidoptera.

ERUCIFORM Caterpillar shaped.

EXOPHAGOUS Feeding externally, not boring into the feeding substrate.

FEMUR Segment of the leg between the coxa/trochanter and the tibia.

FLAGELLUM, FLAGELLAR The slender, distal segments comprising the antennal shaft.

FLAGELLOMERE Segment of the antenna.

FOVEA A pit, depression, or hollow.

FRASS Excretory pellets produced by external-feeding caterpillars, or liquid by sap-feeding miners; from the German verb *fressen*, "to eat."

FRENULUM Process from the HW costa basally, usually single and spinelike in males, two or three separate setae in females, which projects under a flap on the FW, serving to couple the wings during flight.

FRONS Frontal sclerite of the head.

GNATHOS In male genitalia, paired, lateroventral structures, usually joined, of the tegumen anterior to the uncus.

HAUSTELLUM (PROBOSCIS) In adult moths, tubular structure formed from the two lacinia of the galea, used to probe nectar and other sources of liquid nourishment.

HETEROIDEUS Mesoseries of crochets having rudimentary crochets at both ends.

HETEROMORPHIC Having more than one larval instar of a different form.

HOLOMETABOLOUS, HOLOMETABOLA Complete metamorphosis, having egg, larva, pupa, and adult; the derived lineage of insect orders having this life cycle.

HOMOIDEUS Crochets all of about the same size.

HYPERMETAMORPHOSIS More than one distinct form of larval stage.

HYPOGNATHOUS Mouthparts directed ventrally.

INSTAR One of a series of stages in the growth of a caterpillar.

JUXTA In male genitalia, sclerotized portion of the diaphragma anterior to the aedeagus.

LABIUM Fused sclerites forming the ventral component of the mouthparts of chewing insects, including caterpillars.

LABIAL PALPUS Paired, segmented appendage of the labium.

LABRUM Sclerotized "lip" above the haustellum, attached to the clypeus.

LARVA (PL. LARVAE, NO EXCEPTIONS) The feeding stage during which all growth occurs in holometabolous insects.

MALPIGHIAN TUBULES Long, slender ducts that conduct urinary waste to the hind gut.

MAXILLAE Paired, lateral structures at the base of the mouthparts, which give rise to appendages, lacinia, galea, and palpus, which vary greatly in form and function.

MEDIAL VEIN The middle wing vein, between the radial and cubital veins.

MESOTHORAX, MESOTHORACIC Refers to the middle segment of the thorax, which bears the FW.

METATHORAX, METATHORACIC Refers to the hind segment of the thorax, which bears the HW.

MICROLEPIDOPTERA A nickname applied to members of the superfamilies retaining primitive features, and the basal ditrysian moths (Tineoidea through Tortricoidea in the present sequence), as contrasted with the pyraloids and macromoths, most of which are defined by different kinds of tympana, implying they evolved with and following the radiation of bats in the early Tertiary. Most microlepidoptera are tiny to small, but some are large (e.g., hepialids, cossids).

MONOPHYLETIC An evolutionary lineage derived from a common ancestor and including all descendants of that ancestor.

MONOTRYSIAN, MONOTRYSIA Primitive moths having a single orifice for mating and oviposition; the lineage defined by the condition.

NOTUM Dorsal plate of a thoracic segment.

OBTECT In Lepidoptera, refers to pupae that have the appendages and abdominal segments sealed, immobile.

OCELLUS (PL. OCELLI) Small, simple eye, located behind the antennal socket in moths.

OSTIUM, OSTIUM BURSAE The female orifice of copulation.

OVIPORE The orifice through which the egg passes during oviposition.

OVIPOSITION Act of depositing eggs externally.

PALPUS (PL. PALPI) Appendage of the maxilla or labium.

PAPILLAE ANALES Setate, padlike lobes of the last abdominal segment of female moths, employed to sense positioning of the eggs.

PARAPHYLETIC An evolutionary lineage descended from a common ancestor but not including all its descendants.

PARTHENOGENESIS, PARTHENOGENETIC Haploid reproduction by females without fertilization.

PATAGIUM (PL. PATAGIA) Paired, small lobes at the base of the FW.

PECTEN (PL. PECTINES) Comblike structure.

PERITREME A sclerite of the body wall containing a spiracle.

PHALLOBASE Basal portion of the phallus.

PHEROMONE Chemical secreted externally to communicate with individuals of the same species. A sex pheromone functions in attraction of members of the opposite sex and in courtship.

PHOTOPERIOD Duration of daily exposure to light.

PHYTOPHAGOUS Plant feeding.

PILLIFER Paired, sensory, posterolateral lobes at the base of the haustellum.

PINACULUM (PL. PINACULA) Of the larval integument, raised disc bearing one or more primary setae.

PLANTA (PL. PLANTAE) Basal segment of the tarsus.

PLESIOMORPHY, PLESIOMORPHIC State of a character that was ancestral.

PRETORNAL Area of the FW basad from the tornus.

POLYPHYLETIC A group or grade of taxa arising from different evolutionary lineages.

PRAECINCTORUM Median expansion of the intersegmental membrane between the pyraloid tympanal lobes.

PROGNATHOUS Mouthparts directed anteriorly.

PROLEGS Fleshy abdominal legs of caterpillars.

PROTHORAX, PROTHORACIC First thoracic segment.

PULVILLUS Paired, padlike structures beneath the tarsal claws.

PUPA (PL. PUPAE) In holometabolous insects, the stage following the last larval instar, usually encased in a sclerotized shell, in which metamorphosis to the adult occurs.

RADIUS, RADIAL VEIN Wing vein between the costal and medial veins.

RAMUS (PL. RAMI) Any branchlike structure.

RETINACULUM Loop of scales (female) or flap of integument (male) on the underside of the FW under which the frenulum hooks.

SACCULUS In male genitalia, the anterior edge of the valva, often ornately sclerotized.

SCLERITE Any plate of the hardened (sclerotized) integument defined by sutures.

SCLEROTIZED Hardened insect exoskeleton produced by polymerization and cross-bonding of protein.

SCOLUS (PL. SCOLI) On macromoth larvae, enlarged, fleshy protuberance with setae or branched spines.

SOCIUS, SOCII In male genitalia, paired membranous lobes, which are setate and presumably sensory, arising from the venter of the tegumen, just below the uncus.

SPERMATOPHORE Sac containing the sperm, secreted by the male and deposited in the female's corpus bursae.

SPINNERET Organ that exudes the silk, usually located behind the larval mouthparts.

SPIRACLE Breathing pore; opening to the trachea, through which oxygen passes to the blood.

SPICULA, SPICULATE Tiny, sclerotized spike; beset with spiculae.

STEMMATA Light-sensitive organs consisting of a convex lens and subtended by three epidermal cells and optic cells; typically six stemmata on each side of the larval head.

STERIGMA Sclerite surrounding the female ostium bursae.

STERNITE Ventral plate defined by sutures.

SUBMENTUM Sclerite on venter of the larval head behind the mouthparts and spinneret.

SYMPLESIOMORPHY, SYMPLESIOMORPHIC Shared ancestral trait.

SYNAPOMORPHY, SYNAPOMORPHIC Shared derived trait.

TARSUS (PL. TARSI) Terminal part of the thoracic leg; on moths usually consisting of five "segments."(tarsomeres).

TEGULA (PL. TEGULAE) Paired, articulated, scaled sclerites at the costal base of the FW.

TERGITE Dorsal sclerite.

TEGUMEN The central framework of the male genitalia in moths, formed from the dorsal half of A9.

THORACIC SHIELD   Sclerotized plate formed of the pronotum of moth larvae.

TIBIA   Leg segment between the femur and tarsi.

TORNUS   FW margin at the distal angle between the terminal and posterior (dorsal) margins.

TRACHEA (PL. TRACHEAE)   Cuticular, internal air tubes.

TRANSTILLA   Sclerite of the diaphragma posterior to the aedeagus.

TROCHANTER   Small leg segment between the coxa and femur.

TYMPANUM   Tightly stretched membrane; in Lepidoptera, the membranes of the auditory systems.

UNCUS   Terminal structure of the male genitalia, usually forming a rod or hood curving ventrad from the tegumen.

UNIVOLTINE, BIVOLTINE, MULTIVOLTINE   Single, two, or more annual generations.

VALVA (PL. VALVAE)   Lateral, usually flaplike lobes, derived from segmental parameres; termed "claspers" in older taxonomic literature.

VENTER, VENTRAL   Underside, or front of an erect animal.

VERRUCA (PL. VERRUCAE)   In larvae, an elevated area of the cuticle bearing numerous setae, flared, like a pincushion.

VESICA   Membranous tube, the ejaculatory duct, which is eversible from within the aedeagus as the intromittent organ; highly variable in form and armature (cornuti).

VINCULUM   Basal, usually ringlike portion of the male genitalia, formed from the ventral half of A9.

INSECT INDEX

Arctonotus, pl. 41, pl. 63
arctostaphylana (Kearfott), *Epinotia*, 139, pl. 16
arctostaphylella (Walsingham), *Ethmia*, 63, 69, pl. 5
arcuata Walker, *Drepana*, 201, pl. 27
ardita Franclemont, *Euclidia*, 259, pl. 44
areli Strecker, *Acontia*, 281, pl. 50
arenaeola Balogh & Wilterding, *Pyla*, 191
areneoviridella Ragonot, *Pyla*, 192, pl. 25
Areniscythris, 78 (figs.), 79 (fig.), 80
Arenochroa, 174, pl. 22
argentana (Clerck), *Eana*, 146, pl. 17
argentata (Packard), *Lophocampa*, 272–273, pl. 48
argentata Barnes & McDunnough, *Provia*, 287, pl. 51
argenticostana (Walsingham), *Phaneta*, 133, pl. 15
argentifera Busck, *Aristotelia*, 86, pl. 7
Argentostiria, 290, pl. 52
argillacea (Hübner), *Alabama*, 256, pl. 43
argillacea Kaila, *Elachista*, 73
arguta Hodges, *Psilocorsis*, 67, pl. 5
argutanus (Clemens), *Episimus*, 130, pl. 14
Argyresthia, 103 (figs.), 105, pl. 11
Argyresthiidae, 47, 103 (figs.), 105
Argyria, 183,
Argyrogrammatini, 277
argyrospilus (Walker), *Archips*, 145, 149, pl. 18
Argyrotaenia, 23, 148, pl. 18
arida (Skinner), *Eudesmia*, 265, pl. 46
aridalis (Barnes & Benjamin), *Noctueliopsis*, 171, pl. 22
aridata Barnes & Benjamin, *Chesiadodes*, 208
arietis (Grote), *Lasionycta*, 305, pl. 56
Aristotelia, 85, 86, pl. 7
Arizona underwing, 262
arizonae Grote, *Catocala junctura*, 262
arizonae Kearfott, *Proteoteras*, 136, pl. 15
arizonaria (Grote), *Nemoria*, 220, pl. 31
arizonella Dietz, *Amydria*, 51, pl. 2
arizonenella Dietz, *Pigritia*, 62
arizonnella Ragonot, *Martia*, 194, pl. 26
arizonensis (Capps), *Ceratonyx*, 214
arizonensis Capps, *Haimbachia*, 183, pl. 24
arizonensis Davis, *Opogona*, 51, pl. 2
arizonensis Heinrich, *Retinia*, pl. 14
arizonensis Munroe, *Caphys*, 185 (figs.), 187, pl. 25, pl. 61
arizonensis Packard, *Gloveria*, 232, 235, pl. 34, pl. 62
arizonica Tarmann, *Neoilliberis*, 160 *arizonicum* Dyar, *Lacosoma*, 232
arizoniensis (Stretch), *Notarctia*, 268
Arla, 89 (figs.), 90
army cutworm, 310
armyworm, 304
arnicella (Clarke), *Scrobipalpopsis*, 99
Aroga, 85, 89 (figs.), 92, 96, pl. 9
Arotrura, 79 (fig.), 78, pl. 6
Arrhenophanidae
Arta, 187, pl. 24
Artace, 260, pl. 38
artemisiae Nickerl, *Depressaria*, 67, pl. 5
artemisiana (Walsingham), *Phaneta*, 133, pl. 15
artemisiella McDunnough, *Depressaria*, 67
artemisiella (Treitschke), *Euscobipalpa*, 99, pl. 10
artichoke plume moth, 119
arvalis Grote, *Axenus*, 289, pl. 52
arvalis Henry Edwards, *Annaphila*, 289, pl. 52
Ascalapha, 255–256, pl. 43
asella (Rothschild & Jordan), *Sphinx*, 243, pl. 39, pl. 63
Aseptis, 299, pl. 54

ash borer, 125
aspen carpenterworm, 127
aspen twoleaf tier, 303
Aspitates, 213, pl. 29
assimilans Walker, *Phragmatobia*, 269, pl. 47
Astalotesia, 210, pl. 28
Asthenini, 228
astricta Morrison, *Eurois*, 315
astrigata Barnes & McDunnough, *Oncocnemis*, 285–286, pl. 51
astrologa Barnes & McDunnough, *Annaphila*, 289, pl. 52
Atethmiina, 294
Athrips, 88, pl. 7
Atlas moth, 236
atomaris (Smith), *Euxoa*, 311, pl. 57
atomaris Hübner, *Phoberia*, 257, pl. 43
atrata Hodges, *Periploca*, 83, pl. 7
atratella (Blanchard & Ferguson), *Quasisalebria*, 191
atrolinea (Barnes & McDunnough), *Chrysoecia*, 288, pl. 52
atrupictella (Dietz), *Coleotechnites*, 87
Attacus, 236
attenuata (Grote), *Agriphila*, 182, pl. 24
Atteriini, 128, 155
Atteva, 105, pl. 10
augur (Fabricius), *Graphiphora*, 315, pl. 56
augustella (Clarke), *Caloreas*, 122
augustipennis (Grote), *Therasea*, 280, pl. 50
aulaea (Clarke), *Filatima*, 96, pl. 9
aurantiaca (Henry Edwards), *Schinia*, 292, pl. 52
aurantiaca (Hübner), *Virbia*, 267
auranticella (Grote), *Dioryctria*, 192, pl. 25
auratella (Clemens), *Vaxi*, 183, pl. 24
aureoalbida (Walsingham), *Phtheochroa*, 156, pl. 20
auricyanea (Walsingham), *Dyseriocrania*, 34, pl. 1
auripennis Lafontaine, *Euxoa*, 312, pl. 57
aurirubra Braun, *Antispila*, 39
aurora (J. E. Smith), *Hyparpax*, 250–251, pl. 42
aurosparsella (Walsingham), *Eriocraniella*, 34, 35, pl. 1
aurostriata Graef, *Hyparpax aurora*, 250–251
Austrandesiina, 307
Autographa, 259 (fig.), 277–278, pl. 49
Automeris, 237, 239–240, pl. 37
Autoplusia, 277, pl. 48
Autoplusiina, 277
autumnata Borkhaus, *Epirrita*, 228, pl. 33
auxiliaris (Grote), *Euxoa*, 310, pl. 57
avalona McDunnough, *Eucosma*, 134, pl. 15
averna Barnes & McDunnough, *Heterocampa*, 250, 253 (fig.), pl. 42
aversa (H. Edwards), *Neoprocris*, 161
avuncularia (Guenée), *Dasyfidonia*, 210, pl. 28
Axenus, 289, pl. 52
Axyliina, 313
azaleella (Brants), *Caloptilia*, 53, pl. 3

baccatella Pellmyr, *Tegeticula*, 42
bacchariella (Keifer), *Coleotechnites*, 87, pl. 7
baccharisella Busck, *Gnorimoschema*, 98, pl. 10
Bactra, 129 (figs.), 130, pl. 14
Bactrini, 129
badia Hodges, *Pyroderces*, 82, pl. 6
badium (Braun), *Helcystogramma*, 102, pl. 10
Bagisara, 280, pl. 50
Bagisarinae, 280
baldiana (Barnes & Busck), *Teliopsis*, 89, pl. 8
balsamorrhizae McDunnough, *Oidaematophorus*, 120
balteata (Smith), *Tripudia*, 280, pl. 50
banded sunflower moth, 158

banded tussock moth, 272
banded woollybear, 269
Bandera, 198, pl. 26
banksaria Sperry, *Chlorosea*, 220, pl. 31, pl. 62
Baptarma, 290, pl. 52
Baptini, 212
Barbara, 131, 132, pl. 14
barberella (Busck), *Apachea*, 67, pl. 5
barberella (Busck), *Ypsolopha*, 106, pl. 10
barberellus (Busck), *Prodoxus*, 43, pl. 1
bardus Hodges, *Chionodes*, 93, pl. 8
barnesi Swett, *Xanthotype*, 214
barnesii (Dyar), *Cisthene*, 265, pl. 46
basilaris (Zeller), *Sciota*, 191, pl. 25
Basilodes, 288–289, pl. 52
basiochrealis Grote, *Cacozelia*, 187, pl. 25, pl. 61
basiplaganus (Walsingham), *Decodes*, 146, pl. 18
Batia, 62, pl. 4
Batrachedra, 74, pl. 6
Batrachedridae, 74
Batteristis, 101, pl. 10
battis Hodges, *Perimede*, 84, pl. 7
baxa Hodges, *Dichomeris*, 102, pl. 10
bayensis Heppner, *Ellabella*, 166, pl. 21
beanii (Neumoegen), *Neoarctia*, 267, pl. 46
bean-leaf skeletonizer, 279
bean pod borer, 190
bear sphinx, 245
beautiful eutelia, 263
Bedellia, 109 (fig.), 112, pl. 11
Bedelliidae, 109 (fig.), 110
bedstraw hawkmoth, 246
beet armyworm, 294
beet webworm, 174
behrensi Stretch, *Paraphymatopus*, 37, pl. 13
behrensaria (Packard), *Pero*, 214, pl. 29
Behrensia, 286, pl. 51
bella Chambers, *Aetole*, 111, pl. 11
bella (Linnaeus), *Utetheisa*, 271
bella moth, 271
bellela (Walker), *Nemophora*, 40, pl. 2
Belonoptera, 167
benesimilis McDunnough, *Euplexia*, 294, pl. 53
bent-winged owlet, 254
berkeleyellus (Powell), *Morophagoides*, 49, 50, pl. 2
bertha armyworm, 304
bertholdi (Grote), *Comadia*, 127, pl. 13
Bertholdia, 273, pl. 48
beskei Hübner, *Crinodes*, 253 (fig.)
Besma, 217, pl. 30
beta Heinrich, *Sarata*, 193
betularia (Linnaeus), *Biston*, 211, pl. 28
bibionipennis (Boisduval), *Synanthedon*, 125, 126, pl. 14, pl. 60
bicava Lafontaine, *Spaelotis* 314–315, pl. 58
bicolor (Grote), *Eilema*, 266, pl. 46
bicolor Clarke, *Chionodes*, 95, pl. 9
bicolor McDunnough, *Behrensia*, 286
bicolor Powell, *Decodes*, 145 (fig.), 146, 147, pl. 18
bicolorago (Guenée), *Agrochola*, 298
bicoloraria Packard, *Chloraspilates*, 213, pl. 29
bicolorata (Grote), *Neleucania*, 307, pl. 56
bicolored moth, 266
bicurvata Blanchard & Knudson, *Astalotesia*, 210
biedermanata Cassino & Swett, *Eupithecia*, 230, pl. 33
biedermani (Skinner), *Crinodes*, 250, pl. 42
biedermani Barnes & McDunnough, *Schizura*, 251, pl. 42, pl. 64
bifasciata Davis, *Neocrania*, 34, pl. 1
bifasciata (Smith), *Euxoa*, 311, pl. 57

dashlined looper, 208
Dasyblemma, 252–253
Dasychira, 253 (fig.), 274–275, pl. 48
Dasyfidonia, 230, pl. 28
Dasylophia, 252, pl. 42
Dasypyga, 189 (figs.), 190, pl. 25
Datana, 248, 249–250, pl. 33, pl. 64
dataria (Hulst), *Cyclophora*, 222, pl. 31
daucella (Denis & Schiffermüller), *Depressaria*, 67, pl. 5
dayi Blackmore, *Tolype*, 232
davidsonii Davis, *Hyaloscotes*, 52
Daviscardia, 50, pl. 3
davisi Barnes & Benjamin, *Coloradia pandora*, 238, pl. 36
dayi Blackmore, *Tolype*, 232
debenedictisi Povolny & Powell, *Gnorimoschema*, 98, pl. 9
Decantha, 62, pl. 4
Decaturia, 186, pl. 24
deceptiva McDunnough, *Feralia*, 287, pl. 51
decipiens (Grote), *Agrochola*, 298, pl. 54
declinata (Grote), *Paectes*, 263, pl. 46
Decodes, 145 (figs.), 146, pl. 18
decolor (Walker), *Enargia*, 300, pl. 54
decorata Smith, *Hypena*, 255, pl. 43
deducta (Morrison), *Bulia*, 258, pl. 44
defectellus (Zeller), *Nemapogon*, 50. pl. 2
defensaria (Guenée), *Xanthorhoe*, 227, pl. 32
definita (Barnes & McDunnough), *Oncocnemis*, 286, pl. 51
deflecta (Busck), *Scodes*, 102, pl. 10
defoliaria (Linnaeus), *Erannis*, 212
Deilephila, 246, pl. 41
Dejongia, 119, p-l. 12
delawaricus (Walsingham), *Oxyptilus*, 118, pl. 12
delicate cycnia, 271–272
delilah ssp *desdemona* Henry Edwards, *Catocala*, 262
delotella (Busck), *Dichomeris*, 102, pl. 10
Delphasidae, 162
delphinii (Boisduval), *Euclea*, 164
demissae (Keifer), *Filatima*, 96, pl. 9
demissaria (Hübner), *Idaea*, 222, pl. 31
dentata (Grote), *Melitara*, 196, pl. 13
denticulata (Ragonot), *Interjectio*, 190, pl. 25
dentiferella (Walsingham), *Ypsolopha*, 107, pl. 11
Deoclona, 74, pl. 5
Deoclonidae, 47, 59, 74
deprecatorius Heinrich, *Olethreutes*, 129, 131, pl. 14
Depressaria, 66, pl. 5
Depressariidae, 47, 59, 60, 63, 65
depuratella (Busck), *Filatima*, 96, pl. 9
derelecta Heinrich, *Eucosma*, 135
deridens (Guenée), *Charadra*, 281, pl. 50
deserta (Felder), *Cisthene*, 265, pl. 46
deserticola Davis & Deschka, *Phyllonorycter*, 56, pl. 3
desertus Henry Edwards, *Lycomorpha*, 265
deshaisiana (Lucas), *Cydia*, 142, pl. 17
designalis Guenée, *Agathodes*, 179, pl. 23
desiliens Meyrick, *Gelechia*, 91, pl. 8
Desmia, 179, pl. 23
desperaria (Hulst), *Ixala*, 213, pl. 29
Destutia, 217, pl. 30, pl. 62
determinata Walker, *Metanema*, 216
Detersa species group, 311
deutschiana (Zetterstedt), *Aethes*, 158, pl. 20
devastator (Brace), *Apamea*, 295, pl. 53
diamondback moth, 107
diana (Hübner), *Choreutis*, 122, pl. 12
Diaphania, 179, pl. 23
Diastictis, 180, pl. 23

Diataga, 50
Diathrausta, 179, pl. 23
diazoma Grote, *Quadrina*, 235, pl. 34
Dichagyris, 308, pl. 56
Dichogaster, 236, pl. 34, pl. 62
Dichomeridinae, 101
Dichomeris, 101, 103 (figs.), pl. 10, pl. 60
Dichorda, 221, pl. 31
Dichozoma, 170, pl. 21
Dichrorampha, 141, pl. 16
Dictyophoridae, 162
Dicymolomia, 169 (figs.), 172, pl. 22
Diedra, 149, pl. 18
diffasciae (Braun), *Stigmella*, 37, pl. 1
diffinis (Boisduval), *Hemaris*, 233 (fig.), 245, pl. 41, pl. 63
diffusa (Barnes), *Thurberiphaga*, 290, pl. 52
diffusa (Walker), *Faronta*, 304, pl. 55
digitana (Walsingham), *Epinotia*, 140, pl. 16
digitula Mustelin & Mikkola, *Apamea*, 295, pl. 53
Digrammia, 205, pl. 27
dilatifasciella (Ragonot), *Laetilia*, 195, pl. 26, pl. 61
dilatocula (Smith), *Lithophane*, 297, pl. 54
dimidiata (Grote), *Plagiomimicus*, 288, pl. 52
dimidiata Herrich-Schaeffer, *Pyromorpha*, 162
dimorphic bomolocha, 255
dimunendis (Barnes & McDunnough), *Idia*, 254
dingy cutworm, 312
Dioptinae, 252
Dioryctria, 192, pl. 25
Diphthera, 281, pl. 50
Diphtherinae, 281
Diploschizia, 110, pl. 11
Diptychophora, 183, pl. 24
dircella Braun, *Leucanthiza*, 56, pl. 3
directana (Walker), *Cenopis*, 154, pl. 19
discistriga (Smith), *Condica*, 293, pl. 52
discoidalis (Grote), *Abagrotis*, 317, pl. 58
disconcolorata Barnes & Benjamin, *Monoleuca*, 164
discospilata (Walker), *Eufidonia*, 210, pl. 28
discostrigella (Chambers) race *subcaerulea* Walsingham, *Ethmia*, 68, pl. 5
discostrigella (Chambers), *Ethmia*, 68, 69 pl. 5
discreta Braun, *Tischeria*, 44
dispar (Linnaeus), *Lymantria*, 274, pl. 48, pl. 63
disputabilis Obraztsov, *Acleris*, 144, pl. 17
dissociarius (Hulst), *Holochroa*, 214, pl. 29
disstria Hübner, *Malacosoma*, 235, pl. 37, pl. 62
disticha (Morrison), *Ulolonche*, 306, pl. 56
distincta Braun, *Tischeria*, 35
distincta French, *Tolype*, 234
distincta Keifer, *Leucogoniella*, 38, pl. 7
Ditula, 152, pl. 19, pl. 60
diva Grote, *Annaphila*, 259 (figs.), 289, pl. 52
divaricata (Braun), *Arotrura*, 79, pl. 6
diversella (Busck), *Arla*, 90, pl. 8
diversilobiella Opler, *Caloptilia*, 54, pl. 3
divesta (Grote), *Chytonix*, 301, pl. 54
dividua (Grote), *Cobubatha*, 280, pl. 50
dnopherella Ragonot, *Sarata*, 193
Doa, 247, pl. 33
dock rustic, 295
Doidae, 201, 247, pl. 33
Doll's sphinx, 243
dolliana Dyar, *Euclea*, 164
dollii (Neumoegen), *Paranthrene*, 124
dollii Neumoegen, *Sphinx*, 243, pl. 40

Dolocucullia, 284
dolosa (Grote), *Egira*, 302, pl. 55, pl. 64
dolosa Franclemont, *Xestia*, 315
dominatrix Rubinoff & Osborne, *Bucculatrix*, 57, pl. 4
Donacaula, 170, pl. 21
dora Neumoegen & Dyar, *Doa*, 247
doris Barnes, *Coloradia*, 238, pl. 36
Dorithia, 145 (figs.), 147, pl. 17
dorothea Dyar, *Macrurocampa*, 250, pl. 42
dorsalana (Dyar), *Argyrotaenia*, 149, pl. 18
dorsimacula (Dyar), *Cisthene*, 265, pl. 46
dorsipunctella (Kearfott), *Pediasia*, 182, pl. 24
dotalis (Hulst), *Ragonotia*, 194, pl. 26
dotella Dyar, *Honora*, 195, pl. 26
Douglas-fir cone moth, 132
Douglas-fir pitch moth, 125
Douglas-fir tussock moth, 276
Douglasiidae, 47, 58, 103 (figs.)
Drasteria, 258, pl. 44
Drepana, 201, pl. 27, pl. 61
Drepanidae, 201, pl. 27, pl. 61
Drepaninae, 201
Drepanoidea, 201
Drepanulatrix, 213, pl. 29, pl. 61
dromiella Busck, *Gelechia*, 91, pl. 8
drucei Schaus, *Epiperola*, 164
drurella (Fabricius), *Chrysoesthia*
Dryadaula, 51, pl. 2
Dryadaulinae, 51
Dryadaulinae, 51
Dryotype, 300, pl. 54
duaria (Guenée), *Metarrhanthis*, 216, pl. 29
ducta (Grote), *Mniotype*, 300, pl. 54
Dudusinae, 250
dulcedo (Hodges), *Stilbosis*, 83, pl. 7
duodecemstriata (Walsingham), *Ofatulena*, 143, pl. 17
duplex (Walsingham), *Pseudosciaphila*, 130, 150, pl. 14
durata (Opler), *Trachycera*, 189
dusca Barnes & McDunnough, *Crambidia*, 266, pl. 46
duskywing skipper, 259
dyari (Jordan), *Pyromorpha*, 162, pl. 21
dyari Cockerell, *Cirrhophanus*, 289, pl. 52
dyari Taylor, *Gabriola*, 214, pl. 29
Dyotopasta, 50, pl. 2
Dyseriocrania, 34, pl. 1, pl. 59
Dysodea, 167, pl. 21, pl. 61
Dysschema, 271, pl. 48
Dysstroma, 223, pl. 31, pl. 62

Eacles, 237, pl. 35, pl. 62
Eana, 146, pl. 17
eburnella (Denis & Schiffermüller), *Mirificarma*, 100, pl. 10
eburnea (Walsingham), *Arotrura*, 79, pl. 6
Ecdytolopha, 143, pl. 16
echo (Dyar), *Eudonia*, 170, pl. 21
Ecliptoptera, 224, pl. 31
ectrapelaria Grossbeck, *Somatolophia*, 218, pl. 30
Ectropis, 210, pl. 28
Ectypia, 272, pl. 48, pl. 64
Edia, 170, pl. 21
editha Busck, *Ellabella*, 166, pl. 21
edmansae (Packard), *Vitula*, 194, pl. 26
edna (Hulst), *Eupithecia*, 229, pl. 33
Edwards' glassywing, 273
edwardsata (Hulst), *Sabulodes*, 219, pl. 31
edwardsata (Packard), *Neoterpes*, 216, pl. 30
edwardsialis (Hulst), *Sarata*, 193, pl. 13
edwardsiana (Kearfott), *Grapholita*, 142, pl. 17, pl. 60 larva
edwardsii (Beutenmüller), *Penstemonia*, 109 (figs.)

edwardsii (Packard), *Hemihyalea*, 273, pl. 48
effrentella Clemens, *Amydria*, 52
egena (Guenée), *Autoplusia*, 277, pl. 48
egenoides Franclemont & Todd, *Autoplusia*, 277
Egira, 302, pl. 55, pl. 64
eglanterina (Boisduval), *Hemileuca*, 233 (fig.), 238–239, pl. 36, pl. 63
eight spot, 280
eightspotted forester, 284
Eilema, 266, pl. 46
Elachista, 71, 72 (figs.), pl. 6
Elachistid lineage, 49, 64
Elachistidae, 47–54, 60, 70–72, 74 (figs.)
Elaphriini, 294
Elasmopalpus, 193, pl. 26
elaboratella (Braun), *Aroga*, 97
elatella Pellmyr, *Tegeticula*, 42, pl. 1
Elatobia, 48, pl. 2
elautalis (Grote), *Pseudoschinia*, 171, pl. 21
elbea Smith, *Raphia*, 282
elcodes (Dyar), *Lineodes*, 178, pl. 23
elder moth, 299–300
eldorada (Keifer), *Aroga*, 96, pl. 9
eldorada Powell, *Adela*, 40, pl. 2
electellum (Hulst), *Homeoesoma*, 193, pl. 26
Electra buck moth, 238–239
electra W. G. Wright, *Hemileuca*, 238–239, pl. 36
elegans Powell, *Lithariapteryx*, 109 (fig.), 110, pl. 11
elegans (Strecker), *Odontosia*, 249, pl. 33, pl. 63
elegans Stretch, *Euchaetes*, 272, pl. 48
elegant prominent, 249
elegant sphinx, 244
elegantalis Warren, *Choristostigma*, 178, pl. 23
elegantella (Chambers), *Aristotelia*, 86, pl. 7
eleonora Obraztsov, *Anopina*, 148, pl. 18
elephant hawkmoth, 246
elisae Lafontaine & Troubridge, *Cosmia*, 299
Ellabella, 166, pl. 21
ello (Linnaeus), *Erinnyis*, 242
elmorei (Keifer), *Keiferia*, 100
eloisella (Clemens), *Mompha*, 77, pl. 6
elpenor (Linnaeus), *Deilephila*, 246, pl. 41
Elpiste species group, 206
elsa (Strecker), *Sagenosoma*, 243, pl. 39, pl. 63
Elsa sphinx, 243
elutella (Hübner), *Ephestia*, 198, pl. 26
emarginana (Walsingham), *Epinotia*, 137, 138, pl. 16, pl. 60
Emarriana, 284
emasculata (Dyar), *Iridopsis*, 208, pl. 28
Ematurga, 212, pl. 28
Embola, 110, pl. 11, pl. 60
Emeralds, 220
emmeli Powell, *Ethmia monticola*, pl. 5
Emmelina, 117 (fig.), 121, pl. 12, pl. 60
Enargia, 300, pl. 54
enceliae Braun, *Bucculatrix*, 58
Enchoria, 227–228, pl. 33
enchrysa Hodges, *Stagmatophora*, 82, pl. 7
Endothenia, 130, pl. 14
Endromidae, 236
Endrosis, 64, pl. 4
engelhardti B. Clark, *Sphinx sequoiae*, 244
Engelhardtia, 307, pl. 56
enigmatica Lafontaine & Crabo, *Cerastis*, 313, pl. 58
Ennominae, 204
Ennomini, 215
Ennomos, 215, pl. 29
enormis Meyrick, *Batrachedra*, 75, pl. 6
Entephria, 225, pl. 32
entima Franclemont, *Caloecia*, 236, pl. 35
Enypia, 220, pl. 31

Eoreuma, 183, pl. 24
Eosphoropteryx, 278, pl. 48, pl. 64
Epermenia, 116, 117 (fig.), pl. 12
Epermeniidae, 115, 116, 117 (fig.)
Epermenoidea, 20, 115
ephelidaria Hulst, *Neoterpes*, 217
ephemeraeformis (Haworth), *Thyridopteryx*, 49
Ephestia, 189 (figs.), 197, pl. 26
Ephestiodes, 197, pl. 26
epia (Dyar), *Hemiplatytes*, 184, pl. 24
Epiblema, 135, pl. 15
epicoenalis Ragonot, *Arta*, 187, pl. 24
Epidemas, 299, pl. 54
Epimartyria, 5 (fig.), 33, 35, pl. 1, pl. 59
Epinotia, 131, 137–140, pl. 16
Epipaschiinae, 185 (figs.), 187
Epiperola, 164
Epiphyas, 153, pl. 19
epiphysaria Dyar, *Glaucina*, 207, pl. 27
Epiplemidae, 203
Epipleminae, 203
Epipyropidae, 115, 160, 162
Epirrhoe, 227, pl. 32
Epirranthini, 215
Epirrita, 228, pl. 33
Episimus, 130, pl. 14
epsilon Heinrich, *Sarata*, 193
equivoca Neunzig, *Salebriaria*, 191, pl. 25
Eralea, 81, pl. 7
Erannis, 212, pl. 28
Erebinae, 255
erechtea (Cramer), *Caenurgina*, 258, pl. 44
erectarius Grossbeck, *Tornos*, 207, 209 (figs.), pl. 27
Eremberga, 196, pl. 26
Eremobina, 295, pl. 53
ericameriae Braun, *Bucculatrix*, 57
ericameriae Keifer, *Gnorimoschema*, 89 (fig.), 98, pl. 9
Eriocottidae, 47
Eriocrania, 34, pl. 1
Eriocraniella, 5, 19, 33–35, pl. 1
Eriocraniidae, 5, 19, 33–35
Eriocranioidea, 33, 34
eriogonella (Clarke), *Aroga*, 96, pl. 9
Eriopygini, 305
Eriplatymetra, 216, pl. 29
erosa Hübner, *Anomis*, 256, pl. 43
erransella Chambers, *Perimede*, 84
Ersephila, 225, pl. 32
erugatus Davis & Deschka, *Phyllonorycter*, 56, pl. 3
Erynnis, 259
erythremaria Guénee, *Cabera*, 213, pl. 28
erythropasa (Dyar), *Dioryctria*, 192
esmeralda (Oberthür), *Polychrysia*, 278, pl. 48
Esperia, 62, pl. 4, pl. 59
essigana (Busck), *Amorbia*, 153
Estigmene, 264, 269, pl. 47
Ethmia, 7, 11, 63 (figs.), 68, 71 (figs.), 79, pl. 5
Ethmiidae, 6, 7, 11, 13, 47, 59, 60, 63 (fig.), 67
ethnica (Smith), *Aseptis*, 299, pl. 54
Etiella, 190, pl. 25
Eubaphe, 229, pl. 33
Eublemma, 254, pl. 42
Eublemminae, 253
Eublemmini, 254
Eucalantica, 108, pl. 11
Eucaterva, 218, pl. 30
Euceratia, 108, pl. 11
Euchaciinae, 277
Euchaetes, 264, 272, pl. 48
Euchlaena, 213, pl. 29
Euchromiina, 267, 274

Euchromius, 183, pl. 24
Euclea, 161 (figs.), 164, pl. 20
Euclidia, 259, 260, pl. 44
Euclidiini, 258
Eucoptocnemis, 309, pl. 56
Eucosma, 129 (figs.), 131, 132, 133–135, pl. 15
Eucosmini, 131
Eudesmia, 265, pl. 46
Eudesmiina, 265
Eudonia, 169, 170, pl. 21
Eudrepanulatrix, 213, pl. 29
Eudulini, 229
Eufidonia, 210, pl. 28
Eugnosta, 158, pl. 20
Euhagena, 124, pl. 12
Eulia, 147, pl. 18
Euliini, 147
Eulithis, 223, pl. 31, pl. 62
Eulithosia, 289, pl. 52
Eumacaria, 205, pl. 27, pl. 61
Eumorpha, 7, 233 (figs.), 245, pl. 41, pl. 63
Eumysia, 194, pl. 26
Eupackardia, 237, 240, 241, pl. 38, pl. 63
Euparthenos, 260, pl. 45
euphemia Beutenmüller, *Catocala neogama*, 261, pl. 45
euphorbia moths, 247
euphorbiae (Linnaeus), *Hyles*, 242, 246, pl. 41, pl. 63
Euphyia, 227, pl. 33
Eupithecia, 14, 223 (figs.), 229–230, pl. 33, pl. 62
eupitheciaria Grote, *Glaucina*, 207, pl. 27
Eupitheciini, 229
Euplexia, 294, pl. 53
Euproctis, 274
Euproserpinus, 246, pl. 41, pl. 63 (larva, adult)
Eupsilia, 298, pl. 54
Eupterotidae, 236
Eurois, 315, pl. 58
European corn borer, 17, 173
European grain moth, 50
European pine shoot moth, 131
euryalus (Boisduval), *Hyalophora*, 237, 241, pl. 38
Eurythmia, 198
Eusarca, 218, pl. 30
Euscirrhopterus, 283, pl. 51
Euscrobipalpa, 89, 99, pl. 10
Eustroma, 224, pl. 31
Eustrotiinae, 280
Eutelia, 263, pl. 46
Euteliinae, 262
euterpe Edwards, *Euproserpinus*, 246, pl. 63 (larva, adult)
Euthyatira, 202, pl. 27, pl. 61
Eutricopis, 290, pl. 52
Euura sawflies, 74, 75
Euxoa, 307, 309–312, pl. 57
Euzophera, 196, pl. 26
evansi Rindge & Smith, *Annaphila*, 290, pl. 52
even-lined sallow, 300
Evergestiinae, 172
Evergestis, 172, pl. 22
everlasting bud moth, 254
Evippe, 86, pl. 7
Exaeretia, 66, pl. 5
excelsa Strecker, *Destutia*, 217, pl. 30, pl. 62
excelsaria (Strecker), *Stenoporpia*, 207, pl. 27
excelsicola Braun, *Elachista*, 73, pl. 6
Exceptia, 100, pl. 10
excurvata (Packard), *Digrammia*, 205
exigua (Henry Edwards), *Fulgoraecia*, 162, pl. 21
exigua (Hübner), *Spodoptera*, 294, pl. 52
exitiosa (Say), *Synanthedon*, 125, pl. 14
Exoporia, 20, 33
exornata (Walker), *Orthofidonia*, 209, pl. 28

Exoteleia, 87, 89 (fig.)
expallidalis (Dyar), *Eudonia*, 170, pl. 21
expolitana (Heinrich), *Pelochrista*, 135, pl. 15
exposed bird-dropping moth, 280
exprimens (Walker), *Pyrrhia*, 290, pl. 52
exsertistigma (Morrison), *Parabagrotis*, 316, pl. 58
exsiccatus (Lederer), *Tathorhynchus*, 257, pl. 43
extensa (Braun), *Stilbosis*, 84, pl. 7
extima (Walker), *Platyperigea*, 294
extranea (Henry Edwards), *Tegeticula maculata*, 42, 43, 73, pl. 1,
extraneella (Walsingham), *Aetole*, 111, pl. 11
exuberans (Smith), *Anicla* (*Euagrotis*), 308, pl. 56
eyespotted budmoth, 132

fabriciana (Linnaeus), *Anthophila*, 122
fabulosa (Ferguson), *Xestia* (*Pachnobia*), 316, pl. 58
Faculta, 97, pl. 10
Fala, 288, pl. 52
falcata Packard, *Eusarca*, 218, pl. 30
falcata Schaus, *Orgyia*, 276
facultaria (Packard), *Drepanulatrix*, 213
falcifera (Kirby), *Anagrapha*, 279, pl. 50, pl. 64
falciferella (Walsingham), *Ypsolopha*, 107, pl. 11
falco (Walker), *Xylophanes*, 246, pl. 41
fall cankerworm, 204, 220
fall webworm, 264, 270
fallax (Hampson), *Pseudobryomima*, 301, pl. 55
fallax (Herrich-Schäffer), *Agriopodes*, 283
falsarius Clemens, *Acoloithus*, 160, pl. 21
false armyworm, 297
false celery leaf-tier, 177
false pinion, 297
false-sphinx, 249
farcta (Grote), *Leucania*, 305, pl. 56
farinalis Linnaeus, *Pyralis*, 184, 185 (figs.), pl. 24
farnhami (Grote), *Anarta*, 303, pl. 55
Faronta, 304, pl. 55
fasciata (Barnes & McDunnough), *Ceranemota*, 202, pl. 27
faustina cleopatra Henry Edwards, *Catocala*, 262, pl. 44
faustinula (Boisduval), *Cisthene*, 265, pl. 46
februalis Grote, *Feralia*, 287, pl. 51
felicella (Walsingham), *Schreckensteinia*, 116, pl. 11, pl. 60
felicita Smith, *Baptarma*, 290, pl. 52
felinella Heinrich, *Phyllonorycter*, 56, pl. 3
felix (Walsingham), *Platphalonidia*, 145, 158, pl. 20
Feltia, 307, 312, pl. 57
fenestralis (Barnes & McDunnough), *Hydropionea*, 178, pl. 23
fenestrella (Packard), *Lipographis*, 193, pl. 26
feniseca Heppner, *Glyphipterix*, 109, pl. 11
fennica (Tauscher), *Actebia*, 308, pl. 56
Feralia, 283, 287, pl. 51
Feraliini, 287
feriella Hulst, *Tacoma*, 190, pl. 25
fernaldana (Grote), *Eucosma*, 134, pl. 15
fernaldella Kearfott, *Thaumatopsis*, 188
fernaldella (Riley), *Rectiostoma*, 65, pl. 4
fernaldialis (Hulst), *Cactobrosis*, 196, pl. 13
ferruginella (Braun), *Caloptilia*, 54
ferruginosa (Walker), *Virbia*, 267
fervida Barnes & McDunnough, *Scotogramma*, 303, pl. 55
fervida (Walker), *Sonorarctia*, 269, pl. 47
fervifactaria (Grote), *Stamnodes*, 226, pl. 32
festaliella (Hübner), *Schreckensteinia*, 115, 117, pl. 11

festiva (Hübner), *Diphthera*, 281, pl. 50
fidelis Meyrick, *Bondia*, 166, pl. 21
fieldi Barnes & Benjamin, *Monoleuca*, 164
fieldi Barnes & McDunnough, *Tallula*, 188, pl. 25
fieldi (W. S. Wright), *Oidaematophorus*, pl. 12, pl. 60
fifia Dyar, *Homohadena*, 285, pl. 51
fig moth, 198
figulilella Gregson, *Cadra*, 198, pl. 26
figurata (Harvey), *Oncocnemis*, 286
figurella (Busck), *Chionodes*, 95, pl. 9
filament bearer, 220
Filatima, 85, 92, 96
filbertworm, 142
filiana (Busck), *Sonia*, 136, pl. 15
filifera Dyar, *Promlimacodes*, 163
fimetaria (Grote & Robinson), *Narraga*, 206, pl. 27
finarfinella Kaila, *Elachista*, 73, pl. 6
fine-lined sallow, 285
finitima (Smith), *Hemeroplanis*, 255, pl. 43
first-born, 204
fiscellaria (Guenée), *Lambdina*, 204, 217, pl. 30
Fishia, 300, pl. 54
five-spotted hawkmoth, 244
fissirostris (Meyrick), *Neoscythris*, 80
flammeusella (Chambers), *Adela*, 40, 41, pl. 2, pl. 59
Flatidae, 162
flava Barnes & McDunnough, *Euclea*, 164
flava (Grote), *Pseudanarta*, 301, pl. 55
flavalis (Fernald), *Arenochroa*, 174, pl. 22
flavidens (Grote), *Pseudanarta*, 301, pl. 55
flavidens (Smith), *Euxoa*, 311, pl. 57
flavigutta (Hulst), *Eupithecia*, 230, pl. 33
flaviguttalis Barnes & McDunnough, *Bleptina*, 254
flavistrigella (Busck), *Ypsolopha*, 106, pl. 10
flavofasciata (Walker), *Proserpinus*, 245, pl. 41
flegax Hodges, *Anoncia*, 83, pl. 7
flexilis Freeman, *Argyresthia*, 103
florestan (Cramer), *Manduca*, 243, 244, pl. 40
floridensis Clark, *Amphion*, 245, pl. 41
floridus Zeller, *Crambus*, 181, pl. 24
fodinalis (Lederer), *Pyrausta*, 175 (fig.), 177, pl. 22
foeminalis (Smith), *Euxoa*, 311
foliana (Walsingham), *Acleris*, 146, pl. 17
forage looper, 258
forbesi (Benjamin), *Abagrotis*, 317, pl. 58
forbesi Obraztsov, *Clepsis persicana*, 149, pl. 19
forest tent caterpillar, 235
forficaria (Guenée), *Prochoerodes*, 219, pl. 31, pl. 62
formosa (Hulst), *Dysstroma*, 223, pl. 31, pl. 62
formosa (Hulst), *Synaxis*, 219, pl. 30
formosata (Strecker), *Stamnodes*, 226, pl. 32
formosella (Murtfeldt), *Chionodes*, 92
Formosella Group, 92
Forsebia, 258, pl. 44
fortis (Grote), *Homoncocnemis*, 285, pl. 51
four-spotted fungus moth, 254
f-pallida (Strecker), *Grammia*, 268, pl. 46
fractilinea (Grote), "*Oligia*", 296,
fracturalis (Zeller), *Diastictis*, 180, pl. 23
fragarianus (Busck), *Decodes*, 145, 146, pl. 18
fragariella (Busck), *Anacampsis*, 101, pl. 10
fragilis (Strecker), *Virbia*, 267, pl. 46
fragilis (Walsingham), *Paraplatyptilia*, 120, pl. 12
francisca Keifer, *Recurvaria*, 87
franciscana (Walsingham), *Argyrotaenia*, 23, 148, pl. 18

franciscana insulana Powell, *Argyrotaenia*, 149, pl. 18
franciscella Busck, *Argyresthia*, 106, pl. 11
franclemonti Hodges, *Mompha*, 72, 77, pl. 6
frater Grote, *Raphia*, 282, pl. 50
fratercula (Pagenstecher), *Belonoptera*, 167
fraterna (Smith), *Heteranassa*, 260, pl. 44
fraxini Linnaeus, *Catocala*, 262
Frechinia, 170, pl. 21
Frederic's underwing, 262
frederici Grote, *Catocala*, 262
frigid owlet, 263
frigidana (Walker), *Nycteola*, 263, pl. 46, pl. 64
fringata (Barnes & McDunnough), *Eupsilia*, 298, pl. 54
fringed looper, 215
fringe-tree sallow, 285
Friseria, 90, pl. 8
fruit-piercers, 256
fruit-tree leaf-roller, 129, 149
fucana (Walsingham), *Clepsis*, 152, pl. 19
fucosa (Hübner), *Hypoprepia*, 266
fulgens (Henry Edwards), *Lycomorpha*, 265
Fulgoraecia, 162, pl. 21
Fulgoridae, 162
fulgoroid leafhoppers, 162
fuliginosa rubricosa (Harris), *Phragmatobia*, 269, pl. 47
fulvicollis (Hübner), *Cisseps*, 274, pl. 48
fulviplicana (Walsingham), *Phtheochroa*, 156, pl. 20
fulvum Stretch, *Kodiosoma*, 269, pl. 47, pl. 64
fumiferana (Clemens), *Choristoneura*, 128, 150
fumosa Butler, *Hyaloscotes*, 52, pl. 2
funebris (Strömberg), *Anania*, 174, pl. 22
funeralis (Hübner), *Desmia*, 179, pl. 23
funeralis Grote & Robinson, *Acronicta*, 282, pl. 50
funerary dagger moth, 282
funerellus (Dyar), *Phobus*, 191, pl. 25
furcata Borgstrom, *Hydriomena*, 223 (fig.), 224, pl. 31
furcata (Walker), *Eutelia*, 263
Furcula, 249, pl. 33
furfurana (Haworth), *Bactra*, 130, pl. 14
furfurata (Grote), *Homorthodes*, 306, pl. 56
fusca (Haworth), *Pyla*, 191, pl. 25
fusca (Henry Edwards), *Neoilliberis*, 160, pl. 21
fuscesens (Haworth), *Borhausenia*, 62
fuscigera (Grote), *Euxoa*, 311, pl. 57
fusciterminella Clarke, *Agonopterix*, 65, pl. 4
fuscodorsanus (Kearfott), *Henricus*, 157, pl. 19
fuscoleuca (Braun), *Tridentaforma*, 41, pl. 1
fuscomaculella (Wright), *Ozamia*, 196, pl. 26
fuscotaeniella (Chambers), *Rifseria*, 91, pl. 8
fuscula (Grote), *Meganola*, 263, pl. 46
fusimacula Smith, *Oxycnemis*, 287–288, pl. 51
futilalis (Lederer), *Saucrobotys*, 173, pl. 22

Gabriola, 214, pl. 29
Galasa, 187, pl. 24, pl. 61
Galenara, 210, pl. 28
Galgula, 294, pl. 53
Galleria, 186, pl. 24
Galleriinae, 186
gallii (Rottenburg), *Hyles*, 246, pl. 41, pl. 63
garden tiger moth, 268
garden tortrix, 151
garden webworm, 174
Gardinia, 267, pl. 46, pl. 64
gargamelle (Strecker), *Gloveria*, 235, pl. 34
gaultheriella (Walsingham), *Cameraria*, 55, pl. 3
gaurae (J. E. Smith), *Schinia*, 292, pl. 52
gausapalis Hulst, *Crambus*, 181, pl. 24

lithosina Henry Edwards, *Annaphila*, 290,
pl. 52
Lithosiina, 266
Lthosiini, 264
Lithostege, 231, pl. 33
Litocala, 257, pl. 43
Litodonta, 250, pl. 42
Litoprosopus, 257, pl. 43
little underwing, 262
little white lichen moth, 266
littoralis (Packard), *Paradiarsia* 314, pl. 58
lixarioides McDunnough *Galenara*, 210, pl. 28
lobatiella Opler & Davis, *Cameraria*, 55, pl. 3
Lobocleta, 222, pl. 31
Lobophora, 231, pl. 33
Lobophorini, 231
locust twig borer, 143
locust underwing, 260
lodgepole needleminer, 87
loexyla Hodges, *Anoncia*, 83, pl. 7
loftini (Dyar), *Eoreuma*, 183, pl. 24
Lomographa, 212, pl. 28
lomonana (Kearfott), *Epinotia*, 137
longa (Meyrick), *Anoncia*, 83, pl. 7
longana (Haworth), *Cnephasia*, 146,
pl. 17
long-horned owlet, 254
longiciliata (Hulst), *Paleacrita*, 212, pl. 28
longidens (Smith), *Dichagyris* (*Mesembragrotis*),
308, pl. 56
longifrons (Walsingham), *Oidaematophorus*,
121, pl. 12
longipennis (Hampson), *Petrophila*, 175
longissima J. F. Landry, *Arotrura*, 79, pl. 6
Longivesica, 310
longula Braun, *Bucculatrix*, 57, pl. 4
Lophocampa, 272, pl. 48, pl. 64
Lophocoronoidea, 20
lophosella (Busck), *Chionodes*, 92, pl. 8
lorata (Grote), *Euthyatira*, 202, pl. 27
lorea (Guenée), *Lacinipolia*, 305, pl. 56
loricaria (Eversmann), *Macaria*, 206
lorquinaria (Guenée), *Macaria*, 206, pl. 27
lotella Wagner, *Micracalyptris*, 38, pl. 1
Lotisma, 166, pl. 21
lotta Barnes & McDunnough, *Givira*, 127
lowriei Barnes & McDunnough, *Tolype*, 232
Loxostege, 174–175, pl. 22
Loxostegopsis, 178, pl. 23
lucidus Boisduval, *Arctonotus*, 245–246, pl. 41,
pl. 63
lugubrella (Fabricius), *Chionodes*, 95, pl. 9
luminosa Heinrich, *Ofatulena*, 143, pl. 17
luna Morrison, *Cucullia*, 285, pl. 51
luna moths, 236
lunaris (Haworth), *Batia*, 62, pl. 4
lunata (Drury), *Zale*, 260, pl. 44
lunata Henry Edwards, *Heterocampa*, 250,
pl. 42
lunata (Smith), *Amphipoea*, 296, pl. 53
lunatana Walsingham, *Grapholita*, 142, pl. 17
lunate zale, 260
luniferella Hulst, *Oneida*, 185 (figs.), 188,
pl. 25
lupini (Grote), *Merolonche*, 283, pl. 50
luscitalis (Barnes & McDunnough), *Mecyna*,
178
luski Barnes & Benjamin, *Coloradia*, 238
lustralis (Grote), *Lacinipolia*, 305, pl. 56
lustrans (Beutenmüller), *Triprocris*, 162
luteola (Grote & Robinson), *Coranarta*, 303,
pl. 55
lutescella (Clarke), *Scrobipalpulopsis*, 99, pl. 10
Lychnosea, 216, pl. 30
Lycia, 211, pl. 28, pl. 61
lycid beetles, 265

Lycomorpha, 264, 265, pl. 46
lycopersicella (Walsingham), *Keiferia*, 100,
pl. 10
Lygephila, 257, pl. 43
Lygropia, 180, pl. 23
Lymantria, 274, pl. 48, pl. 63
Lymantriinae, 13, 274
Lymantriini, 274
lynceella Zeller, *Gelechia*, 91, pl. 8
lynosiridella Walsingham, *Coleophora*, 76, pl. 6
lynosyrana Walsingham, *Synnoma*, 145, 155,
pl. 19, pl. 61
Lyonetia, 112, pl. 11
Lyonetiidae, 47, 112
lyonothamnae (Powell), *Ypsolopha*, 103 (figs.),
107, pl. 11

Macalla, 187, pl. 25
Macaria, 205–206, pl. 27, pl. 61
Macariini, 205
maccullochii (Kirby), *Androloma*, 284, pl. 51
macdunnoughi Guedet, *Dasyfidonia*, 210
macdunnoughi Swett, *Hydriomena*, 224,
pl. 31
machimiana (Barnes & Busck),
Sparganothoides, 154, pl. 19
mackiei (Keifer), *Coleotechnites*, 87, 89 (figs.),
pl. 17
macneilli Munroe, *Orenaia*, 173, pl. 22
macneilli Powell, *Epiblema*, 136, pl. 15
macneilli Powell, *Ethmia*, 70, pl. 5
macrocarpanus (Walsingham), *Henricus*, 157,
pl. 19
macrocarpella (Frey & Boll), *Cameraria*, 56
Macroglossinae, 244
Macromphalinae, 234
Macrothecini, 186
Macrurocampa, 250, pl. 42
mactata (Guenée), "*Oligia*", 296, pl. 53
macula (Druce), *Schrankia*, 253, pl. 42
macularia Harris, *Sicya*, 218, pl. 30
maculata Harris, *Lophocampa*, 273, pl. 48,
pl. 64
maculata Harris, *Thyris*, 167, pl. 21
maculata (Riley), *Tegeticula*, 42, pl. 1
maculatana (Walsingham), *Eucosma*, 134, pl. 15
maculatella (Busck), *Ypsolopha*, 107, pl. 11,
pl. 60
maderae Pellmyr, *Tegeticula*, 42, pl. 1
madiae Povolný & Powell, *Scrobipalpopsis*, 99
madusaria (Walker), *Euchlaena*, 214,
magdalena Hulst, *Melemaea*, 216, pl. 20
magna Ferguson, *Orgyia*, 276
magna, *Orgyia*, 276
magnaria Guenée, *Ennomos*, 215, pl. 29
magnella (Busck), *Isophrictis*, 85, pl. 7
magnifica Grote, *Spragueia*, 280, pl. 50
magnimacula Handfield & Handfield, *Plusia*,
279
magnoliata Guenée, *Spargania*, 226, pl. 32,
pl. 62
magnoliatoidata (Dyar), *Lobophora*, 231, pl. 33
Magusa, 301, pl. 55
maia (Drury), *Hemileuca*, 238
maida (Dyar), *Hillia*, 298–299, pl. 54
major Smith, *Acontia*, 281, pl. 50
majuscula Herrich-Schäffer, *Orthodes*, 306,
pl. 56
Malacosoma, 232, 233 (fig.), 234, pl. 37, pl. 62
malefida Guenée, *Agrotis* 313, pl. 58
Mamestra, 304, pl. 55
mancalis (Lederer), *Hahncappsia*, 174, pl. 22
Manduca, 242, 243, 244, pl. 40, pl. 63
mansueta Smith, *Acronicta*, 283, pl. 50
many-lined wainscot, 305
manzanita (Braun), *Phyllonorycter*, 56, pl. 3

manzanita Taylor, *Hydriomena*, 225, pl. 32
manzanitae Keifer, *Antaeotricha*, 65, pl. 4
manzanitae (Keifer), *Pseudochelaria*, 88, pl. 8
maple spanworm, 215
mappa (Grote & Robinson), *Autographa*, 278,
pl. 49
marachella Kaila, *Elachista*, 73, pl. 6
marcesaria (Packard), *Macaria*, 206
marculenta (Grote & Robinson), *Hahncappsia*,
174, pl. 22
marga Barnes & McDunnough, *Givira*, 126,
pl. 13
marginata (Harris), *Pennisetia*, 123, pl. 12
marginatus Riley, *Prodoxus*, 43, pl. 1
marginella (Denis & Schiffermüller),
Dichomeris, 102, pl. 10
marginimaculella (Chambers), *Homosetia*, 50,
pl. 2
mariona (Beutenmüller), *Carmenta*, 125,
pl. 14
mariposa Grote & Robinson, *Alypia*, 284,
pl. 51
maritella McDunnough, *Coleophora*, 75, pl. 6
Marmara, 55, pl. 3, pl. 59
marmarodactyla (Dyar), *Anstenoptilia*, 117,
119, pl. 12, pl. 60
Marmopterix, 226
marmorata (Packard), *Stamnodes*, 226, pl. 32
marmorata Smith, *Acronicta*, 283, pl. 50
marmorea (Walsingham), *Ethmia*, 63, 70, pl. 5
marmoreana (Heinrich), *Catastega*, 140
martenii (French), *Neoilliberis*, 160
Martia, 194, pl. 26
martini CB Knowlton, *Cisthene*, 253 (figs.),
265, pl. 46
Marythyssa, 263, pl. 46
masoni (Smith), *Schinia*, 292, pl. 52, pl. 64
matilda (Dyar), *Protitame*, 204, pl. 27
matuta Henry Edwards, *Alypia octomaculata*,
284, pl. 51
matutella, *Landryia*, 78
maxima Dyar, *Apoda*, 163
maximana (Barnes & Busck), *Acleris*, 144, pl. 17
maximella (Fernald), *Donacaula*, 170, pl. 21
mayrella (Hübner), *Coleophora*, 75, pl. 6
maysi Schaus, *Parasa*, 164
mcglashani Henry Edwards, *Gazoryctra*, 37,
pl. 13
meadi (Grote), *Schinia*, 292, pl. 52
meadi (Packard), *Coryphista*, 225, pl. 32
meadii Henry Edwards, *Thyridopteryx*, 52, pl. 2
meadowsi Buckett, *Feralia*, 287
meal moth, 184
meal snout moth, 184
mealybugs, 195
Mecyna, 178, pl. 23
mediodentata (Barnes & McDunnough),
Hymenodria, 225, pl. 32
mediofasciana (Clemens), *Ancylis*, 140, pl. 16
mediofuscella (Clemens), *Chionodes*, 94, pl. 9
mediostriata Braun, *Tischeria*, 44, pl. 1
Mediterranean flour moth, 17, 197
medusa (Strecker), *Gloveria*, 235, pl. 34
Meessiinae, 50
Megachile, 197
Megalographa, 278, pl. 50
Megalopyge, 165, pl. 20, pl. 61
Megalopygidae, 115, 160, 161 (figs.), 165
Meganola, 263, pl. 46
Megasema, 315, pl. 58, pl. 64
Melanchra, 303, pl. 55
melanifera (Keifer), *Caloreas*, 122, pl. 12
Melanolophia, 210, 211, pl. 28
Melanolophiini, 207
melanopa (Thunberg), *Anarta*, 303
Melemaea, 216, pl. 30, pl. 62

Sematurinae, 203
semiatratra (Hulst), *Eustroma*, 223, pl. 31
semicirculana (Fernald), *Dorithia*, 145, 147, pl. 17
semicircularis (Grote), *Euthyatira*, 202, pl. 27
semiclarata Walker, *Lomographa*, 212, pl. 28
semicollaris Smith, *Oncocnemis*, 286
semicrocea (Guenée), *Exyra*, 280
semiferanus (Walker), *Archips*, 150
semiflava (Guenée), *Tarachidia*, 280, pl. 50
semifuneralis (Walker), *Euzophera*, 196, pl. 26
semilugens (Zeller), *Ethmia*, 70, pl. 5
semiluna (Smith), *Edia*, 170, pl. 21
Semioscopis, 66, pl. 5
semipurpurella (Stephens), *Eriocrania*, 34, pl. 1
semirelict underwing, 262
semirelicta Grote *Catocala*, 262, pl. 45
semirubralis (Packard), *Pyrausta*, 177, pl. 22
semitenebrella Dyar, *Ethmia*, 69, pl. 5
senecionana (Walsingham), *Sparganothis*, 153, pl. 19
senescens Zeller, *Acleris*, 145, pl. 17
senta (Strecker), *Hemaris*, 245, pl. 41, pl. 63
sentinaria (Geyer), *Scopula*, 222, pl. 31
seorsa Braun, *Bucculatrix*, 57
separabilis Braun, *Bucculatrix*, 57, pl. 3, pl. 4
separataria (Grote), *Stenoporpia*, 207, 207
septentrionalis Barnes & McDunnough, *Agylla*, 267, pl. 46
septentrionella Busck, *Glyphidocera*, 61, pl. 4
septentrionella Walsingham, *Adela*, 40, pl. 2, pl. 59
septentrionis (Walker), *Gluphisia*, 249, pl. 33
Sequoia sphinx, 243
sequoiae Boisduval, *Sphinx*, 243, pl. 40
sequoiae (Henry Edwards), *Synanthedon*, pl. 60
seraphica (Dyar), *Olceclostera*, 236, pl. 35
Sericosema, 212, pl. 28
serrata (Grote), *Hydroeciodes*, 307, pl. 56
serratilineella Ragonot, *Vitula*, 197
Sesia, 109 (fig.), 124, pl. 14
Sesiidae, 7, 11, 109 (figs.), 115, 123
Sesiinae, 123
Sesioidea, 20, 115, 123
Setagrotis, 316, pl. 58
setonana (McDunnough), *Digrammia*, 205, pl. 27
severa Henry Edwards, *Gluphisia*, 249, pl. 33
sevorsa (Grote), *Gerra*, 283, pl. 51
sexmaculata (Packard), *Macaria*, 206, pl. 27, pl. 61
sexnotata Braun, *Bucculatrix*, 57, pl. 4
sexpunctata Grote, *Spargaloma*, 255, pl. 43
sexpunctella (Fabricius), *Lita*, 90, pl. 8
sexsignata (Harvey), *Litocala*, 257, pl. 43
sexstrigella Braun, *Mompha*, 77
sexta (Linnaeus), *Manduca*, 242, 244, pl. 40
shastana Davis, *Bondia*, 166, pl. 21
Sicya, 218, pl. 30
sidalceae Englehardt, *Zenodoxus*, 123, pl. 12
siderana (Treitschke), *Olethreutes*, 131
sideraria (Guenée), *Scopula*, 223 (figs.)
Sideridis, 304, pl. 55
sierralis Munroe, *Loxostege*, 175, pl. 22
sierralis Munroe, *Orenaia*, 173, pl. 22
Sierran pericopid, 271
sigmoid prominent, 248
signaria (Walker), *Macaria*, 206, pl. 27
signatalis (Walker), *Pyrausta*, 176, pl. 23
signatella (Herrich-Schäffer), *Symmoca*, 60, pl. 4
signiferana Heinrich, *Epinotia*, 139, pl. 16
silaceata (Denis & Schiffermüller), *Ecliptoptera*, 224, pl. 31

silens (Grote), *Euxoa*, 311, pl. 57
Silens species group, 311
silkworm, 236
silverspotted tiger moth, 272
silvertonana Obraztsov, *Anopina*, 148
simalis Grote, *Prorasea*, 173
similaris Richards, *Bulia*, 258, 259 (figs.)
similaris Barnes, *Admetovis*, 304
similis (Stainton), *Bryotropha*, 91
simona McDunnough, *Euxoa*, 311, pl. 57
simplex (Walker), *Egira*, 302, pl. 55
simpliciella (Walsingham), *Cauchas*, 41, pl. 2
simulana (Clemens), *Dichrorampha*, 141, pl. 16
simuloides (McDunnough), *Ancylis*, 140, pl. 16
Simyra, 282, pl. 50
sinaloensis Hoffman, *Citheronia splendens*, 237
singularis Barnes & McDunnough, *Oncocnemis*, 286, pl. 51
singulella Walsingham, *Adela*, 40, pl. 2, pl. 59
sinuaria (Barnes & McDunnough), *Cochisea*, 211, pl. 28
sironae Hodges, *Lita*, 90, pl. 8
siskiyouana (Kearfott), *Eucosma*, 134, pl. 15
siskiyouensis Heinrich, *Epinotia*, 138, pl. 16
sisterina Povolny & Powell, *Exceptia*, 100, pl. 10
sistrella (Busck), *Chionodes*, 93, pl. 8
Sitochroa, 174, pl. 22
Sitotroga, 85, 102, pl. 10
slosseri Peigler & Stone, *Hemileuca*, 238
Slossonia, 215, pl. 29
small heterocampa, 250
small mocis, 260
small necklace moth, 256
small tabby, 185
small-eyed sphinx, 243
smeathmanniana (Fabricius), *Aethes*, 158, pl. 20
Smerinthinae, 242
Smerinthus, 243, pl. 39, pl. 63
Smith's dart, 315
smithii (Pearsall), *Animomyia*, 214, pl. 29
smithii (Snellen), *Xestia*, 315, pl. 58
smithsoniana (Clemens), *Triprocris*, 162, pl. 21
smoky moth, 266
snaky arches, 305
snellenella (Walsingham), *Neodactylota*, 90, pl. 8
snout moths, 255
snowberry clearwing, 245
snowi Henry Edwards, *Melittia*, 124, pl. 13
Snowia, 217, pl. 30
sobrina (Stretch), *Lophocampa*, 272, pl. 64
solandriana (Linnaeus), *Epinotia*, 138, pl. 16
Somabrachyidae, 160
Somatolophia, 218, pl. 30
somnulentella (Zeller), *Bedellia*, 109, 112, pl. 11
Sonia, 101, 132, 136, pl. 15
sonoma (McDunnough), *Pseudocopivaleria*, 287
sonomana Kearfott, *Eucosma*, 133, pl. 15
Sonoran sphinx, 243
Sonoran tent caterpillar, 235
Sonorarctia, 269, pl. 47
sonorensis Hodges, *Ceratomia*, 243
sonorensis Pellmyr & Balcázar-Lara, *Prodoxus*, 43
sordens (Hufnagel), *Apamea*, 295, pl. 53
sordidus Riley, *Prodoxus*, 43, pl. 1
Sorhagenia, 83, 84, pl. 7
sororculella (Dyar), *Prochoreutis*, 122, pl. 12
sosana (Kearfott), *Epiblema*, 135, pl. 15
Sosipatra, 197, pl. 26
sospeta (Drury), *Xanthotype*, 213, pl. 29
southwestern pine tip moth, 131
southwestern tent caterpillar, 235

soybean looper, 277
spadicus Grossbeck, *Monoleuca*, 160, 165
Spaelotis, 314, pl. 58
Spargaloma, 255, pl. 43
Spargania, 226, pl. 32, pl. 62
Sparganothini, 128, 153
Sparganothis, 153, pl. 19
Sparganothoides, 154, pl. 19
spartifoliella (Hübner), *Leucoptera*, 112
spearmarked black moth, 225
specca Davis, *Mesepiola*, 43, pl. 2
speciosa (Hübner), *Xestia (Pachnobia)*, 316, pl. 58
speciosa (Hulst), *Nemeris*, 217, pl. 30
speckled cutworm moth, 303
speckled green fruitworm, 302
speculella Clemens, *Lyonetia*, 112, pl. 11
spenceri Munroe, *Eudonia*, 170, pl. 21
sperryellus Klots, *Crambus*, 175 (figs.), 181, pl. 24
sperryellus Klots, *Diastictis*, 180, pl. 23
sperryorum Munroe, *Diastictis*, 180, pl. 23
sphacelina (Keifer), *Anoncia*, 83, pl. 7
Sphenarches, 118
sphenisca Powell, *Ethmia*, 70, pl. 5
Sphingicampa, 233 (fig.), 237, pl. 36, pl. 63
Sphingidae, 7, 17, 201, 233, 242, pl. 39–41, pl. 63
Sphinginae, 242, 243
Sphinx, 243, pl. 39–40, pl. 63
spila Rindge & Smith, *Annaphila*, 290
spider nests, 74
Spilomelinae, 168, 177
Spilomelini, 168
Spilonota, 132, pl. 15
Spilosoma, 264, 270, pl. 47, pl. 64
spilotus Davis, *Acrolophus*, 49
spiritum Crabo & Fauske, *Copablepharon*, 309, pl. 57
spissicornis Coleophora, 75
splendens Barnes & McDunnough, *Hemihyalea*, 273, pl. 48
splendens Barnes & McDunnough, *Lycomorpha*, 266, pl. 46
splendens sinaloaensis Hoffman, *Citheronia*, 237, pl. 35, pl. 62
splendida Braun, *Tischeria*, 44, pl. 1
spododea (Hulst), *Merisma*, 210, pl. 28
Spodolepis, 215, pl. 29, pl. 62
Spodoptera, 293–294, pl. 52
Spoladea, 179, pl. 23
sponsella (Busck), *Arotrura*, 79, pl. 6
spotted cutworm, 315
spotted datana, 249
spotted straw, 291
spotted tussock moth, 273
Spragueia, 280, pl. 50
spruce bud moth, 137
spruce budworm, 17
spruce coneworm, 192
spruce harlequin, 255
spruce-fir looper, 206
spumosum (Grote), *Plagiomimicus*, 288, pl. 52
spurge hawkmoth, 246
St. Lawrence tiger moth, 268
staghorn cholla moth, 283
Stagmatophora, 81–82, pl. 7
stalachtaria (Strecker), *Narraga*, 206–207
Stamnoctenis, 226
Stamnodes, 226, pl. 32
Stamnodini, 226
starki (Freeman), *Coleotechnites*, 87, 89 (fig.)
statalis Grote, *Arta*, 187
stathmopoda, 74
Staudingeria, 194, pl. 26
Stegea, 172, pl. 22
Stellata species group, 229
Stenomatidae, 47, 63 (fig.), 64

PLANT INDEX

Abies, 87, 132, 137, 139, 145, 149, 150, 151, 153, 154, 157, 167, 192, 206, 207, 210, 216, 217, 220, 231, 272, 275, 276, 279, 316
  amabilis, 147
  balsamea, 150, 208, 279
  concolor, 87, 92, 107, 134, 149, 151, 159, 276, 299
  grandis, 87, 276, 302
  lasiocarpa, 96
  magnifica, 209
Abietoideae, 192
Abronia, 110, 227, 305
  latifolia, 110
  umbellata, 111
  maritima, 227
Abutilon, 171
Acacia, 165, 190, 237, 238, 258
  angustissima, 237
  dealbata, 256
  greggii, 238, 260
  greggii var. *wrightii*, 237
Acamptopappus, 98
  sphaerocephalus, 98, 285
Acanthaceae, 280
Acer, 53, 163, 202, 209, 211, 215, 217, 219, 220, 227, 228, 240, 273, 282, 298
  circinatum, 53
  grandidentata, 56
  macrophyllum, 53, 94
  negundo, 136, 150
Aceraceae, 53, 163, 202, 211, 219, 227, 228, 240, 248, 250, 273, 281, 282, 298, 315
Achillea millefolium, 80 (nectar), 92, 119, 135, 158
Acourtia microcephala, 163
Adenostoma fasciculatum, 86, 149, 209, 210–212, 214, 231
Aesculus, 131, 137
  hippocastanum, 136
Agaricaceae, 50
Agavaceae, 43, 128, 271
Agave, 43, 158
  deserti, 43, 158
  palmeri, 43, 158
  shawii, 158
Ageratum, 119
Ailanthus altissima, 105
Aizoaceae, 111
Albizzia, 257

Alcea (=*Althaea*), 58, 104, 256, 281
  rosea, 281
Alchemilla, 190
alder (see *Alnus*)
alfalfa (see *Medicago*)
algae, 253, 264, 267
Alismataceae, 180
alkali health, 97, 118 (see *Frankenia*)
Allionia nyctaginea, 162
almond, 125, 190, 198 (see *Prunus amygdalus*)
desert almond, 239 (see *Prunus fasciculata*)
Alnus, 39, 53, 105, 123, 137, 138, 202, 204, 206–209, 210, 215, 217, 219, 224–227, 230, 251, 268, 270, 272, 273, 278, 281, 282, 297, 300 301, 306, 316
  rhombifolia, 140
  rubra, 202
  viridis var. *sinuata*, 202
alumroot, 137, 140 (see *Heuchera*)
Amaranthaceae, 79, 179, 280, 308, 317
Amaranthus, 80, 174
Ambrosia, 102, 120, 135, 174, 208, 276, 280
  acanthicarpa, 170
  artemisiella, 102
  artemisiifolia var. *elatior*, 280
  chamissonis (=*bipinnatifida*), 98, 194, 312
  confertifolia, 98
  dumosa, 215
  psilostachya, ,66, 81, 94, 98
Amelanchier, 54, 202, 209, 212, 213, 243, 297, 301, 316, 317
  alnifolia, 202
Amoracia rusticana, 172
Amorpha, 96, 155
  californica, 155, 191
  fruticosa, 155
Ampelopsis, 245, 284,
Amsinckia, 79, 125, 268
  tessellata, 69
Anacardiaceae, 37, 38, 54, 89, 130, 155, 190, 191, 211, 221, 235, 237, 239, 241, 242, 248, 249, 262, 263
Anaphalis, 91, 99, 194
  margaritacea, 122
Annonaceae, 242, 256
Antennaria, 177, 191, 290
Anthemis, 158
Antirrhinum, 119, 130, 286
Anulocaulis annulatus, 111
Apache plume, 239 (see *Fallugia*)

Apiaceae, 41, 42, 65, 66, 108, 116, 154, 160
Apiastrum, 66 (suspected)
Apium graveolens (celery), 177, 279
Apocynaceae, 245, 255, 266, 271, 274, 284
Apocynum, 154, 173, 242, 245, 255, 266, 271, 272, 274, 284
  androsaemifolium, 274
apple (see *Malus*)
apricot (see *Prunus armeniaca*)
Aquilegia, 278, 290, 296
Arabis, 41
Araliaceae, 116, 219
Aralia californica, 116
Arbutus, 39, 211, 219, 225, 229, 230, 273, 250, 299, 302
  arizonica, 55, 230, 250,
  menziesii, 39, 55, 56, 76, 88, 91, 138–140, 153, 166, 212, 219, 225, 229, 240, 273, 299, 302
Arceuthobium, 190
Arctium, 119, 280
  lapella, 85
Arctostaphylos, 39, 41 (adult assoc.), 52, 56, 65, 75, 87, 88, 91, 94, 118, 137–140, 148, 153, 154, 163, 165, 166, 205 (adult assoc.), 209, 211, 214, 225, 229, 234, 237, 240, 241, 243, 244, 257, 287, 298, 299, 302
  auriculata, 87
  columbiana, 298
  glandulosa, 76
  glauca, 76, 140, 287, 299
  hooveri, 140, 147
  insularis, 76, 94, 139
  manzanita, 76, 88, 140
  nevadensis, 298
  patula, 139, 140
  pungens, 165, 234, 237, 243
  uva-ursi, 192, 245, 303
  virgata, 147
Arecaceae, 74, 190, 257
Argemone, 155
Aristolochiaceae, 180, 248
Armillaria mellea, 50
Arnica, 99
Artemisia, 67, 98, 99, 119, 133, 206, 208, 210, 215, 239, 268, 291, 293, 312, 317
  californica, 57, 133, 134, 218, 307
  douglasiana, 55, 81, 99, 122, 133
  dracunculoides, 67

GENERAL INDEX

Composition: Aptara, Inc.

Text and Display: Stone Serif